MW00849090

ENCYCLOPEDIA OF MATERIALS SCIENCE AND ENGINEERING

SUPPLEMENTARY VOLUME 2

ADVANCES IN MATERIALS SCIENCE AND ENGINEERING

This is a new series of Pergamon scientific reference works, each volume providing comprehensive, self-contained and up-to-date coverage of a selected area in the field of materials science and engineering. The series is being developed primarily from the highly acclaimed *Encyclopedia of Materials Science and Engineering*, published in 1986. Titles in the series are listed below.

BEVER (ed.)
Concise Encyclopedia of Materials Economics, Policy & Management

BROOK (ed.)
Concise Encyclopedia of Advanced Ceramic Materials

CARR & HERZ (eds.)
Concise Encyclopedia of Mineral Resources

CORISH (ed.)
Concise Encyclopedia of Polymer Processing & Applications

EVETTS (ed.)
Concise Encyclopedia of Magnetic & Superconducting Materials

KELLY (ed.)
Concise Encyclopedia of Composite Materials

KIMERLING (ed.)
Concise Encyclopedia of Electronic & Optoelectronic Materials

MOAVENZADEH (ed.)
Concise Encyclopedia of Building & Construction Materials

SCHNIEWIND (ed.)
Concise Encyclopedia of Wood & Wood-Based Materials

WILLIAMS (ed.)
Concise Encyclopedia of Medical & Dental Materials

Also available to complement the *Encyclopedia of Materials Science and Engineering*:

CAHN (ed.)
Encyclopedia of Materials Science and Engineering, Supplementary Volume 1

NOTICE TO READERS

Dear Reader
If your library is not already a standing order/continuation order customer to the series **Advances in Materials Science and Engineering**, may we recommend that you place a standing/continuation order to receive immediately upon publication all new volumes. Should you find that these volumes no longer serve your needs, your order can be cancelled at any time without notice.

ROBERT MAXWELL
Publisher at Pergamon Press

ENCYCLOPEDIA OF MATERIALS SCIENCE AND ENGINEERING

Supplementary
Volume 2

Editor
ROBERT W CAHN
University of Cambridge, UK

Senior Advisory Editor
MICHAEL B BEVER
Massachusetts Institute of Technology
Cambridge, MA, USA

PERGAMON PRESS
Member of Maxwell Macmillan Pergamon Publishing Corporation
OXFORD • BEIJING • FRANKFURT
SYDNEY • TOKYO

THE MIT PRESS
CAMBRIDGE
MASSACHUSETTS

Distributed exclusively in North and South America by The MIT Press, Cambridge, Massachusetts, USA

Published and distributed exclusively throughout the rest of the world by Pergamon Press:

U.K.	Pergamon Press plc, Headington Hill Hall, Oxford OX3 0BW, England
PEOPLE'S REPUBLIC OF CHINA	Pergamon Press, Room 4037, Qianmen Hotel, Beijing, People's Republic of China
FEDERAL REPUBLIC OF GERMANY	Pergamon Press GmbH, Hammerweg 6, D-6242 Kronberg, Federal Republic of Germany
AUSTRALIA	Pergamon Press Australia Pty Ltd., P.O. Box 544, Potts Point, N.S.W. 2011, Australia
JAPAN	Pergamon Press, 5th Floor, Matsuoka Central Building, 1-7-1 Nishishinjuku, Shinjuku-ku, Tokyo 160, Japan

First edition 1990

Library of Congress Cataloging in Publication Data
(Revised for vol. 2)

Encyclopedia of materials science and engineering.
Supplementary volume.

(Advances in materials science and engineering)
Includes bibliographies and indexes.
1. Materials—Encyclopedias. I. Cahn, R. W.
(Robert W.), 1924– . II. Bever, Michael B.
(Michael Berliner). III. Encyclopedia of materials
science and engineering. IV. Series.
TA402.E53 1986 Suppl. 620.1'03 88–12518

ISBN 0–262–03173–6 (The MIT Press)

British Library Cataloguing in Publication Data
Encyclopedia of materials science and engineering.
Supplementary volume 2
1. Materials science
I. Cahn, R. W. (Robert Wolfgang), *1924–* II. Series
620.11

ISBN 0–08–036196–X (Pergamon Press)

Printed in Great Britain by BPCC Wheatons Ltd, Exeter

CONTENTS

HONORARY EDITORIAL ADVISORY BOARD

FOREWORD

In the short time since its publication, the *Encyclopedia of Materials Science and Engineering* has been accepted throughout the world as the standard reference about all aspects of materials. This is a well-deserved tribute to the scholarship and dedication of the Editor-in-Chief, Professor Michael Bever, the Subject Editors and the numerous contributors.

During its preparation, it soon became clear that change in some areas is so rapid that publication would have to be a continuing activity if the Encyclopedia were to retain its position as an authoritative and up-to-date systematic compilation of our knowledge and understanding of materials in all their diversity and complexity. Thus, the need for some form of supplementary publication was recognized at the outset. The Publisher has met this challenge most handsomely: both a continuing series of Supplementary Volumes to the main work and a number of smaller encyclopedias, each covering a selected area of materials science and engineering, will be published in the next few years.

Professor Robert Cahn, the Executive Editor, was previously the editor of an important subject area of the main work, and many other people associated with the Encyclopedia will contribute to its Supplementary Volumes and derived Concise Encyclopedias. Thus, continuity of style and respect for the high standards set by the *Encyclopedia of Materials Science and Engineering* are assured. They have been joined by some new editors and contributors with knowledge and experience of important subject areas of particular interest at the present time. Thus, the Advisory Board is confident that the new publications will significantly add to the understanding of emerging topics wherever they may appear in the vast tapestry of knowledge about materials.

The appearance of Supplementary Volumes and the new series *Advances in Materials Science and Engineering* is an event that will be welcomed by scientists and engineers throughout the world. We are sure that it will add still more luster to a most important enterprise.

Walter S Owen
Chairman
Honorary Editorial Advisory Board

EXECUTIVE EDITOR'S PREFACE

The objectives of the Supplementary Volumes and their relation to the parallel series of Concise Encyclopedias were fully set out in the Preface to *Supplementary Volume 1*; these have not changed in the interim.

Supplementary Volume 2, now presented to the readership of the *Encyclopedia of Materials Science and Engineering,* contains, in addition to articles directly commissioned by the Executive Editor, a proportion of articles that were commissioned by Editors of the various Concise Encyclopedias (in consultation with the Executive Editor). Some of these articles derive from Concise Encyclopedias that have already been published, others from Concise Encyclopedias that have not yet appeared. This volume contains 130 articles in all, compared with 113 in *Supplementary Volume 1.* I am most grateful to the Editors concerned for their sterling work in putting together the Concise Encyclopedias.

The Systematic Outline, the List of Contributors and the Subject Index are all cumulated to include information in respect of both Supplementary Volumes, and this practice will be continued for *Supplementary Volume 3*.

The sources of advice used by the Executive Editor have been as before, except that on this occasion, no additional Special Advisers were appointed.

Again, I wish to express my gratitude to Dr Colin Drayton, Mr Michael Mabe and Mr Peter Frank of the editorial staff of Pergamon Press, and to their efficient commissioning and production staff, for their highly professional work (in the face of unusual practical obstacles) in bringing this volume into print.

Robert W Cahn
Executive Editor

GUIDE TO USE OF THE ENCYCLOPEDIA

Readers of this Supplementary Volume are referred to the "Guide to Use of the Encyclopedia" printed on p. xxv of Volume 1 of the *Encyclopedia of Materials Science and Engineering*. Guidance given here is in addition to that already given in the Main Encyclopedia.

A virtually identical titling policy has been adopted with minor adjustments to take into account the updating nature of this publication. Completely new articles have been given titles which accord with the policy of the Main Encyclopedia. Some articles have inevitably become out of date since the publication of the Main Encyclopedia, and two classes of article were commissioned to update them: those intended to be read in conjunction with the original article and those intended to completely replace it. In the case of the former, the title of the original article has been used with some qualifying phrase indicating that it is updated or expands the scope of the original. In the case of the latter, the title of the original article has been used unchanged. For both these categories of article, a statement appears in the first paragraph indicating whether the article supplements or replaces an existing article in the Main Encyclopedia.

The existing system of cross-referencing has been adopted and expanded. As this Supplementary Volume is intended to sit alongside the Main Encyclopedia as an extra volume, cross-references to articles within the Main Encyclopedia are given as before and without qualification. Cross-references to articles within the Supplementary Volumes are qualified by the addition of "Suppl. 1" or "Suppl. 2" to the title. The following forms thus appear:

> The effect of this phenomenon on the creep curve is to initiate a new primary regime at a dramatically higher creep rate (see *Dynamic Recovery and Recrystallization of Metals*).

> The unusual mechanical strength of the intermetallic compound Ni_3Al ($L1_2$ structure) is caused by the atomic long-range order (see *Aluminides for Structural Use*, Suppl. 1).

As before, a high degree of uniformity in terminology and notation has been imposed upon the text. In general, the International System of Units and the recommendations on *Quantities, Units and Symbols* of the Royal Society of London have been used. Where this has not been possible, notation has been standardized throughout a particular field. In all cases, symbols are defined at their first appearance in the text of an article.

Following the Main Encyclopedia, the name/date system of bibliographic citation has been used. The contributor's name appears at the end of each article. All contributors are listed along with their correspondence address and the titles of the articles that they submitted in the alphabetical List of Contributors.

The Subject Index follows the same principles as in the Main Encyclopedia and, for the convenience of the reader, it is cumulative for the Supplementary Volumes. It should be noted that the page numbering of *Supplementary Volume 2* follows on from that of *Supplementary Volume 1*, with the first article beginning on page 655. This numbering system will aid the reader to locate references in the Subject Index.

An addendum to the Systematic Outline of the Encyclopedia, which appears in Volume 8, is contained in this Supplementary Volume and, like the index, this contains information in respect of both Supplementary Volumes. This information will also be cumulated for *Supplementary Volume 3*, so that the reader has only to look in two places, namely the Main Encyclopedia index or Systematic Outline, and the index or Systematic Outline in the latest Supplementary Volume, to find the information required.

An alphabetical list of all the articles contained in this Supplementary Volume is also provided.

ALPHABETICAL LIST OF ARTICLES

Articles appearing in *Supplementary Volume 2* are listed in alphabetical order, as they appear in the volume. A title followed by (R) indicates an article that replaces one with the same or a similar title appearing in the Main Encyclopedia, while an (S) indicates an article that is to be read as a supplement to a similarly titled article in the Main Encyclopedia.

Adhesives for Wood: An Update

This article supplements the article *Adhesives for Wood* in the Main Encyclopedia.

The first evidence of adhesive-bonded wood products can be traced back over 3500 years, when the joining of wood was undertaken primarily for decorative purposes. The discovery of improved adhesives led to more sophisticated developments in wood composites and greater utilization of these products in structural applications. Until this century most of these adhesives were derived from animal, vegetable or mineral sources.

Projected limitations in the available harvest of wood resources and the overall decrease in wood quality emphasize the importance of adhesives and the science of adhesion in improving wood utilization by allowing smaller trees and present waste wood to be restructured into useful wood products. By proper choice of wood component, wood orientation, adhesive and manufacturing conditions, a variety of engineered composites are possible; these can have greater mechanical strength, decreased variability and improved stress-distributing properties compared to conventional solid wood.

This article outlines the mechanism of wood bonding and pertinent physical and chemical factors involved in achieving good bonds. Both wood-related and adhesive-related characteristics influencing the formation and performance of bonds in products such as laminated lumber, plywood, flakeboard and particleboard are considered.

1. Nature of Adhesion

Bond formation depends upon the development of physical and chemical interactions both within the bulk adhesive polymer and at the interface between adhesive and wood. Interactions within the adhesive accumulate to give cohesive strength while the forces between adhesive and wood provide adhesive strength. Both should exceed the strength of the wood, allowing substantial wood failure during destructive testing of high-quality bonds.

Most wood adhesives are liquids consisting of either 100% polymer material or more frequently an aqueous solution, emulsion or dispersion, wherein one half to one third of the mixture is polymer. Other solvents such as volatile alcohols, esters or aromatics may sometimes be used to replace all or part of the water. For more specialized bonding requirements, adhesives in a solid state (i.e., powders, hot melts) may be utilized with heat applied to transform the polymer into a flowable material.

The adhesive must accomplish three distinct stages during bonding processes: it must wet the wood substrate, flow in a controllable manner during pressing and finally set to a solid phase. Failure of the adhesive to accomplish any of these stages often results in reduced bond quality. The extent of wetting depends on the physical–chemical nature of both adhesive and the wood surface. Solvent molecules and low-molecular-weight polymer molecules, being smaller and more mobile, tend to wet and penetrate into the wood substrate most rapidly. Adsorption and diffusion of these liquids into the cell wall often leads to swelling of the lignocellulose wood substance. The porous structure of wood and defect areas created by machining processes facilitate the transport of these materials into the wood. In contrast, higher-molecular-weight polymer material would wet and penetrate the wood substrate at a significantly slower rate, remaining largely as a concentrated polymer on the wood surface. Ultimately, the bonding area should consist of a steadily decreasing concentration of adhesive polymer as one proceeds from the surface to a few micrometers within the wood. Achieving this condition will depend on polymer physical properties, thermodynamic considerations and wood-surface properties. With solid adhesives, an intermediate heating stage which melts or softens the adhesive is necessary to accomplish wetting of the wood surface.

Optimum bond formation requires intimate contact between adhesive and wood substrates to ensure molecular interaction over a large area. This is accomplished using pressure and heat which causes viscous adhesive components to transfer and flow throughout the bonding sites while deforming the wood to achieve better contact between wood surfaces of different texture. Adhesive penetration into the wood structure must not be so great that inadequate amounts of adhesive remain in the glueline (starved glueline) or too little so that fibers and microdefects just below the wood surface are insufficiently bonded together.

Solidification and cure of the wood–adhesive system occurs either by solvent loss, polymerization and cross-linking reactions, or both. This process is time, temperature and pressure dependent with bond quality being a reflection of how well the final solidification stage is accomplished in the particular adhesive and wood substrate present.

2. Mechanism of Adhesion

A number of theories have been proposed to explain bonding mechanisms in materials science. The unique heterogeneous characteristics of wood limit the relevance of any one particular theory, leading instead to wood bonding being rationalized in terms of a combination of mechanisms involving mechanical, physical and chemical parameters.

The ability of mechanical interlocking to occur readily is a function of the porous nature of most wood surfaces and the ease with which the adhesive can flow into these microcavities under the action of heat and external pressure. The mechanical anchoring of adhesive into wood occurs when the adhesive polymer hardens. This bonding mechanism is of obvious importance in hot melts and does increase the contact area available for bond formation. It cannot, however, account for the documented phenomena of bond-strength reduction with aging of wood surfaces and the limited correlation between adhesive penetration depth and bond strength development.

Adsorption and diffusion mechanisms allow for intimate molecular contact between adhesive and wood substance over a large area. The accumulative physical interaction arising from attractive Van der Waals, hydrogen bonding and dispersion forces between adhesive polymer and wood components can result in strong bond formation which many consider a primary bonding mechanism. Complete wetting of the adhered surface enhances bonding potential. Both adhesive surface tension and wood surface tension have some bearing on wetting effectiveness. Adhesive spreading methods are additional factors of this mechanism since the mechanical energy input during glue application may overcome some surface force resistance to wetting.

Both adhesive and wood reactive functional groups allow for the high probability of covalent bond formation. In cross-linkable systems these chemical interactions can produce a strongly bonded wood–adhesive matrix capable of maintaining long-term stability under a variety of heat, moisture and stress conditions. While indications of covalent bonding have been found in model systems, conclusive evidence of their presence in wood-bonded products has not yet been made.

3. Wood-Related Factors Affecting Adhesion

Wood, being a complex, three-dimensional tissue, has a distinct but variable structural and chemical makeup which must be allowed for in adhesive formulation. The heterogeneous character of wood is reflected in both its bulk and surface properties.

3.1 Bulk Features

Anatomical characteristics, permeability, density and moisture content are among the more important interrelated factors influencing wood–bond formation.

Fiber orientation influences the ease with which the adhesive polymer can wet wood while at the same time retaining sufficient polymer on the bonding surface to form a glueline interface. Rapid adhesive penetration is possible along the grain because of open lumen structures, while penetration across the grain is highly restricted by fiber walls. Anatomical differences between species will affect the permeability of wood to liquids and gases, thus influencing the degree to which adhesive can successfully intermix with wood fibers. In softwoods, fiber lumina are laterally interconnected by small capillaries which allow the ready passage of air and liquid to various depths below the wood surface. In hardwoods, many of these interconnecting routes are blocked or absent, resulting in more restricted flow of the adhesive into bonding sites. Further variations such as knots or grain distortions associated with tree growth patterns can provide additional impediments to achieving bonding potential. Other structural features of importance are variations imposed by the presence of juvenile and mature wood and earlywood and latewood effects within one piece of wood. These features are interrelated to wood density.

The density of wood relates directly to its strength and has primarily a physical influence on adhesion. The stronger the wood, the stronger the adhesive must be in order to utilize the full strength of wood. Density variations are associated with the growth ring structure of wood, particularly in the ratio of fiber-wall to fiber-lumen volume. In dense wood adhesive penetration is restricted, while in lower density woods care must be taken to prevent over-penetration. Where large differences in density can occur (i.e., earlywood–latewood zones) the adhesive must be designed to adapt adequately to both of these conditions. In general wood density in the range 250–450 kg m^{-3} (i.e., spruce, pine, fir, aspen) are easier to bond than woods of density greater than 600 kg m^{-3} (i.e., birch, oak, maple).

Wood physical properties are highly dependent on the interaction of wood density and moisture content which in turn affect gluing behavior. High-moisture-content conditions often result in increased adhesive flow and penetration while retarding polymer cure by limiting solvent loss into the wood or restricting condensation polymerization reactions. Excessive water at the bonding interface can also act as a barrier to bond formation, especially with hot-melt systems. Typical moisture content ranges for bonding are 2–16% although successful bond formation has been reported at up to 40% moisture content, with some adhesives curing under ambient or slightly elevated temperature conditions. The likelihood of steam formation at higher moisture contents, with its destructive possibilities in the glueline, necessitates wood to be dried to 2–6% moisture content when using high-temperature-curing adhesives. Wood volumetric changes occurring between fiber saturation and oven-dry conditions result in directional shrinkage and swelling forces, directly proportional to density, which can place considerable stress on the glueline. Cured gluelines should partially maintain bond integrity with dimensional-change-induced stresses, even during continual cycling between wet and dry conditions, because they may weaken the wood–adhesive interface considerably in high-density woods. Moisture and density also affect wood deformational properties requiring some adjustments in bonding pressure (i.e., lower density or higher moisture content requires less bonding pressure).

3.2 Surface Features

Both physical and chemical properties of the wood surface can profoundly affect bond-formation potential. Their primary influence impacts on the accessibility of adhesives to wood-fiber bonding sites. From a microscopic perspective, wood surfaces have a rough physical appearance due to anatomy, in particular grain orientation and fiber size. Surfaces produced from transverse cuts (i.e., cut across the tracheids) expose end-grain areas whose large lumen volume often results in substantial overpenetration of adhesive. Consequently, direct bonding of end-grain surfaces is difficult and not commonly attempted. Cutting along tangential and radial surfaces inevitably excises some fiber walls, exposing different amounts and parts of fiber lumina. Surface topography may vary appreciably between species since lumen diameters may range from as small as 1 μm in fine-grained woods to as large as 200 μm in coarse-grained woods. Presence of resin ducts, vessel elements or knots gives added variations to the wood surface. Machine operations during surface formation often result in further irregularities which can be associated with anatomical features. These can include saw-tooth abrasions, indentations from planer or shaper knives, lathe checks and roughness resulting from the peeling of veneer, torn and broken fibers caused during flake cutting and other splits or checks. Although these features may increase available bonding area, they hinder good bond formation by weakening the wood surface. Heat and pressure can be utilized to compress and plasticize the interface to improve contact in lower-density species, but high-density woods must rely on smooth machining operations to achieve the intimate contact necessary for bonding. Mechanically weakened wood surfaces can be repaired during gluing by the action of adhesive penetration between and into the wood fiber.

Wood surfaces are chemically heterogeneous. Their composition does not necessarily correspond directly to the chemical makeup of bulk wood but rather is more a function of the conditions and methods of surface formation. Since wood fibers are composed of strands of oriented cellulose molecules intertwined within hemicellulose and lignin–polymer matrices, cleavage of a fiber wall may expose varying amounts of each chemical component at the wood surface layer. The carbohydrate components are highly polar, with fractions existing in both crystalline and amorphous forms and containing numerous functional groups available for reaction. Lignin, an amorphous polymer, offers additional reactive sites on its phenolic components. Both carbohydrate and lignin fractions are present in a range of molecular sizes and polarities which allow for bond formation with polar or nonpolar adhesives.

Extractives, although normally a minor component of the wood, can play a significant role in bond formation. These relatively low-molecular-weight materials are often highly mobile, nonpolar substances of limited reactivity which inhibit bond formation by blocking adhesive accessibility to reactive wood sites. This is accomplished by restricting wetting, altering the wood surface pH or forming complexes with adhesive or wood bonding sites.

Wood-surface history is an additional chemical feature affecting bonding potential. A smooth, freshly cut wood surface dried at moderate temperatures offers the optimum conditions for bonding. As aging occurs, the surface is chemically modified at a rate depending on temperature and atmospheric conditions. Oxidation or contamination of the wood surface by nonpolar, mobile extractives and airborne pollutants are the primary phenomena occurring during this time. Oxidation reduces the number of active hydroxyl groups available for bonding. Under high-temperature drying conditions the chance of wood-surface oxidative degradation becomes highly probable. Sometimes chemical and thermal treatments can be utilized to change wood-surface chemistry and enhance bonding potential. Strong acids and oxidizing agents are common activators. They create new functional groups through depolymerization and ring cleavage reactions of lignin and carbohydrate molecules, or generate free radicals which can lead to bond formation by oxidative coupling. Resulting gluelines tend, however, to be brittle and have highly variable strength because wood can be degraded by many of the chemicals used.

4. Wood Adhesives

Adhesives can be classified as either natural or synthetic. Typical natural adhesives are carbohydrate or protein based, derived from either animal or vegetable sources, or may be an inorganic material. Synthetic adhesives are produced by the controlled polymerization of various monomeric organic molecules. These synthetics can be further divided into thermosetting (most common for wood adhesives) and thermoplastic categories. They can function to produce structural or nonstructural bonds. The present discussion deals only with major adhesives used in the wood industry. The majority of these adhesives are formed by condensation polymerization processes involving the use of formaldehyde as a cross-linking agent. Details of their chemistry have been presented elsewhere (Skeist 1977).

Phenol–formaldehyde (PF) resin is the primary wood adhesive used in the production of strong, durable exterior products. These adhesives are low-to-moderate molecular weight thermosetting polymers, manufactured in aqueous solutions with acid or alkaline catalysts. By varying reaction time and temperature, catalyst type and amounts and ratio of reactants, a variety of different adhesives can be prepared. Resoles are a class of PF resins produced under alkaline conditions with a molar excess of formaldehyde to phenol. These reactions produce highly branched polymeric structures while remaining soluble in strongly alkaline solutions. Sufficient formaldehyde is present to allow a highly cross-linked network to be formed during cure. Novolacs are PF resins synthesized under mild acid or neutral conditions with a molar excess of phenol to formaldehyde. These polymers have a linear structure, relatively low molecular weight and maintain thermoplastic behavior unless additional formaldehyde is added when curing is undertaken. These are referred to as two-stage resin systems. Plywood, oriented strandboard, particleboard and fiberboard intended for exterior service are manufactured using aqueous resoles tailored to give particular application, flow and cure properties for a specific product. A typical aqueous resole contains a phenol to formaldehyde to sodium hydroxide molar ratio of 1:(1.8–2.2):(0.1–0.8), with a resin-solids range from 35–50% and averaging 5–40 phenolic units linked together. These alkaline resoles require cure temperatures greater than 100 °C to attain adequate cross-linking. Fast cure at ambient temperatures can be achieved by addition of strong acids. Equipment corrosion and long-term acid degradation of the wood interface are, however, deterrents to this procedure.

For waferboard applications, the need for good adhesive distribution on large wafers and the limitations on moisture level in the glueline during high-temperature pressing require the use of powder PF resins. These are produced by spray drying or other solidification techniques.

Significant increases in phenolic resin reactivity are possible by copolymerizing with resorcinol to produce a phenol–resorcinol–formaldehyde (PRF) system. These are often used in specialized applications (i.e., glued laminated timbers, fingerjointed structures) where durable bonds and ambient cure conditions are needed. To ensure adequate pot life, the PRF is synthesized as a novolac system. Additional formaldehyde needed for cure is mixed in just prior to adhesive application. The high cost of resorcinol, relative to phenol, restricts most PRF applications to higher value and large structural products.

Aqueous urea–formaldehyde (UF) polymers are low-cost, light-color, fast-curing adhesives with poor weather resistance. Their use is limited to products such as decorative plywood panels, particleboard subfloors or fiberboard for furniture or interior applications. Resins are prepared in molar ratios of formaldehyde to urea of (1–2):1 with the higher ratios usually utilized in solid wood or veneer bonding where higher strengths are required. Cure takes place at moderate acid conditions either by addition of acid salts or by use of the inherent acidity of wood species such as oak or cedar. Heat accelerates cure speed; polymer cross-linking, however, can be achieved at ambient temperatures. Although a thermosetting polymer, UF is sensitive to heat and moisture with formaldehyde slowly released as decomposition occurs. Requirements exist restricting formaldehyde emissions in wood products to low concentrations in air. To help control these emissions most particleboard is manufactured with formaldehyde to urea molar ratios less than 1.2:1. Alternatively, fortification of UF to improve its durability and strength is possible by copolymerizing with melamine. This modification adds increased chemical costs while requiring higher cure temperatures.

Polyvinyl acetates (PVAs) are thermoplastic, aqueous emulsions used primarily for furniture assembly and other nonstructural applications. Bond strength develops from water loss into the wood from the adhesive, with a resultant increase in polymer concentration in the glueline. While providing excellent dry adhesion strength and good gap-filling properties, the tendency of PVAs to cold-flow or creep when subjected to sustained loads is a primary disadvantage. Modification of PVA with thermosetting phenolic resin results in cross-linkable PVAs with improved heat and moisture resistance but only moderate alterations in creep behavior.

Other adhesive systems such as epoxy, polyurethanes and polymeric isocyanates have limited applications to wood bonding at the present time. While these adhesives provide improved bond strength with no formaldehyde emissions, they have disadvantages of greater cost and incompatibility with water, which requires the use of organic solvents for applicator cleanup. Application of polymeric isocyanate in particleboard, waferboard and oriented strandboard have recently gained more favor due to their faster cure times and greater tolerance to moisture compared to conventional thermosets.

Demand for durable, fireproof wood panels also has resulted in the application of inorganic cement adhesives in particleboard, although cure sensitivity to certain wood species and substantial increase in panel density limits their usage.

5. Adhesive Factors Influencing Bond Formation

Adhesive physical and chemical properties play as important a role as wood properties in determining bond performance. Strength, durability, working properties and cost are primary considerations governing adhesive selection. Interrelated polymer parameters which influence the overall performance of an adhesive include reactive functionality, molecular weight distribution, solubility, rheology and cure rate.

Number and reactivity of adhesive functional groups determine how many active sites are available for further polymerization. These affect directly the reaction speed and degree of cure possible during bond formation. The functionality of the adhesive components must be greater than 2.0 to achieve a cross-linkable system.

Molecular weight and molecular weight distribution are polymer properties affecting viscosity, flow and cure speed of the adhesive. Provided sufficient functional-group reactivity remains, higher-molecular-weight components can more rapidly attain cure but exhibit significantly greater solution viscosities and poorer flow properties than low-molecular-weight components. Optimum adhesive molecular weight distribution, however, depends upon the wood composite. Solid wood products such as glued laminated lumber and plywood utilize a major portion of higher-molecular-weight components (i.e., $\bar{M}_N$ of 1500–3500) to avoid overpenetration of polymer into wood. Adhesives for waferboard and particleboard have their molecular weight distribution shifted to lower values (i.e., $\bar{M}_N$ of 500–1200). This allows efficient and homogeneous resin application while improving adhesive mobility to facilitate better resin transfer between wood particles during blending operations.

Polymer solubility or dispersibility in solution is an important factor affecting uniform application and cleanup of the adhesive. Improved solubility allows more concentrated adhesive solutions to be used which reduces the amount of solvent to be driven off during cure. Adhesive rheology and cure development, while related to polymer molecular weight and reactive groups present, are highly responsive to heat, pressure and wood moisture content. Various segments of the polymer molecule achieve mobility as plasticization occurs with moisture, or temperature increases to the glass transition (T_g) or softening temperature. Cure results in an increase of T_g. Often compromises are needed to ensure that heat and pressure are applied in a controlled sequence such that the adhesive adequately wets the wood surface and cures before resin overpenetration occurs.

6. Bond Requirements for Composites

Performance criteria for wood composites relate directly to product end use. This requires consideration of engineering strength needs, safety and short- and long-term response of the material to the service environment. Structural, exterior-grade products have the most demanding bond-quality requirements, since glueline failure could be catastrophic to these structures. For these situations glueline strength, durability and reliability must be assured, usually by extensive bond-quality testing programs. While conventional bond strength and short-term durability evaluation methodology is well established, there is at present only a limited understanding of the effect of long-term aging on bond integrity. Insight on how product dimensional change and sustained loading affect time-dependent performance properties of wood composites is developed by cyclic testing through simulated environments using variables of heat, moisture and loading stress. The nonlinear deterioration behavior of many adhesive bonds, together with our lack of knowledge of the aging process, are serious obstacles to assessing long-term performance of new wood adhesives.

For conditions where bond performance is less crucial, cost may be the overriding factor—not only in terms of adhesive price but also of the impact adhesive costs have on overall production expenses. With more automated wood-composite manufacturing systems, compatibility of the adhesive with the assembly system often is the major consideration.

7. Future Direction

Continuing reductions in available harvest volumes and lower wood quality will lead to increased reliance on wood composites and the adhesives used to manufacture them. For commodity products, the importance of manufacturing efficiency will place greater emphasis on adhesives capable of tolerating variable wood properties and moisture contents while having rapid cure speed and reasonable cost. Adhesives will gain prominence in higher-valued wood composites by allowing products of varying shapes, improved engineering strength and greater aesthetic value to be more easily manufactured. Expanded use of continuous pressing techniques and heating methods such as steam pressing or microwave use will create added challenges for adhesive formulation development and wood-handling concepts. Improved knowledge of chemical factors influencing the reactivity of the wood-bonding surface would help enhance functional group reactivity, perhaps leading to reductions in amounts of adhesive required or possibly improved binderless wood composites. Advanced, high-strength composites are likely increasingly to utilize wood fiber or cellulose fiber as a directional, reinforcing matrix surrounded by adhesive polymers which may comprise up to 50% of the composite weight.

Future adhesive improvements will be concentrated in areas of durable, faster curing systems, capable of enduring moderate to long assembly times. Copolymer systems will increase in prominence with emphasis being directed toward multicomponent, cold-setting adhesives possibly applicable to on-site assembly. Isocyanates, because of their extensive reactivity and ability to bond at higher moisture contents are likely to be a component of these newer wood-bonding systems. Reactive hot melts, capable of rapidly curing to thermosetting polymers, will also be a development having a significant impact on future bonding systems.

See also: Glued Joints in Wood

Bibliography

Collett B M 1972 A review of surface and interfacial adhesion in wood science and related fields. *Wood Sci. Technol.* 6: 1–42

Koch G S, Klareich F, Exstrum B 1987 *Adhesives for the Composite Wood Panel Industry*. Noyes Data Corporation, NJ

Kollmann F F 1975 *Principles of Wood Science and Technology,* Vol.II: *Wood Based Materials*. Springer, New York

Marian J E 1967 Wood, reconstituted wood and glued laminated structures. In: Houwink R, Salomon G (eds.) 1967 *Adhesion and Adhesives*, Vol. II: *Applications*. Elsevier, New York

Oliver J F 1981 *Adhesion in Cellulosic and Wood-based Composites*. Plenum, New York

Pizzi A 1983 *Wood Adhesives Chemistry and Technology*. Dekker, New York

Skeist I 1977 *Handbook of Adhesives* 2nd edn. Van Nostrand Reinhold, New York

Subramanian R V 1984 Chemistry of adhesion. In: Rowell R (ed.) 1984 *The Chemistry of Solid Wood*, Advances in Chemistry Series 207. American Chemical Society, Washington, DC

Wellons J D 1983 The adherends and their preparation for bonding. In: Blomquist R F, Christiansen A W, Gillespie R H, Myers G E (eds.) 1981 *Adhesive Bonding of Wood and Other Structural Materials,* Vol. 3. Clark C Heritage Series on Wood, Pennsylvania State University, University Park, PA

P. R. Steiner
[University of British Columbia, Vancouver, British Columbia, Canada]

Aluminum-Based Glassy Alloys

Since a glassy alloy in the Au–Si system was produced for the first time by liquid quenching in 1960 (Klement et al. 1960), a great number of glassy alloys have been developed successively in noble metal, transition metal, refractory metal and rare earth metal-based systems. Amorphization has been achieved in alloy systems of engineering importance containing iron, cobalt, nickel, copper or titanium as a main component and some of these glasses have seen practical use through the utilization of their good mechanical, physical and chemical properties. In addition to the above-described alloy systems, amorphization of aluminum-based alloys has been actively pursued because of the expectation of obtaining high strength materials with low density. However, the first successful data on the formation of aluminum-based glassy alloys exhibiting good ductility and high strength were presented only very recently (Inoue et al. 1987). This article briefly reviews the recent progress of high-strength aluminum-based glassy alloys, based mainly on work by the present authors, although there has also been some recent work in France and the USA (Dubois and Le Caer 1985; Dubois et al. 1985, 1987; He et al. 1988). A brief historical account has been published by Cahn (1989).

1. History of Aluminum-Based Glassy Alloys

The formation of aluminum-based glassy alloys by liquid quenching was first tried in binary systems of aluminum–metalloid and aluminum–transition metal (TM) alloys. As a result, it was found in Al–Si (Predecki et al. 1965), Al–Ge (Ramachandrarao et al. 1972) and Al–TM (TM = Cu) (Davies and Hall 1972), Ni (Chattopadyhay et al. 1976), Cr (Furrer and Warlimont 1977) or Pd (Sastry et al. 1978) alloys that glassy and crystalline phases coexist but only near the holes in thin foils prepared by the gun quenching technique in which the cooling rate is higher than that for the melt spinning method (see *Rapid Quenching from the Melt*). However, no glassy phase without an accompanying crystalline phase has been prepared by melt spinning or by the gun- or piston-anvil methods. The first formation of a glassy single phase in aluminum-based alloys containing more than 50 at.% aluminum was observed in 1981 in Al–Fe–B and Al–Co–B ternary alloys (Inoue et al. 1981). However, these glassy alloys are extremely brittle and hence have not attracted much attention. Subsequently, a glassy phase was found in melt-spun Al–Fe–Si, Al–Fe–Ge and Al–Mn–Si alloys, but they were also very brittle like the Al–(Fe or Co)–B glasses. It was consequently believed that the brittleness might be an inherent property for aluminum-based glassy alloys. In 1987, a glassy phase with good bending ductility was discovered to be formed at compositions above about

80 at.% aluminum in the Al–Ni–Si system (Inoue et al. 1987). Since this discovery, ductile aluminum-based glassy alloys have successively been found in a number of ternary alloys consisting of aluminum on early transition metal (ETM) and a late transition metal (LTM) (Tsai et al. 1988) exemplified by Al–Zr–Cu, Al–Zr–Ni and Al–Nb–Ni, followed by aluminum–lanthanide metal (Ln)–LTM ternary alloys (Inoue et al. 1988a) in which the ETM is substituted by a lanthanide metal and finally in Al–Ln binary alloys without TM elements (Inoue et al. 1988h).

2. Alloy System and Compositional Range

Aluminum-based glassy alloys are divided into metal–metalloid and metal–metal types. The former type consists of multi-component alloys containing three elements such Al–B–TM (Inoue et al. 1981), Al–Si–TM and Al–Ge–TM (Inoue et al. 1988a) (TM = V, Nb, Cr, Mo, Mn, Fe, Co, Ni or Cu); no glassy phase is formed in aluminum–metalloid binary alloys. These glasses form in the range of 12–43 at.% silicon and 5–23 at.% TM for the Al–Si–TM system and 12–52 at.% germanium and 5–23 at.% TM for the Al–Ge–TM system. On the other hand, the metal–metal glasses are formed in three different types of systems: Al–Ln (Inoue et al. 1988c, 1988d) binary, Al–Ln–TM (Inoue et al. 1988g, 1988h) and Al–ETM–LTM (Tsai et al. 1988) ternary alloys. The binary glassy alloys are obtained in Al–Ln (Ln = Y, La, Ce, Pr, Nd, Sm, Eu, Gd, Tb, Dy, Ho, Er, Tm or Yb) systems. The glass-forming ranges in melt-spun Al–Ln alloys are greatly extended by the addition of TM (Fe, Co, Ni, Cu, etc.) for Al–Y–TM, Al–La–TM and Al–Ce–TM ternary alloys, and nickel is the most effective element for the extension, as shown in Fig. 1. The glassy phase is also formed in the ternary alloys containing the transition metals vanadium, niobium, tantalum, chromium, molybdenum or manganese. Glass formation in Al–Ln and Al–Ln–TM alloys has been presumed to result from the simultaneous satisfaction of the two factors of a strong attractive interaction of the constituent elements and a large disparity of atom sizes (atomic size ratio near 0.8). In addition, considering Al–ETM–LTM ternary glasses, we note that most corresponding binary alloys consisting of ETM and LTM without aluminum can be amorphized by melt spinning, suggesting that the interaction between ETM and LTM as well as between Al and ETM or LTM also plays an important role in the glass formation of the aluminum-based ternary alloys.

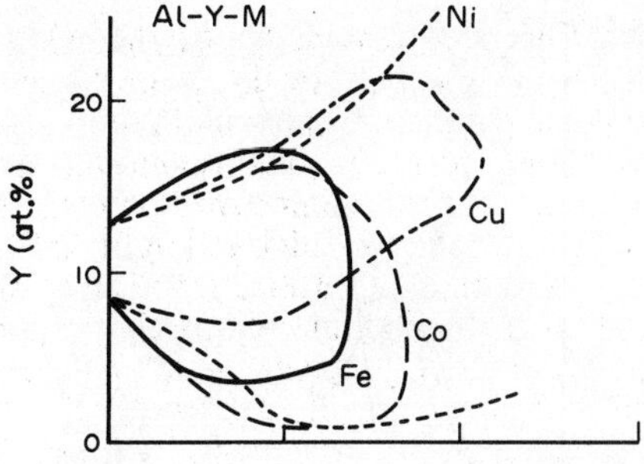

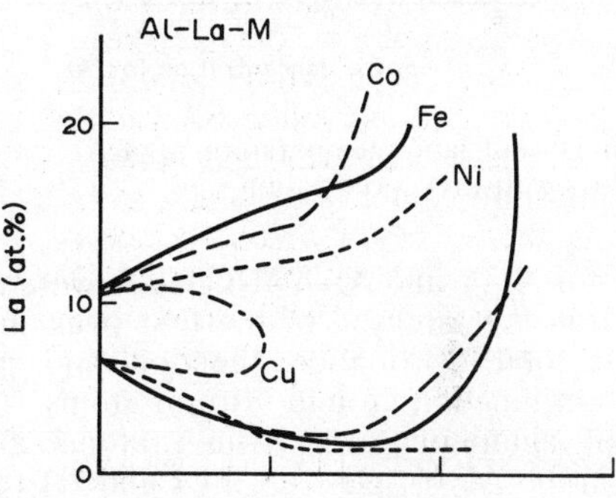

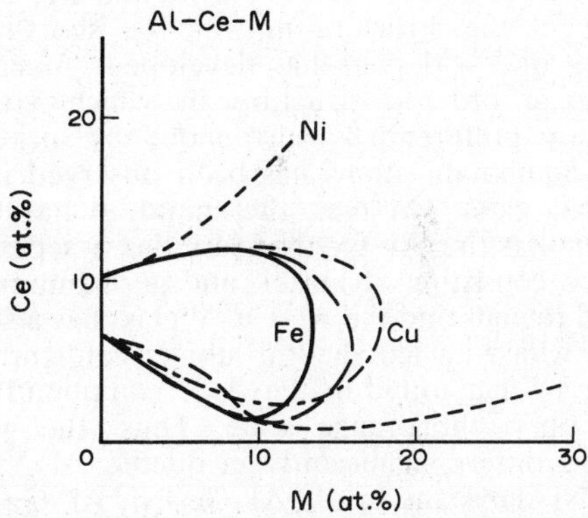

Figure 1
Compositional range for formation of glassy phase in Al–Y–M, Al–La–M and Al–Ce–M (M = Fe, Co, Ni or Cu) systems

3. Glassy Structure

The structure of aluminum-based glassy alloys in the Al–Y, Al–Y–Ni, Al–Ge–(Mn or Ni) and Al–Cu–V systems has been examined by the anomalous x-ray scattering method (Matsubara et al. 1988a, 1988b, 1989). These alloys can be divided into two groups of alloys—one, $Al_{90}Y_{10}$ and $Al_{87}Y_8Ni_5$ exhibiting high strength and good bending ductility and the other $Al_{50}Ge_{40}Ni_{10}$ and $Al_{75}Cu_{15}V_{10}$ with extreme brittleness. Clarification of the structure of these alloys was expected to shed some light on the origin of the ductile and brittle natures. The coordination number *N* and first-neighbor atomic

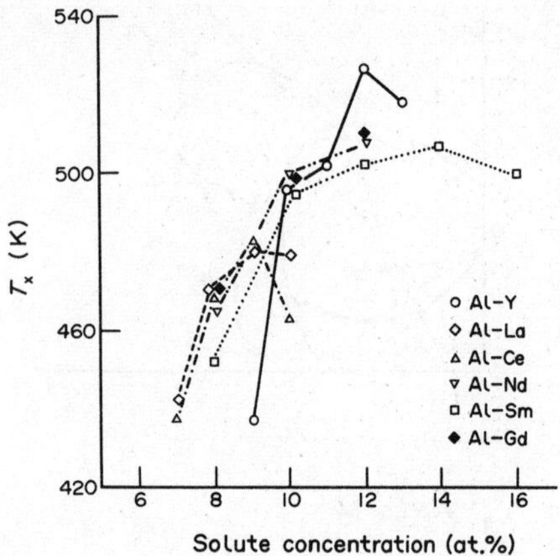

Figure 2
Change in crystallization temperature T_x of Al–Ln glassy alloys with Ln concentration

distance for Al–Y and Al–Y–Ni glasses were determined from the intensity of the first peak in conventional total distribution function and partial distribution function around yttrium atoms. The N values of aluminum and yttrium atoms around yttrium in the $Al_{90}Y_{10}$ glass are 14.1 and 1.1 respectively, indicating that the yttrium atoms are surrounded mainly by aluminum atoms. The alloy does not have a random distribution of yttrium atoms, but consists of the structure in which a short-range ordering of Y–Al pairs has developed. A similar short-range ordered structure in which yttrium atoms are preferentially surrounded by nickel as well as aluminum atoms has been observed in an $Al_{87}Y_8Ni_5$ glass. On the other hand, it has been clarified that the Al–Ge–Ni glass has a separated structure consisting of nickel- and germanium-rich ordered regions and the Al–Cu–V glass has a structure in which an icosahedral atomic configuration (similar to that found in the Al_7V compound) develops on a short-range scale. Thus, the glassy structure differs significantly in ductile Al–Y and Al–Y–Ni alloys and brittle Al–Ge–Ni, Al–Ge–Mn and Al–Cu–V alloys. The appearance of high strength combined with ductility for the former glasses is probably due to the structure in which the short-range ordering of Y–Al and Y–Al–Ni pairs developed homogeneously.

4. *Crystallization Behavior*

It has been mentioned in Sects. 2 and 3 that the large attractive interaction between aluminum and solute atoms is a dominating factor for the formation of the glassy phase and has a significant effect on the atomic configuration in the glassy structure. It is therefore presumed that the glass transition temperature, T_g, and crystallization temperature, T_x, also increase with an increase of the number and force of the bondings with attractive interaction. Figure 2 shows that T_x of glassy Al–Ln binary alloys increases significantly in the low solute concentration range and then becomes almost constant (Inoue et al. 1988c, 1988d). The change of T_x with lanthanide content is closely related to the crystallization behavior. In the lower solute concentration range where the crystallization occurs through the sequence

$$\text{glass} \rightarrow \text{Al} + \text{glass} \rightarrow \text{Al} + \text{Al}_{11}\text{Ln}_3$$

the precipitation of the primary aluminum phase is suppressed with increasing Ln content, resulting in the increase of T_x. In the higher solute concentration range where a compound phase appears directly from the glassy phase, T_x becomes almost constant. It is also seen in Fig. 2 that T_x has nearly the same values for the Al–Ln binary alloys with the same lanthanide concentrations.

The T_x values of Al–10 at.% Ln alloys are increased from 460–500 K to about 800 K by the addition of TM as shown in Fig. 3 (Inoue et al. 1988a, 1988b). The effect of TM on T_x seems to reflect the difference of the bonding force between aluminum and TM atoms. A similar change of T_x with TM content has also been observed for Al–ETM–LTM ternary alloys; that is, for corresponding concentrations, T_x increases from 531 K to 825 K in the order Cu<Ni<Co<Fe for LTM belonging to the same period and Ti<Zr<Hf for ETM of the same periodic group. Since the melting temperature of Al_3M binary compounds increases in the same order, the bonding force of Al–TM and ETM–LTM pairs seems to be a dominant factor for the change of T_x with TM.

5. *Glass Transition Behavior*

A distinct glass transition has been found to appear in the temperature range just below T_x for Al–Ln–Ni and Al–Ln–Co alloys (Inoue et al. 1988e). The alloy composition where the glass transition can be observed is limited to about 10 at.% yttrium and 3–10 at.% nickel for the Al–Y–Ni system and about 6 at.% lanthanum or cerium and 5–12 at.% nickel for Al–La–Ni and Al–Ce–Ni systems. No distinct glass transition has been observed in other aluminum-based alloys. The T_g increases almost linearly from 490 K to 582 K with increasing nickel content (with similar changes for yttrium and cerium) as shown in Fig. 4, and the compositional dependence is similar to that for T_x. Thus, the alloy composition for the appearance of a glass transition is limited to a part region in their glass formation ranges and the limitation appears to be closely related to the crystallization behavior: the glass transition is observed only in the vicinity of the

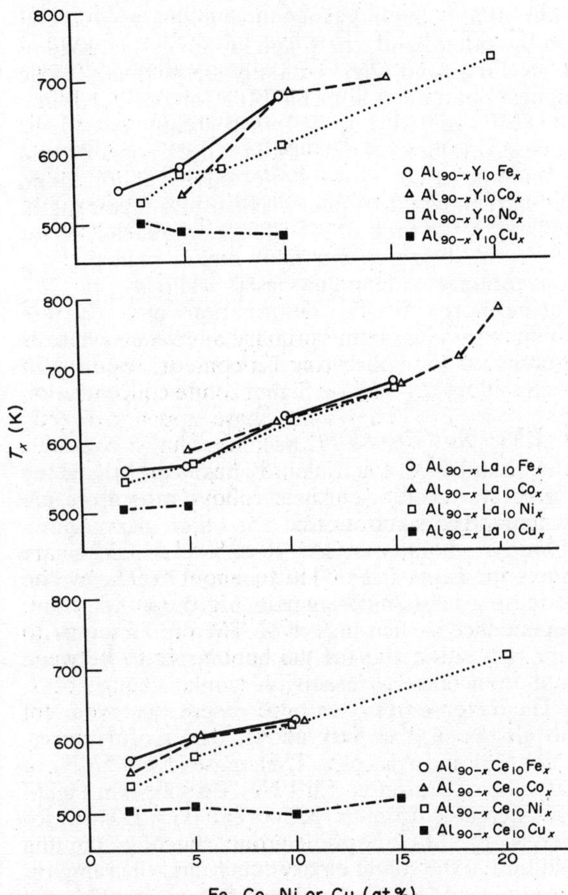

Figure 3
Change in T_x of $Al_{90-x}Y_{10}M_x$, $Al_{90-x}La_{10}M_x$ and $Al_{90-x}Ce_{10}M_x$ glassy alloys with transition metal M concentration

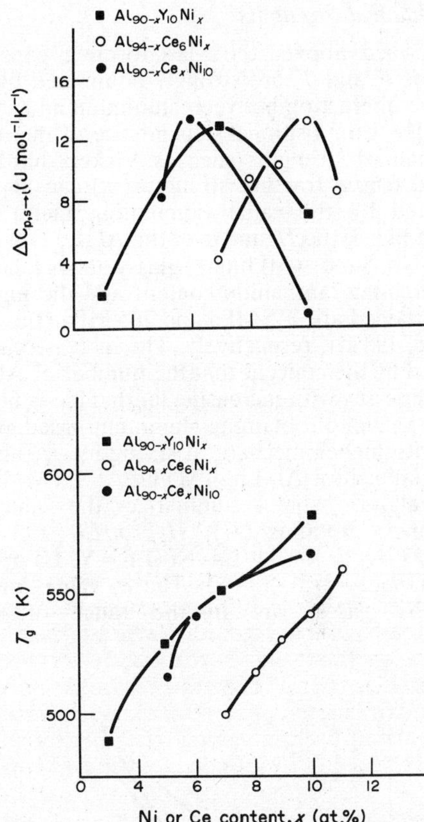

Figure 4
Glass transition temperature T_g and the difference of the specific heat between amorphous solid and supercooled liquid $\Delta C_{p,s\to l}$ as a function of nickel or cerium concentration for Al–Y–Ni and Al–Ce–Ni glassy alloys

compositions at which the crystallization process changes from

$$\text{glass} \rightarrow \text{Al} + \text{glass} \rightarrow \text{Al} + \text{compound}$$

to

$$\text{glass} \rightarrow \text{Al} + \text{compound and/or glass} \rightarrow \text{compound}$$

The close relation between the appearance of T_g and the crystallization process allows us to derive the inference that T_g appears only in the compositional range where almost all the aluminum atoms combine with the solute atoms through the formation of Al–Ln, Al–Ni and Al–Co pairs. The largest temperature span between T_g and T_x is 30 K and the reduced glass transition temperature ($T_{rg} = T_g/T_m$) is 0.58–0.59 for the Al–Ln–Ni (Ln = Y, La or Ce) alloys.

The difference in specific heat between the glassy solid and supercooled liquid, $\Delta C_{p,s\to l}$, has been measured to be 9–10 J mol^{-1} K^{-1} for the Al–Ln–Ni glasses and the largest $\Delta C_{p,s\to l}$ value is obtained for the alloys near the centers of the compositional ranges where the alloys with glass transition are obtained (Inoue et al. 1988f). The viscosities η as a function of temperature have been evaluated from thermal dilatometric curves under different applied stresses (Inoue et al. 1988e). The η values of $Al_{85}Y_{10}Ni_5$ and $Al_{84}Ce_6Ni_{10}$ glassy alloys decrease significantly from 2×10^{14} N s m^{-2} at 488 K to 3×10^{12} N s m^{-2} at 521 K, and no distinct difference in the temperature dependence, η (T) is seen for the two alloys. The η value at 521 K is nearly equal to 10^{12} N s m^{-2} which has been estimated to be the viscosity for a supercooled liquid near T_g, indicating that the glassy solid heated at 521 K changes to a near-equilibrium supercooled liquid state.

6. *Mechanical Strengths*

As described above, the glass-forming capacity, structure, T_x and T_g are strongly dominated by the attractive interaction between aluminum and solute atoms. The compositional dependence of measures of mechanical strengths such as Vickers hardness (H_v) and tensile fracture strength (σ_f) can also be interpreted by the same dominating factor. As shown in Fig. 5, the H_v and σ_f of the Al–Ln (Ln = Y, La, Ce, Nd, Sm or Gd) binary glassy alloys increase with increasing lanthanide content and the highest values attained are 245 MPa and 765 MPa (Inoue et al. 1988c, 1988d), respectively. The increase can be explained by the concept that the number of Al–Ln bonds increases with increasing lanthanide content.

With the aim of obtaining aluminum-based glassy alloys with higher mechanical strengths, H_v and σ_f were examined for Al–Ln–TM and Al–ETM–LTM ternary alloys. Table 1 summarizes the values of σ_f, Young's modulus (E), H_v, $\sigma_f/E(=\varepsilon_{t,f})$ and $9.8H_v/3E(\simeq\varepsilon_{c,y})$ for Al–Ln–Ni (Ln = Y, La or Ce) glasses (Inoue et al. 1988f). The σ_f values of the Al–Ln–Ni glasses are in the range of 900–1140 MPa, being independent of alloy composition. On the other hand, the E and H_v are in the range of 60–89 GPa and 260–380 DPN respectively. The highest σ_f attained are 1140 MPa for $Al_{87}Y_8Ni_5$ and 1080 MPa for $Al_{87}La_8Ni_5$, which greatly exceed the highest value of 550 MPa for conventional aluminum-based alloys subjected to an optimum age-hardening treatment. The specific strength, defined by the ratio of σ_f to density ρ, reaches 38 for $Al_{87}Y_8Ni_5$, being considerably higher than that for conventional aluminum-based alloys at 20. Furthermore, tensile deformation and fracture behavior of aluminum-based glassy alloys have been established to be the same as those for other ductile glassy alloys.

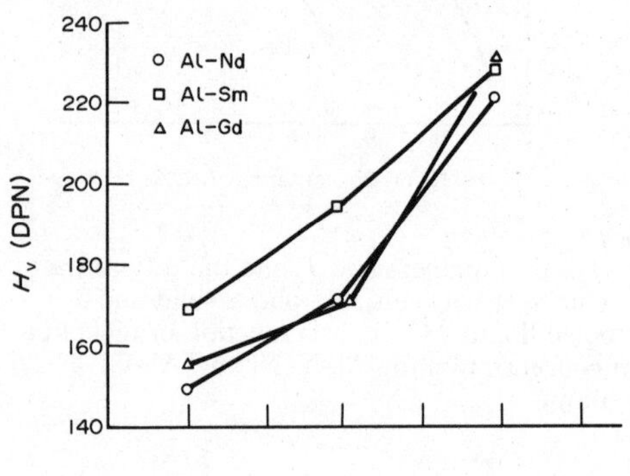

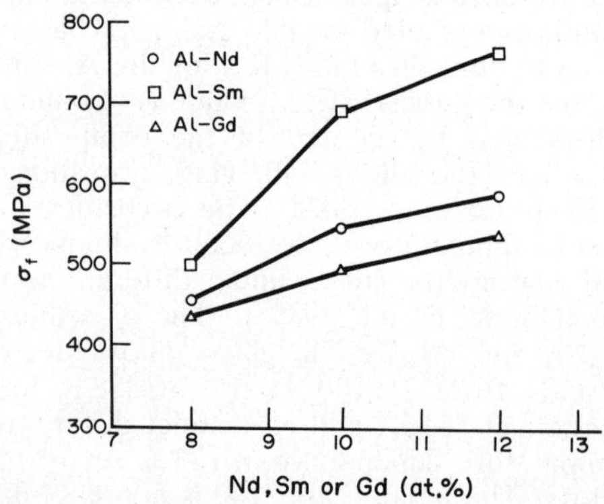

Figure 5
Changes in tensile fracture strength σ_f and Vickers hardness H_v of Al–Ln glassy alloys with Ln concentration

7. *Electronic Properties*

The feature of electrical resistivity $\rho(T)$ and its temperature coefficient near room temperature ρ_{RT} is nearly the same as that for other glassy alloys (Inoue et al. 1988i). The ρ_{RT} for the Al–Ln binary alloys increases from 54 μΩ cm to 170 μΩ cm with increasing lanthanide content and the temperature dependence in the range of 100 K to room temperature is positive and the gradient tends to decrease with an increase of resistivity.

The increase of ρ_{RT} with increasing solute content has also been observed for $Al_{90-x}Ln_xCo_{10}$ (Ln = Y, La or Ce) and $Al_{90-x}Y_{10}(TM)_x$ (TM = Fe, Co, Ni or Cu) alloys (Gonzalo et al. 1989). As shown in Fig. 6, ρ_{RT} increases from 62 μΩ cm to 230 μΩ cm with increasing TM contents. The increase of ρ_{RT} by the addition of the solute elements is nearly the same for yttrium and lanthanum but differs significantly between the TM elements, that is, the increase becomes remarkable in the order of Cu < Ni < Co < Fe. The electrical conductivity is presumed to be mainly dominated by the *s* and *p* electrons because of the aluminum-rich compositions and hence the change in ρ_{RT} for the Al–Ln–TM alloys may be interpreted as follows: The number of the outer electrons for the yttrium and lanthanum is the same ($d^1\,s^2$), resulting in nearly the same $\rho(T)$ values; on the other hand, the number of 3*d* electrons in the TM elements decreases in the order of Cu < Ni < Co < Fe and hence the decrease of electrical conductivity occurs in the same order, being consistent with the experimental result. However, the result cannot be explained by the scattering of conduction electrons caused by the difference in atomic size. Furthermore, the increase of ρ_{RT} with increasing lanthanide and TM contents is probably because the number of the *s* and *p* electrons in aluminum leading to electrical conductivity, decreases through the spend for the bonding to lanthanide and TM elements. This assumption is also supported by the result (Fig. 6) that the Hall coefficient is negative and the magnitude increases with increasing solute content.

Table 1
Mechanical properties of aluminum-based glassy alloys

Alloy (at.%)	σ_f (MPa)	E (GPa)	H_v (DPN)	$\varepsilon_{t,f}=\sigma_f/E$	$\varepsilon_{c,y}\simeq 9.8H_v/3E$	$\sigma_{c,y}\simeq 9.8H_v/3$ (MPa)
$Al_{88}Y_2Ni_{10}$	920	71	340	0.013	0.016	1110
$Al_{87}Y_8Ni_5$	1140	71	300	0.016	0.014	980
$Al_{85}Y_{10}Ni_5$	920	63	380	0.015	0.020	1240
$Al_{87}La_8Ni_5$	1080	89	260	0.012	0.010	850
$Al_{84}La_6Ni_{10}$	1010	84	280	0.012	0.011	915
$Al_{82}La_8Ni_{10}$	900	79	320	0.011	0.013	1045
$Al_{85}Ce_5Ni_{10}$	935	60	320	0.016	0.018	1045

8. *Change in Mechanical Properties by Glass Transition*

It is generally known that the glass transition gives rise to significant changes of E, σ_f and deformation behavior, in addition to the changes in specific heat and viscosity. It was important for the present aluminum-based glassy alloys to establish the changes in E, σ_f, ε_f and fracture behavior caused by the transition from glassy solid to supercooled liquid. Figure 7 shows the temperature dependences of E, σ_f and ε_f for an $Al_{85}Y_{10}Ni_5$ alloy (Inouc ct al. 1988f). With increasing temperature, the E values decrease gradually from 73 GPa to 55 GPa in the amorphous solid, decrease rapidly to 21 GPa in the supercooled liquid and then increase steeply upon crystallization. The σ_f decreases gradually from 920 MPa to 700 MPa in the glassy solid at temperatures below 453 K and rapidly to 240 MPa in the supercooled

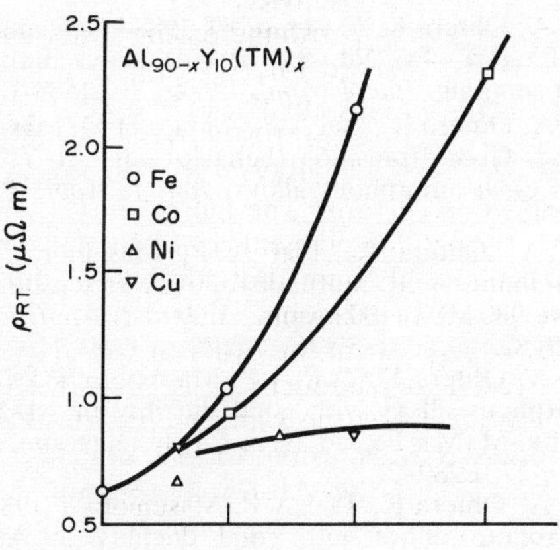

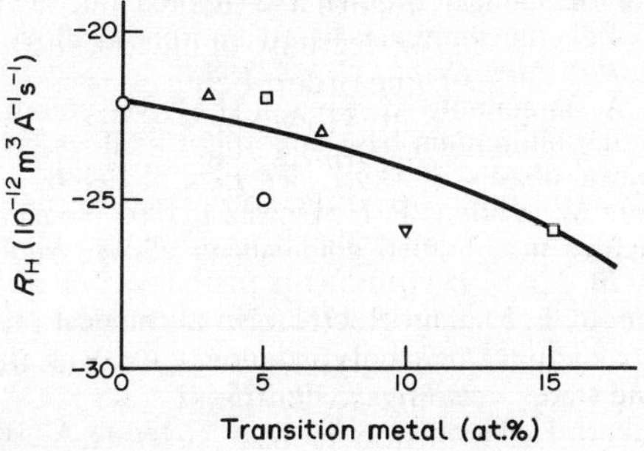

Figure 6
Changes in electrical resistivity ρ_{RT} and Hall coefficient R_H at room temperature with TM concentration for $Al_{90-x}Y_{10}M_x$ glassy alloys

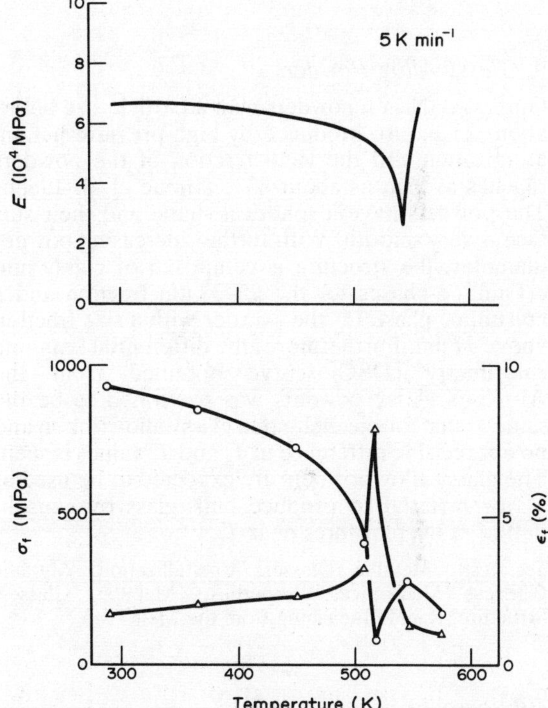

Figure 7
Young's modulus E, tensile fracture strength σ_f and fracture elongation ε_f as a function of temperature for an $Al_{85}Y_{10}Ni_5$ glassy alloy

liquid at 508 K, accompanied by a significant increase of elongation. Further increase of temperature gives an increase of σ_f and a decrease of ε_f due to the crystallization. The fracture surface at room temperature consists of clearly distinguishable regions of smooth and veined patterns. On the other hand, an alloy tested in the supercooled liquid range undergoes severe homogeneous necking by viscous flow. With further increase in testing temperature, the alloy crystallizes and the fracture surface consists of an embrittled pattern which is typical for the crystallized alloys. The feature of deformation and fracture behavior shown in Fig. 7 agrees well with that (Masumoto and Maddin 1971) for palladium-based glassy alloys.

9. Glassy Alloy Powders

Glassy $Al_{85}Ni_5Y_{10}$ powders with a particle size below about 25 μm are produced by high-pressure helium atomization and the yield fraction of the powders reaches as high as about 83% (Inoue et al. 1988b) The powders have a spherical shape and their surface is very smooth. With further increasing powder diameter, the structure is composed of glassy and crystalline phases for the 25–37 μm fraction and a crystalline phase for the powder with a size fraction above 37 μm. Furthermore, the differential scanning calorimetry (DSC) curve obtained from the Al–Ni–Y glassy powders was confirmed to be the same as that for the melt-spun glassy alloy ribbon and no appreciable difference in T_g and T_x values is seen. The glassy alloy powders are expected to be used as a raw material to produce bulk glass by consolidation at temperatures near T_g.

See also: Metallic Glasses: Crystallization; Metallic Glasses: Mechanical Properties; Metallic Glasses: Structure; Rapid Quenching from the Melt

Bibliography

Cahn R W 1989 Aluminium-based glassy alloys. *Nature* 341: 183–4

Chattopadhyay K, Ramachandrarao R, Lele S, Anantharaman T R 1976 Crystal structure of a metastable aluminum–nickel phase obtained by splat cooling. In: Grant N J, Giessen B C (eds.) *Proc. 2nd Int. Conf. on Rapidly Quenched Metals*. MIT Press, Cambridge, MA, pp. 157–61

Davies H A, Hull J B 1972 An amorphous phase in a splat-quenched Al–17.3 at% Cu alloy. *Scripta Metall.* 6: 241–6

Dubois J-M, Dehghan K, Janot C, Chieux P, Chenal B 1985 Small angle scattering and structure factor neutron measurements of an amorphous $Al_{70}Si_{17}Fe_{13}$ alloy. *J. Physique* 46(C–8): 461–6

Dubois J-M, Le Caer G 1985 Tendency to vitrification and physical properties of amorphous alloys based on aluminum. *C. R. Acad. Sc. Paris* 301: 73–8

Dubois J-M, Malaman B, Chenal B, Venturini G 1987 Crystallization and structure of an amorphous alloy based on aluminum. *C.R. Acad. Sci. Paris* 304: 641–6

Furrer P, Warlimont H 1977 Crystalline and amorphous structures of rapidly solidified Al–Cr alloys. *Mater. Sci. Eng.* 28: 127–37

Gonzalo J, Pont M, Rao K V, Inoue A, Masumoto T 1989 Electronic properties of a new class of aluminum-based metallic glasses: $Al_{100-x-y}(TM)_x(La, Y)_y$ with TM = Fe, Co, Ni or Cu. *Phys. Rev. B* 40(3): 1345–8

He Y, Poon S J, Shiflet G J 1988 Synthesis and properties of metallic glasses that contain aluminum. *Science* 241: 1640–2

Inoue A, Bizen Y, Kimura H M, Masumoto T, Sakamoto M 1988a Compositional range, thermal stability, hardness and electrical resistivity of amorphous alloys in Al–Si (or Al–Ge)–transition metal systems. *J. Mater. Sci.* 23: 3640–7

Inoue A, Kita K, Ohtera K, Masumoto T 1988b Al–Y–Ni amorphous powders prepared by high-pressure gas atomization. *J. Mater. Sci. Lett.* 7: 1287–90

Inoue A, Kitamura A, Masumoto T 1981 The effect of aluminium on mechanical properties and thermal stability of (Fe, Co, Ni)–Al–B ternary amorphous alloys. *J. Mater. Sci.* 16: 1895–1908

Inoue A, Ohtera K, Masumoto T 1988c New amorphous Al–Y, Al–La and Al–Ce alloys prepared by melt-spinning. *Jpn. J. Appl. Phys.* 27: L 736–9

Inoue A, Ohtera K, Masumoto T 1988d New amorphous Al–La (La = Pr, Nd, Sm or Gd) alloys prepared by melt-spinning. *Jpn. J. Appl. Phys.* 27: L1583–6

Inoue A, Ohtera K, Tsai A P, Kimura H M, Masumoto T 1988e Glass transition behavior of Al–Y–Ni and Al–Ce–Ni amorphous alloys. *Jpn. J. Appl. Phys.* 27: L1579–82

Inoue A, Ohtera K, Tsai A P, Masumoto T 1988f Aluminum-based amorphous alloys with tensile strength above 980 MPa (100 kg mm^{-2}). *Jpn. J. Appl. Phys.* 27: L479–82

Inoue A, Ohtera K, Tsai A P, Masumoto T 1988g New amorphous alloys with good ductility in Al–Y–M or Al–La–M (M = Fe, Co, Ni or Cu) systems. *Jpn. J. Appl. Phys.* 27: L280–2

Inoue A, Ohtera K, Tsai A P, Masumoto T 1988h New amorphous alloys with good ductility in Al–Ce–M (M = Nb, Fe, Co, Ni or Cu) systems. *Jpn. J. Appl. Phys.* 27: L1796–9

Inoue A, Tsai A P, Bechet D, Matsuzaki K, Masumoto T 1988i Mechanical, thermal and superconducting properties of ductile aluminum-based amorphous alloys. *Trans. Jpn. Inst. Met. (Suppl.)* 29: 325–8

Inoue A, Yamamoto M, Kimura H M, Masumoto T 1987 Ductile aluminium-base amorphous alloys with two separate phases. *J. Mater. Sci. Lett.* 6: 194–6

Klement W, Willens R H, Duwez P 1960 Noncrystalline structure in solidified gold–silicon alloys. *Nature* 187: 869–70

Masumoto T, Maddin R 1971 The mechanical properties of palladium 20a/o silicon alloy quenched from the liquid state. *Acta Metall.* 19: 725–41

Matsubara E, Harada K, Waseda Y, Inoue A, Bizen Y, Masumoto T 1988a X-ray diffraction study of an amorphous $Al_{60}Ge_{30}Ni_{10}$ alloy. *J. Mater. Sci.* 23: 3485–9

Matsubara E, Waseda Y, Inoue A, Ohtera K, Masumoto T 1988b On the structural evolution of the amorphous

$Al_{75}Cu_{15}V_{10}$ alloy to the icosahedral phase. *Z. Naturforsch.* 43(5): 505–6
Matsubara E, Waseda Y, Inoue A, Ohtera K, Masumoto T 1989 Anomalous x-ray scattering study of the structure of amorphous $Al_{87}Y_8Ni_5$and Al_{90}–Y_{10} alloys. *Z. Naturforsch.* 44(a): 814–20
Predecki P, Giessen B C, Grant N J 1965 New metastable alloy phases of gold, silver and aluminum. *Trans. Metall. Soc. AIME* 233: 1438–9
Ramachandrarao P, Laridjani M, Cahn R W 1972 Diamond as a splat-cooling substrate. *Z. Metallkd.* 63: 43–9
Sastry G V S, Suryanarayana C, Srivastava O N, Davies H A 1978 Crystallization of an amorphous Al–Pd alloy. *Trans. Indian Inst. Metals* 31: 292–4
Tsai A P, Inoue A, Masumoto T 1988 Formation of metal–metal type aluminum-based amorphous alloys. *Metall. Trans., A* 19: 1369–71

A. Inoue and T. Masumoto
[Tohoku University, Sendai, Japan]

Aluminum–Lithium Based Alloys

Lithium is not only one of the few elements that is significantly soluble in aluminum but it also has the merits of reducing density by approximately 3% and increasing stiffness by approximately 6% per weight percentage addition to the aluminum alloy. Consequently, since the early 1980s, the development of aluminum–lithium based alloys for aerospace applications has become one of the biggest and most intensive alloy development programs ever conducted.

Initially, this development was spurred by the search for higher performance alloys for military aircraft but it was then given considerably greater impetus by the concern at the escalation in fuel prices resulting from the oil crises of the 1970s and the consequent demand for more fuel efficient civil aircraft.

However, interest in lithium-containing alloys actually started very much earlier, certainly well before 1920, and has been virtually continuous since the 1940s. While this article concentrates on the recent phases of development, it begins with a brief description of the historical background.

1. Background

The initial interest in lithium-containing aluminum alloys seems to have been centered in Germany around 1920 and was probably driven by a wish to improve upon the early heat treatable aluminum–copper alloys developed prior to World War I. The addition of lithium was intended simply to improve the strength of the alloys and there was probably no appreciation of the potential of lithium for reducing density or increasing stiffness in aluminum alloys. It is interesting to note that the addition of up to about 6% lithium to aluminum alloys was regarded (Mond 1921) as "comparatively easy" relative to the addition of magnesium, a view unlikely to be shared by those currently involved in the manufacture of these alloys. Applications for these early alloys were seen in general construction work and early mentions of uses for aircraft were dismissive (Mortimer 1929).

By the 1940s the demands upon aluminum alloys for aircraft were increasing greatly, particularly as gas turbine engines and supersonic flight imposed requirements for higher operating temperatures. It was recognized that additions of lithium to aluminum–copper alloys, particularly in conjunction with other minor additions such as cadmium, could result in very strong alloys (Le Baron 1942) and this led to the development of the alloy designated 2020, the first commercial lithium-containing aluminum aircraft alloy. This alloy contained a nominal 1 wt% lithium and was capable of operating at modestly greater temperatures than had been previously possible. It was also some 3% lighter and 8% stiffer than the established high strength aircraft alloy 7075.

More fundamental metallurgical investigations had continued in the UK and these demonstrated that the role of the lithium in the basically aluminum–copper alloy was to refine and stabilize the θ' precipitates (Hardy 1955, Hardy and Silcock 1955). It had, by then, been recognized that lithium precipitates from supersaturated aluminum–lithium in the form of an ordered, spherical Al_3Li precipitate, δ', analogous to the Cu_3Au structure. Other contemporary work in the UK had led to the development of the creep-resistant aluminum base alloy 2618 (the so-called Concorde alloy). The similarity between the Ni_3Al precipitate structure in the creep-resistant, nickel-based superalloys and the Al_3Li precipitate structure led M. A. P. Dewey to the idea that an alloy with superior creep resistance to 2618, based upon aluminum–lithium, might be possible. It was demonstrated that good creep resistance could be obtained but that the basic aluminum–lithium alloy was too weak. The addition of magnesium improved the strength but also destroyed the creep resistance, leading to the termination of the search for a creep-resistant alloy and the initiation of the development of a low density, high stiffness Al–Mg–Li alloy containing additions of zirconium and/or manganese (Brook 1976).

Simultaneously, and independent of the UK development, work in the USSR had led to the development of the Al–Mg–Li alloy known as 01420 (Fridlyander et al. 1967). This alloy has the nominal composition Al–5%Mg–2%Li–0.5%Mn. (All compositions in this article are cited in wt%.) The alloy has a very low density (2.47 g cm^3), has excellent fatigue and corrosion behavior, and is reported as being weldable. However, its strength is only modest and its toughness is rather low.

Nevertheless, it was probably the potential of this alloy that renewed enthusiasm for aluminum–lithium based alloys and resulted in the very intense development activity of the 1980s.

During the 1980s, five international conferences were devoted exclusively to the topic of aluminum–lithium alloy development and the majority of papers cited in this review have been published in the conference proceedings (Sanders and Starke 1981, 1984, 1989, Baker et al. 1986, Champier et al. 1987).

2. *Metallurgy of Aluminum–Lithium Based Alloys*

The binary alloy system aluminum–lithium produces alloys too weak for general aerospace applications. Additionally, the shearable nature of the δ' precipitate tends to create localized planar slip that can result in limited ductility. Attempts have, therefore, been made to improve these properties by ternary or quaternary additions. As previously mentioned, the first really lightweight alloys were based on the Al–Mg–Li (–Zr or –Mn) system but their limited strength and poor fracture behavior were perceived as serious limitations to their application. In the USA, probably following on from the earlier Al–Cu–Li–Cd alloy 2020, most effort was devoted to the Al–Cu–Li system: reducing the copper content, increasing the lithium addition and omitting cadmium. This led to alloys capable of very high strength although, so far, "through thickness" properties in thick section products remain an unresolved problem. In the UK, work initiated by the Royal Aerospace Establishment concentrated upon alloys of the Al–Li–Cu–Mg–Zr system. The initial alloy was reasonably light, much easier to manufacture and tougher than the Al–Mg–Li alloys but, while stronger than the Al–Mg–Li alloys, it was not as strong as those from the Al–Cu–Li series.

While other alloy systems have also been considered, and may well have development potential, all of the aluminum–lithium based alloys currently in, or nearing, commercial production are based upon the alloy systems mentioned above. The microstructural features of these alloys are summarized below.

2.1 *Aluminum–Lithium*

After quenching from the single phase region and aging, the precipitation sequence is

$$\alpha \text{ (supersaturated solid solution)} \rightarrow \delta'(\mathrm{Al_3Li}) \rightarrow \delta(\mathrm{AlLi})$$

The δ' (see Fig. 1), as mentioned above, forms as coherent, ordered, spherical particles and it is probably not possible to achieve quench rates fast enough to suppress nucleation of the δ' precipitates. The mechanism by which δ' forms during quenching has not been conclusively established; the alternatives being classical coherent nucleation and growth (Williams and Edington 1975) or spinodal decomposition (Nozato and Nakai 1977). However, subsequent artificial aging has shown growth of δ' precipitates to follow classical $t^{1/3}$ coarsening kinetics. In the presence of grain boundaries and dislocations, coarsening of the δ' is accelerated. Precipitate-free zones (PFZs) are formed around these and also around grain-boundary precipitates of the phase.

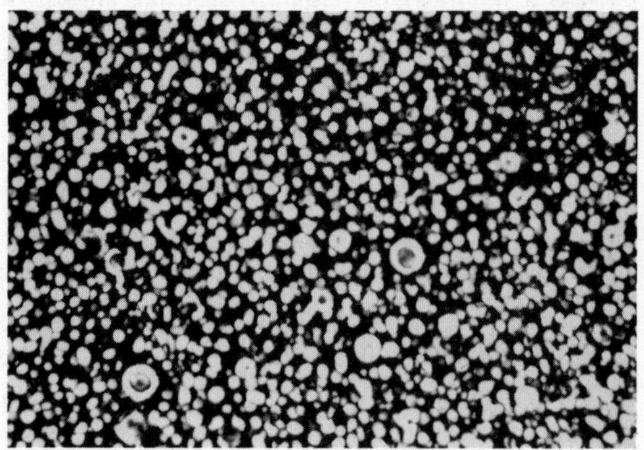

Figure 1
Precipitates of δ' and δ' on $\mathrm{ZrAl_3}$ in an Al–Li–Zr alloy (×50 000)

If, after formation of the metastable δ' precipitates, aging is continued below the δ' solvus, the equilibrium δ phase forms. This occurs more rapidly in the grain-boundary regions but also occurs within the matrix. The δ phase has a highly reactive nature and can form as relatively coarse particles that are generally detrimental to service properties.

2.2 *Aluminum–Lithium–Zirconium*

All the aluminum–lithium based alloys in, or near to, commercial use contain dispersoid-forming additions to control recrystallization behavior. Most development has involved either manganese or zirconium but the latter has found the greatest favor and is present in all three alloys currently near to application on a significant scale.

The role of the manganese appears relatively simple as the manganese-containing intermetallics do not incorporate lithium, are incoherent and, if present in a relatively coarse form, can encourage recrystallization. This may be useful in those products where recrystallization is beneficial to properties; for example, in damage-tolerant sheets.

The role of zirconium is considerably more complex. Zirconium additions are used to inhibit recrystallization in several established, high strength alloys (e.g., 7010 and 7150), the mechanism being the provision of small, coherent, spherical Al_3Zr particles that are very effective in retarding subgrain boundary migration and coalescence. In lithium-bearing alloys, the Al_3Zr particles frequently act as nuclei for precipitation of the isostructural Al_3Li.

Thus a shell of δ' forms around the Al_3Zr particle (Gu et al. 1986). It has been suggested that the precipitate phase stable at normal solution treatment temperatures is not Al_3Zr, but a complex $Al_3(Li, Zr)$ phase. However, attempts to confirm this suggest that the "dark imaging" centers are merely Al_3Zr particles of slightly different orientation from the δ' shell. In terms of practical properties the $ZrAl_3/\delta'$ particles are more resistant to planar slip than δ' and so improve the strength of the alloy. However, the major effect of the particles is in inhibiting recrystallization and thus making a significant strength contribution. Aluminum–lithium alloys containing zirconium age more rapidly and achieve higher strength than either the binary alloy or alloys using manganese to provide the dispersoid. It is suggested (Gu et al. 1986) that the more rapid aging of the zirconium–containing alloy is a consequence of the presence of Al_3Zr particles with a diameter of approximately 200 Å which would grow reasonably rapidly to the optimum diameter for strength of 300 Å. Thus, the Al–Li–Zr alloy after aging contains separate particle size distributions for the Al_3Li that has not nucleated on the Al_3Zr (finer particles) and for the Al_3Li that has formed as a shell around the Al_3Zr (coarser particles). The Al–Li–Zr achieves a peak yield strength of about 360 MPa in 2–3 h at 200 °C whereas a corresponding Al–Li–Mn alloy shows a maximum strength approximately 250 MPa after about 50 h at 200 °C.

2.3 *Aluminum–Lithium–Magnesium*

The initial precipitation sequence from supersaturated solid solution for Al–Li–Mg alloys is very little different from that for the binary aluminum–lithium alloy system. The supersaturated solid solution breaks down to δ' and the main initial difference is that the presence of magnesium reduces the solubility of lithium so that, for a given lithium level, an increased volume of δ' results (Thompson and Noble 1973). Later in the aging process magnesium enters the precipitation reaction to form Al_2MgLi, although some magnesium remains in solid solution. The Al_2MgLi is incoherent and forms as rods with a ⟨110⟩ growth direction, either by overaging or by heterogeneous precipitation, primarily at grain boundaries, and resulting in significant deleterious effects upon ductility and toughness.

2.4 *Aluminum–Copper–Lithium*

In effect, the aluminum–copper–lithium system operates as a combination of two binary systems, at least until the later stages of aging when the T_1 phase forms (Noble and Thompson 1972). The precipitation reaction is as follows:

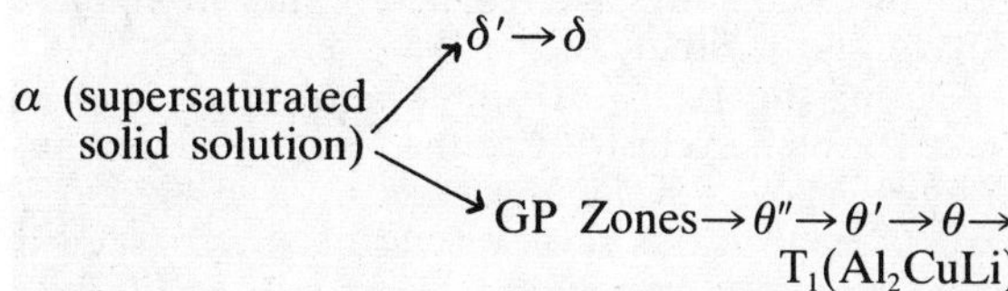

As with magnesium, the presence of copper reduces the solubility of lithium. By appropriate choice of heat-treatment time and temperature the precipitates and their volumes can be controlled. However, precipitation of the copper-containing phases is heterogeneous so that the introduction of a dislocation structure by cold working after solution treatment is necessary if a reasonably uniform distribution of the copper-containing phases is to be achieved.

Basically, this alloy system gives the possibility for the formation of three precipitates: δ', θ' and T_1. The balance between these will be determined by such factors as copper content and stretching, high copper content tending to increase θ' precipitation while stretching and low copper content will favor T_1 at the expense of δ'.

While the above precipitation sequence shows the traditional steps for a binary aluminum–copper alloy, there is also some debate as to whether GP zones actually form in Al–Cu–Li alloys. It is argued that the high lithium-vacancy binding energy results in the presence of too few vacancies to allow the formation of GP zones.

2.5 *Aluminum–Lithium–Copper–Magnesium*

No phase diagram exists for the Al–Li–Cu–Mg system, although several of the more promising alloys lie within it. Despite this lack of the basic phase diagram, a large number of investigations have considered, in detail, the metallurgy of particular alloys. In general, the alloys show the precipitate structures characteristic of the Al–Li–Cu, Al–Li–Zr and Al–Cu–Mg systems. Indication of the precipitation reactions in the first two of these has been given above and the decomposition of supersaturated solid solutions of the last is as follows:

$$\alpha(\text{supersaturated solid solution}) \rightarrow \text{GP zones} \rightarrow S'' \rightarrow S' \rightarrow S\ (Al_2CuMg)$$

The structures and orientations of the intermediate phased S″ and S′ are similar to those of S. S′ forms as rods that then widen into laths with an orthorhombic structure. The equilibrium S phase forms either by loss of coherency of the S′ or by heterogeneous nucleation. As with binary aluminum–copper alloys, a small cold strain after

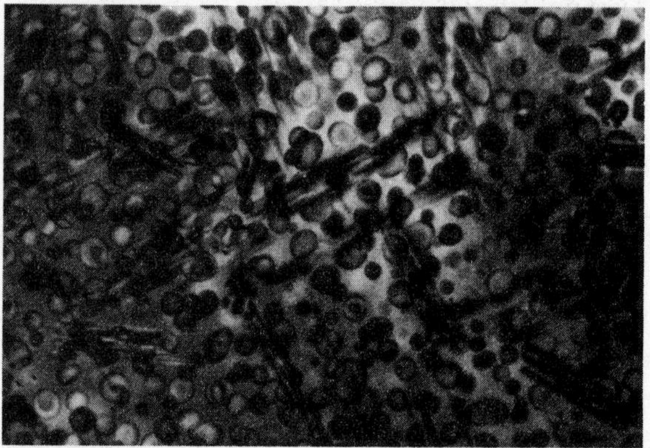

Figure 2
The coexistence of δ', T_1 and S precipitation in Al–Li–Cu–Mg–Zr alloy 8090 (×75 000) (δ' are the circular precipitates; T_1 and S are the plate-shaped features)

solution treatment is very effective in refining and in improving the distribution of the precipitate.

In the quaternary Al–Li–Cu–Mg alloys the $T_1(Al_2CuLi)$ phase still precipitates at low levels of magnesium. However, as the magnesium-to-copper ratio increases, the T_1 phase is progressively replaced by $S(Al_2CuMg)$. The protagonists for this alloy system (e.g., Miller et al. 1987) argue that this gives an advantage over the Al–Li–Cu system since the S phase forms without altering δ' precipitation. In the Al–Li–Cu system, precipitation of the T_1 phase takes up part of the lithium and further δ' nucleation on θ' occurs so that the volume fraction of δ' available for strengthening is significantly reduced. However, some T_1 precipitation occurs (see Fig. 2) even in the preferred compositional region for alloys in this system and the balance of δ' relative to T_1 is critically dependent upon the relative concentrations of the three main alloying additions. With a typical composition of lithium 2.5 wt% and copper 1.4 wt%, a magnesium content of greater than 0.5 wt% is required to give S′ dominance.

2.6 The Icosahedral Phase

Although the T_2 phase was described in the mid-1950s (Hardy and Silcock 1955), its importance in aluminum–lithium based alloys and its five-fold icosahedral symmetry have only been appreciated since the late 1980s. Nevertheless, its detrimental influence on service properties is such as to justify the inclusion of this section. The phase was described as being close to Al_6CuLi_3 and is the equilibrium phase in the Al–Cu–Li system; however, it can also form as $Al_6Cu(Li, Mg)_3$ in the Al–Li–Cu–Mg system (Cassada et al. 1986). The icosahedral phase structural unit is the 105-atom truncated icosahedron (Gayle 1987). The quasicrystalline phase forms during slow solidification but in this situation can be taken back into solid solution by an elevated-temperature homogenization treatment subsequent to solidification. More seriously, from the service property viewpoint, the icosahedral phase can form on grain boundaries during the quench of heavy section products, resulting in significant quench sensitivity, particularly in high copper alloys. It can grow rapidly along the grain boundaries (see Fig. 3) and has a seriously damaging effect on properties such as fracture toughness and stress corrosion.

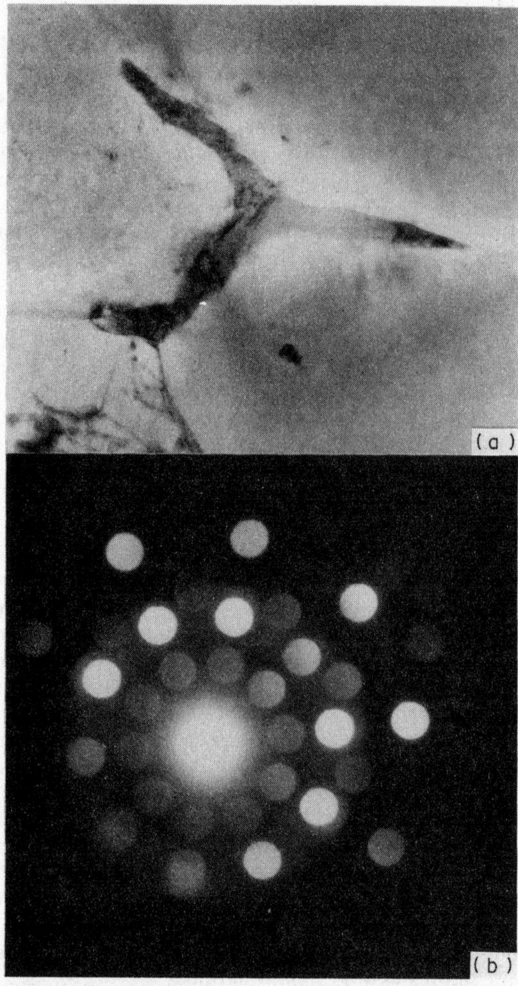

Figure 3
The icosahedral phase: (a) growing at grain boundary in thick plate (×40 000) and (b) an electron diffraction pattern illustrating the five fold symmetry

Additionally, the icosahedral phase forms during elevated temperature aging (Lewis et al. 1987), although the susceptibility to its formation diminishes as the aging temperature is reduced.

While the majority of investigators (e.g., Gayle 1987, Lapasset and Loiseau 1987) believe that T_2 and the icosahedral phase are the same, it has been suggested (Cassada et al. 1986) that T_2 is body-centered cubic with an aluminum-to-copper ratio of approximately six to one compared with an aluminum-to-copper ratio of seven to three for the icosahedral phase. In any event, it is clear that several fairly complex grain boundary phases can exist in addition to the icosahedral phase (Ball and Lagace 1986, Dubost et al. 1987) and, in practical terms, all are almost certainly damaging to the service properties of the alloys in which they form.

3. *Manufacture*

From the points of view of both the aluminum alloy producers and of the eventual aerospace users, one of the great attractions of aluminum–lithium based alloys is that, for the most part, they can be manufactured using facilities and procedures virtually identical to those used for conventional aluminum aerospace alloys. However, because lithium is an extremely reactive element, there are a number of areas in which different procedures have to be adopted and these are summarized below.

The aluminum companies can employ almost identical procedures of hot and cold rolling, extrusion and forging. Indeed, it is probably true that in the majority of hot-working situations the lithium-containing alloys can be worked more readily than the conventional alloys that they are designed to replace. During cold-working processes, some greater difficulties may be encountered because of the high work-hardening rates and the presence of oxide and other surface films that can interfere with the operation of conventional lubricants. The aerospace companies can, in turn, machine heavy section material and form thin sections with the same equipment and near-to-standard procedures while the processes for finishing (anodizing, conversion coatings, etc.), joining, chemical milling and adhesive bonding need be only marginally modified.

3.1 *Melting and Casting*

Towards the end of the 1970s the view seemed to be developing that the conventional "ingot metallurgy" route would not be good enough for aluminum–lithium based alloys and, particularly in the USA, there were suggestions that it would be necessary to adopt a powder metallurgy route in order to achieve a satisfactory starting product. However, the Al–Li–Cu–Mg–Zr alloys developed in Europe and the Al–Cu–Li alloys developed in the USA in the early 1980s proved to be considerably easier to cast and fabricate than had the alloys based upon the Soviet Al–Mg–Li system, and from the early 1980s onwards all of the major aluminum–lithium programs have concentrated upon the ingot metallurgy approach. For the future, it remains possible that processes based upon spray deposition of a solid ingot may have greater potential for the production of a high quality starting ingot than either conventional ingot or powder metallurgy (Kojima et al. 1989).

Conventional large-scale manufacture of aluminum alloys employs large reverberatory melting furnaces (generally gas or oil fired) in which little or no attempt is made to control furnace atmosphere and in which the products of combustion come into contact with the molten metal. These furnaces cannot, of course, be used with aluminum–lithium based alloys because the very expensive lithium would be rapidly burnt off. Thus indirectly heated, generally electrically powered, furnaces with very careful atmosphere control have to be used and, once the lithium-containing alloy is molten, all metal-transfer operations have to be conducted under equally carefully atmosphere-controlled conditions.

None of the manufacturers of aluminum–lithium based alloys has disclosed any details of the actual casting molds and hardware involved, although it is assumed that all are using virtually conventional (other than for atmosphere control with an inert gas over the molten alloy) vertical semicontinuous direct chill (DC) casting systems. With conventional alloys, the DC mold is mounted over a deep, water-containing pit and the ingot is lowered into this pit as casting proceeds. In certain fault circumstances, molten metal can run from the casting head into the pit water and operational codes of practice have been developed over the years to ensure that such "run-outs" occur safely. An early report (Long 1957) had suggested that molten aluminum–lithium alloys in contact with water were more likely to explode than pure aluminum under the same conditions and this led to trials to assess the explosion potential of the newer generation of alloys in contact with water. Qualitatively these trials demonstrated that very violent explosions could occur between a molten aluminum–lithium alloy and water and also suggested that spontaneous explosions between the alloy and the water would occur if the lithium content significantly exceeded 2 wt%. More quantitative trials (Page et al. 1987) confirmed the increased violence with increased lithium content (see Fig. 4) but did not confirm the spontaneous explosion possibility. Nevertheless, the substantially greater energy release that could accompany an explosion occurring due to molten aluminum–lithium

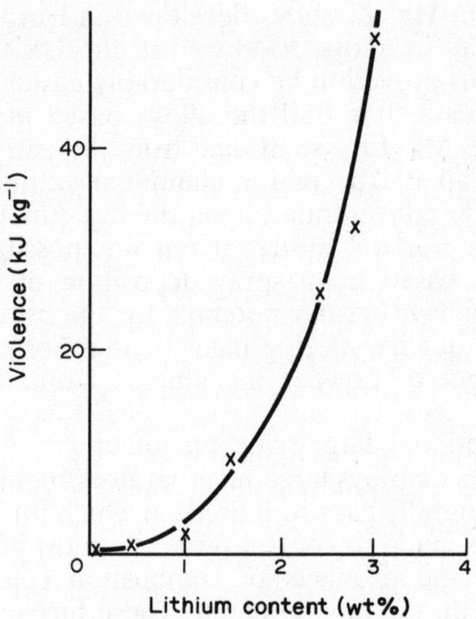

Figure 4
Increase in energy release with lithium content in explosions between molten binary aluminum–lithium alloys and water

alloy and water represents an industrially unacceptable risk and, it is believed that all the major manufacturers have adopted practices to prevent this from happening. These have included substituting ethylene glycol for water as the coolant (Jacoby et al. 1986). While the use of organic coolants does not eliminate the explosion risk it does substantially reduce the energy released in any explosion, particularly if a high boiling point coolant is chosen (Page et al. 1987).

No entirely satisfactory explanation has yet been given for the increased explosion risk with aluminum–lithium alloys, although it has been suggested that the release of hydrogen that occurs when the molten alloy comes into contact with water plays a significant role. The quantity of hydrogen released is too small to contribute significantly to the energy release. However, hydrogen has a thermal conductivity approximately ten times higher than that of steam so that the presence of significant quantities of of hydrogen in the steam blanket that normally separates the molten metal from the water can greatly increase the efficiency of heat transfer from the molten metal to the water as the explosion is detonated.

3.2 Heat Treatment

In the course of the aluminum–lithium based alloy development, two issues in particular have given cause for concern. These were, first, the possibility that heat treatment in conventional nitrate–nitrite salt baths might be hazardous and, second, the loss of lithium from the metal surface during high-temperature heat treatment. Investigation has proved that the former should not be a cause for concern while the latter is almost inevitably present but can be confined within reasonable bounds by correct care during manufacture. Both issues are briefly considered below.

Salt baths are still widely used for intermediate heat treatments by the airframe constructors and are also, but to a lesser extent, used by the aluminum companies for solution heat treatment. Within the aluminum industry there has been a long-standing belief that aluminum–magnesium alloys containing high levels of magnesium should not be heat treated in nitrate salt baths (Clark et al. 1986). It has been suggested that the highly exothermic reaction

$$5\,Mg + 2\,NaNO_3 \rightarrow Na_2O + 5\,MgO + O_2$$

can create a violent explosion. It was felt that lithium-containing alloys could well react in the same way and, since the lithium atom is considerably smaller than the magnesium atom, there were fears that the permissible maximum lithium content might be even lower than the 3 wt% magnesium traditionally permitted in the UK.

The work by Clark et al. (1986) demonstrated that: first, there appeared to be no historical factual basis for the ban on the use of nitrate baths for aluminum–magnesium alloys; second, violent reactions could only be generated in aluminum–magnesium alloys when the combination of composition and temperature was such that at least partial liquation occurred; and third, the aluminum–lithium based alloys also failed to show any significant reaction until partial liquation occurred. Thus the original fears proved unsubstantiated and in the intervening years substantial quantities of aluminum–lithium based alloys have been safely solution treated in salt baths. Such comments have to carry the caution that any salt bath is potentially dangerous unless the normal precautions are observed; obviously such precautions still apply with aluminum–lithium based alloys.

On the issue of the lithium-depleted surface layer, its existence is very well established. At temperatures around those used for solution treatment of the alloys, lithium migrates to the surface where it reacts to form a surface layer largely consisting of Li_2CO_3. Since this layer does not provide a protective film, the bulk of the alloy continues to act as a lithium reservoir and the reaction proceeds. A surface-depleted layer of up to, perhaps, 100 μm in thickness can be produced during the periods of exposure to elevated temperature that would be normal during the production of such materials (Thorne et al. 1987). However, if the exposure to elevated temperature is prolonged (16 h at 500 °C) it has been shown (Papazian et al. 1987) that the depth of denudation can approach 400 μm and is accompanied by the formation of a layer of subsurface porosity. It is suggested that this porosity forms as a consequence of the Kirkendall effect by the annihilation of vacancies created to compensate for the unequal fluxes of lithium and aluminum. Later work (Dickenson et al. 1988) suggests that the Kirkendall effect actually results in the formation of LiH particles and that the pores reported by earlier workers may have been cavities resulting from dissolution of the LiH during metallographic preparation.

In practical terms, the lithium-depleted layer is not generally perceived as presenting a problem although it has been shown to result in a small reduction in the fatigue performance of sheet alloy (Fox et al. 1986). In thicker section aerospace products such as plate, heavy forgings and heavy extrusions, it would be normal for far more metal to be machined from the original surfaces than was originally present as a lithium-depleted zone. It is only in thin extruded sections and in thin sheet that difficulties can arise, particularly in the latter case where the forming of complex sheet components may require multiple solution treatments in order to achieve the required shape. Even then, the time necessary to complete solution treatment of these alloys in thin sheet form is only a very few minutes so that careful control of manufacturing conditions should contain the problem. Certainly it has been demonstrated (Peel et al. 1983) that only very modest strength loss need occur in sheet alloy which has been solution treated three times. There is, however, a small but detectable reduction in the maximum strength attainable in thin sheet (of gauge <1.2 mm) and this strength reduction would, presumably, become larger if very thin gauges (<0.5 mm) of aluminum–lithium alloy sheet were produced.

4. The Development of Commercial Alloys

During the 1980s there have been three dominant programs in the West aimed at producing aluminum–lithium based alloys in the major product forms suitable for substitution in all applications using conventional alloys. The major manufacturing activity, and funding, has been provided by aluminum companies (such as Alcan, Alcoa and Pechiney) but numerous airframe constructors, universities, national aerospace laboratories and research institutes have also been heavily involved. UK developers were the first to state publicly the program targets for this alloy development and these were (Peel and Evans 1983):

(a) for medium strength alloy "A" to be equivalent to alloy 2014 in the form of T6 sheet, plate, extrusions and forgings;

(b) for high strength alloy "B" to be equivalent to alloy 7075 in the form of T6 sheet, plate, extrusions and forgings; and

(c) for damage tolerant alloy "C" to be equivalent to alloy 2024 in the form of T3 sheet, plate and extrusions.

Simultaneously with achieving properties equalling (or exceeding) those of the corresponding baseline alloys, the new alloys were to achieve a density reduction of 10% and a stiffness increase of 10%. Subsequently the other two principal developers announced very similar targets (Meyer and Dubost 1986, Bretz and Sawtell 1986). Probably because the initial impetus for this development was for military aircraft, most of the early activity concentrated upon the medium and/or high strength alloy developments. By so doing it is likely that the potentially larger volume market for damage tolerant materials was neglected. The concept of damage tolerant materials is not simply defined. Materials for use in applications requiring high damage tolerance, for example, the sheet used for the skinning of pressure cabins or the plate used for lower wing skins of large passenger aircraft, require a combination of good fatigue behavior and good fracture behavior. The typical combinations of properties required are indicated in the Damage Tolerant Design Handbook (1983). It is also probable that undue effort was devoted to such difficult issues as fracture toughness in the short transverse orientation of thick plate whereas more recent work on damage tolerant tempers (Gatenby et al. 1989) suggests that the property combinations required in damage tolerant tempers of aluminum–lithium based alloys may well be rather easier to achieve.

In the following sections the processing requirements for achieving the required performance in particular products and tempers will be discussed. They are generally described for particular alloys because information on the specified alloy or product has been made publicly available. It is believed, however, that the general metallurgical principles involved would apply with only minor modifications to most of the aluminum–lithium alloys currently under development.

Table 1 contains a summary of the chemical compositions and other details of the conventional ingot metallurgy aluminum–lithium based alloys on which significant development has been done. Of these alloys, the overwhelming majority of effort has been concentrated upon 8090, 2090 and 2091. Not included in this review are the mechanically alloyed lithium-containing alloys, such as IN905XL (Al–4%Mg–1.5%Li–1.2%C–0.4%O) (Gilman et al. 1986), because the principles involved in their manufacture are so different from those of the alloys of Table 1. Nevertheless, the mechanically alloyed aluminum–lithium based alloys do appear capable of very attractive property combinations, albeit at price premiums that may prove to be prohibitive. Also not included in Table 1 is the more recently

Table 1
Aluminum Association registered compositions for aluminum–lithium based alloys

Aluminum association designation	Company of origin	Trade name	Chemical composition (wt%)										
			Li	Cu	Mg	Zr	Si	Fe	Cr	Ti	Zn	Mn	Al
8090	Alcan	Lital A	2.20	1.00	0.60	0.04	0	0	0	0	0	0	
	Pechiney	Lital C	to	to	to	to	to	to	to	to	to	to	bal.
		CP 271	2.70	1.60	1.30	0.16	0.20	0.30	0.10	0.10	0.25	0.10	
X8090A	Alcoa	Alithalite A	2.10	1.10	0.80	0.08	0	0	0	0	0	0	
			to	to	to	to	to	to	to	to	to	to	bal.
			2.70	1.60	1.40	0.15	0.10	0.15	0.05	0.15	0.10	0.05	
8091	Alcan	Lital B	2.40	0.60	0.50	0.08	0	0	0	0	0	0	
			to	to	to	to	to	to	to	to	to	to	bal.
			2.80	2.20	1.20	0.16	0.20	0.30	0.10	0.10	0.25	0.10	
X8192	Alcoa	Alithalite C	2.30	0.40	0.90	0.08	0	0	0	0	0	0	
			to	to	to	to	to	to	to	to	to	to	bal.
			2.90	0.70	1.40	0.15	0.10	0.15	0.05	0.15	0.10	0.05	
X8092	Alcoa	Alithalite D	2.10	0.50	0.90	0.08	0	0	0	0	0	0	
			to	to	to	to	to	to	to	to	to	to	bal.
			2.70	0.80	1.40	0.15	0.10	0.15	0.05	0.15	0.10	0.05	
2090	Alcoa	Alithalite B	1.90	2.40	0	0.08	0	0	0	0	0	0	
			to	to	to	to	to	to	to	to	to	to	bal.
			2.60	3.00	0.25	0.15	0.10	0.12	0.50	0.15	0.10	0.05	
2091	Pechiney	CP 274	1.70	1.80	1.10	0.04	0	0	0	0	0	0	
			to	to	to	to	to	to	to	to	to	to	bal.
			2.30	2.50	1.90	0.16	0.20	0.30	0.10	0.10	0.25	0.10	
	Pechiney	CP 276	1.90	2.50	0.20	0.04	0	0	0	0	0	0	
			to	to	to	to	to	to	to	to	to	to	bal.
			2.60	3.30	0.80	0.16	0.20	0.30	0.10	0.10	0.25	0.10	

developed alloy designated commercially as Weldalite 049 (Pickens et al. 1989) (see *Aluminum–Lithium Based Alloys: Weldability*, Suppl. 2). This alloy has been developed for rather different applications from the preceding alloys, being particularly intended for cryogenic tank applications in space launch systems. This alloy contains less lithium than the other contemporary aluminum–lithium based alloys and is, therefore, heavier but it is very readily weldable and is capable of very high strengths, even in T6 tempers.

4.1 Plate

The major requirement in aluminum alloys for aerospace applications is to achieve a good strength/toughness relationship, preferably simultaneously with good corrosion/stress-corrosion resistance. During the recent developments, the retention of acceptable fracture toughness at high strength levels has proved to be one of the most difficult targets to achieve and numerous investigations have sought to define the optimum structure and the optimum processing route to achieve this structure.

It should be noted that the presence of zirconium in all of the developed aluminum–lithium based alloys results in an unrecrystallized product in which, very largely, the initial cast grains persist into the final product. Very little recrystallization occurs during processing and such that does is disadvantageous. Nearly all the tempers investigated so far have involved the use of either underaged or, at most, peak aged treatments. The alloys require excessively long overaging treatments in order to achieve matrix softening and any recovery of fracture toughness. Fracture behavior in aluminum–lithium based alloys is dominated by grain boundary failure (Miller et al. 1987) and this can be attributed either to the grain boundary structure itself or to the deformation behavior within the matrix. In the former case, precipitation of large phases (such as the icosahedral phase) together with the development of the attendant precipitate-free zone lead to low grain-boundary strengths. The argument in the latter situation is that intense planar slip, resulting from the presence of the coherent, ordered δ', lead to stress concentration at the high-angle grain boundaries. In order to optimize fracture behavior both the matrix and the grain boundary situations have to receive attention.

Consider, then, the grain boundary situation. In order to avoid formation of the icosahedral phase it is important to achieve a sufficiently rapid quench after solution treatment, and the ability to do this will be determined by the thickness of the plate and by the composition of the alloy. In general, the higher copper alloys of the Al–Cu–Li system will have a

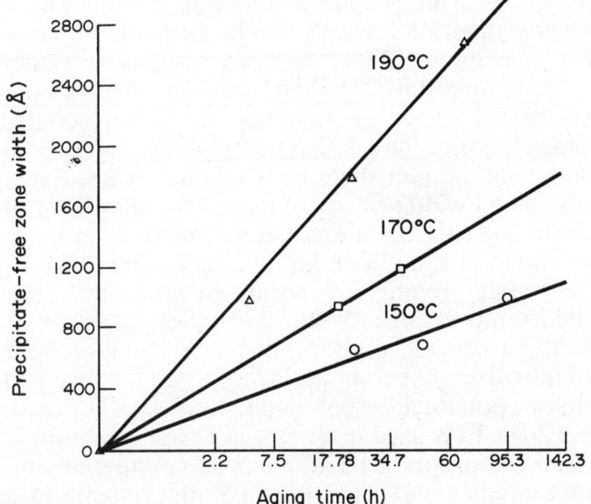

Figure 5
Influence of aging temperature and time on precipitate-free zone width in aluminum–lithium plate

higher susceptibility to icosahedral phase precipitation and the maximum plate thickness in which precipitation can be suppressed will be reduced accordingly. Even in the lower copper alloys of the Al–Li–Cu–Mg system some icosahedral phase precipitation will occur in a plate thicker than about 50 mm, although at thicknesses of 100 mm the extent is sufficiently small for the fracture behavior in the fully aged condition to be dominated by the aging treatment rather than by the quench inefficiency.

As indicated in Sect. 2.6, the icosahedral phase can also form during precipitation treatment so that the use of a low aging temperature is preferred. This has the added advantage (see Fig. 5) of reducing the precipitate-free zone width. However, with aging temperatures of 150 °C or less, the duration of treatment necessary to achieve peak strength can be greatly prolonged. This difficulty can be overcome by the use of stretching after solution treatment and this produces three benefits: first, a dislocation structure is introduced into the matrix, greatly improving the distribution of the heterogeneously precipated strengthening and slip-dispersing phases; second, the aging process is considerably accelerated; and third, the relative volume fraction of precipitates on high-angle boundaries is reduced compared with unstretched material (White and Miller 1987). With increasing stretch the precipitation process is accelerated further and the strength/toughness relationship improved (see Fig. 6). The improvements in strength and toughness that can be achieved by applying the above reasoning are illustrated in Fig. 7 for 50 mm 8090 plate made using an early manufacturing route compared with material made using optimized processing.

4.2 Sheet

Whereas the plate product is invariably unrecrystallized, with sheet products there are arguments in favor of an unrecrystallized product for some types of application and a recrystallized product for others. Unless special processing techniques are adopted, the sheet product will be very largely unrecrystallized (see Fig. 8a). If this kind of structure is achieved by a conventional strip-rolling route, the product will be capable of quite high strengths, particularly in a T8 temper, but will also exhibit a high degree of anisotropy (see Fig. 9). Considerable reductions in this anisotropy can be achieved by encouraging copious S phase precipitation in the 8090 alloy (reducing the variation in tensile strength with rolling direction to about 20 MPa from the 100 MPa shown in Fig. 9) (Peel and McDarmaid 1988), but this is only effective in stretched T8 sheet. The unrecrystallized product will frequently be required for high strength applications involving difficult forming operations and, therefore, multiple solution treatments. In such situations a combination of a modified rolling route with an optimized aging treatment is likely to be more effective, but the presence of matrix substructures will always be important in achieving high strength levels.

For damage tolerant applications, sheet strength becomes relatively unimportant and plane stress fracture toughness and fatigue performance dominate the property requirements (always, of course, accompanied by acceptable corrosion performance). It has been pointed out (Miller et al. 1987) that plane stress fracture toughness is dependent upon the amount of deformation occurring in the plastic zone and this will be strongly influenced by the nature of the grain structure. It is a relatively simple

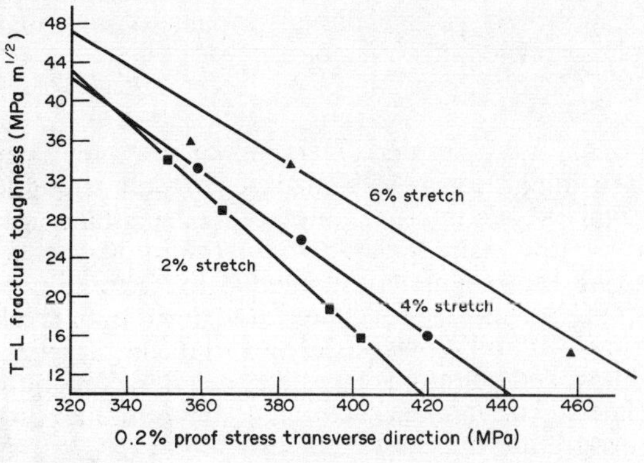

Figure 6
Relationship between strength (0.2% proof stress) and toughness at different levels of post solution treatment stretch in Al–Li–Cu–Mg–Zr plate

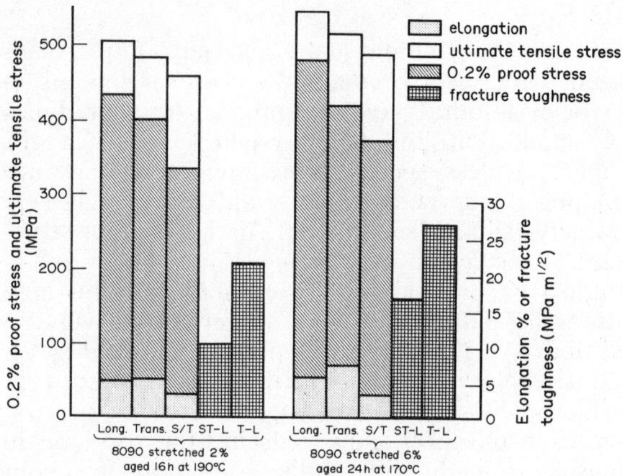

Figure 7
Mechanical properties for 8090 plate processed by early manufacturing route (left) and after optimization (right)

matter to achieve the transition from the unrecrystallized structure of Fig. 8a to the laminar structure of Fig. 8b. However, to achieve a fine-grained recrystallized structure as shown in Fig. 8c requires detailed attention to all aspects of composition, intermediate thermal treatments, cold rolling reductions and rate of heating to solution treatment temperature. The fracture toughness of 8090 sheets processed to possess this range of microstructures is shown in Fig. 10, and, at least at damage tolerant strength levels (0.2% proof stress ≈290 MPa), the fine-grained recrystallized material exhibits considerably superior fracture toughness to the alternative structures. These fracture toughness values compare quite favorably with those for the traditional 2024 T3 fuselage skinning sheet, but detailed comparisons of the relative fracture performances of aluminum–lithium sheet and 2024 T3 sheet at very wide panel widths is continuing. The other issue of continuing concern in this area relates to the relative stabilities of aluminum-lithium sheet and 2024 T3 sheet after prolonged exposure to moderately elevated temperatures. While the typical aging treatment for damage tolerant aluminum–lithium sheet would currently be about 24 h at 150 °C, which still leaves a potential for precipitation, it is also the case that 2024 T3 is used in a naturally aged condition in which the properties could change substantially in, for example, a 20 year life operating largely in a desert environment. Work continues to clarify these issues.

4.3 Extrusions

In heavy extrusions the manufacturing principles involved are very similar to those involved in plate manufacture. Thus, quench rates from solution treatment to avoid icosahedral phase formation are important as is stretching to maximize matrix precipitation and minimize grain boundary precipitation. The extent of stretching, however, is probably best confined to the conventional 2–3% level since the high degree of crystallographic preferred orientation that is retained in the unrecrystallized extrusion structure produces high properties in the longitudinal direction (it should be noted that extrusions in

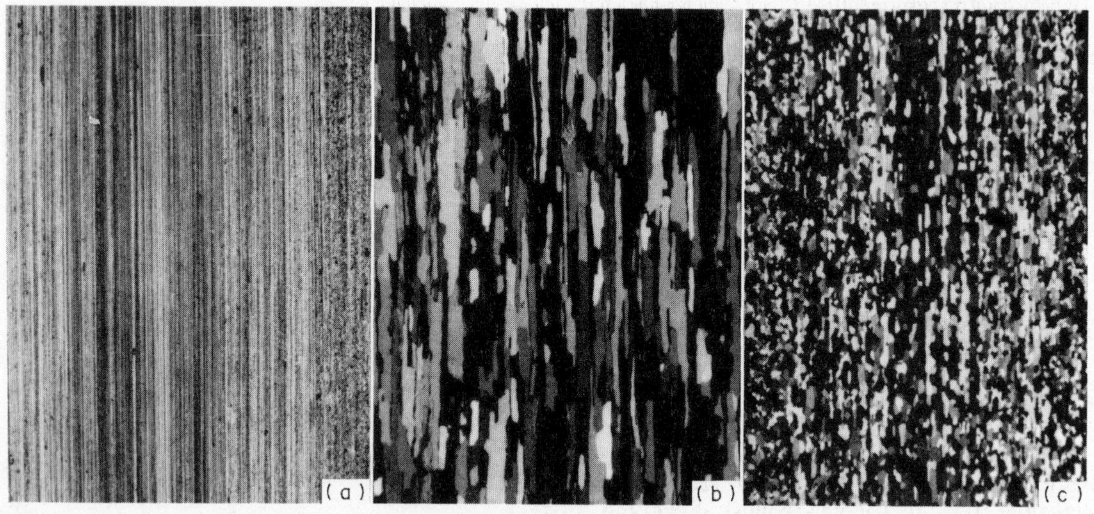

Figure 8
Optical micrographs of the range of structures possible in solution-treated aluminum–lithium sheet (8090): (a) unrecrystallized; (b) laminar recrystallized; (c) fine grained recrystallized (×50)

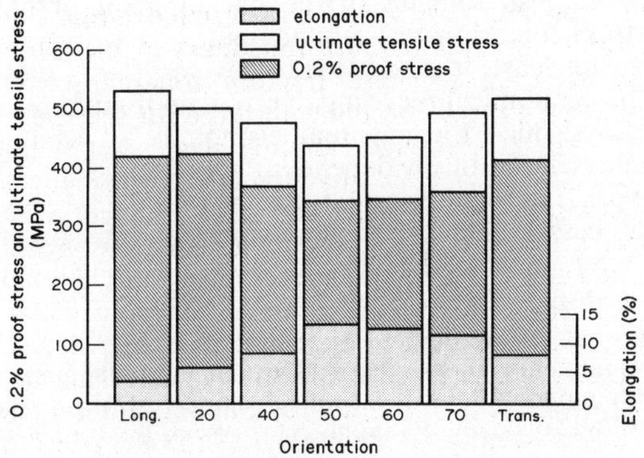

Figure 9
Anisotropy of properties in strip-rolled, unrecrystallized aluminum–lithium sheet in the T8 temper, tested in different directions relative to the rolling direction

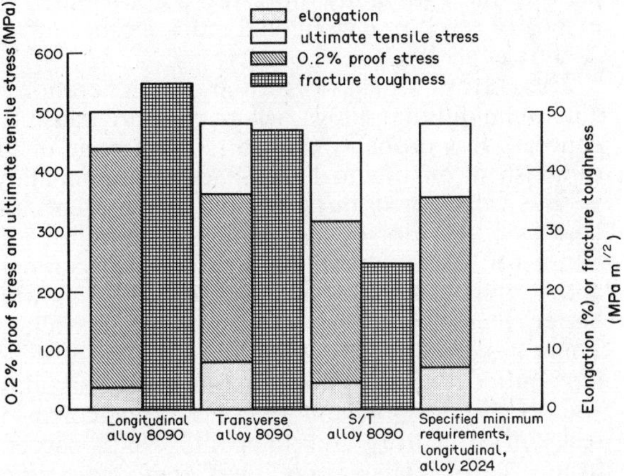

Figure 11
Strength and toughness properties for a large, damage tolerant, wing stringer extrusion in 8090 alloy compared with the 2024 specification in temper T351 (the 2024 specification has no toughness requirements)

the baseline alloys are also highly anisotropic). Higher degrees of stretch would be likely to worsen the balance between longitudinal and long-transverse properties. Aging conditions, again, should be chosen to minimize formation of the icosahedral phase. In general there are probably fewer problems in achieving extrusion properties that match, or exceed, those of the baseline alloys than with any other product form. As an example, Fig. 11 shows the properties of a large stringer in 8090 alloy in a damage tolerant temper, compared with the properties required from alloy 2024 T351.

With thin extrusions, the problem is to suppress recrystallization and this has to be achieved by producing a cast billet with an optimum dispersion of Al_3Zr for inhibiting recrystallization together with the selection of appropriate extrusion conditions.

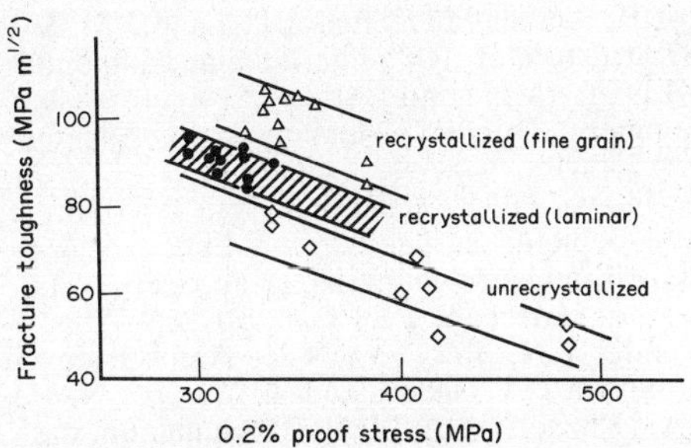

Figure 10
Strength/toughness relationship for sheet with the microstructures illustrated in Fig. 8 (500 mm centre notched panels, LT orientation)

4.4 Forgings

With hand forgings, where cold compression after solution treatment would normally be possible, the principles involved in achieving the optimum combination of properties are identical with the principles for plate. Applying these principles, it has been demonstrated (Lewis et al. 1987) that it is possible to achieve properties suitable for high performance aerospace applications from the 8090 alloy.

5. *Fatigue and Corrosion Behavior*

The previous sections have covered the development of alloys and of processing routes for optimum strength/toughness combinations. For aerospace applications both fatigue performance and corrosion performance are also, of course, of great importance.

As far as fatigue is concerned, little or no effort has been devoted to the development of fatigue resistant compositions or structures, probably because all of the alloys developed so far have demonstrated fatigue performance at least as good as the baseline alloys. However, substantial effort has been applied to explaining the fatigue behavior, including sixteen papers on various aspects of the topic in the Proceedings of the Fourth International Aluminum–Lithium Conference (Champier et al. 1987). In very simple terms, the low fatigue crack growth rate generally observed is attributed to extensive crack branching and crack closure effects (Peters et al. 1987).

The corrosion behavior of the new generation of aluminum–lithium alloys has given more cause for concern. It is probably true to say that none of the established conventional aerospace aluminum alloys is free from corrosion problems. At present, the corrosion behavior of the aluminum–lithium based alloys (in tempers developed to optimize strength and fracture toughness) is the same as, or somewhat better than, that of the equivalent baseline alloys. There has certainly been effort devoted to the development of generally more corrosion resistant alloys and to the development of more stress-corrosion resistant structures and tempers. While success, from a corrosion viewpoint, has been achieved in both these endeavors, the resulting materials have not yet simultaneously achieved the desired density reductions, strength and toughness requirements, and corrosion resistance. A detailed review of the published data on this topic is beyond the scope of this article but, again, a large number of papers on the topic have been presented in the three most recent International Aluminum–Lithium conferences (Baker et al. 1986, Champier et al. 1987, Sanders and Starke 1989).

6. The Future

This paper summarizes in a few thousand words a development that has consumed millions of dollars over a period of a decade. The alloys concerned have, so far, been employed in a number of "demonstrator" aircraft applications and substantial quantities have also been supplied for use in classified space applications. For true success to be achieved the materials will have to be used in the main structure of large civil aircraft, and test/qualification programs are currently underway which, it is hoped, will enable this to happen.

Despite the effort expended so far, there remain significant development tasks if major substitutions of the existing aluminum aerospace alloys are to be achieved. None of the currently proposed high strength aluminum–lithium based alloys can be regarded as wholly satisfactory, particularly because of problems associated with quench sensitivity in thick sections. The alloys are, necessarily, expensive and this will encourage the use of near net-shape forming processes such as precision forging or superplastic sheet forming. Considerable success has been achieved with the latter but with both precision forgings and superplastically formed components the finished article cannot take advantage of cold working after solution treatment so that the strength potential is restricted. Development of an alloy giving high T6 properties would be highly desirable. Work also remains in the development of highly stress-corrosion resistant tempers. Last, but by no means least, it is essential to develop techniques for the economical recycling of scrap aluminum–lithium alloys since the eventual economics of using the alloys are critically dependent upon this issue. While work has demonstrated the potential of recovery processes, such as vacuum distillation (Wilson et al. 1989), more work is certainly required on techniques of scrap segregation and scrap handling.

See also: Aerospace Materials; Airframe Materials; Aluminum Alloys: Heat Treatment; Aluminum and Aluminum Alloys: Selection; Aluminum–Lithium Based Alloys: Weldability (Suppl. 2)

Bibliography

Baker C, Gregson P J, Harris S J, Peel C J (eds.) 1986 *Aluminum–Lithium Alloys* III. Institute of Metals, London

Ball M D, Lagace H 1986 Characterisation of coarse precipitates in an overaged Al–Li–Cu–Mg alloy. In: Baker, Gregson, Harris and Peel 1986, pp. 555–64

Bretz P E, Sawtell R R 1986 "Alithalite" alloys; progress, products and properties. In: Baker, Gregson, Harris and Peel 1986, pp. 47–56

Brook G B 1976 Influence of trace elements on the control of properties of high strength creep resistant alloys. In: *Aluminium Alloys in Aircraft Industry*. Technicopy, Gloucestershire, UK, pp 185–93

Cassada W A, Shiflet G J, Starke E A 1986 Characterisation of two grain boundary precipitates in Al–Li–Cu alloys with electron micro diffraction. In: Starke E A Jr, Sanders TH (eds.) 1986 *Aluminium Alloys—Their Physical and Mechanical Properties,* Vol. II. Chamelion Press, London, pp. 695–710

Champier G, Dubost B, Miannay D, Sabetay L (eds.) 1987 *4th Int. Aluminium–Lithium Conf.* Editions de Physique, Les Ulis, France

Clark E R, Gillespie P, Page F M 1986 Heat treatment of Li/Al alloys in salt baths. In: Baker, Gregson, Harris and Peel 1986, pp. 159–63

Damage Tolerant Design Handbook 1983. Metals and Ceramics Information Center, Columbus, OH

Dickenson R C, Lawless K R, Wefers K 1988 Internal LiH and hydrogen porosity in solutionized Al–Li alloys. *Scr. Metall.* 22: 917–22

Dubost B, Audier M, Jeanmart P, Lang J M, Sainfort P 1987 Structure of stable intermetallic compounds of the Al–Li–Cu–(Mg) and Al–Li–Zn–(Cu) systems. In: Champier, Dubost, Miannay and Sabetay 1987, pp. 497–504

Fox S, Flower H M, McDarmaid D S 1986 Formation of solute depleted surfaces in Al–Li–Cu–Mg–Zr alloys and their influence on mechancial properties. In: Baker, Gregson, Harris and Peel 1986, pp. 263–72

Fridlyander I N, Shiryaeva N V, Ambortsumyan S M, Gorokhova T A, Gabidullin R M, Sidorov N G, Sorokin N A, Kuznetsov A N 1967 Aluminium-base alloy. British Patent No. 1, 172, 736

Gatenby K M, Reynolds M A, White J, Palmer I G 1989 The role of microstructure in controlling the mechanical properties of 8090 in damage tolerant tempers. In: Sanders and Starke 1989, pp. 909–19

Gayle F W 1987 The icosahedral Al–Li–Cu phase. In: Champier, Dubost, Miannay and Sabetay 1987, pp. 481–8

Gilman P S, Brooks J W, Bridges P J 1986 High temperature tensile properties of mechanically alloyed Al–Mg–Li alloys. In: Baker, Gregson, Harris and Peel 1986, pp. 112–20

Gu B P, Mahalingham K, Liedl G L, Sanders T H 1986 The δ' (Al_3Li) particle size distributions in a variety of Al–Li alloys. In: Baker, Gregson, Harris and Peel 1986, pp. 360–8

Hardy H K 1955 Trace element effects in some precipitation hardening aluminium alloys. *J. Inst. Met.* 84: 429–39

Hardy H K, Silcock J M 1955 The phase sections at 500 °C and 350 °C of aluminium rich aluminium–copper–lithium alloys. *J. Inst. Met.* 84: 423–8

Jacoby J E, Yu H, Ramser R A 1986 Direct chill casting of aluminum–lithium alloys. US Patent No. 4, 610, 295

Kojima K A, Lewis R E, Kaufmann B 1989 Microstructural characterisation and mechanical properties of a spray cast Al–Li–Cu–Mg–Zr alloy. In: Sanders and Starke 1989, pp. 85–91

Lapasset G, Loiseau A 1987 A TEM study of icosahedral and near icosahedral phases in 8090 alloy. In: Champier, Dubost, Miannay and Sabetay 1987, pp. 489–95

Le Baron I M 1942 Aluminum alloy. US Patent No. 2, 381, 219

Lewis R E, Starke E A, Coons W C, Shiflet G J, Willner E, Bjeletich J G, Mills C H, Harrington R M, Petrakis D N 1987 Microstructure and properties of Al–Li–Cu–Mg–Zr (8090) heavy section forgings. In: Champier, Dubost, Miannay and Sabetay 1987, pp. 643–52

Long G 1957 Explosions of molten aluminum in water and their prevention. *Met. Prog.* 71: 107–12

Meyer P, Dubost B 1986 Production of aluminium–lithium alloy with high specific properties. In: Baker, Gregson, Harris and Peel 1986, pp. 37–46

Miller W S, White J, Lloyd D J 1987 The physical metallurgy of aluminium–lithium–copper–magnesium–zirconium alloys 8090 and 8091. In: Champier, Dubost, Miannay and Sabetay 1987, pp. 139–49

Mond A L 1921 Alloys of aluminium with lithium. British Patent No. 147, 903

Mortimer, G. 1929. *Aircraft Engineering* (September)

Noble B, Thompson G E 1972 T_1 (Al_2CuLi) precipitation in Al–Cu–Li alloys. *Met. Sci. J.* 6: 167–74

Nozato R, Nakai G R 1977 Thermal analysis of precipitation in Al–Li alloys. *Trans. Jpn. Inst. Met.* 18: 679–89

Page F M, Chamberlain A T, Grimes R 1987 The safety of molten aluminium–lithium alloys in the presence of coolants. In: Champier, Dubost, Miannay and Sabetay 1987, pp. 63–73

Papazian J M, Wagncr J P W, Rooney W D 1987 Porosity development during heat treatment of aluminium–lithium alloys. In: Champier, Dubost, Miannay and Sabetay 1987, pp. 513–9

Peel C J, Evans B 1983 The philosophy of the development of improved Al–Li alloys for use in aerospace structures. In: *The Metallurgy of Light Alloys*. Institution of Metallurgists, London, pp. 32–47

Peel C J, Evans B, Baker C, Bennett D A, Gregson P J, Flower H M 1983 The development and application of improved aluminium–lithium alloys. In: Sanders and Starke 1984, pp. 363–92

Peel C J, McDarmaid D S 1988 The present status of the development and application of 8090 and 8091 alloys. Presented at A.S.M. Westec Aluminium–Lithium Symposium, Los Angeles, March 1988

Peters M, Bachmann V, Welpmann K 1987 Fatigue crack propagation behaviour of the Al–Li alloy 8090 compared to 2024. In: Champier, Dubost, Miannay and Sabetay 1987, pp. 785–91

Pickens J R, Heubaum F H, Langan T J, Kramer L S 1989 Al–(4.5–6.3) Cu–1.3 Li–0.4 Ag–0.4 Mg–0.14 Zr Alloy Weldalite 049. In: Sanders and Starke 1989, pp. 1397–414

Sanders T H, Starke E A Jr (eds.) 1981 *Aluminum–Lithium Alloys*. Metallurgic Society of American Institute of Mining, Metallurgical and Petroleum Engineers, Warrendale, PA

Sanders T H, Starke E A Jr (eds.) 1984 *Aluminum–Lithium Alloys* II. Metallurgical Society of American Institute of Mining, Metallurgical and Petroleum Engineers, Warrendale, PA

Sanders T H, Starke E A Jr (eds.) 1989 *Aluminum–Lithium Alloys* V. MCE Publications, Birmingham, UK

Thompson G E, Noble B 1973 Precipitation characteristics of Al–Li alloys containing Mg. *J. Inst. Met.* 101: 111–5

Thorne N, Dubas A, Lang J M, Degreve F, Meyer P 1987 SIMS determination of the surface lithium depletion zone in Al–Li alloys by quantitative image analysis. In: Champier, Dubost, Miannay and Sabetay 1987, pp. 521–6

White J, Miller W S 1987 The effect of FTMT on the grain boundary microstructure of aluminium–lithium–copper–magnesium–zirconium alloys. In: Champier, Dubost, Miannay and Sabetay 1987, pp. 425–31

Williams D B, Edington J W 1975 The precipitation of delta-prime (Al_3Li) in dilute aluminium–lithium alloys. *Met. Sci.* 9: 529–32

Wilson W R, Allan D J, Dalmijn W L, Brassinga R D 1989 Segregation of Al–Li from mixed alloy turnings. In: Sanders and Starke 1989, pp. 497–518

Wilson W R, Allan D J, Stenzel O, Lorke M, Krone K W, Seebauer C 1989 Al–Li scrap recycling by vacuum distillation. In: Sanders and Starke 1989, pp. 473–96

R. Grimes
[British Alcan Aluminium, Solihull, UK]

Aluminum–Lithium Based Alloys: Weldability

Lithium is an extremely beneficial alloying addition to aluminum because it reduces density and increases stiffness (elastic modulus). In addition, lithium can induce potent strengthening in aluminum, particularly when combined with copper or copper and magnesium alloying additions. Unfortunately, the cost premium for aluminum–lithium alloys (three to four times the cost of conventional aluminum aircraft alloys) has been a major obstacle to widespread acceptance of these alloys.

Table 1
Nominal compositions of conventional aluminum-base filler alloys used in aluminum–lithium welding studies (wt%)

Alloy	Cu	Mg	Si	Mn	Ti	Zr	V	Cr	Al
1100									>99
2319	6.3			0.3	0.15	0.18	0.10		92.97
4043			5.2						94.8
4047			12.0						88.0
4145	4.0		10.0						86.0
5356		5.0		0.12			0.12	0.12	94.64
5556		5.1		0.8	0.12		0.12	0.12	93.74

However, the high cost to put a kilogram of payload into low earth orbit (US$7900 kg^{-1}) has created potential applications in space launch systems where reduced weight is critical and the cost premium could be tolerated. Since space launch systems are most often fabricated by welding, to effectively contain propellants under pressure, the weldability of aluminum–lithium alloys has assumed increasing importance in the late 1980s.

1. Historical Perspective of Aluminum–Lithium Alloy Weldability

Weldability is the resistance of an alloy to hot cracking during welding—a condition caused by the inability of the solidifying metal to withstand the strains associated with solidification shrinkage and the stresses imposed by the structure that is "fit up" to be welded. An excellent review of the hot cracking of aluminum and aluminum–lithium alloys was compiled by Cross (1986). The weldability of aluminum–lithium alloys was first emphasized in the USSR in the mid-1960s by Fridlyander and his coworkers who developed the Al–5%Mg–2%Li (wt%) alloy 01420 for welded applications. Alloy 01420 exists in several variants, which differ in grain-refining additions; 0.1% zirconium, 0.4% manganese, and 0.1% zirconium and 0.4% manganese are the most common. Alloy 01420 was shown to be weldable by virtually every major technique, but the alloy was particularly sensitive to weld-zone porosity because of the affinity of aluminum–lithium alloy surfaces for moisture. However, porosity was reduced to tolerable levels by either mechanical milling or chemical milling in a 200 $g\,l^{-1}$ NaOH aqueous solution to remove ~0.3 mm from the surface prior to welding. Although little information is available concerning the maximum allowable time interval between pretreatment and welding of aluminum–lithium alloys, the interval probably depends upon ambient humidity. It is generally advisable to weld soon after pretreatment. The weldability of 01420 and the effectiveness of the Soviet pretreatments in reducing porosity were demonstrated in the West (Pickens et al. 1983). Furthermore, the weldability of 01420 was extensively addressed in a review paper by Pickens (1985), which was updated by Pickens et al. (1989).

With the renewed interest in the weldability of aluminum–lithium alloys, which was stimulated by the desire of the US Government to develop alternative launch systems to the Space Shuttle, several studies were performed to assess the weldability of the leading aluminum–lithium alloys—Al–2.6%Li–1.2%Cu–0.7%Mg–0.12%Zr (alloy 8090) and Al–2.7%Cu–2.2%Li–0.12%Zr (alloy 2090). Furthermore, a new alloy, Weldalite 049, was designed specifically to be weldable and thus competitive for space launch applications.

2. Weldability of Alloy 8090

Gittos (1987) studied the gas tungsten arc (GTA) and gas metal arc (GMA) weldability of 8090 using conventional and parent alloy fillers. Hot-cracking susceptibility was observed with the aluminum–silicon filler 4043, the aluminum–copper filler 2319 and the parent alloy filler (filler compositions are shown in Table 1); the most severe hot cracking occurred with the parent alloy filler. The aluminum–magnesium filler alloy 5556 produced hot-crack-free weldments. Gittos also used the Houlcroft test (a welding sheet that contains transverse slots of increasing length to vary restraint) to assess the inherent susceptibility of 8090 and found it to be similar to that of the aluminum–magnesium–silicon alloy 6082, a hot-crack-sensitive alloy. Edwards and Stoneham (1987) also used the Houlcroft test and determined that the hot-cracking susceptibility of 8090 is similar to that of the Al–4.4%Cu–0.8%Mn–0.8%Si–0.5%Mg alloy 2014. This conventional alloy is also susceptible to hot cracking and is generally welded with a rather forgiving filler, such as 4043, to alleviate hot-cracking problems—but at the expense of weldment strength.

Skillingberg (1986) assessed the GTA and GMA

weldability of 8090 with 1100, 4043, 5356, 2319 and parent alloy fillers. Crack-free weldments were fabricated with 1100 and 4043, fillers that often reduce hot cracking, as well as with 5356, but significant cracking was observed with 2319 and, to a lesser extent, with the parent alloy filler. In these studies, as-welded tensile strengths of 8090 ranged from about 200–300 MPa with the better conventional fillers. Higher weldment strengths were reported with the parent alloy filler; for example, 349 MPa by Gittos (1987), but hot-cracking susceptibility will probably reduce the viability of 8090 weldments made with parent alloy filler in a production environment.

3. *Weldability of Alloy 2090*

Martukanitz et al. (1987) investigated the GTA weldability of alloy 2090 in the T8E41 temper, using conventional fillers and an inverted T joint to assess hot-cracking susceptibility and butt weldments to assess joint strength. The plates were prepared for welding by either mechanical milling of 0.08–0.23 mm from the surface or chemical milling of 0.04–0.15 mm from the surface in a 5 wt% NaOH aqueous solution at 49 °C, followed by desmutting in a chromic–sulfuric acid solution. Prior to welding, the edges adjacent to all surfaces were mechanically scraped. Weldment strength increased linearly with the amount of material removed over the ranges investigated. Hot-cracking susceptibility of the fillers, ranked in increasing order, was: 4047, 4145, 2319, 4043, 5356. Only 5356 displayed sufficiently severe hot-cracking susceptibility to be considered beyond the limit for commercial weldability. The as-welded strength of 2090 T8E41 welded with 2319 was 232 MPa; re-solution, heat treatment, quenching and aging raised this strength to 386 MPa. A strength of 322 MPa was obtained from an electron beam (EB) weldment in the as-welded condition.

Skillingberg (1986) also assessed the GTA and GMA weldability of 2090 using parent alloy and 5356 fillers. He selected a single V groove with a 60° included angle and filed and degreased the edges of the plates shortly before welding. Skillingberg observed hot cracking in one of the 2090/5356 weldments and attained as-welded strengths of 253 MPa with the 5356 filler and 230 MPa with the 2090 filler.

The weldment strengths attained with 8090 are generally higher than those with 2090. Both alloys are weldable, although the susceptibility to hot cracking reported for each alloy suggests that care must be exercised in selecting the appropriate filler alloy for each situation. It is significant to note that the pretreatments used by the various investigators, mostly mechanical milling, were successful in reducing weld-zone porosity.

4. *Weldalite 049*

After conducting the 1985 review, Pickens endeavored to design a weldable aluminum–lithium alloy to replace mainstay alloys 2219 and 2014 in space launch applications (see Pickens et al. 1989). The alloy is a lithium-containing modification of 2219, to which silver and magnesium have been added to stimulate precipitation. The composition of the alloy is Al–(4.0–6.3%)Cu–1.3%Li–0.4%Ag–0.4%Mg–0.14%Zr; a copper content in the lower part of the range was selected for commercial scale-up of sheet and of plate. The alloy is unique in that it attains high strength in several useful tempers, for example, 700 MPa in either the T6 or T8 artificially aged tempers, and 590 MPa in the naturally aged, T4 temper.

The alloy is readily weldable by GTA, GMA, and EB welding. Tack and Loechel (1989) obtained a mean as-welded strength of 311 ± 9 MPa using the 2319 filler while developing GTA welding parameters to fabricate a subscale cryogenic Weldalite 049 tank (see Fig. 1). Other GTA welding studies by Kramer et al. (1989) routinely demonstrated 340 MPa strengths using the 2319 filler. Autogenous EB weldment strengths of 435 MPa were obtained in the as-welded condition.

Using a proprietary, specifically matched filler wire designed for Weldalite 049, Martin Marietta Manned Space Systems engineers performed variable polarity plasma arc (VPPA) welding studies on 2219, 2090, and Weldalite 049. The Weldalite filler increased the strength of 2219 and 2090 significantly compared with weldments made using 2319 (see Table 2). Most noteworthy of all, as-welded VPPA weldments using the 049 filler had a mean strength of 370 MPa, as high as any reported on similar VPPA aluminum alloy weldments. No hot-cracking susceptibility was observed on any Weldalite 049 weldment, despite the severe restraints imposed. This work emphasizes the need for specifically designed fillers for welding high-performance aluminum alloys.

5. *Conclusion*

Interest in the weldability of aluminum–lithium alloys has increased dramatically in the mid-1980s because of the need to reduce the weight of space launch systems, where the price premium of such alloys can probably be tolerated. Aluminum–lithium alloys can indeed be weldable, and leading alloys such as 8090 and 2090 have been welded by several techniques. Hot cracking has been observed for these two alloys with certain filler alloys commonly used in existing applications. Consequently, care must be exercised in selecting an appropriate filler. The susceptibility of aluminum–lithium alloys to weld-zone porosity is controllable

Figure 1
Subscale, prototype cryogenic tank made using Weldalite 049 welded by GTA using 2319 filler

using appropriate pretreatment. This most often entails mechanical milling of up to ~0.3 mm from the alloy prior to welding, although chemical milling techniques in sodium hydroxide are also successful. To further reduce porosity, root-side shielding is often required to fabricate good quality, large-scale welded structures. This complicates welding and increases fabrication costs.

A new aluminum–copper–lithium alloy, Weldalite 049, has been specifically designed to be weldable, and produces extremely high joint strengths. Moreover, a specifically designed Weldalite filler alloy has been developed that improves the weldment strength of 2219, as well as that of aluminum–copper–lithium alloys.

No commercial welding parameters for aluminum–lithium alloys had been reported by 1989. It would be useful to document correlations between welding parameters and resulting weldment microstructures and properties. Furthermore, much work needs to be performed on weldment toughness, fatigue and corrosion susceptibility, leading to weldment design tolerances, before aluminum–lithium alloys are widely used in welded applications.

Table 2
Mean VPPA, square-butt, as-welded tensile properties of Weldalite 049, 2090 and 2219 weldments made with conventional and Weldalite filler

Base metal/ filler	Thickness (cm)	Ultimate tensile strength (MPa)	Apparent yield strength (MPa)	Apparent elongation in 2.54 cm (%)
2219/2319	0.95	273	141	7.9
2219/049	0.50	325	161	9.0
2090/2319	1.27	252	156	8.6
2090/049	0.65	285	147	7.1
049/2319	0.95	274	248	1.5
049/049	0.95	372	290	3.0

See also: Aluminum: Alloying; Aluminum–Lithium Based Alloys (Suppl. 2)

Bibliography

Cross C E 1986 Weldability of aluminum lithium alloys: An investigation of hot tearing mechanisms. Ph.D. thesis, Colorado School of Mines, Golden, CO

Edwards M R, Stoneham V E 1987 The fusion welding of Al–Li–Cu–Mg (8090) alloy. *J. Physique* 48(C-3): 293–9

Gittos M F 1987 Gas shielded arc welding of the Al–Li alloy 8090. Report No. 7944. 01/87/556.2, The Welding Institute, Abington, UK

Kramer L S, Heubaum F H, Pickens J R 1989 The weldability of high-strength Al–Cu–Li alloys. In: Starke E A Jr, Sanders T H (eds.) *Proc. 5th Int. Aluminum–Lithium Conf.* Materials and Components Engineering Publications, Warley, UK, pp. 1415–24

Martukanitz R P, Natalie C A, Knoefel J O 1987 The weldability of an Al–Cu–Li alloy. *J. Met.* 39 (November): 38–42

Pickens J R 1985 A review of the weldability of lithium-containing aluminum alloys. *J. Mater. Sci.* 20: 4247–58

Pickens J R, Heubaum F H, Langan T J, Kramer L S 1989 Al–(4.5–6.3Cu)–1.3Li—0.4Ag—0.4Mg—0.14Zr alloy Weldalite™ 049. In: Starke E A Jr, Sanders T H (eds.) *Proc. 5th Int. Aluminum–Lithium Conf.* Materials and Components Engineering Publications, Warley, UK, pp. 1397–414

Pickens J R, Langan T J, Barta E 1986 Weldability of Al–5Mg–2Li–0.1Zr alloy 01420. In: Baker C, Gregson

P J, Harris S J, Peel C J (eds.) *Aluminum Lithium III.* The Institute of Metals, London, pp. 137–47
Skillingberg M H 1986 Fusion welding of Al–Li–Cu–(Mg)–Zr plate. In: *Aluminum Tech. 86, Proc. Int. Conf.* pp. 509–15
Tack W T, Loechel L W 1989 Weldalite™ 049: Applicability of a new high-strength weldable Al–Li–Cu alloy. In: Starke E A Jr, Sanders T H (eds.) *Int. Aluminum–Lithium Conf.* Materials and Components Engineering Publications, Warley, UK, pp. 1457–67

J. R. Pickens
[Martin Marietta Corporation, Baltimore, Maryland, USA]

Aluminum Nitride

The development of aluminum nitride ceramic materials has been driven largely by the demand for a high thermal conductivity material which is also an electrical insulator, for use in applications such as substrates, packaging, power electronics or microwave tubes. The self-heating property of densely arrayed electronic devices is one of the problems that has to be solved to improve the reliability of electronic systems. The heat removal from very large scale integrated (VLSI) circuits is determined by the heat transfer from the silicon chip via the packaging material to the environment and one factor which influences the transfer rate is the thermal conductivity of the packaging material. This is true for power electronics, in VLSI and in LSI circuits.

Commercial aluminum nitride materials are available with a thermal conductivity greater than $150\,W\,m^{-1}\,K^{-1}$ at a temperature of 300 K. In comparison, iron has the value of $78\,W\,m^{-1}\,K^{-1}$ at the same temperature; however, as in the case of all metals, it is also a good electrical conductor with a resistivity of $8.6\times10^{-8}\,\Omega\,m$. Compared with other potential substrate materials, aluminum nitride is the only one which combines high thermal conductivity, excellent electrical insulation and nontoxicity (see Table 1). The specific resistance of aluminum nitride materials is between 10^{19} and 10^{21} times higher than that of iron. As in all materials with more than 95 vol.% of one phase, the electrical and thermophysical properties are dominated by the major phase; that is, by the character of the aluminum nitride single crystal.

Table 1
Properties of commercial aluminum nitride ceramics at room temperature

Property	Value
Density ($kg\,m^{-3}$)	3255
Bending strength (MPa)	300–500
Fracture toughness ($Mpa\,m^{1/2}$)	3–4
Thermal conductivity ($W\,m^{-1}\,K^{-1}$)	100–200
Electrical conductivity ($\Omega^{-1}\,m^{-1}$)	10^{-11}–10^{-13}
Relative dielectric constant	8–8.5
Dielectric strength ($kV\,mm^{-1}$)	20–25
Dielectric loss tangent at 1 MHz	<0.001

1. Single Crystal Aluminum Nitride

Field properties such as the dielectric constant and the thermal conductivity are mainly determined by the properties and the partial volume of the phases existing in the bulk material. Highly thermally conductive aluminum nitride materials include more than 95 vol.% aluminum nitride. It is important, therefore, to know the single crystal properties.

The symmetry of the aluminum nitride lattice is hexagonal and it belongs to the space group $P6_3mc$. The lattice constants at room temperature are $a = 311.1\pm0.1$ pm and $c = 498.0\pm0.1$ pm. The most important property of aluminum nitride is the high value of thermal conductivity, which has been theoretically estimated by Slack to be $320\,W\,m^{-1}\,K^{-1}$. The experimentally estimated value of the thermal conductivity parallel to the [001] direction of a single crystal contaminated with 343 ppm oxygen and 50 ppm tungsten by weight is $285\,W\,m^{-1}\,K^{-1}$ (Slack et al. 1987). The value of the thermal conductivity of a single crystal is mainly influenced by the anharmonicity of the lattice. The anharmonicity increases with increasing mass differences between occupants of the lattice sites and with increasing anisotropy of the lattice binding forces. The main impurities in aluminum nitride are oxygen and carbon. Iron, magnesium and silicon can also be dissolved in the aluminum nitride lattice at higher temperatures.

In commercial aluminum nitride ceramic materials, the content of secondary phases such as glass or crystallized sintering additives is always below 2–3 vol.%. If the orientation of the aluminum nitride grains is random, the dielectric properties of the material reflect a mixing of the tensor properties of the single crystal, which have not yet been measured. The dielectric constant of the polycrystalline material is given in Table 1.

The thermal expansion of aluminum nitride ($3.9\,MK^{-1}$ from room temperature to 500 K) is very similar to that of silicon ($3\,MK^{-1}$) over the same temperature interval, which obviates cracking from thermal misfit stresses in layered AlN–Si materials.

In air, oxidation of the material starts at 1250 K. In pure nitrogen at 100 kPa pressure, the material is stable up to 2790 ± 50 K; it decomposes at higher temperatures.

2. Preparation of Aluminum Nitride

The main commercial production route for aluminum nitride powder is the carbo-nitridation of aluminum oxide, Al_2O_3. There are two problems with this reaction: first, the high temperature of 1884 K and second, the fact that all impurities which are dissolved in the carbon or in the aluminum oxide can contaminate the final aluminum nitride powder product. Residual oxygen from the unreacted aluminum oxide and carbon will be found in the powder. The impurities of commercial powders are oxygen ($\approx$1 wt%), iron ($<$0.1 wt%), carbon ($<$0.2 wt%) and other metallic impurities ($<$100 ppm). At higher temperatures during the densification process, these impurities can be dissolved in the aluminum nitride lattice; this will decrease the thermal conductivity of the material.

Reactions between pure gas species allow the preparation of very pure aluminum nitride. The preferred route is the reaction between aluminum chloride ($AlCl_3$) and ammonia, (NH_3), because of the low temperature required the possibility of cleaning the chloride by sublimation.

$$AlCl_3(g) + NH_3(g) \xrightarrow{873\text{–}1973\,K} AlN(s) + 3HCl(g)$$

Aluminum nitride powders are always hydrophilic and react with water to form amorphous aluminum hydroxide ($Al(OH)_3$). During tempering of $Al(OH)_3$ up to 1320 K, it loses water and forms τ-AlOOH below 700 K, which finally reacts to form α–Al_2O_3. If the aluminum oxide is not dissolved in another phase it will react with the aluminum nitride to form mixed crystals with low thermal conductivity during further thermal treatment. For these reasons, the reaction, the storage and the handling of the material should be conducted under controlled atmospheric conditions.

3. Densification of Aluminum Nitride Powders

Aluminum nitride powders are densified by pressureless sintering or by hot pressing. From the the commercial point of view, pressureless densification of aluminum nitride is more important because it allows the sintering of complex shapes and confers consequent economic advantages. The mechanism of densification by pressureless sintering is influenced by additive melt formation and by the solution-reprecipitation of the aluminum nitride. The mobility and the density of lattice vacancies determine the mechanism of densification during hot pressing. Sintering additives are lanthanide oxides, yttrium oxide (Y_2O_3) and alkaline-earth oxides. The melt incorporates impurities such as metals and oxygen and therefore a "gettering" effect, cleaning the aluminum nitride grains, is observed. Densities higher than 99% of the theoretical density and a high thermal conductivity ($>$150 W m^{-1} K^{-1}) are observed with the additions of yttrium oxide or calcium oxide (see Fig. 1). The sintering temperatures are in the ranges 2050–2100 K and 2080–2130 K, respectively.

Figure 1
Scanning electron micrograph of fracture surface of aluminum nitride pressureless sintered with 1 wt% CaO at 2090 K for 15 min

4. Applications

Because aluminum nitride is not poisonous, it will replace beryllium oxide in applications such as heat-sinks in electronic power devices. Semiconductor multilayer substrates, heat exchangers, crucibles for molten metals and fast-heating elements are other potential applications.

Bibliography

Shinozaki K, Iwase N, Tsuge A 1986 High thermal conductive aluminium nitride (AlN) substrates. Japanese Fine Ceramics Association Annual Report: 16–22

Slack G A, McNelly T F 1976 Growth of high purity AlN crystals. *J. Cryst. Growth* 34: 263–79

Slack G A, Tanzili R A, Pohl R O, Vandersande J W 1987 The intrinsic thermal conductivity of AlN. *J. Phys. Chem. Solids* 48: 641–7

A. Kranzmann
[Max-Planck-Institut, Stuttgart, FRG]

Aluminum Oxide for Prosthetic Devices

This article supplements the article *Aluminum Oxide in Biomedical Applications* in the Main Encyclopedia.

From the large number of ceramic materials generally available only a few have been found to possess the combination of properties from which improvements in prosthetic devices can be expected. Among the ceramics used in reconstructive surgery, the aluminum oxide ceramic (alumina ceramic) is the most chemically inactive. After its early evaluation, the alumina ceramic was soon regarded as the prototype of the so-called bioinert materials.

Due to its mechanical properties, especially its rigidity and hardness, the alumina ceramic has found its main uses in hard-tissue, structural applications in orthopedic surgery and dentistry. Most of these applications are based on new aspects of bone remodelling resulting from observations of the tissue reactions in the vicinity of such implants. As a consequence of its inertness, the remodelling of bony tissue adjacent to alumina implants is not disturbed biochemically (by ions or other matter going into solution) or by immune reactions. The interface reactions have, subsequently, been found to be controlled solely by the stress and strain fields created inside the bony tissue by the insertion of the implant. Using this knowledge, rules have been derived for the design of joint and dental implants so that the remodelling results in close bone contact and thus "osseo-integration." Many years of experience with these implant systems has allowed for some judgement about the validity of these rules and the information on which they are based.

In addition, the bioinertness of these ceramics has led to improved devices in nearly load-free applications in ear, nose and throat surgery and in some soft tissue replacements.

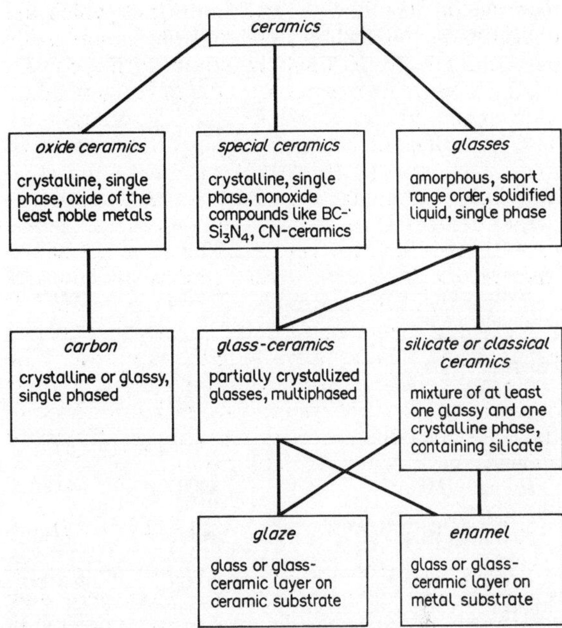

Figure 1
Survey of ceramic materials according to crystallinity and number of phases

1. *Alumina Ceramic*

The alumina ceramic is the main representative of the oxide ceramics. Their definition and position within the realm of ceramic materials can be seen in Fig. 1.

From the structural point of view, ceramics are clearly different from metals and plastics (Heimke 1984). Generally, homopolar and heteropolar compounds and solids are distinguished because either they share some of their electrons which are delocalized from the parent atoms, or they have the electrons clearly bound to the atoms concerned; that is, they are ions. In metals, the freely moving electrons are an essential feature controlling most of their basic properties. In ceramics, and particularly in all oxide ceramics, the ionic nature of the bond is predominant. This implies that ceramics always contain more than one kind of atom (pure carbon is the exception to this), often with several types of both positive and negative species. Oxide ceramics usually consist of one or very few metal oxides.

The different groups of ceramic materials can also be distinguished from their manufacturing processes. All classical ceramics contain silicates, a proportion of which are introduced by the use of clays, which allows the shaping of the desired parts when mixed with the correct amount of water. The oxide and the special ceramics start from industrially prepared mostly pure powders to which some plastics are temporarily added to allow approximate shaping. However, in some ceramics prepared from industrially made raw materials the final compound may not be formed until the "firing" process. (Firing is the process for all ceramics in which the shaped or molded raw piece is transformed into a solid.)

There is also another difference between silicate and oxide ceramics. In the silicate-containing ceramics one or more melts are formed during the firing cycle that solidify and partially crystallize during the process. Thus, the final product is always a mixture of different phases. However, in oxide ceramics, the shaped agglomeration of powder particles is transformed into a polycrystalline solid by solid-state reactions only, mostly by diffusion.

The essential properties of alumina ceramics for all load-bearing applications are summarized in Table 1. Included here are the specifications according to the first generation of standards as well as the actual values for one of the most widely used commercial materials. The differences between these two sets of values result from experiences which indicate the necessity for an extension of the safety margins in some high load-bearing applications (Griss and Heimke 1981a, Heimke and Griss 1981); these improvements will also be included in forthcoming revisions of the standards. For implants in

Table 1
Properties of medical-grade alumina ceramics[a]

Property	Ceramic requirements according to ISO 6474 ASTM F603-83 DIN 58 8353	Frialit bioceramic
Density (g cm^{-3})	>3.9	>3.98
Alumina content (%)	>99.5	>99.9
SiO_2 and alkali metal oxides (%)	<0.1	<0.05
Microstructure, average grain size (μm)	<7	<2.5
Microhardness (MPa)	23 000	23 000
Compressive strength (MPa)	4000	4000
Flexural strength (MPa)	>400	>450
Young's modulus (MPa)	380 000	380 000
Impact strength (cm MPa)	>40	>40
Wear resistance[b] (mm^3 hr^{-1})	0.01	0.001
Corrosion resistance[b] (mg m^{-2} day^{-1})	<0.1	<0.1

a Many of these values are not average but maximum or minimum requirements to be met by all test pieces b The wear and corrosion resistance values refer to particular test arrangements

load-free (or nearly load-free) situations, for example in ear, nose and throat surgery or ophthalmology, somewhat reduced values for the mechanical strength will be stated in the revised versions of some of these standards.

2. *Reasons for Using Alumina Ceramics in Medicine*

The first large scale clinical testing and subsequent applications of alumina ceramics were in total hip replacements and in dentistry.

In the 1960s, total hip replacement became a standard treatment in orthopedic surgery after the introduction of polymethylmethacrylate (PMMA) as the so-called bone cement for the fixation of the polyethylene acetubular socket and the metal stem in the medullary canal of the femur. By 1970, increasing numbers of patients requiring repeat operations stimulated effort to find the causes for these failures. In the early 1970s, it was widely agreed that the soft tissue layer, which separates the surface of the PMMA along most parts of the interface from the normally proliferating bony tissue, was the major cause of the problem. It was also recognized that the polyethylene wear particles contributed, among other things, to a thickening of the soft tissue layer and, therefore, to implant loosening. In dentistry, a direct correlation exists between the thickness of the soft tissue interlayer, the mobility of the implant, the pocket depth of the gingiva surrounding the implant, and the probability of implant failure (Spiekermann 1980). At that time, it was generally assumed that a basic cause for the formation of the soft tissue interlayer was the chemical instability of the materials used, mainly the stainless steels, the cobalt-based alloys, and the PMMA. This instability resulted in products (metal ions or monomers) disturbing the highly sensitive differentiation processes necessary for undisturbed bone formation. The particular type of biocompatibility of some ceramics raised hopes of avoiding such soft tissue interlayers and thereby achieving direct anchorage of implant devices. This soft-tissue-free implant fixation was termed osseo-integration. For joint replacements, the low friction and high wear resistance of alumina ceramics offered additional advantages.

The early compatibility studies of alumina ceramics, including sarcoma rate testing (Griss and Heimke 1981b), had confirmed the expected biological inertness of this material as well as its tribological advantages along the articulating surfaces. The results of other scientists contradicted these findings by showing that soft tissue interlayers between the alumina implants and the surrounding bone could be related to the mechanical effects (Heimke et al. 1981).

3. *Osseo-Integration by Biomechanically Controlled Tissue Remodelling*

Since a bioinert material by definition (summarized in Table 2) does not react with the surrounding tissue, it cannot form any bond with the adjacent

Table 2
The concept of a bioinert material

Requirement	Result
"Nothing goes into solution": leakage of ions or other matter from the implant into the surrounding tissue is below detectability by the cells and without any systemic effect	No biochemical influence on cell differentiation and proliferation. No biochemical information to the cells about the presence of the implant
Strong and fast adsorption of molecules contained in body fluid so that the surface of the implant is covered completely (coated) by the body's own matter	No enzyme reactions: the implant is "camouflaged" against the host's immune system. No foreign body reactions

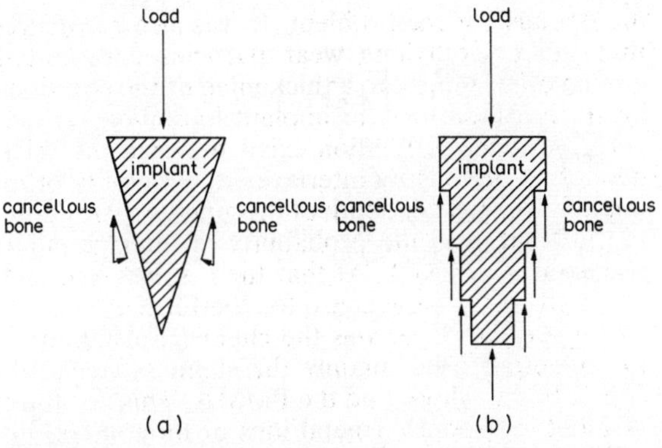

Figure 2
The influence of implant design (shape) on the interfacial stresses (small arrows): (a) large tangential component of force, resulting in shear movement and (b) mainly pressure along the surfaces of the steps, thus avoiding relative movements along these parts of the interface

bone. Thus, only a purely mechanical fixation could be anticipated.

In the initial experiments, in which the test pieces were placed in the bony tissue under nearly load-free conditions, it was recognized that the reorganization of the bony tissue adjacent to an alumina implant follows exactly the same sequence of reactions characteristic of fracture healing (Griss et al. 1975). Further experiments with fully functional total hip replacements in sheep and dogs made it possible to define the design criteria, allowing for a stable and reliable anchorage of implants in the adjacent bony tissue (Griss et al. 1976).

The essential biomechanical requirement for fracture healing is to avoid any shear motion along the interface. In addition to fracture healing, however, it was found that, for maintaining the close bone contact with a bioinert prosthesis, it was essential to preserve this condition of motionlessness along the interface concerned for the entire lifetime of the prosthesis.

Basically, there are only two mechanical situations allowing for a motionless contact: the load-free situation and the situation where the interface is oriented perpendicular to the force acting along the interface. The first, trivial, situation is of no interest for load-bearing implants. Thus, all load-bearing implants must offer sufficiently large interfaces to the adjacent bony tissue so that they are transmitters of mostly pure pressure (forces perpendicular to the surface). Surfaces along which the forces are mainly transmitted parallel (resulting in shear along the interface) are separated by a soft tissue interlayer from the surrounding bond and therefore cannot contribute to osseo-integration and direct load transmission. Figure 2 summarizes these considerations. Clinical experience has shown that the requirement of perpendicularity of the forces meeting the interface allows for a deviation of up to 15°.

The validity of the concept of purely stress-and-strain-field controlled tissue reactions along the interface between bone and an implant of a bioinert material such as alumina ceramic was confirmed by the discovery of the "load-line shadow effect" (Heimke et al. 1982). The details of this effect are shown schematically in Fig. 3. Immediately after pressfitting the implant, the lacunae are filled with blood clots (Fig. 3a). If the condition of motionlessness is maintained during the healing-in period, the blood clot serves as a scaffold for the formation of new bone, filling the lacunae primarily homogeneously (Fig. 3b). After load application, the bony tissue reorganizes to adjust to the stresses acting, resulting in an area of reduced calcification where no stresses are acting—the area shadowed from the load lines (Fig. 3c).

According to the previously described requirements for close bone contact, such contact could not be expected along mainly tangentially loaded interfaces such as the cylindrical surface of the Tübingen dental implant shown in Fig. 4. However, the evaluation of the tissue surrounding two implants, which became available for histology because of an accident suffered by a patient, revealed a close bone apposition not only along the pressure-transmitting interfaces (the surfaces of the steps connecting the cylinders of different diameters) but also along the cylindrical, mainly tangentially loaded areas. Due to this interlocking effect (Heimke et al. 1982), the relative movements created by all load changes resulting from the much higher stiffness of the implant compared with the surroundings (mostly cancellous bone) are realized by the deformation of bony structures some distance away from the surface of the implant rather than by shear movements along the interface. Thus, an interpendence exists between the shape of the implant and the reactions of the bony tissue surrounding it. To achieve a stable and reliable osseo-integration, the shape of the implant must be chosen so that the load pattern (the stress and strain field) created by its insertion allows for remodelling reactions resulting in a close bone contact along as many interfaces as possible.

4. *Alumina Ceramic Implants in Orthopedic Surgery*

The first implant design to follow the above conclusions was that of the cylindrical socket of Frialit total hip replacement (see Fig. 5). This is combined with a ball of the same ceramic fixed on a metal stem. Clinical trials started in September 1974 and the reports of many clinics, up to the late 1980s, show

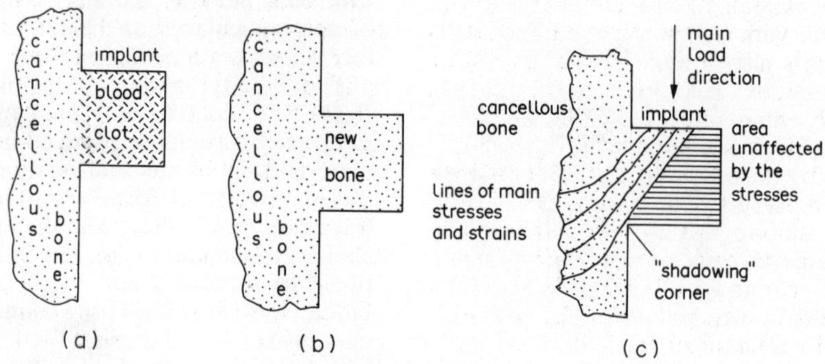

Figure 3
The load-line shadow effect: (a) immediately after pressfitting the implant; (b) the formation of new bone; and (c) the area shadowed from the load lines

they have endured the test of clinical applications. The results of the first five years had been reported in detail and have led to some improvements which became possible because of further progress in the science of the alumina ceramic concerned. In the first generation the ceramic sockets and balls had been combined with femoral components designed for cement fixation. Results of further work led to the design of stems with specially shaped steps and lacunae made from a titanium alloy. This now completely cement-free implantable system stood the test of clinical applications throughout the late 1980s.

The other total hip replacement system using a bioinert alumina ceramic and based approximately on the biomechanical ideas described above was introduced into clinical trials by Mittelmaier in 1974 and this has also withstood widespread clinical tests.

Some years earlier Boutin had already started the clinical use of alumina ceramics in hip surgery. His design, however, closely followed the shape of plastic cups designed for cement implantation and did not take account of the aforementioned biomechanical stress and strain field considerations. Within the field of orthopedics, relatively early alumina ceramic implants have also been tested clinically in tumor surgery of the proximal femur and of the upper arm, but there have been no convincing results.

In knee surgery, early attempts to use alumina ceramic components have not led to large scale applications. Extended experimental work with ceramic-on-ceramic articulating total knee replacements has never entered the state of clinical trials (Geduldig et al. 1976). It was not until the introduction of knee endoprostheses with anchoring portions, femoral condyles of alumina ceramic and an

Figure 4
Alumina ceramic Frialit dental implant, Tübingen type. The implant body is cylindrical with decreasing diameters connected by steps. The surface of the cylinders carry lacunae for interlocking with the bony tissue. The groove in the coronal portion receives the gingiva

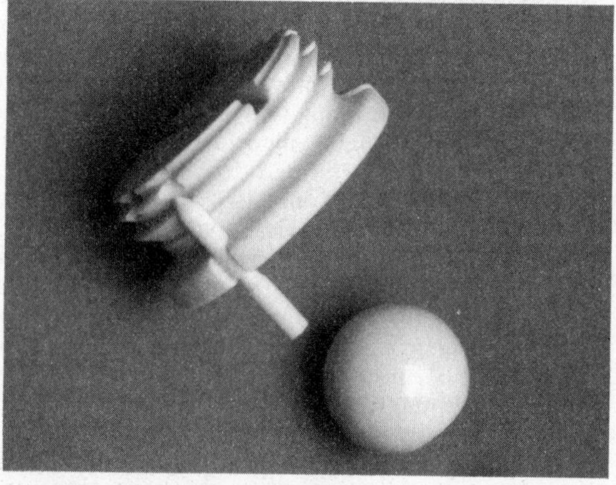

Figure 5
Frialit cylindrical sockets with asymmetric threads for biomechanically favorable load transmission, made of dense, high purity alumina ceramic

articulating tibial plateau of polyethylene, and of ceramic ankle joints with a similar combination, that ceramic components also found application within the lower extremities (Oonishi et al. 1983). However, the loosening rate of the tibial components of the knee prostheses appears to be too high.

In the late 1970s, considerable reduction in wear was reported when polyethylene sockets of hip prostheses were articulating against alumina ceramic balls rather than metal balls. After some exaggerated initial hopes, it is now well estabished that this combination results in a reduction of the polyethylene wear debris by a factor of approximately two.

5. *Alumina Ceramic Dental Implants*

The application of dense alumina ceramics for dental implants was suggested in 1964, but the first ceramic dental implant system fully accounting for the biomechanical rules discussed, the Tübingen dental implant (see Fig. 4), was not designed until 1975 and it was subsequently tested on a strictly experimental basis with a complete follow-up of each single implant for the next five years. Only after the results of this survey became available (Schulte 1984) were these implants made available to dental practitioners.

These implants can be used to replace teeth immediately or shortly after extraction (and after a reshaping of the alveoli to achieve the press-fit necessary for an undisturbed period of bone healing and osseo-integration). They are also used as so-called late implants in edentulous regions of the jaw. In 1983 a modification of the Tübingen implant was introduced after careful clinical tests (Nentwig 1985). Its thinner shape underneath the groove into which the gingiva is positioned allows for its implantation even in cases where the alveolar ridge is too small for the original Tübingen implants.

Scientifically, all the different studies (e.g., histology) on retrieved implants, the different types of mobility tests, and the detailed studies of the gingival attachment, either confirmed the initial design rules or, like the discovery of the load-line shadow effect, further contributed to a more detailed understanding of the load-pattern-controlled tissue reactions around bioinert implants. In addition, the observation of the full preservation of the shape of the alveolar ridge around such immediate implants after more than twelve years opens a new regime of dental care with the possibility of improving public health considerably.

A surprising discovery is also worth mentioning, amplifying the statement about dental care. From the definition of the term "bioinert" (see Table 1), and considering the adsorption behavior with respect to macromolecules in tissue, one would have expected a tendency of increased plaque formation with such ceramic implants. However, experience has shown beyond any doubt that the opposite is true: these implants cause much less plaque formation than metal implants or natural teeth.

The follow-up of all implants inserted in the Tübingen clinic (now more than 1000) has yielded more than 1 000 000 sets of data which have been computerized and evaluated after five and ten years (d'Hoedt et al. 1987) from all relevant points of view. Preliminary results of the ten-year analysis are given in Table 3.

The comparison of the two different follow-up periods (ten years in the left-hand and three years in the right-hand column) is justified by the fact that implant losses of this system are confined nearly completely to the healing-in and early loading period; that is, to the first year. This had already been shown in the five-year statistical evaluation and is clearly confirmed by the ten-year statistics. From the more detailed evaluations of all the data, including those treatments which deviated from the standard, the following additional results are worth mentioning.

(a) The success rate does not depend on the location of the implant.

(b) The success rate is reduced considerably for any deviation from the standard procedure, in particular if no initial press-fit can be achieved or if the implants are loaded within three months of the operation.

Table 3
Summary of ten years follow-up evaluation of all Frialit Tübingen implants inserted in the Tübingen Clinic by the standard treatment (including reimplantation)

	All standard treatments 1975–1985	Treatments commenced after reaching final routine, 1982–1985
Total number of treatments commenced	610	352
Disappeared in follow-up	18	5
Treatments remaining in follow-up	592 (100%)	347 (100%)
Failed treatments	92 (15.5%)	26 (7.5%)
Attempts of reimplantation contained in these failures	14	5
Successful treatments	500 (84.5%)	321 (92.5%)
Reimplantation contained in the successful treatments	28	10

(c) The sulcus fluid flow rates are identical with those mentioned for teeth of the same patients, and they are constant over time as are the pocket depths.

(d) There are indications of a markedly reduced plaque adhesion to alumina ceramic implants as compared to metal implants or natural teeth.

(e) The observations indicate that the shape of the alveolar ridge is maintained around these implants as it is maintained around natural teeth.

Several other implant systems use either alumina ceramics or alumina single crystals (Kawahara et al. 1980). These do not, however, take account of the biomechanical requirements for the true osseointegration of bioinert implants, but rather duplicate more or less closely metallic implant designs. Therefore, all of these implants are separated from the surrounding bony tissue by a soft tissue interlayer (Ehrl and Frenkel 1980, Kawahara 1983). This, of course, results in some mobility of the implant which, in turn, creates some mechanical irritation of the gingival attachment.

6. Alumina Ceramics in Ear, Nose and Throat

Extended animal experiments have shown that the middle ear mucosa proliferates normally on the surface of dense, pure alumina ceramic (Plester and Jahnke 1981). Total and partial oscicular replacements made of alumina ceramic (see Fig. 6) have, since the early 1980s, stood the test of clinical experience showing an improved success rate as compared to plastic parts. More recently, trachea-supporting rings have passed the clinical testing period and have been introduced into clinical application. Orbital support plates have also been tested clinically and have proved successful.

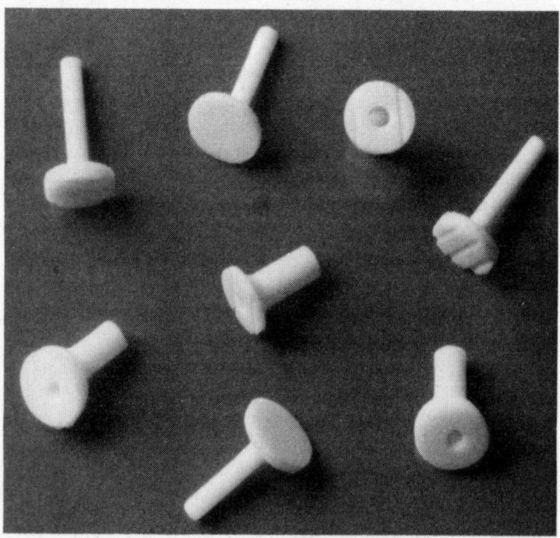

Figure 6
Frialit ossicular replacements, Tübingen type

7. Alumina Keratoprostheses

Another completely soft tissue application of dense, pure alumina ceramics is the Frialit Keratoprosthesis. It consists of a corundum single crystal as its optical part and an alumina ceramic holding ring. It has since the early 1980s stood the test of clinical application in the case of implantation without perforating the lid (Polack 1983). The attempts at using this implant in the "through the lid technique" have shown that in this application mechanical irritation prevents a close integration of the corneal tissue.

8. Outlook

The highly pure, dense alumina ceramic is the main bioinert material and has withstood long-term applications in hip surgery and dental implantology. For other joint replacements, the combination with polyethylene may enable certain improvements.

The interrelation between the implant shape, the stress and strain field created by the implant in the surrounding tissue, and the remodelling reactions has only recently been qualitatively understood. However, combining this knowledge with recent more detailed information on the basic process of bone remodelling can result in implant designs from which an improved osseo-integration can be expected. The succcess rates achieved and documented with the already well established systems must be regarded as the standard that any new system must surpass. Any judgement about the success of an intended improvement can only be given after at least five years of carefully documented clinical experience.

See also: Aluminum Oxide in Biomedical Applications; Biomedical Materials: An Overview; Dental Implants (Suppl. 1)

Bibliography

d'Hoedt B, Heimke G, Schulte W 1987 Bioinert ceramics in dental implantology. In: Vincencini P (ed) 1987 *High Tech Ceramics*. Elsevier, Amsterdam, pp. 219–33

Ehrl P A, Frenkel G 1980 Experimental and clinical experiences with a blade vent-abutment of Al_2O_3-ceramic in the shortened dental row-situation of the mandible. In: Heimke G (ed.) 1980 *Dental Implants*. Hanser, Munich, pp. 63–7

Geduldig D, Lade R, Prüssner P, Willert H G, Zichner L, Dörre E 1976 Experimental investigations of dense alumina ceramic for hip and knee joint replacements. In: Schaldach M, Hohmann D (eds.) 1976 *Artificial Hip and Knee Joint Technology*. Springer, Berlin pp. 434–45

Griss P, Heimke G 1981a Five years' experience with ceramic-metal-composite hip endoprostheses, I. Clinical evaluation. *Arch. Orthop. Traumat. Surg.* 98: 157–63

Griss P, Heimke G 1981b Biocompatibility of high density alumina and its application in orthopedic surgery. In: Williams D F (ed.) 1981 *Biocompatibility of Clinical Implant Materials*. CRC Press, Boca Raton, FL, pp. 155–98

Griss P, Heimke G, von Adrian-Werburg H, Krempien B, Reipa S, Lauterbach H J, Wartung H J 1975 Morphological and biomechanical aspects of Al_2O_3 ceramic joint replacement. Experimental results and design considerations for human endoprostheses. *J. Biomed. Mater. Res. Symp.* 6: 177–88

Griss P, Heimke G, Krempien B, Jentschura G 1976 Ceramic hip joint replacement—Experimental results and early clinical experience. In: Schaldach M, Hohmann D (eds.) 1976 *Advances in Hip and Knee Joint Technology*, Engineering in Medicine, Vol. 2. Springer, Berlin, pp. 446–55

Heimke G 1984 Structural characteristics of metal and ceramics. In: Ducheyene P, Hastings G W (eds.) 1984 *Metal and Ceramic Biomaterials* Vol. 1. CRC Press, Boca Raton, FL, pp. 7–61

Heimke G, Griss P 1981 Five years' experience with ceramic-metal-composite hip endoprostheses, II. Mechanical evaluations and improvements. *Arch. Orthop. Traumat. Surg.* 98: 165–71

Heimke G, Griss P, Werner E, Jentschura G 1981 The effect of mechanical factors on biocompatibility test. *J. Biomed. Mater. Res.* 15: 209–13

Heimke G, Schulte W, d'Hoedt B, Griss P, Büsing C M, Stock D 1982 The influence of fine surface structures on the osseo-integration of implants. *J. Artif. Organs* 5: 207–12

Kawahara H 1983 Cellular responses to implant materials: biological, physical and chemical factors. *Int. Dent. J.* 33:(4) 350–75

Kawahara H, Hirabayashi M, Shikita T 1980 Single crystal alumina for dental implants and bone screws. *J. Biomed. Mater. Res.* 14: 597–605

Nentwig G H 1985 Late implantation with the Frialit implant. *Type Munich Quintessenz* 36: Report 6772

Oonishi H, Hamaguchi T, Okabe N, Nabeshima T, Hasegawa T, Kitamura Y 1983 Cementless alumina ceramic artificial ankle joint. In: Ducheyene P, Van der Perre G, Aubert A E (eds.) 1983 *Biomaterials and Biomechanics*, Advances in Biomaterials, Vol. 5, Wiley, Chichester, UK, pp. 85–90

Plester D, Jahnke K 1981 Ceramic implants in otologic surgery. *Am. J. Otology* 3: 104–8

Polack F M 1983 Clinical results with a ceramic keratoprosthesis. *Cornea* 2: 185–96

Schulte W 1984 The intra-osseous Al_2O_3 (Frialit) Tübingen implant; Development status after eight years. *Quintessenz* 15: (I III) Report 2267

Schulte W, Heimke G 1976 The Tübingen dental implant. *Quintessenz* 6: 17–23

Spiekermann H 1980 Clinical and animal experiences with endosseus implants. In: Heimke G (ed.) 1980 *Dental Implants*. Hanser, Munich, pp. 49–54

G. Heimke
[Clemson University, Clemson, South Carolina, USA]

Aluminum–Silicon: Cast Alloys Modification

The aluminum–silicon is of simple eutectic form (see Fig. 1). Alloys in this system are attractive since in addition to magnesium and lithium, silicon is one of the few elements which can be added to aluminum without loss of the weight advantage that the latter offers. Commercial alloys based on aluminum–silicon are in the range up to ~20 wt% silicon and find wide application as castings for automotive parts, such as pistons, engine blocks, component housings and wheels. The advantages of these alloys, especially those around the eutectic composition of 12.7 wt% silicon, gained greater realization with the accidental discovery by Pacz (1920) that a change in flux composition caused a dramatic improvement in mechanical properties and in particular a more reproducible ductility and improved toughness. Pacz observed that after using a different flux, the previously brittle fracture features had changed to those of a more ductile failure and he declared that the structure had been "modified." This discovery received immediate attention from the aluminum industry and it was shown that the active ingredient was sodium, which could also be effective when added in metallic form to the melt prior to casting; final analyzed concentrations as low as 0.01 wt% sodium were sufficient to effect modification (Edwards et al. 1926). As prospectors have faith in a lucky strike, so foundrymen hanker after that trick or magical additive which will transform a casting to new levels of quality: perhaps the modification of the aluminum–silicon system represents the first established example. Subsequent examples include the modification of gray cast irons and the

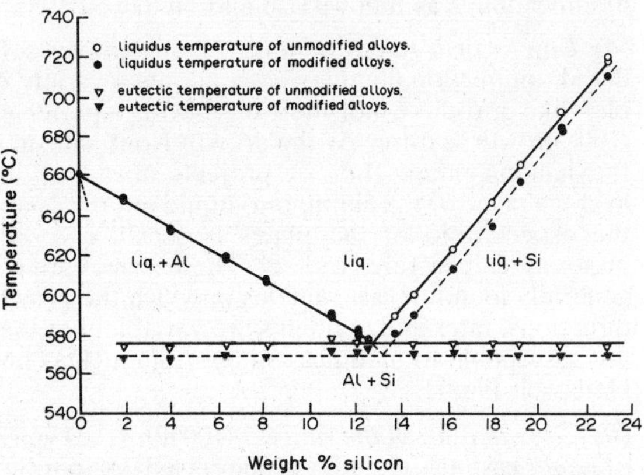

Figure 1
The aluminum–silicon phase diagram based on thermal analysis, cooling at 4 K min^{-1}: full lines indicate unmodified alloy, broken lines indicate sodium modified alloy

widespread use of grain refiners, notably in aluminum foundry practice.

Since Pacz's discovery there have been hundreds of publications concerned with the modification of aluminum–silicon alloys, either seeking to explain how it occurs or to find alternatives to sodium as an innoculant. One of the practical difficulties in using sodium is that its retention time in the melt is limited to some 5–15 minutes, after which time the effect "fades". It is now known that some other metallic additions produce a similar, if less dramatic modification; these include some alkaline earths (notably strontium) and some rare earth metals. Traces of other elements such as phosphorus also change the microstructure but in a different way, affecting the nucleation of primary silicon, which will not be considered in this article. Understanding how the process works has required detailed studies of the morphology and internal structure of the phases, particularly of silicon, and these have progressed significantly with the development and application of electron optical and microanalytical techniques in recent years. It is necessary to describe some of these more important details before considering what may be the mechanism(s) for modification.

1. Microstructure

The aluminum–silicon system is an example of one in which the two eutectic phases solidify from the melt with essentially different morphologies: the major aluminum phase crystallizes in a nonfaceted manner, while the minor silicon phase develops a faceted morphology except at rapid freezing rates ($\gtrsim 1\ \mathrm{mm\,s^{-1}}$) when it may be described as "quench modified." With impurity modification it is the morphology of the silicon rather than the aluminum which changes. The structure of the silicon without modification is as follows (Lu and Hellawell 1987):

(*a*) *Low growth rates* ($\lesssim 5\ \mu\mathrm{m\,s^{-1}}$). In samples solidified undirectionally the silicon adopts a variety of platelike or rodlike morphologies having a preferred ⟨100⟩ growth texture. At the growth front, silicon is the leading phase (i.e., it projects ahead of the average plane of the aluminum–liquid interface) and the aspect ratio of the plates is sensitive to the imposed temperature gradient. These shapes are not generally found in cast samples in which the growth directions, rates and gradients are variable, but they are susceptible to modification by sodium (Day and Hellawell 1968).

(*b*) *Growth rates in the range* $5\text{–}100\ \mu\mathrm{m\,s^{-1}}$. Typical of larger castings, the silicon adopts a flaky morphology in the plane of the {111} close packing of the diamond cubic crystal structure. In the eutectic the silicon consists of interconnected flakes (see Fig. 2), exhibiting a variety of branching modes and containing a few growth twins, most of them in the plane of the {111} habit. Primary silicon, in hypereutectic alloys, consists of coarser, generally twinned flakes and occasionally untwinned octahedra.

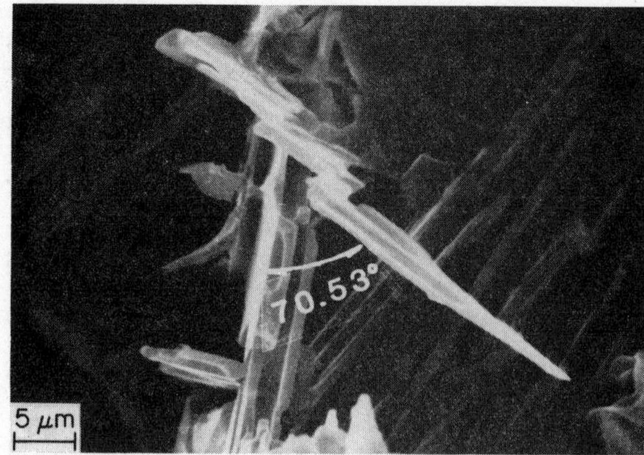

Figure 2
Eutectic silicon flake structure of unmodified eutectic alloy, SEM with aluminum matrix dissolved away

(*c*) *Rapid rates* ($\gtrsim 1\ \mathrm{mm\,s^{-1}}$). As the rate increases from the flake regime, the faceted morphology becomes less sharply defined and the scale finer, until, at around $1\ \mathrm{mm\,s^{-1}}$ or above, the eutectic morphology is one of fine nonfaceted fibers which are occasionally twinned (see Fig. 3). This is a structure occurring in small chilled castings and which may be called "quench modified."

(*d*) *Sodium modified.* In the range of growth rates over which faceted flakes occur in the unmodified alloys, sodium changes the structure to an essentially fibrous morphology in which the interfiber spacing is

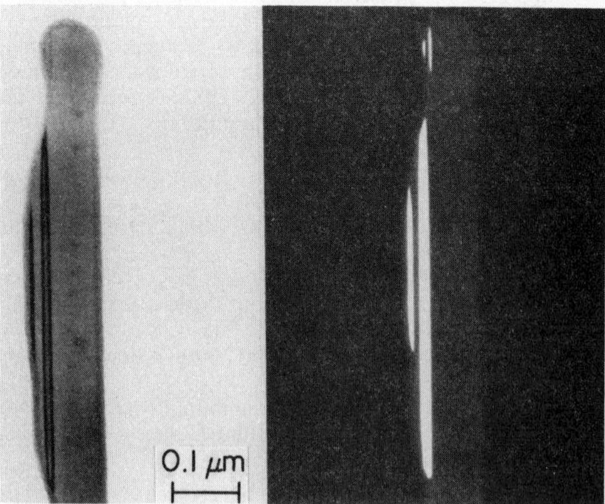

Figure 3
Quench modified silicon fibers of eutectic alloy showing rare twin, TEM, light and dark field

not very different from the former interflake separation and, therefore, much coarser than the quench modified fibers. However, the impurity modified fibers (see Fig. 4) differ significantly in that they are very heavily twinned on all four possible {111} systems and, in addition, on a fine scale can be seen to be externally rough or microfaceted. Taking into account the estimated average growth rate of such fibers, combined with measured twin densities (separations ≲10 nm or 20–30 interplanar spacings), reveals that during growth the twinning frequency was of the order of $10^4\ s^{-1}$. The consequence of this multiple twinning is a corresponding selection of locally preferred growth directions and microfacets, so that taken collectively the growth characteristics appear to be more isotropic—although in sharp contrast to the smooth and rarely twinned fibers which form on quenching the pure binary alloy. If the melt is both treated with sodium and quenched (doubly modified) the silicon is finer, fibrous and very heavily twinned. Primary silicon is also modified by sodium at slightly higher concentrations, becoming heavily twinned and more spherical in shape (Hellawell 1970).

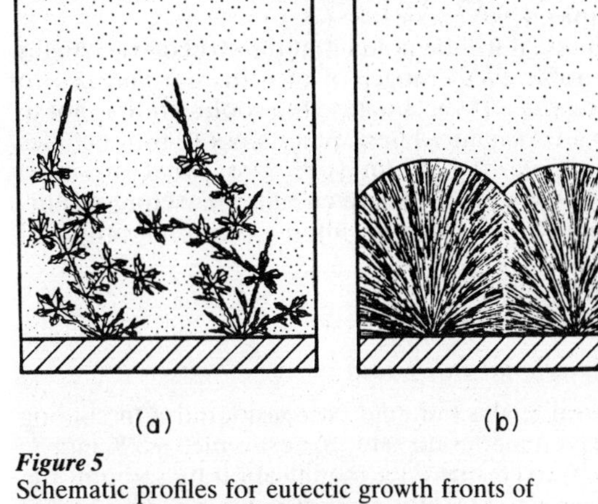

Figure 5
Schematic profiles for eutectic growth fronts of (a) unmodified and (b) modified alloy

With the change from a flake to impurity modified fibrous morphology there is a striking change in the profile of the growth front, as revealed by quenching samples during growth. The flakes form at a front where the silicon projects forward into the melt at various angles, while the fibrous silicon of the modified alloy advances at a relatively closely coupled planar or cellular front (see Fig. 5).

Whether quench and/or impurity modified, the silicon shape change is from flake to fiber (with the internal difference). The fracture plane for silicon is {111} coincident with the flake habit, so it will be understood that with a semicontinuous flake structure there are numerous potential fracture paths through the material. When modified, the continuity of possible brittle failure paths is lost or much reduced. There are no data available to compare the strength of twinned with untwinned fibers, but intuitively it might be anticipated that the former will be less prone to fracture, the quenched material showing lower elongation in tension. As noted previously, the structure of the aluminum matrix is not a significant variable, consisting of many subgrains of a few micrometers width and having 1–2° misorientation. The only change, with growth rate or with the better aligned coupled growth of the sodium modified structure, is a tendency for subgrains to be elongated along the growth axis and to develop a loose ⟨100⟩ texture. There is some uncertainty as to whether the two phases exhibit preferred epitaxial orientations but with twinning in the silicon this seems improbable. It seems that the most highly preferred orientations are separate, single-fiber

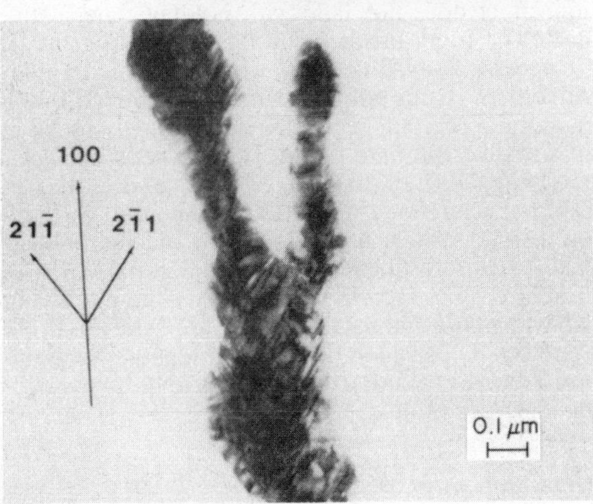

Figure 4
Sodium modified silicon of eutectic alloy showing multiple twinning

textures developed independently in each phase, for example, ⟨100⟩ (Si) ∥⟨100⟩ (Al).

Finally, it will be noted that the structural changes were referred to ranges of growth rate and not to cooling rate. This is because the cooling rate is not in itself an adequate description unless it also includes an estimate of how far (i.e., over what distance) growth took place in a given time, therefore requiring more information about the separation of nucleation sites.

2. Additional Information

Thermal analysis should be considered. On cooling at a given moderate rate, for example, ~5 K min^{-1}, it is observed that with modification by sodium, the eutectic arrest is depressed by some 6 K (see Fig. 6) but that the corresponding heating arrest is lower by only ~1 K. Cooling arrests for the primary (aluminum) liquids are not noticeably different, but those for silicon in hypereutectic alloys are depressed similarly to those for the eutectic. These data can be summarized on the equilibrium phase diagram in Fig. 1. A consequence of the freezing point changes is that the modified material of nominally eutectic composition contains primary (aluminum) dendrites as the eutectic point is effectively displaced, on cooling only, to a higher silicon concentration. In the modified alloy a small arrest may be identified, prior to the eutectic arrest, corresponding to the primary (aluminum) precipitation. It is important to note also that it is the horizontal part of the eutectic arrest which is depressed as well as the initial supercooling. This indicates that it is the growth kinetics in addition to those for nucleation which have been modified (Plumb and Lewis 1957, Day 1968, Lu et al. 1984).

Other useful observations include the epitaxial growth of silicon from eutectic liquid onto a silicon

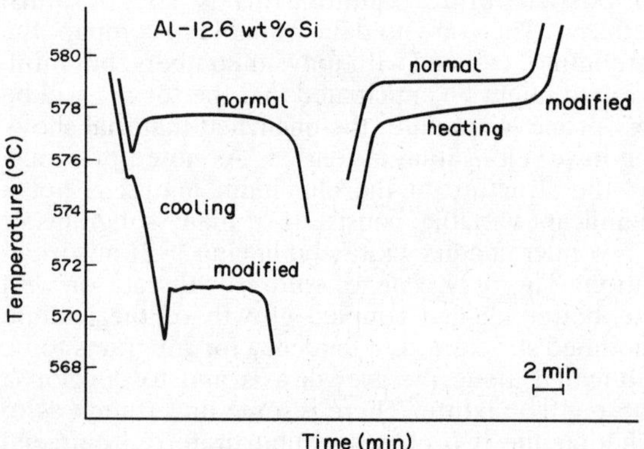

Figure 6
Cooling and heating curves for eutectic alloy, unmodified and sodium modified

substrate (Davies and West 1963). In the presence of sodium this deposition is inhibited and negligible growth occurs. There is, therefore, a strong indication that the modification is brought about by interaction of sodium with the silicon–liquid growth front, as might already have been concluded from the changed internal structure in impurity modified fibers. This view is also supported by measurements of silicon activity in aluminum melts (Brisby and Fray 1983) which show that this is sharply reduced by sodium, implying association of the two elements and probably adsorption of sodium onto silicon in contact with the melt. Finally, although chemical analysis at low concentrations on a fine scale microstructure is difficult, Auger analysis of fractured surfaces in sodium modified material indicates that the sodium is associated with the silicon rather than with the aluminum matrix.

3. Other Elements

Most elements and many compounds have been used to explore the possibility of modification in aluminum–silicon alloys. In addition to sodium, the other alkalis, alkaline earths and more recently rare earths, have received much attention. On an optical basis, the evidence for similar (i.e., to sodium) modification is somewhat ambiguous, but it is reasonably clear that calcium, strontium, barium and certain rare earths exert the comparable effect of "rounding off" the silicon flake morphology. The effects are much less dramatic than those promoted by sodium but it has been established that the three alkaline earths and the rare earth ytterbium all combine this shape change with an increase in twin density (Lu and Hellawell 1987, Shamsuzzoha and Hogan 1985). As noted, strontium is now widely used in foundry practice because its milder modification effect does not fade with holding time in the melt. The literature is inconsistent but, in general, the necessary levels of concentration are higher than for sodium, being around 0.1 wt% (although lower in atomic percentage). It is not easy to make a quantitative estimate of the silicon shape changes, but twin densities have been estimated, listed in Table 1, where they may also be compared with the low twin densities to be found in the unmodified flake or purely quench modified material. Clearly there are other factors which modify the modifying behavior itself and some of these are listed in Table 1 and briefly considered in Sect. 4. together with the promotion of twinning which seems to be an overriding consideration.

4. Modification Mechanism

The bulk of available evidence concerning aluminum–silicon modification relates to sodium and suggests that only the silicon phase is affected,

Table 1
Elements reported to cause modification in order of radius ratios to silicon

Element	Melting point (K)	Vapor pressure at 1000 K (atm)	r/r_{Si}	Twin spacing (nm)	Bulk concentration
Ba	998	5.0×10^{-5}	1.85	30–100	~0.05
Sr	1042	1.0×10^{-3}	1.84	30–100	~0.05
Eu	1095	2.0×10^{-4}	1.72		
Ca	1112	2.0×10^{-4}	1.68	100–200	~0.05
Yb	1097	6.0×10^{-3}	1.65	50–150	~0.05
La	1193	6.0×10^{-6}	1.59		
Na	371	0.2	1.58	5–12	0.01
Ce	1071	1.0×10^{-16}	1.56		
Pr	1204	1.0×10^{-13}	1.55		
Nd	1283	1.0×10^{-11}	1.55		
Al–Si (comparison)	850	Normal flake structure twin spacing 400–1000 nm			

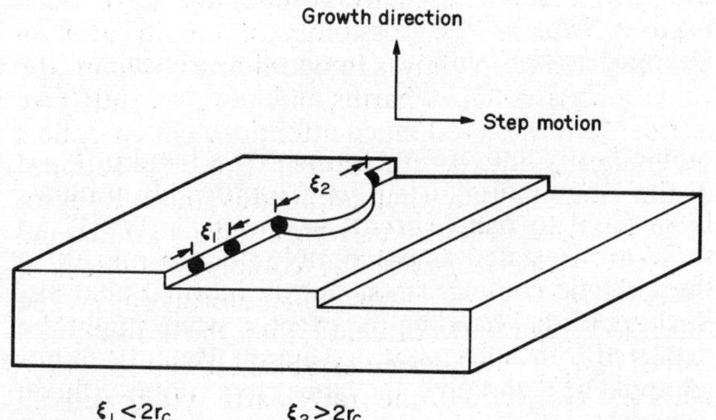

Figure 7
Schematic diagram to illustrate monolayer step growth with impurity adsorption along the step

the principal effect being one of shape change coupled with an increase in twin density. It seems that the minor additive is somehow incorporated in, or adsorbed on, to the silicon crystal, thereby promoting twinning during growth. It is thus necessary to explain (a) the reason why particular elements induce this twinning and (b) why sodium is so much more efficient in doing so than other additives. Possible explanations that have been offered include those by Lu and Hellawell (1987).

4.1 Induced Twinning

The classical model for faceted crystal growth is one of monolayer step advance across the slower growing (generally close packed) crystal faces (see Fig. 7), in this case the {111} faces of silicon. Adsorption of impurity atoms, if it occurs, is preferred at step sites. If this occurs in sufficient concentration, such adsorption can retard or inhibit the growth, requiring overgrowth by monolayers in order to continue across the surface.

Considering now the elements which promote modification, one noticeable common feature is a much larger atomic radius than that of silicon, typically lying with a radius ratio between ≃1.6 to 1.8 (see Table 1). This could be a clue to a mechanism for twin formation. Referring to the diamond cubic structure viewed in 011 plane projection in Fig. 8, with monolayer step motion proceeding from left to right, atoms normally fall in the A–B–C stacking sequence (actually –AA–BB–CC– in diamond cubic). Should the step include a row of ad-atoms of suitable size, however, any further continuation of the step will require a displacement in the stacking sequence to –CC–BB–AA–BB–CC–, thereby creating a twin. Geometrically, using a hard sphere

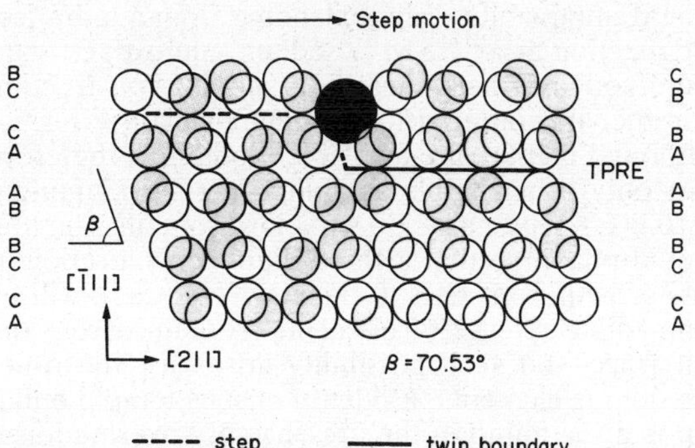

Figure 8
011 plane projection of diamond cubic lattice to show how an impurity atom of "suitable" size could cause displacement of the attacking sequence to induce a twin

model, an exact radius ratio may be calculated and in the diamond cubic structure this would be ~1.6 or ~1.8, depending on how the atoms were arranged. Of course, it may be coincidence that this geometrical model provides a ratio within the observed range for modifiers as the adoption of a hard sphere model is subject to some skepticism, but it does offer a viable mechanism for twinning during growth of a faceted phase, given sufficient concentration of adsorbed species.

In addition there will be qualifying factors other than the prerequisite adsorption and size effect, thus the modifier should dissolve fairly rapidly in the molten alloy. As in the case of sodium it should not have too high a vapor pressure and it should presumably be in elemental rather than a combined form, such as the oxide. It also follows that if the model is correct for silicon in the aluminum–silicon system it should apply to the silver–silicon system or to germanium, with the same structure in, for example, the aluminum–germanium system—given the same prerequisite condition of adsorption. There is optical evidence that this is the case.

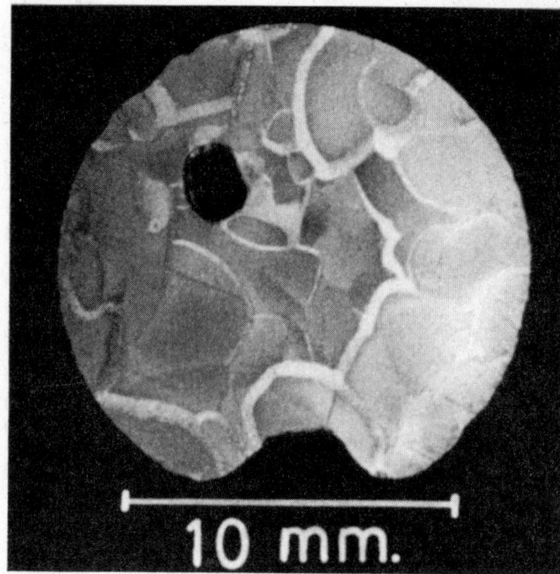

Figure 9
Overmodification bands in eutectic alloy (sodium)

4.2 Sodium

Comparison of the twin spacings (see Table 1) shows that the density induced by sodium is at least an order of magnitude greater than that for the other relevant elements for which measurements exist. One criterion for a sort of specific "modifying power" might be based on a ratio of induced twin density to adsorbed concentration (if it were possible to measure the latter). Certainly, sodium is effective at lower bulk analyzed concentrations than the other metals, so that on that basis it would appear to be even more powerful than the measured twin densities alone would imply. An important consideration for sodium is the form of the aluminum–sodium phase diagram, which is different from those with the other elements, involving a wide liquid immiscibility gap, extending from a monotectic reaction at ~0.15 wt% sodium, almost across to pure sodium; there must be a similar construction for the aluminum rich aluminum–silicon eutectic, although no data are available. Sodium is therefore not only of negligible solubility in solid aluminum (<0.003%) but also of very low solubility in the liquid phase, so that if the sodium concentration at the growth front rises to this low level there will be immediate release of elemental sodium across the interface—no such possibility arises for the other modifying elements. Evidence of such a rapid build-up is to be found in the presence of "overmodification" bands (see Fig. 9) which are unique to sodium modification. When the bulk sodium concentration exceeds ~0.14 wt% sodium, it is found that the duplex eutectic growth front becomes temporarily halted and the silicon phase fibers terminate momentarily and are overgrown by a band or layer of the aluminum matrix. Growth proceeds in bursts from band to band (Fredricksson et al. 1973) and these are revealed as perceptible fluctuations along the eutectic cooling arrest during thermal analysis. Such periodic banding is exactly what might be expected from such a phase diagram when the minor component (sodium) is able to inhibit silicon growth.

5. Comparisons with Cast Iron Modification

Graphitic cast irons solidify under normal foundry conditions as gray irons having the minor graphite phase in a flake form in a nonfaceted austenitic or ferritic matrix. The material is sensitive to brittle failure in the (0001) plane of the flakes, thus there is an obvious parallel with aluminum–silicon alloys in the unmodified conditions and the consequent expectation that such a structure be susceptible to modification. There is no reason to expect that sodium or other elements which modify aluminum–silicon alloys will behave similarly in cast irons. First, the former would boil off from a molten iron immediately and, second, there is no crystallographic similarity between the diamond cubic structure and the open hexagonal structure of graphite. However, with systematic additions, the search for one or more additives which would modify flake graphite was brought to a successful conclusion during World War II with the discovery that magnesium and rare earths will promote the spheroidization of graphite. The search was conducted in the UK, Canada and Germany over a similar period, but the program organized by Morrogh and Williams (1947) is generally regarded as definitive.

In the case of silicon modification, there appears to be a direct interaction between it and the addition, involving some form of incorporation into the silicon phase with the promotion of internal defects. In the case of graphite, the structural change involves the basal layer planes of the hexagonal lattice rotating from the flake habit into a segmented spherical morphology. The modification is, in some ways, more striking than that of silicon, but it is not clear whether this arises because of incorporation of magnesium into the graphite lattice and there is some evidence that the effect is indirect and depends upon the removal or neutralization of a third element or elements by magnesium, etc. Oxygen, sulfur and possibly phosphorus are candidates. It has been shown (Sadocha and Gruzleski 1974) that prolonged vacuum melting of a gray cast iron with a gradual reduction of the more volatile nonmetallic impurities results in a transition from flake to approximately spheroidal shapes (although the latter are less perfect than the spherulites commonly promoted by addition of elements such as magnesium). Further, Auger analysis of fractured flake and magnesium innoculated spheroidal graphite irons shows that at flake–metal interfaces, sulfur and oxygen are to be found close to the metal surfaces. These elements cannot be detected at spherulite–metal interfaces, being then localized in combined form with magnesium (Johnson and Smartt 1979). Information of this type suggests that rather than spherulitic graphite being the modified form, it is actually the normal morphology forming from a melt which has been cleaned by evaporation or scavenging with a reactive addition, the flake form being the modified product promoted by elements such as oxygen or sulfur. A possible explanation might be that the (0001) graphite–metal interface is stabilized by adsorption but that in a clean melt the interfacial energy is higher and a spherical shape having minimum ratio of surface area to volume is preferred. This is still speculative, however, and the overall problem has many aspects which are ill-understood. It is clear that the growth (and perhaps nucleation) characteristics of a faceting phase are idiosyncratic for the crystal structure and environment concerned and there is little justification for trying to extrapolate from one example to another without extreme caution.

See also: Aluminum: Alloying; Spheroidal Graphite Cast Irons

Bibliography

Brisby R J, Fray D J 1983 *Metall. Trans., B* 14: 435

Davies V de L, West J M 1963 *J. Inst. Met.* 92: 175

Day M G 1968 *Nature* 219: 1357

Day M G, Hellawell A 1968 *Proc. R. Soc. London, Ser. A* 305: 473

Edwards J D, Frary F C, Churchill H V 1926 US Patent No. 1410461. *J. Inst. Met.* 36: 283

Fredrickson H, Hillert M, Lange N 1973 *J. Inst. Met.* 101: 285

Hellawell A 1970 *Prog. Mat. Sci.* 15: 1

Johnson W C, Smartt H B 1979 Solidification and casting of metals, *The Metals Society Handbook 192*, pp. 125

Lu S-Z, Hanna M D, Hellawell A 1984 Modification in the aluminum–silicon system. *Metall. Trans. A* 15(3): 459–69

Lu S-Z, Hellawell A 1987 The mechanism of silicon modification in aluminum–silicon alloys—impurity induced twinning. *Metall. Trans. A* 18(10): 1721–33

Morrogh H, Williams W J 1947 *J. Iron Steel Inst.* 155: 321

Pacz A 1920 US Patent No. 1387900

Plumb R C, Lewis J E 1957 *J. Inst. Met.* 86: 393

Sadocha J P, Gruzleski J E 1974 *Proc. Int. Symp. on the Metallurgy of Cast Irons.* Georgi, pp. 443

Shamsuzzoha M, Hogan L M 1985 Twinning in fibrous eutectic silicon in modified Al–Si alloys. *J. Cryst. Growth* 72: 735–7

A. Hellawell
[Michigan Technological University, Houghton, Michigan, USA]

Armor, Ceramic

In 1918 it was reported that a 1/16 in layer of hard enamel could improve the resistance to bullet penetration of a steel sheet. However, it was not until World War Two that the potential advantages of including a hard, ceramic component in an armor system were generally recognized. Plate glass, backed by fiberglass-filled nylon (known as Doron), was introduced in 1945. This lightweight composite armor combination was shown to be capable of stopping rifle bullets. In 1962 Doron-backed alumina panels were found to stop 0.30 in caliber armor-piercing projectiles at half the armor weight per unit area which was required for conventional steel armor. A systematic study of the ballistic performance of a wide range of ceramics subsequently led to the introduction of boron carbide armor components during the Vietnam conflict. This was used for the protection of low-flying helicopters and their crews. The search for more effective ceramic composite armors is continuing still.

1. Ballistic Requirements

The performance of an armor system is measured in terms of the weight per unit area required to defeat a defined threat from a projectile. The level of threat varies, from low-velocity shell fragments and projectiles, through armor-piercing high-velocity bullets, to shaped-charge weapons (which generate a fine,

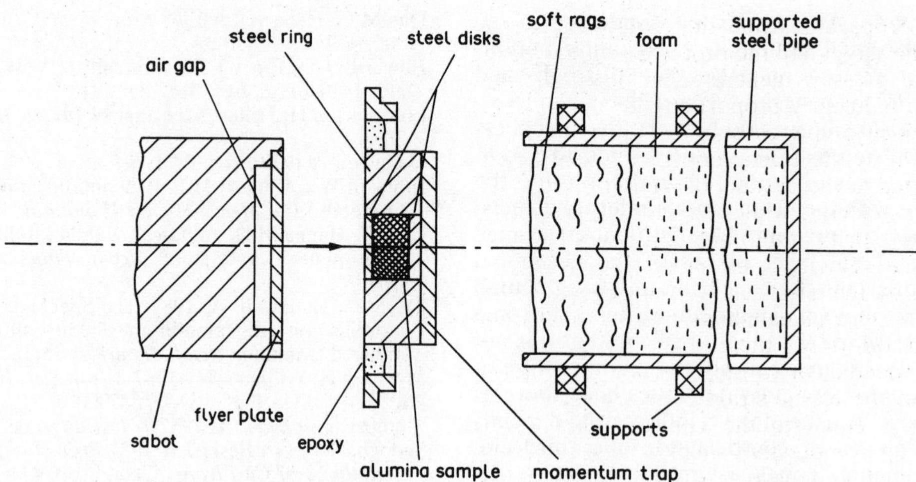

Figure 1
Projectile assembly, target assembly and recovery system for a typical planar impact experiment on a ceramic sample

high-velocity jet of metal) and long heavy-metal projectiles. These levels of threat are termed light, medium and heavy respectively, corresponding to the increasing kinetic energy of the projectiles. Typically, armor protection for individuals (personnel armor) is only intended to defeat light threats, since the weight of armor that an individual can carry comfortably is limited. On the other hand, the armor carried by military vehicles (armored personnel carriers, for example) is intended to defeat medium threats, such as the 0.50 in caliber armor-piercing projectile. Finally, the armor for main battle-line tanks is intended to withstand the heavy threats posed by shaped-charge weapons, such as the various forms of rocket-propelled grenades (RPG), and long-rod, heavy-metal penetrators (which may travel at velocities exceeding $1500\,m\,s^{-1}$).

The measure of performance generally used to rank armor systems is the v_{50} ballistic test. v_{50} is the velocity of the projectile which will have a 50% probability of penetrating the armor at a fixed angle of incidence. Typically the requirement is for a weight per unit area of armor which will guarantee a given minimum value of v_{50} when tested against a specific projectile. While the v_{50} test serves well to evaluate the effectiveness of an armor system, it does not differentiate well between armor materials. In particular, the effectiveness of a ceramic armor is strongly dependent on the backing provided for the ceramic. One method of ranking the ballistic performance of ceramic materials is to mount them on a thick metallic block (usually an aluminum alloy) and to measure the depth of penetration of a test projectile travelling at a standard velocity. The thick metal block ensures that the ceramic does not fail by bending prematurely during penetration, so that the reduction in kinetic energy of the projectile by the ceramic is maximized. Erosion of both the projectile and the ceramic during penetration is the primary mechanism responsible for reducing the kinetic energy of the projectile. Plots of the residual penetration into the metal backing block as a function of the thickness of the ceramic front plate are generally straight lines, the slope of which can be normalized with respect to density to yield a figure for ballistic efficiency.

2. *Dynamic Properties*

Three main damage regimes exist in materials subjected to impact. In the first regime, at low projectile velocities, fracture in brittle materials is well described by linear elastic fracture mechanics. In the second regime, as the projectile velocity approaches the sonic velocity (of the order of $500\,m\,s^{-1}$), the loading history becomes dominated by shock wave propagation. This transition occurs at strain rates of the order of $10^3\,s^{-1}$. The transition to the third regime occurs when the damage is dominated by inertial effects and the adiabatic response of the target leads to melting. In this regime, increasing the projectile energy eventually results in vaporization at the point of impact.

In assessing the suitability of ceramics for armor applications, it is the response to a shock wave which is decisive in most cases. To determine this response, a lapped target plate of the ceramic under test is subjected to planar impact by a projectile accelerated to a known velocity (Fig. 1). The planar geometry allows the material response to be plotted in two dimensions on a time vs distance plot (Fig. 2). At the moment of impact planar compressive shock waves are propagated from the surface of contact into both the target and the projectile. In the elastic regime, the pressure generated in a compressive shock wave is linearly proportional to the material particle velocity, the constant of proportionality being known as the elastic impedance. The impedance is equal to the product of the density and the shock wave velocity, while the particle velocity is equal to the velocity of the contact surface. The ratio of the particle velocity to the shock wave velocity is thus equal to the elastic strain generated by the shock wave. Since planar shock waves propagate in one dimension only, there are no dimensional changes perpendicular to the direction of propagation, so that the elastic strain perpendicular to the wave front is equal to the volume strain in the material. The compressive shock waves are eventually reflected as tensile release waves from the back surfaces of both target and projectile. These cancel the compressive stress in the material. A tensile stress is only produced if the release waves from the target and the projectile can act to reinforce each other. This is the situation in Fig. 2, for which the target thickness is much greater than that of the projectile. The tensile stress can give rise to fracture by delamination in the plane in which the release waves first overlap—a failure mode which is called spalling and is also illustrated in Fig. 2.

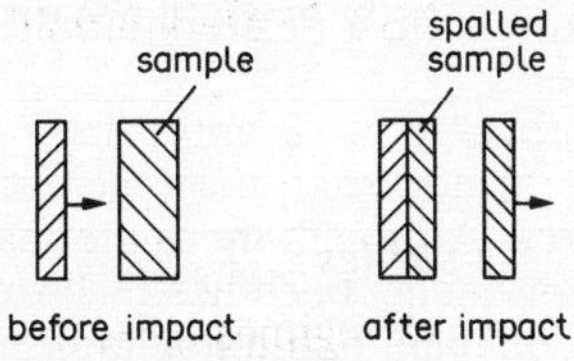

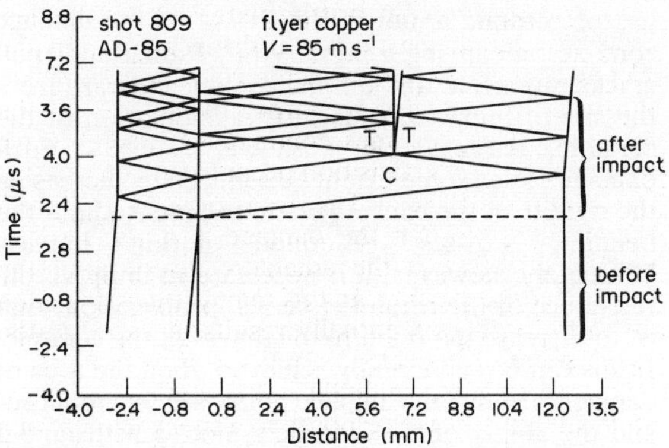

Figure 2
Calculated strain history for a simple planar impact experiment plotted on a time vs distance diagram. The target is under compression in the region C. Spalling is initiated in the region T, where the tensile release waves first overlap. Subsequent shock wave reverberations are also shown

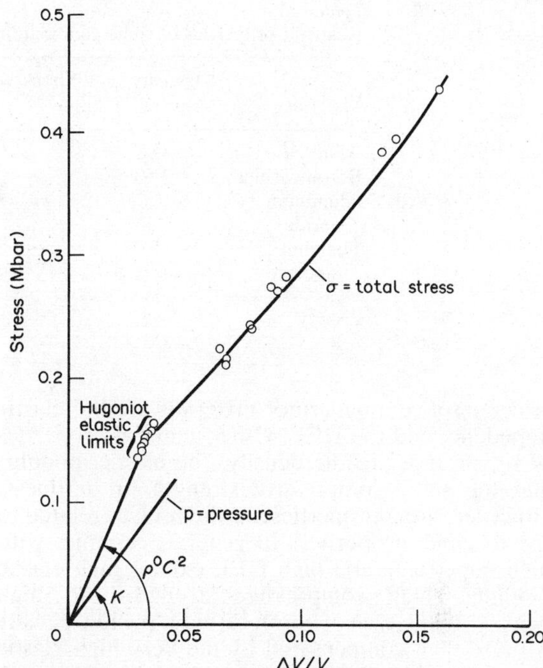

Figure 3
Hugoniot curve for boron carbide plotted as pulse pressure against volume strain (the position of the hydrostat is also shown) $\rho^0 = 2.50$ g cm^{-3}, $K = 2.63$ Mbar, $\rho^0 C^2 = 4.83$ Mbar

The dependence of the material particle velocity on the compressive impact pressure is termed the Hugoniot function and is a characteristic of the material (see *Hugoniot Curve*, Suppl.1). A typical curve for a ballistic ceramic is shown in Fig. 3. The curve is the dynamic equivalent of a static stress–strain curve, in which the elastic impedance governs the initial slope of the curve in place of the elastic modulus. Like a stress–strain curve, the region of elastic response is terminated at a critical pressure, termed the Hugoniot elastic limit (HEL), at which the material starts to undergo irreversible shear. Above the HEL, the Hugoniot curve approaches and becomes parallel to a pressure response curve, the hydrostat, typical of the liquid state, which has no shear strength. In metals the HEL is the critical pressure for the onset of plastic yielding, and is generally low compared with the spall strength of the metal (usually determined from the material response to the subsequent tensile history of the stress pulse). In ceramics, however, the HEL is very much higher than the spall strength, which appears to fall to zero as the compressive pulse pressure approaches the HEL. This is in line with the static behavior of ceramics, which have excellent compressive strengths but are very weak in tension. The dynamic properties of major importance in the

Table 1
Relevant properties of some candidate materials for ceramic armor

Material	Density, ρ (g cm^{-3})	Shock wave velocity, C_l (cm µs^{-1})	HEL (Mbar)	Elastic impedance, ρC_l
Alumina	3.92	1.07	0.140	4.19
Boron carbide	2.50	1.33	0.150	3.33
Aluminum boride	2.58	1.31	0.096	3.38
Titanium diboride	4.52	1.13		5.11

selection of ceramic armor materials are the elastic impedance and the HEL. Other material properties of significance are the density, the elastic modulus and the static compressive strength or hardness. However, these properties are themselves related to the dynamic properties. In general, ceramics with high impedance and high HEL exhibit high elastic modulus, high compressive strength and high density. In the case of boron carbide, the low density is more than compensated by the very high elastic modulus and hardness, to give a material with excellent ballistic properties.

3. Ceramics for Ballistic Applications

The relevant properties of some of the ceramics that have been proposed for armor applications are listed in Table 1. Nearly all successful ceramic armors have been dense, single phase, monolithic materials. Two-phase mixtures and porous ceramics have proved markedly less successful. This is assumed to be connected with the high stresses generated at the interface between two phases of different impedance. Applications in personnel armor have been dominated by economic considerations. Although boron carbide is the ceramic of first choice, the high cost of boron carbide plates has limited their use to critical components, such as the pilot's seat of an attack helicopter. The boron carbide plates are backed by an aramide fiber-reinforced plastic (Kevlar), which prevents fragmentation. The much cheaper alumina plates are used to provide protection in armored vests. Typically, the plates are offered as an optional extra which can be placed in the pockets of a multiweave Kevlar vest whenever additional protection is desired and the extra weight is justified.

Figure 4
A ceramic tile after impact: although the projectile has not penetrated into the backing plate, the tile has been fragmented by the intersection of radial and spall cracks at distances far from the point of impact

The introduction of ceramics into medium- and heavy-threat armors has been generally a closely guarded secret. Although there are many reports of ceramic armor having been used in armored personnel carriers and main battle-line tanks, these reports have not been confirmed. A major problem in the use of ceramic armor is the extent of the damage zone accompanying a primary hit. Radial and spall cracks propagate through the ceramic outward from the site of impact, leading to fragmentation of the ceramic. An example is shown in Fig. 4. Such damage can to some extent be reduced by increasing the rigidity of the backing plate, in order to limit the bending stresses generated during impact. Ultimately, however, it is necessary to improve the resistance of the ceramic to crack propagation; that is, to increase the fracture toughness of the ceramic. If this can be successfully achieved, then the area of damage around the impact point will be reduced, and the ability of the ceramic armor to withstand a second hit can be improved. In the 1980s, the accepted method of limiting the size of the damage zone has been to reduce the initial size of the tiles. Since this introduces edge effects, which reduce the ballistic efficiency, an improvement in the toughness of the ceramic would be very desirable. It is not at all clear whether this can be achieved without drastically reducing the resistance of the ceramic to projectile penetration.

See also: Armor, Composite (Suppl. 2); Failure Analysis of Ceramics; Fracture Toughness Testing of Brittle Materials

Bibliography

Blazynski T Z 1987 *Materials at High Strain Rates.* Elsevier, London

Wong A K, Berman I 1971 *Lightweight Ceramic Armor—A Review.* AMMRC, MS 71-1

Viechnicki D, Blumenthal W, Slavin M, Tracy C, Skeele H 1987 Armor ceramics. *Proc. 3rd TACOM Armor Coordinating Conf.*

D. G. Brandon
[Technion, Haifa, Israel]

Armor, Composite

A list of possible composite armor materials may include traditional materials such as aluminum and steel or less traditional ones such as silk and leather. In the past, many combinations have been applied and found satisfactory, usually over a narrow range of applicability. Recent efforts to increase ballistic protection and reduce the weight necessary for this protection have led to a new focus in the armor community. Two classes of materials have emerged as having great potential over a wide range: fibrous composites and ceramics. However, a vital aspect of armor design is an understanding of the properties which control ballistic performance. Unfortunately, for these new materials, the fundamental research necessary for this understanding has only recently received attention. This is mainly a consequence of the fact that the processes occurring during the projectile penetration event occur at extremely high strain rates. Few materials have been fully characterized under these conditions, least of all ceramics. In this article the material properties important in armor design are examined, briefly for fibrous composites, and in greater detail for ceramics. Ceramics will receive more focused attention because of the lack of published information on their behavior at high strain rates.

Before proceeding further, it is necessary to describe how armors are rated for performance. Two fundamental quantities always considered in describing the performance of any armor system are weight and ballistic protection. Ballistic protection is characterized in terms of the projectile used and the maximum projectile velocity withstood by the armor. A popular method is the v_{50} method, in which v_{50} is defined as the velocity at which 50% of the projectiles are defeated, and 50% penetrate completely through the armor. Many other methods are also used to determine when the armor has been defeated. Most often, a thin aluminum plate is placed behind the armor, and when either the projectile or debris from the armor penetrate the plate, the armor is considered to have been defeated. Weight efficiency is determined relative to an accepted standard material. The efficiency is expressed as the ratio of standard material weight necessary to defeat the projectile to the required weight of the new armor required to defeat the same projectile. The weight is usually computed per unit area or areal density of the armor. The standard material used by the US military is rolled homogeneous armor (RHA), a high strength steel defined by US specification MIL-A-12560.

The important point to be emphasized here is that many methods are used in armor performance characterization. One should be certain of the method used before making direct comparisons.

1. Fibrous Composite Armors

The use of fibers or fabrics as armor dates from the sixth century AD when the Chinese used silk as padding for heavy metal outer armors. It soon became apparent, however, that such metal armor was so encumbering that a lightly armored, more mobile adversary had a tactical advantage. An efficient armor composed entirely of lightweight materials was needed. The use of silk as protection against handguns was first introduced in Russia in around 1914. Unfortunately, widespread use of composites for body armor did not occur until the Korean War when the US Army introduced a protective vest composed of hardened nylon plates for use by ground troops. This vest was only effective against low-velocity fragmentation and could not stop projectiles fired from rifles. The Vietnam War brought about a new vest capable of defeating high-energy rifle bullets. This vest consisted of ceramic plates backed by a fiberglass laminate. Its 60 kg m^{-2} weight was soon found to be an unacceptable burden for the average infantryman, so a lighter vest composed entirely of Kevlar was produced in the late 1970s and is still in use today. In addition to body armor for the military, many law enforcement and private security agencies use fibrous composites for protection. In this application, the threat is usually limited to low-energy handgun projectiles. The vest has the additional requirements of concealment and comfort. The role of fibrous composites as armor will be expanded in years to come.

As armored-vehicle designers are forced to reduce weight and increase protection, fibers will be combined with ceramics, metals and plastics to create more affective armors. These new armor systems will come in two forms, either as structural load-bearing components or as nonstructural parasitic appliqués.

The principal advantage of fibers is their high strength-to-weight ratio. Kevlar, for example, approaches a tensile strength of 2.75 GPa. Other properties such as tensile modulus, elongation, tenacity, energy absorption and sound speed have proved to be important in characterizing impact. Impact, however, is a complex problem and one should not rely on any single property as a predictor; often combinations of properties are the key. For example, glass fibers with relatively low impact strength, and polyster resins with poor tenacity, can be combined to produce a laminate of high impact strength. In addition, properties can be severely affected by strain rate. Most fibers show increasing strength and modulus with increasing strain rate, but elongation to break varies. Examination of fiber properties at the strain rate expected is important in ballistic design.

The application of fibers to ballistic protection generally takes two forms. Fabric or soft armor is constructed from multiple layers of woven fabric without a resin binder. The layers are sewn together using a common sewing thread in a straight or zigzag pattern. Composite laminate armors consist of multilayered fabrics combined with a resin binder. The application of the resin greatly affects the performance of the fabric and must be chosen to meet the given application. Depending on fiber and resin content, the laminate may be classified as either structural or nonstructrual. Structural laminates usually contain a greater percentage of resin than of fiber.

Most ballistic threats which fiber composites are designed to resist are either soft lead projectiles or fragment-simulating projectiles (FSP). Ball ammunition common to handguns is typically fabricated from lead and easily "mushrooms" when striking a composite target. FSPs, on the other hand, are machined from hard steel and are used to simulate the effects of blunt-ended fragments propelled from larger exploding warheads. Fabric construction is used for the soft lead projectiles, whereas fragmentation is generally stopped using a laminate. Neither construction has proved particularly effective against the sharp point typical of armor-piercing projectiles. The sharp point can easily push aside the woven fibers and penetrate the composite.

1.1 Failure of Fibrous Composites

As a first step in the study of failure mechanisms in composites, it is instructive to examine the response of a single fiber to impact. In this discussion, it is assumed the fiber is oriented such that its longitudinal axis is perpendicular to the path of the projectile. At the moment of impact, a compressive wave is propagated outward along the longitudinal axis of the fiber. The speed of this wave, c, is defined by

$$c=\left(\frac{E}{\rho}\right)^{1/2} \tag{1}$$

where E is the fiber modulus and ρ is the density. The wave reaches the end of the fiber and is reflected as a tensile wave. Reflection of the wave as a tensile wave is necessary in order to satisfy the boundary condition of zero strain at the fiber ends. The resulting tensile strain is now experienced by the fiber in contact with the projectile, and the magnitude of strain is dictated by the impact velocity of the projectile. Continued wave reflections intensify the tensile strain, with energy absorbed by the fiber being proportional to this tensile strain. At the same time, a second wave propagates along the transverse axis (across the diameter) of the fiber moving parallel to the projectile and at the same velocity. Because of this transverse wave, the fiber deflects in the direction of the projectile path and as a result increases the energy absorbed by the fiber. The fiber continues to deflect and absorb energy until the projectile is decelerated to a stop or the fiber strains past its dynamic yield point and breaks. If the impact velocity is sufficiently high, the fiber cannot respond fast enough to exhibit a strain. The lowest velocity for which this occurs is the critical velocity and is defined as the velocity for which the fiber will break without straining. Critical velocity is a function of wave speed and in turn a function of fiber modulus and density. Increasing the wave speed of a fiber will allow quicker strain response and thus increase its critical velocity.

The mode by which the fiber separates is also important in determining its resistance to penetration. Three principal modes have been identified: melting, brittle fracture and plastic deformation with longitudinal splitting. Nylon fibers melt but exhibit large deformations before breaking. High values of elongation equate to large work-to-rupture values and thus more energy absorption by the fiber. Brittle fracture with low elongation-to-break is typical of glass fibers. Kevlar fibers do not show as high a value of elongation as nylon, but have proved to be ballistically superior. This may seem to be a contradiction until one examines the fiber breaking mode. Kevlar fibers show some degree of plastic deformation before rupturing. More important, however, is the fibers' ability to split along the longitudinal axis. Longitudinal splitting not only absorbs energy but acts as a crack arrester as well.

Failure in composite laminates is a far more complex process. The process is three-dimensional in nature, characterized by interactions occurring in the individual fabric planes and across those planes to adjacent fabric layers. Additional strain-wave reflections at fiber crossovers serve to distribute impact energy over a larger area, thus increasing ballistic resistance. Because of the numerous crossovers found in woven fabrics, the energy is quickly dissipated as the wave encounters more and more fiber crossovers, and the impact damage is usually restricted to a small area adjacent to the point of impact. Energy transfer to adjacent layers (through the thickness) is facilitated by use of a resin binder. Strain waves are transmitted from the fabric layer to the resin matrix and then to adjacent fabric layers.

At the same time, waves are reflected back into the fabric or matrix at each interface. The amplitudes of these waves are determined by a quantity called mechanical impedance I, defined as:

$$I=\rho c \tag{2}$$

where ρ is the density and c is the wave speed defined in Eqn. (1). Waves may be reflected or transmitted according to the following relationships:

$$P_{\mathrm{t}}=\left(\frac{2I_{\mathrm{t}}}{I_{\mathrm{t}}+I_{\mathrm{o}}}\right)P_{\mathrm{o}} \tag{3}$$

$$P_{\mathrm{r}}=\left(\frac{I_{\mathrm{t}}-I_{\mathrm{o}}}{I_{\mathrm{t}}+I_{\mathrm{o}}}\right)P_{0} \tag{4}$$

where P_{o}, P_{r} and P_{t} are the pressure amplitudes of the incident wave, reflected wave and transmitted wave, respectively. I_{o} and I_{t} are the impedance of the incident and acceptor materials, respectively. As an example, suppose that Kevlar with an impedance of 1784 g cm^{-3} s^{-1} is used with an epoxy resin with an impedance of 320 g cm^{-3} s^{-1}. Further suppose that the wave is travelling from the Kevlar to the epoxy. At the interface, a wave of amplitude -0.696 P_{o} is reflected back into the glass and a wave of amplitude 0.304 P_{o} is transmitted to the epoxy matrix (a negative sign indicates a tensile wave).

Transmission of waves to the resin matrix increases the energy-absorption capability of the laminate. Cracks are formed in the matrix which propagate parallel to the fabric layers. This process is called delamination and is characterized by separation of the fabric layers. Impact energy is dissipated in the formation and advancement of these cracks and in the deflection caused by separation of the individual fabric layers.

1.2 Designing with Fibrous Composites

No armor design will be applicable to all situations. The designer is often faced with compromises and must tailor the armor to the specific application. The ballistic performance of the system often depends on the interaction of its various components. It is difficult to separate out the characteristics important in each case but some generalizations can be made. The following paragraphs will examine some of the properties important in selecting components in a fibrous armor design. The properties to be examined include fiber selection, yarn denier, weave, resin binder and environmental susceptibility.

As was seen earlier, ballistic efficiency resulting from individual fiber properties is difficult to characterize. Modulus, wave speed and elongation to break, however, tend to be the predominant factors involved. Increasing these properties will result in a greater energy absorption capability of the armor and thus an increase in protection.

Yarn denier is a measure of the diameter of the spun yarn used to produce the fabric. Denier numbers increase with increasing yarn diameters. Fabrics woven from high-denier yarns will have fewer fiber crossovers and less ability to distribute energy. Ballistic protection will usually decrease with increasing yarn denier with all other factors held constant.

Following the same logic, a close weave with many crossovers will be more efficient than a loose weave. In fabric armors, this seems to hold for handgun projectiles, but fragmentation-type projectiles show a lesser dependence, the difference being the area involved in the impact. Armors made with resin binders do not show a strong dependence on weave construction. The effect of weave is probably masked by the resin–fiber bonding which distributes the impact energy more efficiently.

Ductile resins such as vinyl esters perform better than more brittle resins like epoxy. Resins showing increased ductility will absorb more energy both in crack initiation and propagation. Evidence suggests that, for the best ballistic efficiency, the resin content should be 20–25 wt%. Laminates with less resin content show increased ballistic protection, but the deformations are severe and unacceptable for most applications. Some applications require the laminate to serve as a structural component as well as to give ballistic protection. The best mechanical properties occur at much higher resin contents, so a compromise must be reached between mechanical and ballistic performance.

Because of environmental factors, the fiber with the best ballistic properties may not make the best armor. Many fibers degrade when exposed to sunlight, moisture, seawater, temperature extremes and petroleum products. Military and, to a lesser extent, civilian armor designs, must take environmental factors into account. Fibers must be protected, or resistant fibers selected, when designing any such armor. In addition, cost, weight and space requirements may determine the design. No military commander will opt for an armor that is too heavy or costly when a lighter, less expensive material will suffice. An armor whose cost does not justify the additional weight saving will not be implemented. For maximum ballistic efficiency, fibrous armors must be allowed to deflect or delaminate. For this reason, fibrous armors must never be combined with rigid components in a way which restricts deflection. When minimal space is allowed for the armor, fibrous armors are not the best choice.

2. *Ceramic Composite Armors*

If the definition of ceramics is limited to refractory materials, the history of their use as armor materials is a short one. A broader definition, however, extends their use to several thousand years. Stone and, more recently, concrete have long been recognized for their effectiveness as defensive materials against fragmentation, projectiles and blast. In the defensive posture, weight is not an issue. Offensive operations, on the other hand, require a high degree of mobility. Materials selected for this kind of mission must be light and effective. This requirement led the US Navy to evaluate ceramics for use on warships during World War II. It was found that plate glass backed by Doron (a glass-reinforced polyester) was an efficient combination against fragmenting warheads and resulted in weight savings over the existing steel armor. The question of structural integrity, however, prevented its acceptance. During the Vietnam War, the search for lightweight armors again turned to ceramics. In the early 1960s, Goodyear Aerospace Corporation and the US Army Materials and Mechanics Research Center investigated the use of ceramics for aircraft and river patrol boats, and for protection of ground combat personnel and aircrews. In 1970, Goodyear was issued a patent for an armor design consisting of a ceramic front face backed by resin-impregnated glass fabric. The bonding agent between the ceramic and glass fabric was identified as an elastomeric adhesive such as polysulfide.

The principal advantages of ceramics over conventional steel are their high compressive strength and hardness. Hardened and sharp-pointed armor-piercing projectiles are effective against softer materials. This effectiveness is lost in impacts with ceramics because the dynamic stress limits in ceramics are much greater than in the projectiles. The sharp point is quickly eroded away leaving a less effective blunt cylinder. The loss of projectile mass reduces the energy of impact and increases the protection afforded by the ceramic system. The system aspect of ceramic armor design is important and must be emphasized. Because of the inherent brittleness of ceramics, they may never be suitable as "stand alone" armor. Ceramics are often combined with other more ductile materials to achieve the required durability and structural integrity. The following section will examine the failure mechanisms in ceramics during impact from a theoretical point of view, followed by a discussion of important aspects of ceramic armor design. Design aspects are the result of the authors' observations and experience and are not founded on strict theoretical principles. Much fundamental research is needed before ceramic armor design is reduced to an exact science.

2.1 *Failure Mechanisms in Ceramics*

Failure of the ceramic during impact is a multistage process. These stages are: (a) erosion of the projectile tip and formation of surface waves; (b) failure of the ceramic in compression; (c) penetration into the ceramic and formation of the fracture conoid; (d) failure of the ceramic in tension; and (e) removal of ceramic debris. Discussion of each stage will be taken in turn. The discussion assumes the projectile velocity is sufficiently small to prevent hydrodynamic flow in the ceramic. Projectile material may, however, fail by flowing if the impact stress is sufficiently high.

When ceramics are confined, their compressive strength will increase, as outlined in the following section. The confinement does not have to be by outside forces but merely by the presence of ceramic material adjacent to the point of impact. The first few microseconds of the impact event are characterized by this increased compressive strength and the projectile is quickly overwhelmed. The projectile tip is eroded away and initially flows radially outward and then backward. Surface waves are propagated radially outward from the point of impact. When these waves reach the lateral boundaries of the ceramic, they are reflected back in the form of relief waves.

Relief waves reach the impact point, relieving the confinement of the ceramic. A decreased compressive yield strength results. The projectile is no longer severely overmatched and begins to penetrate into the ceramic. At the same time, combinations of compressive waves and relief waves begin to form the fracture conoid. In its final form, the fracture conoid will form a cone of fracture surfaces and debris with an apex angle of the order of 60° and a base on the rear surface of the ceramic opposite the point of impact.

The compressive wave mentioned earlier forms a spherical front which travels ahead of the projectile parallel to the direction of impact, again propagating at the wave speed. When this wave reaches the rear surface of the ceramic it is reflected as a tensile wave as described earlier in Eqn. (4). The ceramic, being very weak in tension, fails; this tensile failure propagates back toward the point of impact. When the tensile fracture zones meet with the fracture conoid propagating rearward, the conoid is complete. The final stage is characterized by the pushing of the remaining projectile and ceramic debris from the conoid out of the rear of the ceramic.

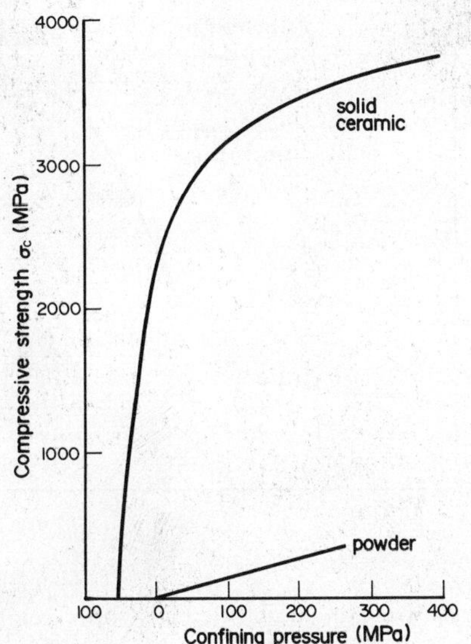

Figure 1
Compressive strength versus confining pressure for AD-85 alumina

2.2 Ceramic Behavior Under Confining Pressure at High Loading Rates

Although the ceramics envisioned for armor use are typically extremely brittle, their mechanical behavior nevertheless is remarkably sensitive to two parameters related to such applications: confining pressure and strain rate. Relevant knowledge regarding these factors can be outlined as follows.

Uniaxial compressive stress–strain curves for ceramics such as alumina show that these materials exhibit no plasticity and fail by rubblization. As increasing confining pressure is applied, the modulus is unaffected, but the failure strength increases significantly; this increase is caused by the superposition of compressive stresses which reduce the opening of microcracks parallel to the nominal loading axis. If the failed (rubblized) ceramic is subsequently tested, increases in confinement raise both the slope of the stress curve (reflecting compaction of the fragments) and also the compressive strength. This behavior is summarized for AD-85 alumina as a function of confining pressure in Fig. 1. The compressive strength of solid alumina contrasts markedly with that of the same material in powder form.

Such pressure dependence is not accounted for by the familiar von Mises criterion which describes the flow of metals, since the criterion is pressure independent. Instead, the Coulomb–Mohr equation (or a higher order variant)

$$(J_2)^{1/2} = k - aP \tag{5}$$

is used, where J_2 is the second invariant of the deviatoric stress tensor, P is the hydrostatic pressure, and k and a are constants.

It should be noted that while confinement in the experiments described above is provided by a hydrostatic test chamber, the impact of a penetrator on a target provides confinement in and of itself because of the transient uniaxial strain conditions which are obtained. The high intrinsic compressive strength of the ceramic, strongly abetted by confinement, severely overmatches the strength of the penetrator, resulting in erosion of the latter as it is turned and flows in the opposite direction (failure). This is true even after the ceramic has failed, as long as the resulting powder is physically confined (Fig. 1).

Compressive strength (σ_c) versus strain-rate ($\dot{\varepsilon}$) data for several technological ceramics are shown in Fig. 2. It is evident that for $\dot{\varepsilon} \gtrsim 100\ \mathrm{s}^{-1}$, some of the materials (SiC and hot-pressed Si_3N_4) exhibit a marked strain-rate strengthening, whereby

$$\sigma_c \propto \dot{\varepsilon}^n \tag{6}$$

where $n = 0.33$, while for others (Al_2O_3, reaction bonded Si_3N_4) no such strengthening is observed. These results suggest that there exists a threshold strain rate ($\dot{\varepsilon}^*$) for dynamic hardening which varies from one material to another (and which for Al_2O_3 and reaction-bonded Si_3N_4 obviously must exceed $2 \times 10^3\ \mathrm{s}^{-1}$). The dynamic fracture analysis of Grady (1982) and Kipp et al. (1980) indicates that this should be so, and predicts Eqn. (6) above. Specifically, it is shown that the failure stress corresponding to the coalescence of a multitude of

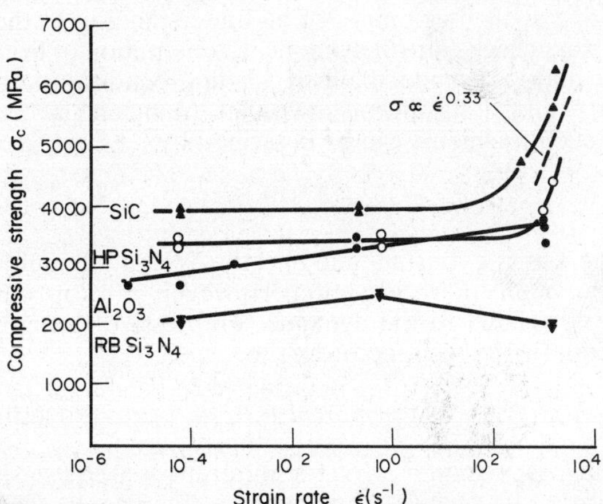

Figure 2
Compressive strength versus strain rate for several ceramics (HP, hot pressed, RB, reaction bonded)

microcracks is given by

$$\sigma_c = (\rho_0 c_0 K_{IC}^2 \dot{\varepsilon})^{1/3} \quad (7)$$

where ρ_0 is the density, c_0 is the wave speed, and K_{IC} is the fracture toughness. Physically, Eqn. (7) reflects the fact that cracks possess "inertia," related to the finite time interval required for a stress wave to travel the length of a crack.

It should be noted that not all strain-rate-hardening brittle materials give evidence that $n = 0.33$. Recent work involving fiber-reinforced ceramic-matrix composites, in which failure occurs through more complex micromechanics (fiber kinking and shear-band propagation) than the simple coalescence of a relatively homogeneous array of microcracks, has shown that the strain rate exponent n can approach unity, depending upon the specific microstructure. This potential variability is important, since it will be shown that fragment size under impulsive loading is a sensitive function of n, and the higher the value of n, the lower the strain rate (or ballistic velocity, for armor applications) at which the ultimate strength is achieved. It thus appears that appropriate reinforcement design would permit the fabrication of composite ceramics with dynamic strengths much higher than their monolithic matrices alone.

Theoretical estimates of the lower bound strain rate $\dot{\varepsilon}^*$ for inertial control of microfracture in monolithic materials yield

$$\dot{\varepsilon}^* = \frac{K_{IC}}{\rho_0 c_0 r_0^{3/2}} \quad (8)$$

which is a function of the initial crack size r_0. Substituting appropriate values of ρ_0, c_0, r_0 and K_{IC} for armor-type ceramics yields $10^3 \lesssim \dot{\varepsilon}^* \lesssim 10^4\ s^{-1}$, in basic agreement with Fig. 2.

It is possible to carry these concepts further, to arrive at the beginnings of an understanding of the factors which control dynamic fragmentation of brittle materials under ballistic loading conditions. In particular, it has been shown that the mean size L_M of such fragments can be described by:

$$L_M = A\dot{\varepsilon}^{(m/m+3)} \quad (9)$$

where A is a constant and m is the Weibull modulus for dynamic flaw activation. However, m turns out to be related to the dynamic compressive strength (Eqn. (6)) as well, according to:

$$n = 3/(m+3) \quad (10)$$

Thus, if n should vary for different materials, so should m, hence the average fragment size. Fragment dimensions are potentially very important in penetrator erosion, since they determine the extent to which the fractured ceramic can behave

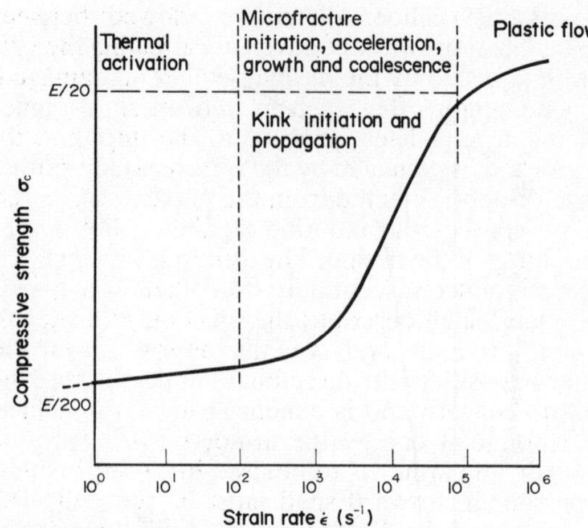

Figure 3
Schematic of strength–strain rate behavior of ceramic composites. E is Young's modulus. By adjusting the material microstructure, the steeply rising portion of the curve may be moved about within the range $10^3 \lesssim \dot{\varepsilon} \lesssim 10^5\ s^{-1}$. The $E/20$ line is not necessarily the HEL; there is evidence that microfracture can occur below and above the HEL for brittle materials

like a fluid in the path of the penetrator as well as the density and surface area of eroding particles.

The fracture behavior of ceramic materials can be summarized as shown in Fig. 3. For $\dot{\varepsilon} \lesssim 10^2\ s^{-1}$, fracture is controlled by thermal activation of microcracks, a process that is unimportant in ballistic penetration. At higher loading rates, inertial and flaw propagation (microcrack growth, fiber kinking) effects provide significant strain-rate hardening. For $\dot{\varepsilon} > 10^4\ s^{-1}$, microfracture is suppressed, and plastic flow begins to occur as the Hugoniot elastic limit (HEL) is exceeded. Recent work on Al_2O_3 suggests that the HEL represents true plastic flow, and that microfracture, although probably plasticity related, is not initiated until stresses exceed at least twice the value of the HEL.

2.3 Designing with Ceramics

A typical ceramic armor design will consist of a forward spall cover, ceramic, adhesive and backing plate. Many variations of this design are possible of course, but a description of this design will aid in illustrating the important aspects of ceramic armor design. Most industrial ceramics suitable for use in armors are available in a square or rectangular configuration. This shape will be assumed in the following discussion and will be referred to as ceramic tiles. Two important criteria not discussed previously apply to ceramic armors more than any other type of armor. These criteria are multihit capability and collateral damage. Multihit requirements are usually specified in terms of the maximum number of projectile caliber diameters allowed between impact locations without adversely affecting the protection afforded by the design. Collateral damage is a phenomenon often seen in improperly designed ceramic armor. Tiles adjacent to the impacted tile are cracked or damaged by the transmission of shock waves or debris ejected from the impact. This effect greatly reduces the multihit capability and overall desirability of the design. The armor designer should strive to reduce susceptibility to collateral damage.

Forward spall describes the ejection of projectile erosion products and ceramic debris out of the armor opposite to the direction of impact. Often this is of no concern and is beneficial in that it reduces the amount of energy the armor must absorb. If, however, the armor is to be used in close proximity to personnel, forward spall must be reduced. This can most easily be done by covering the ceramic with a flexible material such as a glass-reinforced plastic.

Ceramic material properties important in armor design include compressive strength, impedance, density and tensile strength. Ballistic efficiency can be correlated with the average of confined and unconfined compressive strengths

$$(Y + l_H)/2 \tag{11}$$

where Y is the unconfined compressive strength and l_H is the Hugoniot elastic limit, a measure of the confined strength. High tensile strength is desirable to delay the onset of tensile failure on the tile rear face described earlier. Delaying this failure only a few microseconds will significantly increase the perforation resistance of the ceramic. Low density allows the use of thicker tiles for the same areal density and reduces the stress on the back plate. Impedance should be high, as this is a measure of distortion produced in the ceramic. In short, the desired ceramic should have high compressive and tensile strength, high impedance and low density.

It has been observed that a single tile performs better than a stack of multiple tiles of equivalent thickness. The reason is that the single tile has fewer free boundaries where tensile cracks can form, particularly with respect to the time for wave travel to the rear surface of the tile. This time difference is large enough to greatly influence penetration resistance. The lateral dimensions of the tile should be as large as is consistent with acceptable multihit criteria. The larger the distance from the impact point to the sides of the tile, the longer it takes for relief waves to reach the point under the impacting projectile. The ceramic is then no longer confined and will fail in compression more readily. This represents a paradox for the ceramic-armor designer. A tile with large lateral dimensions is ballistically more efficient but has less capability for multihit. The thickness needed is dependent on the type of projectile and velocity. Generally, since lateral dimensions are larger than the thickness, thickness will be the predominant design factor. The final design thickness is often dictated by the corners and intersections of adjacent tiles. At corners where four tiles intersect, the ballistic efficiency may be reduced by as much as 30% compared to center-of-tile impacts. This reduction is a function of the gap size, so gap size should be reduced to the minimum value that does not cause collateral damage.

As mentioned earlier, collateral damage is a tremendous problem in ceramic armors. It has been observed that this problem can virtually be eliminated by placing an elastomer in the lateral gaps between tiles. The elastomer helps to dissipate the shock transmitted between tiles, but more importantly it prevents debris from impacting into adjacent tiles. A thin layer of elastomer between tiles and the back plate will also aid in mitigating transmitted shock waves and further reduce collateral damage. This aspect is more important if a metallic back plate is used, and less so if fibrous composites or other soft materials are used.

Backing plates are necessary in ceramic armors because of the inability of ceramics to provide structural integrity. In addition to a structural role, the backing plate catches the residual debris that is pushed out of the rear of the fracture conoid. In order to perform this task, the back plate must be a ductile material. Popular choices are aluminum and fibrous composites. Care must be taken not to allow too much deflection in the backing plate. Severe distortion can affect the armor's ability to sustain multiple impacts and remain structurally sound. The ratio of ceramic to back plate thickness should be kept between 50% and 60%.

Adhesives are often used to secure the ceramic to the backing plate. In general, the adhesive has little effect on the ballistic performance of thick tiles, but has significant effect on thin tiles. In thin tiles, the adhesive thickness should be kept to a minimum to secure the tile to the backing plate. Typical thicknesses are in the order of 0.5 mm. Care should be taken to ensure the adhesive is not too spongy and that it adequately transfers stress from the ceramic to the back plate. In other words, the adhesive must not sacrifice the rigid support provided by the back plate.

Bibliography

Agarwal B D, Broutman L J, 1980 *Analysis and Performance of Fiber Composites*. Wiley, New York

Grady D E 1982 Local inertial effects in dynamic fragmentation *J. Appl. Phys.* 53: 322–5

Grady D E, Lipkin J 1980 Criteria for impulsive rock fracture. *Geophys. Res. Lett.* 7: 255–8

Kipp M E, Grady D E, Chen E P 1980 Strain-rate dependent fracture initiation. *Int. J. Fract.* 16: 471–8

Kolsky H 1963 *Stress Waves in Solids*. Dover, New York

Lankford J 1987 Temperature, strain-rate, and fiber orientation effects in the compressive fracture of SiC fiber-reinforced glass-matrix composites. *Composites* 18: 145–52
Liable R C (ed.) 1980 *Methods and Phenomena* 5: *Ballistic Materials and Penetration Mechanics*. Elsevier, New York
Lindholm U S, Yeakley L M, Nagy A 1974 The dynamic strength and fracture properties of Dresser basalt. *Int. J. Rock. Mech. Min. Sci. Geomech. Abstr.* 11: 181–91
Zukas J A, Nicholas T, Swift H F, Greszczk L B, Curran D R 1982 *Impact Dynamics*. Wiley, New York

J. Lankford and W. Gray
[Southwest Research Institute, San Antonio, Texas, USA]

Automobile Tires

Although automobile and truck tires are manufactured in vast quantities—over 200 000 000 per year—it is not generally recognized that the tire is a highly sophisticated composite structure. Layers of rubber and rubber-impregnated sheets of parallel fiber bundles (cords) are combined to make what is basically a toroidal laminate with different properties in each of its three principal directions. Indeed, it is this carefully chosen anisotropy that enables the tire to perform diverse and often conflicting functions. Tire mechanics, outlined here, deals with the appropriate choice of materials and designs in order to achieve five main goals:

(a) high sliding friction in wet and dry conditions, over a wide range of temperatures and with varied road surfaces, to provide adequate acceleration, braking and steering;

(b) low rolling resistance under the same circumstances;

(c) a low vertical stiffness to cushion the ride;

(d) high longitudinal and lateral stiffnesses to minimize sliding motions in the contact patch (see Sect. 3); and

(e) resistance to cutting, puncturing, abrasion, etc.

How these diverse requirements are met is discussed below, after a description of the materials of which the tire is composed.

1. Tire Materials

R. W. Thomson, in 1845, was the first to employ air to provide pneumatic cushioning in a rubber tire, but the first practical use was developed in 1888 by Dr J. B. Dunlop. In modern tires, loss of air is minimized by a special rubber inner lining, usually of butyl rubber, chosen for its low permeability (Fig. 1).

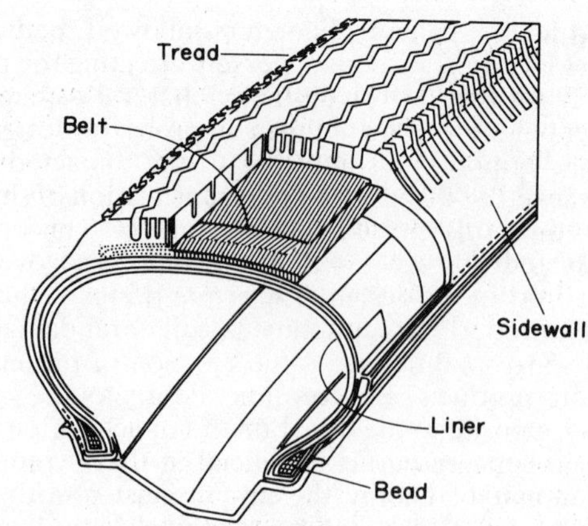

Figure 1
Structure of a radial tire

The tire must be reinforced by layers of relatively inextensible cords to contain the air pressure and to restrict longitudinal and lateral deformations and growth of the tire in service. Such cord layers have been made of a variety of materials over the years: cotton, rayon, nylon, polyester, glass, polyaramide and steel, the latter two materials being the dominant ones at the present time. The choice of cord material is based on the need for high stiffness, good resistance to repeated flexing, high strength-to-weight ratio, good adhesion to rubber and low cost.

Rubber sidewalls of the tire (Fig. 1) must resist scraping, flexing, and the attack of ozone in the air. Rubber treads must primarily resist abrasive wear. Indeed, this is one of the most severe applications of rubber, and the continuous improvement in wear resistance by improved formulations has been a remarkable technological achievement. Typical formulations for tire sidewalls and treads are given in Table 1.

It must be emphasized, however, that rubber formulation is still more art than science. For example, the reason why a tread recipe based on polybutadiene is superior in abrasion resistance to one based on polyisoprene or on a butadiene–styrene copolymer is not known. It should also be pointed out that strong rubber compounds are themselves composite in nature. Typically, about 30% by volume of a fine particulate solid filler, usually carbon black, is included in the formulation. Without this ingredient, the abrasion resistance of rubber would be much lower. Indeed, the introduction of carbon black reinforcement in rubber about 1910 was a major factor in prolonging tire life, from about 8000 km to about 16 000 km. But, again, the mechanism of reinforcement of rubber by fillers is still obscure.

2. Tire Structure

The most important feature of tire structures is the arrangement of the reinforcing cords, which determines the degree of anisotropy. At first they were placed to minimize distortion. For a straight tube, alternate layers would then be arranged at a cord (crown) angle to the tube axis of $\alpha = 57°$ (Fig. 2) so that the line tensions along and across the tube, given by $nt\cos^2\alpha$ and $nt\sin^2\alpha$, would meet the

Table 1
Representative formulations for passenger car tires in parts by weight. The mixes are converted into strong elastic solids by chemically interlinking the rubber molecules ("vulcanization"), on heating for about 20 min at a temperature of about 150 °C in a tire mold

	Compound	
Ingredient	Sidewall	Tread
Natural rubber	50	–
Butadiene–styrene copolymer (75/25)	–	65
Cis-1, 4 polybutadiene	50	35
Heavy aromatic oil	–	25
Carbon black (HAF)	50	65
Processing oil	10	15
Zinc oxide	4	4
Stearic acid	1	1
Sulfur	2	2
Vulcanization accelerator[a]	1	1
Antioxidant[b]	3	2
Protective wax	2	2

a Example: *N*-cyclohexyl-2-benzothiazolesulfenamide
b Example: *N*-(1, 3 dimethylbutyl)-*N*′-phenyl-paraphenylenediamine

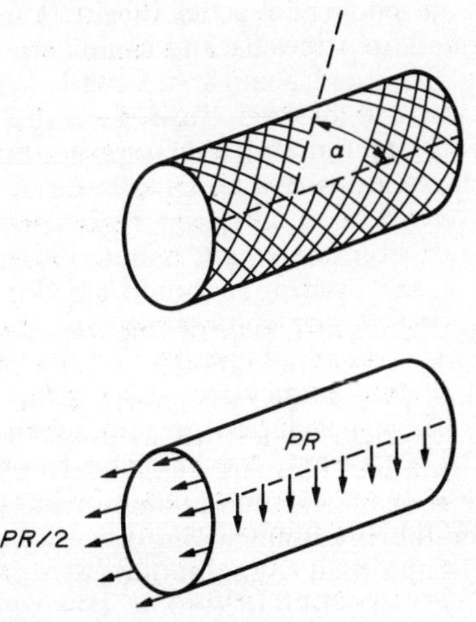

Figure 2
Cord angle α

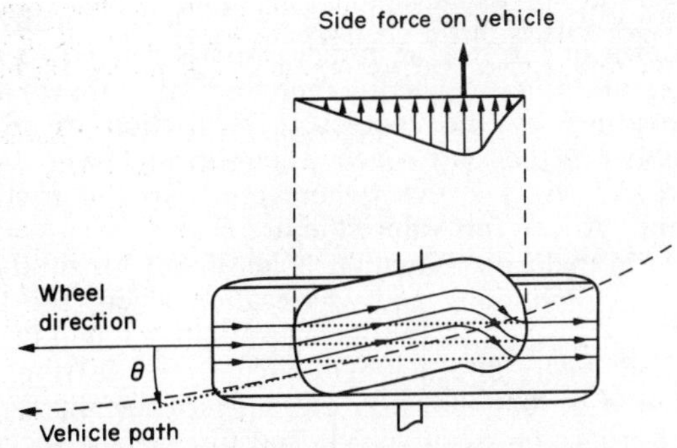

Figure 3
Tire deformation and ground stresses set up in cornering

inflation requirements, $PR/2$ and PR, respectively, where P is the inflating pressure, R is the tube radius, t is the cord tension and n is the number of cords per unit of width normal to the cord direction. In the early 1950s, however, Michelin introduced the radial tire (patented by Gray and Sloper in 1913), in which one set of cords is arranged to lie almost parallel to the tire circumference, forming a circumferential belt (Fig. 1). In practice, because the tire is a toroid rather than a cylindrical tube, the belt cords are made to lie at a crown angle of about 11° rather than at 0°. This arrangement maximizes the line tension in the circumferential direction for a given value of the inflation pressure, and hence gives greater resistance to lateral distortion of the tire on cornering. This feature provides improved steering and wear resistance (see Sect. 4).

3. Tire Traction

Consider a steadily applied traction or vehicle/maneuvering force, as in acceleration, braking or cornering. The tire footprint takes up a steady-state condition, in which tread rubber, entering and advancing through the footprint, is gripped fixedly by the ground while the interior of the tire body moves or "slips" with respect to the ground. Thus the tread is progressively strained under the ground and wheel reactions. Figure 3 illustrates vehicle cornering, where the tread is displaced laterally from the plane of wheel rotation, set (by the driver) at a "slip" angle θ to the direction of vehicle motion. A point is reached, usually near the exit from the footprint, where the ground can no longer exert a sufficiently large stress to maintain the deformation because the local friction has been exceeded. The deformation of the tire then recovers by sliding of the surface over the ground, back to the plane of the wheel.

The total lateral force exerted on the wheel is the summation of the ground stress distribution over the footprint. It varies with the slip angle θ typically as shown in Fig. 4. The initial slope is related to the lateral stiffness of the tire, and the maximum value is governed by the coefficient of friction of tread rubber on the road, reached at a slip angle where the ground stress is everywhere equal to the friction limit. Above this value, the tire skids.

An analogous situation holds under longitudinal forces of braking and acceleration where the tire slips by "windup," producing a strain gradient in the wheel plane and a longitudinal stress distribution similar to that shown in Fig. 3. Here the slip is the difference between vehicle and tire surface speeds, relative to vehicle speed. However, in braking, the tire gets so hot that the effective coefficient of friction of rubber against the road decreases, and Fig. 4 becomes modified as shown in Fig. 5. A so-called friction peak is encountered, which antiskid devices try to maintain.

The initial slopes of curves like those shown in Figs. 4 and 5 are steeper for the stiffer radial construction so that more rapid responses and higher levels of tractive force are obtained for the same steering or braking (or accelerative) inputs.

Because the resultant of lateral and longitudinal forces cannot exceed the coefficient of friction times the weight on the tire, when both act together, as in a braked turn, the resistance to skidding is diminished.

Traction is maximized by the choice of tread materials having high hysteresis or energy dissipation, which raises the coefficient of friction of rubber. Addition of carbon black and the use of high styrene–butadiene copolymers have this effect.

4. Tire Wear

Life expectancy of tires has increased to approach 150 000 km because of advances in rubber compounding and tire design. Use of improved vulcanization systems and blending of elastomers, including

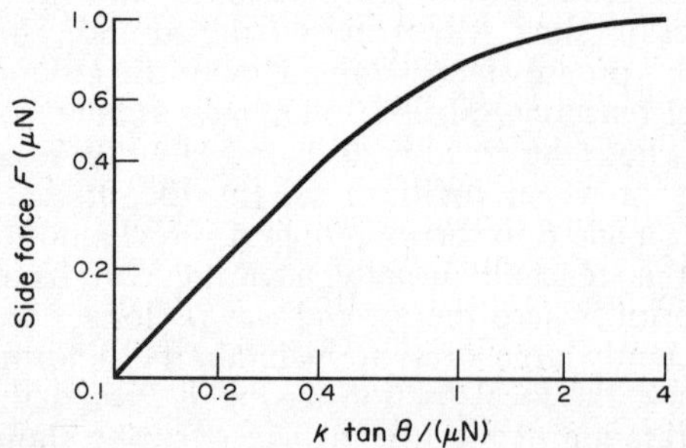

Figure 4
Side force F vs slip angle θ. N is the normal load on the tire, μ the coefficient of friction, and k the lateral stiffness × the half-length of contact patch

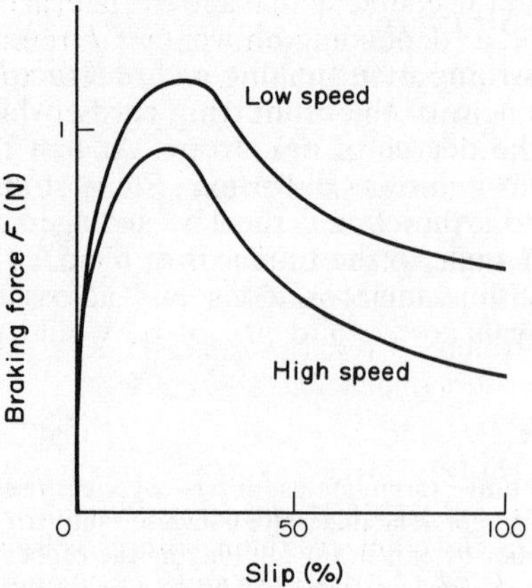

Figure 5
Braking force F vs slip (windup), expressed as the slip velocity relative to the circumferential velocity of the wheel

the abrasion resistant *cis*-polybutadiene, has greatly improved the wear resistance of the tread. The almost universal adoption of the superior radial construction is another important factor.

However, radial tires are more prone than bias tires to uneven wear, in which parts of the tread pattern tend to wear away at a greater rate than their surroundings. When such concentrated wear reaches the vicinity of the belt, the tire must be replaced even though most of the tread remains. Uneven wear is minimized by empirical design of tread elements and by ensuring that the vehicle suspension does not impose severe camber or slip angles on the tire.

Wear proceeds by the mechanism mentioned in Sect. 3. At some point in the footprint the cumulative stress on the tread reaches the friction limit and the deformed tire slides back to its undistorted shape scrubbing the tread against the road. When wear occurs evenly over the tread face, the rate R ($m^3\,m^{-1}$) of material abraded from the tread in unit distance travelled is

$$R = \rho\gamma F^2/k$$

where ρ is the tire resilience or fraction of energy not dissipated in a loading–unloading cycle and therefore available to produce abrasion; γ is the abradability ($m^3\,J^{-1}$), defined as the amount of material abraded for a unit of energy expended in sliding; F (N) is the root mean square value of the distortion force over the tire run, arising from accelerations imposed by the road configuration, by the necessity to stop and start, by the vehicle speed and driver habit; and k is the appropriate stiffness coefficient in the ground plane, defined as the product of the half-length of the contact patch and the lateral or longitudinal force (depending on whether F refers to wear in cornering, or in braking and acceleration) for a unit of displacement of the contact patch with respect to the wheel rim.

Higher k means smaller excursions of the tread in slip and, therefore, shorter sliding paths and reduced wear. This explains the enhanced tread life of the stiffer radial over the softer bias construction.

See also: Rubber Tires; Tire Adhesion: Role of Elastomer Characteristics (Suppl. 2)

Bibliography

Clark S K (ed.) 1981 *Mechanics of Pneumatic Tires*, 2nd edn. US Government Printing Office, Washington, DC

Eirich F R (ed.) 1978 *Science and Technology of Rubber*. Academic Press, New York

Hays D F, Browne A L (eds.) 1974 *The Physics of Tire Traction*. Plenum, New York

Pearson H C 1906 *Rubber Tires and All About Them*. India Rubber Publishing Company, New York

Tompkins E 1981 *The History of the Pneumatic Tire*. Lavenham Press, Lavenham, UK

A. N. Gent
[University of Akron, Akron, Ohio, USA]

D. J. Livingston
[Livingston Associates, Akron, Ohio, USA]

Automotive Composite Components: Fabrication

There is a fundamental difference in the strategy of application of composites between the aerospace industry and the automobile industry. This is primarily due to the volume requirements of the two businesses. In aerospace and defense, the design of the structure is optimized to provide the required functionality and performance, and the manufacturing process (and associated cost) is subsequently selected on the basis that the process is capable of achieving the desired design. In direct contrast, in high-volume production industries such as the automotive industry, the rate of manufacture is critical to satisfying the economics of this consumer industry. Thus, manufacturing processes which are capable of satisfying production output are the primary consideration, and design of a component or structure must be within the boundary constraints of the selected fabrication process. It is extremely important to appreciate this philosophical difference of approach in order to understand the difficulty of translating the extensive design and fabrication experience in aerospace and defense into the automotive industry. Perhaps the clearest example is the use of prepreg materials and hand layup procedures, common in aerospace and amenable to optimal design, which is clearly unacceptable in an industry requiring high manufacturing output but with a quality level of the same degree.

The use of plastic-based composite materials in automotive applications has gradually evolved during the 1970s and 1980s. Virtually all uses of these materials in high-volume vehicles are limited to decorative or semistructural applications. Sheet-molding component (SMC) materials are the highest-performance composites in general automotive use today, and the most widely used SMC materials consist of approximately 25 wt% chopped glass fibers in a polyester matrix and so cannot really be classified as high-performance composites. Typically, SMC materials are used for grille opening panels on many car lines, and closures panels (hoods, deck lids and doors) on a few select models. A characteristic molding time for SMC is of the order of two minutes, which is on the borderline of viability for automotive production rates. A recent breakthrough by one manufacturer has reduced this cycle time to one minute.

The next major step for composites in the automotive business is the extension of usage into truly structural applications such as primary body structure and chassis/suspension systems. These structures have to sustain the major road load inputs and crash loads and, in addition, must deliver an acceptable level of vehicle dynamics so that passengers enjoy a comfortable ride. These functional requirements must be totally satisfied for any new material to achieve extensive application in body structures and must do so in a cost-effective manner. Appropriate composite fabrication procedures must be applied or developed which satisfy high production rates but still maintain the critical control of fiber placement and distribution.

1. Composite Materials for Automotive Usage

By far the most comprehensive property data have been developed on aerospace composites, in particular carbon-fiber-reinforced epoxy designed for fabrication by hand layup from prepreg materials. Relatively extensive databases are available on these materials and it would be very convenient to be able to build from this database for less esoteric applications such as automotive structures. Carbon fibers are the preferred material in aerospace applications because of the superior combination of stiffness, strength and fatigue resistance exhibited by these fibers. Unfortunately for cost-sensitive mass-production industries, these properties are only attained at significant expense (typically carbon fibers cost US$25 or more per pound at 1988 prices). Intensive research efforts are being devoted to reducing these costs by utilizing a pitch-based precursor for production of these fibers, but the most optimistic cost predictions are around $10 per pound, which would severely limit the potential of

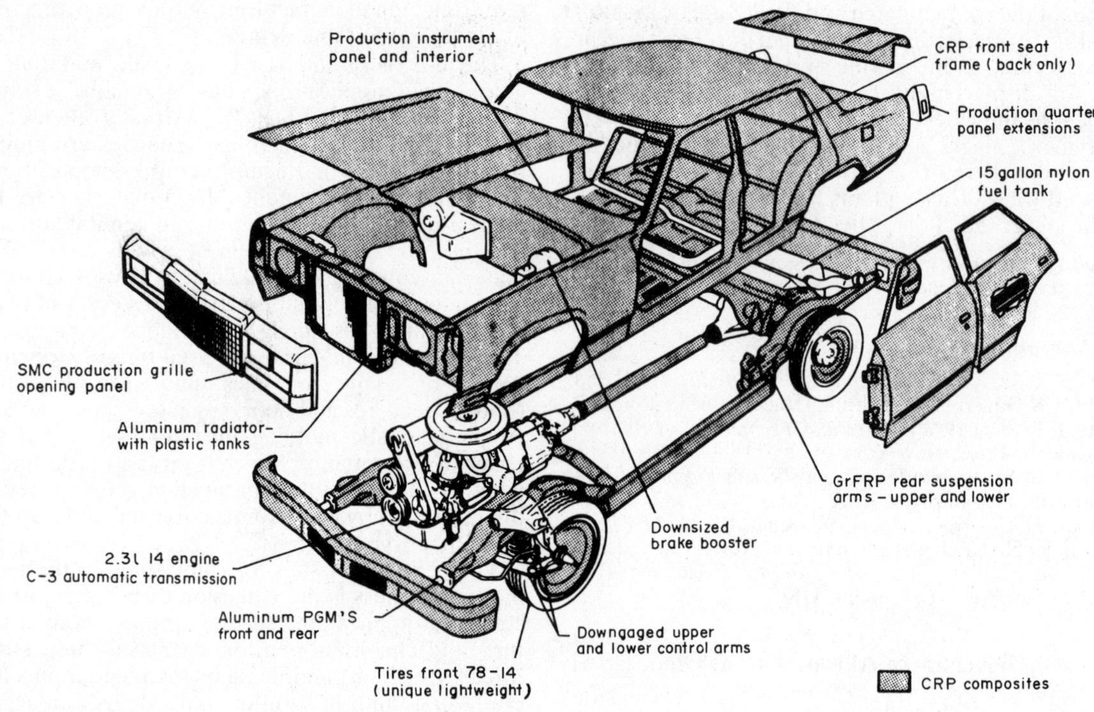

Figure 1
Exploded schematic view of a composite vehicle, with delineation of the composite components. GrFRP, graphite-fiber-reinforced plastic

these fibers for use in consumer-oriented industries. To illustrate the potential that carbon fiber could offer the automobile industry if a breakthrough ever occurred in the reduction of cost, it is interesting to summarize the data on the prototype carbon fiber LTD built by Ford to compare directly with a production steel vehicle (Beardmore et al. 1980, Kulkarni and Beardmore 1980). Although the vehicle was fabricated by hand-layup procedures several interesting features were evaluated. An exploded schematic showing the composite parts of the carbon fiber composite car, made of carbon-fiber-reinforced plastic (CFRP), is shown in Fig. 1. Virtually all structures were designed using aerospace techniques, utilizing 0°/90° and ±45° prepreg materials. The weight savings for the various structures are given in Table 1. While these weight savings (of the order of 55–65%) might be considered optimal because of the use of carbon fibers, other more cost-effective fibers can achieve a major proportion of these weight savings (see Sect. 2.3). Although the CFRP vehicle weighed 1138 kg compared to a similar production steel vehicle of 1705 kg, vehicle evaluation tests indicated no perceptible differences between the vehicles. Ride quality and vehicle dynamics were judged at least equal to top-quality production steel LTD cars. Thus, on a direct comparison basis, a vehicle with a structure made entirely of fiber-reinforced plastic (FRP) was proven at least equivalent to a steel vehicle from a vehicle dynamics viewpoint, at a weight of only 67% of that of the steel vehicle.

The CFRP car clearly showed that high-cost fibers (carbon) and high-cost fabrication techniques (hand layup) can yield a perfectly acceptable vehicle based

Table 1
Major weight savings in a CFRP vehicle. Figures in brackets are weights in pounds

	Weight (kg)					
Component	Steel		CFRP		Reduction	
Body-in-white	192.3	(423.0)	72.7	(160.0)	115.0	(253.0)
Front end	43.2	(95.0)	13.6	(30.0)	29.5	(65.0)
Frame	128.6	(283.0)	93.6	(206.0)	35.0	(77.0)
Wheels (5)	41.7	(91.7)	22.3	(49.0)	19.4	(42.7)
Hood	22.3	(49.0)	7.8	(17.2)	14.7	(32.3)
Decklid	19.5	(42.8)	6.5	(14.3)	13.1	(28.9)
Doors (4)	64.1	(141.0)	25.2	(55.5)	38.9	(85.5)
Bumpers (2)	55.9	(123.0)	20.0	(44.0)	35.9	(79.0)
Driveshaft	9.6	(21.1)	6.8	(14.9)	2.8	(6.2)
Total vehicle	1705	(3750)	1138	(2504)	566	(1246)

on handling, performance and vehicle dynamics criteria. However, crash and durability performance were not demonstrated and these will need serious development work. An even bigger challenge is to translate that performance into realistic economics by the use of cost-effective fibers, resins and fabrication procedures.

The fiber with the greatest potential for automobile structural applications, based on optimal combination of cost and performance, is E-glass fiber (costing approximately US$0.80 per pound at 1988 prices). Likewise, the resin systems likely to dominate, at least in the near term, are polyester and vinyl-ester resins based primarily on a cost–processibility trade-off. High-performance resins will find only specialized applications (in much the same way as carbon fibers) even though their ultimate properties may be somewhat superior.

The form of the glass fiber will be very application-specific and both chopped and continuous glass fibers will find extensive use. It is expected that most of the structural applications involving significant load inputs will utilize a combination of both chopped and continuous glass fiber with the particular proportions of each depending on the component or structure. Since all the fabrication processes anticipated to play a significant role in automotive production are capable of handling mixtures of continuous and chopped glass, this requirement should not present major restrictions. One potential development which is likely if glass fiber composites come to occupy a significant portion of the structural content of an automobile is the tailoring of glass fiber and corresponding specialty-resin development. Approximately 35 million vehicles per year are produced worldwide and consequently each pound of composite per vehicle implies the usage of 35 million pounds in the industry as a whole. This dictates that it should be economically feasible to have fiber and resin production tailored exclusively for the automobile industry. The advantage of such an approach is that these developments will lead to incremental improvements in specific composite materials, which in turn can promote usage.

Glass-fiber-reinforced composites must be capable of satisfying the functional requirements of the various elements of the structure in an automobile. The three primary criteria are fatigue (durability), energy absorption and ride quality (vehicle dynamics). Provided sufficient data are available on all three characteristics, it should then be possible to utilize these to satisfy the design requirements of the automobile. For example, unidirectional glass FRP materials typically have a well-defined fatigue limit of the order of 35–40% of the ultimate strength (Dharan 1975). By contrast the chopped-glass composite would have a fatigue limit closer to 25% of the ultimate strength (Smith and Owen 1969, Aotem and Hashin 1976) and would exhibit much greater scatter in properties. Clearly components and structures composed of a mixture of continuous and chopped fibers would have fatigue resistance specifically related to the proportions of each type of fiber. There is clear evidence that glass-fiber-reinforced composites can be designed to withstand the rigorous fatigue loads experienced under vehicle operating conditions. It must be emphasized, however, that fatigue design data are a critical function of the relevant manufacturing procedure.

In a similar manner, evidence is accumulating that fiber-reinforced-plastic composites can be efficient energy absorbing materials. Relative data from the collapse of tubes is given in Table 2 to illustrate this point and is based on the extensive work of Thornton (1979), Thornton and Edwards (1982) and Thornton et al. (1985). The fragmentation–fracture mechanism of energy absorption typical of glass-fiber-reinforced composites, as compared to the plastic deformation mechanism in materials, is actually very weight-effective. The real issue is the translation of this effective fracture mechanism into complex structures.

The third and less quantifiable requirement for composites in vehicle applications is vehicle dynamics. Glass-fiber-reinforced composites are inherently less stiff than steel, typically by factors of up to 10 (Table 3). However, there are two offsetting factors to compensate for these apparent deficiencies. First, an increase in section thickness can be used to partially offset the decreased material stiffness; also, the flexibility of composite fabrication processes allows the thickening of local areas as is required to optimize properties. Since the composite has a density approximately one-third that of steel, a significant increase in thickness can be achieved while maintaining an appreciable weight reduction. The second, and perhaps major, compensating factor is the additional stiffness attained in composite structures by virtue of part integration. This integration leads directly to the elimination of joints, which results in a significant increase in effective stiffness. It is becoming increasingly evident that this synergism is such that structures of acceptable stiffness and considerably reduced weight are feasible in glass-fiber-reinforced composites. As a rule of thumb, a glass FRP structure with significant part integration relative to the steel structure being replaced can be designed for a nominal stiffness of

Table 2
Energy absorption (typical properties)

Material	Relative energy absorption per unit weight
High-performance composites	100
Commercial composites	60–75
Mild steel	40

Table 3
Typical stiffness of composites: XMC is a sheet molding compound containing oriented continuous glass fiber; SMC-R50 is a sheet molding compound containing 50% chopped glass fiber

Material	Modulus (KPa)
Unidirectional CFRP	137.8
Unidirectional GFRP	41.3
Unidirectional Kevlar	75.8
XMC	31.0
SMC-R50	15.8
SMC-25	9.0

50–60% of that of the steel structure. Such a design procedure should lead to adequate stiffness and typical weight reductions of 30–50%.

2. Potential Applications

The potential use of composites in structural applications in automobiles can be designated into two categories: the direct replacement of existing components and the integration of multiple steel components into one composite component or structure. The second category, involving part integration, is by far the most cost-effective and ultimately will predominate. In the shorter term, however, much experience and confidence in the capabilities of composites is being generated by the singular component programs currently in production.

2.1 One-On-One Component Substitution

The significant differences in material costs between glass FRP and steel (approximately US$0.80 per pound versus US$0.25 per pound at 1988 prices) means that, even with significant weight saving, direct substitution in a singular component can rarely be cost effective. There are two examples currently in production which can be cited in the direct substitution category, namely driveshafts and leaf springs.

Composite driveshafts have been utilized in production on at least two vehicles in North America. One example is in the special drive line configuration of the Ford Econoline Van. These vehicles would normally utilize a two-piece steel driveshaft incorporating a connecting center bearing. The total length of the drive line dictates that a two-piece steel shaft must be used because unacceptable vibrations occur in a one-piece steel driveshaft. In contrast, a one-piece composite driveshaft will provide satisfactory vibratory characteristics since the lower weight combined with high stiffness satisfies the bending-frequency requirements. The driveshaft is fabricated by filament winding on a continuous machine operating at speeds up to $2\,\mathrm{m\,min^{-1}}$. The longitudinal fibers (0°) are carbon (220 GPa modulus), to generate the required bending stiffness, and the ±45° fibers are E-glass, to provide the torsional strength. The resin used is a vinyl ester which gives the appropriate combination of properties, processability and cost. This particular driveshaft is economically feasible only because it results in the elimination of the center bearing—such cost-effective usage is impossible in the case of single-piece steel (or aluminum) driveshafts.

The leaf spring is an example of a singular component utilizing glass-fiber-reinforced epoxy which has found a niche in the North American vehicle market. Current production in the USA is greater than 500 000 FRP springs per annum and their use to date has been a reliable demonstration of the feasibility and durability of composites. FRP leaf springs can be mass-produced by three basic techniques, although there may be several variants on each particular process. The basic materials are always the same, namely continuous E-glass fiber and epoxy resin. (Attempts to substitute more economical resins have all been unsuccessful because of creep requirements.) The three potential techniques are illustrated in Fig. 2. The compression-molding technique involves development of a preform (usually by a filament-winding procedure) followed by molding, and is the process currently used in the USA. An alternative procedure involves the automated layup of a wide slab of material, followed by molding under pressure and subsequent slitting of the slab to form springs. The third, least developed, technique is the pulforming procedure which has perhaps the greatest long-term potential for minimizing cost.

One-for-one replacements of steel components with FRP components will be exceedingly rare, primarily because of economic considerations but also because of complex attachment considerations. As a general principle, therefore, multiple-part integration to reduce assembly costs will be necessary to make FRP structures cost-effective relative to steel.

A simple example of this principle is demonstrated by the prototype composite integrated rear suspension for a Ford Escort (Morris 1986). The basic steel rear suspension and the replacement FRP integrated suspension are shown in Fig. 3. A simple extension of FRP leaf-spring design methodology allowed substitution for the complete steel rear suspension by allowing the FRP spring to perform the dual function of spring and suspension arms. The function and part integration resulted in a weight saving of approximately 4 kg (about 50% weight reduction). In addition to the weight saving and cost-effectiveness of this type of synergism, there are obvious improvements in package space, which is a major consideration in vehicle design.

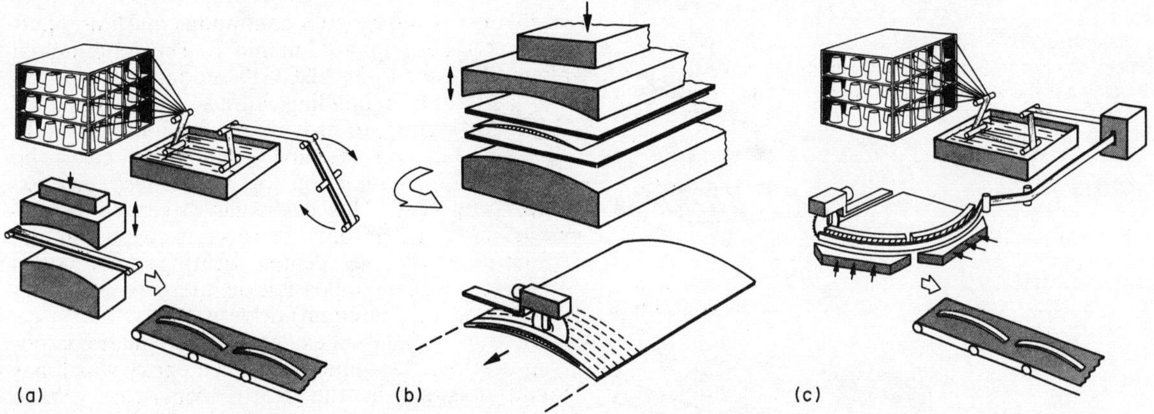

Figure 2
Different fabrication techniques for FRP leaf springs: (a) filament winding–compression molding, (b) compression molding and (c) pultrusion

2.2 Large Integrated Structures

The successful application of structural composites to large integrated automotive structures is more dependent on the ability to use rapid and economic fabrication processes than on any other single factor. The fabrication process must also be capable of close control of composite properties to achieve lightweight, efficient structures. Currently, the only commercial process which comes close to satisfying these requirements is compression molding of sheet molding components (SMC) or some variant of the process. There are, however, processes still at the development stage which hold distinct potential for the future in terms of combining high production rates, precise fiber control and high degrees of part integration. In particular, high speed resin-transfer molding (HSRTM) offers these potential benefits, provided technical developments can be achieved. The following examples demonstrate these fabrication approaches and part-integration concepts which can be utilized in automotive production and compare the utilization of both SMC and HSRTM procedures. It is not the intent here to describe these two processes in detail but for illustrative purposes the techniques are summarized in Fig. 4. Further details on these and other fabrication procedures may be obtained from the ASM Composite Materials Handbook (1988).

2.3 Primary Body Structure

The primary body structure of an automobile consists of approximately 250–350 major steel parts and utilizes approximately 300 assembly robots. For comparison of the two alternate composite construction techniques, consider the body side assembly. In Fig. 5a, a typical SMC approach is illustrated in which the complete body side consists of two moldings which would be bonded together. The HSRTM procedure for the same structure is shown in Fig. 5b and the major difference is the elimination of the adhesive bonding and the incorporation of a foam core. Utilizing either of these composite fabrication techniques for the other major segments of the body shell would lead to a typical body construction

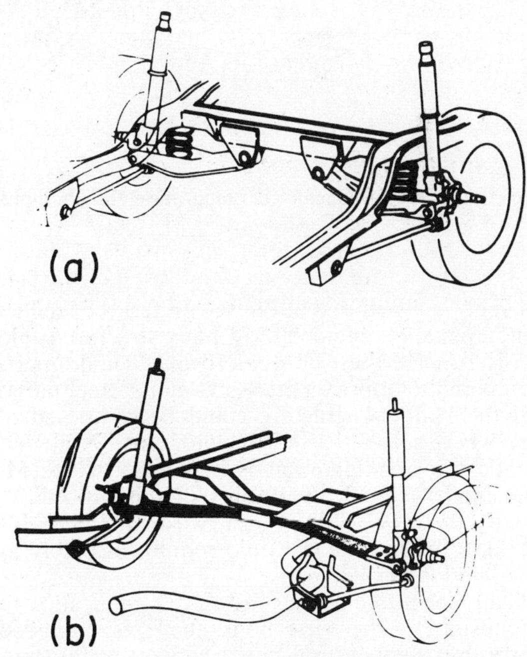

Figure 3
Comparison of (a) production Ford Escort steel rear suspension and (b) corresponding integrated composite suspension

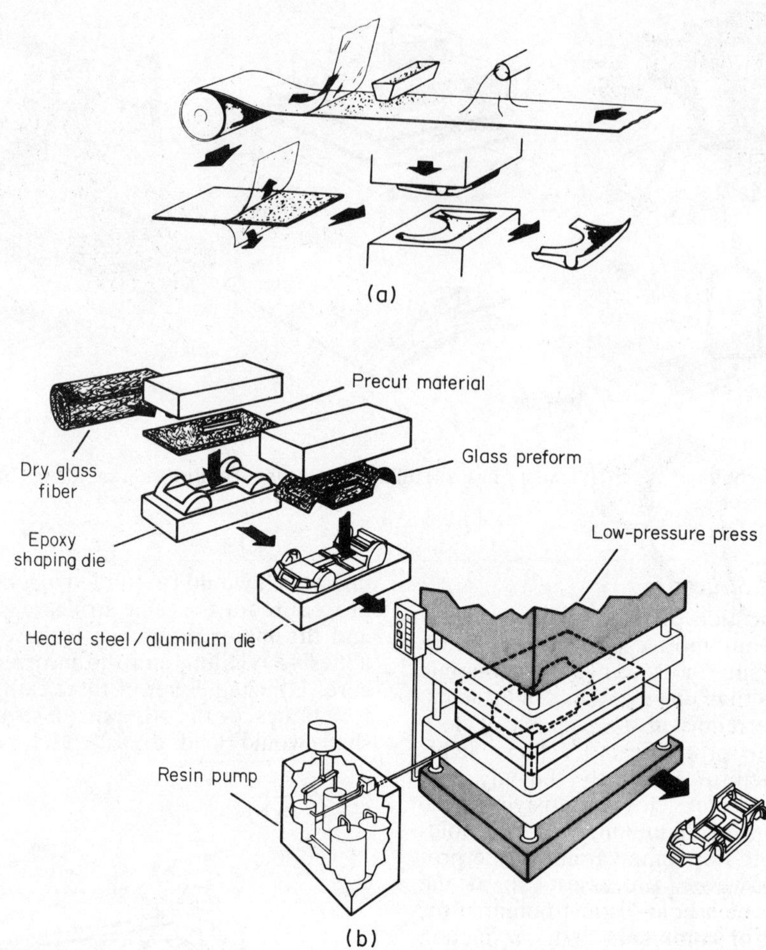

Figure 4
Depiction of (a) SMC material preparation and component fabrication and (b) high-speed resin-transfer molding process

assembly as illustrated in Fig. 6a. In terms of reduction in parts, the molded SMC body structure could consist of somewhere between 10 and 20 major parts (compared to approximately 300 major steel parts) and the HSRTM structure could be composed of somewhere between 2 and 10 major parts. Note that the degree of integration is higher for HSRTM, reflecting the greater versatility of this procedure. Clearly the lower limit of just two major parts for the HSRTM technique would require assembly as illustrated in Fig 6b.

While the attraction for the all-out application of composites to the total body shell, as discussed above, is appealing from the viewpoint of the materials connoisseur, reality dictates that the usage of the composites in automobile structures will be evolutionary and progressive rather than revolutionary in nature. Therefore, it is to be expected that structural segments of vehicles will be first to utilize these procedures and there are several examples of prototype structures where these techniques have been used. Complex, high-load cross members have been developed for evaluation by both Chrysler (Farris 1987) and Ford (Johnson et al. 1987). The cross member is fabricated from glass-fiber-reinforced vinylester utilizing the resin-transfer molding (RTM) process. The unique cross member is projected to save approximately 30% in weight relative to the comparable steel member. The fabrication and testing of these high load, critical members is part of the gathering of detailed information on the realistic expectations that glass-fiber-reinforced plastic composite can satisfy the functionality required. Concurrently, the fabrication technique (RTM) has to be further developed to satisfy the economic constraints of the industry. The success of these

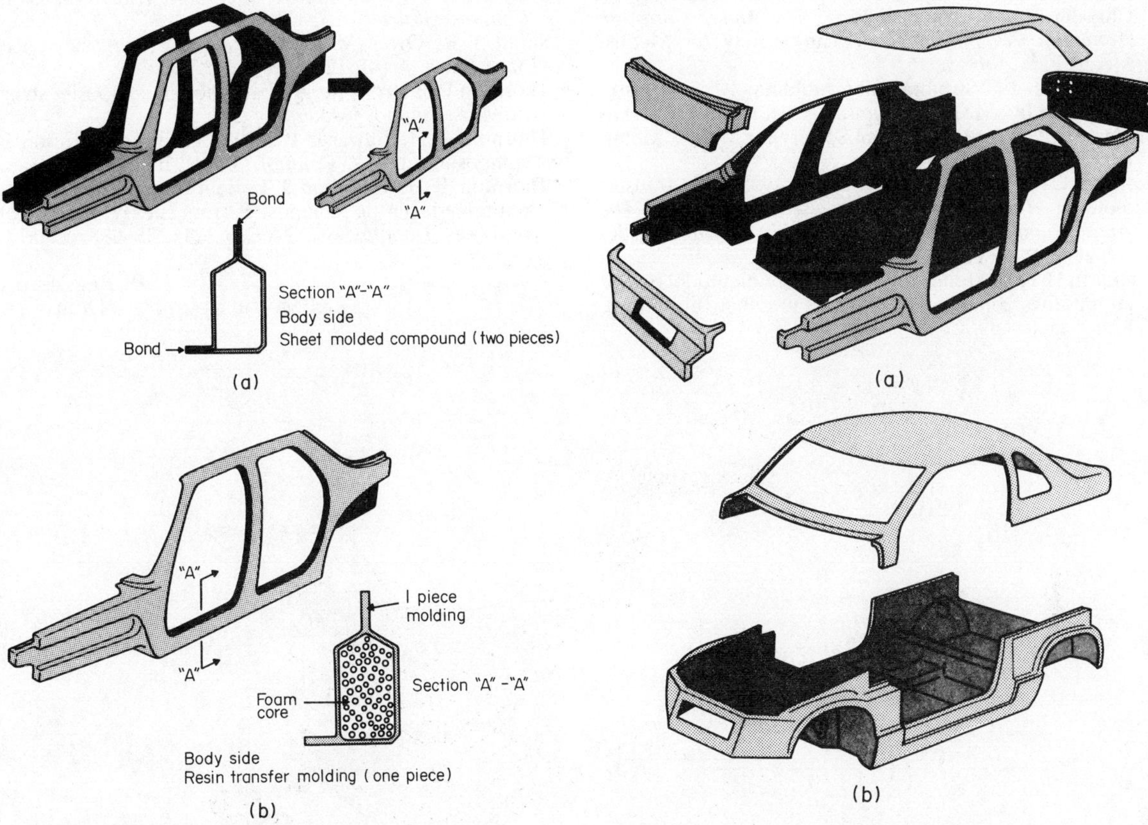

Figure 5
Typical assembly of composite body side panel by (a) adhesive bonding of inner and outer SMC molded panels and (b) HSRTM panels with foam core

Figure 6
Composite body shell: (a) typical assembly, (b) assembly using two major moldings

prototypes is providing the impetus to drive the fabrication developments towards the required goal. A similar example concerning the fabrication of a highly integrated composite Escort front structure by RTM (Johnson et al. 1985) has been well documented and reported. In this prototype, 42 steel parts were replaced by one RTM molding resulting in a 30% reduction in weight, and structural stiffness in excess of that of the steel structure. This structure is currently being utilized to develop energy absorption capability in a typical vehicle design.

3. Summary

The extension of composites use to automotive integrated structures will require an expanded knowledge of the design parameters for glass-fiber-reinforced plastic materials together with major innovations in fabrication techniques. There is accumulating evidence, both in the laboratory and in prototype vehicles, which strongly indicates that glass-fiber-reinforced composites are capable of meeting the functional requirements of highly loaded automotive structures. The more imperative requirement is the cost-effective fabrication advancements that appear necessary to justify such increased use of composites. High-volume, less stringent performance components can be manufactured by variations on compression-molding techniques. However, the high volume, high-performance manufacturing techniques still need development and improved SMC materials and processes, and the HSRTM process holds promise in these areas.

Bibliography

Aotem A, Hashin Z 1976 Fatigue failure of angle ply laminates. *AIAA J.* 14: 868–72

Beardmore P, Harwood J J, Horton E J 1980 Design and manufacture of a GrFRP concept automobile. *Proc. Int. Conf. on Composite Materials, Paris,* August 1980

Dharan C K H 1975 Fatigue failure in graphite fiber and glass fiber–polymer composites. *J. Mater. Sci.* 10: 1665–70

Farris R D 1987 Composite front crossmember for the Chrysler T-115 mini-van. *Proc. 3rd Annu. Conf. on Advanced Composites*. American Society for Metals, Metals Park, OH, pp. 63–73

Johnson C F 1988 Compression molding; Resin transfer molding. In: *ASM Composite Engineered Materials Handbook,* Vol. 1. American Society for Metals, Metals Park, OH pp. 559–68

Johnson C F, Chavka N G, Jeryan R J 1985 Resin transfer molding of complex automotive structures. *Proc. 41st Annu. Conf. SPI*, Paper 12a. Society of the Plastics Industry, New York

Kulkarni H T, Beardmore P 1980 Design methodology for automotive components using continuous fiber reinforced materials. *Composites* 12: 225–35

Morris C J 1986 Composite integrated rear suspension. *Compos. Struct.* 5: 233–42

Smith T R, Owen M J 1969 Fatigue properties of RP. *Mod. Plast.* 46(4): 124–9

Thornton P H 1979 Energy absorption in composite structures. *J. Compos. Mater.* 13: 262–74

Thornton P H, Edwards P J 1982 Energy absorption in composite tubes. *J. Compos. Mater.* 16: 521–45

Thornton P H, Harwood J J, Beardmore P 1985 Fiber reinforced plastic composites for energy absorption purposes. *Compos. Sci. Technol.* 24: 275–98

P. Beardmore
[Ford Motor Company, Dearborn, Michigan, USA]

B

Bainite—The Current Situation

This article updates and extends the article *Bainite Reaction* in the Main Encyclopedia.

Bainite is a nonlamellar mixture of ferrite and carbides that can be obtained in steels by the transformation of austenite in a temperature range above the martensite-start temperature but below that at which fine pearlite can grow at a reasonable rate. During the diffusionless formation of martensite, the change in crystal structure is in effect achieved by a physical deformation of the parent phase. In contrast, a diffusion reaction requires mass transport in order to facilitate the change in structure without entailing the sort of strains associated with martensitic transformations (Bhadeshia 1985a). The awkward ranking of bainite between the obviously diffusionless martensitic reaction, and pearlite, which is a clear example of a diffusional transformation, has in the past led to difficulties, but a clearer picture is beginning to emerge (Christian and Edmonds 1984, Bhadeshia 1988, Bhadeshia and Christian 1990). This is fortunate because several major commercial developments involving the use of bainitic steels in large quantities are imminent. Examples include the high-toughness "ultra low carbon bainitic steels" for use primarily in the oil and gas industries (Nakasugi et al. 1983), "silicon-enriched high-strength, directly transformed" steels as crash reinforcement bars in the automobile industry and the "austempered ductile cast irons" (Dorazil et al. 1962) which in the bainitic condition show remarkable combinations of strength, toughness and wear resistance. As will become evident, the so-called "acicular ferrite" that imparts good toughness in steel weld deposits is also closely related to bainite. These and other developments stand to benefit from the improved understanding of the bainite reaction. Indeed, the subject has now advanced sufficiently to enable quantitative alloy design and heat-treatment procedures to be defined with reasonable confidence.

This article is intended to present a concise account of the mechanism of the bainite transformation in steels which is consistent with all the available data. Some attempts have previously been made to include certain transformations in nonferrous metals within a generalized definition of bainite, but these are fraught with difficulties since the experimental data on which they rely are, in general, ambiguous and incomplete. It is for this reason that no attempt is made to include transformations in nonferrous alloys in this article.

1. Morphology and Carbide Precipitation

Bainite consists of a nonlamellar aggregate of ferrite and carbides; the ferrite is in the form of clusters of thin platelets (Fig. 1), commonly referred to as "sheaves" of bainite (Hehemann 1970). The platelets grow to a limiting size which is less than that of the parent austenite grains and have a relatively high dislocation density.

The formation of bainitic ferrite leads to an increase in the carbon concentration of the remaining austenite. In upper bainite, which is obtained by transformation at relatively high temperatures, the cementite is found to precipitate from the films of carbon-enriched austenite which separate the platelets of bainitic ferrite. In lower bainite, the cementite also precipitates inside the plates of ferrite. There are, therefore, two kinds of cementite precipitates: those which grow from carbon-enriched austenite, and those which precipitate from supersaturated ferrite. The latter particles thus exhibit a "tempering" orientation relationship—the crystallographic orientation relationship observed between carbides and martensite when the carbides precipitate from supersaturated martensite during tempering (Bhadeshia 1988)—with the parent ferrite.

The precipitation of cementite usually occurs as a secondary reaction after the growth of bainitic ferrite (Hehemann 1970, Bhadeshia 1988). In some

Figure 1
Transmission electron micrograph illustrating a sheaf of upper bainite in a matrix of martensite: the sheaf consists of many smaller platelets (subunits) of bainitic ferrite

alloys, especially those containing high concentrations of silicon or aluminum, the cementite precipitation reaction is so sluggish that for all practical purposes the bainite consists of a mixture of just bainitic ferrite and residual austenite. The carbides that form may not always be cementite, their detailed crystallography, morphology and chemical composition depending on the alloy chemistry and transformation conditions.

2. *Shape Change and its Implications*

Like martensite, the growth of bainitic ferrite is accompanied by an invariant-plane strain (IPS) shape change (Ko and Cottrell 1952) which has a large shear component of approximately 0.24 and a dilatational component (~0.04) corresponding to the volume change on transformation. Such a shape change, together with high resolution observations that substitutional solute atoms do not diffuse even on the finest conceivable scale at the interface (Bhadeshia and Waugh 1982), demonstrate the existence of an atomic correspondence between the parent and product crystals for the substitutional solute and iron atoms (Fig. 2).

The shape change implies a coordinated movement of atoms which causes considerable strains in conditions of constrained transformation. If it is elastically accommodated, then the strain energy of bainitic ferrite amounts to about 400 J mol^{-1}. However, because the transformation occurs at relatively high temperatures where the yield strength of both the parent and product phases is rather low, the shape change is partly accommodated by plastic deformation. The dislocation debris associated with this plastic flow is responsible for stifling the growth of individual platelets, which as a consequence grow to a limiting size smaller than that permitted by the austenite grain structure. The plastic relaxation also has the effect of reducing the strain energy due to the shape change. The platelets within a given sheaf all have an identical orientation in space and the same shape deformation. The ferrite plates always have a crystallographic orientation relationship with the parent austenite which is similar to that found between austenite and martensite. These observations at the very least imply that the growth of bainite involves a coordinated movement of the substitutional solute and iron atoms across a glissile transformation interface. Such coordinated movements cannot, in general, be sustained across austenite grains which are randomly oriented. As a consequence, bainite growth is impeded by austenite grain boundaries; this constrasts with diffusional transformations in which the product phase may readily grow across grains of the parent phase that are in different orientations.

Although it has been suggested that an IPS shape change can arise when a sessile semicoherent interface is displaced by the motion of incoherent steps (Kinsman et al. 1975), there is no mechanism that can explain how an uncoordinated transfer of atoms across the steps can be responsible for the systematic displacements implied by the IPS shape change. The hypothesis also contradicts the fact that allotriomorphic ferrite, which often grows by a step mechanism and which has the necessary semicoherent interface, does not show an IPS surface relief. The concept also violates the idea that the atomic correspondence implied by the shape deformation is a property of a particle as a whole, irrespective of the interface orientation (Christian and Edmonds 1984).

3. *The Role of Carbon*

It is much more difficult to determine the precise role of carbon during the growth of bainitic ferrite (Christian and Edmonds 1984, Bhadeshia 1988). Bainite forms below the T_0 temperature where austenite and bainitic ferrite of the same chemical composition have equal free energy. This makes it thermodynamically possible for the transformation to be diffusionless. An alternative possibility is of paraequilibrium transformation, in which the substitutional lattice is configurationally frozen, but subject to that constraint, the carbon redistributes to an extent which allows its chemical potential to be equal in all phases. A degree of partitioning of carbon which is between paraequilibrium transformation and diffusionless growth is also a possibility.

The diffusivity of interstitial carbon is typically many orders of magnitude larger than that of substitutional solutes in iron. Early work indicated that the time t_d required to decarburize a plate of supersaturated ferrite is of the order of a few milliseconds; however, a corrected version of the theory (Bhadeshia 1988) gives t_d as

$$t_d \sim \frac{t_h^2(\bar{x} - x^{\alpha\gamma})^2 \pi}{16\bar{D}(x^{\gamma\alpha} - \bar{x})^2} \tag{1}$$

where t_h is the thickness of the ferrite plate, $\bar{D}$ is the weighted average diffusivity of carbon in the austenite, $x^{\alpha\gamma}$ and $x^{\gamma\alpha}$ are the paraequilibrium carbon concentrations in the ferrite (α) and austenite (γ), respectively, and $\bar{x}$ is the average carbon concentration in the alloy as a whole. If growth involves diffusionless transformation, then any excess carbon in the bainitic ferrite can within a matter of seconds partition into the residual austenite. This makes it impossible to determine directly the concentration of carbon in the ferrite during its growth.

There is, however, an indirect method of assessing the conditions that exist during transformation. On an iron–carbon phase diagram it is possible to plot the T_0 curve which defines (as a function of temperature) the locus of all carbon concentrations where austenite and ferrite of identical compositions have

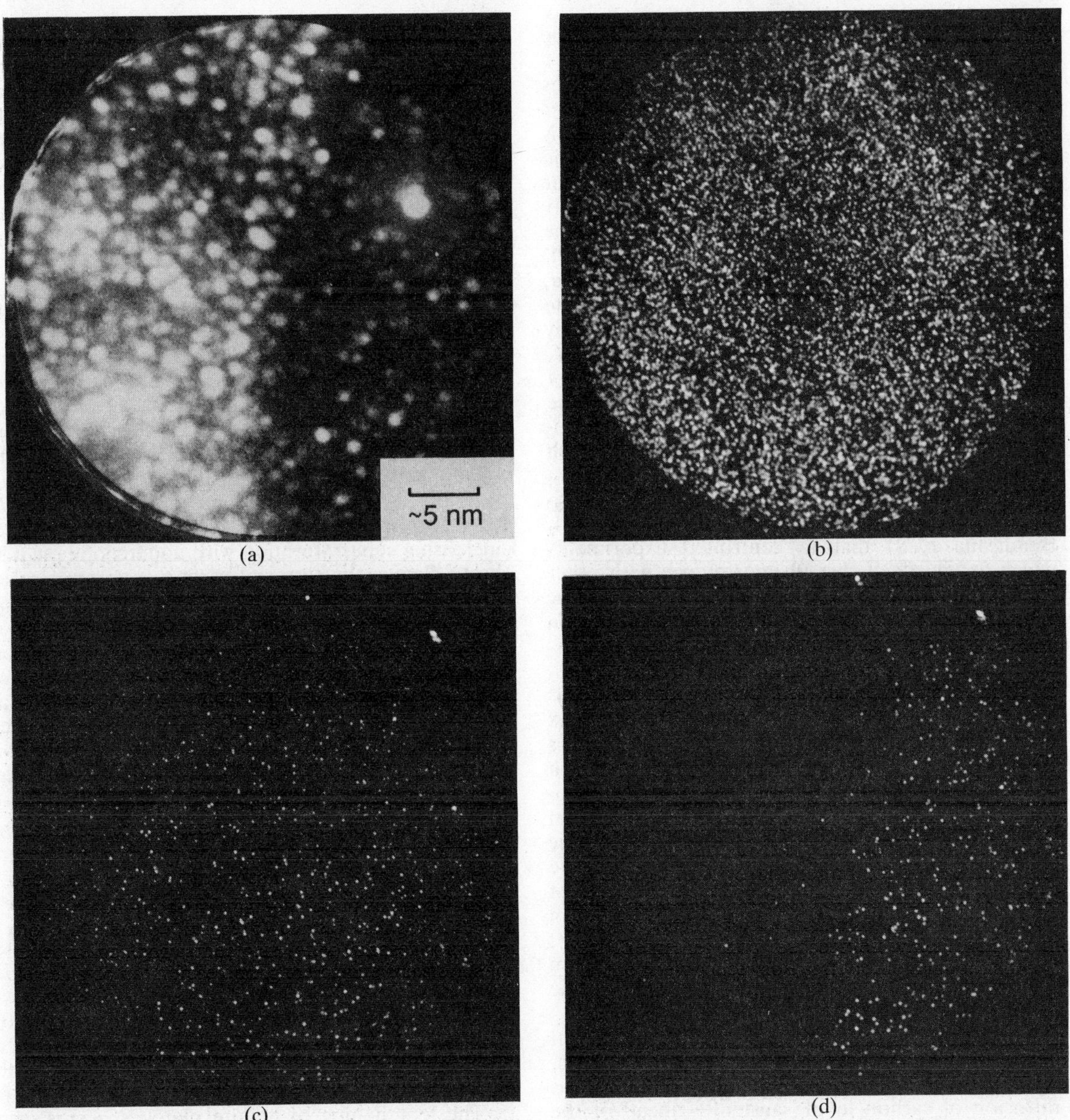

Figure 2
Imaging atom-probe micrographs taken across an austenite–bainitic ferrite interface in an Fe–0.43C–2.20Si–3Mn wt% alloy: (a) field-ion image; (b) corresponding iron atom map; (c) corresponding silicon atom map; and (d) corresponding carbon map (after Bhadeshia and Waugh 1982). The images show that substitutional alloying elements do not diffuse during the bainite transformation, even on the finest conceivable scale at the transformation interface

equal free energy. If the carbon concentration of the austenite (x_γ) exceeds that given by the T_0 curve, then its subsequent diffusionless transformation becomes thermodynamically impossible. However, transformation involving the equilibrium partitioning of carbon during growth can continue until x_γ reaches $(\alpha+\gamma)/\gamma$ (i.e., the Ae_3) phase boundary. The same considerations apply to substitutionally alloyed steels, in which case the relevant $(\alpha+\gamma)/\gamma$ phase boundary is the paraequilibrium Ae_3' curve since substitutional solutes do not diffuse during the bainite reaction. Since the transformation to bainite is displacive, these curves have to be further modified to the T_0' and Ae_3'' curves which include the effects of the strain energy as a result of the shape change.

It can be hypothesized that transformation begins with the diffusionless growth of a plate of upper bainite but that shortly afterwards its growth is stifled and its excess carbon partitioned into the residual austenite. A process like this is realistic for bainite because, unlike martensite, the relatively higher temperatures involved make the time required for the partitioning process rather small. As a consequence of the redistribution of carbon, the next plate of bainite has to grow from austenite enriched in carbon. Growth in this manner can only continue if x_γ is less than the value given by the T_0' curve. As noted earlier, if the ferrite at all stages of growth is not supersaturated with carbon, then the reaction may continue until the larger paraequilibrium volume fraction of ferrite is reached, at which stage x_γ is given by the Ae_3'' phase boundary.

It has been verified using many different experimental techniques (Christian and Edmonds 1984, Bhadeshia 1988) that in controlled experiments which avoid effects due to interference with other overlapping transformations, the growth of bainitic ferrite stops when the diffusionless transformation of carbon-enriched retained austenite becomes thermodynamically impossible (i.e., $x_\gamma \sim x_{T_0'}$). It seems, therefore, that the growth of bainitic ferrite is initially diffusionless but unlike martensite the plates then have the opportunity to reject their excess carbon into the residual austenite. The fact that with this mechanism of overall transformation the austenite is to some extent enriched in carbon, in spite of the intially diffusionless growth of the bainitic ferrite, leads to an effect called the "incomplete reaction phenomenon," in which the degree of transformation to bainite is always far less than demanded by equilibrium. Furthermore, the extent of transformation increases with undercooling from zero at the bainite start (B_s) temperature to a value determined by the T_0' phase boundary.

There is no fundamental reason why growth might not occur with a partial supersaturation of carbon, even though this would involve the simultaneous diffusion of carbon in the austenite ahead of the interface, but at a rate low enough to permit some of the carbon to be trapped by the advancing interface (trapping implies an increase in chemical potential on transfer across the boundary). Such a state of affairs can only lead to steady-state growth (as observed experimentally) if the diffusion field can somehow be stabilized, since random perturbations in the concentration at the interface should otherwise lead to a collapse towards equilibrium. The driving force available for transformation can be partitioned into that dissipated in the diffusion of carbon ahead of the interface, and a quantity dissipated in the transfer of atoms across the interface. Each of these dissipation terms can, with an appropriate model, be related to interface velocity functions, one being the diffusion velocity and the other a velocity determined by the mobility of the interface. The intersection of these functions gives an actual solution of the inerfacial velocity. By imposing different levels of supersaturation in the ferrite, a series of solutions can be calculated as a function of supersaturation. Some criterion (such as maximum velocity) can then be used to select the most likely solution and hence the most likely supersaturation. In this way, it has been demonstrated theoretically that growth involving partial supersaturation can be stable (Olson et al. 1988).

There are, nevertheless, difficulties with the concept of growth with partial supersaturation in the context of bainite in steels. The models all predict that the degree of supersaturation rises with undercooling; this is inconsistent with the fact that the bainite reaction stops when the carbon concentration of the residual austenite approaches the T_0 curve. It might be thought that consequence of increasing supersaturation with undercooling would be to change the terminal carbon concentration of the austenite from the Ae_3'' down to T_0' at full supersaturation as the transformation temperature is reduced. Alternatively, for a given transformation temperature it might be expected that the level of supersaturation would decrease as the transformation progresses (since the austenite becomes enriched in carbon) so that the terminal carbon concentration may then always be given by the Ae_3'' curve.

4. Kinetics

Little is known about the nucleation of bainite except that the activation energy for nucleation is directly proportional to the driving force for transformation (Bhadeshia 1981). This is consistent with the theory for martensite nucleation, although it is a requirement that carbon must partition into the austenite during bainite nucleation.

The lengthening of bainite platelets occurs at a rate much faster than would be expected from growth being controlled by the diffusion of carbon in the austenite ahead of the interface (Bhadeshia 1984). If a sheaf forms by the repeated nucleation and propagation of individual platelets, then the overall lengthening rate of a sheaf is expected to lower since some time will be consumed in each nucleation event. Nevertheless, sheaf lengthening rates are generally found to be higher than expected from carbon diffusion-controlled growth (Bhadeshia 1985b).

Although it has been suggested that substitutional alloying elements, especially those that are strong carbide formers, sometimes exert a drag on the transformation interface, there is no evidence for this (Bhadeshia 1983). Indeed, direct observations of the interface using the atom-probe technique have demonstrated that substitutional solute atoms such as manganese, silicon, chromium, nickel and molybdenum do not partition between the phases during the bainite transformation, nor do they segregate to the transformation interface. The substitutional lattice is therefore configurationally frozen during transformation (Bhadeshia and Christian 1990). The solute-drag hypothesis is also supposed to explain the existence of a "bay" in the time–temperature–transformation diagram of certain steels (see *Bainite Reaction*), but this explanation does not stand up to detailed examination (Bhadeshia 1983); the bay can be explained naturally if the diagram is considered to consist of two overlapping *C*-curves: the one at higher temperatures representing diffusional transformations such as allotriomorphic ferrite and pearlite, the lower *C*-curve representing displacive transformations such as Widmanstätten ferrite, bainite and acicular ferrite (Bhadeshia 1988).

5. *Acicular Ferrite*

Acicular ferrite is a plate-shaped phase formed by the transformation of austenite during cooling of low-alloy steel weld deposits (Grong and Matlock 1986, Abson and Pargeter 1986). In the early stages of transformation the plates nucleate intragranularly on nonmetallic inclusions that are present as impurities in the weld metal. Plates of acicular ferrite are then found to radiate from inclusions, giving an overall morphology of nonparallel plates dispersed throughout the austenite grain structure. The reaction has otherwise been shown to have all of the major characteristics of bainite: it exhibits the incomplete-reaction phenomenon, there is an IPS shape deformation, and so on (Bhadeshia 1988). The detailed morphology differs from that of conventional bainite because the former nucleates intragranularly at inclusions within relatively large austenite grains, whereas in wrought steels which are essentially free of nonmetallic inclusions, bainite nucleates initially at the austenite grain surfaces. The bainite then continues growth by the repeated formation of subunits to generate the classical sheaf morphology. Acicular ferrite does not normally grow in sheaves because the development of sheaves is stifled by hard impingement between plates nucleated at adjacent sites. Indeed, conventional bainite or acicular ferrite can be obtained under identical isothermal transformation conditions in the same (inclusion-rich) steel; in the former case, the austenite grain size has to be small in order that nucleation from grain surfaces dominates and subsequent growth then swamps the interiors of the austenite grains. For the same reasons, acicular ferrite is not usually obtained in relatively clean wrought steels, although titanium oxide inoculated steels in which the oxide particles are intended to induce the formation of acicular ferrite for improved toughness are now commercially available (Nishioka and Tamehiro 1988).

6. *Bainite in the Context of Other Transformations*

There are several displacive transformation products in wrought steels such as Widmanstätten ferrite, bainite and martensite (Bhadeshia 1981). Widmanstätten ferrite is obtained at relatively low undercoolings and generally involves the cooperative formation of adjacent and mutually accommodating crystallographic variants which are usually (though not always) similarly oriented in space. Because the adjacent variants have slightly different habit planes with the austenite, the plate as a whole has a thin-wedge morphology. Hence, what appears in a light microscope to be a single plate of Widmanstätten ferrite is in reality a composite of two adjacent and mutually accommodating plates. During its growth, Widdmanstätten ferrite has an equilibrium or paraequilibrium carbon concentration, lengthening occurring approximately at a rate controlled by the diffusion of carbon in the austenite ahead of the interface. Carbon is partitioned into the austenite during the nucleation of both bainite and Widmanstätten ferrite. However, the activation free energy of nucleation is directly proportional to the chemical driving force so that the nucleation process is, in this sense, similar to that in martensitic transformations. Martensite nucleation (and growth) is, however, diffusionless.

Bainite, therefore, differs from Widmanstätten ferrite in two important respects: its growth is diffusionless and its stored energy term is much higher since, in the case of Widmanstätten ferrite, much of the strain energy due to the shape change is cancelled by the simultaneous growth of mutually accommodating plates.

Bainite differs from martensite in that the nucleation of bainite involves the partitioning of carbon, and because its morphology (i.e., sheaf) is significantly influenced by the fact that it forms at a relatively lower driving force and higher temperature. More detailed comparisons between the different kinds of transformation products can be found elsewhere (Bhadeshia 1988).

7. *Summary*

Bainite seems to grow by a diffusionless transformation mechanism. The excess carbon trapped in the bainitic ferrite is, subsequent to growth, removed by a combination of diffusion into the residual austenite and by the precipitation of carbides within the ferrite. Upper bainite is obtained if the time taken for the diffusion process is small when compared with that required for the precipitation carbides in the bainitic ferrite. Otherwise the transformation product is conventionally classified as lower bainite, with carbide particles being present inside the bainitic ferrite. The residual austenite, which is enriched with carbon, eventually decomposes by diffusional transformation into a mixture of more carbides and ferrite. The morphology of individual plates of bainite is determined by the need to minimize the strain energy owing to the IPS shape change associated with the displacive mechanism of transformation. The fact that the plates grow to a limiting size, and that they cluster into sheaves of platelets in the same orientation, is ultimately a consequence of the high transformation temperatures involved and of the relatively low driving force that is available at the temperatures where bainite forms. The kinetics of the bainite transformation are consistent with the proposed diffusionless, displacive transformation mechanism.

See also: Bainite Reaction; Martensitic Transformations: Nucleation; Martensitic Transformations: Phenomenology; Phase Transformations: Classification; Steels: Physical Metallurgy Principles

Bibliography

Abson D J, Pargeter R J 1986 Factors influencing the as-deposited strength, microstructure and toughness of manual metal arc welds suitable for C–Mn steel fabrications. *Int. Met. Rev.* 31: 141–94

Bhadeshia H K D H 1981 A rationalisation of shear transformations in steels. *Acta Metall.* 29: 1117–30

Bhadeshia H K D H 1983 Considerations of solute-drag in relation to transformations in steels. *J. Mater. Sci.* 18: 335–40

Bhadeshia H K D H 1984 Solute-drag, kinetics and the mechanism of the bainite reaction in steels. *Phase Transformations in Ferrous Alloys*, American Society for Metals, Metals Park, Ohio, pp. 335–40

Bhadeshia H K D H 1985a Diffusional formation of ferrite in iron and its alloys. *Progress in Materials Science*, Vol. 29. Pergamon, Oxford, pp. 321–86

Bhadeshia H K D H 1985b Critical assessment: Diffusion-controlled growth of ferrite plates in plain-carbon steels. *Mater. Sci. Technol.* 1(7): 497–504

Bhadeshia H K D H 1988 Bainite in steels. In: Lorimer G W (ed.) 1988 *Phase Transformations '87* Institute of Metals, London, pp. 309–14

Bhadeshia H K D H, Christian J W 1990 Bainite in steels. *Metall. Trans. A* (in press)

Bhadeshia H K D H, Waugh A R 1982 Bainite: An atom-probe study of the incomplete reaction phenomenon. *Acta. Metall.* 30: 775–84

Christian J W, Edmonds D V 1984 The bainite transformation. *Phase Transformations in Ferrous Alloys*, American Society for Metals, Metals Park, Ohio, pp. 293–326

Dorazil E, Barta B, Munsterova E, Stransky L, Huvar A 1962 High-strength bainitic ductile cast iron. *AFS Int. Cast Met. J.* (January): 52–62

Grong O, Matlock D K 1986 Microstructural development in mild and low-alloy steel weld metals *Int. Met. Rev.* 31: 27–48

Hehemann R F 1970 The bainite transformation. *Phase Transformations*, American Society for Metals, Metals Park, Ohio, pp. 397–432

Kinsman K R, Eichen E, Aaronson H I 1975 Thickening kinetics of proeutectoid ferrite plates in Fe–C alloys. *Metall. Trans. A* 6: 303–17

Ko T, Cottrell S A 1952 The formation of bainite. *J. Iron Steel Inst.* 184: 307–13

Nakasugi N, Matsuda H, Tamehiro H 1983 Ultra-low carbon bainitic steel for line pipes. *Steels for Line Pipe and Pipeline Fittings*, Metals Society, London, pp. 90–5

Nishioka K, Tamehiro H 1988 High-strength titanium-oxide bearing line pipe steel for low-temperature service. *Microalloying '88*, American Society for Metals, Metals Park, Ohio, pp. 1–9

Olson G B, Bhadeshia H K D H, Cohen M 1988 Coupled diffusional/displacive transformations. In: Lorimer G W (ed.) 1988 *Phase Transformations '87* Institute of Metals, London, pp. 322–5

H. K. D. H. Bhadeshia
[University of Cambridge, Cambridge, UK]

Bauschinger Effect in Metals and Composites

Johann Bauschinger brought a rigorous analytical approach to the mechanical testing of materials, and contributed much both to the development of instrumentation, and to the standardization of testing procedures (Mughrabi 1987). In his pioneering paper of 1886, Bauschinger dealt with the mechanical response of a variety of ferrous materials, and and outlined the effect of strain path dependence which is now termed the Bauschinger effect. The essential observation is summarized in Fig. 1, and indicates that if a material is strained beyond its elastic limit in tension or compression, then on reversing the direction of deformation a decrease in elastic limit will be observed. In some cases this will be accompanied by an offset of the stress–strain curves, which is termed permanent softening. The observation that the state resulting from plastic flow is sensitive to the subsequent direction of plastic strain is of major importance both from the fundamental viewpoint of understanding the micromechanics of single and multiphase materials and from the more practical viewpoint of predicting the response of materials subject to operations involving complex changes in strain path.

1. Modelling the Bauschinger Effect

The magnitude of the Bauschinger effect can be estimated in a number of ways. The quantification of the Bauschinger effect in terms of stress, strain or an energy parameter has been discussed at length by Abel and Muir (1973) and Abel (1987), and can be summarized as shown in Fig. 2. It must be emphasized that a variety of stress and strain related parameters have been used in the literature to define the Bauschinger effect, and extreme care must be taken in comparing the results of different authors.

In the century since Bauschinger's original paper, there have been many conflicting models of both the origin and the scale of flow events which give rise to the Bauschinger effect. Heyn (1918) developed a view which ascribed the effect to the action of internal stress which arose due to the difference in elastic and plastic response of neighboring grains in

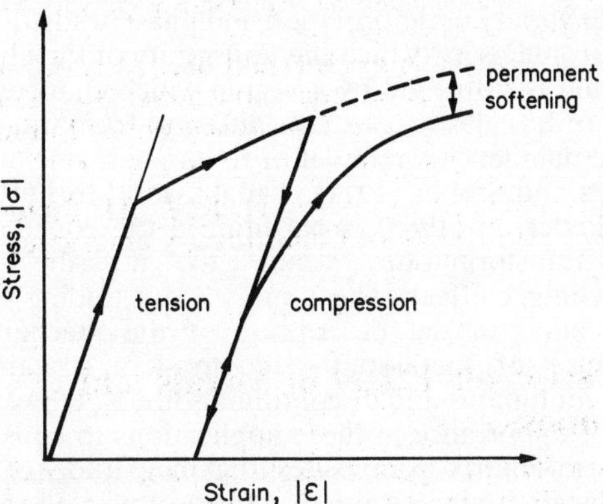

Figure 1
A schematic representation of the Bauschinger effect in a stress–strain diagram. The sequence of loading can be either tension, compression or compression, tension (Mughrabi 1987).

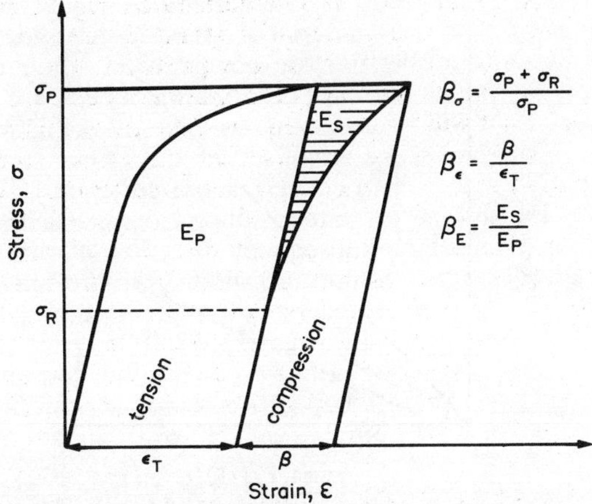

Figure 2
A schematic diagram showing the deformation of the magnitude of the Bauschinger effect in terms of stress β_σ, strain β_ε and energy β_E (Abel and Muir 1973)

a polycrystal. However, the numerous careful studies of the Bauschinger effect in single crystals by Sachs and Shoji (1927), Edwards and Washburn (1953), Patterson (1955) and Buckley and Entwistle (1956), clearly indicate that at least one source of the Bauschinger effect must lie in the reverse motion of dislocations from their aggregation in the work-hardened state. Orowan's (1958) hypothesis based on the anisotropy of the driving force for dislocation motion in the work-hardened state accounts for a number of features of the plastic hysteresis.

Interest in the detailed micromechanical features that give rise to the Bauschinger effect was intensified by studies of two-phase materials. The pioneering work of Wilson and Konnan (1964) showed that x-ray diffraction methods could be used to estimate the elastic stresses carried by hard second phase particles such as Fe_3C and to correlate these stresses with the back stresses deduced from the mechanical response. An example of the dependence of the elastic strains in Fe_3C on the macroscopic pre-strain is shown in Fig. 3.

In two-phase systems containing hard particles the flow stress σ_f can be considered as the sum of three terms: the initial flow stress σ_0, a nondirectional hardening term σ_s and a component that can be characterized as an internal stress, and is directional in character $\langle\sigma_m\rangle$. Thus the flow stress in the forward direction is:

$$\sigma_f = \sigma_0 + \sigma_s + \langle\sigma_m\rangle$$

while in the reverse direction it is:

$$\sigma_r = \sigma_0 + \sigma_s - \langle\sigma_m\rangle$$

and thus the internal stress can be expressed in a variety of ways by comparing σ_f and σ_r.

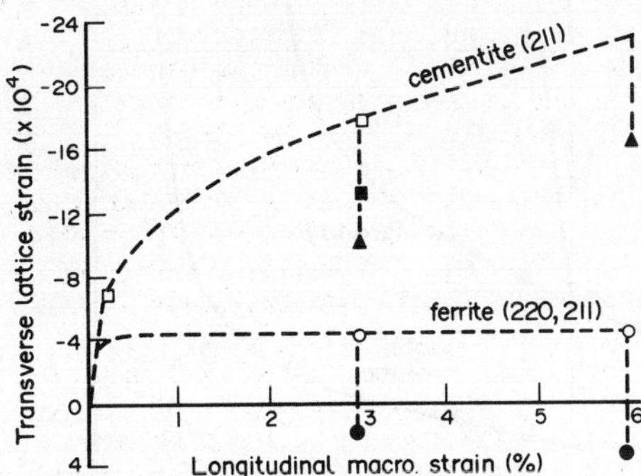

Figure 3
A diagram showing data on the lattice strains measured by x rays in cementite and ferrite as a function of pre-strain. The open points show the strains measured with the sample under load, and the closed points show data for the unloaded condition (Wilson and Konnan 1964)

The resistance of the second phase particles is a function of size, shape and structure, and measurements of the Bauschinger effect can be combined with detailed studies of the work-hardening in two-phase systems to determine whether the particles are sheared noncommittantly with the matrix or whether they act as rigid obstacles which both support large local elastic stresses and force complex local relaxations in the vicinity of the particles. An example of this difference in response is shown in Fig. 4 for the Al–3.6%Cu system.

In simple dispersion-hardened systems such as Cu–SiO_2, Atkinson et al. (1974) showed that the curvature of the reverse stress–strain curve could be related to the degree of permanent softening. It has also been shown that the ability to support local elastic stresses at second phase particles is very temperature dependent and various forms of relaxation can occur because the distances involved in the relaxation events are only of the order of the diameter of these second phase particles. The concept of the mean stress $\langle\sigma_m\rangle$ acting to oppose flow in the matrix has been developed by Brown and coworkers in a number of papers: Brown and Clarke (1975), Brown and Stobbs (1975) and Brown (1984). The magnitude of $\langle\sigma_m\rangle$ may be expressed in terms of the unrelaxed strain ε_p^* which is a measure of the incompatibility of the particle, and the matrix as:

$$\langle\sigma_m\rangle = 2D\gamma\mu f\varepsilon_p^*$$

where f is the volume fraction of second phase particles, D is a function of the elastic constants of the particle, the matrix μ is the shear modulus of the matrix and γ is a factor dependent on particle geometry. As the average internal stress in the body is equal to zero, the mean stress $\langle\sigma_m\rangle$ is balanced by a stress $\langle\sigma_i\rangle$ inside the particles such that

$$f\langle\sigma_i\rangle + (1-f)\langle\sigma_m\rangle = 0$$

In 1957 Eshelby introduced the important concept of the transformation strain which can be applied to a wide variety of deformation and phase transformation problems in "which the uniformity of the elastic medium is disturbed by a region which changed its form or has elastic constants different from those of the remainder." A number of researchers, including Brown and Stobbs (1971), Tanaka and Mori (1971) and Pederson (1983), have utilized the concept of the transformation strain to describe the Bauschinger effect. These models are finding wide application in the description of the mechanical response of metal-matrix composites containing both continuous and discontinuous fibers. However, it is of importance in these applications to consider in metal-matrix composites the magnitude of the unrelaxed strains ε_p^* relative to the imposed strain ε_p^*. An example is shown for Cu–W and Al–B composites in Fig. 5.

The mechanical response of metal-matrix composites containing continuous or discontinuous ceramic phases is very complex because the transformation strains can arise both from thermal history, due to

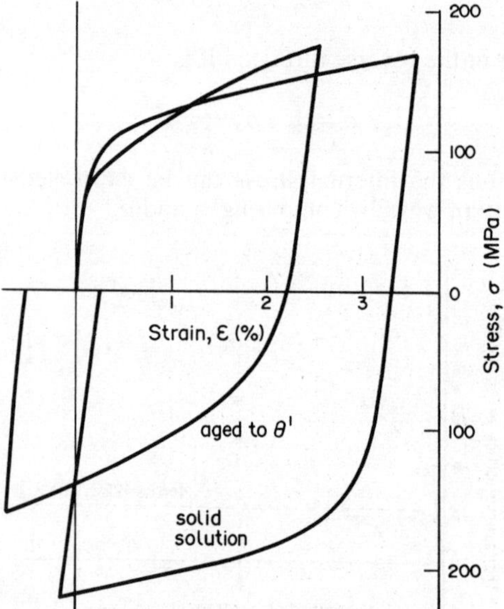

Figure 4
A diagram showing the difference in mechanical response for Al–3.6%Cu in the form of (a) solid solution and (b) aged to a two-phase system containing θ^1 particles. The diagram shows that the Bauschinger effect is much larger when the hard θ^1 particles are present (Moan and Embury 1979)

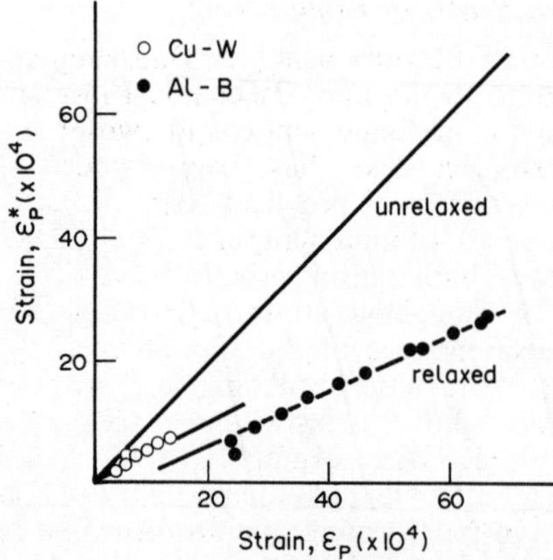

Figure 5
A diagram showing the relationship between the unrelaxed strain ε_p^* measured by x-ray diffraction, and the relaxed plastic strain in fiber composite materials (Pederson 1984)

the difference in thermal expansion coefficient between the matrix phase and the strengthening phase, and from differences in elastic and plastic properties of the phases.

It is difficult to define *a priori* the relaxation mechanisms which are operative in relation to these sources of local incompatibility, and thus a variety of experiments are needed that consider strain history, the temperature dependence of the mechanical response including the Bauschinger effect and the processing history of the material. Further, in many composite materials the volume fractions of strengthening phases are large (of the order of 10–20%), and hence it may no longer be possible to determine the response based on models of isolated inclusions, but instead it may be determined on the basis of the interactions between both elastic and plastic inclusions that depend on the local volume fraction and detailed particle distribution.

2. *Microscopic Studies*

In addition to studies of the Bauschinger effect in materials subject to macroscopic strains, considerable effort has been made to understand the forward and reverse motion of dislocations in the microplastic range with strains of the order of 10^{-4} to 10^{-7} (Marschall and Maringer 1977). In this regime the dislocations move over much smaller distances, and hence sample both obstacles and distributions of residual stress in a different manner than pertinent to macroscopic flow.

Thus it is clear that the Bauschinger effect remains a fruitful field for study in relation to the micromechanical modelling of many advanced materials.

3. *Uses in Practical Engineering*

In addition to the work using the Bauschinger effect to contribute to the micromechanical modelling of flow behavior in single phase and two-phase materials, extensive use has been made of the Bauschinger effect in practical engineering predictions. These are of major importance because many forming and fabrication processes involve strain reversals or changes in strain path. Thus there has been widespread use of the Bauschinger effect in relation to kinematic hardening and the shape of subsequent yield surfaces in plasticity theory (Prager 1955). Other examples of the use of the Bauschinger effect include the prediction of the flow stress loss in pipeforming operations which involve reverse bending and expansion (Uko et al. 1980), the dependence of the Bauschinger effect on strain aging (Wilson and Bate 1987) and its correlation with low temperature creep events and dimensional instabilities in spring steels (Furr 1972, Brownrigg and Sritharan 1987). Thus Johann Bauschinger's discovery of the asymmetry of plastic response following initial plastic flow has served to initiate a wide range of both basic and applied research aimed at elucidating the strain path dependence of the work-hardened state. This will have continued value to the development and utilization of a broad spectrum of modern materials.

Bibliography

Abel A 1987 Historical perspectives and some of the main features of the Bauschinger effect. *Mater. Forum* 10: 11–26

Abel A, Muir H 1973 The effect of cyclic loading on subsequent yielding. *Acta Metall.* 21: 93–102

Atkinson J D, Brown L M, Stobbs W M 1974 The work hardening of copper–silica IV. *Philos. Mag.* 30: 1247

Bauschinger J 1886 *Mitteilung XV, Technische Hochschule München* 13: 1–61

Brown L M 1984 Fundamentals of deformation and fracture. *Proc. Eshelby Memorial Symposium*. Cambridge University Press, Cambridge, pp. 357–67

Brown L M, Clarke D R 1975 Work hardening due to internal stress in composite materials. *Acta Metall.* 23: 821–30

Brown L M, Stobbs W M 1971 The work hardening of copper–silica I—A model based on internal stresses with no plastic relaxation. *Philos. Mag.* 23: 1185–99

Brown L M, Stobbs W M 1975 In: Argon A S (ed.) 1975 *Constructive Equations in Plasticity*. MIT Press, Cambridge, MA, pp. 387–429

Brownrigg A Sritharan T 1987 Spring steel hysteresis. *Mater. Forum* 10: 58–63

Buckley S N, Entwistle K M 1956 The Bauschinger effect in super purity Al single crystals and polycrystals. *Acta Metall.* 4: 352

Edwards E H, Washburn J 1953 Some observations on the work hardening of metals. *Trans. AIME* 197: 1525

Eshelby J D 1957 The determination of the elastic stress field of an ellipsoidal inclusion and related problems. *Proc. R. Soc. London, Ser A* 241: 376–96

Furr S T J 1972 Development of a new laboratory test method for spring steels *Trans. ASME* 94: 223

Heyn E 1918 *Metall und Erz* 22, pp. 436–442

Marschall C W, Maringer R E 1977 *Dimensional Instability—An Introduction*. Pergamon, Oxford, pp. 63–137

Moan G D, Embury J D 1979 *Acta Metall.* 27: 903

Mughrabi H 1987 Johann Bauschinger: Pioneer of materials testing. *Mater. Forum* 10: 5–10

Orowan E 1958 *Symp. on Internal Stresses in Metals and Alloys*. Institute of Metals, London, p. 47

Patterson M S 1955 Plastic deformation of copper crystals under alternating tension and compression. *Acta Metall.* 3: 491

Pederson O B 1983 Thermo-elasticity and plasticity of composites—Mean field theory. *Acta Metall.* 31: 1795–808

Pederson O B 1984 Fundamentals of deformation and fracture. *Proc. Eshelby Memorial Symposium*. Cambridge University Press, Cambridge, p. 139

Prager W 1955 The theory of plasticity—Survey of recent developments. *Proc. Inst. Mech. Eng.* 169: 41

Sachs G, Shoji H 1927 Zug—Druckversuche an Messingkristallen. *Zeit für Physik* 45: 775–81
Tanaka K, Mori T 1971 The hardening of crystals by non-deforming particles and fibres. *Acta Metall.* 21: 571–4
Uko D, Sowerby R, Embury J D 1980 The Bauschinger effect in structural steels, and its role in the fabrication of line pipe. *Met. Tech.* 7: 359
Wilson D V, Bate P S 1987 The Bauschinger effect in strain aged steels. *Mater. Forum* 10: 33–42
Wilson D V, Konnan Y A 1964 Work hardening in a steel containing a course dispersion of cementite particles. *Acta Metall.* 12: 617

J. D. Embury
[McMaster University, Hamilton, Ontario, Canada]

Biomaterial–Blood Interactions

Since 1940, and chiefly since 1950, numerous materials and devices have been introduced into the circulatory system, such as artificial heart valves, circulatory assist devices, totally artificial hearts, oxygenators, liver support systems, renal dialysers, blood vessel substitutes and catheters.

Whenever materials are introduced, severe problems can arise, most of which are related to the interaction between the blood and the devices in use. When blood comes into contact with an artificial surface adsorption and/or degradation of blood components, consumption of coagulation factors, and adhesion and aggregation of platelets induce the generation of microthrombi and occasionally life-threatening thromboembolism. In addition, device dysfunction, impairment of blood flow to distal organs, embolization of thrombus and removal or modification of circulating blood elements are all possible and may be described as "artificial surfaces hemopathy" or "post-perfusion syndrome." Therefore, pharmacological inhibition of the activation of hemostasis must often be employed when cardiovascular biomaterials are used. After heart–lung bypass, for example, the result may be a prolonged incidence of post-perfusion bleeding by 5% to 25%, with a general blood cell count decrease sometimes associated with functional activation and/or impairment. The amount of the coagulation substrate fibrinogen decreases with initiation of extracorporeal oxygenation and levels of fibrinolytic activity are low.

After the insertion of mechanical heart valves, local thrombotic occlusion and thromboembolism caused by the prostheses, associated with the anticoagulant therapy intended to prevent these complications, account for the majority of prosthesis-related deaths. The incidence of thromboembolism with mitral prostheses ranges from 1–11% per patient year. The incidence of significant hemorrhage is 4% per patient year, and up to 10% of these will be fatal. "Blood compatibility is therefore of the utmost importance in determining the performance of devices within the cardiovascular system" (Williams 1987).

The sequence of events leading to thrombus formation when blood comes into contact with an artificial surface is complex due to the many plasma components involved, the nonspecific and competitive nature of protein adsorption, the infinite possibilities for cooperative interactions between proteins and cells at the interface, and the continuously transient state of the interface. Further, physicochemical reactions involved in the adsorption of a mixture of proteins, cells and other nonproteic plasma components remain difficult to analyze and to correlate with the known superficial physicochemical structures of biomaterials. Nevertheless, an understanding of the origin of the surface activity of proteins, the multiple states of adsorbed protein and the competitive adsorption behavior of proteins is a requisite for the description of interfacial phenomena.

1. The Blood–Material Interface

The surface of a biomaterial is an area of transition; chemical constituents of the superficial layer a few nanometers from the surface are responsible for the forces accounting for the interfacial phenomena of surface tension, adsorption and adhesion.

Surface energy can be used to classify materials approximately and correlates directly with the nature and concentration of superficial chemical groups. Contact angle measurement of a drop of fluid deposited on the surface is a classical method of determining surface energy (γ_s) which is, for example, 17×10^{-6} J cm^{-2} for glass, 100×10^{-6} J cm^{-2} for pure metals and $1.5–5 \times 10^{-6}$ J cm^{-2} for polymers. Values of γ_s may be enhanced by the introduction of polar groups (OH, Cl, C≡N). Relative wettability based on contact angle measurements has been used to predict blood compatibility of materials but such a single criterion is inadequate.

The first occurrence is the adsorption of inorganic ions and water molecules. On contact with water the superficial structure of hydrophobic materials is not changed: it is the structure of the water itself that is modified. An intermediary layer a few molecules thick in which the molecules have a more ordered structure than in pure liquid water is formed by a dynamic process with constant rearrangement of ordered microdomains. However, the surface of hydrophilic materials is greatly modified by water. Water molecules become enmeshed in the macromolecular network and break certain interactions between the polymer chains.

2. Protein Adsorption: An Early Event

Proteins adsorb to any surface rapidly and selectively to form a layer about 100 nm thick. Adsorption of proteins and cells on the surface of biomaterials occurs in three steps: transport to the interface, the reaction of adsorption, and the conformational rearrangement of proteins and changes in shapes of the adherent cells. The kinetics of adsorption are dependent on the rate at which the first two steps occur. When a wet biomaterial comes into contact with a flowing solution of protein an intermediary layer is formed, which does not contain protein. The transfer of proteins to the surface through the intermediary layer is due to molecular diffusion which may be supplemented by convective transport in flow situations.

The most obvious means of binding protein to a surface is via an electrostatic bond formation between charged groups on the protein and oppositely charged surface sites. On glass, electrostatic adsorption is high, shows considerable reversibility and desorption occurs in response to electrolytes. Other possible protein–surface interactions are hydrogen bonding, probably for relatively polar substrates, and β-turns. These areas of the protein where the polypeptide chain folds back on itself by nearly 180° lie in the surface region of the protein and have been incriminated in adsorption. However, the most important binding mechanism is probably the hydrophobic reaction defined as the interaction of nonpolar groups in aqueous media. Proteins that contain hydrophobic patches on their surface are able to interact with hydrophobic-contacting surfaces. Proteins probably bind in this way to surfaces such as polyethylene, silicon and certain segmented polyurethanes. On such surfaces compact monolayers are formed, there is no desorption and only a slow partial exchange between adsorbed and dissolved protein occurs. The character of the surface, whether it is hydrophobic or hydrophilic, charged or neutral, polar or nonpolar, determines the degree to which any specific process can occur. Quantities adsorbed are generally greater for hydrophobic than for hydrophilic substrates, where desorption or ion exchange may take place.

The spontaneity of protein adsorption to artificial surfaces is promoted by:

(a) their size, since larger molecules may have multiple contact points with a surface (although this is not a determining factor);

(b) their charge, since molecules nearer their isoelectric pH may adsorb more easily;

(c) their structure, since more stable molecules, and covalent and sulfur–sulfur bridging reduce surface activity; and

(d) their chemical properties, related to their amphipathic nature, hydrophobicity and low solubility resulting from their high molecular weight.

The composition of the protein layer which has a controlling influence on the subsequent biomaterial interactions depends on the concentration of proteins, on their relative mobility in plasma and on their affinity for the surface. The smoothness of the surface and the heterogeneity of the distribution of reactive groups are partly responsible for the selectivity of composition of the adsorbed plasma proteins.

Adsorption is relatively rapid, but variable, being as much as 75% complete in 5 min and it reaches an equilibrium in about 1 h. The amount of fibrinogen adsorbed is independent of surface shear rate from $0\,s^{-1}$ to $2100\,s^{-1}$ (transport is not the controlling step under these conditions). Real-time studies have shown that in adsorption to quartz from an equimolar solution of bovine serum albumin (BSA) and γ-globulin, BSA reaches an equilibrium in 8 s and γ-globulin in 40 min.

The isotherms of adsorption are generally of the Langmuir type. Most protein surfaces exhibit a limited surface concentration, and adsorption isotherms showing surface concentration as a function of solution concentration usually approach a plateau beginning at a moderate solution concentration and reaching the concentrations found in blood. Although many studies used this simple Langmuir model to describe protein adsorption, more recent models suggest that adsorbed proteins may exist in more than one state. The mechanisms that could lead to multiple states of adsorbed proteins include the "occupancy" effect owing to the unavailability for adsorption of surface sites that are already occupied, the conformational changes before, during or after adsorption and the multiplicity of binding models due to amphipathicity of the molecules and the multiple site nature of the surface.

Biological indicators suggest the existence of different states of adsorbed proteins; for example, enzymes such as thrombin may lose some of their activity and proteins such as fibronectin and fibrinogen will exhibit different reactivity towards subsequent cell events. Conformational changes of protein molecules during the adsorption–desorption process may be either positive (surface passivation with an irreversible adsorbed protein layer) or negative (initiation of blood-enzyme systems, in particular the activation of the coagulation and complement system because of the desorption of the surface-activated protein components).

Adsorption is most often described as being irreversible on an ordinary time scale on hydrophobic solid surfaces while for hydrophilic surfaces both reversibility and irreversibility have been found. While desorption is either nonexistent or extremely low, there is a measurable exchange of proteins between the surface and the solution as has been clearly demonstrated by the use of labelled proteins. The extent of such an exchange varies with the surface, being greater for hydrophilic than for hydrophobic surfaces, and depends on protein concentration and flow. The rates of exchanges also vary from days (polyethylene–albumin) to hours (glass–fibrinogen) exhibiting slow-exchanging and fast-exchanging population of proteic molecules.

Although the amounts adsorbed at the plateau vary from one surface to another, the surface concentrations at these plateaux are of the order of 0.1–1.0 $\mu g\, cm^{-2}$ and correspond to monomolecular layers. Proof of the existence of multilayer adsorption has been obtained for a number of polymers. Apparent exceptions to the "monolayer rule" are hydrogel-like materials such as segmented polyurethanes which essentially adsorb no protein, but the rapid reversibility of adsorption on these materials has to be considered.

2.1 *Adsorption from Solution of Single Proteins*

Protein adsorption has been studied experimentally by measuring the adsorption isotherms of single purified proteins involved in hemostasis and thrombus formation (fibrinogen, fibrin, factor XII, high molecular weight kininogen (HMWK), factor VIII (Von Willebrand factor, VWF) and thrombin), in inflammatory and immunological reactions (immunoglobulin G, immunoglobulin M, immune complexes, C5a and C3a), and in cell adhesion (albumin, fibronection and collagens).

Measurements of equilibrium layer thickness, performed on a fibrinogen–glass system, support the fact that adsorption occurs at the monolayer level. The molecules in the layer do not seem to be drastically altered since their average dimensions are similar to their dimensions in solution.

For the proteins studied, as the isoelectric pH increases so does the adsorption of the proteins. Until the early 1990s, there have been no attempts to determine the properties of protein–surface systems that govern the relative surface affinity of different proteins. On a hydrophobic substrate, increased surface concentration of the four proteins tested (immunoglobulin G, immunoglobulin M, human serum albumin and α_2 macroglobulin) correspond to increased protein hydrophobicity. For polyethylene, plasma protein affinity varies as follows:

$$\text{hemoglobin} \gg \text{fibrinogen} > \text{albumin} \simeq \gamma\text{-globulin}$$

Small proteins, by virtue of their higher diffusion coefficients, will initially adsorb faster than large proteins during competitive adsorption. The relative extent of protein adsorption onto polymer surfaces is influenced by the surface tensions of the substrate material, the suspension liquid and the proteins themselves. The more hydrophobic proteins will adsorb to the largest extent with one substrate material. Protein adsorption changes the surface properties of the "naked" surface markedly, and most polymers are rendered more hydrophilic as a result of protein adsorption.

Mixture studies using a two- or three-protein system on a variety of surfaces with various mixture compositions and various total concentrations demonstrate that fibrinogen is preferentially adsorbed when it is in solution with albumin and immunoglobulin G. For glass, and for several other surfaces, the surface enrichment of fibrinogen (between 10- and 200-fold) has been demonstrated. However, many of these measurements refer to equilibrium and involve long adsorption times; the relative quantities adsorbed in multiprotein systems are relatively time independent. One of the few studies showed that, in a one-to-one mixture of albumin and fibrinogen, albumin predominated in the first seven minutes and was then gradually displaced by fibrinogen. The faster diffusion of albumin will delay but will not halt the subsequent adsorption of fibrinogen.

In a multiprotein system the competitive adsorption behavior of proteins at interfaces is of great importance. The factors influencing competitive adsorption may be, in order of importance, the chemical surface properties (in terms of charge, hydrophobicity and chemical functional groups) and the protein properties (such as electrical charge, hydrophobicity, available chemical functional groups, stability of the protein structure, interactions between proteins in the adsorbed layer, relative concentration in the bulk phase and molecular size). The problem of how to combine all of the possible factors into a quantitative model of competitive adsorption is complex since data for such models are unavailable.

2.2 *Adsorption of Proteins from Plasma*

Since blood is a multicomponent system where the influence of one protein on another is an important factor, only studies using blood or plasma can provide definitive answers regarding blood–material interactions.

In general, it has been found that preferential or selective adsorption occurs so that certain proteins may be enriched in the surface and vice versa. Nevertheless, the currently available data concern only major plasma proteins (fibrinogen, albumin and immunoglobulin G) and proteins involved in hemostasis and cell adhesion. Surface chemistry, time of adsorption and protein type are major factors in determining the composition of the adsorbed layer. This layer is formed from the mixture and thus contains a rather complex and changeable combination of proteins.

Adsorption of proteins from plasma to various hydrophilic–hydrophobic copolymers showed that nine or more protein peaks can be observed after gel electrophoresis of the sodium dodecyl sulfate (SDS) eluates from the surfaces. The adsorption of each protein varies in a characteristic way with copolymer composition.

A typical hydrophilic surface like glass shows not only some albumin, immunoglobulin G and fibrinogen and some plasminogen, but also some fibrin degradation products (FDP), suggesting surface activation of plasminogen, which is even reduced when plasminogen-deficient plasma or factor-XII-deficient plasma is used. Contact factors of blood coagulation are known activators of the fibrinolytic system but many other still unidentified species are also present in the glass eluates. Activation of fibrinolysis attested by the formation of FDP is a common feature of all hydrophilic materials.

On hydrophobic materials such as polystyrene, fibrinogen is the major constituent of the layer. FDP are also observed but in smaller amounts than for glass. The eluates are in general less complex than those for glass and this reflects the increased difficulty of eluting proteins from hydrophobic surfaces.

On polyethylene and siliconized glass, fibrinogen adsorption is greatest at 2 min and then decreases to zero. Immunoglobulin G is detected on all surfaces though in relatively low surface concentrations. More polar polymers have low initial adsorption *in vitro* which increases with time, but *in vivo* fibrinogen deposition is characterized by a second stage of greatly increased adsorption that can be inhibited by heparin. These polymers also cause enhanced platelet consumption. Gas discharge treatment on (Dacron) poly(ester terephthalate) reduces the initial adsorption of fibrinogen.

For hydrogels the most frequent observation is a quantitative reduction in the amount of a particular protein adsorbed to the hydrogel compared with the hydrophobic surface. Desorption and exchange studies indicate that proteins are less tightly bound to hydrogels than to nonhydrogels. A qualitative difference in protein adsorption is the apparent preference of certain proteins, especially albumin, for hydrogel surfaces.

Much work remains to be done to establish the exact composition of proteins adsorbed from plasma and blood, particularly for trace proteins. The major constituents of the adsorbed protein layer are fibrinogen, γ-globulin and albumin. Other proteins adsorbed in smaller quantities include VWF, fibronectin, α- and β-globulins, transferrin, caeruloplasmin, the coagulation factors XI and XII, and HMWK.

The spectra and amount of protein adsorbed vary with time. The adsorption of hemoglobin and immunoglobulin G to different polymers increases with time although the degree of adsorption depends on the polymer, whereas fibrinogen and albumin adsorption increase initially with time but then decrease.

Enrichment of the proteins, calculated as the ratio of the weight fraction of each protein in the surface to the weight fraction of the protein in the bulk, varies with the polymer composition and is for fibrinogen 0.8–3.4, for immunoglobulin G 0.8–1.5, for albumin 0.3–0.8 and for hemoglobin 150–400. Thus, the surface enrichment relative to the bulk phase is not very great for immunoglobulin G, albumin or fibrinogen but hemoglobin, which is not considered as a normal plasma constituent (with a concentration of 0.003 mg ml^{-1}), is an exception to this rule.

Generally, however, plasma proteins appear to adsorb in proportion to their bulk concentration as might be expected from mass action. Since the major plasma proteins are only minimally adsorbed on surfaces that are known to adsorb compact monolayers of proteins from single solutions, it has been suggested that in the plasma there are trace proteins that are strongly surface active and that these are adsorbed in large quantities or that the plasma itself restricts adsorption. The ability of hemoglobin to inhibit fibrinogen adsorption depends on the ratio of the proteins but it is also strongly dependent on the total concentration at which the competition takes place. For fibrinogen the maximum adsorption occurs at concentrations in plasma of 0.1% for glass, 1% for polyethylene and about 10% for polytetrafluoroethylene (PTFE). Both time and concentration effects on competitive adsorption of fibrinogen from hemoglobin solution are consistent with a mass action model for competitive effectiveness. Dilution of the plasma reduces the rate of adsorption of any protein as well as its rate of replacement by another protein and this is referred to as the Vroman effect, manifested either as a maximum in fibrinogen adsorbed to a surface as a function of time or as a maximum in fibrinogen adsorbed as a function of plasma concentration. The residence time of fibrinogen on the surface varies with the material. Fibrinogen is displaced within 2 s of contact by HMWK and is also displaced to some degree by factor XII, often within 1 min. Since there is no change in the protein film thickness and since fibrinogen is no longer reactive to antifibrinogen antiserum, it is assumed that fibrinogen is not covered but replaced by other proteins.

The competitiveness of protein adsorptions in the context of blood led Vroman (1987) to postulate that a rapid sequence of adsorption and displacement occurs with time, by which more abundant proteins are displaced by less abundant ones. Albumin is adsorbed and displaced in a fraction of a second (it is not often observed on a surface after plasma contact), and the absence of the Vroman effect for hydrophilic polyurethanes may reflect a very rapid sequence. This competitive adsorption is probably controlled not only by concentration but also by protein affinity for the surface (free energy of adsorption), protein–protein interactions in the layer and kinetics factors. It must be stressed that all studies of the Vroman effect were performed under static conditions.

The replacing factors are HMWK, kallikrein, factor XIa and/or high density lipoproteins. On activating surfaces the sequence appears to be albumin followed by the immunoglobulins, fibrinogen and fibronectin, and lastly HMWK. Plasmin does not affect the Vroman effect which is, however, dependent on an unidentified protein that binds tightly to lysine-agarose. All the surfaces studied except for some hydrophilic polyurethanes displayed the Vroman effect. However, quantitative differences (peak height, peak position) were found among surfaces, and there is some evidence that the height of the peak in the adsorption versus plasma concentration relation is correlated with gross thrombogenicity. It is at present difficult to understand the Vroman effect in relation to surface thrombogenesis, since contact phase activation occurs where fibrinogen is displaced from the surface whereas platelets will adhere only where fibrinogen remains on the surface.

2.3 Consequences of Protein Adsorption on Blood–Material Interactions

It is now accepted that when albumin layering occurs in preference to fibrinogen layering, thromboresistance is favored since albumin minimizes platelet adhesion and thrombogenesis. On the contrary, if fibrinogen and γ-globulin are preferentially adsorbed thrombosis may ensue, causing enhanced platelet interactions and platelet activation. Fibrinogen is also believed to interact with leucocytes. γ-globulin coatings on the surface not only enhance platelet and granulocyte adhesion but they also stimulate platelet release. After the initial layer of plasma proteins is adsorbed onto a surface, a crosslinked fibrin network may develop over a period of minutes or hours. The formed blood elements will be entrapped within this fibrin network. If the rate of thrombus formation is slow, the process of fibrinolysis may prevent gross thrombus formation.

2.4 Fate of Adsorbed Proteins

The local concentration of adsorbed proteins is extremely high (about 1000 mg ml^{-1}) under a variety of adsorbed states (but the structure of adsorbed protein is not well understood, including whether proteins are denatured at the solid–liquid interface) thus modifying cellular interactions controlled by the presence of specific proteins on the foreign surface at a sufficiently high surface density and degree of reactivity.

The organization of adsorbed protein films suggests that on polyvinylchloride (PVC), polytetrafluoroethylene (PTFE) and polyvinyldifluoride (PVDF) bare areas of polymer remain after adsorption, and the degree of coverage of the polymer surface with adsorbed proteins increases with increasing critical surface tension of the polymer. Protein films form in "islands." Albumin exhibits a globular surface deposition pattern with low coverage of the surface, influenced by fluid shear rate. In contrast, the glycoproteins, fibrinogen, γ-globulin and plasma fibronectin exhibit a more extensive coating of the surface with a characteristic reticulated pattern. The extent of surface coverage depends on the roughness of the surface; preferential binding of fibrinogen and, possibly, fibrinogen clusters is observed in surface cracks of the order of 1 μm. These results obtained by partial gold decoration transmission electron microscopy support the finding that smooth surfaces are less thrombogenic than rough ones.

Fibrinogen molecules are visible, and side-on binding configurations as well as side-to-side and end-to-end alignments are observed. Since the recent understanding of the activity of the fibrinogen D region in the end-on configuration, which is more reactive and prone to interact with cell membrane receptors, the side-on configuration of fibrinogen observed for hydrophobic surfaces suggests less reactivity. Albumin pretreatment significantly reduces the total amount of protein visualized on the surface. The conformational changes of protein are observed to a greater extent on hydrophobic than on hydrophilic surfaces.

Denaturation of adsorbed proteins is another matter for debate. Some systems have shown evidence of denaturation and others have not. For example, factor XII adsorbed on quartz was shown to undergo a conformational change while thrombin adsorbed on cuprophan and PVC retained most of its biological activity. Denaturation should be evaluated on the surface, which would be the best method but a difficult aim to achieve, or on eluted proteins. Fibrinogen eluted from glass tubing was examined by circular dichroism and a loss of alpha-helix content of the order of 50% relative to the native fibrinogen was observed. However, a different behavior of fibrinogen molecules initially eluted (less firmly bonded) with considerable chain degradation and of later-eluting fractions hardly different from "native" fibrinogen has been reported. The gel band patterns of the degraded fractions resemble those of early plasmin-induced fibrinogen degradation products (FDP). Plasminogen activation and subsequent fibrinogen partial degradation are non-negligible components of blood–material interactions.

3. Coagulation Activation

When normal plasma comes into contact with negatively charged surfaces, such as glass, kaolin or connective tissue, the intrinsic pathway of blood coagulation is activated.

3.1 Proteins Involved in the Contact Phase

The proteins that have been identified as being involved in the contact activation reaction are factor XII, prekallikrein, HMWK and factor XI.

(*a*) *Factor XII.* Human factor XII is a single-chain glycoprotein with a molecular weight of about 80 000 D, and it is almost entirely in the zymogen form. Conversion of the zymogen into an active serine protease is accomplished by cleavage of a single Arg–Val bond, producing the so-called α-factor XIIa possessing amidolytic and esterase activity. The three domains recognizable in the molecule are the amino-terminal domain (40 000 D), which contains the structural information for the binding of factor XIIa to negatively charged surfaces, followed by a 12 000 D connecting region and by the 28 000 D light chain, which contains the active site.

A number of additional peptide bond cleavages can occur in the human α-factor XIIa, in which case the amino-terminal region is lost, and the resulting enzyme β-factor XIIa (or Hageman factor fragment) lacks surface-dependent coagulant activity but it does contain the active site.

(*b*) *Factor XI.* This is an α-glycoprotein that contains two identical glycoproteins (80 000 D each) that are held together by disulfide bonds. Factor XI is present in plasma in a complex with HMWK and it is activated to factor XIa by limited proteolysis by factor XIIa.

Factor XIa is capable of propagating the intrinsic coagulation pathway by activating factor IX and it is also capable of proteolytically activating factor XII, factor VII (involved in the extrinsic pathway) and plasminogen. Factor XIa is mainly inhibited by α_1-antitrypsin. It can also be inhibited by antithrombin III, and heparin stimulates this inhibition.

(*c*) *Plasma prekallikrein.* Prekallikrein is a glycoprotein that exists as a single polypeptide chain with a molecular weight of approximately 80 000 D. It is a serine protease zymogen that is activated by limited proteolysis by factor XIIa. Kallikrein exhibits homology to factor XIa. As a protease, kallikrein is capable of liberating kinins from kininogens, of activating factors XII (particularly surface-bound factor XII) and IX, and of activating plasminogen. Inactivation of plasma kallikrein is accomplished by the C1 inhibitor and α_2-macroglobulin; less potent in this inhibition is antithrombin III.

(*d*) *High molecular weight kininogen.* HMWK contains approximately one-fifth of the kinin content of plasma and it exists as a single protein chain with a molecular weight of approximately 105 000 D. It is a nonenzymatic cofactor that is central to contact activation reactions. It has been shown to contain a most unusual region of amino acid sequences that is rich in histidine, lysine and glycine. This region is essential for the contact activation cofactor activity, and it is tempting to speculate that this highly positively charged region of the molecule is responsible for the critical binding of the molecule to negatively charged surfaces.

3.2 The Contact Activation Reaction

Data have been accumulated during the 1980s to explain the contributions of negatively charged surfaces to the activation of factor XII and to the expression of the activities of factor XIIa. Available evidence indicates that the three major roles of negatively charged surfaces are:

(a) to induce a structural change in factor XII such that it becomes highly susceptible to proteolytic activation,

(b) to promote HMWK-dependent interactions between factor XII and prekallikrein that result in reciprocal proteolytic activations of each molecule, and

(c) to promote the HMWK-dependent activation of factor XI by surface-bound α-factor XIIa.

However, the actual trigger event that initiates contact activation, after the exposure of a suitable surface on which the activation of factor XII, prekallikrein and factor XI can take place, is a catalytic activity responsible for the initial conversion of these contact activation zymogens into active serine proteases. Four different hypotheses have been put forward for the initiation of contact activation reactions once a negatively charged surface is exposed.

(a) The zymogens, factor XII and prekallikrein may assemble on the negatively charged surface and slowly generate the first active enzymes that subsequently account for the burst of activity.

(b) The interaction of surface bound factor XII with its natural substrates (factor XII and/or prekallikrein) may result in the expression of full proteolytic activity toward these molecules without requiring the formation of two α-factor XIIa chains.

(c) Low levels of factor XIIa and/or kallikrein may be permanently circulating in the blood.

(d) The protease not involved in the extraneous enzyme (coagulation system) may be responsible for the initiation of contact activation reactions.

3.3 Interactions Between Coagulation Contact Activation and other Biological Systems

Many interactions have been demonstrated between coagulation, fibrinolysis and the kallikrein–kinin complement systems during *in vitro* studies. However, the *in vivo* significance of many of the these recorded observations is not established.

Clearly, however, the potential for interactions between these and other systems in the control of physiological and pathological processes, such as thrombus formation on artificial surfaces, is considerable.

4. Complement Activation

The complement activation process is similar to the clotting system of blood in that restricted proteolysis of some of the factors is one of the main regulation mechanisms and activation is of the cascade type. The system consists of more than 19 components and regulatory plasma proteins. C3 is the quantitatively dominating complement protein and it has a central position in the complement activation cascade. Most of the methods used are based on the analysis of complement degrading factors but some of these factors, especially C3, are also deposited on the solid surface. Contact of blood with an activating surface triggers complement activation through one of two pathways: the classical pathway or the alternative pathway. These cleave the third component of complement C3 and generate the anaphylatoxin C3a and a major cleavage fragment C3b. An amplification pathway consisting of alternative pathway proteins augments C3 cleavage once C3b has been generated. A common effector sequence generates the vasoactive, chemotactic, leucocyte-stimulating immune regulatory and cytolytic activities of the activated complement. The classical pathway, initiated by the binding activation of C1 to the modified F_C part of the immunoglobulin G– or immunoglobulin M–antigen complex, can also be induced by the surfaces of viruses, by lipopolysaccharides, by certain artificial surfaces or by immunoglobulin G adsorbed on artificial surfaces. The alternative pathway represents a recognition mechanism for foreign surfaces in a nonimmune host.

The alternative pathway is activated by surfaces which exhibit specific biochemical characteristics allowing bound C3b to initiate the assembly of the amplification C3 convertase on that surface. The activation of the alternative pathway is as follows:

(a) binding to the activating surface of C3b molecules formed continuously in plasma,

(b) Mg^{2+}-dependent binding of factor B to surface-bound C3b,

(c) cleavage of B within the C3b, B complex by a serine protease factor, and

(d) formation of C3 by the surface-bound C3b, Bb and deposition of additional C3b.

Complex C3b, Bb rapidly loses its convertase activity by spontaneous dissociation (half-life of 3 min at 37 °C). This dissociation of Bb is retarded by P which binds to C3b and increases by five-fold the half-life of the convertase. However, the enhancing effect of P is counterbalanced by H which binds to C3b and impairs the binding of B to bound C3b, actively dissociates Bb from C3b, Bb and facilitates proteolytic inactivation of bound C3b by I, yielding C3bi, an inactive form of C3b that cannot bind B.

The competition between H and B for binding to surface-bound C3b discriminates between activating and nonactivating surfaces. On a nonactivating surface fixed C3b binds H with an affinity almost 100-fold greater than that with which it binds B. On an activating surface C3b binds H less effectively, and C3b is relatively resistant to inactivation by H and I. A protected formation and the expression of the amplification convertase on an activating surface result in enhanced C3 cleavage and the deposition of additional molecules of C3b.

Covalent coupling of heparin, and heparin in the fluid phase, inhibit alternative pathway activation. The site of the molecule which is responsible for the anticomplementary effect of heparin is not the same site involved in binding antithrombin III. With increasing amounts of C3b, molecules are deposited on a surface, binding of C5 leads to C5 cleavage by both the alternative and classical pathway convertases, and cleavage of C5 releases the anaphylatoxin C5a and a major fragment C5b which will give rise to the subsequent cytolytic complexes C5b–9. Complement activation also liberates the anaphylatoxins C3a, C5a and C4a which are of similar molecular structure and have similar biological properties, such as contraction of smooth muscle cells, chemotaxis for neutrophils, eosinophils and monocytes, histamin release, oedema, and granulocyte aggregating ability.

The interactions between C5a and specific high affinity receptors on neutrophils (2×10^5 receptors) play an important role in the deleterious effects of the activation of the complement system. The embolization of granulocyte aggregates leads to leucostasis, the increased expression of C3 receptors on the cells, lysosomal enzyme release, the generation of toxin oxygen products, damage to endothelial cells, and the enhancement of granulocyte adhesion to endothelial cells which results in increased capillary permeability. C3b on activating surfaces may induce the release of lysosomal enzymes or derivatives from arachidonic acid metabolism and trigger superoxide formation from phagocytic cells.

Both the deposition reaction from serum and the subsequent reaction with C3 are smaller on hydrophilic surfaces than on hydrophobic surfaces. It is likely that the interaction of C3 with the surface is one of the recognition mechanisms that leads to surface-induced complement activation. C3 undergoes conformational changes on the hydrophobic surface similar to SDS denaturation or biological activation. The affinity of H for surface-bound C3b is also influenced by biochemical characteristics of the particle surface. A nonactivating surface of the alternative pathway should either be incapable of covalently binding C3b, possess a surface with preferential binding of H over that of B to surface-bound C3b, or express few antigenic sites recognized by natural or acquired antibodies.

5. Blood Cell–Surface Interactions

5.1 Platelet–Surface Interactions

The adhesion and aggregation of platelets are an invariable accompaniment of the exposure of blood to an artificial surface (see Fig. 1). The adhesion is linked to protein adsorption and the interaction of platelets with adsorbed fibrinogen or γ-globulin has been attributed to the formation of a complex between glycosyltransferases located in the platelet membrane and incomplete heterosaccharides in the protein layer. The adhesive proteins for platelets should be present both in the plasma and in secretory pools within the platelets. In this light, the α-granule of the platelet contains several candidates: fibronectin, VWF, fibrinogen and thrombospondin.

It is accepted that if a polymer or glass surface is precoated with albumin platelet, adhesion is less than if the platelets were exposed to the bare polymer or glass surface, and it has been claimed that platelet adhesion to surfaces preadsorbed with proteins parallels the saccharide content of the adsorbed proteins. Fibrinogen precoating even at low levels causes increased platelet adhesion of about 40-fold that of albumin coating and lies in the same range as γ-globulin and fibronectin. Greater platelet adhesion than with fibrinogen is obtained with preadsorption of VWF, collagen and thrombospondin. An α-globulin coating on an artificial surface not only enhances platelet adhesion but it stimulates the platelet release reaction as well.

However, platelet adhesion does not correlate with fibrinogen adsorption to glass pretreated with a series of plasma dilutions. Other plasma proteins binding fibrinogen or acting in concert with fibrinogen may be important in influencing platelet adhesion. Alternatively, the state of the adsorbed fibrinogen may be different at different plasma concentrations, that is, how fibrinogen is adsorbed rather than how much is adsorbed may be most important. Platelet and macrophage responses do not correlate with the amount of adsorption of proteins for which they have receptors. *In vitro* platelets do not adhere to "converted" (i.e., having lost its antigenicity) fibrinogen, so the shorter the time for transition of native to converted form, the lower the probability of platelet adhesion.

Following platelet adhesion, the platelet release reaction takes place in the adhering platelets, and platelet aggregation then occurs on the surface. Among the different platelet constituents released, serotonin is less dependent than platelet adhesion on surface properties and it is different from platelet adhesion in that the amount of serotonin released increases with hydrophilicity.

The release of B thromboglobulin, platelet factor IV and thromboxane B2 (TXB2) is related to the implantation of prosthetic valves used in cardiopulmonary bypasses. Both platelets and white cells (monocytes and granulocytes) liberate tissue factor III (FTIII), which will then activate the intrinsic coagulation system. As blood clotting on the artificial surface proceeds, an interaction between platelets and the intrinsic pathway is likely. Thus, the generation of thrombin causes the parallel platelet release reaction and aggregation. Platelet aggregation leads to the secretion from the alpha granules of thrombospondin which is then partially bound to the platelet membrane and may play an important role in mediating platelet aggregation on artificial surfaces.

Hemodynamic conditions are of major importance in the adhesion of platelets to the vessel wall and in determining localization, growth and fragmentation of thrombi. Platelet adhesion to a surface is governed by two independent mechanisms, the transport of platelets to the surface, which depends on the flow conditions, and the reaction of platelets with the surface, which depends on the nature of the surface and of the adsorbed protein layers.

Platelet response following the contact of blood with an artificial surface is also influenced by diffusion and shear forces. Single cells nonuniformly distributed on the surface appear in conditions of low flow rate (0.3 ml s^{-1}) while multiple cell aggregates form at higher shear rates. The adhesion and diffusion of platelets onto surfaces are markedly increased by the presence of red cells until the hematocrit reading reaches about 40%.

Taking into account the multitude of tests performed, no firm conclusions can be drawn as to which materials are the most passive to platelets although, in general, polyurethanes, polyethylene oxide, silicone rubber and PTFE tend to be "good" materials. Materials recognized to be reactive to platelets include polystyrene, PVC, polyethylene and polymethylmethacrylate. Platelet adhesion and activation by polyalkylacrylates have been found to increase with the length of the alkyl side chains and the resultant hydrophobicity. It has been suggested that clusters of surface-bound fibrinogen molecules with minimal conformational alterations may provide a stimulus for platelet activation by surfaces. The alteration of surface-bound protein through adherence and detachment of platelets is one way in which cells can alter a protein substrate.

5.2 Red Cell–Surface Interactions

Red cells, the most numerous cells in blood, do not adhere to endothelium, but they can bind weakly to

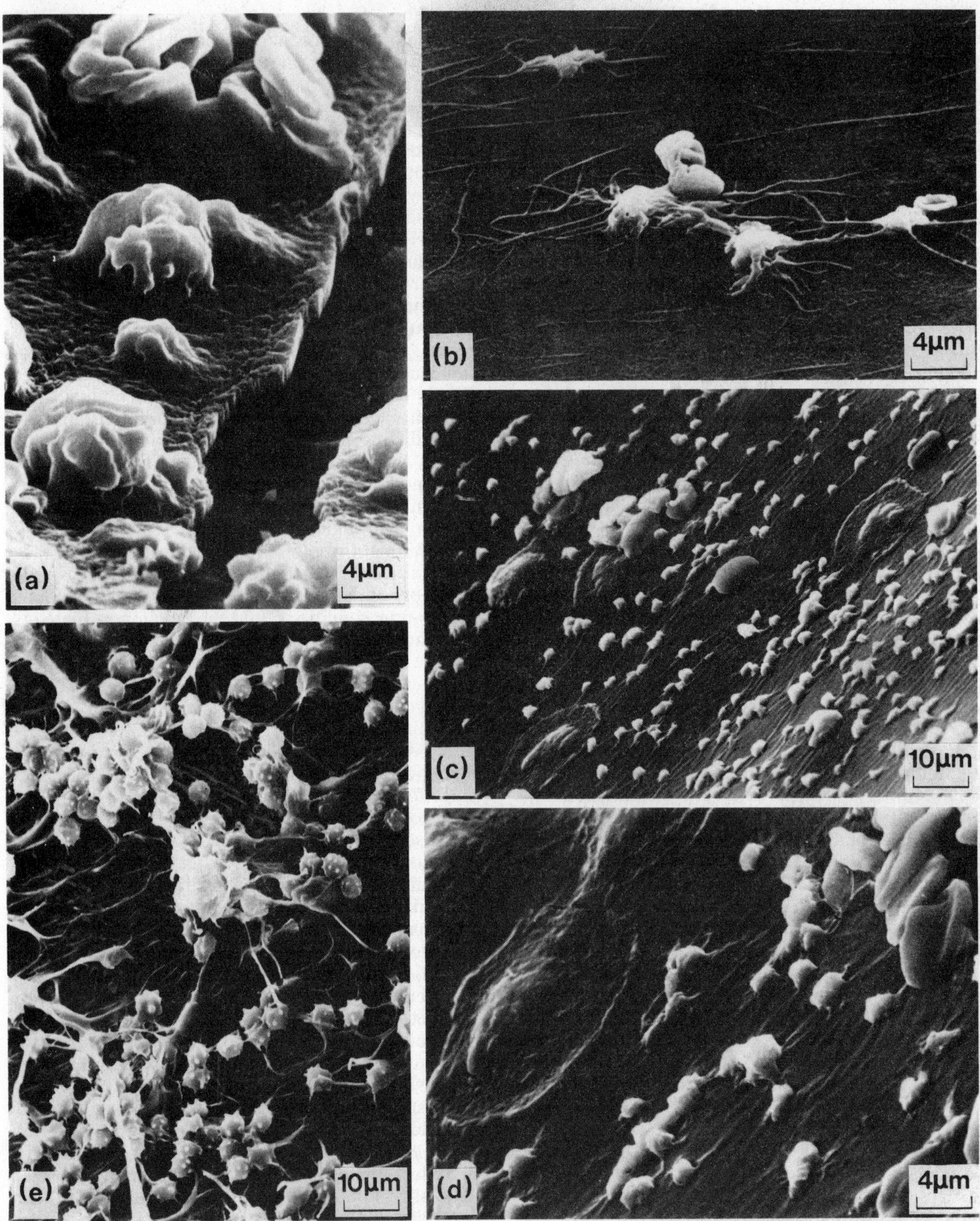

Figure 1
Blood–material interactions: (a) platelet adhesion on a proteic layer deposited on the material; (b) platelet spreading and pseudopodes emission in an early interaction; (c) and (d) large mononuclear cells spreading and platelet adhesion; and (e) thrombus formation with fibrin mesh development and erythrocytes engulfment (by permission of Editions Medicales Internationales, Paris)

some nonendothelial materials without spreading. They contribute significantly to blood–surface interactions in several ways. Rheological effects of red cells have been described. Collision with red cells increases the force of diffusion which brings other blood cells and plasma proteins towards the vessel surface. The addition of erythrocytes to protein solution reduces the amount of protein adsorbed. Red cells do not interact directly with biomaterials to a great extent, but they do promote the adhesion of other blood cells, particularly platelets, due to the hemodynamic behavior of red cells, a reduction in the adsorption of platelet protective proteins and even the deposition of an adhesive substance by red cells.

The function and morphology of red cells can be affected by repeated contacts with artificial materials. First, the turbulence associated with prosthetic and extracorporeal devices can produce the shearing stresses necessary to damage red cells with the loss of ADP (a powerful vasoconstrictor and platelet aggregating agent) and of a procoagulant material. Both high shear stresses (300 Pa) and low shear stresses (150 Pa) cause overt hemolysis. The hemoglobin level released from red cells *in vitro* is related to shear stress, time and the ratio of the surface area of contact to the blood volume and varies with different plastic surfaces. Although controverted, the nature of the material may be most important in low shear stress conditions. Second, the impairment of binding sites of C3b and PGI_2 (a platelet antiaggregating agent), acting effectively as inhibitors of C3b and PGI_2, can lead to significant complications for complement activation and platelet deposition.

Exposure of blood to biomaterials does not usually have measurable effects on red cells but occasionally, during the use of cardiac assist devices, serious anemia and red cell fragmentation can occur. Low-grade hemolysis associated with cardiopulmonary bypass circuits may contribute to platelet fibrin deposition and to the post-perfusion syndrome.

5.3 Leucocyte–Surface Interactions

Leucocytes, particularly neutrophils and monocytes, have a strong tendency to adhere to surfaces; they react with platelets, complement, coagulation and fibrinolytic systems, and they are major mediators of the inflammatory response. The half-life of neutrophils in the blood is about 7 h, irrespective of the age of the cells. Many paths lead to white cell adhesion and aggregation; kallikrein formation and complement activation, such as the formation of C5a, can be expected to link the process of surface-activated clotting to that of immune adhesion. Immunoglobulin G deposited onto hydrophobic rather than onto hydrophilic surfaces will irreversibly bind one or more complement components, thus localizing chemotactic activity at the surface and causing granulocyte adhesion and spreading.

As a result of leucocyte adhesion, several reactions are initiated. These include platelet–platelet and platelet–leucocyte interactions, the detachment of adherent thrombi by the action of leucocyte proteases, the detachment of adherent platelets and adsorbed proteins by leucocytes, the release of leucocyte products (e.g., leukotriene B4) which may give rise to both local and systemic vascular reactions, inflammatory reactions which are leucocyte dependent, and the promotion of fibroblast ingrowths into prosthetic material.

Further, chemoattractants derived from platelets or released during blood coagulation contribute to the adherence of leucocytes to the edge of developing thrombi. Significant quantities of procoagulant activity are generated, at least by the provision of tissue thromboplastin. Leucocytic proteases can also initiate intrinsic coagulation and cleavage of C3. Sensitized basophils and macrophages can elaborate the platelet activating factor, which is a potent inflammatory agent.

Fibronectin and immunoglobulin G preadsorptions promote macrophage attachment while all other proteins (albumin, hemoglobin, fibrinogen) decrease or prevent attachment. Receptors for fibronectin and immunoglobulin G on macrophages are probably responsible for the effects of these proteins on attachment.

5.4 The Adhesion of Blood Cells to Artificial Substrates

It is now accepted that blood cell adhesion to physiological substrates (e.g., collagen, immunoglobulin G, endothelial cells and bacteria) is mediated by cytoadhesins, a family of molecules present in cell membranes or penetrating the cytoskeleton (integrins). These molecules are membrane receptors of glycoproteic nature (glycoprotein Ib, IIb/IIIa, Ia complexes) for plasma proteins (VWF, thrombospondin, fibronectin, vitronectin, immunoglobulins and activated components of the complement system: that is, C3b). They are present on platelets, leukocytes, monocytes and lymphocytes T and B.

These plasma proteins might, on an artificial surface, behave as substrate adhesion molecules (SAMs) and exhibit by adsorption on a surface the characteristic amino acid sequence recognized by cytoadhesins; that is, the arginine–glycine–aspartate–serine (RGDs) sequence. This sequence is localized in the NH^2 or COOH terminal domain for VWF, in the α-and γ-chain for fibrinogen, and in fragment IV for fibronectin. Vitronectin and human thrombin also contain RGDs sequences. The precise analysis of the behavior of SAMs on polymeric substrates is thus considered as an essential step in the understanding of cell–substrate interactions.

6. Conclusions

Large numbers of people have been treated with biomaterials and devices introduced within the cardiovascular system that are able to both prolong life and improve its quality. Progress is not so much hampered by the difficulties and constraints related to the functional requirements of the devices as by our lack of understanding of the interaction between the blood and the devices. In spite of a large body of knowledge relating material characteristics to blood compatibility, it remains unclear which materials are supportive of compatibility with blood. Further, specific materials do not display the same interactions in all situations, in particular, low and high blood flow conditions have different mechanisms of interaction.

In physiological conditions the blood interface is the vascular endothelium, a unique monocellular layer which by active and/or passive participation maintains a nonthrombotic state. Endothelial cells are significantly involved in all the major processes associated with hemostasis and thrombosis (vasoregulation, platelet reactivity, coagulation and fibrinolysis). Moreover, they exhibit highly specialized functions, according to their distribution in the vascular system, as required by the hemodynamic conditions and/or locoregional regulatory systems. These cells are able to synthesize in the pulmonary artery PGI_2, a potent vasodilator and platelet antiaggregating agent, whereas at the venous level endothelial cells synthesize the tissue type plasminogen activator (tpA) thus regulating fibrin deposition in venous vessels. In addition, these polarized cells possess on their luminal membrane specialized microdomains and, in particular, receptors for thrombin. Clearance is in fact an essential step since this protease activates protein C in the presence of thrombomodulin, protein C being essential in the proteolytic inactivation of clotting factors, induces synthesis and excretion of tpA, and stimulates PGI_2 production.

Therefore, mimicking the normal endothelium will represent a major advance in blood-contacting materials research. Until this can be achieved, partial improvements have been found:

(a) by developing new materials of reduced surface energy with sequential hydrophobic and hydrophilic microdomains, or with relative mobility of chains bearing anionic sites at the interface;

(b) by improving the physical surface of existing materials or by modifying their chemical surface, by coating or grafting proteins, as well as by plasma surface treatment;

(c) by introducing pharmacological agents at the interface (e.g., heparin, antiaggregants or fibrinolytic molecules) covalent or ionic binding may be used and result in a short or long term efficacy; and

(d) by introducing into the polymeric backbone chains that mimic glycosaminoglycans of endothelial cells or of heparin.

Bibliography

Anderson J M, Kottke-Marchant K 1987 Platelet interactions with biomaterials and artificial devices. In: Williams D F (ed.) 1987, pp. 103–50

Brash J L 1983 Mechanism of adsorption of proteins to solid surfaces and its relationship to blood compatibility. In: Szycher M (ed.) 1983 *Biocompatible Polymers, Metals and Composites.* Technomic, Lancaster, PA, pp. 35–52

Brash J L, Horbett T A 1987 *Proteins at Interfaces: Physicochemical and Biochemical Studies.* American Chemical Society, Washington, DC, pp. 706

Brash J L, Uniyal S 1979 Dependence of albumin–fibrinogen simple and competitive adsorption on surface properties of biomaterials. *J. Polym. Sci.* 66: 377–89

Cazenave J P, Davies J A, Kazatchkine M D, Van Aken W G (eds.) 1986 *Blood–Surface Interactions.* Elsevier, Amsterdam

Forbes C D, Courtney J M 1987 Thrombosis and artificial surfaces. In: Bloom A L, Thomas D (eds.) 1987 *Haemostasis and Thrombosis*, 2nd edn. Churchill Livingstone, Edinburgh, pp. 902–21

Kazatchkine M D, Carreno M P 1988 Activation of the complement system at the interface between blood and artificial surfaces. *Biomaterials* 9: 30–5

Salzman E W, Merrill E W 1987 Interaction of blood with artificial surfaces. In: Colman R W, Hirsh J, Marder V J, Salzman E W (eds.) 1987 *Hemostasis and Thrombosis Basic Principles and Clinical Practice*, 2nd edn. Lippincott, Philadelphia, PA, pp. 1335–47

Sevastianov V I 1988 Role of protein adsorption in blood compatibility of polymers. *CRC Crit. Rev. Biocompatibility* 4(2): 109–54

Van Aken G G, Davies J A 1986 Interaction of leucocytes and red cells with surfaces. In: Cazenave J P, Davies J A, Kazatchkine M D, Van Aken W G (eds.) 1986, pp. 107–24

Van Mourik J A, Van Aken W G 1986 Initiation of blood coagulation and fibrinolysis. In: Cazenave J P, Davies J A, Kazatchkine M D, Van Aken W G (eds.) 1986, pp. 61–74

Vroman L 1987 The importance of surfaces in contact phase reactions. *Semin. Thromb. Hemostasis* 13(1): 79–85

Vroman L, Adams A L, Fischer G C, Munoz P C 1980 Interactions of high molecular weight kininogen factor XII and fibrinogen in plasma at interfaces. *Blood* 55: 156–9

Williams D F (ed.) 1987 *Blood Compatibility*, Vol. 1. CRC Press, Boca Raton, FL

R. Eloy and J. Belleville
[Inserm, Bron, France]

Biomaterials: Surface Structure and Properties

Surface structure directs or strongly influences the biological reaction to materials. However, in order to observe unambiguously the influence of surfaces on biological reactions, materials must be free from leachable substances; where leachables are present, the biological response might be controlled by the pharmacologic or toxicologic potential of those leachables. Biomaterials are expected not to leach substances except when specifically designed to do so (e.g., a drug release system). Other factors that influence the performance of biomaterials include the mechanical properties of the materials and the design of the implant fabricated from them. The focus of this review is on surfaces and their ability to induce biological reactions, and does not specifically address these other factors.

1. The Nature of a Surface

The surface zone of a solid has been described as a unique state of matter (Duke 1984). This is a reasonable assessment considering that the chemistry and structure of the outermost layers of a material are different from that in the bulk of the material, and that the particular combination of chemistry and structure found at the surface can exist nowhere else but at the surface.

The unique properties of surfaces can be attributed to three factors. First, there is a thermodynamic driving force to minimize the high energy situation created in simply producing a surface (Fig. 1). This minimization of surface energy occurs by translating and/or rotating lower energy molecular structures toward the outside (a number of examples of this are presented in this article). Second, the outside atoms are more exposed to the external world and, therefore, more subject to reaction (e.g., with oxygen) or to contamination (e.g., with hydrocarbons). Third, if the material contains diffusible components that are of lower energy than the bulk of the surface, these components (generally additives or contaminants) will migrate to the surface.

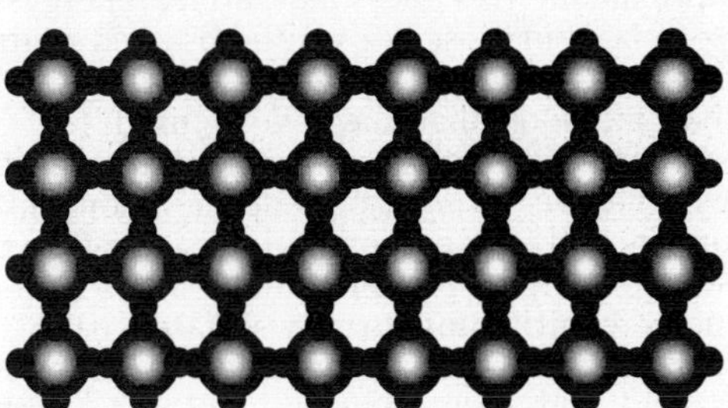

Figure 1
The creation of a surface leads to unfulfilled bonding. Note that in this two-dimensional represention of a crystal, the atoms at the surface can only participate in trivalent or divalent bonding, while those in the bulk are each bound to four other atoms

There are a number of factors that must be identified to adequately describe a surface. These factors fall into two categories: chemical and morphological. Figure 2 outlines some of the considerations involved in a complete description of a surface.

2. Surfaces as the Trigger of Biological Reactions

Surfaces, partly because of their intrinsic energy, can effectuate chemical reactions. Synthetic catalysts (e.g., platinum and zeolites) are being exhaustively studied and are of industrial importance. The surfaces of these materials can lower reaction energy barriers, thereby raising reaction rates. Similarly, examples of biological materials acting as catalysts (i.e., exhibiting high specificity and speeding reactions) are plentiful (enzymes, antibodies, lectins, etc.).

Contemporary synthetic biomaterials do not generally exhibit the chemical specificity associated with synthetic or biological catalysts. However, processes do occur at their surfaces. These processes are often associated with adsorption of components. Upon placement of a material into a biological environment, rapid adsorption of water, ions and proteins to the surface takes place. The nature of the surface will dictate the materials that will adsorb and how they adsorb. Thus, different surfaces can fractionate the complex mixture of proteins that exist in the biological environment in unique ways, leading to differing combinations of proteins at the surface. The surface can also denature or alter the conformation of proteins adsorbed to it (Horbett and Brash 1987). Since cells can respond with high specificity and reactivity (i.e., catalystlike behavior) to certain proteins, the nature of the proteins at the surface will control the biological reaction to the surface. Furthermore, since cellular reactions can be related to the tissue reactions observed by a physician, the biological activity of a surface can be described in terms of a hierarchical series of events: the surface influences the adsorbed protein layer, which influences the cellular reaction which can be related to the ultimate tissue reaction. Thus, the reactions observed by the physician can be attributed to the nature of the surface.

3. Surface Measurement

A wide variety of powerful tools are available to measure surfaces (Ratner et al. 1987, Briggs and Seah 1983). Great demands are placed upon these

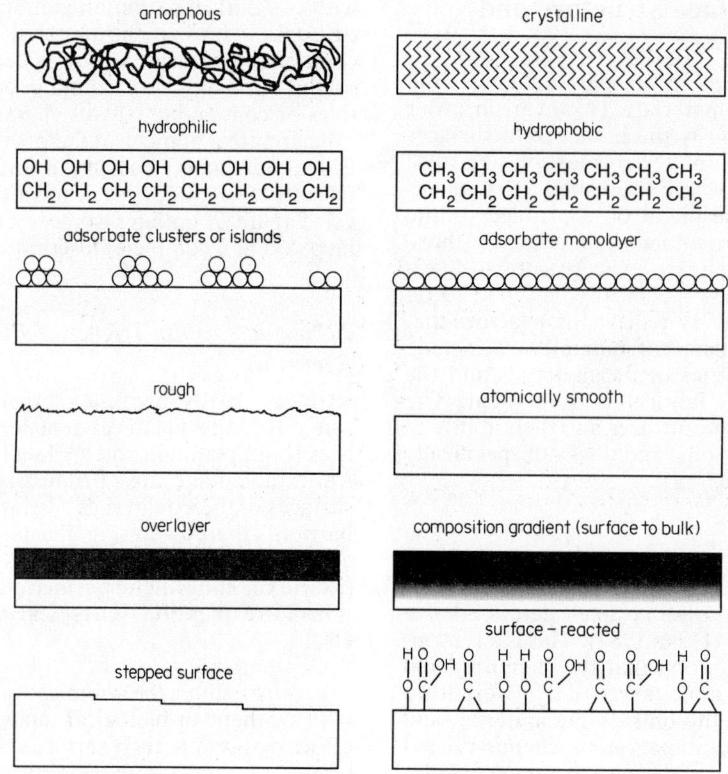

Figure 2
Polymer surfaces: some considerations

techniques because of the small total mass of material associated with surfaces (approximately 10^{14} atoms per cm^2) and because of the requirement that the surface atoms be uniquely resolved while in close proximity to an immensely larger population of atoms in the bulk phase. A few of the more important methods that are applicable to biomaterials are briefly described in this section.

3.1 Contact Angle Methods

Measurement of the angle of contact of a drop of liquid with a solid surface (Fig. 3) is one of the oldest, least expensive, and most informative methods available to probe the surface properties of a solid. The Young equation describes the equilibrium between the surface tension components interacting at the rim of any drop of liquid on a surface limiting its spread (Fig. 4). The Young equation is written as

$$\gamma_{SV} = \gamma_{SL} + \gamma_{LV} \cos \theta \qquad (1)$$

where γ_{SV} is the surface tension between the solid and the vapor, γ_{SL} is the surface tension between the solid and the liquid, γ_{LV} is the surface tension between the liquid and the vapor and θ is the contact angle in degrees. Ideally, one wants to know γ_{SV}, which is directly related to the wettability of the material. The contact angle and the liquid–vapor surface tension are readily measured. This leaves two unknown quantities in this relationship. From this starting point, a number of schemes have been developed to calculate or approximate γ_{SV} from contact angle data. Certainly the most widely used is the scheme developed by Zisman, which allows the computation of a critical surface tension, γ_C; that is, an approximation to the solid surface energy and that correlates in most cases with observed wettability (Zisman 1964).

Other methods that have been used (or considered) for calculating an approximate surface energy value, or that have provided additional information on surface energetics, include underwater contact angle methods, computation of a dispersive and polar contribution to the surface energy, an equation of state approach and the calculation of a Lewis acid and base contribution to the surface energy. Much controversy exists over which method is most appropriate for a given situation, and also the measurements themselves and the meaning of the results are subject to artifact and interpretational difficulties. Therefore, the reader is directed to the many review articles and monographs published on

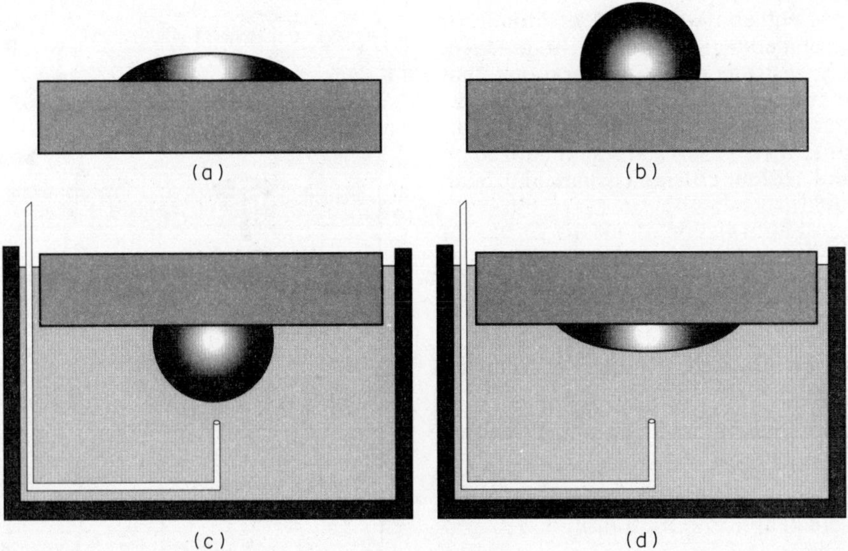

Figure 3
Possibilities observed during contact angle measurement: (a) hydrophilic surface, water drop in air, (b) hydrophobic surface, water drop in air, (c) hydrophilic surface, *n*-octane drop under water, and (d) hydrophobic surface, *n*-octane drop under water

this subject (e.g., Good 1979, Sacher 1988). Still, it is clear that relationships do exist between the biological reaction to surfaces and surface wettability.

3.2 Electron Spectroscopy for Chemical Analysis (ESCA)

ESCA (also called x-ray photoelectron spectroscopy or XPS) has rapidly become a key method for the surface characterization of materials intended for biomedical application. This is because of its information-rich character, its specificity in chemical identification, and its proven value in optimizing biomedical devices and in correlating surface properties to biological responses.

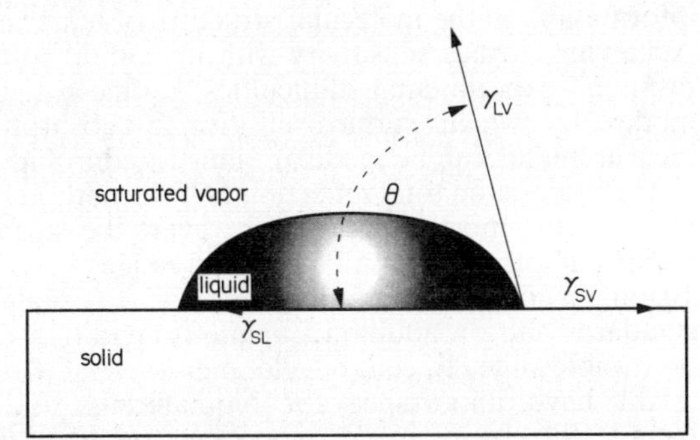

Figure 4
The set-up describing Young's equation by showing the equilibrium between surface tension components

The ESCA method is based upon the photoelectric effect, the nature of which was explained by Einstein in 1905. When a surface is exposed to photons of sufficiently high energy (x rays, in this case), electrons are released. The energy of these electrons is directly related to the atomic or molecular environment from which they originated, while the number of emitted electrons is proportional to the surface concentration of the species from which the electrons originate. The electrons have a poor ability to penetrate through matter without losing energy to inelastic collisions. Therefore, only those electrons in the outermost surface of the material can escape from the surface without energy loss and be detected, hence the surface sensitivity of ESCA. ESCA usually samples a surface to a depth of approximately 100 Å. However, by using specialized x-ray anodes in the ESCA instrument, or by varying the angle of the sample with regard to the analyzer lens in the ESCA spectrometer, the effective depth sampled by ESCA might be expanded over a 10 Å to 250 Å range.

The basic equation used for ESCA qualitative surface analysis is

$$\mathrm{BE} = \mathrm{h}\nu - \mathrm{KE} \tag{2}$$

where BE is the electron binding energy in the atom of origin, $\mathrm{h}\nu$ is the energy of the x-ray source (a known value) and KE is the kinetic energy of the electron that is measured by the electron spectrometer. Thus, in principle, BE can be simply calculated, and from this value one can learn which elements are present and what kind of atomic or molecular environments they are in. In practice, there are many instrumental considerations that complicate the interpretation of these BE values. These, and other considerations important to qualitative and quantitative ESCA, are summarized in many books and review articles (Briggs and Seah 1983, Ratner 1983).

A wide range of information about the surface of a sample can be obtained from ESCA including

(a) which elements are present (except hydrogen and helium),

(b) approximate surface concentrations of elements (±10%),

(c) bonding state (molecular environment) and/or oxidation level of most atoms,

(d) information on aromatic or unsaturated structures from shakeup ($\pi^* \leftarrow \pi$) transitions,

(e) information on surface electrical properties from charging studies,

(f) nondestructive depth profile and surface heterogeneity assessment using
  (i) photoelectrons with differing escape depths, and
  (ii) angular dependent ESCA studies.

(g) destructive depth profile using argon etching (for inorganics),

(h) positive identification of functional groups using derivitization reactions, and

(i) "fingerprinting" materials using valence band spectra.

There are many application areas in biomedical and biomaterials studies for which ESCA is useful, and these are discussed in the literature.

A diagram illustrating the components of an ESCA spectrometer is presented in Fig. 5.

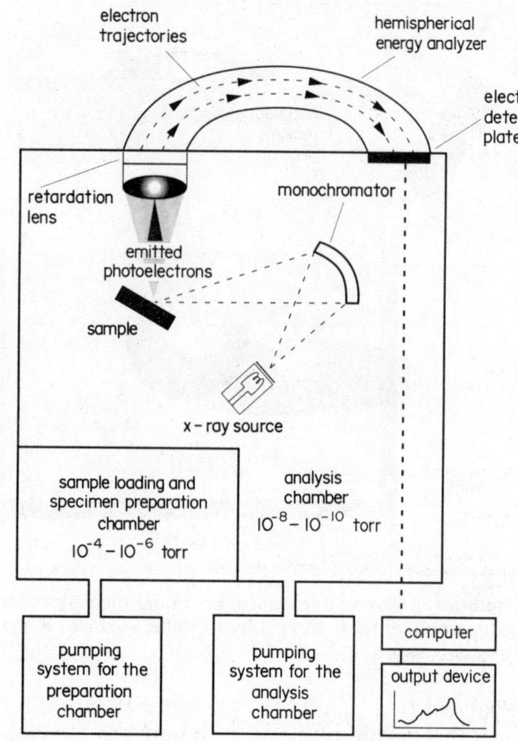

Figure 5
The components of an ESCA spectrometer system

3.3 Secondary Ion Mass Spectrometry (SIMS)

The SIMS method has only recently been applied to biomaterials problems, largely because of the advent of static SIMS, which permits highly surface localized (uppermost 10 Å) analysis of polymeric materials without massive surface damage.

The SIMS method, in essence, allows the acquisition of a mass spectrum of the outermost region of a material. Since mass spectra are rich in structural information, static SIMS has the potential to provide much previously unobtainable information about a surface. The principle of the SIMS method is illustrated in Fig. 6.

The number of publications describing the application of static SIMS to biomaterials problems has, to date, been limited. Static SIMS has shown particular utility for analyzing methacrylate and acrylate polymers, for detecting contaminants and surface localized additives and for analyzing the surface composition of polyurethanes. The potential to perform semiquantitative SIMS analysis has also been demonstrated. General review articles on static SIMS, particularly as applied to polymeric systems, have been published (e.g., Castner and Ratner 1988).

3.4 Surface Infrared Techniques

Infrared (ir) absorption spectroscopy provides much information on the molecular structure of materials. Achieving surface sensitivity with ir methods often presents experimental difficulties. Some of the modes by which surface-sensitive ir absorption measurements can be made are illustrated in Fig. 7.

The attenuated total reflection (ATR) method has received the most attention. However, the surface region sampled by the ATR method ranges from 1–10 μm, a deep penetration by surface analysis standards and, in addition, sample preparation can be troublesome. Recent developments in ATR that might have importance for biomaterials studies include the use of metal depositions on the ATR element to enhance sensitivity, the application of ATR to study surface crystallinity and the use of ATR to study the solid–aqueous interface.

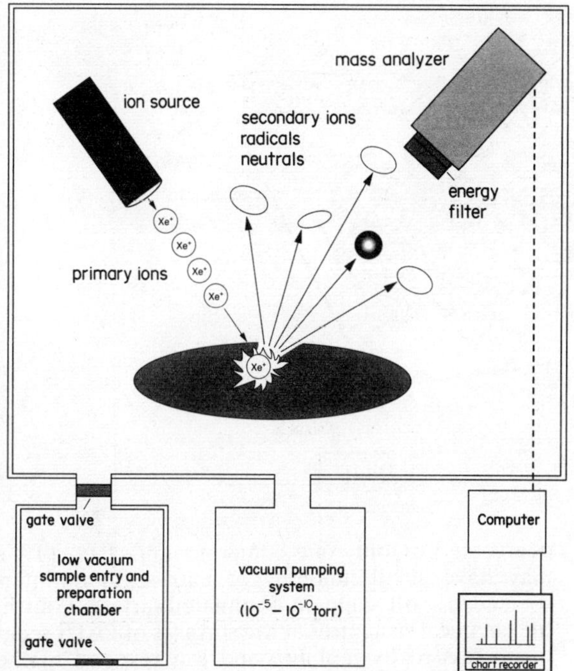

Figure 6
The SIMS technique measures the molecular weight of atoms sputtered from a surface by an energetic, focused ion beam. The experiment is conducted in a high vacuum chamber

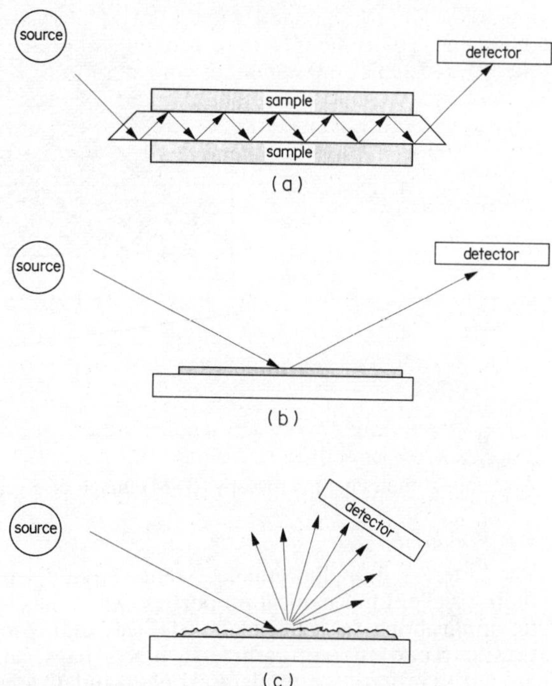

Figure 7
Modes for measuring surface-localized infrared absorption

The external reflection ir method (also called infrared reflection absorption spectroscopy) provides extreme surface sensitivity and also offers information on the orientation of molecules at the surface. However, this method requires a highly reflective metal substrate for the absorbing molecules and often suffers from a poor signal-to-noise ratio.

The diffuse reflectance infrared technique requires little or no sample preparation prior to analysis. However, with this method, the sampling depth is poorly defined.

3.5 New Developments in Surface Characterization

There are a number of new developments that might contribute strongly to the surface characterization of biomaterials. Foremost among these are scanning tunnelling microscopy (STM) and atomic force microscopy (AFM). These methods have the potential to image surfaces and adsorbates on surfaces at the atomic level of resolution (Binnig and Rohrer 1985, Binnig et al. 1987). An STM image of a graphite surface is shown in Fig. 8. Atomic level imaging using STM has also been demonstrated under water.

Surface-enhanced Raman spectroscopy (SERS) may permit high sensitivity Raman measurements of phenomena occurring at interfaces (Allara et al. 1981). However, a rough silver surface must be in close proximity to the interface being studied, a difficult requirement in many analysis situations relevant to biomaterials problems.

High-resolution electron energy loss spectroscopy (HREELS) has recently been applied to polymer surfaces (Pireaux et al. 1986). This method is highly surface sensitive and information-rich. However, even with new high-sensitivity detectors, there is concern over sample damage from the impinging electron source.

New developments in microscopy have also been applied to surface studies. In particular, the potential of transmission electron microscopy and low-voltage scanning electron microscopy for observing biomaterial surfaces is being explored (Goodman et al. 1988).

4. The Nature of Biomaterial Surfaces

A number of specific classes of biomaterials have been subjected to exhaustive surface characterization. This section will review some of the observations made in the study of these systems.

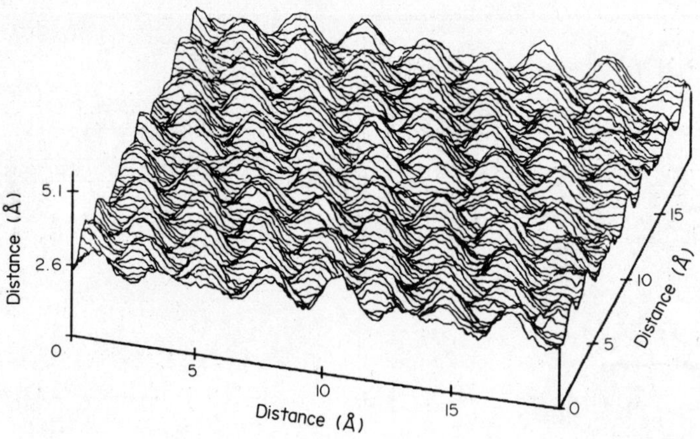

Figure 8
A scanning tunnelling microscopy (STM) image of a graphite surface

4.1 Polyurethanes

The interest in polyurethanes stems largely from their excellent mechanical properties, which has led to applications in artificial hearts, vascular prostheses, blood tubing, catheters, blood bags, and numerous soft tissue implants (Lelah and Cooper 1986). The strong interest in the surfaces of this class of polymers has its origins in two observations. First, the chemistry of the surfaces of polyurethanes has been shown to directly correlate with blood–polymer reactivity (Hanson et al. 1982). Second, the surface composition and structure of polyurethanes are generally different from those observed in the bulk of the material. A number of papers have been published that define the basic principles of polyurethane surface analysis (e.g., Ratner and McElroy 1986).

There are five important surface observations made on polyurethanes. These observations also define the primary areas of study of polyurethane surfaces. First, polyurethane surfaces are often compositionally enriched with low molecular weight, surface active additives that migrate from the bulk of the polymers. Thus, low molecular weight extrusion lubricants (*bis*-stearamides) and silicones are often found dominating or overcoating polyurethane surfaces. Second, in the absence of these additives, the surface of block copolyurethanes (i.e., the poly(ether urethanes) (PEUs) favored for medical applications) are dominated by the soft polyether segment. This is illustrated in the ESCA depth profile diagram in Fig. 9. Third, the interest in the relationship between the domains (100 Å to 200 Å phase separated regions) found in the bulk of PEUs and such structures at the surfaces of PEUs has been a fertile area for study. Some groups argue that hard-segment and soft-segment domains exist at PEU surfaces and can influence biological interactions, while others claim that the surface is totally overcoated by only one component. In fact, PEUs may have small amounts of hard-segment interspersed in a soft-segment-dominated surface. Fourth, the chemical structures at the surfaces of PEUs seem to possess ready mobility and can respond to the external environment by moving to or from the surface. The facile surface mobility of PEUs may account for many of the contradictory observations on phase separation at polyurethane surfaces. Finally, there is a great interest in surface modification of polyurethanes to improve blood compatibility. Such modifications have been made by the

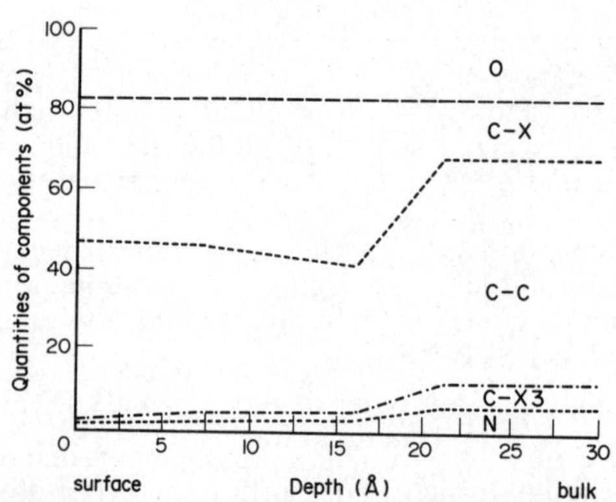

Figure 9
A depth profile diagram based upon angular dependent ESCA data taken from a polyurethane surface. C–C represents hydrocarbon-like functionalities resolved from the carbon 1s spectrum. C–X represents carbon species singly bound to oxygen. C–X3 represents carbon species bound with two or more bonds to oxygen or nitrogen. The high concentration of C–X near the outermost surface suggests polyether surface localization (depth profile constructed by R. Paynter)

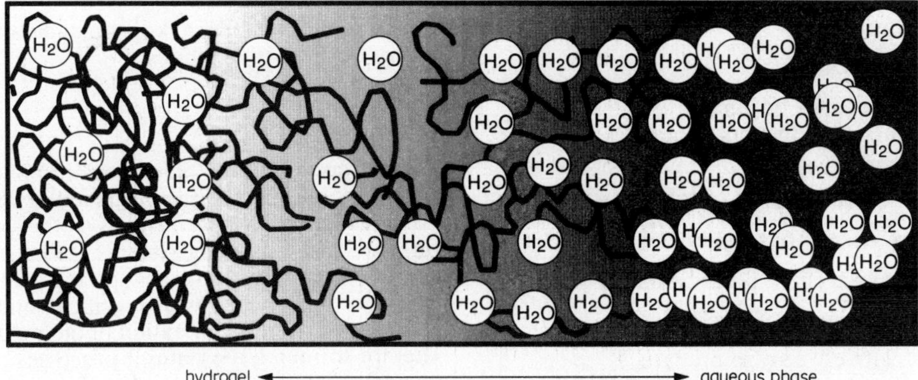

Figure 10
Hydrogels present a diffuse, gradient boundary between the solid phase and the liquid phase

addition of specially engineered oligomeric components and by covalently incorporating specific side chain or block components expected to be surface active.

4.2 Hydrogel Surfaces

Hydrogel surfaces present a diffuse gradient boundary between the solid phase and the liquid phase (Fig. 10). These surfaces are primarily characterized by (a) their high polymer chain flexibility (rotational and translational), (b) gradients of diffusible, lower molecular weight components from the bulk, through the surface zone, and to the fluid phase and (c) their relatively low adhesiveness to proteins and cells (perhaps due to polymer chain flexibility). Review articles on hydrogel surfaces have been published (Ratner 1986, Silverberg 1976), and their surface mobility has been studied. The extremely low interaction of proteins and cells with poly(ethylene oxide) hydrogel surfaces has been the focus of much study in recent years (Gombotz et al. 1988).

4.3 Plasma-Deposited Surfaces

The plasma-deposition method has many advantages for synthesizing surfaces useful for biomedical materials and medical devices (Yasuda and Gazicki 1982, Ratner et al. 1989). In this method, a gas (the monomer) is introduced into a rapidly oscillating electrical field, leading to fragmentation and ionization. The highly reactive species formed in this ionized gas phase environment can, under appropriate conditions, deposit on surfaces and recombine (or vice versa), leading to a surface film with unique characteristics. This is illustrated in the ESCA spectra in Fig. 11, which show the massive molecular rearrangement that occurs when tetrafluoroethylene gas is deposited as a film from a plasma environment. Surface characterization methods are important for understanding these films since (a) their chemistry is different from that of the starting gas and (b) they are so thin (often in the range 50–1000 Å) that the high sensitivity of surface methods is needed. The unique biological interactions observed with the surfaces of these films are of special relevance to this article. Thus, various plasma-deposited layers have been found to exhibit low thrombogenicity, a high affinity for albumin, a low cell adhesiveness and an ability to serve as effective substrates for cell growth. The nature of these unique surfaces has been under study using such methods as ESCA, chemical derivitization and SIMS. Some of the specific advantages of plasma-deposited thin films for biomedical applications are

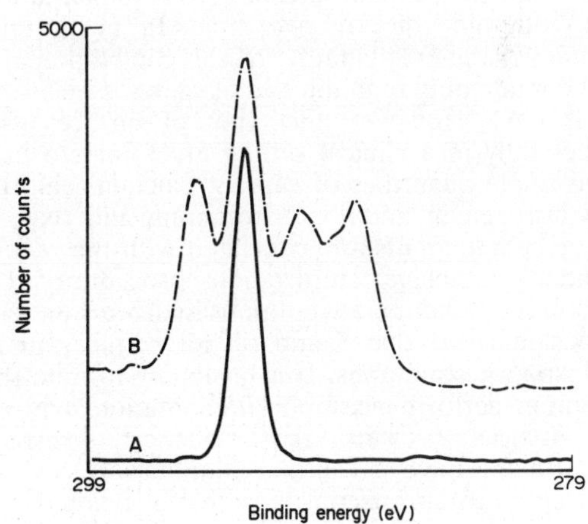

Figure 11
The plasma environment alters the nature of a monomer introduced into it, and yields a polymeric film with a structure different from that obtained by conventional polymerization methods. Lower curve: the carbon 1s ESCA spectrum of polytetrafluoroethylene (Teflon, PTFE), upper curve: the carbon 1s spectrum of an RF plasma-deposited film prepared using tetrafluoroethylene gas

(a) they are conformal,

(b) they are pinhole free,

(c) they can be coated on unique substrates (e.g., metals, glasses, polymers, ceramics and carbons),

(d) they show good adhesion to substrates,

(e) unique film chemistries can be achieved,

(f) they serve as excellent permeation barriers,

(g) they show low levels of leachables,

(h) they are relatively easy to prepare,

(i) a well-developed technology for performing these depositions already exists,

(j) they can be characterized, and

(k) plasma-deposited layers are sterile upon preparation.

4.4 Titanium Surfaces

Titanium surfaces, when prepared using specifically defined procedures, have been reported to bond directly to bone and tissue (Kasemo and Lausmaa 1988). Titanium is the only metal that does not become encapsulated by collagen and that exhibits the interfacially intimate bone-bonding property. Thus, by studying titanium surfaces, important insights might be gained into the nature of biomaterial interactions.

Titanium surfaces are dominated by a thin layer that is mostly titanium dioxide. After implantation, this oxide film thickens over time. In an ultrahigh vacuum chamber, titanium surfaces can be prepared that contain only titanium and oxygen (as measured by ESCA). However, the types of surfaces used successfully in a clinical setting are found to have measurable quantities of carbon, calcium, chlorine and nitrogen in addition to titanium and oxygen. These contaminants are associated with the recommended autoclave sterilization procedure. The carbon component is adventitious hydrocarbon, possibly similar to that found on all surfaces under atmospheric conditions. It is intriguing to note that optimum performance upon implantation may not be associated with the "cleanest" surface. Experiments have shown that titanium surfaces may pick up more fibronectin, a protein associated with cell adhesion, when hydrocarbon contamination levels are high (Lew 1986).

Because of the unique (and clinically efficacious) performance of titanium surfaces, these materials will remain an area for serious study for some time. Since titanium shows a clear correlation between surface preparation and clinical performance, it represents a prime example of the importance of surface properties for biomedical material performance and, possibly, a key to understanding the nature of these interactions.

5. New Developments

The surfaces used for biomaterials applications are generally amorphous or polycrystalline and exhibit little regularity or specificity. On the other hand, surfaces found in living systems (cell surfaces, protein surfaces and antibody surfaces) have a high degree of order, leading to specificity and biological recognition. A new generation of biomaterials might attempt to mimic the natural processes and exploit this specificity so that interfacial reactions can be engineered to give the most appropriate results.

One method used to accomplish this involves immobilizing biomolecules that produce desired reactions to surfaces. Although the biospecificity of surfaces can be vastly enhanced by these methods, the long-term stability of the biomolecules at surfaces and the immunologic reactions to those molecules can impose limitations on their use. Another approach to engineering surfaces to give desired responses uses chemical surface modification methods. Surface modification has achieved significant success, but, in most cases, the surfaces still exhibit the nonspecific response associated with synthetic materials.

Many synthetic molecular systems have been shown to self-assemble or organize on surfaces and this may present new possibilities for developing biologically recognizable surfaces (Ringsdorf et al. 1988). Thus, molecular-level organized structures can be fabricated that have many of the desirable properties of nonbiological materials with the geometric organization of natural biological surfaces. Future biomaterials might be fabricated or surface-modified using Langmuir–Blodgett structures or surface-localized, self-assembled nanostructures (Lehn 1988). In addition, vesicles and microtubules might also assume increasing importance in biomaterial fabrication.

6. Conclusions

This article is centered upon the premise that the surface properties of materials govern biomedical performance. This thesis is still strongly held and ample evidence exists that surface properties can directly affect biological system–biomaterial interactions. However, a clear, predictive relationship that can be used in an engineering sense to design new materials has not been established. New developments in well-defined surfaces and in methods to characterize those surfaces (Ratner 1988) may lead to this understanding, which in turn should lead to rationally designed clinical prostheses that invoke specific, desired biological responses.

Acknowledgement

Support was received from NIH grants HL25951 and RR01296 during the preparation of this article and for some of the studies described herein.

Bibliography

Allara D L, Murray C A, Bodoff S 1981 Spectroscopy of polymer surfaces using the surface enhanced Raman effect. In: Mittal K L (ed.) *Physicochemical Aspects of Polymer Surfaces*. Plenum, New York

Binnig G, Gerber C, Stoll E, Albrecht T R, Quate C F 1987 Atomic resolution with atomic force microscope. *Euorphys. Lett.* 3: 1281–6

Binnig G, Rohrer H 1985 The scanning tunneling microscope. *Sci. Am.* 253: 50–6

Briggs D, Seah M P 1983 *Practical Surface Analysis*. Wiley, Chichester, UK

Castner D G, Ratner B D 1988 Static secondary ion mass spectroscopy: A new technique for the characterization of biomedical polymer surfaces. In: Ratner B D (ed.) *Surface Characterization of Biomaterials*. Elsevier, Amsterdam

Duke C B 1984 Atoms and electrons at surfaces: A modern scientific revolution. *J. Vac. Sci. Technol.* 2: 139–43

Gombotz W R, Guanghui W, Hoffman A S 1988 Immobilization of poly(ethylene oxide) on poly(ethylene terephthalate) using a plasma polymerization process. *J. Appl. Polym. Sci.* 35: 1–17

Good R J 1979 Contact angles and the surface free energy of solids. In: Good R J, Stromberg R R (eds.) 1979 *Surface and Colloid Science*. Plenum, New York

Goodman S L, Pawley J B, Cooper S L, Albrecht R M 1988 Surface and bulk morphology of polyurethanes by electron microscopies. In: Ratner B D (ed.) 1988 *Surface Characterization of Biomaterials*. Elsevier, Amsterdam

Hanson S R, Harker L A, Ratner B D, Hoffman A S 1982 Evaluation of artificial surfaces using baboon arteriovenous shunt model. In: Winter G D, Gibbons D F, Plenk H Jr (eds.) 1982 *Biomaterials 1980; Advances in Biomaterials*. Wiley, Chichester, UK

Horbett T A, Brash J L 1987 Proteins at interfaces: Current issues and future prospects. In: Horbett T A, Brash J L (eds.) *Proteins at Interfaces: Physicochemical and Biochemical Studies,* ACS Symposium Series. American Chemical Society, Washington, DC

Kasemo B, Lausmaa J 1988 Biomaterial and implant surfaces: On the role of cleanliness, contamination, and preparation procedures. *J. Biomed. Mater. Res: Appl. Biomat.* 22: 145–58

Lehn J M 1988 Supramolecular chemistry—scope and perspectives: Molecules, supermolecules, and molecular devices (Nobel lecture). *Angew. Chem. Int. Ed. Engl.* 27: 89–112

Lelah M D, Cooper S L 1986 *Polyurethanes in Medicine*. CRC Press, Boca Raton, FL

Lew P J 1986 Titanium Surfaces: ESCA analysis, protein adsorption, and cell interaction. M.S. Thesis, University of Washington, Seattle, WA

Pireaux J J, Thiry P A, Caudano R 1986 Surface analysis of polyethylene and hexatriacontane by high resolution electron energy loss spectroscopy. *J. Chem. Phys.* 84: 6452–7

Ratner B D 1983 Analysis of surface contaminants on intraocular lenses. *Arch Ophthal.* 101: 1434–8

Ratner B D 1986 Hydrogel surfaces. In: Peppas N A (ed.) *Hydrogels in Medicine and Pharmacy*. CRC Press, Boca Raton, FL

Ratner B D 1988 The surface characterization of biomedical materials: How finely can we resolve surface structure? In: Ratner B D (ed.) *Surface Characterization of Biomaterials*. Elsevier, Amsterdam

Ratner B D, Chilkoti A, Lopez G P 1990 Plasma deposition and treatment for biomaterial applications. In: D'Agostino R (ed.) *Plasma Deposition of Polymer Films*. Academic Press, New York

Ratner B D, Johnston A B, Lenk T J 1987 Biomaterial surfaces. *J. Biomed. Mater. Res: Appl. Biomat.* 21: 59–90

Ratner B D, McElroy B J 1986 Electron spectroscopy for chemical analysis: Applications in the biomedical sciences. In: Gendreau R M (ed.) *Spectroscopy in the Biomedical Sciences*. CRC Press, Boca Raton, FL

Ringsdorf H, Schlarb B, Venzmer J 1988 Molecular architecture and function of polymeric oriented systems: Models for the study of organization, surface recognition, and dynamics of biomembranes. *Angew. Chem. Int. Ed. Engl.* 27: 113–58

Sacher E 1988 The determination of the surface tensions of solid films. In: Ratner B D (ed.) *Surface Characterization of Biomaterials*. Elsevier, Amsterdam

Silverberg A 1976 The hydrogel–water interface. In: Andrade J D (ed.) *Hydrogels for Medical and Related Applications*. American Chemical Society, Washington, DC

Yasuda H, Gazicki M 1982 Biomedical applications of plasma polymerization and plasma treatment of polymer surfaces. *Biomaterials* 3: 68–77

Zisman W A 1964 Relation of the equilibrium contact angle to liquid and solid constitution. In: Fowkes F M (ed.) *Contact Angle, Wettability and Adhesion*, ACS Advances in Chemistry Series. American Chemical Society, Washington, DC

B. D. Ratner
[University of Washington, Seattle, Washington, USA]

C

Carbon-Fiber-Reinforced Plastics

To utilize the high stiffness and strength that carbon fibers possess it is necessary to combine them with a matrix material that bonds well to the fiber surface and which transfers stress efficiently between the fibers. Fiber alignment, fiber content and the strength of the fiber–matrix interface all influence the performance of the composite. Furthermore, highly specialized processing techniques are necessary which take account of the handling characteristics of carbon fibers, which are different from those of other reinforcements. It is the processibility of polymers that makes them particularly attractive for use in a whole range of composite materials. For example, one form of carbon-fiber-reinforced plastic (CFRP) contains continuous carbon fibers, possibly in the form of a unidirectional or woven mat, which are impregnated by a thermosetting resin such as an epoxy. Prepreg material of this type can then be layed up to ensure the appropriate orientation of the fiber before curing of the resin is undertaken. Alternatively, if fiber orientation does not have to be so rigorously defined, then the processability of thermoplastic polymers means that, with the reinforcement in the form of short fiber (typically 0.2–10 μm), a molding compound of fiber and thermoplastic can be injection molded to produce directly finished components.

From the perspective of materials performance, the potential of CFRP is enormous: a civil aircraft constructed, where possible, entirely of CFRP instead of aluminum alloy would have a total weight reduction of 40%; a communications satellite built from CFRP can carry more telecommunications channels than permitted by previously specified materials; and a new generation of tactical aircraft can be planned, which will have a versatility regarded as impossible prior to the advent of CFRP, e.g., forward swept wings for low stalling speeds, and very small turning-circle radii. However, the high cost of the fiber and the lack of cost-effective manufacturing processes are still inhibiting the more general use of CFRP, as is the newness of the art of composites design.

1. Matrix Materials

1.1 Thermosets

Examples of thermoset resins commonly used for carbon-fiber composites are listed in Table 1, together with their advantages and disadvantages. Long shelf life is an important advantage for prepregs, while low viscosity is vital to processing techniques such as pultrusion, filament winding and rapid resin injection molding (RRIM). Cure of the resin should not take long times or require extreme tempertures, while a high value of the glass transition temperature (T_g) (the temperature of the glassy–rubbery transition) is necessary if the

Table 1
Examples of thermosetting resins used for CFRP

Resin	Advantages	Disadvantages
Epoxides		
DGEBA[a] (aliphatic amine curing agent)	very low viscosity; room-temperature cure	very low T_g; short-shelf life; flammability
DGEBA[a] (aromatic diamine curing agent)	medium viscosity; long shelf life	medium T_g; difficult cure
TGMDA[b] (aromatic diamine curing agent)	high T_g; long shelf life; toughness	high viscosity; difficult cure; moisture uptake; flammability
Novolac (phenolic)	nonflammability; high T_g	difficult cure; shrinkage; brittleness
PMR 15 (polyimide)	very high T_g; low moisture uptake, high strength	difficult cure; thermal fatigue

a diglycidylether of bisphenol A b tetraglycidylmethylenedianiline composite is to be used at elevated temperatures. Carbon-fiber–epoxy composites represent about 90% of CFRP production. As can be seen in Table 1 the properties of an epoxy are dependent on the curing agent used and the functionality of the epoxy. The attractions of epoxy resins are that (a) they polymerize without the generation of condensation products which can cause porosity, (b) they exhibit little volumetric shrinkage during cure, which reduces internal stresses, and (c) they are resistant to most chemical environments.

The major disadvantages of epoxy resins are their flammability and their vulnerability to degradation in aggressive media. Phenolic resins are, however, quite stable in both respects, although as structural composite materials, phenolic CFRPs are quite uninteresting, as their fatigue and impact properties are very poor. As indicated earlier, the prospect of using CFRPs as a substitute for light metal alloys is particularly attractive but there are problems to be overcome at high temperatures. Using polyimides, peak service temperatures of 400 °C can be reached. However, conventional polyimides present significant processing problems. Toxic solvents have to be evaporated and, during curing, condensation products have to be vented from the tooling. Cure cycles are also prohibitively long. PMR–15 is a polyimide developed to surmount the above problems. Three monomer reactants are mixed in an alcohol, the solution being used to impregnate the fiber. The alcohol is then evaporated, during which time a thermoplastic prepolymer is formed. This prepreg can then be cured using tooling identical to that used with epoxies, although much higher temperatures (~380 °C) are required.

Another compromise using polyimide chemistry to enhance CFRP performance is to form copolymers with epoxy resins. In this way flame retardancy and reduced moisture absorption are realized, but there is no substantial increase in high-temperature performance. Polybismaleimides represent a very good compromise in many respects and should command the applications arena between the epoxies and the polyimides.

1.2 Thermoplastics

A considerable number of thermoplastics are currently being used in CFRP. They may be either semicrystalline or amorphous; semicrystalline thermoplastics offer greater solvent resistance, but are generally more difficult to process.

In load-bearing applications, carbon fibers are most widely used in combination with nylon 66 which gives the highest strength and modulus over a fairly wide temperature range. A thermoplastic that is currently of considerable interest to fabricators looking to mass-produce articles is poly-1,4 butanediol terephthalate (PBT). It has a melting point of 224 °C and, if a low molecular weight grade is used, it can be injection molded at 250 °C. It crystallizes very rapidly so very short cycle times can be employed. CFRP based upon PBT exhibits a heat distortion temperature of 200 °C with adequate resistance to creep up to 120 °C.

Table 2
High-temperature engineering thermoplastics

Generic name (trade name; manufacturer)	Heat distortion temperature at 1.82 MPa (°C)
Polycarbonate (Lexan, Merlon; GE, Mobay)	133
Polyphenylene sulfide (Ryton; Phillips)	137
Polyether ether ketone (Victrex PEEK; ICI)	148
Polysulfone (Udel; Union Carbide)	174
Polyetherimide (Ultem; GE)	204
Polyethersulfone (Victrex PES; ICI)	204

The prospects of using high-temperature engineering thermoplastics, examples of which are shown in Table 2, with continuous-fiber reinforcements instead of thermosetting resins look particularly exciting. The advantages of thermoplastics are lower production costs, improved toughness and impact resistance, and lower susceptibility to moisture uptake, which is found to reduce the value of T_g for epoxy matrices. Additional benefits are the ease of repair of the composites and the recovery of waste water or used material.

In Table 3 the properties of injection-molded nylon 66, polyethersulphone (PES) and polyether ether ketone (PEEK), each with 40 wt% of carbon fiber are compared. The nylon 66 compound exhibits superior properties at normal ambient temperature. It is with properties such as creep modulus (stress to induce 0.1% strain within 100 s) and tensile strength at elevated temperatures that the high-temperature thermoplastics reveal their merit.

2. Fiber–Matrix Interphase

The properties of the material between the fiber reinforcement and the plastic matrix have a strong influence on the mechanical properties of the CFRP. To produce a high-modulus composite requires maximizing adhesion between the fiber and matrix. However, if the polymer matrix is brittle (e.g., epoxies) there is a corresponding reduction in impact strength as mechanisms of energy dissipation such as debonding and fiber pullout become suppressed. If a tough polymer matrix is to be used (e.g., PEEK), where under impact the principal mechanism of energy dissipation is yielding within the polymer, good adhesion between the fiber and matrix is required.

Table 3
Properties of three thermoplastics reinforced with carbon fiber (CF)[a]

Property	Material		
	40% CF–nylon 66	40% CF–PES	40% CF–PEEK
Density (gm cm^{-3})	1.34	1.52	1.45
Tensile strength (MPa)			
at −55 °C			275
24 °C	246	176	227
50 °C	168	168	208
100 °C	108	140	165
140 °C	75	112	129
180 °C			72
Tensile elongation (%)	1.65	1.1	1.35
Tensile creep modulus (GPa)			
at 24 °C	26.2	22.1	26.5
120 °C	11.5	21.4	23.9
160 °C	9.0	21.1	10.5
200 °C		17.5	6.6
Flexural strength (MPa)	413	244	338
Flexural modulus (GPa)	23.4	16.8	20.8
Compressive strength (MPa)	240	216	
Compressive modulus (GPa)	20.2	17.2	

a "Grafil" product, reference RG40

Unfortunately this region of the composite, sometimes known as the interphase, is extremely complex. Constituents of the interphase might include a modified fiber surface, the treatment being undertaken to promote better adhesion, or a coating of polymer present to ensure compatibility with the matrix polymer. In addition, the presence of the fiber surface may affect the cure of thermosets, resulting in chemically different polymer, or may affect the morphology of semicrystalline thermoplastics.

2.1 *Surface Treatment*

Surface treatment processes are now established as an integral part of fiber manufacture. Most are either wet oxidative treatments using solutions of, for example, sodium hypochlorite, sodium bicarbonate or chromic acid, or dry treatments in which the oxidizing agent is ozone. The precise action of the treatment is undoubtedly complex, but is currently thought to include the introduction of chemically active sites onto which the polymer can graft, as well as chemical cleaning, an increase in rugosity of the fiber surface and the removal of any weak surface layer on the fiber.

A plethora of techniques has been proposed to measure fiber–polymer adhesion. Such techniques are experimentally tedious but, more significantly, the assumptions necessary to deduce adhesive strengths from the various types of measurement make them only comparative. The techniques include measurement of the composite short-beam interlaminar shear strength, the distribution of fiber lengths in a fully fragmented single fiber, the stress required to pull out a fiber embedded in the polymer and the stress required for a bead of polymer to become detached from a single fiber.

2.2 *Surface Coating and Sizing*

Polymer coatings and sizes are frequently applied to fibers immediately after manufacture. The light application of resin binds individual filaments together and maintains a compact bundle for processing into an intermediate product or final composite form. Size compositions applied to the surface can afford protection against filament damage and provide lubrication for a smooth delivery of fibers in continuous processing. The sizing or coating may be bulk polymer or resin specially functionalized to improve fiber wetting and adhesion while still retaining compatibility with the bulk polymer.

3. *Properties of CFRP*

The main points to appreciate when considering a carbon-fiber composite as compared to a conventional material such as a metal alloy are that the composite is heterogeneous and anisotropic. By heterogeneous it is meant that the properties vary from point to point within the material and by anisotropic it is meant that the properties depend on the direction in which they are measured. These two factors are important when considering not only mechanical properties (such as stiffness and strength) but also the physical, electrical and thermal characteristics of the composite.

The physical and mechanical properties of a CFRP are strongly dependent on the properties of

Table 4
Types of commercially available carbon fibers

Carbon fiber type	Tensile strength (GPa)	Elastic modulus (GPa)	Breaking strain (%)
High strength (PAN)	3.5	240	1.5
High strain (PAN)	5.5	240	2.2
Intermediate modulus (PAN)	4.5	285	1.6
High modulus (PAN)	2.5	350	0.7
Ultrahigh modulus (PAN)	2.0	450	0.5
Isotropic (pitch)	1.0	35	2.5
Ultrahigh modulus (pitch)	2.0	550	0.4

the particular fiber used. There are a variety of fiber types available and the properties of fibers are being advanced continually by the development of new grades (see Table 4).

The section below outlines some of the main features of the mechanical, thermal and electrical properties of CFRP.

3.1 Mechanical Properties

CFRP containing continuous aligned fibers (which would typically have a volume fraction of 60%) are essentially elastic to failure (when loaded in tension parallel to the fibers) and exhibit no yield or plasticity region. There is, however, a slight nonlinearity in the stress–strain curve, with the stiffness increasing slightly with applied strain, as a result of improved orientation of the graphitic planes within the fibers. Although the failure strains in tension for CFRP may seem small when compared with those of other structural materials, the predictably elastic behavior under load allows a high proportion of the ultimate strength to be utilized in practice. Consequently, strain levels at useful working stresses are comparable with those of metals, alloys and glass-fiber composites. The properties of unidirectional CFRP are compared with similar performance characteristics of other commonly used structural materials in Table 5. A range of material properties typical of those required for design calculations is given in Table 6. These data confirm the anisotropy of the unidirectional material (which is much more marked in CFRP than in glass or aramid composite); the stiffness and strength of the materials are very much lower when they are loaded perpendicular to the fibers than when they are loaded parallel to the fibers.

It should also be noted that the unidirectional compression strength of the lamina is only about two-thirds of the tensile strength. In early carbon-fiber composite materials (from the early 1970s) the tensile strength and compression strength were approximately equal, with the failure mode in compression being shear failure. Improved fiber tensile properties have been achieved with a reduction in fiber diameter and, during this time, more ductile resins have come into favor. These factors have tended to promote fiber instability or microbuckling failure when unidirectional materials are loaded in compression. Hence the compression strength of the material is not a true measure of the fiber strength, but more of the ability of the matrix to support the fiber against buckling, and the integrity of the fiber–matrix interface. As a result of these developments, compressive performance of CFRP can be a limiting design condition.

There are very few practical applications for CFRP which require only unidirectional reinforcement. Laminates are assembled from unidirectional layers stacked at various orientations to one another. Layers are included at 90° to give adequate transverse properties and at 45° to give good shear properties. A laminate construction of 0°/90°/±45° plies has negligible anisotropy when loaded in-plane and is known as quasi-isotropic. Typical data for such a laminate are shown in Table 7. The mechanics of laminated composites are quite complicated

Table 5
Some properties of fiber-reinforced composites (55–70 vol.% fiber) and metals

Material	Density ($g\,cm^{-3}$)	Ultimate tensile strength (GPa)	Young's modulus (GPa)	Specific tensile strength (GPa)	Specific Young's modulus (GPa)
CF–Epoxy					
high strength	1.5	1.9	130	1.27	87
high modulus	1.6	1.2	210	0.94	119
E-glass–epoxy	2.0	1.0	42	0.50	21
Aramid–epoxy	1.4	1.8	77	1.30	56
Steel	7.8	1.0	210	0.13	27
Titanium	4.5	1.0	110	0.21	25
Aluminum L65	2.8	0.5	75	0.17	26

Table 6
Typical performance data for unidirectional laminates

Property	High strength	High modulus
0° Tensile strength (GPa)	1.90	1.15
0° Tensile modulus (GPa)	137	200
Elongation (%)	1.4	0.6
Interlaminar shear strength (MPa)	90	70
0° Flexural strength (GPa)	1.80	1.10
0° Flexural modulus (GPa)	130	190
0° Compression strength (GPa)	1.25	0.70
0° Compression modulus (GPa)	140	210
Density ($g\,cm^{-3}$)	1.55	1.63
90° Tensile strength (MPa)	50	35
90° Tensile modulus (GPa)	10	7

Table 7
Performance data for multidirectional laminates with 60 vol.% fiber and 0°/90°/±45° laminate construction

Laminate	Tensile strength (GPa)	Tensile modulus (GPa)
High strength	0.6	50
High modulus	0.35	75

and the interested reader is referred to the article *Elastic Properties of Laminates*.

As a result of the marked anisotropy, CFRP can experience large stress concentration factors at cutouts or mechanically fastened joints. Given that the material is predominantly elastic to failure, this might be expected to present serious practical problems. Fortunately, certain damage mechanisms (such as matrix cracking parallel to the fibers and delamination cracking between layers) can occur in laminates so that the stress concentrations do not have the deleterious effects that might be expected. Even so, the designer must take care when dealing with cutouts or joints. Bonded joints must also be designed carefully to avoid failure of the laminate by interlaminar (through-thickness) failure.

The range of basic mechanical properties of CFRP is enhanced by excellent resistance to creep rupture and fatigue. Creep rupture times at equivalent stress levels are several orders of magnitude greater for CFRPs compared with those for glass-fiber-reinforced composites (GFRP), and also significantly better than for aluminum. CFRP is also capable of operating well under conditions of alternating stress. For high strength fibers, S–N fatigue curves are much flatter for CFRP compared with those for GFRP or aluminum, and a higher ratio of working stress to ultimate stress can be utilized. However, the newer fiber-matrix systems are not yet characterized fully and there are signs that, with these materials, fatigue may be more of a consideration than in the past. In passing it should be noted that fatigue failure in composite materials is not caused by the initiation and growth of a dominant crack as it is in metallic materials; instead there is an accumulation of damage (matrix cracks, delamination, fiber breaks) which may result in a lowering of the stiffness and the residual strength. Much more work is needed to develop damage-tolerant design for CFRP composites.

The impact resistance of CFRP is generally considered low, and in fact the post-impact compression strength is considered to be a design-limiting factor. There is, however, scope for the modification of impact performance and fracture toughness by fiber hybridization with glass or aramid reinforcements leading to structures with much greater damage tolerance. Moreover, impact performance will improve as strain-to-failure of the fiber continues to increase.

3.2 Thermal Properties

CFRP is thermally conductive along the direction of the fibers. Conductivity perpendicular to the fibers is much less as it is dominated by the polymer matrix. The ability to dissipate heat is thought to contribute to the very good fatigue properties of CFRP. The conductivity of fibers increases with graphite content and highly graphitized fibers, such as high-modulus fibers, have thermal conductivity values of $700\,W\,m^{-1}\,K^{-1}$ in the longitudinal direction which surpass the value for steel ($50\,W\,m^{-1}\,K^{-1}$).

Coefficients of thermal expansion for carbon-fiber composites are dependent on the fiber type and on temperature. The fibers themselves have small negative coefficients of thermal expansion and this leads to the possibility of designing composites with almost zero overall thermal expansion. This dimensional stability, which can be available over extremes of temperature, is very attractive in applications such as antennae for aerospace vehicles. As with the thermal conductivity, the coefficient of thermal expansion is anisotropic—the value measured perpendicular to the fibers is dominated by the polymer matrix and is usually much higher (see Table 8) than the value measured parallel to the fibers. This can have important practical consequences in that residual stresses are left trapped in a laminate after it has cooled from the processing temperature. These stresses can be large enough to promote matrix cracking, especially in systems such as PMR-15 which, as described previously, is processed at high temperatures. For such systems to achieve wider use this problem has to be overcome either by toughening the resin further, or by further development of a damage-tolerant approach to design.

Table 8
Typical thermal expansion data for CFRP with 60 vol.% fiber

CFRP	Coefficient of thermal expansion[a] (10^{-6} K^{-1}) Longitudinal	Transverse
High modulus–epoxy	−0.25 to −0.60	20–65
High strength–epoxy	0.30 to −0.30	20–65

a over the temperature range 100–400 K

3.3 Electrical Properties

CFRP is electrically conductive along the direction of the fiber. As with thermal conductivity, electrical conductivity increases with the level of graphitization of the fiber.

There are practical applications which make use of the electrical properties of carbon fiber, notably to provide resistance heating effects (in nonmetallic tooling, for example) or static charge dissipation in composite structures.

The electrochemical behavior of graphite can lead to problems when carbon-fiber composites are in contact with metal alloys, as they may be in an aircraft structure, for example. Graphite is cathodic to most metal alloys and this can lead to appreciable corrosion (especially in mechanically fastened assemblies) unless adequate precautions are taken.

4. Forms of Carbon Fibers

Continuous filament fibers represent the primary production route for manufacture and other products are derived from this form by secondary processes. The conversion of fibers into some conveniently handleable intermediate product plays an important part in the effective utilization of the fibers as a reinforcement. Specific products are needed to meet the individual processing requirements of different fabrication techniques.

4.1 Continuous-Filament Tow

Continuous fibers are grouped together in tow bundles, each containing a specified number of individual parallel filaments. The tows are produced in different bundle sizes, ranging in discrete steps from 400 to 320 000 filaments per tow. The main sizes used are in the 3000–12 000 range; finer tows are employed only for specialized applications. Heavier tows offer some advantages for rapid volume conversion to composite and can lead to economies of scale for both fiber production and composite manufacture by certain processes.

Twisted tows, containing typically 15 turns per meter, arise from processes employing twisted precursor fibers. The presence of twist reduces the efficiency of the fiber reinforcement but can enhance the handleability of the tow, aiding conversion to secondary products such as woven fabrics and braids.

4.2 Woven Fabrics

Carbon fibers can be woven into a variety of unidirectional and bidirectional fabrics capable of being processed efficiently into composite materials. The weave construction (e.g., plain, satin, twill) determines the handle of the fabric and controls its ability to conform to contoured shapes.

4.3 Preimpregnated Products

Forms of preimpregnated materials (prepregs) include continuous tows, unidirectional tapes and sheets, and woven fabrics. A resin is applied to the fibers using a solvent, melt or film impregnation technique. The resins used are specially formulated with a latent cure, so that at room temperature the cure reaction proceeds only very slowly and the material can be handled and stored at ambient temperature for a reasonable length of time. Curing temperatures are typically between 100 °C and 180 °C. A choice can be made from the wide range of commercial prepregs to suit the end use and fabrication technique employed.

4.4 Nonwoven Fabrics

Carbon-fiber mats, felts and papers manufactured by conventional nonwoven processes from randomly distributed short fibers (about 6 mm in length) provide multidirectional reinforcement but convert inefficiently into composite form. New products containing longer fibers (about 25 mm) are now being developed, exhibiting good handling characteristics and generating improved composite performance.

4.5 Molding Materials

Short-length fibers chopped from continuous tows are widely used for the manufacture of molding compounds in both thermoplastic and thermosetting matrices. The application of special resin sizes and coatings facilitates processing into molding materials and other products such as papers and felts.

Thermoplastic molding compounds are supplied as granules containing short-length (0.2 mm) fiber reinforcements for processing on conventional injection-molding equipment.

Sheet molding compounds (SMCs) and dough molding compounds (DMCs) are already well established with glass fibers. The use of a small specific quantity of carbon-fiber reinforcement is envisaged in SMC with careful placement of aligned continuous fibers for optimum reinforcing effect at minimum additional cost. Carbon-fiber-reinforced DMCs in epoxy and polyester systems achieve excellent stiffness but strength levels achieved so far are disappointing.

Aligned short fibers in the form of prepreg sheets (ASSM) are available from special processes developed in the UK and Germany. The products can be molded to generate high performance in composites with highly curved or contoured shapes by virtue of the aligned fiber reinforcement coupled with the ability to promote controlled fiber movement for mold conformity.

5. *Fabrication Techniques*

Fabrication techniques have an important bearing on the properties obtained from carbon-fiber composites, and efficient processes must be devised for handling the fibers to ensure the maximum contribution to mechanical performance. The manufacture of GFRP provides some useful experience although two major points of difference arise. First, the handling characteristics of the two fibers are different; secondly, the aim in the manufacture of carbon-fiber composites is to obtain the best possible mechanical performance, requiring the fiber to be incorporated in the precise direction and in the exact amount necessary to meet the stresses applied.

The basic principles of the manufacture of carbon-fiber composites are similar to those used for GFRP (see *Glass-Reinforced Plastics: Thermosetting Resins*), although changes in the actual techniques are necessary in most cases. Processes unique to CFRP production are also being developed.

5.1 Compression Molding

Three suitable material forms containing carbon-fiber reinforcement are available: DMCs, SMCs, and pre-impregnated tapes, sheets and fabrics.

Certain considerations which affect plant and mold tool design and process procedures are necessary for the manufacture of CFRPs. The thermal conductivity and low thermal expansion characteristics of carbon fibers introduce the need for carefully controlled heating and cooling cycles and special attention to molding shrinkage (and expansion) to achieve dimensional accuracy in the final part. SMCs usually comprise hybrid reinforcements, mainly glass, with a small amount of carbon included for specific stiffening. The effectiveness of the carbon fiber is influenced by careful placement for maximum reinforcing effect and for minimum risk of misalignment caused by resin and glass-fiber flow during molding. Alignment accuracy, parallelism of mold-tool platens and additional considerations arise with the use of prepreg materials.

5.2 Autoclave and Vacuum-Bag Molding

The production of high-quality composites from prepreg materials (tapes and fabrics) is achieved by the consolidation on a single-sided tool of a laminated preform. The method makes use of a flexible bag or blanket through which pressure is applied to the prepreg layup on the mold. The mold is supported on a platen inside an autoclave which is pressurized with gas heated to the required cure temperature.

Unidirectional and woven materials can be molded equally successfully and the choice depends on the end use requirements and the layup techniques employed. Highly automated procedures for the precise placements of prepreg layers in laminate preforms are now well developed and are being introduced on a production basis in aircraft component manufacture. Computer-controlled tape-laying and ply-profiling machines are a major step towards a fully automated integrated composite-manufacturing scheme.

Current trends are towards prepregs with simplified cure cycles and zero resin-bleed characteristics. The careful use of viscosity modifiers in resin formulation offers greater flow control, which is important for large or thick composite moldings in which significant temperature variations may exist.

Vacuum molding represents a simplified version of the autoclave technique where the pressure chamber is substituted by an oven and consolidation is achieved by virtue of the vacuum drawn inside the flexible membrane. The relatively low net pressure applied requires the resin to have easy flow characteristics.

5.3 Filament Winding

The filament winding process is appropriate for the production of hollow carbon-fiber components which have an axis of rotation. Excellent mechanical properties are achieved by virtue of the accurate positioning of the fibers and the high volume fractions possible in the composite. The process is highly automated and can be programmed to lay down specific reinforcement patterns to comply with complex design requirements in multistressed components.

Carbon fibers can be processed successfully by both wet-winding and prepreg-winding techniques. Special attention to the gradual buildup of winding tensions and to the choice of smooth-surfaced materials for guide points is necessary in order to promote even, high-speed delivery of carbon fibers with minimum risk of abrasion and filamentation.

5.4 Mandrel Wrapping

Prepreg wrapping is used widely in the manufacture of sports goods, for the production of parallel and taped tubular shapes. Special purpose-built equipment is available for semiautomated production. Mechanical performance is determined by the layup pattern of the tube, although simple configurations are required for successful volume throughput. Unidirectional tapes and woven fabrics are equally applicable. Good material handleability is essential; unidirectional prepregs are often combined with a light scrim support to improve transverse properties.

5.5 Pultrusion

Pultrusion is a well-established technique for the production of profiles of glass-reinforced polyester composites and offers scope for CFRP manufacture. The process is simple, continuous, can be highly automated and does not require a high level of financial or technological investment. The use of carbon-fiber reinforcement with high-performance systems, such as epoxies, presents a completely new set of considerations for a continuous molding operation. The interacting effects of thermal expansion and resin shrinkage impose strict requirements for the tool design both in terms of dimensional accuracy and material of construction.

High-quality, high-performance pultruded stock is commercially available. Hybrid reinforcement configurations, new matrix materials (e.g., thermoplastics offering postformability) and methods for introducing off-axis reinforcement are being developed to extend the capabilities of the process.

5.6 Wet Layup

The contact molding method is simple and is a low cost method, but is prone to variability in quality and performance. The method is therefore not attractive for the widespread use of carbon fibers. However, the addition of a small proportion of carbon fibers as a specific reinforcement for GFRP structures can be both cost and performance effective. For example, unidirectional woven tapes strategically incorporated into GFRP laminates promote maximum stiffness for minimum added cost, and nonwoven fabrics applied as surfacing tissues for GFRP structures provide chemical resistance and electrical conductivity.

5.7 Other Methods

Thermoplastic molding compounds are processed by injection molding. Long-fiber reinforcements, combined with polyethersulfone, polyether ether ketone and other polymers can be converted to composites by thermoforming operations (see *Glass-Reinforced Plastics: Thermoplastic Resins*).

6. Applications

Applications for CFRPs fall generally into three broad categories: aerospace (aircraft, space and satellite structures), sports goods, and engineering and industrial equipment. The overriding application is that of metals substitution for improved mechanical performance and weight reduction. The stiffness-to-weight ratio for high-modulus CFRPs is the highest available for any structural material and hence carbon-fiber composites are applied where stiffness, strength and light weight are essential.

7. Future Perspective

Today, about 4000 t of carbon fiber per year are consumed worldwide. Compared to glass fiber used in composite materials, estimated at five to six million tonnes per year, this is indeed a modest market. The latter is now a mature product and advances at the approximate rate of world economic growth. Before carbon fiber arrives at this phase of its history, it must first establish a concrete role in the engineering consciousness and achieve an economic scale of production. Carbon fiber costs US$30 a kilogram at 1988 prices. Potentially, using polyacrylonitrile (PAN) produced on a commercial scale, it could reduce in price to US$10 a kilogram. On the basis of modulus, volume for volume, this would make CFRP less expensive than GFRP. If the early promise of using pitch as a precursor to carbon fiber should be fulfilled, the cost expectations would exceed this prediction by far.

The most important factor influencing the future of CFRP is the cost of energy. By the late 1980s, earlier anxieties about energy supplies had abated, but ultimately the driving force in the materials economy will be the energy cost of materials plus the energy cost of their use. CFRP offers considerable advantages in both respects compared to conventional materials.

Bibliography

Anon 1981 *Processing and Uses of Carbon Fiber Reinforced Plastics*. VDI, Dusseldorf

Anon 1983 Fiber composites; Design, manufacture and performance. *Composites* 14(2): 87–139

Anon 1986 *Carbon Fibers, Uses and Prospects*. Noyes Data Corporation, Park Ridge, NJ

Clegg D W 1986 *Mechanical Properties of Reinforced Thermoplastics*. Elsevier, London

Goodman S 1986 *Handbook of Thermoset Plastics*. Noyes Data Corporation, Park Ridge, NJ

Kelly A, Mileiko S T 1983 *Fabrication of Composites*. North-Holland, Amsterdam

P. J. Mills and P. A. Smith
[University of Surrey, Guildford, UK]

Cellular Materials in Construction

Many materials have a cellular structure, composed of an interconnected network of struts or plates. Such cellular materials occur in nature as wood, cancellous bone and cork. Recently, humankind has begun making its own cellular materials, in the form of honeycomb-like materials and foams. Although initially honeycombs and foams could only be made from a limited range of materials, both can now be made from a wide range of polymers, metals, ceramics and glasses. Micrographs of several natural and man-made cellular materials are shown in Fig. 1.

The cellular structure of these materials gives rise to unique properties which make them useful in many applications. Since they can undergo large deformations at relatively low loads, they are widely used in the packaging industry for providing protection from impacts. Their low volume fraction of solids and small cell size give them a low thermal conductivity, making them ideal thermal insulators. Their low weight is exploited in structural sandwich panels, where they are used as a lightweight core to separate two stiff, strong faces; the resulting sandwich gives an efficient member for resisting bending and buckling loads.

In this article, the structure and mechanical behavior of cellular materials are reviewed, and their use in construction is described.

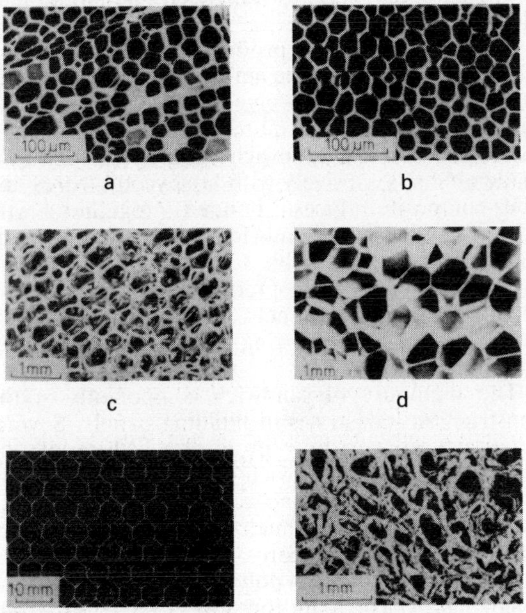

Figure 1
Micrographs of cellular materials: (a) balsa wood, (b) cork, (c) polyurethane foam, (d) polyethylene foam, (e) aluminum honeycomb, (f) copper foam

1. Structure and Mechanical Behavior

Several features characterize the structure of a cellular material. The volume fraction of solids in the material can be described by its relative density, that is, the density of the cellular material divided by that of the solid cell wall material from which it is made. The cells may extend in one plane only (as they do in honeycombs), so that the normals to all of the cell walls lie in a single plane, producing a 'two-dimensional' cellular structure. Alternatively, they may extend in any direction (as they do in foams), giving a less restricted, 'three-dimensional' structure. The cells may be open or closed, depending on whether or not the faces of the cells are empty or covered by thin membranes. The cells may be geometrically anisotropic, with different mean intercept lengths in different directions. Anisotropy is often the result of the rise of the foaming gas in the liquid matrix which produces cells which are elongated in the rise direction and roughly equiaxed in the plane normal to it. The mechanical properties of foams are often anisotropic as a result of anisotropy in the cell geometry. Each of these structural features is shown in Fig. 1.

A typical stress–strain curve for a cellular material is shown in Fig. 2. There are three distinct regimes of behavior. At relatively low loads, the material is linear elastic. At some critical level of load, the cells begin to collapse by elastic buckling, plastic yielding or brittle fracture, depending on the nature of the cell wall properties. Cell collapse then progresses at a roughly constant load, producing a roughly horizontal stress plateau, until, at relatively large strains (typically about 0.8), the opposing cell walls begin to meet and touch. The stress then rises steeply as the material densifies (Maiti et al. 1984).

Each of these regimes of behavior is related to the mechanism by which the cells deform within that

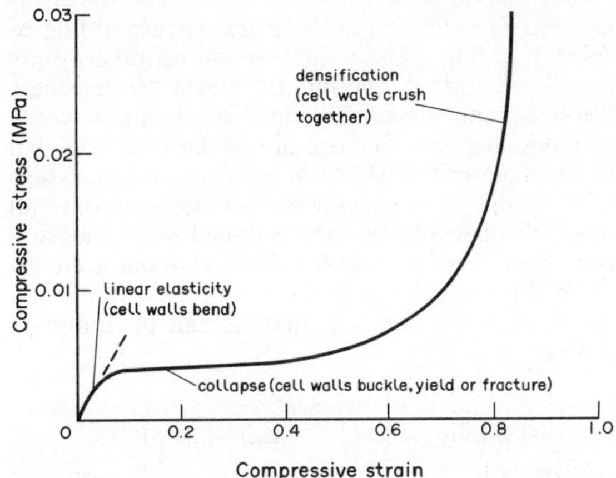

Figure 2
A typical stress–strain curve for a cellular material in uniaxial compression

Table 1
Property–density relationships for cellular materials

Property	Open celled foams	Honeycombs In-plane	Honeycombs Axial
E	$C(\rho/\rho_s)^2 E_s$	$C(\rho/\rho_s)^3 E_s$	$C(\rho/\rho_s) E_s$
ν	C	C	C
G	$C(\rho/\rho_s)^2 E_s$	$C(\rho/\rho_s)^3 E_s$	$C(\rho/\rho_s) E_s$
σ^*_{el}	$C(\rho/\rho_s)^2 E_s$	$C(\rho/\rho_s)^3 E_s$	$C(\rho/\rho_s)^3 E_s$
σ^*_{pl}	$C(\rho/\rho_s)^{3/2} \sigma_{ys}$	$C(\rho/\rho_s)^2 \sigma_{ys}$	$C(\rho/\rho_s)^2 \sigma_{ys}$
σ^*_{cr}	$C(\rho/\rho_s)^{3/2} \sigma_{fs}$	$C(\rho/\rho_s)^2 \sigma_{fs}$	$C(\rho/\rho_s) \sigma_{fs}$
K^*_{Ic}	$C(\rho/\rho_s)^{3/2} \sigma_{fs}(\pi l)^{1/2}$	$C(\rho/\rho_s)^2 \sigma_{fs}(\pi l)^{1/2}$	$C(\rho/\rho_s) \sigma_{fs}(\pi l)^{1/2}$

a The constants of proportionality are different for each property and each case (foams, honeycomb-in-plane and honeycomb-axial). b E_s, σ_{ys} and σ_{fs} are the Young's modulus, yield strength and modulus of rupture, respectively, of the solid cell-wall material. c The characteristic length of the cell size is l

regime. Observation of model honeycomb materials and of foams in a scanning electron microscope reveals that the initial linear elasticity is produced by bending in the cell walls. Cell collapse is caused by elastic buckling in elastomeric honeycombs and foams, by plastic yielding in materials which yield, and by brittle crushing in materials which fracture (Gibson and Ashby 1982, Maiti et al. 1984). The strain at which the honeycomb or foam begins to densify depends on its relative density.

The tensile behavior of cellular materials is similar to the compressive behavior with the following two exceptions. First, elastomeric honeycombs and foams do not buckle so that the stress plateau between the linear elastic regime and densification is eliminated in tension. Second, the tensile stress–strain curves of brittle cellular solids are terminated by brittle fracture at the end of the linear elastic regime.

This stress–strain behavior can be characterized by several mechanical properties. The linear elastic regime is described by a set of elastic moduli such as Young's modulus E, Poisson's ratio ν and the shear modulus G. The plateau stresses corresponding to elastic buckling, plastic yielding and brittle crushing are σ^*_{el}, σ^*_{pl} and σ^*_{cr}, the limiting strain for densification is ε_D, and the tensile fracture toughness is K_{Ic}. By modelling the structure of cellular materials and the mechanisms by which they deform, relationships between the properties of the cellular material and its relative density, its solid cell wall properties and its geometry can be found. Table 1 contains a summary of these relationships for both honeycombs and foams. A more detailed analysis can be found in Maiti et al. (1984).

2. Applications of Cellular Materials in Construction

2.1 Lightweight Structural Sandwich Panels

Structural panels composed of two stiff, strong faces separated by a lightweight core are known as sandwich panels. The core acts to separate the faces, increasing the moment of inertia of the panel, with little increase in weight, producing an efficient structure for resisting bending and buckling loads. Usually, the core is made of a lightweight cellular material. In aerospace applications, aluminum or paper–resin honeycomb cores are preferred for their extremely low weight. In building applications foam cores, which are typically somewhat heavier than honeycomb cores, are preferred as they offer low thermal conductivity in addition to relatively low weight. Typical face materials include fiber-reinforced composites, steel, aluminum and plywood.

The technology for producing sandwich panels was first developed in the aerospace industry, beginning with the balsa core–plywood face sandwiches of the World War II Mosquito aircraft (Allen 1969). Modern aerospace sandwich components are now made almost exclusively with honeycomb cores and fiber-composite faces bonded together with advanced adhesives. Sandwich panels are now made routinely and are reliable for components such as the rotor blades of helicopters, wing-tip sections, tail sections and flooring panels. This technology is now well developed and is being transferred to the construction industry.

The main use of sandwich components in the construction industry is in building panels. Several manufacturers are now producing wall, roof and door panels using sandwich construction. Different combinations of face and core materials are used; typically the faces are made of steel, aluminum or waferboard and the cores are made of foamed polyurethane, foamed polystyrene beadboard or foamed glass. The foam core gives excellent thermal insulation while the faces provide the structural stiffness and strength of the panel. The panels are used in housing and in low-rise, nonresidential buildings, and as fascia panels in commercial buildings.

The reduced weight of sandwich construction is also useful in applications where the component must be lifted into place by crane or helicopter. For example, a recent EPA ruling required the addition of roofing covers to exposed water distribution reservoirs. Domed covers made with fiberglass-reinforced plastic faces and a paper–resin honeycomb core were assembled adjacent to the reservoir sites and then lifted into place by crane. The low weight of the sandwich reduced the installation costs. In another application, modular sandwich lighthouse structures with fiber-reinforced plastic faces and balsa cores were installed using helicopters; again, the low weight of the units reduced installation costs.

Sandwich construction is also used for portable structures, such as the folding bridges being developed by the US Army. The deck of these bridges is doubly sandwiched, with a top face of high-strength aluminum and a bottom face of a graphite-epoxy–aluminum composite. Between these two faces is a webbed core which is itself a sandwich with aluminum faces and a honeycomb core. The low weight of the sandwich structure allows the bridge to be transported easily. The bridges are hinged so that they fold during transport and can be deployed by a hydraulic actuator for use.

The weight of sandwich panels can be minimized by appropriate optimization analyses. Wittrick (1945), Ackers (1945) and Kuenzi (1965) have developed minimum weight analyses for pin-ended columns subject to an end load. Huang and Alspaugh (1974) and Ueng and Liu (1979) have minimized the weight of a sandwich panel subject to multiple constraints. More recently, Gibson (1984) and Demsetz and Gibson (1987) have found the core density as well as the face and core thickness which minimize the weight of a sandwich beam or plate for a given stiffness. These optimization analyses are useful in applications where the weight of the component is critical.

2.2 Wood

Wood is still one of the most widely used structural materials throughout the world. The present world production is about 10^9 t per year, roughly equal to that of iron and steel (Dinwoodie 1981). In North America more wood products are used for construction than all other construction materials combined (Bodig and Jayne 1982).

The cellular structure of wood is shown in Fig. 1 (a). Softwoods used in construction have two main types of cells: tracheids and parenchyma. The tracheids are the long, narrow cells (aspect ratio about 100:1) which make up the bulk of the tree. They conduct fluids and nutrients within the tree and give it its structural support. The parenchyma cells, found in the rays, store sugars.

At a finer level, the walls of the cells have a fiber-composite structure with partially crystallized cellulose molecules forming stiff, strong, fiber-like chains helically wound in a matrix of amorphous lignin and semicrystalline hemicellulose. Each cell wall is made up of four layers, each with a different orientation of cellulose fibers.

Mechanically, wood behaves very much like other cellular materials. Stress–strain curves for several types of wood in uniaxial compression along and across the grain are shown in Figs. 3 and 4. These curves show that wood is highly anisotropic; it is much stiffer and stronger when loaded along the grain than across. Part of this anisotropy results from the different mechanisms by which wood deforms when loaded along and across the grain, and part comes from the anisotropy in the cell wall properties. The general shape of the curves resembles that of Fig. 2.

The mechanical behavior of wood can be modelled by analyzing the mechanisms by which it deforms (Easterling et al. 1982). When loaded across the grain, the cell walls bend, in the same way that a honeycomb does when loaded in the plane of the honeycomb and, because of this, the Young's moduli and compressive strength of woods loaded across the grain depend on the cube and the square of the relative density of the wood (see Table 1). When loaded along the grain, the cell walls compress uniaxially, as a honeycomb does if loaded normal to the plane of the hexagonal cells and, because of this, the Young's modulus along the grain depends linearly on the relative density of the wood. Finally, the axial compressive strength of the wood is related to yielding in the cell walls, again giving a

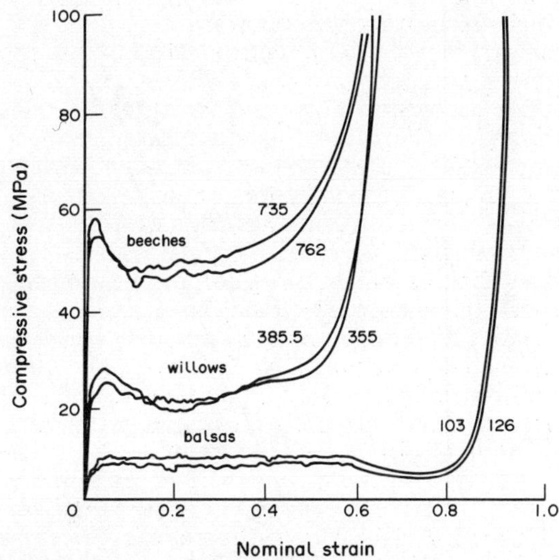

Figure 3
Stress–strain curves for wood loaded along the grain. Figures by curves indicate density ($kg\,m^{-3}$); axial compression = $1.25–1.5 \times 10^{-5}\,s^{-1}$

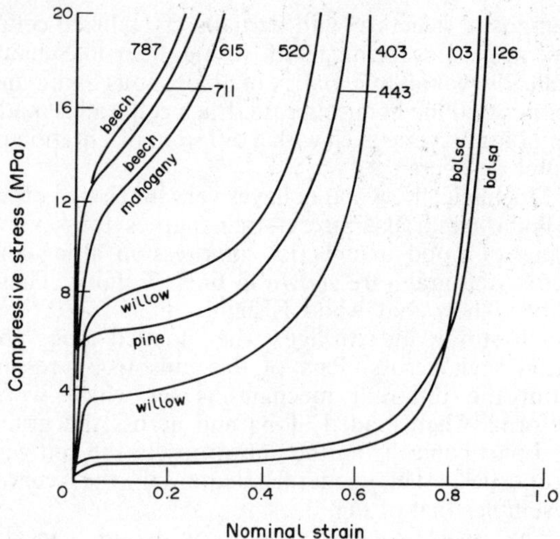

Figure 4
Stress–strain curves for wood loaded along the grain. Figures by curves indicate density ($\mathrm{kg\,m^{-3}}$); tangential compression $= 1.6 \times 10^{-3}\,\mathrm{s}^{-1}$

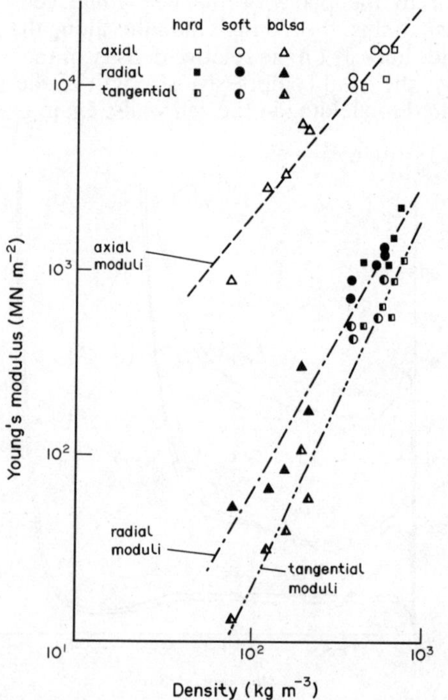

Figure 5
Young's modulus for woods, plotted against density. The lines indicate the theoretical dependencies

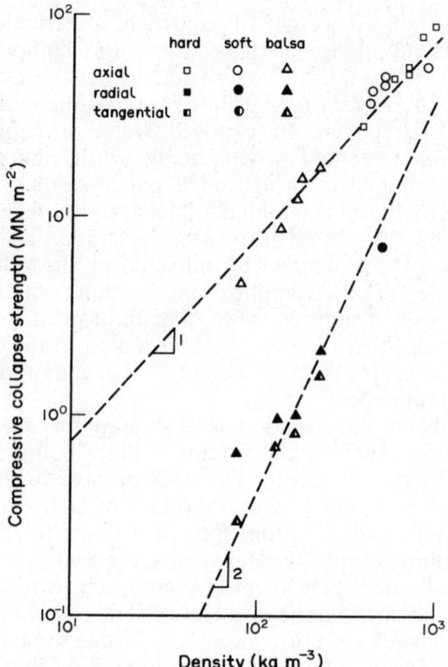

Figure 6
Compressive strength for woods, plotted against density. The lines indicate the theoretical dependencies

linear dependence on relative density. These results are plotted in Figs. 5 and 6, along with data for wood. This modelling begins to explain the dependence of the properties of wood on the loading direction and the relative density of the wood. The tensile fracture toughness of wood has been explained in a similar way; see, for example, Ashby et al. (1985).

The mechanical properties of wood tend to be variable, due to the presence of defects in the wood, for example, nonparallel grain and knots. Recently, glue-laminated wood members (glulam) have been developed to minimize the effect of such defects. Glue-laminated members are made by gluing together thin strips of wood from which the defects have been cut out. The properties of such members are more uniform than those of dimensioned lumber. Glulam sections also have the advantage that they can be curved and built up to any size. The efficiency of wood members can also be improved by using them in sandwich constructions. Typically, plywood faces are used in conjunction with honeycomb, foam or balsa cores.

2.3 Thermal Insulation

The thermal conductivity of cellular solids is typically low, partly because of their low volume fraction of solids (about 10%) and partly because of their small cell size (usually between about 0.04 mm and 1 mm). This makes them ideal thermal insulators; one of the most prevalent applications of foams is in thermal insulation.

In Arctic construction, icing can be a problem; foams are sometimes used as thermal insulation to reduce this. For example, polyethylene foam sheets have recently been used to insulate railway tunnels outside Anchorage, Alaska, and in British Columbia, Canada. Near the portals of the tunnels ice build-up on the crown and sidewalls of the tunnels and on the track caused severe icing problems. Insulation shields, made of polyethylene foam faced with galvanized steel plates, installed at the portals of the tunnels, reduced icing to a tolerable level.

Foam sheet has also been used to insulate foundations from permafrost in the Arctic. In this case, the foam is used to prevent the permafrost melting from the heat of the building since melted permafrost can cause excessive sliding and settlement of the foundations.

Bibliography

Ackers P 1945 The efficiency of sandwich struts utilizing calcium alginate core, R & M 2015. Aeronautical Research Council, Farnborough, UK

Allen H G 1969 *Analysis and Design of Structural Sandwich Panels*. Pergamon, Oxford

Ashby M F, Easterling K E, Harrysson R 1985 The fracture toughness of woods. *Proc. R. Soc. London. Ser. A*. 398: 261–80

Bodig J, Jayne B A 1982 *Mechanics of Wood and Wood Composites*. Van Nostrand Reinhold, New York

Demsetz L A, Gibson L J 1987 Minimum weight design for stiffness in sandwich plates with rigid foam cores. *Mater. Sci. Eng*. 85: 33–42

Dinwoodie J M 1981 *Timber: Its Nature and Behaviour*. Van Nostrand Reinhold, New York

Easterling K E, Harrysson R, Gibson L J, Ashby M F 1982 On the mechanics of balsa and other woods. *Proc. R. Soc. London. Ser. A*. 383: 31–41

Gibson L J 1984 Optimization of stiffness in sandwich beams with rigid foam cores. *Mater. Sci. Eng*. 67: 125–35

Gibson L J, Ashby M F 1982 The mechanics of three-dimensional cellular materials. *Proc. R. Soc. London. Ser. A*. 382: 43–59

Huang S N, Alspaugh D W 1974 Minimum weight sandwich beam design. *AIAA J*. 12: 1617–18

Kuenzi E W 1965 Minimum weight structural sandwich. US Forest Service, Research Note FPL-086. Forest Products Laboratory, Madison, WI

Maiti S K, Gibson L J, Ashby M F 1984 Deformation and energy absorption diagrams for cellular solids. *Acta Metall*. 32: 1963–75

Ueng C E S, Liu T L 1979 Least weight of a sandwich panel. In: Craig R R (ed.) 1979 *Proc. ASCE Engineering Mechanics Division 3rd Specialty Conf*. University of Texas at Austin, pp. 41–4

Wittrick W H 1945 A theoretical analysis of the efficiency of sandwich construction under compressive end load, R & M 2016. Aeronautical Research Council, Farnborough, UK

L. J. Gibson
[Massachusetts Institute of Technology, Cambridge, Massachusetts, USA]

Ceramic Cutting Tools

The history and evolution of humanity has always been associated with tools. Tools were a prerequisite for the early development of hunting and earth moving equipment. The first tools were made from ceramic materials, mainly because their brittleness made it easy to form a sharp cutting edge. The first simple machine which used ceramic tools, the bow drill, dates back as far as the Upper Paleolithic period.

More recently, during and after the industrial revolution in Britain, the introduction of new tools for the chip-forming metal-working industry has been an endless process, the pace of which has steadily increased. As a result, the rate of increase of productivity has continuously risen. The development of new tool materials and machines has, on average, led to a doubling of productivity every tenth year (see Fig. 1) during the twentieth century.

The contribution of ceramic tools to this rapid development was very small until the 1980s. This can be attributed mainly to the failure, by brittle fracture, of early ceramic tools because of insufficient strength, toughness and thermal shock resistance and to the successful development of other tool materials, especially cemented carbides. With a coating of a ceramic material (e.g., alumina) on a cemented carbide substrate, the chemical inertness of ceramics can be combined with the favorable combination of hardness and toughness of cemented carbides to give a product which has taken a very large part of the cutting tool business.

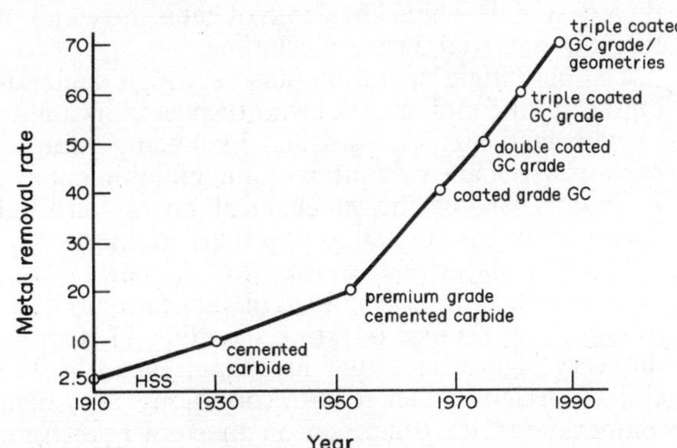

Figure 1
The effect of new tool materials on the metal removal rate

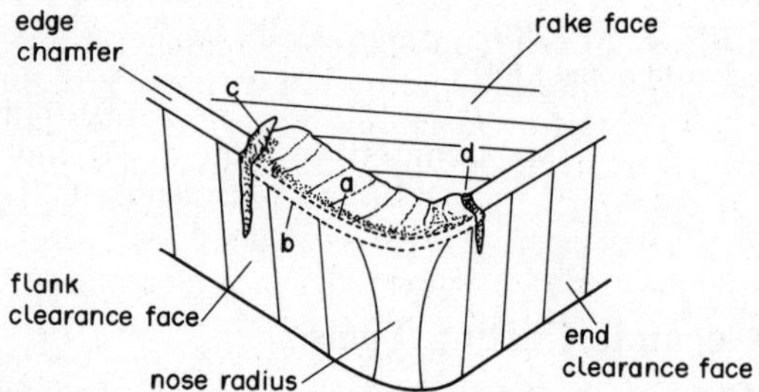

Figure 2
Regions of tool wear: (a) crater wear, (b) flank wear, (c) depth of cut notch wear and (d) trailing edge notch wear

Two factors have contributed to a renewed interest in ceramic cutting tools. First, the development of machine tools with increased power and stability has decreased the need for toughness and has also made it possible to perform machining with cutting conditions well suited to ceramics; the versatility offered by computer numerical control (CNC) programming has made it possible to adapt the cutting cycle more precisely to the workpiece geometry in question. Second, large international investments in research and development have produced new ceramic materials with improved properties and a greater consistency due to better processing techniques.

1. *Metal Cutting*

Metal cutting is an operation where a thin layer of metal, the chip, is removed from the workpiece with a tool. The tool material is, in the majority of operations, in the form of an indexable insert (typically about 13 mm across and 5 mm thick) mechanically clamped to a tool holder so that when one edge is worn it can be indexed to bring another edge into operation. The cutting speed is the rate at which the workpiece material moves over the cutting edge. The feed is the distance moved by the tool along the workpiece per revolution and it is related to the thickness of the chip. The depth of cut is the width of the chip removed during machining.

The machining operation puts very high demands on the cutting tool material with respect to its ability to withstand high stresses and high temperatures, both of which are very often of a fluctuating nature. A major part of the mechanical energy utilized during cutting is converted into thermal energy.

The two major heat sources for the cutting tool are located along the chip–tool interface and the tool-flank–workpiece interface (see Fig. 2). During the very high strains that are experienced by the chip, especially under seizure conditions, very high temperatures are generated on the tool rake face; temperatures up to 1600 °C have been measured. The temperatures on the flank face are normally lower and are caused by the rubbing action of the machined workpiece on the tool flank. However, on a heavily worn insert, flank face temperatures may reach values similar to those on the rake face.

During the cutting process, the cutting edge undergoes a continuous change in geometry due to several wear mechanisms which act simultaneously (see Fig. 2). These mechanisms are generally classified as either mechanical or chemical. Mechanical wear can be caused by the abrasive action of hard inclusions in the workpiece on the tool. Chemical wear, including diffusion/solution of the tool material into the hot chip, depends mainly on the high temperatures which exist during the cutting. The dominant wear mechanism for a specific cutting operation is a function of the tool and workpiece materials, the geometry of the cutting process, and the machining conditions with respect to speed, feed and depth of cut.

The two different material categories that have evolved during the development of ceramic materials for metal machining are alumina-based and silicon nitride-based ceramics.

2. *Production of Ceramic Cutting Tool Materials*

The fabrication of ceramic cutting tools is normally performed using the same methods as those used in the powder metallurgy industry. Suitable powders are mixed and milled in a suspension and, after drying, the powder to which suitable binders have normally been added is pressed into the desired shapes and sintered.

Pure oxide compositions are normally sintered in air whereas ceramics containing carbides or nitrides require other furnace atmospheres to avoid unwanted chemical reactions. Ceramics for tool materials normally contain very small amounts of the liquid phase during sintering, which explains why high temperatures are required to obtain a pore-free material and avoid grain growth and reduced mechanical strength. Some ceramic compositions also tend to thermally decompose at the high temperatures required for densification since diffusion is small at lower temperatures. For these reasons alternative sintering methods have been developed for the fabrication of some ceramic cutting tools, and these are described next.

Uniaxial hot pressing differs from normal sintering in that pressure is applied during the consolidation process. A powder or presintered blank is placed in a die made from graphite and the pressing operation is carried out at a pressure of about 25 MPa and at temperatures of about 1500–1800 °C for 1 h. The temperatures and times required for densification are lower than for conventional sintering and consequently finer grain sizes can be achieved. The hot-pressing method is commonly used for alumina-based materials and for silicon nitride compositions.

Composites based on alumina produced by hot pressing normally show different mechanical and physical properties in the directions parallel and perpendicular to the hot-pressing direction. The material has been pressed in one direction only and as a result the microstructural components tend to be preferentially aligned. Another process using an applied pressure is hot isostatic pressing (HIP). This process uses an inert gas to apply a uniform pressure to the blank during the sintering. The blank must be encapsulated in a medium which transmits the pressure to the individual particles. This medium can either be metallic or a glass. HIP can be performed at pressures up to 200 MPa and temperatures up to 2000 °C. Due to the encapsulation the HIP process is normally a more costly production process than sintering.

More recently, however, a new technique which combines the advantages of both sintering and the HIP process has been developed. This sinter–HIP process is based on conventional sintering but the last part of the consolidation process is performed at a higher pressure. Pressure is not applied until the sintering process has led to the formation of closed porosity so that canning is not necessary. The applied pressure is normally about 20 MPa. For reactive materials like Si_3N_4 not only can inert gases be used but also active gases like nitrogen can promote densification due to nitrogen pickup.

3. *Alumina-Based Ceramics*

3.1 *Pure Oxide Ceramics*

Pure alumina (Al_2O_3) ceramics were already considered as a suitable cutting tool material at the beginning of the twentieth century and the first patents were issued in about 1910 prior to the introduction of cemented carbides. The commercial introduction of alumina ceramics had to wait, however, until World War II and occurred primarily because of the lack of strategic raw materials, like tungsten, for the production of cemented carbides.

These early tools were prone to failure by brittle fracture. Their low fracture strength was directly attributed to large grain size and enclosed porosity. Further development led to higher-quality alumina-based tools mainly as the result of using more finely grained raw materials and improved processing. The use of sintering aids also enabled the fabrication of fine-grained materials with a high density. The most common addition is magnesia which is added to commercial alumina tools at the level of a few tenths of a percent.

Pure alumina tools have relatively low strength and toughness properties and also have a low thermal conductivity, which is why the inserts are prone to fracture under conditions of rapid stress and temperature variations. In the early 1970s it was discovered that additions of zirconia to alumina could significantly improve all of these properties. The reasons for the improvements are related to the properties of zirconia itself. Pure zirconia transforms spontaneously upon cooling from a tetragonal to a monoclinic crystal structure with an associated volume increase. If the zirconia grains are embedded in an alumina matrix the transformation normally leads to microcracking in the alumina. However, if the zirconia grains are small enough (about 0.3 μm), they are kept in a metastable tetragonal form due to the hydrostatic compressive forces exerted on the zirconia grains in the alumina matrix. Zirconia in its high temperature form can also be stabilized to varying degrees by additions of yttria, magnesia or calcia. When a crack forms and propagates in a tool in use the tensile stresses associated with the crack propagation allow the metastable zirconia to transform causing the formation of both compressive stresses and microcracking ahead of the crack tip. These mechanisms effectively increase the fracture toughness of the material.

Materials containing transformed zirconia particles (microcracking mechanism) or metastable zirconia (transformation toughening mechanism) show almost the same fracture toughness at a given level of zirconia. However, the occurrence of microcracks lowers the strength and can also cause accelerated wear during metal cutting; thus ceramic tools usually contain metastable zirconia.

In alumina tools used for machining cast iron, the zirconia content is about 4 wt%. When machining steel the demands for toughness and thermal shock resistance are higher and the optimum zirconia content is 10–20 wt%.

3.2 *Mixed Ceramics*

One of the reasons for the very poor thermal shock resistance of alumina-based tools is their low thermal conductivity. The thermal shock resistance can be improved by adding a metallic phase. The most common additives are titanium carbide and titanium nitride at a level of 20–40 wt% as these both give improved thermal conductivity, strength, hardness and toughness compared with pure alumina ceramics.

These types of ceramics, usually referred to as mixed ceramics, were introduced in the late 1960s. The most common additive is titanium carbide and the material is normally hot pressed as reactions between aluminum and titanium carbide generally prohibit pressureless sintering.

3.3 *Whisker-Reinforced Ceramics*

A whisker is a single crystal fiber. Whiskers used for reinforcing ceramic cutting tools have a diameter of approximately 1 μm and a length of 10–30 μm. Due to their crystalline perfection and the very small diameter they have a very high strength. Silicon carbide whiskers are at present used to reinforce alumina-based tools, as they increase strength, fracture toughness and thermal shock resistance. Whisker reinforcement also leads to a large increase in the Weibull modulus which means that variations in strength are reduced.

The exact causes of these improvements are still under discussion but the following mechanisms have been identified. Crack bridging occurs when crack propagation is hindered by those whiskers in the crack plane oriented perpendicular to the crack. Whisker pullout makes a major contribution to the fracture energy and is favored by a level of interfacial bonding which is not too strong and by a high transverse fracture toughness in the whisker. Crack deflection processes further increase the toughness when cracking of the relatively weak interface causes a blunting or redirection of the crack.

Alumina–silicon-carbide inserts generally contain 25–35 wt% silicon carbide whiskers. This addition results in a two-fold increase of the fracture toughness and in an increase in the thermal shock resistance so that the use of coolants in the metal cutting process now becomes possible.

Alumina–silicon-carbide composites are normally produced by hot pressing at a temperature of about 1800 °C. The hot-pressing route gives rise to an anisotropic whisker distribution with whiskers preferably oriented in a plane perpendicular to the hot-pressing direction. When machining nickel-based alloys the optimum tool performance is observed when the insert is oriented so that the hot-pressing direction is parallel to the flank face.

4. Si_3N_4 and Sialons

Si_3N_4 is a covalent compound with a number of interesting engineering properties. The most significant property for metal cutting is a very low thermal expansion coefficient which makes the material very resistant to thermal shocks.

Unlike alumina-based tools which are normally sintered with very little liquid phase, silicon nitride cannot be consolidated by solid-state sintering. Sintering additives such as Y_2O_3, MgO and Al_2O_3 are used to form a liquid with the SiO_2 which is always present on the surface of the Si_3N_4 grains. The resulting microstructure is a two-phase material consisting of crystalline silicon nitride embedded in an intergranular bonding phase. Upon cooling the liquid forms a glass which can be more or less devitrified during the cooling cycle. The devitrification can be further controlled by slow cooling or subsequent heat treatment. The properties of silicon nitride tools are to a large extent controlled by the amount and composition of the intergranular phase especially at high temperatures. Materials prepared by hot pressing or by the HIP process normally require lower amounts of sintering additives than materials prepared by pressureless sintering. Si_3N_4 cutting tools from different suppliers can therefore show significant differences in metal cutting behavior.

In the early 1970s it was discovered that oxygen (O^{2-}) could be substituted for nitrogen (N^{3-}) in the β-Si_3N_4 crystal with silicon (Si^{4+}) being substituted by aluminum (Al^{3+}) to maintain electrical charge neutrality. The solid solution is denoted $Si_{6-z}Al_zO_zN_{6-z}$ where $0<z\leq 4.2$ and belongs to a group of engineering ceramics called sialons. Like Si_3N_4, sialon materials have to be sintered with a fairly large amount of liquid phase. The silicon nitride raw material α-Si_3N_4 dissolves during sintering in the oxynitride glass and is precipitated as β'-sialon. If AlN and Y_2O_3 are simultaneously added it is also possible to form α-sialon with the general formula $M_x(SiAl)_{12}(ON)_{16}$ where $x\leq 2$ and M can be lithium, magnesium, calcium or yttrium. α-sialon has a higher hot hardness than β-sialon. In the sialon system it is generally possible to partly devitrify the glassy phase to form various crystalline phases; for example, yttrium–aluminum–garnet (YAG). The microstructure and consequently the cutting properties can be varied within large limits.

5. Application of Ceramic Cutting Tools

5.1 Cast Iron Machining

The machining of cast iron brake drums and disks for the automotive industry is the largest application area for ceramic cutting tools and all of the different types of ceramic grades can be used.

Pure oxide ceramics are preferred when castings are of high quality; that is, with no slag inclusions and with no 'as cast skin' which gives superior wear behavior to other ceramic grades. Si_3N_4-based tools are more resistant to the chemical wear caused by SiO_2 inclusions and by sand. They show a superior tool life in rough-turning operations with poor quality castings. Variable depths of cut and short machining times often lead to thermal cracking and Si_3N_4-based tools perform better than oxide ceramics under such circumstances.

Mixed ceramic grades show the best performance in finishing operations where tool life is determined by the surface finish required on the workpiece. Localized wear on the trailing edge is critical under such circumstances and the better attrition wear of mixed ceramics compared with pure ceramics makes it perform better under these conditions.

5.2 Steel Machining

The machining of steel represents the major part of machining operations. Ceramic cutting tool materials have not yet penetrated this area to any large extent.

Alumina-based materials lack adequate toughness and thermal shock resistance to be able to compete successfully with the materials in use in the early 1990s (mainly coated cemented carbides). At low cutting speeds tool failure is often experienced due to lack of strength and toughness whereas at higher cutting speeds tool life is often limited by the initiation and growth of thermal cracks.

Si_3N_4-based tools cannot be used as they are chemically unstable in the presence of iron-based materials. Diffusion and solution of Si_3N_4 into the chip leads to rapid crater formation and weakening of the cutting edge leading to catastrophic tool failure. The low chemical stability of silicon carbide also makes it impossible to use whisker-reinforced materials in steel machining operations. Hardened steels, however, represent an area where ceramic materials, especially mixed ceramics and whisker-reinforced ceramics, show a good performance due to a high hot hardness and a relatively good thermal shock resistance. When machining hardened steels cutting temperatures are relatively low so that solution/diffusion wear plays a less significant role.

5.3 Heat-Resistant Alloys

The machining of heat-resistant alloys is the area where the greatest benefits of ceramic cutting tools have been achieved. These workpiece materials have high temperature strengths and generate highly segmented chips which cause severe notch wear due to attrition at the depth of cut positions (see Fig. 2). The outstanding increase in productivity due to the introduction of new ceramic cutting tool materials is shown in Fig. 3 which plots tool life as a function of cutting speed when machining a nickel-based alloy, Inconel 718.

Materials of this type have commonly been machined with uncoated cemented carbides at low cutting speeds ($25–50\,m\,min^{-1}$). Mixed ceramics permit higher cutting speeds, but tool life is determined by localized depth-of-cut (DOC) notching.

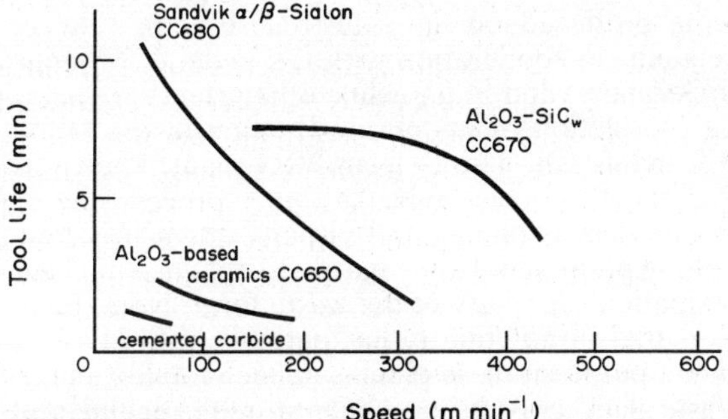

Figure 3
Tool life of ceramic and cemented carbide materials when machining Inconel 718 (feed $0.2\,mm\,rev^{-1}$, depth of cut 2 mm)

Sialon materials show a much better resistance to this type of wear, generally failing by uniform flank wear, and they permit cutting speeds up to ten times higher than those for cemented carbides. Silicon-carbide-whisker-reinforced ceramics can be used at cutting speeds up to 20 times those of cemented carbides. Tool life is generally determined by DOC localized notching. At these high cutting speeds the cutting tool temperatures generated are so high that a considerable softening of the workpiece material occurs, which decreases the stresses on the tool material.

6. Conclusion

At present ceramic cutting tools have a fairly low share of the total indexable insert market (approximately 4%). However, newly developed ceramic materials have been able to significantly increase their productivity in specific machining applications; for example, in the machining of cast iron and heat-resistant alloys.

The usage of ceramic cutting tools will increase owing to the development of more suitable machine tools but a rapid development towards a more general utilization is not anticipated unless new and much improved ceramic tool materials can be developed.

See also: Aluminum Oxide; Cutting-Tool Materials; Sialons

Bibliography

Baldoni J G, Buljan S T 1988 Ceramics for machining. *Am. Ceram. S.* 67: 381–7

Ezugwu E O, Wallbank J 1987 Manufacture and properties of ceramic cutting tools: A review. *Mater. Sci. T.* 3: 881–6

King A G, Wheildon W M 1966 *Ceramics in Machining Processes*. Academic Press, London

Trent E M 1984 *Metal Cutting*. Butterworth, London

G. Brandt
[AB Sandvik Coromant, Stockholm, Sweden]

Ceramic Materials for Nuclear Waste Storage

The operation of a nuclear power reactor producing 1300 MW of electrical power typically results in the accumulation of around 30 t of spent fuel per annum. The fuel elements are intact but highly radioactive. Many of the isotopic species comprising the one tonne or so of fission products in the spent fuel are short-, medium- and/or long-lived β or γ-emitters. In addition, various isotopes of the transuranic elements neptunium, plutonium, americium and curium—typically amounting to 250 kg in the same mass of fuel—have grown in by various nuclear reactions which follow the absorption of fast neutrons by ^{238}U. Most of these transuranics are medium- to long-lived α-emitters.

The management of spent fuel is technically and socially challenging in both the short and long term, because of the initially very high and then eventually much lower but very long-lived radioactivity. In the short term (i.e., for several decades) spent fuel is routinely stored in water-filled pools, and later it may be transferred to air-cooled dry storage vaults. Further management depends on whether or not the fuel is reprocessed to remove reusable uranium and plutonium. Some spent fuel may eventually be disposed of in deep geological repositories without ever having been reprocessed, although this has not yet happened in any country. In the reprocessing option, the nuclear waste contains only the residual fission products, the transuranics neptunium, americium and curium and a very small fraction of the uranium and plutonium. It is widely accepted that this high-level waste (HLW)—which is produced as a corrosive nitrate solution—must be solidified, perhaps then stored in air-cooled vaults for up to 50–100 years and eventually disposed of by deep geological burial. It is also accepted that the solidified waste form will be, in the broad sense of the term, a ceramic material. The ceramic may be either crystalline, partly crystalline (glass-ceramic) or non crystalline (glass).

1. Requirements of a Nuclear Waste Form

The prime requirements of a nuclear waste form are as follows.

(a) It must encapsulate the individual waste species either physically or in solid solution in such a way that their dispersal in ground water under geological disposal conditions by aqueous leaching will be acceptably slow. This leach resistance must be maintained for thousands—preferably hundreds of thousands—of years without significant deterioration from the effects of radiation damage resulting from the decay of contained radionuclides.

(b) It must be capable of fabrication into large (250–400 mm diameter) cylindrical monoliths with minimal internal cracking; these monoliths, which may be initially supported by metal cladding, must withstand routine transport and handling operations during storage and eventual geological disposal, and in a maximum credible accident must not produce an unacceptably high fraction of respirable fines.

(c) The cost of fabrication, storage and disposal of the waste form must be acceptable in the context of the other costs associated with electricity generation by nuclear power.

The range of ceramic nuclear waste forms accepted or proposed for future use was discussed recently in a comprehensive survey (Lutze and Ewing 1988).

2. Borosilicate Glasses

Borosilicate glasses (Lutze 1988) are accepted as the first-generation form for HLW solidification and storage in virtually all western nuclear countries. Waste oxides are immobilized partly in the glass structure and partly by physical encapsulation. Their aqueous leach resistance (requirement (a)) varies with glass composition, waste loading, temperature, water chemistry and flow rate but is reasonably low up to around, but not much above, 100 °C. As their structures are basically amorphous, internal actinide decay has a minor effect on their structure, density and leach resistance. They also appear to satisfy requirements (b) and (c).

Basic borosilicate glasses typically consist of 45–68 wt% SiO_2, 10–33 wt% B_2O_3, 1.4–18 wt% Na_2O, 0–7 wt% Li_2O, 0–6.5 wt% CaO and sometimes lesser amounts of TiO_2, MgO, Al_2O_3 and ZnO. Waste loadings up to 33 wt% (on an oxide basis) have been considered. However, even at low waste loadings, fission products such as palladium, rhodium, ruthenium and technetium do not dissolve in the glass, and other crystalline phases are rarely absent. As the waste loading is increased, so is the occurrence of crystalline precipitates, which may contain fission product elements which adversely affect leach resistance. This means that the choice of waste loading in any particular case will be a complex decision based on many factors, including the age of the waste and the specified waste glass properties. The practical upper limit of waste loading is likely to be around 25 wt%.

Fully active borosilicate waste glasses (in 250–400 mm diameter cylindrical metal canisters) are being produced on an industrial scale in France, Belgium (in cooperation with the Federal Republic of Germany) and India, while other plants are nearing completion or under construction in the USA, UK and Japan. The Belgian/FRG plant, known as 'Pamela', also contains a Vitromet process line to produce glass spheres and disperse them in a low-melting point metal alloy matrix to enhance the heat dissipation capability of the waste form. Note, however, that in no country has permanent disposal of the output from these plants yet been implemented. Waste glass canisters have been stored in engineered air-cooled vaults in France since production began at the Atelier de Vitrification de Marcoule (AVM) in 1978, and this same practice will be followed in the other countries until geological repositories gain technical and public acceptance. Glass storage times before disposal are likely to be a minimum of 30 years and may be as long as 100 years.

3. *Synroc*

Of all the potential second-generation improved alternatives to borosilicate glass for HLW solidification, storage and ultimate disposal, Synroc (synthetic rock) (Ringwood et al. 1979, 1988, Reeve 1988) is the most advanced in its stage of development. It is being developed mainly in Australia, with cooperation on specific aspects from the UK, Japan and Italy. The Australian program is centered at the Australian Nuclear Science and Technology Organisation (ANSTO). Various Synroc compositions are possible, but all are titanate-mineral phase assemblages. The reference form, Synroc-C, consists of zirconolite ($CaZrTi_2O_7$), hollandite ($Ba_{1.1}\ (Al, Ti)_{2.3}^{3+}\ Ti_{5.7}^{4+}\ O_{16}$), perovskite ($CaTiO_3$) and rutile phases. The first three have the capacity to accept nearly all the elements present in HLW into their crystal structures at regular lattice sites, while rutile is a matrix and chemical buffer phase with a very minor HLW content. Similar minerals occur in nature, where they have survived in a wide range of geochemical–geological environments for periods up to two billion years. This evidence suggests that Synroc should be a superior waste form which should help to make HLW disposal more acceptable to the sophisticated public of the twenty-first century.

The composition of Synroc-C is typically 71.1 wt% TiO_2, 6.8 wt% ZrO_2, 5.5 wt% Al_2O_3, 5.5 wt% BaO and 11.1 wt% CaO. HLW is added at a level of up to 20 wt%. Synroc is made by mixing the basic ingredients, such as oxides or gel-like hydrolysates of the appropriate metal alkoxides, with simulated or actual HLW, drying and calcining the slurry under reducing conditions, and uniaxially or isostatically hot pressing the calcine in graphite dies, stainless steel or inconel bellows or bellowslike containers.

In the Australian program at ANSTO, simulated HLW and simulated HLW spiked separately with mixed fission products and individual transuranic actinides (Reeve et al. 1987) have been used to demonstrate the intrinsic superior aqueous leach resistance of Synroc compared with borosilicate glass. Under dynamic leaching conditions at 90 °C, elemental leach rates were factors of 5×10^2 to 5×10^4 lower than expected values (Lutze 1988) for borosilicate glass. For fully active Synroc, made at the Harwell Laboratory in the UK in collaboration with Australia, the leach rates of various elements at 90 °C were lower by factors of 15–750 than expected for borosilicate glass under exactly the same testing conditions (Boult et al. 1988).

From evidence of the structural integrity of very old metamict samples of natural zirconolite and perovskite, and from the results of as yet incomplete studies of accelerated radiation damage in Synroc and its constituent minerals by means of fast neutron irradiation and actinide doping, it is argued (Ringwood et al. 1988) that although the Synroc minerals will eventually expand linearly by up to several per cent and become metamict, the aqueous leach resistance will be affected to only a small extent over geological time.

The production technology of Synroc has been demonstrated in a full-scale nonradioactive plant at ANSTO, which is further committed to the production of a conceptual design for a fully active Synroc production plant together with a comprehensive Synroc database by mid-1991. This will make possible a more firmly based economic assessment of the Synroc process for second-generation HLW immobilization than is possible at present.

4. *Other Ceramic and Glass Waste Forms*

4.1 *Tailored Ceramics*

Synroc (see Sect. 3) is one example of a tailored ceramic waste form. Harker (1988) has discussed other examples of this approach, the progenitor of which was the 'Supercalcine' concept (McCarthy 1977). In Supercalcine, the complex mixture of elements in HLW was modified by minimum chemical additions to encourage the formation of an assemblage of crystalline phases such as monazite, pollucite, apatite, fluorites, scheelite, corundum, spinels and various perovskites. However, because of the presence of silicates, an undesirable glass phase was usually present.

Harker discusses a range of durable polyphase ceramics tailored for commercial HLW and for several US defense wastes whose composition is dominated by the presence of either aluminum or zirconium arising from complete dissolution of the fuel cladding during reprocessing. Because the defense wastes are relatively dilute in fission products, economic considerations require relatively high waste loadings. Harker's listed compositions achieve waste loadings of up to 80 wt% by means of relatively small phase-tailoring additions of oxides such as TiO_2, ZrO_2, rare earth oxides, CaO, Al_2O_3 and SiO_2. Major phases formed include magnetoplumbites, spinels, uraninite (UO_2) solid solutions, zirconolite, perovskite, pseudobrookite, loveringite, rutile, nepheline and corundum. The proposed fabrication process is based on rotary or fluidized bed calcination followed by hot isostatic pressing. However, in the late 1980s work on tailored ceramics other than Synroc has been limited to a small effort in the USA with the Idaho Chemical Processing Plant waste.

4.2 *TiO_2-Matrix Ceramics*

Adelhelm et al. (1988) discuss an alternative approach applicable to commercial high-level wastes, which do not contain dissolved fuel cladding and where, as in the Synroc concept, much lower waste loadings can be economically accepted. In this approach, a continuous rutile (TiO_2) matrix is used to physically encapsulate waste particles, which are embedded heterogeneously rather than being incorporated in host phases. Waste loadings up to 12 wt% are possible. The proposed consolidation process, which involves in-can hot pressing, has been demonstrated nonradioactively up to a diameter of 100 mm. From nonactive tests, the product has a significantly lower leach rate at 100 °C than borosilicate glass, by a factor of up to 50 (for cesium). Much further work needs to be done to characterize this waste form fully and to develop a process amendable to hot-cell operation.

4.3 Single-Phase Ceramics

The potential of synthetic monazite (basically a lanthanide orthophosphate) for HLW immobilization, particularly for wastes rich in actinides, has been discussed by Boatner and Sales (1988). Waste loadings up to 20 wt% of US defense waste or 10 wt% of commercial HLW were investigated. Monoclinic monazite, which was shown to be capable of incorporating all the actinides, rare earths, strontium and barium in HLW in its structure, was the only crystalline phase seen by x-ray diffraction. The leach rates of uranium, cesium and strontium from the monazite ceramic were found to be at least a factor of 20 below the corresponding leach rates from borosilicate glass at 90 °C and the monazite matrix itself is approximately one thousand times more durable than glass. The structure retains a very high leach resistance even after radiation damage has made it metamict, but no radiation damage tests have been carried out on monazite containing actual HLW.

Ewing (1988) discusses the potential of sodium zirconium phosphate and a range of low-temperature hydroxylated ceramics including micas, apatite and nepheline for the incorporation of HLW elements. However, research to determine this potential fully is at a very early stage.

4.4 Glass Ceramics

The largest effort under this heading has been in Canada, where intensive development of a glass-ceramic waste form based on sphene ($CaTiSiO_5$) commenced in 1981 (Hayward 1988). The basic composition (6.6 mol.% Na_2O, 5.1 mol.% Al_2O_3, 16.5 mol.% CaO, 14.8 mol.% TiO_2, 57 mol.% SiO_2) forms a fluid melt at ~1300 °C which rapidly unmixes during cooling and which may be converted to a glass ceramic by reheating for one hour at ~1050 °C. The sole crystalline phase, sphene, amounts to 40 vol.% while the glass is an alumino-silicate which is more durable than typical borosilicate waste glasses by one to two orders of magnitude. Waste ions partition between the sphene and glass phases. For example, rare earths, yttrium and zirconium are preferentially partitioned into the sphene; strontium exhibits little preference for either phase; while barium, cesium and hexavalent uranium partition strongly into the glass. The waste loading limit for sphene to be the sole crystalline phase is 15 wt%. A major advantage claimed for sphene is its thermodynamic compatibility with groundwaters of the deep Canadian granitic shield.

Leaching tests on sphene glass ceramics containing simulated waste, including some with trace-active doping, have confirmed their predicted superiority, by at least one to two orders of magnitude, over borosilicate glass. In the associated radiation damage assessment program, severe radiation damage produced by argon ion bombardment did not increase the leach rate by more than a factor of five. No tests have been carried out with fully active waste. The largest scale demonstrated, nonradioactively, has involved a 3 kw glass melter from which the molten glass can be poured into 81, 25 mm internal diameter stainless steel containers.

Other glass ceramics which have been studied as nuclear waste forms have included (Hayward 1988): celsian glass ceramics (Lutze et al. 1979); fresnoite glass ceramics (Lutze et al. 1979) and basalt glass ceramics (Welch et al. 1982).

4.5 Miscellaneous Waste Form Proposals

Only brief mention can be made in this article of the following proposals, most of which are at an early stage of development.

(a) *Sintered glass* (Gahlert and Ondracek 1988). Glass obtained by a process for hot pressing (rather than melting) borosilicate glass plus HLW at a temperature of only 750 °C has been demonstrated in a full-scale nonradioactive process line.

(b) *Lead–iron–phosphate glasses* (Sales and Boatner 1988). This is a range of glasses with greatly improved leach resistance compared with borosilicate glass at temperatures below 150 °C. Their fabrication has been demonstrated only nonradioactively, in sizes up to 6 cm diameter.

(c) *High-silica porous glass matrix* (Ewing 1988). Glass obtained by a process in which a porous high-silica glass is infiltrated with HLW solution, calcined to 700 °C and sintered at 900 °C. Improved leach resistance compared with borosilicate glass has been demonstrated on a nonradioactive, small laboratory scale.

(d) *Multibarrier concepts* (Ewing 1988). These include the application of ceramic coatings (e.g., of Al_2O_3, SiC or pyrolytic carbon) to ceramic or glass particles containing HLW and the incorporation of these in a glass or metal matrix, and the related Vitromet process (see Sect. 2). The latter is the only one of these processes to have passed the laboratory research and development stage.

(e) *FUETAP* (McDaniel and Delzer 1988). This is the process in which concrete containing US defense waste is Formed Under Elevated Temperature and Pressure. This process is apparently not amenable to commercial HLW. The FUETAP program at Oak Ridge National Laboratory in the USA was terminated in 1980, but could generate interest at some future time.

5. Conclusion

Borosilicate glass is the generally accepted first-generation form for high-level nuclear waste solidification and storage. Second-generation waste forms designed to increase public acceptability of nuclear waste disposal in the twenty-first century should have greatly increased leach resistance in ground water. Of the alternatives to borosilicate glass covered in this survey, the crystalline titanate assemblage Synroc is the most advanced in its stage of development.

See also: High-level Radioactive Waste Disposal: Safety (Suppl. 2); Radioactive Waste Disposal

Bibliography

Adelhelm C, Bauer C, Gahlert S, Ondracek G 1988 TiO_2—A ceramic matrix. In: Lutze and Ewing 1988, Chap. 6

Boatner L A, Sales B C 1988 Monazite. In: Lutze and Ewing 1988, Chap. 8

Boult K A, Dalton J T, Evans J P, Hall A R, Inns A J, Marples J A C, Paige E L 1988 The preparation of fully-active Synroc and its radiation stability. Report AERE R 13318

Dé A K, Luckscheiter B, Lutze W, Malow G, Schiewer E 1976 Development of glass ceramics for the incorporation of fission products. *Am. Ceram. Soc. Bull.* 55: 500–3

Ewing R C 1988 Novel waste forms. In: Lutze and Ewing 1988, Chap. 10

Gahlert S, Ondracek G 1988 Sintered glass. In: Lutze and Ewing 1988, Chap. 2

Harker A B 1988 Tailored ceramics. In: Lutze and Ewing 1988, Chap. 5

Hayward P J 1988 Glass-ceramics. In: Lutze and Ewing 1988, Chap. 7

Lutze W 1988 Silicate glasses. In: Lutze and Ewing 1988, Chap. 1

Lutze W, Borchardt J, Dé A K 1979 Characterization of glass and glass-ceramic nuclear waste forms. In: McCarthy G J (ed.) *Scientific Basis for Nuclear Waste Management*, Vol. 1. Plenum, New York, pp. 69–81

Lutze W, Ewing R C (eds.) 1988 *Radioactive Waste Forms for the Future*. North-Holland, Amsterdam

McCarthy G J 1977 High-level waste ceramics: Materials considerations, process simulation and product characterisation. *Nucl. Technol.* 32: 92–105

McDaniel E W, Delzer D B 1988 FUETAP concrete. In: Lutze and Ewing 1988, Chap. 9

Reeve K D 1988 Synroc—A high quality ceramic nuclear waste form. *Mater. Sci. Forum* 34–36: 567–70

Reeve K D, Levins D M, Seatonberry B W, Ryan R K, Hart K P, Stevens G T 1987 Final report on fabrication and study of Synroc containing radioactive waste elements. Australian Nuclear Science and Technology Organisation Rept AAEC/C60

Ringwood A E, Kelly P M 1986 Immobilisation of high-level waste in ceramic waste forms. *Philos. Trans. R. Soc. London, Ser. A* 319: 63–82

Ringwood A E, Kesson S E, Reeve K D, Levins D M, Ramm E J 1988 Synroc. In: Lutze and Ewing 1988, Chap. 4

Ringwood A E, Kesson S E, Ware N G, Hibberson W D, Major A 1979 The Synroc process: A geochemical approach to nuclear waste immobilisation. *Geochem. J.* 13: 141–65

Roy R 1982 *Radioactive Waste Disposal*, Vol. 1: The Waste Package. Pergamon, New York

Sales B C, Boatner L A 1988 Lead–iron phosphate glass. In: Lutze and Ewing 1988, Chap. 3

Welch J M, Schuman R P, Flinn J E 1982 Iron-enriched basalt for containment of nuclear wastes. In: Topp S V (ed.) *Scientific Basis for Nuclear Waste Management*, Vol. 6. North-Holland, Amsterdam, pp. 23–30

Wicks G G, Ross W A (eds.) 1984 *Nuclear Waste Management—Advances in Ceramics*, Vol. 8. The American Ceramic Society, Columbus, OH

K. D. Reeve
[Australian Nuclear Science and Technology Organisation, Menai, New South Wales, Australia]

Ceramic Membranes, Porous

Research and development of thermostable ceramic membranes is a new field which has received much interest since 1980. It is rapidly developing since it has become obvious that defect and crack-free membranes can be produced.

Membranes were originally studied for separation of mixtures of gases or liquids. More recently it has become obvious that their separative properties can be used for the manipulation of chemical reactions resulting in increased yields or selectivities and an improved energy balance of chemical reactors. Finally, a number of nonseparation applications are emerging, such as the immobilization of enzymes and cells in bioreactors or in diagnostic kits, or their use in sensors and in photocatalysis, and so on.

1. General Considerations

A schematic overview of the several types of membranes is given in Table 1. In 1988, supported porous, asymmetric ceramic membranes were the most important class of inorganic membranes. This section covers some basic points for such a porous membrane system.

Table 1
Schematic of inorganic membrane types

	Main characteristics	Notes
Dense	metal foil oxidic solid electrolyte liquid immobilized (LIM)	solution/diffusion of atomic or ionic species
Dynamic	permanent nonpermanent[a]	ion exchange in hydroxide layers on a support
Porous metal or non metallic inorganic	symmetric, asymmetric supported, nonsupported pore shape, morphology and size; chemical nature of pore surface	perm-selective diffusion affected by the pore characteristics
Composite	two phase particle mixture	
Modified	pores in matrix partial filled with 2nd phase sandwich structures	distribution important, microparticles, props

a nonpermanent means: separation layer is formed during the preparation process *in situ* on a porous support and is only present then in a stable form

1.1 Structure

The essential structural features of the membrane system are shown in Fig. 1. The system consists of a porous support with a thickness of a few millimeters and with pores with diameters in the range of 1–10 μm, a porous intermediate layer of thickness 10–100 μm with pores of 0.05–0.5 μm diameter and a top layer with a thickness of 1–5 μm and pores of 2–50 nm diameter. The intermediate layer must prevent the penetration of the precursor of the top layer material into the pores of the support during the synthesis and the collapse of the finished top layer into the large pores of the support. Furthermore, it enables the regulation of the pressure drop across the top layer of the membrane in service.

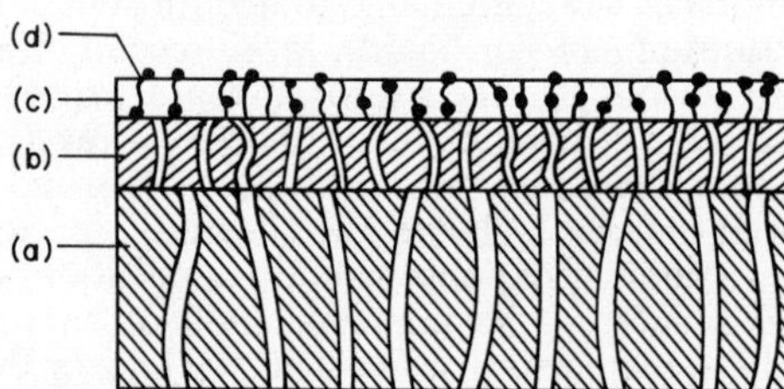

Figure 1
Schematic representation of an asymmetric composite membrane: (a) porous support (1–15 μm pores); (b) intermediate layer(s) (100–1500 nm pores); (c) separation top layer (3–100 nm pores); and (d) modification of separation layer. (a) + (b) microfiltration or "primary" membrane, (a) + (b) + (c) ultrafiltration or "secondary" membrane, and (a) + (b) + (c) + (d) hyperfiltration and/or gas separation membrane

1.2 Function

A membrane can be described as a semipermeable barrier between two phases that prevents intimate contact. The barrier must be perm-selective, which means that it restricts the movement of molecules within it in a specific way. The barrier can be solid, liquid or gas. The perm-selectivity for mixtures of different molecules can be obtained by one of several mechanisms:

(a) by size exclusion or molecular sieving,
(b) by differences in diffusion coefficients (bulk as well as surface),
(c) electric charge,
(d) solubility, or
(e) adsorption and reactivity on internal surfaces.

The pore system of the top layer can be modified further, mainly by precipitation of a new phase from liquids or gases. This results in a further decrease of the pore diameter to below 2 nm and/or in a change of the chemical nature of the surface. The object of this is to produce a further increase in the separation effectiveness and/or to introduce catalytic activity. The support, or the combination of both support and intermediate layer, can also be used in certain applications (e.g., microfiltration). It is only in the very small pores of the top layer that interaction of the gas or liquid mixture with the modified internal surface of the membrane can play a dominant role in the membrane properties (e.g., gas separation, catalytically active membranes).

1.3 Inorganic Membranes

Dense, inorganic membranes consist of solid layers of metals (palladium, silver, alloys) or oxidic solid electrolytes which allow selective diffusion of hydrogen or oxygen atoms and ions. An interesting category of dense materials consists of a porous support in which a liquid is immobilized (so-called LIM). Interesting examples are molten salts immobilized in in a porous support and semipermeable with very high selectivity for oxygen or ammonia and with permeabilities not far below that of the porous membranes.

1.4 Advantages and Disadvantages

The main advantages of ceramic membranes are: (a) high-temperature thermal stability and chemical stability are high, (b) mechanical stability in large pressure gradients is good, (c) rigorous cleaning operations are possible (steam sterilization or backflush), and (d) catalytic activity is easily realizable.

Their disadvantages are their brittle character, relatively high costs and complicated sealing technology for high-temperature applications.

2. Basic Principles of Membrane Synthesis

The effectiveness of a membrane in a certain application depends on the detailed morphology and microstructure of the membrane. These are critically determined by the synthesis process. The most important and well-developed processes are (a) packing of particles from dispersions, (b) phase separation and leaching methods, (c) anodic oxidation (of aluminum), (d) pyrolysis, (e) track-etch methods, and (f) modification methods of primary membranes.

A review of these methods is given by Burggraaf and Keizer (1990). Phase separation of borosilicate glasses followed by leaching of the boron-rich phase, drying and calcination result in a porous silica-rich glass with pores as small as 4 nm diameter. The material can be produced in hollow fiber form. The main disadvantage is the relatively poor chemical stability. Straight pores can be produced by anodic oxidation of an aluminum foil in an acid electrolyte. The unaffected portion of the foil is subsequently etched away. The resultant structure of the nonsupported ceramic membranes has distinctive conical pores perpendicular to the macroscopic surface. Pore sizes are obtainable ranging from 25 nm to 200 nm. To get sufficient mechanical stability, these membranes must be supported in some way. Regular, linear-shaped pores can be made by bombarding a thin layer of material with highly energetic particles and etching away the residual track (the track-etch method) (see *Solid-State Nuclear Track Detectors: Applications*, Suppl. 1). Pore diameters ranging from 6 nm to 1200 nm can be obtained with porosities of 2–5%. Membranes of this type are attractive for model systems in fundamental studies.

Pyrolysis of thermosetting polymers or silicone rubbers can result in 'molecular sieve' carbon or silicon membranes with pore diameters of $\leqslant 1.0$ nm.

By far the most frequently used and promising method is the formation of a layer consisting of a packing of well-ordered, uniformly sized particles. The size and shape of these particles determine the minimum obtainable mean pore diameter and the pore size distribution. These parameters, as well as the porosity, can be changed by further heat treatment or by modification procedures.

The main process for making ceramic membranes is to prepare a dispersion of fine particles (called slip) and then to deposit the particles in the slip onto a porous support by a slip casting or by a film-coating. The capillary pressure drop created by contacting the slip with the macroporous support forces the dispersion medium of the slip to flow into the dry pores of the support. The slip particles are concentrated at the pore entrance to form a gel layer. Organic binders, plasticizers and viscosity modifiers may be added to the precursor slips to enhance the binding between layer and support, to control the consistency of the slip and to prevent cracking of the layer during drying and further processing.

For making the top layers with very small pores, colloidal dispersions are used. Here nanometer-sized particles are needed which are stabilized in a liquid dispersion medium by colloid chemical methods. The colloidal dispersions are obtained by so-called sol-gel methods. The essential features of these methods are the controlled hydrolysis of a metal-organic compound or salt and its subsequent peptization or polymerization.

If the colloidal dispersion with particles of between 5 nm and 15 nm makes contact with the porous support, the subtraction of water at the interface into the support pores (by capillary action) causes the formation of a *gel* layer at this interface. The thickness of this gel layer increases with the square root of time in a slipcasting process.

The gel layer is dried to form a xerogel. This drying step is critical because the large capillary pressures acting in this process step can easily cause cracking of the layer. Important controlling parameters are particle shape and concentration, agglomeration degree, binders/plasticizers, roughness and pore size of the support, and so on.

After drying, the xerogel is calcined to form the final oxidic structure. The minimum diameter of pores obtainable in this way is about 2.5–3 nm with porosities up to 50%. To further reduce the pore size the pore structure must be modified. (Heat treatment at higher temperatures increases the mean pore size).

Commercially available in 1988 were γ- and α-alumina and zirconia membranes on alumina supports, zirconia membranes on carbon supports, porous glass membranes and membranes produced by track-etch methods. Laboratory-scale results have been published for titania and silica membranes on alumina and metal supports, zirconia on zirconia supports, unsupported and supported carbon membranes and membranes consisting of binary mixtures of alumina with titania, zirconia or ceria on alumina supports. For dense membranes results have been published for metal membranes (silver, palladium) and solid electrolyte membranes (oxygen-ion-conducting solid solutions of ZrO_2 or Bi_2O_3 with stabilizing oxides like Y_2O_3 or rare earth oxides).

3. *Methods for Modification of Membranes*

The pores of a top layer of an existing membrane system can be modified by deposition of a new phase from liquids or gases. Several possible morphologies are schematically shown in Fig. 2, but in principle a large variety of techniques can be used. Interesting results were obtained by homogeneous precipitation *in situ* within the pore structure from mixtures of a metal salt and urea in water. The liquid solution was brought into the pores by impregnation. In this way magnesia or silver could be deposited, filling up to 50% of the pore volume. By careful regulation of the evaporation of the solvent and the temperature it seems possible to control the concentration profile of the deposit. Instead of precipitation, material can be deposited by adsorption from, for example, metal acetyl acetate or alkoxide solutions. In this way, vanadia or iron oxide monolayers are produced in alumina or titania membranes.

Controlled hydrolysis of alcoholic tetraorthosilicate solutions on or in alumina membranes followed by calcination results in silica deposits that decrease the effective pore diameter to between 1 nm and 2 nm (Uhlhorn et al. 1989a). Finally, the synthesis of zeolite membranes is observed in the reaction of alumina membranes with a range of reagents in a complex series of steps, including hydrothermal treatment (Burggraaf and Keizer, 1990).

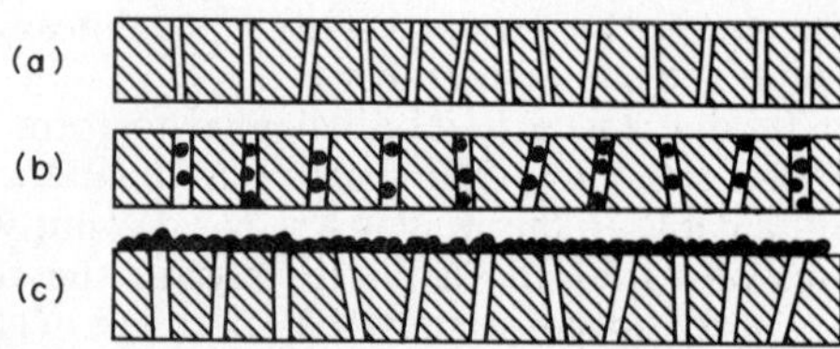

Figure 2
Schematic representation of the microstructure of modified membrane top layers: (a) homogeneous multilayers in the pores; (b) plugs in the pores (constrictions); and (c) plugs/layers on top of the pores

4. *Separation and Reactions with Membranes*

4.1 *Separation*

In both liquid and gas separation a mixture of compounds permeates through the membrane, usually driven by a pressure gradient but in some cases by an electrochemical or electric field gradient. In all cases the aim is to obtain a large permeability (flux) in combination with a large selectivity (separation factor). Separations in liquids can take place between solvents and solutes, or between species in liquid media due to the effect of size exclusion which predominates in pressure driven porous ceramic membranes (micro- and ultrafiltration). Additional interactions between feed components and the membrane pore surface invariably exist and these become significant for pores with a diameter of a few nanometers. The membrane pore surface can play an important role in separating a given liquid system depending on the surface charge and preferential adsorption effects.

Applications of these separations include: treating oil–water and oil–latex emulsions in waste water, recovery of textile sizing agents, extraction of proteins from whey, classification of beverages and a number of biotechnological applications. High temperature applications include the processing of high molecular weight mixtures in the petrochemical industry.

Serious problems associated with the separations are concentration polarization and fouling, which are are partially solved by application of favorable dynamic flow conditions and by regeneration operations like steam sterilization and periodical application of back-flush pulses.

The flux J under laminar flow conditions can be described by a Kozeny–Carman-type equation:

$$J=\frac{\varepsilon^3\Delta P}{Kt_m S_i^2(1-\varepsilon)^2\eta} \tag{1}$$

where J is the permeate flux ($m\,s^{-1}$), ΔP is the applied transmembrane pressure difference (Pa), K is the dimensionless Kozeny–Carman constant which incorporates the tortuosity of the membrane structure, η is the permeate viscosity (Pa s), ε is the membrane porosity, t_m is membrane thickness (m), S_i is the internal surface area per unit volume of membrane ($m^2\,m^{-3}$). Reported flux values are between $0.0003\,m\,s^{-1}$ and $0.02\,m\,s^{-1}$ depending on the conditions, and it is obvious that asymmetric structures (small thickness of the separating layer) promote a large permeability.

By the late 1980s gas separation had not been used industrially except in isotope enrichment. However, a significant volume of research shows its potential as a gas separation tool and recently the industrial use of vapor separation is reported. For a given membrane system the degree of separation α_{ij} between gases depends on the relative permeabilities of the gases to be separated.

This separation α_{ij} is defined as:

$$\alpha_{ij} = \frac{y_i/x_i}{y_j/x_j} \tag{2}$$

where i represents the more permeable gas, j the less permeable gas, and y and x are the mole-fractions of the gas species downstream and upstream of the membrane respectively.

For a binary mixture of gases α can be expressed as:

$$\alpha_{12} = a_{12}^{*} F_1(P_d/P_u, y_1/x_1)F_2 \tag{3}$$

with P_d and P_u are the pressures downstream and upstream and α_{12}^{*} is the ideal separation factor given by the ratio of the individual Knudsen permeabilities (ϕ_1 and ϕ_2) of the two permeating gases. This gives the so called perm-selectivity. F_2 represents a correction factor which incorporates process effects such as back-diffusion and concentration polarization. For a large value of P_d/P_u, Eqn. (3) reduces to $\alpha_{12} = \alpha_{12}^{*}$ provided that no concentration polarization and laminar flow occur.

Useful transport mechanisms for gas separation are (a) size exclusion and molecular sieving (in very small pores), (b) Knudsen diffusion, (c) surface diffusion, and (d) capillary condensation and multilayer diffusion. Laminar and turbulent flows which occur in larger pores must be eliminated because they decrease the separative action of the membrane. Equation (4) gives the permeability F_0 for some mechanisms (Uhlhorn et al. 1989b).

$$F_0 = \left(\frac{\varepsilon}{\tau}\right) c_1 M^{-1/2} r + c_2 P \eta r^2 + c_3 D_s \frac{dx_s}{dp} r^{-1} \tag{4}$$

where c_1, c_2 and c_3 are constants, ε is the membrane porosity, τ is the tortuosity, M is the molecular mass, P is the pressure, η is the gas viscosity, D_s is the surface diffusion coefficient of adsorbed gases, x_s is the fraction of occupied sites (by adsorption) on the membrane internal surface compared with a monolayer and r is the pore radius.

With decreasing pore size, the laminar flow contribution (the second term in Eqn. 4) vanishes and Knudsen diffusion (the first term in Eqn. 4) becomes the predominant mechanism. This happens when the pore diameter measures approximately 5–50 nm depending on the pressure. The separation factor is limited by the square root of the molecular weight ratios of the gases being separated. With a further decrease in pore size, the surface diffusion flux (the third term in Eqn. 4) can become predominant. Surface diffusion is important when one component is preferentially adsorbed. As it accumulates, the adsorbed component diffuses faster than the other nonadsorbed component and an additional surface flux parallel to the flux through the gas phase causes a difference in the permeability and consequently in the separation. This phenomenon becomes important at pore diameters between 1 nm and 3 nm and especially when the specific surface area is large. This is expressed by the term r^{-1} in Eqn. 4. The effect of the surface concentration and of details of the interaction are incorporated in the parameters c_1, c_2 and c_3 and in dx_s/dP.

With multilayer adsorption and capillary condensation the pores are gradually filled when pressure increases (multilayer adsorption) and the effective pore size decreases. This results in a decreasing flux of the nonadsorbing component and in an increasing flux of the adsorbing specimen. Finally the pores are blocked either by multilayer adsorption or by capillary condensation.

When other gases do not dissolve or dissolve only in very small amounts in the condensed component, separation occurs. This phenomenon occurs especially with easily condensable gases (and with vapors). Examples are the alcohol–water, water–air and SO_2–H_2 systems.

Modification of the membrane pores, resulting in a pore size decrease or in a change of chemical character, can considerably change the interaction parameters and hence change the perm-selectivity of the membrane. With Knudsen diffusion, only separation factors of 2–5 are reported; these can be increased by surface diffusion and modification to 10–20 in materials with pores larger than 2 nm in diameter (Hsieh 1988, Uhlhorn et al. 1989b, Catalytica Studies Division 1988). With pore diameters smaller than 2 nm, separation of the flux into its different contributions has not been reported. Separation factors in the range of 7–500 have been reported for H_2O–alcohol and H_2–SO_2 mixtures, probably by a capillary condensation mechanism in modified alumina and carbon membranes. Perm-selectivities of 10–40 are reported for H_2–N_2 in carbon and porous glass membranes (Hsieh 1988, Catalytica Studies Division 1988) and for mixtures of a number of hydrocarbons and nitrogen or helium in silica modified alumina (Hsieh 1988, Zaspalis et al. 1989a, Uhlhorn et al. 1989a) with very small pores.

For dense membranes consisting of a porous steel gauze filled with molten salts, a separation factor of 172 for O_2–N_2 combined with a high permeability (2000 Barrer) has been reported.

4.2 Membrane Reactors

Ceramic membranes can be combined in several ways with reactors:

(a) membranes physically separated from reactors,

(b) membranes as part of the reactors containing catalysts, and

(c) membranes as catalysts or catalyst carriers loaded with a catalyst on their internal surface. This last category of membranes are called catalytic membranes.

Incorporating membranes in reactors can increase the conversion of reaction yield and/or enhance the reaction selectivity towards a desired product. Catalytic action is not always necessary. An increase in the yield of product C can be obtained when a reaction product B is continuously removed by the membrane from the gaseous equilibrium reaction $A \rightleftharpoons B + C$.

It is obvious that dehydrogenation reactions especially can be affected in this way (relatively large separation is not necessary but is obtainable). Improved selectivity in catalytic reactions can be obtained in different ways. First, precise control of the contact time and its distribution is possible by adjusting membrane thickness and pore size, pressure and partial-pressure gradient and by altering the catalyst distribution in the membrane. In this way unwanted side-reactions can be partially eliminated. Second, a controlled injection of a reactant B through the membrane into the other reactant C is possible in the reaction $A \rightleftharpoons B + C$, instead of contacting a gas mixture with the catalyst. In this case the membrane is used to separate two reaction chambers each of which contains one of the reactants and the product A of the reaction can be rapidly removed. Typical possibilities here are selective oxidation or oxidative dehydrogenation reactions.

Investigations with dense metal membranes in the early 1980s showed interesting results, but the low permeability of these membranes hampers further development (Burggraaf and Keizer 1990, Catalytica Studies Division 1988).

The use of porous glass membranes is more recent while the uses of porous ceramic and especially modified porous ceramic membranes are just becoming apparent (Burggraaf and Keizer 1990, Catalytica Studies Division 1988, Zaspalis et al. 1989a). An interesting illustration is the oxidative dehydrogenation of methanol to formaldehyde with silver-modified γ-alumina membranes (Zaspalis et al. 1989a).

Recently, chemical reactors based on electrically or electrochemically driven dense ceramic membranes (solid electrolytes) and oxygen pumps were reported (Catalytica Studies Division 1988). As with metals, they have a relatively low permeability which, however, can be improved.

Reported examples of the use of membrane reactors are some dehydrogenation reactions; for example, the conversion of cyclohexane to benzene and the decomposition of H_2S (Burggraaf and Keizer 1990, Hsieh 1988, Catalytica Studies Division 1988).

The large potential of membrane reactors is just emerging, and besides increased conversion and selectivity, other benefits may be possible. Calculations show that an improved energy balance of the total reactor system can be obtained with membrane reactors.

See also: Gas Sensors, Solid State (Suppl. 2)

Bibliography

Burggraaf A J, Keizer K 1990 Synthesis of inorganic membranes. In: Bhave R (ed.) 1990 *Inorganic Membranes: Synthesis, Characteristics and Applications*. Van Nostrand Reinhold, New York, Chap. 2.

Catalytica Studies Division 1988 Catalytic membrane reactors: Concepts and applications, Catalytica Study No. 4187 MR 1988. CSD, Mountain View, CA

Hsieh H S 1988 In: Sirkar K K, Lloyd D R (eds.) 1988 *New Membrane Materials and Processes for Separation*, AIChE Symp. Series 261 84: 1–8

Uhlhorn R J R, Huis in't Veld M J, Keizer K, Burggraaf A J 1989a High perm-selectivities of microporous modified γ-alumina membranes. *J. Mater. Sci. Lett.* 8: 1135

Uhlhorn R J R, Keizer K, Burggraaf A J 1989b Gas and surface diffusion in (modified) γ–Al_2O_3. *J. Membrane Science* 46: 225

Uhlhorn R J R, Keizer K, Burggraaf A J 1990 Gas separation with inorganic membranes. In: Bhave R (ed.) 1990 *Inorganic Membranes: Synthesis, Characteristics and Applications*. Van Nostrand Reinhold, New York, Chap. 4. 2

Zaspalis V, Keizer K, van Ommen J G, Ross J R H, Burggraaf A J 1989a Ceramic membranes as catalytic active materials in selective (oxidative) dehydrogenation reactions. *Proc. Br. Ceram. Soc.* 43: 103

Zaspalis V, Keizer K, van Ommen J G, Ross J R H, Burggraaf A J 1989b *Porous Ceramic Membranes as Catalytic Active Materials*. AIChE Symp. Series, in press

A. J. Burggraaf and K. Keizer
[University of Twente, Enschede,
The Netherlands]

Ceramic Powders: Chemical Preparation

Successful economic exploitation of advanced ceramic materials requires that components can be reproduced from powders and function reliably in various working environments. Worldwide research and development activities have resulted in an increasing awareness that the chemical synthesis or processing methods used for ceramic materials allow control of powder properties important for fabrication of monolithic components. While progress in materials science has contributions from many disciplines, it is chemistry which occupies a pivotal role in the early 1990s for preparation of ceramic powders. However, surveys of the scientific literature and attendance at conferences on materials science show an increasing chemical input to powder synthesis since the 1970s rather than a sudden emergence of this chemical role.

1. Conventional Ceramic Synthesis

Three conventional techniques are used for preparation of ceramic powders, both multicomponent oxides and nonoxide materials.

1.1 Powder Mixing

Oxides, carbonates and hydroxides are blended together and then either wet or dry milled, after which the powder mixture undergoes repeated calcination and mechanical grinding for completion of the solid-state reaction (the rate of which is increased by compaction before firing). Powder mixing is used widely for advanced ceramic manufacture, particularly for electroceramics such as barium titanates, and has also attracted considerable attention recently for synthesis of high-temperature oxide superconductors. Ceramic synthesis does not often receive widespread publicity, but since oxide superconductors were discovered (Bednorz and Müller 1986), its profile has been raised not only within scientific circles, but also among the general public. The powder mixing method does, however, have disadvantages. First, chemical impurities are introduced into the powders from the grinding media, often alumina, zirconia and stainless steel, used for comminution. Second, it is difficult to achieve an intimate mixture and maintain chemical homogeneity on the molecular level in the final product, especially when one reactant is present in minor quantities as frequently occurs in electroceramics. Third, powders made in solid-state reactions are often aggregated, an undesirable property for fabrication of high-strength components, while volatile compounds such as lead oxide can be removed by evaporation at the high temperatures, typically 1373 K or more, which are encountered in these reactions.

Conventional synthesis for nonoxides involves carbothermic reduction of carbon–oxide mixtures to the metal which is then nitrided. Thus, silicon nitride is manufactured according to the overall equation

$$3SiO_2 + 6C + 2N_2 \rightarrow Si_3N_4 + 6CO \qquad (1)$$

As in the case of oxide ceramics, mechanical grinding introduces impurities but, additionally, residual oxygen present as silica has deleterious effects on the mechanical behavior of fabricated components.

1.2 Precipitation from Solution

Precipitation of hydroxides and oxalates from solution is the second conventional method of ceramic synthesis. The Bayer process for manufacture of alumina involves dissolution in alkali of bauxite, a naturally occurring mineral, followed by bayerite precipitation from the resulting sodium aluminate solution, although it should be appreciated that Bayer alumina for advanced ceramic applications constitutes only about 5% (by weight) of the total alumina output. Also, yttrium and zirconium hydroxides are coprecipitated from mixed salt solutions during production of partially stabilized zirconia. While precipitation and powder mixing share the same disadvantages, careful control of solution conditions (e.g., pH) is essential in order to deposit all components as finely divided precipitates, although the latter can be leached or peptized during washing for removal of entrained electrolyte.

1.3 Fusion

In the fusion route, oxides and carbonates are melted together and, after solidification, lumps are comminuted to ceramic powders by mechanical grinding. Fusion is used for both traditional and advanced ceramic synthesis. The first category involves architectural glass, while alumina-based abrasive, glass ceramics and glasses (e.g., calcia–silica–alumina compositions of ionomer cements used as dental restorative materials) are manufactured by this process. However, the melting point of reactants, together with their volatility, limits the versatility of the fusion method. While conventional syntheses have been applied to laboratory-scale preparations and manufacturing processes, the requirement for better control of impurity levels, particle diameters and particle aggregation, as well as enhanced chemical homogeneity, has promoted a variety of nonconventional syntheses (Segal 1989).

2. Nonconventional Ceramic Synthesis

2.1 Sol–Gel Processing

The phrase "sol–gel" refers to two processing techniques when discussing advanced ceramic synthesis. In the first (Segal 1989), colloidal dispersions (i.e., sols) of oxides or hydroxides are prepared either by peptization of coarse precipitates or by growth of polymeric cationic species from salt solutions; particle size in these sols is by definition in the range 1 nm–1 μm. Dispersions are derived from hydrolyzable cations, in practice tri- and tetravalent species (e.g., Al^{3+}, Cr^{3+}, Ce^{4+} and Zr^{4+}), and are reversibly dehydrated to gel powders which can be produced in a spherical form by spray-drying. Oxide powders are obtained on gel calcination and, when spherical, have diameters in the range 5–100 μm. Sol–gel powders have good flow properties compared with angular-shaped material produced in conventional syntheses, and this is an advantage when they are used in thermal-spraying processes for deposition of ceramic coatings.

Multicomponent oxides are prepared by mixing different sols at the colloidal level, which corresponds to a smaller particle size than is encountered in conventional syntheses. The result is lower reaction temperatures and better homogeneity than is attainable in conventional syntheses. In addition, oxide porosity is determined by the aggregated nature of colloidal units; that is, whether primary or aggregated particles are present. This sol-gel process is suitable for preparation of "high value, low volume" products, but economic considerations prevent it from replacing conventional syntheses for traditional ceramics such as tableware. Developed initially as a dust-free route to microspherical powders containing mixed uranium–thorium oxides for nuclear fuels, sol–gel has been widely applied to other ceramic powders (Segal 1989) including glass compositions, chromatographic adsorbents, catalyst supports, alumina-based abrasives and stabilized zirconia, as well as many electroceramics such as ferrites and zinc oxide.

Sol–gel processing of metal organic compounds (Thomas 1988) involves irreversible hydrolysis of alkoxide solutions (also called sols), a reaction which is both acid- and base-catalyzed. An electrophilic reaction mechanism in acid catalysis yields rigid monolithic gels that are calcined to oxide powders, whereas base hydrolysis proceeds by nucleophilic substitution which results in direct precipitation of oxide particles. While alkoxides have been prepared from most elements in the Periodic Table, this sol–gel process has mainly involved studies of alkoxides of silicon, zirconium, aluminum and titanium, which reflects in part their commercial availability. Extensive work has been carried out for glass compositions based on silicon esters such as tetraethoxysilane, $Si(OC_2H_5)_4$, whereby hydrolysis involves formation of silanol groups ($\equiv$Si—OH) followed by monomer condensation and polymer growth. Homogeneous nucleation at high pH yields approximately submicrometer unaggregated spherical powders with a narrow size distribution. Oxides prepared by this method have attracted considerable attention (Barringer et al. 1984) in recent years as they are often considered to have the characteristics of ideal sinterable powders from which strong monolithic components can be reproducibly fabricated at temperatures lower than required for powders made by conventional synthesis. Submicrometer silica, alumina, titania and zirconia particles have been synthesized, although large-scale production has not yet been achieved. Homogeneous nucleation in alkoxide solutions involves stages of particle nucleation, growth and aggregation. The first two stages may be controlled by the reaction kinetics, whereas particle aggregation can be considered in terms of attractive and repulsive forces acting between particles in lyophobic colloidal systems (Hiemenz 1986).

Powders derived from alkoxides contain few impurities as the starting reagents are often liquids or volatile solids which can be purified by distillation. In contrast with the processing of colloids, alkoxides interact at the molecular level in solution rather than at the colloidal level with a resulting enhanced chemical homogeneity and lower reaction temperatures. The technique is associated with nonfusion routes to glasses, but it is particularly useful for high-purity compositions that are difficult to make by conventional synthesis. These advantages have been appreciated in the increasing use of this technique for the synthesis of superconducting ceramic powders, although the method has disadvantages, namely cost and availability of alkoxides. Sol–gel processing of alkoxides is not restricted to electrophilic and nucleophilic mechanisms involving water, as hydrogen sulfide has been reacted with germanium alkoxides yielding amorphous germanium sulfide.

2.2 Gas-Phase Reactions

Reactions between gases, or between solids and gases, are used for the synthesis of ceramics, particularly nonoxides. An example from the latter category is given in Eqn. (1), although this reaction mechanism is complex and involves formation of gaseous silicon monoxide. A gas-phase reaction which has been used for manufacture of silicon nitride is given by the overall equation

$$3SiCl_4 + 4NH_3 \rightarrow Si_3N_4 + 12HCl \qquad (2)$$

where the heat source is a conventional resistance furnace. Residual chlorine in the product can reduce high-temperature strength, but exceptionally pure silicon nitride has been made by utilizing thermodynamic equilibria in the Si–S–N system whereby silicon disulfide, SiS_2, is reacted with gaseous ammonia. Fumed or flame-hydrolyzed oxides are prepared by passage of anhydrous metal halide vapor through an oxygen–hydrogen stationary flame, hence for silica

$$SiCl_4 + 2H_2 + O_2 \rightarrow SiO_2 + 4HCl \qquad (3)$$

where the aggregated oxide powder is formed by coalescence of liquid droplets.

Recent experimental procedures have concentrated on developing heating methods that allow control of nucleation, growth and particle aggregation from supersaturated vapors, an analogous approach to that used for alkoxide hydrolysis described in Sect. 2.1. For example, silicon carbide has been synthesized by heating silane–ethylene mixtures with a high-powered (150 W) CO_2 laser. These laser-driven reactions are characterized by rapid heating and cooling rates (about 10^5 K s^{-1}) and fast reaction times (around 10^{-3} s) while absence of wall heating eliminates heterogeneous nucleation. In addition, inductively coupled radio-frequency (RF) sources have been used (Fauchais et al. 1983) for heating gases and generating plasmas from which oxide powders such as SnO_2–Al_2O_3 have been condensed and, as for laser-driven reactions, fast heating and cooling rates encourage homogeneous nucleation. High purity ceramics are obtained because gaseous reactants are readily purified and do not come into contact with electrode material, while powders made in laser-driven and plasma reactions are often considered ideal.

The growth of experimental procedures, highlighted for gas-phase reactions, has been accompanied by an increasing number of theoretical attempts to describe the complex aggregated structures found in ceramic systems. The branch of mathematics known as fractal geometry (Dauod 1987) which has been used for modelling macroscopic random structures, including geographic coastlines, has gained wide popularity since the mid-1980s for analysis of the microscopic morphologies generated in ceramic systems (see *Fractals*, Suppl. 1). Its use has been aided by advances in computer technology. Particle aggregates in fumed material, as well as gel structures in hydrolyzed alkoxide solutions, have been subjected to analysis and their fractal dimensions determined.

2.3 Hydrothermal Synthesis

Monodispersed ceramic powders can be obtained by heating salt solutions under hydrothermal conditions (Matijević 1987) at temperatures of approximately 600 K and pressures of approximately 100 MPa. Although the detailed chemistry of particle nucleation, growth and aggregation in solutions at elevated temperatures has not been fully clarified, cation hydrolysis can occur in the absence of an added base according to the representative equation

$$a[M(H_2O)_x]^{z+} \rightarrow [M_a(H_2O)_{ax-b}(OH)_b]^{(az-b)+} + bH^+ \quad (4)$$

where z is the valency of a cation M, and a, b, x are molar quantities. The hydrolyzed species shown in Eqn. (4) can be considered a source for particle nucleation. Monodispersed Cr_2O_3 and TiO_2 have been synthesized using this technique, and while its advantage is as a low-temperature processing route to oxides, the drawback, as for homogeneous nucleation, is that dilute cation concentrations about 10^{-2} M are essential for formation of monosized particles. However, hydrothermal synthesis also applies to thermal treatment of oxide and hydroxide powders which undergo recrystallization and dissolution processes in an aqueous environment at elevated temperatures, and is under consideration as a potential route for manufacture of stabilized zirconia.

2.4 Nonaqueous Liquid-Phase Reactions

Nonaqueous liquid-phase reactions were mentioned in Sect. 2.1 in connection with preparation of germanium sulfide, although nowadays they are associated with synthesis of silicon nitride (Kohtoku et al. 1986). An insoluble solid containing the polymeric species, silicon diimide $Si(NH)_2$, is deposited from the interaction of silicon tetrachloride liquid with ammonia, either as a liquefied gas or when dissolved in an inert solvent. Silicon diimide decomposes on heating and crystallizes to α–Si_3N_4 at temperatures up to 1800 K. Submicrometer powder can be obtained with high chemical purity and low oxygen content compared with material prepared by carbothermic reduction (see Sect. 1.1) and this reaction forms the basis for a manufacturing route to silicon nitride powder. The increasing variety of nonaqueous liquid-phase reactions is illustrated by formation of submicrometer zinc sulfide powder, through reaction of H_2S with zinc diethyl and silicon carbide by dechlorination of a carbon tetrachloride–silicon tetrachloride mixture with sodium which yields a solid ceramic precursor. The latter crystallizes to silicon carbide on calcination and, while the mechanism of many liquid-phase reactions has not been fully established, nucleophilic species such as $SiCl_3^-$ have been postulated as chemical intermediates in this reaction.

3. Future Developments in Powder Preparation

A major driving force for the preparation of ceramic powders is specification of powder properties which give rise to dense, strong ceramics on fabrication. However, there is no intrinsic reason why only powder sources should be used for fabrication, and increasing attention is being focused on the direct conversion of liquid ceramic precursors to monolithic components. This approach eliminates sintering aids and any deleterious effects they may have on mechanical properties, as well as the need to define the characteristics for an ideal sinterable powder. However, use of liquid ceramic precursors has a disadvantage: shrinkage on solvent evaporation gives rise to stresses that are responsible for crack formation in the monoliths, although the drying process can minimize the latter effect. Hence monolithic gels derived from metal alkoxides yield monolithic aerogels when subjected to supercritical drying. In addition, the presence of additives such as formamide in alkoxide solutions before gelation has been shown to minimize crack formation on drying gel monoliths at ambient temperatures. Chlorosilanes react with amines to form silylamines, silazanes and polysilazanes, some of which are viscous, oily liquids that have been pyrolyzed to monolithic pieces of silicon nitride, thus illustrating the potential of this approach for nonoxide ceramics. In a recent innovation, composite materials have been prepared using the Lanxide process which involves oxidation of molten metal (e.g., aluminum) containing fibrous filler, whereby the ceramic matrix is grown directly through the fibers. In conclusion, direct conversion of liquid ceramic precursors to monolithic ceramics is an active research area at the present time, and may have a significant influence on future developments concerned with the chemical preparation of ceramic powders.

See also: Ceramic Powders: Packing Characterization; Ceramics Process Engineering: An Overview; Colloid Chemistry; Fractals (Suppl. 1); Mixing of Particulate Solids; Oxidation of Molten Metals, Directed (Suppl. 2); Sintering of Ceramics

Bibliography

Barringer E, Jubb N, Fegley B, Pober R L, Bowen H K 1984 Processing monosized powders. In: Hench L L, Ulrich D R (eds.) *Ultrastructure Processing of Ceramics, Glasses and Composites*. Wiley, New York, pp. 315–33

Bednorz J G, Müller K A 1986 Possible high-T_c superconductivity in the Ba–La–Cu–O system. *Z. Phys. B* 64: 189–93

Dauod M 1987 Polymerization and aggregation during gelation. In: Freeman G R (ed.) *Kinetics of Nonhomogeneous Processes: A Practical Introduction for Chemists, Biologists, Physicists and Materials Scientists*. Wiley, New York, pp. 651–723

Fauchais P, Boudrin E, Caudert J F, McPherson R 1983 High pressure plasmas and their application to ceramic technology. In: Veprek S, Venugoplan M (eds.) *Topics in Current Chemistry*, Vol. 107. Springer, Berlin, pp. 59–183

Hiemenz P C 1986 *Principles of Colloid and Surface Chemistry*, 2nd edn. Dekker, New York

Kohtoku Y, Yamada T, Miyazaki H, Iwai T 1986 The development of ceramics from amorphous silicon nitride. In: Bunk W, Hausner H (eds.) *Proc. 2nd Int. Symp. on Ceramic Materials and Components for Engines*. Deutsche Keramische Gesellschaft, FRG, pp. 101–8

Matijević E 1987 Colloid science in ceramic powders preparation. In: Vincenzini P (ed.) *Materials Science Monographs: High Tech Ceramics*, Vol. 38A. Elsevier, Amsterdam, pp. 441–58

Segal D L 1989 *Chemical Synthesis of Advanced Ceramic Materials*. Cambridge University Press, Cambridge

Thomas I M 1988 Multicomponent glasses from the sol-gel process. In: Klein L C (ed.) *Sol-Gel Technology for Thin Films, Fibers, Preforms, Electronics and Specialty Shapes*. Noyes Publications, NJ, pp. 1–15

D. L. Segal
[UKAEA Harwell Laboratory, Harwell, Oxfordshire, UK]

Ceramic Transformations, Seeding

Transformations in ceramics play a major role in microstructure and property development. Examples include solid-state reactions (e.g., $MgO + Al_2O_3 \rightarrow MgAl_2O_4$), phase transformations (e.g., γ-$Al_2O_3 \rightarrow \alpha$-$Al_2O_3$) and decompositions (e.g., $MgCO_3 \rightarrow MgO + CO_2$). Common to most ceramic transformations, variations in nucleation and growth determine the process temperature and time, the product particle or grain size and the chemical phase development. Since the classic studies of Zsigmondy and Turnbull and Vonnegut, seeding has been widely accepted as a means of controlling nucleation and growth and is firmly ingrained in such diverse fields as cloud seeding, bone mineralization, semiconductor processing and chemical synthesis (Matthews 1975).

A seed is defined as a separate and distinct substance added to a system for the purposeful manipulation of nucleation. The scientific principles governing the requisite seed charcteristics are related to the ability of the seed crystal to induce oriented overgrowth (i.e., epitaxy) in the reacting system as a consequence of their crystallographic matching. Solid-phase epitaxy, or the introduction of solid seed particles to a system undergoing a solid-state reaction, as a means of controlling ceramic transformations has been demonstrated in the late 1980s as a means of influencing solid-state transformation kinetics, microstructure development and phase development (Messing et al. 1986).

1. Epitaxy

Epitaxy concerns the degree of lattice parameter matching δ during growth at the interface between a reacting system or matrix (a_α) and the seed crystal or substrate (a_β) as follows:

$$\delta = \frac{a_\alpha - a_\beta}{a_\beta} \qquad (1)$$

The nucleation effectiveness of a seed crystal (α) is related to the seed–matrix interfacial energy ($\gamma_{\alpha\beta}$) and the seed–matrix contact angle (θ) by

$$\cos\theta = \frac{\gamma_{\beta\psi} - \gamma_{\alpha\beta}}{\gamma_{\alpha\psi}} \qquad (2)$$

For low values of δ the seed and host matrix are coherently oriented at their interface and $\gamma_{\alpha\beta}$ ranges from 1–200 mJ m^{-2}. For larger degrees of lattice disregistry the strained, incoherent interface is accommodated by dislocations and $\gamma_{\alpha\beta}$ ranges from 200–500 mJ m^{-2}. For values of δ greater than $\pm15\%$, the incoherent interphase interface is ~1000 mJ m^{-2}

and there is no effect on nucleation. Although most seed crystals have a lattice parameter mismatch with the transformed product of less than ±15%, there are many examples of nucleation catalysts having either no crystallographic relation or a large lattice parameter difference with the matrix (Turnbull and Vonnegut 1952).

The addition of seed crystals lowers the free energy for nucleation to less than that required for homogeneous nucleation (ΔG_{homo}) through its effect on the interphase interfacial energy and the seed–matrix contact angle

$$\Delta G_{het} = \Delta G_{homo}(2 + \cos\theta)(1 - \cos\theta)^2/4 \qquad (3)$$

For example, when $\theta = 10°$, $\Delta G_{het} \simeq 10^{-4} \Delta G_{homo}$, demonstrating the significant reduction in nucleation energy afforded by seeding.

2. Seeding of Ceramic Reactions

Seeding of ceramic transformations has been demonstrated in a variety of systems (Messing et al. 1986). Homoepitaxy, in which the seed and desired phase have identical lattice parameters and crystal structures, is the most likely prospect for nucleation manipulation in ceramic systems. In some cases, the seed surface is sufficiently different (i.e., distorted or hydrated) from the bulk crystal so that its potency for initiating nucleation is diminished. Besides selection by seed crystal structure and lattice parameter, it is also essential to know the intrinsic nucleation density of the system being seeded. For ceramic systems of the type mentioned above, intrinsic nucleation densities range from 10^9–10^{15} nuclei cm^{-3}. To effect the nucleation of a reaction, it is necessary to exceed the intrinsic nucleation densities. Seed crystals may act as sites for multiple nucleation, making exact estimations of extrinsic nucleation density difficult.

An example of a seeded transformation is the introduction of α-Al_2O_3 or α-Fe_2O_3 crystals into an AlOOH gel that transforms to α-Al_2O_3 on heating to 1200 °C. The α-Fe_2O_3 seed crystals are heteroepitaxial sites for crystallization and, with $\delta = 5.5\%$, fulfill the general seed selection rule of $\delta < 15\%$. Assuming a single grain develops from each nucleus and the nuclei are uniformly distributed in the transforming matrix, the grain size (GS) of the transformed material after seeding at a density (f) is given by

$$\mathrm{GS} = kf^{-1/3} \qquad (4)$$

where k is a correction factor for grain shape and the molar volume difference between the untransformed and transformed materials. Thus, assuming no volume contributed by the nuclei, the α-Al_2O_3 grain size obtained by seeding AlOOH with 2.4×10^{14} α-Al_2O_3 seed particles cm^{-3} is 0.18 μm. The intrinsic nucleation density of the AlOOH before seeding is 10^{11} nuclei cm^{-3} which corresponds to an α-Al_2O_3 grain size of 2.2 μm after transformation. This example illustrates that much smaller grain sizes can be obtained by seeding. However, a significant technical challenge is posed by the need to obtain extremely fine particles or nuclei to achieve nucleation densities of this magnitude. (Obviously, the seed crystals cannot be larger than the target grain size).

It should be noted that as seed crystals become smaller their reactivity and propensity for solid solution with the matrix increases. If a chemical reaction occurs below the nucleation temperature, there will be no seed crystals remaining for nucleation.

3. Kinetics of Seeded Transformations

Nucleation and growth transformations of unseeded systems are characterized by S-shaped kinetics preceded by an incubation period of no reaction. When seeded and heated at the same temperature, the transformation is almost instantaneous and the transformation temperature is significantly lowered. By increasing the seeding (nucleation) density, the transformation rate increases in proportion to the number of transformation growth centers. In this manner the nucleation-limited transformation of θ-alumina to α-alumina has been lowered by as much as 200 °C. At this temperature the transformation is kinetics- or growth-limited by the intrinsic diffusion processes of the system, and further temperature reductions can only be obtained by changing the transport process. For example, by changing the diffusion path to a liquid or vapor, the enhanced rate of diffusion relative to solid-state diffusion can cause the lowering of the α-alumina transformation temperature by an additional 200 °C (Shelleman and Messing 1988).

4. Nucleation Strategy

Given the tremendous potential for manipulation of ceramic transformations, seeding will become an increasingly important part of ceramic synthesis. However, seeding by the physical addition of seed crystals is limited by the unavailability of sufficiently small particles to realize high nucleation densities. Therefore, an important prospect is the development of *in situ* processes for nucleation control (see *Glass Ceramics*). McArdle and Messing (1988) demonstrated that by introducing an iron salt into an AlOOH gel, subsequent heating at low temperature resulted in the precipitation of α-Fe_2O_3 crystallites within the matrix and thus in nucleation sites for subsequent transformation. Despite the 5.5% lattice parameter mismatch, α-Fe_2O_3 seeding was as effective as α-Al_2O_3 seeding. Huling and Messing mixed molecular and colloidal gels to form a hybrid gel for the *in situ* nucleation and transformation of mullite ($3Al_2O_3.2SiO_2$). Upon heating, the molecular gel transformed to mullite crystallites in the colloidal gel matrix, providing nuclei for the bulk transformation of the colloidal gel to mullite. Both of these examples illustrate the potential for the control of solid-state transformations by controlling the original chemistry of the system so as to yield the *in situ* growth of seed crystals as a means of controlling nucleation of the bulk material.

To date, the seeding of ceramic transformations has received little attention and has been studied in only a few systems. However, the tremendous control it allows over reaction temperatures, reaction rates, microstructure development and phase development suggests that its importance in the ceramics field will increase dramatically in the future. One potential benefit is the ability to control the growth process, which should permit greater control of final microstructure and properties. The inherent attractiveness and technological simplicity of *in situ* chemical seeding promises to provide a variety of new seeded ceramic materials and to establish epitaxy as an important step in processing control, as it already is in other advanced materials applications.

Bibliography

Gebhardt M 1973 Epitaxy. In: Hartman P (ed.) 1973 *Crystal Growth: An Introduction*. North-Holland, Amsterdam

McArdle J L, Messing G L 1988 Transformation and microstructure control in boehmite-derived alumina by ferric oxide seeding. *Adv. Ceram. Mater.* 3: 387–92

Matthews J W (ed.) 1975 *Epitaxial Growth*. Academic Press, New York

Messing G L, McArdle J L, Shelleman R A 1986 The need for controlled heterogeneous nucleation in ceramic processing. In: Brinker C J, Clark D E, Ulrich D R (eds.) 1986 *Better Ceramics Through Chemistry*. Materials Research Society, Pittsburgh, PA, pp. 471–80

Shelleman R A, Messing G L 1988 Liquid phase assisted transformation of seeded γ-alumina. *J. Am. Ceram. Soc.* 71: 317–22

Turnbull D, Vonnegut B 1952 Nucleation catalysis. *Ind. Eng. Chem.* 44: 1292–8

G. L. Messing

[Pennsylvania State University, University Park, Pennsylvania, USA]

Ceramics: Construction Applications

Ceramics, materials whose essential ingredients have inorganic or nonmetallic bonds, are among the oldest and most enduring of construction materials. Brick and tile have a history of at least 100 000 years' use, and there is evidence that sheet glass was cast for windows as long ago as the time of Christ. Very early applications of cements and concrete-like materials took place in the Middle East thousands of years ago, and the ancient Egyptians set stone with mortars derived from both gypsum and lime. The Greeks and Romans added volcanic ash or ground burnt clay tiles (pozzolanic minerals) to render lime cements hydraulic (i.e., mixed with water to cure, and resistant to water when hardened). By further adding sand, broken tiles and brick, and crushed stone, the Romans also produced a concrete, the technology and quality of which were not surpassed until the eighteenth century.

Today, ceramics encompass a wide selection of materials with many different uses in construction. These materials can be grouped within three general categories: structural clay products (e.g., brick, tile, porcelain fixtures, sewer and drain pipe), glass, and cement-based products (e.g., mortar, plaster, wallboard, concrete). While the raw ingredients and basic products of construction ceramics have remained largely the same for centuries, many advances and improvements have been introduced to gain better performance, reduce costs and extend the range of applications of these products. Thus, for example, in addition to its predominant use in windows, glass is also applied in construction as glass blocks, curtain walls and, as fibers, in insulation and for reinforcing plastics. Similarly, with advances in its chemistry and methods of production, delivery and placement, concrete has become the most widely used construction material (by weight, constituting about 75% of all construction materials used annually) with many different applications in buildings, roads, tunnels, foundations, linings and subaqueous structures. Most ceramic products used in construction are composed of materials containing silica (SiO_2) and thus are included in the segment of the ceramics industry known as the silicate industries.

1. Entering the Modern Era

Many of the advances in both structural clay products and glass products have been in their production processes. Structural clay products have seen improvements in the form of more efficient use of plant and ease of installation. From the early centuries of the last millennium when window glass was made by blowing a long cylinder of glass, cutting it open and unrolling it, the glass industry has developed increasingly better methods of producing flat glass, leading to the current float glass method wherein a sheet of glass is produced by floating molten glass on a bed of molten metal (see *Float Glass Process*, Suppl. 1). Other improvements have affected the materials themselves or produced wholly new products. For instance, through more careful selection of constituents and control in firing, bricks and tiles have been made stronger and, when necessary, lighter.

Modern advances in cementitious materials occurred in the eighteenth and nineteenth centuries, beginning in England and spreading to other European countries and the USA. It was during this period that interest in hydraulic cement and concrete was revived, accompanied by a growing understanding of both the raw materials and the manufacturing processes needed to produce higher quality cements. In a series of experimental developments begun by Smeaton in 1756 (with the use of a mortar derived from clayey limestone and a pozzolanic material) improvements in cements were gained by investigators such as Parker, Frost, Aspdin and Johnson in the first half of the nineteenth century. A patent for "Portland cement" manufactured by heating finely ground limestone and clay was issued to Aspdin in 1824, but the first cement factory was established by Frost (to produce his "British cement") in 1825. Subsequent discoveries by Johnson improved the technology of cement production and stressed the importance of higher, controlled temperatures in heating the raw materials to produce cement clinker. By the latter half of the nineteenth century, cement industries were established in England, Belgium, Germany and France; in the 1890s the US industry grew very rapidly. The twentieth century has seen further improvements in the scientific understanding of the chemistry of cement and of concrete batching and placement; in the development of new cements and additives to alter the characteristics of cements; and in the development of new products for construction. Today, concrete and other cementitious products are used virtually worldwide in construction.

2. *Structural Clay Products*

2.1 *Industry Characteristics*

Since the beginning of the 1960s the importance of these products, as measured by their intensity of use (the ratio of the quantity consumed to the GNP), has shown contrasting trends. On the one hand, the intensity of use of brick has steadily declined in the face of competition from other exterior building materials while, on the other, the intensity of use of tile has greatly increased, due to a surge in popularity. In 1982, 4.8 billion clay bricks and about $2.75 \times 10^7\,m^2$ of floor and wall tile were shipped from US plants. Altogether, the value of structural clay products (brick, tile and pipe) shipped in that year from the USA amounted to about $1.2 billion.

The raw materials for clay products are abundant and available. Often production plants are situated close to the raw material source (or the market) to avoid transporting clay products long distances, since transportation costs of the relatively inexpensive but heavy products would constitute a major share of the final product cost. Energy costs are also important to this industry since the processing of clay products requires preparation of the clay powder using mechanical means as well as firing of the end product at fairly high temperatures.

Common brick is probably the most important product of the brick and structural clay tile industry. Glazed floor and wall tile and clay sewer pipe are also major products of the industries providing clay products to construction. Clay products have encountered much competition from other materials. For instance, brick must compete with siding materials like wood and aluminum which are easier to install; wall and floor tiles, popular for bathrooms, compete with plastic tiles and fiberglass-reinforced showerstalls and tub units; and clay sewer pipe must compete with concrete, cast iron and plastics. A recent countertrend is the increased popularity of ceramic floor and wall tiles. For this product, as well as other clay products, competition exists internationally. The USA is a net importer of these products which include floor and wall tile and unglazed brick.

2.2 *Materials Properties*

Structural clay products are composed of a combination of three types of minerals. The first is a hydrated clay which is composed of layers of silicon bonded to oxygen (with a general chemical formula of $(Si_2O_5)_n$) and aluminum bonded to oxygen and hydroxide ions ($AlO(OH)_2$). Typically, this alumina layer will be bonded to one or two silica layers to form one layer. Between these thicker layers the bonding is very weak (van der Waals bonds) which gives these clays a platelike character and highly directional properties. The second constituent is a fluxing material, like feldspar ($K_2(Na_2)O \cdot A_{12}O_3 \cdot 6SiO_2$). The third component is an inert filler like quartz.

Within this general composition there are certain variations and additives that make a particular material suitable for particular products. For instance, clays for building and paving bricks are usually rich in fluxes, or materials that lower the firing temperature of the clay. Ceramics used for wall and roof tiles are usually made from low-melting clays whereas facing ceramics can be made from low- or high-melting clays. Certain minerals, such as chamotte or sand, can lower the plasticity of the clay and decrease shrinkage on firing. Combustible materials (like wood shavings, ashes or fuel slag) can be added as extenders which burn out, producing a lighter, more porous product.

Sometimes structural clay products, like floor and wall tiles, are glazed. A glaze is a smooth, glossy coating that may be applied by painting, spraying or dipping. Glazing materials can be varied to achieve a certain color, gloss or texture.

Once the clay mixture has been formed and fired, the product consists of a crystalline phase embedded in a glassy (noncrystalline) matrix. The crystals are strongly bonded by ionic–covalent bonds, the strength of which give rise to many of the properties attributed to traditional ceramic products.

The strong bonding of structural clay products makes them fairly strong (up to 138 MPa compressive strength for facing bricks) but very brittle since planes of atoms do not easily slide past each other under a load to produce deformation. The water absorption of bricks is preferred to be 5–8%, although it can be as high as 25–30% for bricks or tiles. Traditional ceramic products are electrical and thermal insulators, again due to the strong bonding at the microstructural level. For this reason also, ceramics have good abrasion, chemical and heat resistance.

2.3 Processing and Manufacturing

The processing of structural clay products follows that of traditional ceramic products, which can be divided into four steps: powder preparation, forming, heat treatment and finishing. Powder preparation includes the raw material selection and the powder production. Once the raw materials (clays, fluxes, etc.) have been selected, usually from whatever local sources are available, they are ground and sieved to form a powder.

Forming of the product can take place in a variety of ways. Sewer and drain pipe can be extruded through a shaping die. Bricks can be formed in sand-coated wooden molds or extruded into long strips which are subsequently cut to size. Floor and wall tiles are formed by dry pressing them into molds. Sanitary ware is formed by slip casting. In this process a slurry of clay particles in water is poured into a porous plaster of Paris mold. The mold absorbs water from the slurry and a layer of compacted clay particles is deposited inside the mold. For hollow pieces this is allowed to proceed until the desired part thickness is achieved, and then the remaining slurry is drained out. For solid pieces the slurry remains in the mold until all of the water is absorbed.

Heat treatment occurs when the formed piece is baked or sintered in a kiln to turn the loosely bonded powder into a dense body. Finishing procedures usually entail grinding to the desired shape.

2.4 Typical Products

Structural clay products fall into several categories. Building brick comes in various colors and shapes. Some are solid, others are hollow or perforated to decrease weight or enhance insulation properties. Bricklaying is a time- and labor-intensive process so prefabricated panels of brick are now manufactured which can be a full story high. Sand-lime brick is often used in Europe, although less so in the USA. Other brick products include paving brick, glazed brick and firebrick. Terracotta, a product made from clay and grog (prefired clay, to reduce shrinkage), is an exterior finish that can be produced in different shapes; a glaze is also applied for a smooth finish.

Several types of tiles are available, both for indoor and outdoor exposure. These also appear in a wide variety of colors and shapes. Quarry tile is a red or brown floor tile usually about 20 cm square. A rectangular hollow tile is used for flue linings. Roofing tile, common in the western USA, is attractive and durable but rather expensive. Drainage tile is used to convey water away from, for example, building foundations.

Other types of structural clay products are pipes and porcelain fixtures. Sewer pipes must be produced from a good grade of clay and glazed on their interior to avoid absorption. Sanitary ware, made from a combination of ball-clay, china clay and other clays, is often included under construction materials since it is an integral part of building and housing construction. Sanitary ware includes all of the ceramic bathroom fixtures: sink, toilet, and tub.

3. Glass

3.1 Industry Characteristics

The US flat glass industry is fairly concentrated, with five firms producing almost all of the flat glass. In 1985 the flat glass industry in the USA shipped about $3.15 \times 10^8 m^2$, amounting to glass worth more than a billion dollars. One-third of the shipments of this product goes to the construction market, with most of the remainder used by the automotive market.

In the early 1980s the USA maintained a positive trade balance in flat glass, although imports rose sharply in 1982 due to a growing domestic demand. Some competition is coming from the use of plastic, especially in residential skylights, two-thirds of which were plastic in 1984. On the other hand, nonresidential skylights maintain a large and growing share of the demand for glass skylights. In addition, glass has been very popular as a design medium, both for exterior architecture and interior design, entailing applications such as glass curtain walls and glass blocks.

3.2 Materials Properties

Glass building products are composed of sand (SiO_2), soda (Na_2O) and lime (CaO). Various other oxides like potash and boric oxide are often used. In addition, coloring and opacifying agents can be added to give a wide variety of looks to common glass. When molten glass cools to its solid state, the molecules do not fall into a crystalline arrangement, as is the case in other types of ceramics, but rather maintain their unordered or amorphous arrangement.

Various properties of glass make it very suitable as a window material. Its transparency allows natural light into a building. It is a thermal insulator, so can be used in fairly thin sheets (or as part of an insulating system of several sheets). Its insulating properties and ability to be processed into fibers has also led to its incorporation into insulating systems for buildings and pipes. Glass does allow some solar radiation to enter a building so windows have been developed which can reflect this heat, preventing it from raising the temperature inside the building during hotter months. Glass is fire resistant although, unless specially laminated, it can transmit radiation from fires, thereby possibly spreading the fire. Window glass is very durable and easy to maintain.

3.3 Processing

The main process used for making flat glass, from which windows are made, is the float process. In this process, molten glass is poured onto a pool of molten tin. At one end of the tin bed the glass layer is pulled off by rollers producing a continuous sheet of glass of uniform thickness and free from distortions. Older, less common methods of producing flat glass are to pull the glass vertically from a furnace (sheet glass) or to draw a thicker layer between rollers and then grind and polish to form plate glass. Glass can also be drawn into tubing or fibers.

3.4 Products

Flat glass comes in many varieties used for windows, skylights and curtain walls. Clear transparent glass is the standard, although various colors and degrees of opacity can be achieved. Laminated glass is composed of two layers of glass bonded with an elastomeric layer to make penetration into the glass difficult. Insulating glass comprises two or more sheets of glass with sealed spaces between them. Tempered glass is strengthened by heating and rapid cooling, and sometimes by chemical means. When broken, tempered glass breaks into small, relatively harmless pieces. Wired glass is also available. It consists of a wire mesh embedded in the sheet of glass. Low emissivity (or "Low-E") glass is coated to reflect radiant heat to reduce heat loss in the winter and heat gain in the summer.

Other glass products used in construction include fibers and process-related equipment. Fibers are used for insulation and to reinforce plastics. Process equipment includes glass pipe and glass-lined steel pipe, pumps and pipe fittings.

4. Cement and Concrete

4.1 Industry Characteristics

The primary cementitious material with respect to ceramics in construction is Portland cement, which accounts for over 90% of the total production of hydraulic cement in the USA. From the start of US cement production in the late 1800s, production capacity of Portland cement clinker grew to a peak of over 90 Mt in 1975, but has since stabilized. This capacity is provided by 58 firms operating 164 Portland cement plants (as of 1980) and although the long-term trend is toward fewer plants in the USA, the number has not changed much since 1974.

The production of Portland cement is energy intensive: costs of fuel for heating raw materials to form clinker, and for electricity to grind and pulverize raw materials and clinker, account for 30–50% of the costs of cement production. The Europeans and Japanese have led the industry worldwide in applying new, more efficient production technology. Challenges facing the industry in the USA, apart from the need to adopt this new technology, include the difficulty in raising the capital needed for costly plant modernization, antipollution and other government regulations, energy availability, and the matching of production capacity to shifting and fluctuating construction demand.

The primary use of Portland cement is in concrete, and several other industries contribute to the production of this construction material; for example, crushed rock and sand and gravel firms which supply the aggregates, and chemical companies and other sources which provide admixtures, coatings, sealants and substitutes for natural aggregate. The major consumer of Portland cement in the USA is the ready-mix concrete industry, comprising over 6000 firms which alone consume more than half of the cement shipped. The concrete masonry industry (bricks, blocks, etc.) accounts for between 5% and 6% of consumption, and other concrete products account for almost 8%. The remainder is used by concrete site plants, with about 1.5% being used by other industries. Among segments of construction output, housing accounts for 30–33% of cement consumption, with the remainder distributed as follows: industrial/commercial, 21–23%; public buildings, 7%; public works, 17–18%; transportation, 17–19%; and miscellaneous uses, 3–4%. These data encompass new construction, maintenance and repair of facilities, and include both batch concrete and concrete products. Besides concrete, Portland cement is also used in construction in mortar and grout.

The gypsum industry is another major dealer in cementitious products related to construction. One major product is drywall or gypsum board, used primarily in housing construction, although nonresidential building markets are also being developed.

The growing volume of building repairs, alterations and additions offers another attractive opportunity for this material. Although the primary constituent of drywall is gypsum, the major costs are associated with the paper covering and the energy of production. Other major products of this industry are plasters, including plaster of Paris, wall plaster and hard-finish plaster.

4.2 *Properties and Processing of Cement*

Portland cement is primarily a hydrated calcium silicate. The raw material for calcium is typically limestone, chalk, marl or seashells. The silica is provided by shale, clay, slate, slag, sand or iron ore. Other ingredients, such as alumina and iron, are also included in cement from readily available sources. The constituent rocks and minerals are carefully ground, proportioned and mixed (using wet or dry processes). The ground material is burned in a kiln at about 1500 °C, causing the formation of clinker. The clinker is then pulverized (with any additional chemicals, such as gypsum) to form Portland cement.

The ultimate properties of concrete are influenced (but not determined) by the chemistry of the cement. Unlike sintered ceramic products (e.g., from clay) the hardened paste in concrete results from a chemical reaction of hydration. Portland cement encompasses four primary compounds: dicalcium silicate ($2CaO.SiO_2$, abbreviated C_2S), tricalcium silicate (C_3S), tricalcium aluminate ($3CaO.Al_2O_3$, or C_3A) and $4CaO.Al_2O_3.Fe_2O_3$, or C_4AF. The short-term and long-term characteristics of the resulting product depend on the relative amounts of these four compounds and how they react during hydration. For example, a greater proportion of C_3S will produce a higher early-strength cement (since C_3S hydrates relatively quickly) but more C_2S results in higher ultimate strength (through a greater production of the hydration product calcium silicate hydrate, or C–S–H). C_3A hydrates rapidly and is, therefore, useful in cements that must set up quickly; however, this reaction is so rapid that it must be retarded by the addition of gypsum during cement manufacture. The hydration reaction is complex, produces heat (which is desirable in some applications, undesirable in others), is influenced by time, temperature, amount of water and other chemicals present, and yields a time-varying series of compounds whose structures and mechanisms in influencing the properties of the hydrated cement paste are still only incompletely understood.

Several types of cement are available, depending on what characteristics are desired in concrete during placement, over the short term, and ultimately. Briefly, these are as follows: Type I—general use; Type II—general use where moderate sulfate resistance and moderate heat of hydration are needed (C_3A and C_3S are limited in such cements, because of their more rapid hydration); Type III—high early strength cement; Type IV—low heat of hydration cement; and Type V—high sulfate resistance cement (C_3A is limited, since its hydration product is unstable in sulfate environments).

There are variations of these cement types to include air entrainment for frost resistance. Not all these cements are manufactured or used widely, however, since it is possible to achieve some of these same objectives more economically and with specific control by using admixtures. In the USA, 90% of the cements produced are of Types I and II; the remainder are Type III and special cements such as the following: expansive cements (used to counteract shrinkage cracking or to induce compressive stresses (chemically prestressing) in the concrete); rapid setting and hardening cements; oil-well cements (used to cement the well casing to the surrounding rock, requiring fluidity during pumping and placement, but rapid hardening thereafter); and calcium aluminate cements or high alumina cements (in which strength is derived from aluminates rather than silicates). In addition, Portland cements may be blended with blast furnace slag or pozzolans for lower heat of hydration and greater economy and durability.

4.3 *Concrete*

Concrete is the product that results from the combination of cement, coarse and fine aggregate, and admixtures. Because of its economy, durability, versatility and use of readily available (or synthetic) constituent materials, concrete has found wide application in virtually all types of construction. Most of the concrete used in construction has hydrated Portland cement as its bonding agent, although polymer concrete (having a polymerized monomer as its bonding matrix) is also used. While the type of Portland cement influences the characteristics of the concrete at all stages of its life, there are other important factors; for example, any admixtures that are included, the size, shape, and porosity of the aggregate, and conditions prevalent during placement of the material.

Important improvements to the workability, control, strength, and durability of concrete, may be achieved by using admixtures. Admixtures are materials other than cement, water and aggregate that are included in the concrete. Several types of admixtures have been developed, for example, in the following areas: air entrainment (for frost protection, increased durability, improved workability), plasticizers or water-reducers (to provide higher strength, more economical use of cement, or greater workability), accelerators or retardants (to control the rate of set for reasons of weather, job requirements, methods of placement (e.g., rapid set needed when spraying shotcrete), and so on), and pozzolans (materials which react with the free lime during hydration to form additional cementitious compounds, saving in cement and providing a more durable concrete). Admixtures often fulfill more than one function simultaneously, and if used in combination may react with each other. Therefore, proposed mixes involving admixtures are tested beforehand for their effect on all concrete properties.

The tensile strength of concrete is about one-tenth its compressive strength. Therefore, concrete is usually reinforced with steel bars or wire mesh to withstand tensile loads. Concrete may also be prestressed (i.e., loaded prior to service in compression, to offset loads that would otherwise cause tension). Fiber-reinforced concrete includes short, discontinuous fibers throughout the cement paste; although strength is not increased, the resulting concrete is tougher. The coarse and fine aggregate bound by the cement matrix influence the weight, elastic modulus and dimensional stability of the concrete. Where higher strength-to-weight ratios are desired, lightweight aggregates (either natural or artificial) are used. Conversely, heavy aggregates are used where a dense concrete is needed (e.g., for radiation protection). Through careful control of constituent materials and production, higher strength concretes are continually being developed (from 35 MPa in the 1950s to 76–96 MPa available in the 1980s).

5. *Advances in Ceramic Construction Products*

Ceramic materials are well suited to the concept of engineering construction materials to meet the requirements of the design, since their macrostructural properties can be well controlled through careful selection of raw materials and precise formation of microstructure in the processing steps. This control of manufacture is what distinguishes advanced or high-performance ceramics from traditional ceramics, many of which have been discussed in this article. The use of such concepts to fashion materials is beginning to occur in construction. For example, computer simulation models are being developed and applied to understand better the internal structure, physics and chemistry of concrete, in order to improve strength, durability and placement during construction. In other cases innovative products have been developed. One example is a high-performance ceramic tile manufactured for use in heavily trafficked, high-maintenance areas on highways and bridges. The product is made from a high alumina (87%) powder which is cold pressed and sintered at 1470 °C. Another example is an opaque glass-ceramic panel developed in Japan as an exterior facing for buildings. (A glass-ceramic is a glass which has been allowed to crystallize.) The panel looks like marble but performs better than marble or granite in weather resistance. It is strong, nonabsorbent, fire-, chemical- and wear-resistant and relatively easy to use and maintain.

See also: Ceramic Materials for Nuclear Waste Storage (Suppl. 2); Ceramic Tile Manufacture; Ceramics in Automotive Applications; Ceramics: Nondestructive Evaluation

Bibliography

Budnikov P P 1964 *The Technology of Ceramics and Refractories*. MIT Press, Cambridge, MA

Jones J T, Berard M F 1972 *Ceramics: Industrial Processing and Testing*. Iowa State University Press, Ames, IA

Kingery W D et al. 1976 *Introduction to Ceramics*. Wiley, New York

Mehta K P 1986 *Concrete: Structure, Properties, and Materials*. Prentice-Hall, Englewood Cliffs, NJ

Newton F H 1974 *Elements of Ceramics*. Addison-Wesley, Reading, MA

Phillips C J 1960 *Glass: Its Industrial Applications*. Reinhold, New York

M. J. Markow and A. M. Brach
[Massachusetts Institute of Technology, Cambridge, Massachusetts, USA]

Channelling-Enhanced Microanalysis

The products of interaction events between an electron beam and a material, such as the characteristic x rays representative of the constituent elements of the material or the electron energy loss corresponding to x-ray production, are commonly used for chemical analysis. In a modern transmission electron microscope equipped with an energy dispersive x-ray detector (EDXS) or an electron energy loss spectrometer (EELS), the ability to produce incident electron probes with diameters of the order of 10 nm easily (with a routinely available image resolution of 2–5 nm) results in a microanalytical technique of high spatial resolution, limited by the interaction volume of the electron probe and the electron transparent sample (approximately 50–200 nm thick). The technique of channelling-enhanced microanalysis further refines the spatial resolution to the level of specific crystallographic site chemistry, with no attendant requirement of increased instrument capability (Spence and Taftø 1983, Krishnan 1988).

1. *Principles*

The principles of channelling-enhanced microanalysis may be understood by examining the solutions for the equations of dynamic scattering describing

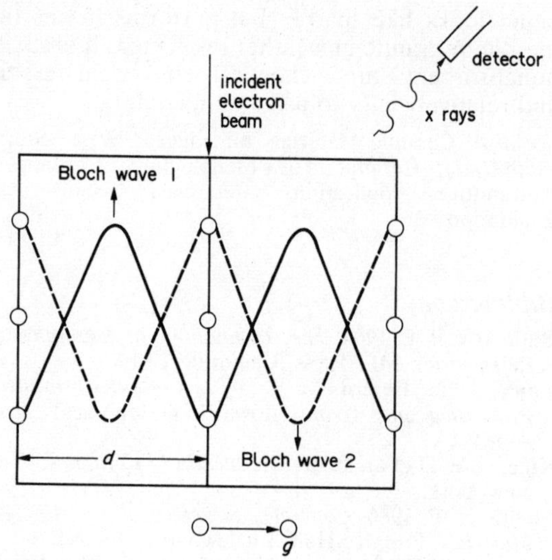

Figure 1
The standing wave pattern of the primary electron beam in relation to the atomic planes giving rise to a single diffracted beam

the propagation of an electron beam through a material. If a thin crystal is oriented in such a way that only one diffracted beam is excited, the solution has two roots leading to the two Bloch wave functions (Hirsch et al. 1965). The physical implication of these solutions is that the electron current flow in the direction parallel to the reflecting planes (i.e., parallel to the incident beam) is modulated laterally as $\sin^2 \pi g$ or $\cos^2 \pi g$ to form standing waves with the periodicity of the set of planes which give rise to the diffracted beam (see Fig. 1). At the exact Bragg condition both waves are excited equally, while for positive or negative deviations from the Bragg orientation either of the two waves may be preferentially excited. Since the wave function is two-dimensional for the two-beam case, that is, constant in the direction normal to the plane of Fig. 1, a selective sampling of the chemistry of specific crystallographic planes is possible by exciting either of the two waves, provided that the energy loss process and the consequent x-ray generation associated with the electron–atom interaction on each set of planes is sufficiently localized. This interaction distance is believed to be of the order of the Bohr radius for x-ray production even for very light elements, while electron energy loss processes are sufficiently localized for losses of greater than 2 keV.

It is clear from Fig. 1 that if the electron wave fields can be theoretically calculated it is possible to quantify in absolute terms the x-ray yield for selected atom layers. However, simple two-beam theory does not suffice and complex many-beam calculations are required since the elimination of multiple scattering is experimentally difficult to attain even in the thinnest of crystals for electrons accelerated through high potentials (100–300 kV). It must be noted that calculation involves the assumption of a delta function for the electron–atom interaction distance which may not always be justified (Roussouw et al. 1988).

2. Applications

2.1 Planar Channelling

The need for calculating the electron wavefield is obviated if a reference x-ray signal from an element whose distribution in the lattice is known can be utilized. This is usually possible for structures that are chemically layered on specific planes. This method, ALCHEMI (atom location by channelling-enhanced microanalysis), is illustrated using the B2 structure (body centered cubic with two sublattices) in Figs. 2a, b. The (002) planes contain alternate layers of only A or only B atom type sublattices while the (110) planes are chemically identical. A planar channelling experiment for this structure would therefore consist of orienting the thin crystal of the phase to be analyzed so that the transmitted beam and the (100) beam are excited. The corresponding Bloch waves have the periodicity of the (100) planes with current density peaks on the A or B atom type layers. If the problem is now to determine the site occupation of a third alloying addition (atom C) to this phase, x-ray spectra are collected with a positive or negative deviation from the exact Bragg condition for the (100) plane and the intensity ratios I_A/I_C and I_B/I_C of the characteristic peaks of these elements are compared for the two incident beam orientations. C atoms occupy A atom sites if I_A/I_C does not vary with orientation and they occupy B atom sites if I_B/I_C does not vary with orientation.

The site occupation can be quantified if an assumption is made that either A or B atoms occupy only one of the sublattices and the phase composition is known. In general the characteristic x-ray intensity (I) for any element X_i (i elements in the phase) for a given orientation η of the electron beam can be expressed as a linear sum of the contributions from each of the parallel planes (A, B, A, B . . .)

$$I^{\eta}_{X_i} = P_{X_i} M_{X_i} f^{A}_{X_i} C_{X_i} I^{\eta}_{A} + P_{X_i} M_{X_i} f^{B}_{X_i} (1 - C_{X_i}) I^{\eta}_{B} \quad (1)$$

Here, $f^{A,B}_{Xi}$ is the fraction of the total number of sites available for element X_i on plane A or B per unit cell, M_{Xi} is the total number of available sites per unit cell, P_{Xi} is a scaling factor including the fluorescence yield, C_{Xi} is the fraction of X_i on A sites and $I^{\eta}_{A,B}$ are the effective Bloch wave intensities averaged over the thickness of the crystal.

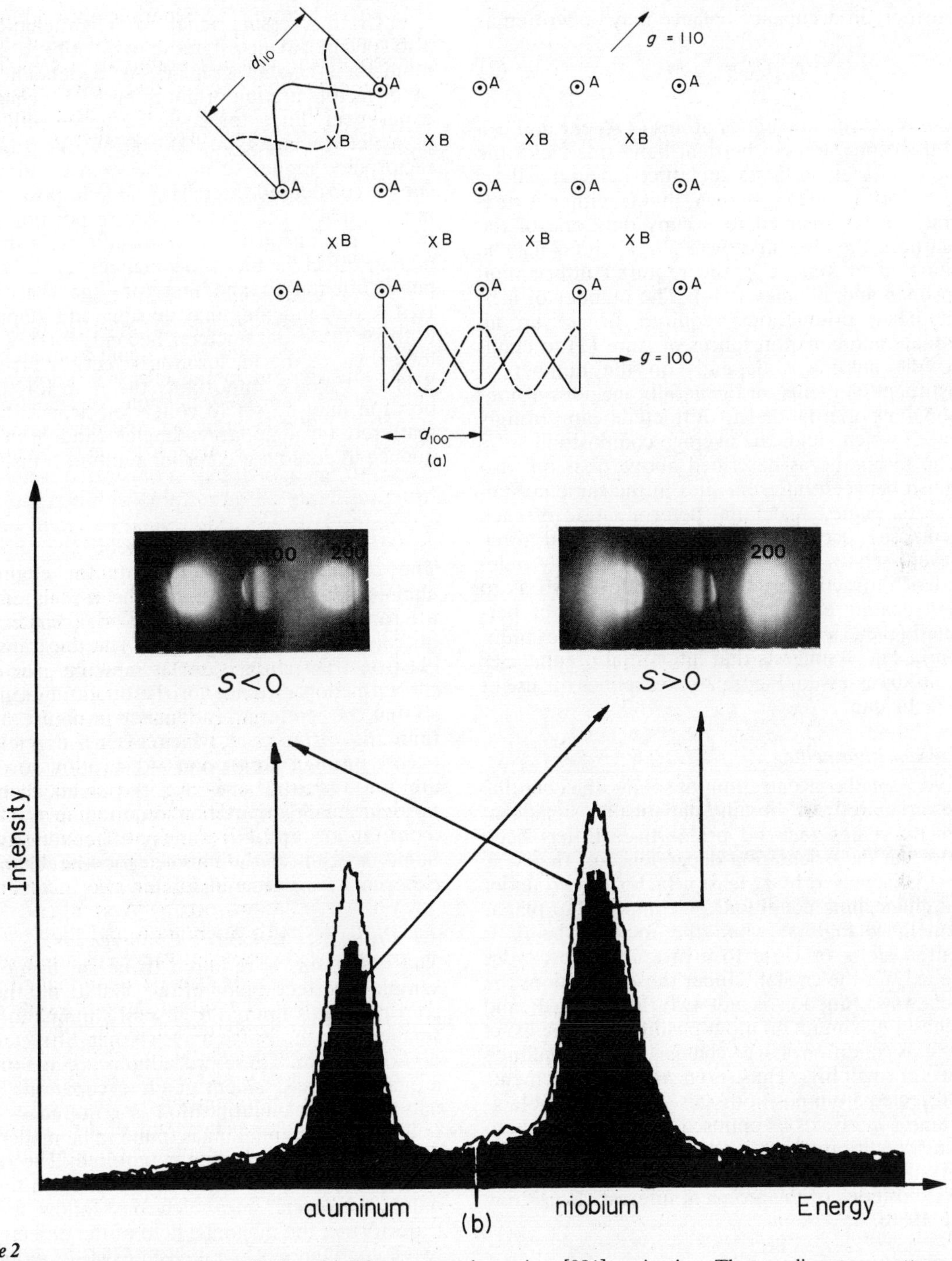

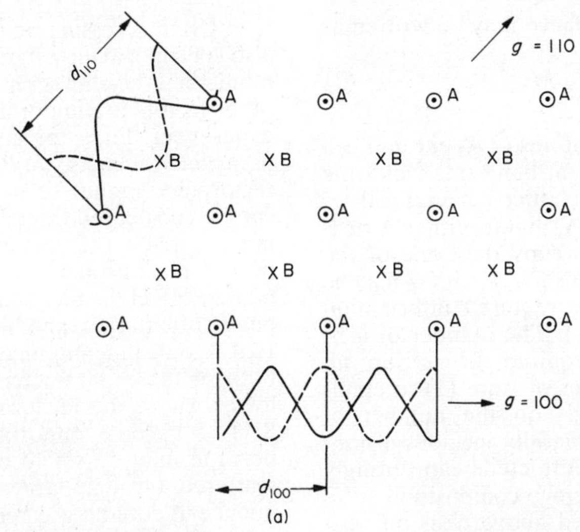

Figure 2
(a) The channelling technique applied to the B2 structure shown in a [001] projection. The standing wave pattern with $g = 100$ has intensity peaks on A planes or B planes. The standing wave pattern with $g = 110$ has intensity peaks on planes which are chemically identical. (b) Experimental spectra from a Ti–Al–Nb alloy with a B2 structure obtained with a negative excitation error (S<O) and a positive excitation error (S>O) for $g = 100$. The spectra are normalized to the titanium peak (not shown) and therefore indicate large changes in I_{Ti}/I_{Nb} but little change in I_{Al}/I_{Nb} for the two orientations. Aluminum and niobium atoms therefore occupy the same sublattice. Note that $g = 200$ is also excited illustrating the difficulty of attaining a strictly two-beam condition, and therefore the standing pattern in (a) will include contributions from other Bloch waves. (For further details see Banerjee et al. 1988)

Further, an occupancy balance may be written as

$$\sum N_{X_i} C_{X_i} = \sum M_{X_i} f_{X_i}^{A} \qquad (2)$$

where N_{Xi} is the number of atoms of X_i per unit cell and the expression on the right-hand side yields the number of sites on the A sublattice per unit cell.

If f_{Xi} for any one i is known, that is, either A or B atoms can be assumed to occupy only one of the sublattices, then the unknowns P_{Xi}, I_A^n and I_B^n may be eliminated to yield C_{Xi}, the required information (Krishnan and Thomas 1984). The number of incident beam orientations required to set up an adequate number of instances of Eqn. (1) to eliminate the unknowns depends on the number of constituent elements and generally includes a non-channelling orientation (no diffracted beam strongly excited) which yields the average composition.

The technique as described above does not distinguish between different sites in the same crystallographic plane. A definite determination of each specific site occupancy can be achieved if three different sets of parallel planes with linearly independent diffraction vectors can be found so as to allow separate experiments for each set to be performed (Banerjee et al. 1988). Further, an examination of Fig. 1 suggests that interstitial occupancies can also be assessed. Figure 2b exemplifies the use of this technique.

2.2 Axial Channelling

For very small concentrations of solute, the counting times required to obtain statistically significant intensity values may be prohibitive. It has been found that the orientation dependence of the x-ray signal is increased by at least a factor of two under axial channelling conditions as compared to planar channelling; that is, when the incident beam is oriented along or close to a prominent low-order zone axis for the crystal. Under these conditions the Bloch wave function is not two-dimensional, and reaches a maximum on atom positions for the exact zone axis condition. Large changes in the amplitude occur for small tilts. Thus, even small displacements in projected atom positions can yield measurable x-ray emission effects as opposed to the planar channelling condition where it is necessary to orient the beam along planes which clearly separate the projected positions of the atoms of interest (Roussouw et al. 1988).

2.3 Site Occupation by Electron Energy Loss Spectroscopy

The principles of channelling are applicable to the primary energy loss event in the electron beam (which leads to x-ray emission) and therefore the advantages of an electron energy loss spectrometer (EELS) over an energy dispersive x-ray detector (EDXS) in accessing signals from light elements and also specific valence states may be utilized to determine their crystallographic site dependence. Two other factors are important in an EELS channelling experiment. First, for single scattering with an incident electron energy of 100–300 KV, the principle of reciprocity applies so that the source and detector can be considered interchangeable in position. This implies that the detector aperture position in relation to the diffraction pattern influences the channelling effect in the same manner as the incident beam orientation and therefore the sensitivity of EELS to channelling may be squared by optimizing both of these parameters. Second, for low-energy losses where the localization effects referred to in Sect. 1 become important, the detector aperture position may be set to collect electrons which are scattered through larger angles (and as a consequence undergo a greater momentum transfer).

3. Experimental Requirements

The experimental requirements for channelling-enhanced microanalysis and the associated errors are similar to those for conventional thin film x-ray microanalysis (Williams 1984). The thickness of the electron transparent samples must be greater than the extinction distance for the operating reflection, so that the scattering is dynamic in nature, but less than the distance at which extensive attenuation occurs through absorption. Absorption corrections are not required since for the small changes in incident beam orientation required in the experiment the absorption paths are effectively constant. Some special considerations for the channelling experiment are described below.

3.1 Probe Size and Beam Convergence

The probe size is required to be small enough to sample a perfect region of the crystal, and the beam convergence is limited to a semi-illumination angle smaller than the Bragg angle so that diffracted disks do not overlap. These precautions are adequate for a qualitative assessment of site occupation. For situations where quantification is required or where sensitivity to channelling is found to be inadequate a more careful optimization is required. The requirement of parallel illumination is implicit in the technique; however, since electrons follow a helical trajectory in the magnetic field of the objective lens two components of parallelism exist—a radial component and a tangential component. Increasing the beam convergence improves the tangential parallelism but results in a poorer radial component. The appropriate experimental conditions vary with the structure, sample and microscope, and may be optimized through a simple set of experiments (Krishnan 1987).

3.2 Orientation and Voltage

The precise value of the positive and negative excitation errors (deviations from the Bragg condition) used for the channelling condition are not critical, and will yield the same fractional occupancy for different sets of orientations since the Bloch currents I_A and I_B are eliminated in the formulation in Sect. 2.1. However, the sensitivity of the experiment, that is, the differences in x-ray emission or energy loss edge height due to channelling at different orientations, is a function of the exact values of excitation used in both planar and axial channelling. In general, the sensitivity increases with increased positive and negative deviation in planar channelling. A limit to the deviation from the Bragg condition exists due to the possible excitation of the next diffracted spot in the systematic row used in planar channelling and will depend on both the interplanar spacing and the extinction distance. Experiments reveal no significant effects due to the accelerating voltage except for a possible inversion in the character of orientation dependence in the zone axis condition above a critical voltage.

It must be pointed out that the reasons that permit the channelling condition to be utilized in determining site occupation are precisely those for which the excitation of diffracted beams are avoided in conventional x-ray microanalysis. Planar channelling and axial channelling in EDXS and EELS have now been applied to determine, qualitatively and quantitatively, site occupations in a very wide variety of materials such as alloying additions to intermetallics, various minerals, nuclear waste ceramics, dopants in semiconductors and magnetic alloys.

See also: Electron Energy-Loss Spectrometry (Suppl. 1); Transmission Electron Microscopy; X-Ray Microanalysis, Quantitative (Suppl. 1)

Bibliography

Banerjee D, Gogia A K, Nandi T K, Joshi V A 1988 A new ordered orthorhombic phase in a Ti_3Al–Nb alloy. *Acta Metall.* 36: 871–82

Hirsch M A, Howie A, Nicholson R B, Pashley D W, Whelan M J 1965 *Electron Microscopy of Thin Crystals*. Butterworth, London

Krishnan K M 1987 When is parallel illumination best for ALCHEMI? *Ultramicroscopy* 23: 199–204

Krishnan K M 1988 Atomic site and species determinations using channelling and related effects in analytical electron microscopy. *Ultramicroscopy* 24: 125–42

Krishnan K M, Thomas G 1984 A generalisation of atom location by channelling enhanced microanalysis. *J. Microscopy* 136: 97–101

Roussouw C J, Turner P S, White T J 1988 Axial electron channelling analysis of perovskite. *Phil. Mag. B* 57: 209–47

Spence J C H, Taftø J 1983 ALCHEMI: A new technique for locating atoms in small crystals. *J. Microscopy* 130: 147–54

Williams D B 1984 *Practical Analytical Electron Microscopy in Materials Science*. Philips Electron Optics Publishing, Mahwah, NJ

D. Banerjee
[Defence Metallurgical Research Laboratory, Hyderabad, India]

Coal: World Resources

Coal deposits occur as layers of variable thickness interbedded with shales, claystones, siltstones, sandstones and, less commonly, limestones. Coal seams originate as peat accumulations in swampy areas (mires). The types and quantities of plants and minerals making up a peat deposit vary greatly, depending on the environments of deposition (facies). Subsequent burial of the peat layer under hundreds to thousands of meters of younger sediments in regions of long-term subsidence (sedimentary basins) results in the conversion of peat into first lignite, then subbituminous and bituminous coal and finally—rather infrequently—into anthracite, owing to the rising overburden pressure and temperature with increasingly deeper burial. The latter sequence is the coalification series. Lignite and subbituminous, bituminous and anthracite coal are termed ranks of coal. The rising pressure and temperature bring about significant changes in the physical and chemical properties of the original plant material in the peat. Coal deposits are also subjected to whatever deformation (folding and faulting) the segment of crust they are part of is subjected to during its geological evolution.

These geological factors determine to a large degree how attractive a given coal deposit is for exploitation and thus the likelihood of it being used for future economic development. Other important factors are size of deposit, distance to potential markets, infrastructure and prevailing, as well as anticipated, market conditions.

1. Coal Resource and Reserve Classification and Databases

1.1 Coal Quality and Bonity

Coal resource and reserve databases, in order to be useful, should contain information on coal quality and bonity (Fettweis 1976, 1979). Coal quality depends on coal facies and rank, and controls the value of any production from a coal deposit. Bonity (derived from the German word "Bonität") encompasses factors that control the cost of mining (e.g. thickness, depth, deformation, location and size of deposit). Quality and bonity together determine the economic feasibility of mining a given deposit.

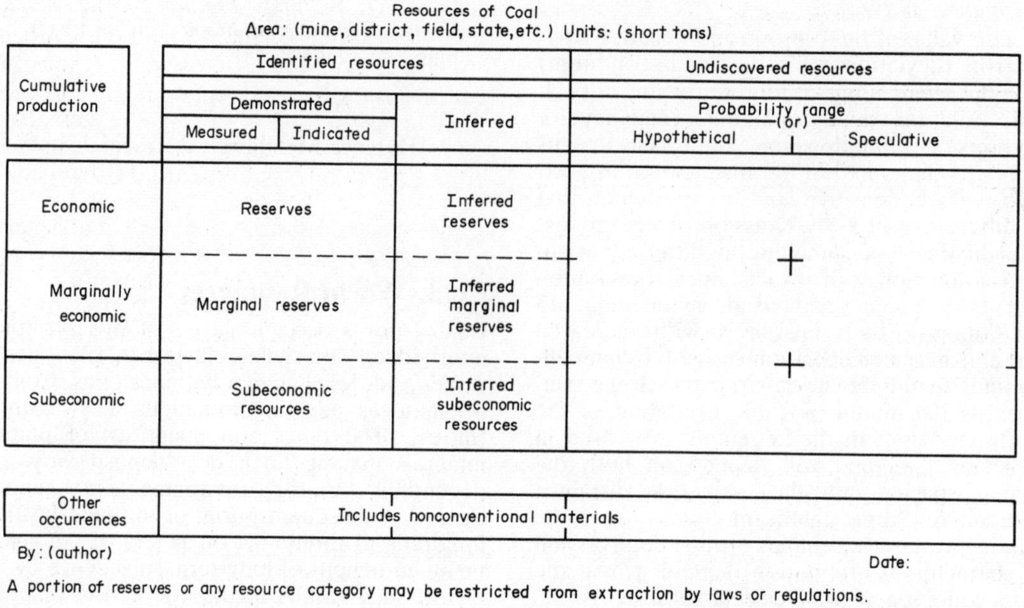

Resources of Coal
Area: (mine, district, field, state, etc.) Units: (short tons)

Cumulative production	Identified resources			Undiscovered resources	
	Demonstrated		Inferred	Probability range (or)	
	Measured	Indicated		Hypothetical	Speculative
Economic	Reserves		Inferred reserves		
Marginally economic	Marginal reserves		Inferred marginal reserves		
Subeconomic	Subeconomic resources		Inferred subeconomic resources		

Other occurrences	Includes nonconventional materials

By: (author) Date:

A portion of reserves or any resource category may be restricted from extraction by laws or regulations.

Figure 1
Classification scheme for coal resources and reserves used in the USA (after Wood et al. 1983)

1.2 Degree of Geological Assurance

Information on quality and bonity is generated through exploration. As the amount of exploration increases (through geological mapping, drilling and trenching, as well as sampling and testing of samples) and increasing knowledge of coal quality and bonity is gained, the "degree of geological assurance" of the existence (or absence) of an economic coal deposit increases. Information on the degree of assurance is expressed through such descriptors as measured, proved, known, assured, demonstrated and discovered resources for high degrees of assurance; indicated, probable, inferred, additional and semiproved for lower degrees of assurance; and possible, undiscovered, hypothetical, speculative and prognostic for lowest degrees of assurance. These terms refer primarily to the degree of geological assurance of the existence of a deposit, rather than the degree of knowledge of its numerous quality and bonity parameters.

1.3 Economic Feasibility

Exploration also generates the information needed to assess the economic minability of coal deposits ("feasibility" to mine for profit). The degree of feasibility is expressed through such terms as economic, marginally economic, subeconomic, uneconomic, or similar terms. Economic minability includes an assessment of the technical minability. However, many coal deposits could be technically mined, yet they could not now be exploited for economic gain. On the other hand, a deposit of economic minability may not be legally minable; for instance, in areas set aside as wilderness or wetlands, or otherwise excluded from mining by law. In modern usage the term reserve always implies that economic minability has been established by considering both technical and legal constraints.

1.4 Classification Schemes

The degrees of economic minability and of geological assurance are combined to classify known and surmised coal deposits within a logical system that permits aggregation into databases for regions, nations, continents and the world. Unfortunately, up to 1988, no universally accepted standard for the estimation of amounts of coal has emerged. Consequently, compilations on a worldwide basis bring together data that are not necessarily comparable between countries (and even within countries). Examples of national classification standards for coal resources and reserves are illustrated in Figs. 1, 2 and 3.

(*a*) *World Energy Conference survey of world coal resources and reserves*. The categories of both degree of geological assurance and degree of economic minability are defined differently in various countries and make worldwide compilations of coal resource data inherently suspect. The World Energy Conference (WEC) has attempted to overcome these disparities through the adoption of precisely worded definitions (see Table 1); however, much judgment is required in applying the WEC definitions to the various national databases in order to

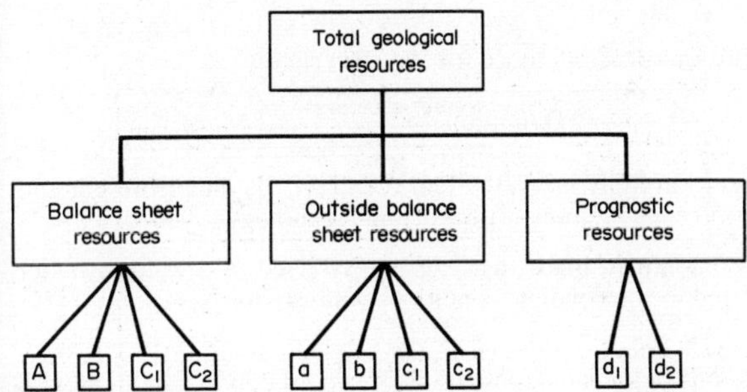

Figure 2
Classification scheme for coal resources and reserves used in the USSR and the GDR (after Fettweis 1979 © Glückauf, Essen. Reproduced with permission)

determine the amounts of coal within each of the WEC categories. The WEC compilation for 1984 shown in Table 2 follows previous surveys, in particular the one for 1974; significant earlier surveys were those for the World Power Conference in 1929 and the most comprehensive worldwide compilation of coal resource data completed for the International Geological Congress in Toronto in 1913. Coal resources, though not evenly distributed throughout the world (see Fig. 4, Table 2), are nevertheless sufficiently widespread, and occur in such abundance, that no shortages of coal supply are expected to occur anywhere in the foreseeable future (World Energy Conference 1986).

(*b*) *Resources versus reserves*. Coal that is currently or will be economically exploitable within a reasonable timescale (e.g., within 25 years in the Canadian scheme) is generally referred to as reserve (amount recoverable) or reserve base (amount *in situ*). Coal in the ground that may in the future become of economic interest, but cannot now or in the foreseeable future be economically recovered is commonly referred to as resource. However, the amount included in reserves and that allocated to resources depends very much on the background and intent of the person or group that promulgated or applied a particular classification scheme. The terminology is not uniform; for instance, the term reserve base has not been generally adopted and is considered confusing by some (van Rensburg 1980). Resources

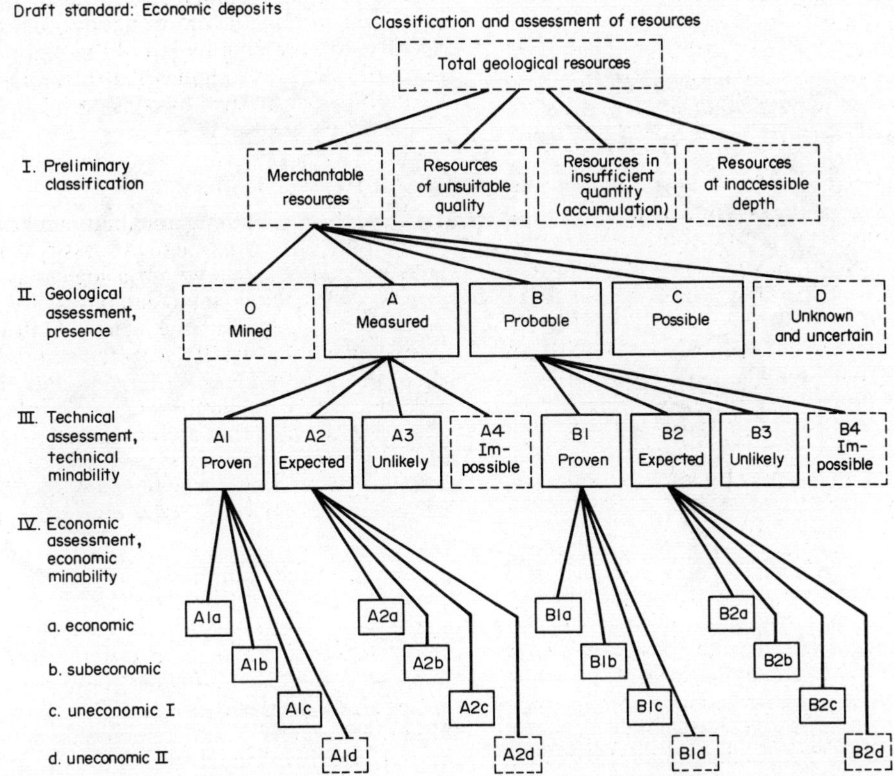

Figure 3
Classification scheme for coal resources and reserves used in the FRG. The dashed outlines indicate resources which are without particular importance for mining (after Fettweis 1979 © Glückauf, Essen: reproduced with permission)

Table 1
Terminology for world resources and reserves adopted by the World Energy Conference for the 1984 survey

Term	Definition
Proved amount in place	Tonnage that has been carefully measured and is exploitable under present and expected local economic conditions with existing technology
Proved recoverable reserves[a]	Tonnage of proved amount in place that can be extracted in raw form under present and expected local economic conditions with existing technology
Estimated additional amount in place	Indicated and inferred tonnage additional to the proved amount in place, e.g., unexplored extensions of known deposits, undiscovered deposits in areas with known coal resources, amounts inferred through knowledge of favorable geological conditions. Deposits whose existence is merely speculative are not included
Estimated additional reserves recoverable	Quantity of the estimated additional amount in place which might become recoverable within foreseeable economic and technological limits
Accessible coal in significant coal fields	Based on figures from the International Energy Agency (IEA), Coal Information 1986; this includes amount of coal likely to be considered for extraction from significant coal fields within the next 20 years. A significant coal field is one whose collective physical characteristics render it likely either to make significant contributions to, or to enter into, the detailed commercial mining and market evaluations required in order to achieve world coal supply over the next 20 years

a in most countries only a fraction of proved recoverable reserves is considered to be accessible coal

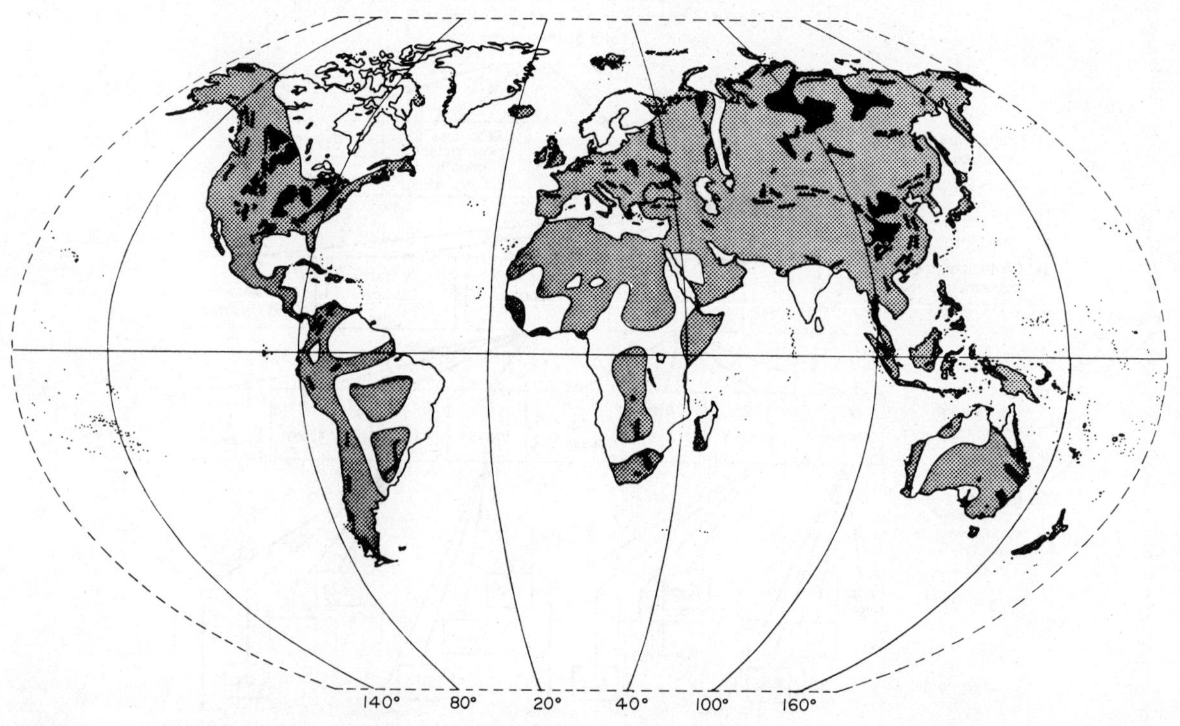

Figure 4
Important coal occurrences of the world: black areas, known and assumed coal occurrences; white areas, nearly coal-free nonsedimentary rocks; shaded areas, sedimentary rocks with potential for coal occurrences (after Fettweis 1979 © Glückauf, Essen. Reproduced with permission)

Table 2
World coal resources and reserves of the 1984 survey by the World Energy Conference (1986)

	Proved amount in place (10^9 t)				Proved recoverable reserves (10^9 t)				Accessible coal in significant coal fields (10^9 t)				Estimated additional amount in place (10^9 t)			
	BT	SB	LN	Total	BT	SB	LN	Total	BT	SB	LN	Total	BT	SB	LN	Total
Africa																
Botswana	7.0			7.0	3.5			3.5	3.0			3.0	100.0			100.0
South Africa	115.5			115.5	58.4			58.4	30.0			30.0	17.1			17.1
Zimbabwe	1.5	1.0		2.5	0.7	0		0.7	1.0			1.0	5.8	U		5.8
other	3.2	0.3	0.1	3.6	3.0	0.2	<0.1	3.2	0.3	0.2	0.1	0.6	3.2	1.0		4.2
Total	127.2	1.3	0.1	128.6	65.6	0.2	<0.1	65.8	34.3	0.2	0.1	34.6	126.1	1.0	0	127.1
North America																
Canada	4.9	1.3	2.8	9.0	3.5	0.9	2.4	6.8	1.4	2.2	1.3	4.9	25.7	27.8	4.2	57.7
Mexico	1.6	0.8	N	2.4	1.3	0.6	N	1.9	1.0			1.0	2.0	0.8		2.8
USA	237.6	164.4	40.9	442.9	132.0	99.2	32.7	263.9	55.0	37.0		92.0	458.1	276.2	392.8	1127.1
Total	244.1	166.5	43.7	454.3	136.8	100.7	35.1	272.6	57.4	39.2	1.3	97.9	485.8	304.8	397.0	1187.6
South and Central America																
Colombia	2.0	<0.1	<0.1	2.0	1.0	<0.1	<0.1	1.0	2.4			2.4	7.2	0.6	0.2	8.0
Venezuela	0.5			0.5	0.4			0.4	1.6		<0.1	1.6	2.2			2.2
other	<0.1	23.2	<0.1	23.2	<0.1	14.3	<0.1	14.3	1.1	5.2		6.3	0.1	28.3	7.3	35.7
Total	2.5	23.2	<0.1	25.7	1.4	14.3	<0.1	15.7	5.1	5.2	<0.1	10.3	10.5	28.9	7.5	46.9
Asia, Eastern Europe																
China	610.6		126.5	737.1	U		U	U	70.7			70.7	1700.0		300.0	2000.0
Czechoslovakia	5.8		7.2	13.0	2.7		2.9	5.6	1.0	5.0		6.0	5.5		1.6	7.2
GDR			47.0	47.0			21.0	21.0			13.0	13.0			0.0	0.0
India	26.3		1.6	27.9	U		1.6	1.6	19.0		0.5	19.5	85.6		1.9	87.5
Japan	8.5		0.2	8.7	1.0		<0.1	1.0	0.8			0.8	U		U	U
Poland	63.0		13.2	76.2	28.3		14.4	42.7	36.5		10.9	47.4	100.5		20.4	120.9
Turkey	0.1		5.3	5.4	0.1		4.8	4.9	0.1		3.0	3.1	1.2		2.8	4.0
USSR	136.0	51.8	105.0	292.8	108.8	41.4	94.5	244.7	73.8	21.0	77.4	172.2	2163.0	2085.9	960.0	5208.9
other Asia	16.8	5.1	30.0	51.9	0.8	0.5	0.9	2.2	2.3	0.4	1.8	4.5	5.1	2.8	1.6	9.5
other Eastern Europe	1.5	4.6	25.9	32.0	0.7	2.5	21.6	24.8	0.1	0.8	0.5	1.4	1.9	2.0	4.0	7.9
Total	868.6	61.5	361.9	1292.0	142.4	44.4	161.7	348.5	204.3	27.2	107.1	338.6	4062.8	2090.7	1292.3	7445.9
Australia and New Zealand																
Australia	48.5	3.0	39.3	90.8	27.4	2.1	36.2	65.7	20.8		35.0	55.8	507.0	100.4	87.0	694.4
New Zealand	<0.1	0.4	1.6	2.0	<0.1	0.2	0.1	0.3		0.2		0.2	0.4	1.4	8.9	10.7
Total	48.5	3.4	40.9	92.8	27.4	2.3	36.3	66.0	20.8	0.2	35.0	56.0	507.4	101.8	95.9	705.1
Western Europe																
Belgium	0.7			0.7	0.4			0.4	0.8			0.8	1.4			1.4
France	0.9	0.2	U	1.1	0.3	<0.1	U	0.3	0.4			0.4	NR	NR	<0.1	<0.1
FRG	44.0		55.0	99.0	23.9		35.2	59.1	13.4		12.0	25.4	186.3		U	186.3
UK	U		0.4	0.4	4.6		U	4.6	14.0			14.0	185.4		U	185.4
Greece			5.3	5.3			3.0	3.0			1.5	1.5			U	U
Spain	0.9	0.5	0.3	1.7	0.4	0.2	0.2	0.9	0.4		1.2	1.6	2.6	0.9	0.2	3.7
other Western Europe	1.4	0.1	0.3	1.8	0.5	0.1	0.1	0.7					0.1	1.4	3.6	5.1
Total	47.9	0.8	61.3	110.0	30.1	0.3	38.5	69.0	29.0		14.7	43.7	375.8	2.3	3.8	381.9
World total (rounded)	1340	257	508	2105	404	162	272	838	351	72	158	581	5568	2530	1797	9895

Key: BT, bituminous and anthracitic coals; SB, subbituminous coal; LN, lignite; U, unknown or not available; C, confidential; N, negligible amount; NR, not reported always reflect coal in the ground, but some schemes include coal, the existence of which may only be surmised by geological inference and the economic minability of which is unknown and may never materialize; for example, hypothetical and speculative resources in the US classification scheme. Other schemes (e.g., the South African scheme) reject the inclusion of any data on undiscovered resources in national inventories (van Rensburg 1980).

The term reserves has often been restricted to the amounts of coal that can be economically recovered that is, after mining losses or after mining and

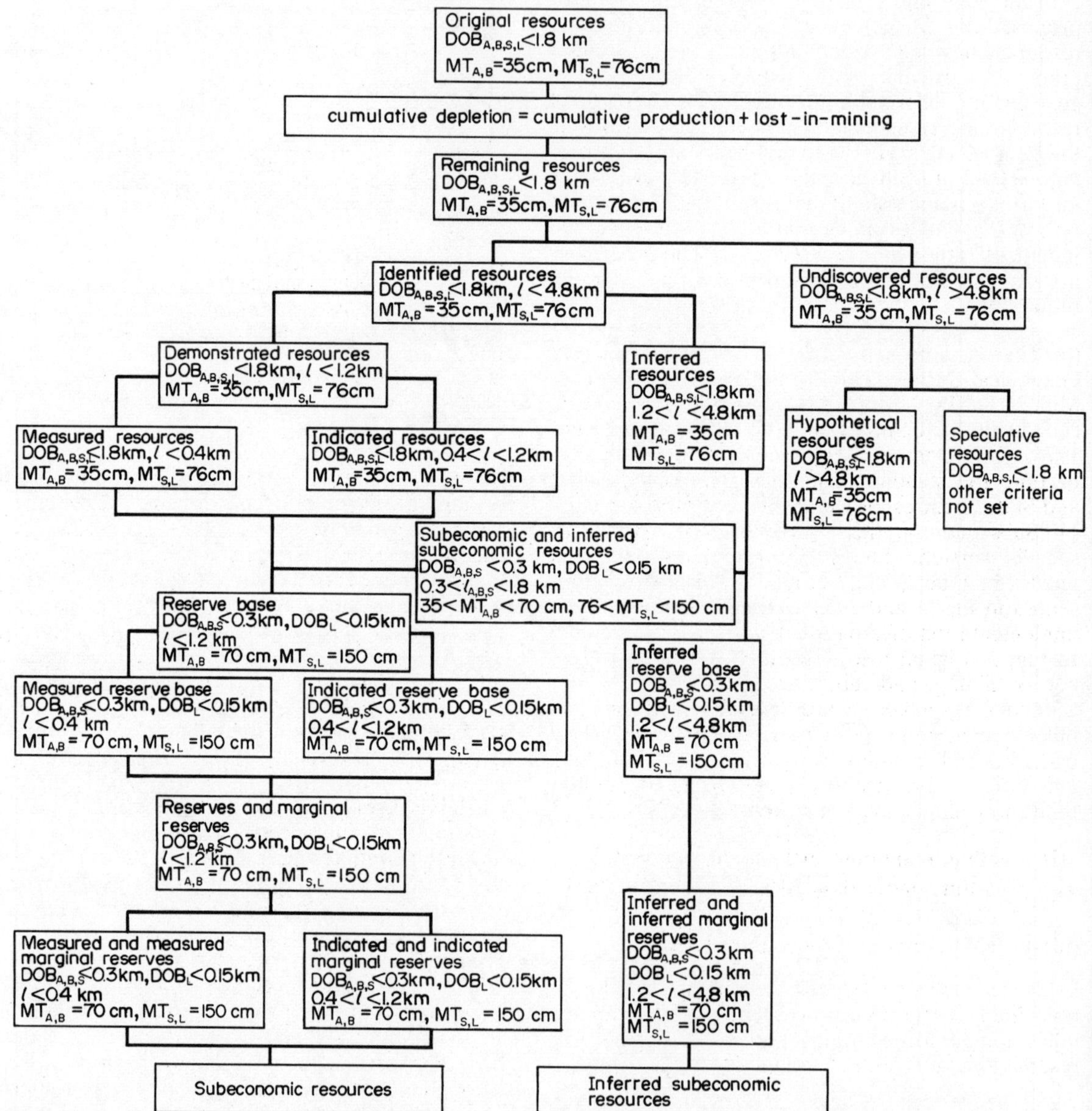

Figure 5
Classification and hierarchy of coal resources as currently used by the US Geological Survey (Wood et al. 1983). Coal resource terms are defined in Table 3. The criteria given in each box are the depth of burial (DOB), the maximum distance l from the point of measurement of coal thickness and the minimum thickness (MT) of the coal. The subscripts A, B, S and L refer to anthracite, bituminous coal, subbituminous coal and lignite

beneficiation losses. To avoid confusion, the terms reserves and resources are sometimes used with qualifying adjectives such as recoverable (out of ground), exploitable (*in situ*), inferred, and so forth. However, Wood et al. (1983) point out that reserves include only recoverable coal and that terms such as extractable reserves are redundant and should be avoided. Unfortunately, usage of the terms reserves and resources has been and still is inconsistent.

In the USA and Canada the classification scheme proposed by McKelvey (1973) for solid mineral resources has been widely adopted (McKelvey diagram), in particular by the agencies of the Federal governments charged with compiling national coal resource and reserve databases (Fig. 1). However, the geological surveys of individual states and provinces have not all adopted these standards. It is important to understand that resulting differences in resource estimates are a matter of preference and judgment rather than 'correctness'. The acceptability of estimates to the user depends on the user's requirements.

(*c*) *Classification standards*. In the USA the US Geological Survey (USGS) and the US Bureau of Mines (USBM) adopted a classification standard (US Geological Survey and US Bureau of Mines 1976, US Geological Survey 1980) which has become the standard reference for many Federal and State agencies, as well as the coal industry (Fig. 1). Subsequently, the USGS adopted a detailed, specific version of this system for its own coal resource studies (Wood et al. 1983) which is represented in Fig. 5 and Table 3 as an example of the implementation of a national classification scheme. In Fig. 5 original resources refers to the resources before mining, and subeconomic and inferred subeconomic resources include coal left in room and pillar mining, in property barriers, coal too thick to be recovered completely by conventional mining, and mine and preparation waste. Occurrences other than those shown in the hierarchy include those:

(i) less than minimum thickness at any depth;
(ii) containing more than 33 wt% ash on dry basis; or
(iii) buried at depths of more than 1.8 km.

Estimated tonnage, where calculated, should be reported as 'other occurrences' and not as resources, unless mined. Mined tonnage quantity is included in reserve base and reserve estimates.

(*d*) *Problems in quantifying the degree of geological assurance*. It will be noted that the degree of geological assurance of the existence of a resource in the 1983 USGS classification system is defined in terms of distance from points of measurement of thickness. Other classification schemes define assurance

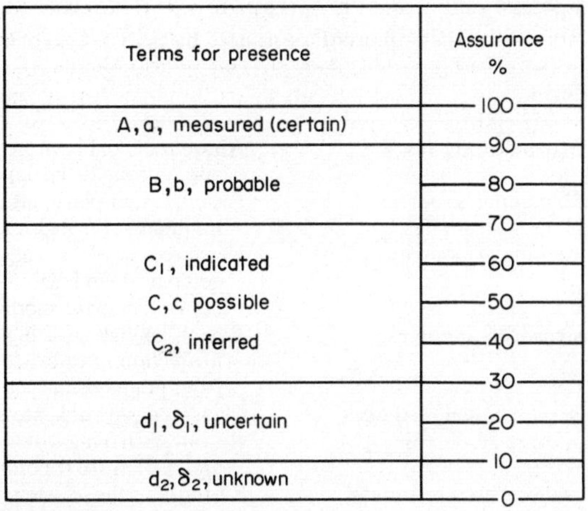

Terms for presence	Assurance %
	100
A, a, measured (certain)	
	90
B, b, probable	80
	70
C_1, indicated	60
C, c possible	50
C_2, inferred	40
	30
d_1, δ_1, uncertain	20
	10
d_2, δ_2, unknown	
	0

Figure 6
Degrees of geological assurance used in the FRG and eastern European mineral classification schemes. 100% assurance indicates that the presence of mineral resources is totally certain and 0% assurance indicates that the presence is totally unknown. Upper-case letters refer to currently minable resources while lower-case letters refer to potentially minable resources (compare with Figs. 2 and 3)

categories in terms of percentage certainty of the resource calculation (Fig. 6). The definitions of assurance categories in the USGS and USBM (1976) classification include a statement of percentage error margin for measured resources (less than 20%), but none for indicated resources, inferred resources, and so forth. Such restraint is not without reason. The less information available on a coal resource, the more difficult it becomes to calculate or estimate the percentage error of a computed tonnage. Tewalt et al. (1983) discussed the difficulties of estimating geological uncertainties for quantitative coal resource evaluations.

Classical statistical and geostatistical analyses can be applied to determine the uncertainties inherent in any calculated resource tonnage (or average quality), but the methods are most meaningful when a sufficently large, closely spaced (say 0.5–1 km grid) database is available. Geostatistical methods (e.g., Kriging) should be, but are not generally, applied where the area distribution of data points is uneven, which is usually the case for exploration data on a regional scale. Generally, only deposits considered or targeted for exploitation are sufficiently explored through a regular grid of exploration wells to permit meaningful statistical analysis. Even then, simple statistical analyses (averages, standard deviations, etc.) which are in common use in the industry may be misleading because they disregard regional and local trends in quality and bonity parameters. Trend surface, regression and other geostatistical analyses

Table 3
Abbreviated definitions of terms used by the US Geological Survey (USGS) to classify coal resources (after Wood et al. 1983)

Term	Definition
Original resources	Amount of coal in-place before production, occurring in coal seams at least 0.35 m (bituminous and anthracitic coals) or 0.75 m (subbituminous and lignitic coals) at depths to 1800 m
Remaining resources	Resources in-place after mining; does not include any amounts left in ground rendered unminable through exploitation
Identified resources	Resources whose location, rank, quality and quantity are known or estimated from specific geological evidence; they include economic, marginally economic and subeconomic components. They are subdivided into measured, indicated and inferred resources
Measured resources	Resources having a high degree of geological assurance; specifically, they are computed by projection of thickness of coal and overburden, rank and quality data for a radius of 0.4 km from points of measurement
Demonstrated resources	Measured plus indicated resources
Inferred resources	Resources having a low degree of geological assurance; specifically, they lie between 1.2 km and 4.8 km from points of measurement
Undiscovered resources	Resources, the existence of which is only postulated, comprising deposits that are either separate from or are extensions of identified resources; they are subdivided into hypothetical and speculative resources
Hypothetical resources	Resources that are either similar to known coal deposits which may be reasonably expected to exist in the same coal field or region under analogous geological conditions or are an extension from inferred resources, beyond 4.8 km of points of measurement
Speculative resources	Undiscovered resources that may occur either in known types of deposits in favorable geological settings where coal deposits have not been discovered or in types of deposits as yet unrecognized for their economic potential. (Note: the USGS has not estimated the amount of speculative coal resources for the USA)
Economic resources	Implies that profitable extraction under defined investment assumptions has been established, analytically demonstrated, or assumed with reasonable certainty
Marginal economic resources	Implies that coal does not meet economic criteria under current costs and prices but may do so under reasonable future projections
Subeconomic resources	Resources which do not meet current or reasonable future criteria for profitable extraction
Reserve base	Those parts of the identified resources that meet specified minimum physical and chemical criteria related to current mining and production practices, including those for quality, depth, thickness, rank and distance from points of measurement. Specifically, the reserve base is the in-place demonstrated (measured plus indicated) resource from which reserves are estimated: coal seams thicker than 0.7 m (bituminous and anthracite) or 1.5 m (subbituminous and lignite), subbituminous coals only to a depth of 300 m and lignites only to a depth of 150 m are included
Reserves	Virgin and (or) accessed parts of coal reserve base which could be economically extracted considering environmental, legal and technological constraints. The term reserves need not signify that extraction facilities are in-place or operative. Reserves include only recoverable coal, not coal lost in mining
Restricted reserves	Those parts of any reserve category that are restricted or prohibited by laws or regulations from extraction

produce more meaningful results, but are not yet commonly used even at the level of reserve blocks for individual mines.

2. Outlook

The introduction of computers and specialized software in the coal-mining and consulting industries and in government agencies is leading to changes in the ways coal deposits are evaluated. Large computerized coal resource databases, which incorporate not only the traditional information on amount (thickness and size), depth and rank, but also many other quality and bonity parameters, have been compiled by governments and industry, particularly since the oil embargo of 1973. These databases and software, written for their analysis, are now maturing in the USA, Canada, the FRG and other countries. They will eventually permit more flexible approaches to coal resource assessments than were possible in the past. Instead of preselecting boundary values (e.g., for thickness, depth, rank) at the beginning of a resource calculation, as was required for 'manual' compilations (to limit the number of resource categories to be carried to a manageable level), the computer permits the manipulation of a vast amount of data for evaluation according to 'what if' type questions. Eventually, these databases will permit re-analysis according to various national and international standards, as well as to any other standards preferred by industry, various agencies or individual researchers.

Bibliography

Averitt P 1975 *Coal Resources of the United States, January 1, 1974*, US Geological Survey Bulletin 1412. US Government Printing Office, Washington, DC
Bielenstein H V, Chrismas L P, Latour B A, Tibbetts T E 1980 Coal resources and reserves of Canada. *Can. Dep. Energy, Mines Resour. Rep.* ER 79–9
Fettweis G B 1976 *Weltkohlenvorräte. Eine vergleichende Analyse ihrer Erfassung und Bewertung*. Glückauf, Essen
Fettweis G B 1979 *World Coal Resources. Methods of Assessment and Results*, Developments in Geology Vol. 10. Elsevier, Amsterdam
International Geological Congress 1913 *The Coal Resources of the World*, Vols. 1, 2, 3. 12th International Geological Congress, Toronto
McKelvey V E 1973 *Mineral Resources Estimates and Public Policy*, US Geological Survey Professional Paper 820. US Government Printing Office, Washington, DC, pp. 9–19
Tewalt S J, Bauer M A, Mathew D, Roberts M P, Ayers W B, Barnes J W, Kaiser W R 1983 Estimation of uncertainty in coal resources. *Rept. Invest. Univ. Tex. Austin. Bur. Econ. Geol.* 136
US Department of Energy 1981 *Demonstrated Reserve Base of Coal in the United States on January 1, 1979*, US DOE/EIA-0280 (79) UC-88. US Department of Energy, Washington, DC
US Geological Survey 1980 *Principles of a Resource/Reserve Classification for Minerals*, US Geological Survey Circular 831. USGS, Washington, DC
US Geological Survey and US Bureau of Mines 1976 *Coal Resource Classification System of the U.S. Bureau of Mines and U.S. Geological Survey*, US Geological Survey Bulletin 1450-B. US Government Printing Office, Washington, DC, pp. B1–B7
van Rensburg W C J 1980 The classification of coal resources and reserves. *Miner. Resour. Circ. (Univ. Tex. Austin. Bur. Econ. Geol.)* 65
Wood G H Jr, Kehn T M, Carter M D, Culbertson W C 1983 *Coal Resource Classification System of the U.S. Geological Survey*, US Geological Survey Circular 891. USGS, Washington, DC
World Energy Conference 1986 *Survey of Energy Resources. Section 1–Coal (Including Lignite)*. WEC, London, pp. 6–28

H. H. Damberger
[Illinois State Geological Survey, Champaign, Illinois, USA]

Colloidal Crystals

Since many industrial processes involve colloids, these have been widely studied since the late 1960s. Any aggregate of 10^6–10^{12} atoms or molecules can be called a colloidal particle, size being typically in the range 0.1–1 μm; that is, on a length scale halfway between atomic and macroscopic dimensions. Any dispersion of such particles can be called a colloid. Colloidal particles usually have various shapes and sizes; that is, they are polydisperse. Important results in the study of colloids have so far been obtained with the use of monodisperse particles. The identity of shape and size of the particles simplifies many physicochemical problems. Monodisperse colloids are, therefore, model systems for colloid science. However, monodispersity is often linked to better macroscopic qualities, for example in ceramic production.

Investigations of monodisperse particles have revealed the existence of a crystallization phenomenon. Under appropriate conditions and depending on their nature, monodisperse colloidal particles behave like identical atoms and organize themselves in long-range ordered crystals called colloidal crystals. Several types of interaction can give rise to crystallization; for example, hard sphere contacts and electrostatic repulsion.

Two situations can occur: either the long-range order can be easily destroyed (for instance, by using mechanical shear in which case crystallization is a reversible process), or particles coming into contact remain stuck to each other, leading to irreversible aggregation. Only the first case will be considered in this article.

Many properties of real crystals are observed in colloidal crystals, although on different scales. Indeed the main difference between the two systems lies in their microscopic size: 1 Å for atoms and approximately 10^3 Å for colloidal particles. The local study of the crystalline state requires x-ray scattering; the equivalent tool to investigate ordering of colloidal crystals is visible light scattering. The colored flecks observed in natural opals as well as in synthetized systems are, in fact, due to light scattering. Another singular property of colloidal crystals is their very low elastic modulus which allows colloidal crystals to be handled as liquids and poured from one vessel to another. This property is also revealed by gravitational compression, propagation of low velocity transverse waves and so on. Another striking consequence of the low elastic modulus of colloidal crystals is their behavior under shear. Above a critical stress value, shear induces first orientation of the crystallites and then, for higher stress, a series of structural transitions leading finally to melting.

1. Preparation of Monodisperse Colloids

1.1 Natural Colloidal Crystals

Natural colloidal crystals have been known of for a considerable time: in 1938 it was proposed that the properties of Schiller layers in iron oxide sols were due to the ordering of quasimonodisperse particles. The discovery in 1935 of tomato and tobacco viruses provided another example of natural monodispersity. Each viral particle is an ingenious arrangement of identical protein bricks. In quasispherical viruses, the subunits are organized in regular polyhedra (icosahedra) whereas helical proteins build helical shells; crystalline order has been observed in such concentrated suspensions of viruses. So far it has been impossible to imitate such an architecture. However, the study of a much simpler natural system, opals, has stimulated research on synthetic colloidal crystals.

Mineralogical and x-ray studies revealed that precious opals are composed of amorphous silica, the most common material on earth. The reason they present such attractive colored flecks was given in 1964 with the use of scanning electron microscopy. Precious opals are composed of tiny spherical silica particles cemented together. These spheres often have a shell structure which probably results from their formation by aggregation of smaller particles. They are disposed in a long-range ordered lattice, the parameter of which is of the order of the wavelength of visible light. The voids between the silica spheres are filled with a strongly hydrated form of amorphous silica; the refractive index of the spheres is therefore different from that of the voids. The splendid visual appearance of opals is simply due to the Bragg scattering of visible light by the microscopic network of silica spheres.

From these results, the probable mechanism of opal genesis can be sketched. First, silica spheres grow in water suspension forming uniform colloidal particles. Under their own weight they sediment very slowly, packing in a regular lattice. The voids between them are filled with highly hydrated amorphous silica. These three distinct processes: (a) formation of monodisperse silica particles, (b) slow sedimentation (one year long) and packing in a crystal lattice and (c) cementing together, have been reproduced in the laboratory to obtain an opal-like material which is now commercially available.

1.2 Synthetic Monodisperse Colloids

Understanding the structure of precious opals and how they are formed has opened the door to the development of simple and low-cost methods for preparing synthetic monodisperse particles. Great efforts have been devoted to this end over the last decades (Overbeek 1982, Sugimoto 1987). Monodisperse particles are produced by chemical reaction, for example, the redox reaction between chloroauric acid and formaldehyde gold sols. At least three general conditions are required in such a reaction to produce monodisperse colloids: the separation between nucleation and growth, the inhibition of coagulation and the choice of growth process (diffusion-controlled growth).

Emulsion polymerization, for example, is a technique used to produce polymeric monodisperse colloids or latex (Pieranski 1983). Many experiments have been performed using suspensions of polystyrene spheres, of various diameter, synthetized by this technique. In most cases, the polymerization reaction is begun by an initiator which gives rise to free radicals; the reaction can then propagate leading to the formation of polymeric chains. The main physicochemical features of the emulsion polymerization can be summarized as follows. The monomer, styrene for example, is nearly insoluble in water. It is mainly contained in droplets which may need to be stabilized by an emulsifier. Only a few molecules are present in the water. An initiator such as potassium persulfate and its free radicals are water soluble, thus the initiation and first steps of propagation take place in water. This is the nucleation stage of emulsion polymerization. The solubility of oligomers decreases rapidly as their size increases, thus they tend to segregate and form micelles, the growth of which is slow. During the growth process, monomers are supplied from the droplets (which act as a monomer reservoir). This decomposition of the monomer micelles free from oligomer radicals is probably the key to producing uniform particles since no nucleation sites appear in the growth stage. The width of statistical size distribution becomes narrower with increasing radius of the particles. To produce large polymer spheres ($\geqslant 0.5\,\mu m$) it is recommended that the seeding technique is used. The mass of a spherical particle $0.1\,\mu m$ in diameter is about 5×10^{-19} kg. As the average mass of polystyrene chains produced by the emulsion polymerization technique is about 5×10^{-22} kg, the number of polymeric chains coiled into one particle is typically 10^3. Each of the polymeric chains begins and ends with the initiator group X, for example $-SO_4K$. In water they can dissociate, providing a large electrostatic charge per particle of the order of $10^3\,e$. The surface area σ per end group is of the order of 3000 Å^2. The particle charge can be determined from titration experiments. The uniformity of the final colloid depends on many parameters, such as the concentration of the different components of the initial solution. Some parameters are still unknown since the same recipe can give different suspensions and, at worst, flocculated ones.

2. *Interactions and Methods of Crystallization*

Suspensions of monodisperse colloidal particles usually have a milky aspect and under appropriate conditions become iridescent. The latter occurs when the particles are organized in regular lattices. The ordering is linked to the types of interaction between the colloidal particles in suspension.

2.1 Steric Repulsion

The simplest case to consider is that of hard spheres with no electric charge. These interact only by steric repulsion when they come into contact. Such a system is theoretically described by two parameters: the volume fraction Φ occupied by the particles, and the reduced pressure $\tilde{P}=P/n_0kT$. The volume fraction Φ is equal to nv where n is the particle number per unit volume and v is the volume of one particle. The symbol n_0 is the density of close-packing limit face-centered-cubic (fcc) or hexagonal-close-packed (hcp): $\Phi_0=74\%$. Numerical simulations of such a hard sphere system (Hoover and Ree 1968) have shown the existence of a transition between a long-range ordered phase and a disordered one. The solid phase is stable at high volume fraction and high pressure. When Φ reaches 50%, crystallization occurs. Between 50% and 55%, the two phases coexist, but beyond 55%, only the ordered phase should exist. An obvious method of crystallization consists of concentrating monodisperse colloidal suspensions—this happens naturally in precious opals under sedimentation. The process usually takes a long time: for polystyrene spheres (density$\simeq 1.06\times10^3\,\mathrm{kg\,m^{-3}}$) in water, the sedimentation velocity is about $2\,\mu\mathrm{m\,min^{-1}}$ when their diameter is $1\,\mu\mathrm{m}$. It decreases as R^2 with decreasing radius R. Up to several months are therefore necessary to attain a sedimentary equilibrium in a typical test tube. This can be greatly reduced (by a factor of 10^3) by the use of centrifugation.

2.2 Electrostatic Interactions

Colloidal particles often carry chemical groups at their surface that tend to dissociate in highly polar solvents such as water. Depending on the environment of the particle (e.g., the pH, temperature, etc.), only a fraction of these groups are dissociated. Each particle, therefore, becomes a macro-ion of charge $\pm Ze$ surrounded by a cloud of counter-ions; for instance, Z monovalent ions. The number Z can be very large ($\sim10^3$ or 10^4). Independent of these, the suspension may contain stray ions from an added salt with concentration c_s. Such a suspension is globally neutral but the particles and ions interact via long-range electrostatic forces. The interaction range is given to a first approximation by the Debye length λ_D.

$$\lambda_D=\left(\varepsilon_0\varepsilon kT\Big/\sum_i c_iZ_i^2\right)^{1/2} \quad (1)$$

where i is the type of ions, c_i their concentration and Z_i their charge. Increasing the salt concentration c_s makes λ_D decrease. In water, the maximum value of λ_D is a few micrometers. Two types of interaction can play a role in the crystallization phenomenon: attractive Van der Waals forces and repulsive electrostatic interactions of range λ_D. It might be expected that the balance between these forces would eventually lead to a minimum energy. This can happen, but the minimum is not deep enough to explain the stability of the ordered phase.

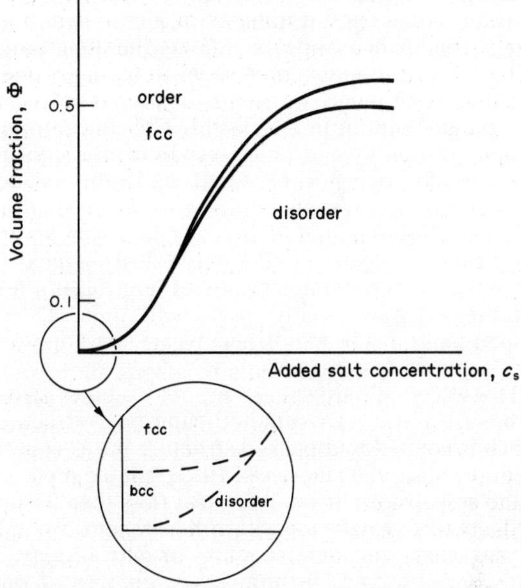

Figure 1
Phase diagram of charged monodisperse spherical colloids. Crystallization occurs at either large volume fraction Φ or low added salt concentration c_s. The structure is usually fcc, but a bcc structure is observed in dilute suspensions of small particles

The simplest model that takes into account the electrostatic interaction is the effective hard-sphere model. When λ_D is shorter than the interparticle distance, each colloidal sphere is surrounded by its own electrical double layer of size λ_D. Thus one can introduce an effective radius $R^*=R+\lambda_D$ and an effective volume fraction $\Phi^*=\Phi(1+\lambda_D/a)^3$. The crystallization transition takes place when $\Phi^*=0.5$; that is, for smaller values of Φ. The higher the interaction range, the smaller the possible volume fraction of the ordered suspension. When λ_D is large enough, colloidal crystals at low volume fraction (a few percent or less) can be obtained.

The over-simplified model is in good qualitative agreement with experimental diagrams (Fig. 1) which have been largely studied in latexes (Hachisu and Kobayashi 1974, Kose and Hachisu 1974); that is, suspensions of polymeric particles. Crystallization appears either at high volume fraction (large Φ) or at low ionic strength (large λ_D). This last condition is achieved in water by removing the stray ions in solution, for example, by dialysis or with the use of ion-exchange resins (Amenati and Fujita 1978). The easiest procedure to perform consists of directly mixing the ion-exchange resin and the suspension. After several minutes of gentle agitation, bright iridescence appears due to the long-range order. The resin grains then seem to be imprisoned in the bulk, although slow centrifugation is sufficient to separate them from the crystal. The concentrated samples are milky and not transparent, the crystallites usually being very small and not visible. Crystallization is revealed only by iridescence at the surface of the container. The very dilute samples are practically transparent and colored flecks appear in the bulk. The crystallites can be of large size (a few millimeters) and usually grow with time. Dilute suspensions must be handled with care since they are very sensitive to any salt impurity.

The effective hard-sphere model excludes structures such as the body-centered-cubic (bcc) structure which is not a closed-packed structure but is experimentally observed (Pieranski 1983), for example, in dilute suspensions of small spheres (less than 0.5 μm in diameter). Analogies with other systems, metals in particular, aid understanding of why bcc structures are observed. In both cases, charges of one sign are localized on a crystal lattice while charges of the opposite sign are delocalized. In colloidal crystals this delocalization has a thermal origin. In metals it has been proved that the bcc structure is energetically more favorable than the fcc, especially at low temperature. This analogy is certainly strong when λ_D is large compared with the interparticle distance: the counter-ions are no longer localized around the colloidal particles. As an example, an fcc–bcc transition has been observed in suspensions of relatively small polystyrene spheres ($2R \sim 0.1$ μm) when the volume fraction is decreased. Indeed, a lower volume fraction induces a longer range of electrostatic interactions.

None of these simple models take into account the size of the particles, which certainly has a great influence on the phase diagram. Nevertheless, they give the basic concepts which enable us to understand the crystallization of monodisperse colloids. Long-range order appears in these systems either as large volume fraction or when repulsive interactions with a long enough range exist between the particles. In the two cases, monodispersity in the size and shape of the particles is indispensable in obtaining colloidal crystals.

3. Characterization of Colloidal Crystals: Light Scattering

Colloidal particles are composed of either metal or dielectric matter (silica, polystyrene, etc.), the refractive index of which is usually different from that of the suspending medium. Each particle consequently behaves like a diffuser. The simplest case is when the incident beam with wavelength λ is parallel and represented by the vector $\boldsymbol{k}_0$, with $k_0 = 2\pi/\lambda$. Assume that each scatterer contributes to the amplitude in the direction $\boldsymbol{k}$ independently (there is no multiple scattering). Each scatterer introduces a phase shift $\phi = (\boldsymbol{k} - \boldsymbol{k}_0).\boldsymbol{r}$, where $\boldsymbol{r}$ represents its position. The maximum intensity is reached when ϕ is equal to $2\pi n$ (n is an integer) for all the diffusers, and this is achieved when one family of crystallographic planes is in the Bragg position:

$$\frac{\lambda}{n_0} = \frac{2d \sin \theta}{p} \tag{2}$$

where d is the distance between two neighboring planes, p is an integer and n_0 is the refractive index of the suspending medium. In colloidal crystals, d is typically of the order of 0.1 μm—the range of visible light wavelengths.

When a sample is lit with white light, crystallites which are in a correct position provide Bragg scattering. The color of the fleck is related to the wavelength λ and depends on both θ and d. The study of the colored flecks can reveal details of the structure of the crystallites, their orientation and the lattice parameter. This is analogous to the Laue method in x-ray crystallography: the scattering of a polychromatic beam enables the symmetry class as well as the crystalline orientation to be determined. In the case of colloidal crystals, additional information is provided by the color of the spots, linked to the interparticle distance, which can be precisely analyzed with the use of a monochromator. Concentrated colloidal crystals are milky and not transparent, and only the crystallites closest to the vessel surface contribute to scattering. It is therefore impossible to study the bulk structure with this method. Dilute suspensions are more transparent, however, and iridescence appears in the bulk. The distribution of the colored flecks indicates, for example, that the crystallites very often have a preferred orientation with respect to the container surfaces. They usually align their high density planes parallel to the glass surfaces. For an fcc structure these are the 111 planes perpendicular to the body diagonal of the cubic cell. For a bcc structure these are the 110 planes perpendicular to a diagonal of a cube face. This orientation involves interactions with the glass surface which are often difficult to analyze. The classical powder method can also be used to determine both the structure and the lattice parameter of dilute crystals. A typical device consists of immersing a sample, held in a glass or quartz cuvette, into a spherical vessel filled with water in order to match the refractive index of glass. A monochromatic beam provided by a laser is incident on the sample through a window. When all the possible orientations of crystallites are present in the suspension, Debye–Scherrer rings are obtained on the sphere (which is used as a screen), the distribution of which gives both the structure and the lattice parameter.

Kossel diagrams are another optical method used to characterize the structure of colloidal crystals (Pieranski 1983). They are composed of lines similar to the Kossel lines in x-ray scattering or to the Kikuchi lines in electron diffraction experiments. Kossel diagrams require a divergent monochromatic beam. A family of crystallographic planes can be considered with a light ray incident on these planes. The incidence angle is $\pi/2-\theta$. If θ satisfies the Bragg condition of Eqn. (2), the ray will be diffracted and therefore the transmitted intensity will be much lower in that direction. This is also the case for all the rays incident at the same angle on the family of planes. The Bragg reflected rays are distributed along a cone, the axis of which is normal to the planes, with an angle equal to $\mu=\pi/2-\theta$. The Kossel lines are the intersection of this cone with the projection screen. This simple explanation of Kossel lines enables their appearance and position to be understood. A rigorous treatment of this phenomenon requires the analysis of multiple scattering.

Kossel diagrams have been used in suspensions of polystyrene spheres to determine the structure and the lattice parameter using, for example, the same setup as in the powder method. The laser beam is made divergent by using a thin sheet of Teflon. The Kossel lines are observed on the translucent sphere; they can also appear when reflecting a parallel laser beam. This happens when scatterers (dust) are present close to the colloidal monocrystal. Interpretation of Kossel diagrams is generally difficult when the structure, orientation and lattice parameter are all unknown. In the case of colloidal crystals, the problem is much simplified since the structure is either bcc or fcc, and crystallites have a preferred orientation with respect to the container surfaces.

4. *Elastic Behavior*

An unusual property of colloidal crystals, owing to their large microscopic size, is their very low elastic modulus. An exact calculation of the elastic tensor is very difficult since many parameters have to be taken into account: volume fraction of the suspension, particle size, dielectric constants of the spheres and of the suspending medium, effective charge and thus the pH and ionic strength. Nevertheless, the order of magnitude of Young's modulus E can be determined by a simple argument. It is given by the ratio of a typical energy U to the cube of the involved length scale. The energy U lies in the range of kT, whereas the length scale is given by the interparticle distance, say 0.3 μm. Therefore E is of the order of a few tenths of a pascal, about ten orders of magnitude smaller than the Young's modulus for metals.

4.1 *Gravitational Compression*

One of the first methods of determining E was proposed by Crandall and Williams (1977) and consists of measuring the deformation e of a colloidal crystal under its own weight, as defined by the relative variation of the lattice parameter a, $e=\delta a/a$. Polystyrene spheres, for example, which have a density of $1.06\times10^3\,\mathrm{kg\,m^{-3}}$, tend to sink in water and exert a downward force proportional to their effective density $\rho_{\mathrm{eff}}=0.05\times10^3\,\mathrm{kg\,m^{-3}}$. The repulsive interaction between the spheres prevents compression. The force balance gives

$$nm_{\mathrm{eff}}g=E\nabla e \tag{3}$$

where n is the particle number per unit volume and m_{eff} is their effective mass; e and ∇e are measured using Bragg scattering of an He–Ne laser beam. As the particle density n is higher at the bottom of the container than at the top, this method also gives the variation of E as a function of n.

The time necessary to attain gravitational compression equilibrium is nevertheless very long. Indeed, when considering displacements of colloidal crystals, two situations can occur: either the fluid and the particles move together, or they do not. In the second case, the friction linked to the Stokes force due to the backflow of the solvent is important and overdamps the relative motion of the two components of the suspension. In sedimentation, this force acts against gravity and leads in a first approximation to a constant velocity $u_{\mathrm{sed}}\simeq2\rho_{\mathrm{eff}}gR^2/9\eta$ where η is the viscosity of the fluid and R is the radius of the spherical particles. As an example, particles of radius $R=0.5$ μm sediment at about 1.5 μm min^{-1}.

4.2 *Propagation of Elastic Waves*

In real crystals, elastic waves (periodic displacements of the atoms around their equilibrium position) can propagate at a velocity in the order of $10^3\,\mathrm{m\,s^{-1}}$. They also exist in colloidal crystals but with a different velocity scale, $u_{\mathrm{el}}\sim10^{-2}\,\mathrm{m\,s^{-1}}$. The other main difference is the presence of the suspending medium which overdamps any relative motion of particles and fluid. This explains why the longitudinal waves (displacement parallel to the wave vector) must be overdamped. In transverse waves (displacement perpendicular to the wave vector), the suspension medium can follow the motion of the sphere since the shear deformation keeps the volume constant. For wavelength λ much larger than the microscopic length, that is for a frequency f much smaller than 10^5 Hz, the difference between the fluid velocity v and the particle velocity $\dot{u}$ must be small (Mitaku et al. 1978). Colloidal crystals thus behave like a viscoelastic medium with a shear modulus E, a shear viscosity η and an average density ρ. The dispersion relation for the transverse waves is

$$\omega^2 + i\eta\rho^{-1}k^2\omega = E\rho^{-1}k^2 = 0 \tag{4}$$

At low frequency ($\omega \ll 10^3$ Hz) $\eta\omega \ll E$, the dissipative term is small and transverse waves can propagate. Several experimental techniques have been developed at low frequency. A surface-loaded resonance oscillator can detect the full mechanical impedence of the suspension (Mitaku et al. 1978, Benzing and Russel 1981, Lindsay and Chaikin 1982) which depends on E. At low frequency (a few Hz) the damping is small enough that the resonance peaks are sharp. Another method consists of establishing standing waves using reflections at the sample boundaries. As the propagation velocity $u_{el} = (E/\rho)^{1/2}$ is about 10^{-1} cm s^{-1}, elastic waves must be excited at a frequency f around 10 Hz to have a wavelength λ in the range of a few centimeters, which is the typical size of the container. A simple device consists of pouring the suspension into a glass cylinder and making it oscillate with the use of a loudspeaker (Dubois-Violette et al. 1980). The crystal deformation for a given excitation frequency can be monitored by observing the displacement of the Kossel lines which change both their position and diameter under shear deformation.

The shear elastic modulus of a colloidal crystal depends on many parameters since it is controlled by interparticle interactions. Some of these parameters can be changed by a few orders of magnitude in a given sample: the value of E was found to decrease by a factor of 200 upon a 60-fold dilution (Benzing and Russel 1981, Lindsay and Chaikin 1982) while the stray ion concentration n_s was held as low as possible by the ionic exchange purification method. The salt content n_S can also be changed at a given volume fraction Φ. When n_s is lower than the concentration of counter-ions necessary to achieve electroneutrality, the elastic modulus does not vary much. When n_s reaches the concentration of electroneutrality, a sharp decrease is noted until melting occurs ($E = 0$).

4.3 Lattice Dynamics

Whereas Bragg spots and Kossel lines are caused by the scattering of light by the well-ordered lattice, thermal elastic vibrations of the particles around their equilibrium position remove intensity from the Bragg spots and distribute it in a very anisotropic pattern. The study of the diffuse background thus gives information, not only on the shear modulus E (as for the mechanical measurements), but on the entire elasticity tensor. In real crystals, phonon dispersion relations $\omega_p(\boldsymbol{q})$ are obtained using neutron scattering. For a given direction of detection $\boldsymbol{K}+\boldsymbol{q}$ where $\boldsymbol{K}$ is a reciprocal Bravais lattice vector, the shift in energy of the scattered beam compared to the incident beam is related to $\omega_p(\boldsymbol{q})$. Such variations cannot be detected using x rays which, unlike neutrons, have a much higher energy than phonons. In colloidal crystals, the wavelength is high enough that such an analysis can be performed in light scattering experiments using autocorrelation spectroscopy (Hurd et al. 1982). Very dilute colloidal crystals (10^{17}–10^{18} particles m^{-3}) of small polymeric particles (0.1 μm in diameter) provide transparent suspensions with quite large crystallites. When placed between two parallel quartz plates (less than a few millimeters apart) they are suitable for such lattice dynamics studies. Unfortunately, when the container is too thin, viscous friction on the walls introduces additional damping even for transverse modes.

5. Colloidal Crystals Under Shear

Another consequence of the value of the elastic modulus E is that even low shear has a great influence on the structure of these suspensions. It has been shown using a sensitive rheometer (Mitaku et al. 1978) that polymeric colloidal crystals start to flow at above a critical shear stress σ_c which is a small fraction of E. This shear stress σ_c is easily reached by flowing the suspension. Different geometries have been used to study shearing flows. The suspension can be poured in between rotating cylinders (Couette flow) or rotating disks or in a pipe and submitted to a pressure gradient (Poiseuille flow).

5.1 Orientation in a Low-Shear Flow

As stated in Sect. 3 the crystallites at the container boundaries tend to align with their high density planes parallel to the surface. This leaves one degree of freedom; that is, the rotation around the axis normal to the planes. Analysis of Bragg-scattered light shows that this degeneracy is removed by a low shear (Pieranski 1983). The high density lines which lie in the high-density planes tend to be parallel to the flow velocity. In the case of a bcc structure, the high-density plane $1\bar{1}0$ contains two high density lines [111] and $[11\bar{1}]$. Two orientations are therefore possible and the crystallites form a twinned texture (Fig. 2) which is generally local. In alternative Poiseuille flow, Dozier and Chaikin (1982) have shown that a periodic pattern of alternate orientations could appear along the tube under appropriate conditions. The wavelength of this pattern is macroscopic (a few millimeters); that is, of the order of the tube radius and much larger than the crystallite size.

5.2 Shear-Induced Melting

With a high shear stress, the colored flecks disappear, the colloidal crystal 'melts', and there is no longer any long-range order in the suspension. This can be easily achieved by rocking a bottle which contains the colloidal suspension. Such shear flow melting is a reversible process. After stopping the shear, the colored flecks reappear a few seconds later. Small crystallites grow with time and become macroscopic in dilute suspensions after several weeks. Ackerson and Clark (1983) have observed a series of structural transitions in sheared bcc colloidal crystals leading finally to melting. The sequence, with increasing shear, is as follows:

(a) bcc twinning;

(b) bcc–hcp phase transition, the hcp layers being parallel to the twinning plane;

(c) free slipping hexagonal layers;

(d) strings of particles running parallel to the velocity; and

(e) disordered state.

A quantitative study of the critical shear stress which induces melting has been carried out using rotating disks geometry (Pieranski 1983). The suspension is poured between two glass disks separated by a few millimeters. One disk rotates with an angular velocity ω (a few Hz). The velocity of the suspension thus increases continuously with the distance r to the disk center. For a distance r_c, one observes that the central part of the sample is crystallized with a twinned structure as deduced from visible light scattering, whereas the outer part is, disordered. This critical radius provides a measurement of the critical shear stress σ_c.

A shear-melted suspension behaves like a Newtonian fluid with a constant viscosity depending on both the volume fraction and the interaction

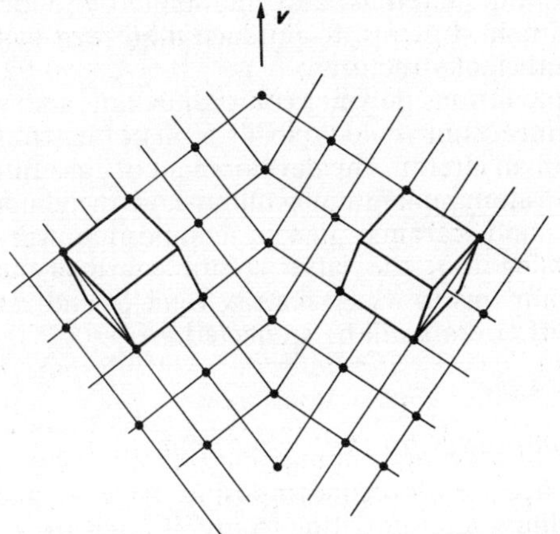

Figure 2
Orientation of a colloidal crystal under low shear: the density rows tend to align parallel to the velocity field v. In a bcc structure these are the body diagonals of the cube and twinning is often observed. Higher shear rates induce structural transitions, leading finally to melting

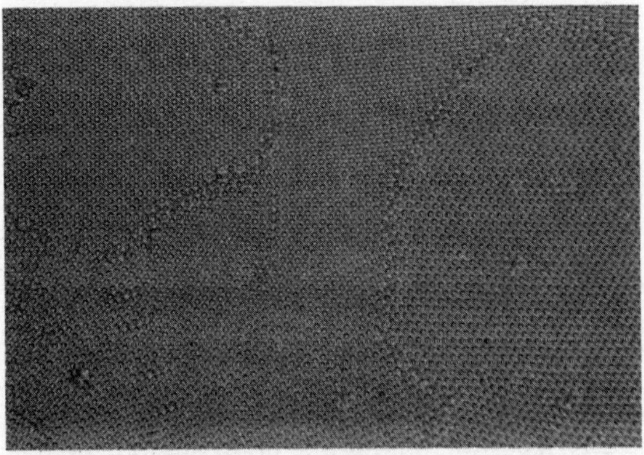

Figure 3
Observation of an ordered suspension of large (1μm in diameter) polystyrene spheres with an optical microscope. The high density planes (with a hexagonal structure in an fcc crystal) are parallel to the glass surface

range. The existence of translational order at low shear may induce interesting effects: for instance, a colloidal suspension exhibits Taylor instabilities (rolls) in a Couette geometry (Joanicot and Pieranski 1985) when the angular velocity of the external cylinder is large enough. For this value of the velocity, the suspension is melted but in the region near the container walls and between two neighboring rolls the shear stress is low enough for crystallization to occur, underlining the presence and structure of the instability. Moreover, the existence of crystallites can influence the establishment of the Taylor rolls.

6. *Observation with an Optical Microscope*

The most fascinating sight offered by colloidal crystals is probably when they are observed through an optical microscope (Fig. 3). The size of the particles and of the lattice parameter is so large compared with that of usual crystals that this observation reveals many details of their local structure. In samples composed of large particles the individual particles are even visible using low magnification objectives (Kose et al. 1973).

Colloidal crystals mimic some of the local behavior of crystals, as they do for macroscopic properties: defects such as grain boundaries, dislocations or vacancies have been observed. Edge dislocations can be created in a well controlled manner by pouring the suspension between a spherical and a flat glass surface (Pieranski 1983). Edge dislocations have to be present to fill the available space with crystal and form concentric loops. This can be seen even with small particles, although they are not individually observed. Indeed, the strain field around the defect modulates both the color and intensity of the scattered light.

This experiment shows that the solid surfaces appear perfectly smooth and rigid. This important detail has made possible the observation and study of two-dimensional colloidal crystals. An fcc colloidal crystal made of large particles (diameter 1 μm), observed through the thin cover glass plate shows the first crystalline plane in the vicinity of this plate. The particles form a two-dimensional hexagonal lattice at some distance from the plate. This indicates that the crystal really aligns the high-density planes parallel to the glass surface and that the interaction between the particles and the glass surface is repulsive. Owing to this repulsion, perfect two-dimensional colloidal crystals have been realized between two glass surfaces a few micrometers (or less) apart. This geometry has been used to study different processes rarely performed with usual crystals, such as crystal growth, passage from two dimensions to three dimensions by a series of structural transitions (Pansu et al. 1983) and two-dimensional melting (Murray and Van Winkle 1987).

See also: Ceramic Powders: Packing Characterization

Bibliography

Ackerson B J, Clark N A 1983 Sheared colloidal suspensions. *Physica A* 118 (1–3): 221–49

Amenati K, Fujita H 1978 Purification of latex suspensions. *Jpn. J. Appl. Phys.* 17: 17–21

Benzing D W, Russel W B 1981 The viscoelastic properties of ordered lattices: Experiments. *J. Colloidal Interface Sci.* 83: 178–90

Crandall R S, Williams R 1977 Gravitational compression of crystallized suspensions of polystyrene spheres. *Science* 198: 293–5

Dozier W D, Chaikin P M 1982 Periodic structures in colloidal crystals with oscillatory flow. *J. Phys. (Paris)* 43: 843–51

Dubois-Violette E, Pieranski P, Rothen F, Strzelecki L 1980 Shear waves in colloidal crystals: I Determination of the elastic modulus. *J. Phys. (Paris)* 41: 369–76

Hachisu S, Kobayashi Y 1974 Kirkwood–Alder transition in monodisperse latexes: Aqueous latexes of high electrolyte concentration. *J. Colloid. Interface Sci.* 46: 470–6

Hoover W G, Ree F M 1968 Melting transition and communal entropy for hard spheres. *J. Chem. Phys.* 49: 3609–17

Hurd A J, Clark N A, Mockler R C, O'Sullivan W J 1982 Lattice dynamics of colloidal crystals. *Phys. Rev. A* 26: 2869–81

Joanicot M, Pieranski P 1985 Taylor instabilities in colloidal crystals. *J. Phys. Lett.* 46: 91–6

Kose A, Hachisu S 1974 Kirkwood–Alder transition in monodisperse latexes: Non aqueous systems. *J. Colloid Interface Sci.* 46: 460–9

Kose A, Ozaki M, Takano K, Kobayashi Y, Hachisu S 1973 Direct observation of ordered latex suspension by metallurgical microscope. *J. Colloid Interface Sci.* 44: 330–8

Lindsay H M, Chaikin P M 1982 Elastic properties of colloidal crystals and glasses. *J. Chem. Phys.* 76: 3774–81

Mitaku S, Ohtsuki T, Enari K, Kishimoto A, Okano K 1978 Studies of ordered monodisperse latexes. *Jpn. J. Appl. Phys.* 17: 305–13

Murray C A, Van Winkle D H 1987 Experimental observation of two stage melting in a classical two-dimensional screened Coulomb system. *Phys. Rev. Lett.* 58: 1200–3

Overbeek J Th G 1982 Monodisperse colloidal systems, fascinating and useful. *Adv. Colloidal Interface Sci.* 15: 251–77

Pansu B, Pieranski P, Strzelecki L 1983 Thin colloidal crystals: A series of structural transitions. *J. Phys. (Paris)* 44: 531–6

Pieranski P 1983 Colloidal crystals. *Contemp. Phys.* 24: 25–73

Sugimoto T 1987 Preparation of monodispersed colloidal particles. *Adv. Colloidal Interface Sci.* 28: 65–108

B. Pansu
[Université Paris-Sud, Paris, France]

Composite Materials: Aerospace Applications

Fibrous polymer composites have found applications in aircraft from the first flight of the Wright Brothers' *Flyer 1*, in North Carolina on December 17th 1903, to the plethora of uses now enjoyed by them on both military and civil aircraft, in addition to the more exotic applications on space launcher vehicles and satellites. Their growing use has arisen from their high specific strength and stiffness, when compared to the more conventional existing and developing materials, and the ability to shape and tailor their structure to produce more aerodynamically efficient structures.

While fibrous polymer composites can, and will in the future, contribute up to 40–50% of the structural mass of an aircraft, the development of conventional alloys, aluminum–lithium alloys and the emerging metal and ceramic matrix composites will find increasing use, the latter in the more aggressive environments to which reentry vehicles and vertical take-off aircraft will be subjected.

1. Composites for Aircraft Applications

Glass-fiber-reinforced composites have been used on military aircraft dating from 1940, but their poor relative specific stiffness has prevented them from extending the foothold they have found on fairings, doors, etc., to the primary structural applications of wings, stabilizers and major fuselage sections. Aramid fibers introduced in the 1960s found parallel applications with glass fibers, but their lack of specific stiffness and poor compressive strength limited their use, despite the tolerance to damage that composites utilizing these fibers can afford.

The adoption of composite materials as a major contribution to aircraft structures followed on from the discovery of carbon fiber at the Royal Aircraft Establishment at Farnborough, UK, in 1964. However, not until the late 1960s did these new composites start to be applied, on a demonstrator basis, to military aircraft. Examples of such demonstrators were trim tabs, spoilers, rudders and doors. With increasing application and experience of their use came improved fibers and matrix materials resulting in composites with improved properties, allowing them to displace the more conventional materials—aluminum and titanium alloys—from primary structures.

1.1 Structural Properties of Composites

Whereas composite strength is primarily a function of fiber properties, the ability of the matrix to both support the fibers and provide out-of-plane strength is, in many load situations, equally important. The aim of the material supplier is to provide a system with a balanced set of properties. While improvements in fiber and matrix properties can lead to improved laminate properties, the all-important field of fiber–matrix interface must not be neglected.

All laminate properties, with the exception of those relying on interlaminar strength and stiffness, are almost directly proportional to the basic strength of the fiber. As laminate stiffness also follows this relationship then clearly any improvement in fiber properties will almost inevitably lead to a product having improved laminate properties with a consequential improvement in their application to aerospace structures.

(a) Effects of fiber properties. The early carbon fibers, following on from the Farnborough discovery, exhibited properties moderate by current standards and improvements have occurred in stiffness, strength and a combination of the two. Figure 1 shows a carpet plot of fiber strength against modulus with an indication of the years in which fibers were developed. While it is difficult to be specific as to the direction in which developments should best take place, the trend of development required for military aircraft would be one in which both properties were increased simultaneously along the "ideal" direction indicated. Satellite applications, in contrast, benefit from the use of high fiber modulus, improving stability and stiffness for reflector dishes, antennas and their supporting structure.

(b) Effects of matrix properties. In order to produce a laminated structural element the fibers have to be bonded one to the other and, while thermoplastic materials are becoming available, the more conventional matrix material is a thermosetting epoxy. The

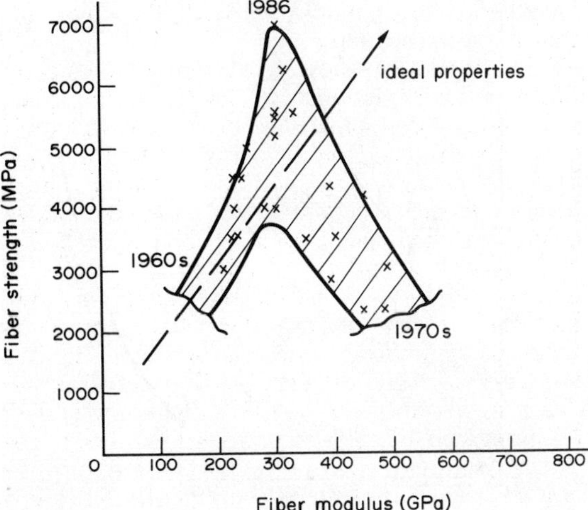

Figure 1
Tensile strength versus tensile modulus for carbon fiber

matrix material is the Achilles' heel of the system and limits the fiber from exhibiting its full potential in terms of laminate properties. The matrix performs a number of functions, amongst which are: stabilizing the fiber in compression, translating the fiber properties into the laminate, minimizing damage due to impact by exhibiting plastic deformation and providing out-of-plane properties to the laminate.

Matrix-dominated properties are reduced when the glass transition temperature is exceeded and whereas with a dry laminate this is close to the cure temperature, the inevitable absorption of moisture reduces this temperature and hence limits the application of most high-temperature-cure thermoset epoxy composites to less than 120 °C.

The first generation of aerospace composites introduced in the 1960s and 1970s utilized brittle-matrix systems leading to laminates with a poor tolerance to low-energy impact. These impacts can be caused by runway debris thrown up by the aircraft wheels, or the impacts occurring during manufacture and subsequent servicing operation. Although the emerging toughened epoxy systems provide improvements in this respect, they fall far short of the damage tolerance provided by the new solvent-resistant thermoplastic materials. A measure of damage tolerance is the laminate compression strength after impact and this property is plotted on the abscissa of Fig. 2 with the laminate notched compression strength on the ordinate. The ideal solution is to provide a material exhibiting equal properties and it can be seen that whereas the thermoplastic systems are tougher they have not capitalized on this by yielding higher notched compression properties.

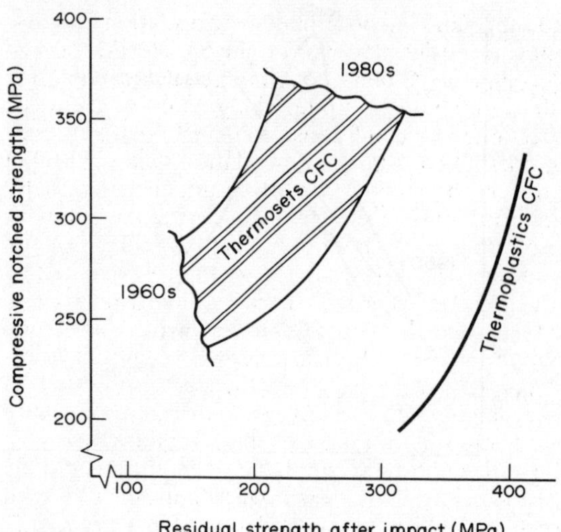

Figure 2
Comparison of notched compression and residual strength after impact of carbon-fiber composite (CFC)

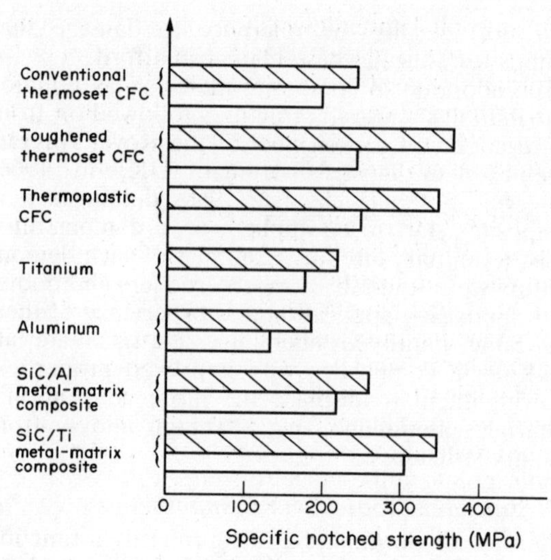

Figure 3
Structural material strength comparison

(c) *The fiber–matrix interface*. This, the third factor governing strength, is considered by many to be crucial in maximizing the degree to which the fiber properties can be translated into the laminate, whether the matrix is thermosetting or thermoplastic. The interface between the fiber and matrix may be less than one micrometer yet excessive bonding between the fiber and matrix will result in poor notched tensile strength while an inadequate bond will result in poor interlaminar properties. Conflict therefore exists and the designer must select the material most nearly meeting his requirements.

2. *Alternative Fabrication Materials*

Of two materials with identical mechanical properties, the one with the lowest density will produce the minimum mass structure if all or part of the material is stability designed. As a consequence, carbon-fiber composites have an advantage over the more conventional materials used in aircraft fabrication. Composite materials are displacing the 2000 and 7000 series aluminum alloys from their dominant position in aerospace but emerging metallic materials have the potential to challenge the growing composite applications. Aluminum–lithium alloys, again a development from the Royal Aircraft Establishment, Farnborough, UK, are offering improvements in strength with a parallel reduction in density. These, in their own right, offer a challenge but when reinforced by ceramic particulates, such as silicon carbide, yield specific properties approaching those of today's composites. However, they are prohibitively expensive and difficult to manipulate and will only find application in niche areas in the near future. A comparison has been made, and shown in Fig. 3, of the specific notched tensile strength of various aerospace materials including both thermoset and thermoplastic materials.

3. *Implications of Composites on Design*

Aircraft design from the 1940s has been based primarily on the use of aluminum alloys and as such a plethora of data and experience exists to facilitate the design process. With the advent of laminated composites exhibiting anisotropic properties the methodology of design had to be reviewed and, in many areas, replaced. It is accepted that designs in composites should not merely replace the metallic alloy but should take advantage of exceptional composite properties if the most efficient designs are to evolve.

Hitherto the material properties for direct use by the designer had been available and his function was, based on a limited number of properties, to ensure that the structure offered sufficient strength and stiffness. With composite materials, however, the extra function of designing the material from a set of laminae properties has to be undertaken. Although in many instances laminae properties are available and documented for strength, stiffness and stability, stacking sequences have to be compromised to ensure continuity and compatibility of structure. An example of the complexity of the process is the derivation of the bending stiffness of an anisotropic laminate which involves the manipulation of a 6×6 matrix, a task which can only be performed by computer. Although these complexities lengthen the design process, they are more than compensated for by the mass savings and improvements in aerodynamic efficiency that result.

3.1 Aerodynamic Improvements

Because of the limitation of metal-manipulation techniques, the majority of aircraft control–lift surfaces produced have a single degree of curvature. Improvements in aerodynamic efficiency can be obtained by moving to double curvature allowing, for example, the production of variable camber, twisted wings. Composites allow the shape to be tailored to meet the required performance targets at various points in the maneuver envelope.

A further and equally important benefit is the ability to tailor the aeroelastics of the surface to further improve the aerodynamic performance. This tailoring can involve adopting laminate configurations which allow the cross-coupling of flexure and torsion such that wing twist can result from bending and vice versa. Modern computational techniques of structural optimization, using finite-element analysis techniques, allow this process of aeroelastic tailoring, along with strength and dynamic stiffness (flutter) requirements, to be performed automatically with a minimum of postanalysis engineering yielding a minimum mass solution.

3.2 Airworthiness Procedures

High material variability and the deleterious effects of temperature and moisture preclude the use of the conventional airworthiness route required for the clearance of aircraft structures. The composite design allowables, particularly in compression, can be as low as 60% of the mean test result obtained from a test conducted at room temperature. As both material variability and degradation are failure-mode dependent, a room-temperature test cannot practically demonstrate the true strength of a structure. This is exacerbated if the structure is hybrid, containing metallic components, as the variability and thermal and moisture degradation of these materials can be negligible.

As the most critical condition, the production of minimum design values, generally takes place at elevated temperature when the laminate has absorbed moisture then the obvious method would be to undertake the testing under these conditions. However, testing in this manner on full-scale flight vehicles can be prohibitive in terms of both cost and timescales and, as a consequence, alternative routes have been found to ensure the competence of the structure.

These routes have as their foundation the collection of data from many thousands of specimens, giving strength for the particular property along with variability and environmental degradation factors; from this, design data are produced. Verification of these data on elements containing structural features is then obtained by undertaking some of the tests under ambient conditions, and other elements are degraded to confirm the degradation factor. Parallel to this is the verification of strain distribution by comparing strain measurements during the test with those obtained from a finite-element analysis of the specimen. The full-scale test is then performed at ambient conditions with each of the critical cases taken to the ultimate design load with the results from the extensively instrumented structure extrapolated to verify that, had the test proceeded to higher loads, both variability and degradation would have been accommodated. Should failure not have occurred at ultimate load then the test would be taken to failure, with this occurring in the area with lowest variability and degradation effects. A metallic component usually shows this property. Figure 4 has been constructed in an attempt to demonstrate graphically the way in which a full-scale static test undertaken at ambient conditions can be read across to provide the effective ambient condition design allowable.

3.3 Test Instrumentation

Instrumentation, particularly strain measurement, is essential to the airworthiness clearance of composite structures. The two major agents for understanding the strain distribution are electrical resistance strain gauges and photoelastic coatings, the former giving field strains while the latter are more useful for indicating the distribution around stress concentrations.

Work is ongoing, taking advantage of the laminated form of composite structures, to embed strain-

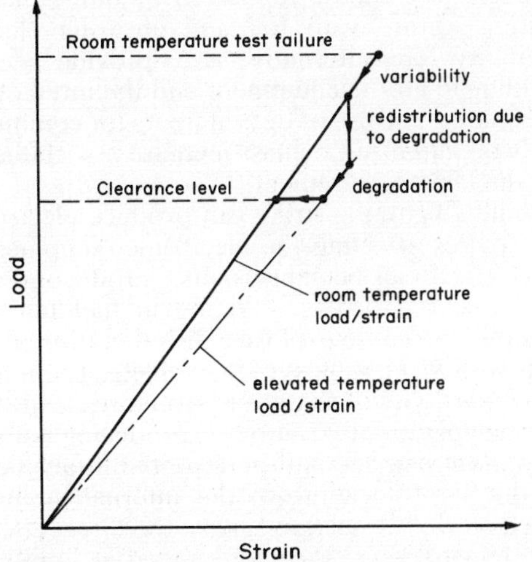

Figure 4
Airworthiness clearance of composite structures

measuring fiber optics into the laminate during manufacture, producing a matrix of multipoint data measurements leading to an "intelligent" structure. Currently, however, only single-point strain measurement along a fiber is possible.

More direct methods of failure prediction have been investigated by many researchers and prominent amongst these methods is acoustic emission (AE). This relies on the ultrasonic emission from structures caused by matrix cracking, fiber failure, joint slip, etc. AE has been used as an inspection tool for such items as pressure vessels where a characteristic pressure–emission profile can be obtained from acceptable cylinders and then used to clear production components. However, on structures where this characteristic has not been determined, the current capability to predict failure is limited. Emission characterization is being developed to identify the impending failure mode and it is considered that this may lead to a more accurate prediction capability.

3.4 Electromagnetic Compatibility

All aircraft contain electronic equipment and with modern "fly-by-wire" aircraft the use of such equipment is increasing rapidly in the critical flight safety area of the systems. On metallic aircraft these delicate systems are in the main adequately protected from electromagnetic interference by the inherent shielding provided by the fuselage skins. Fibrous polymer composites unfortunately do not provide this protection and, as a consequence, special action has to be taken to ensure that this deficiency is overcome.

Electromagnetic interference can result from natural phenomena: lightning strike, radio transmitters, etc., or in the case of military aircraft from malicious intent. The effect of the traditional aluminum structure can be replicated by lining the composite structure with foil and ensuring electrical continuity. The alternative is to provide adequate shielding to all the equipment and the interconnecting cabling. The use of optical fibers for communication will minimize the mass increase associated with this shielding requirement.

While a lightning strike can produce electromagnetic pulses affecting the electronic equipment, its direct effect can be catastrophic, producing significant structural damage. Arcing in fuel tanks, for example, can cause explosive disintegration of these tanks with potential loss of the vehicle. The inherent low conductivity of composite structures leads to the poor dissipation of arc current, producing hot spots, mechanical damage and preferential attachment to metallic items, leading to the internal arcing discussed above. Protection against the direct effects of lightning strike are therefore aimed at limiting the physical damage to an acceptable level, particularly over fuel.

The above-mentioned internal foil is inadequate for system protection, and external foils or meshes have to be provided on laminates in the more critical zones of the aircraft. Meshes can be cocured along with the laminate, as can foils, but an attractive protection capable of easy repair is that of flame or plasma spraying metallic coatings.

A side issue that must be addressed is the conflicting requirements of adequate electrical continuity and corrosion. In the presence of aluminum, carbon-fiber composites act as noble metals with the former corroding in their presence. Careful design is therefore essential to prevent the creation of galvanic cells and the resulting corrosion.

4. Manufacture

The aerospace industry has, over the past decades, geared itself to the batch manufacture of metallic parts for aircraft. This has led to sophisticated methods of manipulating these metals with numerically controlled machines for metal removal, routing, forming etc. Further improvements, with the introduction of flexible manufacturing systems, ensure the optimum availability of part, tool and machine, producing a system capable of manufacturing a number of different components with minimal interference to efficiency.

With the extensive introduction of fibrous composites in the 1970s, new skills had to be developed in both the production engineering of the design and the subsequent manufacture of the component.

Mold tools capable of operating at the cure and postcure temperatures (200 °C) while remaining stable and durable had to be designed and, while this can be readily effected in steel, the mismatch in coefficient of thermal expansion between the two materials requires careful consideration. Early composite mold tools were neither durable nor stable but modern tooling materials, usually in preimpregnated form, are proving more reliable. Although the steel thermal mismatch can be overcome on laminate molding tools, the resolution of the problem is more difficult if this material is to be used in assembly tools requiring thermal cycling.

4.1 Laminate Manufacture

The largest proportion of carbon-fiber composites used on primary class-one structures is fabricated by placing layer upon layer of unidirectional material to the designers requirement in terms of ply profile and fiber orientation. On less critical items, woven fabrics very often replace the prime unidirectional form.

A number of techniques have been developed for the accurate placement of the material, ranging from labor intensive hand layup techniques to those requiring high capital investment in automated tape layers. These latter machines can require, in certain

Table 1
CFC layup rate comparison

Method of manufacture	Deposition rate ($kg\,h^{-1}$)
Foil transfer	0.1–0.25
Broadgoods	0.7–1.25
Tape layer	5
Filament winding	20

instances, investments of up to US$3 million (at 1988 prices).

Processes adopted by the aerospace industry range from hand layup using foil-transfer techniques, through the use of numerically controlled ply profiling machines to numerically controlled tape laying or filament winding. The first of these is an obsolescent process and has, in general, been replaced by the latter two. It involves the hand profiling of plies and their subsequent location on the mold tool, again by hand, using draughting foils for their location.

Use has been made in both the tailoring and shoe industries of automated profiling equipment and out of these has arisen the broadgoods system in which numerically controlled machines under two-axis control profile the plies to the appropriate shape. Three cutting media have been used on these machines: high pressure water, reciprocating knife cutting and laser cutting, with reciprocating knife becoming prominent because of the moisture uptake with the former and charring of ply ends with the latter. Although the use of broadgoods systems reduces the effort in profiling the plies, their manual laying remains a costly aspect of the laminate production. A number of attempts have been made to automate this ply-placement process, with success only being found on small components with limited complexity and curvature.

An obvious development is to combine the ply profiling and laying into one operation and this has been accomplished with limited success in automatic tape-laying machines operating under numerical control. These machines are currently limited in production applications to flat layup, and significant effort is being directed by machine manufacturers at overcoming the many problems associated with laying on contoured surfaces.

Cost reductions in the manufacture of carbon-fiber composite components have been obtained by the progression from hand layup to the use of tape-laying machines. Table 1 shows the corresponding increase in productivity.

4.2 Assembly of Aerospace Components

Early composite designs were replicas of those which employed metallic materials and, as a consequence, the high material cost and man-hour-

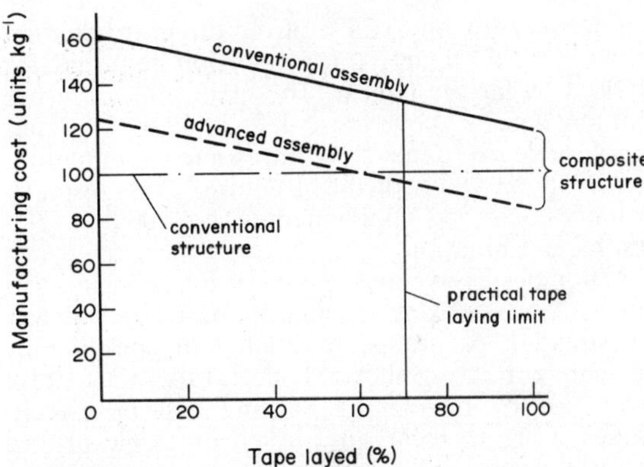

Figure 5
Manufacturing cost of conventional and composite structures

intensive laminate production jeopardized their acceptance. This was compounded by the increase in assembly costs due to the initial difficulties of machining and hole production. Evidence exists to support the argument that cost is directly proportional to the number of parts in the assembly and, as a consequence, design and manufacturing techniques had to be modified to integrate parts, thereby reducing the number of associated fasteners.

A number of avenues are available for reducing the parts count, among which are the use of honeycomb sandwich panels—sometimes frowned upon in metallic structures due to corrosion—the use of integrally stiffened structures and, more ambitiously, the cocuring or cobonding of substructures onto lift surfaces such as wings and stabilizers. Figure 5 has been constructed to demonstrate the cost of manufacture of aircraft structure in terms of units of cost per kilogram of structure using the conventional aluminum structure as the datum. Hand layup techniques and conventional assembly result in manufacturing costs 60% higher than the datum and only with the progressive introduction of automated layup and advanced assembly techniques can composites compete with their metallic predecessors.

4.3 Nondestructive Examination

All aerospace materials are quality assured to a high standard with the bulk of metallic materials being examined by the suppliers using ultrasonics, x ray and other appropriate processes. The carbon-fiber preimpregnated material supplier quality-assures his product by various techniques, but the verification of a sound cured structural element is now the responsibility of the aircraft fabricator.

Composites attract a high level of both conventional inspection and the more sophisticated nondestructive examination (NDE) and these combined can account for up to 25% of the direct production effort. Of this figure, the proportion required for NDE is by far the largest. The main tool to ensure that any defects above those defined by the designer as tolerable are detected is ultrasonic examination, invariably on 100% of the structure. This, in many instances, can be supplemented by 100% radiographic examination.

A number of avenues are open for reducing this burden, by relaxing standards based on defect investigation programs, by zoning of components through correlating allowable defect to stress–strain levels, or by developing existing and new techniques. Current techniques based on single probes or multiprobes are time consuming and attention is being directed at phased arrays, thermography, etc. (see *Composite Materials: Nondestructive Evaluation*).

5. *Aircraft and Aerospace Applications*

From the cautious application of composites in the early 1970s has developed the more confident use of these materials in the primary load-carrying structures of both military and civil aircraft. Although military applications are far in excess of civil aircraft applications, the latter are growing rapidly.

5.1 *Civil Aircraft Applications*

These have concentrated on replacing the secondary structure, bounding major structural elements, with fibrous composites where the reinforcing media has either been carbon, glass, Kevlar, or hybrids of these. The matrix material, a thermosetting epoxy system, is either a 125 °C or 175 °C curing system with the latter becoming dominant because of its greater tolerance to environmental degradation. Typical examples of the extensive application of composites in this manner are the later aircraft from Boeing (the 757 and 767) and from Europe, the aircraft produced by Airbus Industrie. This consortium of aircraft companies is currently producing the A320 which, like the later versions of the A310, carries a vertical stabilizer, a primary aerodynamic and structural member, fabricated almost in its entirety from carbon composite. In addition, the A320 has extended the use of composites to the horizontal stabilizer in addition to the plethora of panels and secondary control surfaces.

In two small transport aircraft attempts have been made to apply composite materials to almost the entire aircraft; these are the Lear Fan fabricated in Belfast, Northern Ireland, and the US Beech Starship. Of these ambitious projects the former failed for technical and commercial reasons although the latter is showing every prospect of entering service.

5.2 *Military Aircraft Applications*

These applications throughout the world, initially funded from national research and development budgets, have enabled composite technology to be developed to its current level. This includes material manufacture, design techniques and product manufacture as well as in-service experience and maintenance.

Without exception all agile fighter aircraft currently being designed throughout the world contain in the region of 40% of composites in the structural mass, covering some 80% of the surface area of the aircraft. Without this degree of compositization the essential agility of the aircraft would be lost because of the consequential mass increase.

The US advanced tactical fighter (ATF) and the European Fighter Aircraft (EFA) are examples of the proposed degree of compositization for aircraft entering service in the 1990s. Examples of current applications in which the structural mass of composites is 25% of the total can be found in the McDonnell–Douglas AV-8B/Harrier II GR Mk. 5 and the European Experimental Aircraft Programme (EAP).

5.3 *Space Applications*

Satellites and space platforms clearly require ultralight structures to minimize the launch costs and, to this end, exotic materials have been developed. These include ultrathin lamina and high-modulus materials, the former to enable precise tailoring to the structural requirement and the latter to provide stability for the antennas and reflectors. With the demand for observation and communication satellites ever increasing, the developments for this niche market will inevitably continue.

Current launch vehicle availability in terms of tonnes per annum is incapable of meeting demand and the available vehicles, the US Space Shuttle and the European Ariane rocket, are prohibitively expensive. Worldwide studies are determining the feasibility of fully reusable launch vehicles requiring minimum turnaround activity, thereby reducing launch costs significantly. Amongst these are the US National Aero-Spaceplane (NASP) and the British Horizontal Take-Off Launch Vehicle (HOTOL). The latter experiences structural temperatures ranging from that of liquid hydrogen (20 K) to those caused by reentry (1750 K). Although materials are available that will perform structurally at these temperatures, their performance, particularly towards the upper temperature limit, is inadequate to meet the demanding specific-strength requirements. These inadequacies in available materials provide the next major challenge to material suppliers, designers and fabricators involved in the introduction of fibrous composites with polymer, ceramic or metal matrices.

See also: Composite Materials: An Overview

Bibliography

Ashton J E, Halpin J C, Petit P H 1969 *Primer on Composite Materials Analysis*. Technomic, Stanford, CT

Curtiss P T (ed.) 1984 Crag test methods for the measurement of the engineering properties of fiber reinforced composites, Royal Aircraft Establishment Report RAE TR84102. RAE, Farnborough, UK

Haresceugh R I 1987 Composites—the way ahead. *Proc. 6th Int. Conf. on Composite Materials, 2nd European Conf. on Composite Materials*, Vol. 5. Elsevier, London

Jones R M 1975 *Mechanics of Composite Materials*. McGraw–Hill, New York

Lechnitski S G 1963 *Theory of Elasticity of an Anisotropic Elastic Body*. Holden–Day, San Francisco, CA

Royal Aeronautical Society 1986 *Materials In Aerospace*, Vols. 1, 2. RAS, London

R. I. Haresceugh
[British Aerospace Military Aircraft Division, Preston, UK]

Composite Materials: Applications Overview

Composite materials now exert an influence on the daily lives of most people in industrialized societies. Much personal transport, and indeed safety, depends upon the cord or fiber reinforced elastomer tire; many commercial airliners fly with some parts of the primary structure and much of the interior furnishings and trim made from composites; many types of sporting and leisure goods, such as boats, gliders, sailboards, skis and racquets, make extensive use of composite materials in their construction; and in the home, most plastic-bodied appliances incorporate reinforcement in the form of short, chopped fibers.

The term composite material can be defined in a number of ways. The most commonly held view is of a material in which strong, high-modulus man-made fibers, with diameters of the order 10–100 μm, are embedded in a matrix material which may be a polymer, a metal, a glass or a ceramic. The objective is to enhance the mechanical or physical properties of the host material. Alternatively, from a mechanical standpoint, the matrix may be regarded simply as a binder whose role is to transfer stress to the reinforcing fibers and ensure their cooperative interaction. The matrix may also fulfill other functions, such as protecting the fine reinforcing filaments from corrosion, oxidation, or other forms of environmental degradation.

Reinforcing agents are not necessarily confined to man-made fibers, although continuous filaments of glass, carbon, ceramics and high-modulus polymers are the most widely used at present. However, vegetable fibers, inorganic whiskers, metal wires and particles of refractory solids all find a place in the spectrum of reinforcing materials. On a somewhat

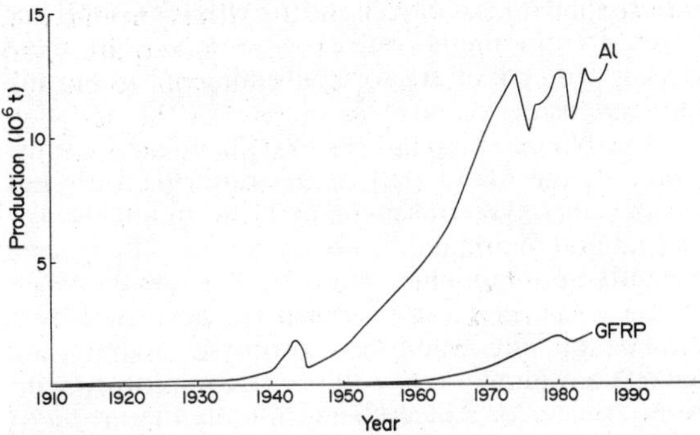

Figure 1
Comparative growth of aluminum and GFRP consumption

different scale, concrete reinforced with steel bars can be regarded as a composite with mechanical behavior analogous to that of fiber composites, but where application in tonnage terms dwarfs that of other constructional materials. However, this article is concerned with the families of materials known as fiber-reinforced plastics (FRPs), metal-matrix composites (MMCs) and ceramic-matrix composites (CMCs) in which at least two of the three dimensions of the reinforcement are typically of the order 100 μm or less.

Of these materials, the group that has reached the most advanced stage of both technical development and market penetration is fiber-reinforced plastics. One of their principal attributes is low density and to indicate the scale of their application, world consumption of glass-fiber-reinforced plastics (GFRP) is compared in Fig. 1 to that of aluminum, a material of roughly comparable cost and density, but established in the market place some 50 years earlier. The data for aluminum are plotted from the time (approximately 1890–1900) when commercial quantities became available following the development of the Hall process for the electrolysis of fused cryolite. Figures for GFRP date from the mid-1940s when significant quantities of glass fiber first became commercially available. Both curves reflect the recession in world trade during the 1980s, and the aluminum plot also illustrates the boost and subsequent fall in demand associated with World War II, and the influence of the 1973 oil crisis.

Although consumption of GFRP in the 1980s was small compared to that of aluminum, it is apparent from Fig. 1 that the rate of growth of consumption over the first 30 years of composites technology, averaging 10–15% per annum, has been considerably greater than in the first 30 years of aluminum technology. It is probably not meaningful to pursue the comparison further in view of the general increase in the pace of technological development that has occurred over the past few decades, except to note that at least part of the growth for GFRP has resulted from metals substitution, and that most market projections are for continuing growth in the applications of composites.

Glass-fiber-reinforced plastics, in volume terms, represent the major part of the reinforced plastics market, mainly on account of their relatively low cost, good corrosion resistance, good specific strength and modulus, and the ease with which complex shapes can be molded or fabricated by a number of well-established processes. The same general advantages apply to plastics reinforced with even stronger or stiffer fibers, notably of carbon or of highly oriented polymers such as the aramids, and although the resulting composites are considerably more expensive than GFRP, they can nevertheless be highly cost-effective in applications where weight reduction gives rise to operating economies, as in aerospace. For example, one source estimates the acceptable cost of weight saving in helicopter construction as US\$250 kg^{-1}, and in a satellite as US\$650 kg^{-1} (Jardon and Costes 1987).

One of the major limitations of polymer-based composites is the decrease in their mechanical properties with increased temperature. A large number of polymer formulations, both thermosetting and thermoplastic, are used as matrices for composites and although a few, presently expensive, materials are suitable for extended service above 300 °C, most reinforced plastics are limited to temperatures below ~200 °C for continuous operation. It is necessary to employ more refractory reinforcements with metal or ceramic matrices in order to exploit the advantages of composite materials for high temperature applications, such as in heat engines. Table 1, from Hancox and Phillips (1985), indicates maximum temperatures for continuous operation of the three principal types of composite material considered in this article.

Reinforcements being explored in the late 1980s with metal and ceramic-matrix composites include continuous fibers of carbon, silicon carbide, boron, aluminum oxide and whiskers of silicon nitride and silicon carbide. Most of the resulting composites are still at the research and development stage (see, for example, Anderson et al. 1988), and to give some indication of the relative scale of applications for polymer-, metal-, and ceramic-matrix composites, Table 2 lists some estimates of approximate consumption levels for the principal reinforcing materials. Despite the fact that carbon fibers are used to reinforce some low-melting-point metals and alloys, it can be assumed that virtually all the glass, carbon and aramid fibers listed in Table 2 are used for plastics reinforcement at the present time. Although the quantities of carbon, aramid and ceramic fibers used in composites are small compared to glass fiber consumption, most market analysts predict growth rates up to ~20% per annum for high-performance polymer-matrix composites, based largely on assessments of the continuing adoption of these materials by the aerospace industry.

In discussing applications for composite materials it is important to recognize that they are generally more expensive than other construction materials. Table 3 gives broad ranges of cost for reinforcements, matrix resins and some conventional construction materials. Exact costs will depend on the form in which the basic materials are to be supplied.

Table 1
Upper temperature limits for continuous operation of composite materials

Composite	Maximum operating temperature (°C)
Polymer matrix	400
Metal matrix	580
Ceramic matrix	1000

Table 2
Approximate production levels for the principal reinforcements for composite materials

Reinforcement	Production capacity/consumption (t year^{-1})
Continuous fibers	
glass	[a]1280×10^3
aramid	3600[a]
carbon	3300[a]
ceramics (B, SiC, Al_2O_3)	few[b]
Short ceramic fibers	~40[b]
Whiskers (SiC, Si_3N_4)	~100[b]

a Brehm and Sprenger 1985 b Feest et al. 1986

Table 3
Approximate price ranges for reinforcements, thermosetting resins and some traditional materials of construction

Material	Approximate price, 1988 (£10^3 t^{-1})
Continuous fibers	
glass	1–2
aramid	20–75
carbon	25–100
SiC	200–400
Thermosetting resins	1–10
Whiskers	50–100
Unreinforced materials	
metals (e.g., steel, aluminum)	0.1–1
timber	0.3–1.5
engineering thermoplastics	2–10

Table 4
European market for glass-resin composites in 1986, breakdown by application (estimated volume 860 000 t)

Application	Distribution (%)
Transport	21
Electrical industry	19
Industrial and agricultural equipment	20
Building and public works	16
Sports and leisure	6
Consumer goods	6
Others	12

In the case of reinforced plastics, for example, cost will be determined by whether the fibers are in the form of rovings or woven fabric, whether preimpregnated with resin, and by the tow size or areal weight of the fabric. The costs of manufactured items will also depend on the nature of the production process and on the complexity of the component. For GFRP, the cost of a finished component may be typically two or three times the materials cost, and the added value may be significantly higher if there is a substantial degree of proof testing and quality assurance involved in the manufacturing process, as with many aerospace parts.

1. *Applications of Fiber-Reinforced Plastics*

In view of the difference between market sizes for GFRP and for composites based on carbon and aramid fibers, it is convenient, for a broad description of applications, to discuss the two areas separately. There is, however, no clear-cut definition of what constitutes a high-performance composite in terms of its composition. Thus, although some types of GFRP fulfill relatively undemanding uses such as decorative panels or cable ducting, other types are used in primary structures of some aircraft designs (Lubin and Donohue 1980) and in the construction of deep-sea submersibles (Oliver 1980).

1.1 *Applications of GFRP*

Table 4 shows a breakdown of the European market in terms of applications for GFRP based mainly on the relatively low-priced polyester resins (Chevalier 1987). Some attempt is made below to give an indication of the types of application in some of the larger sectors.

(*a*) *Transport.* In Western Europe, some 170 000 t of GFRP in its various forms is used in land transportation applications, and in the USA, total consumption in this sector is substantially greater. The incentive to introduce reinforced plastics into land-based transport is mainly to take advantage of their corrosion resistance and of manufacturing economies. Weight reduction, although leading to improved fuel economy or performance, is unlikely to be an important factor in new designs unless it can be achieved without a cost penalty.

The use of reinforced plastic panels for car bodies in the 1980s has been mainly restricted to small-volume production of specialist vehicles for which the versatility of GFRP fabrication methods enables design changes to be made fairly easily, and where the high capital cost of large metal press tools is not justified. One of the largest production runs has been General Motors' Fiero car which employs body panels produced by press-molding a sheet molding compound (SMC) and by the reinforced reaction injection molding (RRIM) process, and these are attached to a steel structural frame (Ferrarini et al. 1984). Similar schemes have been used for many years for the production of cabs for heavy commercial vehicles (Seamark 1981).

The adoption of molded reinforced plastics for a wide variety of panels, bumpers and interior fittings appears to be growing steadily, at least in terms of the average quantity per car (Waterman 1986). In the USA, reinforced plastic front and rear ends have almost completely superseded die-cast zinc because a single molding can replace an assembly of many individually cast items. In Europe, Renault were producing 9600 GFRP bumpers per day in 1983, and the Citroen BX features an SMC bonnet (hood) and tailgate, produced at a rate of the order of 1000 per day (Buisson 1983). Attention has also been directed towards underbonnet components, of which one of the most ambitious has been the molding, by a lost-wax-type process, of a dough molding compound (DMC) inlet manifold (Suthurst and Rowbotham 1980).

The use of composites for structural automotive components has not been overlooked, and there are numerous references in the literature to the development of stressed components such as drive shafts and GFRP springs. Leaf springs make use of the excellent fatigue properties of fiber composites and can give extended lives in comparison to steel springs, particularly for commercial vehicles (Lea and Dimmock 1984, Jardon and Costes 1987). Figure 2 shows a GFRP spring undergoing fatigue testing. Such springs provide a substantial reduction in mass over steel springs, as well as providing an improved ride and a decrease in the noise transmitted to the driving compartment.

The use of composites is not confined to road transport, however. Applications in rolling stock on European railways have been reviewed (Anon 1980) and the construction of the cabs on British Rail's Intercity vehicles has been described by Gotch and Plowman (1978). In the USA a rail-borne grain transporter, 15 m long and 4.6 m in diameter, has been produced by filament winding (Ruhmann et al. 1982).

(*b*) *Electrical*. The most common type of glass fiber used for reinforcement purposes is E-glass, which was originally developed for electrical insulation purposes, so it is not surprising that, when combined with organic resins, a class of materials results which can readily be molded into complex shapes possessing excellent insulating properties and a high dielectric strength. Thus switch casings, junction boxes, cable and distribution cabinets, relay components and lamp housings, to name just a few examples, are produced from a variety of glass-fiber molding compounds. Cable ducts and standard cross-sectional shapes for transformer insulation are produced by the pultrusion process, and reinforcing bands for rapidly rotating electrical machinery are often filament wound. Mathweb, a proprietary process for winding lattice-type structural beams (Preedy and Kelly 1982), is now used by British Rail to provide emergency support to overhead power lines because the light weight simplifies replacement of damaged gantries. Pultruded composites are also used in the construction of pantograph arms for current pickup on electric traction vehicles.

(*c*) *Industrial and agricultural*. Generally speaking, applications in agriculture, in chemical process plants and in tanks and pipes for industrial purposes rely upon the combination of low density and good corrosion resistance of GFRP. One of the advantages of the material is the relative ease with which large and complex-shaped components can be fabricated on a one-off basis. Typical applications include vats and silos for the storage of aggressive chemicals, pipelines for the transport of water and sewerage, water storage tanks, wine vats and the construction of certain types of chemical process plants as, for example, in the manufacture of chlorine. A particular advantage of the low density is that the cost of transporting components is minimized and this particularly facilitates the installation of pipelines in remote or rugged terrain. It is also possible to transport and handle longer lengths of pipe, thus

Figure 2
Fatigue test of a GFRP automotive leaf spring (photograph courtesy of GKN Technology Ltd.)

Figure 3
GFRP tank lined with PVC for storage of HCl (photograph courtesy of Plastics Design and Engineering Ltd.)

minimizing the number of joints and hence the assembly costs. Figure 3 illustrates a PVC-lined GFRP storage vessel for hydrochloric acid, of dimensions 10.5 m in length and 3.5 m in diameter.

(*d*) *Building*. The corrosion resistance and lightness, and hence ease of transport and installation of GFRP, coupled with the ability to mold surfaces with decorative textures and color, has led to extensive use of this material as cladding for prestige buildings (Jaafari et al. 1976). Considerable quantities of GFRP in the form of flat and corrugated translucent sheeting are also used for roofing purposes, particularly in outdoor canopies of large structures such as stadia, where the reduced weight of the roofing panels enables economies to be made in the design of the roof support structure. In the USA, the development of prefabricated bathroom units comprising sanitary ware, furniture and partitioning has generated a significant market (Seymour and Tompkins 1975).

(*e*) *Marine*. This area provides a good example of how a number of attributes of GFRP in combination have resulted in almost complete substitution for traditional materials, particularly timber. Low density, corrosion resistance and the ability to produce complete hull and deck moldings repetitively from a single set of molds have revolutionized the leisure and small workboat building industries, and reduced maintenance costs have played a large part in gaining widespread consumer acceptance of GFRP. Of less general applicability, but of particular importance to naval mine-clearance operations, is the nonmagnetic character of GFRP. The UK pioneered the introduction of GFRP minesweepers with the launch, in 1972, of HMS Wilton with an overall length of 46 m (Dixon et al. 1973), and vessels up to 60 m in length and with displacements of ~600 t are being built in a number of countries (Anon 1987).

1.2 Applications of High-Performance Composites

As indicated earlier, the term high-performance composites is somewhat loosely employed to describe composites based primarily on the more expensive carbon and aramid reinforcements, although it also reflects the improved performance to be gained in the two major market sectors to have developed thus far—aerospace and sports goods.

(*a*) *Aerospace*. The use of lighter materials of construction means either that a greater payload can be carried or that operating costs can be reduced for the same payload. The outstanding example of the cost-effectiveness of lightweight composites is in spacecraft technology where the cost of launching may often exceed the cost of designing and building the spacecraft (see, for example, Zweben 1981, Burke 1986). Consequently, carbon- and aramid-fiber-reinforced plastics are frequently used for such items as the basic structure of a spacecraft, as supports for solar arrays, for the construction of dish antennas, in pressure vessels for propellant gases and in apogee motor cases and thrust cones. In addition to weight savings, the low axial thermal expansivity of these materials enables very high dimensional stability to be achieved in antennas, microwave resonant cavities and filters and in telescope support structures (see, for example, Reibaldi 1985). Such components may experience temperature excursions in orbit within the range −100 °C to approximately 100 °C depending on whether the component is exposed to full sunlight or is in the shadow of the spacecraft.

In military aircraft, composites offer improved performance, and a number of aircraft design and construction programmes are in progress involving major use of advanced composites (Hadcock 1986). For example, approximately 35% of the structural weight of the projected European Fighter Aircraft (EFA) is likely to be built from composite materials, including the main wing, the forward fuselage and the fin and rudder. Figure 4 illustrates the use of composites in the construction of the AV-8B Mk. 5 Harrier II airframe giving rise to a 25% weight saving in the airframe compared to all-metal construction (Riley 1986). In commercial aircraft, composites have for some time been used to reduce weight in internal fittings such as cargo floors, galley furniture, luggage lockers and for internal trim. Seat frames are now beginning to be built in carbon-fiber-reinforced plastics (CFRP) since they not only lead to weight savings but occupy less space. With regard to the use of composites in primary aircraft structures, considerable caution has been, and still is, exercised in proving such materials before flight certification is granted, but nevertheless major structural items are now being flown on passenger aircraft. In the A310 Airbus, for example, the vertical stabilizer makes substantial use of composites construction involving glass, carbon and aramid fibers, with a total weight saving of 397 kg (Pinzelli 1983). The A320 Airbus will use CFRP in the fin and tailplane leading to a weight saving of 800 kg over aluminum-alloy skin construction (Anon 1986). As an indication of the benefit of such weight savings, it has been estimated that a 1 kg weight reduction on a DC-18 saves over 2900 l of fuel per year (Zweben 1981).

On a smaller scale, a number of designs of light aircraft and gliders (Riddell 1985), built primarily of GFRP, have been flying for a number of years and attention is now being directed to the development of all-composite aircraft in the executive jet category. In this class, two prototypes of the Lear Fan 2100 aircraft with 70% of structural weight in CFRP were flown before the company went into liquidation for reasons unconnected with the use of composites (Noyes 1983). The Beech Starship, under development in the late 1980s, is also constructed largely from composite materials and one phase of the program involves a comparison between hand-lay and filament-winding techniques for the construction of the fuselage (Wood 1986).

As frequently happens when an existing technology is threatened by innovation, the aluminum industry has responded to the introduction of composites with the development of alloys, such as the aluminum–lithium series, with improved strength and stiffness (Clementson 1985). Although their specific properties do not match those of the advanced composites, they possess the advantage that established metal skin design and fabrication procedures can be applied, obviating the need for investment in new skills to handle design with anisotropic materials and the very different manufacturing techniques involved with composites. Nevertheless, whatever the outcome of the trade-off between metal or composite properties, one important economic advantage of composites remains, namely the ability to fabricate large and complex parts as a single molding, as opposed to assembly of a number of individual metal pressings.

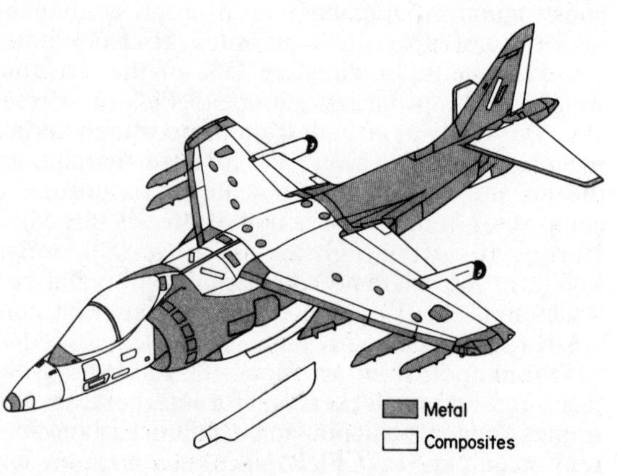

Thus far, little has been said of the excellent fatigue behavior of composites in fiber-dominated directions. An outstanding example of a component whose life is governed by fatigue damage is the helicopter rotor blade which experiences complex dynamic stresses including torsion, together with bending in, and perpendicular to, the plane of the rotor (Goddard 1985). Composite blades are now fitted to most major helicopter designs on the basis that they give virtually unlimited fatigue life in relation to the lifetime of the aircraft, reduce manufacturing costs and enable novel and more efficient aerofoil structures to be created. Most helicopter manufacturers also plan to use composites extensively in future airframes. For example, the Boeing 360 demonstrator has an all-composite fuselage consisting of CFRP frames to which are bolted and bonded honeycomb sandwich panels with aramid-fiber-reinforced skins. This form of construction gives a 25% weight saving in the fuselage and substantial cost reductions related to parts integration and the shorter time required for assembly (Anon 1987a). As with helicopter rotor blades, aircraft propellor blades made from composites show excellent fatigue and corrosion resistance compared to metals, and at reduced weight. McCarthy (1986) has described the design and construction of composite blades and has illustrated the weight savings compared to metal blades.

(b) Leisure and sporting goods. Generally speaking, the production of aerospace structures and components is a low-volume manufacturing operation measured typically in terms of a few tens or hundreds of parts per year. The second largest market sector for composites based on carbon and aramid fibers, and to some extent glass fibers, has in recent years been the leisure goods industry, where production rates have been measured in thousands or tens of thousands of units per year. However, in items like tennis and squash racquets, fishing rods, skis and ski poles, golf shafts, windsurfers and other small sailing craft, the quantities of expensive reinforcements are small, but aggregate to approximately one-third of present carbon and aramid fiber consumption.

Reference to the use of composites in sailplanes and small boats has already been made in this article. A feature of the use of such materials in high-performance sailing craft is that the advanced

Figure 4
The extent of composites construction in the AV-8B Mk. 5 Harrier II aircraft (from Riley 1986)

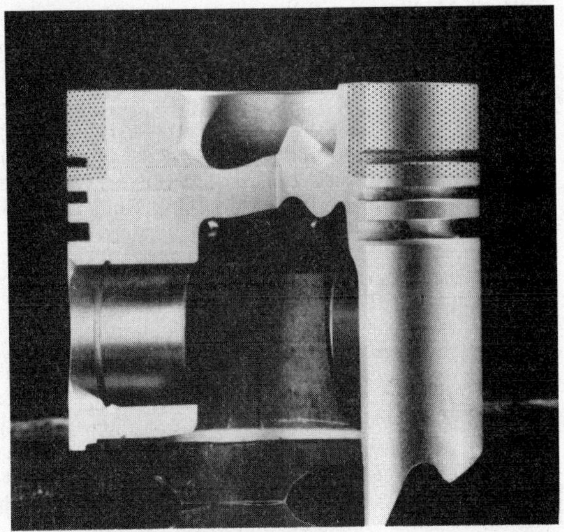

Figure 5
Toyota diesel engine piston. The metal-matrix composite insert can be seen as the darker region incorporating the upper piston ring groove

materials and techniques developed for the aerospace industry are being adapted to produce racing vessels with strong rigid hulls constructed from honeycomb sandwich panels, giving substantial reductions in weight, and increases in performance. For similar reasons, many racing cars now employ a composite chassis consisting of CFRP skins bonded to a honeycomb core (Clarke 1985).

2. *Metal-Matrix Composites*

Among the attributes of whisker reinforced MMCs, in addition to improved strength and stiffness, are increased creep and fatigue resistance, giving rise to higher operating temperatures than those of the unreinforced metal (Feest 1986). The presence of ceramic reinforcements, whether in fibrous or particulate form, can also provide increased hardness, wear and abrasion resistance. These properties are potentially exploitable in many aspects of pump and engine technology including compressor bodies, vanes and rotors, piston sleeves and inserts, connecting rods, cylinder heads and clutch components. Many such components are under development (Feest et al. 1986) but few have yet entered commercial production. One exception is the use by the Toyota Motor Corporation (Donomoto et al. 1983) of aluminum-alloy diesel engine pistons incorporating a ceramic-fiber-reinforced top land and piston ring groove (Fig. 5). Honda have also engaged in a limited production run of aluminum-alloy connecting rods reinforced with stainless steel wire for their 1.2 l City vehicle in Japan. On a much smaller scale of application, boron-fiber-reinforced aluminum-alloy struts are used in the midsection framework of the US space shuttle fuselages (Irving 1983). An application that exploits the thermal properties of aluminum reinforced with silicon carbide particles is the packaging of microwave circuits (Thaw et al. 1987). In this respect, the presence of the silicon carbide reduces the thermal expansion coefficient of the packaging material, providing a better match to the ceramic circuit substrate without significantly decreasing the high thermal conductivity required for good heat dissipation from the device.

3. *Ceramic-Matrix Composites*

The present picture for CMCs is similar in many ways to that of MMCs in that considerable research and development activity exists (Phillips 1985) but there are few established applications. Potential applications for CMCs include high-temperature seals, bearings, blades and rotors, particularly in gas turbine engines, piston crowns in reciprocating engines, and wear-resistant duties such as valve bodies, cutting and forming tools. In the latter context, can-forming punches made from alumina reinforced with silicon carbide whiskers are reported to show better performance than tungsten carbide punches in machines that produce 400 cans per minute round the clock (Anon 1988).

The availability of strong, tough, lightweight materials with good thermal shock resistance, capable of continuous operation in air at temperatures above 1200 °C, would make possible significant increases in the efficiency of gas turbine engines. Phillips (1987) and Jamet (1987) have reviewed the prospects of developing ceramic-matrix composites to meet this type of requirement. Existing ceramic reinforcements suffer degradation in strength or undergo creep in the range 1000–1200 °C and the only material available in the late 1980s with attractive mechanical properties above 1200 °C is carbon-fiber-reinforced carbon, operating in a nonoxidizing atmosphere (Hill et al. 1974). Carbon–carbon materials are currently used for ablative purposes on those surfaces of spacecraft and reentry vehicles subject to severe aerodynamic heating, and considerable research is being devoted to identifying ways of increasing oxidation resistance, particularly for reusable space vehicles.

At lower temperatures, and for short periods of time at higher temperatures, carbon–carbon materials have proved suitable in disk brakes for aircraft and racing cars, where sufficiently large amounts of energy are generated in emergency braking to cause conventional steel disk brakes to fade. For a review of carbon–carbon technology, see Fitzer (1987).

See also: Automobile Tires (Suppl. 2); Automotive Composite Components: Fabrication (Suppl. 2); Composite Materials: Aerospace Applications (Suppl. 2); Helicopter Materials

Bibliography

Anderson L I, Lilholt H, Pederson O B (eds.) 1988 Mechanical and physical behaviour of metallic and ceramic composites. *Proc. 9th Risø Int. Symp. on Metallurgy and Materials Science*. Risø National Laboratory, Denmark

Anon 1980 GRP for rolling stock on European railways. *Vetrotex Fibreworld* 9: 11–18

Anon 1986 A320–Fly-by-wire airliner. *Flight Int.* 130 (4026): 86–94

Anon 1987a Boeing 360—Helicopter hi-tech. *Flight Int.* 131(4058): 22–7

Anon 1987b Growth of composites in military naval construction: Slowly but surely. *Vetrotex Fibreworld* 24: 6–7

Anon 1988 ACMC's ceramic composites—Tooling up for a first in cans. *Mater. Edge* 7: 10

Brehm B, Sprenger K H 1985 Modern fibre reinforced materials. *Sprechsaal* 118 (3): 253

Buisson M 1983 SMC and the car. *Vetrotex Fibreworld* 17: 10–12

Burke W R (ed.) 1986 Composites design for space applications. *Proc. Workshop ESA SP-243*. European Space Agency Publications Division, Noordwijk, The Netherlands

Chevalier A 1987 Glass/resin composite market *Vetrotex Fibreworld* 24: 8

Clarke G P 1985 The use of composite materials in racing car design. *Proc. 3rd Int. Conf. on Carbon Fibres—Uses and Prospects*, Paper 18. Plastics and Rubber Institute, London

Clementson A 1985 Materials and manufacturing in aerospace. *Proc. 2nd Conf. on Materials Engineering*. Institute of Mechanical Engineers, London, pp. 189–94

Dixon R H, Ramsey B W, Usher P J 1973 Design and build of the GRP hull of HMS Wilton. *Proc. Symp. on GRP Ship Construction*. Royal Institute of Naval Architects, London, pp. 1–32

Donomoto T, Funatani K, Miura N, Miyake N 1983 *Ceramic Fiber Reinforced Piston for High Performance Diesel Engines*, SAE Trans. 830252. Society of Automotive Engineers, Warrendale, PA

Feest E A 1986 Metal matrix composites for industrial application. *Mater. Des.* 7(2): 58–64

Feest E A, Ball M J, Begg A R, Biggs D A 1986 *Metal Matrix Composites Development in Japan*, Report on OSTEM visit to Japan, Oct. 1986. Harwell Laboratory, Didcot, UK

Ferrarini L J, Spence D H, Walker M G 1984 Broadening the limits of reinforced polyurethane RIM. *Proc. 1st Int. Conf. on Fibre Reinforced Composites*, Paper 25. Plastics and Rubber Institute, London

Fitzer E 1987 The future of carbon–carbon composites. *Carbon* 25 (2): 163–90

Goddard P N 1985 The use of new materials in helicopter load bearing structures—Present trends and future predictions. *Proc. 2nd Conf. on Materials Engineering*. Institute of Mechanical Engineers, London, pp. 243–51

Gotch T M, Plowman P E R 1978 Improved production processes for manufacture of GRP on British Rail. *Proc. Reinforced Plastics Congr.*, Paper 4. British Plastics Federation, London

Hadcock R N 1986 Design of advanced composite aircraft structures. In: Dorgham M A (ed.) 1986 *Designing with Plastics and Advanced Plastic Composites. Proc. Int. Association for Vehicle Design*. Interscience, Geneva

Hancox N L, Phillips D C 1985 Fibre composites for intermediate and high temperature applications. *Proc. 2nd Conf. on Materials Engineering*. Institute of Mechanical Engineers, London, pp. 139–44

Hill J, Thomas C R, Walker E J 1974 Advanced carbon–carbon composites for structural applications. *Proc. Conf. on Carbon Fibers, Their Place in Modern Technology,* Paper 19. Plastics Institute, London

Irving R R 1983 Metal matrix composites pose a big challenge to conventional alloys. *Iron Age* 226(2): 35–9

Jaafari A, Holloway L, Burstell M L 1976 Analysis of the use of glass reinforced plastics in the construction industry. *Proc. Reinforced Plastics Congr.*, Paper 27. British Plastics Federation, London

Jamet J F 1987 *Ceramic—Ceramic Composites for Use at High Temperature, New Materials and Their Applications*, Inst. Phys. Conf. Ser. No. 89. Institute of Physics, Bristol, pp. 63–75

Jardon A, Costes M 1987 Mass production composites. *Proc. 6th Int. Conf. on Composite Materials*, Vol. 1. Elsevier, London, pp. 1.1–1.4

Lea M, Dimmock J 1984 The development of leaf springs for commercial vehicle applications. *Proc. 1st Int. Conf. on Fibre Reinforced Composites*. Paper 10. Plastics and Rubber Institute, London

Lubin G, Donohue P 1980 Real life ageing properties of composites. *Proc. 35th Annual Conf. SPI*, Paper 17-E. Society for Plastics Industry, New York

McCarthy R 1986 *Manufacture of Composite Propellor Blades for Commuter Aircraft*, SAE Trans. 850875. Society of Automotive Engineers, Warrendale, PA, pp. 4.606–4.613

Noyes J V 1983 Composites in the construction of the Lear Fan 2100 aircraft. *Composites* 14(2): 129–39

Oliver P C 1980 Applications for composites in the offshore environment. *Proc. 3rd Int. Conf. on Composite Materials*. Pergamon, Oxford, p. 2395

Phillips D C 1985 Fibre reinforced ceramics. In: Davidge R W (ed.) *Ceramic Composites for High Temperature Engineering Applications*. CEC Publication EUR 9565EN. Commission of the European Communities, Luxembourg

Phillips D C 1987 High temperature fibre composites. In: Matthews F L, Buskell N C R, Hodgkinson J M, Morton J (eds.) 1987 *Proc. 6th Int. Conf. on Composite Materials*, Vol. 2. Elsevier, London pp. 1–32

Pinzelli R 1983 Concept, benefits and applications of aramid fiber in hybrid composites. *Proc. 4th Int. Conf. SAMPE European Chapter*. Comité Pour la Développement des Matériaux Composites, Bordeaux, pp. 63–77

Preedy J, Kelly J F 1982 An application for Mathweb structures—Development of a temporary overhead electrical gantry by British Rail. *Proc. Reinforced Plastics Congr.*, Paper 23. British Plastics Federation, London

Riddell J C 1985 Composite materials and the sailplane market. *Proc. 2nd Conf. on Materials Engineering*.

Institute of Mechanical Engineers, London, pp. 253–8
Riebaldi G G 1985 Dimensional stability of CFRP tubes for space structures. *Proc. Workshop on Composites Design for Space Applications*, ESA-SP243. European Space Agency Publications Division, Noordwijk, The Netherlands
Riley B L 1986 AV-8B/GR Mk. 5 airframe composite applications. *Proc. Inst. Mech. Eng.* 200(50): 1–17
Ruhmann D C, Mundlock J D, Britton R A 1982 Glasshopper—The fiberglass reinforced polyester covered hopper car. *Proc. 37th Annual Conf. SPI*, Paper 4-E. Society for Plastics Industry, New York
Seamark M J 1981 Facelifting the world's first all-SMC clad truck cab after 5 years—A unique case study. *Proc. 36th Annual Conf. SPI*, Paper 11-C. Society for Plastics Industry, New York
Seymour M W, Tompkins D D 1975 Design approaches to the fibrous glass reinforced polyester bathroom as related to market needs. *Proc. 30th Annual Conf. SPI*, Paper 3-A. Society for Plastics Industry, New York
Suthurst G D, Rowbotham E M 1980 Achieving the impossible—Plastic intake manifold. *Proc. Reinforced Plastics Congr.*, Paper 33. British Plastics Federation, London
Thaw C, Minet R, Zemany J, Zweben C 1987 Metal matrix composite microwave packaging components. *SAMPE J.* 23(6): 40–3
Waterman N A 1986 The economic case for plastics. In: Dorgham M A (ed.) 1986 *Designing with Plastics and Advanced Plastic Composites*, Proc. Int. Association for Vehicle Design. Interscience, Geneva, pp. 1–15
Wood A S 1986 The majors are taking over in advanced composites. *Mod. Plast. Int.* 16(4): 40–3
Zweben C 1981 Advanced composites for aerospace applications. *Composites* 12(4): 235–40

D. H. Bowen
[UKAEA Harwell Laboratory, Didcot, UK]

Composite Materials: Fatigue

Fatigue of metals has been studied for over a century and despite significant advances it remains a major cause of catastrophic failure of structures. Composites, on the other hand, have high potential for fatigue resistance and can, in certain cases, be designed to eliminate the fatigue problem. The fatigue properties of composites are anisotropic, that is, directionally dependent, and can be dangerously low in some directions. This warrants careful use of composites based on proper understanding of the mechanisms that govern the fatigue behavior. The mechanisms are admittedly complex, but once analyzed and understood can provide a key to developing a new generation of engineered materials.

The composites considered here are continuous-fiber laminates having fibers of glass or carbon in a thermosetting polymeric matrix. The fiber orientations considered are the unidirectional, the bidirectional (angle-plied and cross-plied laminates) and some combinations of these.

1. Unidirectional Composites

Mechanisms of fatigue damage in unidirectional composites depend on the loading mode, for example tensile or compressive, and on whether the loading is parallel to or inclined to the fiber direction. For illustration, only tensile loads are considered here and the mechanisms for parallel loading and for inclined loading are described separately.

1.1 Loading Parallel to Fibers

The mechanisms may be divided into three types (see Fig. 1). Fiber breakage (Fig. 1a) occurs at a local stress exceeding the strength of the weakest fiber in the composite. An isolated fiber break causes shear-stress concentration at the fiber–matrix interface near the broken fiber tip. The interface may then fail, leading to debonding of the fiber from the surrounding matrix. The debond length depends on the shear strength of the interface and is usually small, of the order of a few fiber diameters. The debonded area acts as a stress concentration site for the longitudinal tensile stress. The magnified stress may exceed the fracture stress of the matrix, leading to a transverse crack in the matrix.

The matrix undergoes a fatigue process of crack initiation and crack propagation and generates cracks normal to the longitudinal tensile stress. These cracks are randomly distributed and initially restricted by fibers (Fig. 1b). If the cyclic strain in the matrix is sufficiently low, the cracks remain arrested by the fibers. When the local strains are higher than a certain threshold, the cracks break the fibers and propagate. In this progressive crack-growth mechanism the fiber–matrix interface will also fail due to severe shear stresses generated at the crack tip (Fig. 1c).

Final failure results when the progressive crack-growth mechanism has generated a sufficiently large crack (which may be only of the order of a millimeter or less for brittle composites). The fracture surface of a specimen looks messy or broom-like if the fiber–matrix interface is weak and is increasingly neat for stronger interfaces.

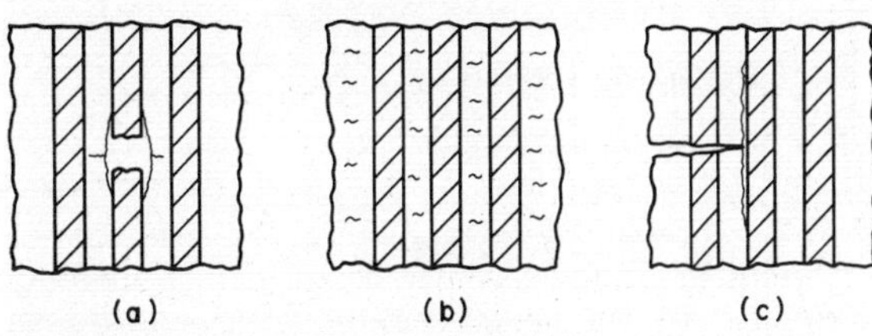

Figure 1
Fatigue damage mechanisms in unidirectional composites under loading parallel to fibers: (a) fiber breakage, interfacial debonding; (b) matrix cracking; and (c) interfacial shear failure

1.2 Fatigue-Life Diagram

The mechanisms of damage described above may operate simultaneously. However, observations indicate that the predominant mechanism leading to failure may be effective in a limited range of the applied cyclic strain. This is illustrated in a fatigue-life diagram shown schematically in Fig. 2. The horizontal axis shows the number of load cycles to failure on a logarithmic scale and the vertical axis plots the maximum strain, i.e., the maximum stress divided by the modulus of elasticity of the composite, applied initially to a test specimen. Strain instead of stress is plotted since strain is roughly the same in fibers and in the matrix while stress differs in the two phases depending on the volume fraction of fibers and the elastic moduli of the two phases.

The lower limit to the diagram is given by the fatigue limit of the matrix ε_m, that is, the threshold strain below which the matrix cracks remain arrested by the fibers. This strain is observed to be approximately the fatigue strain limit of the unreinforced matrix material. The upper limit to the diagram is given by the strain-to-failure of the composite ε_c, which is also the strain-to-failure of fibers in a composite reinforced by stiff fibers. The diagram shows a scatter band on the failure strain since this quantity is usually subjected to significant scatter. The mechanism governing static failure, i.e., failure not preceded by significant cycle-dependent growth process, is fiber breakage with associated interfacial debonding. This mechanism is indicated in the scatter band on the failure strain.

The progressive damage mechanism is matrix cracking with associated interfacial shear failure, as described above, and this governs fatigue life. The sloping band of scatter on fatigue life in the diagram is due to this mechanism.

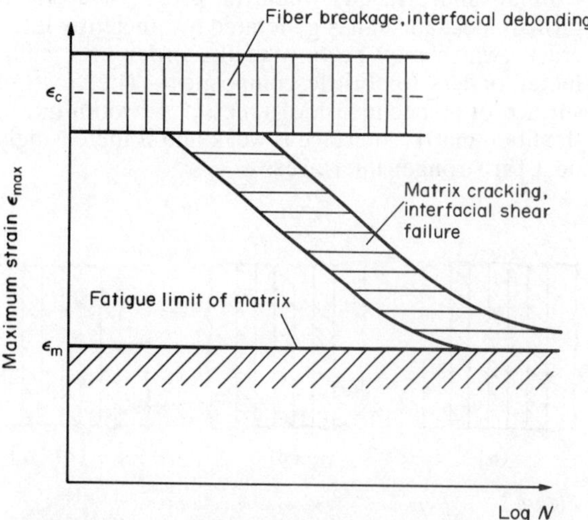

Figure 2
Fatigue-life diagram for unidirectional composites under loading parallel to fibers

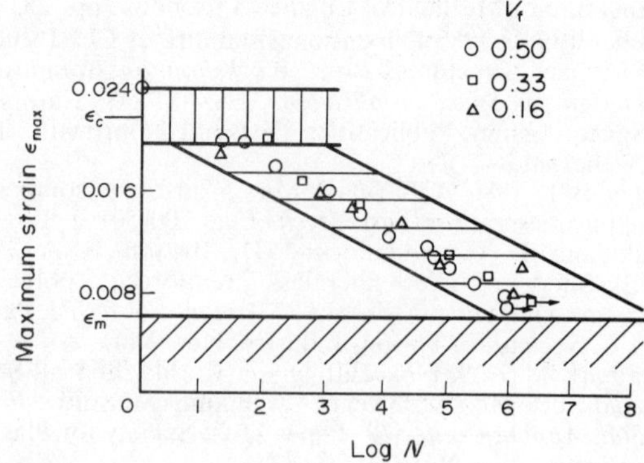

Figure 3
Fatigue-life diagram for a glass–epoxy under loading parallel to fibers. V_f is the volume fraction of fibers

1.3 Effect of Fiber Stiffness

Consider two unidirectional composites with the same matrix and different fibers. The fatigue limit of the two composites will be the same and is given by the fatigue limit of the matrix ε_m. The upper limit of the fatigue diagram, given by the composite failure strain, which is equal to the fiber failure strain, will be different for the two composites. Thus the range of strain in which progressive fatigue damage occurs will be different in the two composites. In a particular case where the composite-failure strain and the fatigue-limit strain are equal, the range of strain with progressive fatigue damage will be zero. In such a case fatigue damage will be absent and only static failure will be possible.

Figures 3 and 4 show data and fatigue life diagrams of unidirectionally reinforced epoxy with glass fibers and with carbon fibers, respectively. The

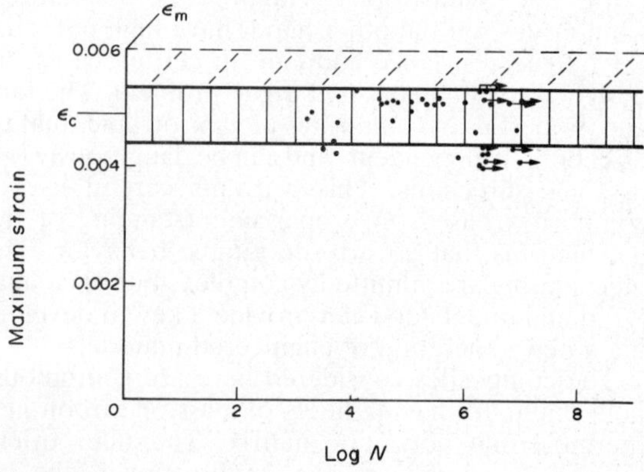

Figure 4
Fatigue-life diagram for a carbon–epoxy under loading parallel to fibers

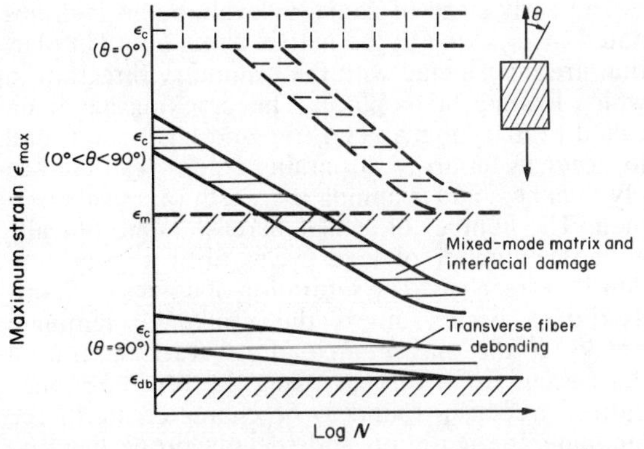

Figure 5
Fatigue-life diagram of unidirectional composites under loading inclined to fibers. ε_{db} is the strain to debonding of fibers from matrix

fatigue-limit strain in both composites is 0.6% while the mean failure strains are 2.20% for glass–epoxy and 0.48% for carbon–epoxy. It is seen that the glass–epoxy composite has a wide range of strain with progressive fatigue damage while the carbon–epoxy composite of highly stiff fibers has its fatigue damage totally suppressed. Other carbon–epoxy composites with less stiff fibers and having failure strain of about 1% show some progressive fatigue damage.

1.4 Loading Inclined to Fibers

When the cyclic loading axis is inclined at angles of more than a few degrees to the fiber axis, the predominant damage mechanism is matrix cracking along the fiber–matrix interface. The static failure band in the fatigue-life diagram is then lost and the fatigue limit drops with increasing off-axis angle. The lowest fatigue limit is given by the strain for transverse fiber debonding, i.e., failure of the fiber–matrix interface by growth of an interfacial crack in opening mode. This occurs at the off-axis angle of 90°; that is, when loading is transverse to the fiber direction. Figure 5 shows a schematic fatigue-life diagram for off-axis fatigue of unidirectional composites.

The anisotropy of fatigue properties of composites is illustrated dramatically by Fig. 6 which shows the decrease of the fatigue limit strain with the off-axis angle. The strain below which a composite is safe against fatigue failure when loaded normal to fibers is only 0.1% for glass–epoxy composites, the data for which has been plotted in Fig. 6. This is one-sixth of the same strain for loading along the fibers. However, the ratio of the allowable stresses in the two directions is 1:24, when the elastic moduli in the two directions differ by a factor of four, a typical value for glass–epoxy composites.

2. Bidirectional Composites

The inferior fatigue properties of unidirectional composites in the direction normal to fibers can be improved by building up laminates with plies of unidirectional composites stacked in two orientations. Such composites, called angle-plied composites, when loaded in direction bisecting the angle between fibers, suffer damage similar to that in a unidirectional composite loaded inclined to fibers. However, the rate of progression of damage is reduced due to the constraint provided by plies of one orientation to cracking in plies of the other orientation. The constraint is highly effective at low angles between fibers but loses effect increasingly with increasing angle. This is illustrated by Fig. 7 where the fatigue-limit strain of angle-plied laminates is plotted against the half-angle between fibers. For comparison the dotted line shows the fatigue limit of the unidirectional composite of the same material, glass–epoxy, against the off-axis angle. Significant improvement in the fatigue limit is seen for angles up to about 45°.

An important class of bidirectional composites is the cross-plied laminates where two orthogonal fiber directions are used. When loaded along one fiber direction a cross-plied laminate develops cracks along fibers that are loaded transversely. These transverse cracks are now constrained by plies with fibers normal to the crack planes, and the degree of the constraint depends on the thickness of the cracked ply (equal to the crack length) and the stiffness properties of the constraining plies. The load shed by a cracked ply is carried by the constraining plies over a distance determined by the constraint conditions. This distance determines the position of another transverse crack. Thus a crack-density progression process occurs, leading to a saturation crack density.

Load cycling beyond attainment of the transverse crack saturation may lead to diversion of the transverse crack tips into the interfaces between plies. An

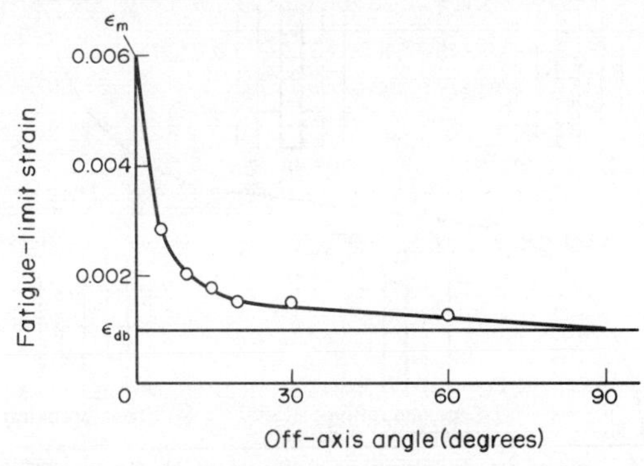

Figure 6
Variation of the fatigue limit with the off-axis angle. ε_{db} is the strain to debonding of fibers from matrix

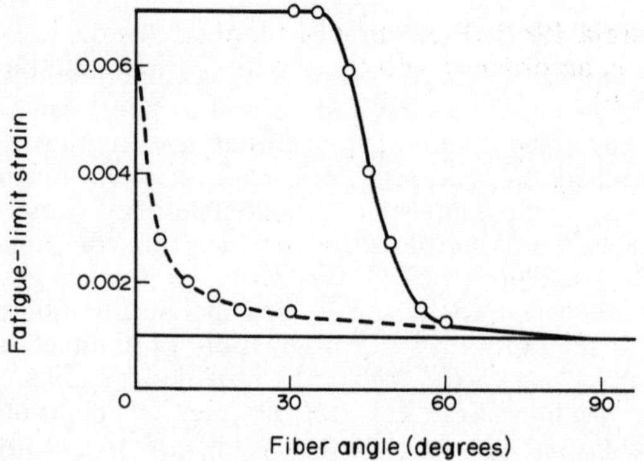

Figure 7
Variation of the fatigue limit with the fiber angle in an angle-plied laminate of glass–epoxy. The dashed line corresponds to the fatigue limit variation of Fig. 6

interlaminar crack may thus form and grow causing an eventual delamination.

The fatigue-life diagram of a cross-plied laminate has the static failure band as the upper limit which is given by the failure strain of fibers. The lower bound, that is, the fatigue limit, is determined by the strain to initiation of transverse cracking.

3. General Laminates

Combinations of the unidirectional, the angle-plied and the cross-plied orientations are used in various configurations to satisfy the performance requirements to which a structure may be subjected. The fatigue properties have been studied primarily under loading along one of the in-plane symmetry directions of laminates. These studies, carried out over a decade, have formed the basis for an understanding of the development of damage pictured schematically in Fig. 8 (Reifsnider et al. 1983).

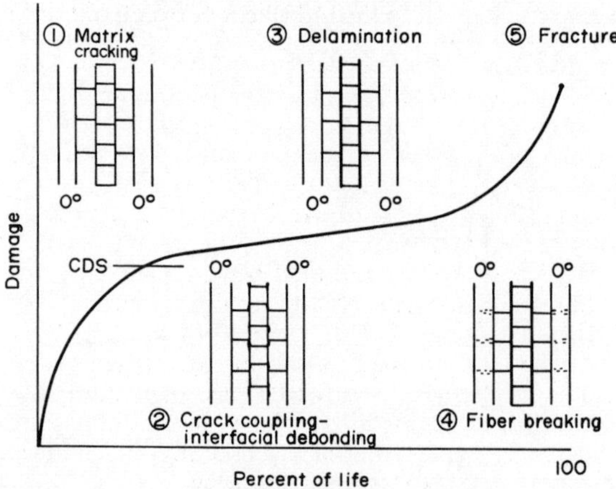

Figure 8
Development of damage in composite laminates

The early stage of damage development is dominated by cracking of the matrix along fibers in plies that are not aligned with the symmetry direction in which loading takes place. This cracking has been called primary matrix cracking and has been found to occur as an array of parallel cracks restricted to ply thickness and spanning the width of a test specimen. The number of cracks increase monotonically with the number of load cycles until a saturation density is reached. The saturation of cracks in all off-axis plies occurs unless the weakened laminate breaks at the maximum load. If crack saturation does occur in all off-axis plies, the resulting crack pattern has been found to be characteristic of the laminate configuration and the ply properties and independent of the load amplitude. The damage state associated with the characteristic crack pattern has been called the characteristic damage state (CDS) and signifies the termination of the first stage of matrix cracking damage. The following stage begins by initiation of cracks transverse to the primary cracks and lying in a ply adjacent to the ply with primary cracks. These cracks, called the secondary matrix cracks, extend short distances and appear to be initiators of the interlaminar cracks. The interlaminar cracks are initially distributed in the interlaminar plane and are confined to small areas. The resulting local delamination spreads with continued loading, merges with neighboring delamination and leads to large-scale delamination. The final stage is dominated by fiber breakage and ultimate failure occurs when the locally failed regions have sufficiently weakened the laminate to cause failure under the maximum load.

4. Degradation of Properties

Mechanisms of fatigue damage in composites result in distribution of cracks of various orientation in the volume of the material. This leads to degradation of the overall material properties. An example is shown in Fig. 9, where the elastic modulus of a composite normalized by its initial value has been plotted against the fatigue cycles. The modulus degradation shows three stages corresponding to the damage development shown schematically in Fig. 8. In stage I extending until CDS the modulus degradation is abrupt. In stage II, which follows CDS and is characterized by a steady damage development, the modulus degradation is gradual. In stage III the damage development is unstable and gives rise to an erratic drop in the modulus.

The overall properties degradation in stages I and II has been modelled by regarding the composite with damage as a continuum with changing microstructure (Talreja 1985). A phenomenological theory of constitutive behavior then provides relationships between the severity of damage and the

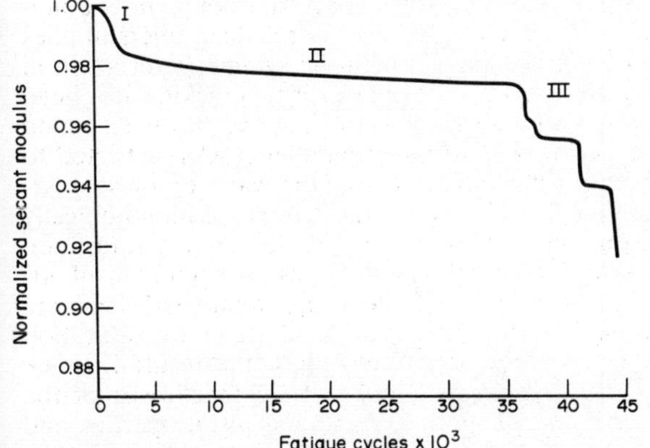

Figure 9
Degradation of the elastic modulus of a cross-plied laminate of carbon–epoxy under fatigue

overall stiffness properties of a composite. An example of such relationships for intralaminar cracking is

$$E_1 = E_1^0 - 2t_c^2/ts[a + b(\nu_{12}^0)^2 - c\nu_{12}^0]$$

where E_1 and E_1^0 are the current and initial values, respectively, of the elastic modulus along a symmetry axis labelled 1 in an orthotropic laminate, ν_{12}^0 is the initial Poisson's ratio for straining in the direction 1 with contraction in the orthogonal direction 2, t_c and t are thicknesses of the cracked ply and of the laminate, respectively, s is the spacing of the intralaminar cracks and a, b and c are the material constants. These constants have been determined for glass–epoxy and for carbon–epoxy laminates

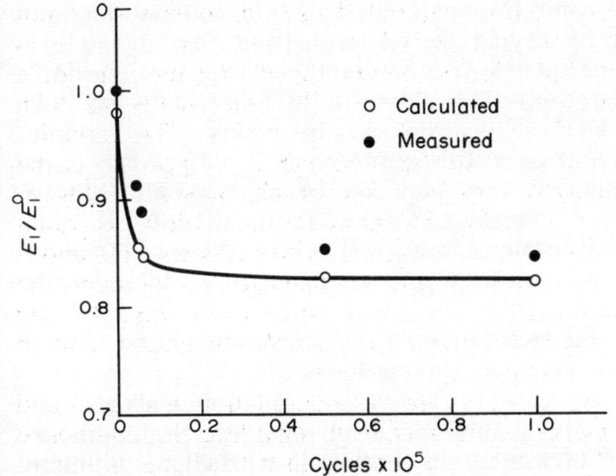

Figure 10
Predicted and measured degradation of the elastic modulus of a glass–epoxy laminate under fatigue

(Talreja 1987). Prediction of the modulus degradation in accordance with the equation is illustrated by Fig. 10.

See also: Failure of Composites: Stress Concentrations, Cracks and Notches

Bibliography

Reifsnider K L, Henneke E G, Stinchcomb W W, Duke J C 1983 Damage mechanics and NDE of composite laminates. In: Hashin Z, Herakovich C T (eds.) 1983 *Mechanics of Composite Materials—Recent Advances*. Pergamon, New York, pp. 399–420

Talreja R 1985 A continuum mechanics characterization of damage in composite materials. *Proc. R. Soc. London. Ser. A* 399: 195–216

Talreja R 1986 Stiffness properties of composite laminates with matrix cracking and interior delamination. *Eng. Fract. Mech.* 25: 751–62

Talreja R 1987 *Fatigue of Composite Materials*. Technomic, Lancaster, UK

R. Talreja
[Technical University of Denmark, Lyngby, Denmark]

Composite Materials: Joining

As fiber-reinforced materials, particularly continuous-fiber-reinforced plastics (FRP), become more widely used an important requirement for their full exploitation is the development of suitable attachment methods and general joint design philosophies.

Unlike isotropic materials, problems of joining FRP arise mainly due to the unique characteristics of the material, that is, the inherent weakness of the material in interlaminar shear, transverse tension and transverse compression. Also, since the material remains elastic to failure (in unidirectional material where fibers are oriented in one direction), there is no plastic stress–strain behavior, and therefore the mechanisms of stress relief around geometric discontinuities, such as holes, require special attention. Clearly all of these problems require full consideration before engineers can embark upon the design of an efficient joint, efficiency being measured in terms of high strength for low weight and/or cost.

The epoxy-resin-matrix FRP (which make up the bulk of the current high-technology matrices for FRP) cannot be welded, soldered or brazed. With the exception of components made using filament winding techniques, where there is usually no significant end attachment problem, most joints are effected by means of adhesive bonding, mechanical fasteners or a combination of the two.

With the recent introduction of thermoplastics into the field of structural plastics, especially for the aerospace industry, other methods of joining are being explored, and one of particular interest that promises future potential is that of welding.

1. Adhesive Joints

The form of adhesive-bonded joints that has probably received most attention is a simple lap joint. This is attractive for two main reasons: it is simple to fabricate, and many of the problems of load introduction such as are found in mechanical joints are largely eliminated. However, even with the use of advanced adhesive technology only relatively low load-transfer rates can be realized because of the low shear strength of both the adhesive and the composite adherend, and the high thermal strains that can exist within the bonded regions due to the high cure temperature of adhesives. Also degradation of the adhesive, and the interface between adhesive and composite, can occur after exposure to certain environments. This is discussed in greater depth later in this section.

Failure of FRP adhesive joints can occur in any of four modes: adhesive failure at the interface between the adhesive and adherend, cohesive failure through shearing of the adhesive, peel failure due to out-of-plane loads tending to pull a joint or adherend apart in transverse tension, and adherend failure in axial tension or compression. Axial failure of the adherend occurs at or close to the end of a lap joint with failure initiating in axial tension or compression of the outer plies. Analysis shows that due to a shear lag phenomenon caused by shear deformation in the composite matrix, axial load distribution through a laminate's thickness is a function of axial distance from the joint. Consequently the most highly loaded plies are those on the face that is bonded, even outside the joint. A solution to the problem is to ensure the outer plies are oriented to give maximum strength, that is, laid parallel to the load direction. Because of this mode of failure, high joint efficiencies are possible only in very thin laminates; for thick laminates, high joint efficiencies are achieved by the use of multistepped or scarf (acute-angled-butt) joints, both of which are more difficult and costly to produce. The configurations of bonded joints available to a designer are numerous; the most common, together with their design limitations, are given in Fig. 1.

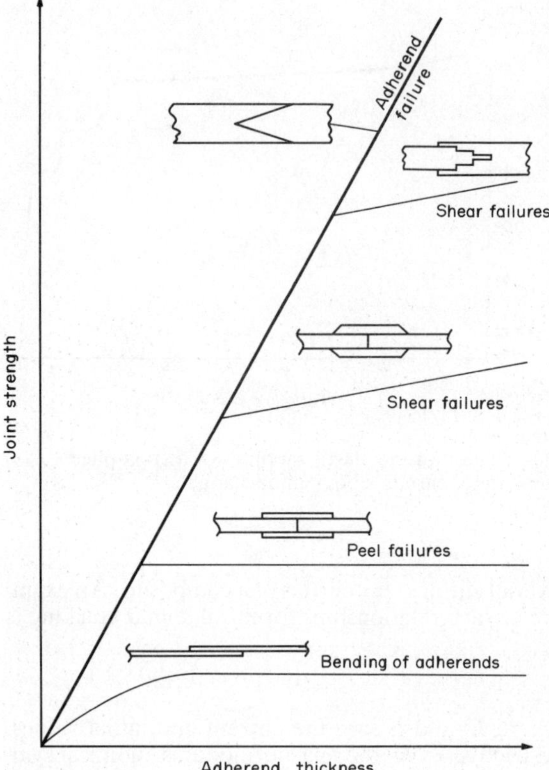

Figure 1
Bonded joint types and their design limitations

One of the problems associated with FRP is that of environmental degradation. This arises because of the readiness of both FRP adherends and adhesive to absorb moisture (water) when exposed to a humid atmosphere. This causes problems during the bonding operation, and also after bonding has been successfully achieved. During the bonding operation, which is usually carried out at an elevated temperature to ensure high-temperature structural properties, moisture diffuses out of the adherend into the adhesive producing high voiding in the adhesive and bond line. As a consequence the adhesive and cohesive properties of the adhesive are significantly reduced. Clearly, therefore, the bonding of 'wet' adherends must be avoided, either by ensuring the bond surfaces do not absorb moisture or by drying the surfaces prior to bonding. The consequence of environmental exposure of assembled joints is discussed in the following paragraphs.

Central to any successful adhesive joint is joint preparation; this ensures that a maximum strength bond will exist between the adherend and adhesive layer interface. Since both FRP-to-FRP and FRP-to-metal joints will need to be used it is worth commenting on the essential differences between them.

The joint between an FRP adherend and an adhesive layer is, for epoxy-matrix FRP, fairly simple since they are essentially similar materials and usually the interface is not a plane of weakness since joint failure more often occurs in the FRP adherend. A major concern in the surface preparation is that of contamination prior to bonding. Provided this is avoided a good joint should be obtained. During the

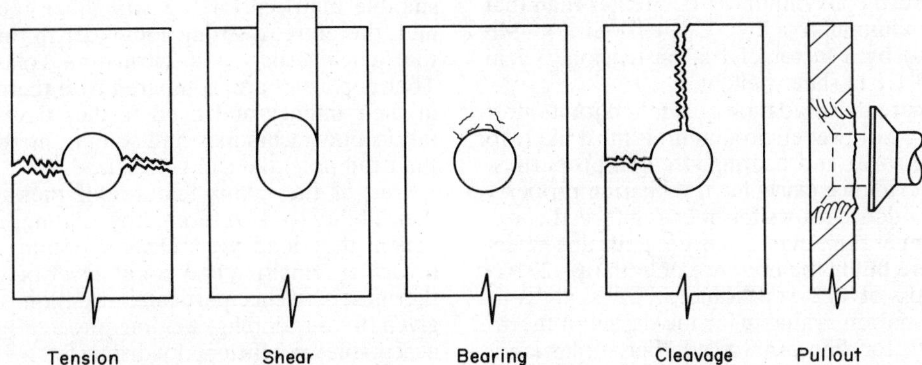

Figure 2
Modes of failure for mechanical joints in FRP

exposure of the joint to a humid environment moisture is absorbed by the FRP adherend, and a plasticization of the matrix and a lowering of the material glass-transition temperature occurs; this directly influences the matrix-dependent strengths properties. A similar effect is observed in most adhesives, causing some reduction in compliance and modulus. Moisture absorption also produces swelling of both adherend and adhesive which can result in differential swelling stresses and possible damage of the bond.

Metallic adherends are the most complex to prepare for bonding since it is necessary to etch and/or anodize the bonding surface to produce an oxide layer. Priming of the surface is often necessary to improve the bonding further and to provide a surface protection for production purposes. One of the main difficulties associated with metallic-to-FRP bonding is that of subsequent environmental exposure. Due to moisture ingress through the FRP and through the adhesive edges, hydrolizing of the metallic bond surface can occur. This is usually brought about when substandard surface treatment has occurred. Unlike the FRP-to-FRP bond the metallic-to-FRP bond can have a major plane of weakness at the metal–adhesive interface.

The adhesive joining of thermoplastics using standard epoxy adhesives is not as simple as joining thermosets, owing to the surface chemistry of the two being so different. The surface pretreatment of a thermoplastic is, therefore, more critical and needs special attention if good adhesion is to be achieved. One method that is currently showing great promise is that of the corona discharge method. This improves the 'wettability' of the surface which gives a correspondingly improved bond over other abrasive type preparations.

2. *Mechanical Joints*

With the increasing interest in the use of mechanical fasteners for joining FRP components a variety of fastening techniques that are used for metals have been successfully applied to many forms of FRP. These range from self-tapping screws which are used for lightly loaded connections, to quality engineering bolted connections used in heavily loaded structures. As with metals, bolted or demountable joints are needed in FRP structures where there is a call for inspection and servicing, modification, repair or replacement of damaged or worn components, or in situations where off-axis loads have to be accounted for and where bonding is uneconomic or impractical. Hence holes or cutouts in FRP laminated composites have to be catered for in design. Like metallic structures, FRP exhibit failure modes in tension, shear and bearing but, because of the complex failure mechanisms of FRP, two further modes are possible, namely cleavage and pullout. Figure 2 shows the location of each of the modes.

It is well known that holes can seriously weaken a laminate, particularly if the laminate is highly anisotropic. This is due in part to the high stress concentrations that occur in the region around such discontinuities, which in tension can be as large as 8 for highly anisotropic carbon-fiber-reinforced plastics (CFRP), compared with a much lower value of 3 normally associated with isotropic materials. Furthermore, isotropic materials yield before failure and the effect of stress concentrations on the net failing stress is small. Since unidirectional FRP do not respond in this way it is not surprising that mechanical fastening of unidirectional-fiber materials is inefficient. However, the problem can be overcome by reducing the degree of anisotropy in the region of a hole and imparting a degree of "plasticity" by the incorporation of fibers oriented in a number of different directions. In fact the failing stress of all three principal modes of failure, that is, tension, bearing and shear pullout, is increased by the inclusion of fibers oriented at $\pm 45°$. Figure 3 shows the change in failure mode of a $0° \pm 45°$ laminate as the contribution of $\pm 45°$ fibers is varied. If we express the efficiency of a joint in terms of its specific strength, then the efficiency of a carbon-fiber-reinforced epoxy joint can be greater than that of an L71 aluminum alloy (Cu $4\frac{3}{4}$%–Si $\frac{3}{4}$%–Mn $\frac{3}{4}$%–Mg $\frac{3}{4}$%) by a factor of 1.32 in tension, 3.7 in bearing and 1.1 in shear pullout.

Environmental degradation of a bolted joint, after exposure to a hot, wet environment, is most likely to occur in the shear and bearing strength properties. At present evidence exists for the bearing property only. The evidence shows for fiber-reinforced epoxies that temperature has a more significant effect than moisture but in the presence of both (at 127 °C) a strength loss of 40% is possible.

The information available for the design of mechanical joints for fiber-reinforced thermoplastics is at present incomplete. The key ingredient still required for establishing a design capability is the characterization of variables such as fiber lay-up and stacking sequence in terms of joint parameters. Despite the incompleteness of this work sufficient evidence exists to suggest the failure behavior of thermoplastics is much the same as for thermoset FRP and that similar high joint efficiencies can be obtained.

3. Welding

The special properties of the currently available thermoplastics, in particular damage tolerance and high resistance to hot wet environments, make them suitable matrices for use with fiber reinforcement and, therefore, a strong contender to thermosets in the engineering and structural plastics field. Thermoplastics are, compared with thermosets, still in their infancy and need further development of fabrication techniques and joining methods before their full potential can be realized.

One of the major features of thermoplastics is their ability to be reformed by heating, and for this reason they lend themselves to joining techniques involving remelting that could never be expected of thermosets. Hence particular attention is now being given to developing welding procedures that are acceptable as efficient load-transfer methods. The experimental work has, so far, examined one particular thermoplastic, a polyether ether ketone (PEEK) manufactured by ICI (UK). Currently several techniques are under review, including the following.

(a) Hot-plate welding—this method gives sound welds and can achieve good lap shear and fracture properties.

(b) Resistance-heating welding—with this method a resistively heated implant technique is needed to produce good properties without disturbance of the fibers. Unfortunately it is limited, subject to further development, to small joint areas.

(c) Induction welding—this technique is capable of continuous welding and produces good mechanical properties. The method shows great promise for the production of large assemblies and potentially of varying joint areas.

All of the three processes show a strong potential for use in production, with induction welding showing the most promise. Further investigation into these techniques is required, together with the development of nondestructive testing techniques to qualify them.

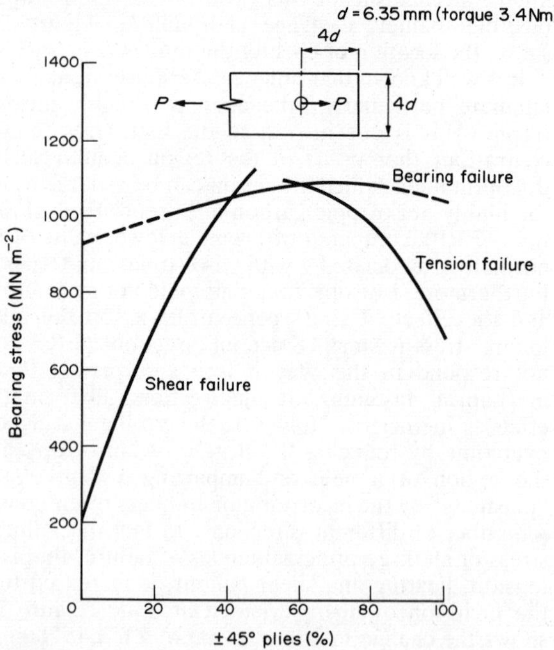

Figure 3
Influence of fiber orientation on failure mode (0°±45° CFRP)

4. Shim Joints

Shim joints use thin metallic interleaved layers to reinforce the composite material in the region of the joint. The action of the shims is to transfer load from the composite material through interlaminar shear between the shim and the composite and then through pin bearing by means of conventional bolts. Failure of this type of joint is much the same as for metals but includes a failure mode due to excessive shear between the composite and metal shims. The multishim joint is a compromise between bonded joints and mechanical joints, and it attempts to make use of the advantages found in each. Unfortunately, because of fabrication difficulties and emphasis on overall cost and weight, shim joints are not always an attractive method of joining.

All of the joining techniques discussed can be used without undue difficulty for FRP, but the behavior of the joints must be properly understood and the limitations realized if structural loads are to be carried safely.

Bibliography

Collings T A 1977 The strength of bolted joints in multi-directional CFRP laminates. *Composites* 7: 43–55

Collings T A 1987 Experimentally determined strength of mechanically fastened joints. In: Matthews F L (ed.) 1987 *Joining Fibre-Reinforced Plastics*. Elsevier, London, pp. 9–63

Erdogan F, Ratwani M 1971 Stress distribution in bonded joints. *J. Compos. Mater.* 5: 378–93

Hart-Smith L J 1987 Design of adhesively bonded joints. In: Matthews F L (ed.) 1987 *Joining Fibre-Reinforced Plastics*. Elsevier, London, pp. 271–311

Parker B M 1978 The effect of hot-humid conditions on adhesive-bonded CFRP-CFRP joints. In: *Symp. on Jointing in Fibre Reinforced Plastics*. IPCS Science and Technology Press, pp. 95–103

Wong J P, Cole B W, Courtney A L 1969 Development of the shim-joint concept for composite structural members. *J. Aircr.* 6(1): 18–24

T. A. Collings

[Royal Aerospace Establishment, Farnborough, UK]

Composite Materials: Manufacturing Overview

Examples of composites involving natural fibers, such as clay bricks reinforced with straw, and plaster reinforced with animal hair, can be traced back over many centuries. For significant engineering purposes, however, fiber composites can be said to originate with the development during the 1940s of processes for spinning glass fibers in commercial quantities. Subsequent development during the 1960s and 1970s of methods of manufacturing strong stiff fibers from carbon, aramid polymers and ceramics such as silicon carbide and aluminum oxide has now made feasible the production of a variety of composite materials in which the fibers are used to enhance the properties of metals and ceramics as well as polymers.

To illustrate the range of possibilities, Fig. 1 plots the strength and modulus of materials of construction in bulk form, and as fibers and whiskers. It can be seen that conventional materials occupy the lower two decades of strength whereas most fibers display strengths greater by one or two orders of magnitude. Whiskers, as a result of their high degree of crystal perfection, show even higher strengths which in some cases approach theoretical values as calculated from interatomic binding energies.

The physical and mechanical properties of a composite can, to a first approximation, be represented

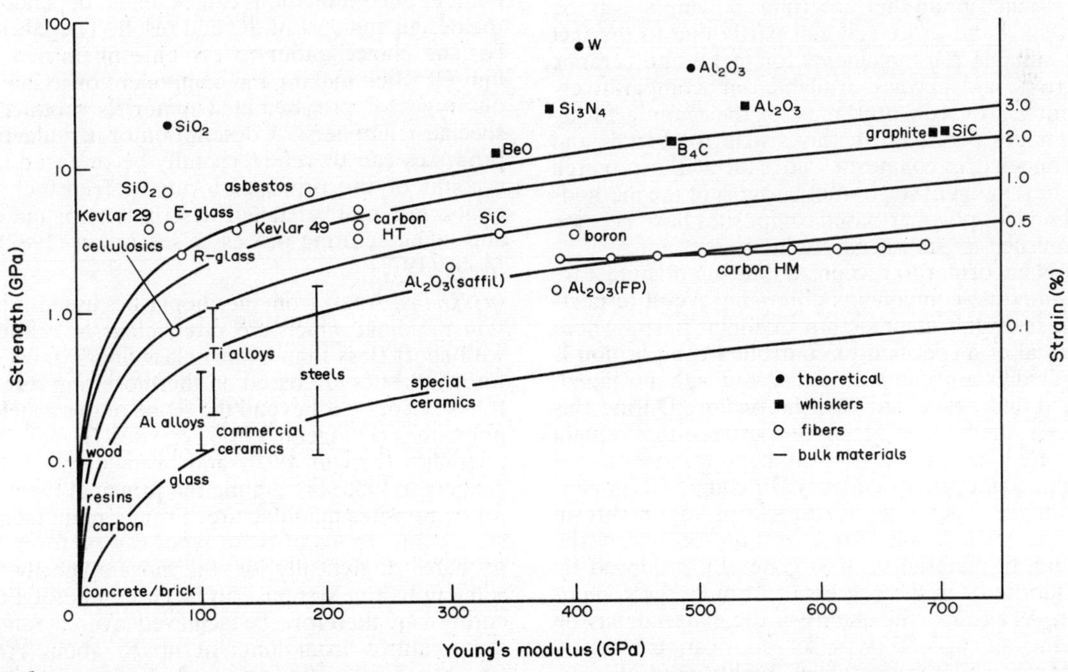

Figure 1
Mechanical properties of various material forms

by a simple mixtures law:

$$P_c = P_f V_f + P_m V_m$$

where P_c represents the composite property (for example tensile strength, Young's modulus or density) and P_f and P_m are the values of the corresponding property for the fibers and matrix, respectively. V_f and V_m are the fractions of the total volume occupied by the fibers and the matrix, respectively. The value of V_f typically lies in the range 0.25–0.65, depending on whether the fibers are aligned and closely packed or randomly oriented with respect to each other. Since most synthetic fibers are spun continuously as multifilament bundles or rovings it is relatively easy to achieve high packing fractions in components. Figure 1 indicates that there is then considerable potential to increase the strength, and to a lesser extent the modulus, of many bulk engineering and constructional materials provided suitable manufacturing methods are available to combine the fibers and the matrix. Unlike spun fibers, whiskers tend to be only a few millimeters in length and their organization into parallel arrays is a slow and expensive process. Consequently, they are more frequently used as randomly oriented mats with low packing fractions, and composite properties are correspondingly low despite the high intrinsic properties of the whiskers themselves.

At the present time only polymer-matrix composites have achieved significant market penetration and this is due partly to the relative ease with which fibers and liquid-thermosetting polymers can be combined and processed and partly due to the fact that suitable reinforcements for metal and ceramic matrices have become available only comparatively recently. Consequently, most of the manufacturing techniques described in this article for metal- and ceramic-matrix composites are still at the research and development stage whereas most of the methods outlined for polymer-based composites have become established for industrial production.

It is important to recognize that the manufacture of composite components differs in several respects from, say, the manufacture of metal parts, where material of a specified or controlled composition is supplied by a primary producer and is then shaped, joined and assembled by a fabricator. During this process, the basic materials properties either remain unaltered or, in the case of heat treatment, are changed in a predictable way. In composites manufacture, although some feedstocks of fiber and resin may be preassembled by a primary producer, the composite material itself is generally produced by the fabricator in the process of forming the component. As a result, the quality of the material may be a function of the type of fabrication process employed and the degree of quality control exercised by the fabricator.

Although the maximum reinforcing effect is achieved with arrays of continuous parallel fibers, the properties in the transverse direction are governed primarily by those of the matrix material and by the strength of the bond between the fiber surface and the matrix. Thus unidirectionally reinforced materials are highly anisotropic in their properties and the anisotropy must be taken into account in the design of a component and may have an influence on its manufacturing conditions. Anisotropy may be reduced, of course, by randomizing fiber directions or by laminating sheets of parallel fibers with the fibers in each sheet oriented in different directions.

1. Reinforced Plastics

1.1 Feedstock Materials

Manufacturing processes for composites, particularly those with polymer matrices, are intimately bound up with the forms in which the fibers and matrices are available to the fabricator. The range of materials is very wide. For instance, there are at least three types of reinforcing fiber of commercial importance—glass, carbon and aramid—each available in different grades, and there are many types and formulations of resin, both thermosetting and thermoplastic. The fiber length, depending on the mechanical duty required of the component, may fall within the range 1–50 mm or may be effectively continuous; the fibers may all be aligned parallel or laminated in sheets at angles to each other or lie in a plane in random orientations. Thus an enormous number of combinations can be made, depending on the design and cost of the end result. The fabricator has the choice either to combine fibers and resin himself when making the component or to use various types of preassembled materials produced by specialist suppliers. A description of manufacturing processes can therefore usefully be prefaced by an account of the principal forms of feedstock materials. For more detailed descriptions of materials and manufacturing processes see Lubin (1982) and Mohr (1973).

(*a*) *Resins*. Apart from thermoplastics used for injection molding, which are often filled or reinforced with short (less than 1 mm) glass fibers, most reinforced plastics are based on thermosetting resins. A full description is beyond the scope of this article but polyesters (Dudgeon 1987), epoxies (Bauer 1980), phenolics (Knight 1980) and bismaleimides (Stenzenberger 1986) are among the principal types used for composites manufacture. From a manufacturing standpoint, resins of these types can be formulated to cure catalytically or thermocatalytically, and some polyester systems can be cured photolytically: curing can therefore be achieved over a range of temperatures from ambient up to about 200 °C. Choice of resin type depends upon a variety of design and manufacturing considerations, not least of which are the viscosity and flow characteristics of the resin as a function of time and temperature. An important step in the production of some types of composite feedstocks is the ability to cure partially (B-stage) a thermosetting resin to convert it from a liquid to a tacky solid, and then to inhibit further curing, usually by refrigeration, until the component has been assembled, when completion of the cure can be achieved by raising the temperature.

Although in terms of volume, thermosetting resins are used predominantly for making composites, interest has developed in recent years in continuous-fiber-reinforced thermoplastics. Candidate materials range from commodity thermoplastics such as polypropylene and nylon to high-performance engineering materials including polysulfone (PS), polyether sulfone (PES), polyetherimide (PEI) and polyether ether ketone (PEEK) (McMahon 1984, Cogswell 1986). Thermoplastic-matrix composites possess certain potential manufacturing advantages over thermosets. For example, feedstocks possess unlimited shelf-life at ambient temperature: the fabricator does not have to be concerned with proportioning and mixing resins, hardeners and accelerators as with thermosets; and the reversible thermal behavior of thermoplastics means that components can be fabricated more quickly because the lengthy cure schedules for thermosets, sometimes extending over several hours, are eliminated.

(*b*) *Fibers*. Rovings are the basic forms in which fibers are supplied, a roving being a number of strands or bundles of filaments wound into a package or creel, the length of the roving being up to several kilometers, depending on the package size. The term rovings is usually applied to glass fibers, whereas bundles of continuous carbon or aramid fibers are often referred to as tows, this name reflecting the textile fiber technology involved in their manufacture. A strand of glass fibers contains typically some 200 filaments, each approximately 10 μm in diameter, and the number of strands in the roving will depend upon the type of application for which the glass is employed (Waring 1970). Carbon fibers, with similar diameters to glass fibers, are produced in a variety of tow sizes ranging from a few hundred filaments per tow to 320 000 per tow. The commonest sizes in most manufacturers' ranges are 3000, 6000, 10 000 and 12 000 ends (Lovell 1988). Fibers are usually sized during production, partly to protect the surface against mechanical damage during subsequent processing or handling, partly to promote wetting by matrix resins and partly to increase adhesion between the fiber and the matrix material. Glass fibers are usually sized with a complex organosilicon compound with other additives; carbon fibers are often treated with a dilute resin compatible with the matrix resin.

Rovings or tows can be woven into fabrics, and a range of fabric constructions are available commercially, such as plain weave, twills and various satin weave styles, woven with a choice of roving or tow size depending on the weight or areal density of fabric required. Fabrics can be woven with different kinds of fiber, for example, carbon in the weft and glass in the warp direction, and this increases the range of properties available to the designer. One advantage of fabrics for reinforcing purposes is their ability to drape, or conform to curved surfaces without wrinkling. Special forms of woven rovings include braids and knitted constructions which can be used to reinforce tubular components and will readily accommodate bends or curves. Indeed it is now possible, with certain types of knitting machine, to produce fiber preforms tailored to the shape of the eventual component. Generally speaking, however, the more highly convoluted each filament becomes, as at crossover points in woven fabrics, or as loops in knitted fabrics, the lower its reinforcing efficiency.

Chopped fibers are used in lengths ranging from less than 1 mm as a reinforcement or filler in thermoplastics for injection-molding purposes, to approximately 50 mm randomly oriented in sheets or mats for the production of laminates with isotropic properties. For handling purposes the fibers in chopped strand mats are held together with a polymeric binder designed to be soluble in the matrix resin. Chopped strand mats composed of glass fibers are one of the most commonly used feedstocks for hand-placement manufacturing processes, being relatively cheap and easy to cut and drape over curved mold surfaces. Thin mats or tissues are often used in the surface layer of moldings to impart a smooth finish.

An alternative use for chopped fibers is in sprayed constructions. Guns are available that simultaneously spray liquid resin and fibers, the latter being derived from a continuous roving that is reduced to appropriate lengths by a chopper within the gun. Spray-up processes are used to build large shell structures on a suitable core and can be controlled robotically to eliminate exposure of operatives to the styrene vapor associated with the polyester resins widely used in glass-reinforced-plastics technology. Spraying of chopped fibers and resin is also sometimes used in conjunction with other manufacturing processes, such as the filament winding of pipes, to introduce a degree of isotropy in the properties of products.

To simplify the handling of fibers and resins for the fabricator, preimpregnated fibers (prepregs) are available from specialist suppliers. The simplest form, continuous warp sheet, consists of an array of parallel rovings or tows in which the filaments have been spread out laterally to produce a uniform distribution within the thickness of the sheet. The sheet is then impregnated with liquid resin to produce a known fiber/resin volume fraction. Prepregs, developed originally with thermosetting resins, are now also available with certain high-melting-point thermoplastics. In the case of thermosets, the resin is partly cured (B-staged) to a slightly tacky condition after impregnation and the sheet surfaces are protected with siliconized paper during storage. Fabrics can also be obtained in the preimpregnated condition and such materials are used typically to build up shell structures by stacking as a preliminary to press or autoclave molding operations. Unidirectional warp sheet is frequently stacked with adjacent layers oriented with respect to each other in such a way as to reduce anisotropy in the final product, rather like the orthogonal grain directions in adjacent veneers in plywood. Warp sheet is also produced in tape of various widths for use in the automated assembly of thin shell structures such as aircraft wing skins and helicopter blades.

Sheet and dough molding compounds are materials which generally comprise polyester resin, an inert particulate filler and chopped fibers, typically in roughly equal proportions by volume. Sheet molding compounds (SMC) are produced by sandwiching a layer of chopped-strand mat between layers of resin mixed with a filler such as talc or chalk, and rolling to ensure penetration of the mixture into the bed of fibers. A material of paste-like consistency results, additives being employed to obtain the appropriate rheological characteristics to ensure the flow of the mixture into mold cavities. Sheets are usually supplied in thicknesses between 3 mm and 10 mm and are protected on each surface until use by a polyethylene film to prevent evaporation of volatile constituents of the resin.

In standard sheet-molding compounds containing 25–35% fiber, the fibers are generally short and are oriented randomly to give isotropic properties and good flow characteristics into deeply drawn sections. Mechanical properties are correspondingly poor but can be improved by increasing the fiber content at the expense of the filler. Fiber contents up to 65% can be achieved, in which case the materials are usually referred to as HMCs. Further improvements in mechanical strength can be attained with XMCs, which embody continuous fibers or rovings in two directions at the surface of the sheet. Clearly the presence of continuous fibers modifies the ability of such compounds to flow readily and to conform to complex surfaces.

Dough molding compounds (DMC) or bulk molding compounds (BMC) differ from SMCs principally in employing a shorter fiber length. The compounds are made by combining the constituents in a high-shear type mechanical mixer and the fibers, although short, may provide a greater degree of three-dimensional reinforcement than in sheet molding compounds.

It may sometimes be desirable to prefabricate the mass of reinforcing fibers to fit a shape to be molded, rather than to rely on flow processes to carry fibers to all parts of the mold. For example, in resin-transfer molding processes, where the mold is first filled with dry reinforcing fibers and the resin is then injected, ideally to flow around the fibers to fill the mold. However, some fibers may be displaced by the flow of the resin. In these circumstances, the fiber mass may first be shaped to fit the mold cavity, for example by spraying on to a porous screen along with a dilute binder, whose function is simply to hold the fibers to the required shape while impregnation with the matrix resin and molding take place. Such a fiber mass is called a preform.

1.2 Manufacturing Methods

A basic aim in almost all reinforced-plastic manufacturing processes is to avoid the entrapment of air or vapor in the form of bubbles or voids. These represent flaws, often located at the fiber–matrix interface or between plies or sheets of material, and generally reduce the strength of the cured composite. A rule of thumb for glass-fiber-reinforced plastics (GRP) is that the interlaminar-shear strength decreases by approximately 7% for every 1% of porosity. The use of low viscosity resins, the vacuum outgassing of resins prior to application and pressure consolidation of the composite are among methods adopted to reduce porosity in the final product.

(*a*) *Hand and spray placement*. In the simplest form, chopped-strand mat or woven rovings are laid over a polished mold surface previously treated with a release or nonstick agent, and liquid thermosetting resin is worked into the reinforcement by hand with the aid of a brush or roller. Polyester resins are most commonly used with glass fibers because of their low cost and good chemical resistance. Resin and curing agent are mixed prior to application and most systems are formulated to cure at ambient temperature. The principal advantages of hand-lay processes are versatility—there is little limitation on the size and complexity of the molding—and low capital cost, the major investment usually being the cost of the tooling. The major disadvantages are low fiber volume fraction and hence low mechanical properties, partly due to the absence of positive consolidation measures. Spray-up methods offer greater rates of material lay-down in addition to the advantages previously outlined. Both methods are well suited to one-offs or short production runs and an outstanding example of the importance of this type of technology is the boat-building industry where GRP has virtually superseded traditional construction materials. Hand-lay methods have been used for the construction of quite large craft such as minesweepers up to approximately 60 m in length (Dixon et al. 1973).

(*b*) *Press molding*. Molding under pressure in heated matched male/female tools is the most widely used process for the volume production of reinforced plastic components. Most of the feedstocks described in Sect. 1.1 can be molded in this way, but the process is particularly appropriate for use with sheet and dough molding compounds. These materials are formulated first to undergo a rapid reduction in viscosity from their paste-like consistency as the charge of compound heats up by contact with the mold surfaces. This enables the compound to flow smoothly, under molding pressures of the order 3–7 MPa, into complex and deeply re-entrant parts of the mold. A rapid gelation and cure then ensues, total cycle times ranging typically from seconds to minutes, depending on the size and thickness of the molding. The process is extensively employed for the production of domestic articles, cabinets and containers for electronic equipment and office machinery, and for automotive parts in sizes up to those of truck doors and cab panels. Figure 2 illustrates schematically a simple press-molding operation.

Press molding is also used for components reinforced with continuous fibers, but the fibers inhibit flow in the mold and such materials are used primarily for the production of parts of constant wall-thickness and of limited curvature.

In production runs involving many thousands of components, steel or plated-steel tools are employed and for large complex parts can represent a substantial capital investment. For short production runs or the development of prototypes, molds can be made by casting a high-strength epoxide resin around a pattern at a relatively modest cost.

(*c*) *Vacuum molding*. This process makes use of atmospheric pressure to consolidate the material while curing, thereby obviating the need for a hydraulic press. The laminate, in the form of preimpregnated fibers or fabric, or as chopped-strand mat

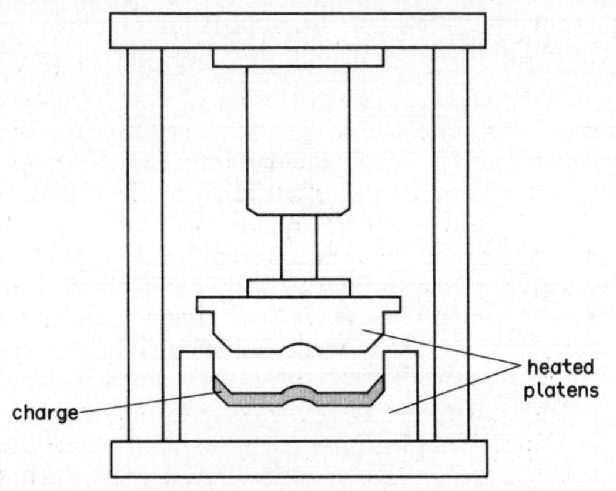

Figure 2
Simple press molding arrangement

or woven rovings impregnated with liquid resin, is placed on a single mold surface and is overlaid by a flexible membrane which is sealed around the edges of the mold by a suitable clamping arrangement. The space between the mold and the membrane is then evacuated and the vacuum maintained until the resin has cured. Quite large, thin shell moldings can be made in this way at low cost.

(*d*) *Autoclave molding*. If a higher density and lower void content are required, a consolidating pressure greater than atmospheric pressure may be necessary. The component, laid up on a mold, is enclosed in a flexible bag tailored approximately to the desired shape and the assembly is enclosed in an autoclave—a pressure vessel designed to contain a gas at pressures generally up to approximately 1.5 MPa and fitted with a means of raising the internal temperature to that required to cure the resin. The flexible bag is first evacuated, thereby removing trapped air and organic vapors from the composite, after which the chamber is pressurized to provide additional consolidation during cure. The process is used extensively in the aircraft industry for the production of thin shell structures of high mechanical integrity, and large autoclaves capable of housing complete wing or tail sections have been installed.

(*e*) *Resin-transfer molding*. The molding processes thus far described start with an open mold or tool which is subsequently closed to surround and compress the charge of fibers and resin. In resin-transfer molding, the fibers only, sometimes as a preform, are enclosed in the mold and the resin is subsequently injected to infiltrate the fibers and fill the cavity. Low-viscosity resins are used to ensure low voidage and good penetration of resin to all parts of the mold, and in vacuum-assisted resin injection, a variant of the process, air in the mold is pumped out prior to injecting the resin. Resin-transfer molding can be used to produce quite large moldings, such as car body-shells, in small-scale production of a few hundred moldings per year. Compared to press-molding of similar-sized parts, the capital cost is low because the low pressures involved enable the molds themselves to be made from GRP.

(*f*) *Reaction injection molding*. Reaction injection molding (RIM) and reinforced reaction injection molding (RRIM) also use relatively low pressures (approximately 0.5–1 MPa) to produce large moldings, principally in polyurethane elastomers (Becker 1979). The principal difference from resin-transfer molding is that, instead of using a precatalyzed resin with a relatively slow cure, the RIM process brings two fast-reacting components together and mixes them just prior to injection into the mold. The mixed system can be tailored to cure in the mold within 30–60 s, thus giving rise to component cycle times of the order 1–2 min. The basis of the process

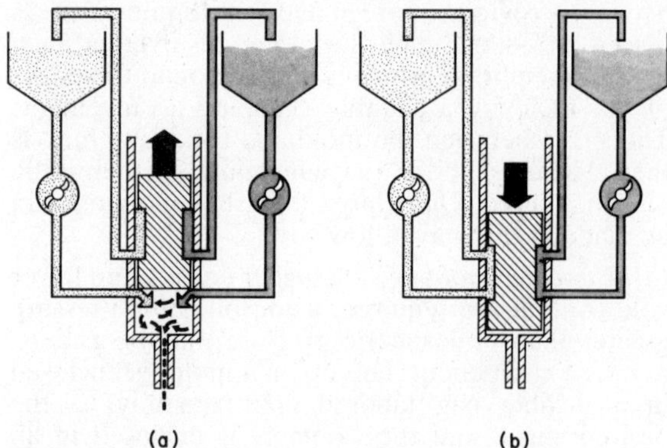

Figure 3
Reaction injection molding process: (a) mixing/injection stage, (b) mixed polymer ejection stage

is illustrated in Fig. 3. The equipment comprises two reservoirs storing, typically, a diisocyanate and a polyol, which polymerize when intimately mixed. Figure 3a shows the mixing/injection stage during which the two reactants are pumped through an impingement chamber in the mixing head, and thence to the mold cavity. The impingement chamber produces highly turbulent flow conditions and hence rapid mixing of the two constituents, and this is followed by rapid gelation and cure in the mold. In the second stage, represented by Fig. 3b, all the mixed polymer is ejected from the mixing head and the two reactants are separated and continuously recirculated ready for the next molding cycle. The mixing head is thus self-cleaning and any cured polymer in the connecting pipe from the mixing head to the mold cavity is arranged to be removed as a piece of flash attached to the molding.

A variety of urethane formulations is available to give polymers with a range of hardness from soft and rubbery to hard and brittle. The moldings can be reinforced to some degree with short glass fibers by including them in one or both of the liquid reactants to form a slurry. The limitation on fiber length and hence reinforcing efficiency, and the quantity of fiber that can be incorporated, is set by the design of the mixing head and its ability to handle the fibers without risk of blockage or of undue agglomeration of the fibers, resulting in a nonuniform distribution in the molding, RIM and RRIM processes compete in many ways with press-molded sheet and dough-molding compounds for markets in office and domestic equipment and for automotive products such as bumpers, body and trim panels, and both materials and equipment are the subject of intensive development at the present time.

(*g*) *Pultrusion*. The processes thus far described produce discrete moldings. Pultrusion enables profiles of constant cross-section to be manufactured continuously. The process is analogous to the extrusion of plastics and nonferrous metals in that the profile is shaped by continuous passage of the feedstock through a forming die, but in the case of pultrusion the reinforcing fibers are used to pull the material through the die. In Fig. 4, continuous rovings are passed through a bath of resin followed by a series of carding plates to encourage complete penetration of resin into the fiber bundles and to remove excess resin. The impregnated fibers are then drawn through a heated die in which the resin first gels and then cures. The finished profile finally passes through a traction mechanism before being cut to length.

Glass fibers and polyester resins are the most commonly employed materials because of their relatively low cost and good corrosion resistance. Preheating by radiofrequency methods is often used to increase the throughput, and production rates of several meters per minute are achieved depending on section thickness. The process is ideally suited to the production of structural elements of high strength and stiffness owing to the presence of the high proportion of axial fibers necessary to sustain large tractive forces, particularly when high volume fractions of fiber (up to approximately 0.7) are required in the cross section. It is, however, possible

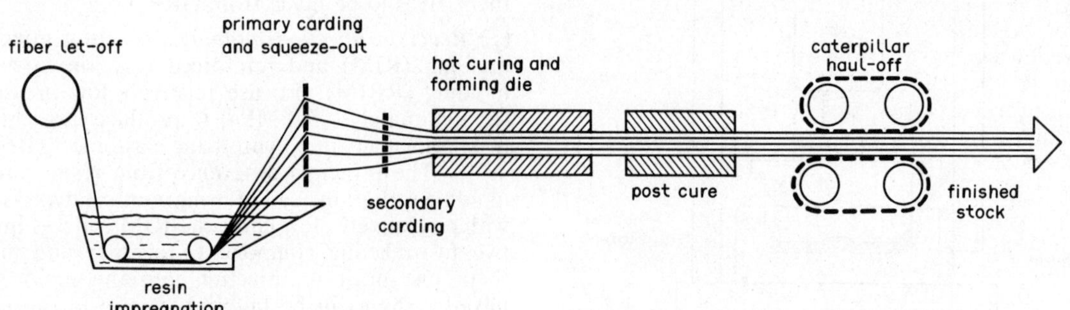

Figure 4
Schematic of pultrusion process

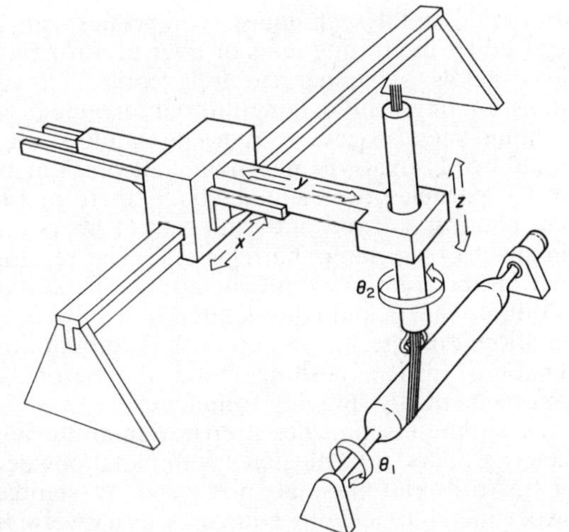

Figure 5
Basic motions of a 5-axis filament winding machine

to incorporate varying cross-sectional geometry although of constant cross-sectional area.

(*h*) *Filament winding*. In this process, strong stiff shell-structures can be made by winding bands of continuous fibers impregnated with resin on to a former or mandrel which can be withdrawn after the resin has been cured. The method is analogous to the technique of wire wrapping for the reinforcement of steel pressure-vessels and gun barrels and is extensively used for the manufacture of reinforced-plastic pipe, and for cylindrical tanks and vessels in GRP for the storage of aggressive chemicals and a wide variety of agricultural products. The size of component that can be manufactured is limited only by the capacity of the winding machine, and methods also exist for the building of large vertical silos *in situ*, involving the construction of a circular railway track upon which a carriage, dispensing fiber and resin, travels continuously around a vertical former.

Modern filament-winding machinery is capable of producing shapes that are much more complex than the simple axisymmetric structures indicated above, largely as a consequence of adopting robotics technology. Figure 5 shows schematically a gantry-type filament-winding machine with five axes or machine motions that are controlled independently by a microprocessor. Two such motions, rotation (θ_1) of a mandrel between head and tailstock, and translation (x) of a carriage backwards and forwards parallel to the axis of mandrel rotation, suffice for the winding of cylindrical shells at helix angles dictated by the relative rates of rotation and translation. The additional facility of movement of the feed eye along the y and z directions, however, enables a band of rovings to be laid in a controlled way on convex surfaces of any given shape, and rotation of the feed eye (θ_2) enables the plane of the band to be matched to the inclination of the mandrel surface at any given point.

For a stable winding it is necessary to identify geodesic paths, defined as the shortest distance between two points on a surface, to avoid the possibility of the band of fibers slipping from the desired trajectory. This condition, coupled with the sheer magnitude of the task of programming a multiaxis robot to traverse a constantly changing and nonrepetitive path, has led to the development of computer-based methods aimed at integrating design and manufacturing procedures for filament-wound components by analogy with CADCAM processes for metal machining (Wells and McAnulty 1987). This type of technology opens up the possibility of winding components as complex as aircraft fuselages, steering wheels and bicycle frames.

2. *Reinforced Metals*

Reference to Fig. 1 shows that the mechanical properties of the nonferrous metals and alloys commonly used for structural purposes are considerably better than those of unreinforced polymers, and the improvements to be obtained by combining them with fibers or whiskers will therefore be less dramatic than with polymers. However, other worthwhile improvements can be achieved by incorporating fibers or even particulates into a metal matrix. For instance, creep resistance at a particular temperature may be improved or the resulting material may be more wear and abrasion resistant, or less susceptible to fatigue failure. Such improvements, even if modest, may be of great technological importance if, for example, the temperature limit for continuous operation of an engine component or an airframe structural part can be increased. Considerable research is now being devoted to both the development of reinforcements suitable for metal matrices, and of manufacturing methods. Much of the drive behind the upsurge in metal-matrix composites development derives from the recent development of processes for producing refractory fibers and whiskers in commercial quantities (Bunsell 1987).

2.1 Feedstock Materials

(*a*) *Matrix materials*. The most common matrices are the low density metals, such as aluminum and aluminum alloys, and magnesium and its alloys. Some work has been carried out on Pb–Sn alloys, mainly for bearing applications, and there is interest in the reinforcement, for example, of titanium-, nickel- and iron-base alloys for higher-temperature performance. However, the problems encountered in achieving the thermodynamic stability of fibers in intimate contact with metals become more severe as the potential service temperature is raised, and the bulk of development work at present rests with the light alloys.

(*b*) *Reinforcements*. The principal reinforcements for metal matrices include continuous fibers of carbon, boron, aluminum oxide, silica, aluminosilicate compositions and silicon carbide. Some ceramic fibers are also available in short staple form, and whiskers of carbon, silicon carbide and silicon nitride can be obtained commercially in limited quantities. There is also interest in the use of refractory particles to modify alloy properties such as wear and abrasion resistance. In this case particle sizes and volume fractions are greater than those developed metallurgically in conventional alloys, and incorporation of the particles into the metal is achieved mechanically rather than by precipitation as a consequence of heat treatment.

Most metal-matrix composites consist of a dispersed reinforcing phase of fibers, whiskers or particles, with each reinforcing element ideally separated from the next by a region of metal. An alternative approach is to reinforce the metal, generally in sheet form, with a fiber-reinforced polymer. An example is ARALL (Vogelesang and Gunnink 1983), a laminate consisting of alternate plies of aluminum alloy bonded to sheets of aramid-fiber-reinforced resin. This has the advantage of a simple, low-temperature production process in which the continuous parallel fibers can display a high reinforcing efficiency. Applications are, however, restricted to thin shell-type structures.

Apart from sheet materials similar to ARALL, the production of feedstocks in which reinforcement and matrix can be purchased preassembled for the convenience of the fabricator is less well advanced than the production of polymer-matrix composites. Wires or tapes of aluminum containing up to 500 continuous silicon carbide filaments are available in development quantities and in some cases reinforcements can be obtained woven into fabrics or agglomerated into randomly oriented mats.

2.2 Manufacturing Methods

Many of the processes used for the production of unreinforced metal parts, or for the deposition of metal layers, have been explored or adapted with the aim of producing a continuous, void-free metal matrix with a controlled distribution of reinforcements. Fabrication methods can involve processing the metal in either the molten or the solid state, and components can be formed either by direct combination of matrix and reinforcement or by the production of a precursor composite which, in the form of composite wires, sheets or plies, is used to build up a component. In the latter case, the assemblage of plies must be consolidated and bonded in a subsequent process.

In liquid-metal techniques, composites can be prepared by infiltrating mats or fiber preforms with liquid metals or, under carefully controlled conditions, by physically mixing the reinforcement and the liquid metal together. A pseudoliquid route is offered by plasma or flame spraying in which metal-powder particles are heated above their melting point and are sprayed onto an array of fibers on a thin sheet of the same matrix metal. The resulting sheet of fiber-reinforced metal can then be stacked with other sheets and consolidated in a subsequent operation. An alternative approach is to build up a solid body by codepositing metal and particulate reinforcements by spraying techniques.

The simplest solid-state preparation route is to mix short fibers or particulates with metal powders. The mixture can then be processed by standard powder-metallurgical procedures. Alternatively, the metal may be coated onto the reinforcement by electrochemical or chemical-vapor deposition methods.

Many of these processes are still under development or evaluation and, as with polymer-matrix composites, the methods of manufacture that will eventually be adopted commercially will depend on production costs as well as technical suitability for particular components. For this reason, considerable attention is being focused on infiltration and casting processes because of their relatively low cost. For detailed reviews of manufacturing techniques see Mileiko (1983) and Chou et al. (1985).

(*a*) *Liquid-metal infiltration*. Direct impregnation of bundles of continuous filaments has been used to produce precursor wires or tapes that can be subsequently consolidated into component forms. A typical scheme involves drawing a bundle of filaments through a bath of molten metal and then through an orifice which shapes the impregnated bundle to a circular or rectangular cross section. For complete impregnation to take place it is important that the metal should wet the fiber surface, and considerable research has been devoted to coatings for fibers or additives to the metal that will reduce the contact angle of the molten metal at the fiber surface and promote wetting. Penetration of fiber bundles may be enhanced by infiltrating under vacuum or using inert-gas pressure over the molten metal (Masur et al. 1987).

(*b*) *Squeeze casting*. In this process a mass or preform of fibers or whiskers is infiltrated with molten metal under a high pressure derived from a hydraulic press. The method is illustrated schematically in Fig. 6 and can result in a fully dense composite with good dispersion of the reinforcement. Whisker concentrations of up to 30% by volume have been achieved, resulting typically in increases in strength and Young's modulus of a factor of two over unreinforced aluminum alloys. The method can be used to produce billets that can then be shaped by conventional metal hot-working techniques such as extrusion or forging, or it can be used directly for near net-shaping of components such as automotive connecting rods. With some materials it may be desirable for the infiltration to be carried out under vacuum, to avoid either the risk of oxidation of the reinforcement or the possibility of entrainment of air bubbles in the matrix.

(*c*) *Stir casting and compo-casting*. In some cases it is possible to introduce particulate or short-fiber reinforcements simply by feeding into the vortex region produced by vigorously stirring the molten metal. Long stirring times may be necessary to develop a good bond between the two components, and the volume fraction of the reinforcement may be limited to about 10% by the rapid increase in the viscosity of the mixture with the further addition of reinforcement.

Differences in density between molten metals and reinforcing fibers or whiskers may mean that separation of the two components will occur due to gravitational force if simple mixing of the fibers in molten metal is attempted. The tendency to separate can be reduced or eliminated in compo-casting whereby stirring takes place as the metal undergoes solidification. Under these conditions, mixed phases of solid and liquid metal can coexist in the form of a slurry in which reinforcement particles can be mixed and become trapped between the solid-phase regions with uniform dispersion.

(*d*) *Precursor production*. Preimpregnated sheet that can be used to build up more complicated structures can be made in a variety of ways. The most common method using continuous monofilament or bundles of filaments is to assemble them with the required spacing by winding on to a drum or around pegs to produce a parallel array on a flat surface. The metal is then coated onto the filaments by plasma spraying, electroplating or by chemical or chemical vapor deposition techniques. Alternatively, bundles or tows of filaments can be coated by electrochemical

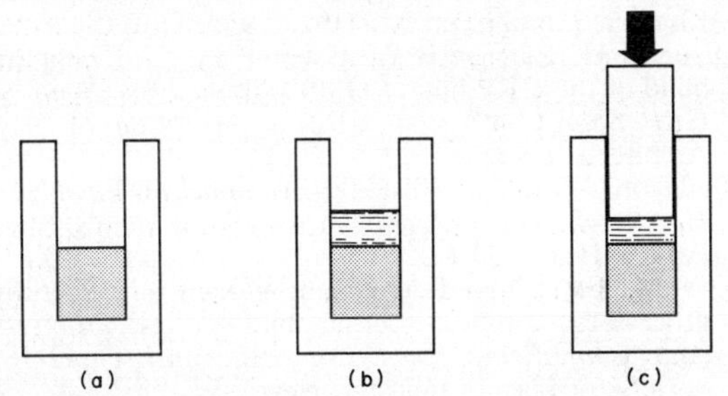

Figure 6
Schematic of squeeze casting process: (a) fiber preform placed in heated mold, (b) liquid metal poured into mold and (c) infiltration under pressure

or plasma-activated deposition processes. These various precursor forms must then be shaped, stacked, consolidated and bonded to form a continuous matrix phase.

(*e*) *Consolidation and bonding methods*. Densification and bonding of sheets of precursor materials or of mixtures of metal powder and reinforcement are usually carried out in the solid state under conditions of temperature and pressure such that consolidation results from a combination of plastic deformation and diffusion in the metal phase. Methods of placing the material under the appropriate conditions of temperature and pressure include conventional hot-pressing, hot and cold isostatic pressing, hot rolling and explosive welding. As an example see Hack and Amateau (1982) who describe diffusion bonding of carbon-fiber-reinforced aluminum alloy wires to aluminum alloy foils in a press at a temperature between 554 °C and 562 °C and a pressure of 24 MPa for a dwell time of 20 minutes. An account of the technology of boron-reinforced aluminum composites, including details of diffusion bonding, has been given by Miller and Robertson (1977).

Many of the composites incorporating short-fiber or particulate reinforcements are produced in the form of billets by the methods described above and may require subsequent processing by standard metallurgical techniques such as extrusion, forging or drawing to achieve their final form.

3. *Reinforced Ceramics*

Research into the properties of fiber-reinforced ceramics has been in progress since the 1950s, but until the advent of refractory, oxidation-resistant fibers such as alumina and silicon carbide in commercial quantities, much of the work was confined to developing an understanding of the principles of reinforcement and mechanisms of failure (Phillips 1983). As with metal-matrix composites, there is currently an increase in research and development activity, although little commercial application of materials thus far. This description of manufacturing methods consequently relates to laboratory-type processes.

3.1. *Feedstock Materials*

(*a*) *Matrix materials*. The choice of matrix depends primarily on the mechanical and thermal duty required of the composite. For potential applications in aerospace and in heat engines, oxidation resistance is of prime importance and most refractory oxides, carbides and borides have been considered as matrices. The most extensively employed, however, have been aluminum oxide, silicon carbide and silicon nitride, mainly because the technology of producing well-consolidated artifacts in the unreinforced state is better established for these materials.

Glasses and silica have also been explored because of the relative ease with which matrix consolidation can be accomplished by processing in the liquid state. Glass–ceramic materials that can be combined with reinforcements while in a liquid state and then devitrified by heat treatment have also been the subject of considerable investigation.

Although not a ceramic, carbon is one of the most refractory materials in nonoxidizing atmospheres, and Fig. 1 indicates that substantial improvement to the mechanical properties of bulk carbon is feasible by fiber reinforcement. In samples unidirectionally reinforced with carbon fibers, tensile strengths of up to 1 GPa have been reported (Hill et al. 1974). For some applications, reinforcement in three dimensions is required and weaving or knitting processes have been developed for producing continuous-carbon-fiber preforms that can subsequently be invested with a carbon matrix.

(*b*) *Reinforcements*. Only a few of the fibers available commercially are suitable for ceramic-matrix composites by virtue of their compatibility and stability with the matrix during fabrication and at intended service temperatures. Continuous fibers include silicon carbide, boron, aluminum oxide, aluminosilicates and carbon (Bunsell 1987). In oxidizing atmospheres, only silicon carbide and the oxide fibers offer the prospect of high-temperature composites and, depending on the stress to which they are subjected, the upper limit of ceramic-matrix composites is currently in the region of 1200 °C.

(*c*) *Manufacturing methods*. Hot-pressing of sheet-material preforms is the route most widely adopted for ceramic-matrix composites, and hot-pressing conditions generally follow established procedures for consolidation and sintering of monolithic ceramics. Early work has been reviewed by Sambell (1970) and a method of producing sheet preforms is described by Sambell et al. (1974). More recently, chemical vapor deposition (CVD) methods have been adopted to infiltrate a fiber preform with, for example, silicon carbide derived from the gaseous-phase cracking of methyl trichlorosilane in hydrogen. Alternatively, the preform may be impregnated with a liquid organometallic compound which is then pyrolyzed to yield a refractory ceramic residue (Bernhart et al. 1985).

Similar methods are used for the manufacture of carbon-fiber-reinforced carbon materials (Fitzer 1987). Preforms are either coated with carbon produced in the vapor phase by reacting, say, methane and hydrogen, or are subjected to impregnation of a polymer possessing a high carbon yield when pyrolyzed. One of the major disadvantages of both CVD and pyrolysis routes is that successive depositions are necessary to produce a fully dense matrix. The processes are therefore slow and correspondingly expensive. Nevertheless, such processes are operated commercially, notably for the manufacture of carbon–carbon disk brakes for aircraft and for racing cars.

4. Conclusion

Within the scope of this article it has been possible to describe only briefly some of the methods used for manufacturing composite materials and components. In practically every case, however, the aim is to control the dispersion of the reinforcement in the matrix in such a way as to maximize the properties of the composite while minimizing the manufacturing cost. The great variety of materials and material forms involved in composites technology provides considerable promise for the evolution of materials with improved performance, but the availability of cost-effective manufacturing methods capable of providing products with a high degree of consistent and reproducible performance is a key factor in determining the rate at which they can compete with more traditional materials of construction.

Bibliography

Bauer R S 1980 The versatile epoxies. *Chem. Technol.* 9: 692–700

Becker W E 1979 *Reaction Injection Moulding*. Van Nostrand Reinhold, New York

Bernhart G, Chateigner S, Heraud L 1985 Les composites ceramique–ceramique, du concept à la piece. In: Bunsell A R, Lamicq P, Massiah A (eds.) 1985 *Developments in the Science and Technology of Composite Materials*. European Association for Composite Materials, Bordeaux, pp. 475–81

Bunsell A R 1987 Fibre reinforcements—past, present and future. In: Matthews F L, Baskell C R, Hodgkinson J M, Morton J (eds.) 1987 *Proc. 6th Int. Conf. on Composite Materials*, Vol. 5. Elsevier, London, pp. 1–13

Chou T W, Kelly A, Okura A 1985 Fibre-reinforced metal-matrix composites. *Composites* 16(3): 187–206

Cogswell F N 1986 Continuous fibre reinforced thermoplastics. In: Collyer A A, Clegg D W (eds.) 1986 *Mechanical Properties of Reinforced Thermoplastics*. Elsevier, London, pp. 83–119

Dixon R H, Ramsay B W, Usher P J 1973 Design and build of the GRP hull of HMS Wilton. *Proc. Symp. on GRP Ship Construction*. Royal Institution of Naval Architects, London, p. 1

Dudgeon C A 1987 Polyester resins. *Materials Engineering Handbook*, Vol. 1. American Society for Metals, Metals Park, OH, pp. 90–6

Ewald G 1981 Curved pulforming—a new manufacturing process for composite automobile springs. *Proc. 36th Ann. Conf. Reinf. Plastic/Composites Inst.* Paper 16C. Society for Plastics Industry, New York

Fitzer E 1987 The future of carbon-carbon composites. *Carbon* 25(2): 163–90

Hack J E, Amateau M F 1977 Mechanical behaviour of metal matrix composites. *Proc. AIME Conf. Dallas,*

Texas. 16–18 Feb. 1982. American Institute of Mining Engineers, New York
Hill J, Thomas C R, Walker E J 1974 Advanced carbon-carbon composites for structural applications. In: *Carbon Fibers, their Place in Modern Technology*, paper 19. The Plastics Institute, London, pp. 122–30
Knight G J 1980 High temperature properties of thermally stable resins. In: Pritchard G (ed.) 1980 *Developments in Reinforced Plastics*, Vol. 1. Applied Science, London, pp. 145–210
Lovell D R 1988 *Carbon and High Performance Fibres Directory*, 4th edn. Pammac Directories Ltd., High Wycombe, UK, pp. 45–74
Lubin G (ed.) 1982 *Handbook of Composites*. Van Nostrand Reinhold, New York
McAdams L V, Gannon J A 1986 Epoxy resins. In: *Encyclopedia of Polymer Science and Engineering*, Vol. 6, 2nd edn. Wiley, New York, pp. 322–82
McMahon P E 1984 Thermoplastic carbon fibre composites. In: Pritchard G (ed.) 1984 *Developments in Reinforced Plastics*, 4. Elsevier, London, pp. 1–30
Masur L J, Mortensen A, Cornie J A, Flemings M C 1987 Pressure casting of fiber-reinforced metals. In: Matthews F L, Buskell N C R, Hodgkinson J M, Morton J (eds.) 1987 *Proc. 6th Int. Conf. on Composite Materials*, Vol. 2. Elsevier, London, pp. 320–9
Mileiko S T 1983 Fabrication of metal-matrix composites. In: Kelly A, Mileiko S T (eds.) 1983 *Handbook of Composites*, Vol. 4. North-Holland, Amsterdam, pp. 221–94
Miller M, Robertson A 1977 Boron/aluminium and borsic/aluminium analysis, design, application and fabrication. In: Renton W J (ed.) 1977 *Hybrid and Select Metal Matrix Composites: A State of the Art Review*. American Institute of Aeronautics and Astronautics, New York, pp. 99–157
Mohr J G (ed.) 1973 *SPI Handbook of Technology and Engineering of Reinforced Plastics/Composites*, 2nd edn. Van Nostrand Reinhold, New York
Phillips D C 1983 Fibre reinforced ceramics. In: Kelly A, Mileiko S T (eds.) 1983 *Fabrication of Composites*, Vol. 4. Elsevier, Amsterdam, pp. 373–428
Sambell R A J 1970 The technology of ceramic-fibre ceramic-matrix composites. *Composites* 1(5): 276–85
Sambell R A J, Phillips D C, Bowen D H 1974 The technology of carbon fibre reinforced glasses and ceramics. In: *Carbon Fibres, their Place in Modern Technology*, paper 16. The Plastics Institute, London, pp. 105–13
Stenzenberger H 1986 Bismaleimide resins. In: Kinloch A J (ed.) 1986 *Structural Adhesives—Developments in Resins and Primers*. Elsevier, London, pp. 77–126.
Vogelesang L B, Gunnink J W 1983 ARALL, a material for the next generation of aircraft—a state of the art. In: Jube G, Massiah A, Naslain R, Popot M (eds.) 1983 *High Performance Composites Materials: Proc. 4th Int. Conf. Society for the Advancement of Materials and Process Engineering, European Chapter*. SAMPE, Bordeaux, pp. 81–92
Waring L A R 1970 Reinforcement. In: Parkyn B (ed.) 1970 *Glass Reinforced Plastics*. Butterworth, London, pp. 121–45
Wells G M, McAnulty K F 1987 Computer aided filament winding using non-geodesic trajectories. In: Matthews F L, Buskell N C R, Hodgkinson J M, Morton J (eds.) 1987 *Proc. Sixth Int. Conf. on Composite Materials*, Vol. 1. Elsevier, London, pp. 161–73

D. H. Bowen
[UKAEA Harwell Laboratory, Didcot, UK]

Composite Materials: Structure–Performance Maps

The purpose of the structure–performance map is to present in a concise manner the effective properties of composites based upon various reinforcement forms and fiber–matrix combinations. The maps are constructed using the results of extensive analytical studies of the thermoelastic properties of unidirectional laminated composites as well as two-dimensional and three-dimensional textile structural composites with polymeric, metal and ceramic matrices. The effectiveness and uniqueness of the reinforcement forms can be evaluated from the maps. Also, these maps can guide engineers in selecting materials for structural design, and researchers in identifying the needs of future work.

1. Material Systems

The primary materials for reinforcement include graphite, boron, glass and aramid fibers in continuous, discontinuous (e.g., short fibers, whiskers and particulates) or textile forms. Other reinforcement materials with a promising potential of development are silicon carbide and alumina fibers. A number of other ceramic fibers and whiskers have also received considerable attention. Table 1 lists the mechanical and physical properties of several fibers. The superior strength and modulus of fibers are also demonstrated in Fig. 1 on the specific basis (strength/density and modulus/density).

In the following discussions, continuous fibers in unidirectional laminae and woven-fabric composites are considered. It should be noted that the term woven-fabric composites is used to encompass composites reinforced with two-dimensional (2-D) and three-dimensional (3-D) textile preforms.

The types of matrix materials along with their mechanical and physical properties are given in Table 2. Polymeric, metal and glass/ceramic matrices for composites are included.

2. Analytical Modelling

The analytical modelling work for constructing the structure–performance map focuses on unidirectional laminae, and 2-D and 3-D woven-fabric composites.

Table 1
Properties of selected fibers

	Density ρ (g cm^{-3})	Longitudinal tensile modulus E_1 (GPa)	Transverse tensile modulus E_2 (GPa)	Poisson's ratio ν_{12}	Shear modulus G_{12} (GPa)	Longitudinal tensile strength σ (MPa)	Longitudinal thermal expansion coefficient α_1 (10^{-6} °C^{-1})	Transverse thermal expansion coefficient α_2 (10^{-6} °C^{-1})
Glass	2.45	71	71	0.22	30	3500	5	5
Kevlar	1.47	154	4.2	0.35	2.9	2800	−4	54
Graphite (AS)	1.75	224	14	0.20	14	2100	−1	10
Graphite (HMS)	1.94	385	6.3	0.20	7.7	1750	−1	10
Boron	2.64	420	420	0.20	170	4200	5.0	5.0
SiC	3.2	406	406	0.20	169	3395	5.2	5.2
Al_2O_3	3.9	385	385	–	154	1400	8.5	8.5

Unidirectional laminae are the building blocks of laminated composites. Various approaches have been successfully developed to predict the thermoelastic properties of unidirectional laminae, such as the mechanics-of-materials approach and the elasticity approach. The off-axis thermoelastic properties of a unidirectional lamina inclined at an angle to the principal material direction can be obtained through proper transformation of the tensorial quantities. The cross-ply laminate, which is a special case of the off-axis laminate, is composed of unidirectional laminae with principal material directions oriented alternately at 0° and 90° to the laminate axis. The laminated angle-ply composite is made up of off-axis unidirectional laminae in $+\theta$ and $-\theta$ orientations.

The effective thermoelastic properties of laminated composites can be predicted from the well-established classical laminated plate theory. The key laminate geometric parameters are fiber orientation, lamina thickness, lamina stacking sequence and thermoelastic properties of the individual laminae. Comprehensive discussion of the analysis for laminated composites can be found in the article *Elastic Properties of Laminates*.

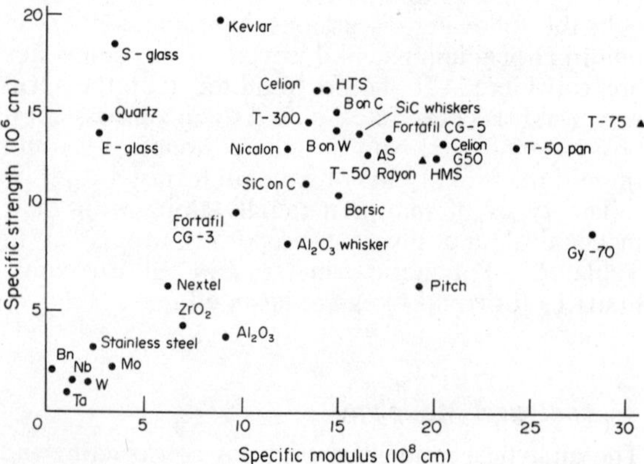

Figure 1
Variation of specific strength with specific volume for several fibrous materials

In the construction of performance maps, a range of fabric materials with increasingly larger size of repeating unit—from plain weave to 8-harness satin weave—are considered.

In the case of 3-D fabric composites, the braided-fiber preform is adopted for modelling purposes. The concepts developed for analysis can be readily applied to other types of 3-D preforms. The "unit cell" approach has been proposed, in which a fiber assumes a position along a diagonal in the unit cell and defines an angle with respect to the braiding axis. Two analytical approaches have been developed: the fiber interlock model and fiber inclination model. These are used for predicting the thermoelastic properties as functions of fiber spatial orientation, fiber volume fraction, and braiding parameters. The fiber inclination model is employed for the performance-map analysis which treats the unit cell of the 3-D braided composite as an assemblage of four inclined unidirectional laminae. The classical laminated plate theory is employed to derive the effective laminate thermoelastic properties. The validity of these analytical methods has been substantiated by experimental characterizations of 3-D braided reinforcements in both polymeric- and metal-matrix composites.

3. Structure–Performance Maps

Four types of reinforcement forms are under consideration: unidirectional laminated angle-plies with the off-axis angle θ ranging from 0° to 90°; a 0°/90°

Table 2
Properties of selected matrices

	Density ρ (g cm^{-3})	Tensile modulus E (GPa)	Poisson's ratio ν	Shear modulus G (GPa)	Tensile strength σ (MPa)	Thermal expansion coefficient α (10^{-6} °C^{-1})
Epoxy	1.246	3.5	0.33	1.25	35	57.5
PEEK	1.30	4	0.37	1.4	70	45
Aluminum	2.71	69	0.33	26	74	23.6
Titanium	4.51	113.8	0.33	43.5	238	8.4
Magnesium	1.74	45.5	0.33	7.5	189	26
Borosilicate glass	2.23	63.7	0.21	28	90	3.25

cross-ply; two-dimensional woven fabric with n_g ranging from 2 (plain weave) to 8 (8-harness satin); and three-dimensional braided composites with the braiding angle between a fiber segment and braiding axis in the range 15–35°.

The results of the parametric studies have been used to construct maps of the following four types:

(a) longitudinal Young's modulus E_x vs transverse Young's modulus E_y;

(b) longitudinal Young's modulus E_x vs in-plane shear modulus G_{xy};

(c) longitudinal Young's modulus E_x vs Poisson's ratio ν_{xy}; and

(d) longitudinal thermal expansion coefficient α_x vs transverse thermal expansion coefficient α_y.

The fiber volume fraction of all the composites presented in the maps is assumed to be 60%.

The structure–performance maps for various combinations of fiber and matrix materials and reinforcement forms are shown in Figs. 2–6. The general characteristics of the maps are summarized below.

The in-plane thermoelastic properties of unidirectional laminae depend strongly on fiber orientation. Unidirectional reinforcement provides the highest elastic stiffness along the fiber direction. The 0°/90° cross-ply yields identical thermoelastic properties in 0° and 90° orientations. Its in-plane shear rigidity is poor. The in-plane thermoelastic properties of angle-ply laminates are also dependent upon the fiber orientations. The longitudinal Young's modulus of an angle-ply laminate is lower than that of a unidirectional lamina. Better transverse elastic properties and in-plane shear resistance can be achieved through stacking of the unidirectional laminae with different fiber orientations. For ±45° angle-ply, the in-plane stiffness is relatively low, while the shear modulus reaches a maximum.

The 2-D biaxial woven-fabric composites can provide balanced in-plane thermoelastic properties within a single ply. They behave similarly to 0°/90° cross-plies, although the fiber "waviness" tends to reduce the in-plane efficiency of the reinforcements. As the fabric construction changes from plain weave to 8-harness satin, the frequency of crimp due to fiber crossover is reduced, and the fabric structure approaches that of 0°/90° cross-ply.

The thermoelastic properties of three-dimensional braided composites also show a strong dependence

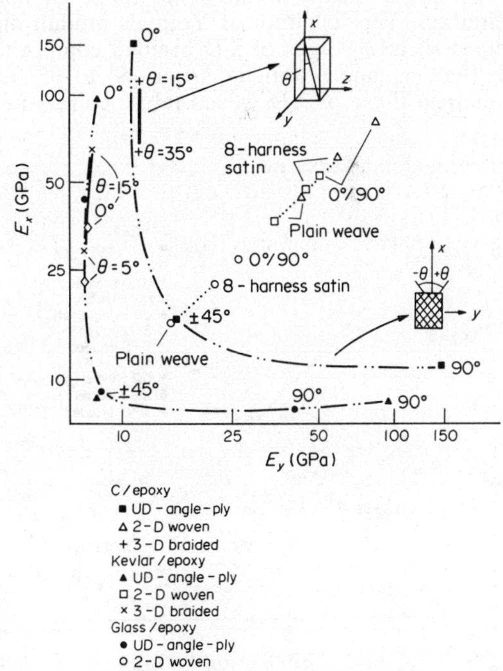

Figure 2
Variation of longitudinal Young's modulus E_x with transverse Young's modulus E_y for C–epoxy, Kevlar–epoxy and glass–epoxy composites

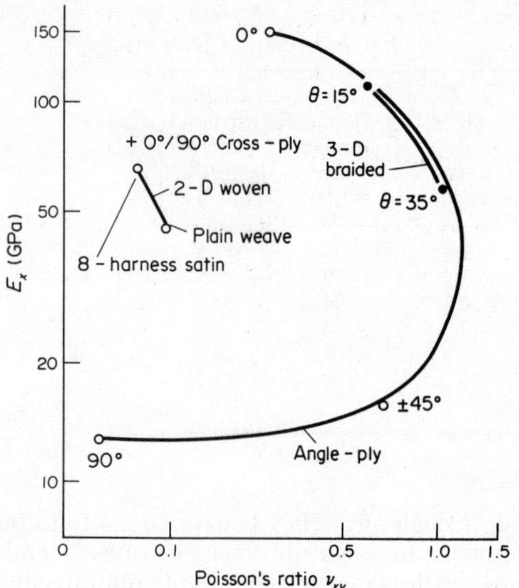

Figure 3
Variation of longitudinal Young's modulus E_x with Poisson's ratio for C–epoxy composites

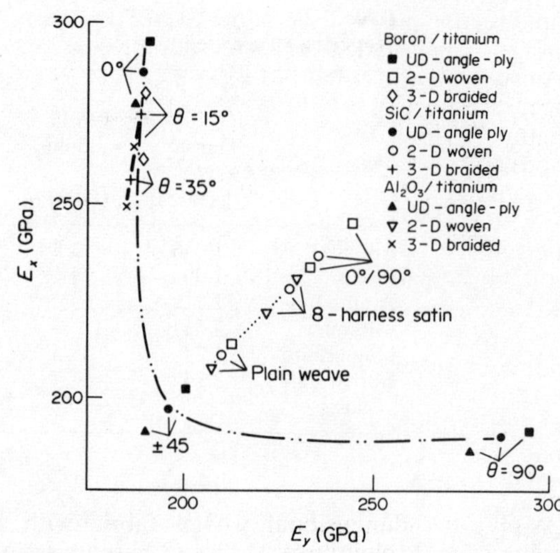

Figure 5
Variation of longitudinal Young's modulus E_x with transverse Young's modulus E_y for B–Ti, SiC–Ti and Al_2O_3–Ti composites

on fiber orientation. Three-dimensionally braided composites have demonstrated good in-plane properties, which are comparable to those of unidirectional angle-plies with the same range of fiber orientation. The longitudinal Young's moduli and in-plane shear rigidities of 3-D braided composites with braiding angles ranging from 15° to 35° are better than those of 2-D woven-fabric composites. However, the transverse Young's moduli are lower and the major Poisson's ratios higher than those of 2-D woven-fabric composites. It is a unique capability of 3-D braided composites that they can provide both stiffness and shear rigidity along the thickness direction. Also, because of the integrated nature of fiber arrangement, there are no interlaminar surfaces in braided composites, such as would be found

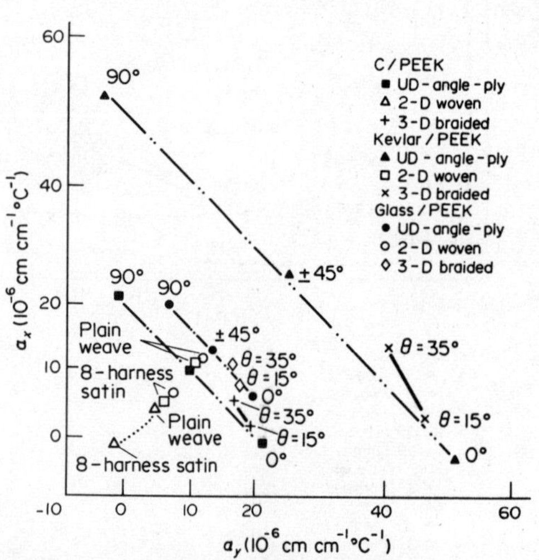

Figure 4
Variation of longitudinal thermal expansion coefficient α_x with transverse thermal expansion coefficient α_y for C–PEEK, Kevlar–PEEK and glass–PEEK composites

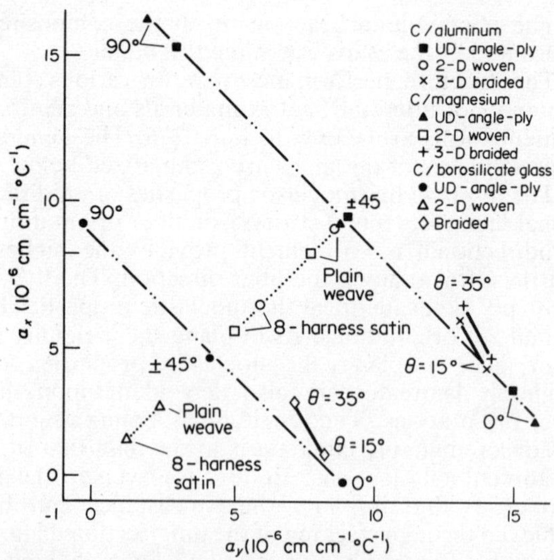

Figure 6
Variation of longitudinal thermal expansion coefficient α_x with transverse thermal expansion coefficient α_y for C–aluminum, C–magnesium and C–borosilicate-glass composites

in unidirectional and 2-D fabric laminate composites. The elimination of interlaminar failure is responsible for the enhanced damage tolerance and impact resistance in 3-D braided composites.

Figure 2 shows the variation of E_x with E_y for composites of carbon, Kevlar and glass-fibers in epoxy matrix. The variation of the major Poisson's ratio with various forms of reinforcement is demonstrated in Fig. 3 for carbon/epoxy composites. The lateral contractions under axial loading of 3-D braided composites are relatively higher. A unique characteristic of Kevlar and graphite fibers is their negative coefficients of thermal expansion along the axial direction. The thermal expansion coefficient map for polyether ether ketone (PEEK) matrix composites is shown in Fig. 4. Composites with zero thermal expansion and hence dimensional stability can be achieved by suitable material selection, geometric design and fiber hybridization.

A wide variety of metal-matrix materials has been adopted, ranging from low melting point, lightweight metal alloys such as those of aluminum and magnesium to various superalloys for application at elevated temperature. The variation of E_x with E_y for titanium matrix composite is shown in Fig. 5. The high specific strength and high resistance to corrosion, as well as low thermal expansion coefficient, make titanium-based composites attractive for aerospace applications.

Figure 6 compares the thermal expansion coefficients α_x and α_y for carbon-fiber-reinforced aluminum, magnesium and borosilicate glass matrices. The low thermal expansion characteristic of borosilicate glass composite is evident.

The advances in analytical modelling have laid the foundation for the understanding of structure–performance relationships of laminated and fabric-based composites. Maps of other mechanical and physical properties of various advanced composites can be constructed.

See also: Elastic Properties of Laminates

Bibliography

Ashby M F, Gandi C, Taplin D M R 1979 Fracture-mechanism maps and their construction for FCC metals and alloys. *Acta Metall.* 27: 699

Chou T-W, Kelly A, Okura A 1985 Fiber-reinforced metal-matrix composites. *Composites* 16: 187

Chou T-W, Yang J M 1986 Structure–performance maps of polymeric, metal and ceramic matrix composites. *Metall. Trans.* 17A: 1547–59

Frost H J, Ashby M F 1982 *Deformation-mechanism Maps*. Pergamon, Oxford

Gandi C, Ashby M F 1979 Fracture-mechanism maps for materials which cleave: FCC, BCC and HCP metals and ceramics. *Acta Metall.* 27: 1565

Ishikawa T, Chou T-W 1982a Stiffness and strength behavior of woven-fabric composites. *J. Mater. Sci.* 17: 3211–20

Ishikawa T, Chou T-W 1982b Elastic behavior of woven-fabric composites. *J. Compos. Mater.* 16: 2–19

Ishikawa T, Chou T-W 1983a One-dimensional micromechanical analysis of woven-fabric composites. *AIAA J.* 21: 1714–21

Ishikawa T, Chou T-W 1983b In-plane thermal expansion and thermal bending coefficients of fabric composites. *J. Compos. Mater.* 17: 92–104

Ishikawa T, Chou T-W 1983c Thermoelastic analysis of hybrid fabric composites. *J. Mater. Sci.* 18: 2260–8

Ishikawa T, Chou T-W 1983d Nonlinear behavior of woven-fabric composites. *J. Compos. Mater.* 17: 399–413

Kelly A 1985 Composites in context. *Compos. Sci. Technol.* 23: 171

Ma C L, Yang J M, Chou T-W 1986 Elastic stiffness of three-dimensional woven-fabric composites, ASTM-STP 893. American Society for Testing and Materials, Philadelphia, PA

Majidi A P, Yang J M, Chou T-W 1986 *Toughness Characteristics of Three-dimensionally Braided Al_2O_3/Al–Li Composites*. The Metallurgical Society, Warrendale, PA

Vinson J R, Chou T-W 1975 *Composite Materials and Their Use in Structures*. Wiley, New York

Yang J M, Ma C L, Chou T-W 1986 Fiber inclination model of 3-dimensional textile structural composites. *J. Compos. Mater.* 20: 472–84

T.-W. Chou
[University of Delaware, Newark, Delaware, USA]

Compton Scattering

The Compton effect describes the inelastic scattering of x rays or γ rays from the electrons in a target material. It has two applications within the area of materials investigation. At a fundamental level, the spectral distribution of the scattered radiation reveals information about the momentum distribution of the electrons in the target. A less profound use follows from the proportionality between the intensity of the scattered signal and the density of electrons from which they are scattered. It results in densitometry and imaging from the backscattered signal in situations where transmission techniques are ineffective for nondestructive examinations.

1. Basic Theory

The scattering interaction is shown schematically in the inset to Fig. 1. Compton's shift formula

$$\Delta\lambda = \frac{h}{mc}(1 - \cos\phi) \tag{1}$$

relates the wavelength shift $\Delta\lambda$ to the scattering angle ϕ where h is Planck's constant, m is the

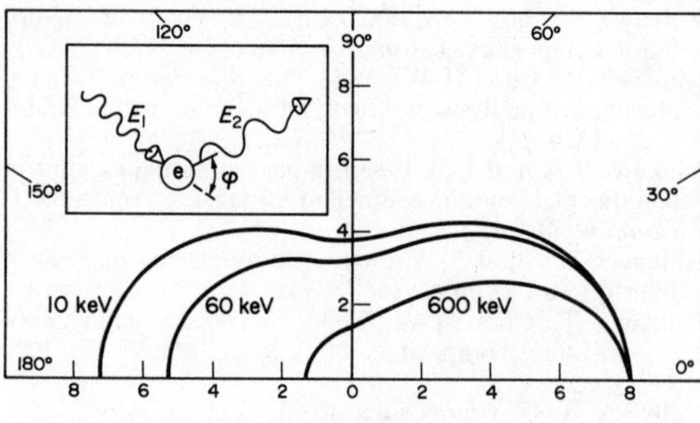

Figure 1
The Klein–Nishina cross section for Compton scattering (as given by Eqn. (3)) plotted as a polar diagram for incident photon energies E_1 of 10, 60 and 600 keV. The energy of the scattered photon E_2 varies with the angle of scattering ϕ according to Eqn. (2)

electron mass and c is the velocity of light. It can also be written more usefully, if less economically, in terms of the photon energies as

$$E_2 = E_1\left[1+\frac{E_1}{mc^2}(1-\cos\phi)\right]^{-1} \tag{2}$$

The differential scattering cross section for an unpolarized beam is given by the Klein–Nishina formula, which can be written as

$$\frac{d\sigma}{d\Omega}=\frac{1}{2}\left(\frac{e^2}{mc^2}\right)^2\left(\frac{E_2}{E_1}\right)^2\left(\frac{E_1}{E_2}+\frac{E_2}{E_1}-\sin^2\frac{\phi}{2}\right) \tag{3}$$

This preserves the symmetry between forward scattering and backscattering predicted by the classical Thomson cross section at low energies, although at higher energies the scattering becomes peaked in the forward direction, as is evident from Fig. 1. The energy loss, given by Eqn. (2), becomes progressively greater as is illustrated in Fig. 2 for a selection of incident energies up to 1 MeV. In the high energy limit, that is, as $E_1 \rightarrow \infty$, the backscattered photon will have an energy of only $\frac{1}{2}mc^2$, which is 256 keV. These fundamental formulae, and others, are detailed by Evans (1958).

The Compton shift formula (Eqn. (1) or (2)) appears to be devoid of interest because it contains no information about the target material. This is because the derivation is based on the assumption of stationary electrons. In reality, electron motion broadens the spectral distribution of the scattered beam appreciably; conduction electrons in a metal travel with speeds of approximately $10^6\,\mathrm{m\,s^{-1}}$ and inner shell electrons in heavy elements are relativistic. This is akin to the Doppler effect, the component of momentum along the x-ray scattering vector (taken as the z axis) causing the additional energy shift. Equation (2) then has an additional term proportional to this component of momentum p_z. Therefore, at any particular scattered beam energy, the line shape is described by the projection of the electron momentum distribution $n(p)$ along the x-ray scattering vector, that is, the probability of observing that component of momentum. This is known as the Compton profile $J(p_z)$ which can be expressed by the equation

$$J(p_z)=\int_{p_x}\int_{p_y} n(p)\,dp_x dp_y \tag{4}$$

It is related in this simple way to the measured spectral distribution when the energy transfer $E_1 - E_2$ clearly exceeds the electron binding energies in the target. This so-called impulse approximation is certainly valid when γ rays with energies greater than 100 keV are backscattered from common elements, but its use is questionable when x-rays from conventional tube sources or synchrotrons (energies less than 30 keV) are used.

The validity of the impulse approximation has rarely been given attention by experimenters since there are more pressing practical problems with

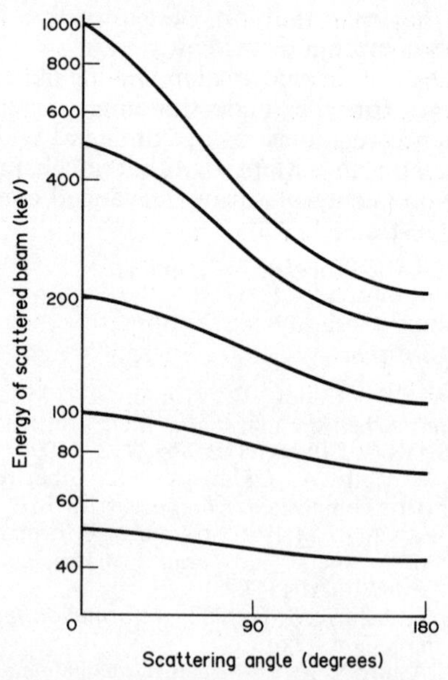

Figure 2
The energy of Compton scattered radiation as a function of the angle of scattering for different incident photon energies. Large energy losses occur for high energy photons at high angles of scattering

softer radiation. These arise from the prohibitive photoelectric absorption in all but the lightest elements (i.e., the first 14 elements) coupled with the fact that the inelastic scattering cross section, described by Eqn. (3) and depicted in Fig. 1, only amounts to a few barns per atom (1 barn = 10^{-28} m^2) whereas the photoelectric absorption cross section may be several orders of magnitude larger at conventional x-ray energies in all but the lightest materials. In a diffraction experiment, the coherence of the Bragg-scattered beam leads to large signals, but in the incoherent Compton scattering process the intensities are much lower; this is the main reason for the comparative neglect of the technique after the pioneering days of DuMond and his co-workers (DuMond 1933). Their measurement of the Compton profile of beryllium provided the first direct evidence for the Fermi–Dirac statistics in an electron gas, but there was little further activity until the 1960s when the work was revived. Recently, line shape analysis has received a fillip from two developments with synchrotron sources. First, high resolution experiments with dispersive crystal spectrometers have become practicable as a consequence of improving flux and brightness and, second, magnetic properties can now be studied because of polarization of the beam.

Despite the weakness of the Compton cross section, the fact remains that it is the dominant interaction in light materials (including biological tissue) at the energies traditionally used for x-ray radiography, which are typically an order of magnitude higher than the energies of the characteristic emissions used in crystallographic studies. This point is emphasized in Fig. 3 where the contributions to the mass attenuation coefficient for concrete are plotted. Scattering as well as absorption processes attenuate the transmitted beam. Both photoelectric absorption and elastic scattering rapidly diminish in importance at higher photon energies; the Compton scattering cross section does not and it becomes the dominant interaction at typical radiographic energies. Thus, Compton scattering is responsible for contrast in many typical radiographic situations and this suggests the direct use of the signal for imaging/densitometry, instead of its indirect use, by virtue of its major contribution to the photon attenuation coefficient.

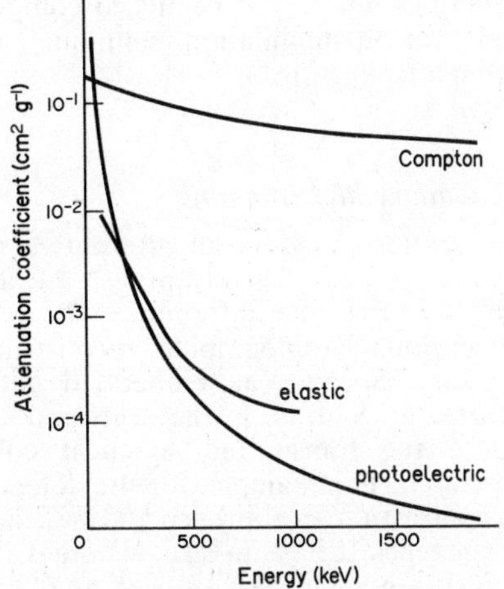

Figure 3
The mass attenuation coefficients for photoelectric absorption, elastic scattering and Compton scattering in concrete as a function of the photon energy. [Source: J H Hubbell (1969) National Bureau of Standards, NSRDS-NBS 29]

2. Line Profile Analysis

This is aimed at providing information about electronic behavior in the bulk material. Defects, inclusions, grain boundaries, impurities and so on have no measurable effect at their normal concentrations. Although the Compton line shape is a projection of the total electron distribution (as shown by Eqn. (4)) it retains a surprising sensitivity to the behavior of the outer, slow-moving electrons involved in solid-state bonding. This research topic has been covered by Williams (1977) and has been updated in review articles (see, for example, Cooper 1985).

Good resolution of the Compton line shape might be thought of as, for example, 10% of the Fermi momentum in a metal. Such resolution has only recently become available from the use of focusing x-ray optics. While the characteristic emission from an x-ray tube anode remained as the source, the flux was too low to make such measurements viable. Instead, the bulk of the work of the 1980s traded resolution for statistical precision by replacing the crystal spectrometer with a semiconductor detector and the x-ray source with a γ-ray isotope emitter of generally higher energy (i.e., greater than 50 keV). In practice this has meant momentum resolution four or five times worse than the target figure mentioned above, but at acceptable count rates (greater than 10^6 counts per day). The favored isotope used has been ^{198}Au with an emission line at 412 keV but with a half-life of less than three days, which perhaps indicates the difficulty of optimizing these experiments. The method has proved a critical test of theories of electrons in solids and has, for example, been used to investigate the behavior of hydrogen in metals and the Coulomb correlations between the electrons in transition metals.

As sensitivity has been compromised by such poor resolution, great emphasis has been placed on the accuracy of the data processing. Fortunately, this will no longer be the case due to the arrival of brighter x-ray synchrotron sources; these make the crystal spectrometer a practical proposition at a respectable resolution that begins to compete with the allied positron annihilation technique (see West 1973 for an introduction).

3. Densitometry and Imaging

Compton scattering offers an alternative to x-ray transmission methods. Its advantages are apparent from Fig. 4. First, the information is essentially three-dimensional; tomographic reconstruction is not necessary. Second, an embedded defect can produce greater contrast in the scattered signal as opposed to the transmitted signal if collimated beams are used. For example, for the detection of a void the contrast C can approach 100% if the size of the void matches the beam size, whereas in transmission it depends on the void size as $C=\mu l$ for a small void. If sensitivity S is defined as the modulation of the signal (scattered or transmitted) by the defect, divided by the statistical noise in the measurement, it can be shown that $S_c/S_t \simeq (\mu L)^{-1/2}$ which implies that the method is competitive for the inspection of thin materials or of near-surface regions of thick objects.

These advantages, which have been summarized by Harding (1985) and Holt (1985), have to be set against adverse considerations of flux; for example, the inspection time for collimated beam imaging is inversely proportional to the square of the volume defined by the incident and scattered beams. Nevertheless, it is an alternative when the object to be inspected is so large or so complex that transmission methods cannot be used.

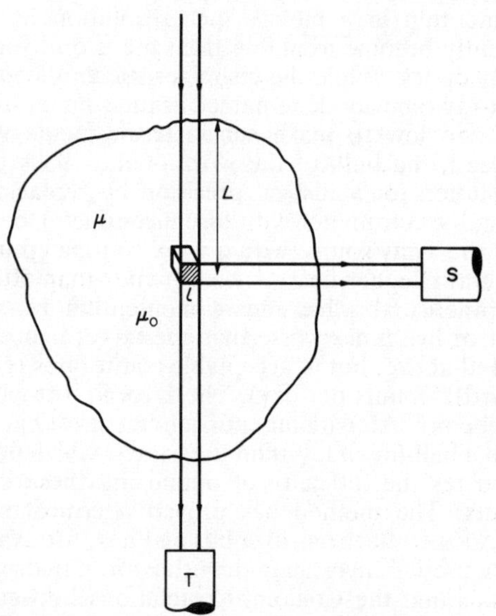

Figure 4
A comparison of scattering (detector at S) and transmission (detector at T) geometries for the detection of a void or inclusion of linear dimension l and attenuation coefficient μ_0 embedded to a distance L in a matrix characterized by an attenuation coefficient μ

The densitometry of light material enclosed in dense material and the imaging of near-surface defects are well suited to a scattering approach. In this category, the characterization of osteoporotic bone and the inspection of explosive fillers in munitions offer two diverse illustrations of the same principle: that real time measurements can be competitive with, or superior to, the traditional approaches. The use of intense white-beam bremsstrahlung sources such as LINACS or high-voltage x-ray units coupled with multidetector configurations has led to the development of practical systems for backscatter imaging of near-surface defects in alloy castings (Stokes et al. 1982), fiber-reinforced composites, air-frame components and fastenings. Recent developments have shown that high-voltage x-ray systems with scanning collimators and multiple detectors can be used to inspect areas of approximately 10^4 mm^2 down to depths of approximately 5–10 mm in scan times of around a minute with a spatial defect resolution approaching 0.5 mm. The effective depth of penetration depends on the tube voltage and the target material, but typical figures of merit indicate that the technique can be applied effectively to a wide range of problems in nondestructive evaluation (Harding and Kozanetzky 1987, 1989) and some commercial development is under way. In many of these cases other superficial techniques (ultrasound, eddy currents, etc.) are not viable.

Bibliography

Cooper M J 1985 Compton scattering and electron momentum. *Rep. Prog. Phys.* 48: 415–81

DuMond J W M 1933 The determination of electron momentum. *Rev. Mod. Phys.* 5: 1–33

Evans R D 1958 The Compton effect. In *Corpuscles and Radiation in Matter II*, Handbuch der Physik, Vol. 34. Springer, Berlin, pp. 218–97

Harding G 1985 X-ray scatter imaging in NDT. *Int. Adv. NDT* 11: 271–95

Harding G, Kosanetzky J 1987 Status and outlook of coherent x-ray scatter imaging. *J. Opt. Soc. Am.* 933–44

Harding G, Kosanetzky J 1989 Scattered x-ray beam NDT. Proc. ISRP-4. *Nucl. Instrum. Methods* A280: 517–28

Holt R S 1985 Compton imaging. *Endeavour* 9: 97–106

Stokes J A, Alvar K R, Corey R L, Costello D G, John J, Kocimski S, Lurie N A, Thayer D D, Tripe A P, Young J C 1982 Some new applications of collimated photon scattering for nondestructive examination. *Nucl. Instrum. Methods* 193: 261–7

West R N 1973 Positron annihilation. *Adv. Phys.* 22: 263–382
Williams B G 1977 (ed.) *Compton scattering*. McGraw-Hill, New York

M. J. Cooper
[University of Warwick, Coventry, UK]

Computer Simulation of Microstructural Evolution

Prediction of the microstructure of a material as a function of processing conditions is a major objective of materials science. Since microstructure may be defined as the spatial arrangement of defects, the existence of microstructure (e.g., the presence and arrangement of grain boundaries, dislocations, etc.) is a consequence of the nonequilibrium nature of technologically interesting materials. This nonequilibrium feature of microstructure leads naturally to the well-known dependence on processing history. Therefore, the understanding of microstructure and processing conditions are, by necessity, intimately entwined.

A further difficulty in describing the formation and evolution of microstructure is the complexity of the arrangements of, and interactions between, the microstructural subconstituents (the grain structure of a polycrystalline material is a good example of this type of complexity). In all but the simplest microstructures, grain boundaries join together to form an intricate, topologically connected network. These networks are far from random: in addition to fulfilling simple topological rules (e.g., Euler relations), such a network has a well-defined grain-size distribution, a distribution of grain faces and edges, angles at which pairs of boundaries meet, and so on, that depend on the physics of the process controlling the dynamics of grain-boundary motion. Two very similar classes of microstructures—grains and soap bubbles—have rather different network properties. Historically, many of the difficulties in understanding such microstructures and their evolution may be traced to problems in visualizing and describing such complex structures.

Until recently, methods which attempted to describe such simple processes as normal grain growth had to simplify greatly either the complexity of the microstructure or the physics which led to its evolution. The first class of simplifications consists of those approaches which may be collectively referred to as mean-field; that is, the actual microstructure is replaced by a single 'typical' grain responding to the average interactions of all other grains in the system. While such approaches have had some limited success in describing a proportion of the cruder features of the microstructure, in general they do not lead to realistic microstructures or accurately predict the distributions of the grain size or topological parameters. The second class of simplifications consists of those which attempt to describe microstructures in terms of simple geometrical models. These include the use of Dirichilet cells or Voronoi polyhedra based on a random distribution of points to create grain structures. Such methods are limited to describing special types of structures and processes (e.g., primary recrystallization in the absence of any grain growth) and have only a minimal physical basis.

In principle, computer simulation provides an ideal method for attacking the complexity inherent in microstructural problems. The major uses of computer simulation in materials science may be divided into three general classes: continuum, atomistic and electronic. Electronic structure calculations are either performed for small clusters of atoms (<100) or for perfect crystals where the number of atoms employed in the calculation is equal to the number of atoms in a unit cell. Atomistic calculations typically employ between 10^2 and 10^4 atoms. While these numbers are increasing with improvements in available computational techniques and computer technology, such simulations cannot be performed on a scale sufficiently large to include much more than a single isolated microstructural feature such as grain boundaries or dislocations. For example, in order to simulate one cubic millimeter of a material (a typical microstructural scale) one must account for $\sim 10^{20}$ atoms. This is beyond the current state of the art by approximately sixteen orders of magnitude. Similarly, dynamical atomistic simulations (e.g., molecular dynamics) are typically performed for nanoseconds while microstructural events occur on time scales ranging from microseconds to hours. This discrepancy is in the range of 10^3–10^{13}. Because of the disparity between the size and time scales on which significant microstructural complexity exists and the inherent size–time limitations of atomistic simulations, microstructural simulation must operate on a nonatomistic basis.

Most simulations of microstructural evolution are based on continuum models (see Srolovitz 1986). These continuum models are generally solved by either (a) numerical integration of the equations of motion describing the evolution of individual defects or defect segments, or (b) Monte Carlo methods which are based upon kinetic rate theory and a description of the energy of the microstructure in terms of the location of the defects. Application of well-established methodologies such as finite-element and finite-difference methods to such problems as plasticity and solidification are not discussed in this article since they have been reviewed well elsewhere (see *Solidification and Casting: Computer Simulation*, Suppl. 1).

1. The Monte Carlo Method

1.1 General Features

The Monte Carlo method (see Binder 1979) employs random numbers to evaluate the probability with which the various configurations of the system occur. The physics of any problem under consideration is contained in the expression for this probability and the variables on which it depends. Since the equilibrium configuration of the system is generally unknown, the probability is commonly written in terms of a transition probability $W(\alpha \to \beta)$ from one configuration of the system, α, to another, β. If the transition probability between states can be identified with a microscopic model, then (with appropriate averaging) the development of the system from its initial configuration to its final state corresponds to the time-dependent approach to equilibrium. In this way the kinetic evolution of complex systems can be modelled.

Two types of transition rates are commonly employed:

$$W(\alpha \Rightarrow \beta) = \begin{cases} \frac{1}{\tau} \exp(-\Delta H/kT) & \text{if } \Delta H > 0 \\ \frac{1}{\tau} & \text{otherwise} \end{cases} \quad (1)$$

and

$$W(\alpha \Rightarrow \beta) = \frac{1}{\tau} \frac{\exp(-\Delta H/kT)}{1+\exp(-\Delta H/kT)} \quad (2)$$

where τ is a positive constant with dimensions of time, $\Delta H = H_\alpha - H_\beta$ is the difference in energy between configurations β and α, and kT is the thermal energy. Equation (2) is generally employed with conserved (Kawasaki) dynamics and is used in cases such as diffusion, where the total number of each species is conserved. Equation (1) is used with nonconserved (Glauber) dynamics and is employed in cases such as grain growth, where the crystallographic orientation of a part of the solid may change. The configuration of the system approaches equilibrium as the number of transition attempts (using Eqn. (1) or (2)) becomes large.

The connection between the Monte Carlo method and real kinetic processes can be demonstrated by simple rate theory. For example, in the case of vacancy diffusion, the net flux of vacancies J from site α to site β is given by

$$J(\alpha \Rightarrow \beta) = \frac{1}{\tau} A\{\exp(-Q/kT) - \exp[-(\Delta H + Q)/kT]\} \quad (3)$$

where A is a positive constant, $1/\tau$ is the attempt frequency and Q is the activation energy. In Eqn. (3) the back jump rate (i.e., from β to α) has been subtracted out. Normalizing Eqn. (3) yields the transition probability

$$W(\alpha \Rightarrow \beta) = \frac{1}{\tau'} \left[\frac{\exp(-\Delta H/kT)}{1+\exp(-\Delta H/kT)} \right] \quad (4)$$

where an effective attempt rate τ' has been defined as $\tau[\exp(Q/kT)]/A$. The final expression of Eqn. (4) is equivalent to the conserved-dynamics Monte Carlo transition rate of Eqn (2). This equivalence establishes the similarity between the Monte Carlo and rate theory approaches.

Q has been assumed to be a constant. This is clearly not true for cases where the diffusing vacancy samples different atomic configurations (e.g., regions adjacent to defects, strained regions, etc.). For atomic systems in general, Q will not be a constant and hence the kinetic (rate theory) view of the Monte Carlo method is inappropriate. However, the simulations described below are not atomistic, but are instead applications of the Monte Carlo technique to continuum systems. This system is represented in a discrete form in which all sites are equivalent, hence the assumption of a constant Q is appropriate. As in the rate theory description of the diffusion process, τ scales the time in the Monte Carlo simulations (the number of Monte Carlo transition attempts is proportional to time). Therefore, provided that Q is known, the relationship between the number of Monte Carlo attempts and real time is specified.

1.2 Grain Growth

In order to simulate microstructural evolution, a model must be identified that describes the appropriate microstructural features. As a simple example, consider the case of normal grain growth (i.e., grain-size evolution driven by grain-boundary curvature) (Anderson et al. 1984). To incorporate the complexity of grain-boundary topology, the microstructure is mapped onto a discrete lattice (Fig. 1). Each lattice site is assigned a number between 1 and Q corresponding to the orientation of the grain in which it is embedded. The value of Q must be chosen to be sufficiently large that impingement of grains with like orientation is infrequent. A grain-boundary segment is defined to lie between two sites of unlike orientation. A site which is surrounded by others of like orientation is in the interior of a grain. The Hamiltonian describing the interaction between

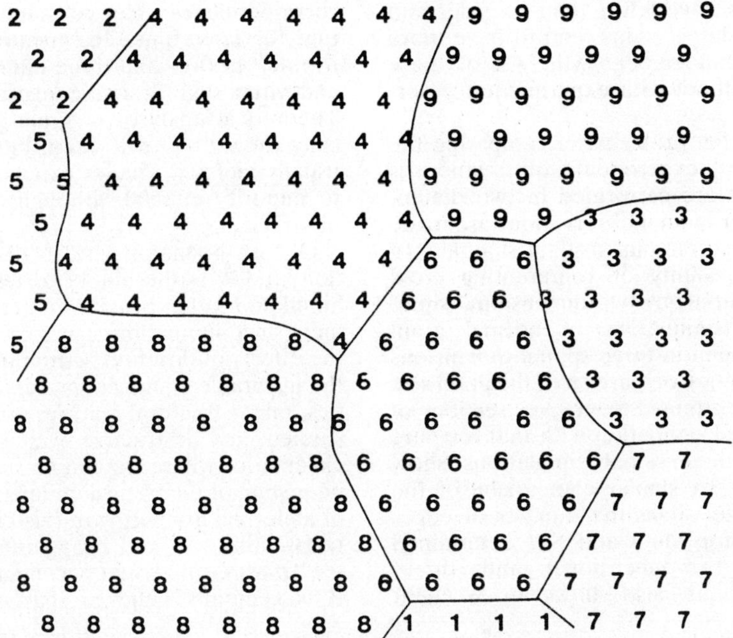

Figure 1
Sample microstructure on a triangular lattice where the integers denote local crystallographic orientation and the lines represent grain boundaries

the lattice sites is

$$H = -J \sum_{\text{nn}} (\delta_{S_i S_j} - 1) \tag{5}$$

where S_i is one of the Q possible grain orientations associated with site i and δ_{ab} is the Kronecker delta. The sum is taken over all nearest-neighbor (nn) pairs of sites. Thus, nearest-neighbor pairs contribute J to the system energy when they are of unlike orientation, and zero otherwise. Therefore, the grain boundary energy is proportional to J. This model is known as the Q-state Potts model in statistical-mechanics literature.

The kinetics of boundary motion are simulated by employing the Monte Carlo method described above. Since grain orientations are not conserved during the process of grain growth, nonconserved dynamics are employed. In the standard Monte Carlo method, a lattice site and a new trial orientation are chosen at random, the latter from one of the other Q-1 possible orientations. The change in the energy of the system due to the reorientation of the site is calculated via Eqn. (5). The probability that the new trial orientation is accepted is τW, where W is defined as in Eqn. (1). Successful transitions of sites at grain boundaries to orientations of nearest-neighbor grains correspond to grain boundary migration. This leads to a description of boundary motion which is equivalent to that derived from classical reaction-rate theory.

In the simulations described below, the system is started at a temperature well in excess of the melting or critical temperature T_c and rapidly quenched to $T \ll T_c$. To reduce finite size effects, very large systems with periodic boundary conditions are usually employed: 40 000 sites on triangular lattices or 10^6 sites on a cubic lattice are typical. The unit of time is defined as 1 Monte Carlo step (MCS) which corresponds to N microtrials or reorientation attempts, where N is the total number of sites in the system. Variations of the procedure are often employed to increase the efficiency of the simulation method.

The main result of any study of microstructural evolution is the actual microstructure: all other information may be extracted from this and its development over time. Figure 2 shows a representative series of micrographs representing different times in a two-dimensional normal grain-growth simulation. The general appearance of the microstructure shows good correspondence with those in polished sections of samples having undergone normal grain growth. More detailed examination shows that grain boundaries tend to meet at 120°;

small grains tend to have fewer than six sides and undergo shrinkage; large grains tend to have more than six sides and undergo growth. All of these results are in agreement with experimental observations.

One of the potential problems of comparing the simulation results with experimental observations is that the simulations are performed in two dimensions while the experimental observations are made on cross sections of three-dimensional samples. In order to test the validity of representing cross sections of real materials by two-dimensional simulations, additional simulations of normal grain growth were performed in three spatial dimensions (Anderson et al. 1989). Comparison of the grain size (R_A) distribution determined from cross sections of the three-dimensional simulation with that for pure iron and the two-dimensional simulations show excellent agreement. A similar comparison of the normalized grain size versus the number of edges per grain (Fig. 3) for MgO and Sn, determined experimentally in two-dimensional and three-dimensional simulations, also shows an excellent correspondence. Recent simulation results show that for later times the mean grain size grows as $R = At^{1/2}$ in two- and three-dimensional simulations, and with a slightly smaller exponent at earlier times. Theoretical analyses of grain-growth kinetics typically show $t^{1/2}$ growth laws, but experimental work usually yields a smaller exponent (presumably due to impurity effects) although $t^{1/2}$ growth has been observed.

One of the major benefits of this class of simulation studies is the ability to test theoretical ideas. Simulation often provides a better test than experiment since simulations can be performed to examine the effects of different assumptions directly without the inevitable complications associated with impurities, microstructural features other than those being studied, and destructive evaluation techniques. An example of where simulation has demonstrated the poor state of theoretical understanding is in the area of abnormal grain growth (also known as secondary recrystallization and exaggerated grain growth)—see *Grain Growth and Secondary Recrystallization*. It is generally believed that abnormal growth is

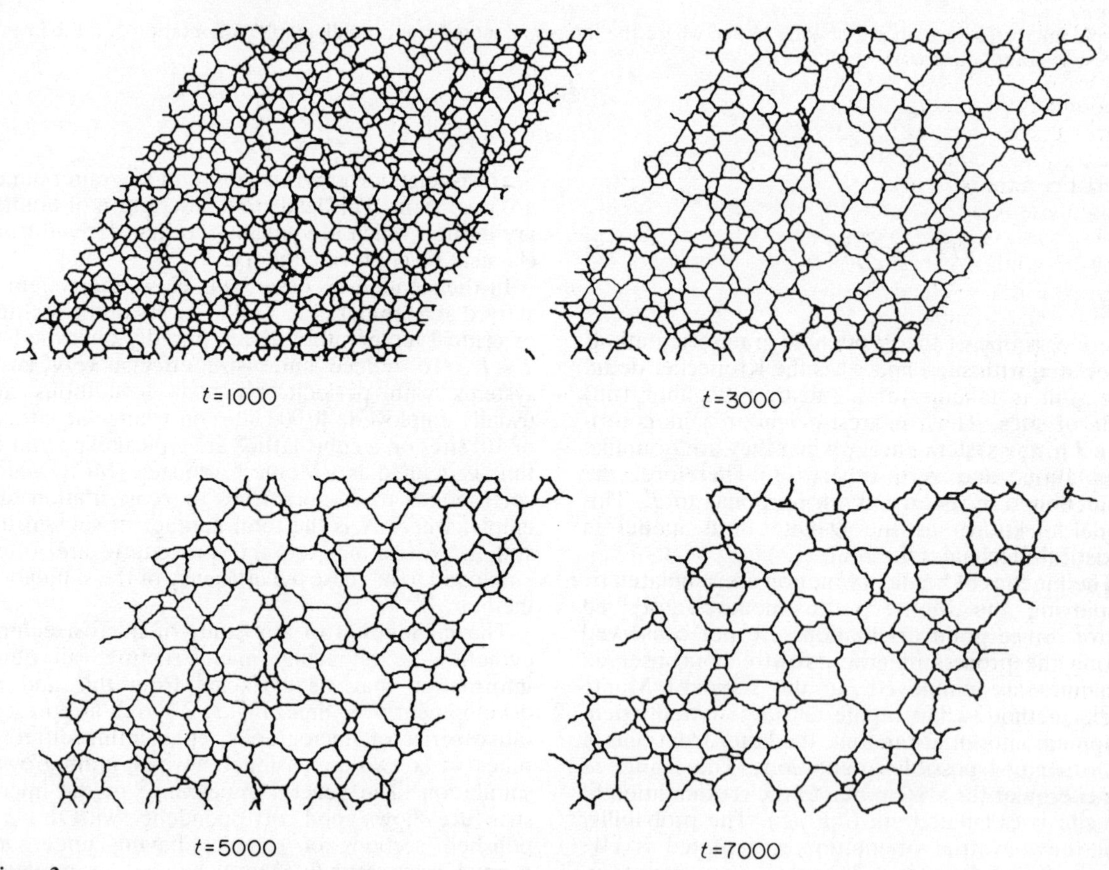

Figure 2
The temporal evolution of a typical microstructure undergoing normal grain growth

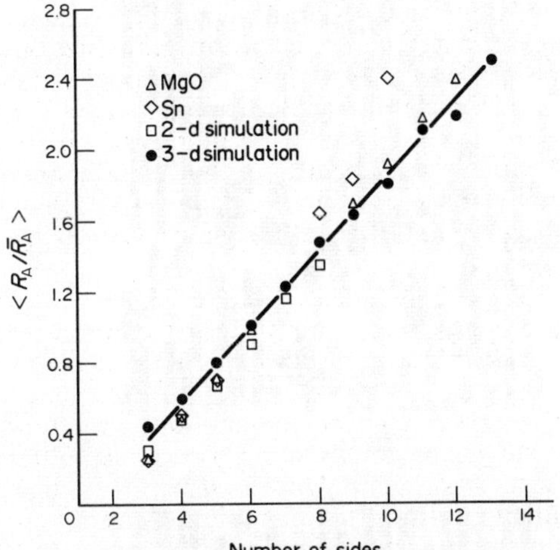

Figure 3
Relationship between the number of sides exhibited by grains when viewed in cross section and the average grain radius determined from the cross-sectional area for the given number of sides

associated with the presence and then dissolution of second-phase particles. Simulations have shown, however, that texture effects and the related grain-boundary mobility anisotropy can lead to abnormal growth (Rollett et al. 1989). A relatively strong texture is a common by-product of processing material containing particles. These texture-controlled simulations lead to microstructures and the presence of a texture change consistent with experimental observations. A theoretical framework for abnormal growth has evolved from the knowledge derived from these simulations.

Generalizations of the grain growth model to include primary recrystallization are made by adding a nucleation mechanism and by the addition of an energetic term which accounts for the energy stored in a material in the form of dislocations (Srolovitz et al. 1986). Primary recrystallization has been simulated as a function of the degree of stored energy, the density of nuclei, the mode of nucleation (site saturated or constant rate), inhomogeneously stored energy (e.g., due to inhomogeneous deformation) and the density of second-phase particles. In addition to microstructures and their temporal evolution, such simulations yield recrystallized fraction as a function of time, Johnson–Mehl–Avrami (JMA) exponents (which indicate the nucleation and growth mode), grain size versus time, grain-size distributions, and so on. In the simplest cases (uniform stored energy, no particles) the simulations yield results that are consistent with theory. However, when the stored energy is nonuniformly distributed through the material (varies on a grain by grain basis) the JMA exponents are much lower than the theoretical values. Such low exponents are the rule in experimental studies of primary recrystallization (see *Recrystallization of Deformed Metals (Primary)*).

Applications of this type of model to other types of microstructures and their evolution require suitable modification of the energetics expression, as in Eqn. (5), to account for the different possible driving forces, and also a suitable discretization of the microstructure. To date, this type of model has been applied to normal grain growth, grain growth in the presence of a dispersed second phase, abnormal grain growth, primary recrystallization, recrystallization in the presence of a dispersed second phase including particle stimulated nucleation, and dynamic recrystallization. Combining this simulation method with a Monte Carlo diffusion technique (based on Kawasaki dynamics and Eqn. (2)) yields a method appropriate for diffusion-limited problems. This hybrid technique has been applied to the study of boundary migration and sintering controlled by a diffusing impurity. Figure 4 shows a late-stage sintering microstructure produced using such a technique. If the Monte Carlo diffusion mechanism is coupled with an energy expression which accounts for both elastic and surface energy, the coarsening of precipitates in a stress field can be simulated. Figure 5 shows the microstructural evolution of the two phase structure (γ and γ') typical of nickel-based superalloys under the influence of an applied stress (see Gayda and Srolovitz 1989). Such a stress field biased γ–γ' morphology is typically observed (see *Phase Equilibria in Alloys: Influence of Stress*, Suppl. 2).

1.3 Ostwald Ripening

This is the process of diffusional coarsening of second-phase particles. Application of the Monte Carlo method to Ostwald ripening is accomplished by assigning $S_i = \pm 1$ to each site on a lattice; $S_i = 1$

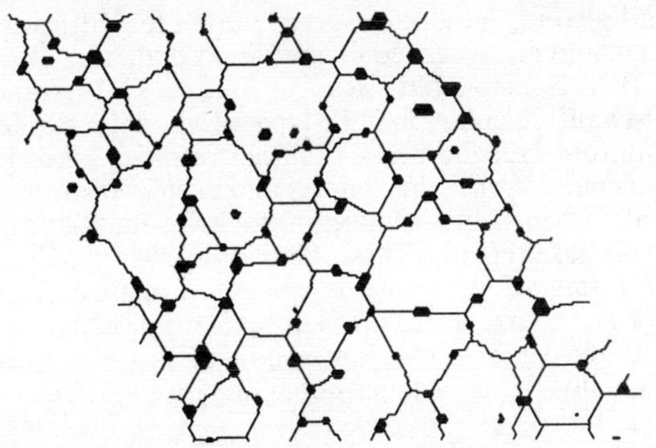

Figure 4
A late stage sintering microstructure: the black regions are voids, the white ones are grains and the lines represent grain boundaries

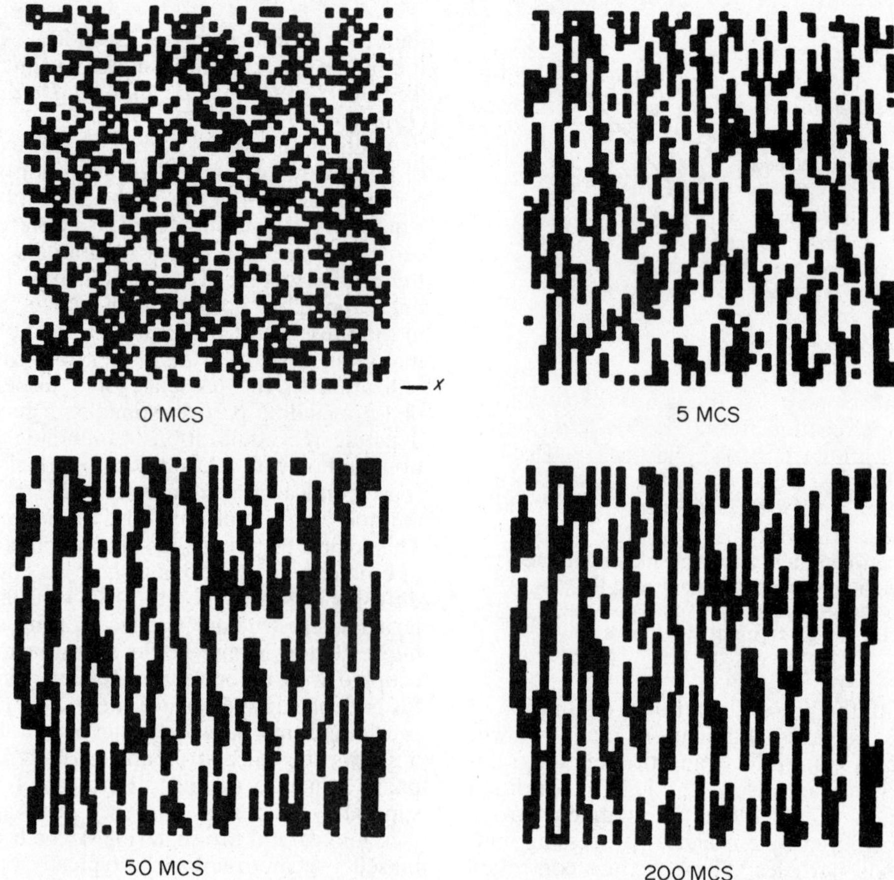

Figure 5
The temporal evolution of a two-phase (misfitting, coherent) microstructure under the influence of a tensile stress applied vertically

corresponds to matrix sites and $S_i = -1$ corresponds to sites occupied by the second phase. A Hamiltonian similar to that given in Eqn. (5) is employed. This model corresponds to the classical Ising model. Since the total amount of both matrix and second phase is conserved during the diffusional coarsening, conserved or Kawasaki dynamics (Eqn. (1)) are employed. Rao et al. (1976) studied Ostwald ripening in this type of model at a few different second-phase concentrations following quenches from the high-temperature disordered state (i.e., solid solution) into lower-temperature two-phase regions. These simulations were analyzed in terms of the time-dependent structure factor $S(k,t)$, the size distribution of second-phase particles and the surface-to-volume ratio of these particles (actually the two-dimensional analog perimeter-to-area ratio).

The microstructure resulting from these simulations consists of both well-defined second-phase particles containing many $S=-1$ sites and a large number of isolated $S=-1$ sites (i.e., surrounded by $S=1$ sites). These small "particles" are the main diffusing species. The densities of these single sites agree with calculations of the "vapor pressure" of the larger second-phase particles which have a finite radius. In addition to the diffusion of single second-phase sites, the larger particles were also seen to migrate, although at a much slower rate than the single sites, as expected. Due to the limited simulation statistics, it was not possible to make an accurate comparison between the simulation and predicted particle size distribution functions. Analysis of the simulated coarsening kinetics showed that the mean particle size increased in time as t^{α}, where $\alpha \simeq 0.2$. This contrasts with the classical Lifshitz–Slyozov–Wagner prediction of $\alpha = 1/3$. However, more recent calculations on a variant of this model (Sadiq and Binder 1983) yielded $\alpha = 0.35$. Recent experience with Monte Carlo models of this general class have shown that power-law kinetics with exponents smaller than theoretically predicted are common, but a crossover to the theoretical exponents eventually occurs.

2. Equation of Motion Methods

The equation of motion methods for simulating microstructural evolution can be divided into two classes: front tracking methods and particle methods. Unlike the Monte Carlo method, which simulates the evolution of the entire solid body, the equation of motion methods focus on only those parts of the microstructure that are evolving. By doing this, these methods are able to simulate microstructural evolution far more efficiently.

2.1 Front Tracking

Front tracking methods have found wide application in the physical sciences, including studies of flame fronts, chemical reaction fronts, clouds, fluid flow, shocks, solidification fronts and plasmas (see Bishop et al. 1984). These methods reduce the physical problem to one which may be solved strictly on a front or interface. The true *d*-dimensional (where *d* is the dimension of the body) dynamical problem is generally transformed into a set of partial differential or integro-differential equations which are numerically solved at a finite set of points on the front or interface. As the interface deforms, expands or contracts, the number of points describing the interface is often changed in order to guarantee numerical accuracy. The actual implementation of this method varies widely: the computational techniques involved are described by Wilson et al. (1978). This section describes the application of such methods to two problems in microstructural evolution, namely to the formation of dendrites during solidification, and to grain growth.

(*a*) *Dendritic growth*. This has received a great deal of attention in recent years, due, in part, to the rich patterns formed during the growth process and our inability to describe theoretically the complete dynamic evolution of these structures. When the melt is undercooled (i.e., below the melting point T_m), the solid–liquid interface is unstable and dendrites form (see *Dendritic Solidification*). The rate-limiting step in such a process is the diffusion of the (latent) heat away from the solidification front. Such a process may be modelled by the (thermal) diffusion equation in the entire body plus an equilibrium condition for the interface. This condition accounts for the melting temperature and its variations with curvature (i.e., the Gibbs–Thompson effect).

Recent phenomenological work on dendritic growth (see Brower et al. 1984) based on a set of simplistic assumptions known as the "geometrical model," reduces the problem from the dimensionality of the sample to that of the interface. Application of this model to dendritic growth in two dimensions yields

$$\hat{\boldsymbol{n}}\cdot\frac{d\boldsymbol{x}}{dt}=\left(\kappa+A\kappa^2-B\kappa^3+\frac{\partial^2\kappa}{\partial s^2}\right)[1+\varepsilon\cos(m\theta)] \qquad (6)$$

In this equation, the interface is represented by the time evolution of the points $\boldsymbol{x}$ on the surface which is parameterized in terms of its normal $\boldsymbol{n}$, arclength s, curvature κ, and orientation angle θ (with respect to the underlying crystal structure). A and B are physical parameters governing the growth, m is the degeneracy of the underlying crystal symmetry and ε is a measure of the anisotropy of the interfacial properties. The linear term in κ accounts for the Gibbs–Thompson effect, the κ^2 term for the undercooling of the melt, the κ^3 term for the finite solid nucleation barrier, and the surface Laplacian term accounts for diffusion. The square bracketed term allows for variations in the interfacial properties with the interface orientation, with respect to the underlying crystal structure.

If the right hand side of Eqn. (6) is denoted by $F(\theta)$, then this equation becomes

$$\left.\begin{aligned}\dot{\theta}&=-\frac{\partial F}{\partial s}\\ \dot{s}&=\int_0^s\kappa(s')F[\theta(s')]\,ds'\end{aligned}\right\} \qquad (7)$$

where both θ and s depend on some time-independent parameterization of the interface (which is assumed to be a closed curve). It is convenient to introduce a relative arclength parameterization $\alpha=s/s_T$ where s_T is the total length of the interface. Equations (7) can then be rewritten as

$$\left.\begin{aligned}\frac{\partial\theta(\alpha)}{\partial t}&=-\frac{1}{s_T}\frac{\partial F(\theta)}{\partial\alpha}\\ &\quad-s_T\left(\int_0^\alpha\kappa F(\theta)\,d\alpha'-\alpha\int^1\kappa F(\theta)\,d\alpha'\right)\\ \frac{\partial s_T}{\partial_t}&=s_T\int_0^1\alpha F(\theta)\,d\alpha\end{aligned}\right\} \qquad (8)$$

where $\kappa=(1/s_T)\partial\theta/\partial\alpha$. Discretization of Eqns. (8) yields a coupled set of ordinary differential equations which can then be integrated using a predictor-corrector ordinary differential equation solver. Finally, the physical interface is reconstructed by integrating

$$\left.\begin{aligned}x(\alpha)&=\int_0^\alpha\cos[\theta(\alpha')]\,d\alpha'\\ y(\alpha)&=y_0-\int_0^\alpha\sin[\theta(\alpha')]\,d\alpha'\end{aligned}\right\} \qquad (9)$$

Figure 6 shows a typical dendrite simulated using this method with 300 points and the parameters

$m=4$, $A=4$, $B=2$, and $\varepsilon=0.15$. The resultant dendritic structure exhibits a four-fold symmetry, side-branches and overall shows an excellent correspondence with experimentally observed dendritic morphologies. Furthermore, this class of simulations has led to a theoretical understanding of open questions such as what determines the growth velocity.

(*b*) *Grain growth*. Front tracking methods have recently been applied to the study of grain growth (see Ceppi and Nasello 1984, Frost et al. 1988). In this case the partial differential equation describing the temporal evolution of the grain boundaries is quite simple. Grain boundaries are assumed to move according to $v_n=\mu\kappa$, where v_n is the velocity of a boundary segment normal to itself, κ is the local boundary curvature and μ is a constant "mobility". An initial two-dimensional structure is constructed, integration points are distributed along the grain boundaries, and each point is moved along the local boundary normal by $\Delta t\mu\kappa$ during each integration time step Δt. Since three boundaries tend to meet at 120° triple points, the positions of the triple points must also be updated. This is accomplished by moving them to such a position, with respect to the three boundaries, that the 120° condition is satisfied. The boundary-migration and triple-point update steps are iterated alternately. Topological changes, such as the shrinking away of a small grain or changes in the number of edges of a grain by vertex impingement, are accounted for by reconnecting the local network in the vicinity of the appropriate vertex.

This method has been applied to both normal and abnormal grain growth. In both cases calculations yield reasonable microstructures, and kinetics are in good agreement with experiment and with those found in the Monte Carlo simulations. However, since this method only requires points on the grain boundaries, the number of points needed in the integration is much less than that required in the Monte Carlo simulations and so it is, in principle, more efficient.

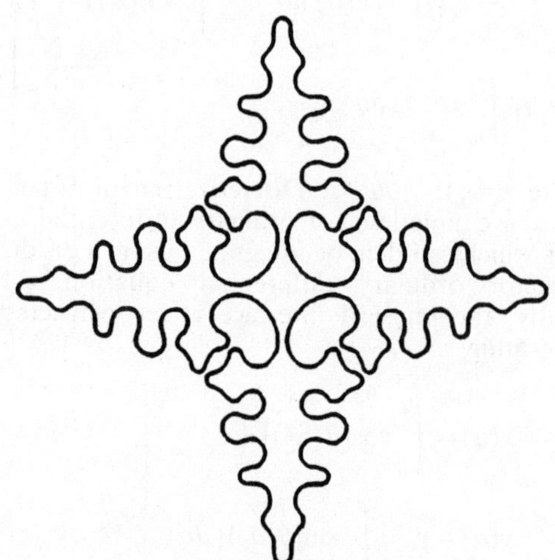

Figure 6
A two-dimensional dendrite simulated within the framework of the geometrical model

(*c*) *Ostwald ripening*. Front tracking methods have also been applied to Ostwald ripening. Voorhees et al. (1988) have developed a theoretical framework which results in an equation of motion for the surfaces of diffusionally interacting second-phase particles. Their method is based on a boundary-integral technique for the solution of the diffusion problem for arbitrary particle shapes. The resultant integral equations were discretized in terms of mesh points on the particle surfaces and integrated using a geometrically convergent iterative procedure. The time evolution of the Ostwald ripening particle shapes was simulated for 2–16 initially circular particles.

These simulations showed a number of interesting features. First, diffusional interactions led to significant distortion of particle shapes, sometimes resulting in the relatively flat particle surfaces often observed in liquid phase sintering. Second, particle migration was observed over distances larger than the initial particle radius. Third, particles that shrink on one side but grow on another were observed (this leads to both particle migration and shape change). Related models which account for stress effects have been shown to lead to the development of spatial correlations between particles during Ostwald ripening (Voorhees and Johnson 1988).

2.2 Particle Methods

Particle methods are very similar to the front tracking methods described in Sect. 2.1. Instead of concentrating on interfaces, however, they focus on the evolution of special points in the microstructure such as the triple points along which grain boundaries meet, or individual defects. The evolution of such points is determined by the solution of their local equations of motion. This section describes the application of such methods to two problems in microstructural evolution; namely, to the evolution of dislocation substructure and to grain growth.

(*a*) *Dislocation substructure*. Dislocations interact by way of their elastic stress fields. Since the force one dislocation exerts on another is well known, it is possible to simulate the evolution of a large number of interacting dislocations (see Ghoniem and Amodeo 1988). However, in order to make the model tractable, it is necessary to restrict attention to the relatively simple case of straight parallel dislocations (i.e., two dimensions with zero-dimensional dislocation lines). The equations of motion of such dislocations are dominated by the observed linear dependence of dislocation velocity

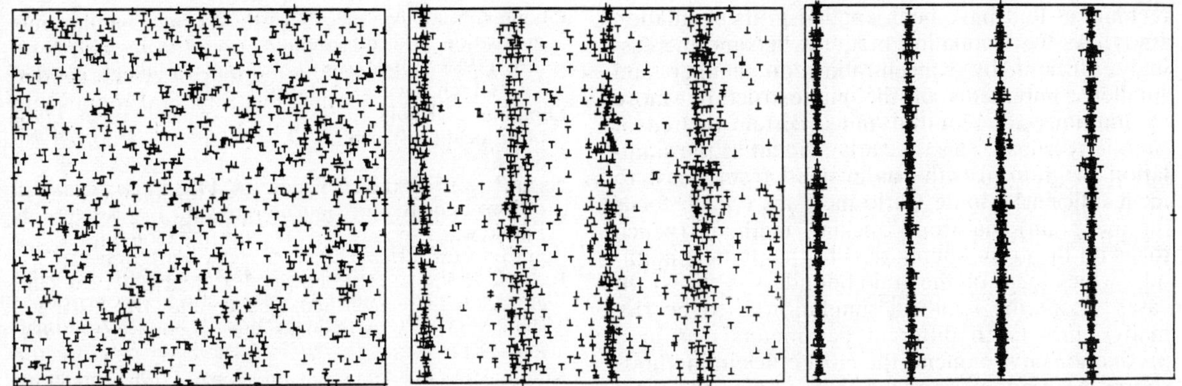

Figure 7
The evolution of a microstructure containing 10^3 dislocations distributed on parallel slip planes: the $\perp$ and $\top$ symbols indicate edge dislocations with Burgers vectors of opposite signs

on driving force; however, for fast-moving dislocations, dislocation momentum may be important. The resultant equation of motion may be written as

$$m^* \ddot{x}_i = F_{\mathrm{ext}} - \Gamma \dot{x}_i + \sum_{j \neq i} F^j_{\mathrm{int}} \qquad (10)$$

where x_i is the coordinate of dislocation i, dots indicate differentiation with respect to time, m^* is the effective dislocation mass, F_{ext} is the force on dislocation i in the x direction due to an external stress, Γ is the damping factor, and the last term accounts for the interaction of dislocation i with all other dislocations. Separate equations must be written for glide and climb, and appropriate damping factors must be determined for each. All of the terms in Eqn. (10) can be calculated directly from the elastic theory of dislocations.

Equation (1) can be integrated for each dislocation by taking small finite time-steps. If dislocations of opposite sign become closer than a cutoff distance (defined in terms of the size of the time-step Δt), they are annihilated. The integration of Eqn. (10) is carried out in the same way as the integration of Newton's equation of motion in normal molecular dynamics calculations. However, unlike molecular dynamics, the particles are not atoms but dislocations. Figure 7 shows the temporal evolution of 10^3 dislocations on the same slip system with either positively or negatively signed Burgers vectors. The initial structure is random, but after a short time it forms oppositely signed dislocation walls. The dislocation density, the strength and density of second-phase particles, the applied stress, the temperature (through the relative values of the glide and climb damping) and the available slip systems can be varied independently in these simulations. Variations of this type of model based on different equations of motion and methods of solution have been developed: for example, cellular automata—see Lepinoux and Kubin (1987).

(*b*) *Grain growth*. The particle method approach to grain growth focuses on the dynamical evolution of the positions of the vertices at which three grain boundaries meet in two dimensions. Soares et al. (1985) and Nagai et al. (1988) independently proposed to write the equation of motion of a vertex (grain boundaries are assumed to be straight and connect two vertices) in two dimensions as

$$\boldsymbol{v}_i = M \sum_{j=1}^{3} \boldsymbol{\varepsilon}_{ij} \qquad (11)$$

where $\boldsymbol{v}_i$ is the velocity of vertex i, M is a constant "mobility", $\boldsymbol{\varepsilon}_{ij}$ is the line tension of the grain boundary connecting sites i and j, and the sum is taken over the three grain boundaries meeting at vertex i. Equation (11) is then integrated with a small time-step Δt. As in the front tracking method, special handling is required to deal with the topological changes that occur when vertices meet.

This model yields microstructures and growth kinetics which are in agreement with those found by the front tracking and Monte Carlo methods and in cross sections of experimental grain-growth samples. Since this method focuses on vertices rather than grain boundaries (as is done in the front tracking methods) or grains (as is done in the Monte Carlo model) it is in principle more efficient.

3. *Conclusions*

While grain growth is not necessarily the most important area of microstructural evolution, it has been emphasized because of the wide variety of techniques that have been applied in its simulation study. As the simulation methods become increasingly efficient by concentrating on smaller and smaller components of the microstructure, more assumptions are required and a certain degree of simplicity is lost. This is clearly evident as the simulation of grain growth has evolved from (least to most efficient) Monte Carlo methods (which focus on the grain) via front tracking methods (which focus on the grain boundaries) to the particle methods (which focus on the grain boundary vertices). In cases where the identical simulation is to be run many times (with different parameters), as in a production environment, the most efficient methods should be used even though they employ the most assumptions and are not easily generalizable. In a research environment, however, generalizability, simplicity and the minimum number of assumptions are necessities and clearly favor the more basic approaches.

The computer simulation methods and applications described above represent only a sampling of the simulation methods currently applied to microstructure-related problems in materials science. Other areas to which microstructural simulations have been applied include directional solidification, film growth, particle coarsening and texture development during deformation. To date, most of the interest in the simulation of microstructural evolution has come from the basic research area of materials science and engineering. However, as the power and availability of computers continues to grow and new simulation methods are developed, computer simulation of microstructural evolution will move increasingly from the hands of researchers to those of engineers.

Bibliography

Anderson M P, Grest G S, Srolovitz D J 1989 Computer simulation of normal grain growth in 3 dimensions. *Phil. Mag. B* 59(3): 293–329

Anderson M P, Srolovitz D J, Grest G S, Sahni P S 1984 Computer simulation of grain growth. 1. Kinetics. *Acta Metall.* 32: 783–91

Binder K 1979 *Monte Carlo Methods in Statistical Physics*. Springer, Berlin

Bishop A R, Campbell L J, Channell P J 1984 *Fronts, Interfaces and Patterns*. North-Holland, Amsterdam

Brower R C, Kessler D A, Koplik J, Levine H 1984 Geometrical models of interface evolution. *Phys. Rev. A* 29: 1335–42

Ceppi E A, Nasello O B 1984 Computer simulation of bidimensional grain growth. *Scr. Metall.* 18: 1221–5

Frost H J, Thompson C V, Howe C L, Whang J 1988 A two-dimensional computer simulation of capillarity-driven grain growth: Preliminary results. *Scr. Metall.* 22: 65–70

Gayda J, Srolovitz D J 1989 A Monte-Carlo finite element model for strain-energy controlled microstructural evolution—Rafting in superalloys. *Acta Metall.* 37(2): 641–50

Ghoniem N M, Amodeo R 1988 Computer simulation of dislocation pattern formation. In: Kubin L P, Martin G (eds.) 1988 *Non-Linear Phenomena in Materials Science*. Trans Tech, Switzerland, pp. 377–88

Lepinoux J, Kubin L P 1987 The dynamic organization of dislocation structures. A simulation. *Scr. Metall.* 21: 833–8

Nagai T, Kawasaki K, Nakamura K 1988 Vertex dynamics of two-dimensional domain growth. In: Komura S, Furukawa H (eds.) 1988 *Dynamics of Ordering Process in Condensed Matter*. Plenum, New York, pp. 121–5

Rao M, Kalos M H, Lebowitz J L, Marro J 1976 Time evolution of a quenched binary alloy: III. Computer simulation of a two-dimensional model system. *Phys. Rev. B* 13: 4328–35

Rollett A D, Srolovitz D J, Anderson M P 1989 Simulation and theory of abnormal grain-growth anisotropic grain-boundary energies and mobilities. *Acta Metall.* 37(4): 1227–40

Sadiq A, Binder K 1983 Kinetics of domain growth in two dimensions. *Phys. Rev. Lett.* 51: 674–8

Soares A, Ferro A C, Fortes M A 1985 Computer simulation of grain growth in a bidimensional polycrystal. *Scr. Metall.* 19: 1491–6

Srolovitz D J (ed.) 1986 *Computer Simulation of Microstructural Evolution*. The Metallurgical Society, Warrendale, PA

Srolovitz D J, Grest G S, Anderson M P 1986 Computer simulation of recrystallization. 1. Homogeneous nucleation and growth. *Acta Metall.* 34: 1833–45

Voorhees P W, Johnson W C 1988 Development of spatial correlation during diffusional late-stage phase transformations in stressed solids. *Phys. Rev. Lett.* 61: 2225–8

Voorhees P W, McFadden G B, Boisvert R F, Meiron D I 1988 Numerical simulation of morphological development during Ostwald ripening. *Acta Metall.* 36: 207–22

Wilson D G, Solomon A D, Boggs P T 1978 *Moving Boundary Problems*. Academic Press, New York

D. J. Srolovitz
[University of Michigan, Ann Arbor, Michigan, USA]

Cork

Cork is a natural material with a remarkable combination of properties. It is light yet resilient; it is an outstanding insulator for heat and sound; it has exceptional qualities for damping vibration; it has a high coefficient of friction; and it is impervious to liquids, is chemically stable and is fire resistant. These attributes have made cork commercially attractive for well over 2000 years.

1. Origins, Chemistry and Structure

Commercial cork is the outer bark of *Quercus suber* L., the cork oak. The tree, first described by Pliny in AD 77, is cultivated for cork production in Portugal, Spain, Algeria, France and parts of California. As Pliny said, the cork oak is relatively small and unattractive; its only useful product is its bark which is extremely thick and which, when cut, grows again.

Cork occupies a special place in the history of microscopy and of plant anatomy. When Robert Hooke perfected his microscope around 1660, one of the first materials he examined was cork. What he saw led him to identify the basic unit of plant and biological structure, which he called "the cell". His book, *Micrographia* (Hooke 1664), contains careful drawings of cork cells showing their hexagonal-prismatic shape, rather like cells in a bee's honeycomb. Subsequent studies by Lewis (1928), Natividade (1938), Eames and MacDaniels (1951), Gibson et al. (1981), Ford (1982) and Pereira et al. (1987) confirm Hooke's picture, and add to it the observation that the cell walls themselves are corrugated like the bellows of a concertina (see Fig. 1), a structure which gives the cork extra resilience. The cells themselves are considerably smaller than those in most foamed polymers: the hexagonal prisms are between 10 μm and 40 μm in height and measure between 10 μm and 15 μm across the base, with a cell-wall thickness of between 1 μm and 1.5 μm. There are between 4×10^4 and 2×10^5 cells per cubic millimeter.

The cell walls of cork are made up of lignin and cellulose, with a thick secondary wall of suberin and waxes (Eames and MacDaniels 1951, Sitte 1962, Esau 1965, Zimmerman and Brown 1971). All trees have a thin layer of cork in their bark. *Quercus suber* is unique in that, at maturity, the cork forms a layer several centimeters thick around the trunk of the tree.

2. Properties

Many of the properties of cork derive from its cellular structure (Gibson et al. 1981, Gibson and Ashby 1988). When cork is loaded in compression or tension the cell walls bend, and this bending allows large elastic change of shape. For this reason, the elastic moduli are low (see Tables 1, 2)—roughly one hundred times lower than those of the solid of which the cell walls are made. When compressed beyond the linear-elastic limit (see Fig. 2) the cell walls buckle giving "resilience", that is, a large nonlinear elastic deformation. The prismatic shape of the cells makes cork anisotropic. Young's modulus E_1 measured in a direction parallel to the prism axis (the "radial" direction of Fig. 1) is almost twice as great as E_2 or E_3, measured in directions normal to this axis (the "axial" and "tangential" directions). The shear moduli are also anisotropic. Most anisotropic of all is Poisson's ratio which, for compression parallel to the prism axis, is almost zero. This means that compression in this direction produces little or no lateral spreading—a property useful in gaskets, and in corks which must seal particularly tightly. The wide range of values for a given property derives from the range of density of natural cork (which depends on growing conditions and the age of the tree) and on its moisture content.

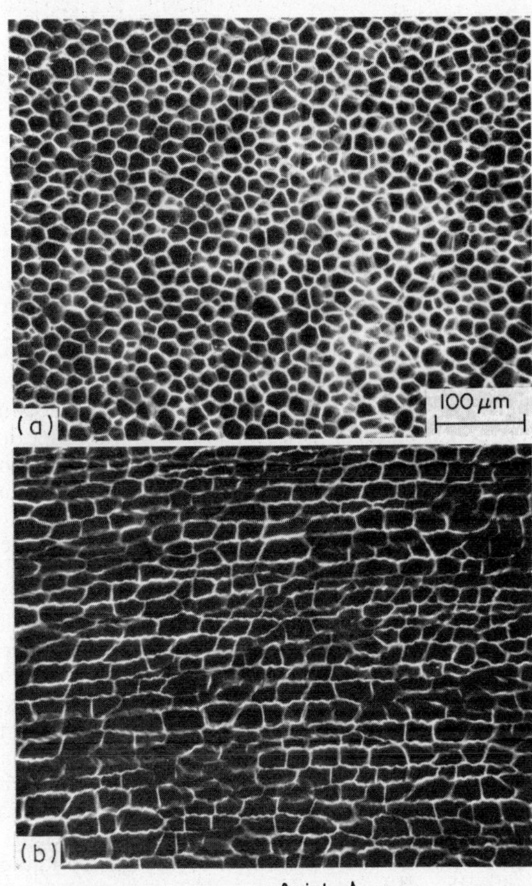

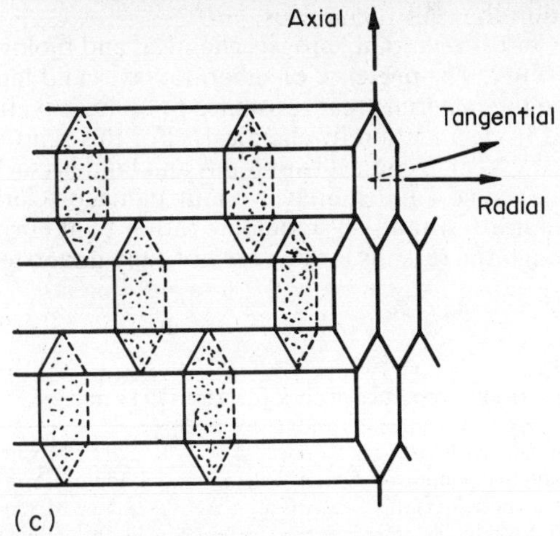

Figure 1
Cells in cork: (a) section normal to the radial direction of the tree, (b) section normal to the axial direction of the tree (i.e., the direction parallel to the trunk), (c) the cells are hexagonal prisms

Table 1
Average properties of cork

Density	120–240 kg m^{-3}
Young's modulus	4–60 MN m^{-2}
Shear modulus	2–30 MN m^{-2}
Poisson's ratio	0.15–0.22
Collapse strength (compression)	0.5–2.5 MN m^{-2}
Fracture strength (tension)	0.8–3.0 MN m^{-2}
Loss coefficient	0.1–0.3
Coefficient of friction	0.2–0.4
Thermal conductivity	0.025–0.028 J mK^{-1}

Cork is a particularly "lossy" material; that is, when it is deformed and then released, a considerable fraction of the work of deformation is dissipated. This fraction is measured by the loss coefficient (see Table 1); for cork it has a value between 0.1 and 0.3, giving the material superior vibration- and acoustic-damping properties. The high coefficient of friction also derives from the ability to dissipate energy. When an object slides or rolls on a cork surface, the cells deform and then recover again as the slider passes; the energy loss results in friction.

Cork has a particularly low thermal conductivity because of its cellular structure. Heat is conducted through a cellular solid by conduction through the cell walls, by conduction through the gas within the cells and by convection of this gas (Gibson and Ashby 1988). The cells in cork are sufficiently small so that convection is suppressed completely, and the thin cell walls contribute little to heat transfer. Therefore, the thermal conductivity is reduced to a value only slightly above that of the gas contained within the cells themselves.

Cork is remarkable for its chemical and biological stability. The presence of suberin, waxes and tannin give the material great resistance to biological attack and to degradation by chemicals. For this reason, it can be kept in contact with foods and liquids (such as wine) almost indefinitely without damage. Cork is also fire-resistant—it smoulders rather than burns—though the reasons for this are not fully understood.

Table 2
Anisotropic properties of cork (density 170 kg m^{-3})

Young's modulus E_1	20±7 MN m^{-2}
Young's modulus E_2, E_3	13±5 MN m^{-2}
Shear moduli G_{12}, G_{21}, G_{13}, G_{31}	2.5±1 MN m^{-2}
Shear moduli G_{23}, G_{32}	4.3±1.5 MN m^{-2}
Poisson's ratio ν_{12}, ν_{13}, ν_{21}, ν_{31}	0.05±0.05
Poisson's ratio ν_{23}, ν_{32}	0.5±0.05
Collapse strength, compression, σ_1^*	0.8±0.2 MN m^{-2}
Collapse strength, compression σ_2^*, σ_3^*	0.7±0.2 MN m^{-2}

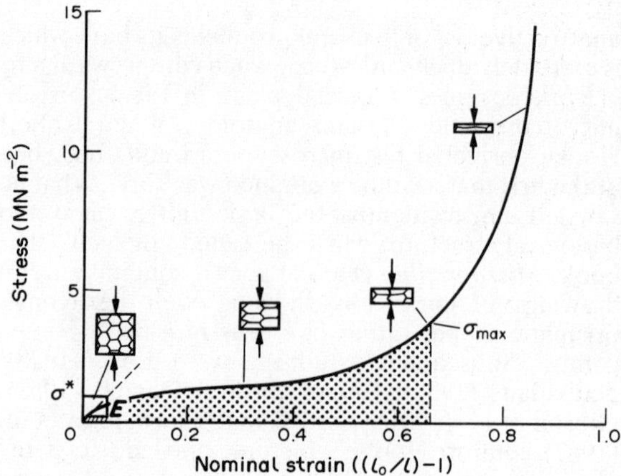

Figure 2
A compression stress–strain curve for cork, showing Young's modulus E and the collapse stress σ^*

3. Uses

For at least 2000 years, cork has been used for a variety of purposes, including the examples given by Pliny in AD 77 of "floats for fishing nets, and bungs for bottles, and also to make the soles for women's winter shoes." Few materials have such a long history or have survived the competition from man-made substitutes so well. The special structure of cork that suits it to its uses is described in this section.

3.1 Bungs for Bottles and Gaskets for Woodwind Instruments

Connoisseurs of wine agree that there is no substitute for corks made of cork. Plastic corks are hard to insert and remove, they do not always give a good seal and they may contaminate the wine. Cork corks are inert and seal well for as long as the wine need to be kept. The excellence of the seal derives from the elastic properties of the cork. It has a low Young's modulus E and, more importantly, it also has a low bulk modulus K. Solid rubber and solid polymers above their glass transition temperature have a low E but a large K, and it is this that makes them hard to insert into a bottle and gives a poor seal when they are in place.

It might be expected that the best seal would be obtained by cutting the axis of the cork parallel to the prism axis of the cork cells; the circular symmetry of the cork and of its properties are then matched. This is correct. However, natural cork contains lenticels—tubular channels that connect the outer surface to the inner surface of the bark, allowing oxygen into, and CO_2 out of, the new cells that grow there. The lenticels lie parallel to the prism axis (the radial direction) and so a cork cut parallel to this axis will leak. This is why almost all commercial corks are cut with the prism axis (and

Figure 3
Ordinary corks (a) have their axis at right angles to the prism axis to prevent leakage through the lenticels (dark lines). A laminated cork (b) makes better use of the symmetry of the cork structure

the lenticels) at right angles to the axis of the bung (see Fig. 3a).

A solution to this problem is shown in Fig. 3b. The base of the cork, where sealing is most critical, is made of two disks cut with the prism axis (and lenticels) parallel to the axis of the bung itself. Leakage is prevented by gluing the two disks together so that the lenticels do not connect. The cork, when forced into the bottle, is then compressed (radially) in the plane in which it is isotropic, exerting a uniform pressure on the inside of the neck of the bottle.

Cork makes good gaskets for the same reason that it makes good bungs: it accommodates large elastic distortion and volume change, and its closed cells are impervious to water and oil. Thin sheets of cork are used, for example, for the joints of woodwind and brass instruments. The sheet is always cut with prism axis (and lenticels) normal to its plane. The sheet is then isotropic in its plane and this may be the reason for cutting it in this way, but it seems more likely that it is cut like this because Poisson's ratio for compression down the prism axis is zero. Thus, when the joints of the instrument are mated, there is no tendency for the sheet to spread in its plane and wrinkle.

3.2 Floor Covering and the Soles of Shoes

Manufacturers who sell cork flooring claim that it retains its friction even when polished or covered with soap, and experiments by Gibson et al. (1981) confirm this.

Friction between a shoe and a cork floor has two origins. One is adhesion: atomic bonds form between the two contacting surfaces and work must be done to break them if the shoe slides. Between a hard slider and a tiled or stone floor this is the only source of friction and, since it is a surface effect, it is destroyed by a film of polish or soap. The second source of friction is due to anelastic loss. When a rough slider moves on a cork floor, the bumps on the slider deform the cork. If cork were perfectly elastic no net work would be done and the work done in deforming the cork would be recovered as the slider moves on. However, if the cork has a high loss coefficient (as it does) then it is like riding a bicycle through sand: the work done in deforming the material ahead of the slider is not recovered as the slider passes on, and a large coefficient of friction results. This anelastic loss is the main source of friction when rough surfaces slide on cork and since it depends on processes taking place below the surface, not on it, it is not affected by films of polish or soap. The same thing happens when a cylinder or sphere rolls on cork, which therefore shows a high coefficient of rolling friction.

3.3 Packaging and Energy Absorption

Many of the uses of cork depend on its capacity to absorb energy. Cork is attractive for the soles of shoes and flooring not only because it has good frictional properties, but because it is resilient under foot, absorbing the shocks of walking. It makes good packaging because it compresses on impact, limiting the stresses to which the contents of the package are exposed. It is used as handles of tools to insulate the hand from the impact loads applied to the tool. In each of these applications it is essential that the stresses generated by the impact are kept low, but that considerable energy is absorbed.

Cellular materials are particularly good at this. The stress–strain curve for cork (see Fig. 2) shows that the collapse strength of the cells (see Table 1) is low, so that the peak stress during impact is limited. Large compressive strains are possible, absorbing energy as the cells progressively collapse. In this regard, its structure and properties resemble polystyrene foam, which has replaced cork (because it is cheap) in many packaging applications.

3.4 Thermal Insulation

Trees are thought to surround themselves with cork to prevent loss of water in the hotter seasons of the year. The two properties involved—low thermal conductivity and low permeability to water—make it an excellent material for insulation against cold and damp (the hermit caves of southern Portugal, for example, are lined with cork). It is for these reasons that crates and boxes are sometimes lined with cork, and the cork tip of a cigarette must appeal to the smoker because it insulates (a little) and prevents the tobacco getting moist.

3.5 Indentation and Bulletin Boards

Cellular materials densify when they are compressed or indented. Therefore, when a sharp object, like a drawing pin, is stuck into cork, the deformation is very localized. A layer of cork cells, occupying a thickness of only about one quarter of the diameter of the indenter, collapses, suffering a large strain. The volume of the indenter is taken up by the collapse of the cells so that no long-range deformation is necessary. For this reason the force needed to push the indenter in is small. Since the deformation is (nonlinear) elastic, the hole closes up when the pin is removed.

Bibliography

Eames A J, MacDaniels L H 1951 *An Introduction to Plant Anatomy*. McGraw-Hill, London

Emilia Rosa M, Fortes M A 1988 Stress relaxation and stress of cork. *J. Mater. Sci.* 23: 35–42

Esau L 1965 *Plant Anatomy*. Wiley, New York, p. 340

Ford B J 1982 The origins of plant anatomy: Leeuwenhoek's cork sections examined. *IAWA Bull.* 3: 7–10

Gibson L J, Ashby M F 1988 *Cellular Solids: Structure and Properties*. Pergamon, Oxford, Chap. 12

Gibson L J, Easterling K E, Ashby M F 1981 The structure and mechanics of cork. *Proc. R. Soc. London. Ser. A.* 377: 99–117

Hooke R 1664 *Micrographia*. Royal Society, London, pp. 112–21

Lewis P T 1928 The typical shape of polyhedral cells in vegetable parenchyma. *Science (N.Y.)* 68: 635–41

Natividade J V 1938 What is cork? *Bol. Junta Nac. da Cortica (Lisboa)* 1: 13–21

Pereira H, Emilia Rosa M, Fortes M A 1987 The cellular structure of cork from *Quercus suber* L. *IAWA Bull.* 8: 213–18

Sitte P 1962 Zum Feinbau der Suberinschichten in Flaschen Kork. *Protoplasma* 54: 55–559

Zimmerman M H, Brown C L 1971 *Trees Structure and Function*. Springer, Berlin, p. 88

M. F. Ashby
[University of Cambridge, Cambridge, UK]

D

Deformation Twinning of Metals and Alloys

Mechanical or deformation twinning, by which a portion of the lattice homogeneously shears in response to an applied stress, can be an important mode of deformation in metals and alloys that lack a sufficient number of independent slip systems. Since homogeneous crack-free deformation of a polycrystal requires five independent shear systems (in the absence of diffusional flow), twinning can provide the necessary additional shear systems in crystals with only a few slip systems, such as hexagonal-close-packed (hcp) and lower-symmetry crystal structures. Twinning differs from slip in which strain is accommodated by translating undeformed blocks of a crystal relative to another across a single lattice plane in multiples of a displacement vector (the Burgers vector). In twinning, the displacement of any plane within the twin is proportional to its distance from the twin-matrix interface. The atoms, therefore, move only a fraction of a lattice spacing relative to each other producing a portion that is a mirror image of the original crystal. Both slip and twinning shear the lattice, but in twinning a small but well-defined volume in the crystal responds to a sudden shear, whereas an individual slip process is two-dimensional in character.

Whether slip or twinning will be the predominant deformation mechanism depends on which requires the least stress to be initiated in the portion of the crystal undergoing shear. The stress levels required for slip or twinning are not constant, but vary with the crystal structure of the metal or alloy, test temperature, strain rate, alloy content and other intrinsic and extrinsic variables. In deformation at low temperature and at high strain rates, twinning is also observed in higher-symmetry crystal structures such as body-centered-cubic (bcc) and face-centered-cubic (fcc). Activation of twinning systems in bcc, and particularly in fcc metals and alloys, occurs because the flow stress for slip increases more rapidly with either decreasing temperature or increasing strain rate (or both) than it does for the flow stress for twinning.

Twinning is not only an alternate deformation mode to slip for shear; it also influences the structure–property behavior of metals because twins can act as barriers to the motion of slip dislocations, particularly in low-symmetry crystals. Such twin-dependent barriers can affect work-hardening and can provide sites for dislocation pileups, which influence cross-slip and consequently fracture nucleation.

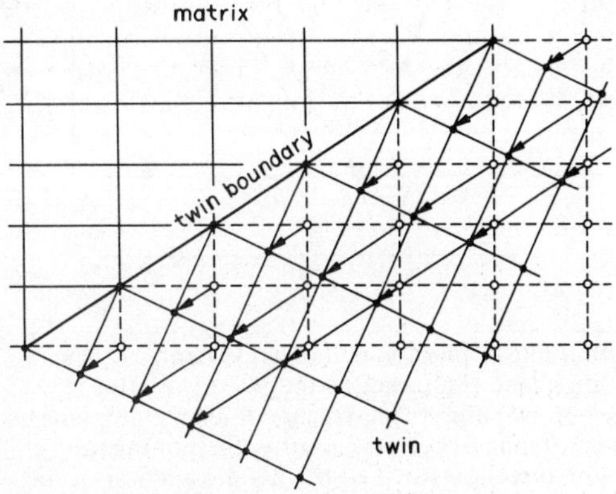

Figure 1
Schematic drawing of twin formation in a tetragonal lattice by uniform shearing of atoms parallel to the twin plane: the dashed lines represent the lattice before twinning; the solid lines represent the lattice after twinning

1. *Crystallography of Twinning*

In this article, twinning will be defined as a homogeneous simple shear in a crystal; a shear which does not change the shape, volume or structure of the crystal, but rather reorients the atoms of the crystal so that a mirror relationship exists between the sheared and unsheared portion of the crystal. The uniform lattice shear which causes the mirror image orientation, such that the displacement of a lattice point in the twinned region is directly proportional to the distance from the boundary and parallel to the boundary, is illustrated in Fig. 1. Any such alteration in the structure is properly classified as a twin-transformation product rather than being a definition of the twin itself. Twins are uniquely defined by four elements, schematically shown in Fig. 2. The invariant (unrotated) plane along which this mirror relationship lies is the composition or twin plane and is designated as K_1. This plane is crystallographically unchanged by the shear and is the identical plane in both the twinned and untwinned crystals. The direction of shear η_1 creates the twin and is in the plane K_1. A second undistorted plane K_2 will also exist in the crystal, which is unchanged in terms of length or shape but which has been rotated about the twin plane. The plane K_2 intersects K_1 in a line perpendicular to the shear direction η_1 and makes equal angles with K_1 before and after the shear. The plane of shear S is the normal to K_1 and contains η_1. The intersection of S

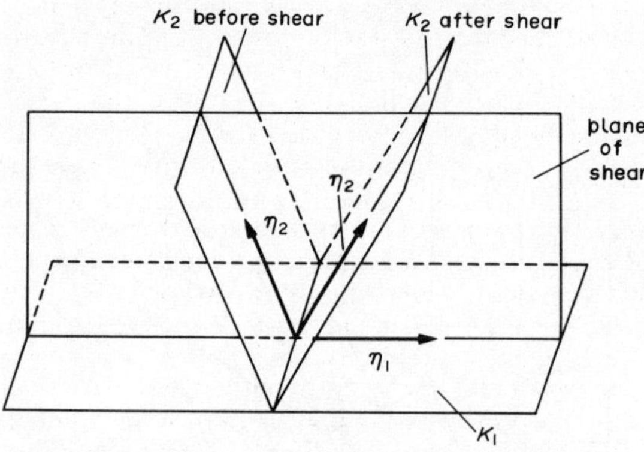

Figure 2
Diagram illustrating the twinning elements: K_1 the twinning plane (first undistorted plane); K_2 the conjugate twinning plane (second undistorted plane); η_1 the twinning–shear direction; η_2 the conjugate twinning direction (defined by the intersection of the plane of shear with K_2); and S the plane of shear

and K_2 is labelled η_2. There are two positions for η_2 corresponding to the positions of K_2 before and after shearing. The magnitude of the twinning shear g can be measured from the twin geometry as follows:

$$g = 2 \cot \Theta$$

where Θ is the angle between K_1 and K_2.

Deformation twinning in a material is completely defined either by K_1 and η_2 or by K_2 and η_1, but it is usual to quote all four parameters; all four elements are normally cited in the order K_1, K_2, η_1 and η_2. In bcc metals, such as iron, niobium or tungsten, the twinning plane K_1 is normally $\{112\}$ and the direction of shear η_1 is $\langle 11\bar{1}\rangle$ (or equivalent planes and directions). In fcc metals, such as silver, copper, aluminum or nickel, the mode is $\{111\}\ \langle 11\bar{2}\rangle$. In the case of hcp metals, the pyramidal $\{10\bar{1}2\}$ planes are the most common twinning system; twin activity on other twin systems varies from metal to metal depending on the c/a (axial) ratio. In rhombohedral structures, such as bismuth, the mode is $\{110\}\ \langle 00\bar{1}\rangle$. Twinning modes in tetragonal, orthorhombic and monoclinic metals and alloys are more complicated, with metal-to-metal variations existing. Table 1 lists crystal structures, some common twin elements and the twin shear for a number of metals.

Since the shape and size of the crystal unit cell is unaltered by twinning, there must be three rational lattice vectors (nonplanar) which have the same magnitudes and the same angular relationships before and after twinning. These vectors, which are not distorted by a twinning deformation, must lie in the two undistorted planes K_1 and K_2. For a simple shear to qualify as a twinning shear it must result in the same crystal structure as the parent lattice but in a different orientation. There are three modes of shear that satisfy these requirements for twinning.

(a) Twins of the first kind: K_1 and η_2 are rational planes and directions.

(b) Twins of the second kind: K_2 and η_1 are both rational.

(c) Rational or compound twins: all four elements K_1, K_2, η_1 and η_2 are rational.

The lattice rotations caused by twinning differ depending on the type of twinning. For twins of the first kind, the orientation relation required to leave the composition plane invariant can be achieved by reflection in the K_1 plane or by a rotation of 180° about the normal to K_1. In twins of the second kind, the orientation relation is created either by a reflection in the plane normal to η_1 or by a rotation of 180°

Table 1
Some commonly observed twin elements in metals

Metal	Crystal	c/a	K_1	K_2	η_1	η_2	g
Ag, Al, Cu, Ni	fcc		$\{111\}$	$\{11\bar{1}\}$	$\langle 11\bar{2}\rangle$	$\langle 112\rangle$	0.707
α-Fe, Nb W, Mo, V	bcc		$\{112\}$	$\{11\bar{2}\}$	$\langle 11\bar{1}\rangle$	$\langle 111\rangle$	0.707
Zn	hcp	1.856	$\{10\bar{1}2\}$	$\{\bar{1}012\}$	$\langle\bar{1}011\rangle$	$\langle 10\bar{1}1\rangle$	0.139
Mg	hcp	1.624	$\{10\bar{1}2\}$	$\{\bar{1}012\}$	$\langle\bar{1}011\rangle$	$\langle 10\bar{1}1\rangle$	−0.131
			$\{11\bar{2}1\}$	$\{0001\}$	$\langle 11\overline{26}\rangle$	$\langle 11\bar{2}0\rangle$	0.640
Ti	hcp	1.587	$\{10\bar{1}2\}$	$\{\bar{1}012\}$	$\langle\bar{1}011\rangle$	$\langle 10\bar{1}1\rangle$	−0.167
			$\{11\bar{2}1\}$	$\{0001\}$	$\langle 11\overline{26}\rangle$	$\langle 11\bar{2}0\rangle$	0.638
			$\{11\bar{2}2\}$	$\{11\overline{24}\}$	$\langle 11\overline{23}\rangle$	$\langle 22\bar{4}3\rangle$	0.225
Bi	rhombohedral		$\{110\}$	$\{001\}$	$\langle 00\bar{1}\rangle$	$\langle 110\rangle$	0.118
β-Sn	tetragonal		$\{301\}$	$\{\bar{1}01\}$	$\langle\bar{1}03\rangle$	$\langle 101\rangle$	0.119
Ge	diamond cubic		$\{111\}$	$\{11\bar{1}\}$	$\langle 11\bar{2}\rangle$	$\langle 112\rangle$	0.707
α-U	orthorhombic		$\{130\}$	$\{1\bar{1}0\}$	$\langle 3\bar{1}0\rangle$	$\langle 110\rangle$	0.299
α-Np	orthorhombic		$\{110\}$	$\{1\bar{1}0\}$	$\langle 1\bar{1}0\rangle$	$\langle 110\rangle$	0.068

about η_1. In crystals of higher symmetry (i.e., cubic or hcp) the reflection and rotation operations are equivalent and therefore twins are generally compound. Twins of the first and second kind are relatively rare in metals. Twins of the second kind in pure metals have only been seen in structures of low symmetry, such as α-uranium, which is orthorhombic, and in crystalline mercury, which is rhombohedral. These kinds of twins are also commonly observed in minerals with low-symmetry crystal structures.

In hcp and other low-symmetry crystals a simple shear will not complete the twin relationship, leaving some atoms slightly out of atomic position. In addition to the shear it then becomes necessary to "shuffle" local arrangements of some of the atom positions or undergo secondary twinning, as in the case of hcp metals, to satisfy the twinning relationship. The reasons that one twinning mode (twin system) operates in preference to another appear to depend on which system best satisfies the following:

(a) the homogeneous shear is minimized (thus minimizing the accommodation slip or twinning necessary in the twin or the parent lattice),

(b) shuffles are unnecessary or minimized,

(c) shuffles should be parallel to the twinning shear direction (to minimize bond breaking),

(d) crystallographic order in an ordered superlattice is undisturbed, and

(e) a convenient dislocation mechanism exists that enables the twin to form.

The last criterion, which is prompted by observation of $\{\bar{1}35\}$ deformation twins in crystalline mercury in preference to twins with smaller shears on the $\{110\}$ and $\{001\}$ planes, suggests that the crystallographic theories of twinning remain incomplete.

2. Twin Formation

The formation of deformation twins is thought to be separable into two parts: (a) the nucleation of a twin nucleus, and (b) development or growth of the twin. Experimental evidence suggests that, while large stresses are necessary to form a twin for a mechanical twin, subsequent twin development occurs under smaller stresses. When this happens, twins grow at very rapid rates, measured to be slightly less than the shear wave velocity, as soon as they are nucleated. In single crystals of zinc, the initial nucleation of twins requires stress levels in excess of $10^{-1}\mu$ (where μ is the shear modulus), while their subsequent development is produced with stresses not much larger than $10^{-4}\mu$. This rapid growth of twins produces two interesting physical phenomena in metals. First, the formation of such twins produces a characteristic yield drop and discontinuous strain increments in the stress–strain curves of single crystals. Second, the rapid growth of twins sets up "plastic" shock waves in the metal, travelling at a velocity slightly less than the speed of sound, which can be heard as audible clicks in some materials when the waves reach the surface of the metal: the "cry" of tin is an example. The large disparity in stress for twin nucleation and growth may result because twinning dislocations have smaller Burgers vectors than normal dislocations. Accordingly, they form "jogs" of lower energy with the dislocations cutting their twin planes, and they are also less strongly bound to impurity atoms, all of which would promote rapid twin growth once the twinning nucleus is present.

Observations of single crystals have shown that, precluding existing twins, plastic deformation by slip is a prerequisite for twinning. Since the experimentally observed shear stress level necessary for twin formation is much smaller than the level for homogeneous twin nucleation (probably a sizable fraction of the theoretical elastic limit 0.05–0.1μ), deformation twins most likely originate in regions of stress concentration. The "high stress" concentration could be a scratch in a sample, a pileup of dislocations or the end of a polygonization subboundary. Detailed transmission electron microscopy (TEM) studies on twin nucleation in dislocation-free zinc platelets have shown that twin nucleation usually occurred at a sample reentrant corner or at a corrosion pit, where the applied stress was locally magnified. Furthermore, wide zinc platelets tended to have more stress raisers and thus twinned at a lower overall applied stress. However, in thin platelets with very smooth edges, when the samples were carefully mounted to achieve uniform loading, the samples always twinned at values approaching the theoretical stress for homogeneous twin nucleation. This observation implies that the energy to form a twin nucleus, or embryo, comes from an aggregate of slip dislocations in some form, so that the release of a few dislocations, which may be so small as to be virtually undetectable, must precede twinning. In turn, this dependency on prior deformation has been linked to the wide variations in resolved shear stress for twinning. Furthermore, the dependence of twinning on stress state is not well established. The scatter in the twinning shear stress data together with the apparent prerequisite of a stress concentration to initiate plastic deformation via a twinning mode make it unlikely that a critical resolved shear stress exists for twinning, as it does for slip.

Several dislocation mechanisms have been modelled to provide the dislocations that must move through a crystal to generate a deformation twin. A dislocation model for twinning requires either that identical dislocations be initiated on every lattice plane or that a single dislocation be able to move from one plane to the next. The first models of a

Figure 3
Optical micrograph of deformation twins in cold-rolled tin

twinning dislocation, independently developed for bcc α-iron and hcp cadmium, postulated a partial dislocation rotating around a sessile dislocation in such a way as to cause a shear on successive planes. A twin can form via this mechanism, called a "pole mechanism," if there is only one twinning dislocation. A "pole dislocation" must exist that has a component of the Burgers vector $\boldsymbol{b}$ normal to the $\{112\}$ plane of a bcc crystal in order for this mechanism to operate. The glissile twinning partial dislocation on the $\{112\}$ plane rotates around this pole with one end attached to it and climbs one layer per revolution. During rotation, successive lattice planes are swept out, displacing the boundary plane normal to itself and thus increasing the layers of twinned material to generate a homogeneous twin. Pole mechanisms have also been proposed for other crystal systems besides bcc and hcp, in particular for fcc crystals.

While there are some objections to this type of mechanism, pole mechanisms for twinning in fcc crystals have wide support in the literature on the basis of TEM observations, lower-energy dislocation dissociation reactions and because they explain the growth of twins in single crystals. Alternate twin nucleation mechanisms include homogeneous nucleation, partial dislocation breakaway or the Lomer–Cottrell barrier mechanism. At present, no single twin nucleation mechanism is considered all encompassing; in fact all twin formation mechanisms still appear unable to explain the experimentally observed high rates of twin formation.

Once nucleated, twins grow in length and thickness in response to the applied stress, and the surrounding lattice accommodates the twinning shear either by slip or by additional twinning. Figure 3 shows an optical micrograph of deformation twins in cold-rolled tin. The twinned regions, as in the case of tin, are often bounded by parallel or nearly parallel sides that correspond to planes of low indices. The lenticular shape, which is characteristic of most deformation twins in metals and alloys, is consistent with the theory that twin shape is governed by the back stress produced by the twinning shear. In hcp metals, regenerative dislocation mechanisms that would allow a twin to grow without nucleating new twinning dislocations appear unlikely. In experimental studies of twin growth in hexagonal zinc platelets, the growth of twins occurs by the repeated nucleation (at the platelet edges) and movement of twinning dislocations, rather than by a pole mechanism. The observation of new twin formation directly adjoining existing twins, termed autocatalytic or polysynthetic twins, further suggests that elastic accommodation stresses in the matrix adjacent to a twin most likely serve to nucleate new twins. Twin growth in thin electropolished tin crystals has also been attributed to repeated nucleation of twinning dislocations. In bulk materials, the question of the exact mechanism of twin growth remains unsettled. However, experiments utilizing nearly perfect crystals have made it clear that investigations of twinning in bulk crystals will probably not reveal the twinning process, but rather are complicated by various internal imperfections.

3. *Metallurgical Factors Influencing Twinning*

3.1 *Alloying*

Alloying additions to pure metals, either as substitutional solid-solution additions or as interstitials, are known to affect twinning behavior. The stress necessary for deformation twin formation in fcc single crystals increases with increasing stacking fault energy (SFE). Studies on solid-solution alloys of fcc metals (e.g., silver–gold, copper–zinc, copper–aluminum, copper–gallium, copper–germanium and copper–indium) show that twinning occurs more readily in alloys than in pure metals under deformation. This conclusion is demonstrated by the lower resolved shear stress necessary to form twins, for example, 40–120 MPa in copper-based solid solutions compared with 150 MPa for pure copper. This effect can be explained by twin embyro studies which suggest that in alloys of low SFE twin growth is propagation controlled while, in contrast, nucleation is the critical event in high-SFE materials. In bcc crystal structure materials, solid-solution alloying increases the incidence of deformation twinning. Twins occur readily at room temperature in solid-solution alloys of molybdenum and tungsten with rhenium, and also in niobium with vanadium additions. The increased frequency of twinning in bcc lattices due to alloying may result from:

(a) differences in the movement of screw dislocations in bcc metals determined by a "restoring force," which acts to eliminate the displacement in the unstable stacking fault (if ever induced);

(b) changes in the core dislocation structure and interatomic forces, causing the retardation of cross-slip and an increase in the lattice friction (as is the case when vanadium is added to niobium); or

(c) a reduced number of mobile dislocations caused by solute additions, thereby enhancing twinning.

In hcp materials, solid-solution additions decrease the propensity for twinning. In titanium–aluminum binary alloys the extent of twinning is greatly decreased when the aluminum content is increased from 2–6 wt% Al.

The effect on twinning of impurity additions and interstitials, such as oxygen, nitrogen or carbon, remains poorly understood. Adding silicon to iron increases deformation twinning but, in contrast, carbon additions to an Fe–4.8 at.% Sn alloy increase the twinning stress. In studies of the twinning behavior of niobium deformed at low rates, twinning was completely suppressed at 2000 ppm of oxygen or 200 ppm of carbon. Other studies on high-purity (low interstitial in particular) niobium and tantalum (both bcc metals) have shown that these metals readily twin at low temperatures and at room temperature during impact loading. In hcp metals, experimental results on the effect of interstitials on twinning are varied. Research on zirconium–oxygen alloys has shown that oxygen apparently does not inhibit deformation twinning more than it does slip. High oxygen concentration (3.6 at.% and 4.2 at.%) Zr–O alloys twin extensively on $\{10\bar{1}2\}$ and $\{11\bar{2}2\}$ twinning systems just above room temperature. In α-titanium, increasing the interstitial oxygen content suppresses $\{11\bar{2}3\}$ twinning in one study and enhances $\{11\bar{2}2\}$ twinning in another, although these differences may be related to grain size variations. Most measurements of twin frequency suggest an overall suppression of twinning in α-titanium with increasing oxygen content.

3.2 Temperature and Strain Rate

Twins in most crystal structures form more readily as the temperature of deformation is decreased or the rate of deformation is increased. In bcc materials, the yield stress for slip increases sharply with decreasing temperature, whereas the twinning stress decreases very slightly with decreasing temperature. While slip is the preferred mode of deformation at most temperatures in bcc materials, at very low temperatures the stress necessary for twinning is less than that required for slip: twinning then becomes the preferred mode of deformation. In fcc metals the temperature dependence of the twinning stress appears to be low; the twin stress increases slightly with increasing temperature. In high-SFE fcc pure metals, such as copper or nickel, deformation twinning only occurs at high stress levels (e.g., 150 MPa for copper and 300 MPa for nickel), which are normally only reached at low temperatures or under heavy deformations. Research on several hcp metals has revealed that the stress for twinning on $\{10\bar{1}2\}$, $\{11\bar{2}1\}$ and $\{11\bar{2}2\}$ planes increases with increasing temperature while that for $\{10\bar{1}1\}$ twinning decreases. The increase in stress for $\{11\bar{2}2\}$ twinning with increasing temperature may suggest that slip is a prerequisite for this twinning mode in α-titanium.

Increasing the strain rate at room temperature, particularly to impact and shock-loading levels, greatly enhances deformation twinning. Many metals and alloys, in particular high-SFE fcc metals that do not exhibit twinning during conventional low strain-rate deformation, readily twin under the severe stress levels imposed by shock deformation. Recent shock experiments on Al–4.8 wt% Mg and 6061-T6 Al alloys produced deformation twins in both alloys. The occurrence of deformation twins in aluminum alloys, which have relatively high SFEs and thus do not exhibit twins under any other observed loading conditions (including low-temperature deformation), illustrates how strain rate can influence twinning in fcc metals.

Some of the increased propensity for twinning in fcc materials may be linked to a decrease in the twinning stress with increasing strain rate, as has been observed in Cu–5 at.% Ge crystals. In shock-loading experiments on hypoeutectoid pearlitic mild steels (bcc), twin lamellae grew without deviation from the proeutectoid ferrite into adjacent pearlitic colonies. This observation is interesting in contrast to conventional low strain-rate experiments in which twins will not go through the pearlite. In hcp metals and alloys, such as zirconium and Ti–6Al–4V, increasing strain rate is known to increase the frequency of deformation twins, in particular $\{11\bar{2}1\}$ twins. Examination by TEM of shock-loaded Ti–6Al–4V revealed $\{11\bar{2}1\}$ twinning to be the sole twinning system activated.

3.3 Microstructure

The starting microstructure of a metal or alloy, including the grain size, texture, the presence of second phase precipitates or dispersoids and prior deformation history, is known to influence the occurrence of deformation twinning. Twins normally form more easily in large grains since twin nucleation is sensitive to restraints that resist the twinning shear. Since twin formation requires accommodation of elastic stresses, which depend on the twinning shear and the twin shape, reducing the grain size increases the twinning stress. For zone-refined titanium of 1–2 μm grain size, no twins were observed for tests at 300 K and above, while for larger grain sizes, heating to 500 K was required before twins were no longer detected. Twin thickness also varies with grain size with thicker twins in coarse-grained specimens.

Experimental studies have ascertained that preferred crystallographic orientation (texture) in a material can influence the prevalence and type of deformation twinning. Studies have shown that once extensive twinning is activated, it can play a major role in subsequent texture development during plastic deformation. Twinning affects texture because unlike slip, for which the forward and backward directions of shear are crystallographically equivalent, there is only one shear direction for twinning on any given twin plane. In hexagonal metals, the correlation between twin activation and mode and the local applied stress direction in a given polycrystal grain is particularly evident.

In hexagonal metals, the mode of deformation, whether tension or compression, plays a role in the occurrence of twins. When twins develop in hcp single crystals, the specimen will either extend or contract depending on the orientation of the crystal axes. This phenomenon is related to the fact that the sense of shear during twinning in hcp metals varies with the c/a ratio. For example, if a single crystal of zinc were oriented with its basal plane parallel to the axis of the specimen, twinning on any $\{10\bar{1}2\}$ twin plane will extend the sample. This sample will therefore twin if deformed in tension but not in compression. In the case of magnesium, the situation is completely reversed: twinning by $\{10\bar{1}2\}$ will occur when compression is applied parallel to the basal plane.

The angular relationship between the texture of an hcp polycrystal and the applied stress direction can similarly influence the mode and extent of twinning. In bcc crystal structure metals, the influence of texture on twinning has also been linked to anisotropic yield behavior. At room temperature, the yield stress of polycrystalline niobium is approximately the same in tension and compression. At 77 K, however, where the compression samples deform predominantly by twinning and the tensile samples deform by slip, the yield stress in compression is lower than that in tension.

The exact influence of second-phase particles on twin formation has not been self-consistently defined. Mechanical twinning in Fe–24.6 at.% Be alloys proceeds almost unimpeded through the $FeBe_2$ particles when the second phase consists of isolated particles. But when the dispersion is in the form of finely distributed solute-atom clusters (as in the initial stages of precipitation) deformation twinning is completely suppressed. In studies of shock-loaded nickel containing thoria particles, the thoria particles prevented the formation of twins or a dislocation cell substructure. The suppression of twinning may be caused by the thoria dispersoids affecting local dislocation motion. However, recent shock-loading studies in which 6061-T6 Al was subjected to a 13 GPa shock found that twins did form in this alloy in the presence of precipitates.

Experiments have shown that prior deformation can suppress deformation twinning during subsequent deformation. Tin crystals deformed extensively by slip at room temperature do not twin upon reloading at room temperature. At lower temperatures, however, extensive slip in tin does not inhibit subsequent twinning. Systematic studies on the twinning response of niobium crystals have similarly found that prestraining, in this case resulting in a substructure of tangled dislocations at cell walls, inhibited twinning during subsequent deformation even at low temperatures. Preexisting dislocation substructure has also been observed to suppress twinning in shock-loaded materials, such as iron and molybdenum.

Collectively, the results of prestraining experiments on twinning pose the question of whether or not the preexisting substructure affects the nucleation or the propagation stage of twin formation. Prestraining will increase the dislocation density and correspondingly the flow stress, making further deformation by slip more difficult. In this regard, the increase in the applied stress necessary to continue deformation will favor twinning. The same dislocation substructure, however, may directly interfere with twin formation mechanisms and suppress twinning, but the evidence is insufficient to differentiate the exact mechanism causing twin suppression.

3.4 Twinning of Ordered Alloys

Deformation twinning of superlattice structures is receiving increasing attention in the literature, both in experimental studies of deformation mechanisms of these materials and in theoretical papers on twinning in ordered structures. Initial theoretical considerations of twinning in ordered cubic crystals predicted that mechanical twins were improbable since the twin would not possess the same crystal structure as the parent lattice. A high shear is necessary in ordered structures to restore the lattice in the twin to that of the original crystal structure. Theoretical discussions of twinning in superlattices have most recently concentrated on differentiating between "true" twins and "pseudo" twinning (or between twinning and "twinning") to describe whether the order and symmetry in twin formation in the superlattice is accomplished solely by shear or if lattice shuffles are required to restore local crystallographic order. Experiments have yielded features thought to be deformation twins in several different superlattice structures deformed via various loading paths. In the $L1_2$ structure, twins (thought to be true twins) have been observed in Ni_3Al as well as in shock-loaded Cu_3Au (in both the ordered and disordered conditions). Twin nucleation in compression specimens of ordered Cu_3Au twinning has only been observed after some plastic deformation by slip.

The three intermetallic compounds of the titanium–aluminum binary system (i.e., Ti_3Al, TiAl and Al_3Ti) are being extensively studied because they have attractive properties as high-temperature structural materials. Ti_3Al has an ordered hexagonal ($D0_{19}$) structure while TiAl is $L1_0$ and Al_3Ti is $D0_{22}$ in structure. Studies of Ti_3Al deformed to 900 °C (still fully ordered) have found no twinning; at 900 °C some extended faults or microtwins were seen. In TiAl, one type of twin $\{11\bar{1}\}\langle 112\rangle$ plays an important role in plastic deformation with the number of operative twinning systems depending on both orientation and load direction. Contrary to the case of disordered crystal structures, increasing the temperature to above 700 °C in TiAl increased the incidence of twinning. This increase occurs because above 700 °C the a/6 [112] partial dislocation (the twinning dislocation) is no longer pinned, so that twinning assumes increasing importance as a deformation mode for TiAl.

Observations of deformed specimens by TEM have revealed that the most notable feature of the deformation structure of Al_3Ti ($D0_{22}$) is an abundance of deformation twins. These twins are always one of four $\{111\}\langle 11\bar{2}\rangle$ systems. Twins of this type are likely to result from the propagation of Shockley partials with Burgers vectors of $1/6\langle 11\bar{2}\rangle$, similar to those seen in twinned Ni_3V. Ternary alloying additions are currently under investigation as a means of increasing the room temperature ductility of intermetallics. Boron, which improves the ductility of nickel-base $L1_2$ compounds, and the group IV and V elements are likely candidates. Ductility of Al_3V, which is isostructural to Al_3Ti, is improved substantially by the partial replacement of vanadium by titanium. The increased ductility of Al_3V with 5% Ti has been attributed to increased activity of "ordered" twinning in Al_3V. Further investigations of single and two-phase superlattice-base alloys will continue to expand our understanding of the twinning behavior of ordered alloys.

4. Influence of Twinning on Plasticity and Fracture

Although many metals and alloys display deformation twinning under various experimental conditions, twinning remains secondary to deformation by slip. The exact extent to which deformation twinning can affect plastic flow in polycrystalline materials appears to depend on the amount of total strain linked to twinning. In high-symmetry crystal structures, where twinning is far less prevalent than in low-symmetry structures, the influence of twinning on flow behavior (i.e., flow stress and work-hardening rate) is usually small. The stress–strain curves of polycrystalline samples of Cu–4.9 at.% Sn, which twins easily at room temperature, exhibit a reduced work-hardening rate related to the onset of twinning. In magnesium (hcp), stress–strain studies have shown that when deformation occurs by a combination of slip and twinning, or by slip in a heavily twinned matrix, a high work-hardening rate can result. Overall, experimental observations on bcc, fcc and hcp metals have shown that when conditions promote deformation *primarily* by twinning, the trend is for the flow stress to have a positive temperature dependence and a negative strain-rate dependence.

The association of twinning with fracture, in particular whether one process can be definitely said to cause the other, remains unclear. When a twin is arrested within a crystal, or within a grain in a polycrystal, a large stress must be relaxed to locally accommodate the twinning shear. Where a twin terminates, such as at twin–twin or twin–grain boundary intersections, a hole or crack may develop if the matrix is unable to accommodate the local shear. Twins can also serve as preferred propagation paths for growing cracks. Experiments on iron–silicon single crystals found that cracks were a preferred accommodation process when nonparallel shear vector twins intersected. Intergranular cracking in tungsten and molybdenum is also linked to twin impingement at grain boundaries. Investigations of the fracture behavior of α-titanium containing 2500 ppm oxygen found that microcracks were initiated at twin boundaries, most commonly at the matrix interface of second-order twins formed in primary $\{11\bar{2}2\}$ twins. But the critical question remains of whether or not twin nucleation of cracks is responsible for the failure of polycrystalline materials. Fracture studies on molybdenum, in both single crystals and polycrystals, have demonstrated that while twin-induced microcracking often causes fracture in the single crystals, intergranular fracture in the polycrystalline molybdenum is slip-induced.

See also: Brittle Fracture: Micromechanics; High-Rate Deformation of Metals (Suppl. 1); Hugoniot Curve (Suppl. 1); Plastic Deformation: Thermally Activated Glide of Dislocations; Time-Dependent Fracture: Mechanisms

Bibliography

Bevis M, Crocker A G 1968 Twinning shears in lattices. *Proc. R. Soc. A* 304: 123–34

Bilby B A, Crocker A G 1965 The theory of the crystallography of deformation twinning. *Proc. R. Soc. A* 288: 240–55

Cahn R W 1954 Twinned crystals. *Adv. Phys.* 3: 363–445

Christian J W 1965 *The Theory of Transformations in Metals and Alloys.* Pergamon, New York, Chap. 20, pp. 743–801

Christian J W, Laughlin D E 1988 The deformation twinning of superlattice structures derived from disordered

B.C.C. or F.C.C. solid solutions. *Acta Metall.* 36: 1617–42
Gray G T III 1988 Deformation twinning in Al–4.8 wt% Mg. *Acta. Metall.* 36: 1745–54
Hirth J P, Lothe J 1968 *Theory of Dislocations*. McGraw-Hill, New York, Chap. 23, pp. 738–60
Klassen-Neklyudova M V 1964 *Mechanical Twinning of Crystals*. Plenum, New York
Mahajan S, Williams D F 1973 Deformation twinning in metals and alloys. *Int. Metall. Rev.* 18: 43–61
Reed-Hill R E, Hirth J P, Rogers H C (eds.) 1963 *Deformation Twinning*. Gordon and Breach, New York
Reid C N 1981 The association of twinning and fracture in BCC metals. *Metall. Trans., A* 12: 371–7
Remy L 1981 The interaction between slip and twinning systems and the influence of twinning on the mechanical behavior of FCC metals and alloys. *Metall. Trans., A* 12: 387–408
Yamaguchi M, Umakoshi Y, Yamane T 1987 Deformation of the intermetallic compound Al_3Ti and some alloys with an Al_3Ti base. In: Stoloff N S, Koch C C, Lin C T, Izumi O (eds.) *High-Temperature Ordered Alloys II*. MRS, Pittsburgh, PA, pp. 275–86
Yoo M H 1981 Slip, twinning, and fracture in hexagonal close-packed metals. *Metall. Trans., A* 12: 409–18

G. T. Gray III
[Los Alamos National Laboratory, New Mexico, USA]

Dental Amalgams: Further Developments

This article supplements the article *Dental Amalgams: Composition, Fabrication and Trituration* in the Main Encyclopedia.

Dental amalgam is prepared by mixing an alloy powder with mercury to produce a plastic mixture which is condensed into a prepared cavity. It has maintained its popularity as a filling material over the years since it is durable, inexpensive and requires only one dental appointment for placement.

1. History of Dental Amalgams

During the T'ang Dynasty, *c.* 700 AD, Chinese physicians used a silver–tin amalgam as a filling material (Hsi-T'ao 1958). Historical accounts of amalgams (Craig 1985, Roggenkamp 1986) state that amalgams were not used in the West until much later. Following the development of a silver–mercury paste by Traveau in France in 1826, amalgams for the restoration of teeth were introduced into the USA by the Crawcour brothers in 1833. Although the use of amalgams has always been controversial, 150 years of research in both fabrication technology and alloy chemistry has greatly improved the quality of the amalgams currently available.

The first systematic investigations of silver–tin alloys and their amalgams were started in the nineteenth century by J. Foster Flagg and were continued by G. V. Black. In 1929, an alloy similar to Black's ternary silver–tin–copper alloy was approved by the American Dental Association (ADA) in the American Dental Association Tentative Specification for Dental Amalgam Alloys. This was revised several times until the current ADA specification was adopted in 1977 and formed the basis for the original International Organization for Standardization (ISO) Standard 1559 adopted in 1978.

The required limits of chemical composition (in wt%) outlined in the previous 1969 ADA specification were similar to those of Black's alloy: a minimum of 65% silver and maximums of 29% tin, 6% copper, 2% zinc and 3% mercury. It was not until the 1960s that formulations that differed from the Black alloy were recognized. The first divergence was high-copper amalgam, produced by mixing a small amount of plasticized copper amalgam into low-copper amalgams. The second clinically significant modification of the amalgam was the development of a high-copper content amalgam by Innes and Youdelis (1963). This discovery did not receive much attention until 1970, when it was shown that amalgam restorations prepared from this high-copper powder were clinically superior to low-copper amalgams, especially in resisting marginal breakdown (Mahler et al. 1970). The work of Innes and Youdelis has since led to the development of numerous high-copper alloys.

2. Alloy Powder

The current ADA Specification No. 1 (1977) requires that the chemical composition of alloy powders consists essentially of silver and tin, but it allows for the inclusion of copper, zinc, gold and/or mercury in amounts smaller than the silver or tin content. Alloy powders containing zinc in excess of 0.01 wt% are designated as "zinc-containing". Those powders containing zinc less than or equal to 0.01 wt% are designated as "nonzinc". The present ADA specification does not list compositional ranges for a low- or high-copper alloy powder. In contrast, the most recently adopted ISO standard (1986) specifies a chemical composition (in wt%) of a minimum of 40% silver and maximums of 32% tin, 30% copper, 2% zinc and 3% mercury. Usually, an alloy powder containing less than 6% copper is classified as a low-copper alloy or traditional powder, whereas an alloy containing greater than 8–10% copper is classified as a high-copper alloy powder. The amalgams made from these two types of alloy powder are called low-copper and high-copper amalgams, respectively; usage of the former is becoming less common in developed countries.

Table 1
Ranges of chemical compositions (wt%) for commercially available copper alloys

	Low-copper alloys	High-copper alloys
Silver	66.7–71.5	39.9–70.1
Tin	24.3–27.6	17.0–30.2
Copper	1.2–5.5	9.5–29.9
Zinc	0–1.5	0–0.7
Mercury	0–4.7	0–0.25

As mentioned in Sect. 1, the first major change in alloy composition occurred in the 1960s when the silver–copper binary eutectic powder was mixed with a low-copper powder. The silver–copper eutectic particles were intended to "dispersion" harden the amalgams produced, though it is now known that, in strict metallurgical terms, this does not occur. The resulting copper-enriched powder was called "high-copper admixed" or "dispersant" powder.

The next major change occurred in 1974 when Asgar developed a new type of alloy for amalgams by preparing a powder from a single silver–copper–tin ternary alloy, rather than by mixing two kinds of powder together, and this was called high-copper single composition powder. The ranges of chemical compositions for commercially available copper alloys are given in Table 1.

2.1 Alloy Phases

Studies of the silver–tin–copper ternary equilibrium phase diagram date back as far as the 1920s. When an alloy is produced by normal melting and heat treatment, the major phase in alloys within the compositional ranges listed in Table 1 is always an intermetallic compound in the silver–tin binary system, Ag_3Sn, called the γ phase (rhombically deformed hexagonal-close-packed). A solid-solution phase of the silver–tin alloy system, the β phase (hexagonal-close-packed), is also sometimes present. Since copper dissolves in Ag_3Sn to approximately 1%, any excess copper will form copper-rich phases, most commonly copper–tin. Depending on the composition of the alloy and its thermal history, other phases which may be present are ε copper–tin (Cu_3Sn, hexagonal), η' copper–tin (Cu_6Sn_5, NiAs-type hexagonal) and/or α tin. Admixed powders contain a silver–copper eutectic (α silver plus α copper) powder and, in some cases, include a hypoeutectic silver–copper alloy powder. In the low-copper powder range, zinc forms Cu_5Zn_8 when the amount of zinc exceeds its solubility in the γ and/or β silver–tin (1.6% and 5.9%, respectively) (Jensen et al. 1973).

2.2 Shape of Particles in the Alloy Powder

The particles, usually 10–30 μm in size, used to prepare dental amalgams may be irregularly shaped powders fabricated by lathe-cutting or ball-milling the alloy ingot, or spherical particles made by atomizing molten metal. The shape of spherical particles depends on the cooling rate during fabrication, ranging from nearly perfect spheres to spheroidal shapes with irregular surfaces. In either case, particles are annealed under inert or reducing gas conditions, followed by acid washing in order to alter the reaction rate with mercury. Powders composed of irregular particles were used until spherical powders were introduced in 1962 (Demaree and Taylor 1962). Alloy powders currently in use may be composed entirely of irregularly shaped particles, entirely of spherical particles, or be a mixture of irregular and spherical particles. Although particle shape does not affect the reaction of the powder with mercury, dramatic differences in shape between lathe-cut and spherical powders can affect the handling characteristics of the dental amalgam.

3. Amalgamation Reaction

Modern dental amalgams are prepared by a process called "trituration": this entails vigorously mixing a powder or tablets with mercury for a short period of time (10–20 s) using a mechanical mixer.

3.1 Low-Copper Alloy

When trituration begins, the silver and tin in the outer surface layers of the particles dissolve into the mercury. Owing to the limited solubility of silver (0.035 wt%) and tin (0.06 wt%) in mercury at room temperature, two kinds of intermetallic compounds, Ag_2Hg_3 (body-centered-cubic) and Sn_8Hg (simple hexagonal) (Fairhurst and Ryge 1962, Fairhurst and Cohen 1972) start to precipitate soon after mixing. These compounds are called γ_1 and γ_2, respectively. In a dissolution–precipitation process, a species must be supersaturated before precipitation can begin. Silver reaches its saturation concentration in mercury, before tin so silver supersaturation results in the precipitation of γ_1 before the precipitation of γ_2. While the reaction takes place, the alloy powder and the precipitates coexist with the mercury, keeping the mix at a plastic consistency for approximately 10–15 min.

The alloy is usually mixed with 40–50 wt% mercury, which is insufficient to consume the alloy particles completely. Consequently, unconsumed particles are present in the hardened amalgam and these become surrounded and bound together by minute equiaxed γ_1 grains (1–3 μm in size) and irregularly shaped, discrete γ_2 crystals. The amalgam is thus a composite in which alloy particles of approximately 30 vol.% are embedded in γ_1 (60%) and γ_2 (10%) phases (Mahler 1988). The structure also includes ε-Cu_3Sn particles and voids in the γ_1 matrix (Phillips 1982).

The copper included in the powder, in the form of either dissolved atoms in the γ-Ag_3Sn lattice or atoms in the ε-Cu_3Sn when copper exceeds its solubility in γ-Ag_3Sn, participates in the reaction in a similar manner to that described in Sect. 3.2 for high-copper alloy. However, copper does not affect the amalgamation reaction to a noticeable level since it is of a low concentration in the alloy.

3.2 High-Copper Alloy

In the case of high-copper admixed alloys, silver and tin enter the mercury from the γ and/or β silver–tin powder, silver also entering the mercury from silver–copper alloy particles. Silver from both sources precipitates as γ_1-Ag_2Hg_3 grains. If sufficient copper is available, the dissolved tin combines with copper to form η'-Cu_6Sn_5 crystals rather than forming γ_2-Sn_8Hg. In most high-copper amalgams, formation of the γ_2 phase is totally suppressed, which is important since γ_2 is the weakest, softest phase of the amalgam phases (Craig 1985). In the admixed powder reaction, the η' crystals form at two different sites: at the surface of the silver–copper eutectic particles and at the γ_1 matrix (Okabe et al. 1979b). Apparently, before supersaturation, the dissolved tin diffuses to the surface of the eutectic particles to combine with copper to form η' crystals. Copper in the high-copper powder acts as a sink for tin migration. As the amalgamation reaction proceeds, the reaction layer (usually 1–2 μm thick) forms on the particle surface. This layer is a mixture of ultrafine η' crystals (approximately 30 nm) (Okabe et al. 1977) and a lesser amount of γ_1 crystals. Saturation of silver occurs after nucleation and growth of η' crystals. The structure of high-copper admixed amalgams includes the unconsumed low-copper alloy particles, unconsumed silver–copper particles, and, sometimes, ε-Cu_3Sn particles embedded in the γ_1 matrix. In addition, individual thin η' rod crystals, which are believed to have formed by the dissolution and precipitation process, are embedded within γ_1 crystals in the high-copper admixed amalgam.

Similarly, in the amalgamation reaction of the high-copper single composition alloy powder (Okabe et al. 1978a, b), silver and tin dissolve from the γ and/or β silver–tin phase in the particle. Concurrently, copper dissolves mainly from the ε-Cu_3Sn phase, which coexists with the silver–tin phases in the silver–tin–copper alloy particles. Tin atoms dissolved in the mercury are attracted to the ε-Cu_3Sn areas located on the alloy powder surfaces where meshes of η'-Cu_6Sn_5 crystals form. The η' crystals in these amalgams are much larger than those found in high-copper admixed amalgams. Included in the structures are unconsumed alloy particles, η' copper–tin crystals and the γ_1 matrix. An important finding is that unconsumed alloy particles are covered by a number of η'-Cu_6Sn_5 rod crystals. Other η' crystals are seen embedded within the γ_1 grains.

4. Mechanical and Physical Properties

4.1 Mechanical Properties

Sufficient strength to resist clinical fracture is a prerequisite for any restorative material. Extensive investigations of mechanical and physical properties have been made with amalgams prepared from experimental and commercial alloy powders, producing a variety of microstructures and phase distributions. Specimens made from these different amalgams have been subjected to many mechanical tests including measurements of tensile, shear, compressive and bending strengths, as well as being subjected to fatigue, creep and, more recently, fracture toughness tests. Thus far, only creep tests have shown a significant correlation with clinical performance (Mahler et al. 1970). The extent of marginal fracture of amalgam restorations increases with creep rate and, consequently, national and international specifications for amalgams now include creep values. For example, the ISO Standard 1559 for creep measures the percentage change in the length of week-old specimens over a three-hour period under a constant compressive stress (36 MPa) at 37 °C. Recent emphasis has been placed on studying amalgam deformation at slow-strain rates because of the established relationship between creep and clinical performance.

Table 2 consists of a summary of the mechanical properties of types of amalgam. The compressive strength ranges given for 30 min, 1 h and 24 h after specimen preparation demonstrate how amalgams strengthen as amalgamation progresses; most commercial amalgams reach their final strength within 24 h. The compressive strength ranges for the three types of amalgam overlap each other. However, the values indicate that high-copper alloys could have higher compressive strengths. Both the diametrical tensile strength and elastic modulus for each type of amalgam fall into a similar range. The much lower values in tensile strength ranges as compared with those of compressive strength reveal the brittleness of these amalgams. On the other hand, there is a dramatic difference in creep rates between low-copper and high-copper amalgams, indicating that the low-copper amalgam is much more plastic.

Published studies (e.g., Mitchell et al. 1987) demonstrate that the fracture mode and strength of the amalgam are strongly sensitive to strain rate. At all strain rates, the mechanical properties of dental amalgams are controlled by the γ_1-based matrix. At a high strain rate, amalgam is very brittle and γ_1 grains undergo intergranular fracture. The plasticity of γ_1 and γ_2 increases as the strain rate is lowered; at extremely low strain rates both γ_1 and γ_2 grains exhibit a high degree of plastic deformation, producing many fine "needles". In addition, significant grain boundary sliding of the γ_1 grains becomes evident (Okabe et al. 1983). The sliding rate of γ_1

Table 2
Ranges of strengths and creep values of low-copper and high-copper amalgams

Amalgam	Compressive strength (MPa)[a]			Diametral tensile strength (MPa)	Elastic modulus[b] (GPa)	ISO Creep (%)
	30 min	1 h	24 h			
Low-copper		124–200	310–390	53–62	20–40	0.70–3.70
High-copper admixed	55–110	80–180	330–500	42–62	35–50	0.05–2.10
High-copper single composition	70–210	100–320	320–510	34–64	30–60	0.03–0.040

a loading rate $=4.2\times10^{-3}$ mm s^{-1} b loading rate $=2.0\times10^{-3}$ mm s^{-1}
Table compiled from data of Osborne et al. (1978), Duke et al. (1982), Bryant (1979, 1980), Bryant and Mahler (1986)

grains increases when the structure contains γ_2, the softest component among the amalgam phases. Therefore, low-copper amalgams, all of which contain γ_2, have lower creep resistance than high-copper amalgams.

A restraint on sliding is operative in high-copper amalgams. When η' crystals span γ_1 grain boundaries, the γ_1 grains must move plastically around much harder η' crystals for sliding to occur; such grain boundaries slide very slowly. Another example of a durable interface is that between the high-copper alloy particles, which are covered by a dense mesh of minute η' crystals, and the γ_1 matrix. The presence of η' crystals, therefore, increases resistance to creep deformation.

Crack nucleation and propagation in amalgams at various strain rates has been studied: at high ($\sim10^{-1}$ mm s^{-1}) and moderate ($\sim10^{-3}$ mm s^{-1}) loading rates, cracks have been found to nucleate at different interfaces between γ_1 and other components and to propagate through the γ_1 matrix. Failure is either by intergranular separation of γ_1 or by cleavage. The tendency for cleavage to occur increases with strain rate and with the percentage of η' crystals.

At low strain rates (10^{-6} mm s^{-1} or less), γ_1 grains fracture transgranularly in a ductile manner. Cracks nucleate near grain boundaries; however, cracks remain within the grain so that individual γ_1 grains fracture by microvoid coalescence and dimpled "cup and cone" rupture surfaces are observed. At higher strain rates, the activation of multiple slip systems required for plastic deformation is blocked by the inability of superdislocations within the γ_1 intermetallic to cross-slip. At low strain rates, dislocation climb reactivates the required slip systems and plastic deformation becomes possible.

4.2 Dimensional Change

The amalgamation reaction continues long after the amalgam is condensed into the prepared cavity. Since the reaction includes the dissolution of alloy particles, the precipitation of crystals and the growth of crystals, dimensional changes occur during hardening. An ideal amalgam for use as a restorative material would neither expand nor contract. In theory, an unconstrained amalgam will first contract due to dissolution and then expand towards the end of the reaction as impinging crystals push one another apart. The dimensional change of the amalgam is the result of the total effects of the type of alloy used and, more importantly, manipulative variables such as mercury percentages, trituration conditions and condensation technique. The ISO specification states that the dimensional change should be measured at 37 °C, between 5 min and 24 h after the preparation of the specimen. Currently, its value is regulated to be within 0±0.20% or 0±20 μm cm^{-1}.

Dimensional change is an important clinical problem, since it is thought to relate to microleakage at the interface which results in postoperative sensitivity and possibly secondary caries. A recent *in vitro* study recommended that the specification be revised to reflect the fact that microleakage involves not only dimensional changes but also the surface texture of the amalgam at the cavity wall. In addition to the evaluations of the dimensional changes of various amalgams *per se*, much research has been devoted to investigating the effects, for example, of cavity varnishes and thermocycling on microleakage. Mahler (1988) summarizes the present limited information.

4.3 Corrosion Resistance

Deterioration of metals in the mouth may be caused by direct chemical action and/or electrolytic processes. Dental amalgam is subject to tarnishing and electrochemical corrosion. Over a period of time, tarnish or surface discoloration may develop into corrosion which reduces the structural integrity of the amalgam. Electrochemical evaluation of various amalgam phases reveals that the γ_2 phase is considerably more anodic than γ_1 and γ-Ag_3Sn. The absence of γ_2 in high-copper amalgams results in a dramatic improvement in their corrosion resistance (see *Corrosion of Dental Materials*). High-copper amalgams also suffer some corrosion, however, since η' is anodic with respect to the surrounding mercury-containing phases.

In most oral environments both types of amalgams are initially in a state of passivity and little corrosion takes place. However, high acidity and a high concentration of chloride ions lead to the development of crevices and pores which destroy passivity and allow corrosion to occur. The surface of amalgam restorations should thus be as free of pores and crevices as possible.

Crevice corrosion can be beneficial, as corrosion products at the interface between the tooth structure and amalgam appear to contribute to the seal that forms between the restorative and the tooth. This seal reduces microleakage. In studies, the following insoluble corrosion products have been identified: SnO, SnO_2, $Sn_4(OH)Cl_2$, $CuCl_2 \cdot 3Cu(OH)_2$, Cu_2O, $CaSn(OH)_6$, $ZnSn(OH)_6$, $Zn_5(OH)_8Cl_2 \cdot H_2O$ and $Zn(OH)Cl$.

A contact between dissimilar metals in the same electrolyte, in this case saliva, results in galvanic corrosion in which the less noble metal suffers more intensive deterioration. Since dental amalgams are less noble than most other dental alloys, amalgam restorations should not be placed in continuous contact with cast restorations, especially alloys of gold and palladium. Even intermittent contacts of dissimilar restorations in opposing teeth will result in the flow of a galvanic current which may cause a discomfort to the patient.

5. *Marginal Fracture and Bulk Fracture*

Many clinical studies show that high-copper amalgams have a significantly lower rate of marginal breakdown than low-copper amalgams. Figure 1 shows surfaces of both low- and high-copper amalgam restorations after four years of service. The photograph clearly shows the difference in their clinical performance: the margins of the low-copper amalgam restoration (right) are severely ditched. In addition, the surface near the interproximal area of the low-copper amalgam has been indented because of the action of the opposing tooth cusp. This is an indication of its low creep resistance. Some studies show that restorations with less breakdown need to be replaced as a result of recurrent caries less often, though not all studies agree with this. Continued research on marginal fractures should clarify the situation.

Research has focused on understanding the mechanism of marginal breakdown. Mahler et al. (1970) demonstrated a correlation between the creep of γ_2-containing amalgams and marginal fracture. There have been many attempts to explain this

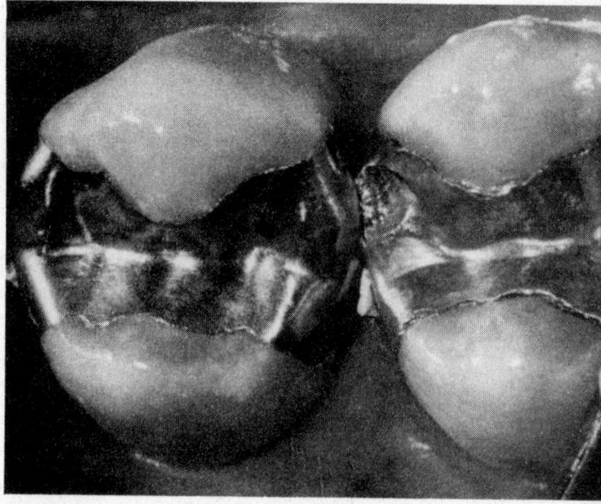

Figure 1
Four-year-old low-copper (right) and high-copper (left) amalgam restorations (courtesy of D. B. Mahler)

relationship, a large number placing primary emphasis on the role of corrosion (e.g., Sarkar 1978). It has been reported that amalgam restorations with the largest quantities of corrosion products had the worst margins; however, other studies have shown that marginal cracks develop in restorations even in the absence of significant corrosion. A recent study of amalgam restorations in denture teeth correlated increasing marginal fracture with the length of time in service (Marker et al. 1987). The mode in the fracturing process appeared to be similar to that reported for natural teeth. Examination of cross sections of these specimens revealed that marginal breakdown occurred even when no defect was apparent near the margins of freshly placed amalgam restorations. Evidence of plastic deformation was observed near crack tips, suggesting that crack nucleation occurred at low strain rates since only at such rates can this plasticity in amalgams occur. Once nucleated, cracks propagate and final fracture occurs at high strain rates.

Explanations of marginal breakdown that rely primarily on mechanical deformation have also been offered. It has been proposed that phase changes in the amalgam or corrosion products in the interface might produce hydrostatic stresses tending to extrude an amalgam, and it has also been suggested that long-term linear expansion might result in extrusion and lead to marginal breakdown. Tooth deflection is another factor that might explain marginal breakdown. Calculations showed that creep resulting from compressive stresses at the margins tended to produce a gap there (Derand 1977). Any change in cavity design tending to increase the deflection of the cusp will increase this gap; for example, deeper or wider preparations would increase the deflection. In this light, it is interesting that significantly greater marginal breakdown has been reported in large restorations. Moreover, it has been found that marginal breakdown increases with biting force.

Several studies have shown that high-copper amalgams undergo less bulk fracture than low-copper amalgams. Since low- and high-copper amalgams are weak in tension, both types of amalgam would be expected to be equally brittle. It is possible that easier crack nucleation at low strain rates could account for the brittleness of γ_2-containing amalgams.

6. *Mercury Release*

Controversy has existed over the use of dental amalgams for the restoration of teeth ever since the technique was introduced. The reason is apparent: mercury without oxide film has a high vapor pressure at room temperature (1.20×10^{-3} torr at 20 °C) and evaporates freely (Okabe 1987). Thus, there has always been a belief that amalgam restorations could cause mercury poisoning by chronically exposing the dental personnel and patient to mercury vapor. In many countries, there has been an increased public awareness and professional concern over the release of mercury from amalgam. This concern has been fuelled by the introduction of highly sensitive mercury vapor detectors with detection limits as low as $1\ \mu g\ m^{-3}$ in air or $0.2\ \mu g\ l^{-1}$ in solution. Measurable amounts of mercury vapor have been reported in the intraoral air of patients with amalgam restorations, and there have been a number of reports of mercury in breath, saliva, blood, nails and hair, resulting from both mercury release and various clinical procedures. Many of these studies have been reviewed in a recent article by Langan et al. (1987). However, current reports reveal that far less mercury is actually released from amalgam restorations than was previously suggested. It is known that mercury is liberated *in vivo* and that the release level temporarily rises when the amalgam is abraded. In *in vitro* studies, mercury release increases when the amalgam is heated in an oxygen-free atmosphere or after it has been corroded. However, in air, processes such as aging, surface oxidation and water film formation reduce mercury vaporization from the amalgam.

An x-ray diffractometry study showed that the rate of reaction of mercury during amalgamation is dependent on the alloy used and that, in a normal situation, all mercury used for trituration is consumed within several hours (Ferracane et al. 1986). However, formation of the solid phases and the consumption of the mercury does not mean that amalgamation has ended. The peritectic isotherms of some of the amalgam phases are near the highest service temperature of the amalgam (60–80 °C), that is, 127 °C for the γ_1-Ag_2Hg_3 phase and 216 °C for the γ_2-Sn_8Hg phase. The homologous temperature (the ratio of the ambient temperature to the melting point temperature, T/T_m) could thus be as high as 0.8. Body temperature is high enough to initiate the solid-state diffusion process necessary to change an amalgam from an initial nonequilibrium structure to a structure that is closer to equilibrium. These diffusional processes result in the decomposition of both the γ_1 and γ_2 phases and the transformation of the γ_1 into the β_1 phase, which contains less mercury than γ_1. While these structural changes are taking place, it is probable that some mercury is liberated by evaporating into mouth air, by dissolving into oral fluids or by diffusing into the tooth structure. However, at the same time, unconsumed alloy particles are likely to be a sink for the mercury produced as a result of such changes.

None of the *in vivo* and *in vitro* corrosion products listed above include mercury in their structure and, consequently, mercury liberated during the corrosion process must either go into solution or reamalgamate with unconsumed alloy particles. Studies of amalgam dissolution into various electrolytes give conflicting results: some investigators report more release of mercury from high-copper amalgams than from low-copper amalgams, but others show the opposite tendency (Okabe 1987). However, almost all of the data are consistent in that the rate of dissolution of mercury decreases drastically with time of immersion. The dissolution rate slows after releasing atoms located near the surface, dissolution becoming a solid-state diffusion control process. Oxide film formation is another factor which slows down the release of metallic ions to the media. There is no doubt that mercury dissolves at different rates from various amalgam phases, dissoluton being found to be greatest from γ_1 though this is reduced in the case of dental amalgams where γ_1 is galvanically coupled with the less noble phases γ and γ_2. This coupling reduces the driving force for mercury release since it keeps the corrosion potential of the system at a low level.

As for the toxicity of mercury and the biocompatibility of amalgams, recent reviews (Craig 1986, Langan et al. 1987, Mahler 1988) summarize the literature published thus far. Although occupational exposure to mercury is a potential hazard for dental personnel, mercury can be handled in the dental office without exceeding limiting values. There is no evidence in the published data that the minute amount of mercury released from amalgam restorations into oral air, saliva or tooth structure can cause mercurialism to patients, except for those who have allergies to mercury, and, furthermore, such allergies are rare.

It has been estimated that the amount of mercury vapor inhaled during the first three-hour period after placement of an amalgam restoration with a surface area of 50 mm^2 is 200 ng, approximately 10% of the normal daily human exposure to mercury through air, water and food intake. Amalgam restorations are unlikely to expose a patient to mercury vapor levels near the threshold limit value (TLV) of 0.05 mg m^{-3} per 40 h working week as set by the US National Institute for Occupational Safety and Health. If several large restorations are releasing enough mercury to keep the mercury vapor concentration at the TLV level, all the mercury in the restoration would evaporate in a few years. Clearly, the life span of amalgam restorations is much longer than this; hence, mercury vapor from restorations cannot exceed the TLV, except very intermittently (Okabe 1987).

7. *Selection of Alloys*

Since numerous alloys for dental amalgams are available, selecting the best amalgam for use can be confusing. Creep and compressive strength are two important factors to consider when making a selection. An amalgam with an ISO creep value of less than 1.0% is likely to show good marginal integrity for a long time span.

The compressive strength one hour after preparation of the amalgam, sometimes called "early strength," is another important factor. Most high-copper single composition amalgams have much higher early strengths than high-copper admixed amalgams. It is generally recommended that impressions of newly carved amalgam posts are not taken; nevertheless, such clinical practices are becoming more widespread. If crown preparations are to be made on freshly placed amalgam, one with high early strength should be selected. Another advantage of high early strength in terms of amalgam polishing has been suggested—that some such high-copper amalgams can be polished 10 min after the end of trituration without harmful effects.

Some generalizations can be made about single composition and admixed high-copper amalgams. Both can have low creep, mainly due to the absence of γ_2 in both types. Since low creep is important, either type of high-copper amalgam will be an improvement over low-copper amalgams. Compressive strengths of single composition amalgams tend to be higher. However, both types of high-copper amalgam are more than strong enough in compression. Admixed amalgams resist condensation well since most admixed amalgams contain irregular particles whereas the totally spherical, single composition amalgams are mushy. Amalgams which resist condensation are generally easier to adapt to cavity walls and may exhibit less leakage. If it is necessary to work with the restoration immediately after placement, single composition amalgams should be used owing to their high early strength. If strength properties are less important, admixed amalgams may be preferred because of their higher resistance to condensation.

Finally, it should be noted that the performance of even the best of these amalgams is influenced by operator technique. High-copper amalgams have been shown to be less technique-sensitive than low-copper amalgams. Nevertheless, careful cavity designing and preparation, condensing, carving and polishing are still essential. High-copper amalgams, placed by skilled dental personnel, allow patients to receive higher quality, possibly longer lasting restorations.

Bibliography

Bryant R W 1979 The strength of fifteen amalgam alloys. *Aust. Dent. J.* 24: 244–52

Bryant R W 1980 The static creep of amalgams from fifteen alloys. *Aust. Dent. J.* 25: 7–11

Bryant R W, Mahler D B 1986 Modulus of elasticity in bending of composites and amalgams. *J. Pros. Dent.* 56: 243–8

Craig R G 1985 *Restorative Dental Materials*. Mosby, St Louis, MO, pp. 198–200

Craig R G 1986 Biocompatibility of mercury derivatives. *Dent. Mater.* 2: 91–6

Demaree N C, Taylor D F 1962 Properties of dental amalgams made from spherical alloy particles. *J. Dent. Res.* 41: 890–906

Derand T 1977 Marginal failure of amalgam class II restorations. *J. Dent. Res.* 56: 481–5

Duke E S, Cochran M A, Moore B K, Clark H E 1982 Laboratory profiles of 30 high-copper amalgam alloys. *J. Amer. Dent. Assoc.* 105: 636–40

Fairhurst C W, Cohen J B 1972 The crystal structures of two compounds found in dental amalgam: Ag_2Hg_3 and Ag_3Sn. *Acta Crystallogr., Sect. B* 28: 371–8

Fairhurst C W, Ryge G 1962 X-ray diffraction investigation of the Sn–Hg phase in dental amalgam. In: Mueller W M (ed.) 1962 *Advances in X-Ray Analysis*. Pergamon, New York, pp. 64–70

Ferracane J, Mafiana P, Spears R, Okabe T 1986 Rate of liquid mercury consumption in freshly made amalgams. *J. Dent. Res.* 65: 192

Hsi-T'ao C 1958 The use of amalgam as filling material in dentistry in ancient China. *Chinese Med. J.* 76: 553–5

Innes D B K, Youdelis W V 1963 Dispersion strengthened amalgams. *Canad. Dent. Assoc. J.* 29: 587–93

Jensen S J, Olsen K B, Utoft L 1973 A zinc-copper phase in dental silver amalgam alloy. *Scand. J. Dent. Res.* 81: 572–6

Langan D A, Fan P L, Hoos A A 1987 The use of mercury in dentistry: A critical review of the recent literature. *J. Amer. Dent. Assoc.* 115: 867–80

Mahler D B 1988 Research on dental amalgam: 1982–1986. In: Reese J A (ed.) 1988 *Advances in Dental Research*, Vol. 2. International Association of Dental Research, Washington, DC, pp. 71–82

Mahler D B, Terkla L G, Van Eysdan J, Reisbick M H 1970 Marginal fracture vs mechanical properties of amalgam. *J. Dent. Res.* 49: 1452–7

Marker V A, McKinney T W, Filler W H, Miller B H, Mitchell R J, Okabe T 1987 A study design for an *in vivo* investigation of marginal fracture in amalgam restorations. *Dent. Mater.* 3: 322–30
Mitchell R J, Ogura H, Nakamura K, Hanawa T, Marker V A, Okabe T 1987 Characterization of fractured surfaces of dental amalgams. *Int. Symp. Testing and Failure-Analysis*. American Society for Metals, Metals Park, OH, pp. 179–87
Okabe T 1987 Mercury in the structure of dental amalgam. *Dent. Mater.* 3: 1–8
Okabe T, Butts M B, Mitchell R J 1983 Changes in the microstructures of silver-tin and admixed high-copper amalgams during creep. *J. Dent. Res.* 62: 37–43
Okabe T, Mitchell R J, Butts M B 1977 Analysis of Asgar–Mahler reaction zone in dispersalloy amalgam by electron diffraction. *J. Dent. Res.* 56: 1037–43
Okabe T, Mitchell R J, Butts M B, Fairhurst C W 1978a A study of high-copper amalgams, III. SEM observation of amalgamation of high-copper powders. *J. Dent. Res.* 57: 975–82
Okabe T, Mitchell R J, Butts M B, Fairhurst C W 1979a A study of high-copper dental amalgams by scanning electron microscopy. In: LeMay I, Fallon P A, McCall J L (eds.) 1979 *Microstructural Science*, Vol. 7. Elsevier, New York, pp. 165–74
Okabe T, Mitchell R J, Butts M B, Wright A H, Fairhurst C W 1978b A study of high-copper amalgams, I: A comparison of amalgamation on high-copper alloy tablets. *J. Dent. Res.* 57: 759–67
Okabe T, Mitchell R J, Fairhurst C W 1979b A study of high-copper amalgams, IV. Formation of η'Cu–Sn (Cu_6Sn_5) crystals in high-copper dispersant amalgam matrix. *J. Dent. Res.* 58: 1087–92
Osborne J W, Gale E N, Chew C L, Rhodes B F, Phillips R W 1978 Clinical performance and physical properties of twelve amalgam alloys. *J. Dent. Res.* 57: 983–8
Phillips R W 1982 *Skinner's Science of Dental Materials*. Saunders, Philadelphia, PA, pp. 311–15
Roggenkamp C L 1986 A history of copper in amalgam and an overview of setting reaction phases. *Quintessence Int.* 17: 129–33
Sarkar N K 1978 Creep, corrosion and marginal fracture of dental amalgams. *J. Oral Rehab.* 5: 413–23

T. Okabe
[Baylor College of Dentistry, Dallas, Texas, USA]

Dental Elastomers: An Overview

The term elastomer embraces those polymers which have the rubberlike long-range extensibility at low forces. They are polymers where intermolecular forces are so low that the energy for extension is used preponderantly to uncoil molecules; in thermodynamic language the energy increase is mostly expended in entropy changes.

Natural rubber was the precursor to the wide and chemically diverse range of synthetic elastomers now available. Until the early 1950s synthetic elastomers were, like natural rubber, thermoplastic solids in the uncross-linked (unvulcanized) state. When thermally softened, such materials are still extremely viscous (~10^{11} Pa s), requiring heavy machinery to mix in compounding ingredients and to shape by extrusion, calendering and molding. Then elevated temperatures (>100 °C) are necessary to effect vulcanization. Indeed, this is still true of the bulk of the rubber industry. However, from about 1950, polymers became available that were mobile, pourable fluids in the uncross-linked state, and which could be cross-linked at room temperature in relatively short times. Major examples are room temperature vulcanizing (RTV) silicone rubbers and fluid polysulfides (Jorczak and Fettes 1950, Nitzsche and Wick 1957). These systems are widely used in the sealants industry, but have also come to be the basis of the major group of dental impression materials, as compliant linings for dentures and as maxillofacial prosthesis materials.

It has already been stated that elastomers owe their characteristics to low intermolecular forces. However, otherwise glassy rigid polymers can be rendered elastomeric by the addition of a plasticizer (a mobile liquid which reduces intermolecular forces to convert the polymer to an elastomer). Such materials are widely used in dentistry as compliant prostheses.

1. Dental Impression Materials

A dentist will need to take an impression of the oral structures in three general circumstances:

(a) an edentulous mouth, for the subsequent fitting of full dentures;

(b) a partially edentulous mouth, including standing teeth, for a partial denture; and

(c) prepared standing teeth in a dentate patient for supplying a crown, a bridge or an inlay.

Impressions of children's teeth are also taken in the course of orthodontic treatment.

From impressions, models are then cast, usually from plaster of Paris. When impressions of a number of teeth and the surrounding oral structures are involved, it is clear that there are very intricate undercuts engaged by the impression material. Hence a fluid RTV elastomer is useful in that it will both flow into the required structures and will have the necessary compliance and elasticity to facilitate withdrawal and faithfully reproduce the structures involved.

These materials are supplied to the dentist as a two-pack system, usually in two tubes (Fig. 1a); one tube contains the polymer, incorporating fillers and other additives, and the other contains the cross-linking (vulcanizing) agent. Equal lengths of each are squeezed out onto a mixing pad and mixed with a spatula. (The example shown is a polysulfide; a

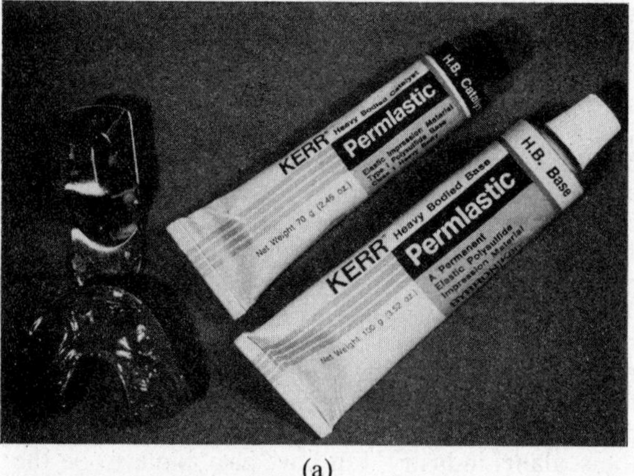

(a)

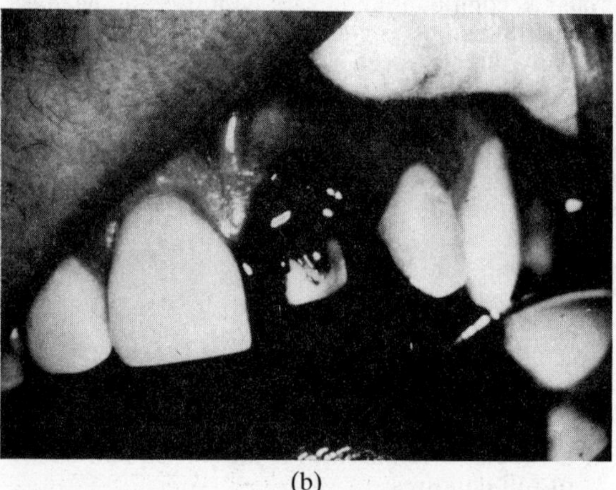

(b)

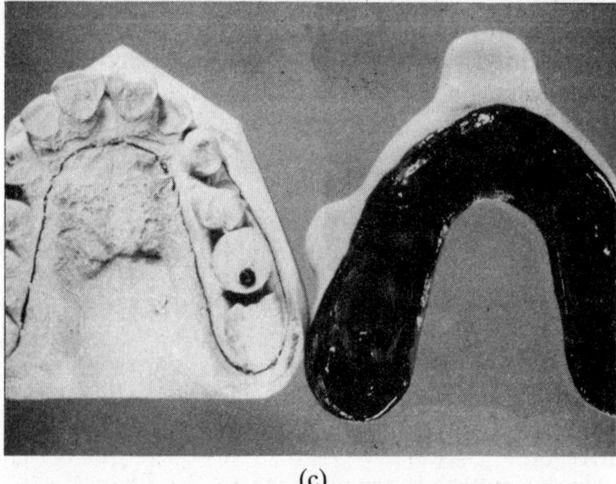

(c)

Figure 1
Polysulfide impression material showing: (a) tubes of the impression material, and the mixed material loaded in the impression tray; (b) the impression tray in the mouth; and (c) a set impression and the mold cast from it

brown paste is the cross-linking system which is lead dioxide in a liquid vehicle.) The mixed material is loaded onto an impression tray (Fig. 1a), inserted in the patient's mouth (Fig. 1b), withdrawn after setting (Fig. 1c) and a plaster mold cast.

Before describing such materials in detail, it is instructive to consider briefly the alternatives. Historically, a form of plaster of Paris was used; when set, this had to be broken in the patient's mouth, and subsequently reassembled jigsaw fashion! The first quasielastomeric material used in dentistry, and still in use, is a thermally reversible Agar–Agar gel (Phillips 1982). This material is an elastic gel at either room or mouth temperature; when heated to about 50 °C it is fluid, and in this state is placed in the mouth in a cored impression tray. When in position, cold water is run through the tray, and the material gels so it can be withdrawn. Such materials are weak, and shrink rapidly due to water evaporation. Hence immediate casting-up of the plaster mold is necessary. A much more common hydrolloid is based on sodium or a related salt of alginic acid (Phillips 1982). This is cross-linked (set) by a divalent salt in the formulation. It is widely used and is suitable for nonprecision work. However, for partial denture and more critical crown and bridge work, elastomers have the necessary strength to avoid tearing on removal from intricate undercuts, the elasticity to recover well and the necessary dimensional stability not to change significantly over several days. A multicomponent bridge is a very expensive item of treatment, and it is clearly necessary that it must fit. This in turn places stringent demands on the impression material that is used in its fabrication.

Elastomeric impression materials fall into three main classes: silicone, polysulfide, and an imine-terminated polyether. The first two are industrially used polymers adapted for dentistry; the third was developed especially for dentistry.

1.1 Silicone Impression Materials

There are a number of RTV silicone rubbers (Watt 1970), two of which are used in dentistry; namely the so-called "condensation" and "addition" types.

(*a*) *Condensation silicones*. These are α, ω-hydroxy terminated poly(dimethyl siloxanes) which are cross-linked by an alkoxy ortho-silicate or related reagent, in conjunction with an organotin activator (Fig. 2). Stannous octoate and dibutyl tin dilaurates are commonly used (Braden and Elliott 1966). The alcohol eliminated will slowly evaporate, giving rise to shrinkage (Fig. 3).

This type of silicone is commonly manufactured in what is termed the "putty" and "wash" system. The putty is a very highly filled system, which is used to make a first impression. The lightly filled very fluid

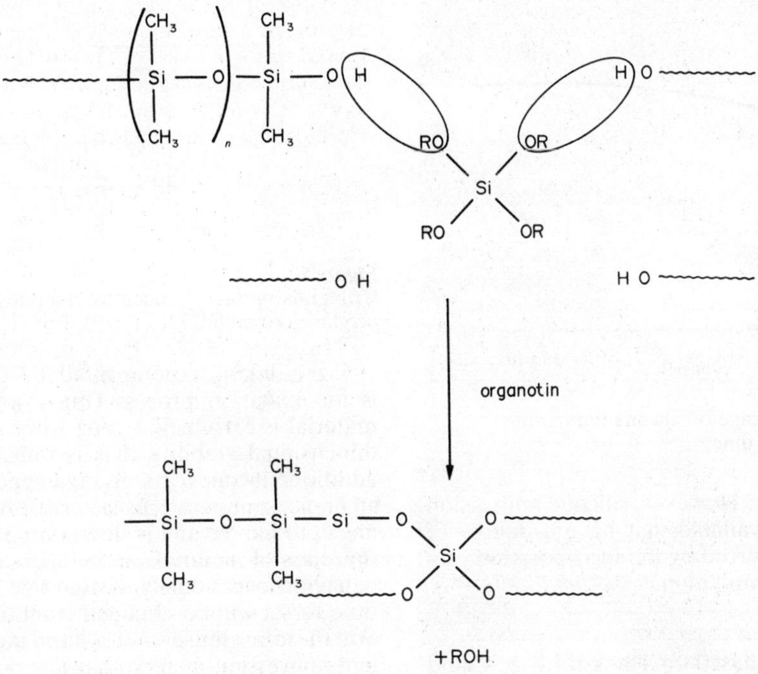

$$\sim\!\!\left(\begin{array}{c}CH_3\\|\\Si\\|\\CH_3\end{array}-O\right)_n\!\!-\begin{array}{c}CH_3\\|\\Si\\|\\CH_3\end{array}\!\!-O\,H \quad + \quad (RO)_4Si \quad + \quad H\,O\!\sim\ \xrightarrow{\text{organotin}}\ \sim\!\!\begin{array}{c}CH_3\\|\\Si\\|\\CH_3\end{array}\!\!-O-\begin{array}{c}CH_3\\|\\Si\\|\\CH_3\end{array}\!\!-Si-O-Si(O\!\sim)_3 \quad +ROH$$

Figure 2
Formation of condensation silicones

system is applied to the surface of the first impression, and reinserted in the mouth. This fills in the very fine detail (Fig. 4).

One disadvantage of this type of material is a limited shelf life. If, as is common, the cross-linking paste contains both the alkoxy ortho-silicate and organotin compound, they react together very slowly, particularly in the presence of tin. The net result is that the impression material will not set.

(*b*) *Addition silicones*. These are vinyl terminated poly(dimethyl siloxanes) and are historically the latest of the impression materials, introduced at the start of the 1980s. They are cross-linked with hydrosilanes as in Eqn. (1).

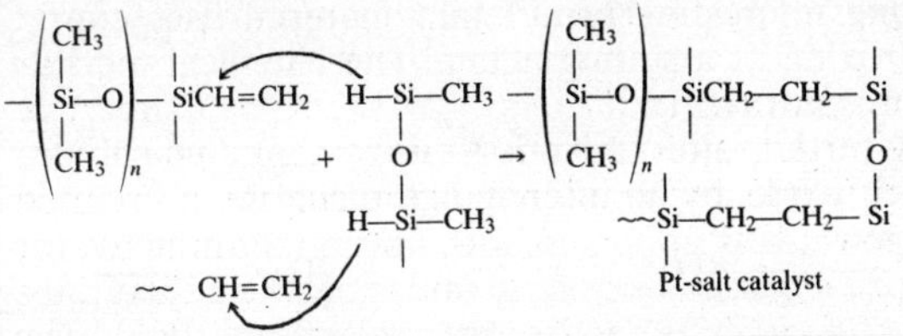

$$-\left(\begin{array}{c}CH_3\\|\\Si-O\\|\\CH_3\end{array}\right)_n\!\!-SiCH{=}CH_2 \quad + \quad \begin{array}{c}H-Si-CH_3\\|\\O\\|\\H-Si-CH_3\end{array} \quad + \sim CH{=}CH_2 \ \xrightarrow{\text{Pt-salt catalyst}}\ -\left(\begin{array}{c}CH_3\\|\\Si-O\\|\\CH_3\end{array}\right)_n\!\!-SiCH_2-CH_2-Si\begin{array}{c}|\\O\\|\end{array}\!\!\sim Si-CH_2-CH_2-Si \quad (1)$$

This type of material, having no by-products eliminated on cross-linking, is dimensionally extremely stable. Shrinkage of the order of only approximately 0.05% over several days is usually found. A number of features seem to have limited their widespread use.

(i) They are expensive.
(ii) If a plaster model is cast immediately with the set impression, residual Si—H groups react with water, with the evolution of hydrogen and the generation of a porous surface to the model. (At the time of writing this has apparently been overcome by incorporating finely divided palladium in the formulation.)
(iii) If the material is handled with rubber gloves, the sulfur in the rubber "poisons" the platinum salt catalyst, and the material does not set. (With the proliferation of the AIDS virus, the use of rubber gloves has become standard clinical practice. In this context it should be noted that sterilization of impressions is also standard practice, usually in glutaraldehyde solution.)

(*c*) *General properties of silicone impression materials*. All set silicone elastomers, with the exception of the highly filled "putty" materials, are extremely elastic, giving rapid and almost complete recovery from deformation. Residual strains of less than 1% are typical.

One disadvantage of silicone elastomers as impression materials is their hydrophobicity; this means that the unset impression wets the oral mucosa with difficulty, particularly if there is any accumulated blood or saliva. It is therefore necessary for the dentist to dry the appropriate area before

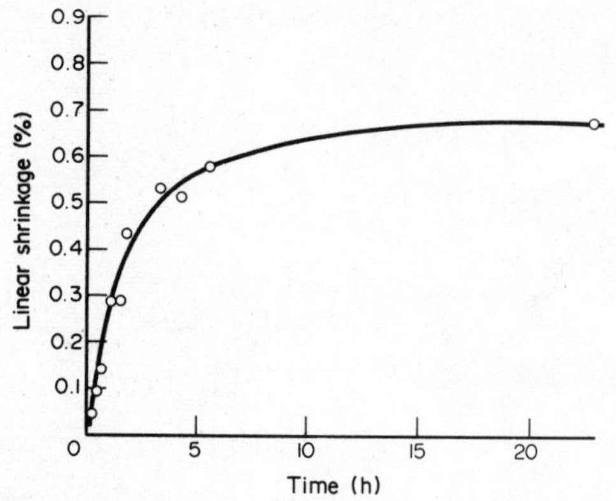

Figure 3
Plot of the linear shrinkage of silicone impression rubber as a function of time

taking the impression. However, silicone impression materials are now available that have a degree of hydrophobicity, conferred by the incorporation of a detergent in their formulation.

1.2 Polysulfides

These are invariably based on Thiokol LP-2, a fluid polysulfide (Braden 1966) having the general structure shown in Eqn. (2).

$$\begin{array}{l} HS{-}(CH_2{-}CH_2{-}O{-}CH_2{-}CH_2{-}S{-}S)_n{-}CH_2 \\ \qquad\qquad\qquad\qquad\qquad\qquad\qquad\quad | \\ \qquad\qquad\quad HS{-}CH_2{-}CH_2{-}O{-}CH_2 \end{array} \quad (2)$$

This also contains 2 mol% of pendant —SH groups as sites for cross-linking. This polymer, a yellow viscous liquid, together with a filler, plasticizer and a small quantity of sulfur (~0.5%) constitutes one of the pastes. The cross-linking paste is usually lead dioxide (PbO_2) in a liquid vehicle, often with a small quantity of oleic acid to control the reactivity of the lead dioxide.

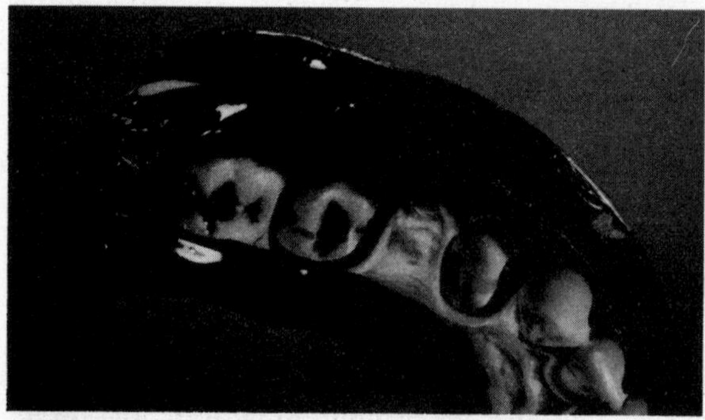

Figure 4
Putty and wash silicone impression: the putty (first impression) is the lighter and the wash (second impression) is the darker

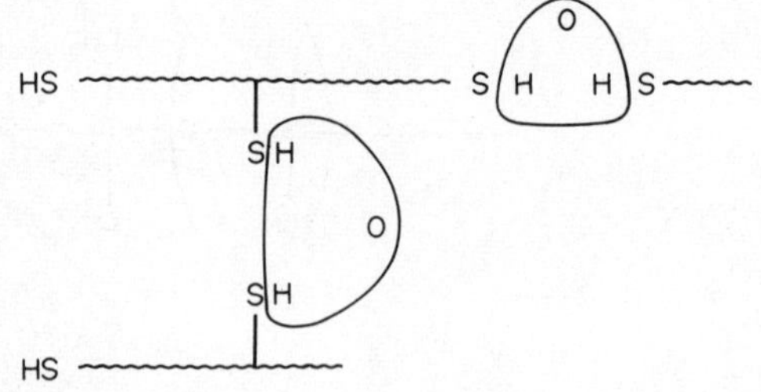

Figure 5
Cross-linking accompanied by chain-lengthening to produce polysulfides

Cross-linking accompanied by chain-lengthening is an oxidation process (Fig. 5). This impression material is extremely strong when set and has good dimensional stability that is only bettered by the addition silicones. Its disadvantages are that it has an unpleasant odor (characteristic of mercaptans), is messy to handle and is slow setting. Unaccountably, this class of material can be capricious in its setting behavior; occasionally a batch that takes a very long time to set will be obtained from the manufacturer. For these reasons, what is fundamentally an excellent impression material has lost popularity in favor of other materials.

Polysulfides are usually sold in three consistency grades. The least viscous grade ("light bodied") is mixed, placed in a syringe and injected round the oral structures of interest; this is then backed up by the most viscous grade ("heavy bodied") in an impression tray.

An obvious question concerns the use of lead dioxide. However, this is a very insoluble lead compound, and there is no evidence to suggest that either patient or dentist is at risk. Nevertheless, the lead dioxide results in an impression which is brown in color, the aesthetics of which combine unfavorably with the odor of the material. Hence various alternatives have been tried. The obvious alternatives are other oxidizing agents, and organic hydroperoxides have been used. Unfortunately, these materials are somewhat volatile, resulting in poor dimensional stability (Braden 1976).

One interesting proprietary material uses cupric hydroxide as a setting agent. The chemical mechanism is not clear.

A further alternative is to use an organic sulfide, which works by an interchange reaction:

$$\begin{array}{c} \sim\!\sim SH \qquad\qquad HS\sim\!\sim \\ X{-}S{-}S{-}X' \\ \downarrow \\ \sim\!\sim S{-}S \sim\!\sim \\ +X{-}SH \\ +X'{-}SH \end{array} \quad (3)$$

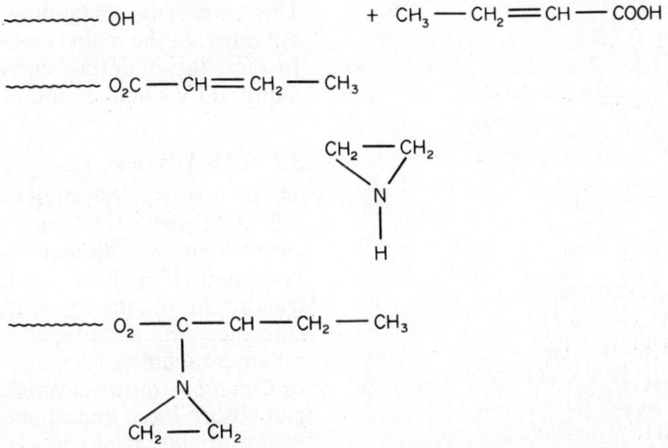

$$\sim\sim OH \quad + \quad CH_3—CH_2{=}CH—COOH$$

$$\sim\sim O_2C—CH{=}CH_2—CH_3$$

$$CH_2—CH_2 \text{ (aziridine ring)}, \; N—H$$

$$\sim\sim O_2—C—CH—CH_2—CH_3 \text{ with aziridine N ring } (CH_2—CH_2)$$

Figure 6
Esterification with an unsaturated acid and addition of aziridine to the double bond of the copolymer of ethylene oxide and tetrahydrofuran

The reverse reaction is quenched by the incorporation of zinc oxide or zinc carbonate.

1.3 Imine-Terminated Polyether

The silicone and polysulfide fluid RTV polymers so far described are industrial systems, adapted for use in dentistry. However, the material now to be described was developed as a dental impression material, and there is probably no industrial equivalent. Because the chemistry is ingenious, it is instructive to review the synthesis of the polymer used (Braden et al. 1972).

The starting material is a polyether comprising a random copolymer of ethylene oxide and tetrahydrofuran:

$$—\{[(CH_2)_2—O—]_n—[(CH_2)_4—O—]_m\}_p \qquad (4)$$

A copolymer is used to avoid the crystallinity which would result from a homopolymer.

This polymer has terminal hydroxy groups, which are esterified with an unsaturated acid, and an aziridine (ethylene imine) ring added to the double bond as in Fig. 6. Hence, the polymer paste of the impression material contains this imine-terminated polyether, together with a filler and plasticizer. The cross-linking paste contains a material capable of donating carbonium ions, and cross-linking is at chain ends via a ring-operating polymerization (Fig. 7). This impression material has the following advantages:

(a) it is rapid setting;

(b) it is clean and easy to handle;

(c) it is hydrophilic, hence readily wets the oral tissues; and

(d) it has excellent dimensional stability.

Indeed, the earlier versions of this material were too hydrophilic, resulting in unwanted dimensional changes if left wet for too long. The disadvantages are that it is very hard when set, necessitating an extra bulk of material in the impression tray to facilitate removal. More serious is the propensity of the cross-linking agent to produce hypersensitivity, notably on the hands of the operator (Nally and Storrs 1973). In the late 1980s a new version of this material appeared, but there is as yet no information on this important aspect.

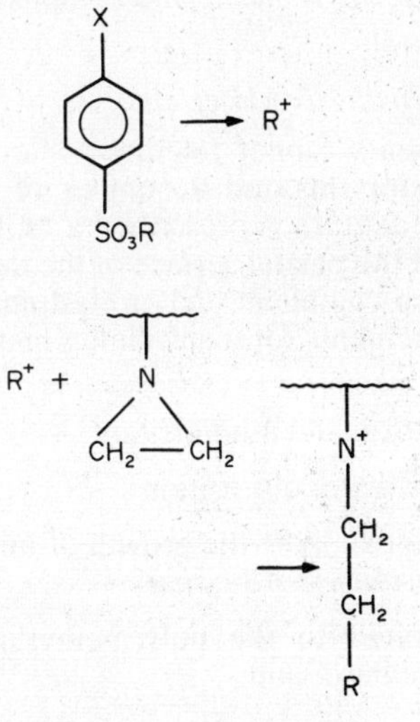

$$X—C_6H_4—SO_3R \longrightarrow R^+$$

$$R^+ + \text{N(aziridine ring, } CH_2—CH_2) \longrightarrow N^+—CH_2—CH_2—R$$

Figure 7
Cross-linking to produce imine-terminated polyether

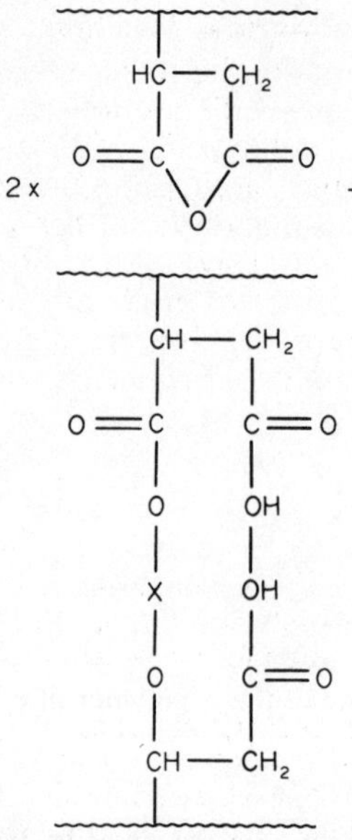

$$2\times \left[\begin{array}{c} HC-CH_2 \\ O=C \quad C=O \\ \backslash O/ \end{array}\right] + X(OH)_2 \rightarrow \begin{array}{c} CH-CH_2 \\ O=C \quad C=O \\ O \quad OH \\ X \quad OH \\ O \quad C=O \\ CH-CH_2 \end{array}$$

Figure 8
Cross-linking of maleinized polybutadiene by diols

1.4 The Future

As stated in Sect. 1, the successful production of highly complex and expensive restorations rests on the accuracy and ease of use. Therefore, the question arises as to whether or not there are any other fluid polymer systems that can combine the advantages of the materials described and eliminate their disadvantages. There is one experimental system that shows some promise in this direction. Maleinized polybutadiene is a fluid polymer that can be cross-linked very rapidly by diols (Fig. 8). This is an addition reaction, in a hydrophilic polymer, and the set material is elastic and compliant.

2. *Soft Prosthesis Materials*

Some patients cannot tolerate a hard acrylic denture, and for this and a number of other clinical reasons, a denture is prescribed a "soft liner." This means that the palatal surface of the denture is to be lined with a compliant and/or elastomeric material. The requirements for a soft lining material are that it:

(a) is nontoxic and a nonirritant;

(b) has low water absorption;

(c) must not support the growth of microorganisms (e.g., *Candida Albicans*);

(d) must bond to the poly(methyl methacrylate) denture base; and

(e) must be amenable to processing by available dental technology techniques.

This extremely demanding list disqualifies many, if not most, of the main classes of industrial elastomer. In fact, the materials currently used are predominantly RTV silicones and plasticized acrylic systems.

2.1 RTV Silicones

These are usually condensation silicones, as described in Sect. 1.1*a*, and are far from ideal. Not surprisingly, it is difficult to form a lasting bond to a poly(methyl methacrylate) denture. An adhesive, usually an α,ω-hydroxy terminated poly(dimethyl siloxane) in a solvent, is necessary. Silicone polymers seem particularly prone to the formation of *Candida Albicans* (Wright 1980), and surprisingly can absorb large quantities of water. The subject of water uptake is discussed in Sect. 2.4. The one advantage (cf. plasticized acrylics) is that they retain their compliance.

2.2 Plasticized Acrylics

Poly(methyl methacrylate) dentures are fabricated from a dough comprising a mixture of methyl methacrylate monomer and poly(methyl methacrylate) powder (see *Acrylic Dental Polymers: Formulation and Synthesis*). Plasticized acrylics follow the same method of fabrication, and Table 1 gives a typical formulation. It will be noted that the polymer powder is poly(ethyl methacrylate) which has a lower glass transition temperature than poly(methyl methacrylate) and so needs less plasticizer to render it compliant. The 2-ethoxyethyl methacrylate, when polymerized, is compliant; the ethylene glycol dimethacrylate is a cross-linking agent; and the phthalate is a plasticizer. The dough made from this is polymerized in contact with the poly(methyl methacrylate) dough, thus ensuring a good bond to the denture base. However, the good bond to the denture is probably the only advantage of this type of soft liner. The final material, while having an elastomeric compliance, is usually very sluggish in elastic recovery. More seriously, plasticizer leaching out in oral fluids results in hardening of the material in a few months.

Table 1
Typical formulation of a plasticized acrylic

Polymer powder	Monomer liquid	Percentage composition
Poly(ethyl methacrylate)	2-ethoxyethyl methacrylate	70
	Ethylene glycol dimethacrylate	5
	Butyl phthalyl butyl glycolate	25

Parker and Braden (1982) have developed materials where the plasticizer does not leach out, but these have other problems referred to in Sect. 2.4.

2.3 A Silicone–Acrylic Copolymer

It is obvious from Sects. 2.1, 2.2 that neither type of material is really suitable for its intended purpose. However, one material offers a reasonable compromise between the above two classes of material.

This material is an adduct of an α,ω-hydroxy terminated poly(dimethyl siloxane) and γ-methacryloxy propyl trimethoxy silane (Eqn. (5)).

$$\mathrm{HO}\left(-\underset{\mathrm{CH_3}}{\overset{\mathrm{CH_3}}{\mathrm{Si}}}-\mathrm{O}\right)_n-\underset{\mathrm{CH_3}}{\overset{\mathrm{CH_3}}{\mathrm{Si}}}-\mathrm{OH}+\mathrm{CH_3O}-\underset{\mathrm{CH_3O}}{\overset{\mathrm{CH_3O}}{\mathrm{Si}}}-\mathrm{O(CH_2)_3}-\mathrm{CO_2}-\overset{\mathrm{CH_3}}{\mathrm{C}}=\mathrm{CH_2}$$
$$\downarrow \qquad (5)$$
$$\sim\sim-\underset{\mathrm{CH_3}}{\overset{\mathrm{CH_3}}{\mathrm{Si}}}-\mathrm{O}-\underset{\mathrm{O}\sim\sim}{\overset{\mathrm{O}\sim\sim}{\mathrm{Si}}}-\mathrm{O}-\mathrm{(CH_2)_3}-\mathrm{CO_2}-\overset{\mathrm{CH_3}}{\mathrm{C}}=\mathrm{CH_2}$$

This is obviously a highly branched silicone polymer, but containing methacrylate groups to achieve bonding to poly(methyl methacrylate). It is cross-linked with a peroxide at 100 °C, usually being processed in contact with the acrylic dough when the denture is made.

2.4 Long-Term Water Absorption of Elastomeric Polymers

When Parker and Braden (1982) developed soft acrylics without extractable plasticizer, the resulting clinical materials failed mechanically after some time. This proved to be because, contrary to expectations, immersion in water resulted in very high prolonged uptake, without equilibrium after many years. This transpired to be a feature of elastomers in general, and in fact Muniandy and Thomas (1984) showed that this was due to the presence in most polymers of water soluble impurities. When water diffuses into an elastomer the water droplets form impurity sites. These grow in size until osmotic and elastic forces balance or, in weak polymers, until fracture occurs.

This almost unavoidable phenomenon makes the development of a long-term soft-lining material extremely difficult.

2.5 Other Materials

A promising group of materials has been described (Gettleman et al. 1987) comprising a polyphosphazine polymer, compounded with a methacrylate monomer, and heat polymerized in contact with the denture. It remains to be seen how such materials perform, particularly with respect to long-term water absorption.

2.6 Maxillofacial Prosthesis Materials

These materials are for the replacement of missing facial tissue associated with a serious accident or the surgical removal of a tumor. Clearly, apart from the physicochemical and biological requirements, the material needs to simulate facial tissue visually to minimize the obviously distressing situation involved. Most of the materials already discussed are used, but much depends on the skill of the surgeon and maxillofacial technician.

Bibliography

Braden M 1966 Characterisation of the setting process in dental polysulphide rubbers. *J. Dent. Res.* 45: 1065–71

Braden M 1976 The quest for a new impression rubber. *J. Dent.* 4: 1–4

Braden M, Causton B E, Clarke R L 1972 A polyether impression rubber. *J. Dent. Res.* 51: 889–96

Braden M, Elliott J C 1966 Characterisation of the setting process in dental silicone rubbers. *J. Dent. Res.* 45: 1016–23

Gettleman L, Vargo J M, Gebert P H, Farris C L, LeBoeuf R J Jr, Rawls H R 1987 PNF as a permanent soft liner for removable dentures. In: Gebelein C G (ed.) 1987 *Advances in Biomedical Polymers*. Plenum, New York, pp. 55–61

Jorczak J S, Fettes E M 1951 Polysulphide polymers. *Ind. Eng. Chem.* 43: 324–8

Muniandy K, Thomas A G 1984 Water absorption in rubbers: Polymers in a marine environment. *Proc. Conf. Institute of Physics and Institute of Marine Engineers*. Institute of Physics, London

Nally F N, Storrs J 1973 Hypersensitivity to a dental impression material. *Br. Dent. J.* 134: 244–6

Nitzsche S, Wick M 1957 Vulcanization systems for silicone rubber (in German). *Kunstoffe* 47: 431–4

Parker S, Braden M 1982 New soft lining materials. *J. Dent.* 10: 149–53

Phillips R W 1982 *Science of Dental Materials*, 7th edn. Saunders, Philadelphia, PA

Watt J A C 1970 RTV silicone rubbers. *Chem. Br.* 6: 519–24

Wright P S 1980 The effect of soft lining materials on the growth of *Candida Albicans*. *J. Dent.* 8: 144–51

M. Braden
[University of London, London, UK]

Dental Gold Alloys: Age-Hardening

Dental gold alloys have been used, not only because their gold color is preferred, but also because they have extremely high chemical stability in the mouth and several desirable mechanical properties such as high strength, ductility and elasticity.

According to American Dental Association Specification No. 5 and International Standard (ISO

1562), dental gold alloys for casting are classified as Type I, II, III and IV depending on their gold and platinum metals contents and their mechanical properties. Among them, the Type IV alloys are designed to be age-hardenable by an appropriate heat treatment. However, the cause of the age-hardening and the nature of the related phase transformations have only recently been extensively studied. This is not surprising because it is difficult to elucidate the hardening mechanisms of alloys as complex as dental gold alloys. Frequently, they contain five or more elements, the essential components of the alloys being gold, copper and silver. There is a consensus among researchers in this field that a more fundamental understanding of phase transformations is necessary for developing new alloys for dentistry, and for explaining the mechanical, chemical and biological properties of dental alloys.

Progress in metallurgy and materials science has been rapid during the 1980s, because transmission electron microscopy (TEM) has made it possible to study directly the relationship between microstructure and phase transformations in alloys. In particular, high-resolution electron microscopic (HREM) observation, coupled with the selected area electron diffraction (SAED) technique, has readily enabled the analysis of age-hardening mechanisms in dental gold alloys.

1. Phase Diagram of the Au–Cu–Ag Ternary System

Although a knowledge of the equilibrium phase diagram is an important aid in predicting the phase transformations resulting from the aging of alloys, the ordering regions in the Au–Cu–Ag ternary system have not yet been established with sufficient accuracy. In 1925, Sterner-Rainer showed that in the Au–Cu–Ag phase diagram, there is a region where a disordered phase coexists with an Au–Cu ordered phase. His phase diagram, however, was too sketchy to be used for quantitative description of phase equilibria in this system.

The Au–Cu–Ag ternary system is characterized by ordering regions and a miscibility gap in which copper-rich and silver-rich face-centered-cubic (fcc) phases coexist. The boundaries of the two-phase region were outlined at 673 K and 987 K using an x-ray diffraction method (Masing and Kloiber 1940). Thereafter, McMullin and Norton (1949) reported the limits of the two-phase region for five different temperatures.

In contrast, information on the stable region of AuCu ordering in the ternary system is much more fragmentary. Raub (1949) found superlattice reflections in the x-ray diffraction patterns of Au–Cu–Ag alloys containing up to 30% silver, while Hultgren and Tarnopol (1939) indicated that the first 5 at.% of silver, when substituted for a corresponding amount of gold in the equiatomic Au–Cu binary alloy, lowered the critical temperature for ordering by as much as 65 K. Hultgren and Tarnopol also reported that the AuCu II orthorhombic ordered phase was stable over a wider range of temperature in the Au–Cu–Ag ternary alloys than in the binary Au–Cu alloys. The occurrence of spinodal decomposition was described by Murakami et al. (1975).

Although the studies by Hultgren and Tarnopol (1939) and by Raub (1949) showed in detail the existing regions and structures of the ordered phases in the Au–Cu–Ag ternary system, Uzuka et al. (1981) were the first to draw phase boundaries involving ordering in the phase diagram. Their experimental results indicated that with increasing silver content the transformation zone of the AuCu I superlattice in the ternary alloys was pushed towards lower temperatures. Moreover, the coexisting region of the AuCu II superlattice and silver-rich disordered phase (designated α_2) was extended towards the side of higher silver concentration. However, their phase diagram was actually a map of their experimental data and was not, apparently, intended to do justice to the phase rule.

Kikuchi et al. (1980) calculated a "coherent" phase diagram for this ternary system, using the interaction-energy parameter determined to fit only the binary phase diagrams of Au–Cu, Cu–Ag and Ag–Au systems. A single parent lattice, (i.e., the fcc lattice) was assumed for all possible phases. Consequently, the AuCu II long period superstructure ($L1_{0-s}$) could not be taken into account, and Cu_3Au ($L1_2$) and AuCu I ($L1_0$) superstructures were considered as well as disordered fcc phases. Eventually, Yamauchi et al. (1980) constructed a plausible "incoherent" phase diagram by superimposing the theoretical "coherent" phase diagram on experimental data of incoherent phases. An isothermal section of their plausible incoherent phase diagram is shown in Fig. 1. This plausible phase diagram explained TEM observations of the AuCu ordering and two-phase decomposition processes in Au–Cu–Ag ternary alloys rationally. However, more experimental work is required to determine the three-phase region topology with certainty in this ternary system. Recently, a coherent phase diagram of the $Au_x(Ag_{0.24}Cu_{0.76})_{1-x}$ pseudobinary system was depicted by Nakagawa and Yasuda (1988) on the basis of TEM and SAED examinations as shown in Fig. 2. Three-phase regions of AuCu II, α_1 (copper-rich fcc) and α_2 phases, AuCu I, AuCu II and α (disordered solid solution of fcc) phases, and Cu_3Au, AuCu II and α_2 phases in the SAED patterns were confirmed. Microstructural features of these regions were also studied in association with phase identifications.

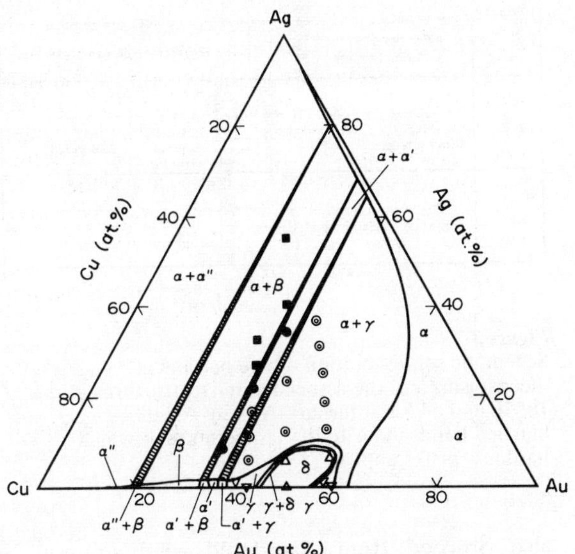

Figure 1
Isothermal section at 573 K of plausible "incoherent" phase diagram of the Au–Cu–Ag ternary system. α: Au or Ag-rich disordered phase; α', α'': Cu-rich disordered phase; β: Au_3Cu type ordered phase; γ: AuCu II type ordered phase; δ: AuCu I type ordered phase (Yamauchi et al. 1980)

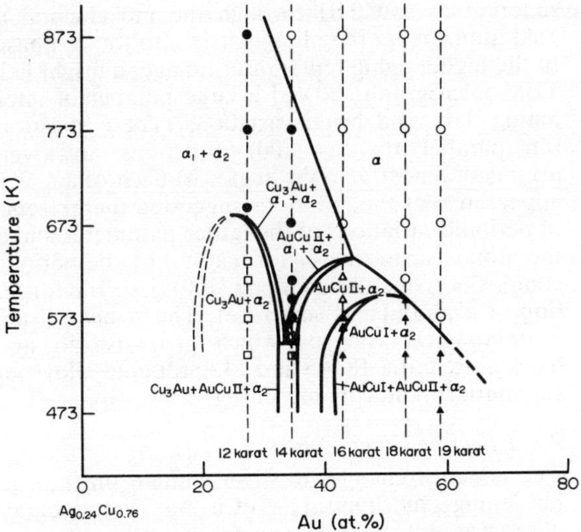

Figure 2
Experimental "coherent" phase diagram of the $Au_x(Ag_{0.24}Cu_{0.76})_{1-x}$ pseudobinary system (Nakagawa and Yasuda 1988)

2. Application of TEM to Studies of Age-Hardening Mechanisms in Au–Cu–Ag Ternary Alloys

TEM has important advantages over other methods of studying the hardening mechanisms in alloys in that:

(a) the results obtained are visual;

(b) it provides high resolution;

(c) additional information about structure and orientation can be obtained by using the SAED technique; and

(d) the dark field image formed using superlattice reflections readily enables ordered regions to be distinguished from disordered regions, since only ordered regions appear bright.

Notwithstanding these advantages, TEM studies of hardening mechanisms in dental gold alloys are relatively recent. The earliest TEM study was carried out by Yasuda's group (Kanzawa et al. 1975) with the object of elucidating the correlation between microstructure and phase transformations in an 18 karat gold dental alloy, Au–35.7 at.% Cu–11.2 at.% Ag. They found that age-hardening in the alloy is brought about by a two-stage process, with initial hardening being due to the formation of coherent AuCu I ordered platelets, and secondary hardening resulting from twinning, as will be shown later. Shortly afterwards, Prasad et al. (1976) studied the age-hardening of a commercial Type III dental gold alloy, using TEM in addition to x-ray diffraction and hardness measurements. They reported that hardening was predominantly due to precipitation. The precipitates, which were formed on the {100} and {111} planes of the matrix, were homogeneously nucleated and coherent with it; ordering also played a role in hardening, but its contribution appeared to be very slight. The above results, however, were not in agreement with those of Yasuda's group. The data may still be insufficient for a definite conclusion to be made as to whether age-hardening in dental gold alloys results from ordering or from precipitation.

As can be seen in Fig. 1, the Au–Cu–Ag ternary system is characterized by ordering regions and a two-phase decomposition region. If the alloy has a composition falling in the ordering region, age-hardening will be due to an order–disorder transformation mechanism. Alternatively, if the alloy is located in the two-phase region, precipitation hardening will occur. Thus, it is supposed that the age-hardening characteristics of these alloys are affected by their composition, especially by the atomic ratio of gold to copper. Because the tendency for two-phase decomposition may markedly increase with decreasing gold content, the effect of the ordered phase on hardening may diminish or disappear altogether. However, it has not been clearly demonstrated that a difference in the age-hardening mechanism of ternary dental Au–Cu–Ag alloys occurs with changes in gold content. Therefore, Yasuda and his coworkers conducted studies to elucidate the differences in hardening mechanisms for 18, 16 and 14 karat gold alloys, in which the ratio of copper to silver was maintained at 65:35 by weight, using SAED and TEM in addition to x-ray diffraction, electrical resistivity measurements and hardness tests.

2.1 The 18 Karat Gold Alloy

Experimental results for the 18 karat gold alloy (Kanzawa et al. 1975), showed that age-hardening was due to the formation of platelets of AuCu I superlattice on the matrix {100} planes. These ordered platelets were arranged in a stepwise fashion that roughly followed ⟨110⟩ trace on the {100} planes, and their c-axes were distributed in the three cube edge directions of the parent disordered matrix. Thus, it was observed that at a later stage of aging, twinning took place on the $(101)_{tet}$ plane to relieve a considerable amount of strain induced in the matrix through changes in crystal symmetry.

2.2 The 16 Karat Gold Alloy

Age-hardening in the 16 karat gold alloy Au–43.2 at.%Cu–13.8 at.%Ag was, however, found to be due to the development of a long period antiphase domain structure in the AuCu II superlattice at lower aging temperatures (Yasuda et al. 1978). The unidirectional, alternating fine lamellar structure, consisting of the AuCu I superlattice, was also found to be present in several different areas of the same specimen. The age-hardening in this temperature range was thought to be due to the development of antiphase domain boundaries in addition to the strain field induced by the structural change. In a higher temperature range, an alternating coarse lamellar structure, consisting of an fcc lattice of copper-rich α_1 and silver-rich α_2 phases, was formed by discontinuous decomposition. These two phases were oriented parallel to one another. No age-hardening was observed in this temperature range. However, considerable age-hardening was found to occur in the middle temperature range, in spite of the similarity between the microstructures after aging at these temperatures and at higher ones. In this respect, it was thought that the ordering and the decomposition of supersaturated solid solution took place simultaneously during the aging process in this temperature range.

2.3 The 14 Karat Gold Alloy

In the 14 karat gold alloy Au–49.7 at.%Cu–15.8 at.%Ag studied by Yasuda and Ohta (1979) which possessed an off-stoichiometric composition of AuCu, it was expected that age-hardening would

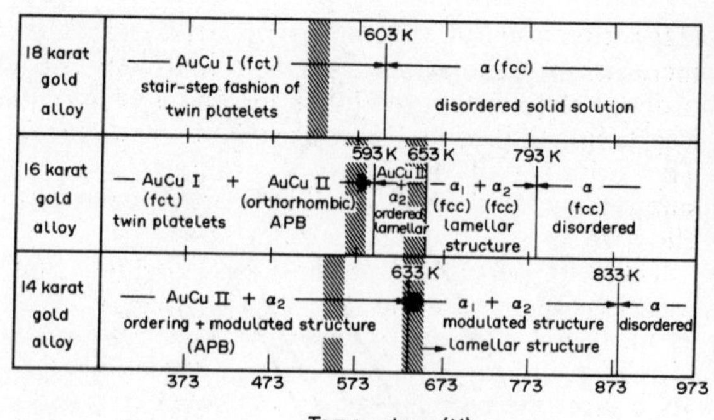

Figure 3
Schematic representation of age-hardening mechanisms and the associated microstructures in the 18, 16 and 14 karat ternary Au–Cu–Ag alloys: hatched bands indicate the temperature at which hardness peaks appear

also proceed from disordered solid solution by AuCu ordering accompanied by the two-phase decomposition. The microstructure of this alloy at lower aging temperature range showed a fine mottled contrast parallel to the ⟨100⟩ directions in the early stages of aging. This fine contrast was due to the formation of a modulated structure induced by spinodal decomposition, which was confirmed by the appearance of sidebands in x-ray diffraction patterns and satellite reflections in SAED patterns. With longer aging periods, the modulated structure changed to a "tweedlike" structure consisting of copper-rich α_1 and silver-rich α_2 phases. The α_1 phase gradually transformed to the long period superlattice AuCu II through the introduction of gold atoms from the alloy matrix into the α_1 phase. In the higher aging temperature range, a bright field TEM micrograph showed a large number of alternating dark and bright striations. These striations ran parallel to the ⟨100⟩ directions and were arranged almost at right angles to each other. The appearance of the striations suggested the presence of periodic variations of the lattice parameter and in the atomic scattering factor, caused by the periodic composition fluctuation along ⟨100⟩ (i.e., the formation of a modulated structure). The structural and morphological changes which give rise to age-hardening in the 18, 16 and 14 karat gold alloys are summarized and collated in Fig. 3.

2.4 Effect of Silver Content

The effect of changes in silver content on the age-hardening mechanism AuCu–Ag pseudobinary alloys was also studied by means of hardness tests, x-ray and SAED as well as TEM (Ohta et al. 1983). Hardening in a low silver-content alloy AuCu–6 at.% Ag aged at 573 K was caused by the formation of AuCu I ordered platelets formed along

⟨110⟩ directions of the matrix and the consequent increase in the elastic strain field. Growth of the ordered platelets caused twinning, which resulted in a softening of the alloy by releasing the elastic strain. The twinning of the AuCu I ordered phase was suppressed by the addition of more silver to the alloy.

The long period antiphase domain configuration was also observed in the alloy aged at 603 K (Yasuda et al. 1987). Fig. 4a shows the dark field image formed using the 110_z superlattice reflection. Figs. 4b, c show the corresponding SAED pattern and its schematic representation, respectively. It can be seen that a large number of the fringes abut each other at an angle which slightly deviates from a right angle, as was reported by Watanabe and Takashima (1975). HREM observation has proved to be an indispensable technique in unravelling the configuration of the long period antiphase domain boundary of AuCu II. Figure 5 shows an HREM image along [001] of periodic antiphase domain boundaries in AuCu–6 at.% Ag alloy. It is clear from this image that the antiphase domain boundaries have a wavy character; they are not strictly confined to {100} planes and their width also varies locally.

In contrast, age-hardening in the high silver-content alloys was attributed to the dual mechanisms of phase decomposition and ordering. In the AuCu–31.1 at.% Ag alloy, for example, aged at 573 K, phase decomposition was thought to proceed by spinodal decomposition, followed by the formation of a modulated structure of increasing periodicity which grows into a characteristic blocklike structure consisting of alternating ordered AuCu II orthorhombic phase and silver-rich α_2 phase. In the initial stages of aging, most of the hardening was caused by the development of the modulated structure. The hardening that occurred upon further aging was attributed to ordering in the copper-rich α_1 phase.

3. *Age-Hardening Mechanisms in Au–Cu–Ag–Pd Quarternary Alloys and Commercial Dental Gold Alloys*

Since the works of Wise et al. (1932) and Wise and Eash (1933) concerning quarternary gold alloys, it has been known that the addition of platinum and/or palladium to ternary Au–Cu–Ag alloys gives rise to conspicuous age-hardening characteristics. To clarify the crystallography and morphology of the hardening phase in these alloys, Yasuda and Kanzawa (1977) carried out TEM studies on a quarternary gold alloy Au–35.4 at.% Cu–17.8 at.% Ag–9.7 at.% Pd which was prepared so as to contain the equivalent of 2 karat of palladium substituted for a corresponding amount of gold in a 16 karat Au–Cu–Ag alloy. The age-hardening observed in this quarternary alloy was due to the formation of AuCu I ordered nuclei on the disordered matrix {100} planes. In adjacent ordered platelets of AuCu I, the c-axes were mutually perpendicular to compensate

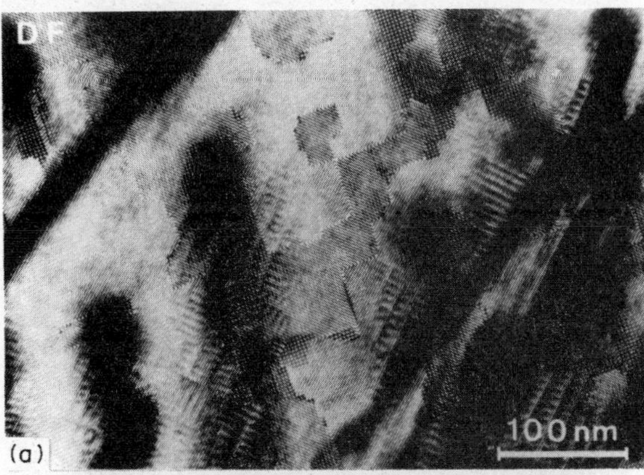

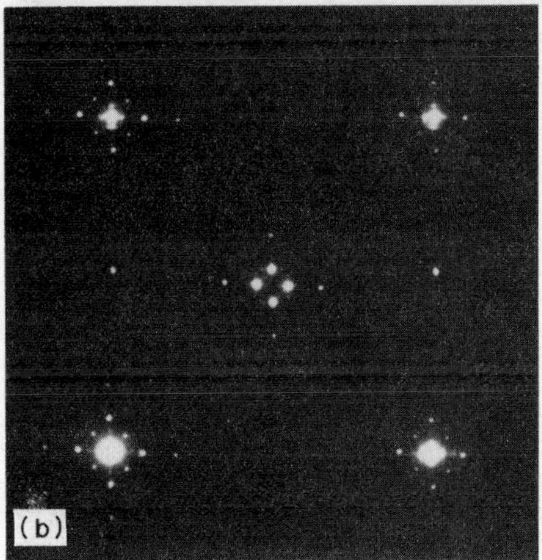

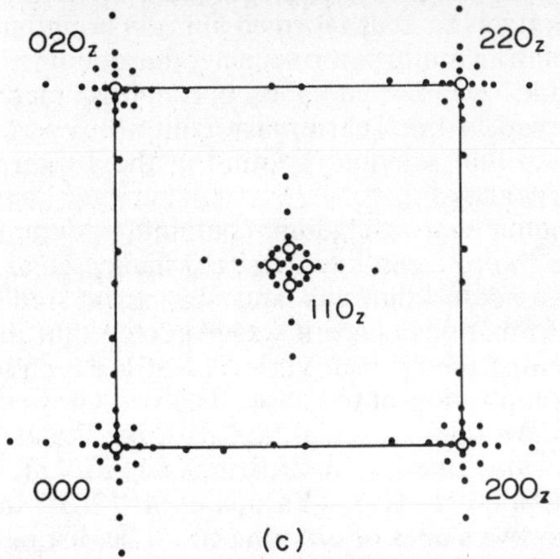

Figure 4
Dark field TEM micrograph (a) produced by using the 110_z superlattice spot and SAED pattern (b) of the AuCu–6 at.% Ag alloy after aging at 613 K for 100 ks

Figure 5
HREM image along [001] of periodic antiphase domain boundaries in AuCu–6 at.% Ag alloy aged at 613 K for 100 ks

for the strain induced by their tetragonality, as was also observed in the 18 karat gold ternary alloy. To relieve the strain caused by the tetragonal distortion in the ordered domains, twinning took place on the $(101)_{tet}$ plane at a later stage of aging. The nodular precipitates that also formed at grain boundaries did not play an important role in age-hardening. Thus, it became clear that the hardening mechanism observed in the quarternary gold alloy was analogous to that previously found in the 18 karat gold ternary alloy.

Commercial palladium-containing dental gold alloys were also studied to make clear their hardening mechanisms and the associated phase transformations, since it was expected that the age-hardening mechanism varied greatly depending on the composition of the alloy. The two alloys studied were Au–26.5 at.% Cu–18.2 at.% Ag–7.9 at.% Pd (16 K–S) and Au–25.3 at.% Cu–27.4 at.% Ag–7.6 at.% Pd (14 K–S) (Yasuda et al. 1983). In these alloys, two stages of ordering could be distinguished by TEM and SAED studies of isochronal aging of a supersaturated solid solution. In the first stage, ordering of a metastable AuCu I′ superlattice occurred as a stepwise arrangement of platelets situated on the {100} planes. This structural feature was also analogous to that found in the 18 karat gold ternary alloy. Prolongation of the aging period caused the formation of a lamellar structure consisting of the equilibrium AuCu I ordered phase and the silver-rich α_2 phase. The metastable AuCu I′ ordered platelets were formed prior to the equilibrium AuCu I ordered phase. Platelet formation was accompanied, in the 16 K–S alloy, by a contraction of about 7% in the surrounding matrix in a direction normal to the plane of the AuCu I′ ordered platelets (i.e., this direction coincided with the c-axis of the AuCu I′ superlattice). In the 14 K–S alloy, the contraction was about 6%. The strain introduced by the tetragonality of the AuCu I′ structure led to substantial hardening in the early stages of aging.

Isothermal age-hardening curves for the two alloys showed that, after a rapid increase in the early stages of aging, the hardness dropped drastically as aging was prolonged (Udoh et al. 1984). This marked decrease in hardness on overaging was due to the formation of the equilibrium AuCu I ordered phase and of the α_2 phase at grain boundaries as a lamellar structure by a heterogeneous mechanism. It was thought that, during lengthy aging, these lamellae would continue to grow at the expense of the metastable AuCu I′ ordered platelets formed inside the grains. The strain field generated by the formation of the AuCu I′ ordered platelets was therefore relieved as a result of the growth of the lamellar structure. Thus, loss of coherency at the interface between the ordered platelets and the matrix caused the hardness decrease.

Recently, a different type of age-hardening mechanism has been found to occur upon appropriate aging in a commercial 18 karat gold dental alloy Au–31.7 at.% Cu–8.1 at.% Pd–5.3 at.% Ag (18 K–S) (Udoh et al. 1986). X-ray and SAED studies as well as TEM showed that a Au_3Cu ($L1_2$) type superlattice and a disordered fcc phase were formed by aging at 673 K. Hardening occurred through the mechanism of the antiphase domain size effect. The $L1_0$ type AuCu I superlattice was also formed in regions adjacent to the Au_3Cu ordered phase in the later stages of aging below 623 K. Thus, it was concluded that the hardening should be attributed not only to the antiphase domain size effect, but also to the elastic strain field induced by the formation of the AuCu I ordered platelets in the aging temperature range below 623 K.

Although the hardening mechanism in dental gold alloys was thought to be attributable to a coherency strain field at the interface between the AuCu I platelets and the surrounding matrix, the configuration of the interface, which could provide direct evidence of the presence of a coherency strain field, could not be deduced by conventional TEM. However, HREM observation coupled with SAED provided direct information on crystal structures down to the scale of interatomic distances, and proved to be extremely effective for elucidating the structural characteristics of interfaces between different phases under suitable imaging conditions. These techniques were therefore employed for studying the configuration of the antiphase domain boundaries and the strain field generated at the interface between the AuCu I and Au_3Cu ordered phases in the 18 K–S commercial dental gold alloy (Yasuda et al. 1986).

Figure 6 shows the dark field image taken from the 18 K–S alloy aged at 673 K for 1.8 Ms by using a 110 superlattice reflection, with the incident electron beam parallel to the [100] direction using conventional TEM techniques. In the micrograph, the dark stripes which run approximately parallel to the ⟨100⟩ directions are antiphase domain boundaries that contribute to the hardening of the alloy. However, the atomic arrangement at the antiphase domain boundary could not be found in the micrographs obtained by the conventional dark field imaging technique. Figure 7 is an HREM image taken from the same region of the crystal as in Fig. 6. In this configuration it is clearly evident that the copper columns are at the correct positions with respect to the basic fcc lattice (i.e., at each corner of the square), and have the same configuration and scale as the projection of the copper columns in the Au_3Cu ordered structure. A shift in the arrays of the bright dots occurs across the interfaces (i.e., at the antiphase domain boundaries as indicated by the arrows in Fig. 7). The presence of antiphase domain boundaries introduces a local change in chemical composition of constituents (i.e., excess atoms with respect to the stoichiometric composition will be segregated along the antiphase domain boundaries). Thus, it is thought that this segregation contributed to the hardening as well as to the higher energy at the antiphase domain boundaries.

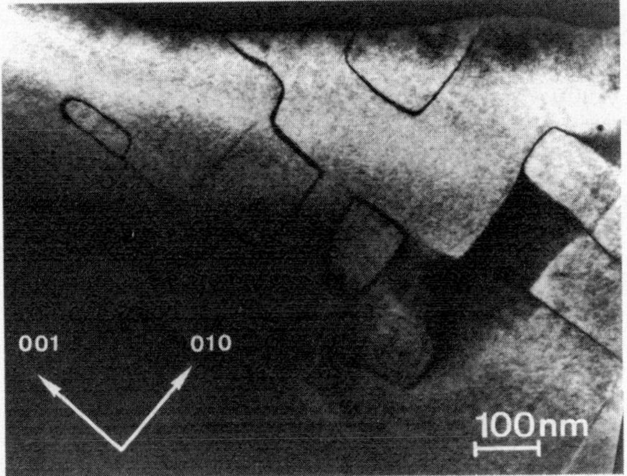

Figure 6
Dark field image of a commercial 18 karat gold dental alloy (18 K–S) aged at 673 K for 1.8 Ms taken by conventional electron microscopy using the 110 superlattice spot

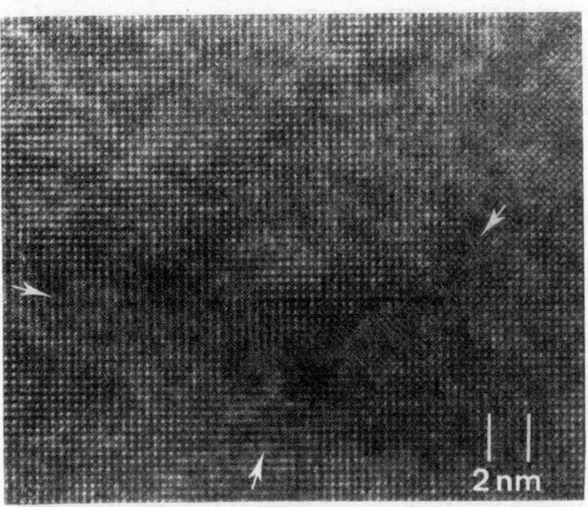

Figure 7
HREM micrograph along the [100] direction of the Au_3Cu superlattice; a triple junction of antiphase domain boundaries (indicated by arrows) is observed (the separation of the bright dots, which represent copper atoms, is 0.4 nm)

When aging was carried out below 623 K, the AuCu I ordered platelet was formed in regions adjacent to the Au_3Cu ordered matrix. Figure 8a is a bright field high resolution image produced using the 001, 100, 110 and equivalent superlattice reflections, the fundamental reflections being excluded, and contains an interface between an AuCu I ordered platelet and the ordered Au_3Cu matrix. In the micrograph, the direction of the c-axis of the AuCu I ordered platelet is indicated by an arrow, while the arrowhead points to an antiphase domain boundary. Figure 8b is also a high resolution image which shows in detail the arrangement of the columns of atoms along the interface between the AuCu I and Au_3Cu ordered phases. In spite of the disturbance caused by the presence of strain and strain contrast, the interface is clearly visible owing to the small difference in interatomic distance between the AuCu I and Au_3Cu phases. These phases were further identified from optical diffraction patterns. In Fig. 8a, it can be seen that the amount of elastic strain increases with the thickness of the AuCu I ordered platelet (i.e., the length along the c-direction within the limited range of the elastic limit). The width, or thickness, of the AuCu I platelet image is certainly limited by the strain contrast associated with the difference in lattice parameters along the c-axis for AuCu I and Au_3Cu ordered phases. If, as shown schematically in Fig. 9, it is assumed that at the top of such a platelet, the pure copper (001) planes in the AuCu I superlattice, coincide with the mixed gold–copper planes in the

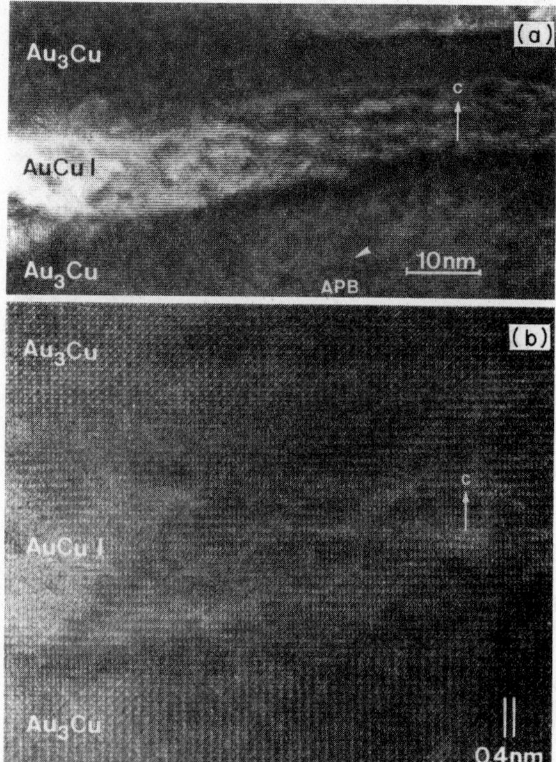

Figure 8
Bright field HREM image of a commercial 18 karat gold dental alloy (18 K–S) aged at 573 K for 1.8 Ms: (a) only the superlattice spots were selected for imaging, and (b) all reflections up to 220_{fcc} were selected for imaging

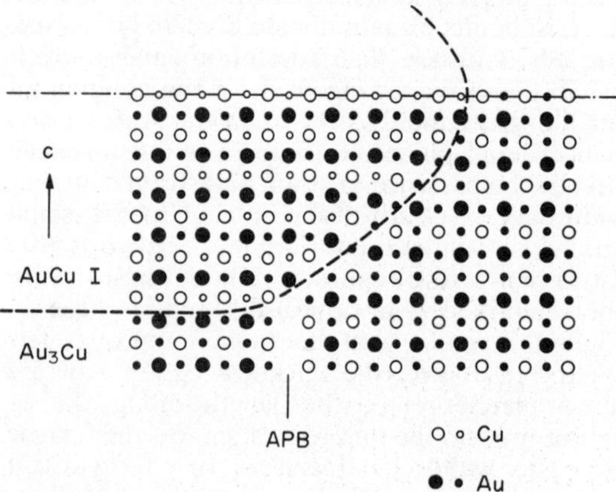

Figure 9
Schematic representation of a AuCu I platelet in an Au_3Cu matrix for axial ratio c/a of 0.9; an antiphase domain boundary (APB) is generated automatically if copper continuity is required along the center of the platelet (large circles are at level 0; small circles are at level 1/2)

Au_3Cu superlattice, then with an axial ratio (c/a) of 0.90 for a platelet with a half-width of 5 AuCu I unit cells, the pure copper planes will now coincide with the pure gold planes. For a c/a ratio of 0.945, which was the measured value from the diffraction patterns, the width of the AuCu I platelet will be 18–19 units thick. This explanation is in good agreement with the experimental observations; indeed, such a mismatch of the antiphase domain boundary is observed in Fig. 8a. Thus, it was concluded that the AuCu I ordered platelets, being coherent with the Au_3Cu ordered phase as matrix, gave rise to considerable amounts of elastic strain which could be a major contribution to the age-hardening of the alloy.

See also: Dental Materials: An Overview; Dental Noble-Metal Casting Alloys: Composition and Properties

Bibliography

Hultgren R, Tarnopol L 1939 Effect of silver on the gold–copper superlattice AuCu. *Trans. Metall. Soc. AIME* 133: 228–38

Kanzawa Y, Yasuda K, Metahi M 1975 Structural changes caused by age-hardening in a dental gold alloy. *J. Less-Common Met.* 43: 121–8

Kikuchi R, Sanchez J, de Fontaine D, Yamauchi H 1980 Theoretical calculation of the Cu–Ag–Au coherent phase diagram. *Acta Metall.* 28: 651–62

McMullin J, Norton J 1949 On the structure of gold–silver–copper alloys. *Metals Trans.* 1: 46–8

Masing G, Kloiber K 1940 Precipitation processes in the copper–silver–gold system (in German). *Z. Metallkd.* 32: 125–32

Murakami M, de Fontaine D, Sanchez J, Fodor J 1975 Ternary diffusion in multilayer Ag–Au–Cu thin films. *Thin Solid Films* 25: 465–82

Nakagawa M, Yasuda K 1988 A coherent phase diagram of the $Au_x(Ag_{0.24}Cu_{0.76})_{1-x}$ section of the Au–Cu–Ag ternary system. *J. Less-Common. Met.* 138: 95–106

Ohta M, Shiraishi T, Yamane M, Yasuda K 1983 Age-hardening mechanism of equiatomic AuCu and AuCu–Ag pseudobinary alloys. *Dent. Mater. J.* 2: 10–17

Prasad A, Eng T, Mukherjee K 1976 Electron microscopic studies of hardening in Type III dental alloy. *Mater. Sci. Eng.* 43: 179–85

Raub E 1949 Effect of silver on transformations in the gold–copper system (in German). *Z. Metallkd.* 40: 46–54

Udoh K, Hisatsune K, Yasuda K, Ohta M 1984 Isothermal age-hardening behaviour in commercial dental gold alloys containing palladium. *Dent. Mater. J.* 3: 253–61

Udoh K, Yasuda K, Ohta M 1986 Age-hardening characteristics in an 18 carat gold commercial dental alloy containing palladium. *J. Less-Common Met.* 118: 249–59

Uzuka T, Kanzawa Y, Yasuda K 1981 Determination of the AuCu superlattice formation region in gold–copper–silver ternary system. *J. Dent. Res.* 60: 883–9

Watanabe D, Takashima K 1975 Periodic antiphase domain structure in the off-stoichiometric CuAu II phase. *J. Appl. Crystallogr.* 8: 598–602

Wise E M, Crowell W, Eash J T 1932 The role of the platinum metals in dental alloys. *Trans. Metall. Soc. AIME* 99: 363–412
Wise E M, Eash J T 1933 The role of the platinum metals in dental alloys. *Trans. Metall. Soc. AIME* 104: 276–307
Yamauchi H, Yoshimatsu H A, Forouhi A R, de Fontaine D 1981 Phase relation in Cu–Ag–Au ternary alloys. In: McGachie R O, Bradley A G (eds.) 1981 *Precious Metals*, Pergamon, Willowdale, Canada, pp. 241–9
Yasuda K 1987 Age-hardening and related phase transformations in dental gold alloys. *Gold Bull.* 20: 90–103
Yasuda K, Kanzawa Y 1977 Electron microscope observation in an age-hardenable dental gold alloy. *Trans. Jpn. Inst. Met.* 18: 46–54
Yasuda K, Metahi H, Kanzawa Y 1978 Structure and morphology of an age-hardened gold–copper–silver dental alloy. *J. Less-Common Met.* 60: 65–78
Yasuda K, Nakagawa M, Van Tendeloo G, Amelinckx S 1987 A high resolution electron microscopy study of pseudobinary AuCu–6 at.% Ag alloy. *J. Less-Common Met.* 135: 169–83
Yasuda K, Ohta M 1979 Age-hardening in a 14 carat dental gold alloy. In: Browning M E, Edson G I, Bubes I S (eds.) 1979 *Proc. 3rd Int. Conf. Precious Metals* International Precious Metals Institute, Chicago, IL, pp. 137–64
Yasuda K, Udoh K, Hisatsune K, Ohta M 1983 Structural changes induced by ageing in commercial dental gold alloys containing palladium. *Dent. Mater. J.* 2: 48–58
Yasuda K, Van Tendeloo G, Van Landuyt J, Amelinckx S 1986 High-resolution electron microscopy study of age-hardening in a commercial dental gold alloy. *J. Dent. Res.* 65: 1179–85

K. Yasuda
[Nagasaki University, Nagasaki, Japan]

Diamondlike and Diamond Thin Films

No class of materials displays such a wide range of properties as the solids of which carbon is a major constituent. These range from plastics to diamond and it has been a longstanding objective to capture the useful properties of these materials in the form of thin films. Thin films are extremely important configurations for many technological applications. Moreover, film deposition techniques often add a degree of control over structure and composition impossible to obtain for bulk materials.

Bridging the areas of organic and inorganic materials, carbon-based films hold not only technological promise but also are of considerable scientific interest. They provide model systems for investigation of questions relating to structure and bonding since both can be changed over wide ranges. Carbon, unlike its isoelectronic cousins silicon and germanium, readily forms double and triple bonds. The promise held by carbon films can only be realized if control over the bonding can be obtained. In certain cases this has been possible; for example, it is now possible to deposit amorphous carbon films with predetermined physical properties such as hardness or transparency. Another example is the deposition of polycrystalline diamond as a thin film, using hydrogen as an intermediate to promote the tetrahedral bonding.

Carbon films, specifically their fabrication, properties and applications, are reviewed here in two distinct parts. This separation, into amorphous and crystalline materials, is for convenience and it should be realized that mixed structural forms, such as microcrystalline regions in an amorphous matrix, do exist. These microcrystallites may be either diamond or graphitic. However, the fabrication as well as the applications of amorphous and crystalline materials are sufficiently different to justify their separate treatment. Notably excluded from the discussion are the graphites, including the intercalated graphites (see *Intercalation Compounds*), since these materials are rarely studied as thin films.

1. Diamondlike Carbon Films

A large amount of literature exists on amorphous carbon films but there is no uniform nomenclature to designate these materials. Terms such as amorphous carbon, diamondlike carbon, DLC films, i-carbon and a-C or a-C:H are all used to designate a class of films which primarily consists of carbon without long range atomic order. The properties of these carbon films which primarily consists of carbon without long-range atomic order. The properties of these carbon films are generally very different from the relatively density of some amorphous carbon films can exceed the value of graphite (2.25×10^3 kg m^{-3}) and be even closer to that for diamond (3.52×10^3 kg m^{-3}).

The addition of hydrogen to amorphous carbon to form hydrogenated amorphous carbon (a-C:H) greatly affects its properties. The film density generally decreases when hydrogen is incorporated and may be as low as 1.2×10^3 kg m^{-3}. Depending on the hydrogen concentration, these hydrogenated amorphous carbon films often do not have the superior mechanical characteristics, such as hardness and wear resistance, that are typical of the unhydrogenated material. However, the incorporation of hydrogen relieves some of the high intrinsic film stress which is typical of an overconstrained unhydrogenated carbon network and allows the deposition of films thicker than 1 μm. A schematic overview which orders different carbon films and other carbon containing materials in terms of the fraction of the carbon sites that have tetrahedral (sp^3) bonding and the hydrogen content is shown in Fig. 1. The dashed curve is the theoretical line for a completely constrained network consisting of sp^2 and sp^3 carbon sites and hydrogen. Materials to the right of this line are underconstrained and are consequently soft. Conventional polymers fall in this category. Materials to the left of the line would be overcon-

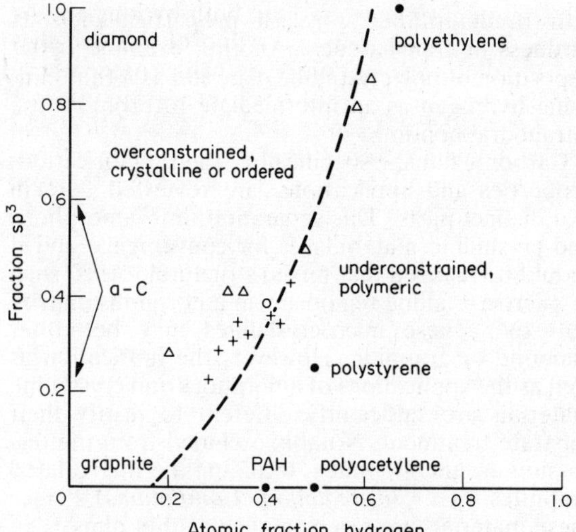

Figure 1
Fraction of carbon sites with sp^3 coordination versus the atomic fraction of hydrogen. The data points are for hydrogenated diamondlike hydrocarbons (a-C:H) where the sp^3 fraction was determined by nuclear magnetic resonance (△ and ○) and electron energy loss spectroscopy (+). Other abbreviations are: a-C for diamondlike amorphous carbon and PAH for polycyclic aromatic hydrocarbons

strained if the network were random. In practice, networks will develop some form of medium or long range order to relieve the constraints.

1.1 Fabrication

The key to depositing dense and hard amorphous carbon films, as opposed to relatively soft graphitic carbon films, is to bombard the surface of the film during growth with energetic ions or atoms. It might be conjectured that the bombardment prevents the formation of extended graphitic regions and promotes the amorphous nature and random bonding in the solid. Although many other deposition techniques have been reported, these films are generally deposited by either a sputtering technique or by plasma-enhanced chemical vapor deposition (PECVD).

Sputtering is a thin film deposition technique in which a condensable vapor stream is created by physically removing atoms and molecules from a solid slab or disk, called a target. In the case of carbon films, the target is often high purity graphite. The target is bombarded by inert gas ions and atoms of a sufficiently high energy, typically 1000 eV, to effect the removal of the target material. The vapor stream from the target is intercepted by a substrate on which the sputtered material condenses and the film is formed. Several different sputtering techniques can be distinguished according to the details of how the energetic inert gas atoms are obtained. Direct current and radiofrequency (rf) sputtering in parallel plate systems (Fig. 2a) have been used but suffer from the problem that the sputter efficiencies for carbon are very low. Magnetic fields can be used to confine the electrons close to the target and to

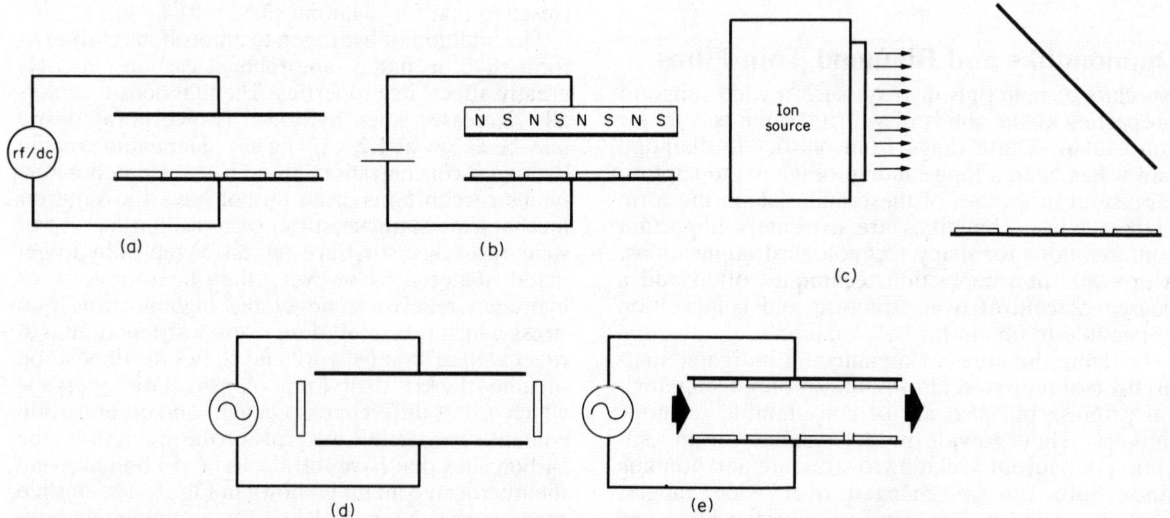

Figure 2
Deposition procedures for amorphous (including diamondlike) carbon films: (a) diode sputtering with target at the upper electrode and substrates at the lower electrode, (b) magnetron sputtering apparatus, (c) ion beam sputtering from an inclined target, (d) symmetrical sputtering apparatus with two targets and cylindrically arranged substrates, and (e) a simple PECVD apparatus

create intense plasmas. Such magnetically enhanced discharges (i.e., magnetron sputtering) allow the deposition of carbon films at higher rates (Fig. 2b). There is evidence that a fraction of the carbon and the inert gas atoms arrive at the substrate with considerable energy, thus creating conditions amenable to the growth of dense amorphous carbon films. In any of these configurations, the substrate may also be biased to achieve further control over the energetic bombardment conditions during growth.

Better control over the ion sputter energy is obtained when ions are generated and accelerated in ion guns. This mode of sputtering is generally referred to as ion beam sputtering (Fig. 2c). The growth environment at the substrate can be changed controllably by allowing ions either from the first or from a second ion gun to bombard the substrate during film growth. A final example of the numerous variations on the basic sputtering technique is shown in Fig. 2d, where both electrodes of a symmetric parallel plate reactor are covered with targets and where the substrates are arranged along a cylindrical circumference. A planetary motion of these substrates provides for uniform coating over large area substrates, for example, computer disks. Again, these substrates may be biased independently of the plasma excitation voltages to control bombardment energies. Any of these sputter techniques can yield films which are composed only of carbon. By sputtering with gases which are reactive with carbon (reactive sputtering) one can change the chemical composition of the films in a controlled manner and therefore change the properties of the carbon films. The element that affects the properties of carbon films the most is hydrogen. Because hydrogen molecules are too light to effect sputtering, hydrogen gas is always admixed with a heavier sputtering gas such as argon.

Hydrogenated amorphous carbon films, a-C:H, can also be deposited by an entirely different technique: plasma-enhanced chemical vapor deposition (PECVD). This method does not use a solid form of carbon as the source material, but rather carbon-containing gases or vapors which are decomposed in a glow discharge (plasma) to form primarily neutral, condensible radicals. These hydrocarbon and carbon radicals react with each other in the gas phase as well as at the surface of the film. Condensation reactions at the surface of the growing film are thermally activated and are also affected by ion bombardment. PECVD reactors are often of the parallel plate type (Fig. 2e) and are excited by rf power, thus closely resembling the rf sputter configuration (Fig. 2a). Carbon films are deposited both on the grounded and the electrically powered electrode. In a symmetric system, the properties of the carbon films deposited on these electrodes are identical. The dc self bias at the electrodes in such a system is one half of the amplitude of the rf voltage. However, bombardment energies of the film, both by ions and by neutral species are only a fraction of this self bias due to collisions and charge exchange processes in the sheaths at the electrodes. Electrodes of different sizes often cause the film properties to differ as a result of the difference in the electrode self bias. These differences can be maximized by the insertion of a blocking capacitor in the electrical supply line to the smaller electrode. In this case, a maximum negative dc bias of approximately the amplitude of the applied, high frequency voltage can be obtained at this electrode whereas the bias at the large electrode is minimal. Harder and less transparent carbon films with a lower hydrogen concentration are obtained at the powered electrode when compared with the film deposited on the large electrode, due to the difference in bombardment conditions of the film during growth.

An interesting and important question is to what extent the properties of the carbon films depend on the nature of the precursor gas or vapor. It has been well established that the deposited film can have some memory of the source gas when deposition energies are low. Spectroscopic methods reveal a correlation between sp^2/sp^3 bonding sites in the source material and in the film as well as the presence of functional groups such as benzene rings in the film if the source material contains these structures. However, most of these results pertain to materials deposited under conditions of minimal ion bombardment, for example in inductively coupled plasmas (sometimes referred to as plasma polymerization). Correlations between the source gas and the film are less apparent when films are grown under the energetic conditions of bombardment to make them diamondlike.

1.2 Properties

The physical properties of amorphous carbon films can be changed to a remarkable extent depending on the method and details of the deposition. At one extreme, carbon films can be made very hard but relatively nontransparent in the visible part of the spectrum. On the other hand, carbon films exist that are transparent but relatively soft.

Films of the first type are likely to contain little or no hydrogen and are often deposited by sputtering in an inert atmosphere or by PECVD under conditions of ion bombardment and heating during growth. These extremely hard carbon films are generally found to be amorphous by x-ray diffraction. However, there is evidence that disordered graphitic regions are often present in the amorphous matrix. The extent of these graphitic regions depends on the details of the deposition process (e.g., ion bombardment, temperature and hydrogen incorporation). The details of these structures and their relationship to the growth conditions have been an active topic of research.

Films which are transparent but not extremely hard typically contain a high percentage (up to 60 at.%) of bonded hydrogen and are deposited under conditions where carbon is not likely to lose hydrogen during the growth process, i.e., minimal ion bombardment and relatively low substrate temperature. The effect of the hydrogen on the structure is to increase the fraction of tetrahedrally (sp^3) coordinated carbon at the expense of the trigonally (sp^2) coordinated carbon atoms.

The chemical properties of amorphous carbon films have been explored in far less detail than the physical properties. Amorphous carbon films are generally inert to both acids and organic solvents, although highly hydrogenated carbon films are soluble in organic solvents. Exposure to oxygen plasmas etches the carbon films and reactive plasma etching can be used to pattern the carbon films when appropriate mask materials are used.

Many other physical properties such as the mass density, electrical conductivity, refractive index, electron spin density, etc., change monotonically with changes in the hydrogen concentration. The challenge to the materials scientist is therefore to find deposition conditions that yield the right combination of properties, keeping in mind that the extreme material characteristics are, to an unknown extent, mutually exclusive. In the following, the mechanical and optical properties will be reviewed in some detail to illustrate the trade-offs that exist with these materials.

The hardness, abrasion and wear resistance of diamondlike carbon films can be outstanding. It is these extreme mechanical properties that caused these films to be referred to as diamondlike carbon (or DLC) films. The hardness of diamondlike films is not surprising when one recognizes that the in-plane yield strength of graphite exceeds that of diamond owing to the strength of the carbon double bond. Graphite is not particularly strong in other directions because the graphitic layers of crystalline graphite are only weakly bonded to each other. In amorphous carbon there is no evidence of extended graphitic sheets and the material can therefore be considered to be a form of intimately cross-linked carbon with no particular directionality. It is only when the network integrity is disrupted (e.g., by the presence of an excess of hydrogen or a high density of dangling bonds) that the amorphous carbon network loses its superior mechanical properties. The incorporation of small amounts of hydrogen into the network can relieve the high intrinsic stresses which exist in the overconstrained carbon network. For unhydrogenated carbon films these stresses are generally so high that even relatively thin films, of the order of one micrometer, fail in adhesion.

The optical properties of the diamondlike carbon films also depend strongly on the hydrogen concentration of the material. The optical absorption, bandgap and the index of refraction all change monotonically with the hydrogen content of the material. These properties are therefore sensitively dependent on the details of the deposition process, particularly the deposition temperature and environment (i.e., ion bombardment). Unhydrogenated amorphous carbon has a bandgap of about 0.5–0.7 eV and is therefore opaque at visible and near infrared wavelengths. The optical absorption is caused by the relatively high density of carbon multiple bonds in the unhydrogenated material. The addition of hydrogen decreases the concentration of these bonds and increases the bandgap and therefore the transparency of the material. Bandgap values as high as 4 eV can be obtained when the incorporated hydrogen approaches 60 at.%. This increase in the transparency of the material is obtained at the cost of its mechanical properties, as discussed earlier. This fundamental trade-off between the optical absorption and the mechanical properties of amorphous carbon films complicates the application of these films as optical protective coatings.

The electrical properties of amorphous carbon films depend primarily on the bandgap of the material and therefore on the concentration of the incorporated hydrogen. High electrical resistivities ($\geqslant 10^{12}\ \Omega\,m$) are measured for the wide band gap material, decreasing to values as low as $1\ \Omega\,m$ for sputtered, unhydrogenated films. There is an indication that the electrical conductivity might be increased by the incorporation of boron or phosphorus, but it is uncertain whether this effect is the result of substitutional doping. Hydrogenated amorphous carbon films do not show comparable sensitivity to the incorporation of these elements as exhibited by hydrogenated amorphous silicon. The prospects of amorphous carbon, either hydrogenated or not, as an extrinsically dopable electronic material are therefore not promising. However, hydrogenated amorphous carbon films have been used as charge transport layers in photoreceptor applications. Here, charge carriers are photogenerated in relatively thin amorphous silicon layers and subsequently electrons are injected into the carbon layer by the high electric fields.

Most other practical applications of amorphous carbon films rely on the exceptional mechanical properties of the material. Diamondlike carbon is presently used as a protective overcoat for thin-film magnetic disks to protect mechanically the stored information from magnetic head bounces and crashes on the disk. Particularly important for this application is the fact that amorphous carbon films with the requisite mechanical properties can be deposited over large areas at temperatures which are sufficiently low to be compatible with the magnetic film. Another commercialized application is the use of diamondlike carbon films as a stiffener on loudspeaker tweeters. Hydrogenated diamondlike carbon films are also being used as a bottom layer for use in the production of etched gratings by photolithography for laser structures. Other proposed uses include protection coatings for magnetic heads, industrial sewing machine bobbins, visible and infrared optics and the inner wall of fusion devices, as well as low index antireflection coatings on solar cells.

2. *Diamond Films*

Diamond is synthesized by either one of two methods: at high temperatures and high pressures where diamond is thermodynamically stable, or at low pressures where diamond is metastable. The high pressure process was developed at General Electric in the 1950s (see *Diamond*) and typically produces individual diamond crystals. The low pressure techniques use some form of the chemical vapor deposition (CVD) method where diamond is grown from an activated carbon-containing gas or vapor. A thermally activated CVD method was originally reported in the US patent and scientific literature and demonstrated in the 1960s. The nucleation of diamond on nondiamond substrates at high growth rates was achieved by Soviet workers in the late 1970s and further developed in the 1980s by Japanese researchers. Presently, the CVD method is capable of producing polycrystalline diamond films on a wide variety of substrates over large areas and at relatively low temperatures. Attempts to deposit diamond films heteroepitaxially as single crystals have been unsuccessful to date, but homoepitaxial deposition, that is the continued growth of a diamond single crystal, has been convincingly demonstrated.

In diamond each carbon atom is tetrahedrally coordinated. As a result of its high atomic density and strong covalent bonding, diamond has the highest hardness and elastic modulus of any material. Additionally, diamond is the least compressible substance known and has the highest known thermal conductivity at room temperature. Diamond is structurally and electronically analogous to silicon and germanium but has a much larger bandgap (5.5 eV) than crystalline silicon and germanium. Although these properties are readily observed in single crystal diamond, the properties of diamond thin films are not necessarily the same because of their lesser degree of crystalline perfection.

2.1 *Deposition*

Many details of the process that allows diamond films to grow from the gas phase are still not fully understood. The growth process of diamond films can be viewed as a balance between the deposition of carbonaceous material and the selective removal, either by etching or bond rearrangement, of the graphitic material. The tetrahedral diamond bonding either survives the reaction with atomic hydrogen or is formed in this process. Thus, depending on the particular point of view, a film that consists primarily of diamond is either left behind or deposited. Some theories emphasize the importance of the species that are formed in the gas phase or the bonding stabilization at the surface of the growing film. However, the details of any of these models are still being refined and experimentally verified.

The medium most often used to suppress the formation of graphitic material is atomic hydrogen. These reactive hydrogen atoms are obtained from the thermal or plasma decomposition of molecular hydrogen. Sometimes oxygen is admixed with the hydrogen or incorporated with the carbonaceous gas. In addition to hydrogen, oxygen and fluorine can also function as effective selective etching media.

The deposition of diamond can be practically realized in many different ways. Three of these processes are schematically illustrated in Fig. 3. In Fig. 3a the gas mixture is thermally decomposed using a hot refractory filament. The substrates may be heated directly by the filament or indirectly, by a heating element incorporated in the substrate table. Filament temperatures are typically in excess of 2000 °C, high enough to decompose 10–20% of the molecular hydrogen into atoms. In the arrangement of Fig. 3b, the gas decomposition is accomplished in a glow discharge plasma. This plasma is often excited at microwave frequencies in order to achieve the high electron energies needed for the electron impact dissociation of hydrogen at a pressure of several tens of torr. The substrate table forms a resonant termination for the wave guide structure. Silicon wafer substrates in the right electrical conductivity range may be directly heated by the plasma. In the dc plasma jet reactor shown in Fig. 3c the gases are activated in an arc maintained between a central cathode and a coaxial anode. When a high gas flow is maintained in the annular space between these electrodes, a plasma jet emerges from the nozzle arrangement. The energy carried by the plasma stream necessitates the cooling of the electrodes as well as the substrate. Variations of this jet structure, such as the hollow cathode discharge and the thermally activated jet (torch) have also been shown to be capable of depositing diamond films at relatively high rates over small areas. Uniform deposition over larger areas by these methods may be obtainable if the substrate is moved with respect to the jet or torch. For all processes, the precursor gas mixture typically contains two orders of magnitude more hydrogen than hydrocarbon molecules. The partial pressure of the carbon-containing gas is therefore not very different from typical plasma deposition processes and deposition rates are com-

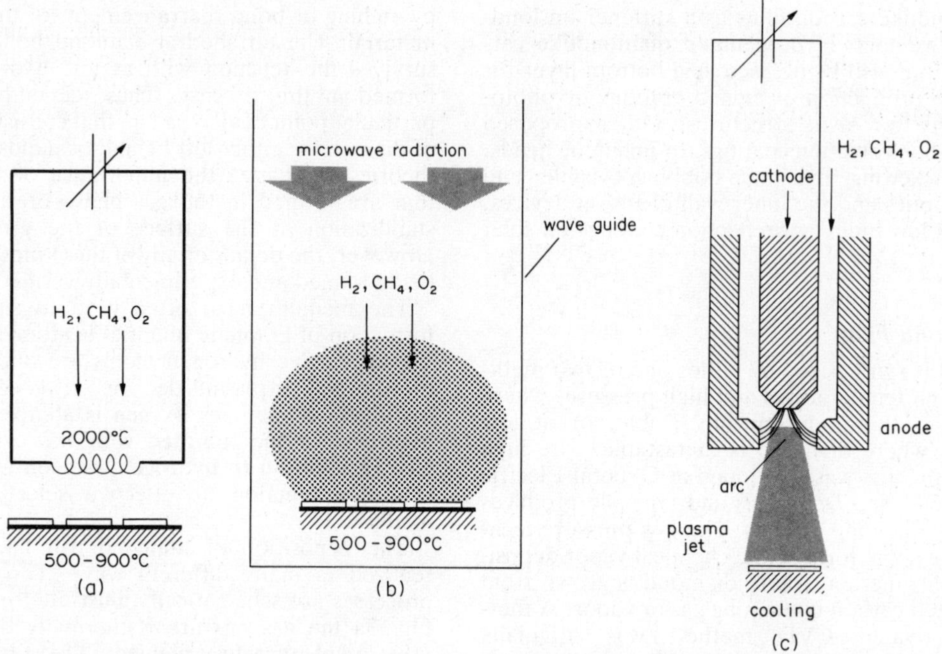

Figure 3
Deposition procedures for diamond films: (a) hot filament deposition, (b) microwave PECVD arrangement, and (c) a dc plasma jet reactor

parably low, of the order of 1 μm h^{-1}. Higher deposition rates can often be obtained by increasing the partial pressure of the carbon-containing radicals, at the expense of incorporating more graphitic material.

The nature of the carbon precursor molecules is not critical and many hydrocarbon gases such as methane, acetylene and ethylene work equally well. The reaction of atomic hydrogen with graphite is a temperature-activated process. The substrate temperature should be high enough for this reaction to proceed at rates that are sufficiently high compared with the deposition rate. Substrate temperatures may range from about 500 °C to 900 °C depending on the formulation of the gas mixture and the desired film quality. As indicated before, it is often advantageous to add oxygen to the gas stream or to incorporate it with the carbon-containing gas as an alcohol or ketone. The specific action of oxygen is not clear, but its addition results in more uniform deposition over larger areas and, importantly, at lower substrate temperatures.

A problem with most low rate diamond deposition methods as illustrated in Fig. 3 is the apparent activation barrier for the nucleation of diamond. Commonly, substrates such as silicon wafers are treated with abrasive powders to create nucleation sites. Diamond from these nucleation centers grows nearly three-dimensionally on the substrate (as distinguished from two-dimensional layer growth) to form heavily twinned crystals and polycrystals before agglomerating to form a continuous film. Depending on the nucleation density, diamond films generally must be thicker than about one micrometer to be substantially free of pin holes. Nucleation centers can also be created by seeding substrates with submicrometer diamond powders in the areas where diamond deposition is required. Typical seeding densities of about one particle per square micrometer result in continuous and pinhole-free films when the film thickness exceeds about two micrometers. With any of the very high deposition rate techniques such as the jet or torch deposition methods, the nucleation of the diamond film seems to be less of a problem. The observation of gas-phase reactions forming diamond microcrystals strongly indicates the possibility of homogeneous (gas phase) nucleation for these cases.

2.2 *Properties*

The crystalline structure of diamond films is readily demonstrated by many spectroscopic and diffraction techniques, including Raman scattering and x-ray diffraction. Grain sizes are between 0.1 and several tens of micrometers. Smaller-grained films are deposited at lower substrate temperatures and higher concentrations of carbon gas. The rough and faceted surface morphology of the film as often observed in a scanning electron microscope reflects the polycrystallinity of the material (Fig. 4a). Films are often composed of columnar grains, which originate at the substrate and increase in diameter towards the surface of the film (Fig. 4b). It is possible to adjust the deposition conditions so that a preferential morphology of these crystallites is obtained. High carbon-containing gas concentrations often result in a predominance of the cubic {100} morphology, whereas lower gas concentrations favor the formation of the octahedral {111} faces. Generally, diamond films do contain some bonded hydrogen, but typically less than 1 at.%. The hydrogen causes increased absorption in the infrared region around $30\pm2\ m^{-1}$. Raman spectroscopy is most often used to characterize diamond films. The Raman line in the first order spectrum at $13.32\ m^{-1}$ is characteristic of tetrahedrally bonded diamond. The width of this peak is generally larger for vapor-grown diamond films than for natural and high pressure diamond and is related to the crystalline domain size. Other, often much broader regions of the Raman spectrum, especially around $15\ m^{-1}$, are indicative of a variety

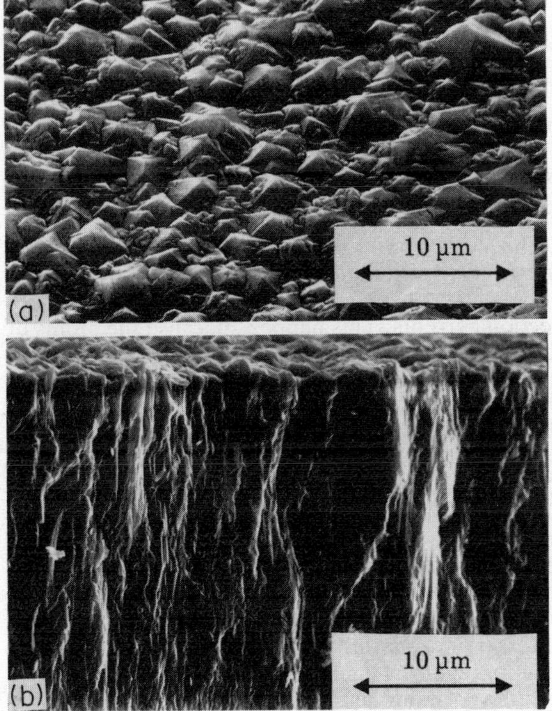

Figure 4
Scanning electron micrographs of (a) the surface of a 20 µm thick diamond film, illustrating the rough, facetted nature of the surface and (b) a cross-sectional fracture surface of the same diamond film, showing the columnar growth pattern. The vertical stride are relatively dense at the film interface (bottom) compared with the surface (top)

of graphitic and disordered bonding configurations. The photoluminescent properties of diamond films often interfere with the Raman spectroscopy and correlate with the presence of defects in the film. Relatively high concentrations of structural defects such as twinned crystals and stacking faults are typically observed with transmission electron microscopy.

Just as in the case of amorphous carbon films, the physical properties of diamond films depend to a remarkable degree on the details of the deposition conditions. In order to have a basis for comparison between bulk single crystalline diamond and its polycrystalline film equivalent, film properties will be discussed which are typical of good quality films as defined by the virtual absence of absorption features other than the $13.32\ m^{-1}$ peak in the Raman spectrum.

Good quality vapor-grown diamond films do not exhibit the same transparency typical for single crystal diamond in the visible wavelength region. The high defect density and the rough surface morphology render the film a dull gray. Films thicker than several micrometers are often only translucent rather than optically transparent. The high defect density does not affect the x-ray transparency and vapor-grown diamond is therefore, as is the single crystalline material, relatively transparent to x-rays. Surprisingly, the extreme hardness and abrasive wear resistance of diamond are also observed for the polycrystalline films, indicating intimate contact between the crystals. The stiffness and cohesive nature of the diamond sheet yields free-standing diamond films that can be easily manipulated when thicker than a few micrometers. Measurements on cantilevered beam specimens of free-standing diamond films yield values for the elastic modulus that are comparable to those of the bulk material. Good quality diamond films have a thermal conductivity comparable to natural diamond. The thermal conductivity is a strong indication of the quality of the film material, as determined by its Raman spectrum, and rapidly decreases when the graphitic content of the film increases. Also, the electrical conductivity of diamond films depends on the film quality. Good quality vapor-grown films have a room temperature resistivity exceeding $10^8\ \Omega\ m$. Large increases in the conductivity of diamond films can be obtained by substitutional doping of the material with boron, which makes it *p*-type. Charge carrier mobilities, which exceed $0.15\ m^2\ V^{-1}\ s^{-1}$ for both electrons and holes in single crystalline diamond, are three orders of magnitude smaller for the polycrystalline material, very likely as a result of structural imperfections. It is for this reason, and because the material is not readily *n*-type dopable, that vapor-grown diamond at the present state of the art is not yet a viable electronic material except for some specialized applications.

The largest application foreseen for vapor-deposited diamond is not as a thin film, however. All kinds of mechanical operations such as polishing, grinding, drilling and sawing rely on the availability of relatively cheap grainy material with the requisite mechanical properties, size and shape. Currently, diamond abrasive grit is primarily derived from the high temperature and pressure growth process, at a rate exceeding 70 t $year^{-1}$. For small crystal sizes and unique crystal shapes, vapor-grown diamond may be cost competitive with the high-pressure material and is predicted to play an important role in this industry. Many of the other real and imagined applications of diamond thin films capitalize on the unique combination of extreme properties that this material offers. Although the field is still in its infancy, such applications have already emerged, not only in the patent literature but also in the market place. The combination of hardness and infrared/x-ray transparency make free-standing diamond films ideal for different window applications. The combination of hardness and wear resistance and high thermal conductivity will find application in printing applications that rely on the transfer of heat, in machine tools and insulating heat sinks. The mechanical stiffness combined with the relatively low specific mass of diamond makes the material in thin film form an excellent candidate for various micromechanical applications and structural reinforcement coatings, as demonstrated by their use on alumina audio diaphragms.

Currently, efforts are under way to synthesize related materials such as cubic boron nitride by vapor deposition methods similar to those used for diamond. Cubic boron nitride, usually obtained at high pressures by the thermal gradient growth method, is slightly less hard than diamond but has some other attractive bulk properties. The bandgap of cubic boron nitride is wider than that of diamond by about 1 eV and the material is readily *p*- and *n*-type dopable. This makes boron nitride an excellent candidate for light emitting devices at short wavelength. Also, for certain mechanical applications, cubic boron nitride holds an edge over diamond. Carbon readily dissolves in ferrous alloys at higher temperatures, severely limiting the life of diamond tools in some high speed machining applications. This problem is reduced with cubic boron nitride. It is because of these properties that this crystalline material is likely to play an important role in selected practical applications. It remains to be determined whether some of these applications can be realized with vapor deposited cubic boron nitride films with properties similar to the high-pressure crystals.

See also: Diamond

Bibliography

Angus J C, Buck F A, Sunkara M, Groth T F, Hayman C C, Gat R 1989 Diamond growth at low pressures. *MRS Bulletin* 14: 38–47

Angus J C, Hayman C C 1988 Low-pressure, metastable growth of diamond and 'diamondlike' phases. *Science*. 241: 913–21

Angus J C, Koidl P, Domitz S 1986 Carbon thin films. In: Mort J, Jansen F (eds.) *Plasma Deposited Thin Films* Chap. 4. CRC Press, Boca Raton, FL

Bar-Yam Y, Moustakas T D 1989 Defect-induced stabilization of diamond films. *Nature* 342: 786–7

De Vries R C 1987 Synthesis of diamond under metastable conditions. *Ann. Rev. Mater. Sci.* 17: 161–87

Dismukes J P, Purdes A J, Meyerson B S, Moustakas T D, Spear K E, Ravi K V, Yoder M (eds.) 1989 *Proc. 1st Int. Symp. on Diamond and Diamond-like Films*. The Electrochemical Society, Pennington, NJ

Robertson J, O'Reilly E P 1987 Electronic and atomic structure of amorphous carbon. *Phys. Rev. B* 35: 2946–57

Spear K E 1989 Diamond–ceramic coating of the future. *J. Am. Ceram. Soc.* 72: 171–91

F. Jansen
[Xerox Corporation, Webster Research Center, Webster, New York, USA]

J. C. Angus
[Case Western Reserve University, Cleveland, Ohio, USA]

Diffusion in Compound Semiconductors

Diffusion occurs in solids, liquids and gases, and is shown most strikingly when two differently colored liquids mix together even though there are no convection currents. This happens because the atoms of the two liquids are in continual random motion, and progressively migrate to cause mixing. The freedom of movement, of course, is much greater in gases than in liquids, and very small in solids, so that diffusion occurs at a much slower rate in solid materials. Nevertheless, diffusion processes in solids are of considerable commercial and scientific significance, and form the basis of many advanced technological developments. Diffusion is used to manufacture solid-state circuits in semiconductor materials, and an appreciation of its importance is necessary to control the growth of single crystals, and to regulate the doping and contamination of multilayer materials.

1. The Diffusion Coefficient

Diffusion in solids can be treated for many purposes by the general diffusion equations developed by Adolf Fick (1855). The first of these equations (called Fick's laws) follows from the observation that, if a material is not uniform, then matter will flow in a manner which will decrease the concentration gradients, particularly if the material is heated or annealed. If the x axis is taken parallel to the concentration gradient, the flux J is given by the equation

$$J = -D\frac{\delta C}{\delta x} \tag{1}$$

where $\delta C/\delta x$ is the concentration gradient and the constant of proportionality D is defined as the diffusion coefficient. Sometimes, of course, the concentration gradient may be changing with time, and continuity leads to Fick's second law, which in three dimensions is

$$\frac{\delta C}{\delta t} = -\nabla J \tag{2}$$

If D does not depend on position, and $J = -D\nabla C$, then

$$\frac{\delta C}{\delta t} = D\nabla^2 C \tag{3}$$

Solutions of this equation for specific diffusion experiments can be used to determine values of the diffusion coefficient for impurity and component atoms in solids, including semiconductors, and the values have been determined in this way for many compound semiconductors. Note particularly, however, the assumptions that are implicit in Fick's laws, and that D is constant with position. It will be seen later that these assumptions are not always true, and neglecting these problems can lead to fundamental misconceptions.

2. Atomic Theory of Diffusion in Solids

Consider in more detail how the diffusion process of the atoms takes place, starting with questions which apply to microscopic particles in a liquid or gas, moving in a random way. For example, how far will a particle be from a given starting point after a large number of random jumps, given the number of jumps per second and the mean jump distance? This is called random walk and was tested first in 1905 by Smoluchowski and later by Einstein in 1956. In crystalline solids, diffusion occurs by the periodic jumping of atoms from one site to another, and so the problem is a similar one. A simple treatment relates the diffusion coefficient to the jump distance α and jump frequency π by the expression

$$D = \tfrac{1}{6}\pi\alpha^2 \tag{4}$$

for jumps in three dimensions.

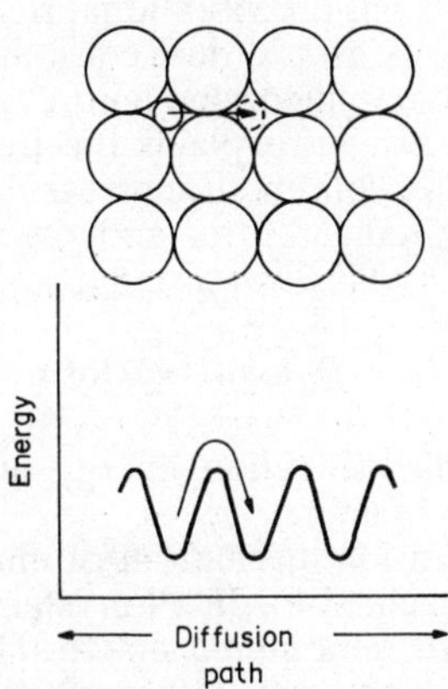

Figure 1
Interstitial diffusion

The atoms vibrate about their equilibrium positions, and occasionally these oscillations become violent enough to allow an atom to change sites. It is these jumps from one site to another which give rise to diffusion in solids. There are various mechanisms by which atoms can diffuse and these are discussed briefly in general terms and then more specifically how they apply to compound semiconductors.

3. Mechanisms of Diffusion

Atoms may move by a variety of mechanisms in solid crystals, mostly involving defects in the crystals. The simplest mechanism, illustrated in Fig. 1, is called the interstitial mechanism, because the atoms are situated in interstices or holes in the lattice, and jump to the next interstitial position. Even if the atom is a very small impurity atom, an activation energy is required to move it, as illustrated in Fig. 1. In the compound semiconductors like gallium arsenide, GaAs, and indium phosphide, InP, which have a structure of the zinc blende or sphalerite type, the interstitials sit in one of two positions, the tetrahedral or hexagonal sites, as shown in Figs. 2a,b. Interstitials in this lattice may be bonded to the lattice, called split interstitials, and may migrate as shown in Fig. 2c. Theoretical calculation of the diffusion coefficient from the size of the energy barrier, or enthalpy of migration ΔH_m, was begun by Zener (1952), who attempted to calculate the jump frequency in Eqn. (4). This simple treatment gives an equation of the form

$$D = \alpha a_0^2 \nu \exp\frac{\Delta S_m}{R} \exp -\frac{\Delta H_m}{RT} \tag{5}$$

where ΔS_m is the entropy of migration, a_0 is the lattice constant, ν is the vibrational frequency of the atom, and α is a geometrical constant. This treatment is highly simplified, and ignores correlation of successive jumps, but explains the frequent observation that diffusion coefficients are exponentially dependent on temperature, and described by the relation

$$D = D_0 \exp(-Q/RT) \quad (6)$$

where Q is the activation energy and D_0 is the pre-exponential factor.

Another diffusion mechanism of much interest is the vacancy mechanism. In all crystals some of the lattice sites are unoccupied above 0 K, and these sites are called vacancies. If one of the atoms on an adjacent site jumps into the vacancy, as shown in Fig. 3, the atom has diffused by the vacancy mechanism. Note that the vacancy moves in the opposite direction to the atoms, and we will see that the diffusion coefficient of the vacancies themselves is the fundamental parameter in describing this diffusion process. With this process, the probability that an atom will jump depends not just on the size of the energy barrier, as for interstitial diffusion, but also on whether the next site is empty. This last probability is related to the equilibrium concentration of vacancies, which in turn depends on the enthalpy of formation of the vacancy ΔH_f. The following equation is then obtained, again ignoring correlation of successive jumps

$$D = a_0^2 \nu \exp \frac{\Delta S_f + \Delta S_m}{R} \exp - \frac{\Delta H_f + \Delta H_m}{RT} \quad (7)$$

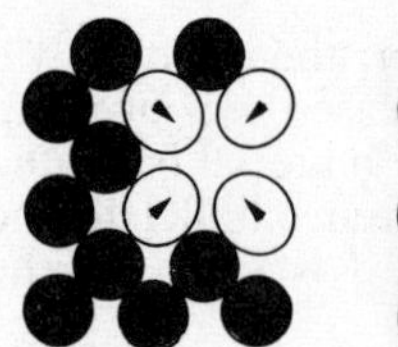
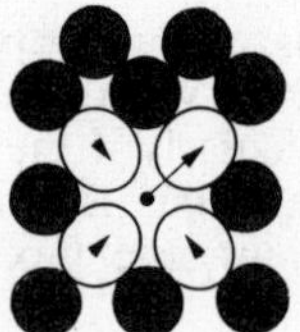
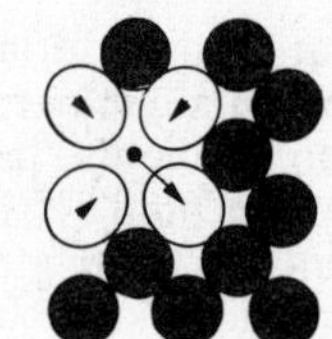

Figure 3
Vacancy diffusion

In the sphalerite lattice of the compound semiconductors illustrated in Fig. 4 we see that, in GaAs, an arsenic atom might diffuse directly to the nearest arsenic vacancy, or possibly to a gallium vacancy site. The second of these two jumps, however, would create another defect—an arsenic atom on a gallium site, which is called an antisite or antistructure defect. Diffusion by this mechanism is more properly described in terms of these defects.

As well as interstitial and vacancy diffusion, many other mechanisms are possible. The diffusing atom may spend part of its time on an interstitial site, and

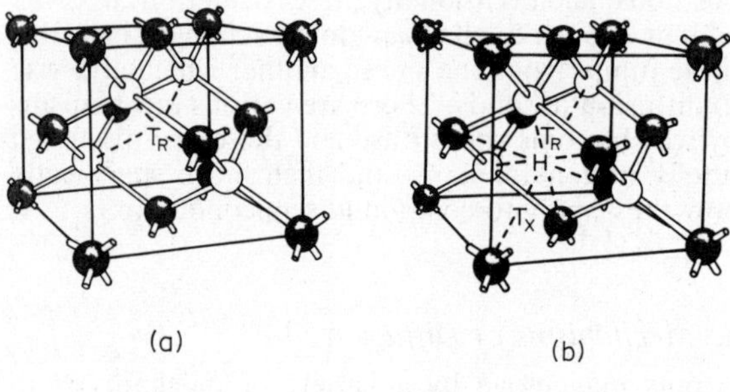

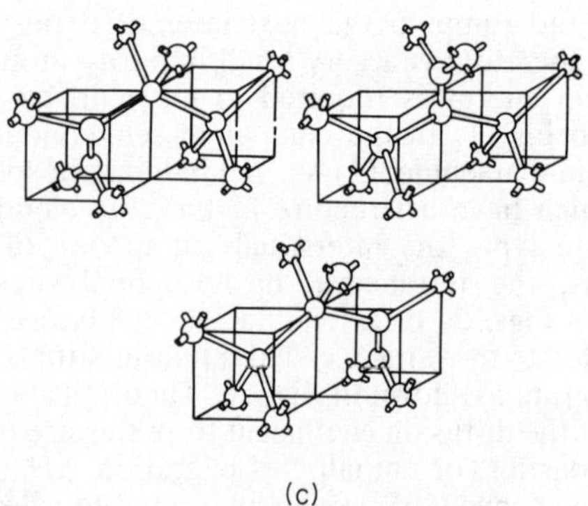

Figure 2
The sphalerite lattice showing (a) the T_R tetrahedral interstitial position, (b) the H hexagonal interstitial site, shown joining a T_R site and a T_X site, and (c) a possible migration mechanism of a split-interstitial (left) via a mixed split-interstitial (middle) to its final position (right)

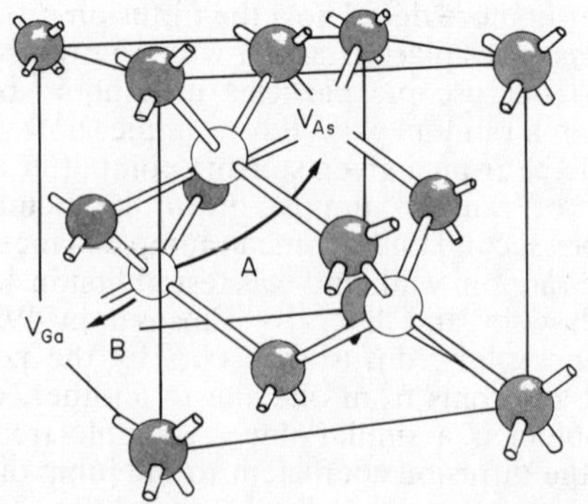

Figure 4
Vacancy diffusion—process A: As diffusion via nearest As vacancy, process B: As diffusion via Ga vacancy

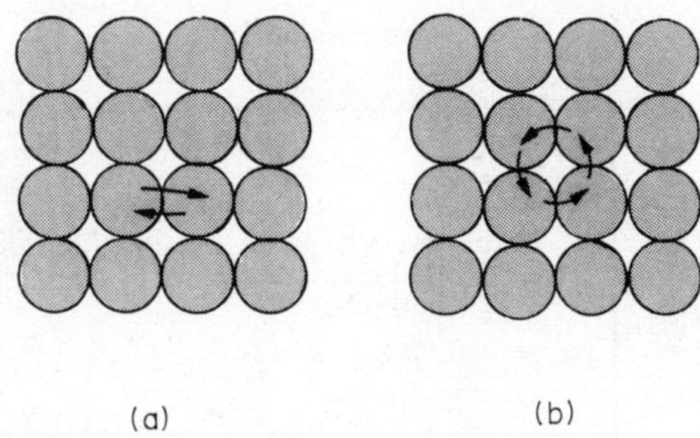

Figure 5
Exchange (a) and ring (b) diffusion mechanisms

part on a lattice or substitutional site, where it has a different electronic activity in the semiconductor. Hence, we encounter two other mechanisms: the interstitial substitutional (or dissociative) mechanism, and the kick-out mechanism. In the first of these the fast-diffusing interstitial atom occasionally moves to a lattice site when it meets a vacancy, the reverse of this process occurring also. In the second mechanism, the interstitial atom displaces, or kicks out, a lattice atom to take up the lattice site. As will be seen, these mechanisms have been invoked for certain impurity atoms in compound semiconductors.

Diffusion mechanisms involving defect pairs, for example divacancies, are also possible, as well as complexes between different defects (e.g., impurity vacancy complexes, etc.). Diffusion, however, may also occur without defects, for example by direct exchange of atoms, or by diffusion in a ring, as illustrated in Fig. 5.

4. *Determination of Diffusion Mechanism*

Diffusion mechanisms are sometimes inferred from the values of D_0 and Q in Eqn. (6) obtained from the temperature dependence of the measured diffusion coefficient. However, this relies on theoretical calculation of quantities such as ΔH_f and ΔH_m, which may be too imprecise to make a certain conclusion. Another approach in compound semiconductors uses the fact, ignored in Eqn. (7), that the point defect concentrations will depend upon the partial pressure of the volatile component (e.g., arsenic over GaAs). A low arsenic pressure will produce more arsenic vacancies than a high arsenic pressure, and so on. The diffusion coefficient of, for example, arsenic depends on the defect concentration in the following way if diffusion is considered via neutral arsenic vacancies and interstitials only:

$$D_{As} = f(V^*_{As})D(V^*_{As})C(V^*_{As}) + f(As^*_i)D(As^*_i)C(As^*_i) \qquad (8)$$

where $D(V^*_{As})$ $C(V^*_{As})$ are the diffusion coefficient and site fraction of arsenic vacancies respectively, and $f(V^*_{As})$ is the correlation factor, which describes the correlation of successive jumps and is related to the likelihood of a reverse jump following a previous jump. Similarly, $D(As^*_i)$ and $C(As^*_i)$ are the diffusion coefficient and site fraction of neutral arsenic interstitials respectively, and $f(As^*_i)$ is the correlation factor. Since the defect concentrations $C(V^*_{As})$ and $C(As^*_i)$ will rise and fall, respectively, if the arsenic partial pressure is lowered during the diffusion anneal, the effect on D_{As} should give a clue to the dominant defect process.

In a similar way, the fact that the defects may be ionized can be used to investigate the diffusion mechanism. If diffusion is considered via neutral and acceptor arsenic vacancies on the arsenic sublattice, the corresponding equation is

$$D_{As} = f(V^*_{As})D(V^*_{As})C(V^*_{As}) + f(V'_{As})D(V'_{As})C(V'_{As}) \qquad (9)$$

where $D(V'_{As})$ is the diffusion coefficient of acceptor arsenic vacancies, and so on. By measuring D_{As} as a function of doping level, conclusions arise about the charge state of the defect involved. A general form of Eqns. (8) and (9) is:

$$D = \sum_j f_j D_j c_j \qquad (10)$$

where f_j, D_j and C_j are respectively the correlation factor, diffusivity and site fraction of the jth native defect (nondefect mechanisms are neglected).

By contrast, the diffusivity arising from a ring or exchange (i.e., nondefect) mechanism (see Fig. 5) will be independent of doping level and of partial pressure. It will be shown how these tests shape up in typical compound semiconductors.

5. *Self-Diffusion in Typical Compound Semiconductors*

Self-diffusion is the diffusion of the component atoms within a compound such as gallium and arsenic in GaAs. It is investigated mainly by the radiotracer method. This involves evaporating a layer of the radioactive isotope of the element on the surface of the compound and, after heating to the required temperature, sectioning to find how far the isotope has penetrated into the solid. Layers are usually removed and counted for radioactivity.

Examples of diffusion profiles obtained by this method are shown in Fig. 6 (Palfrey et al. 1983) for the self-diffusion of arsenic in GaAs. The number of arsenic atoms (on a logarithmic scale) is plotted vs depth and the curves conform closely to an ideal solution of Eqn. (2), called a complementary error function curve (ERFC). This leads directly to values of diffusion coefficient shown in the figure, and the fit confirms that the assumptions of constant D and so on, are correct. Also note that increasing the arsenic pressure during diffusion decreases the diffusion coefficient. This suggests that the most likely diffusion mechanism for arsenic diffusion in GaAs is the vacancy mechanism, process A in Fig. 4.

This type of study has only been carried out over a limited temperature range for GaAs. For another compound semiconductor, GaSb, however, it has been carried out for both gallium and antimony over a wide temperature range. The diffusion coefficients are plotted in Fig. 7 vs reciprocal temperature, showing the effect of changing from gallium-saturated conditions to antimony-saturated conditions. Figure 7 shows that both components diffuse faster in the antimony-saturated material than in the gallium-saturated material, and both diffuse at a similar rate. These measurements, by Mehrer and Weiler (1985), led to the conclusion that the same defect acts as the diffusion vehicle for both atomic species. This defect was suggested to be either a V_{Ga}–V_{Sb} vacancy pair or a triple defect consisting of two gallium vacancies and one gallium antisite defect (see *Constitutional Vacancies*). Note also in Fig. 7 that the temperature dependence is not described simply by Eqn. (6) as in the elemental semiconductors. Limited measurements are also available for some of the other compound semiconductors (e.g., indium phosphide and indium antimonide in Fig. 8), but the information is not sufficient to make firm conclusions about the diffusion mechanisms.

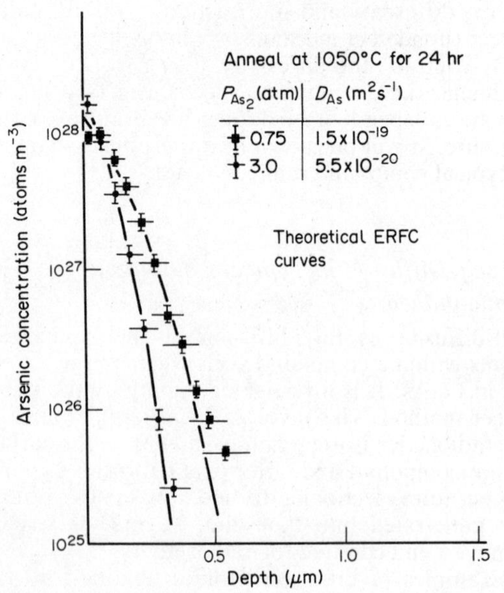

Figure 6
Arsenic diffusion profiles for GaAs (Palfrey et al. 1983)

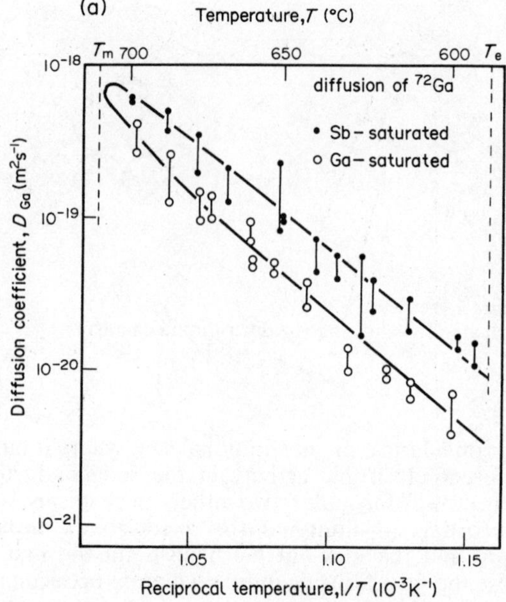

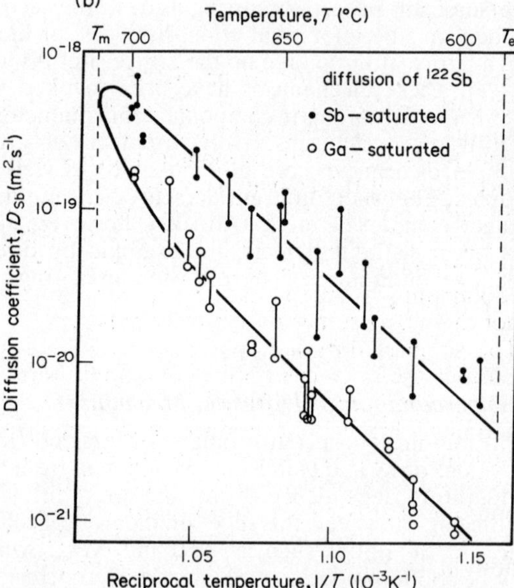

Figure 7
Diffusion of Ga and Sb in GaSb (Mehrer and Weiler 1985)

6. *Impurity Diffusion in Compound Semiconductors*

If diffusion data is collected for impurities in a typical compound semiconductor such as gallium arsenide, an approximately linear relation for $\log_e D$

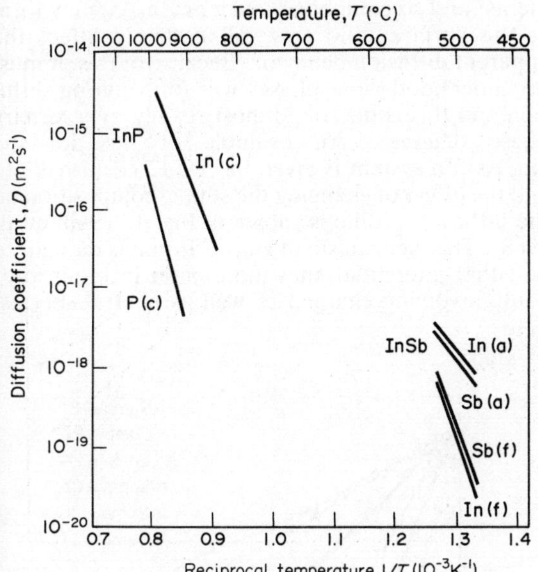

Figure 8
Self-diffusion in InP and InSb

versus $1/T$ (K) is found as predicted in Eqn. (6). Such data is shown in Fig. 9 and the values range from the fast-diffusing elements like copper and lithium to the slow-diffusing elements such as sulfur. Notice also that the activation energy Q for copper and lithium is much less (~0.5 eV) than for most of the slower diffusing impurities (~2.5 eV for sulfur). This is probably because copper and lithium diffuse predominantly as interstitials. Such data, however, is not the whole story in these compounds.

The simplest diffusion conditions to analyze for impurities are isoconcentration conditions, where a radiotracer impurity is diffused into material already containing a higher concentration of the same (non-radioactive) impurity. Most diffusion experiments, however, are carried out under chemical diffusion conditions; that is, under conditions of considerable concentration gradient of the impurity. The last condition often gives profiles that are quite different from the appropriate ideal solution of Eqn. (3), as in the example for zinc in GaAs shown in Fig. 10 where it is compared with an isoconcentration diffusion showing an ideal profile (Chang and Pearson 1964). Practical device processing, however, normally uses chemical diffusion conditions and so an attempt must be made to model this as well.

Like self-diffusion, impurity diffusion often depends on the partial pressure of the volatile component in the compound. This is shown in Fig. 11 for indium in GaSb where Mathiot and Edelin (1980) found that indium diffuses faster in antimony-rich crystals than in gallium-rich crystals. This led them to conclude that indium diffuses by a vacancy

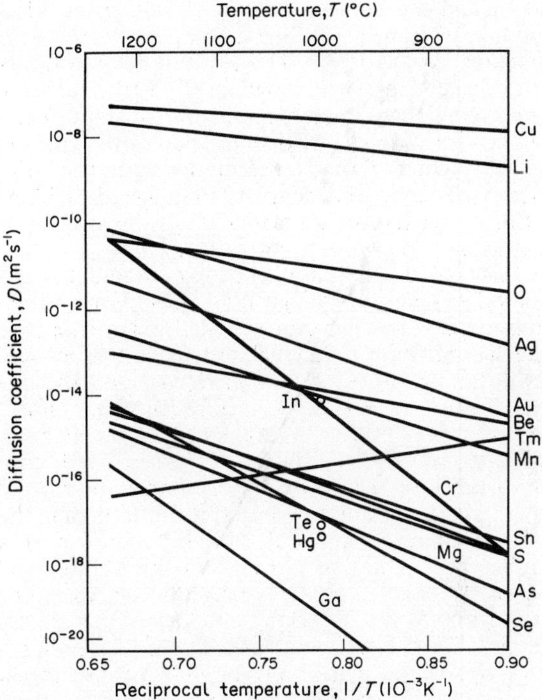

Figure 9
Impurity diffusion in GaAs (Shaw 1973)

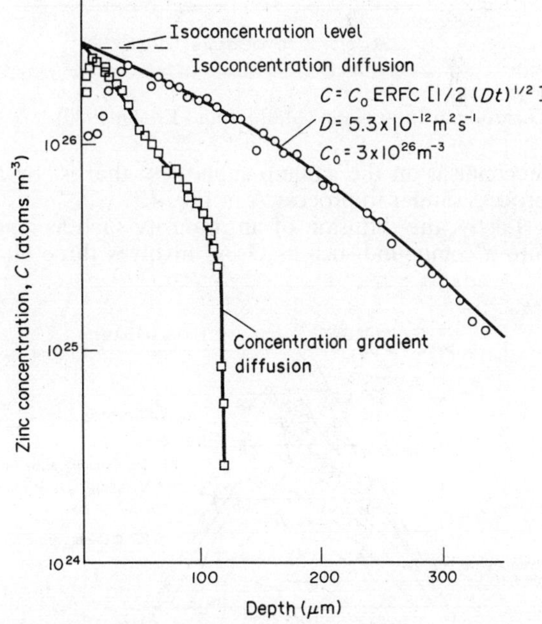

Figure 10
Composition of isoconcentration and chemical diffusion of Zn in GaAs (Chang and Pearson 1960)

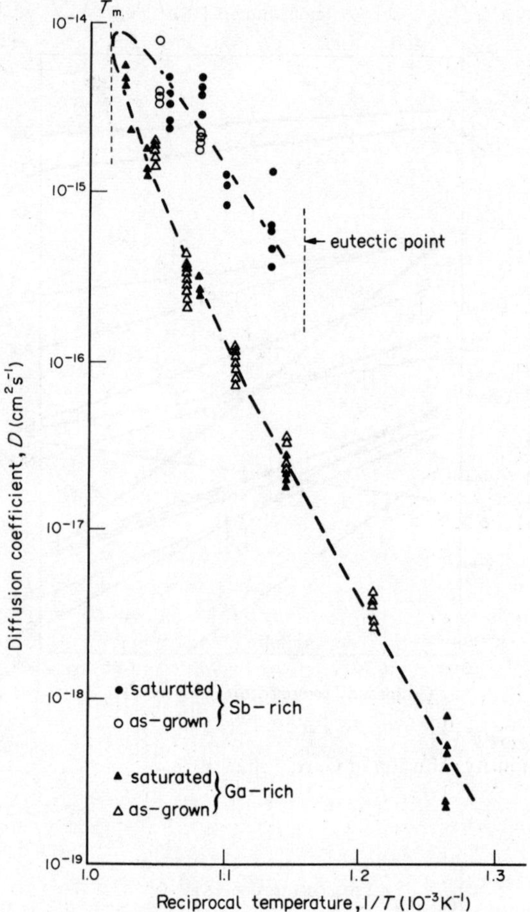

Figure 11
Diffusion of In in GaSb (Mathiot and Edelin 1980)

mechanism on the gallium sublattice; that is, by a process similar to process A in Fig. 4.

Lastly, the diffusion of an impurity such as zinc into a compound such as GaAs involves three elements, and so compounds such as Zn_3As_2 may form at the surface, and this will certainly affect the apparent diffusion behavior. Because of this, it must be understood what phases will form during diffusion and this is described most readily by a ternary phase diagram. An example of this for the Ga–As–Zn system is given in Fig. 12 (Panish 1966) and the effect of changing the source composition on the diffusion profile is shown in Fig. 13 (Shih et al. 1968). The mechanism of zinc diffusion is thought to be either interstitial–substitutional or kick-out, probably involving charged as well as neutral species,

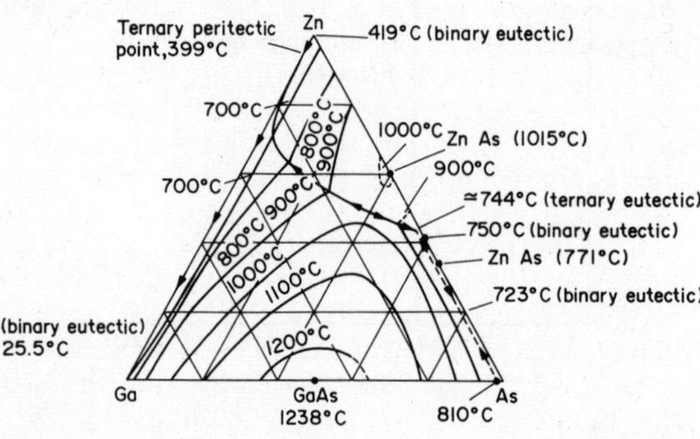

Figure 12
The Ga–As–Zn ternary phase diagram (Panish 1966)

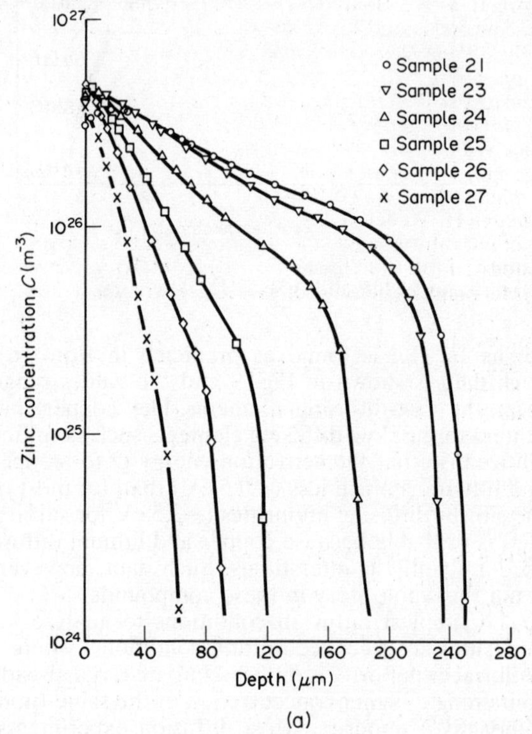

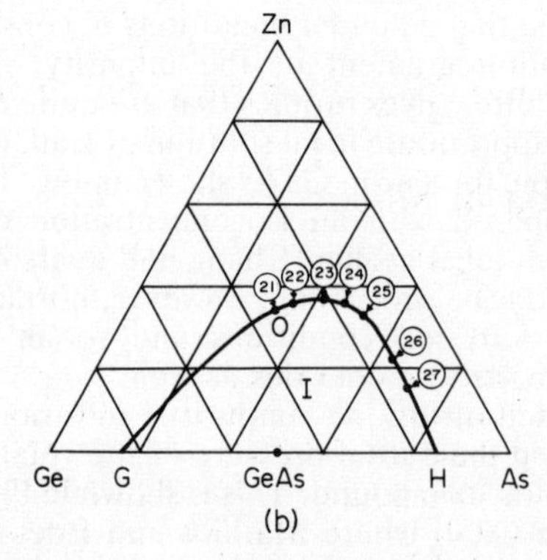

Figure 13
Effect of source compositions in (b) on the diffusion profiles of Zn in GaAs at 1050 °C (Shih et al. 1968)

but the ternary composition of the diffusion source and arsenic pressure are of great importance. This is one of the most thoroughly measured systems, and is an example of how it is hoped the other elements will be understood in future.

See also: Diffusion in Crystalline Metals: Atomic Mechanisms; Diffusion in Crystalline Metals: Phenomenology

Bibliography

Chang L L, Pearson G L 1964 Diffusion mechanism of Zn in GaAs and GaP based on isoconcentration diffusion experiments. *J. Appl. Phys.* 35: 1960–5

Corbett J W, Bourgoin J C 1975 Defect creation in semiconductors. In: Crawford J H, Slifkin L M (eds.) 1975 *Point Defects in Solids*, Vol. 2. Plenum, New York, pp. 1–161

Einstein A 1956 *Investigations on the Theory of Brownian Movement*. Dover, New York

Fick A 1955 *Ann. Phys.* Leipzig 94: 59

Mathiot D, Edelin G 1980 Diffusion of indium in GaSb. *Philos. Mag. A.* 41: 447–58

Mehrer H, Weiler D 1985 Self-diffusion and mechanisms of self-diffusion in GaSb. In: Kimerling L C, Parsey J M (eds.) *Proc. 13th Conf. on Defects in Semiconductors*. Metallurgical Society of AIME, Warrendale, PA, pp. 309–15

Palfrey H D, Brown M, Willoughby A F W 1983 Self-diffusion in gallium arsenide. *J. Electron. Mater.* 12: 863–77

Panish M B 1966 The arsenic-rich region of the Ga–As–Zn ternary phase system. *J. Electrochem. Soc.* 113: 861

Shaw D 1973 *Atomic Diffusion in Semiconductors*. Plenum, London

Shewmon P G 1963 *Diffusion in Solids*. McGraw-Hill, New York

Shih K K, Allen J W, Pearson G L 1968 Diffusion of zinc in gallium arsenide under excess arsenic pressure. *J. Phys. Chem. Solids*. 29: 379–86

Van Vlack L H 1970 *Materials Science for Engineers*. Addison-Wesley, Reading, MA

Willoughby A F W 1990 In: Schulz M (ed.) *Impurities and Defects in Semiconductors*, Vol. 22b. Landolt-Bornstein

Zener C 1952 In: Shockley W (ed.) *Imperfections in Nearly Perfect Crystals*. Wiley, New York, pp. 289

A. F. W. Willoughby
[Southampton University, Southampton, UK]

Diffusion in Silicon

The processing of silicon microelectronic devices has since the early 1950s depended on impurity diffusion techniques to form *p–n* junctions. During this time, numerous approaches have been studied on how to introduce dopants into silicon with the goal of controlling junction electrical properties, concentrations of dopants, doping uniformity and reproducibility, and cost of manufacturing.

The use of diffusion in fabricating modern, very large-scale integration (VLSI) circuits is still important today. Scaling of integrated devices requires *p–n* junctions of the order of 1000 Å below the silicon surface. Such a requirement makes it even more necessary for semiconductor scientists and engineers to refine further their understanding of dopant diffusion. Short-time or low-temperature diffusion steps are needed to produce 1000 Å deep junctions. However, this new processing regime has uncovered complex diffusion phenomena which result in time-dependent diffusion coefficients whose peak values at a given temperature may be many times the steady-state values. In order to understand and to model these effects it is necessary to start with the fundamental basis for diffusion in silicon.

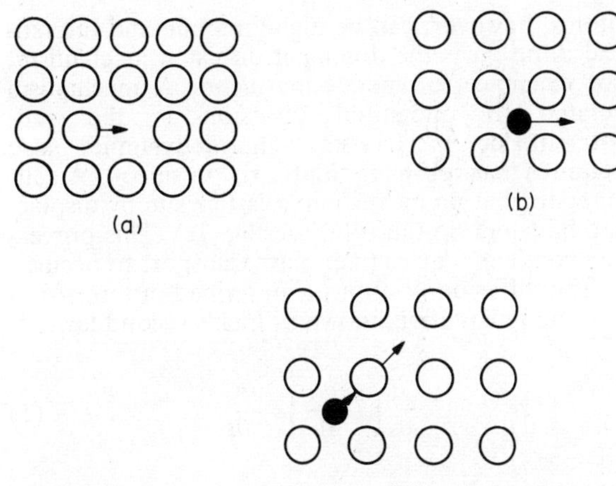

Figure 1
Dominant diffusion mechanisms in silicon:
(a) vacancy diffusion, (b) interstitial diffusion, and
(c) the self-interstitialcy mechanism

1. Diffusion Fundamentals

Dopant diffusion in silicon occurs primarily by processes involving lattice point defects: monovacancies and silicon self-interstitial atoms. Therefore, from an atomic view, diffusion involves the transport of atoms from one part of the silicon crystal to another by the interaction of atoms with these point defects.

Thermodynamic considerations require that some of the lattice sites in the crystal are vacant and that the number of vacant lattice sites is generally a function of temperature. When a lattice atom moves into an adjacent vacant site, the process is called the vacancy diffusion mechanism (see Fig. 1a). In addition to occupying lattice sites, atoms can reside in the space between the lattice sites. These interstitial atoms can readily move to adjacent interstitial sites without displacing the lattice atoms (see Fig. 1b). The interstitial atoms may be impurity atoms or atoms of the host lattice, but in either case they are generally present only in very small numbers. These atoms, however, can be highly mobile and interstitial diffusion is the dominant diffusion mechanism, for example, for metals in silicon. A mechanism related to interstitial diffusion is the self-interstitialcy mechanism. Thermodynamics also requires that self-interstitials exist in silicon. A self-interstitial atom moves into a lattice site by displacing the atom on that site (see Fig. 1c). This process can continue, causing net mass transport to occur.

The diffusion of atoms is described in terms of a continuity equation known as Fick's second law:

$$\frac{\partial}{\partial x}\left(D\frac{\partial C}{\partial x}\right)=\frac{\partial C}{\partial t} \tag{1}$$

This describes the rate of change of the atomic concentration, C, with time. The diffusion coefficient D is expressed in units of $cm^2\,s^{-1}$ and the concentration C is in units of atoms cm^{-3}.

The diffusion coefficient can be shown to be basically the product of the lattice vibration frequency, γ, the atomic jump distance, a_0, the point defect fraction, X_{pd}, and the fraction of atoms that are activated to make a jump, X_m (Fair 1988):

$$D=\tfrac{1}{2}a_0^2X_{pd}X_m\gamma \tag{2}$$

Thus, the activation energy for diffusion consists of the energy necessary to form the point defect and the migration energy of the diffusing atom.

2. Point Defects

It can be seen from Eqn. (2) that the process of diffusion depends upon the concentration of point defects (such as vacancies or self-interstitials) in the crystal. Therefore, if ways can be found of raising or lowering the point defect concentration then diffusion coefficients can be affected.

For the vacancy mechanism (Fair 1981), the single vacancy in silicon is believed to exist in four charge states: $V^+, V^\times, V^-$ and $V^=$ where + refers to a donor level, × a neutral species and − an acceptor level. The creation of a vacancy introduces a new lattice site, and thus four new valence band states in the crystal. These states are available as acceptors but are not shallow. The lattice distortion associated with the vacancy will split states from the valence and conduction bands of the surrounding atoms a few tenths of an electron volt into the forbidden gap. States split from the valence band will become donors and those split from the conduction band will become acceptors. At low temperatures there should be one deep donor level V^+ a few tenths of an electron volt above the valence band edge, a single acceptor level V^- near midgap, and a double acceptor level $V^=$ very near the condition band edge (see Fig. 2).

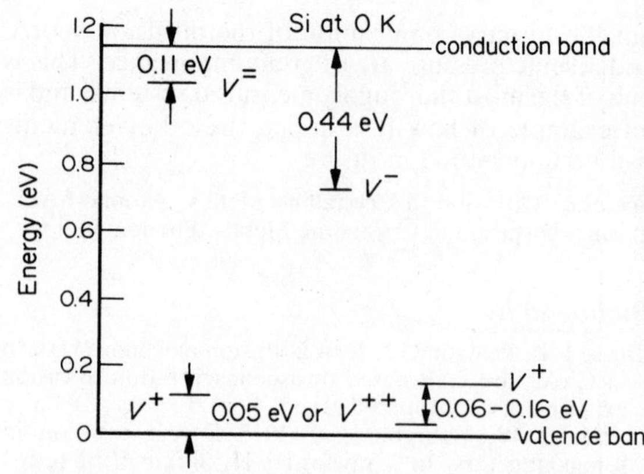

Figure 2
Estimated vacancy energy levels in the silicon band gap at 0 K (based on best guess from experiment)

It has been experimentally verified that both silicon self-diffusion and the diffusion of Group III and V impurities in silicon depend upon the Fermi-level energy, E_f. The initial assumption in the vacancy diffusion model of self-diffusion is that an observed diffusivity arises from the simultaneous movement of neutral and ionized vacancies. Each charge type vacancy has a diffusivity whose value depends upon the charge state, and the relative concentrations of vacancies depend upon the Fermi level. Whereas at low temperature $V^\times$ will be the dominant species in intrinsic silicon, at high temperatures both V^+ and V^- would be more numerous, and there is no value of E_f for which $V^\times$ dominates. Another important concept is that every time an ionized vacancy is formed the silicon crystal must return the neutral vacancy population back to equilibrium by generating an additional vacancy. This is due to the fact that the concentration of uncharged vacancies is considered to be an intrinsic property of the crystal. Hence, this concentration will only be a function of temperature. However, ionized vacancies can be controlled through the law of mass action. In this way, as the doping becomes more n-type or more p-type, the total vacancy concentration will increase as the population of ionized vacancies increases. Since impurity and self-diffusion coefficients depend upon the concentration of vacancies, the diffusion coefficients will also increase with doping. Such concentration-dependent diffusion can occur when the doping level exceeds the intrinsic electron concentration, n_i, at the diffusion temperature. An illustration of concentration-dependent diffusion is shown in Fig. 3. Similar arguments hold for the concentration dependence of diffusion by the self-interstitialcy mechanism which depends on self-interstitials existing in multiple charge states.

In developing dopant diffusion models one must decide whether a vacancy mechanism or a self-

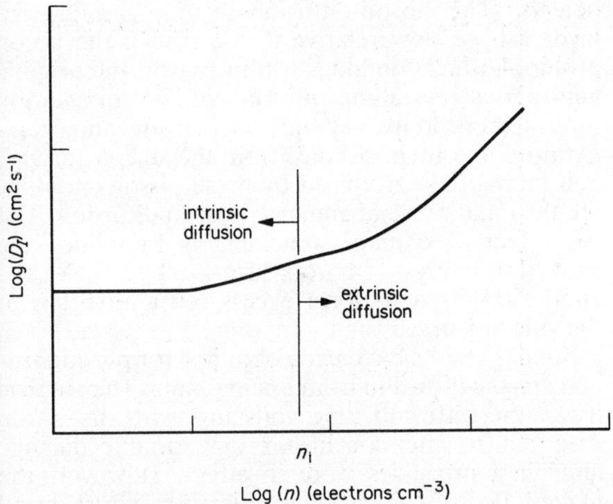

Figure 3
Donor impurity diffusion coefficient vs electron concentration showing regions of intrinsic and doping-dependent diffusion

interstitialcy mechanism, or both, apply. It must also be decided which defect is responsible for dislocation and other defect growth in silicon. The majority opinion is currently that both types of point defects are important. Thermal equilibrium concentrations of point defects at the melting point are orders of magnitude lower in silicon than in metals. Therefore, a direct determination of their nature by volumetric change experiments is not possible. The accuracy of calculated formation and migration enthalpies appears to be within ±1 eV, but these do not help in distinguishing whether it is vacancies or self-interstitials that are the dominant presence in diffusion. The interpretation of low-temperature experiments on the migration of irradiation-induced point defects is complicated by the occurrence of the radiation-induced migration of self-interstitials. In addition, there are indications that the structure and properties of point defects may be different at low and high temperatures.

In view of the uncertainties regarding the native point defect in silicon, it is necessary in discussions of dopant and self-diffusion to take account of both types of defects.

3. Point Defect Models of Diffusion in Silicon

Under thermal equilibrium conditions a silicon crystal will contain a certain equilibrium concentration of vacancies, C_V^O, and a certain equilibrium concentration of silicon self-interstitials, C_I^O. In diffusion models based on vacancy domination ($C_V^O \gg C_I^O$) the dopant as well as self-diffusion coefficients can be described by

$$D_i = D_i^{\times} + D_i^{-} + D_i^{=} + D_i^{+} \quad (3)$$

where D_i is the measured diffusivity and $D_i^{\times}$, D_i^{-}, $D_i^{=}$ and D_i^{+} are the intrinsic diffusivities of the species through interactions with vacancies in the neutral, single acceptor, double acceptor and donor charge states respectively (Fair 1981). These individual contributions to the total measured diffusivity were described in Sect. 2.

Analogous to the vacancy model, self-interstitials can be assumed to be dominant such that $C_I^O \gg C_V^O$. For such a model, dopant and self-diffusion are assumed to occur via an interstitialcy mechanism. Mobile complexes consisting of self-interstitials in various charge states and impurities are assumed to exist.

In principle, both vacancies and self-interstitials may occur simultaneously and somewhat independently. Indeed, any relationship that may exist between C_V^O and C_I^O may be dominated by the silicon surface that can act as a source or as a sink for either species. If a local dynamic equilibrium exists between recombination and spontaneous bulk generation, vacancies and self-interstitials would react according to

$$V + I \rightleftharpoons O \quad (4)$$

where O denotes the undistributed lattice. The law of mass action under equilibrium for this reaction is

$$C_I C_V = C_I^O C_V^O \quad (5)$$

For sufficiently long times and high temperatures Eqn. (5) is fulfilled. However, a substantial amount of time may be required for dynamic equilibrium to occur. This would make vacancy–self-interstitial recombination an activated process. In addition, under point-defect injection conditions, Eqn. (5) may no longer be valid.

If both types of point defects are important, diffusion processes may involve both types:

$$D_i = D_i^I + D_i^V \quad (6)$$

where D_i^I is the interstitialcy contribution and D_i^V is the vacancy contribution to the total measured diffusivity, D_i. To learn about the relative importance of D_i^I and D_i^V, numerous experiments have been conducted, but these have only provided indirect evidence. For example, experiments which provide an external source of either self-interstitials or vacancies allow one to observe enhanced or retarded diffusion of dopant profiles in the presence of excess point defect concentrations. The oxidation of silicon is known to produce an excess of self-interstitial flux into the silicon bulk (Hu 1974). Other surface reactions such as the direct nitridation of silicon with NH_3 gas, oxidation in the presence of chlorine, or silicidation reactions involving metals like titanium or tantalum all produce excess vacancy fluxes into the silicon. Verification of these point defect fluxes is made by observing the growth or shrinkage of interstitial-type defects in the silicon during the experiments. The results of these experiments show that in silicon both boron and phosphorus diffuse primarily by a self-interstitialcy mechanism, arsenic diffuses with a mixed vacancy and self-interstitialcy process, and antimony is exclusively a vacancy diffuser.

4. Issues in Shallow Junction Formation for VLSI

VLSI technologies with circuit feature sizes that are patterned at submicrometer dimensions require *p–n* junctions that are of the order of 1000 Å deep. To achieve such shallow junctions requires sophisticated methods of introducing dopants into the silicon, such as ion implantation. In addition, diffusion of dopants from their initial deposited positions in the silicon must be limited by reducing the time or temperature of processing. However, when these methods are applied in conjunction with ion implantation, enhanced dopant diffusion occurs (Michel 1986).

Ion implantation of atoms into a crystal produces damage in the form of displaced lattice atoms (interstitials), vacancies and even the complete destruction of crystalline order (amorphization). Annealing the implanted silicon will start the processes of damage removal, the regrowth of amorphous layers to a single crystal, or the establishment of extended dislocations and dislocation loops. These processes can produce excess point defects for varying periods of time. Dopant atoms will experience different amounts of enhanced or retarded diffusion during annealing depending on the diffusion mechanism, the spatial relationship of the damage to the dopant distribution and whether the damage produces vacancies or self-interstitials.

Boron diffusion occurs by an interstitialcy mechanism. Boron implantation produces isolated clusters of self-interstitials. At very high implantation doses, extended dislocations will be formed near the peak of the implanted profile where the boron concentration exceeds solid solubility. The annealing of the isolated clusters occurs rapidly—in 1 s at 1000 °C and in 25 min at 750 °C. Self-interstitials produced during this time will cause enhanced, transient boron diffusion. However, the shrinkage of the extended dislocations in the profile peak is very slow. The annihilation of these defects can take hours at 950 °C, causing enhanced boron diffusion to occur during this time.

If an amorphous layer is produced in the silicon surface, a shallow vacancy-rich layer can be produced in addition to regions rich in interstitial-type defects. The boron-diffusion in the vacancy-rich layer will be slow relative to the rest of the boron profile. Dislocation loops will form near the original amorphous–crystalline interface. The dissolution rate of these loops depends on the annealing temperature and their distance from the silicon surface. Self-interstitials produced by these dissolving loops are the main cause of enhanced boron diffusion. The net effect of damage annealing is to reduce the activation energy of boron diffusion by 2.5 eV, the energy required to form excess self-interstitials at the silicon surface.

Similar results have been obtained for phosphorus and arsenic diffusion from ion implants. The similarity of the diffusion time constants with the boron case led to the conclusion that similar damage annealing processes were in effect. However, the magnitude of enhanced arsenic diffusion is small relative to boron and phosphorus. This is due to the fact that arsenic diffuses with a mixed vacancy—the self-interstitialcy mechanism with the vacancy component dominating.

Low-temperature or short-time diffusion technology is far more complex than previously believed, since ion-implantation-related point defects and extended defects dissolve over significant portions of the annealing time. Enhanced or retarded diffusion of dopants is, therefore, dependent upon the amounts and types of defects produced by ion implantation, the annealing procedure, and whether or not the silicon is amorphized. In view of these facts, no first-principle model exists to account for these variables. For silicon technologists the challenge is to control and reproduce these variables or else to develop new dopant deposition methods that eliminate them.

See also: Diffusion in Crystalline Metals: Atomic Mechanisms; Diffusion in Crystalline Metals: Phenomenology; Transition Metal Silicides (Suppl. 2)

Bibliography

Fair R B 1981 Physics and chemistry of impurity diffusion and oxidation of silicon. In: Kahng D (ed.) 1981 *Applied Solid State Science*. Academic Press, New York

Fair R B 1988 Diffusion and ion implantation in silicon. In: McGuire G E (ed.) 1988 *Semiconductor Materials and Process Technology Handbook*. Noyes Publications, Park Ridge, IL

Hu S M 1974 Formation of stacking faults and enhanced diffusion in the oxidation of silicon. *J. Appl. Phys.* 45: 1567–76

Michel A E 1986 Diffusion modeling of the redistribution of ion implanted impurities. In: Sedgwick T O, Seidel T E, Tsaur B (eds.) 1986 *Rapid Thermal Processing*. Materials Research Society, Pittsburgh, PA, pp. 3–13

R. B. Fair
[Microelectronics Center of North Carolina, North Carolina, USA]

E

Endodontic Materials

The term "endodontic" literally means "within the tooth" but is used customarily to refer to therapy of the tooth when the dental pulp is involved directly. In the most conservative of these treatments, the pulp is treated to keep it alive and the interface between the therapeutic material and the pulp may be only micrometers across. In the most radical, not only the pulp but a substantial portion of the tooth root itself may be replaced. The material involved in the former procedure falls very much into the category of pharmacological materials, while those used in the latter are distinctly biomaterials. For materials used in all other endodontic procedures, it may be a point of argument whether the nature of the material is pharmacological or whether it falls into the category of biomaterials science. For completeness, the whole range of endodontic materials will be described in this article. First, however, it is appropriate to look at the structure of the system being treated: a diagram of the pulpal structure of a typical tooth is shown in Fig. 1.

1. *Structural Elements of the Tooth*

1.1 *Enamel*

Dental enamel is composed of large crystals of hydroxyapatite. It is permeated by micropores and there is a movement of fluid through it from the underlying dentine. Although relatively brittle, it is supported by a firm bond to the underlying dentine at the amelo–dentinal junction. The integrity of the surface enamel is related, therefore, to the mechanical properties of the underlying dentine which may change for a number of reasons, described later. Cracks are frequently observed at the enamel surface but these rarely propagate into the dentine, although when this occurs it can cause extreme pain. The enamel forms a relatively impermeable barrier which protects the underlying dentine.

1.2 *Dentine*

Dentine consists of microtubules of mineralized tissue within an organic matrix. In the newly erupted tooth, these tubules are occupied by the long cellular processes of the odontoblasts, the cells which form the dentine. Although there are very few nerve endings within the dentine, the whole structure right up to the enamel is sensitive to pain. The fluid content of the dentine is maintained to a degree by the odontoblastic processes, and this will be lost in favor of simple diffusion when the pulp is removed. This may severely affect the mechanical properties of the dentine. Young dentine is relatively elastic but this diminishes as the tooth ages or when the pulp is lost. As the tooth ages, further deposition of

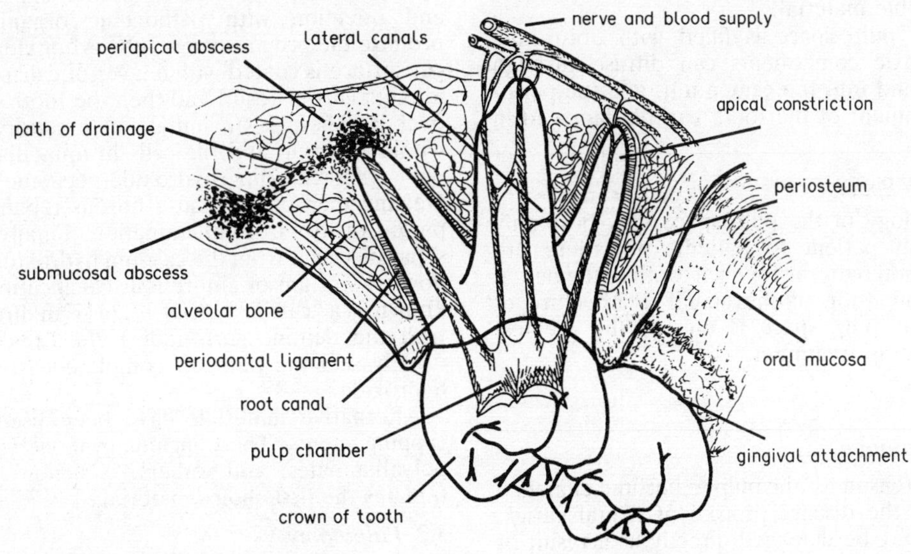

Figure 1
Pulpal structure of a typical tooth

mineral within the tubules may occur and occlude them to a greater or lesser degree. These combined effects will alter the physical properties of the dentine and in consequence the prognosis for the tooth after it has been treated endodontically.

Permeability. Because of its microtubular nature, dentine is extremely permeable. Furthermore, the large surface area within the tubules is relatively active chemically. Damage can occur to the pulp via the permeable dentine. When the pulp is removed and replaced with a chemical substance to obturate the space, irritant substances may penetrate the dentine in a reverse direction and affect adversely the cells of the periodontal ligament which is attached to the cementum surface of the tooth.

The permeability of the dentine can be affected in a number of ways. Most significantly, the permeability can be reduced greatly when a smear layer is produced by the dentist's instruments at the cut surface of the dentine. Chemicals used later may remove this smear layer and again increase the permeability. The presence of a smear layer may partly protect the dentine and associated cellular elements from the effects of applied chemicals such as sealers or adhesives.

1.3 Cementum

Cementum is a thin coat of mineralized tissue that covers the outer surface of the root dentine. The gingival tissue is attached to it by hemidesmosomal junctions and by the penetration of fibers from the periodontal ligament. Cementum is produced continuously but slowly throughout life and this can be an advantage. After root treatment, it is possible for cementum to be laid down over the apertures into the pulp space provided these have been sealed with a biocompatible material.

When the pulp space is filled with obturating materials, toxic components can diffuse through the dentine and into the cementum, and the periodontal attachment of the tooth can be damaged in consequence.

1.4 The Pulp Space

The morphology of the pulp space may be highly variable from patient to patient, but there are certain common patterns for each tooth of the dental arch. Determination of the actual architecture of any particular pulp space is one of the greatest problems of pulp treatment.

2. Pulpal Damage

The primary reason for the pulp to become damaged is as part of the disease process of dental caries, although it may be damaged directly as a result of trauma or via a range of other mechanisms including direct infection via communicating links through the dentine. These links include lateral canals and furcation communications. These may become exposed to the oral environment by the loss of attachment at the neck of the tooth due to periodontal disease. Where this attachment is lost, pockets form and fill with oral bacteria or dental plaque. These bacteria may penetrate into the pulp directly through these channels and lead to pulp death.

The tooth has a number of defense reactions to the carious process. To what extent each of these may be activated depends on the speed and severity of the carious attack. These defences include the laying down of a reactive calcific barrier along the wall of the pulp chamber adjacent to the carious lesion. Such a barrier can make access to the pulp for endodontic treatment very difficult.

3. Endodontic Treatment

From the most conservative to the most radical, the range of endodontic procedures performed are: direct pulp-capping, pulpotomy, pulpectomy, apicectomy, hemisection, radicectomy and diadontic implant. The most commonly performed is the pulpectomy, with the others combined making up only a few percent of the total.

3.1 Pulp-Capping

The objective of pulp-capping is to maintain the viability of a dental pulp which has become exposed by loss or removal of dentine during an operative intervention or due to traumatic breakage of the crown of a tooth in an accident. The key to success lies with the status of the pulp prior to the exposure and the level of infection with microorganisms following the exposure. Existing inflammatory change and infection with pathogenic organisms compromise the prognosis severely. After cleansing, the cut surface is coated with a layer of calcium hydroxide in a carrier resin, and then the tooth is restored over it. There is an initial inflammatory response within the pulp and the cells in immediate contact with the calcium hydroxide become necrotic. Healing then occurs and a fibrous repair follows a period of localized inflammation. Finally, the pulp separates itself from the calcium hydroxide by laying down a stratum of amorphous calcification which is known as a "calcific barrier." This is in direct contact with the dentine surrounding the break and will usually seal the pulp off completely from calcium hydroxide.

Alternative materials have been used as pulp-capping agents. These include cyanoacrylates, glass polyalkanoates and others. Calcium hydroxide remains the first choice material.

3.2 Pulpotomy

The pulpotomy procedure is carried out when the coronal pulp is beyond rescue but the pulp in the root may be kept. This is usually in the case of damage to an immature tooth where the root apex has not fully formed. An open apex prevents a build-up of intracoronal pressure due to inflammation of the pulpal tissue, and pulp death due to ischemic necrosis is less likely.

Access is cut into the pulp chamber and the coronal pulp is cut out using a sharp rotary bur, a dental excavator or a diathermy instrument. The severed end of the radicular pulp is treated in the manner of a direct pulp cap; after cleansing, it is sealed with a layer of calcium hydroxide and the tooth is restored. Success of the procedure is judged primarily by radiographic evidence of the continuation of apex formation. A calcific barrier should also form between the calcium hydroxide layer and the dental pulp in the manner of healing after a direct pulp cap.

3.3 Pulpectomy

When the pulp is damaged more severely or is already necrotic, it is necessary to remove it in its entirety. This leaves a space which has to be filled with a suitable obturating material to prevent microorganisms from propagating in the avascular void of the root canal system and leading to chronic sepsis around the apical opening of the root. The obturating material should be biocompatible, because it may come into contact with the tissues around the tooth root. This contact may be direct where the material closes off apertures into the tooth at the apical foramen, lateral canals and furcation communications. Indirect contact occurs due to the tubular nature of dentine. Dentine permeability allows chemical constituents of the obturating material to diffuse through the tooth root and affect the supporting periodontium and its attachment to the root.

A further requirement of the obturating material is that it should form a total seal with the walls of the pulp canal against the ingress of microorganisms. The formation of this seal requires the material to take on the shape of the prepared canal space. This shape may be both complex and compound, so the material must deform easily in order to adapt to it. It is preferable that the obturating material is radio-opaque so that its presence and position can be confirmed radiographically after, or sometimes during, placement.

The conventional method of sealing a tooth after removal of the pulp is to shape the canal to remove as much of the complexity of shape as possible and open and widen it to make insertion of the sealing material easier. This process also cleans the canal if it is infected and reduces considerably the microorganism count. The process of filing the walls of the canal system and the removal of debris by irrigation is known as the 'biomechanical preparation' of the root canal. Irrigation of the canal during preparation is an essential part of the process and serves both to lubricate the cutting edge of the files in the canal as well as cleaning out dentine fillings. Sterile water will achieve these objectives, although sodium hypochlorite solution is used by many practitioners because of its ability to dissolve organic debris and its bleaching action on the dentine.

The canal system is then sealed temporarily to allow the resolution of apical inflammation prior to the placement of the definitive canal seal. During this period, which should be about three to seven days, the growth of microorganisms in the canal must be suppressed. This is achieved by incorporating an antiseptic into the dressing. A pledget of cotton wool is lightly moistened with a solution of *p*-monochlorophenol and placed into the canal opening before sealing temporarily with a zinc oxide–eugenol cement. Antibiotic pastes have been used but have little advantage over the method described and may in fact be irritant to the apical tissues. A further problem with such pastes is that they have to be completely removed from the canal structure prior to the placement of the definitive root filling, and this may be very difficult to achieve in practice due to the insoluble nature of many of the paste vehicles.

The most satisfactory method of sealing the canal system definitively is known as "lateral condensation." In this technique, a deformable plastic point is coated with a semifluid sealer and inserted to the full length of the canal. With the forcible insertion of pointed instruments called "spreaders," this point is squashed laterally against the sides of the canal. A further point is inserted into the space created and the process repeated until the canal is filled. The repeated squashing of the plastic points forces the sealer between the points and the canal walls, and between the points themselves, thus creating a tight-sealing laminate within the canal. This system of lateral condensation is the standard against which all other systems of canal obturation are compared. The most successful material used for constructing the plastic point is a composite of gutta percha and zinc oxide powder. Gutta percha is a natural product derived from the coagulated exudate of mazer trees, which are native to Malaysia. It is an isomer of natural rubber and has the same basic unit, isoprene. Other additives include plasticizing waxes, radioopacifiers such as barium sulfate and colorants. Gutta percha is remarkably low in cytotoxicity, even in tissue culture studies, and this contributes to its universal acceptability in this role. However, it should be remembered that a sealer paste will usually intervene between the surface of the gutta percha and the tissues and it is this that will determine to a greater extent the biocompatibility of the obturating system.

In the past, solvents such as chloroform have been used during the placement of gutta percha root fillings to help adapt the material to the shape of the canal. Several methods were used, ranging from injecting directly a paste of gutta percha dissolved in chloroform, coating a gutta percha point with such a paste as a sealer, or just dipping the gutta percha into the solvent before placing it into the canal. However, as the solvent evaporates, shrinkage occurs and the vital seal between the gutta percha point and the dentine of the root canal walls is lost. There is also a safety aspect in that chloroform is no longer considered acceptable as a material for direct clinical use. A recent innovation has been to melt the gutta percha in a specially designed heat gun and then inject it into the canal in the molten state. While this system works acceptably in skilled hands, the depth of penetration of the molten gutta percha can be impossible to determine during placement.

Synthetic polymers such as polystyrene have been used, but these are generally less deformable and therefore rely on the sealer paste to guarantee a seal. The use of silicone points was examined more than a decade ago, but has not proved advantageous over gutta percha and in consequence has not been developed.

Nondeformable points, such as those made of silver, were popular in the past but are now not much used. Titanium points are now available commercially and may have limited application in canals which can be shaped to a circular cross section, but such cases are few. A further disadvantage of metal points is that they are easily displaced with loss of the required seal during the subsequent restoration of the crown of the tooth.

The importance of the sealer paste is obvious. The success of root canal treatment is probably dictated more by this than by the obturating points that are used. Ideally, this material should be biocompatible, adhesive and highly thixotropic. The most popular sealers used today are based on zinc oxide–eugenol systems. These are acceptably biocompatible although they show no adhesive behavior. Various formulations of zinc oxide sealers have been promoted. One of the most popular is Grossman's formula which contains barium sulfate as a radioopacifier. Anhydrous sodium borate in the powder removes water from the system and this slows down the setting reaction to a more workable level.

Epoxide resins have been used successfully and adhesion has been claimed even against wet canal walls. One such product uses a base paste of bisphenol diglycidyl ether which is mixed with a hardener containing hexamethylene tetramine. The material has an extremely long working time, measured in hours rather than minutes, and this may be advantageous in the time-consuming lateral condensation of multiple complex canals in molar teeth.

Several sealers based on calcium hydroxide have appeared on the market in the late 1980s. Excellent results from the use of these materials have been reported in the literature.

Glass polyalkanoate materials are being developed and show great promise. Sealers made of this material should adhere well to the dentine by the formation of chemical adhesive bonds.

Paste-fillers alone have been used and there are many permutations that have been marketed. These range from thermoplastic resins that are injected into the tooth in the molten state to two-part resins which set after injection. Many of the paste-fillers contain anti-inflammatory drugs such as the steroids dexamethasone or triamcinolone in order to damp down the response of the apical tissues to the material. Antiseptics are sometimes incorporated as well and may be highly irritant to the tissues, although the full force of the inflammatory response may be damped down and delayed by the presence of the anti-inflammatory steroids.

The major drawback with paste-fillers is the difficulty in establishing the seal at the correct place in the tooth root. By its very nature, the canal is an open-ended cylinder. A constriction at the apex may block successfully the passage of a semirigid point, but it may not limit the flow of a paste. The possibility of the paste being extruded into the tissues beyond the apex of the tooth limits both the range of materials which can be used and the technique and circumstances of placement.

Occasions arise when a tooth has to be root-treated when its apex is not fully formed or has been opened up by root resorption. In such cases, it is inevitable that a large area of root-sealing material is going to come into contact directly with the apical bone. The most desirable technique in such cases is to pack the apical aperture with calcium hydroxide powder to form an interface against which an insoluble and nonresorbable sealer may be packed. There are several commercial preparations based on calcium hydroxide that may be selected from with caution but offer little advantage over plain pharmaceutical grade calcium hydroxide powder. Such preparations contain a range of additives such as radioopacifiers, antiseptics and even antibiotics.

Whichever technique is used for obturating a root canal, it should be obvious that the primary criterion of success is the integrity of seal that has been achieved at every opening of the pulp space on the root surface. A wide range of scientific techniques has been developed to examine this seal, but there is no agreement on the best method. Without a standard assessment it is difficult to compare results for different materials. *In vitro* leakage techniques that have been reported have used dyes, radioisotopes, bacteria, scanning electron microscopic examination and electrical impedance measurement.

3.4 Apicectomy

When other, more conservative methods of sealing the apical tissues from the pulp space of the tooth have failed, a direct approach is sometimes undertaken and is known as an apicectomy. This procedure involves gaining direct access to the apex of the tooth by removing the overlying bone after a mucoperiosteal flap has been lifted. The apex of the tooth contains the often troublesome delta of apical canals and is resected to expose the main trunk of the root canal. This enables the direct placement of an apical seal. Traditionally, dental amalgam has been the chosen material for this seal and, perhaps surprisingly, it is still the most commonly used. When set, dental amalgam is well tolerated by the tissues. However, great care has to be taken to restrict the material to the prepared root canal and to avoid overspill of amalgam into the surrounding bone.

Plastic resins and composite materials have been used in this application, but generally they do not form an acceptable seal with the dentine walls in the compromised conditions of a surgical procedure. margins of the retrograde seal to microleakage.

Of all other materials that may be used to achieve the apical seal, probably the most promising are the glass polyalkanoates. These materials adhere to dentine and are relatively biocompatible, but few have the desired radioopacity necessary for postoperative radiographic examination. Lanthanum is added to some glass polyalkanoate cements to make them radioopaque, but in the volume used to fill a deftly prepared apical cavity they may be barely visible. For radioopacity, amalgam has no competitor.

An apicectomy causes great damage to the mechanical and structural integrity of a tooth and is a procedure to be avoided whenever possible. By the time an apicectomy is completed, the root has been hollowed out, had the apex cut off and is embrittled by the original removal of the pulp anyway. It is obvious that its strength must suffer considerably and fracture of the remaining root mass is not unusual in the long term.

3.5 Diadontic Implant

For a number of reasons, the area of root surface supporting a tooth may become reduced to the point where the tooth becomes nonviable. For example, the apical part of the root may be fractured away from the coronal part due to a blow. Root resorption can occur and reduce root support, or the level of the periodontal attachment can be reduced severely due to periodontal disease. In all these cases it may become necessary to supplement the support which the root provides. This can be achieved with a diadontic implant. The diadontic implant is a rod which can be cemented into the pulp space of the crown of the tooth and extended into the alveolar bone through what remains of the tooth root to provide increased radicular support. The implant is usually about 1–2 mm in diameter and made of surgical-grade cobalt–chromium alloy. The length of implant that may be employed depends on the anatomical structure in the vicinity of the tooth root.

Provided the implant remains sealed from the oral environment, its prognosis remains good. However, if contact with the mouth is established by the formation or extension of a periodontal pocket, then failure follows rapidly.

4. Restoration of Root-Filled Teeth

One common feature associated with pulp death and subsequent endodontic treatment is staining of the enamel and dentine by blood and tissue breakdown products, or even by endodontic materials themselves. Staining in anterior teeth is unacceptable and various techniques have been devised for bleaching the dentine and enamel. Bleaching agents include strong solutions of hydrogen peroxide, sodium hypochlorite and various acids. The decomposition of hydrogen peroxide to produce nascent oxygen is often accelerated by the application of heat or strong light to the outer surface of the tooth being bleached. Often, several lengthy visits are required to bleach a tooth to an acceptable color.

The vast majority of root-filled teeth have crowns that are damaged very severely. This is due in part to the original damage which was deep enough to destroy the dental pulp and in part to the clinician having to cut down into the pulp space to place the root filling. This means that, frequently, there is insufficient remaining crown mass to support a restoration. Furthermore, the mechanical properties of the remaining tissue are so compromised that fracture under loads that are normal in the mouth may be expected. In the few cases where a substantial bulk of coronal tissue remains it is possible to place an amalgam filling or a gold inlay, but the design of the restoration must be modified to take into account the reduced structural properties of the tooth. The entire load-bearing surface of the tooth is covered by an extension of the restoration known as "capping the cusps" so that occlusal loads are not concentrated in one weakened area such as a cusp tip. It has been demonstrated recently that a bracing effect can be achieved by employing an adhesive-bonded composite restoration. Instead of capping the cusps, the restoration is bonded firmly to acid-etched enamel and with dentine adhesives to the underlying dentine. However, the long-term strength of such bonds remains to be established.

When the loss of coronal tissue is greater it is necessary to provide a replacement crown of metal or porcelain. In cases of moderate loss it is possible to build up a core of replacement dentine using adhesively bonded composites or glass ionomer materials. Where loss is severe, gaining retention for a replacement crown is achieved by enlarging the radicular pulp space and inserting a metal post which is held in place with a cement. The prognosis for such restorations is not good and loss of the restoration due to fracture of the remaining root, breakage of the cement seal or recurrence of dental caries may occur.

The distribution of stress in post crowns has been examined by polarimetric analysis and by finite-element computer modelling methods. These methods have highlighted the critical importance of the cement layer in the system. Both the retention of the post and resistance of the root to fracture are markedly affected by modelling a different boundary layer.

Recent work has shown that adhesion to dentine can be gained by etching the dentine surface in the post hole and then washing out the dentinal tubules before permeating the tissue with a plastic resin. The usual resin used is bisphenol-A glycidyl dimethacrylate. The supporting post is constructed from an alloy that can be etched electrolytically in an acid bath to create a porous surface that can be permeated with the same resin. A filled composite resin may be used in the cement layer between the treated post and the treated dentine. In this way, a strong adhesive bond can be achieved between the post and the supporting dentine. This not only increases the retention of the system substantially, but also distributes stress better and reduces the possibility of root fracture. Furthermore, it has been used successfully to repair fractured roots.

Bibliography

Anon 1988. In vitro assessment of the biocompatibility of dental materials. Report of an international conference 1987. *International Endodontic Journal* 21: 49–187

Cohen S, Burns R C 1987 *Pathways of the Pulp*. Mosby, St Louis, MN

Mumford J M, Jedynakiewicz N M 1988 *Principles of Endodontics*. Quintessence, London

Schilder H 1967 Filling root canals in three dimensions. *Dent. Clin. North Am.* 11: 723–44

N. M. Jedynakiewicz
[University of Liverpool, Liverpool, UK]

F

Fiber-Reinforced Ceramics

This article replaces the article of the same title in the Main Encyclopedia.

The development of fiber-reinforced ceramics has been prompted by the need for materials with the advantages of ceramics combined with increased toughness and strength, and a reduced variability of strength. For a variety of technological and commercial reasons, a very considerable increase in the amount of research and development carried out on ceramic-matrix fiber composites (CMC) began in the early 1980s. The increase in effort and rate of development made this one of the most rapidly advancing fields in composite technology by the mid-1980s, a field which is continuing to develop rapidly. By the late 1980s, three different generic types of CMC had been developed, characterized by the methods used to incorporate fibers or whiskers into the matrix. These are: continuous-fiber reinforced glass-ceramic and glass systems produced by a solid-state slurry impregnation route followed by hot-pressing; CMC produced from continuous woven-fiber preforms infiltrated with a ceramic matrix by gas or liquid phase routes; and hot-pressed whisker toughened ceramics (Phillips 1987). In addition there are other promising fabrication routes which are at a less well developed stage. The advantages and disadvantages of the different fabrication routes and the mechanical properties of these different generic types of material differ, as do, to some extent, the applications of the materials.

The useful properties of ceramics include retention of strength at high temperature, chemical inertness, low density, hardness and high electrical resistance. Their principal disadvantage is their brittleness: failure strain, fracture energy and fracture toughness (K_{IC}) being low compared with tough plastics and metals. This renders them susceptible to damage by thermal or mechanical shock, makes them easily weakened by damage introduced during service, and causes them to have a large variability in strength. Fiber reinforcement can increase their fracture energies by several orders of magnitude, to values approaching $10^5\,J\,m^{-2}$, through energy-absorbing mechanisms similar to those in reinforced polymers, with consequent increases in apparent values of K_{IC} to ~30 MPa $m^{1/2}$, as shown in Table 1. Reductions in variability of strength, as characterized by the Weibull parameter m, are typically from an m of less than 10 for unreinforced ceramics to 20–30 for the best CMC. This is equivalent, very approximately, to reducing the coefficient of variation (CV) in strength from more than 0.12 down to 0.04. This is important for the design of highly stressed engineering components, as an increase in m and a reduction in CV reduce the required safety factors and thus enable the component to be designed to operate at higher stresses.

Table 1
Representative strength and toughness data for monolithic and some fiber-reinforced ceramics

	Strength (MPa)	Fracture energy ($J\,m^{-2}$)	K_{IC} (MPa $m^{1/2}$)
Glasses	100	2–4	0.5
Engineering ceramics (untoughened)	500–1200	40–100	3–5
Zirconia toughened	600–2000		6–12
Whisker toughened	300–800		6–9
Short-fiber reinforced glass	50–150	600–800	7
Continuous-fiber reinforced glass and glass-ceramic	1600	10^4–10^5	20–30

1. Fabrication Routes

Currently, all successful techniques for the manufacture of CMC require processing at temperatures of the order of 1000 °C and upwards. Chemical and thermal expansion compatibility between fibers and matrix are therefore important. Since strains induced by thermal expansion differences as the composite cools can cause the composite to crack and fragment, the fibers and matrices which can be combined successfully are limited. In general a fiber with the same or a higher thermal expansion coefficient than the matrix is preferred. High temperature chemical reactions between the fiber and matrix during fabrication can have significant effects on the properties of the composite: the most severe are either the degradation of the fibers or production of too strong a fiber–matrix bond resulting in a brittle, low-strength composite. To control the bond strength and improve toughness an interlayer between the fiber and matrix may be necessary; for example, the fibers may be coated with graphite prior to composite manufacture.

Glass and glass-ceramic matrix composites are most successfully manufactured by a slurry impregnation route (Phillips 1983, 1985). Fibers, in the form of multifilament continuous tow, are passed through a slurry of finely powdered matrix material in a mixture of solvent and binder. As the tow passes out of the slurry, the solvent evaporates and the resulting tape can be wound onto a drum to produce handleable prepreg sheets consisting of intimately mixed fibers and powder held together by the binder. The prepreg sheets are then stacked as a series of plies, in a similar way to conventional polymer prepreg technology, and hot-pressed at temperatures of about 1000 °C or more and pressures of about 5–10 MPa, to produce a laminate with a low-porosity matrix. This process has produced the highest strength CMC and has enabled the manufacture of composites containing up to 60 vol% of unidirectional fibers with strengths equal to those predicted by the law of mixtures. The most successful composites produced in this way to date consist of multifilament carbon (graphite) or silicon carbide (e.g., Nicalon) fibers in borosilicate glass or lithium aluminosilicate (LAS) glass-ceramic matrices. The carbon fiber composites oxidize in air above about 450 °C while the SiC fiber composites can be employed to around 1100 °C. Softening of borosilicate glass restricts the maximum temperature of this matrix to around 580 °C while LAS glass-ceramics can, in principle, operate to around 1000–1200 °C. The glass-ceramic composite can be produced either from a glass-ceramic powder or from a glass powder which is then heat treated to devitrify it. The latter process offers the prospect of lower-temperature processing than the former, but the fabrication process is more difficult to optimize because of the dimensional changes which occur when the glass crystallizes. The principal advantages of the slurry impregnation manufacturing process are the high strengths achievable and the relatively short manufacturing times; the main disadvantages are the need for high temperatures and pressures and the difficulty of manufacturing complicated shapes.

In the gas- and liquid-phase filtration fabrication routes, a fiber preform is infiltrated with a medium which deposits the ceramic matrix. The preform may be made by stacking together sheets of woven fabric cut to the required shape, or may consist of a multidimensional woven or knitted structure. The most mature infiltration technique is through chemical vapor infiltration (CVI) (Naslain and Langlais 1985). The preform is heated in a chemical reactor vessel and a gas or mixture of gases is passed through the vessel to cause the deposition of the ceramic matrix within the fiber network, resulting in the slow growth of the matrix as an interpenetrating network. A typical example of such a reaction is:

$$CH_3SiCl_3(g) + H_2(g) \rightarrow SiC(s) + HCl(g)$$

Other ceramics which have been deposited in this way include B_4C, TiC, BN, Si_3N_4, Al_2O_3 and ZrO_2. Typical temperatures required for this process are around 1000 °C.

In order to achieve low matrix porosity it is necessary to maintain an open-pore network for as long as possible as the ceramic matrix grows. In practice this is difficult to achieve and it is necessary to carry out the impregnation several times, with the surface of the composite material having to be removed by machining between each stage, to permit gas to continue to enter the composite. In practice, the minimum porosities which are achieved are between 10 and 20 vol.% of the matrix volume. Because of matrix porosity the strength of these composites tends to be around half of that expected from the law of mixtures. Figure 1 shows how the strength of a ceramic composite can be affected by porosity. The advantages of this fabrication route are: lower temperatures than those needed for powder hot-pressing routes; the potential for the easier production of more complex shapes by near-net shape processing; and the production materials which are more nearly isotropic (because of their use of multidirectional preforms) and which are very tough. A major disadvantage is the long time required to manufacture a component and the necessity for intermediate stages of machining: it can take several weeks or even months to manufacture a component the size of a gas turbine blade. The long fabrication times can be mitigated against by using a batch production process in which a large number of components are manufactured simultaneously in one reaction vessel. Another disadvantage is the relatively poor utilization of fiber strength because of the high porosity of the final product. The most important material produced by this route is SiC-fiber-reinforced SiC. The procedure has also been used to modify carbon–carbon composites to produce a composite in which the outer surface is a hybrid matrix with improved oxidation resistance. Examples of such mixed matrices are SiC/C and B_4C/C.

Similar composites have been manufactured at the laboratory scale by liquid-phase infiltration employing sol-gel and polymer pyrolysis processes. Both of these require the initial synthesis of liquid organic compounds which can be converted to a ceramic after infiltrating a fiber preform. The sol-gel route requires the synthesis of an alkoxide—an organic salt of an alcohol and a metal. After hydrolysis and polycondensation, the resulting gel is dehydrated, leaving a porous ceramic of high chemical reactivity and sintering ability. An example of polymer pyrolysis is the conversion of polycarbosilane to SiC.

Whisker-toughened ceramics are the most successful category of discontinuous-fiber-reinforced ceramics. In principle discontinuous-fiber-reinforced ceramics are the simplest CMC to manufacture. Short fibers or whiskers are mixed intimately

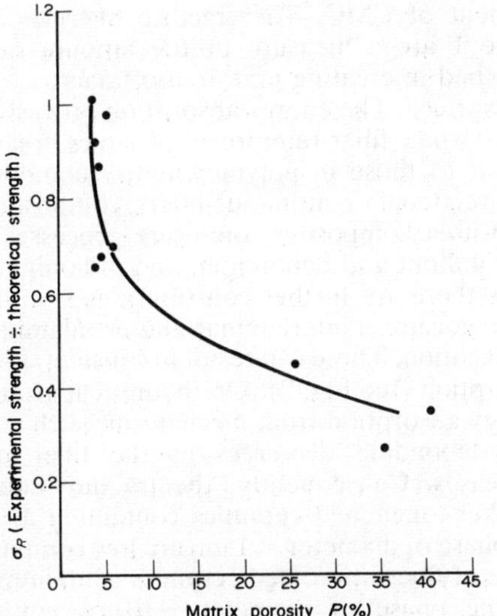

Figure 1
The variation of experimental strength, expressed as a function of theoretical strength, with matrix porosity for a ceramic-matrix composite (Phillips 1983)

with powdered matrix material and then either hot-pressed uniaxially or hot-isostatically pressed after isostatic cold-pressing. Temperatures in excess of 1500 °C and pressures in excess of 100 MPa are required for consolidation of ceramic matrices such as alumina. Following this process, the reinforcing phase in the resulting composite is randomly oriented in the pressing plane. Agglomeration of the fibers or whiskers makes it difficult to produce homogeneous materials with reinforcing volume fractions greater than about 30%.

Composites of this type, containing short random fibers as opposed to whiskers, have tended to be of lower strength than the unreinforced matrix but can have substantially greater toughness: techniques exist for aligning the fibers before consolidation and composites produced in this way can have higher strengths than the matrix. Whisker-toughened ceramics containing randomly oriented whiskers can be substantially stronger than the unreinforced matrix and have greater toughness. Successful systems of this type are SiC whiskers in Si_3N_4 and SiC whiskers in Al_2O_3. As an example of the improvement which can be obtained by whisker toughening, an unreinforced Al_2O_3 had a strength of 500 MPa and a K_{IC} of 4.5 MPa $m^{1/2}$ while 30 vol.% of SiC whiskers increased these to 650 MPa and 9.0 MPa $m^{1/2}$ respectively (Wei and Becher 1985).

The development of whisker-toughened ceramics has occurred at the same time as the development and exploitation of a completely different but very important competing method of toughening ceramics, which employs the martensitic transformation of zirconia (Butler 1985). There is promising research aimed at combining whisker toughening and zirconia toughening and recent results appear to indicate synergistic effects, the combined toughening effect being greater than the sum of the individual effects. Advantages of whisker toughening over zirconia toughening include the lower densities of whisker-toughened materials and the retention of the toughening mechanisms to high temperatures—a major disadvantage of zirconia toughening is that the contribution to toughness from the martensitic transformation decreases as temperature increases. However, during the late 1980s a potential problem arose for whisker-reinforced materials. Serious concerns have been expressed about the possible toxicity of whiskers with diameters less than 1 μm. It is unclear at present whether this will be confirmed as a serious problem for the whisker toughening of ceramics.

Other promising fabrication techniques, but at a less advanced stage, include the manufacture of fiber-reinforced silicon nitride by the nitridation of a silicon matrix to silicon nitride, and the oxidation of an aluminum matrix to alumina. Interesting composites have also been produced by the *in situ* growth of a whisker-like phase, for example the growth of needle-like β-Si_3N_4 crystals in Si_3N_4 (Suzuki 1987).

2. *Applications*

The main demand for the development of CMC has been the requirement for higher-temperature materials, particularly in aerospace as components of gas turbine engines, for rocket motors and for hot spots on spacecraft reentering the atmosphere. There are, however, potentially important applications where extreme temperature capability is not required. For example, fiber-reinforced glasses and glass ceramics have mechanical properties comparable with high-performance fiber-reinforced polymers, and could be used at intermediate temperatures higher than attainable with polymeric systems. They are more stable to ionizing radiations than fiber-reinforced plastics and this could make them attractive for nuclear and space environments. They are less susceptible to hygrothermal effects and therefore could be attractive for aircraft or marine applications. Their hardness could make them more erosion resistant than fiber-reinforced plastics and therefore attractive as radome materials, while hardness combined with a tailorable toughness could make them useful in armor. The chemical inertness of CMC could make them suitable in specialized chemical plants and they are already marketed as biomedical implant materials. Whisker-toughened ceramics are already marketed as cutting tool and die materials.

3. Properties: Matrix Microcracking and Toughness

The maximum strength, toughness and anisotropy of properties of the different classes of CMC differ widely and a detailed discussion of their properties is not appropriate here. However, a key question concerning the higher strength systems is the importance of matrix microcracking.

The strain to failure of a ceramic matrix is usually less than that of the reinforcing fiber and, on loading, the matrix of the composite can crack at loads lower than the ultimate load. For a unidirectional composite stressed in the fiber direction this is manifested as an array of regularly spaced cracks. To a first approximation, the stress at which matrix cracking would be expected to occur on the basis of a simple isostrain model is

$$(\sigma_m)_u\left[1+V_f\left(\frac{E_f}{E_m}-1\right)\right]$$

where $(\sigma_m)_u$ is the strength of the unreinforced matrix, V_f is the fiber volume fraction, and E_f and E_m are the fiber and matrix Young's moduli, respectively. A composite with a multiply-cracked matrix can retain useful properties even under fatigue conditions, but the cracks lead to easier ingress of aggressive environments. However, suppression of matrix cracking to higher stresses can occur and the theories of Aveston et al. (1971), Marshall et al. (1985) and McCartney and Kelly (McCartney 1986) provide an explanation of crack inhibition. In practice the use of higher fiber volume fractions and, in the case of laminates, thin plies provide the best methods of minimizing multiple matrix cracking.

The importance of matrix microcracking in determining the life of a composite under service conditions has not yet been established. Matrix microcracking provides paths for the easier ingress of aggressive environments and hence more severe attack on fibers. By analogy with advanced polymer composite systems it can also be expected to initiate a mechanism of fatigue failure, in which matrix cracks in one ply initiate orthogonal interlaminar cracks. Current data indicate weakening effects in SiC/LAS composites under both static and fatigue conditions at modest temperatures (~800 °C) due to matrix microcracking (Prewo 1986, 1987), although strengths increase again at higher temperatures. Further research is needed to understand and resolve this problem since, if it becomes apparent that the SiC/glass-ceramic systems can be operated only at stresses below the matrix microcracking stress, it will imply a substantial decrease in their acceptable operating stresses.

Another issue concerns the toughness of fiber-reinforced ceramics. Increased toughness is one of the most important driving forces behind the development of CMC. The fracture energies listed in Table 1 are a measure of the amount of energy absorbed in creating unit area of macroscopic fracture surface. The energy absorption processes which occur when fiber-reinforced ceramics fracture are similar to those in polymer matrix composites. In unidirectional continuous-fiber composites, and short-fiber composites, these are processes such as fiber pullout and debonding, and in laminated materials there are further contributions from multiple and delocalized interlaminar and intralaminar crack propagation. These can result in considerable energy absorption (see Fig. 2). On theoretical grounds, the energy absorption from mechanisms such as pullout and debonding decreases as the fiber diameter decreases. Consequently, the fracture energies of whisker-toughened ceramics containing a reinforcing phase of diameter <1 μm are low compared with those of fiber-reinforced ceramics containing a reinforcing phase of diameter ~10 μm upwards. In whisker-toughened ceramics a substantial contribution to toughness derives from crack deflection processes and the conventional mechanisms of pullout and debonding may play a relatively minor role, although this still remains an issue requiring further research. In continuous-fiber systems the fracture process is complex and may be delocalized with extended damage (see Fig. 3). Under these circumstances, linear elastic fracture mechanics is inapplicable and the values of K_{IC} in Table 1 for continuous-fiber systems should be regarded, as for similar polymer matrix composites, as a measure of damage tolerance rather than a design quantity.

4. Temperature Limitations

Currently the main practical temperature limitation on CMC is due to the lack of available ceramic fibers with good properties above 1000–1200 °C (Mah et

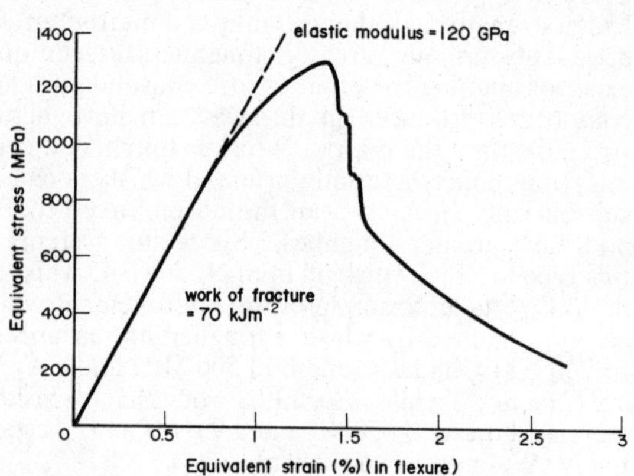

Figure 2
Load–extension behavior in flexure of a SiC fiber-glass matrix composite (courtesy D. M. Dawson and R. W. Davidge)

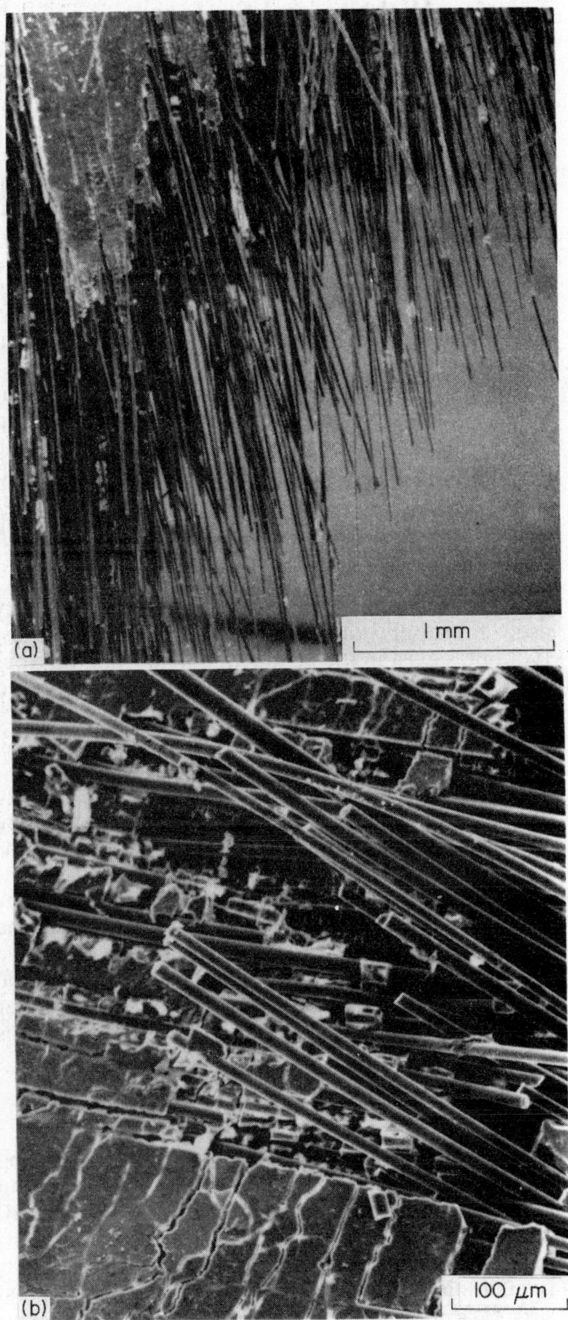

Figure 3
Examples of fracture morphology of a SiC fiber-glass matrix composite (courtesy of D. M. Dawson and R. W. Davidge)

al. 1987). Above this temperature range, degradation or creep of existing ceramic fibers becomes excessive. Carbon (graphite) fibers provide a higher temperature capability, maintaining their strengths in inert atmospheres to temperatures in excess of 2000 °C, but successful high-temperature ceramic-matrix composites utilizing carbon fibers and able to operate to high temperatures in oxidizing atmospheres have not yet been developed. At present carbon–carbon composites offer the best prospect of very high temperature capability, and much research is under way to enhance oxidation resistance. Promising approaches include the use of oxidation inhibitors such as borates incorporated in the carbon matrix for use at temperatures up to around 1000–1100 °C, and oxidation barriers such as borate glasses, Al_2O_3 and SiC coatings for higher temperatures.

The prospects for developing ultrahigh temperature composites, materials which can operate in air for long periods of time at temperatures above 1200 °C, have been assessed by Hillig (1985) who considered the necessary criteria which must be satisfied for a composite to survive and have useful properties under stress at high temperatures in air. These are, as a minimum: stability with respect to volatilization; low internal chemical reactivity; retention of stiffness; and a creep rate for the fiber of less than $10^{-7}\,s^{-1}$. He assumed that only an oxide matrix would have sufficient stability against oxidation; a nonoxide fiber would be necessary on grounds of stiffness and bonding; and that the matrix would provide a measure of oxidation protection for the fiber. He also pointed out that a nonoxide matrix might be suitable with an oxide coating to protect it; and that a further barrier between fibers and matrix may be necessary even for an oxide matrix. Although there is a dearth of good quality thermochemical data he was able to compile some, and calculate or infer others, and concluded that there are a number of oxides, carbides, borides and nitrides, as well as carbon, which might function together to temperatures in the range 1700–2100 °C. Of the nonoxides only carbon, Si_3N_4 and SiC are currently available as fibers or whiskers. A conclusion which can be drawn from this survey is that SiC whisker-reinforced Al_2O_3 is one of the most promising practical systems. Such a material already exists but displays a disappointing decrease in strength at 1200 °C. This highlights the uncertainty in predicting the high-temperature performance of ceramic composites. Much work is currently under way throughout the world to attempt to develop new fibers and extend the temperature range of ceramic-matrix–fiber composites.

Bibliography

Aveston J, Cooper G A, Kelly A 1971 Single and multiple fracture. *Proc. Conf. on Properties of Fiber Composites*. IPC, Guildford, UK, pp. 15–24

Butler E P 1985 Transformation-toughened zirconia ceramics. *Mater. Sci. Technol.* 1: 417–32
Ceramic Bulletin 65(2), 1986 Issue devoted to papers on ceramic matrix composites
Hillig W B 1986 Prospects for ultra-high-temperature ceramic composites. In: Tessler R T, Messing G L, Pantano C G, Newnham R E (eds.) 1986 *Proc. Conf. on Tailoring Multiphase and Composite Ceramics*. Plenum, New York, pp. 697–712
Hillig W B 1987 Strength and toughness of ceramic matrix composites. *Annu. Rev. Mater. Sci.* 17: 341–83
McCartney L N 1987 Mechanics of matrix cracking in brittle-matrix fibre-reinforced composites. *Proc. R. Soc. Lon. Ser. A* 409: 329–50
Mah T, Mendiratta M G, Katz A P, Mazdiyasni K S 1987 Recent developments in fibre-reinforced high temperature ceramic composites. *Ceram. Bull.* 66: 304–8
Marshall D B, Cox B N, Evans A G 1985 The mechanics of matrix cracking in brittle-matrix fiber composites. *Acta. Metall.* 33: 2013–21
Naslain R, Langlais F 1986 CVD processing of ceramic–ceramic composite materials. In: Tessler R T, Messing G L, Pantano C G, Newnham R E (eds.) 1986 *Proc. Conf. on Tailoring Multiphase and Composite Ceramics*. Plenum, New York, pp. 145–64
Phillips D C 1983 Fibre reinforced ceramics. In: Kelly A, Mileiko S T (eds.) 1983 *Handbook of Composites*, Vol. 4: *Fabrication of Composites*. North-Holland, Amsterdam, pp. 373–428
Phillips D C 1985 Fibre reinforced ceramics. In: Davidge R W 1985 *Survey of the Technological Requirements for High Temperature Materials Research and Development*. Sect. 3: *Ceramic Composites for High Temperature Engineering Applications*. Commission of the European Communities, EUR 9565, pp. 48–73
Phillips D C 1987 High temperature fibre composites. In: Matthews F L, Buskell N C R, Hodgkinson J M, Morton J (eds.) 1987 *Proc. 6th Int. Conf. on Composite Materials/2nd European Conf. on Composite Materials*, Vol. 2. Elsevier, London, pp. 2.1–2.32
Prewo K M 1986 Tension and flexural strength of silicon carbide fiber reinforced glass ceramics. *J. Mater. Sci.* 21: 3590–600
Prewo K M 1987 Fatigue and stress rupture of silicon carbide fiber-reinforced glass ceramics. *J. Mater. Sci.* 22: 2695–2701
Suzuki H 1987 A perspective on new ceramics and ceramic composites. *Philos. Trans. R. Soc. Lon. Ser. A* 322: 465–8
Wei G C, Becher P F 1985 Development of SiC-whisker-reinforced ceramics. *Am. Ceram. Soc. Bull.* 64(2): 298–304

One of the leading forums for presentation of recent work on ceramic composites is the Annual Conference on Composites and Advanced Ceramic Materials organized by the American Ceramic Society. The proceedings of these conferences are published each year as a volume of *Ceramic Engineering and Science Proceedings* and provide a useful overview of the current state of this rapidly advancing field.

D. C. Phillips
[UKAEA Harwell Laboratory, Didcot, UK]

Friction Welding and Surfacing

One dictionary defines friction as "the resistance which a moving body meets with from the surface on which it moves." The resistance referred to can result in the development of thermal energy. As a result, relative movement between two surfaces while pressed together under an applied load will generate thermomechanical conditions to permit welding. In reality this happens all too frequently in, for example, engineering machinery and internal combustion engines, when lubrication to the moving parts fails and seizure results with catastrophic consequences. Friction welding has been known since as early as the 1890s (Bevington 1891), but it was not put into practice in a usable form, particularly for joining metals, until the 1950s (Bishop 1960, Vill 1962). One unique feature of this joining method is that welds can be made in the solid phase—that is to say, no macromelting takes place. The self-regulating heating phase allows the joining of a wide range of materials from plastics to nickel-based superalloys.

parts to be welded is the approach predominantly used for applications in a variety of industries. Examples exist in the production field, ranging from the automotive industry to the highly sophisticated aerospace and nuclear industries. Other developments which also use rotary motion (such as radial friction welding for joining pipes, metal deposition by friction, and recent developments leading to the provision of lightweight, portable stud welding equipment) have further advanced the applications potential of the process.

In general, rotary motion joins parts which by necessity are of circular section, so the scope for the method is restricted. In order, therefore, to remove this restriction, alternative motions (which include orbital, linear and angular reciprocation) are under investigation and their status will be reviewed in this article.

1. The Process

To provide the two main requisites for generating heat it is necessary to introduce relative motion between the parts while an end thrust is applied. This arrangement is conveniently obtained by rotating one part about its axis while the other, fixed part, is maintained in contact under a preset axial pressure. After a suitable period of heat generation, during which the interface temperature rises and the material in the weld zone reaches a plastic state, rotation is terminated while the axial pressure is either maintained or increased to complete bond formation (Anon 1970). As a result of the thermomechanical conditions developed during the weld sequence, axial shortening of the parts occurs with the formation of an extruded metal collar in the weld region.

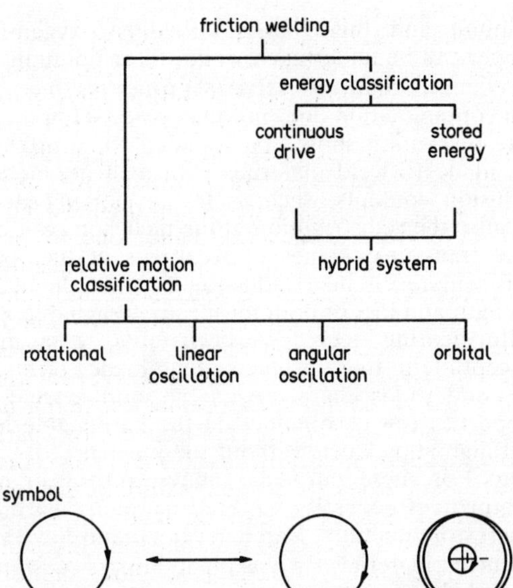

Figure 1
Classification for friction welding

In view of the continuing independent development of the friction welding method worldwide, it is worthwhile to refer to Fig. 1 which provides a classification for friction welding. Two process variants appear under the energy classification. First, there is continuous drive, the most well known and practised: this is the process variant where power or energy is provided by an infinite duration source and maintained for a preset period. It is primarily used in Europe, the USSR and Japan. Second, there is the stored energy, often known as the inertia or flywheel, method (Kiwalle 1968): this is the process variant where the energy for welding is supplied by the kinetic energy stored in a rotating system or fluid storage system. It is used chiefly in the USA. Both variants are capable of providing high-integrity joints at production rates higher than the alternative joining processes of flash and resistance butt welding. Significant cost savings are made as a result of lower power consumption, reduced pre- and post-weld machining and greater material conservation. There is also a hybrid system combining some features of both methods.

Reference is made in Fig. 1 to four relative motion categories:

(a) rotational motion, in which one component is rotated relative to and in contact with the mating face of another component;

(b) linear oscillation, in which one component is moved in a linear oscillating motion relative to and in contact with the mating face of another component;

(c) angular oscillation, in which one component is moved in an angular oscillating motion about a common component axis relative to and in contact with the mating face of another component; and

(d) orbital motion, in which one component is moved in a small circular motion relative to and in contact with the mating face of another component, either with neither component about its own central axis, or with both components rotating in the same direction about their own axes at the same speed but with their axes displaced.

Conventional rotation has predominated since the upsurge in development of friction welding in the mid-1950s. The major limitations of this motion are that only parts of essentially circular cross section are suitable and precise angular orientation between parts cannot be guaranteed. Angular, linear and orbital motion systems have, however, overcome to some extent these limitations, although it is fair to say that the two former motions have only been successfully applied as yet for welding plastics (Anon 1978). (See *Welding of Thermoplastics*, Suppl. 2.)

A British company has devised a mechanical arrangement for orbital motion (Searle 1971), which is incorporated in a production machine capable of welding circular and noncircular metal and plastic parts with precise orientation up to 2000 mm^2 cross section. Additionally, orbital motion is characterized by a constant surface velocity as distinct from a velocity gradient that exists with simple rotary motion. This offers benefits in terms of shorter weld times and the capability to produce a multiplicity of welds in one heating sequence. This latter feature allows complex weldments to be fabricated and also provides for a significant increase in production rates.

2. *Welding Variables*

The major welding variables that influence the metallurgical and mechanical properties as well as productivity for all friction welding systems are relative surface velocity, heating pressure (applied during forced motion), forging pressure (applied during or after deceleration) and duration of heating (Jennings 1970, Ellis 1972, Wang 1975). The choice of variables used must be considered together with secondary factors such as pre-weld surface conditions and geometry and the thermal and physical properties of the materials being welded. The tolerant nature of the process allows a wide spectrum of velocities (0.3–7 m s^{-1}) to be used. A useful guide when operating with continuous-drive rotary friction welding is to aim for a peripheral velocity of 2 m s^{-1} for solid bars and slightly higher for tubular sections.

Depending on diameter, this will give a very wide range of rotational speeds: 16 000 revolutions per minute (rpm) for a 2 mm diameter rod, and 480 rpm for a 76 mm diameter bar. Care must be taken if low velocities are chosen since high torques are developed; these can affect work holding and produce nonuniform deformation which may lead to defects at the weld interface and/or severe deformation in the weld region. High suface velocities can lead to wide heat-affected zones and overheated structures in ferrous metals and the formation of undesirable intermetallic compounds when welding certain dissimilar metals such as copper–aluminum and aluminum–steel.

The applied pressure is most important since it influences the driving torque (transmission power) and determines the metal displacement and temperature characteristics. Material properties and joint configuration (e.g., bar to bar, bar or tube to plate) determine the values of heating and forging pressures adopted. When chosen, they should ensure uniform heating and maintain the two surfaces in intimate contact to prevent atmospheric contamination. Typical values used for steel are 31–60 MPa and 77–150 MPa for the friction and forging pressures, respectively. For stainless steel and nickel-based alloys with high hot strength properties, greater pressures are required.

The duration of heating can be controlled by either an electronic timer or burnoff (length loss during rotation), since both are to some extent complementary. The control must ensure that the heating duration is sufficient to reproduce similar thermal conditions in the material from weld to weld.

3. Materials

One of the inherent features of friction welding is the efficient utilization of the thermal energy developed. With careful control of the main welding variables, the weld region does not suffer from excessive heat buildup. This is particularly important when welding mild, low-alloy steels and stainless steels, since overheated microstructures are to be avoided. Moreover, the process is readily amenable to join widely dissimilar metal combinations such as aluminum to stainless steel, copper to steel, and aluminum to titanium. For guidance, the weldability chart (Table 1) indicates pairs of materials that have been joined at the Welding Institute to yield high integrity welds, and others that cannot be welded. Further information regarding materials weldability is given by Wang (1975) for both continuous drive and stored energy processes.

Exclusion of the atmosphere during welding allows the formation of sound high-strength bonds in the more highly reactive metals such as zirconium, titanium and their alloys. Welding oxygen-free copper can be undertaken while still maintaining its excellent thermal and electrical properties since oxygen contamination does not take place. However, a note of caution should be sounded when attempts are made to weld materials with high nonmetallic inclusion contents, such as free-machining steels, because the redistribution of the inclusions will produce transverse planes of weakness at the weld. This will result in a reduction of ultimate tensile strength and loss of both tensile and notch ductility. Sulfur-bearing steels have been found to be more susceptible to these problems than leaded steels.

Sound welds characterized by good mechanical properties can be obtained in the hardenable low- and high-alloy steels without the incidence of cracking. For these particular alloys, post-weld heat treatment is generally required to reduce hardness and restore ductility. Rapid developments have seen advanced materials such as oxide dispersion strengthened alloys, metal matrix composites, aluminum–lithium alloys, rapidly solidified alloys and ceramics become available for industrial application. (See *Aluminum–Lithium Alloys: Weldability*, Suppl. 2.) Preliminary friction welding trials have established that sound bonds of acceptable strength can be achieved with these materials.

Most of the methods of motion used under continuous drive conditions have been successfully applied to join various types of thermoplastic. Linear, angular and orbital motions have been particularly successful so that equipment based on the first two is now used in production. Experience has shown that the following plastics are readily weldable by friction: monocast nylon, glass-fiber-reinforced nylon, polycarbonate and polyvinyl chloride. Many other plastics have been reported as successfully joined by other researchers (see *Welding of Thermoplastics*, Suppl. 2).

4. Applications

The applications for friction welding processes (Welding Institute 1979) can be conveniently classified into four principal groups.

4.1 Mass Production

The process is particularly suitable for quantity production where the relatively high capital cost of fully automated equipment can be justified by high throughput and low labor involvement. The automobile industry is by far the greatest user of the process: examples include engine exhaust valves, propeller shafts, steering columns and rear axle tube assemblies. It is interesting to note that, in the case of the latter, the Ford Motor Company alone have made tens of millions of welds since introducing the process in the late 1960s.

Table 1
Material combinations friction welded at the Welding Institute

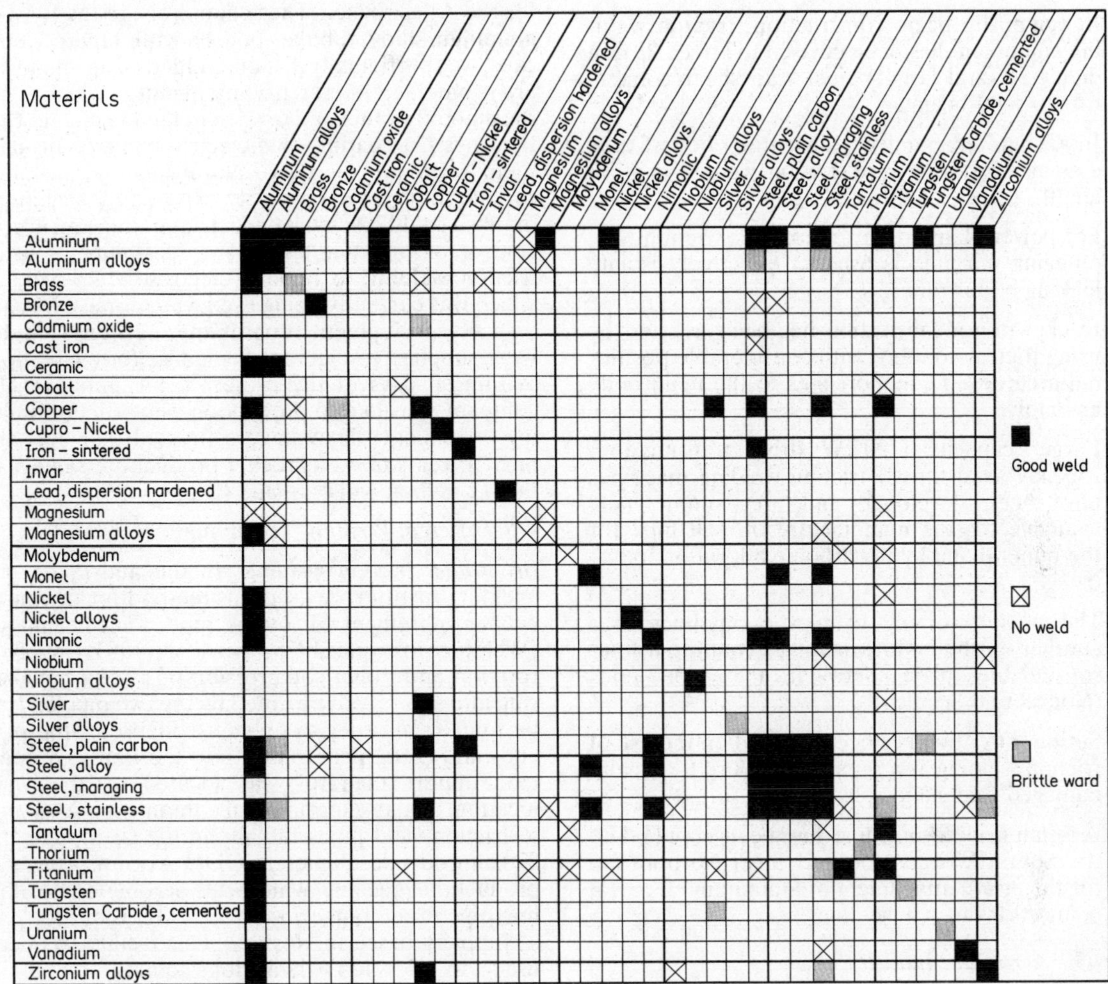

4.2 Short-Run (Jobbing) Production

The versatility of the process and the ease of machine adjustment also make the welding of small batches attractive. This is particularly useful where the cost of expensive dies or molds cannot be justified. The fabrication of nonstandard bolts in expensive materials is an example where friction welding eliminates die costs and economizes on material.

4.3 Dissimilar Metal Welding

The process is used where a direct or indirect bond is required between two widely dissimilar metals. Alternatively the welding method is chosen to make short transition pieces in materials which are otherwise impossible to join by conventional processes. Tubular joints between aluminum and stainless steel are used in the cryogenic and nuclear industries, while aluminum alloy (He9)–mild steel, aluminum–titanium and aluminum–copper welds are required in electrolytic smelting of aluminum, electrolytic extraction of chlorine from brine and electrical power distribution (connectors), respectively.

4.4 Specialized Requirements

Under this classification the following applications are worthy of note.

(a) The attachment of flanges to pipes—a specially designed machine has already been constructed for this purpose.

(b) The "on-site" coupling of small bimetal connectors to live conductor cables using a portable machine.

(c) The joining of stub tubes to headers in the manufacture of steam generator plants.

Burmeister and Wain, a leading Danish company, have built such a machine.

(d) In composite beam construction, friction welding equipment has been devised to attach 25 mm diameter (and larger) steel stud shear connectors to steel beams.

(e) In China, economizer serpentine boiler tube elements (straight) are friction welded in 12 m lengths prior to final bending.

(f) For power plant boilers, a portable friction tube plugging machine is used to effectively isolate leaking boiler tubes.

(g) Rocket motor fabrication has been assisted by using friction to joint, with considerable positional accuracy, 12 support pins to the main body assembly.

(h) Larger capacity (190 kW drive transmission, 2500 kN axial thrust) friction-welding machines have been developed and are commercially available for the manufacture of drill pipe for the mineral and hydrocarbon exploration industry.

(i) The on-site joining of plastic pipelines uses another method which involves rotating a nonconsumable plate between the pipe ends (Nockel and Schuld 1975).

(j) Serious attention has been focused on the use of friction welding as a repair procedure to reclaim damaged shafts and rolls, and so on.

(k) A semiportable machine has been supplied to the South African Iron and Steel Corporation for the *in situ* attachment of aluminum electrical connections to railway lines.

4.5 The Aerospace Industry

The Rocketdyne Division of Rockwell International designed, developed and built the main engines for the space shuttle (Schaffer 1981). These engines were considered the most advanced liquid-propulsion rockets ever developed. At the heart of the engine lies the main-injector assembly which mixes fuel and oxidizer. In simple terms, the assembly takes the appearance of a tube sheet with 600 upstands attached to it. Oxidizer passes through the hollow injector posts while hydrogen is swirled by their spiral-fluted exteriors. Because of their close proximity, attachment of these posts by conventional arc welding was considered impossible, thus inertia friction welding was proposed as the joining process. The choice for a solid phase method was further justified as the ultimate requirement was to join Haynes 188 (cobalt alloy) post material to the Inconel 718 (nickel-based alloy) base plate.

Yet another development involving friction welding for the aerospace industry came after high reject rates (85%) were being experienced by General Dynamics during the manufacture of fuel mandrels for the Atlas rocket. These mandrels, in the form of aluminum-alloy tubular bodies with tapered end-caps, were originally fusion welded. The mandrels are submerged in mercury manometers which sense the amount of fuel and oxygen in the tanks, and thus they need a coating of dielectric paint to insulate them from the mercury. Unfortunately, just one surface spot of weld porosity exposed by machining off flash could crack the dielectric coating and impair control of the fuel level. The feasibility of using friction welding to replace electron beam welding which was currently being used was appraised. After a short development programme, 12 sample welds were supplied to General Dynamics for evaluation. Within 90 days of the project being initiated, the company moved into production using friction welding technology. They have estimated that over the next 10 years of Atlas rocket production, savings of US$12 million will be made (Bargs 1986).

4.6 Different Welding Techniques

(*a*) *Radial friction welding*. In the mid-1970s, the Welding Institute devised this method for the particular requirement of joining pipes. The technique (Welding Institute 1978, Nicholas 1983) involved rotation and radial compression of a solid bevelled ring into a 'V' groove formed by the two pipe ends to be joined. In order to prevent collapse of the pipe ends and metal penetration into the bore, an internal support mandrel was located at the weld location. Today experimental equipment is available to friction-weld pipes radially in the size range 50–270 mm outside diameter (OD). Also, a prototype production machine which can accommodate flow line pipe in the range 75–176 mm OD is available at a company based in Norway. This facility has been used to develop data for joining nominally 100 m OD × 12.7 mm wall API 5LX–X52 grade steel and 2205 duplex stainless steel (Dunkerton et al. 1987).

After this technology had been developed, another potential application of totally different character, namely that of attaching soft metal driving bands to projectiles, was found suited to the process. In this context a solid ring of copper alloy (gilding metal) is rotated and radially compressed onto the external surface of a steel shell body, thus generating a complete circumferential bond (Peters 1981). Trials with shell bodies in the range 105–155 mm diameter have shown that the welds can be made in 10–15 s. Subsequent firing trials have demonstrated that bond integrity is equal to, if not better than, the conventional method of band attachment, in which the bands are cold crimped into complex machined recesses.

It is also possible to reverse the deformation direction; that is, to expand a ring outwards into a hollow cylinder. Applications such as locating bearing inserts, internal upstands and attachment of tubes to tube sheet can be considered.

(*b*) *Friction stud welding*. Since the late 1960s many machines that could carry the tag "portable" have been produced. Such machines, developed for attaching shear connectors to girders (Gilmour 1974), electrical studs to railway lines (Anon 1980) and electrically coupling sacrificial anodes to offshore structures underwater (Nicholas 1984), have utilized a hydraulic system to provide both rotation and welding force. Unfortunately, all machines needed a number of hydraulic umbilicals to connect between the hydraulic power pack and the welding head. These machines can still be considered portable but not exactly "lightweight."

Continuing developments have led to systems which use electromechanical and pneumatic means for providing the necessary requirements for friction stud welding. Novel designs involving screw force actuation (Needham and Ellis 1971) and the centrifugal action of bob-weights to develop axial force (King and Sproulle 1986) are at present being evaluated.

(*c*) *Orbital motion*. By using this type of relative motion, both round and nonround parts can be welded with precise angular orientation. Also, as the motion generated will exhibit uniform surface velocity, it will be possible to joint complex parts and a number of separate parts in one friction weld sequence. Unfortunately, the mechanical arrangement to provide orbital movement of one part about the face of a stationary part has not been developed. However, using the arrangement of rotating both parts in the same direction with their axes offset (Searle 1971), it is possible to generate orbital motion. Using such a machine, experimental trials have been undertaken to develop information pertaining to the influence of welding machine settings on the mechanical properties of welds made in 25 mm diameter cold-drawn mild steel. In brief, it was possible to achieve high-quality welds using lower than normal welding forces and surface velocities, and also to produce weld-zone microstructures which performed far better when subjected to impact loading (Cousans and Camping 1985).

(*d*) *Angular motion*. This arcuate reciprocating motion has been successfully applied to the joining of plastics. Examples of its use extend to the manufacture of thermostat housings for the heating–cooling systems of automobiles (Anon 1973a) and petrol tanks for mopeds. It was necessary to use this motion for these circular parts because angular alignment was required.

(*e*) *Linear motion*. Successful exploitation of this form of reciprocating motion system has again been seen in the plastics industry. Applications include the front bumper assemblies (Murray and Rusch 1984) and manifold units for certain models of Ford cars. Another reported use was for the manufacture of emission control canisters, again in the automotive industry.

In order to investigate the potential of this form of motion for joining metals, the orbital friction welder referred to earlier was interfaced with special drive plates and a cruciform to provide linear reciprocating motion. Preliminary welding trials with rectangular sections of 25 × 10 mm and 22 × 6 mm in mild steel, austenitic stainless steel, titanium alloy and aluminum alloy were carried out with varying degrees of success (Nicholas 1987). This exciting new development will provide many more areas of application where friction welding can be introduced (Barrett 1989).

5. *Advantages and Limitations of Friction Welding*

5.1 Advantages

The attributes associated with friction welding can be summarized as follows.

(a) The process is inherently simple and reliable, producing welds of high mechanical strength and sound metallurgical properties.

(b) A wide range of materials are weldable to themselves or in combinations which are not normally weldable by conventional methods.

(c) Little or no end preparation of surfaces is necessary for similar materials since the process is to some extent self-cleaning.

(d) There is no intense ultraviolet light, or hot metal ejected from the weld zone.

(e) No external consumables are needed and the atmosphere is excluded from the welding surfaces. As a result, highly reactive metals such as zirconium alloys and titanium can be welded.

(f) The process is efficient since there is direct conversion of mechanical energy to thermal energy at the rubbing surfaces, and a balanced three-phase load of approximately 0.9 power factor is demanded from the electrical supply.

(g) There are numerous geometries of parts that can be joined, for example, bar to bar, tube to tube, bar or tube to plate in wrought, forging or casting stock.

(h) The equipment can be regarded as a machine tool, and is readily adaptable to automatic or semiautomatic control, thus producing increased output, reduced costs and close control of dimensional tolerances.

(i) Complex shapes can be fabricated from many small components.

Table 2
Friction surfacing applications evaluated

Requirements	Deposit metal	Substrate metal
Experience	Mild steel	Mild steel
Repair	Mild steel	0.4% C, 1.5% Ni, 1.2% Cr, 0.25% Mo alloy steel
Corrosion protection	316S16 austenitic stainless steel	Mild steel
Repair	316S16 stainless steel	316S16 stainless steel
Corrosion protection	Ni-based alloy (alloy 625)	Mild steel
Corrosion protection	Ni-based alloy (alloy 625)	316S16 stainless steel
Wear resistance	Ni-based alloy (Hastelloy CW-12M-1	316S16 stainless steel
Wear resistance	Co-based alloy (Stellite grade 6)	316S16 stainless steel
Repair	Aluminum alloy (4% Cu)	Aluminum alloy (nominally 4% Cu)

(j) As there is no liquation, the resulting weld does not suffer defects normally encountered in fusion-welding processes, such as porosity, shrinkage cracking and hydrogen embrittlement.

(k) The process can also be operated in adverse environments (underwater, oil, etc.) and still provide sound high strength properties.

5.2 Limitations

(a) During the early years the process could not provide part alignment and was restricted to circular parts. However, other forms of motion and the use of synchronized alignment are now available (Anon 1973b, Dunkerton 1983).

(b) The process does not lend itself readily to the welding of cast irons in general. However, reasonable success as defined by good mechanical properties has been achieved with the austenitic spheroidal graphite grade.

(c) Considerable caution must be exercised when the process is considered for joining materials which contain an abundance of nonmetallic inclusions. For example, when welding sulfur-bearing free cutting steels it is unlikely that better than 85% of the parent metal tensile strength will be obtained from the joint due to unfavorable distribution of the inclusions.

(d) There is (in theory) no limitation as to the size of part that can be welded provided a machine of sufficient drive power, force and rigidity can be manufactured. However, such a machine will involve considerable capital investment. Welds have been made in the range 0.75–250 mm diameter.

(e) Unfortunately, the nature of a solid phase weld does not lend itself to satisfactory inspection by the existing conventional nondestructive test methods available. However, batch destructive testing, in-process monitoring and proof load tests in some way attempt to provide quality assurance during production. This feature, perhaps more than any other, can be held responsible for limiting the rate of commercial exploitation.

6. Friction Surfacing

Klopstock and Neelands (1941) indicated that frictional heating could be utilized to deposit one metal onto another. In brief, a rotating consumable rod is pressed onto the substrate material under an applied axial pressure, while the latter is traversed across the consumable rod to deposit a layer of metal. Deposition in longitudinal, circumferential and annular rings has been successfully achieved (Nicholas and Thomas 1986).

It has been determined that the metal laid down is in the hot worked condition, with no evidence of melting; thus dilution (i.e., pick-up from the base material) is for all intents and purposes zero. Table 2 identifies some of the applications and deposit–substrate combinations that have been successfully investigated.

See also: Welding of Thermoplastics (Suppl. 2)

Bibliography

Anon 1970 Friction welding: Special feature. *Met. Constr.* 2(5): 185–8

Anon 1973a Angular welding simplifies design of symmetric plaster parts. *Des. Eng.* February, 99

Anon 1973b Friction welding machines in Japan (1972). International Institute of Welding, Document III-469-73

Anon 1978 Vibration welding (of plastics) permits novel designs. *Des. Eng.* January, 47–8

Anon 1980 Welding unit made in South Africa. *ISCOR News* October, 18–20

Bargs S 1986 Inertia welding for fuel mandrels. *Weld. Des. Fabr.* June, 37–9

Barrett J 1989 Linear friction enables non-circular metal welding. *Eureka* March, 75–9
Bevington J 1981 Spinning Tubes. US Patent No. 444, 721
Bishop E 1960 Friction welding in the Soviet Union. *Weld. Met. Fabr.* October, 408–10
Cousans A K, Camping M J 1985 An initial investigation of the orbital friction welding of mild steel. Welding Institute Research Report 270
Dawes C J 1970 Development of micro-friction welding. *Proc. 2nd Int. Conf. Advances in Welding Processes*, pp. 57–62
Dunkerton S B 1983 Recent innovations in friction welding. *Proc. 1st Int. Conf. Developments and Innovation for Improved Welding Production*
Dunkerton S B, Johansen A, Frich S 1987 Radial friction welding for offshore pipelines. *Weld. J.* 66(7): 40–7
Ellis C R G 1972 Continuous drive function welding of mild steel. *Weld. J.* 51: 183–97
Gilmour R S 1974 Friction welding stud shear connections to steel beams. *Met. Constr.* 6(5): 150–2
Jennings P 1970 Some properties of dissimilar metal joints made by friction welding. *Proc. 2nd Int. Conf. Advances in Welding Processes*, pp. 147–52
King C G, Sproulle R J 1986 Is this the start of something small? *Welding Institute Res. Bull.* 27(6): 190–2
Kiwalle J 1968 Designing for inertia welding. *Mach. Des.* November 7 40(26): 161–6
Klopstock H, Needlands A R 1941 An improved method of joining and welding metals. UK Patent No. 572,849
Murray A D, Rusch K C 1984 SP-566 plastics in passenger cars. *Proc. Int. Congr. Society of Automotive Engineers*, pp. 13–20
Needham J C, Ellis C R G 1971 Automation and quality control in friction welding. *Welding Institute Res. Bull.* 12: 333–9
Nicholas E D 1983 Radial friction welding. *Weld. J.* 62(7): 17–29
Nicholas E D 1984 Underwater friction welding for electrical coupling of sacrificial anodes. *Proc. 16th Annual Offshore Technology Conf.*
Nicholas E D 1987 Friction welding non circular sections with linear motion—A preliminary study. Welding Institute Research Report 332
Nicholas E D, Thomas W M 1986 Metal-deposition by friction welding. *Weld. J.* 65(8): 17–27
Nockel S, Schuld N 1975 Friction welding of plastic pipes (in German). *Schweisstechnik (Berlin)* 25(4): 164–6
Peters T J 1981 Ultrasonic inspection locates discontinuities in inertia welded rotating bands. *Mater. Eval.* 39(11): 1143–6
Schaffer G 1981 Inertia welds launch shuttle. *Am. Mach.* 123–7
Searle J G 1971 Friction welding non-circular components using orbital motion. *Weld. Met. Fabr.* 39(8): 294–7
Vill V I 1962 *Friction Welding of Metals*. American Welding Society
Wang K K 1975 Friction welding. Welding Research Council (UK) Bulletin 204
Welding Institute 1978 Friction welding methods and apparatus. UK Patent No. 1,505,832
Welding Institute 1979 *Exploiting Friction Welding in Production*, Information Package Series. WI, Abington, UK

E. D. Nicholas
[Welding Research Institute, Abington, UK]

Fumigation of Wood

Damage from decay and insect attack is a major problem of wood in service. Even extremely dry wood can be vulnerable to attack by certain insects, and wood is often used in places where it cannot be kept dry. Decay fungi and insects can colonize the wood and rapidly reduce its service life. Certain chemicals have commonly been used to control surface deterioration, but they generally cannot penetrate beyond the surface to control internal decay. However, in the late 1960s it was found that even though wood is impermeable to liquids, common agricultural fumigants can migrate through it. This discovery opened new avenues for controlling the deterioration of wood.

1. *Fumigants*

Fumigants used to protect wood fall into two broad categories: those that rapidly eliminate established fungi or insects, usually within 48 hours, but do not remain in the wood for long periods; and those that move more slowly through the wood but remain for between three and 17 years (see Table 1). The short-term chemicals are used to kill insects in buildings and to eliminate plant pathogens from logs destined for export, while the long-term chemicals are used to protect wood from decay for ten or more years.

All fumigants are toxic chemicals that can be hazardous to humans when misapplied. Fumigant use is regulated in many states and countries. Before using any fumigants, it is advisable to consult the appropriate government authorities for restrictions that may apply in a particular area.

1.1 Short-Term Fumigation

Short-term fumigant treatments generally employ highly volatile chemicals that rapidly penetrate the wood, eliminate insects and decay fungi, and escape into the atmosphere. They provide rapid control but no long-term protection against reinvasion. Short-term fumigation is typically applied to small wood pieces, such as historic wood artifacts in museums, to eliminate insect pests. Because these pieces often go unobserved for long periods, powderpost beetle infestations of the dry wood can cause substantial damage. Short-term fumigation is also effective against insects and fungi in logs destined for export.

Occasionally the need arises to fumigate large structures to eliminate internal insect pests, such as old house borers or drywood termites. This space fumigation, as it is called, is commonly used in agricultural storage buildings to eliminate insect infestations in grain.

The two chemicals most commonly used for short-term fumigation are methyl bromide and sulfuryl

Table 1
Characteristics of currently registered wood fumigants or their active ingredients

Chemical	Trade names	Persistence in wood	Application method	Target	Uses
Trichloronitromethane	chloropicrin Pic-clor Timber-Fume	>18 year	internal	fungi	poles piles timbers
Methylisothiocyanate	MIT	>9 year	internal	fungi	—[a]
MIT + chlorinated C_3 hydrocarbons	Vorlex Di-Trapex	>18 year	internal	fungi insects	poles piles
Methyl bromide	Bromo-Gas Meth-o-Gas Celfume	<48 h	space	insects fungi	logs
Sodium *n*-methyl-dithiocarbamate	Vapam, Wood-Fume, Metam Sodium	<10 year	internal	fungi	poles piles chips
Sulfuryl fluoride	Vikane	<48 h	space	insects	buildings furniture

a Not currently registered, but shows great promise

fluoride (Vikane). Both these compounds are colorless, odorless gases. They are often mixed with low levels of chloropicrin, which serves as an indicator gas. Methyl bromide is used to eliminate the oak wilt fungus (*Ceratocystis fagacearum*) from oak logs prior to export. Sulfuryl fluoride is most often used to eliminate drywood termite nests in wood structures.

Short-term fumigation is done under an airtight seal. Small objects are treated in airtight bags. For space fumigation, the area to be treated is cleared and sealed off with tarpaulins. A fixed amount of chemical is then pumped under pressure into the treatment space. A series of fans, strategically located inside the structure, mixes the chemical, which remains in the building for 20–24 hours. Since both chemicals are highly toxic, it is imperative that personnel performing the treatment wear respirators. After treatment the building is ventilated, and may then be safely entered.

The chemical dosage and length of treatment depend on the area to be treated, the type of seal used to contain the chemical, the temperature, and the wind speed, if any. These factors are used to express a dosage in grams of fumigant per hour, which is converted to grams of fumigant per 1000 liters multiplied by the duration of the treatment. Fortunately, several convenient slide rules, or Fumiguides, are available, which may be used to calculate these values quickly.

Short-term treatments usually do not alter the appearance of the wood. However, when fumigating historic wood, it is advisable to expose a small section before treating the whole object, to ensure that the treatment does not damage the color or finish of the wood.

1.2 Long-Term Fumigation

For some wood members it is expedient to provide slower, longer-term control of decay. This type of protection is particularly needed for large-dimension timber or roundwood that is decaying internally in areas that initial preservative treatments cannot reach. For such long-term protection, the chemical in its liquid form is poured into steep, angled holes drilled into the wood. These holes are then plugged with tight-fitting wooden dowels. The liquid chemical volatilizes and migrates through the wood, eliminating established decay fungi up to 4 m from the point of application. Fumigants applied in this way can arrest decay within one year and protect the wood from reinvasion for up to 17 years (see Fig. 1).

Three currently registered fumigants are commonly used for long-term protection of wood. They are Vapam (32.1% sodium *n*-methyldithiocarbamate), Vorlex (20% methylisothiocyanate in chlorinated C_3 hydrocarbons) and chloropicrin (trichloronitromethane). Of these three, Vapam is the most commonly used and the easiest to handle. Vapam decomposes slowly in the presence of organic matter to release a mixture of 14 different compounds, including carbon disulfide, carbonyl sulfide and methylisothiocyanate (MIT). MIT is believed to be the main fungitoxic component of Vapam, but recent studies suggest that other compounds may also provide some long-term protection. Of the three chemicals, Vapam remains in the wood for the shortest time—its volatile fungitoxic compounds cannot be detected two years after treatment. Yet poles treated with Vapam have remained free of decay fungi for up to ten years.

One drawback of treatment with Vapam is its relatively low concentration of active ingredient per liter of formulated chemical. Vapam is a 32.1% solution that decomposes to produce MIT at a 40% conversion rate. As a result, only 12–16% of the applied chemical is released as MIT. Moreover, in many structures to be treated, the decay is so far

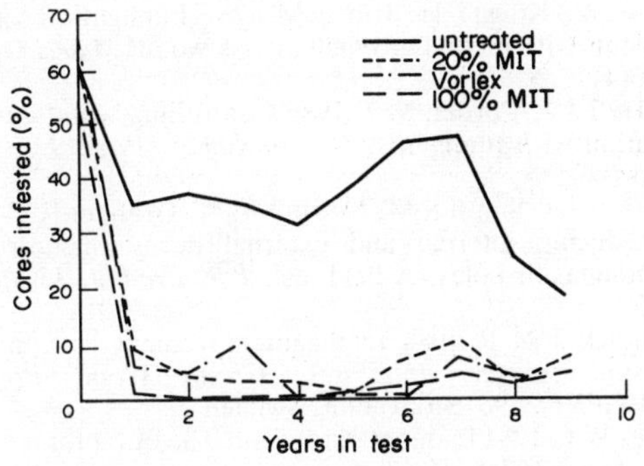

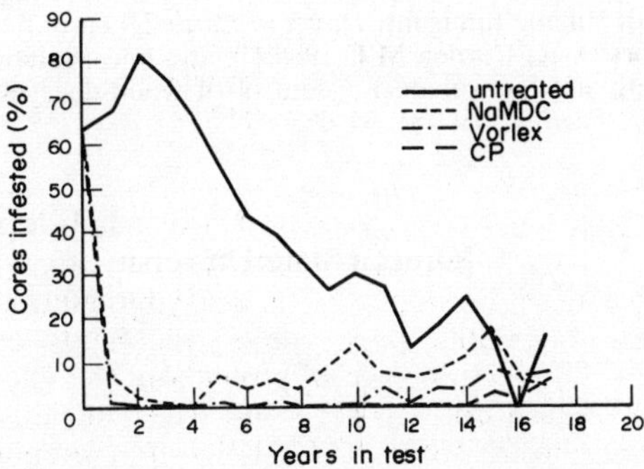

Figure 1
Ability of Vapam (NaMDC), Vorlex, chloropicrin (CP) and methylisothiocyanate (MIT) to eliminate and prevent Basidiomycete colonization of Douglas-fir poles, as measured by culturing core samples removed from each pole at yearly intervals

advanced that the number of treatment holes must be limited. For such cases, chloropicrin and Vorlex, which have relatively high levels of active ingredient, may provide both more rapid control and better long-term control; residues of chloropicrin and Vorlex, which have relatively high levels of active ingredients, may provide both more rapid control and better long-term control; residues of chloropicrin and Vorlex have been shown to persist in wood for more than 18 years. However, both these chemicals are highly volatile and highly toxic. They are best applied to wood located in areas where there is ample ventilation and where direct human contact is unlikely to occur—for example, on utility poles, pier pilings and bridge timbers.

Tests indicate that most wood species can be treated with fumigants, but that chemicals will move faster through some woods than through others. For example, fumigant movement is slower through lodgepole pine than through southern pine; thus, southern pine will be more likely to require more frequent retreatments than lodgepole pine. Any species of wood to be fumigated should be tested for fumigant movement first, if this has not already been done.

2. *Fumigant Applications*

Fumigants have been most widely used to control internal decay in utility poles. They have also helped to preserve laminated building timbers, marine pilings and bridge timbers. They have slowed the attack of marine borers for up to three years; further studies are now underway on full-sized marine pilings. Fumigants have also successfully controlled root rot in living trees and in freshly cut stumps. Vapam is used to eliminate the pine wood nematode from wood chips destined for export.

In summary, fumigants can provide long-term protection against decay in most large poles and timbers. The only fumigant application that does not work well by itself is the remedial treatment of wood in contact with the ground. Such treatment alone does not keep subterranean termites and soft-rot fungi from attacking the surface of the wood. Consequently, fumigant treatment in these cases is recommended only if some type of preservative wrap is applied at the same time.

3. *The Future of Wood Fumigation*

The fumigation of wood is a relatively recent method for arresting fungal and insect attack. Its growing popularity parallels an increasing demand, prompted by environmental concerns, for safer chemicals. Several research projects have been undertaken to learn more about how fumigants function in wood and what their effect is on the environment, with the aim of developing safer formulations and improved methods of application without sacrificing effectiveness.

Two chemical formulations have recently been identified as safer than those now used, yet still effective for long-term remedial fumigation of wood. They are solid MIT and Mylone. MIT is a crystalline solid which sublimes at room temperature to rapidly eliminate decay fungi established in the wood. It remains in sufficient concentrations to limit reinvasion for up to ten years. This chemical must be encapsulated for safe use; however, studies indicate that encapsulation does not alter its effectiveness. Several field tests of MIT are underway. Mylone is a crystalline powder which slowly decomposes to produce MIT and several other compounds. The natural slow rate of decomposition can be altered by adding certain buffers. High pH (basic) buffers markedly accelerate decomposition, while low pH (acidic) buffers inhibit it. These buffers may be useful for selectively treating wood in varying stages of decay. For example, Mylone with a high pH buffer might be applied to actively decaying wood, while the same chemical without a buffer might be applied to protect a decay-free piece of wood. Another potentially safer fumigant is Vapam in its solid formulation. In its sodium salt form, Vapam produces a powder that, when hydrated, rapidly eliminates fungi established in wood. Studies are now underway to better develop this formulation for use in the field.

Fumigation is increasingly becoming a part of routine wood maintenance programs because of its effectiveness in lengthening the service life of wood. While fumigants should never replace the careful design and construction practices that help ensure long service life, they improve the performance of wood used under adverse conditions.

See also: Wood: Decay During Use; Wood: Deterioration by Insects and Other Animals During Use; Preservative-Treated Wood

Bibliography

Graham R D 1973 Preventing and stopping internal decay of Douglas fir poles. *Holzforschung* 27: 168–73

Kinn D N, Springer E L 1985 Using sodium N-methyldithiocarbamate to exterminate the pine wood nematode in wood chips. *Tappi* 68(12): 88

Liese W, Knigge H, Rütze M 1981 Fumigation experiments with methyl bromide on oak wood. *Mater. Organ.* 16(4): 265–80

Morrell J J, Corden M E 1986 Controlling wood deterioration with fumigants: A review. *For. Prod. J.* 36(10): 26–34

Morrell J J, Smith S M, Newbill M A, Graham R D 1986 Reducing internal and external decay of untreated Douglas-fir-poles: A field test. *For. Prod. J.* 36(4): 47–52

Ruddick J N R 1984 Fumigant movement in Canadian wood species. *Int. Res. Group Wood Preserv.* IRG/WP/3296. Stockholm, Sweden

Thies W G 1984 Laminated roof rot: The quest for control. *J. For.* 82: 345–56

Young E D 1972 Short exposures in structural fumigation with Vikane fumigant. *Down to Earth* 27(4): 6–7

Zahora A R, Corden M E 1985 Gelatin encapsulation of methylisothiocyanate for control of wood-decay fungi. *For. Prod. J.* 35(7/8): 64–9

J. J. Morrell
[Oregon State University, Corvallis, Oregon, USA]

G

Gas Sensors, Solid State

In recent years, there has been an explosion in the application of computing and microelectronics with the overall aim of monitoring, controlling, optimizing and automating industrial processes. The latter application can be of particular importance to industries where working conditions are either hazardous or environmentally unacceptable. However, in many cases, the computer applications are restricted by the lack of reliable sensors to detect changes in the processes and their environment, particularly with gases. Conventional gas analysis using chromatographic or spectroscopic methods is not always suitable due to expense, the need for continuous monitoring and poor sensitivity at low concentrations. The development of new and more reliable sensors is, therefore, of paramount importance for the future control of many processes, especially those operating at high temperatures.

The development of chemical sensors to monitor gaseous species falls into two main categories. First, at elevated temperatures, sensors are generally based upon ionically conducting ceramics. These devices may be operated in the potentiometric or coulometric modes. Second, sensors can be based upon materials which rely upon changes in physical effects such as the Curie point for piezoelectric-based and Hall-effect devices, adsorption bands for optical devices and conduction band saturation for semiconducting materials. This article reviews these two major groups of sensors which, if successfully applied, allow the control of a diverse range of processes, from gas combustion in the internal combustion engine through pyrometallurgical processes to microwave cooking.

1. Solid Electrolyte Sensors

These sensors are based upon ceramic materials which are, almost exclusively, ionic conductors. Most ceramics are insulators, both electronically and ionically, but there are several groups which conduct ions because of doping of the material which induces defects in the structure (CaO, ZrO_2), because of order–disorder phase transformations which cause the cation lattice to disorder ($RbAg_4I_5$), or because there are planes or tunnels in the material through which ionic species may move ($Na_20.11Al_2O_3$ or $Na_{1+x}Zr_2Si_xP_{3-x}O_{12}$). Other solid electrolytes are based upon polymeric materials (polyethylene oxide and perflurosulfonic acid). In the potentiometric mode these sensors are based upon the very simple thermodynamic concept that a concentration gradient of a species across an ionic conductor results in a potential which can be detected by a high impedance voltmeter. An oxygen sensor can be represented by

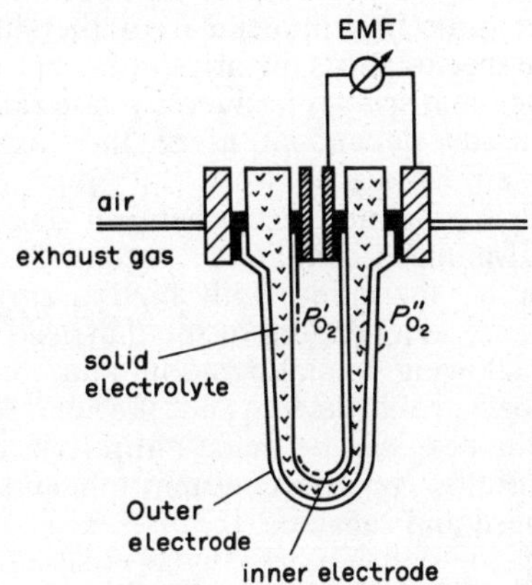

Figure 1
Gas sensor to monitor oxygen content of exhaust gases from car engines

$$\text{Pt, } O_2'|O^{2-} \text{ conductor}|O_2'', \text{ Pt}$$

where "O^{2-} conductor" is an ionic conductor of oxygen ions, and O_2' and O_2'' denote different partial pressures of oxygen. The physical arrangement for measuring the concentration of oxygen is shown in Fig. 1. The potential (E) is given by the Nernst equation

$$-ZEF = RT \ln \frac{P'_{O_2}}{P''_{O_2}} \tag{1}$$

where P'_{O2}, P''_{O2} refer to the partial pressures of fixed. Sensors based on this concept are used to temperature, R is the gas constant, F is Faraday's constant and Z is the unit of charge carried. The main requirements for developing a successful sensor are an ionic conductor for the species of interest and a reference material which gives a constant partial pressure of the species.

Examination of Eqn. (1) shows that the potential change is the same for a concentration change of 1 ppm to 10 ppm as it is for a concentration change of 10 000 ppm to 100 000 ppm. Sensors based upon the Nernst equation are particularly sensitive at low concentrations but they are relatively insensitive at high concentrations. If, instead of measuring the potential, a potential is applied across the electrolyte, the current flow will be directly related to the flux of ions through the electrolyte and this, in turn, will depend upon the rate of arrival of the species at the surface of the electrolyte. The current flow is, therefore, directly proportional to the diffusional flux of the species and is linearly dependent upon the concentration of species in the gas phase rather than logarithmically dependent as in the case of the potentiometric mode.

Overall, solid-state electrochemical sensors have several advantages in that the potentials and currents can be measured with high accuracy, the potentials are independent of the dimensions of the material allowing miniaturization and integration into microelectronic devices and, if a solid reference is used, the cells can be made entirely in the solid state, have low energy consumption and can be made rugged and reliable.

2. *Oxygen Sensors*

Most oxygen sensors use the cubic form of zirconia which has been stabilized with calcium oxide or yttria. In some cases, it is possible to use air or a mixture of gases as the reference but, frequently, the reference is a mixture of a metal and its oxide which gives a fixed oxygen partial pressure at a given temperature through the following equilibrium:

$$2\mathrm{MO} \rightleftharpoons 2\mathrm{M} + \tfrac{1}{2}\mathrm{O_2}, \qquad K = \frac{a_{\mathrm{M}}^2 P_{\mathrm{O_2}}^{1/2}}{a_{\mathrm{MO}}} \tag{2}$$

where K is the equilibrium constant, P_{O_2} is the partial pressure of the oxygen and a_M and a_{MO} are the activities of the metal and metal oxide, respectively. As the metal and metal oxide are present as pure phases, the activities are unity, K is constant and, therefore, the partial pressure of the oxygen is fixed. Sensors based on this concept are used to measure the oxygen content of gases, control furnace atmospheres, measure dissolved oxygen in molten copper and steel, and control the fuel-to-air ratio in internal combustion engines, especially where catalytic converters are fitted. Changing from a fuel-rich to an air-rich gas causes the oxygen partial pressure to change dramatically resulting in a potential change of about 0.8 V, which is more than sufficient for a microprocessor to monitor and make the necessary adjustments to the carburetor. In the case of lean-burn engines, the gas is always oxidizing and, therefore, measurements of potential are not sensitive to changes in the oxygen content. In this case, limiting-current devices are used, an example of which is shown in Fig. 2.

3. *Hydrogen Sensors*

The quantitative detection of hydrogen is of considerable importance in a variety of industrial situations, such as the synthesis of ammonia, methanol and other chemicals, petroleum refining, semiconductor manufacture, chemical reduction and corrosion control. Unfortunately, there are no satisfactory high-temperature proton conductors and the main approach has been to use materials such as Nafion (perfluorosulfonic acid) and HUP (hydrogen uranyl phosphate) which become proton conductors when hydrated. The reference material used has been either hydrogen gas, a mixture of hydrides or a redox mixture. These sensors have been used to measure hydrogen in gases and hydrogen in solid steel generated by corrosion reactions.

4. *Sulfur and Sulfur-Containing Gases*

The electrochemical determination of sulfur and sulfur dioxide is not as straightforward as the detection of oxygen or hydrogen, since the electrolytes are neither as stable nor as conducting. Calcium sulfide has been used as an electrolyte in the following cell:

$$\mathrm{Fe,\ FeS|CaS(Ca^{2+}\ conductor)|H_2,\ H_2S}$$

Since the electrolyte is a calcium ion conductor, the changes in sulfur potential are related to a change in the calcium activity through the relationship

$$\mathrm{Ca} + \tfrac{1}{2}\mathrm{S_2} \rightleftharpoons \mathrm{CaS}, \qquad K = \frac{a_{\mathrm{CaS}}}{a_{\mathrm{Ca}} P_{\mathrm{S_2}}^{1/2}} \tag{3}$$

More recently, in order to combat the poor ionic conductivity of calcium sulfide, mixed electrolytes of calcium sulfide and stabilized zirconia have been used in which the reaction of the working electrode is

$$\mathrm{CaS} + \tfrac{1}{2}\mathrm{O_2} \rightleftharpoons \mathrm{CaO(ZrO_2)} + \tfrac{1}{2}\mathrm{S_2}, \qquad K = \frac{a_{\mathrm{CaO}} P_{\mathrm{S_2}}^{1/2}}{a_{\mathrm{CaS}} P_{\mathrm{O_2}}^{1/2}} \tag{4}$$

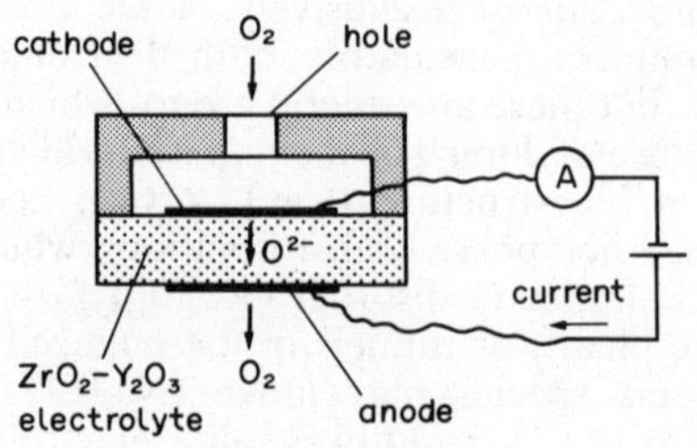

Figure 2
Limiting-current-type oxygen sensor

As the activities of the solid phases are fixed, the oxygen pressure must be proportional to the sulfur pressure. Unfortunately, these sensors only function in reducing atmospheres owing to the instability of calcium sulfide. At higher oxygen pressures, sulfates are thermodynamically stable and may be used to detect sulfur dioxide. The overall reaction is

$$SO_4^{2-} \rightleftharpoons SO_2 + O_2 + 2e^- \qquad (5)$$

and the potential can be expressed by

$$E = \frac{RT}{ZF}\ln\frac{P_{O_2''}}{P_{O_2'}} + \frac{RT}{ZF}\ln\frac{P_{SO_2''}}{P_{SO_2'}} \qquad (6)$$

and if the oxygen potential is fixed at both electrodes the potential is given by the ratio of the partial pressures of the sulfur dioxide.

5. *Gas Sensors Dependent upon Changes in Electrical Conductivity*

These sensors depend upon the electronic conductivity of certain semiconducting oxides, such as TiO_2, Nb_2O_5 and CeO_2, changing with the oxygen partial pressure. In the case of TiO_2, the intrinsic electrical conductivity is mainly due to oxygen lattice defects and can be represented by the following equation:

$$O_o^x \quad \tfrac{1}{2}O_2 + V_o - 2e^- \qquad (7)$$

where O_o^x is an oxygen ion on an anionic site of the crystal, V_o is a positive vacancy in the anion sublattice and e^- represents an electron donated to the conduction band. From Eqn. (7), it can be seen that decreasing the oxygen partial pressure leads to an increase in vacancies and electrons. Using the electroneutrality relationship and assuming thermodynamic equilibrium, the following equation can be obtained:

$$n = P_{O_2}^{-1/6}\exp(-E/kT) \qquad (8)$$

where n is the number of free electrons generated according to Eqn. (7), E is the energy of formation of an oxygen vacancy and k is the Boltzmann constant. In fact, experimental results show a $P_{O_2}^{-1/4}$ relationship with the conductivity owing to the presence of impurities in the crystal which cause a relatively high concentration of oxygen vacancies. As a consequence, the vacancy concentration is virtually independent of the oxygen partial pressure and the dependence of the electrical conductivity is proportional to $P_{O_2}^{1/4}$.

According to Eqn. (8), it can be estimated that a TiO_2 resistor will exhibit a change in conductivity of four orders of magnitude as the air-to-fuel ratio in an internal combustion engine passes from being reducing to oxidizing. Superimposed upon this conductivity change, there is also a conductivity change due to the variation in temperature, which can be overcome either by maintaining the resistor at a constant temperature by means of a heater or by interfacing a very porous TiO_2 pellet with a dense TiO_2 pellet. The resistivity of both pellets changes with temperature but only the resistance of the porous pellet changes with oxygen pressure due to the much larger surface area exposed to the gas. The resistance change with oxygen pressure can be accurately determined by, in essence, subtracting the variation of resistance for the dense pellet from that of the porous pellet.

For application in lean-burn engines, the quarter power dependence of the conductivity on oxygen partial pressure gives a low sensitivity. However, other semiconducting oxides, such as $CoO.Co_{1-x}Mg_xO$ and $SrMg_xTi_{1-x}O_3$, are being investigated to overcome this problem.

6. *Humidity Sensors*

There is an increasing demand for humidity sensors in automatic cooking by microwave ovens, in fridges, in air conditioners, in driers and in dew sensing on the cylinder heads of videotape recorders. Humidity sensors are also made from semiconductors, but the mechanism by which the sensor functions is not related to the structure of the bulk material, but is due more to unsatisfied bonds at the surface of the material which promote the adsorption of hydroxyl groups which then bond to form water molecules. The surface conductivity increases as the humidity increases. For $MgCr_2O_4$–TiO_2 materials, the adsorbed water molecules are easily desorbed by either decreased pressure or by temperatures slightly higher than the ambient temperature. These sensors are used in the control of microwave cooking.

7. *Gas-Sensitive Resistors*

Similar sensors have been developed to detect other gases but the exact mechanism of operation is not fully understood though it is thought that chemisorbed oxygen must be present. In the case of tin oxide sputtered onto an alumina substrate, oxygen is thought to deplete the valence band, which lowers the conductivity. When carbon monoxide is present, this is also adsorbed, increasing the conductivity of the tin oxide. However, the relationship between the gas concentration and the resistance is nonlinear and the selectivity is poor. In spite of these disadvantages, however, metal oxide gas sensors have been

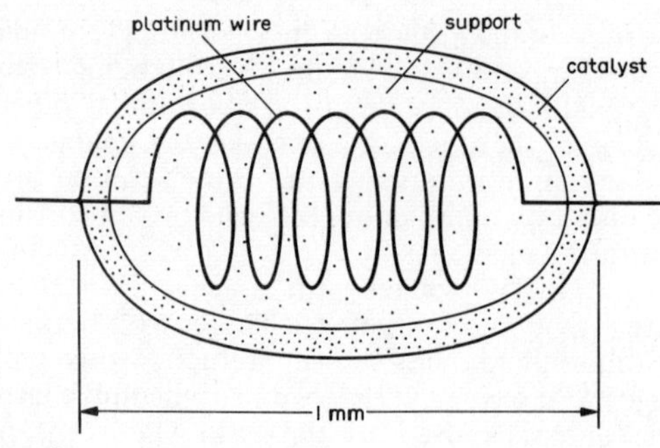

Figure 3
Catalytic gas sensor bead

used for monitoring combustible gases and as gas leakage detectors, as the construction and electronic circuitry is relatively simple.

8. Surface Acoustic Wave Sensors

The strong piezoelectricity and low acoustic losses of lithium niobate ($LiNbO_3$) have laid the basis of surface acoustic wave sensors. A set of electrodes generates an acoustic wave which propagates along the surface and is detected by a second set of electrodes. The velocity of the surface acoustic can be influenced by a deposited coating which, in turn, can be altered by absorbed gases. For example, a thin palladium film can absorb and desorb hydrogen and this can be detected by phase shifts in the acoustic wave. Similarly, a lead phthalocyanate coating has been used to detect nitric oxide.

9. Catalytic Gas Detectors

Inflammable gases can be monitored by catalytic gas detectors which are made from a platinum coil upon which is deposited a porous oxide material, such as alumina or thoria. A catalytic gas sensor bead is shown in Fig. 3. The actual detector consists of two such beads, one of which is activated with catalyst while the other has no catalytic activity. In the presence of inflammable gases and at a temperature of approximately 600 °C, the catalytic bead causes the inflammable gases to react with the oxygen in air and this raises the temperature and, thus, the resistance of the platinum coil. The inert bead, having no catalytic activity, does not increase in temperature and, therefore, the coil resistance of this bead remains constant. The difference between the resistances of the two beads is related to the concentration of the inflammable gases present.

10. Semiconductor Junction Devices

These sensors rely on the adsorption of a gas that modifies the electron distribution in the surface region of a semiconductor. This can occur either by direct interaction with the gas or by indirect interaction with oxygen which will be displaced by the gas being measured. Instead of measuring the surface conductivity, the conduction across the interface or the capacitance of the surface space-charge layer is measured. A typical metal oxide semiconductor field effect transistor (MOSFET) is shown in Fig. 4.

Initial attempts to develop a hydrogen-sensitive palladium–silicon Schottky diode failed because of the formation of an intermediate silicide layer of Pd_2Si. This problem was solved by deliberately growing a very thin layer of silica prior to the deposition of the palladium. This very thin layer does not affect either the diode characteristics of the structure or the hydrogen sensitivity. It is likely that as well as the formation of a dipole layer at the palladium–silica interface, the modulation of the surface states at the silicon–silica interface and the corresponding tunnelling of electrons between these surface states and the palladium may also play an important role.

Hydrogen atoms diffuse through the thin palladium layer and are adsorbed at the metal–insulator interface. The hydrogen atoms are polarized and give rise to a dipole layer resulting in a voltage drop which is added to the externally applied voltage. This voltage drop is dependent upon the concentration and it should be noted that this concentration is not only a function of the hydrogen pressure, but can also be influenced by the presence of oxidizing gases which can remove absorbed hydrogen. For nonporous palladium films, the devices only respond to gases which can dissociate into hydrogen.

Langmuir–Blodgett films, which are homogeneous organic films, one molecule thick, can be used as an insulating layer between the gate and the channel of a silicon MOSFET or between the electrodes of a MOS diode. Redox or complexation reactions with the active molecules of the Langmuir–Blodgett film result in a shift of the flat band voltage and a change in the low-frequency

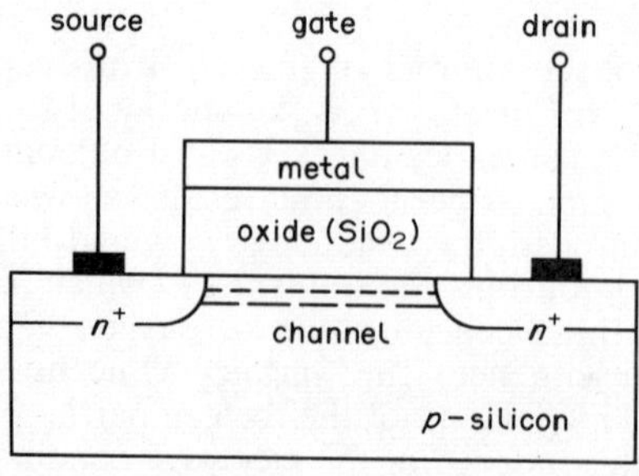

Figure 4
Metal oxide semiconductor field effect transistor (MOSFET)

dielectric losses of the device. It has been found that MOS structures containing a Langmuir–Blodgett film of phthalocyanine respond quickly to a few parts per million of NO_x at room temperature.

11. Conclusions

From an industrial viewpoint, there is a considerable need for control in order to maintain a permanent analysis of the physicochemical parameters of many processes with the aim of achieving quality and reproducibility. Furthermore, with the ever-increasing pressure to minimize atmospheric pollution, the monitoring of effluents is becoming of paramount importance. The increase in consumer goods, such as cars and air conditioners, is responsible not only for direct pollution of the environment but also for indirect pollution by consuming electricity produced by the combustion of fossil fuels. Any improvements in efficiency made possible by control will pay dividends both in terms of improved production values and decreased atmospheric pollution.

In this article it has been demonstrated that there are a wide variety of sensors based on different physical and chemical effects. The integration of chemical sensors, such as humidity sensors, is much more difficult than with physical sensors because chemical sensors must be directly exposed to the various environments and, therefore, they cannot be totally sealed.

Another problem for the integration of chemical sensors is the incompatibility of the materials of the sensing element and the integrated circuits. However, considerable effort is being put into developing thin film devices using sputtering, vapor deposition and silicon technology with a view to improving sensor characteristics and lowering the overall cost.

Finally, it can be confidently stated that the demand for gas sensors will grow and that research will develop new sensors and make existing sensors more reliable.

See also: Zirconia and Hafnia

Bibliography

Aucouturier J-L, Cauhape J-S, Destriau M, Hagenmuller P, Lucat C, Menil F, Portier J, Salardenne J (eds.) 1986 *Proc. 2nd Int. Meeting Chemical Sensors*, Bordeaux, France

Edmonds T E (ed.) 1988 *Chemical Sensors*. Blackie, Glasgow

Fisher G 1986 Ceramic sensors: Providing control through chemical reactions. *Ceram. Bull.* 65: 622–9

Goto K S 1988 *Solid State Electrochemistry and its Applications to Sensors and Electronic Devices*. Elsevier, Amsterdam

Moseley P T, Tofield B C (eds.) 1987 *Solid State Gas Sensors*. Hilger, Bristol

Williams D E, McGeehin P 1984 Solid state sensors and monitors. In: Pletcher D (ed.) 1984 *Electrochemistry*, Vol. 9. The Royal Society of Chemistry, London, pp. 246–90

D. J. Fray
[University of Cambridge, Cambridge, UK]

Germanium

This article replaces the article of the same title in the Main Encyclopedia.

The early history of the semiconductor revolution in electronics is dominated by the study of germanium. In this respect, germanium is the prototypical semiconductor. Germanium is most appropriately treated as a historical material since its major contribution to electronics was in the proof of semiconductor concepts and their demonstration in useful devices. Many concepts from the quantum theory of solids (Seitz 1940) such as hole conduction, effective mass, band gaps, forbidden-gap impurity states, and so on, were first studied and reduced to routine engineering practice using germanium. The reasons for the dominance of germanium in the early successes in semiconductor electronics is due to its unique chemical, physical and electrical characteristics.

1. History

Until recently, the only significant uses for germanium were for the electrical properties of the element. The novel electrical properties of germanium (and silicon) were found to make the best "cat's whisker" or "crystal rectifier" detectors (Torrey and Whitmar 1948) for radar signals. Intensive efforts during World War II to understand and predict the performance of these detectors led, by the end of the war, to the concept of the p–n junction and an awareness of the crucial importance of the control of minute concentrations of impurities and defects. The required impurity control and measurement was not achieved during the war, and even the concept of a crystal dislocation was widely disbelieved—a dislocation had not yet been seen in any solid (Bardeen 1987).

This activity during the war interested many solid state physicists and convinced them of the need for further investigation. In 1945, Bell Telephone Laboratories established a solid state division under the joint leadership of W. Shockley and W. Brattain with the directive to organize a concentrated effort of physicists, chemists and engineers to study the properties of germanium and silicon with the goal of producing a solid state amplifier. It soon became apparent that silicon was a very intractable material to purify and the initial efforts were concentrated on germanium. These studies led to methods for measuring resistivity, mobility, carrier type, lifetime, diffusion length and dislocation density, and to techniques for impurity control by melt segregation, diffusion and zone refining (Pfann 1966). The birth of the semiconductor age came with the invention in 1947 of the point contact transistor (Bardeen and Brattain 1948) and the field effect transistor (Shockley and Pearson 1948), and also the junction transistor (Shockley 1949). The Nobel prize for physics was given to Bardeen, Brattain and Shockley in 1956 for these discoveries.

A symposium was held at Bell Laboratories in 1952 to present the results of the transistor work to a selected audience and the proceedings, although classified by the military, were widely distributed. These proceedings were eventually declassified and published (Bridgers et al. 1958) together with two volumes (Biondi 1958) of results obtained since the initial symposium.

2. Properties

Germanium is a hard brittle element with a silvery metallic luster. It is one of only two elements—the other being silicon—that crystallize from the melt in the diamond cubic structure (diamond and gray tin are metastable phases). As is characteristic of semiconductors, the electrical conductivity of germanium increases with temperature, and at liquid helium temperature it is an electrical insulator. Germanium is not attacked by hydrochloric or hydrofluoric acids, or by strong alkalis, but is dissolved slowly by nitric acid and aqua regia. Mixtures of an oxidant and HF readily attack germanium, for example, HNO_3-HF, H_2O_2-HF; and it is dissolved by H_2O_2 and halides in solution at room temperature. Germanium is stable in air to 500 °C.

The diamond cubic lattice is a very open structure, and consequently elements and compounds with this structure have a melt density greater than the solid. Germanium expands by 4.9% on freezing. The crust of the earth contains about 8×10^{-4} wt% germanium, but because there are no rich ores the supply is very price sensitive. A typical price during the past 30 years has been about US\$600 kg^{-1} (1988 US\$).

The reasons for the dominance of germanium over silicon in the first decade of the semiconductor age are mostly chemical. A first requirement for proving device concepts was to obtain high-purity structurally perfect crystals. It transpired that germanium is probably the easiest solid element of the periodic table to purify. Germanium can be melted in carbon or fused silica containers with negligible reaction, and its purity can be demonstrated immediately by simply measuring its resistance vs temperature. This simplicity of analysis led to the rapid advances in semiconductor purification so that germanium became and still is demonstrably the purest solid known (Haller et al. 1981). Large, highly perfect crystals of germanium can be obtained commercially with a total electrically active impurity concentration of less than 10^{10} cm^{-3} (or about one impurity atom in roughly 10^{13} germanium host atoms).

A summary of the physical properties of germanium is given in Table 1, and a more extensive listing can be found in Madelung et al. (1982).

3. Applications

The great failing of germanium as an electronic device material when compared with silicon is that it does not have a stable oxide (or coating) which can simultaneously act as a surface passivant and as an impurity diffusion mask. As soon as high-purity silicon technology had been developed, germanium was replaced in all applications except those few where some unique property of germanium was required. The principal uses today are for photodetectors, nuclear radiation detectors and optics.

3.1 Photodetectors

Semiconductor photodetectors are conveniently classified as either intrinsic or extrinsic, depending on the mechanism of photo charge carrier generation (Sze 1981). Intrinsic detectors rely on band-to-band transitions produced by exciting carriers across the full band gap, and extrinsic detectors use ionization of forbidden-gap energy levels. In either case, the long wavelength cutoff λ_c (μm) is:

$$\lambda_c = \frac{hc}{E} = \frac{1.24}{E} \tag{1}$$

where E(eV) is the energy gap for intrinsic detectors or the ionization energy of forbidden-gap impurities for extrinsic detectors, and h and c are Planks constant and the velocity of light, respectively. For germanium intrinsic detectors, $E_g = 0.66$ eV so that $\lambda_c = 1.8$ μm, which therefore includes the technically important 1.3 μm and 1.5 μm low-loss windows in optical-fiber communications. These detectors are usually operated at room temperature, although cooling will improve the signal-to-noise ratio by suppressing thermally generated carriers.

The group III and V impurities in germanium have an ionization energy E_i of about 0.01 eV so that $\lambda_c = 120$ μm. Impurities with larger ionization energy such as Be, Zn, Ni, Hg and Cu can be selected so as to optimize the sensitivity to particular shorter wavelengths. Also E_i may be reduced by uniaxial stress to extend the long wavelength cutoff to 240 μm. Extrinsic germanium detectors must always be used at low temperature to prevent thermal ionization of the impurities.

Intrinsic germanium photodetectors are normally reverse biased p–n junction diodes to minimize both response time ($\sim10^{-11}$ s) and dark current. Extrinsic detectors are usually used as low bias voltage photoconductors to prevent electric field ionization of

Table 1
Properties of germanium

Property	Value
Atomic number	32
Atomic weight	72.59
Natural isotopic composition (%)	
70	20.4
72	27.4
73	7.8
74	36.6
76	7.8
Density (25 °C, g cm^{-3})	5.3234
Atomic density (atoms cm^{-3})	4.416×10^{22}
Crystal structure	diamond cubic
Lattice parameter (300 K)	5.6579060
Mohs hardness (300 K)	6.3
Melting point (mp) (°C)	937.4
Boiling point (°C)	2830
Vapor pressure at mp (atm)	2×10^{-13}
Volume contraction on melting (%)	4.7
Heat of fusion (kJ mol^{-1})	36.945
Heat of sublimation (kJ mol^{-1})	374.5
Thermal expansion coefficient at 300 K (10^{-6} K^{-1})	5.9
Thermal conductivity at 300 K (W cm^{-1} K^{-1})	0.6
Energy gap at 300 K (eV)	0.664
Energy gap at 0 K (eV)	0.785
Intrinsic resistivity at 25 °C (Ω cm)	50
Intrinsic electron concentration at 300 K (cm^{-3})	2.33×10^{13}
Temperature dependence of intrinsic concentration (cm^{-3})	$[3.1 \times 10^{32} T^3 \exp(-0.785/kT)]^{1/2}$
Intrinsic electron mobility at 300 K (cm^2 V^{-1} s^{-1})	3900
Intrinsic hole mobility at 300 K (cm^2 V^{-1} s^{-1})	1800
Effective mass, m^*/m_0	
electrons; longitudinal	1.64
transverse	0.082
holes; light	0.044
heavy	0.28
Refractive index, 2.5–150 μm	4.0
Dielectric constant	16.0

impurities (avalanche breakdown) and as a result are slower than intrinsic detectors. For photoconductors, gain is simply the ratio of carrier lifetime to carrier transit time and gain may be traded for speed of response. In the case of germanium, carrier lifetime is readily adjustable over a wide range so that devices can be designed for specific applications.

The principal application of germanium photoconductors is in infrared astronomy where in the longer wavelength range (30–200 μm) they have the highest detectivity of any device. For low background, space-borne astronomy where detectivity is of paramount importance (Wang et al. 1986), germanium photoconductors are used as unity gain detectors at the lowest possible temperature (2 K).

3.2 Nuclear Radiation Detectors

Germanium nuclear radiation detectors (Haller and Goulding 1981) are widely used for γ-ray spectroscopy where their high-energy resolution allows analysis of complex γ-ray spectra from nuclear power, fuel-cell processing and environmental contaminants. These detectors are essentially giant volume p–i–n diodes in which the i-layer forms the bulk of the detector and the p- and n-layers form blocking contacts. In this terminology, the i-layer is so lightly doped with electrically active impurities that it is totally depleted of free charge carriers at the operating bias voltage. The material requirements for these detectors are that the semiconductor has an atomic number high enough for significant ionization cross section for high-energy γ-rays, a high purity to form a thick depletion layer, and has very small trapping of charge carriers. Germanium is the only material to meet these requirements.

The thickness W of the depletion layer, which must be totally depleted of charge, is given by:

$$W^2 = \frac{2\varepsilon\varepsilon_0 V}{q|N_A - N_D|} \qquad (2)$$

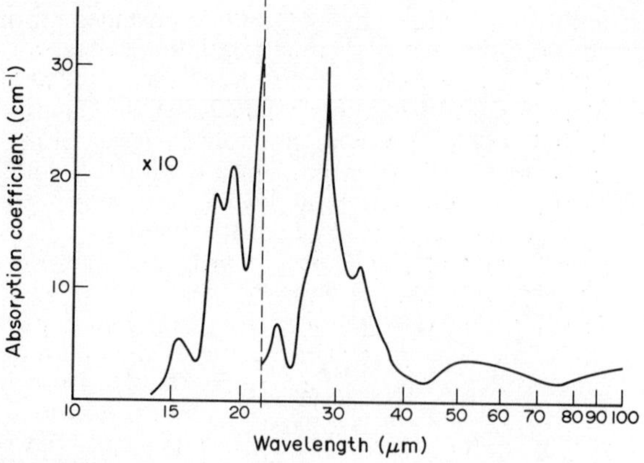

Figure 1
Optical absorption coefficient of germanium in the region of the phonon resonances—adapted from data in Potter (1985)

where ε is the dielectric constant (16 for Ge), ε_0 is the permittivity of vacuum, q is the electronic charge, V is the applied voltage and $|N_A - N_D|$ is the concentration of net electrically active impurity centers. For example, for a depletion layer thickness of 1 cm at 1000 V bias, $N_A - N_D$ must be less than 2×10^{10} cm^{-3}. In practice, detectors with a totally depleted i-layer volume of greater than 100 cm^3 are available commercially.

These detectors are used as single photon or particle energy spectrometers in which 3 eV of incident energy is required to produce each electron–hole pair and 2.3 eV is dissipated as phonons (3 eV $-E_g$). The energy resolution in the higher-energy ranges is limited by the statistics of electron–hole pair generation, and is typically 1.7 keV for a 1 MeV photon.

In an attempt to measure the generation current in germanium nuclear radiation detectors (Hall 1976), "monodes" of 167 cm^3 volume were constructed which, having only one external electrode, had zero surface-leakage current. These radiation-shielded monodes had a bulk generation current of about 10^{-14} A cm^{-3} at temperatures below 100 K, which proved to be electron–hole pair generation due to high-energy cosmic radiation. Thus it has proven impossible to measure such traditional parameters as carrier diffusion length and lifetime in such high-purity germanium.

3.3 Infrared Optics

Germanium has many characteristics which make it useful as an infrared optical material (Potter 1985), especially in the militarily important 8–15 μm range (see Fig. 1). As an optical material, germanium is transparent from 2 μm to 15 μm, has good mechanical properties, moderate price and extremely low dispersion. Germanium has significant optical attenuation in the 20–40 μm range due to intrinsic crystal lattice phonon absorption. The volume of germanium used in optics exceeds that for all other applications combined.

3.4 Miscellaneous Uses and Prospects

Germanium is used in low-voltage high-current power supplies (mainframe computers) where its higher efficiency compared with silicon gives it some economic advantage. Although costly, gallium arsenide on germanium makes the highest efficiency solar cells, which may find application on spacecraft. Graded index optical fibers are made using a high index germanium-doped core. Impatt diodes, tunnel diodes (Sze 1981) and phonon drag detectors (Gibson and Kimmitt 1980) are made using germanium. The heat capacity of germanium at low temperatures is very small and so makes excellent bolometers (a type of thermal detector). A germanium bolometer when attached to a large piece of high-purity germanium and cooled to near 0.1 K can be used as an x-ray photon or nuclear radiation detector with very low noise.

There has been considerable progress recently toward the development of Ge–Si quantum-well strained-layer superlattices for use as photodetectors and transistors (Kaspar and Bean 1988). Theoretically, germanium should make very fast cryogenic electronic devices, but the lack of a good surface passivant with a low density of interface states has inhibited progress in this area. With the discovery of a good surface passivant, fast symmetrical CMOS could be developed because below 100 K the mobility of holes and electrons are nearly equal and at 77 K $\mu_p = \mu_e = 4\times10^4$ cm^2 V^{-1} s^{-1}.

See also: Electronic Materials: An Overview

Bibliography

Bardeen J 1987 Solid state physics—1947. *Solid State Technol.* 30(12): 69–71

Bardeen J, Brattain W H 1948 The transistor, a semiconductor triode. *Phys. Rev.* 74: 230

Biondi F J (ed.) 1958 *Transistor Technology*, Vols. II and III. Van Nostrand, Princeton, NJ

Bridgers H E, Scaff J H, Shive J H (eds.) 1958 *Transistor Technology*, Vol. I. Van Nostrand, Princeton, NJ

Gibson A F, Kimmitt M F 1980 In: Button K J (ed.) *Infrared and Millimeter Waves*, Vol. 3. Academic Press, New York

Hall R N 1976 Bulk generation current in depleted germanium junctions. *Appl. Phys. Lett.* 29: 202–4

Haller E E, Goulding F S 1981 Nuclear radiation detectors. In: Hilsum C (ed.) *Handbook on Semiconductors*. North-Holland, Amsterdam, pp. 799–828

Haller E E, Hansen W L, Goulding F S 1981 Physics of ultra-pure germanium. *Adv. Phys.* 30: 93–138

Kaspar E, Bean J C (eds.) 1988 *Silicon Molecular Beam Epitaxy*. CRC Press, Boca Raton, FL

Madelung O (ed.) 1982 *Semiconductors: Physics of Group IV Elements & III–V Compounds*. Landolt-Boernstein Series, Vol. 17a. Springer, Berlin
Pfann W G 1966 *Zone Melting*. Wiley, New York
Potter R F 1985 Germanium. In: Palik E D (ed.) *Handbook of Optical Constants of Solids*. Academic Press, New York, pp. 465–78
Seitz F 1940 *The Modern Theory of Solids*. McGraw–Hill, New York
Shockley W 1949 The theory of p–n junctions in semiconductors and p–n junction transistors. *Bell Syst. Tech. J.* 28: 435
Shockley W, Pearson G L 1948 Modulation of conductance of thin films of semiconductors by surface charges. *Phys. Rev.* 74: 232–3
Sze S M 1981 *Physics of Semiconductor Devices*. Wiley, New York, pp. 743–89
Torrey H C, Whitmar C A 1948 *Crystal Rectifiers*. McGraw-Hill, New York
Wang J Q, Richards P L, Beeman J W, Haegel N M, Haller E E 1986 Optical efficiency of far-infrared photoconductors. *Appl. Optics* 25: 4127–34

W. L. Hansen
[Lawrence Berkeley Laboratory,
Berkeley, California, USA]

Gold–Titanium Alloys

Gold as an element and in alloy form has been used for nearly 6000 years as a decoration. Today titanium is considered the archetypal space-age metal, but only very recently have both been combined for decorative purposes in the form of gold-rich jewelry alloys.

Gold has the disadvantage that it is rather soft and deforms easily. Therefore, for jewelry the quarternary system gold–silver–copper–zinc (and nickel and palladium for white gold) remained the basic combination for gold alloys from 8–18 karats or from 333‰ to 750‰ fineness (33.3–75 wt% of gold) alloys.

However, the market has increasingly called for an alloy close to fine gold with the same (or even better) properties as the existing alloyed forms of "karat gold."

In a detailed investigation into the influence of the various alloying elements on the properties of karat gold, it has been found that additions of approximately 1 wt% titanium, and/or zirconium and hafnium, together with even smaller amounts of other elements, lead to totally new jewelry alloys containing more than 95 wt% gold. It is this range of compositions where the influence of titanium is truly unique.

1. The Gold–Titanium Phase Diagram

In the beginning of the work on gold–titanium jewelry alloys in 1980 there was little detailed and reliable information available on the gold-rich part of the phase diagram (Fig. 1). Of importance for our development is the terminal solubility of titanium in gold which shows a maximum near 3 wt% (12 at.%) titanium at the peritectic line (Au) L + $TiAu_4$ at 1210 °C. It is, however, not certain if the composition of the phase in equilibrium with the gold is $TiAu_4$ or $TiAu_6$. The solubility of titanium in gold is strongly temperature dependent: at 800 °C it is near 3 at.%, and at 500 °C less than 1 at.%. At lower temperatures the titanium will precipitate as the $TiAu_4$ intermetallic phase, causing age-hardening of the alloy. The preferable technical age-hardening temperature is in the range of 500–600 °C immediately after casting or after a homogenization at 800 °C. In these gold-rich alloys and at this temperature, however, a compromise has to be reached between excessive grain growth and homogenization, especially if the metal cannot be cold worked, as in cast objects.

The large grain growth can be taken care of by adding grain refiners in small amounts. Overaging also has to be avoided.

2. Preparation of the Material

2.1 Melting and Casting

Due to the reactivity of titanium with oxides and carbon, as well as with nitrogen and oxygen, melting in an argon-arc furnace on a water-cooled hearth is best. A master alloy of about 10 wt% titanium as well as pure titanium may be used. If, however, the surface of a carbon crucible exposed to the gold–titanium melt is not too large and melting time (after premelting the gold and introducing the titanium into the molten gold) is not too long, even carbon crucibles (coated with refractory or uncoated) may be used for melting in inert gas induction furnaces. Titanium sponge as a starting material is not recommended since it often contains fairly high amounts of impurities. Melting under glassy (borate-based) fluxes is possible, but can be difficult because inclusions may be formed.

Casting should be done either in a vacuum or in an inert-gas atmosphere, otherwise inclusions of titanium oxynitrides may cause embrittlement or inclusions in the material, generating cracking during deformation or "commas" during polishing. The true binary gold–titanium alloy tends to be fairly coarse-grained and irregular in grain size after being cast. This can be improved by adding the well known grain-fining additions. The use of grain-finers also improves the annealing behavior, especially at the homogenization temperature (~800 °C).

According to ongoing work, gold–titanium based alloys can be cast in phosphate-bound investment material, as is used for platinum-based alloys. However, care has to be taken that the

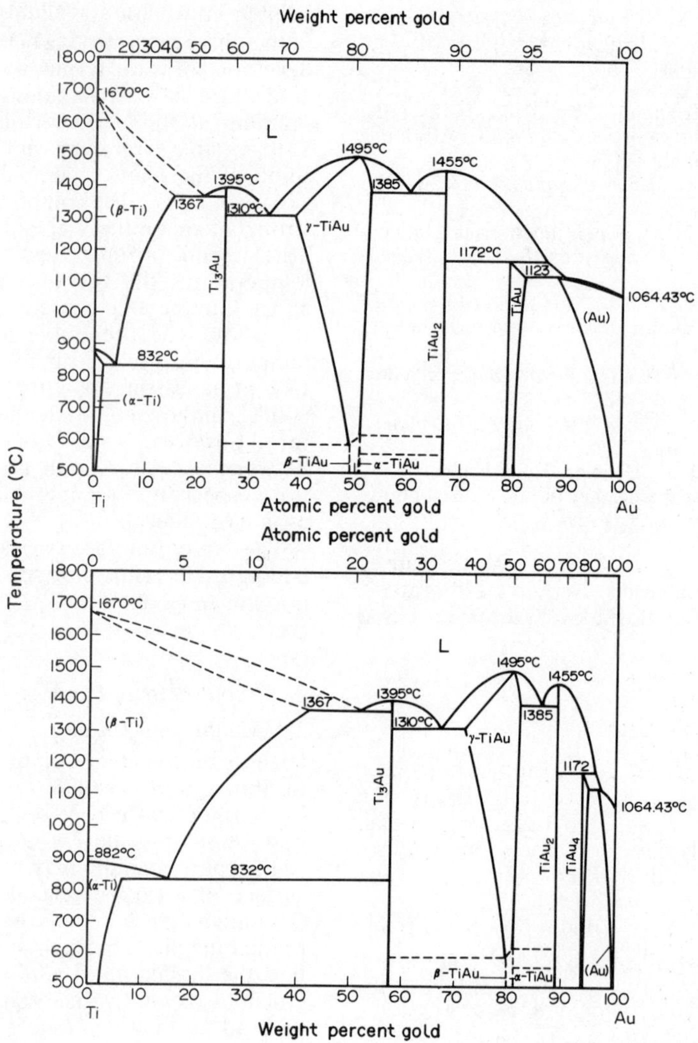

Figure 1
The gold–titanium phase diagram (Massalski 1986)

surface-to-weight and the melt-to-flask temperature ratios of parts to be cast are properly adjusted.

The absence of air during all these operations is critical otherwise a tarnishing layer forms which has to be removed by leaching in salt melts. Simple $KHSO_4$ melts, as well as commercial soldering fluxes, prove quite effective. These can also serve as fluxes during melting or soldering. If the time of flux melting is too long, however, the titanium content of the alloy may be reduced appreciably.

2.2 Alloy Composition

It was found that a concentration between 0.7 wt% and 1.5 wt% titanium gives the best compromise between age-hardening behavior, deformation characteristics, color and in-service behavior. Furthermore, hallmarking regulations may have to be taken into consideration. Below about 0.7 wt%, little age-hardening is observed; above 2 wt%, titanium alloys tend to become brittle, even in the as-cast state. The optimum concentration seems to change drastically to lower values if titanium is added to lower karat alloys based on the elements Au–Cu–Ag–Zn. It should be mentioned in this connection that zirconium and hafnium also cause age-hardening of gold alloys, but at higher concentrations. These higher concentrations generate more production problems; for example, too whitish a color, easy embrittlement in a small concentration range, and so on. The coloring problem (the alloy with 1 wt% titanium is a shade whitish) can be overcome if, for example, an alloy with 970‰ (i.e.,

Table 1
Mechanical properties of variously age-hardened gold–titanium alloys

	Pretreatment Annealed at 800 °C for 30 min	Annealed at 800 °C for 30 min and cold deformed 70%	Annealed at 600 °C for 30 min and cold deformed 70%
Hardness (HV 0.5)	160	180	190
Tensile strength (MPa)	490	580	680
Elongation (%)	12	8	8

97 wt%) gold, 0.8 % titanium, 1.3 wt% copper and the remainder silver, is used. This alloy has in every aspect practically the same properties as the 990‰ (99 wt% gold) alloy; it is, however, more golden-reddish in color.

2.3 Deformation Characteristics

In the as-cast condition, the alloy containing 1 wt% titanium has a hardness of about 80 HV, corresponding closely to that of fine gold. If the alloy is deformed by 70%, after homogenization at 800 °C, the hardness rises to 180 HV. If it is only partially homogenized, at 600 °C for 30 min, before deformation, hardness is near 180 HV. As cast, a tensile strength of about 320 MPa and an elongation of about 20% are obtained. Other data are: 320 MPa and 3% (800 °C annealed, 70% deformed) and 500 MPa and 5% (600 °C annealed, 70% deformed)—as tested on wires. At 800 °C, 30 min is long enough to achieve a homogeneous structure. Age-hardening can be done between 400 °C and 600 °C. At 500 °C, 45–60 min is best. The highest hardness values during age-hardening may be reached by a 500 °C heat treatment of a "semi-hard" metal (as cast, deformed 80%, annealed 600 °C for 30 min, age-hardened at 500 °C). Table 1 shows typical values for an alloy with 1 wt% titanium age-hardened at various conditions. Annealing times vary between 10 min and 60 min according to the pretreatment of the alloy. Longer annealing times or higher hardening temperatures cause "over-hardening" or softening of the alloy. Figure 2 shows a typical hardness vs annealing temperature curve, annealing time being 30 min in all cases. Annealing should be done under inert atmospheres or in vacuum. If the surface of the alloy is tarnished by reaction with air and the surface is treated in a salt melt, the fine gold layer formed on top of the alloy generates a color change towards the typical color of fine gold (dark, warm, yellow brown). On the other hand, the oxynitride skin disappears if the alloy undergoes a fairly long heat treatment above 500 °C. Apparently, this oxynitride layer at first diffuses (and/or decomposes) into the core of the material, leaving a skin of fine gold at the surface which again prevents further oxidation. This very interesting behavior, which can be used intentionally to form a hard dispersion of titanium nitride or other refractory titanium compounds near the surface, is presently under study.

The alloy can be clad (i.e., onto copper and copper-base alloys) in the soft and in the age-hardened condition. Together with the deformation-induced hardness increase, final values of HV near 180 are reached, which are excellent for practical jewelry applications.

For coin blanking, the alloy can be stamped in the soft state and age-hardened afterwards, or may be age-hardened to a certain hardness and then pressed. In both techniques, excellent and sharp

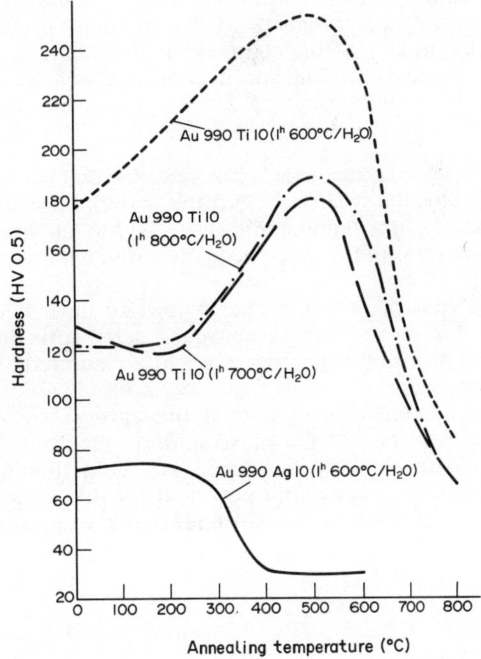

Figure 2
Hardness vs annealing temperature curves for a gold–titanium 1 wt% alloy. Pretreatment of the alloy: soft annealed at the temperatures listed on the curves (600–800 °C), 70% cold deformed by rolling and then annealed for 1 h at the relevant temperature

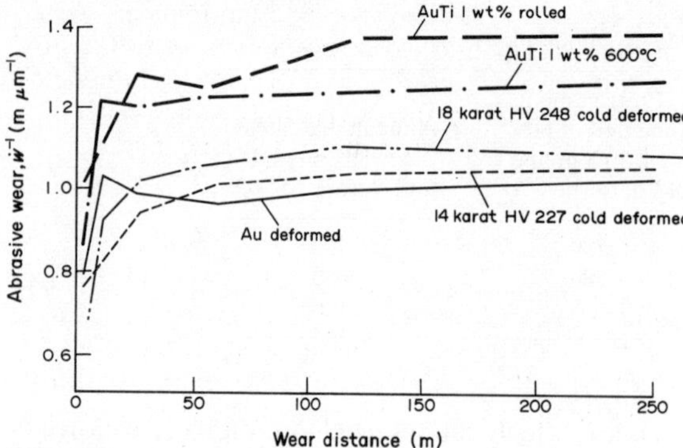

Figure 3
Abrasive wear tests on various gold jewelry alloys according to Clasing and Heidsick

contours are obtained without intermediate annealing.

2.4 Wear

Corrosion and wear resistance are critical for use as a jewelry material. As is to be expected for an alloy with such a high gold content, the corrosion resistance in all accelerated tests used for jewelry is excellent and comparable to that of fine gold. Clasing and Heidsick (1983) developed a method for accelerated wear tests of jewelry alloys. As shown in Fig. 3 the age-hardened 1 wt% titanium proved superior to all existing commercial 14 and 18 karat alloys by 30–40%. The specific abrasive wear resistance $\dot{\omega}^{-1}$ is given by

$$\dot{\omega}^{-1} = sA\delta/\Delta m$$

where $\dot{\omega}$ is the wear rate, Δm the material loss due to wear, s the sliding distance, A the apparent contact area and δ the specific density of the worn metal.

The first pieces of 23 karat jewelry have been worn for more than three years under practical conditions, confirming the excellent accelerated test results.

The combination of one of the earliest known metals with one of the most modern, producing a new alloy for jewelry production—the gold–titanium 1 wt% alloy—opens up a new field for the modern use of gold for both investment and decorative purposes.

See also: Gold and Gold Alloys

Bibliography

Clasing M, Heidsick H 1983 The abrasive wear of gold jewelry alloys. *Gold Bull.* 16: 76–81
Gafner G 1989 The development of 990 gold–titanium: Its production, use and properties. *Gold Bull.* 22: 112–22
Massalski T B (ed.) *1986 Binary Alloy Phase Diagrams.* American Society for Metals, Metals Park, OH
Ott D, Raub Ch J 1982 Influence of small additions on the properties of gold and gold alloys (in German). *Metall.* 36: 150–7
Ott D, Raub Ch J 1986 Developments in modern gold jewellery alloys: Gold 100 3. *Proc. Symp. Industrial Uses of Gold*, SAIMM, Johannesburg, pp. 65–71
Rapson W S, Groenewald T 1978 *Gold Usage*. Academic Press, London
Tasker A M 1988 The promise of 990 gold. *Aurum* 34: 62–7

Ch. J. Raub
[Forschungsinstitut für Edelmetalle und Metallchemie, Schwäbisch Gmünd, FRG]

Gypsum and Anhydrite

Gypsum, the dihydrate form of calcium sulfate ($CaSO_4.2H_2O$), and anhydrite, the anhydrous form ($CaSO_4$), are found in close association in many parts of the earth's surface. Anhydrite has minor economic importance, but gypsum is a widely used industrial mineral owing to its unique ability to give up and retain water of crystallization at moderate temperatures. To understand the position of gypsum and anhydrite in the present and future world economy, it is appropriate to examine their uses, the structure of the industries and the properties of the materials.

1. Uses

The majority of uses of gypsum are based on the reversible reaction

$$CaSO_4 \cdot 2H_2O \rightleftharpoons CaSO_4 \cdot \tfrac{1}{2}H_2O + 1\tfrac{1}{2}H_2O$$

and gypsum's modest solubility of 0.24 g per 100 cm^3 water at 0 °C. Anhydrite reacts exothermically with water to form gypsum, but, except for a few specialized uses, the reaction rate is too slow to be of practical value.

Stucco, the hemihydrate of calcium sulfate ($CaSO_4 \cdot \tfrac{1}{2}H_2O$), is also known as plaster of paris. When it is mixed with water and various additives it forms a fluid that can be molded or shaped into various products. With time the stucco and water recrystallize to form gypsum, a process known as setting. When gypsum is heated, the water of crystallization that is driven off provides fireproofing, a desirable property for construction uses. Building materials is the main area of application of gypsum.

Gypsum panels and plaster are major products. Paper-faced gypsum panels or wallboard, invented about 1890, is the main gypsum-based product in countries where building systems are based on lumber or formed-metal alternatives. Major plaster usage dates from the 1870s when set-control additives were discovered. In countries where building partitions are primarily masonry, plaster is the common gypsum-based building material, competing with cement. Plaster is applied directly to masonry walls, but is also used for decorative ceiling and wall panels.

In cement, gypsum and anhydrite act as set-controlling agents; about 5% of the finished weight of cement is gypsum or anhydrite ground with the cement clinker.

Landplaster is natural gypsum used as a soil amendment, supplying calcium and sulfur. Calcium replaces sodium in bentonitic soils, permanently swelling the clay mineral lattice and making the soil amenable to agriculture.

Although specialty plasters and fillers are a minor use of gypsum in terms of volume, they are an important application. Specialty plasters include molding plasters and castings, including medical and dental plasters, oil-well cements and absorbents. High-purity gypsum is used as a calcium supplement in food and for fillers, filters, glass manufacture and coal-mine rock dust.

In European coal mines, some use is made of anhydrite to build mine pillars with the help of accelerators to convert anhydrite to gypsum. Small amounts of alabaster gypsum are used for artistic carvings and other *objets d'art*. Gypsum or anhydrite can also be used as a raw material for the manufacture of lime, sulfur, sulfuric acid and cement. These processes have high energy requirements and this factor together with wide variations in the sulfur market have resulted in a decrease of this use for gypsum and anhydrite.

2. *Extraction Methods*

Most of the world's gypsum is produced by quarrying, or surface mining, owing to the fact that most gypsum deposits are the result of near-surface hydration of anhydrite. Underground mines are developed in regions where groundwater has dissolved outcrop gypsum or glaciation has covered it and quarrying is not economic. Most underground mines are room-and-pillar construction with extraction ratios between 70% and 85% and pillar sizes dependent on overburden load and the strength of the roof. Local geology dictates mining height. Roof bolting is commonly practised and both quarries and mines are highly mechanized in Europe, Canada, the USA and other developed countries.

In areas where labor is cheap and abundant, contrasting extraction styles are common; local markets are usually served by small operations with minimal equipment and export markets are supplied by highly mechanized quarries. Production varies widely between 1 t and 100 t per person-day.

Gypsum rock is crushed and generally screened to a size suitable for calcining. This size depends on the calcining method, 90% under No. 100 screen to 15 cm maximum size is usual. Ground gypsum, ready for calcining, is also called landplaster. After crushing, impurities are sometimes removed by screening or hand picking, but mechanical beneficiation to upgrade gypsum is generally not practised in preference to selective mining or quarrying.

3. *Manufacturing Technology*

At atmospheric pressure the calcining of gypsum begins at 43 °C; formation of the beta hemihydrate, stucco, is essentially complete at about 120 °C. Calcining with steam produces alpha hemihydrate. These hemihydrates are only two phases of the system $CaO–SO_3–H_2O$. Other phases, soluble anhydrite and insoluble anhydrite, along with alpha hemihydrate, are used for specialty products and plasters where a high-strength, hard, dense material is desirable.

Uncalcined gypsum uses are based on gypsum purity, lack of deleterious impurities and particle size. Processing is usually limited to crushing, grinding, screening or air separation.

Calcined gypsum is made by several methods. The most common method is in a kettle, an enclosed pot-like vessel with a central shaft for stirring the landplaster while the water boils off. Other calcining methods include: hollow flights, a hollow-core auger-like device that has hot oil circulating through the core; immersed combustion conical kettles; and rotary, beehive and Moorish kilns.

Moorish kilns remain unchanged from when they were first used in the 1400s or earlier. They are simple half ellipsoidal (long axis being vertical) hillside excavations lined with coarse noncalcining rocks. An oval doorway allows entry to stack coarse rock, introduce fuel and remove calcined rock, which is then ground into plaster. Stucco produced from kettles generally does not need regrinding before use in plaster or wallboard; however, regrinding stucco for plaster will produce a denser plaster.

Setting or rehydration time of plaster and stucco is controlled by additives. Organic compounds and mineral salts will increase and decrease setting times, respectively. In less technologically developed countries where setting control additives are generally not employed, only as much plaster as can be used within the natural setting time of 15 to 30 min is mixed at any one time.

Wallboard is made by creating a sandwich of two continuous sheets of paper with a layer of stucco–water slurry between, laid out continuously on a conveyor belt; the wallboard thickness is controlled by using a master roll or forming plate. The stucco slurry rehydrates as the newly formed board travels on the belt line away from the stucco–water slurry mixer. Retarders and accelerators are used so that the setting time is closely controlled and generally ranges between three and seven minutes to accommodate boardline speeds in the range 20–100 m min^{-1}. Wallboard weight is an important market factor related to gypsum purity.

4. Geology

4.1 Mineralogy

The mineralogy of the natural calcium-sulfate minerals is relatively simple, consisting of three minerals: anhydrite, bassanite and gypsum. The mineral species that exists at any particular location is dependent on pressure and availability of water, which is itself related to a wide range of variables, including pressure, temperature and other constituent mineral salts.

Gypsum is the common calcium-sulfate mineral in the near-surface environment. The common rock gypsum is usually easily distinguished from other minerals by its low hardness of 2 on Mohs scale, low specific gravity of 2.2–2.4, and its ability to give up water easily. Selenite is the very coarsely crystalline variety; alabaster is compact and very finely crystalline; and satin spar is a variety that crystallizes parallel to an elongated *c* axis and is generally found as vein-type fracture filling. Gypsum is commonly white, pink or tan. Gypsum is an impure variety of gypsum deposit which is gray or brown in color, depending in part on impurities. Anhydrite is harder, rating 3 to 3.5 on Mohs scale, and heavier than gypsum with a specific gravity of 2.7–3.0, and it is commonly brown, blue, gray or tan. Bassanite, $CaSO_4.\frac{1}{2}H_2O$, is distinguishable only by x-ray diffraction and occurs in nature only in trace amounts.

4.2 Deposits

In the classical evaporite precipitation sequence, gypsum precipitates from saturated brines after calcium carbonate and before halite. Owing to the fact that concentrated brines are widely distributed on the earth's surface, depositional environments for gypsum are numerous and consist of playa lakes, brine pans, sabkhas and enclosed or restricted marine basins. Most gypsum deposits are depositionally associated with a basin or basin margin where circulation of seawater is restricted. Studies of modern deposits of anhydrite and gypsum on sabkhas along the Trucial Coast of the Sheikdom of Abu Dhabi have been useful in the understanding of ancient depositional environments. An example of ancient sabkha deposits in which anhydrite and gypsum, but few other evaporites, occur is the Illinois Basin. In contrast, the Michigan Basin contains a nearly complete suite of evaporites and has anhydrite commingled with halite. Lithologic textures and sedimentary structures of sabkhas, typified by nodular structures, allow definition of the depositional environment of many gypsum deposits.

A class of deposits not extensively used for the gypsum itself, but economically important, is salt-dome caps. As salt goes into solution in groundwater near the top of the dome, residual anhydrite collects and is converted to gypsum. Bacterial activity based on hydrocarbons results in the gypsum being converted to calcium carbonate and native sulfur. The sulfur-reducing bacteria utilize the lighter hydrocarbons, preferentially resulting in heavy oil and asphaltic accumulations being associated with salt domes that contain sulfur deposits. Gypsum deposits are associated with hydrocarbons in many deposits.

Similarly, dissolution of salt in bedded deposits can result in gypsum deposits that have characteristics of both shallow- and deep-water deposits, making the original depositional environment difficult to determine.

Solution of existing gypsum deposits by groundwater and subsequent deposition by evaporation results in gypsite deposits. Gypsite usually contains considerable quantities, up to 50%, of the original near-surface materials.

Commercial gypsum deposits are found in sediments of Ordovician age and younger. Owing to the relatively easy destruction of gypsum deposits by solution, Cambrian and older rocks contain few deposits and no known commercially exploited ones. The geological age and location of commercial gypsum deposits are given in Table 1.

Regardless of the original deposition, most actively exploited gypsum deposits are the result of hydration of anhydrite as it becomes exposed to groundwater. Exceptions are found in arid tropical or subtropical climates where Tertiary-age gypsum has not been buried deeply enough to be converted to anhydrite.

As hydration of anhydrite occurs, impurities that are contained in the anhydrite lattice and associated with anhydrite surfaces are rejected and not included in the gypsum crystal lattice. The first product of hydration of anhydrite is likely to be bassanite, which is metastable, and hydrates to gypsum if water is available.

In many places gypsum associated with anhydrite is very hard and dense, but recrystallization appears

Table 1
Geological age and location of main commercial sources of gypsum deposits

Geological age	Area
Tertiary	California, Caribbean, Europe, Mexico, Nevada
Jurassic	Arkansas, Colorado, Europe, Iowa, Nevada, Utah
Permian	Kansas, New Mexico, Oklahoma, Texas
Carboniferous	Europe, Indiana, Nova Scotia, Virginia
Devonian	British Columbia, Iowa, Manitoba
Silurian	Manitoba, New York, Ohio, Ontario

to continue in many deposits, resulting in coarsely crystallized, nearly selenitic, gypsum. Regardless of whether or not the originally deposited calcium sulfate mineral was anhydrite or gypsum, as gypsum becomes buried, it dehydrates to anhydrite at pressures and temperatures approximating those at 1300 m below the land surface.

4.3 Impurities

Impurities are important considerations in the utilization of gypsum deposits. Deposits that originate in low-energy basins are usually very pure, having less than 5% carbonate and fine clastics as impurities. Deposits that originate in sabkha environments also can be pure, but they commonly contain materials that were part of the original sediment of the coastal plain. Carbonates, mainly dolomite, are very common impurities. Clastic impurities are generally little affected by brines, with the exception of smectites which may be altered from sodium-based clays to calcium-based clays.

Soluble salts that may be deposited in small amounts include other sulfates, especially those of potassium, sodium and magnesium, and chlorides. Soluble salts are especially important in gypsum and anhydrite manufacturing technology because they affect water vapor pressure, important in calcining, and paper bonding to the gypsum core in wallboard manufacture.

4.4 Prospecting

Gypsum deposits are relatively easy to find when they crop out on the earth's surface. Blind deposits, those that do not crop out, are generally discovered by core drilling in favorable areas. Geological and hydrological studies, literature searches and examination of existing drill cuttings and cores in repositories are among the methods used to find gypsum deposits. Geophysical methods have little applicability. Gypsum is a low-cost bulk material that has place value; however, a more general rule of gypsum value is that the lowest production cost in the market area has the greatest value. Thus, the value of a gypsum deposit is a combination of place value and other market factors.

4.5 Synthetic Gypsum

Synthetic gypsum is a term applied to gypsum that is the by-product of another process. Several chemical processes result in synthetic gypsum, the most common ones being the manufacture of phosphate fertilizers, acetic acid and titanium dioxide pigments. Gypsum is also a common product of acid neutralization; for example, pickling liquor. Flue-gas desulfurization at power plants produces large quantities of gypsum, second only to phosphogypsum resulting from fertilizer manufacture. Desulfurization gypsum generally contains soluble salts from the original coal: fly-ash, if not collected separately; calcium sulfite, a result of incomplete oxidation; and unreacted calcium carbonate or other desulfurizing agent. Phosphogypsum commonly is very fine, contains a high moisture content and, owing to decay of the radium in the original phosphate rock, gives up radon daughter products.

The amount of synthetic gypsum produced worldwide is approximately twice the amount of natural gypsum that is mined or quarried. Usage of synthetic gypsum, however, has been slow to increase for the reasons below.

(a) Most synthetic gypsums are of lower quality or contain deleterious impurities that limit their utility. The cost to process synthetic gypsum, especially phosphogypsum and desulfurization gypsum, is usually prohibitive.

(b) Sources of synthetic gypsum are not often close to preexisting manufacturing facilities that use gypsum. The manufacturing facilities usually are capital intensive and not often constructed or moved.

(c) Most gypsum-producing mines and quarries are captive operations owned by the manufacturing companies which are sensitive to raw-material supply.

(d) Many synthetic gypsums are not amenable to the material handling systems that are present at existing operations, resulting in prohibitive cost penalties if used. High moisture contents of up to 25% and fine particle size are examples of this problem.

5. Industry

5.1 Structure

In most western countries the individual markets for manufactured gypsum products are generally dominated by three or four companies. One example is Canada where over 95% of all gypsum products are manufactured by Domtar Gypsum Company, Canadian Gypsum Company and Westroc Gypsum Company.

In the USA the two largest manufacturers, United States Gypsum Company and National Gypsum Company, satisfy over half of the market demand, and the next three largest competitors, Domtar Gypsum, Celotex and Georgia–Pacific, supply approximately 35% more. The remaining 10–15% of the manufactured product market is split among ten, one- and two-plant, operating companies. The exception to a very competitive industry is in the UK where British Plasterboard Ltd is a regulated monopoly. In countries with a gypsum industry based on plaster usage, industry is structured to localized supplies and usage, with two or more producers commonly sharing the local market.

Gypsum products are heavy and can normally stand long-distance transportation costs only in times of good markets. Recent building markets in the USA have brought about some changes in product distribution that would have been uneconomic only a few years ago. Examples of such changes are wallboard manufactured in Montana being shipped by truck to the East Coast markets and Norwegian wallboard being imported into Florida.

Most leading North American companies maintain product development facilities to improve product quality and manufacturing techniques, and have technical agreements with European counterparts.

Raw gypsum markets in the USA are very different from the markets of the manufactured products. The major producers of agricultural gypsum and cement rock include several producers which have large local markets and no facilities for making wallboard.

5.2 *Distribution of Operations*

Manufactured gypsum products are heavy and so many gypsum-product operations are located at deposits that are controlled by the producing companies. Historic major gypsum districts in the USA and Canada are in upstate New York, northern Ohio, Michigan, Iowa, Ontario and Nova Scotia. Of these, Michigan, Iowa, Ontario and Nova Scotia are still leading producing areas. Major producing states in the USA are Texas, California, Michigan, Iowa, Nevada, Indiana, Oklahoma, Arizona, Ohio, New York, Colorado, Arkansas, Kansas and Virginia. There are no major gypsum deposits on any coasts of the USA. The three coasts, Atlantic, Gulf and Pacific, are supplied to some extent from interior deposits, but most of the gypsum in coastal manufacturing facilities is imported from Canada or Mexico.

In Canada gypsum is produced in Newfoundland, Nova Scotia, Ontario, Manitoba and British Columbia. Nova Scotia is the major producer, supplying both internal Canadian markets and much of the East Coast of the USA.

In Mexico, gypsum is common and occurs in many states. Markets in the gulf coastal plain, which lacks gypsum deposits, are supplied by internal deposits.

Most European countries have a good distribution of deposits with the exception of the Scandinavian countries, which have none. The gypsum deposits of the Paris Basin and the Midlands of England are well-known historic producers that continue to be major producing areas. Spain is believed to have the largest reserves of high-purity gypsum in Europe.

The Middle Eastern countries have enormous reserves, but they are mostly undeveloped. Australia has large reserves as does Thailand. Japan's gypsum industry is based on synthetic gypsum as it has no natural gypsum deposits.

5.3 *World Production and Trade*

The USA is the largest producer and consumer of gypsum, producing about 15 million tonnes and consuming an additional 7 to 8 million tonnes of imported rock. Of the gypsum imported into the USA, 4 to 5 million tonnes is supplied by Canada, mostly from Nova Scotia, about 2 million tonnes is from Mexico and approximately one million tonnes is from Spain. Major producers of gypsum are shown in Table 2.

Major importers are the USA, Japan, the Scandinavian countries and Taiwan. The main exporters are Canada, Spain, Australia and Thailand. Many countries export smaller amounts, a few hundred thousand tonnes or less, and many countries import roughly the same quantities. Most of this smaller quantity trade is for Portland cement and plaster use.

An important change in the world trade of gypsum has come about since the late 1970s with the availability of low-cost high-purity gypsum supplies from Spain. This gypsum enjoys a very low-cost freight rate, returning to North America in ships that had carried grain or other bulk cargo to Europe. Spanish gypsum now dominates the cement-rock business on the East and Gulf Coasts of the USA, and a wallboard plant being constructed in the New York area will use Spanish gypsum as a raw material.

Monetary devaluation is also a strong factor in gypsum trade. In the late 1980s, gypsum from Mexico was delivered to ports in the USA at a lower dollar cost per tonne than at any time since 1960, and during that time devaluation of the Spanish

Table 2
Major producers of gypsum in 1986

Country	Amount produced (10^6 t)
Australia	1.2
Canada	8.9
People's Republic of China	5.0
UK	3.1
France	5.5
FRG	2.3
Italy	1.4
Iran	4.8
Japan	6.2
Mexico	3.2
Poland	1.3
Romania	1.6
USSR	4.9
Spain	5.4
Thailand	1.4
USA	14.8

Source: US Bureau of Mines

peseta has roughly paralleled the devaluation of the dollar.

5.4 Industry Changes

Over the long term, gypsum usage is likely to continue growing at a few percent annually. Business cycles that affect the gypsum industries result in increases of as much as 50% of the low production figures. Because of the large investment required to build most gypsum facilities, and other factors, the break-even capacity of many gypsum operations is 75% to 80% of capacity; profitability is achieved only when operating at relatively high operating capacities.

In the USA, gypsum usage has historically been related to new home construction. Commercial construction, whose building cycle is different from the housing cycle, an enlarging repair and remodelling market and the proliferation of home mortgage financing has greatly changed the historic building cycle, resulting in levelling out of gypsum usage. Future annual changes in gypsum usage are likely to be smaller than in the past as a result of these moderating factors.

Manufacturing technology will continue to improve, resulting in lighter-weight wallboard and higher boardline speeds. Board weights of 8.54 kg m^{-2} per 1.27 cm thick board are now common compared with 9.3–9.8 kg m^{-2} per 1.27 cm thick board in the early 1960s. Wallboard imports into the USA are dependent on monetary exchange rates and are not likely to grow significantly. Wallboard usage is likely to grow slowly in countries now using gypsum plaster as acceptance of metal framing and labor costs increase. Use of synthetic gypsum will grow and displace some natural gypsums as better quality synthetic gypsum becomes available, historic gypsum deposits become depleted and disposal of synthetic gypsum becomes increasingly problematic.

Bibliography

Appleyard F C 1983 Construction materials, gypsum and anhydrite. In: LeFond S J (ed.) 1983 *Industrial Minerals and Rocks*, 5th edn. American Institute of Mining, Metallurgical and Petroleum Engineers, New York, pp. 183–97

Appleyard F C 1983 Gypsum and anhydrite. In: LeFond S J (ed.) 1983 *Industrial Minerals and Rocks*, 5th edn. American Institute of Mining, Metallurgical and Petroleum Engineers, New York, pp. 775–92

Kelly K K et al. 1941 *Thermodynamic Properties of Gypsum and its Dehydration Products*, Technical Paper 625. US Bureau of Mines, Washington, DC

Kinsman D J J 1969 Modes of formation, sedimentary associations and diagenetic features of shallow water and supratidal evaporites. *Am. Assoc. Pet. Geol. Bull.* 53: 830–40

MacDonald G F J 1953 Anhydrite–gypsum equilibrium relations. *Am. J. Sci.* 251: 883–98

US Bureau of Mines 1985 *Mineral Facts and Problems*, US Bureau of Mines Bulletin 675. USBM, Washington, DC

D. B. Jorgensen
[US Gypsum Co., Chicago, Illinois, USA]

H

Health Hazards in Wood Processing

Wood is a very versatile material that can present risks to the health of woodworkers during processing. Woodworkers include the following occupations: saw-mill workers, loggers, furniture makers and carpenters. This article examines these health risks, including known and presumed causes of the conditions linked with woodworking. Several acute and chronic reactions to wood-dust exposure have been recorded in the medical literature. Acute conditions include dermatitis (skin eruptions) and respiratory ailments such as asthma or pneumonites. Chronic disease responses include long-term respiratory conditions, impaired nasal clearance and fatal cancers. Of particular concern is the very high risk of nasal cancer among European furniture makers. Monitoring the working environment for elevated levels of wood dust is an important means of preventing some of the health hazards described here.

1. Dermatitis and Other Systemic Effects

Exposure to irritating woods can produce lesions on the skin, irritation of mucosal membranes and other clinical reactions; responses that have been described anecdotally for centuries but only recently annotated clinically. These responses are initiated by reaction to sawdust, by mechanical actions of hairs and bristles or by chemical irritation or sensitization.

1.1 Irritant Dermatitis

Skin lesions are produced by contact with the sap or powders in the bark of certain tropical woods. These trees contain irritating chemicals that cause reddening and blistering of exposed skin of loggers and timber handlers. The irritant dermatitis may be exacerbated by high temperature, humidity and a lack of protective clothes or bathing facilities; initial lesions may be accompanied by secondary infections and other skin eruptions. Table 1 lists some of the irritating woods and the endogenous chemicals that produce skin lesions.

1.2 Atopic or Sensitizing Dermatitis

Atopic or sensitizing dermatitis arises when the body produces antibodies to the extracts of heartwood, and with repeated contact, a dermal reaction (including eruptions, redness, itching and scaliness) to the extract usually affecting hands, forearms, face, eyes, neck, scalp and the groin is produced. To differentiate irritant from atopic dermatitis, a physician (often a dermatologist) determines where

Table 1
Toxic substances found in woods (adapted from Woods and Calnan (1976), and Hausen (1981))

Substance	Woods	Organs affected	Comment
Alkaloids and glycosides	Pigeon plum, greenheart, mesquite, ferro, baralocus, wenge, quatambu, Ceylon satinwood, avodire, red peroba, Knysna boxwood, opepe, aburo	Skin	Many structures and compounds are yet undetermined
Saponins	Satine, albiza, vinhaticco, black bean, mahua, makore	Skin, mucous membranes	Large quantities in satine; bassic acid from makoresaponin used to make poison arrows
Phenols, catecols	Poison ivy, grevillea, ginkgo (maidenhair)	Skin	
Quinones	Lapacho, peroba, teak, angelim, partridge, tagayoson, walnut shells, ebony, mansonia, rosewood, purpleheart (possibly yellow poplar, sucupira, cocuswood, quassia, salmwood)	Skin, eyes	Mansonia and rosewood contain dalbergiones; strong sensitizers
Strilbenes	Iroko, Peru balsam	Skin, respiratory tract	
Terpenes	Pines, latex from *Euphorbia* spp.	Skin	Liverworts growing on bark account for woodcutters' eczema
Furocoumarins	Satinwoods, Bergamot orange, latex from figs	Skin	Furocoumarins are sensitive to light

Table 2
Pulmonary symptoms and effects from reactant woods

Observed effects	Woods
Asthma and shortness of breath	Redwood (*Sequoia sempervirens*)
	Cedar of Lebanon (*Cedrus libani*)
	Iroko (*Chlorophora excelsa*)
	Abiruana (*Lucuma* spp.)
	Zebrawood (*Microberlinia* spp.)
	Tanganyika aningre (*Aningeria* spp.)
	Central American walnut (*Juglans olanchana*)
	African maple (*Triplochiton scleroxylon*)
	Western red cedar (*Thuja plicata*)
Decline in FEV_1 or FVC	Redwood (*Sequoia sempervirens*)
	Oak
	Mahogany
	Cedar
	Iroko (*Chlorophora excelsa*)
	Abiruana (*Lucuma* spp.)
	Ramin (*Gonystylus bancanus*)
	Zebrawood (*Microberlinia* spp.)
	Teak (*Tectonia grandis*)
	Okume (*Aucoumea klaineana*)
	Tanganyika aningre (*Aningeria* spp.)
	Central American walnut (*Juglans olanchana*)
	African maple (*Triplochiton scleroxylon*)
	Western red cedar (*Thuja plicata*)

the patient's blood or skin shows a characteristic reaction to an extract of the wood (Woods and Calnan 1976). Both types of dermatitis may have a latent period lasting from hours to two weeks after first (or repeated) contact. Some more common tropical woods producing either sensitizing or irritant dermatitis include African mahagony (*Khaya ivorensis*), rosewoods (*Machaerium* spp and *Dalbergia* spp.), cocobolo (*Dalbergia retusa*), iroko (*Chlorophora excelsa*), teak (*Tectonia grandis*) and mansonia (*Mansonia altissima*). The US National Institute for Occupational Safety and Health (NIOSH) estimates that 1% of workers may develop skin sensitization to sawdust of American woods (NIOSH 1987).

Occasionally, hairs or bristles from tree bark, leaves or other parts of the tree can produce skin irritations among loggers. Another condition, called "woodcutters' eczema" or "wood poisoning," can be produced by skin contact with epiphytes, lichens or liverworts growing on tree bark or surrounding shrubs. A common cause of this lesion is the leafy liverwort, *Frullania*, whose sesquiterpene lactones have been found to be skin sensitizers. Research has also shown that furocoumarins, which become skin sensitizers when exposed to sunlight, may be common in satinwoods (*Fagara flava* and *Chloroxylon swietenia*), Bergamot orange (*Citrus aurantium* var. *bergamia*), and in the latex of fig trees (*Ficus carica*).

2. *Acute Respiratory Ailments*

Exposure to wood dust can produce two acute conditions, either allergy to the wood (or its extracts) or pneumonites caused by fungal spores in the wood. Since the symptoms are similar, patients must be tested by skin patching or by inhalation challenge tests to determine whether they react to the wood or to the spores.

2.1 Woodworkers' Asthma

The clinical definition of woodworkers' asthma is the immediate or delayed hypersensitivity reaction among workers when exposed to the wood dust from exotic or hardwood lumber. After some period without exposure, the constricted breathing returns to normal. Confirmation of the reactivity of the worker to the wood dust is usually obtained by inhalation-testing the patient with an extract of the dust or the dust itself. If the woodworker responds to the dust, a decline in the patient's forced vital capacity (FVC) or forced expiratory volume in one second (FEV_1) confirms reactivity to the wood. Skin patch tests for antibody reactions to specific woods often show little correlation with the suspect woods; declines in either FEV_1 or FVC are a much clearer means to establish individual reactivity. Table 2 shows the symptoms and/or declines in pulmonary function associated with specific woods. Woodworkers' reactivity to some timbers has been reported only in case studies and no information was presented regarding the smoking status of the patients. Some of the more common woods include redwood (*Sequoia sempervirens*), teak (*Tectonia grandis*), oak, mahogany and cedar (Goldsmith and Shy 1988).

Western red cedar or Canadian red cedar asthma is the allergic condition that has been studied most thoroughly by clinicians and epidemiologists. This condition has been confirmed to be caused by exposure to sawdust of western red cedar (*Thuja plicata*). The clinical picture is as follows: woodworkers without familial or personal history of asthma or allergy suffer asthma symptoms when western red cedar is machined, although they work without reacting to other timbers. Initial complaints are of nasal and eye irritation, later becoming nasal obstruction with cough. The symptoms increase in severity with loss of exercise tolerance, development of breathlessness, wheezing and nocturnal cough including phlegm production. These symptoms become chronic and may persist for weeks after cessation of exposure to red cedar. Western red cedar asthma affects both smokers and nonsmokers alike. Bronchial challenge tests with red cedar dust or extracts of the dust produce immediate, delayed and dual hypersensitivity reactions, including loss of up to 50% of normal FEV_1 or FVC. Dr Moira Chan-Yeung from the University of British Columbia showed that Western red cedar asthma is the specific result of pulmonary response to plicatic acid, the largest fraction in the nonvolatile extract of the timber (Chan-Yeung et al. 1973). Plicatic acid has also been implicated in the cause of asthma from Eastern white cedar (*Thuja occidentalis*) (NIOSH 1987). Chan-Yeung's work suggests that approximately 5% of mill workers handling Western red cedar have red cedar asthma (summarized in Goldsmith and Shy 1988 and NIOSH 1987).

2.2 Pneumonites and Wood Dusts

Some timbers, sawdust and wood bark may be come contaminated with mold and fungal spores and produce allergic alveolitis similar to woodworkers' asthma. The symptoms and clinical signs include malaise, sweating, chills, fever, loss of appetite, breathlessness and chest tightness. In addition to positive dermal and pulmonary reactions to the spores there may be x-ray evidence of alveolitis, which is absent in cases of woodworkers' asthma. Some of the identified pneumonites and their assocaited woods are shown in Table 3.

Table 3
Pneumonites and wood dusts

Type of wood	Possible agent(s)
Maple (bark)	*Cryptostroma corticale*
	Saccharomonospora virdis
Cork (suberosis)	*Penicillium frequentans*
Wood chips or pulp (wood trimmers' disease)	*Aspergillus fumigatus*
	Thermoactinomyces vulgaris
	Alternaria
Redwood dust	*Graphium*
	Pullalaria

3. Chronic Health Risks

There are some case reports of nonmalignant pulmonary disease among occupations having wood-dust exposures (NIOSH 1987). However, the greatest concern in the medical literature has been on premalignant and malignant lesions in the nose.

3.1 Premalignant Nasal Conditions

As European wood-furniture workers have been known to be at very high risk from nasal cancer since the 1960s (IARC 1981), several epidemiological studies have sought to elucidate the stages of premalignancy. Goldsmith and Shy (1988) reviewed the studies of nasal symptoms among European woodworkers and showed the following: British and Danish furniture workers had impaired nasal clearance, including greater symptoms of rhinitis, middle-ear inflammation, sinusitis and bloody noses. Tissue samples from the middle nasal turbinates (where nasal tumors arise) of furniture workers from Sweden and Norway showed that they had greater dysplasia and metaplasia than controls who worked in other occupations. Furthermore, among nasal cancer patients who were woodworkers, researchers found similar irregular tissues adjacent to the tumors. Thus, it appears that, in parallel with impaired nasal clearance, there is a sequence of cellular events (requiring several decades) from tissue injury to adenocarcinoma of the nasal sinuses.

3.2 Cancer Mortality Risks

Epidemiologists showed in the mid-1960s that nasal cancers (specifically adenocarcinomas of the nasal turbinates) were strongly related to the manufacture of wood furniture in the High Wycombe area of the UK (reviewed by IARC 1981). This relationship has been confirmed in Denmark, Canada, France, the USA, Sweden, Finland, Norway, Italy, the FRG, the GDR, the Netherlands, Belgium, Austria, Switzerland and Australia. Table 4 shows the range of cancer mortality risks for woodworkers by tumor site. (If the observed risk was the same as the expected risk, the value would equal 100.) None of these studies was able to adjust for differences in cigarette smoking habits or other confounding factors. It is possible that the elevated cancer risks for lymphoma, Hodgkin's disease and leukemia may be related to interactive exposure of wood dust and either solvents or paints applied to wood as preservatives or possibly it is related to contact with

Table 4
Cancer mortality risks for woodworkers

Cancer	Sawmills logging	Furniture makers	Carpenters
Stomach	190–98	107–75	128–74
Nasal sinus	886–76	1318	124–81
Lung	138–45	130–67	120–79
Bladder	109–62		160–100
Lymphoma	100–73		113–99
Hodgkin's disease	193–67		162–58
Leukemia	193–67	133–100	134–38

herbicides having potential carcinogenic properties. The International Agency for Research on Cancer evaluated the scientific evidence and concluded the following (IARC 1981):

> There is *sufficient evidence* that nasal adenocarcinomas have been caused by employment in the furniture-making industry. The excess risk occurs mainly among those exposed to wood dust. Although adenocarcinomas predominate, an increased risk of other nasal cancers among furniture workers is also suggested .... No evaluation of the risk of lung cancer is possible .... The epidemiologic data are not sufficient to make a definitive assessment of the carcinogenic risk of employment as a carpenter or joiner. A number of studies, however raise the possibility of an increased risk of Hodgkin's disease .... The evidence suggesting increased risk of lung, bladder, and stomach cancer comes from large population-based occupational mortality statistical studies and is inadequate to allow an evaluation of risks for these tumors.

4. Preventing Disease Risks Among Woodworkers

There are two means of reducing the risk of disease in woodworking industries: good housekeeping and medical surveillance of employees. Good housekeeping means controlling all wood dusts at or below the occupational health standards prevalent in the country where the plant operates. NIOSH (1987) describes the wood dust exposure regulations in force in many countries at the time the summary of the literature was written. It is important to recognize that toxicity of exposure depends on whether the timber is hardwood or softwood, and the specific toxic reaction for each timber, especially imported or exotic varieties. Appropriate medical surveillance includes baseline medical information when a worker begins employment and periodic evaluation of health in order to determine any clinical change in status since the previous contact. Essential in medical surveillance of woodworkers is a thorough examination of the upper respiratory tract, especially the nasal sinuses, and an estimate of the cumulative exposure to wood dust.

Bibliography

Chan-Yeung M, Barton G M, MacLean L, Grzybowski S 1973 Occupational asthma and rhinitis due to Western red cedar (*Thuja plicata*). *Am. Rev. Resp. Dis.* 108: 1094–102

Goldsmith D F, Shy C M 1988 Respiratory health effects from occupational exposure to wood dusts. *Scan. J. Work Environ. Health* 14:1–15

Hausen B M 1981 *Woods Injurious to Human Health*. De Gruyter, Hawthorne, NY

IARC 1981 *Monographs on the Evaluation of the Carcinogenic Risk of Chemicals to Humans*, Vol. 25: *Wood, Leather and some Associated Industries*. World Health Organization International Agency for Research on Cancer, Lyon

NIOSH 1987 *Health Effects of Exposure to Wood Dust: A Summary of the Literature*. U.S. Department of Health and Human Services, National Institute for Occupational Safety and Health

Woods B, Calnan C D 1976 Toxic woods. *Br. J. Dermatol.* 94 (suppl.): 1–97

D. F. Goldsmith
[University of California, Davis, California, USA]

Heart Valve Replacement Materials

The four natural cardiac valves (i.e., the tricuspid, pulmonic, mitral and aortic) allow unobstructed unidirectional blood flow through the heart. These delicate but complex and highly specialized structures can become distorted by diseases (e.g., rheumatic heart disease or infection) or degeneration ("wear and tear") thereby either obstructing forward motion of blood (stenosis) or allowing reverse flow (regurgitation or insufficiency). Damaged valves can contribute to impaired cardiac function, and in many cases, long-term survival is impossible unless the value is repaired or replaced. Since valve repair is usually difficult and often not possible, frequently the only option available is surgical replacement by a prosthesis.

A successful heart valve replacement should

(a) be easily implantable,

(b) evoke no thrombus formation,

(c) be quickly and permanently healed by host tissues,

(d) present minimal resistance to forward flow,

(e) allow insignificant retrograde flow,

(f) not damage cellular or molecular blood elements,

(g) undergo minimal wear or degenerative changes over extended intervals, and

(h) not generate excessive noise or other discomfort to the patient.

No available prosthetic heart valve completely satisfies these criteria.

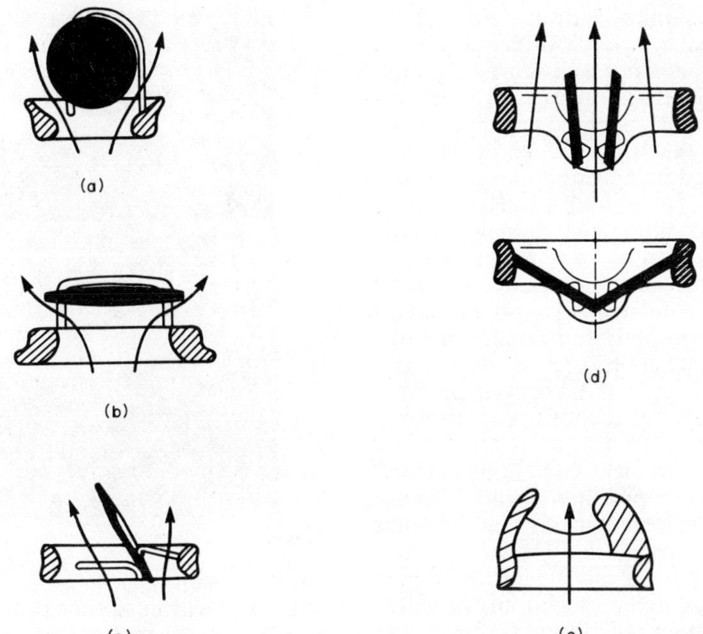

Figure 1
Designs and flow patterns of major types of prosthetic heart valves: (a) caged-ball, (b) caged-disk, (c) tilting-disk, (d) bileaflet tilting-disk, and (e) bioprosthetic (tissue) valves (in part © 1982, Pergamon Press Ltd; reproduced by permission from Schoen F J et al. 1982 Bioengineering aspects of heart valve replacement. *Ann. Biomed. Eng.* 10: 97–128 and in part © 1985, Springer; reproduced by permission from Schoen F J et al. 1985 *Guide to Prosthetic Heart Valves*. Springer, New York, pp. 209–38)

1. *Types of Replacement Valves*

Heart valve prostheses are of two general types—mechanical and tissue. Several models of each are available (see Fig. 1). Mechanical valves generally have three essential components that are fabricated from rigid, nonphysiologic biomaterials:

(a) the flow occluder around both sides of which the blood must flow,

(b) the cagelike superstructure that guides and restricts occluder motion, and

(c) the valve body or base.

The poppet occluder moves passively, responding to pressure and flow changes within the heart. Caged-ball, caged-disk and tilting-disk mechanical valves have been used most widely.

Tissue valves have a flexible trileaflet anatomy, somewhat similar to that of natural valves; they also function passively. Historically, tissue valves have included homografts/allografts (i.e., human cadaver valves), heterografts/xenografts (i.e., porcine (pig) aortic valve or bovine (cow) pericardial tissue) or autografts (fabricated, for example, from the patient's own pulmonary valve, thigh connective tissue (fascia lata) or pericardium). The major advantages of tissue valves relative to mechanical prostheses are their pseudoanatomic central flow and relative freedom from surface-induced thrombus formation, usually without anticoagulation therapy. However, most tissue valves used today are glutaraldehyde-preserved (cross-linked/tanned) and are thereby nonviable altered biological (bioprosthetic) materials, which undergo durability-limiting progressive degenerative processes.

All prostheses, mechanical and tissue alike, have a fabric (Teflon, Dacron or polypropene) sewing ring that surrounds the valve orifice at the base and is used for suturing the device into the surgically prepared implantation site, the anulus. Since the anular anatomy is different for mitral and aortic valves (the valves most frequently diseased) sewing ring configurations vary slightly for prostheses which are to be used at each of these sites. The junction of the sewing ring with the anular tissues is intended to be the major area of direct patient–prosthesis interaction.

2. *Hemodynamic Performance*

All prosthetic valves present some degree of obstruction to forward flow, and thus the effective orifice area of almost all types of devices is less than that of properly functioning natural heart valves. Some functional prosthetic valve orifice areas are sufficiently low that they approach those measured in patients with moderate-to-severe valve stenosis who have not had surgery. Hemodynamic obstruction is accentuated in small sizes, even in the most efficient designs. Caged-ball prostheses are the most obstructive valves, while tilting-disk valves, especially the bileaflet tilting-disk designs, are the least obstructive. The central flow pattern of bioprosthetic heart valves generally yields enhanced hemodynamic function relative to most mechnical prostheses, but even properly functioning bioprostheses can cause significant obstruction, particularly in small sizes, when the bulk of the struts supporting the tissue component is not reduced proportionally to the size of the anulus.

Due in large part to turbulent flow, some destruction of red blood cells by prosthetic heart valves is common. Nevertheless, hemolysis is generally slight and hemolytic anemia is unusual. However, materials degradation or paravalvular leak (i.e., a partial separation of the suture line anchoring the valve, leading to regurgitation of blood around the prosthesis) can cause turbulent blood flow through irregular small spaces, and may precipitate significant hemolysis.

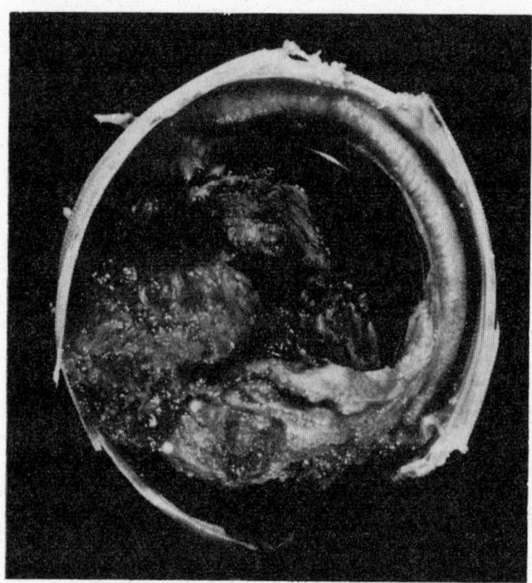

Figure 2
Thrombotic occlusion of tilting-disk prosthesis: the thrombus, initiated in a region of flow stasis, caused near-total occluder immobility (© 1987 W. B. Saunders Inc.; reproduced by permission from Schoen F J 1987 Surgical pathology of removed natural and prosthetic heart valves. *Hum. Pathol.* 18: 558–68)

3. *Complications of Artificial Valves*

3.1 *Thromboembolism*

In most clinical series using mechanical prostheses, the most frequent valve-related complications are thromboembolic problems, including gross thrombosis, causing local occlusion of the prosthesis (Fig. 2) or distant thromboemboli that may compromise blood flow to the heart, brain or kidneys. Thrombosis on artificial valves is potentiated by surface thrombogenicity, blood hypercoagulability and locally static blood flow. Sites of thrombus formation are predicted at areas of turbulence (e.g., the cage apex of caged-ball prostheses and the minor orifice of the outflow region of a tilting-disk prosthesis). The rate of thromboembolic complications with contemporary porcine bioprosthetic valves is approximately 1–2% per patient year, without anticoagulant therapy; this is less frequent than that with most mechanical valves, which is up to 4% per patient year, despite anticoagulant therapy. Thus, tissue valves, with low intrinsic thrombogenicity, are an attractive alternative to mechanical prosthetic valves when anticoagulation is undesirable. Thrombotic occlusion of bioprosthetic valves occurs rarely, and thromboemboli are unusual.

3.2 *Infection*

Infection associated with a valve prosthesis (endocarditis) is a devastating and often fatal complication. Bioprosthetic valve infective endocarditis occurs with approximately the same incidence as that with mechanical valves. However, although endocarditis associated with a mechanical prosthetic valve is localized to the prosthesis–tissue interface at the sewing ring (since the valve biomaterials cannot themselves support the growth of microorganisms) bioprosthetic valve infections can involve, and indeed can be limited to, the cuspal tissue. Cusp-limited infections are probably more readily treated by antibiotic therapy than most prosthetic valve endocarditis.

Infection is one of the most frequent clinically important complications of implanted biomaterials in general, occurring in approximately 1–10% of patients with a variety of prosthetic devices. Several potential mechanisms may be important. Microorganisms can be introduced inadvertently into a patient by supposedly sterile medical devices contaminated during manufacture or implantation. In addition, microorganisms are provided access to the circulation and to deeper tissue by damage to natural infection barriers during implantation surgery or subsequent device function. Moreover, implanted biomaterials and medical devices impede the free migration of inflammatory cells to the infected area and possibly interfere with the local bacteriocidal mechanisms of inflammatory cells.

Thus, the presence of any foreign material in tissues impairs the local resistance to infection, and the frequency and pathology of prosthetic valve endocarditis are almost independent of specific valve materials and design considerations. In contrast, thromboembolic considerations and durability considerations (discussed later) are clearly both materials- and design-dependent.

3.3 Structural Failure

Limitations to valve durability have been an important deterrent to the long-term success of cardiac valve replacement. Among patients who require reoperation for prosthetic valve failure or who have autopsy following prosthesis-associated death, materials degradation is frequently observed. Durability considerations vary widely among mechanical valves and bioprostheses, among different prosthesis configurations, among different models of a specific prosthesis type incorporating different design features and materials, and in some cases for the same model prosthesis placed in the aortic versus the mitral site. Since materials properties play a critical role in valve durability, degenerative dysfunction is discussed in detail below.

4. Degeneration of Mechanical and Tissue Valves

4.1 Mechanical Prostheses

The silicone ball occluder of early caged-ball aortic prostheses absorbed blood lipids, which caused distortion, cracking, embolization of poppet material or abnormal poppet movement, a spectrum of damage known as "ball variance." Mitral ball variance was distinctly less common than aortic, probably due to lower stresses incurred during mitral valve function, and changes in elastomer fabrication in 1964 have virtually eliminated lipid-related ball variance in subsequently implanted valves. Ball variance, although infrequently encountered today, exemplifies the complex interactions among mechanical, chemical and materials factors in degenerative valve failure.

In order to minimize thromboembolism, the cages of some caged-ball valves were covered with cloth to provide a scaffold on which recipient tissue could grow to yield a smooth, natural blood-contacting surface. However, cloth-covered caged-ball valves with metal or polymeric poppets suffered cloth abrasion and fragmentation by the occluder, and fibrous overgrowth with ball entrapment and resultant stenosis. Early tilting-disk designs had polymeric disks in which free rotation of the disk was not possible. These valves failed due to concentrated abrasive wear. Caged-disk valves with plastic disks developed severe disk wear with notching and reduction in diameter. Abrasive wear of mechanical components can interfere with local valve function, or can cause distal migration (embolization) of fragments of material.

Most mechanical heart-valve prostheses currently in use have pyrolytic carbon occluders, and some have both carbon occluders and carbon cage components. Pyrolytic carbon, an advantageous material for construction of heart-valve components owing to its thromboresistance, high strength, wear resistance and ability to be fabricated into a wide variety of shapes, is applied to a preshaped graphite or metallic substrate using a fluidized bed coating process. Tilting-disk designs with pyrolytic carbon disks have good durability, with minimal abrasion of metallic cage components by carbon occluders. The wide use of pyrolytic carbon as an occluder and strut-covering material for mechanical valve prostheses has virtually eliminated abrasive wear as a long-term complication of valve replacement. Nevertheless, catastrophic fractures of both metallic and carbon-coated valve components of some designs have been reported. Fractures are a particularly troublesome feature of mechanical valve prostheses used in circulatory support devices, probably due to the nonphysiological mechanical environment of such valves.

4.2 Bioprostheses

Deterioration of glutaraldehyde-treated porcine aortic valve bioprostheses by calcification or cuspal tearing (primary tissue failure), which causes stenosis or regurgitation, or both, now commonly necessitates reoperation following several years of function. The failure rate in adults is approximately 25% or higher, ten years after implantation, and recent evidence suggests that failure occurs in over 50% of cases at 15 years. Thus, deterioration of bioprosthetic valves is frequent and strongly time dependent; the low rate of valve failure earlier than five years following surgery (1% per year) accelerates to approximately 5% per year thereafter.

Calcification of the cusps is the most important pathologic process contributing to bioprosthetic valve failure. Regurgitation through tears, secondary to calcification, accounts for approximately three-quarters of porcine valve primary tissue failures (Fig. 3). Pure stenosis due to calcific cuspal stiffening and cuspal tears or perforations unrelated to calcification are less frequent. Noncalcific cuspal tears reflect direct mechanical destruction of the valve structure during function, with fraying and disruption of collagen.

Bioprosthetic valve calcification increases with time after implantation, but patients vary widely in their propensity to calcify valves. Calcific deposits accumulate insidiously, and are present in the majority of bioprostheses examined three or more years after implantation. Nevertheless, some patients with bioprostheses functioning for ten years or more have no detectable calcific deposits. Specific factors which control which patients and which valves will have problems are largely unknown, except for age effects. Calcification is more likely to become severe and clinically significant in children and young adults than in older patients. Some data suggest that the increased risk of calcific failure may continue to age 35. The work on Hancock and

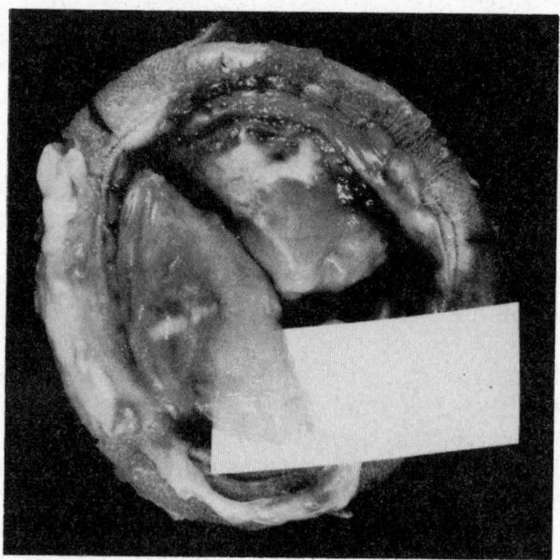

Figure 3
Porcine valve primary tissue failure due to calcification with secondary cuspal tear leading to severe regurgitation (© 1985 W. B. Saunders Inc.; reproduced by permission from Schoen F J and Hobson C E 1985 Anatomic analysis of removed bioprosthetic heart valves: Causes of failure of 33 mechanical valves and 58 bioprostheses, 1980–1983. *Hum. Pathol.* 16: 549–59)

Carpentier-Edwards has shown that there are no apparent differences in durability between these two most popular porcine bioprostheses.

Degenerative intrinsic calcification begins primarily at the cuspal attachments, which are the sites of greatest cuspal mechanical stresses. Mineral deposits are usually related to connective tissue cells and collagen of the transplanted valve (intrinsic calcification). Surface thrombi or vegetations also calcify (extrinsic calcification) but extrinsic mineralization is not usually clinically important. In contrast with the slow accumulation of intrinsic calcific deposits over several years, extrinsic calcification can occur within several days following implantation.

Bioprostheses fabricated from bovine pericardium may have a greater propensity to late primary tissue failure than porcine aortic valves. Calcification of pericardial valves, with or without cuspal tears, and large noncalcific cuspal disruptions are frequent. Cuspal tears often originate at the most peripheral portion of the cusps (i.e., the commissures).

5. *Additional Considerations for Tissue Heart Valves*

The natural aortic valve has optimal hemodynamics (no obstruction or regurgitation) and does not cause areas of localized stress concentration within the tissue, trauma to molecular or formed blood elements or thrombotic complications. The aortic valve achieves the basic requirements of valvular function by means of a highly specialized and complex structure. Although the pressure differential across the closed valve (approximately 80 mm Hg in the aorta, but less than 10 mm Hg in the left ventricle) induces a large load on the leaflets, the fibrous arrangement within the cusps assists in transferring the resultant stresses to the aortic anulus. In particular, aortic valve cusps have highly anisotropic material properties in the plane of the tissue; this reflects an oriented leaflet tissue architecture in which collagenous struts transfer stresses to the anulus.

Pericardial tissue, used as the material for the construction of some bioprosthetic heart valve leaflets, has approximately isotropic elastic properties. Since highly anisotropic elasticity is the major structural property of valve leaflets which promotes uniform distribution of stress during valve function, stresses in pericardial leaflet valves will be directed toward the natural mechanical focal points, the commissures, leading to high stress concentrations which are not encountered in the natural valve. Therefore, it is not surprising that cuspal tearing at the top of the stent posts is a common mode of failure of pericardial and other isotropic tissue valves. Also, repeated loading of pericardium causes dynamic creep of this viscoelastic material, progressively stretching the tissue and leading to permanent leaflet deformation.

The goals of aldehyde pretreatment of bioprosthetic tissue include enhanced material stability and decreased immunological reactivity, with maintained thromboresistance and preserved antimicrobial sterility. Glutaraldehyde is a dialdehyde which has been used since antiquity for the tanning of leather, and more recently as a fixative in the preparation of tissues for electron microscopy. Glutaraldehyde treatment cross-links proteins, especially collagen, the most abundant structural protein of bioprosthetic tissue. Interestingly, glutaraldehyde pretreatment of bioprosthetic tissue also appears to potentiate degenerative calcification. The specific alterations in the functional/mechanical characteristics of the natural aortic valve and other tissues used in bioprostheses by glutaraldehyde cross-linking procedures are controversial. Moreover, tissue mechanical properties may be further altered during valve function, but the exact nature and significance of the changes have yet to be elucidated.

Because of the problems encountered with mechanical prosthetic and bioprosthetic valves, there is considerable enthusiasm for use of sterilized aortic valve homografts derived from human cadavers. Like other tissue valves, homografts have central, unimpeded blood flow. Patients with these valves have almost complete freedom from thromboembolism without anticoagulant therapy. The disadvantages of homograft valves are their relatively more difficult technique of insertion, their limited availability, that their use is restricted to aortic (and not mitral) valve replacement, and that they exhibit a high incidence of late deterioration by tearing. Nevertheless, functional deterioration of the valve due to degeneration occurs slowly and usually can be managed successfully. Moreover, with the development of techniques for frozen storage of such valves to minimize logistical constraints, and novel procedures to use homografts as mitral replacements, it is likely that the clinical use of valve homografts will broaden in the future.

6. *Considerations in Valve Selection*

With all mechanical valve prostheses, lifetime anticoagulation is necessary to minimize the risk of thromboembolic complications. Nevertheless, anticoagulation therapy imposes a significant risk of hemorrhage. Bioprostheses have the important advantage that chronic anticoagulant therapy is generally not required, but their long-term function is often compromised by degeneration. Therefore, the major consideration surgeons use to select the type of prosthesis to implant is the trade-off between risks of thromboembolic complications and hemorrhage related to anticoagulation use with mechanical prostheses (favoring bioprostheses) and the limited durability of bioprostheses (favoring mechnical prostheses). However, extraordinary clinical circumstances may prevail in selected patients. For example, bioprostheses are generally used in young women wishing to become pregnant, in order to avoid chronic anticoagulant drugs, which cross the placental circulation and potentially cause serious problems for the fetus. In contrast, mechanical valves may be preferable in children and young adults, despite the need for anticoagulation, because failure of bioprostheses within several years is almost certain in this population. Moreover, primary tissue degeneration of bioprosthetic valves, the overwhelming mode of valve failure, usually leads to slow progressive development of cardiac dysfunction, which can allow recognition of the problem and elective reoperation. This contrasts with the frequently catastrophic valve failure and subsequent rapid deterioration of patients with mechanical valves who have thromboembolic complications or structural valve failure. The most important advantages and disadvantages of the various types of replacement valves are summarized in Table 1.

7. *Future Directions*

Problems encountered with the clinical use of bioprostheses have stimulated several lines of investigation. Alterations in tissue preparation techniques for bioprosthetic valves and radically altered design configurations are under consideration. Since calcification is the major pathologic feature associated with bioprosthetic valve failure, considerable work is being directed toward elucidation of mechanisms of bioprosthetic valve mineralization and strategies to obviate this problem. The most promising approaches to effective prevention of bioprosthetic valve calcification are empirical and include pretreatment of valve leaflets to remove a structural component which initiates calcification (such as cell membranes) and the use of specific inhibitors of calcification (such as the diphosphonate drugs, which are potent inhibitors of minimal crystal growth used in the therapy of metabolic bone disease). Drugs could be administered either systemically, by valve pretreatment or through sustained local controlled release. The best approach seems to include pretreatment of the valve cusps, either by diphosphonates or some treatment which removes crystal nucleation sites, in conjunction with controlled drug delivery into bioprosthetic valves that have a modified sewing-ring configuration that contains a diphosphonate-containing polymeric matrix.

Although modest development of mechanical poppet prostheses with new designs, particularly those using pyrolytic carbon occluders, has continued, it is unlikely that radical design changes will be forthcoming. Modest modification of tilting-disk configurations has concentrated on enhancing opening to improve hemodynamics and eliminating metallic struts from regions of stasis to reduce thromboembolic risk. Fixed pivot-point bileaflet tilting-disk configurations have achieved wide acceptability over the past several years with generally favorable clinical results. Major developments in the technology of polymeric materials have allowed recent progress in trileaflet flexible prostheses fabricated from polytetrafluoroethylene, polyurethane or silicones to mimic the natural aortic valve anatomy and function. However, long-term durability limitations remain the major obstacle to clinical application.

Table 1
The most important advantages and disadvantages of contemporary valve substitutes

Valve type	Advantages	Disadvantages
Mehanical	Durable	Thrombogenic
Bioprosthetic	Thromboresistant	Propensity to deteriorate
Homograft	Near-normal hemodynamics	Limited availability
	Thromboresistant	Technically demanding insertion
		Late tearing

Bibliography

Ferrans V J, Tomita Y, Hilbert S L, Jones M, Roberts W C 1987 Pathology of bioprosthetic cardiac valves. *Hum. Pathol.* 18: 586–95

Jamieson W R E, Rosado L J, Munro A I, Gerein A N, Burr L H, Miyagishima R T, Janusz M T, Tyers G F O 1988 Carpentier-Edwards standard porcine bioprosthesis: Primary tissue failure (structural valve deterioration) by age groups. *Ann. Thorac. Surg.* 46: 155–62
Matsuki O, Robles A, Gibbs S, Bodnar E, Ross D N 1988 Long-term performance of 555 aortic homografts in the aortic position. *Ann. Thorac. Surg.* 46: 187–91
Milano A D, Bortolotti U, Mazzucco A, Guerra F, Stellin G, Talenti E, Thiene G, Gallucci V 1988 Performance of the Hancock porcine bioprosthesis following aortic valve replacement: Considerations based on a 15-year experience. *Ann. Thorac. Surg.* 46: 216–22
Morse D, Steiner R M, Fernandez J 1985 *Guide to Prosthetic Cardiac Valves*. Springer, New York
Rose A G 1987 *Pathology of Heart Valve Replacement*. MTP Press, Lancaster
Schoen F J 1987a Cardiac valve prostheses: Pathological and bioengineering considerations. *J. Cardiac Surg.* 2: 65–108
Schoen F J 1987b Cardiac valve prostheses: Clinical status and contemporary biomaterials issues. *J. Biomed. Mater. Res.* 21(A1): 91–117
Schoen F J 1989 *Interventional and Surgical Cardiovascular Pathology: Clinical Correlations and Basic Principles*. Saunders, Philadelphia, PA
Schoen F J 1990 Mode of failure and other pathology of mechanical prosthetic and bioprosthetic heart valves. In: Bodnar E, Frater R (eds.) 1990 *Replacement Cardiac Valves*. Pergamon, Oxford
Schoen F J, Fernandez J, Gonzalez-Lavin L, Cernaianu A 1987 Causes of failure and pathologic finding in surgically-removed Ionescu–Shiley standard bovine pericardial heart valve bioprostheses: Emphasis on progressive structural deterioration. *Circulation* 76: 618–27
Schoen F J, Harasaki H, Kim K M, Anderson H C, Levy R J 1988 Biomaterial-associated calcification: Pathology, mechanisms, and strategies for prevention. *J. Biomed. Mater. Res.* 22(A1): 11–36
Schoen F J, Kujovich J L, Levy R J, St John Sutton M 1988 Bioprosthetic heart valve pathology: Clinicopathologic features of valve failure and pathobiology of calcification. *Cardiovasc. Clin.* 18(2): 289–317
Schoen F J, Levy R J, Ratner B D, Lelah M D, Christie G 1990 Materials considerations for improved cardiac valve prostheses. In: Bodnar E, Frater R (eds.) 1990 *Replacement Cardiac Valves*. Pergamon, Oxford
Silver M D, Butany J 1987 Mechanical heart valves: Methods of examination, complications, and modes of failure. *Hum. Pathol.* 18: 577–85

F. J. Schoen
[Brigham and Women's Hospital and Harvard Medical School, Boston, Massachusetts, USA]

Heat-Sensitive Papers

Heat-sensitive papers, commonly called thermal papers, have grown to be a sizable proportion of the specialty papers market. It is important to define what is meant by heat-sensitive papers because a number of different types have been developed, employing a variety of technologies. The four types that can be distinguished are:

(a) type I—iron-stearate-phenolic paper (Clark 1957),

(b) type II—thermal diazo paper (Sato et al. 1984),

(c) type III—thermal paper based on color formers (Sodeyama 1985), and

(d) type IV—thermal paper based on microcapsules containing diazonium salts (Nakamura 1988).

Type I is sold by the 3M Corporation under the trade name ScotchMark. It is manufactured by an old-fashioned technology based on the reaction of iron-stearate with phenolic compounds. It is used in the industrial labels market, for example, in bar codes production where it competes with ink jet, laser and thermal transfer technologies.

Type II is a variation on the well-known diazo-paper still used for drafting in engineering and architecture and it is of minor importance.

Type III is used for telefax, scientific recorders, cash registers, bar code labels and so on. Important manufacturers of type III thermal paper are Jujo, Kanzaki, Mitsubishi, Ricoh and Honshu of Japan; Appleton, Ricoh and Labelon of the USA; and Jointine/Wiggins Teape of the UK. The Japanese provided approximately 80% of world production in 1988. More than 95% of the thermal paper produced worldwide develops a black print on heating to temperatures between 60 °C and 90 °C; a small proportion are colored red or black, depending on the temperature, the remainder are blue.

Type IV is a relatively recent development by the Fuji Photo Film Company of Japan. The advantage of this paper is its indelible print. This is made possible because the reactive diazonium ion is dissolved in a waxy medium and encapsulated ; it can be photolyzed and thus rendered unreactive with the couplers that are also present on the paper. It is only used in speciali⁻ed (niche) markets because of its high cost.

By far the most important of these thermal papers is type III, and its growing importance is documented in Fig. 1 (Anon 1987), and the remainder of this article will therefore refer to this type.

1. Thermal Paper Based on Color Formers

Type III thermal papers were invented in the late 1960s. The basic technology involved was first described by Henry H. Baum, while working for the National Cash Register Company of Dayton, Ohio. In this patent, the thermal reaction of a phenolic compound (e.g., bisphenol A) with crystal violet lactone (a colorless color former) was demonstrated. Following this, formulations to produce thermal paper have become more and more sophisticated, though the basic principle has remained the same.

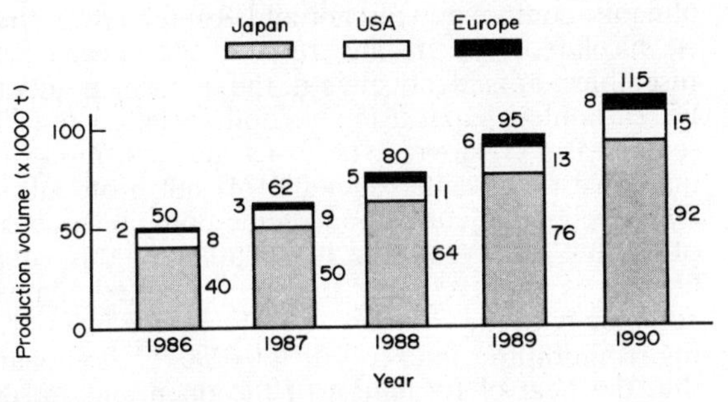

Figure 1
Estimated worldwide production of type III thermal paper 1986–1990

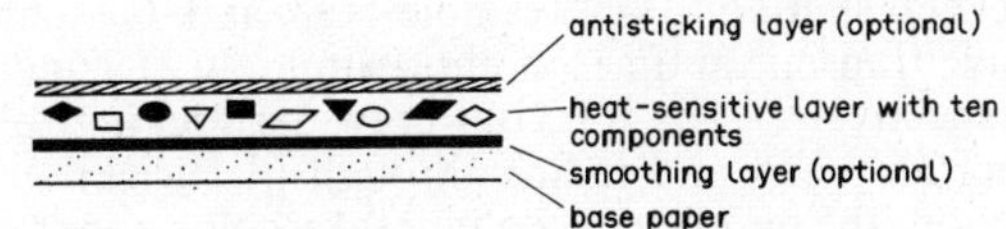

Figure 2
Components of a commercial thermal paper

Figure 2 shows the layers of a commercial thermal paper and Table 1 gives details of the components of the heat-sensitive layer of this paper. All components are usually employed as aqueous dispersions except for the binder, the optical brightener and the dispersant, which are used as aqueous solutions. Thermal papers employing color formers and/or developers as solutions are also known.

In principle only one coat is required in order to produce a simple fax paper. However, the high-performance fax papers G3 and G4 contain a number of different coats. (G stands for group and the higher the number, the faster a page of A4 can

Table 1
Common components of a heat-sensitive layer

Component	Trade name or chemical name	Producer	Function
Color former	PSD-150 ODB-2 S-205	Nippon Soda Hodogaya Yamada	colorant
Developer	bisphenol A = DPP = diphenylol-propane;	Shell, Dow Chem., Rhône Poulenc	reacts with color former to form color
	Paraben = 4-hydroxi, benzylbenzoate	Nagase Europe	
Sensitizer	*p*-benzyl-diphenyl	Nippon Steel Chemical	solvent for color former and developer
Coating pigment	kaolin, $CaCO_3$, talc, SiO_2	Many	keeps thermal print head clean
Binder	polyvinylalcohol	Rhône Poulenc, Wacker, Hoechst, Kuraray	dispersion stabilizer, binds disperse particles to paper
	PVA 105 styrene–acrylicacid copolymers	BASF, others	
Dispersant	Surfynol Suprapal	Air Products BASF	dispersion stabilizer, flow improver
Fluorescent brightener	Blankophor Pinopal Leukophor	Bayer Ciba–Geigy Sandoz	improves whiteness of paper
Antioxidant	Irganox	Ciba–Geigy	improves contrast
Slipping agent	zinc stearate	Akzo Metallgesellschaft Riedel de Haen	provides smooth surface
Wax	Dewahil	BASF	improves whiteness of paper

be transmitted. For example, G3 and G4 papers have transmittal times of approximately 21 s and 9 s, respectively.) The top coat prevents sticking to the thermal head and fading through plasticizer or fat action, the middle coat contains the color-generating layer and the bottom coat is the levelling coat which provides an optimal base for the two previous layers and insulates the base paper from the thermal layer.

The improvements in quality achieved by certain paper manufacturers are best illustrated by the scanning electron micrographs shown in Fig.3. In 1985 the recrystallization of color formers, bisphenol A and sensitizers was a problem whereas in 1988 the new formulations with a top coat prevented this, as can clearly be seen. The top coat also reduces the tendency of the paper surface to stick to the thermal print head. (For a non-Japanese language review see Schüler 1987.)

The essential chemical reaction leading to the dyestuff is the equilibrium between a colorless leucoform (Fig.4a) and the colored open form (Fig.4b) of a fluorane (the color formers commonly used to make thermal papers are summarized in Table 2) or other reactive chemical through the action of a phenolic compound (Fischer and Römer 1909), that is, bisphenol A. In this respect the mechanism resembles, or is identical with, the one which underlies carbonless paper using phenolic-resin-coated CF (coated front) papers. The exact mechanism is still unknown (Saito and Tanaka 1974) but protonation by the phenolic compound seems unlikely because of its low acidity. A puzzling feature of this equilibrium is the fact that it can be shifted almost completely to the colorless-leucoform side by applying temperatures in excess of 150–200 °C. It appears that the heat of formation of the open and closed species in this system are very similar, thus any changes in the system which shift the equilibrium towards the dye-side are highly desirable. By far the most important color formers are the black fluoranes, that is, the structure of Fig. 4a with R^1, R^2 and R^3 as alkyls, and R^4 as chlorine or hydrogen.

2. Manufacture

The color former is dispersed in an approximately 10% aqueous polyvinyl alcohol (or other water soluble polymer) solution with a solids content of 20–

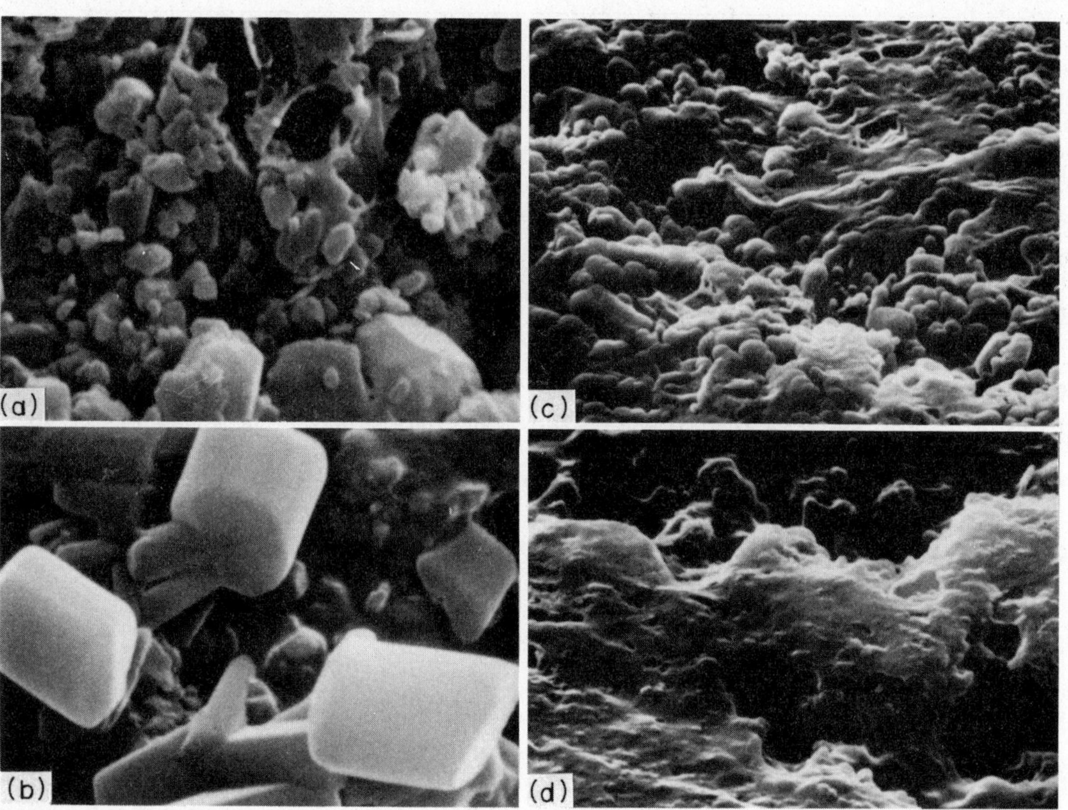

Figure 3
Scanning electron micrographs (×10 000) of thermal papers without a top coat, (a) before and (b) after printing; and with a top coat, (c) before and (d) after printing

Table 2
Fluorane color formers used in black thermal papers

Trade name	Producer	Remarks
ODB-2	Hodogaya (Japan)	no premature reaction, fast development
S 205	Yamada Chem. (Japan)	improved N 102, slower
PSD-150	Nippon Sóda (Japan) Great Lakes Chemical (USA)	similar to S 205
TH 107	Hodogaya (Japan)	no premature reaction, slow development

40%, to a particle size of 1–3 μm, using a ball mill (e.g., Fryma CoBall-Mill MS 32) or other suitable dispersion equipment. Bisphenol A and the coreactant are dispersed in the same way, the former needing more passes through the mill because of its hardness.

Clay, kaolin, $CaCO_3$, amorphous SiO_2, talc or other filler pigments are dispersed in water or ground wet to sizes of the order of 1μm. For faster dispersion, special surfactants are usually added. Special micronized wax is also dispersed in water using surfactants and binders. The aqueous color former, developer, filler, dispersant and additive dispersions are mixed just before the heat sensitive second coat is applied on top of the dried first coat.

It is important to keep the contact time between color former and developer dispersions (ratio one-to-three or higher) as short as possible in order to obtain a high degree of whiteness. The addition of fluorescent brighteners also helps. The heat-sensitive dispersion is then applied to the paper by a so-called airknife at speeds of 200–250 m min^{-1}.

The drying of the paper has to be done at temperatures below 60 °C to prevent premature coloration of the paper. Optionally, a third coat of an aqueous micronized wax can be applied and dried. The final step is calendering with steel rolls applying a well-defined pressure, but care has to be taken since excess pressure leads to a discoloration of the paper due to the chemical reaction taking place prematurely. Such paper is of low quality because of its grayish appearance (background). For continued smooth performance in telefax machines a very smooth surface (Bekk values exceeding 200 s) is very important, because the thermal print head touches the paper during the printing process.

3. Applications

The only applications of heat-sensitive papers are in thermal printers and copiers. The essential component of a modern thermal printer is the thermal print head which is a miniature matrix of heating elements. Each heating element has an approximate size of 0.2 × 0.2 mm and can be switched on and off independently of the other elements. The electric pulse usually lasts several milliseconds, and heats the surface of the element to over 300 °C.

In scientific recorders (e.g., ultraviolet spectrometers), the rate-determining step is the data collection

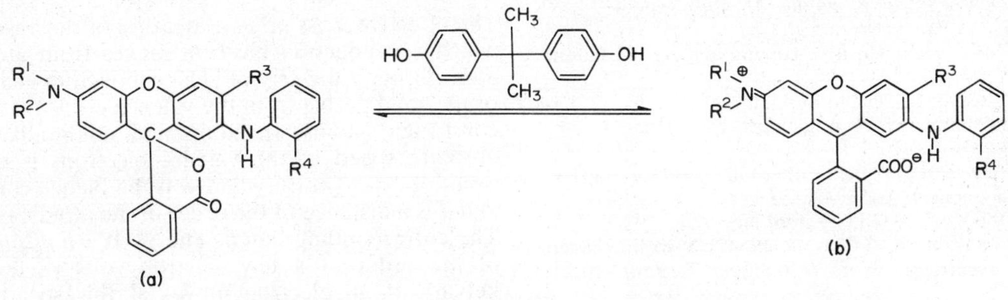

Figure 4
Equilibrium of (a) colorless leucoform and (b) colored open form of a fluorane

and processing and not the printing process, thus slow thermal paper (one coat) can be used. In the selfadhesive label industry for food packaging, use is also made of relatively slow thermal paper with a top coat. Here bar codes (machine readable information) and expiry dates are printed before the label is stuck onto the polythene wrapper of the food package.

The same type of thermal paper is employed in cash registers and in coin operated machines that print parking tickets. Telefax machines have recently become available from Japanese electronic companies at much lower prices (around US$ 600) creating a need for faster fax paper. Since the cost determining factor for long distance messages is the development speed of the thermal paper, fast paper is required to minimize costs. The fastest papers are produced in Japan employing complex multicoat technologies.

4. Conclusion

The great advantage of thermal printing at its present state of development is its low cost and high reliability, with regard to the instrument and the thermal paper. The disadvantages are the rough texture of the paper and the poor fastness properties of the print. Although the light fastness of the black print is good, the white background turns brownish after exposure to sunlight, so lowering the contrast. Some chemicals, for example, oils, solvents, greases, fats and fruit juices, also lower the contrast, thus making the image less legible. However, under normal storage conditions, for example, in a file, the information can be preserved for several years. Thermal printing on heat-sensitive papers will prove to be a durable technology as it fulfills an ever increasing need for fast information exchange around the globe at competitive prices.

Bibliography

Anon 1987 Facsimile drives exploding thermal paper market. *Datek Imaging Supplies Monthly* June 4–7

Baum H H US Patent No. 3,539,375

Clark L C 1957 Heat sensitive copying-paper. US Patent No. 2,813,043

Fischer O, Römer F 1909 Notes on dimethylaniline-phthaleine and similar basic phthaleines. *Ber. Dsch. Chem. Ges.* 42: 2934–5

Nakamura K 1988 UV fixable thermal recording papers. *Kami Pa Gikyoshi* 42(5): 425–32

Saito T, Tanaka D 1974 US Patent No. 3,983,292

Sato H, Sukegawa K, Oobra Y 1984 UV-fixable diazo thermal recording paper. *J. Imaging Technol.* 10(2): 74–9

Schüler D 1987 Heat reactive paper coatings. *Farbe Lack* 93(12): 984–6

Sodeyama H 1985 Thermal paper and its application. *Finat News* 1:

R. Dyllick-Brenzinger
[BASF AG, Ludwigshafen, Rhein, FRG]

Heavy Fermion Metals: Magnetic and Superconducting Properties

Heavy fermion metals are a subclass of the valence fluctuation metals, a class of intermetallic compounds with rare earth or actinide component, which show certain drastic anomalies in their thermal, magnetic and transport properties. In particular, at the Fermi energy of these metals, the effective mass m^* of the conduction electrons is from a few tens to a few thousand times larger than the free electron mass m_0. The heavy fermion subclass is defined at the upper end of this range, at $m^* > 400m_0$.

The large effective masses of the valence fluctuation metals are due to the incipient delocalization of the *f* electrons of the rare earth or actinide component. These *f* electrons are valence electrons, like the outer (*sd*) electrons, but spatially they are closer to the nucleus. Therefore they usually remain localized, even in metals, and then the conduction electrons derive from the outer (*sd*) electrons only. In the heavy fermion metals, the *f* electrons begin to delocalize noticeably for the first time.

Heavy fermion metals were discovered in the mid-1970s. Several years later a few were found to be superconductors below 1K. All previous experience had shown that conventional Bardeen–Cooper–Schrieffer (BCS) superconductivity does not survive the interaction with even minute concentrations of delocalizing *f* electrons. This prospect of a new mechanism for superconductivity has triggered intensive experimental and theoretical work on heavy fermion metals.

1. Effective Mass and Bandwidth

The effective mass m^* is a measure of the ease with which a conduction electron moves from atom to atom through the crystal. The conduction electrons of the solid derive from the valence electrons of the constituent atoms. In atoms and solids alike, the absolute speed of the valence electrons is of the order of $v = h/(m_0 d)$, where h is the Planck constant and d is a distance of the order of the atomic radius. The corresponding kinetic energy $W = h^2/(2m_0 d^2)$ is of the order of a few electron volts (a few 10^4 kelvin). If an electron moves at this speed in a straight line through the solid (e.g., as in copper) its *translational* (effective) mass is close to the free electron value m_0. If, however, it circles at this speed $N \gg 1$ times around a given atom before moving on to the next, its translational (effective) speed and its translational kinetic energy (the "bandwidth") are reduced to $v^* = v/N$ and to $W^* = W/N$, respectively, and its translational (effective) mass is increased to $m^* = Nm_0$. For electrons that stay forever at a given atom $N \rightarrow \infty$ (e.g., those electrons in the completed inner atomic shells) the effective mass is infinite and the bandwidth is zero. One then speaks of localized electrons. In a heavy fermion metal with $m^* = 10^3 m_0$, one has a nearly localized electron with a bandwidth W^* of only a few tens of kelvin. Such metals can be studied experimentally at thermal energies *larger* than the bandwidth; that is, far into the nondegenerate (classical) regime of the Fermi gas, in contrast to the usual metals with $m^* \approx m_0$, which are accessible to the experiment only in the degenerate (quantum) regime, at $T \ll W/k_B \approx 10^4$, where K_B is Boltzmann's constant. The bandwidths W^* of the heavy fermion metals also lie well below the energy of the average phonon; that is, in these metals the effective speed v^* of the heavy electrons is smaller than that of the nuclei, rather than the other way around as is usual. Therefore, the elementary excitations of such a solid can no longer in general be divided into separate electronic and phonic excitations.

2. Valence Fluctuations, Fractional Valence and Lifetime Widths

The delocalization of f electrons in metals is viewed more concretely in terms of a fluctuation of the rare earth or actinide atom between two states with adjacent valences q and $q+1$ in the configurations $f^{n+1}(sd)^q$ and $f^n(sd)^{q+1}$. In both configurations the atom is actually electrically neutral, as it should be in a metal. In $f^{n+1}(sd)^q$, an electron is in the inner f shell (i.e., localized), while in $f^{n+1}(sd)^{q+1}$ it has moved to the outer (sd) shell, where it is then one of the conduction electrons. Averaging over time, this fluctuation implies a nonintegral (fractional) occupation of both the f and the (sd) shells, namely $0<\nu<1$ for $f^n(sd)^{q+1}$ and $1-\nu$ for $f^{n+1}(sd)^q$. The "fractional valence" $q+\nu$, and its change with temperature and pressure, can be detected by several techniques, such as measurements of the lattice constant, of the Curie constant, of the Mössbauer isomer shift and of the structure of the L_{III} x-ray absorption edge. Fractional valence has been detected in all valence fluctuation metals, including the heavy fermion metals. The observed values of $q+\nu$ tend to cluster near those values which one expects from the degeneracies of the ground states of the f^n and f^{n+1} shells at equipartition. The equipartition values are 3.14 for cerium, 2.89 for ytterbium and 3.53 for uranium. In other words, in most valence fluctuation metals with cerium and ytterbium, the fractional valence is close to three (see Table 1). Usually the fractional valences are weak functions of temperature but strong functions of pressure.

With the onset of delocalization of the f shell, the configurations $f^{n+1}(sd)^q$ and $f^n(sd)^{q+1}$ acquire finite lifetimes τ_{n+1} and τ_n, which cause finite lifetime widths $\Delta\varepsilon_{n+1} = h/\tau_{n+1}$ *and* $\Delta\varepsilon_n = h/\tau_n$ for all states of the f^{n+1} and f^n configurations. The longer lifetime belongs to the configuration with q, if $\nu < \nu_e$, and to that with $q+1$, if $\nu_e < \nu$, where ν_e is the equipartition value of ν. Since in the configuration $f^{n+1}(sd)^q$ the fluctuating electron circles the rare earth or actinide atom as a local f electron, while in the configuration $f^n(sd)^{q+1}$ it moves as an ordinary conduction electron with $m^* \approx m_0$, one may identify the lifetime widths with the bandwidths W^*:

$$\left.\begin{aligned} \Delta\varepsilon_{n+1} &= h/\tau_{n+1} = W^*_{n+1} \\ \Delta\varepsilon_n &= h/\tau_n = W^*_n \end{aligned}\right\} \quad (1)$$

These lifetime widths can be detected directly by inelastic magnetic neutron scattering and indirectly in the temperature dependence of the static susceptibility and of the resistivity.

3. Thermal Properties

Experimentally the effective mass m^* at the Fermi level is extracted from the coefficient γ_0 of the linear term of the specific heat $C(T)$ measured at helium temperatures, via

$$\left.\begin{aligned} \Delta(C/T) &\equiv C(T)/T - \beta T^2 = \gamma_0 \\ &= (gk_B)/W^* \\ &= (gk_B m^*)/(m_0 W) \quad (T/W^* \ll 1) \end{aligned}\right\} \quad (2)$$

where βT^3 is the low temperature contribution of the phonons to the specific heat, W^* is the f-bandwidth (in kelvin) and g is the degeneracy of the groundstate of the local f shell. Table 1 exhibits the full range of γ_0 values observed in valence fluctuation metals. In the classical regime $(T/W^* > 1)$, $\Delta(C/T)$ goes to zero rapidly. The integral over $\Delta(C/T)$ from $T/W^* \ll 1$ to $T/W^* \gg 1$ gives the entropy $\Delta S = k_B \ln g$ of the narrow band. One finds, for example, $g=2$ for some cerium heavy fermion compounds such as $CeCu_6$, which have sufficiently large splittings of the $J=5/2$ Hund's rule multiplet of the local f^1 shell caused by the crystalline electric fields of the solid. The linear specific heat coefficient γ_0 decreases rapidly in a sufficiently strong magnetic field H, namely when the Zeeman splitting μH becomes larger than $W^* = gk_B/\gamma_0$, where μ is the magnetic moment of the crystal field ground state of the local f^1 shell. Then a Schottky anomaly develops instead, which corresponds to a two-state system

Table 1
Magnetic ordering temperatures T_M, superconducting transition temperatures T_c, linear specific heat coefficients γ_0, fractional occupations of the higher valence state ν, and magnetic bandwidths W^* at 300 K and near $T=0$ for some representative metals with cerium, ytterbium and uranium with delocalizing f shells

System	T_M (K)	T_c (K)	γ_0 (mJ mol^{-1} K^2)	ν	$W^*(300)$ (K)	$W^*(0)$ (K)
CeB_6	3.15			0.03	40	
$CeAl_2$	3.8			0.04	60	
$CeAl_3$			1700	0.12	100	10
$CeCu_6$			1500	0.12	120	8
$CeCu_2Si_2$			600–900	0.06	≈100	10
$CeCu_{2.2}Si_2$		0.9	700–1100	0.11	≈100	130
$CeBe_{13}$			130	0.12	170	280
$CePd_3$			36	0.29	200	300
$CeSn_3$			40	0.02	250	400
αCe			11	0.20	2300	2300
YbPd	0.5			0.8	130	
YbCuAl			260	0.8	50	30
$YbCu_2Si_2$			125	0.82	50	40
$YbAl_2$			12	0.4	700	700
URu_2Si_2	17	1.5	40		300	15
UBe_{13}		0.9	700	≈0.5	70	100
UPt_3		0.45	415	≈0.5	70	100
UAl_2			90	≈0.5	300	250
USn_3			170			
$UAuPt_4$			720			

with energy splitting $2\mu H$. This proves that the large γ_0 is indeed due to a very narrow doubly-degenerate band of conduction electrons that is intersected by the Fermi level and that derives from the magnetic-doublet crystal-field groundstate of the local f electrons. (For the exceptional behavior of the heavy fermion superconductors see Sect. 7.)

4. Magnetic Properties

The magnetic properties of heavy fermion metals are dominated by the f shell of the rare earth or actinide atoms. They differ significantly from those at strictly local f shell only at low temperatures, at $T \leqslant W^*$. At integral valence, the magnetic properties of rare earth and actinide metals are well understood. First, the magnetic moments seen in the high-temperature Curie constants are almost identical with those of the Hund's rule ground state of the stable f^n shell in isolated atoms, except for a small permanent change of the moment caused by a local magnetic polarization of the (sd) conduction electrons. Second, at temperatures of the order of 100K and below, the susceptibility reflects the splitting of the Hund's rule ground state into crystal-field multiplets by the electric field of the environment. Finally, the interaction between the f^n magnetic moments on neighboring atoms via the polarization of the (sd) conduction electrons induced by the f electrons leads to magnetic order below a critical temperature T_M, which ranges from a few 100 K to a few tenths of a kelvin, depending on n and on the structure of the environment of f^n.

The main effect of the onset of delocalization of the f electron on the magnetic properties is a strong tendency towards a nonmagnetic groundstate of the metal; this is caused by the now finite lifetime width of all its magnetic Zeeman levels. While a lifetime width does not lift any Zeeman degeneracies and does not prevent the usual Zeeman splitting in an external magnetic field or in the exchange field of the neighbor, it does reduce the resulting magnetic polarization (i.e., the susceptibility). Schematically, the temperature dependence of the susceptibility of a heavy fermion metal with cerium in low fields may be written as

$$\chi(T) = \frac{(1-\nu)\mu_1^2}{3k_B(T+T_M+W_1^*(T))}$$
$$\chi_0 \rightarrow \frac{(1-\nu)\mu_1^2}{3k_B W_1^*} \quad (T, T_M \ll 1) \tag{3}$$

Here μ_1 is the magnetic moment of the delocalizing f^1 shell and W_1^* is its lifetime width (μ_0 of f^0 happens to be zero). At high temperatures, in the classical regime, $\chi(T)$ is close to the Curie–Weiss behavior of the local f^1 shell, but with a reduced effective moment, and with a Curie–Weiss temperature which is increased by W_1^* with respect to the T_M (in Eqn. (3), the usually observed antiferromagnetism is assumed). Both $1-\nu$ and W_1^* are therefore measurable by $\chi(T)$. At low temperatures, χ becomes a constant, similar to the Pauli–Sommerfeld theory of ordinary metals, provided $T_M \ll W^*$. The magnetic order is then suppressed, and such a metal is said to have a nonmagnetic groundstate. χ_0 then gives another measurement of the heavy fermion bandwidth W^*. Indeed, the ratio $(\chi_0/\gamma_0)(\mu_i^2/3k_B^2 g_i)$ is found to be close to unity for most valence fluctuation metals, including the nonsuperconducting heavy fermion metals. This tendency can, for example, be read from the constancy of the product $\gamma_0 W^*(0)$. (In Table 1 $W^*(0)$ is taken either from the static susceptibility via Eqn. (3) or from the quasielastic linewidth of the magnetic neutron spectra, or from both.) However, for the heavy fermion superconductors, this ratio is closer to 0.2 (see Sect. 7).

It can happen that the magnetic ordering temperature T_M is not small compared with the bandwidth W^*; that is, T_M and W^* are comparable, or even cross each other as functions of pressure or concentration. (For a rare earth valence fluctuator, T_M can be estimated by deGennes scaling from isostructural reference metals with other rare earth atoms with strictly localized f electrons.) This can lead to very complex behavior and is more likely to occur in the heavy fermion metals than in the valence fluctuation metals with smaller m^* (larger W^*). Some such cases are listed in Table 1. The tendency for magnetic order is seen to decrease from uranium to cerium to ytterbium. This is consistent with the decreasing size of the f electron in this order (i.e., with the decreasing size of the exchange integral between the local f electrons and the light (sd) conduction electrons). If there is magnetic order, the specific heat no longer varies linearly with temperature near $T=0$. No value can then be given for γ_0. This does not mean that there is then no narrow f band, but merely that $W^* < T_M$ (note that CeB_6 and $CeAl_2$ have slightly fractional ν; that is, here the f shells are delocalizing, although only weakly, with $W^*>0$ nevertheless).

W^* can also be measured directly by inelastic magnetic neutron scattering, in terms of the energy width of the Zeeman levels. It was by this powerful technique (which also measures the crystal field splittings) that the key feature was found which distinguishes heavy fermion metals from valence fluctuation metals with similar fractional valence but smaller effective mass: in both types of system W^* is of similar magnitude (a few ten to a few hundred kelvin) at temperatures of order 100 K and above (see W^* (300) in Table 1) and is then roughly independent of temperature, but when the temperature crosses the crystal field splittings on its way to zero, W^* *decreases* in the heavy fermion metals, by a factor of up to ten, while it stays roughly *constant* in the others (compare W^* (300) with W^* (0) in Table 1). Apparently, the fluctuation rate from and to the other valence state depends on the f crystal field level; in the heavy fermion metals, this rate happens to be much smaller in the crystal field groundstate than in the excited crystal field states, but not in the other valence fluctuation metals.

5. *Transport Properties*

For $T \to 0$, the resistivity $\rho(T)$ of valence fluctuation compounds approaches zero, as it should in a metal (if the lattice is sufficiently free of defects). However, when the temperature approaches W^* from below, $\rho(T)$ increases much more rapidly than the usual phonon contribution $\rho_p(T)$ observed in isostructural reference compounds with stable (strictly local) f electrons. The so-called resistivity anomaly $\Delta\rho(T) \equiv \rho(T) - \rho_p(T)$ rises proportional to T^2 up to $T \approx W^*$, then goes through a point of inflection, and then usually saturates above $T \approx 2W^*$ at values of order $100\,\mu\Omega$ cm. Thereafter it remains constant up to the melting point. In some cases, especially in the heavy fermion metals, $\Delta\rho(T)$ goes through a maximum at $W^* \leq T \leq 2W^*$, before saturating at temperatures above about 100 K. This behavior can be understood in the simple picture of conduction electron scattering in a dynamic alloy: the valence fluctuation metal with the f carrying atoms in two configurations $f^n(sd)^{q+1}$ and $f^{n+1}\,(sd)^q$ with concentrations ν and $(1-\nu)$ respectively may be regarded in first approximation as a metallic alloy (a solid solution) of two different types of atoms, whose valence differs by one. In such alloys, the resistivity anomaly follows the relation

$$\Delta\rho(\nu) = A\nu(1\text{-}\nu) \tag{4}$$

where A is a constant, which increases with the difference in the charge density of the outer valence electrons (here with the difference between $(sd)^{q+1}$ and $(sd)^q$). *With* ν from the fractional valence, as determined by the techniques mentioned in Sect. 2, $A \approx 1$ mΩ cm is found from a large set of data. The high temperature saturation of $\Delta\rho(T)$ is simply a consequence of the well established temperature independence of $\nu(T)$ at high temperatures, and the occasional increase at $\Delta\rho(T)$ with decreasing T in the case of the maxima of some heavy fermion metals is consistent with a temperature dependence of $\nu(T)$, as found experimentally in the same region of temperature. This shows that at temperatures above W^*, the valence fluctuation compound exhibits the resistivity of an alloy, on top of the normal phonon contribution. In a static alloy, the resistivity anomaly would remain high and constant for $T \to 0$.

The actually observed steep drop of $\Delta\rho(T)$ towards zero at $T<W^*$ (e.g., by more than 100 $\mu\Omega$ cm within the last few kelvin in the case of $CeCu_2Si_2$) is a consequence of the temporal valence fluctuation; that is, the fact that the alloy is actually dynamic, on a time scale of order $\tau\approx h/W^*$. The charge density on the atoms performing valence fluctuations on the time scale τ is identical on all atoms if averaged over a time $t\gg\tau$. In a pure crystal, the conduction electrons occupy a given Bloch state with momentum $\hbar k$ for a time which has a lower limit of $t_T>h/(k_BT)$. Therefore, in the region $T<W^*$, the conduction electrons exist in Bloch states for times which are longer than the valence fluctuation time τ, the more so the lower the temperature. As the temperature falls the charge density of the valence fluctuating atoms becomes ever closer to the mean charge density, until at $T=0$ a perfect crystal is obtained with the same mean charge on each *f*-carrying atom. There is no conduction electron scattering in a perfect crystal (i.e., $\Delta\rho(T)\rightarrow 0$ for $T\rightarrow 0$).

There are also large anomalies in the thermopower, the Lorentz ratio, the Hall effect and the magnetoresistance (which all behave differently at $T<W^*$ and at $T>W^*$) but these are too numerous and too complicated to enumerate here.

6. *Superconducting Properties*

Four heavy fermion metals are superconductors, with critical temperatures T_c of order 1 K: $CeCu_{2.2}Si_2$, UBe_{13}, UPt_3 and URu_2Si_2 (see Table 1). URu_2Si_2 also exhibits antiferromagnetic order at $T_M=17$ K and does not really qualify as a heavy fermion metal, judging from γ_0. However, as mentioned in Sect. 4, in the case of sufficiently strong magnetic order (at $W^*\leq T_M$) there is still a bandwidth due to the delocalizing *f* electron, and this width is found to be about 15 K just above T_M by neutron scattering. In this sense URu_2Si_2 qualifies. At T_c the specific heat shows jumps of order $\gamma_0 T_c$. This proves not only that the superconductivity is a bulk property, but also that it is a property of the heavy electrons, and not just of the ordinary (*sd*) conduction electrons with their much lower mass. On the other hand, in a magnetic field the static susceptibility with the magnetic formfactor of the *f* electrons does not decrease below T_c, as it should if the superconducting electrons were of *f* character. A smaller but finite linear specific heat coefficient remains near $T=0$; that is, a superconducting gap seems to develop over only part of the Fermi surface. The superconductivity is of extreme type II, with a very large ratio $\kappa=H_{c2}/H_{c1}$ of the upper and lower critical fields. In UBe_{13} the derivative dH_{c2}/dT is nearly infinite at T_c.

The first three heavy fermion superconductors have very large γ_0, but their W^* does not decrease significantly between room and helium temperature, in contrast to the nonsuperconducting heavy fermion metals (see Table 1); the product $\gamma_0 W^*$ is nearly an order of magnitude smaller than in the nonsuperconducting heavy fermion metals. Also, for the superconductors $CeCu_{2.2}Si_2$ and UBe_{13}, $\Delta S\leq 0.5k_B \ln 2$ (i.e., $g<2$). Thirdly, contrary to the nonsuperconducting heavy fermion metals, the specific heat is insensitive to large magnetic fields for the superconductors (see Sect. 3). The distinction between nonsuperconducting and superconducting heavy fermion metals is clearest in $CeCu_{2+\varepsilon}Si_2$, which exists in both versions (superconducting at $\varepsilon=0.2$ and nonsuperconducting at $\varepsilon=0$). For $\varepsilon=0.2$, $W^*(T)$ behaves as in UBe_{13} and UPt_3, but also as in $CeBe_{13}$ and in $CePd_3$, while for $\varepsilon=0$ it behaves as in $CeCu_6$ and $CeAl_3$. It is worth noting that the superconducting transition temperature of $CeCu_{2.2}Si_2$ jumps to about 2 K at a pressure near 4 GPa, while at the same time W^* increases strongly, as seen in the temperature dependence of the resistivity. All this suggests that the large linear specific heat coefficients γ_0 of the heavy fermion superconductors do not originate from narrow *f*-bands; the magnetically determined bandwidth of the *f* levels is nearly an order of magnitude larger than suggested by Eqn. (1). The origin of their large γ_0 (or the large effective mass of the fermions) remains uncertain, and so does the mechanism of the heavy fermion superconductivity, in spite of the intense experimental and theoretical activity of recent years.

Bibliography

Assmus W, Fulde P, Lüthi B, Steglich F (eds.) 1988 *Crystal Field Effects and Heavy Fermion Physics*. North-Holland, Amsterdam

Gupta L C, Malik S K (eds.) 1987 *Theoretical and Experimental Aspects of Valence Functions*. Plenum, New York

D. K. Wohlleben
[Universität zu Köln, Köln, FRG]

High-Level Radioactive Waste Disposal: Safety

Of the various types of radioactive waste (see *Radioactive Waste Disposal*) high-level waste (HLW) presents the highest risk. It is not surprising then that a large amount of theoretical and experimental work has been devoted to evaluating the safety of its disposal.

The analysis of the risk associated with the final disposal of HLW is usually dealt with using the concept of the multiple barrier (see *Radioactive*

Waste Management: A Systems Approach). In this concept the path of the radioactivity released from the waste form toward a human is subdivided into successive barriers, which are subsequently analyzed. The risk is considered acceptable if the dose to the most exposed population groups is within the limit imposed by the International Commission on Radiological Protection. In addition, in the USA the Nuclear Regulatory Commission (NRC) specifies that the engineered barrier system shall be designed so that the containment of HLW (within the waste packages) will be substantially complete during the period when radiation and thermal conditions are dominated by fission product decay (300–1000 years). It is also specified that the release rate of any radionuclide from the engineered barrier system following the containment period shall not exceed one part in 100 000 year^{-1} of the inventory of that radionuclide calculated to be present after 1000 years of permanent closure.

The NRC requirement puts emphasis on the safety of the waste package. In particular, the requirement of a substantially complete containment means a detailed evaluation of the corrosion resistance of the canister surrounding the waste form. Moreover, in order to fulfill the second condition, it is necessary to be able to predict the long-term leaching and corrosion of the waste form.

1. *Waste Form Preparation and Properties*

High-level radioactive waste consists of the spent fuel discharged from reactors and/or the solutions generated in the spent fuel reprocessing. In this section, the solidification of the radioactive solution in an appropriate matrix is considered. The leading method is based on borosilicate glass of specific composition. Some effort, however, is being devoted to advanced matrices such as ceramics, lead-containing glasses and so on.

1.1 *Vitrification of HLW Solutions*

In the preparation of a glass starting from a highly radioactive solution it is important to have simple and reliable equipment in order to minimize the maintenance of the hot cell. Moreover, it is necessary to control the effluent strictly so that no secondary radioactive wastes are produced. Two types of plant have been developed: the first uses a melting pot made from a refractory alloy, the second uses a ceramic melter.

Typical of the first type of plant is the Atelier Vitrification Marcoule (AVM) facility operated in Marcoule (France) since 1978. The plant consists of a calciner, a glass melter and an off-gas treatment. The nitric acid solution and some calcination additives are introduced continuously into a rotating tube heated between 600 °C and 900 °C. In this tube the solution is evaporated and the nitrates are decomposed to oxides. The calcined powder is introduced, together with the glass frit promoter, into a melting pot where the glass is formed. The melting pot, which is made of Inconcel 601, is kept at 1100 °C using induction heating. The glass formed is poured every eight hours into a refractory steel container. Each container holds about 360 kg of glass. When the filling is complete a lid is welded on with a plasma torch. The containers after decontamination are conveyed to a storage where they are cooled by air circulation. The gases produced by evaporation and calcination are purified successively in a cyclone separator, a condenser, a recombination column and a scrubber. Uncondensable gases pass through a series of very high efficiency filters before discharge through the stack. Gas treatment serves to recycle the more volatile elements in the calcinate such as ruthenium. The very high efficiency filtration system installed in this facility helps to achieve a gas decontamination factor of between 500 and 1000.

The second type of plant, a ceramic melter, is used both in the Defence Waste Processing facility at the Savannah River Plant and in the PAMELA vitrification plant at the Eurochemic site in Mol, Belgium. The ceramic melter takes advantage of the fact that the glass at temperatures higher than 750 °C has an electrical resistance sufficiently low to permit direct heating by passing electric current through the molten glass. Usually Inconel electrodes are used. At the start of the process, a preheating system is needed to raise the temperature to 750 °C. Direct heating allows injection into the system of a large amount of heating power so that no calcination unit is needed. The solution, to which the glass frit is added, is fed directly into the melter where the evaporation of the water content, the denitration, the calcination of the solid waste residue to oxide and, finally, the simultaneous melting and homogenization of the glass take place. In some cases formic acid is added to the solution in order to facilitate the denitration and to form the glass in a reducing condition.

1.2 *Choice of Glass Composition*

The first requirement that must be met by an HLW glass composition is that it must have an acceptable melting temperature. In the glass literature, melting temperature denotes not only the temperature of formation and homogenization of the glass but also the temperature at which there is a possibility of pouring the formed glass. Usually, at the melting temperature the glass must have a viscosity of 10–30 N s m^{-2}. For most glasses, or products derived from glasses, the effect of composition on formation temperature is far too complex to be analyzed in this article. In general, elements such as boron, sodium, lithium, potassium and phosphorus are used to reduce melting temperature whereas silicon, aluminum and iron increase the melting temperature. Effects are less pronounced with most other elements of interest, particularly as they are not normally present in significant quantitites. The type of plant defines the maximum attainable temperature. As has been shown in the preceeding subsection, the ceramic melter allows a higher maximum temperature and, therefore, a higher freedom in the composition choice.

A second requirement is a good resistance to leaching and corrosion by the underground water which will be present in the final repository. Many attempts have been made to quantify the relation between the chemical durability of a glass and its composition. The main difficulty for a rigorous thermodynamic approach is to define thermodynamically a multicomponent amorphous material such as glass. Moreover, this approach requires a complete description of the solution and the new phases that can be formed during the corrosion process and their Gibbs free energies. A semi-empirical approach is commonly used and will be described below in connection with the prediction of long-term leaching evaluation.

Finally, the stability of the glass has to be verified. When a glass is cooled from approximately 800 °C down to less than 400 °C, nucleation of crystal growth can take place, depending on the rate of cooling and hence on the dimensions of the block. The rate of crystallization will be a maximum somewhere above the transformation point but, although the rate may be low at lower temperatures, crystallization can still occur over long periods of time, depending on the glass composition. It has been shown that sometimes a recrystallized glass shows a sharp increase in the leaching rate. Crystallization is never complete so that in a partly recrystallized sample at least two phases, glass and crystal, are present. When this point is reached in the crystallization process, the remaining glass is enriched in alkali and the leaching of the glass increases. To avoid this problem, two methods are possible. Either a glass formulation exhibiting a negligible crystallization can be used, or a glass that presents the same leaching resistance in the glassy and in the devitrified form must be found. Owing to the fact that the time in which the glass stays at a high temperature can be of the order of hundreds of years, because of radioactive self-heating, the second method is usually preferred. Table 1 shows a list of the most commonly used glass compositions.

Table 1
Simplified glass composition (in wt%)

Oxide	PNL 76–68	SRL 131	DWRG	SON 68	UK 189
SiO_2	42.8	40.3	49.0	45.5	41.5
B_2O_3	9.2	9.9	6.8	14.2	21.9
TiO_2	3.1	0.8	0.1	0.0	0.0
Fe_2O_3	10.8	13.6	10.1	2.9	2.9
Al_2O_3	0.5	3.1	6.6	4.9	5.0
Nd_2O_3	4.1	0.5	0.0	1.6	1.8
ZrO_2	1.7	0.3	0.9	2.7	1.4
MoO_3	1.9	0.0	0.0	1.7	1.7
ZnO	4.8	0.0	0.0	2.5	0.4
MgO	0.0	1.2	0.7	0.0	6.2
CaO	2.2	1.0	2.0	4.0	0.0
BaO	0.5	0.0	0.0	0.6	0.4
Li_2O	0.0	3.9	4.9	2.0	3.7
Na_2O	15.0	14.8	9.2	9.9	7.7
K_2O	0.0	0.1	0.0	0.0	0.0
Cs_2O	0.8	0.3	0.3	1.3	0.8
rest	2.6	10.2	9.4	6.2	4.8

1.3 Glass Corrosion

The corrosion of glass containing HLW is a rather complex phenomenon. Three main corrosion mechanisms can be identified: the dissolution of the glass network, the alteration of the subsurface layer due to an ion-exchange mechanism and the formation at the surface of a complex layer of insoluble compounds.

The dissolution of the glass network is the main mechanism that gives rise to the release of radioactive materials. The basic reaction is the hydrolysis of silicon dioxide forming monosilicic acid ($Si(OH)_4$) which dissolves progressively. The dissolution rate depends on many physical and chemical parameters, of which the concentration of silicon in the solution is one of the most important. Early work identified the degradation rate as

$$k = k^{+}(1 - a/a_{sat})$$

where a is the activity of the $Si(OH)_4$ in the solution and a_{sat} is the activity under conditions of saturation. It has to be noted, however, that owing to the amorphous nature of the glass a real equilibrium is not possible. Even in a condition of pseudoequilibrium a small but finite difference in activities will be present, giving rise to a continuous dissolution of the glass and a reprecipitation of some specific crystals. When the pH of the solution increases the monosilicic acid is ionized. At pH 9 it is ionized first to $(HO)_3SiO^-$ and at still higher pH values even bivalent ions can be formed. Therefore, with the increase of the pH the leaching rate increases. Temperature has a very large influence: between 25 °C and 90 °C the initial hydrolysis rate increases by a factor of greater than 100. Moreover, the saturation concentration increases for the same range of temperatures by a factor of about four. As a consequence, the long-term corrosion will also increase.

The second mechanism which is commonly observed is ionic exchange between alkali ions in the glass and H^+ (or H_3O^+) ions coming from the solution, which gives rise to a progressive diffusion of alkali from the subsurface layer to the solution. At the beginning of the processes of corrosion and leaching there is competition between dissolution and diffusion. If dissolution is prevalent the zone in the glass that is depleted in alkali is very thin. However, in a near saturation condition, dissolution being very slow, the ionic exchange diffusion can influence a large zone. In a closed system the alkali released induces an increase of the pH of the solution which, as noted above, gives rise to an increase in the leaching rate.

The third mechanism involves the formation of a layer of a very complex nature at the surface of the glass. When leaching is conducted at low temperatures this layer is usually in the form of a gel. Increasing the temperature and the leaching time produces an increased concentration of crystals in the layer. The formation of the layer is due to precipitation from the leaching solution of compounds which have reached the saturation concentration. They can be either simple hydroxides of the elements comprising the glass or compounds of a more complex formulation. In the latter case, through the leaching solution, even the composition of the surrounding rocks and, in general, of all the materials contained in the repository can influence the composition of the layer.

This surface layer, covering the glass surface uniformly, can act as a diffusion barrier and as an absorbing medium for some elements released from the glass. It can be considered as the zone in which all the equilibria of adsorption and dissolution take place. The chemical and crystallographic analysis of the composition is of paramount importance in determining the variation in time of the leaching rate. As noted previously, the dissolution rate of the glass depends on the concentration of silica in the leaching solution. This concentration can be determined from the solubility product of the compounds that precipitate from the leaching solution. Knowledge of the compounds formed on the leaching layer and of the composition of the solution permits calculation of the silica concentration and, therefore, of the leaching rate.

It is clear that a problem so complex can be solved only by the use of computer codes and in the 1980s an increasing number of such codes have been proposed. A typical code is that described by Grambow et al. (1987). Its central part is composed of a geochemical code that, from the composition of the solution contacting the glass, defines the type and the mass of the altered surface products. Therefore, the code is able to deduce the concentration of silica in the solution and the corresponding degradation rate. In a closed system, the leaching solution is continuously enriched in the elements leached from the glass and the surrounding rocks. The compounds which are in equilibrium with the saturated solution will concentrate with time. For example, for an MCC 76–68 glass leached in a closed system with a solution 0.001 molar in $MgCl_2$, $Fe(OH)_3$ will precipitate first, followed by $Nd(OH)_3$, sepiolite (a magnesium silicate), $Zn(OH)_2$, $CaCO_3$ and finally amorphous SiO_2. If the system is open the code can take into account the water flow rate and the corresponding variation in the leaching solution composition. Equally, if an absorbing backfilling is present the code can evaluate absorption and diffusion. A major difficulty in the use of such a code is the very large amount of basic data which are required. Also, data on amorphous systems are difficult to evaluate. The use of such a complex computer model requires a series of experiments intended specifically to validate the model. Using experimental data on different glass compositions existing in the literature, it is found that a good agreement is obtained for the most soluble elements. However, some improvement is needed for the less soluble elements such as iron and magnesium for which thermodynamic data requires further refinement.

1.4 Ceramic Waste Forms

Even though borosilicate glass is generally accepted as a reliable material to use as a matrix for conditioning HLW, a consistent research effort has been devoted to the evaluation of possible alternatives. Among these new matrices the one under the most investigation is probably SYNROC (SYNthetic ROC) which was first proposed in Australia by A. E. Ringwood.

SYNROC is a titanate ceramic composed of three major phases plus minor metal alloys. It has been demonstrated that most of the elements in HLWs are readily accommodated as dilute solid solutions in the crystal lattices of the major constituent phases of SYNROC, that is, hollandite ($BaAl_2Ti_6O_{16}$), zirconolite ($CaZrTi_2O_7$) and perovskite ($CaTiO_3$). The process technology is typical of high-temperature ceramics. The precursor of SYNROC is prepared as a powder with a high specific surface. After mixture with the HLW solution these are dried to a free-flowing powder. The powders, after calcining in a reducing atmosphere at 750 °C, are sintered in a bellow by uniaxial hot pressing at temperatures around 1200 °C. The resulting material presents a leaching behavior which is decidedly better than that of the commonly used borosilicate glasses. At 90 °C it is claimed that in distilled water the leaching coefficient is 100 times lower than that of borosilicate glass. Moreover, its dependence on temperature is lower than that of borosilicate glass. It has to be noted that the minerals are not in equilibrium with the system; in contact with the underground water only hydroxylated minerals, often of the clay type, are in thermodynamic equilibrium. The evolution of the leaching will then be similar to that discussed previously for borosilicate glass, with successive formations at the surface of the minerals of compounds in equilibrium with the solution.

The main obstacle to the use of SYNROC, as well as other types of ceramics, is in the fabrication process. A glass melter has high throughput so that the active part of the production plant is relatively small. Hot pressing, however, is a slow procedure and, in order to have the same production as a glass melter, a large assembly of hot cells is required.

2. *Canister and Overpack Integrity*

In this section a canister is defined as the stainless steel or refractory metal mold used to form the vitrified HLW block. Moreover, the term "overpack" will be used to indicate the metal (or ceramic) shell surrounding the canister. Usually, it is not considered advisable to assume that the canister could have an appreciable resistance to corrosion. This is because the canister, during the pouring of the glass and the interim storage, undergoes a very complex thermal history which does not allow evaluation of its corrosion resistance, particularly its resistance to localized corrosion. Therefore, protection for the so-called thermal period (300–1000 years), as required by the NRC regulation, has to be provided by the overpack.

Two main possibilities have been investigated. The first is the use of materials highly resistant to corrosion while the second is the use of material which corrodes at a predictable rate which is sufficiently low to make it possible to provide an adequate thickness to allow for the corrosion attack.

2.1 *Corrosion Resistant Materials*

The high resistance of some materials to homogeneous corrosion is due to the formation at the surface of the alloy of an oxide passive film. Therefore, the presence of oxygen or oxidizing species in the underground repository is particularly important. Throughout the various types of repository a slightly reducing condition is usually predominant.

A particular case is represented by the site used for the Nevada Nuclear Waste Storage. In this case the repository is situated 300 m below the ground surface but 200 m above the static water table. Owing to the heat released by the waste, the corroding agent will be a mixture of steam and air. In these conditions stainless steel presents an adequate resistance to corrosion. In view of the necessity for long-term integrity, the phase stabilities of austenitic materials 304L and 316L and of alloy 825 have also been investigated. Carbide precipitation and formation of the σ phase have been observed for the austenitic materials. No intermetallic phase formation was documented for alloy 825.

For nonoxidizing conditions a large number of materials have been investigated. Good results have been obtained using Hastelloy C4 and titanium alloys such as Ti–O.2%Pd and Ticode 12. In clay, salt and granite they have a corrosion rate typically in the range 0.1–0.3 μm year^{-1} at 150 °C. The main problem for the use of these materials is the difficulty of extrapolating in a reliable way the results of corrosion tests to time periods that greatly exceed those possible for direct testing. In particular there is a danger that a mechanism of secondary importance becomes, by accumulating with time, the main cause of failure. For example, with titanium alloys the main danger is possible damage due to adsorbed hydrogen. Hydrogen embrittlement may occur if the solubility of hydrogen (20 ppm at room temperature) is exceeded and titanium hydride forms. Significant effects will be obtained, however, only at hydrogen contents of some hundreds of ppm or more. Another type of damage is delayed fracture at stresses below the ultimate strength, owing to the enrichment of hydrogen at defects and other stress raisers.

2.2 *Corrosion Allowance Materials*

The use of materials which have a low, predictable corrosion rate has been particularly studied for repositories which present reducing conditions. Mild carbon steel has been taken into consideration particularly for clay and granite repositories. In anoxic conditions, iron corrodes very slowly generating bivalent iron ions and hydrogen. Tests performed by Marsh (1987) in synthetic granite groundwater have shown a corrosion rate after 500 d of roughly 5 μm year at 25 °C and 30 μm year^{-1} at 90 °C.

Obviously, even in this case the problem of extrapolation of experimental data to very long times is fundamental. In the case of the corrosion allowance materials, however, the main failure mechanism is always homogeneous corrosion, even if in some cases pitting can also develop. Marsh (1987) has developed a mechanistically based mathematical model to predict the rate of attack as a function of time. Starting from electrochemical data and taking into account diffusion of the produced species in the surrounding media it is possible to evaluate the corrosion rate. This model may be considered conservative as it assumes that the principal electrochemical corrosion reaction is unimpeded by the buildup of corrosion products on the metal surface. However, tests performed to validate the model showed that, at least for temperatures lower than

100 °C, the model overestimates the rate of attack. Owing to the thickness of the shell, the influence of γ rays produced in the glass can be considered negligible. For underground water compositions giving rise to pitting attack, an additional shell thickness evaluated using statistical models has to be planned to take the localized corrosion into account.

Copper has also been considered as a possible corrosion allowance material. The Swedish group KBS has proposed that spent unreprocessed nuclear fuel may be disposed of by encapsulation in copper canisters. The canisters should be placed in vertical boreholes in igneous rock and embedded in a buffer of compacted bentonite. The bentonite buffer will possess a very low permeability (10^{-12} m s^{-1}) so that for mass transport through the buffer only diffusion need be considered. The thermodynamic possibilities of various corrosion reactions on copper under the prevailing conditions have been studied. Copper can be attacked by water containing dissolved oxygen, but only at comparatively high redox potential—above 50 mV SH at 25–100 °C. The corrosion products will be Cu_2O or CuO. The maximum corrosion rate possible will then be determined by the supply of oxygen to the canister surface. Oxygen entrapped in the buffer material at the time of closing of the repository was found to be an important source. Sulfide in the groundwater was found to be another important reactant, since in the presence of SH^- copper can be oxidized even by H^+, for example in water. Sulfide can also be formed by bacterial reduction of sulfate. Assuming that there is an excess of sulfate, the limiting factor will be the supply of organic matter that is required by the bacteria for their life process. The evaluation of the rate of attack is performed considering that the supply of aggressive species is limited by diffusion through the backfilling material (compacted bentonite). Again, for copper, an additional thickness has to be planned to take pitting into account.

3. *Conclusions*

The evaluation of the risk associated with a repository containing conditioned HLW necessitates calculation in a reliable way of the amount of the various radioactive elements released. The requirement of the NRC regulation stipulates a complete containment of the radioactive products for at least 300 years.

Direct extrapolation of experimental data does not give sufficient reliability. The problem has been solved using a mechanistic approach, both for the glass corrosion and leaching and for the overpack corrosion, by coupling simple kinetics with a large amount of basic thermodynamic data, through computer codes. The validation of these codes has become the central problem. In this respect the possibility of testing these codes with natural analogues (e.g., archeological remnants or natural glasses) is particularly interesting.

Bibliography

Clarke D H 1983 Ceramic materials for the immobilization of nuclear waste. *Ann. Rev. Mater. Sci.* 13: 191–218

Grambow B, Lutze W, Ewing R C, Warme L O 1987 Performance assessment of glass as a long term barrier to the release of radionuclides into the environment. In: Apted M J, Westerman R F (eds.) 1987 *Scientific Basis for Nuclear Waste Management XI*. Materials Research Society, Pittsburgh, PA, pp.531, 543

International Atomic Energy Agency 1984 Handling, treatment and conditioning of waste from reprocessing plant. In: 1984 Radioactive waste management. IAEA, Vienna, pp. 209, 293

Lutze W, Ewing R C (eds.) 1988 *Radioactive Waste Forms for the Future*. North-Holland, Amsterdam

Marsh G P 1987 Predicting the long term corrosion of metal containers for nuclear waste disposal. In: Apted M J, Westerman R E (eds.) 1987 *Scientific Basis for Nuclear Waste Management XI*. Materials Research Society, Pittsburgh, PA, pp. 85, 99

Thorne M C (ed.) 1985 *Radioactive Protection Principles for the Disposal of Solid Radioactive Wastes,* ICRP Publication 46. Pergamon, Oxford

US Nuclear Regulatory Commission 1983 Disposal of high level radioactive wastes in geological repositories; Technical criteria (10 CFR Part 60), vol. 48, no. 120. Federal Register, Washington, DC

F. Lanza
[Joint Research Centre—CEC, Ispra, Italy]

High-Performance Composites with Thermoplastic Matrices

Throughout the 1970s and 1980s, thermosetting resins have been the principal type of matrix used for high-performance fiber composites. The main disadvantages of these resins are: a low strain to failure and brittleness, lengthy cure cycles, moisture absorption and its deleterious effects on mechanical properties, and the preclusion of forming operations and certain kinds of repair once the matrix has been fully cured.

Some of these problems can be overcome by using a thermoplastic polymer for the matrix. Compared with thermosets these materials usually have a higher strain to failure and a lower moisture absorption, they can be shaped when heated and repaired, do not require a lengthy cure cycle and have a very long storage life at ambient temperature. Their principal disadvantage has been the difficulty in fabricating fiber-reinforced composites. This arises because, at a given temperature, the viscosity of a thermoplastic is much greater than that of an uncured thermoset. It may not be possible to reduce this sufficiently for fiber impregnation, by raising the temperature, before the polymer degrades. There are also problems of bonding to the fibrous reinforcement. Early composites with thermoplastic matrices contained an excessive void volume, and damaged or misaligned fibers, and were inferior in performance to thermoset-based specimens. The approach to the fabrication of thermoplastic-matrix composites that has evolved has been to concentrate on producing fiber-reinforced sheets and tapes by, for instance, solvent and hot-melt impregnation methods, aided by modification of the basic polymer. These feedstocks can then be turned into artifacts by hot pressing, film stacking and hot pressing, tape winding and other techniques. Short-fiber-reinforced molding granules have been developed, but, because of the reduced fiber orientation in the final product, composites based on these materials, although having a better degree of isotropy, have poorer properties in a specific direction than unidirectionally reinforced composites.

Because of the difficulties initially encountered in fabricating thermoplastic-matrix composites the tendency has been, for high-performance continuous-fiber composites, to concentrate on thermoplastics which give maximum resistance to temperature rather than on the use of the commoner, commodity thermoplastics.

1. Types of Thermoplastic Used

Ideally the thermoplastic polymer matrix should have a high softening point and resistance against pyrolysis and chemical degradation. These are properties that may be shown by crystalline polymers with a rigid backbone chain, regularly spaced substituents and strong interchain forces due to van der Waals or polar interactions. Polymer properties usually show a marked reduction at the glass transition temperature T_g, but crystalline materials have useful property retention until close to T_m, the crystalline metling point, as the crystalline phase holds the molecular chains together. Practically, it is necessary to compromise between the desired properties and the ability to process the polymer without causing degradation. Aromatic segments in the main polymer chain can be used to increase chain stiffness and reduce susceptibility to pyrolysis while ether linkages increase chain flexibility. Further details are given by Marks and Atlas (1965), Rose (1984) and Woodhams (1985).

Among the polymers used as matrices are polysulfones, aromatic polyketones, modified polyimides, polysulfides and liquid crystal systems. To obtain a processible polysulfone, polyether sulfone (PES) or polyaryl sulfone, the backbone polymer chain of the simplest aromatic polysulfone, is modified by the incorporation of ether linkages. Commercial polymers of this type include Udel and Radel (Amoco Chemicals), Ultrason S (BASF) and the polyether sulfones Victrex and Ultrason E, produced by ICI and BASF, respectively. Despite their regular structure, polysulfones are essentially amorphous. These materials are tough, resistant to nuclear but not ultraviolet radiation and are subject to environmental stress cracking. The resistance to most acids, alkalis and aliphatic hydrocarbons is good but the polymers are attacked by concentrated sulfuric acid and aircraft hydraulic fluids. Polysulfones are soluble in dimethyl formamide, dimethyl acetamide, dichloromethane and *N*-methyl pyrrolidone. Temperatures in excess of 300 °C are required for their processing.

Polyketones are represented by ICI's polyether ether ketone, Victrex PEEK, and their polyether ketone, Victrex PEK. BASF and Hoechst also manufacture polyether ketones known as Ultrapek and Hostatec, respectively. Unlike the polysulfones, polyketones are crystalline. Both PEEK and PEK have excellent resistance to nuclear radiation and a wide range of chemcials and solvents, though they are attacked by concentrated sulfuric and hydrofluoric acids. The materials are tough and not susceptible to environmental stress cracking. Processing is more difficult than for polysulfones and a temperature of approximately 390 °C is required. In addition it may be necessary to control the degree of crystallinity in the final product, to give the desired properties, by cooling at a specified correct rate. There is evidence that reprocessed PEEK may undergo some thermal cross-linking.

Modified polyimides were developed in attempts to retain the heat resistance of polyimides while rendering their processing more tractable. Torlon, produced by Amoco Chemicals, is a polyamide imide (PAI), which is not attacked by aliphatic, aromatic chlorinated or fluorinated hydrocarbons, dilute acids, aldehydes, ketones, esters or ethers, and is very resistant to nuclear and ultraviolet radiation. It is soluble in *N*-methyl pyrrolidone, is attacked by alkalis and is notch sensitive. Its melt temperature is 355 °C and the processed polymer needs a prolonged period of heat treatment to promote chain extension. The time involved can be up to 100 hours or more. Polyether imide (PEI) (General Electric Plastics), in which imide groups are interspersed with ether linkages, is sold under the name of Ultem. It has a good resistance to nuclear and ultraviolet radiation and to mineral acids, dilute bases and aliphatic hydrocarbons, but is soluble in halogenated hydrocarbons. It is notch sensitive and subject to environmental stress cracking. When PEI is burnt, the smoke generation is low and the gases produced nontoxic. The processing temperature is in the range 350–425 °C.

Probably the most widely produced high-performance thermoplastic is polyphenylene sulfide (PPS). Phillips Petroleum market the material under the name of Ryton. Bayer also provide the polymer,

Ciba Geigy have introduced a variant called Craston, while Solvay in conjunction with Tohpren, and Celanese together with Kureha Chemicals, are entering the market. The Celanese product is more linear and purer than Ryton. Idemitsu have recently developed a very high molecular weight, ultrapure product. The linear version of the polymer is highly crystalline and may cross-link oxidatively. The resistance to acids, aqueous bases, halogenated hydrocarbons and alcohols is good but the polymer is attacked by concentrated sulfuric acid and some amines. It is soluble in chloronaphthalene at elevated temperatures. Though brittle, its environmental stress resistance is good, as is its resistance to nuclear, but not ultraviolet, radiation. Smoke generation is low and nontoxic gases are produced on burning. Polyphenylene sulfide is the simplest member of the polyarylene family. Recently two new members known as polyarylene sulfide polymers 1 and 2 (PAS 1 and PAS 2) have been produced by Phillips Petroleum. The first is a semicrystalline material and the second is amorphous in nature. Mechanical properties appear similar to those of polyphenylene sulfide though the strain to failure of PAS 2 is higher, at 8%. The glass transition temperatures of PAS 1 and PAS 2 are 145 °C and 215 °C, respectively and the crystalline melting point of PAS 1 is 340 °C.

Liquid-crystal polymers are based on aromatic thermoplastic polyesters. Varieties available include Xydar, produced by Dartco, Vectra made by Celanese and Ultrax (BASF). The polymers are self-reinforcing structures in which molecular alignment is frozen-in on cooling, giving a significantly higher modulus and strength than those of the other polymers considered here. To ease processing, the molecular backbone chain is kinked by inserting hydroxy naphthoic acid so that the molecules do not fit together too well. The strain to failure is low, presumably because of the fibrous nature of the materials, and behavior is anisotropic. The heat and chemical resistance are excellent and in organic and chlorinated solvents the behavior of Vectra and Xydar is superior to that of the amorphous polymers such as polysulfone, polyether sulfone and polyether imide. In addition the materials are not attacked by acids, bases, alcohols or esters. The resistance to environmental stress cracking and nuclear radiation is good. The ability of the highly ordered molecular chains to slide over one another in the molten state while retaining their relative orientation enables these materials to be processed relatively easily. The melt temperature of a typical injection or extrusion grade of Vectra is 275–330 °C. For Xydar the figure is about 400 °C and the melt viscosity is approximately two orders of magnitude greater than that of Vectra under similar conditions.

Further details of the various polymers discussed here, except the liquid-crystal systems, are available in Brydson (1982). Liquid-crystal systems are discussed by Collyer and Clegg (1986a). The basic repeat units of many of the systems are illustrated in Fig. 1.

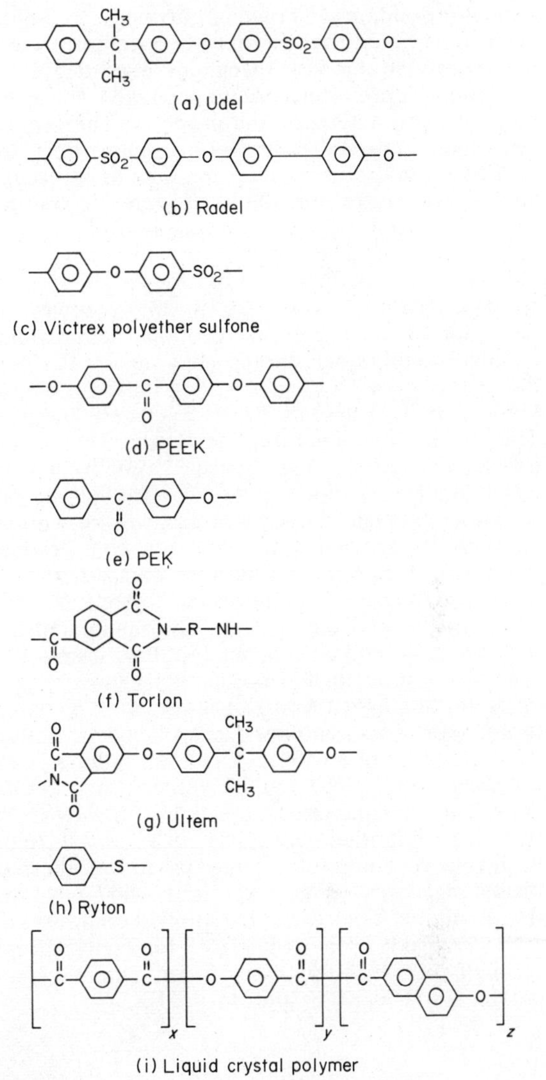

(a) Udel

(b) Radel

(c) Victrex polyether sulfone

(d) PEEK

(e) PEK

(f) Torlon

(g) Ultem

(h) Ryton

(i) Liquid crystal polymer

Figure 1
Basic structural repeat units of thermoplastic polymers

2. *Properties of High-Performance Thermoplastics*

Some typical room-temperature properties of unreinforced grades of various polymers are listed in Table 1. The information has been taken from trade data and the other sources indicated. The values quoted depend on the grade of polymer (e.g., molecular weight, additives to control crystallinity, etc.), the nature of the test specimen (e.g., degree of crystallinity, preferred orientation) and the technique used to determine the property. The properties of liquid-crystal polymers are anisotropic and here the data refer to the direction of molecular alignment, except for the coefficient of thermal expansion where the larger value is for the transverse direction. The limiting oxygen index (LOI) indicates the percentage of oxygen in an oxygen–nitrogen mixture necessary to sustain combustion. Since air contains 22% oxygen all the polymers detailed would be self-extinguishing in air. The continuous service temperature is based on the Underwriters' Laboratories (USA) test method. This gives the temperature at which 50% of the measured property (e.g., tensile strength) will be retained after a period of up to 100 000 hours exposure under static conditions. Stress, with or without an aggressive chemical environment, could reduce this figure. It is not possible to condense details of resistance to specific chemicals or, where it is available, fatigue and creep data. The manufacturers' data sheets should be consulted in these cases.

In 1984 the estimated production figures for the West for some of the high-performance polymers listed here were as follows: polyphenylene sulfide 8200 t, polysulfone 6500 t, polyether sulfone 700 t, polyether imide 300 t and polyether ether ketone 100 t. The total market is projected to grow by 460% by 1995. For the year 1984, 8% of all high-performance composites were based on thermoplastic matrices. This represents 400–600 t of thermoplastic. The cost of the unfilled grades (early 1987) ranged from £4000 t^{-1} for polysulfone, through £15 000–£30 000 t^{-1} for liquid-crystal polymers to £30 000 t^{-1} plus for PEEK.

3. *Fabrication of Thermoplastic Preforms*

Mineral-filled, short-glass and carbon-fiber-reinforced thermoplastic materials suitable for injection molding, or bulk materials from which components can be machined, have been available for some years. Though the strength and modulus are increased, and thermal expansion and shrinkage reduced, these types of material are not high-performance composites. To qualify for inclusion in this category reinforcement with continuous, aligned or woven glass, carbon or aramid fibers, or possibly a fiber mat, is required. Usually the fabricator buys reinforced thermoplastic preform (analogous to a thermoset prepreg) in the form of a sheet, tape or pultruded section and uses this to produce the final artifact, though some manufacturers make the final, shaped product from separate fibers and polymer. Developments involving liquid-crystal polymers and long, but not continuous, fiber reinforcement and discussed by Cogswell (1987).

In making a preform the aim is to produce a sheet or tape of constant thickness in which the fibers are aligned and uniformly distributed and individually wetted by the matrix. Since they are amorphous and hence more readily soluble, solvent impregnation was first used with polysulfones, polyether sulfone and later polyamide imide. The sheet material produced must be thin so that the solvent can be removed before any further fabrication operations. Crystalline or semicrystalline polymers tend to have better solvent resistance and melt impregnation is used to fabricate feedstock. Sometimes, to improve alignment, unidirectional fibers are stitched together with thermoplastic fibers based on, for instance, PEEK. Intimate blends of carbon fibers and PEEK, PEK or polyether sulfone fibers can be knitted, woven or combined together as filaments. This leads to better fiber wet-out and an absence of polymer-rich areas. Another method of preparing a fabric or sheet is to impregnate with a suspension of fine (>5 μm) polymer particles suspended in water or another inert liquid. This method is applicable to polysulfones and polyether ketones as well as polyether imide and polyphenylene sulfide, but not to liquid-crystal polymers since these cannot be cryogenically ground to produce fine particles, because of their inherent toughness and fibrous nature. Pultruded sections or tape may be prepared using a similar approach. Fibers are pulled through powdered resin (30–250 μm) in a coating unit and then shaped and heated in a die. The production rate can be as high as 1 m s^{-1} with the fiber volume loading not deviating by more than ±1 vol%. The method has been used with polysulfones, polyphenylene sulfide, PEEK, polyether imide and polyamide imide. Finally the fibers may be coated with a prepolymer which is then reacted to give a linear thermoplastic. Commercial manufacturers frequently refer to their process as proprietary.

It has been suggested that reinforced thermoplastic sheet and pultruded sections have a lower fiber loading (~55 vol%) and poorer fiber alignment than their thermoset counterparts. The preform products should be stored in clean, dry conditions, since the matrices will pick up small amounts of water (see Table 1). Unlike thermoset prepregs, which become tacky at room temperature, thermoplastic-based materials are stiff and do not become drapable until above their T_g or, for crystalline polymers, above T_m. If the material is unidirectionally reinforced it will easily split unless handled carefully. A soldering iron is recommended for joining pieces of preform before solidification.

Starting materials based on all the polymers listed in Table 1 except PEK, which is a very recent product, and liquid-crystal polymers, are in principle available, though most commercial emphasis

Table 1
Properties of unreinforced polymers: *indicates notched specimens, + indicates specimens exposed for 100 000 h

	Polysulfone	Polyether sulfone	Polyether ether ketone	Polyether ketone	Polyether imide	Polyamide imide	Polyphenylene sulfide	Vectra (LCP)	Xydar (LCP)
Crystalline	no	no	semi-		no	no	semi-	liquid crystal	liquid crystal
Density (Mg m^{-3})	1.25[a]	1.37	1.26–1.32		1.27	1.4[d]	1.36[i]	1.37–1.4	1.35–1.4[j]
σ_f(MPa)	106	129	170		145	197	154[i]	169–245	110–134[j]
E_f(GPa)	2.7	2.6	3.6		3.3	4.8[d]	3.5[i]	9–15.2	11–13[j]
σ_t(MPa)	60–75[a]	84	93	110	105	93[d]	84[i]	165–188	80–123[j]
E_t(MPa)	2.5[a]	3.2[e]	3.2[e]		3.0		3.3[e]	9.7–19.3	
ε_f(%)	50–100	40–80	50		60	17[f]	4[i]	1.3–3.0	3.3–4.9[j]
Izod impact (J m^{-1})	70	76–84*	83*		50*	53[f]	21*[i]	45–530	75–210[j]
Fracture toughness (MN m$^{-3/2}$)	2.2[b]		7.5[b]						
G_{Ic}(KJ m^{-2})			6.6[l]			1.1	1.4		
LOI	32	36	35		47	43	44[i]	35–50	42–47[j]
HDT (1.8 MPa)(°C)	174[a]	200	150	165	200	273[f]	136[e]	222	316–355[j]
T_g(°C)	190[c]	230[c]	143	165	217		93[c]		
T_m(°C)			334	365			285	280	423[j]
Continuous service temp.(°C)	150	180$^+$	250$^+$		170	230–260[g]	200–240[h]		240[k]
α(°C^{-1} × 10^{-6})	55.8	55	47		62	63[d]	54	−5– +75	
Water absorption in 24 h at RT (%)	0.2[d]	0.43	0.1[d]		0.25[d]	0.3[d]	0.2[d]	0.02–0.04	

HDT: heat distortion temperature a Pye 1982 b Gotham and Hough 1984 c Muzzy and Kays 1984 d English 1985 e Specmat (UK) Ltd trade data 1987 f Collyer and Clegg 1984 g Collyer and Clegg 1985 h Woodhams 1985 i O'Connor 1987 j Collyer and Clegg 1986a k Collyer and Clegg 1986b 1 Leach and Moore 1985
All other entries are taken from trade data

Table 2
Room-temperature properties of thermoplastic composites based on polysulfone, polyether sulfone and PEEK: *indicates notched specimens

	Polysulfone 40 vol% glass cloth[a]	Polysulfone 55 vol% TOHO HTA7[b]	Polyether sulfone 40 vol% glass fiber[a]	Polyether sulfone 60 vol% carbon fiber[d]	PEEK APC 2 61 vol% AS-4[e]	PEEK APC 2 61 vol% 1M-6[e]	PEEK 61 vol% Celion G30-50[f]	PEEK T 300[g]
Density (Mg m^{-3})					1.657			
σ_f(MPa)	172	1560	214	1551	1880	2170	2160	1796
E_f(GPa)	83	112	11	94	121	151	131	139
ILSS (MPa)		75		79			90	107
σ_t(MPa)	131		159	1232	2130	2700	1840	
E_t(GPa)	11.7		13.8	153	134	176	145	
ε(%)		1.45			1.45	1.48		1.29
σ_c(MPa)	166		152		1100	1100	1075	
σ_t transverse (MPa)					80			
E_t transverse (GPa)					8.9			
σ_f transverse (MPa)					137	160		171
E_f transverse (GPa)					8.9	9.3		
Izod impact (j m^{-1})	84.6*		79*					
Instrumented impact (KJ m^{-2})		60[c], 62[c]		78[c], 114[c]				75[c] (melt)
G_{Ic}(KJ m^{-2})					2.4	2.5		
HDT (°C)	185		216					
α (°C$^{-1}\times10^{-6}$)	2.34		2.52					

ILSS: interlaminar shear strength; G_{Ic}: critical work of fracture a English 1987 b Weiss and Huttner 1987 c Stori and Magnus 1983 (first reading is for a solution-impregnated specimen, second for a melt impregntated one, both of which have 50 vol% fiber loading) d McMahon 1984 e Anon 1986 f Clemans et al. 1987 g Owens and Lind 1984

appears to be on reinforced PEEK, polyphenylene sulfide and polyether imide products. Phillips Petroleum produce pultruded structural sections and an aramid-fiber tape, both with a polyphenylene sulfide matrix, but the most widely used system appears to be ICI's APC 2 based on carbon fibers in a PEEK matrix. At least one UK manufacturer produces finished articles based on liquid-crystal polymers.

4. Fabrication of Thermoplastic Composite Artifacts

Given a supply of preform the final step is to convert the material into the finished article. One method of doing this involves stacking layers of fiber-reinforced preform with the desired orientation, possibly interspersed with very fine layers of polymer film of either the same type as used in the

Table 3
Room-temperature properties of thermoplastic composites based on polyether imide and polyamide imide

	Polyether imide 55 vol% Carbon fiber[a]	Polyether imide 54 vol% glass fiber[a]	Polyether imide 52 vol% Kevlar fiber[a]	Polyamide imide 0/90 T300[b]	Polyamide imide 0/90 Celion 3000[b]
σ_f(MPa)	711	643	253	1000	1010
E_f(GPa)	51	26	54	63	59
ILSS (MPa)				75	100
σ_t(MPa)	527	355	396		680
E_t(GPa)	57	24	41		62
ε(%)	1.6				
σ_c(MPa)					560
Inst. impact (KJ m^{-2})	28[c] (sol.)				

a Specmat (UK) Ltd. trade data 1987 b McMahon 1984 c Stori and Magnus 1983

Table 4
Room-temperature properties of thermoplastic composites based on polyphenylene sulfide

	70 vol% glass fiber[d]	66 vol% carbon fiber[a]	55 vol% pultruded[b] carbon fiber[b]	56 vol% carbon fiber[c]	61 vol% Celion G30–50[d]
Density ($Mg\,m^{-3}$)	2	1.6	1.52		
σ_f(MPa)	1159	1310	1173	1366	1770
E_f(GPa)	44.2	124	96.9	124	131
ILSS (MPa)		70			62
σ_t(MPa)	911	1656	1380	1172	1820
E_t(MPa)	49.7	135	117.3	130	117
ε(%)			1.1	0.8	
σ_c(MPa)	759	655	586		634
Izod impact ($J\,m^{-1}$)		1586	2078		

a Phillips Petroleum trade data b O'Connor 1987 c Specmat (UK) Ltd. trade data 1987 d McMahon 1984

preform or another kind. The stack is processed by pressing in a matched mold at a pressure of 1–10 MPa and at a temperature in excess of T_g, or T_m if appropriate. Processing times vary from 30–300 s while it has been suggested that 30 min may be required. The exact time will depend on the thickness of material used, and laying-up and cooling may increase the process time very considerably. Care must be taken in molding large, shaped components with shallow curvature because of the long slip path between adjacent plies which can result in fiber wrinkling when a compressive force is applied. In this case, shaping individual preform sheets before fusing them is recommended. As with thermoset fabrication, it is necessary to remove air and any solvents that have been used in preparing the preform before final consolidation. To do this, a properly designed mold with adequate venting or the application of a vacuum is required. Another method is to vacuum-form using an autoclave in a manner similar to that used for a thermoset prepreg. A disadvantage of this method is the high temperature that is required.

Certain sheet-metal-forming techniques can be used to form thermoplastic-composite artifacts. In hydroforming a preheated, consolidated blank is formed to the shape of the mold by using a hydraulically pressurized flexible diaphragm or block of rubber. Because of the hydraulic nature of the loading, an even pressure is applied over the whole surface of the charge, or blank, minimizing fiber damage. A typical forming time of 15 s is claimed. Another method which can be used to produce complex shapes is diaphragm forming. This is analogous to the superplastic forming of metal alloys. A charge of, for instance, APC 2 is laid up in the requisite way and encased between two sheets of Supral aluminum. Air is evacuated from the enclosed charge and the temperature raised to 380 °C when the aluminum sheets become superplastic. Differential pressure and a shaping tool can be used to shape the sandwich structure. The method can presumably be used with other types of thermoplastic, provided that the polymer can withstand the temperature at which the aluminum becomes superplastic. This process can be used to produce complex shapes and is very competitive for short runs. The cycle time can be as low as 20–40 min. One other technique, borrowed from the metal-forming industry, is roll forming. Continuous lengths of preform are heated, consolidated and formed, and further consolidated using shaped rollers. In this way corrugated sheet and "top hat" sections, among others, can be made. The throughput speed is up to $0{\cdot}17\,m\,s^{-1}$.

Other means of fabrication available are filament winding and tape laying. Filament winding is widely used in thermoset fabrication. The main differences when it is applied to thermoplastics are that a filament already impregnated with the matrix is used and additional heating, possibly of both the mandrel and filament, is required to fuse the filaments together. Methods of heating include infrared, lasers, microwaves, induction, contact, ultrasonic welding, flame and nonoxidizing hot gas. Advantages of thermoplastic filament-winding include the ability to deviate markedly from geodesic paths and to fabricate structures with reentrant surfaces. Tape laying is used to fabricate large, planar, or low curvature components. Because of the geometry, it is not possible to consolidate by applying tension to the tape. Instead, an external head must be used. In another method, preform tape is spirally wound between two parallel man-

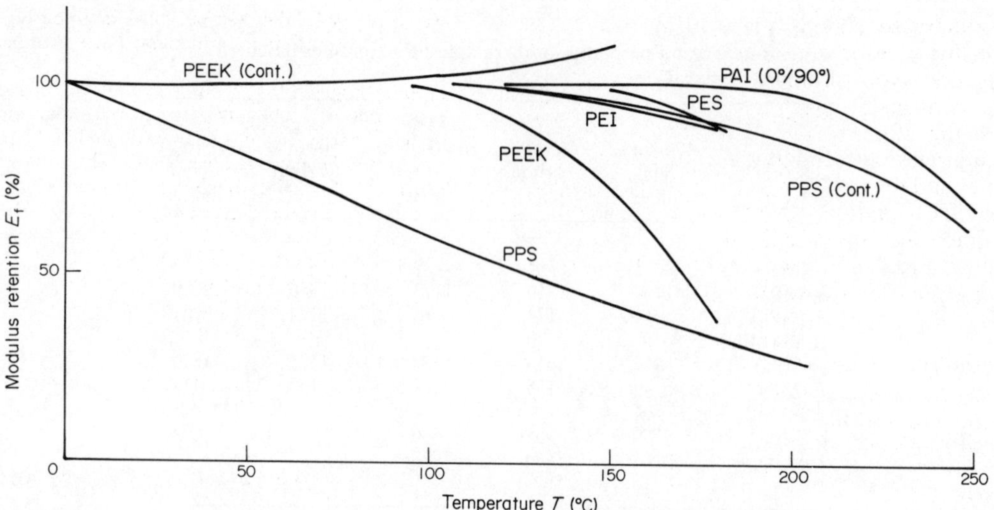

Figure 2
Percentage modulus retention of thermoplastic-matrix composites plotted against temperature. Continuous and 0°/90° reinforcement is carbon fiber; in other cases the reinforcement is 30% short glass fiber

drels with continuous butt welding to create a single-ply tube with any fiber orientation between 0° and 90°. The welding speed for PEEK is $0{\cdot}03\ \mathrm{m\,s^{-1}}$.

Stamping has been suggested as a means of forming components. The tooling costs are moderate and the process suitable for long runs and large, simply shaped components. Phillips Petroleum have developed a polyphenylene sulfide reinforced with chopped or swirl mat glass or woven and nonwoven glass fiber. A temperature of 315–343 °C and a pressure of 14–42 MPa are required for stamping. The cycle time is 20–120 s.

Further details and references on all methods of fabrication, especially when applied to reinforced PEEK (APC 2), are given by Cattanach and Cogswell (1986).

5. *Composite Properties*

Some room-temperature thermomechanical properties of carbon-, glass- and aramid-fiber composites with thermoplastic matrices are given in Tables 2–4.

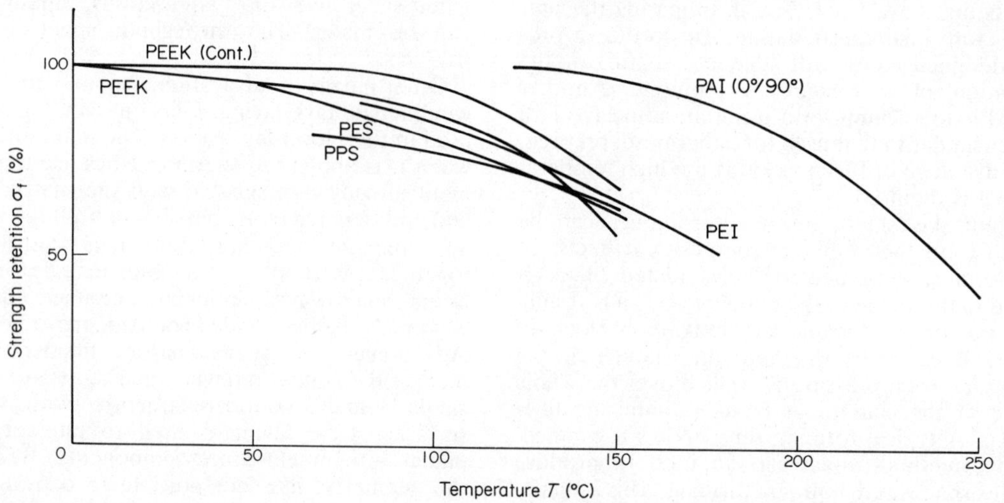

Figure 3
Percentage strength retention of thermoplastic-matrix composites plotted against temperature. Continuous and 0°/90° reinforcement is carbon fiber; in other cases the reinforcement is 30% short glass fiber

Unless otherwise stated, the specimens are unidirectionally reinforced. Unfortunately no data covering one type of fiber with all the different matrices is available and thus it is not possible to compare the effectiveness of the various thermoplastic matrices. Information on high-temperature properties and the performance of laminates is available for PEEK and polyphenylene sulfide materials in the trade and open literature. The percentage retention of modulus and strength with temperature for some systems is shown in Figs. 2 and 3.

Bibliography

Anon 1986 Development of thermoplastic composites for airframe construction. *Aerospace Design and Components* Oct–Nov 1986, 16–21

Brydson J A 1982 *Plastics Materials*, 4th edn. Butterworth, London

Cattanach J B, Cogswell F N 1986 Processing with aromatic polymer composites. In: Pritchard G (ed.) 1986 *Developments in Reinforced Plastics*, 5. Elsevier, London

Clemans S R, Western E D, Handermann A C 1987 *Hybrid Yarns for High Performance Thermoplastic Composites*, Materials Science Monographs 41. Society for the Advancement of Materials and Process Engineering, Azusa, CA

Cogswell F N 1987 The next generation of injection moulding materials. *Plast. Rubber Int.* 12:36–9

Collyer A A, Clegg D W 1985a Polyimides—mechanical properties versus processability. *High Performance Plastics* 2(7): 1–5

Collyer A A, Clegg D W 1985b Radiation resistant polymers. *High Performance Plastics* 2(2): 2

Collyer A A, Clegg D W 1986a Self reinforcing polymers. *High Performance Plastics* 4(1): 2–7

Collyer A A, Clegg D W 1986b Xydar electrical component. *High Performance Plastics* 3(5): 8

English L K 1985 Aerospace composites. *Mater. Eng.* 102(4): 32–6

English L K 1987 Fabricating the future with composite materials, part 3: Matrix resins. *Mater. Eng.* 104(2): 33–7

Gotham K V, Hough M C 1984 *Durability of High Temperature Thermoplastics.* RAPRA, Shrewsbury, UK

Leach D C, Moore D R 1985 Toughness of aromatic polymer composites reinforced with carbon fibers. *Compos. Sci. Technol.* 23: 131–61

McMahon P E 1984 Thermoplastic fiber composites. In: Pritchard G (ed.) 1984 *Developments in Reinforced Plastics,* 4. Elsevier, London, Chap. 1

Marks H F, Atlas S H 1965 Principles of polymer stability. *Polym. Eng. Sci.* 5: 204–7

Muzzy J D, Kays A O 1984 Thermoplastic versus thermosetting structural composites. *Polym. Comp.* 5: 169–72

O'Connor J E 1987 Polyphenylene sulfide pultruded type composite structures. *SAMPE Q.* 18: 32–8

Owens G A, Lind D J 1984 *Fibre Reinforced Composites,* Conf. Proc. Paper 12. Plastics and Rubber Institute, London

Pye A M 1982 High performance engineering plastics. *Mater. Eng.* 3: 407–9

Rose J B 1984 High temperature engineering thermoplastics from aromatic polymers. *Plast. Rubber Int.* 10: 11–15

Stori A, Magnus E 1983 An evaluation of the impact properties of carbon fibre reinforced composites with various matrix materials. In: Marshall I H (ed.) 1983 *Composite Structures*. Elsevier, London, paper 24

Weiss R, Huttner W 1987 *High Performance Carbon Fiber Reinforced Polysulfone,* Materials Science Monographs 41. Society for the Advancement of Materials and Process Engineering, AZUSA, CA

Woodhams R T 1985 History and development of engineering resins. *Polym. Eng. Sci.* 25: 446–52

N. L. Hancox
[UKAEA Harwell Laboratory, Didcot, UK]

High-Pressure Gas Containers: Safety and Efficiency

Currently there are well over 100 million high-pressure gas cylinders known to exist worldwide, of which about a half are used in the USA, with smaller countries such as the UK responsible for about 5% of the total. This latter figure of about five million cylinders is to be compared with a volume of less than 250 000 cylinders in the UK in the late nineteenth century. The tremendous growth in the cylinder industry was essentially launched in the 1870s with the birth of the compressed gases industry, which today forms the basis of numerous industrial, medical and leisure activities.

However, even the very large numbers mentioned above appear insignificant when the vast volumes of low-pressure cylinders used for transporting liquefied petroleum gas products are considered. This article will concentrate only on high-pressure cylinders, primarily of seamless construction, which are filled, in the main, with permanent gases, dominated by three components of the atmosphere (namely oxyen, nitrogen and argon) and carbon dioxide.

Small low-pressure copper vessels were used as early as the 1850s although the first gas containers, as they are known today, were manufactured in 1874 for transporting liquefied carbon dioxide for the aerated drinks industry. Then, as now, it was for a military application that a major breakthrough was achieved in the early 1880s, when wrought-iron welded vessels were used for transporting hydrogen by British expeditionary troops to inflate balloons in Asia and Africa. However, the low tensile strength of the wrought iron meant the walls of the vessel were thick, resulting in a heavy container. In fact only 25 m^3 of gas (reckoned at 15 °C and 1.013×10^5 Pa) was transported in a vessel weighing in excess of 500 kg.

In more recent times, aluminum alloys have also been used for transporting compressed gases. Their

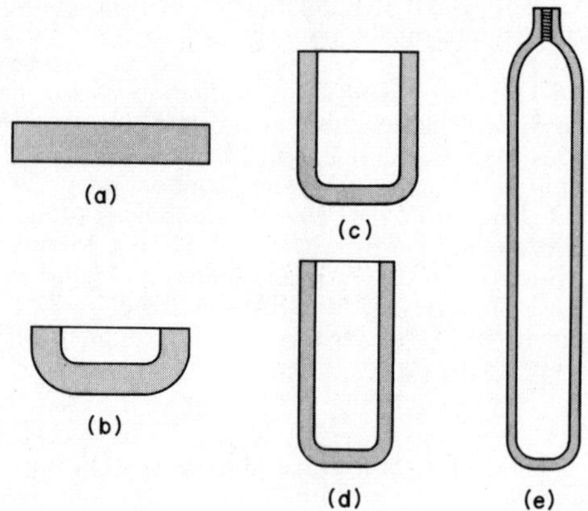

Figure 1
Stages of cylinder production in Lane and Taunton's original patent application: (a) original plate stock, (b) plate "cupping," (c)–(d) further cold reduction and (e) final completed cylinder

lower density compared with steels, coupled with the corrosion resistance of the material, makes them suitable for certain applications, particularly for the containment of high-purity gases.

A forerunner of a popular present-day technique for manufacturing steel cylinders was patented in 1885 by Lane and Taunton in Birmingham, UK. Using a steel plate as their starting stock, this being assisted by concurrent developments in steelmaking processes, a cylinder was produced via a cold-drawing route (see Fig. 1). Such a seamless cylinder rendered the cold route obsolete and before the beginning of the twentieth century seamless cylinders were being manufactured from not only plate, but also hot billet forging and by spinning steel tubes. All three processes are widely used today.

1. Safety Considerations

A modern-day cylinder, when fully charged with gas, stores the equivalent of almost 0.5 kg of TNT if allowed to rupture in an uncontrolled manner. This fact, coupled with the consideration that as a developing technology the industry was faced with some unfortunate incidents in its infancy, meant that cylinders needed to be controlled by a regulatory authority. Since the majority of gas cylinders were transported by rail, particularly in Europe, many of the incidents were in the vicinity of railway stations. Hence as early as 1890, the conditions for the safe transportation of dangerous goods by rail, the "Convention de Berne," were formulated. This regulation was the forerunner of the current RID regulations and was later extended to cover the international carriage of dangerous goods by road, by the implementation of the ADR regulations.

These international regulations paved the way and often formed the basis for different countries to implement their own national codes. Within the UK, it was the publication of the gas cylinders (conveyance) regulations in 1931 that was the first major instrument of legislative control. These regulations are still in force today, although the imminent "pressure systems and gas container regulations" which have now been finalized, will supersede those of 1931 in 1991. In the USA, the Department of Transportation lays down the regulations for the design, testing and maintenance of high-pressure cylinders.

The regulations in force worldwide ensure that all relevant technological requirements are met in such a way that a cost-effective component will result. Having obtained a safe cylinder at the time of manufacture, the user of that cylinder must then maintain the vessel in a satisfactory condition for its intended service, for as long as possible. This is because the product being conveyed is, in the vast majority of cases, considerably cheaper than the cost of the vessel. (The exceptions are some recent special gas mixtures primarily intended for standardization and high-technology applications.) Thus, in order for a cylinder to be cost effective with respect to its high initial investment, it is essential that the cylinder has to be capable of withstanding many years of continuous use.

The cylinder is often used under quite demanding environments (e.g., in civil engineering projects, off-shore exploration and heavy engineering) where either singly or in combination, extremes of temperature, high corrosion rates and rough handling are frequently experienced. Even after encountering a certain degree of misuse, the design and construction of the cylinder should be such as to maintain the package in a safe and integral form. These considerations presuppose the primary objectives of a cylinder which are for it to be a gas-tight package whose material is chemically compatible with the contents; for it to be capable of withstanding the stresses generated by its contents; and for it to permit a large number of repeated pressurization cycles. The transportation of gas cylinders across national boundaries is becoming widely accepted. Hence, after many years of effort and negotiations an international specification ISO4705 has finally been agreed. Also, in March 1986, the European Community published three directives relating to the design and manufacture of seamless steel, seamless aluminum and welded steel cylinders.

The above specifications and the various national specifications pertaining to gas cylinders are evolving as a result of the acceptance of a particular design philosophy, while additional requirements that reflect the consequences of cylinder handling,

Table 1
Improvements in the technical efficiency of seamless steel cylinders

Year	Steel type	Technical efficiency (O_2, $m^3\,kg^{-1}$)	Improvement over 1930 (%)
1930	High-C	0.04	
1931	Low-C	0.07	75
1942	C–Mn	0.09	125
1960	Cr–Mo	0.14	250
1982	Cr–Mo	0.16	300
1986	Cr–Mo	0.18	350

distribution systems, filling considerations and demands from users of compressed gases are, in places, still to be agreed.

2. Materials

As mentioned in Sect. 2, the vast majority of high-pressure cylinders are now manufactured from essentially two families of alloy steels. In the past, both high- and low-carbon steel cylinders were manufactured but in the pursuit of more efficient gas cylinders (i.e., the volume of gas at normal temperature and pressure carried per unit weight of the cylinder) the carbon steels have been replaced by carbon–manganese steels and, since the 1960s, by steels containing chromium–molybdenum additions. The latter category forms the basis of the vast majority of cylinders being manufactured today.

The improvements in the technical efficiency of steel cylinders are shown in Table 1, where the role of additional alloying elements can be clearly seen. This improvement is a direct consequence of superior mechanical properties (Table 2) of the steel being used, although some of the smaller benefits have been from changes in design considerations.

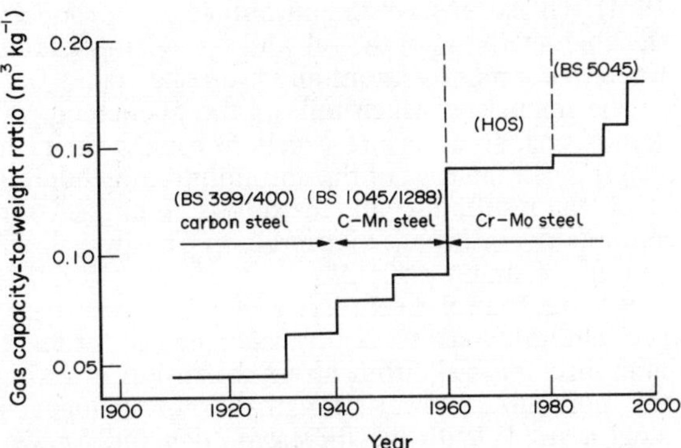

Figure 2
Progressive changes in efficiency of seamless steel cylinders in the UK

This almost 250% improvement in cylinder efficiency (in the case of the UK) achieved since the 1920s is schematically shown in Fig. 2.

As well as the higher strength and toughness properties of the carbon–manganese and chromium–molybdenum steels, modern steelmaking practice has led to a more reliable and consistent product. This, in turn, has led to design considerations being realistically reflected in actual cylinders rather than building into the component a high degree of safety by gross over-design.

The highest mechanical strengths are obtained by quenching and tempering both the chromium–molybdenum and sometimes the carbon–manganese steel cylinders once the latter have been manufactured to the required shape. The quenching is achieved by immersing the cylinder from above its austenitizing temperature into a quench tank, thus achieving a martensitic microstructure. The quenchant is either mineral oil or, increasingly, advanced polymer-based solutions are used (Irani and Hayes

Table 2
Composition, heat treatment and mechanical propeties of seamless steel gas cylinders

	Chemical composition (wt%)									Mechanical properties		
Steel	C	Si	Mn	P	S	Cr	Mo	Ni	Heat treatment[a]	Yield stress[b] (MPa)	Tensile stress (MPa)	Elongation (%)
Low-C	0.15–0.25	0.05–0.35	0.4–0.9	0.05 max	0.05 max				N or N+T	250	430–510	22
High-C	0.35–0.45	0.05–0.35	0.6–1.0	0.05 max	0.05 max				N or N+T	310	570–680	19
C–Mn	0.40 max	0.10–0.35	1.3–1.7	0.05 max	0.05 max				N or N+T	445	650–760	20
Cr–Mo	0.37 max	0.10–0.35	0.4–0.9	0.05 max	0.05 max	0.8–1.2	0.15–0.25	0.50 max (residual)	Q+T	755	890–1030	14

a N = normalized, Q = quenched, T = tempered b Maximum design value; minimum actual value

1984). The latter have the advantage of reproducing the characteristics of the oil while providing a fume-free environment without any fire risks.

The tempering which follows the quenching process is such as to ensure excellent toughness in the material. In the case of the chromium–molybdenum steel, the resultant cylinder exhibits an impact transition temperature, as measured on Charpy V-notch testing, of about −50 °C.

Whenever a relatively corrosive gas such as hydrogen chloride has to be transported, in order to minimize the risk from stress–corrosion cracking, the normalized lower-strength carbon–manganese steel is used. With the increasing demand for very closely specified specialty gases, such as those essential to the production of silicon chips in the semiconductor industry, the use of stainless-steel cylinders is increasing. However, the latter material results in a heavy cylinder on account of the low strength of the alloy, and this, coupled with its high initial cost, is deterring wider application.

In certain cases, the exact choice of cylinder materials relies on the purity of the contained gas(es) and the need to maintain the purity. In particular, the presence of trace impurities of moisture, although within product specification, may dictate the type of cylinder for its safe operation.

A mention must here be made of the significant number of failures during the 1970s of steel cylinders in hydrogen trailer service, although none in a catastrophic manner. Harris and Priest (1984) established that such cylinders, which are subject to a high number of filling cycles on account of their intended service, were most prone to hydrogen-assisted cracking if the Brinell hardness of the cylinders exceeded 290 HB (the then British Standard permitted hardness for hydrogen cylinders up to 312 HB). Additionally, as would be expected, the cracks were in the "knuckle" zone of the cylinder, where there is a sudden change in the wall thickness, and were generally initiated from surface flaws. These results were incorporated into Appendix E of the British Standard Specification BS 5045 part 1 (1982) where a maximum hardness of 290 HB, a tighter ultrasonic inspection and specific heat treatment procedures are stipulated.

A number of different routes and alloys are used for cylinders manufactured from aluminum alloys. In the USA, UK, Australia and many other parts of the world, the 6000 series of alloys is the preferred choice. This is primarily due to this family of alloys not exhibiting stress–corrosion cracking, a very important consideration in view of the vast range of gases currently filled in aluminum-alloy cylinders. Their relatively inert and smooth internal surface makes them particularly suitable for special gas mixtures applications where product stability is a major consideration. For very stringent applications, even this high-quality surface finish needs to be further treated by the gas company before it is acceptable.

Recently, however, a failure mechanism involving room temperature grain-boundary creep cracking (in the neck and shoulder regions of the cylinder) has been identified in some of the 6000 series alloys (Guttman 1983). Ways of minimizing the failure have been identified which include a reduction in the lead content of the alloy and refining the grain structure in the uppermost part of the cylinder.

In recent years, composite cylinders have been manufactured and used in situations where the lightness of the cylinder is of paramount importance. These cylinders are made from a thin metallic liner which is overwrapped with strands of fibers embedded in an epoxy resin. The liner is often an aluminum alloy or a high-strength steel, while the fibers used vary widely (e.g., carbon, glass and Kevlar). Often, the vessel is a conventionally shaped cylinder with either the parallel section fiber-reinforced by means of hoop wrapping or the entire surface overwrapped (i.e., the parallel section and the ends). Less conventional are the overwrapped oblate spheroid and (when the maximum technical efficiency is required) a totally overwrapped spherical liner.

After cylinders have been wound, the epoxy resin has to gel and cure in order to achieve a hard, wear-resistant and chemically impervious layer over the filaments. This is performed by a series of curing operations until the cylinder can be safely handled. Next comes a crucial part of the overall manufacturing cycle, whereby the cylinder is subjected to an autofrettage process. The hydrostatic pressure applied during the autofrettage stage is high enough for the liner to deform plastically, so that when this pressure is removed the liner is under a compressive stress, thus permitting the liner and the fibers to operate as a truly composite component.

Though not yet a competitor for industrial use, composite cylinders have applications where the lightness of a vessel, and not its cost, is important.

A project in the UK on 'advanced gas containers' (Irani and Oldfield 1985) has demonstrated the importance of control over the microstructural characteristics of the steel in order to obtain optimum strength and toughness parameters. The experiences with hydrogen trailer cylinders also demonstrated the role of flaws and defects in the finished product as a major rate-controlling factor in the life of a cylinder.

However, with gas users requiring ever increasing purity and consistency of product over the entire life of the contents of a cylinder, attention is now being focused on the quality of the internal surface of a cylinder which is being constantly upgraded. As an example, a minority gas at the parts per billion concentration is now expected to be retained in a parent gas for several months without degradation.

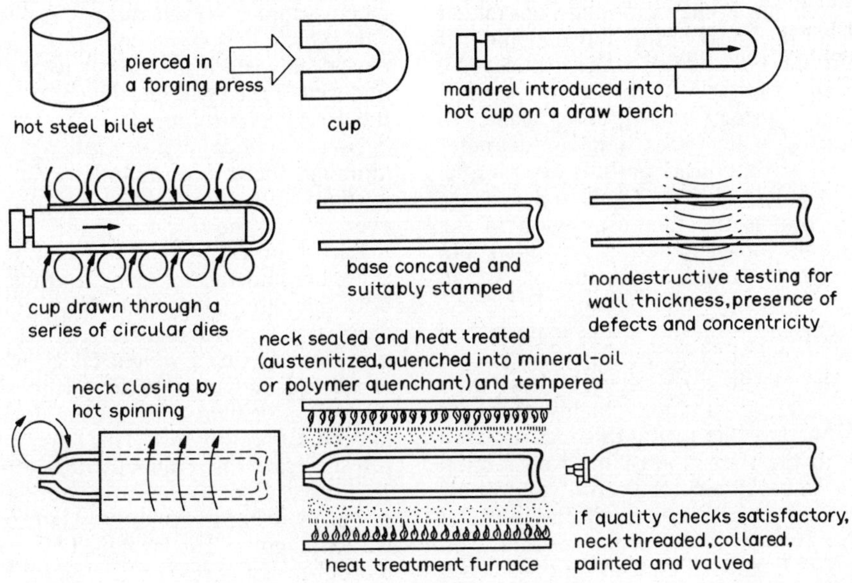

Figure 3
Principal steps in seamless steel cylinder production from a billet

3. *Manufacture*

Seamless steel cylinders can be manufactured involving a plate, billet or a tubular product as the starting stock material. A schematic of the billet piercing process, in its present-day form, incorporating the principal features, is shown in Fig. 3. With the introduction of standards and specifications, there was an international demand for quality products to be manufactured. Hence stringent controls were introduced involving rigorous hydraulic pressure testing, cyclic pressure testing, use of automatic ultrasonic inspection methods (to detect, locate and facilitate the correction of defects), computer-aided design techniques incorporating finite-element analysis, and so on.

In the UK the present specification for transportable gas containers is the BS 5045 series, part 1 of which relates to seamless steel cylinders and was last revised in August 1986.

This particular series was initially launched in 1974 by the publication of a specification relating to welded cylinders with only circumferential welds, with the seamless steel standard appearing in 1976. The formulation of the various parts of BS 5045 was largely the result of the implications of the report of the Home Office Gas Cylinders and Containers Committee published in 1969.

The current design equation used in the UK for seamless steel cylinders is based on the Lamé–von Mises formulae which result in a wall thickness t(m) given by

$$t=\frac{3\times10^{-3}P(\mathrm{OD})}{7\times10^{-5}f_e-4\times10^{-6}\,\mathrm{P}} \tag{1}$$

where P and OD represent test pressure (Pa) and outside diameter (m), respectively. The standard term f_e, which represents the maximum permissible equivalent stress (Pa) at test pressure, is an interesting one and one over which there is considerable debate on the international scene. In the UK a value of $0.75Y$ is used for most cylinders, where Y is the minimum specified yield stress, although a value of $0.875Y$ can be substituted for certain portable containers. This higher value for f_e, as can be seen from Eqn. (1), has the effect of reducing the wall thickness t, enabling a lighter cylinder to be produced for these "special" containers which have applications such as resuscitation apparatus, underwater breathing apparatus, fire extinguishers, and so on.

For a common industrial chromium–molybdenum steel cylinder (OD = 230 mm) intended to transport oxygen at 20 MPa (test pressure 30 MPa; that is, 50% higher than the settled pressure) substituting into Eqn. (1) gives:

$$t=\frac{(3\times10^{-3})(3\times10^{7})0.23}{(7\times10^{-5})0.75(7.55\times10^{7})-(4\times10^{-6})(3\times10^{7})}$$

$$=5.39\ \mathrm{mm}$$

This figure of 5.39 mm is the minimum guaranteed value for the wall thickness of the above cylinder if manufactured to the 1986 British Standard BS 5045/AMD 5145. In contrast, if the corresponding cylinder is designed using similar f_e and Y values to the EEC directive, which uses a mean diameter formula, the result is a considerably heavier vessel with a minimum wall thickness of 5.49 mm.

The cylinder wall is the thinnest region in the vessel and hence the most stressed. If a cylinder is to burst (e.g., under an extreme fire situation) it will normally rupture in its parallel portion. However, implicit in the design is that the strains experienced by the cylinder, under normal operating conditions, are always in the elastic zone. Nevertheless, even under normal operating conditions, high stresses are generated near the knuckle region of concave-based cylinders, due to the transition of the wall section into the thicker base. An essential prototype test for cylinders, involving pulsation testing, has revealed this region of a cylinder to be prone to failure on account of possible stress concentrations. With advances in finite-element analysis techniques, modern cylinders are designed to minimize such stress concentration effects.

Regardless of the choice of alloys or the manufacturing process, the finished cylinder must be capable of withstanding repeated pressure cycles, not only from the filling events (which could number about 5–10 per year) but also from the day-to-day pressure fluctuations resulting from changes in room temperature. Hence the chemistry and integrity of the alloy and the subsequent manufacture, testing and inspection of the cylinder must be such that the vessel, as manufactured, can be used for many decades, often under adverse environmental and service conditions.

4. The Role of Users

Upon receipt, it is the owner's responsibility to ensure that at all times the cylinder is maintained in a safe and satisfactory condition to enable it to be used for a further term of service. When cylinders are to be used in more than one country, additional inspection bodies are involved. It is hoped that the advent of European and international specifications will eliminate this unnecessary duplication, thus reducing the costs of inspection for such cylinders.

Almost always the filler of a gas cylinder is its owner. When, exceptionally, a supplier agrees to fill a cylinder belonging to someone else, whether or not a particular cylinder can be recharged is ultimately the responsibility of the filler. If there are any doubts regarding the integrity or suitability of a cylinder, it is subjected to a retest procedure. In the UK the BS 5430 series, which involves the "periodic inspection, testing and maintenance of transportable gas containers" is applied. This standard is also used when cylinders reach their reinspection time intervals which are stipulated in BS 5430. The actual period between inspections varies depending on the material of the cylinder, the gas being contained and the service environment.

No gas cylinder can be refilled if it is "out of test," although there is no mandatory requirement for a partially full cylinder to be returned for retesting even though the retest date may have passed.

Part 1 of BS 5430 is relevant to seamless steel cylinders and incorporates a visual examination of both the external and internal surfaces, looking for defects such as dents, bulges, fire damage, corrosion, and so on; a weight check which detects any anomalous reduction in the thickness of the steel; an inspection of the quality of the neck threads; and a hydraulic pressure test. This latter crucial part of the retest cycle was originally developed by Brier in 1886 (see Fig. 4) and is essentially still in use, primarily in North America. Many other parts of the world including Europe have been successfully using a proof-pressure test rather than the more elaborate water-jacket method proposed by Brier. This wide use of the proof-pressure test has been the result of considerable improvements in the quality of materials used, a more accurate understanding of the design and manufacturing processes, a refined

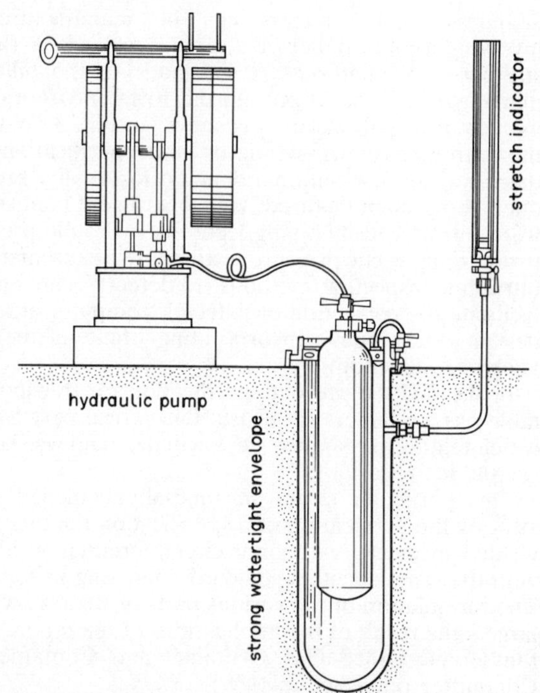

Figure 4
Water jacket as proposed by Brier (1886) for pressure-testing cylinders

Figure 5
A modern proof-pressure testing arrangement

inspection procedure and improved transportation methods in the gas industry. The worthiness of the proof-pressure testing has been justified by the excellent safety record of cylinders so tested. In fact, the main reason why so few cylinders fail at the hydraulic stage (see Fig. 5 for a modern proof-pressure testing arrangement) is that potentially suspect cylinders have been rejected on visual inspection, well before they reach the proof-pressure stage.

Often, individual gas companies have their own additional requirements to those in BS 5430, since such national documents are published on a consensus basis and do not necessarily meet all the requirements of all the parties concerned. Some of the testing under BS 5430, and other international reinspection specifications, is somewhat subjective and needs to be undertaken by competent staff. The time intervals betwen tests are also empirical and reflect experience rather than any solid technical grounds. Nevertheless, the effectiveness of the retesting procedures has proved itself over many years as a sound and economic route for returning cylinders for a further term of service.

If cylinders are found to be satisfactory, they are repainted (including any additional external corrosion protection such as metal-spraying), revalved, certified and refilled with the intended gas.

5. The Total Package

A large range of gas quantities, from about 0.7 m^3 of oxygen in a cylinder contained in a DIY welding kit to about 30 HCM (1 HCM = 100 m^3) in a trailer of hydrogen, can be conveniently transported in gas cylinders. Typically, a hydrogen trailer consists of about 250 large cylinders (say 70 l in water capacity and 20 MPa charging pressure) packaged and manifolded in appropriate bundles to meet the end user's requirements. Requirements for intermediate applications ranging from, for example, 10 m^3 to 150 m^3 can be catered for by a single cylinder or a package of manifolded cylinders convenient for lifting by a fork-lift truck.

Regardless of the number of cylinders in the package, the whole must be fitted with an appropriate valving system and all associated components such as fusible plugs (for acetylene cylinders where required), dip tubes (for liquid withdrawal systems), test-date rings (which signify date of next test), and so on.

In the case of a single cylinder, there are essentially three broad categories for its ease of mobility. These are listed in Table 3. Unlike the earlier discussion, which centered around the technical efficiency of a cylinder, here the important consideration is the ease with which the gas user can

Table 3
Portability aspects of various sizes of gas cylinders

Water capacity (l)	Portability	Implications
0–10	truly portable	low weight essential
10–20	semiportable	keep weight as low as possible
20–50	not portable but must be transportable	avoid excessive weight; marginal cost savings

conveniently move the package. The 10–20 l water capacity category is particularly interesting for, by reducing the weight of the cylinder, it may be possible to reclassify this category under the "truly portable" range rather than its present "semiportable" grouping. In fact, a major research and development project in the UK will make a significant contribution in this area of improved portability (Irani and Oldfield 1985). By careful development of the optimum mechanical properties, it has been proved that a cylinder whose wall thickness is less than those currently in use could be safely used.

When a cylinder is specified and eventually purchased, several of its dimensional features are fixed for its lifetime. Depending on the service conditions of the cylinder this may extend from a few years to decades. Hence it is essential, both for the gas companies (who have automated handling and filling capabilities), and the users of compressed gases (who possess auxiliary equipment such as trolleys or who are accustomed to fitting cylinders in limited space), that the external dimensions of a cylinder remain approximately constant. Hence maintaining the outside diameter and height of a cylinder are of primary importance and need to be reviewed with changes in national and international specifications. To this end, the early cylinders took a common pipe diameter of 9 in (230 mm approximately). Since there are now many millions of cylinders in international traffic, this feature is unlikely to change for a large proportion of high-pressure cylinders. Nevertheless, as new products are marketed, or for reasons of gas users' convenience, a number of diameters exist, particularly for cylinders of less than 20 l water capacity.

With this requirement for external dimensional standardization, extreme care must be exercised in the interest of safety by users of gas cylinders that changes in specifications that permit the use of a thinner-walled vessel are clearly recognizable. In the UK, every permanent gas cylinder is hard stamped in its shoulder region with several requirements including its maximum settled pressure. However, since the latter can at times be difficult to identify conveniently, a more noticeable device is employed. One method is to rivet a particular shaped collar, for a given pressure, around the neck of the cylinder.

As well as for the reasons mentioned above, the height of a cylinder needs to be controlled, for having fixed the diameter and wall thickness, the height must be fixed in order to achieve a constant volume. This parameter, apart from assisting marketing considerations for ease of invoicing, is of particular relevance when dealing with liquefied gases. If the latter are filled into cylinders with a lower volume than expected, then overpressurization of the cylinders could occur, leading to a possible rupture. Also, the need for manifolded packages presupposes a constant height as this considerably assists the construction and ease of transportation of such packages.

6. *Future Considerations*

Work will continue to improve the portability and efficiency of cylinders. Higher-strength steels and aluminum alloys will be developed while fiber-wrapped composite cylinders, which are already being marketed for specific applications, will gain wider acceptance. Enhancement of our understanding of key design features by means of finite-element analysis and more sophisticated stress-analysis models will enable better prediction of the mode and location of potential failures. Improvements in manufacturing technology, including superior heat treatment methods such as the use of advanced quenchants (Irani and Hayes 1984) or even air-cooling grades of high-strength alloys, will further cylinder technology.

Stricter cleanliness of the internal surfaces including closer attention to surface finish will be increasingly demanded by gas users. In this respect, alternative methods for periodically inspecting cylinders, which eliminate the introduction of water as part of the hydraulic test, will be developed. Already ultrasonic testing is a well-respected technique. The latter, and rival approaches such as acoustic-emission testing, currently being developed in conjunction with tube trailers (Harman and Kelly 1984), with further development and in-depth understanding could provide an alternative.

Despite over a century of experience and exhaustive prototype testing, the failures involving hydrogen trailer cylinders illustrate how certain combinations of operating conditions are only really exposed when high-pressure cylinders are put into service. A new development, once successfully through its set of formal acceptance tests, needs to be evaluated in a controlled field-trial. If felt relevant, findings from the field-trials can then be incorporated into a manufacturing and/or periodic inspection specification.

See also: High-Pressure Gas Cylinders: Integrity

Bibliography

Gordon R 1984 *IOMA Broadcaster*. July, pp. 7–11

Guttman M 1983 *Metal Service* 17: 123–40

Harris D, Priest A H 1984 Hydrogen-assisted cracking of gas container steels. *Integrity of Gas Containers: Materials Technology*. National Physical Laboratory, Teddington, UK, pp. 69–82

Hartman W F, Kelly M 1984 *IOMA Broadcaster*. May, pp. 9–12

Irani R S, Hayes D M 1984 Assessing aqueous quenchants for seamless steel gas cylinders. *Integrity of Gas Containers: Materials Technology*. National Physical Laboratory, Teddington, UK, pp. 90–106

Irani R S, Oldfield F K 1985 The objectives of the advanced gas container project. *Proc. Seminar Pressurised Transport Containers*. Institute of Mechanical Engineers, Northwich, UK, pp. 1–5

R. S. Irani
[BOC, Guildford, UK]

High-Resolution Electron Microscopy

In the technique of high-resolution transmission electron microscopy two or more beams of the electron diffraction pattern located in the back focal plane of the objective lens of a transmission electron microscope recombine to form an image in which there is fine detail at an atomic level. In certain favorable circumstances such images can be interpreted directly in terms of the projected potential of the material under investigation, but ambiguities in the interpretation of high-resolution images can arise if the material is too thick or if it contains defects, even if the interpretation is backed up with detailed image simulations based on the dynamic scattering behavior of high-energy electrons. The technique has wide applications in materials science and, for example, has been used to obtain valuable information about the atomic scale structure of interphase interfaces, the surface structures of catalytic particles, the layered structures of ceramic superconductors and the crystal structures of microcrystalline zeolite molecular sieves.

The current generation of commercial high-resolution electron microscopes use high-energy electrons accelerated through potential differences of 200–1000 keV and have interpretable resolutions approaching 0.15 nm which, although far superior to the resolution limit of visible light, is nevertheless orders of magnitude greater than the wavelength of the electrons. This is because the resolution is limited by aberrations in the electron lenses, the two most important of which are spherical aberration and chromatic aberration. Although there are still some gains to be made in improving voltage and current stabilities and in using sophisticated electron lens designs, it is likely that the most important advances in high-resolution electron microscopy will come from improvements in the specimen environment and in the range of auxiliary equipment that can be used to extract chemical information at the atomic level.

1. Instrumentation

Commercial high-resolution electron microscopes (HREMs) in the 200–400 keV potential difference range have the same external appearance as conventional transmission electron microscopes (see *Transmission Electron Microscopy*) and it is the differences in their internal instrumentation which allow high resolutions to be attained. Electron sources in HREMs tend to be either pointed tungsten filaments or lanthanum hexaboride (LaB_6) filaments, both of which give high brightness and enable direct magnifications of over $\times 500\,000$ to be usefully employed. An alternative to these two electron sources is the field emission gun with a tungsten cathode. This combines a low energy spread with a high brightness, both of which are very desirable characteristics, but the drawbacks of a field emission gun are that it needs a high vacuum of the order of 10^{-10} torr and that it is difficult for it to provide a uniform illumination of samples over reasonably sized areas.

The other main difference in the instrumentation between conventional and high-resolution transmission electron microscopes is in the design of the objective lenses. In HREMs the focal length of the objective lens is kept as short as possible (typically less than 2 mm) to minimize lens aberrations (see Sect. 2). In practice this means that there is little room in which to put the specimen and, in the effort to attain high resolution, compromises are usually made in the design of electron lenses that limit the range of specimen tilting to typically $\pm 15°$. It is also vitally important in an HREM to maintain a clean environment around the specimen to minimize specimen contamination over prolonged periods of microscope usage. This means that effective anticontamination shields are needed close to the sample, which also places limitations on the design of the objective lens. It is also not uncommon to have a top entry stage rather than a side entry stage (possibly even with a 2.3 mm diameter specimen holder rather than a 3 mm diameter specimen holder) and it is not unusual to dispense with microanalytical facilities if there is a need to place detectors near the specimen. However, instrumentation of all transmission electron microscopes has improved over the years and the current 400 keV microscopes available commercially tend to be either an HREM version with an interpretable resolution of around 0.18 nm, possibly with microanalytical facilities but with limited tilting facilities, or a version with an interpretable resolution of around 0.25 nm with large scale tilting of ±45° and with the option of incorporating extensive microanalytical facilities.

It has also been increasingly common to attach an image intensifier to the base of the lens column and to send the image to a TV screen, so that the microscope user is able to carry out alignment procedures and astigmatism corrections on images at magnifications of the order of ×20 000 000. Further instrumentation developments in this area have also meant that HREM images can be fed to a computer while the user is still at the microscope, and this has led to the development of systems for correcting specimen alignment and astigmatism on-line. However, it is still the case for the most part that HREM images are recorded on conventional electron microscope fast film and it is the skill of the operator which determines the quality of experimental work produced from high-resolution electron microscopes.

2. *Image Formation*

There are two principal imaging modes which are used to form HREM images: the many-beam axial-illumination mode, in which the undiffracted electron beam is coincident with the optic axis of the microscope, and the two-beam tilted-illumination mode, in which the undiffracted electron beam and a diffracted beam are symmetrically disposed either side of the optic axis (see Fig. 1). In general, the axial mode of high resolution imaging is the one used on the current generation of HREMs. The tilted-beam mode, which was used in the development of the technique of high-resolution electron microscopy, is now seldom used, although if care is taken this mode can give considerable success.

Image formation in the HREM can be treated conveniently by separating the physical process of forming an image into two separate stages mathematically. The first stage is the scattering by the specimen of the high-energy electrons produced by the electron source. This can be described by the dynamical theory of electron diffraction in which the

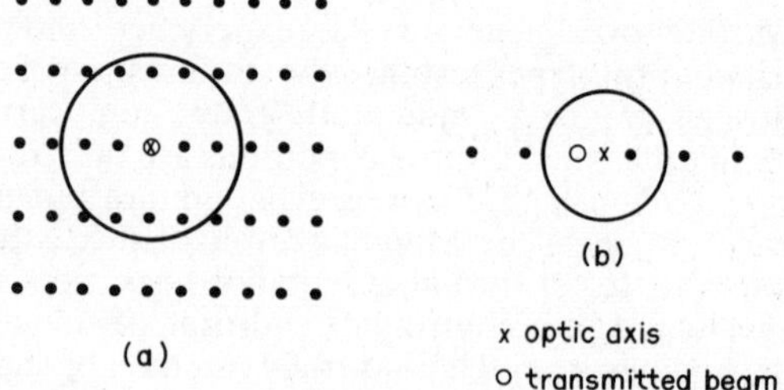

Figure 1
Imaging modes in an HREM: (a) the axial-illumination mode and (b) the two-beam tilted-illumination mode. In both (a) and (b) the outline of a suitable objective aperture is shown

wave function of an electron is divided into a linear sum of Bloch waves, but it is more usual in HREM image simulation computer programs for the scattering of the high-energy electrons to be described using the multiscale theory of electron diffraction. In the multislice approach, the specimen is divided into slices, each of which modifies the wave function of the electron by acting as a phase object, and the electron wave propagates from one slice to the next by Fresnel diffraction. The amplitude distribution, $F(\boldsymbol{u})=F(u, v)$, in the diffraction pattern of the specimen is then the Fourier transform $\mathcal{F}[\Psi(x, y)]$ of the amplitude distribution of the wave function $\Psi(x, y)$ of electrons after they have been transmitted through the specimen:

$$F(\boldsymbol{u})=F(u, v)=\mathcal{F}[\Psi(x,y)] = \int\int \Psi(x,y)\exp[2\pi i(ux+vy)]\,dxdy \quad (1)$$

where $\boldsymbol{u}$ is the spatial frequency in the specimen in the plane normal to the optic axis and has the dimensions of reciprocal space. The second stage in the image formation process is the transfer of information from the back focal plane of the objective lens, in which the diffraction pattern of the specimen is located, through to the final image. It is at this stage that the imperfections of the imaging system are taken into account. The function $F(\boldsymbol{u})$ is multiplied by a function $T(\boldsymbol{u})$, so that the amplitude distribution in the back focal plane of the objective lens is not $F(\boldsymbol{u})$, but $F(\boldsymbol{u})T(\boldsymbol{u})$. $T(\boldsymbol{u})$ modulates the amplitudes and phases within the electron diffraction pattern, taking into account the phase shifts caused by the spherical aberration C_s of the objective lens, the defocus ΔF of the objective lens and the attenuation of high spatial frequencies in the amplitude distribution caused by finite beam divergence, by chromatic aberrations in the objective lens and by instabilities in the high voltage and in the lens currents. A form of $T(\boldsymbol{u})$ commonly used in HREM calculations is:

$$T(\boldsymbol{u}) = \exp[i\chi(\boldsymbol{u})]\, E(\boldsymbol{u}) \quad (2)$$

where

$$\chi(\boldsymbol{u}) = \frac{\pi}{2} C_s \lambda^3 u^4 + \pi \Delta f \lambda u^2 \quad (3)$$

and the envelope function $E(\boldsymbol{u})$ is given by an expression such as

$$E(\boldsymbol{u}) = \exp(-\tfrac{1}{2}\pi\lambda^2\Delta^2 u^4) \quad (4)$$

where C_s is the spherical aberration coefficent, λ the wavelength of the electrons, ΔF is the defocus of the objective lens and Δ is termed the width of spread of defocus which is a consequence of the chromatic aberration of the objective lens and the current and voltage instabilities. To take account of the beam divergence α, $E(\boldsymbol{u})$ can be multiplied further by a function $J_1(\xi)/\xi$, where $\xi = 2\pi\alpha u(\Delta F + C_s\lambda^2)$ and J_1 is the first-order Bessel function. $T(\boldsymbol{u})$ can be multiplied by a suitable aperture function if an objective aperture is used to select beams within the diffraction pattern and can be further modified to take account of residual astigmatism in the objective lens which has not been corrected by the microscope user. The amplitude distribution within the image can then be described as the Fourier transform of $F(\boldsymbol{u})T(\boldsymbol{u})$:

$$\Psi_I(x, y) = \mathscr{F}[F(\boldsymbol{u})T(\boldsymbol{u})] \quad (5)$$

Under certain suitable circumstances Eqn. (5) can be considerably simplified. If the specimen under observation is very thin it can be considered to be simply a phase object, modifying the wave function $\Psi_O(x, y)$ of an electron incident on the specimen, so that the wave function $\Psi(x,y)$ on exit from the specimen is of the form

$$\Psi(x,y) = \Psi_O(x,y)\left[1 - \frac{i\pi}{\lambda E}\phi(x,y)\right]$$
$$= \Psi_O(x, y)[1 - i\sigma\phi(x, y)] \quad (6)$$

where $\phi(x,y)$ is the projected potential of the specimen and E, to a good approximation, is the accelerating voltage of the microscope. If the amplitude $\Psi_O(x, y)$ is taken to be unity, the intensity distribution within a high-resolution image $I(x, y) = \Psi_I(x, y)\Psi_I^*(x, y)$ becomes

$$I(x, y) \simeq 1 + 2\sigma\phi(x, y) * \mathscr{F}\{\sin(T(\boldsymbol{u})]\} \quad (7)$$

neglecting terms of the order of $\sigma^2\phi^2(x, y)$, and where $*$ in Eqn. (7) denotes a convolution operation. Therefore, if $\sin[T(\boldsymbol{u})]$ can be regarded as a constant (up to the limiting resolution of the microscope), the image intensity $I(x, y)$ will be proportional to the projected potential $\phi(x, y)$ to that resolution. This simple result is known as the weak phase object approximation. Thus, under these circumstances, the regions where atoms are present are imaged as dark contrast regions and regions between atom columns are imaged as bright contrast regions.

In practice, however, the weak phase object approximation is of limited use. Although there is a great temptation to interpret high-resolution images intuitively in terms of projected potentials and under circumstances of contrast reversal (where atoms are imaged as bright regions) there is no real substitute for detailed image simulations as a function of thickness and lens defocus. Even under such conditions ambiguities can still arise in the interpretation of high-resolution images if the material is too thick (in which case inelastic scattering events may significantly affect the form of the image) or if the material contains nonperiodic detail.

3. *Microscope Resolution*

From consideration of how the image is formed in a transmission electron microscope it is clear that the function $T(\boldsymbol{u})$ will contain terms that are dependent on the electron optical performance of an HREM. Hence the behavior of $T(\boldsymbol{u})$ as a function of $\boldsymbol{u}$ for different defocus values can be used to characterize the electron optical performance of a particular HREM. In practice, it is common to plot the function $\sin\chi(\boldsymbol{u})$ as a function of $\boldsymbol{u}$ for different defocus values ΔF, with or without the envelope function $E(\boldsymbol{u})$. The curves produced in this manner are termed contrast transfer functions (CTFs). The optimum operating conditions for a particular HREM are usually taken to be at values of defocus where the CTF has a reasonably constant value over a wide range of spatial frequencies. Of particular importance are the values of defocus known as "extended Scherzer defocus," at which $\Delta F = -1.5^{1/2} C_s^{1/2} \lambda^{1/2}$, and "extended second broad band defocus," at which $\Delta F = -3.5^{1/2} C_s^{1/2} \lambda^{1/2}$. The term "Scherzer defocus" denotes a value of defocus ΔF of $-C_s^{1/2}\lambda^{1/2}$. Typical CTFs for a state-of-the-art 400 keV HREM are shown in Fig. 2.

There is no simple definition of the 'point-to-point' resolution of an HREM. The *instrumental* resolution can be taken to be the inverse of the spatial frequency at which the value of the envelope function $E(\boldsymbol{u})$ falls to e^{-2}. This instrumental resolution will often be far superior to $\delta \simeq 0.64 C_s^{1/4}\lambda^{3/4}$, the inverse of the spatial frequency at which the CTF crosses the horizontal axis at extended Scherzer defocus which is often taken as a measure of the *interpretable* resolution of a microscope, because of the difficulty of interpreting accurately information within the strong oscillations of the CTF at high spatial frequencies. Other definitions of interpretable resolution use the value $\delta \sim 0.707 C_s^{1/4} \lambda^{3/4}$ at which the CTF crosses the horizontal axis at Scherzer defocus or the value $\delta \sim 0.67 C_s^{1/4} \lambda^{3/4}$ corresponding to the higher spatial frequency of the "passband" at extended Scherzer defocus at which $\sin \chi$ has a value of $1/\sqrt{2}$.

4. Experimental Conditions

The conditions currently used in most HREM experiments are those of axial illumination. In this mode the specimen is tilted in such a way that the area under investigation is oriented so that the incident electron beam is accurately aligned along a low index direction. The beam tilt is then adjusted to make the incident electron beam come down the optic axis of the microscope. Failure to do this adjustment as precisely as possible will have the effect of introducing antisymmetric phase shifts into the diffracted beams and this can introduce considerable spurious detail into an HREM image as well as lowering the expected symmetry of the image. It is usual to do the beam alignment in conjunction with the astigmatism correction for the objective lens so that when these have been done correctly the contrast from any thin layers of amorphous contamination is extremely hard to observe at Gaussian focus (for which $\Delta F = 0$), even if an image is seen on a TV screen through an image intensifier system. It is also standard practice, particularly for beam-sensitive materials, to do these tilt and astigmatism corrections on an area somewhat away from the area to be photographed to minimize specimen contamination and degradation. Specimens used for HREM examination have to be thin, even by transmission electron microscopy standards, and although it is often possible to obtain good quality HREM images from specimens 100 nm thick, it is more usual to try and examine specimens less than 20 nm thick, if only because of the difficulties in interpreting HREM images from thicker specimens. In practice, this means that the edges of specimens are used for HREM examination, and hence care must be taken not to introduce artifacts from specimen preparation or specimen examination into these areas.

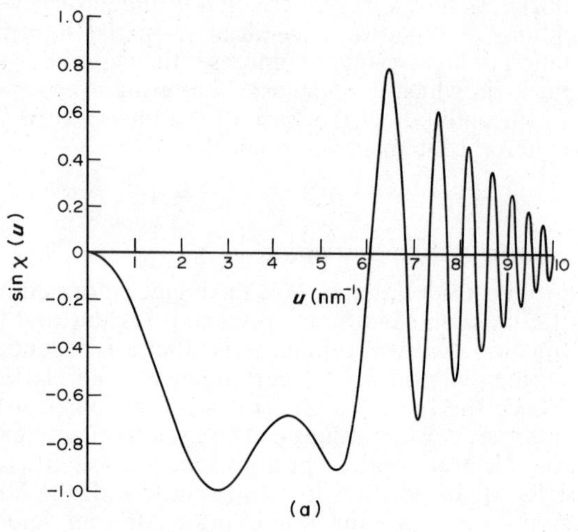

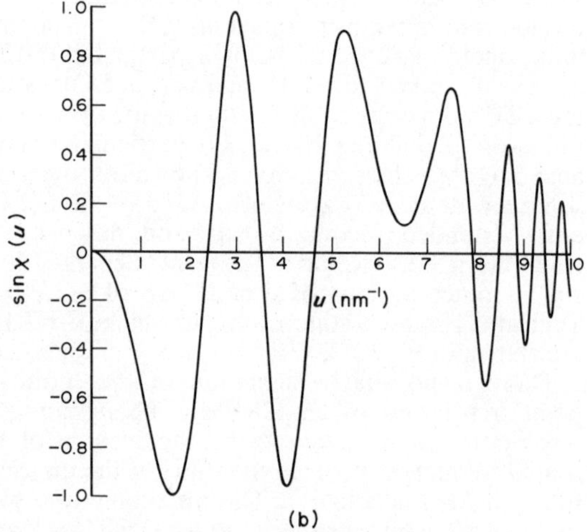

Figure 2
CTFs for a 400 keV microscope: (a) at extended Scherzer defocus and (b) at a defocus of -114 nm. The microscope parameters used were a C_s of 1 mm, Δ of 3 nm and α of 20 mrad, for which extended Scherzer defocus is -49.6 nm

5. Applications

From the preceding discussion it is apparent that transmission electron microscopes can readily produce images which contain information at the atomic level, that is, of the order of atomic spacings within both crystalline and amorphous material. However, it is also apparent that this information need not be readily interpretable in terms of projected potentials of crystal structures, because of the way in which the instrumental parameters of the microscope affect the intensity distribution within the HREM image.

A good example of this difficulty is the interpretation of HREM images from [110] zones in silicon. Silicon has the diamond cubic structure, in which atoms at (0,0,0) and $(\frac{1}{4}, \frac{1}{4}, \frac{1}{4})$ are 0.235 nm apart. In a [110] zone this means that the most closely spaced atom columns are 1.36 Å apart. In the majority of published [110] HREM images of silicon these atom column pairs appear together either as bright blobs (under conditions of reversed contrast) or as the dark regions between bright blobs. However, in the late 1970s [110] images from silicon were published

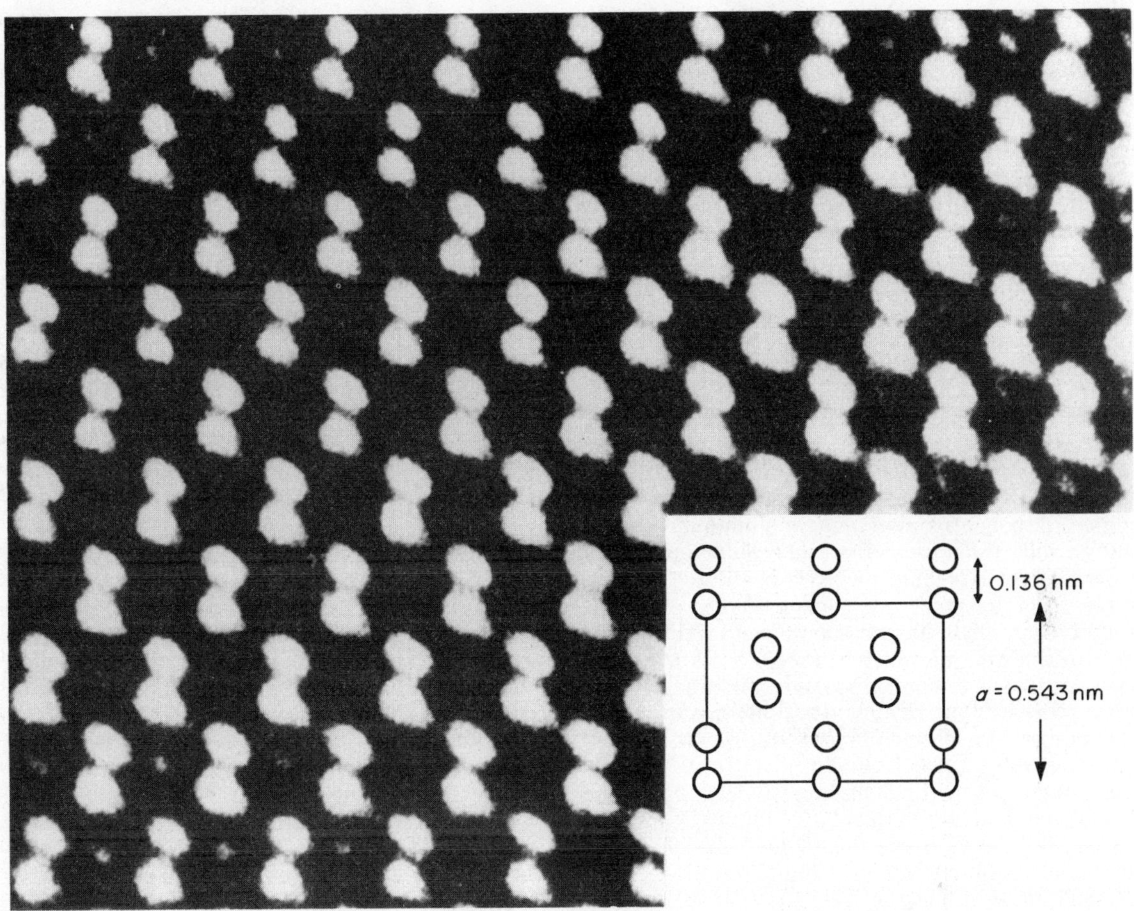

Figure 3
[110] "atomic" image from silicon taken with a 400 keV HREM. This photograph has been enhanced by photographic averaging in order to eliminate random noise from the image. The specimen thickness was 15 nm and the objective lens defocus −114 nm with a C_s of 1 mm, Δ of −3 nm and α of 20 mrad (courtesy of J. L. Hutchison)

in which these white blobs appeared as dumbells and since the distance between the centers of the two parts of the dumbells appeared to be 0.136 nm, these images were interpreted as being true "atomic" images of silicon in the [110] direction, even though they were produced in microscopes which had interpretable point-to-point resolutions of between 0.2 nm and 0.3 nm. It was soon realized that the distance between the centers of the dumbells varied as a function of specimen thickness and objective lens defocus and that the distances seen in the published micrographs were actually significantly larger than 0.136 nm. The dumbells were therefore the result of an imaging artifact. However, with the advent of 400 keV HREMs with resolutions at Scherzer defocus of the order of 0.16 nm, it has been possible to obtain true "atomic" images from silicon in the [110] direction by carefully selecting defocus values in which the important reflections in the HREM image all have sin χ positive, in which case the dumbells can indeed correspond to atomic columns under conditions of reversed contrast. An example of such an image is shown in Fig. 3.

This relatively simple example therefore demonstrates the power of the HREM in obtaining images with atomic level information, but also shows the limitations of the interpretation of these images in terms of projected potentials of even perfect crystalline materials under conditions of perfect axial illumination. However, if due caution is taken in image interpretation, HREMs can be used to great advantage to obtain structural information which is almost impossible to obtain directly by other means. A good example of this is the way in which HREMs have been used in conjunction with techniques such as nuclear magnetic resonance spectroscopy and synchrotron x-ray diffraction to investigate the atomic structure of zeolites. These are a family of materials which can separate complex molecular mixtures by acting as sieves through which molecules below a certain size can pass by diffusing down open channels in the zeolite structure (see *Zeolite Minerals*). The size of these channels is therefore of critical importance in allowing the zeolites to separate particular molecular mixtures and since the size and frequency of these channels can be markedly affected by defects and intergrowths in the zeolites, it is vitally important to be able to characterize fully new synthetic molecular sieves developed by inorganic chemists. In this area the use of HREMs is a valuable technique for establishing the size of channels and the effect of intergrowths on the catalytic ability of a particular zeolite.

Further examples of the use of HREMs can be illustrated by considering the information which can be usefully obtained from micrographs such as those in Figs. 4, 5. Thin plate-like Ω precipitates on {111} planes of the matrix aluminum in aluminum–copper alloys with trace elements magnesium and silver cause strong streaking in electron diffraction patterns from these alloys, similar to the streaking found from plate-like precipitates of Al_2CuLi in Al–Li–Cu–Mg alloys (see *Electron Diffraction*). Al_2CuLi has a hexagonal crystal structure and this gives rise to four distinct precipitate variants in Al–Li–Cu–Mg alloys, one variant on each set of {111} planes. Thus, from the electron diffraction information alone it is tempting to assign a hexagonal crystal structure to the Ω precipitates. However, use of an HREM shows unambiguously that the Ω precipitates cannot be hexagonal because of the distinct forms of image which can occur from different precipitates on the same set of {111} planes (see Fig. 4). Simple image simulations confirm that the

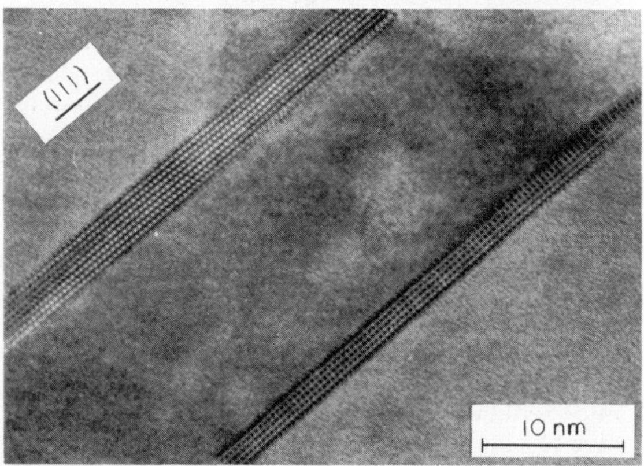

Figure 4
High-resolution electron micrograph of two plate-like Ω precipitates on (111) planes of the matrix aluminum in an Al–Cu–Mg–Ag alloy. The electron beam is parallel to [$\bar{2}$11] of the face-centered-cubic aluminum

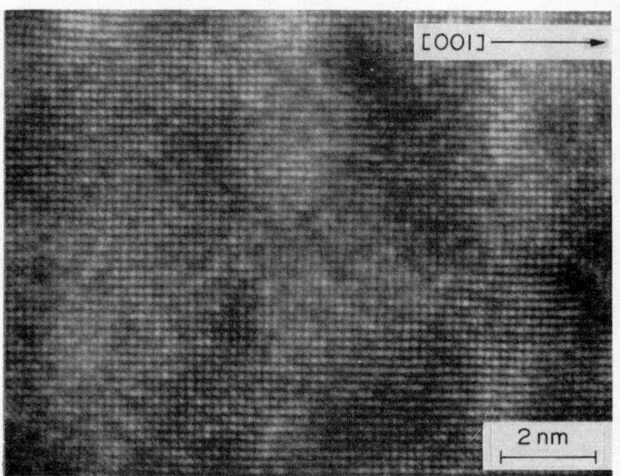

Figure 5
A nonaxial high-resolution image of a Cu–NiPd multilayer sample showing modulated fringe spacings (courtesy of C. S. Baxter)

precipitates have an orthorhombic structure which is a distorted form of the crystal structure of Al_2Cu. However, to be able to assess the role of both magnesium and silver in determining precipitation of Ω precipitates and to be able to fully characterize the precipitate–matrix interfacial structure is much more of a challenge. The use of HREMs is not sufficiently chemically sensitive to be able to take account of trace atom concentrations, and the very existence of an interface and its associated structural relaxations introduces additional information in reciprocal space which will modulate each diffraction spot present. As a result, it is almost impossible to interpret the information present in a through-focal sequence of HREM images in terms of a unique structural model of the interface.

The structure of modulated materials (see *Multilayers, Metallic*, Suppl. 2) can be usefully observed in an HREM by deliberately using nonaxial bright field imaging conditions, in which the beams used to form the image are symmetrically disposed about the optic axis. In the example shown in Fig. 5, the 000, 002, 020 and 022 beams from Cu–NiPd multilayers of 4.4 nm wavelength have been symmetrically disposed about the optic axis. This results in a sequence of micrographs in which there are clear modulated fringe spacings in the [001] direction, whereas for axial illumination the periodicity of the modulations is hidden in the rapid oscillations of the CTF. Intensity profiles from nonaxial micrographs such as Fig. 5 can be used to assess the variability of fringe spacings parallel and perpendicular to the [001] growth direction in these multilayers (see Fig. 6). By examining experimentally observed fringe spacings from multilayers of different wavelengths and correlating these spacings with spacings inferred from image simulations it is

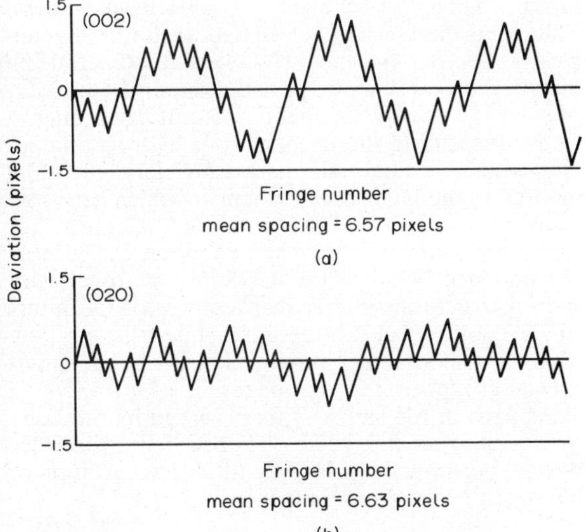

Figure 6
Deviations from the line of least-squares fit of fringe positions measured from intensity profiles of the specimen shown in Fig. 5: (a) parallel to the [001] growth direction, and (b) perpendicular to the [001] growth direction. The mean spacings given in pixels are those of the (002) and (020) lattice planes and would correspond to approximately 0.18 nm in pure copper (courtesy of C. S. Baxter)

possible to deduce the existence of wavelength-dependent structural changes in these materials. Again, it is necessary to do detailed image simulations, because there is not a one-to-one correlation between the fringe spacings measured on HREM micrographs and true interplanar spacings, but when this has been done, as in this example, the interpretation can be used with confidence.

These examples therefore demonstrate the way in which high-resolution electron microscopy can be used for materials problems, but they also reinforce the need to do detailed image simulations to support any image interpretation. Most image simulation packages currently available cater only for periodic boundary conditions, and so they cannot readily take account of nonperiodic detail, such as that found in amorphous materials, and in the vicinity of interfaces and line defects, without lengthy calculations to avoid "edge effects" in the simulations. Furthermore, they do not deal adequately with the effect of inelastic scattering on HREM images and this means in particular that HREM images from thicker crystals tend to contain information that is uninterpretable.

6. *Future Developments*

It is likely that the most important advances in high-resolution electron microscopy will come not from improvements in voltage and current stabilities, nor from improvements in electron lens designs because the designs of lenses are already very sophisticated. Instead, improvements will come from the development of ultra-high vacuum specimen environments and from the increased range of auxiliary equipment that can be used, such as the routine use of on-line image processing and the use of spectrometers for energy filtering of high-resolution images so that the resultant image only includes contributions from elastically scattered electrons. Further developments can also be anticipated in the understanding of how inelastically scattered electrons can be used profitably to obtain chemical information at an atomic level; at present such electrons contribute to the blurring of detail in HREM images, but their precise contribution is still poorly understood.

See also: Electron Diffraction; High-Voltage Electron Microscopy (Suppl. 1); Transmission Electron Microscopy; Transmission Electron Microscopy: Convergent-Beam and Microdiffraction Techniques (Suppl. 1)

Bibliography

Baxter C S, Stobbs W M 1986 High-resolution lattice imaging reveals a "phase transition" in Cu/NiPd multilayers. *Nature (London)* 322:814–16

Bovin J–O, Wallenberg R, Smith D J 1985 Imaging of atomic clouds outside the surface of gold crystals by electron microscopy. *Nature (London)* 317:47–9

Clarke D R 1979 High-resolution techniques and applications to nonoxide ceramics. *J. Am. Ceram. Soc.* 62:236–46

Cowley J M, Smith D J 1987 The present and future of high-resolution electron microscopy. *Acta Crystallogr. Sect. A* 43:737–51

Fujita F E, Hirabayashi M 1986 High-resolution electron microscopy. In: Gonser U (ed.) 1986 *Microscopic Methods in Metals*. Springer, Berlin, pp. 29–74

Goodman P, Moodie A F 1974 Numerical evaluation of N-beam wave functions in electron scattering by the multi-slice method. *Acta Crystallogr. Sect. A* 30:280–90

Kilaas R, Gronsky R 1985 The effect of amorphous surface layers on images of crystals in high-resolution electron microscopy. *Ultramicroscopy* 16:193–202

Self P G, O'Keefe M A, Buseck P R, Spargo A E C 1983 Practical computation of amplitudes and phases in electron diffraction. *Ultramicroscopy* 11:35–52

Smith D J 1989 Instrumentation and operation for high-resolution electron microscopy. In: Mulvey T, Sheppard C J R (eds.) 1989 *Advances in Optical and Electron Microscopy*, Vol. 11. Academic Press, London, pp. 1–55

Smith D J, Saxton W O, O'Keefe M A, Wood G J, Stobbs W M 1983 The importance of beam alignment and crystal tilt in high-resolution electron microscopy. *Ultramicroscopy* 11:263–82

Spence J C H 1988 *Experimental High-Resolution Electron Microscopy*, 2nd edn. Oxford University Press, Oxford

Stobbs W M, Saxton W O 1988 Quantitative high-resolution transmission electron microscopy: The need for energy filtering and the advantages of energy-loss imaging. *J. Microsc. (Oxford)* 151:171–84

Stobbs W M, Wood G J, Smith D J 1985 The measurement of boundary displacements in metals. *Ultramicroscopy* 14:145–54

Thomas J M, Vaughan D E W 1989 Methodologies to establish the structure and composition of new zeolitic molecular sieves. *J. Phys. Chem. Solids* 50:449–67

Ultramicroscopy 18 1985 Proceedings of the Arizona State University Centennial Symposium on High-Resolution Electron Microscopy

K. M. Knowles
[University of Cambridge, Cambridge, UK]

Hydrogen as a Metallurgical Probe

Hydrogen has the smallest atoms of all the elements, which gives rise to a variety of properties that make it especially favorable for use as a metallurgical probe. First, its diffusivity is very high. Thus special structural units or areas (very often defects) in the metal can be reached by the probe within reasonable times and at temperatures at which these units are still stable. Second, hydrogen is dissolved as an atom within interstices of the metal lattice without changing the metal structure. Even after the formation of a hydride the metal lattice remains almost exactly the same, being expanded by only a few percent. Third, its solubility in many metals is high and, therefore, defects strongly interacting with hydrogen can be saturated first. Other structural entities with a lower binding energy can then be studied by adding more hydrogen. The latter property is peculiar to hydrogen, whereas the first two properties are even more pronounced for other light particles such as positrons and muons which are frequently used as probes. The ability of hydrogen to probe different interstices according to their binding energies may be described by the term hydrogen spectrometry (Feenstra et al. 1988). The strong interaction of hydrogen with defects in metals, namely dislocations, grain boundaries and cracks, has deleterious effects on the mechanical properties of some metals as well, this being termed hydrogen embrittlement (Bernstein and Thompson 1981). On the other hand, its high solubility may be beneficial to using metals for hydrogen storage (Alefeld and Völkl 1978).

1. Density of Sites and Hydrogen Distribution

Even in a perfect lattice a discrete spectrum of interstitial sites (i.e., tetrahedral and octahedral sites) is available to be occupied by hydrogen. However, the difference in site energy is usually very large compared to the thermal energy kT and, therefore, only one of these sites is filled with hydrogen atoms. This degeneracy of site energies vanishes whenever a local change of atomic distances (strain) or a change of chemical composition occurs. Following the concepts of statistical thermodynamics, a density of sites (DOS) function can be introduced (Kirchheim 1982) that gives the fraction of sites $n(E)$ within an energy interval dE at energy E. By measuring the properties of hydrogen being dissolved in a material, reasonably detailed information on the DOS can be obtained which has to be transformed into structural and/or chemical information about the corresponding sites. The correspondence between the DOS for hydrogen and special structural or chemical features of the interstices in a material is often unambiguous, especially if the DOS contains pronounced peaks belonging to special groups of interstitial sites.

As each of the sites can be occupied by only one hydrogen atom, the rules of Fermi–Dirac statistics have to be applied yielding the total concentration of hydrogen c at a given temperature T:

$$c = \int_{-\infty}^{\infty} \frac{n(E)}{1+\exp\left(\dfrac{E-\mu}{kT}\right)} dE \qquad (1)$$

where μ is the chemical potential of hydrogen, also called the Fermi level of hydrogen. Thus by measuring the chemical potential of hydrogen at different hydrogen concentrations the DOS can be evaluated from Eqn. (1). Owing to the steplike behavior of the Fermi–Dirac function (the reciprocal of the denominator in Eqn. (1)) hydrogen atoms are usually filled in sites below the Fermi level ($E \leqslant \mu$) and Eqn. (1) is simplified to

$$c = \int_{-\infty}^{\mu} n(E)\, dE \quad \text{or} \quad \frac{\partial c}{\partial \mu} = n(E) \qquad (2)$$

However, the application of Eqn. (2) is subject to conditions on temperature, concentration and DOS, $n(E)$ (Kirchheim and Stolz 1985). In most cases the sites are filled according to their energy beginning with the sites of lowest energy. However, if there is only a very small fraction of low-energy sites, hydrogen atoms populate mostly normal sites, thus decreasing the free energy by an increase of configurational entropy. It then requires more time for the hydrogen atoms to find the traps compared to their residence time within the traps. The partitioning of hydrogen atoms is described totally by Eqn. (1) and the integrand in Eqn. (1) determines the occupancy in different energy intervals of the DOS.

During a gradual filling of sites of different energies other properties such as sample volume, resistivity and hydrogen diffusivity may also change, providing useful information on the nature of the sites.

2. *Interaction with Vacancies and Dislocations*

Vacancies in metals interact strongly with dissolved hydrogen atoms mainly due to a release of the elastic energy that builds up during the squeezing of the hydrogen atoms into the interstices of the metal lattice. Theoretical calculations show that the distance between hydrogen and metal atoms in normal interstitial sites is so small that they repel each other, whereas this distance in a vacancy is large and leads to an attraction. Therefore, a positive volume change occurs during dissolution of hydrogen in normal sites and a negative volume change has been observed for the first 50 at.ppm of hydrogen in amorphous $Pd_{80}Si_{20}$ (Stolz et al. 1984). In the latter case hydrogen was used as a probe to detect the presence and number of vacancylike defects in a metallic glass. This was especially helpful as other techniques developed for crystalline metals cannot be applied to amorphous metals. The negative volume change was also found in heavily cold-rolled crystalline palladium, where it is well known that vacancies are produced by dislocation climb.

The attractive interaction between dislocations and hydrogen is also mainly due to the elastic strain around the hydrogen atom, because the total elastic energy can be reduced by moving the hydrogen atom in an appropriate part of the stress field of a dislocation. There is experimental evidence that the strain field of hydrogen atoms in metals has cubic symmetry and, therefore, the interaction is strongest when there are tensile stresses of an edge dislocation variety.

In a well-annealed metal the density of dislocations is about $10^7\ cm^{-2}$, where the number of sites in the dislocation core is very small and corresponds to concentrations of about 0.01 at.ppm. In order to study the interaction with hydrogen, higher dislocation densities are required. In cold-rolled metals the dislocation densities are as high as 10^9–$10^{11}\ cm^{-2}$ and the concentration of trap sites provided by the cores is of the order of 10 at.ppm. Very sensitive techniques, such as electrochemical methods and the measurement of internal friction and resistivity, are necessary to detect these small amounts of trapped hydrogen. In the case of iron and palladium, site energies of $-50\ kJ\ mol^{-1}$ of hydrogen to $-60\ kJ\ mol^{-1}$ of hydrogen, with respect to normal sites far away from the dislocation, have been determined for sites in the dislocation cores.

Increasing the hydrogen concentration results in an occupation of sites below the glide plane of edge dislocations. The chemical potential of hydrogen then depends linearly on the square root of the hydrogen concentration (Kirchheim 1982) which is a direct consequence of the decrease of hydrostatic tension by $1/r$, where r is the distance from the dislocation core. Using this, hydrogen has been used to probe parts of the stress field around the dislocations.

Sites of constant hydrostatic pressure are on cylindrical loci which are at tangents to the glide plane. With respect to hydrogen these sites have the same site energy and, under special conditions (i.e., low concentrations and/or low temperatures), Eqn. (2) can be used instead of Eqn. (1), which corresponds to an occupation of all the sites within the cylinder by hydrogen atoms. The diameter of this cylinder can be calculated from the total amount of hydrogen, and the site energy on the curved cylindrical surface is equal to the chemical potential of hydrogen (besides an additional constant term). Therefore, an increase of the chemical potential causes an increase of the diameter of the cylinder. As the site energies on the cylindrical surfaces decrease with increasing diameter, a point will be reached where partitioning between sites in the cylinder and normal lattice sites far away from the dislocation occurs, that is, additional hydrogen atoms are used in part to increase the size of the cylinders and in part to fill normal sites. This partitioning can be easily observed by measuring resistivity, because the resistivity increment of hydrogen atoms in the cylinder is small compared to an isolated hydrogen atom in a normal site. The results of such measurements as well as emf results (Kirchheim 1982) and small-angle neutron scattering (Kirchheim et al. 1987) are in agreement with the concept of solute–dislocation interaction previously described. Again, the exceptional properties of hydrogen have enabled the study of this solute–defect interaction in more detail than would have been possible with other solute atoms.

So far it has been tacitly assumed that the nature of the interaction energy is purely elastic. However, it can be shown (Kirchheim 1982) that the negative hydrogen–hydrogen interaction energy plays an important role, because the elastic energy alone would not lead to the observed extended segregation of hydrogen at dislocations.

3. *Interaction with Internal Interfaces*

Grain boundaries in pure metals do not interact very strongly with dissolved hydrogen. However, in the presence of other segregated impurities (e.g., phosphorus in iron) a strong interaction will occur, because the embrittlement effects of some elements is enhanced remarkably in the presence of hydrogen. The usual procedure to measure grain boundary segregation by surface analytical techniques after intergranular fracture fails for hydrogen because common techniques such as Auger electron spectroscopy, electron spectroscopy for chemical analysis and x-ray photoelectron spectroscopy do not detect hydrogen or because the previous grain boundary coverage by hydrogen will be changed after the exposure of the boundary to the high vacuum due to the attainment of a new equilibrium corresponding to surface segregation.

Nanocrystalline metals (i.e., polycrystalline metals with grain diameters of the order of nanometers) contain a large fraction of grain boundaries and they are, therefore, most suitable for studying the interaction of hydrogen with grain boundaries. It has been shown for nanocrystalline palladium (Mütschele and Kirchheim 1987) that the grain boundaries provide a broad spectrum of segregation energies (DOS) as one would expect from simple geometrical considerations. The diffusion behavior is peculiar because the diffusion coefficient in grain boundaries is lower than that within the grains due to a preferred occupation of low-energy sites at low hydrogen concentrations. The diffusivity increases as sites of higher energy have to be occupied by increasing the hydrogen concentration. Although the width of the DOS is about the same as in metallic glasses (see Sect. 4), the average activation energy, calculated as the difference between average site and saddle-point energies, is rather small in nanocrystalline palladium when compared with metallic glasses, indicating a more open structure of the grain boundaries in nanocrystalline materials.

The α–β phase transformation in the system of hydrogen and nanocrystalline palladium differs remarkably from that in normal polycrystalline palladium (20 μm grain diameter) and it can be used to determine the average thickness of grain boundaries. In polycrystalline palladium the volume fraction of grain boundaries and its influence on the phase transformation is negligible. In nanocrystalline palladium the same β phase is formed within the grains but much less hydrogen is needed for the completion of the α phase to β phase transformation because the grain boundaries do not participate in the phase formation. From the difference between the amount of hydrogen used for the β phase formation in poly- and nanocrystalline palladium the volume fraction of the grain boundaries is calculated straightforwardly. Thus a volume fraction ν of 0.27 in a nanocrystalline palladium sample with an average grain diameter d of 8 nm (from transmission electron micrographs) was determined which yields an average thickness of the grain boundaries $\delta = \nu d/3 =$ 0.7 nm. This relationship is valid for grains of spherical or cubic shape which approximates well to the case of nanocrystalline palladium.

From the interaction of hydrogen with Pd–Al_2O_3 (or Pd–ZnO) interfaces information on the phase boundary itself can be extracted. The interfaces are prepared by internal oxidation of palladium–aluminum (palladium–zinc) alloys at 1000 °C in air with leads to the formation of prismatic precipitates of Al_2O_3 (ZnO) having edge lengths within the submicrometer range. The diffusion coefficients of hydrogen at 295 K are shown in Fig. 1 as a function of hydrogen concentration. During the first run a larger trap concentration is observed compared to the following runs. Between each run hydrogen is removed by anodic polarization. The difference in trap concentrations is attributed to irreversible trapping of hydrogen at the palladium–oxide phase boundary, where the number of irreversible trap sites corresponds to a coverage of 1.5×10^{15} hydrogen atoms cm^{-2} (about one monolayer) of the boundary. These hydrogen atoms cannot be removed either by anodic polarization or by annealing at 800 °C in high vacuum. Thus, during the first run of doping the samples, both kinds of trap sites are saturated while in the following runs only reversible sites are filled.

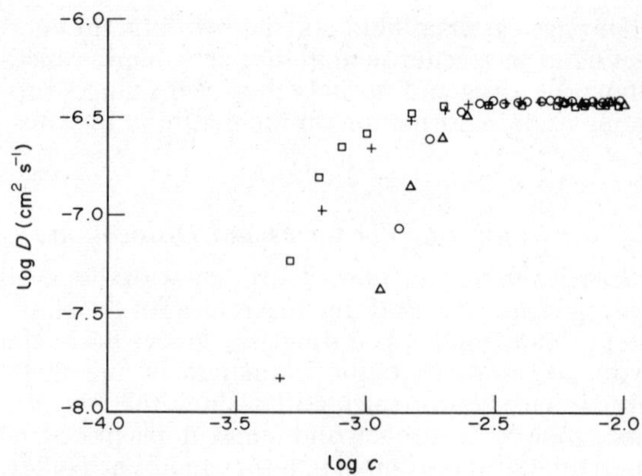

Figure 1
Diffusion coefficient of hydrogen in palladium with Al_2O_3 precipitates D as a function of hydrogen concentration c: saturation of irreversible traps (free oxygen bonds at the Pd–Al_2O_3 interface) and reversible traps (strained interstices of palladium at the interface (○), only reversible trapping occurs (□), after annealing in aluminum vapor the oxygen bonds at the interface are saturated with aluminum and no irreversible trapping is observed (+), and after a final annealing treatment in air the oxygen atoms saturate the free aluminum bonds at the interfaces but still have free bonds to cause irreversible trapping of hydrogen (△). After saturation of all kinds of traps the diffusivity of hydrogen in pure palladium ($\log D = -6.4$) is measured

The irreversible trapping was attributed to the formation of oxygen–hydrogen bonds at the palladium–oxide interface assuming that unsaturated oxygen atoms form the terminating layer of the oxide precipitates. This assumption is supported by direct observations in a high-resolution electron microscope (Necker and Mader 1988). By annealing treatments in zinc or aluminum vapor these oxygen atoms are saturated with the corresponding metal atoms. These form the terminating layer of the oxide and have a lower affinity to hydrogen; therefore, the irreversible trapping of hydrogen vanishes (see Fig. 1). The irreversible traps recover if the samples are reannealed in air. The reversible trapping of the hydrogen can be described quantitatively by a two-level trapping model (Kirchheim 1982) with a free energy of trapping $\Delta G = -44\,000 + 50T$ kJ mol^{-1} of hydrogen. This trapping may be due to the strain field around the precipitates because of misfit dislocations and/or the different thermal expansion coefficients of the palladium and the oxides.

During the internal oxidation of Pd–3 at.% Al alloys at 700 °C in air, no precipitates were detected by conventional transmission electron microscopy and only streaks were observed in the diffraction pattern. According to chemical analysis the sample contained 3.0±0.05 at.% of aluminum and 4.9±0.15 at.% of oxygen, which corresponds to an excess of 0.4±0.2 at.% of oxygen. This value is in good agreement with the 0.5 at.% of irreversible traps that was determined for hydrogen in this alloy. The concentration of irreversible traps is larger when compared with the samples oxidized at 1000 °C because the interfacial area is larger for the small precipitates formed at 700 °C. Thus, hydrogen provides information on the interfacial area or the number of unsaturated oxygen bonds. The interfacial area can be calculated, assuming the same coverage as for irreversible trapping of 1.5×10^{15} hydrogen atoms cm^{-2}.

4. *Behavior in Crystalline and Amorphous Alloys*

For an alloy $A_{1-x}B_x$, which is either amorphous or crystalline and where hydrogen occupies tetrahedral sites, an appropriate DOS function is

$$n(E) = f \sum_{i=1}^{4} \binom{4}{i} x^i (1-x)^{4-i} n_i(E-E_i) \qquad (3)$$

where the factor f is theoretically equal to the number of tetrahedral sites per metal atom which can be occupied by hydrogen. However, due to the repulsive interaction between hydrogen atoms some of the sites remain empty and, therefore, actual values of f are less than the total number of tetrahedral sites per metal atom. The five different types of tetrahedral sites (A_4, A_3B, A_2B_2, AB_3 and B_4) are present in concentrations of $f\binom{4}{i}x^i(1-x)^{4-i}$ and their site energies for hydrogen may be distributed around an average energy E_i according to the functions $n_i(E-E_i)$, which may be Dirac delta functions in the case of disordered crystalline alloys or Gaussian functions in the case of amorphous alloys. The experimental determination of $n(E)$ (e.g., by measurements of the chemical potential $\mu(c)$) and a comparison with Eqn. (3) allows the assessment of whether A and B atoms are randomly distributed or whether a chemical short-range order causes deviations from the binomial distribution. The distribution functions $n_i(E-E_i)$ and/or the average energies E_i can provide information on the structural short-range order (Kirchheim 1982, Richards 1983, Feenstra et al. 1988).

The concept of evaluating the DOS from measurements of hydrogen solubility and drawing conclusions on the properties of the alloy itself has been successfully applied to disordered, crystalline niobium–vanadium alloys (Feenstra et al. 1988) and various amorphous alloys containing an early and a late transition metal (Harris et al. 1987). For the crystalline $Nb_{1-x}V_x$ alloys the results were analyzed (Feenstra et al. 1988) with respect to the site energies E_i of the different tetrahedra as a function of the alloy composition. E_i changes with x because the enthalpies of hydrogen dissolution depend on the atomic volume of the alloy, which also changes as a function of x in the case of niobium–vanadium alloys. Different from an ideal lattice of a pure metal, the atomic distances in an A–B alloy are distributed around an average value, that is, for those cases where A and B have different atomic radii. Thus an Nb_4 tetrahedron in the alloy has a smaller volume than pure niobium whereas a V_4 tetrahedron in the niobium–vanadium alloy has a larger volume than pure vanadium. Therefore, V_4 sites in the alloy are more favorable for hydrogen than the corresponding sites in pure vanadium. From analysis of the site energies the average atomic distances of A–A, A–B and B–B pairs can be calculated.

Measurements of the chemical potential of hydrogen in amorphous nickel–zirconium alloys have shown (Harris et al. 1987) that the distribution of nickel and zirconium atoms over the corners of the tetrahedral sites follows the binomial distribution as given by Eqn. (3) with f determined as having a value of 1.9. This is in agreement with theoretical considerations assuming an infinitely strong repulsion for hydrogen–hydrogen distances less than 0.21 nm and a reasonable structure for the amorphous alloy.

By replacing the natural oxide films on an amorphous sample emf measurements are possible at low concentrations and using Eqn. (2) this yields the distribution of site energies DOS (Kim and Stevenson 1988) as shown for amorphous nickel–titanium and nickel–zirconium alloys in Fig. 2. The measured DOS shows two separated peaks caused by the "chemical disorder" which gives rise to the appearance of interstices with a variety of different chemical environments. Contrary to the crystalline metals, the energies for each type of tetrahedron are distributed due to the structural disorder in amorphous metals. Using Gaussian distributions for $n_i(E\text{-}E_i)$ and $f=1.3$ in Eqn. (3) yields the solid line shown in Fig. 2. The width of the separate Gaussian functions is about 10 kJ mol^{-1} of hydrogen.

The width of the energy distribution for a special interstice contains information on the structural disorder. Neglecting topological short-range order leads to a simple relationship between the site energy distribution for hydrogen and the distribution of atomic distances for the matrix atoms (Kirchheim 1982, Richards 1983). The measured widths of the DOS for hydrogen are usually smaller which may be due to topological short-range order.

Among the amorphous alloys the palladium–silicon alloys are mostly used to measure hydrogen solubility and diffusivity. The nonmetal seems not to play an important role and the DOS for hydrogen can be described by a single Gaussian distribution, which can explain the concentration dependence of the chemical potential and of the diffusion coefficient (Kirchheim and Stolz 1985). In modelling the long-range diffusion behavior the assumption is made that the saddle-point energies are constant which is sufficient to give the correct dependence of the diffusion coefficient on concentration (see Fig. 3). However, the magnitudes of the long-range diffusion and especially the short-range diffusion of hydrogen are strongly determined by the distribution of saddle-point energies as can be shown by Monte Carlo simulations (Kirchheim and Stolz 1987). However, the theory on diffusion in disordered media has to be improved considerably in order to use data on the short- and long-range diffusion of hydrogen for probing saddle-point configurations. An example of the sensitivity of this diffusivity on structural differences is presented in Fig. 3, where the diffusion coefficient of hydrogen in amorphous $Pd_{80}Si_{20}$ is shown as a function of hydrogen concentration (Stolz et al. 1984). The values differ by up to two orders of magnitude for samples prepared by different techniques, although the radial distribution function as measured with x rays reveals no differences. The results of Fig. 3 can be explained partly by different DOS but they are also affected by the distribution of saddle-point energies (Kirchheim and Stolz 1987).

The properties of hydrogen (chemical potential, diffusivity, resistivity increment) were also determined in cold-rolled amorphous $Pd_{80}Si_{20}$ (50% reduction of cross section). No changes have been observed compared to the as-quenched samples (Kirchheim et al. 1985), although all quantities are changed remarkably in the presence of edge dislocations. Thus hydrogen is not able to detect edge type

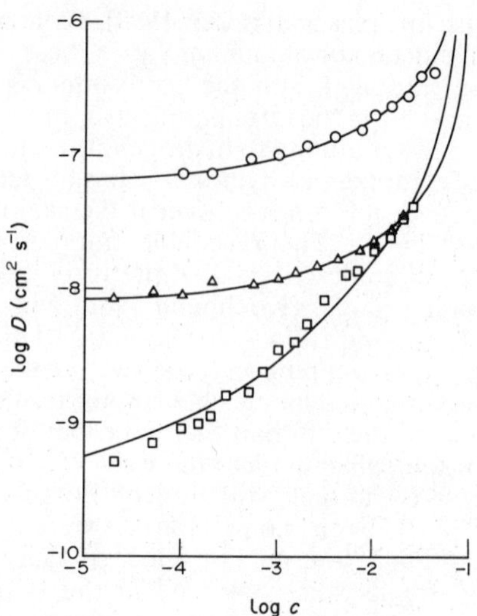

Figure 3
Diffusion coefficient of hydrogen in amorphous $Pd_{80}Si_{20}$ prepared by melt spinning (△), sputtering (□) and the double piston technique (○). The lines were calculated from the DOS

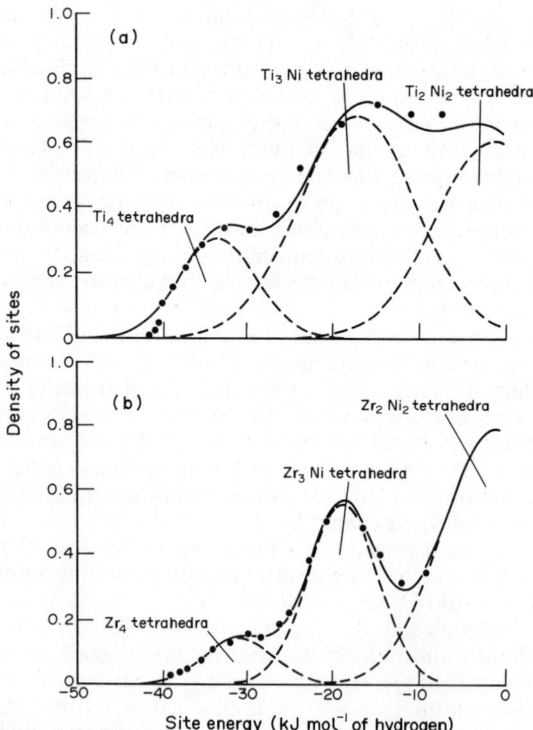

Figure 2
Measured density of sites $\partial c/\partial\mu$ for (a) amorphous $Ni_{35}Ti_{65}$ and (b) $Ni_{50}Zr_{50}$ revealing the concentration and binding energies of different tetrahedral sites

dislocations in plastically deformed, amorphous palladium–silicon alloys.

See also: Defects in Metallic Glasses

Bibliography

Alefeld G, Völkl J 1980 *Hydrogen in Metals*, Topics in Applied Physics, Vol. 29. Springer, Berlin, pp. 201–42

Bernstein I M, Thomson A W (eds.) 1981 *Proc 3rd Int. Conf. Effects of Hydrogen on Behavior of Materials*. Metallurgical Society of AIME, New York

Feenstra R, Brower R, Griessen R 1988 Hydrogen as a local probe of alloys: $Nb_{1-y}V_y$. *Europhys. Lett.* 7(5):425–30

Harris J H, Curtin W A, Tenhover M A 1987 Universal features of hydrogen absorption in amorphous transition metal alloys. *Phys. Rev. B* 36:5784–97

Kim J J, Stevenson D A 1988 Hydrogen permeation studies of amorphous and crystallized Ni–Ti alloys. *J. Non-Cryst. Solids* 101:187–97

Kirchheim R 1982 Solubility, diffusivity and trapping of hydrogen in dilute alloys, deformed and amorphous metals. *Acta Metall.* 30:1069

Kirchheim R, Huang X Y, Carstanjen H-D, Rush J J 1987 Inelastic neutron scattering and resistivity of hydrogen in cold-worked palladium. In: Latanison R M, Jones R H (eds.) 1987 *Chemistry and Physics of Fracture*, NATO ASI series. NATO, Brussels, p.580

Kirchheim R, Stolz U 1985 Modelling tracer diffusion and mobility of interstitials in disordered materials. *J. Non-Cryst. Solids* 70:323

Kirchheim R, Stolz U 1987 Monte-Carlo simulations of interstitial diffusion and trapping II. Amorphous metals. *Acta Metall.* 35:281

Kirchheim R, Szökefalvi-Nagy A, Stolz U, Speitling A 1985 Hydrogen in deformed and amorphous $Pd_{80}Si_{20}$ compared to hydrogen in deformed and crystalline palladium. *Scripta Metall.* 19:843

Mütschele T, Kirchheim R 1987 Hydrogen as a probe for the average thickness of a grain boundary. *Scripta Metall.* 21:1101

Necker G, Mader W 1988 Characterization of Ag/CdO interfaces. *Phil. Mag. Lett.* 58(4):205–12

Richards P 1983 Distribution of activation energies for impurity hopping in amorphous metals. *Phys. Rev. B* 27:2059

Stolz U, Kirchheim R, Wildermuth A 1984 Hydrogen as a probe in amorphous metals. *Proc 5th Int. Conf. Rapidly Quenched Metals*. Würzburg, p.1537

Stolz U, Nagorny U, Kirchheim R 1984 Volume changes during dissolution of hydrogen in metallic glasses. *Scripta Metall.* 18:347

R. Kirchheim
[Max-Planck-Institut für Metallforschung, Stuttgart, FRG]

Hydrogenated Amorphous Silicon

In amorphous silicon (a-Si), almost every silicon atom is tetrahedrally bonded to four nearest-neighbor atoms–just as in crystalline silicon (c-Si). However, in the case of more distant neighbors, due to slight stretching and twisting of bonds, long-range order is lost. In contrast, crystals are characterized by structural periodicity over a long range. A consequence of this disorder is the presence of a great number of dangling bonds. Dangling bonds form states in the energy gap that act as recombination–generation centers that are detrimental to device performance. Fortunately, in hydrogenated amorphous silicon (a-Si:H), most of the silicon dangling bonds which might have remained dangling are terminated by a hydrogen atom. This has two effects: (a) the density of gap states is greatly reduced, and (b) the energy gap, reflecting the average binding energy of the material, is larger than that of c-Si.

The ability to dope a-Si:H either *n*-type or *p*-type stimulated the search for devices that could be mass produced more cheaply than could their c-Si counterparts. Since a-Si:H can be deposited at low temperature, it appeared that almost any substrate would be useable and large area devices or large arrays of devices could be made. Since the material is available as a thin film, little material is consumed and, best of all, the great cost of producing a good-quality single crystal vanishes in the amorphous material.

1. Material Synthesis

The most common method of preparing a-Si:H is the glow-discharge decomposition of silane. An electric field (dc, ac, or radio-frequency (rf)) produces a plasma containing ions and other reactive species that condense on a heated substrate (200–400 °C, typically) to form an amorphous solid, still rich in hydrogen. A dc magnetic field can be used to confine the plasma, increasing its density. In a dc discharge, ion bombardment of the film is reduced by a "proximity electrode" that sets an equipotential plane at the same potential as the substrate but away from the surface of the film.

Sputtering is another method of preparing a-Si:H. A controlled amount of hydrogen is added to the sputtering gas to react with the sputtered silicon. Silicon deposited by evaporation produces an amorphous film that can be subsequently hydrogenated by exposure to atomic hydrogen.

Silane may be directly decomposed by ultraviolet photolysis. Infrared photolysis of silane is possible with the 10.59 mm emission line from a carbon dioxide laser.

Doping is obtained by adding either phosphine or diborane to silane to produce *n*- or *p*-type a-Si:H, respectively.

In hydrogenated amorphous germanium–silicon alloys, the energy gap decreases with increasing germanium content. Conversely, hydrogenated amorphous silicon–carbon alloys result in increasing bandgap as the carbon concentration is increased.

High-resolution transmission electron microscopy and hydrogen-specific neutron scattering have yielded information on the structure of a-Si:H. The types of clustering and the presence of voids in the material have important consequences on the properties of the film. An amazingly large variety of defects, mostly strain related, can be frozen into the structure during deposition, forming lifetime-reducing localized states that lie deep inside the energy gap. Adding impurities such as phosphorus, boron or oxygen further complicates the variety of possible defects.

"Superlattices" have been made that consist of layers, 10–1000 Å thick, of a-Si:H alternating with layers of a-Ge:H, a-$Si_{1-x}C_x$:H or a-Si:N_x:H. The films were deposited by a glow-discharge technique in which the composition of the reactive gases was changed periodically. Atomically sharp interfaces were obtained by exchanging the gases in the plasma reactor in a time that was short compared with the time it takes to grow a monolayer.

2. Optical Properties

Whereas in c-Si optical transitions must conserve both energy and crystal momentum, in a-Si crystal momentum is not a valid quantum number. Hence in a-Si all optical transitions conserve momentum. Therefore, allowed transitions are more efficient in a-Si than in c-Si. If parabolic densities of states are assumed for valence and conduction bands, it would be expected that for photon energies $h\nu$ (where ν is the frequency and h is Planck's constant) greater than that of the energy gap E_g the absorption coefficient would vary as $(1/h\nu)(h\nu-E_g)^2$, an empirical relation first derived by Tauc to determine E_g (Fig. 1). All amorphous semiconductors appear to obey Tauc's relation. The absorption-coefficient data used to determine the Tauc gap are larger than $10^4\,cm^{-1}$. At lower values of absorption coefficient, an exponential dependence, the Urbach edge, is found that has been associated with the perturbation of band edges.

Alloys of a-Si:H with various other group IV elements have energy gaps that can be adjusted by the alloy composition. The absorption edge correspondingly shifts with composition. In superlattices consisting of alternating layers of a-Si:H and a-Si:N_x:H, the bandgap increases and the exponential bandtail broadens with decreasing a-Si:H layer thickness. Quantum-size effects resulting from the two-dimensional confinement of electrons and holes in the a-Si:H layers have been observed.

Below the exponential absorption edge, there are transitions that involve deep levels with such low absorption coefficients that their study requires very sensitive techniques such as photothermal deflection spectroscopy (PDS) that can detect an absorption coefficient as low as $10^{-2}\,cm^{-1}$. Using PDS, it was found that the absorption between 0.8 eV and 1.3 eV is proportional to the number of dangling bonds determined by electron spin resonance.

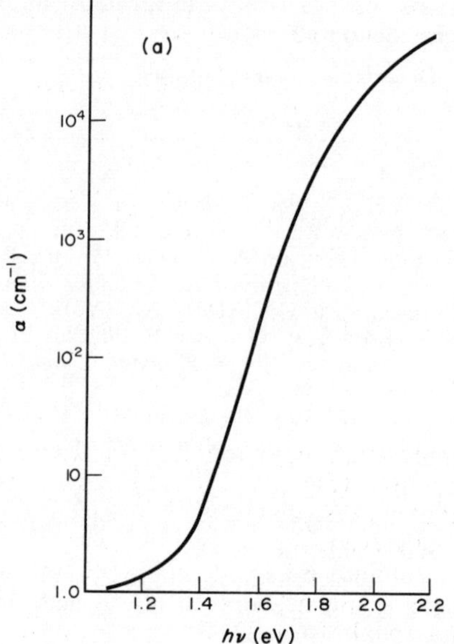

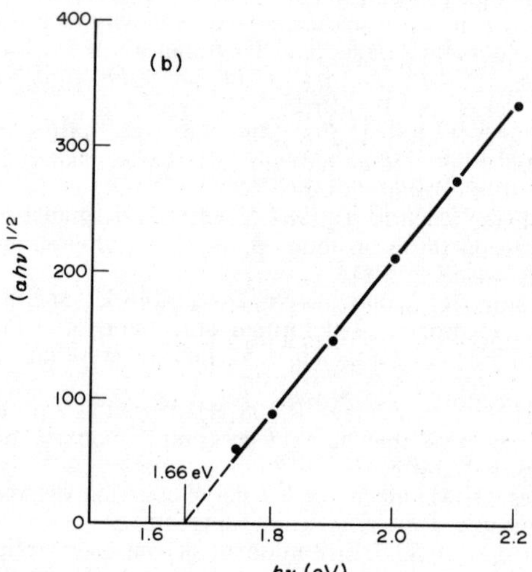

Figure 1
Typical absorption edge of a-Si:H with (a) usual plot and (b) Tauc plot

At still lower photon energies, the realm of photon–phonon interaction is entered, probing atomic vibrational modes, particularly those involving hydrogen. Thus, SiH, SiH_2, SiH_3, and $(SiH_2)_n$ can be distinguished between (see Table 1), and also SiF, SiO, and so on.

The absence of crystal momentum makes radiative transitions more probable, resulting in efficient

Table 1
Vibrational frequencies for SiH_x ($x = 1$–3)

	Vibrational frequencies (cm^{-1})		
	Mode		
Group	Stretching	Bending	Rocking
SiH	2000		630
SiH_2	2090	880	630
$(SiH_2)_n$	2090–2100	890, 845	630
SiH_3	2140	905, 860	630

low-temperature emission that peaks in the 1.2–1.4 eV range (Fig. 2). The carriers thermalize to traps that are distributed in energy. A strong Stokes shift is observed.

The luminescence efficiency decreases with increasing dangling-bond concentration. Another emission peak (usually weaker) appears at 0.8–0.9 eV and depends on defects, such as dangling bonds, following either dehydrogenation, doping or exposure to radiation.

Time-resolved spectroscopy (TRS) follows the decay of luminescence over a broad distribution of decay times, from 10 ns to 10 ms. TRS demonstrates the tunnelling-assisted transitions between electrons and holes located in distant traps or between donors and acceptors.

Photoconductivity involves two phenomena: photon absorption and charge transport (which can be very complex). The primary photocurrent is due to the motion of photogenerated carriers because blocking contacts prevent the influx of external charges. In the secondary photocurrent, the contacts are ohmic (or forward-biased) so that the photogenerated carriers modulate the conductivity of the photoconductor and, depending on the relative trapping time and transit time, considerable gain can be obtained. Thus, traps play a crucial role in secondary photocurrents. Recombination centers, especially dangling bonds, are also important because they determine the survival of photogenerated carriers.

Metastable states can be induced in a-Si:H by irradiation with light or x rays, by bombardment with electrons or ions, or by injection of either carrier. These metastable states appear to result from electron–hole pair recombination because, in the light-induced case, when an electric field is simultaneously applied to sweep away the photogenerated carriers, metastable states do not form. Metastable states persist after the irradiation has ceased and the resulting changes in electrical and optical properties can be monitored: the dark resistivity increases (Staebler–Wronski effect), the photoluminescense efficiency at approximately 1.3 eV

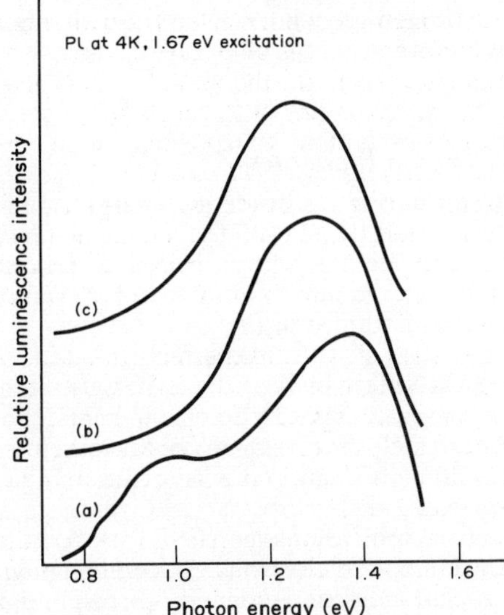

Figure 2
Luminescence spectra of a-Si:H deposited by: (a) diode sputtering, (b) magnetron sputtering, (c) glow discharge (courtesy of R.A. Street)

decreases, luminescence at approximately 0.8 eV increases, and the spin resonance of dangling bonds increases. All of these changes can be reversed by annealing at roughly 200 °C, and some of the changes are removed by applying an electric field in the dark. Metastable states are formed when weak Si–Si bonds are broken by the energy released by electron–hole recombination or trapping. Metastable states may also result from a light-induced charge transfer between already existing dangling bonds.

Light exposure produces two kinds of defects, one responsible for a drop in the mobility-lifetime product $\mu\tau$ and the other responsible for subbandgap absorption. The metastable states that reduce $\mu\tau$ can be created as efficiently at low temperature as at room temperature, whereas the metastable states that increase absorption at around 1 eV seem temperature activated. The two defects anneal at slightly different temperatures. Thus, by controlling the temperature during irradiation, it is possible to create and saturate the $\mu\tau$-killing defect first and then to produce, at a higher temperature, the one-electron-volt-absorbing centers. Later, it is possible to anneal the $\mu\tau$-affecting centers before annealing the one-electron-volt-absorbing centers. The resulting plot of $\mu\tau$ versus absorption at 1 eV forms a roughly rectangular hysteresis loop, demonstrating the separate accessibility of the two types of defects. A dual-beam photoconductivity experiment (infrared quenching or enhancement) supports the hypothesized independence of the two metastable states.

3. Electronic Properties

The density of states inside the energy bandgap can be determined by field-effect and capacitance techniques and by deep-level transient spectroscopy (DLTS). An example of density of states from DLTS data is shown in Fig. 3.

In the Staebler–Wronski effect, the dark conductivity of a-Si:H can drop by orders of magnitude after prolonged exposure to visible light. The Fermi level moves close to midgap, because light-soaking creates 10^{16}–10^{17} new states near the middle of the energy gap.

Electron spin resonance (ESR) probes the resonance of unpaired electrons. Silicon dangling bonds have a characteristic resonance corresponding to a gyromagnetic ratio $g=2.0055$. They number about $10^{19}\ cm^{-3}$ in a-Si; their concentration drops by several orders of magnitude in a-Si:H, becoming as low as $10^{15}\ cm^{-3}$. Phosphorus doping yields a resonance at $g=2.004$, whereas boron doping yields a resonance at $g=2.013$.

Time-of-flight experiments follow the motion of electrons or holes after a short flash of light in a strong electric field. The current decays in time according to two different power laws (Fig. 4), indicating dispersive transport. Holes have a drift mobility two orders of magnitude lower than electrons. The model that acounts for the observed time-

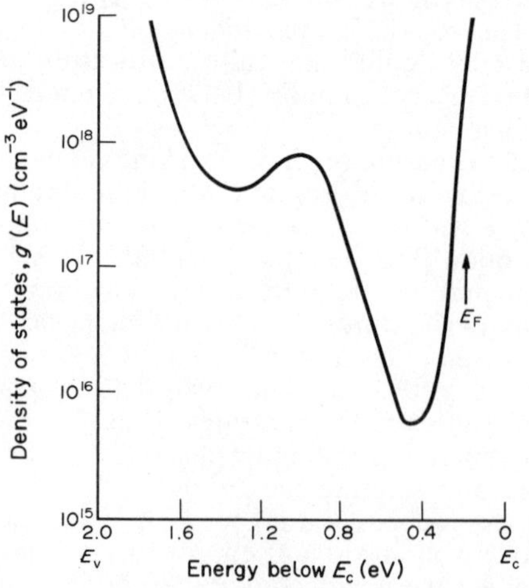

Figure 3
Density of states in a-Si:H derived from DLTS spectra: E_F is the Fermi energy, E_c the conduction-band energy and E_v the valence-band energy (courtesy of J.D. Cohen)

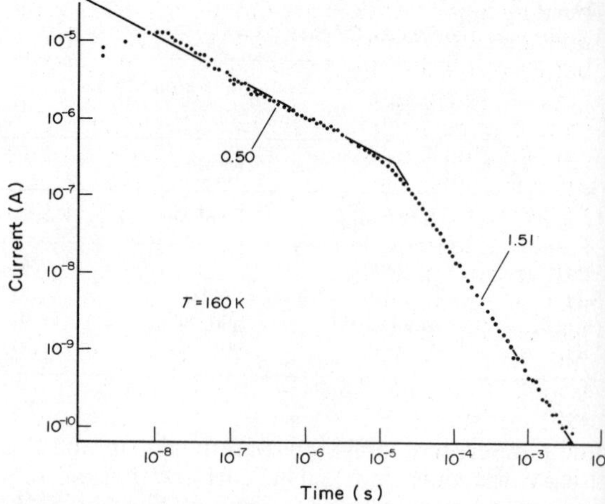

Figure 4
Electron photocurrent decay at 160 K in 3.8 μm-thick a-Si:H under 16 V applied bias: the numbers indicate the slope of the curve (courtesy of T. Tiedje)

dependence leads to values of 42 meV and 27 meV for the characteristic shapes of the exponential valence bandtail and conduction bandtail, respectively.

The diffusion length of carriers in a-Si:H is determined by the "surface photovoltage method." The photovoltage results from photogenerated carriers diffusing to the surface, where an electric field separates the electron–hole pairs. The analysis of the functional dependence of the surface photovoltage on the absorption coefficient α yields the diffusion length L_D. It is found that in *p–i–n* solar cells, most of the photovoltage is generated by the *p–i* transition.

4. Applications

The most important application of a-Si:H and the first commercially successful device made of a-Si:H is the solar cell, widely used to power pocket calculators. Alloying silicon with other elements such as carbon and germanium affects the absorption edge, thus allowing a stacking of cells with successively narrower energy bandgaps to utilize the solar spectrum more efficiently.

a-Si:H is a good photoreceptor for transferring an image by electrophotography. a-Si:H is highly resistive, has a slow charge-decay time, an excellent spectral sensitivity in the visible, and a good chemcial inertness. Excellent image reproducibility with high resolution and good contrast have been demonstrated.

a-Si:H imaging tubes were made with a spatial resolution comparable to that of commercial vidicons. They exhibit neither blooming nor image burning under strong illumination. When used to intesify x-ray images, the a-Si:H imaging tube has a better resolution than conventional image intensifiers. A single-tube color camera has been demonstrated using striped color filters on the face plate.

Field-effect transistors (FETs) made of a-Si:H offer the advantage that very large arrays of a-Si:H FETs are feasible. Hence, even if their performance level is much lower than that of c-Si FETs, there are still many applications where their characteristics may be adequate. a-Si:H FETs have been used to address liquid crystal displays (LCD), including a full-color LCD television. The low-mobility a-Si:H is ideally suited for making large FETs having a high OFF-resistance. Furthermore, a-Si:H has the advantage of low-temperature processing, so that it can be deposited onto inexpensive glass that serves as a transparent substrate.

Image sensors suitable for facsimile transmission or for optical character recognition have been made using a-Si:H.

a-Si:H is a good material for information storage. For example, laser-induced dehydrogenation of a-Si:H forms a microscopic blister or a crater that can be easily read optically against a smooth background.

a-Si:H fast detectors take advantage of the short transit time of photogenerated carriers in very small structures in the presence of an electric field. Response times in the tens of picoseconds have been obtained. a-Si:H can also be used as a fast electrooptic modulator.

An intriguing switching phenomenon has been discovered that may be suitable for memory applications. The device is a diode that can be switched to the conducting state by biasing beyond a threshold with one polarity and switched back to the resistive state by exceeding a threshold with opposite polarity.

Bibliography

Brodsky M H (ed.) 1979 *Amorphous Semiconductors*, Topics in Applied Physics, Vol. 36. Springer, Berlin

Hamakawa Y (ed.) 1982 *Amorphous Semiconductor Technologies and Devices*. North-Holland, Amsterdam

Joannopoulos J D, Lucovsky G (eds.) 1984 *The Physics of Hydrogenated Amorphous Silicon*, Topics in Applied Physics, Vol. 55–56. Springer, Berlin

Pankove J I (ed.) 1984 *Hydrogenated Amorphous Silicon*, Semiconductors and Semimetals, Vol. 21. Academic Press, New York

J. I. Pankove
[University of Colorado, Boulder, Colorado, USA]

I

Ice and Frozen Earth as Construction Materials

The cold regions of the world, characterized by the presence of ice and frozen earth, are centered around the poles and extend to about the 40th parallel in both the northern and southern hemispheres. Ice covers between 9% and 15% of the earth's surface during the year, while nearly one-fifth of all the land surface of the earth is underlain by perennially frozen earth or permafrost. The polar zone consists of continuous masses of perennial ice and frozen earth, while the subpolar zone consists of discontinuous or broken masses of ice and frozen earth, both perennial and seasonal in nature. The temperate zone consists mainly of seasonal ice and frozen earth.

Development and construction, particularly in the polar and extreme subpolar zones, require innovative engineering solutions. Transportation of personnel, materials and equipment as well as communications pose complex and expensive logistical problems. The environmental impact of development on people and wildlife is a problem of public concern. These problems have been compounded in recent years by the rapid pace of development imposed by the need of society to exploit scarce natural resources and to maintain the security of sovereign nations.

The ability to work with ice and frozen earth as construction materials is central to cold-regions engineering. Both materials are plentiful in supply and inexpensive. Specialized knowledge concerning their many unique mechanical and thermal characteristics is essential for developing safe and economical, yet innovative, engineering solutions. Examples of innovations in recent decades include the development of:

(a) ice roads and bridges for aircraft and ground transport;

(b) ice platforms/islands for working space offshore;

(c) ice barriers for protecting offshore structures from moving ice;

(d) small embankment dams built with permafrost;

(e) large dams and dams in warm permafrost built with thaw-stable and artificially thawed soil; and

(f) constructed facilities with passive and active methods of control for the frozen earth in foundations.

1. Structure and Forms of Ice

Ice can be either a naturally occurring or an 'engineered' construction material. Naturally occurring ice is found in many forms and varying morphology (e.g., as river and lake ice, ice islands, glacier ice and icebergs, as well as sea ice sheets, floes and ridges). Engineered ice may be formed either by controlled flooding of an area and waiting for the water to freeze, or by spraying water into freezing air so that the vapor instantly crystallizes and falls back to accumulate as a sheet or mound. In certain applications, such as in the construction of ice roads, natural ice may be strengthened by compaction and by additives such as wood shavings and sawdust.

Naturally occurring and engineered ice are generally crystalline in nature. This type of ice is known to possess ten polymorphic forms or stable structures under differing temperature and pressure conditions. The most prevalent form is ice I_h, which has a hexagonal lattice structure. The oxygen atoms are all concentrated along basal planes that are perpendicular to the principal hexagonal axis, called the *c* axis. The H_2O molecule consists of an oxygen atom covalently bonded to two hydrogen atoms. Each H_2O molecule is hydrogen-bonded to four other molecules in an approximately tetrahedral coordination.

Grain boundaries develop when growing ice crystals intermingle and achieve an equilibrium configuration (associated with a minimum in surface energy). Grain sizes are typically in the 1–20 mm range, although much larger sizes can exist. During the growth and formation of ice masses, the crystals may be granular, columnar or in one of many generally less important forms. The *c* axes of different crystals in an ice mass can exhibit various degrees of alignment, thereby imparting a definite crystallographic texture (or fabric) to the solid.

Ice crystals reject impurities such as gas bubbles, salts and organic or inorganic matter during the freezing process. However, large amounts of impurities get trapped within the ice, especially when freezing is fast. Consequently, most ice has a porous structure. In freshwater ice, the impurities tend to concentrate along grain boundaries, together with a thin water film. In saltwater ice, on the other hand, the brine is mostly contained in platelets or cells within the ice crystals and normal to the *c* axis. The platelets, with typical spacings of 0.1–1.0 mm, impart a cellular substructure to the material. Gravitational forces cause the formation of large tubular brine drainage channels in natural sea ice features. Typically, the brine channels have a diameter of 10 mm and a spacing of 200 mm.

2. Structure and Forms of Frozen Earth

The thermal regime of earth in cold regions depends on its surface temperature, thermal properties and the geothermal gradient. At the top there is the active layer, typically extending from 5 m to 15 m, where seasonal freezing and thawing occurs. Heat and moisture movements between the atmosphere and the earth below take place in this layer.

Below the active layer, the temperature increases steadily under the influence of heat generated deep in the earth. This heat flows upward at a rate dependent on the geothermal gradient, typically 2–3 °C per 100 m, and achieves equilibrium with the mean ground surface temperature. At this depth, the earth is either thawed or may include a layer of permafrost. Sometimes, layers of unfrozen ground may exist within the permafrost or between active layer and permafrost. The depth of permafrost ranges from a few centimeters at the edge of the subpolar zone to about 60–100 m at the boundary with the polar zone. The thickness of the permafrost layer can reach several hundreds of meters in the polar zone. Ice wedge polygons and other frost features may be present in permafrost regions. The wedges form due to the infiltration of cracks in frozen earth by snow and frozen meltwater. Such cracks can endure as long as there is no thawing.

Frozen earth is a multiphase material consisting of solid mineral particles, gaseous inclusions (vapors and gases), liquid (unfrozen and strongly bound) water and ice inclusions. When earth is frozen, its complexity is increased by the appearance of the new solid phase, by the formation of new interfaces and by the random appearance of air bubbles and brine pockets. Ice is the most important component of frozen earth. Frozen water in earth or constitutional ice may exist in the form of large crystals and lenses, or in the form of pore ice which bonds the mineral particles. The frozen pore moisture is termed "ice-cement." Ice-cement bonds depend on many factors, including the degree of negative temperature, the total volume content of ice or iciness, the structure and coarseness of ice inclusions and the content of unfrozen water.

If freezing is rapid, ice formation occurs only in the pores and is characterized by a fairly uniform distribution. This type of constitutional ice imparts a fused or massive cryogenic texture to frozen earth. If freezing is slower, continuous ice interlayers and lenses can form which additionally impart laminar and cellular textures to frozen earth.

Unfrozen water, which is present in frozen earth at temperatures down to about −70 °C, generally exists in two states: (a) water strongly bonded by the surfaces of the mineral particles that cannot crystallize, even at very low temperatures, due to the electromolecular forces of the surface; and (b) loosely bound water of variable-phase composition, between the layer of bound water and free water, that freezes at temperatures below 0 °C. The unfrozen water separates ice from the mineral grains in frozen earth. The amount of unfrozen water in frozen earth decreases as the temperature falls below 0 °C. Depending on the unfrozen water content, frozen earth can be classified as either hard-frozen or plastic-frozen. Hard-frozen earth is firmly cemented by pore ice and contains very little unfrozen water. In most cases this occurs at temperatures below −0.3 °C to −1.5 °C, although for certain very fine mineral particles the limit can be as low as −5 °C to −7 °C. Plastic-frozen earth, which occurs at higher temperatures, has a larger unfrozen water content (e.g., as much as half the total pore water).

3. Structure–Property Considerations for Ice

The structure of polycrystalline ice governs its macroscopic behavior, both when used as a construction material and when present as the most important constituent of frozen earth.

The hexagonal lattice structure makes ice an intrinsically anisotropic material. Single ice crystals are transversely isotropic because of their hexagonal symmetry. Five independent constants are necessary to characterize their elastic behavior. In spite of its open hexagonal structure, the H_2O molecules are relatively close packed in the basal plane. Consequently, ice crystals deform (slip) easily, and may also crack between basal planes when subjected to shear stresses. The resistance against deformation and cracking is much higher on other planes.

The behavior of polycrystalline ice is complicated by the presence of grain boundaries and by the degree of grain alignment or orientation. Grain boundaries can impede slip and cause dislocation pile-ups. Since grain-boundary planes are weakened by the concentration of impurities during freezing and the resulting increase in defect density, intergranular sliding may occur in polycrystalline ice. In addition, stress concentrations can cause crack nucleation at irregularities in the grain boundary and at grain-boundary junctions or triple points.

The degree of grain alignment is generally expressed in terms of the relative orientation of the *c* axis in ice crystals. If the *c* axes are randomly oriented, the resulting ice is isotropic. If the *c* axes lie along a plane but are randomly oriented, the resulting ice is transversely isotropic. If the *c* axes are all aligned in one direction, the resulting ice is truly anisotropic. The latter two types of mechanical behavior represent texture or material anisotropy.

Ice is generally very close to its melting temperature in construction applications; that is, it occurs at a homologous temperature (the ratio of ambient temperature to melting temperature with both measured in kelvins) greater than about 0.8. The mechanical behavior of crystalline solids at such temperatures is controlled by thermally activated rate processes. Thus, ice behavior is very sensitive to rate and temperature variations.

4. *Deformation and Failure of Ice*

The deformation and progressive failure behavior of ice is governed by three primary mechanisms: flow, distributed cracking and localized cracking. In particular, ice may display purely ductile, purely brittle or combined behavior, depending on the temperature and conditions of loading. The primary mechanism associated with flow and the constitutive framework of rate theory are appropriate for characterizing purely ductile behavior. The primary mechanism associated with distributed cracking and the constitutive framework of damage theory are appropriate for characterizing deformations during ductile-to-brittle transition or when the material is purely brittle. Such deformations are accompanied by the formation and stable growth of multiple cracks and/or voids. The primary mechanism associated with localized cracking and fracture mechanics theory are appropriate for predicting the failure strength corresponding to the onset of material instability.

The effect of the evolution of material structure on the macroscopic flow behavior of polycrystalline ice is only beginning to be understood. There is general agreement, based on theoretical and experimental work, that at least two thermally activated deformation systems, a "soft" system and a "hard" system, must be present for flow to occur. They may be either grain-boundary sliding (with diffusional accommodation) and basal slip or basal slip and slip on a nonbasal plane. A combination of these processes could be present as well.

Initially, the solid resists the applied stresses in an elastic manner and then flow begins on the soft and hard systems. However, flow, particularly on the easy soft system, causes the build-up of internal elastic stresses. This may occur as a result of grain-boundary sliding next to grains poorly aligned for deformation or dislocation pile-ups at the boundaries of such grains. Dislocation pile-ups at grain boundaries have been observed in ice through scanning electron microscopy. The internal elastic stresses, termed "back" or "rest" stresses, resist flow. In addition, internal drag stresses which resist dislocation fluxes are generated in annealed materials undergoing flow. These drag stresses are the outcome of creep-resistant substructures; for example, subgrains and cells, formed by grain-boundary sliding, and of dislocation entanglement, dipole formation and kink band formation during slip (particularly on the basal plane).

An increasing characteristic drag stress contributes to isotropic hardening, while an increasing rest stress contributes to kinematic hardening. In isotropic hardening, material properties are independent of the direction of straining. On the other hand, kinematic hardening induces directionally dependent material properties, referred to as deformation or stress-induced anisotropy. The Bauschinger effect in metals is an example of kinematic hardening.

The deformations resulting from the interactions between the soft and hard systems can be decomposed into two components: a transient-flow component and a steady-state flow component. Steady-state flow, representing a balance between work-hardening and recovery processes, is associated with viscous (irrecoverable) strains and obeys an incompressible Norton-type power law. Isotropic and kinematic hardening phenomena are active during transient flow and give rise to anelastic strains. These strains are recoverable on unloading since equilibrium requires the internal elastic back stress to reduce to zero. The transient strain rate can be taken to follow an incompressible Norton-type power law driven by a reduced stress equal to the applied stress minus the rest stress and with the viscous resistance defined in terms of the drag stress. The time-dependent elastic strains defining transient deformation represent the phenomenon of delayed elasticity or anelasticity. The time-temperature equivalence for both deformation systems is given by the Arrhenius law, with similar activation energies for each system.

As the rate of deformation is increased, flow is accompanied by crack formation in ice that is initially crack free. Sudden changes in temperature (i.e., thermal shocks) can also induce cracking. Less is known about cracking in ice than is known about its flow behavior. Cracks generally tend to nucleate in regions of high defect density, which for polycrystalline freshwater ice occur at grain boundaries. Elastic stress concentrations are severe at irregularities in grain boundaries and particularly at grain-boundary junctions, where the lenticular equilibrium shape of a triple point void is similar to the shape of a stressed crack. Once growth initiation occurs, the crack may follow either an intergranular or transgranular path.

At fast loading rates or under thermal shocks, transgranular cracking dominates. Observations indicate that these cracks are characterized by the smooth and often striated surface features of cleavage cracks. In addition they often change direction abruptly on intersecting a boundary in order to be parallel or perpendicular to the basal plane.

At slower loading rates, intergranular cracking tends to dominate. The back stresses associated with anelasticity significantly contribute to the (recoverable) strain energy for crack growth at these rates of deformation. These stresses tend to concentrate on grain-boundary planes associated with crystals poorly oriented for deformation, particularly on the soft system; hence, the preference for intergranular cracking. In addition, the stress required to cause growth initiation (or the stress for crack nucleation) is smaller when anelasticity is present since at any given stress level more elastic strain energy is stored in the material.

Commencement of crack growth does not necessarily lead to instability if toughening mechanisms are active. Crack deflection at grain boundaries and voids provides one form of toughening. If the applied loading is inadequate to overcome these barriers, further growth is arrested. This is typically the case under stress states involving pure compression. The length of the arrested cracks is typically a fraction (0.6) times the grain diameter. The first crack which forms under compressive loading tends to be stable. The material sustains additional compressive stress prior to reaching its ultimate strength. The development of multiple stable cracks leads to a distributed cracking phenomenon which governs the ductile-to-brittle transition or purely brittle behavior in ice. This type of cracking weakens ice; under compressive creep stresses the phenomenon leads to "tertiary" or accelerated creep and under constant strain-rate loading in compression to "strain softening". At relatively fast rates of loading involving brittle behavior, ultimate failure is by splitting or slabbing, while in the transition range of loading rates a shear mode of failure dominates. The application of a low-to-moderate level of confining pressure tends to suppress the development of stable cracks, which in turn allows a higher shear/distortional stress to be sustained. Ultimate failure generally is in the shearing mode. As the confining pressure is further increased the material displays pressure-insensitive ductile flow, and eventually at very high confining pressures pressure-melting is induced and the material can no longer sustain shear/distortional stress. Theoretical models which integrate the flow and damage behavior of ice under multiaxial loading histories are under active development.

States of stress involving tension typically provide adequate energy for unstable crack propagation. In ice with grain diameters larger than a critical value (typical of most engineering applications) the first crack which forms (nucleates) propagates in an unstable manner. The stress at which this crack forms defines the tensile strength. On the other hand, when the grain diameter is smaller than the critical value, crack growth is arrested. Additional stress is necessary for unstable crack propagation. Although viscous flow often accompanies ice deformation prior to the onset of instability or localization of cracking, it is the stored elastic energy in the material which defines the onset of instability.

The energy or stress intensity required for growth initiation and unstable crack propagation in ice are not the same. Growth initiation is associated with a preexisting defect which travels along an intergranular or transgranular path. The resistance to its growth is provided by the surface energy (or energies) appropriate for the particular crack path (or paths). Crack propagation in ice is governed by toughening mechanisms, one of which is crack deflection. Since crack-tip plasticity is generally absent in ice, the development of a process zone containing discontiguous cracks can also toughen the material. The amount of toughening clearly depends on the structure and form of the polycrystal. Experimental determination of the critical fracture energy involving such toughening mechanisms is sensitive to specimen geometry and the amount of crack-tip damage introduced when precracking the specimen.

5. *Deformation and Failure of Frozen Earth*

Engineering problems related to frozen earth are caused by the freezing process, thawing, and the steady-state frozen condition. The freezing of earth causes frost heaving, a phenomenon of special importance in the temperate zone. Three conditions are required for frost heaving to occur:

(a) a cold surface must exist to propagate freezing;

(b) a source of water must exist to feed ice growth; and

(c) the physical composition of the soil must promote the migration of moisture to the freezing front.

When all three conditions are satisfied, free moisture from the earth below migrates along the thermal gradient toward the colder surface on top. The moisture freezes preferentially to existing ice grains when it reaches the frost line, forming small ice lenses. The lenses form normal to the direction of heat flow and are therefore approximately parallel to the ground surface. As these lenses grow and expand, the surface of the earth moves upward. This movement often is nonuniform due to variations in the soil profile, drainage pattern, surface cover or soil conductivity. These differential movements produce undesirable effects on constructed facilities.

Although the ice segregation process is quite complex, it is known that growth occurs as long as moisture can be attracted from the adjacent soil at the rate of freezing. When the heat removal rate exceeds the moisture supply, the freezing front advances and a new ice lens forms. The total amount of heave is often close to the thickness of segregated ice layers in noncompressible earth. Between 2 cm and 5 cm of heaving in a single season is common and movements of 15–20 cm are not unusual. The temperate and subpolar regions characterized by seasonal frost have greater potential for frost heaving, unlike polar regions with permafrost, due to the almost unlimited availability of free moisture from groundwater or earth near the freezing front.

Experimental studies show that particle size provides a valid criterion of frost susceptibility in frozen earth. Although there is no sharp dividing line between frost-susceptible and nonsusceptible materials, smaller particles seem to have a dominating influence on the maximum heaving pressures that are developed in a particular material. Both the rate of heave and the heaving pressure increase with constraint provided by confining pressure.

Two basic countermeasures against frost heaving are available: the first involves earth improvement measures, while the second involves stabilization of constructed facilities. Earth improvement measures include: mechanical techniques such as changes in the composition and density of earth or loads on earth; thermophysical techniques such as changes to the temperature and humidity conditions to control migration of moisture; and physicochemical techniques such as artificial salinization to prevent freezing, the introduction of inorganic compounds to change filtration and capillary properties, the introduction of less wettable compounds, and electrochemical treatment. Stabilization measures include: improved ground drainage for preventing moisture migration to the freezing front; heating of the earth in constructed facilities to prevent freezing using heat insulated screens or artificial methods of heating; the application of counteracting forces; and the use of non-frost-susceptible materials in construction.

Thawing of earth in the active layer causes melting of the ice lenses. The resulting water escapes and voids are left, increasing the porosity of the soil. As melting progresses downward, the meltwater cannot penetrate the frozen earth below and often is unable to dissipate laterally. The trapped water induces excess pore pressures and as a consequence reduces the effective strength of the earth. This phenomenon is termed "thaw weakening". If consolidation occurs subsequently, pore pressures dissipate and the material recovers part of its strength. The dissipation of excess pore pressures is termed "thaw consolidation," while additional dissipation of pore pressure represents ordinary consolidation. Part of the recovery may also be associated with desaturation of the earth. Thawing causes uniform and differential settlements as well as stability problems in earth masses. The movement of mineral particles together with water on thawing can give rise to serious erosion problems. Thawing may be accompanied by a lowering of the permafrost table. Although some variation can occur due to changes in the natural weather pattern, larger shifts are caused by modifications to the surface cover or introduction of artificial heating sources. In ice-rich earth, the potential for settlement associated with the thawing of permafrost is substantial.

The initial effective stress in earth thawed under undrained conditions, called the "residual" stress, governs the magnitude of settlements in the earth. When the residual stress is high, the subsequent consolidation settlements and pore pressures generated during thaw are smaller while the undrained shear strength is higher.

Three major principles of construction are available for dealing with thawing earth. These are:

(a) to preserve the frozen state of the earth;

(b) to design for settlements due to thawing earth; and

(c) to improve the earth by preconstruction thawing.

The first principle is appropriate in the polar, and extreme latitudes of the subpolar, zones and when the constructed facilities do not release substantial amounts of heat and do not occupy large areas. The second principle requires a predictive theory of settlements in thawing earth. If the settlements are excessive, an alternative principle must be adopted. The last principle is adopted when it is necessary to reduce (a) future settlements due to thawing and thawed earth, and (b) differential settlements in earth of highly nonuniform compressibility in the frozen and thawed states. Methods for preserving the frozen state of earth attempt to remove heat. One approach is to provide ventilation between the heat source and the frozen earth through physical separation. A second approach involves the use of special pipes with natural or forced ventilation or coarse-pored ventilated rock fill between the heat source and frozen earth. Artificial cooling of thaw-susceptible soils may also be considered. Preconstruction thawing methods rely on natural solar heat or on artificial heat sources. Examples of artificial heat sources include: hydraulic sources such as hot and cold water; steam; electric heating with alternating current; and thermochemical methods. Preconstruction thawing can be augmented by mechanical, physical and chemical methods for compacting and artificially strengthening the earth.

Frozen earth in its steady state, particularly permafrost, is both stable and strong due to the development of ice-cement bonds. However, the stresses and strains in frozen earth vary with time since the ice cement and ice interlayers flow when loaded. Thus frozen earth displays creep, stress relaxation and recovery behavior. When the degree of ice saturation is large, the mechanical behavior may approximate that for ice and the long-term strength is very small. On the other hand, when the degree of ice saturation is small and interparticle forces begin to contribute, the mechanical behavior may approximate that for unfrozen soil. However, the presence of the unfrozen water film surrounding the mineral particles which restricts interparticle contact complicates mechanical behavior. In addition, pressure melting occurs due to stress concentrations on the ice component between soil particles and from hydrostatic pressure on the ice. The migration of unfrozen water under stress gradients also influences the mechanical behavior of frozen earth.

Two general approaches exist for predicting the deformation and compressibility of frozen earth: (a) to consider frozen earth to be a quasi-single-phase medium with mathematically well-defined properties, neglecting the fact that one portion of the deformation is due to volume changes; or (b) to consider consolidation and flow as two simultaneous but separate phenomena, whose relative amounts of deformation depend on the applied loads and elapsed time. The first approach is easier and more practical, while the second approach is complex but more realistic.

Flow of frozen earth can be divided into three stages: transient flow, steady flow and accelerating flow. Transient flow is characterized by the closure of microcracks, healing of structural defects by moisture that has been pressed out of overstressed zones and refrozen, a decrease in free porosity due to particle dislocation, and partial closure or size reduction of macrocracks. All of these factors cause rheological compacting of the frozen earth. Compressibility of frozen earth can be significant during transient flow, although the actual amount depends on the pressure. This may occur due to instantaneous compression of the gaseous phase, creep of the ice-cement due to shear stresses at the grain contacts, and hydrodynamic consolidation due to the expulsion of unfrozen water under stress. The transition from transient flow to steady flow is characterized by microcrack closure, but the decrease in free porosity is offset by the formation of new (mainly microscopic) cracks. Essentially incompressible steady flow results when equilibrium is established between healing of existing structural defects and the generation of new defects. Finally, during accelerated flow new microcracks are generated at an increasing rate which eventually develop into macrocracks. In addition, the recrystallization and reorientation of ice inclusions cause a decrease in their shear resistance and, consequently, of the frozen earth. In practical applications, it is often sufficient to consider only transient flow in ice-poor earth and only steady flow when excessive ice is present.

See also: Ice: Mechanical Properties (Suppl. 1)

Bibliography

Andersland O B, Anderson D M 1978 *Geotechnical Engineering for Cold Regions*. McGraw-Hill, New York

Ashton G D 1986 *River and Lake Ice Engineering*. Water Resources Publications, Littleton, CO

Barnes P, Tabor D, Walker J C F 1971 The friction and creep of polycrystalline ice. *Proc. R. Soc. London. Ser A* 324: 127–55

Chamberlain E J 1981 *Frost Susceptibility of Soil: Review of Index Tests*, Monograph 81-2. US Army Cold Regions Research and Engineering Laboratory, Hanover, NH

Chung J S, Hallam S D, Maatanen M, Sinha N K, Sodhi D S 1987 *Advances in Ice Mechanics—1987*. American Society of Mechanical Engineers, New York

Eranti E, Lee G C 1986 *Cold Region Structural Engineering*. McGraw-Hill, New York

Farouki O T 1981 *Thermal Properties of Soils*, Monograph 81-1. US Army Cold Regions Research and Engineering Laboratory, Hanover, NH

Gandhi C, Ashby M F 1979 Fracture-mechanism maps for materials which cleave: F.C.C., B.C.C. and H.C.P. metals and ceramics. *Acta Metall.* 27: 1565–602

Glen J W 1975 *The Mechanics of Ice*, Monograph 11-C 2b. US Army Cold Regions Research and Engineering Laboratory, Hanover, NH

Gold L W 1963 Crack formation in ice plates by thermal shock. *Can. J. Phys.* 41: 1712–28

Goodman D J, Frost H J, Ashby M F 1981 The plasticity of polycrystalline ice. *Philos. Mag. A* 43(3): 665–95

Hobbs P V 1974 *Ice Physics*. Clarendon, Oxford

Johnston G H (ed.) 1981 *Permafrost Engineering Design and Construction*. Wiley, Toronto

Ketcham W M, Hobbs P V 1969 An experimental determination of the surface energies of ice. *Philos. Mag.* 19(162): 1161–73

Mellor M 1983 *Mechanical Behavior of Sea Ice*, Monograph 83-1. US Army Cold Regions Research and Engineering Laboratory, Hanover, NH

Michel B 1978 *Ice Mechanics*. University of Laval Press, Quebec City

Shyam Sunder S, Wu M S 1989 A multiaxial differential flow model for polycrystalline ice. *Cold Regions Sci. Technol.* 16: 45–62

Tsytovich N A 1975 *The Mechanics of Frozen Ground*. Scripta Book Company, Washington, DC

Weeks W F, Ackley S F 1982 *The Growth, Structure, and Properties of Sea Ice*, Monograph 82-1. US Army Cold Regions Research and Engineering Laboratory, Hanover, NH

Weertman J 1983 Creep deformation of ice. *Annu. Rev. Earth Planet. Sci.* 11: 215–40

In addition, several scientific journals and international conferences, listed below, emphasize the study of these materials.

Scientific journals:

Annals of Glaciology
Canadian Geotechnical Journal
Cold Regions Science and Technology
Journal of Glaciology
Journal of Offshore Mechanics and Arctic Engineering

Proceedings of international conferences:

IAHR Symposium on Ice

International Permafrost Conference
International Symposium on Ground Freezing (ISGF)
IUTAM Symposium on Ice
Offshore Mechanics and Arctic Engineering (OMAE)
Port and Ocean Engineering Under Arctic Conditions (POAC)

S. Shyam Sunder
[Massachusetts Institute of Technology,
Cambridge, Massachusetts, USA]

Incandescent Lamp Filaments

The filament is the main component of an incandescent lamp. Many substances have been tried as filaments: carbon; the refractory metals tungsten, osmium, tantalum and rhenium; and refractory compounds. Tungsten is the choice material and is used exclusively in the incandescent lamp industry. The attributes of tungsten are detailed in this article.

1. High Melting Point

The melting point of tungsten, 3680 K, is the highest of any metal. The higher the operating temperature, the greater is the light efficiency or "efficacy" in lumens per watt. At the melting point the efficacy of tungsten is 54 lm W^{-1}. In practice, at the typical operating temperature of 2850 K for a standard 100 W lamp, the efficacy is 17 lm W^{-1}. The reduction in lumens per watt is caused by the loss of a large fraction of the input energy as heat by radiation in the infrared region, by gas convection and by conduction through the connecting lead wires. If the filament is surrounded by a halogen gas, as in the so-called tungsten–halogen lamps, the operating temperature is typically 3000–3200 K and the light efficacy is over 33 lm W^{-1}. Tungsten light has a particularly favorable spectrum of wavelengths that are pleasing to the human eye; it is perceived as white light with a warm tint.

2. Low Blackening Rate of the Envelope

Tungsten has a low rate of evaporation at operating temperature which keeps blackening of the envelope at a tolerable level and prevents early filament burnout. Evaporation rates in vacuum at 2800 K are about 10^{-6} kg m^{-2} s^{-1} and about 1.6×10^{-4} kg m^{-2} s^{-1} at 3200 K (Smithells 1952). The vapor pressure at 2800 K is about 10^{-5} torr and at 3000 K, the vapor pressure is about 10^{-4} torr (Langmuir 1913). In contrast, the vapor pressure of carbon at 3000 K is about 10^{-2} torr (Weast 1970).

With the filament operating in inert fill gases such as argon or nitrogen, the evaporation rate of tungsten is substantially reduced (Langmuir 1913). Thus, evacuated lamps usually exhibit noticeable blackening at end of life, whereas gas-filled lamps remain essentially free of blackening. Halogen atmospheres, which consist of bromine or a gaseous iodine compound in an inert gas, are especially effective at preventing blackening. As the tungsten evaporates from the filament, it travels through the fill gas to the bulb wall where it reacts with the halogen or oxyhalides. The reaction products diffuse to the vicinity of the filament where they dissociate and the tungsten is deposited on the cooler parts of the filament. The bulb wall remains clean and the light output is near 100% over the life of the filament. Since the deposition of the tungsten does not occur at the hotter parts of the filament where the evaporation rate is the greatest, the filament life is still limited.

3. Strength at High Temperature

Tungsten wire filaments have good mechanical strength at high temperatures and are able to resist creep deformation or filament sag under the effects of gravity, vibration or sudden accelerations. Of greatest concern are shear stresses in the transverse direction to the axis of the coiled-coil filament, an example of which is shown in Fig. 1. The shear stresses are caused by torque of the helical segments of a tensile-loaded filament. The shear stresses range from about 1 MPa to over 10 MPa in magnitude (Briant and Walter 1988). The most prevalent mechanical failure at high temperatures results from cavitation and shear along transverse grain boundaries. Mechanical failure at low temperatures is caused by brittle fracture along transverse grain boundaries. The strength of the tungsten wire at high temperatures is the result of the formation and stabilization of a unique recrystallized grain morphology in which the grains overlap along the wire axis and the boundaries contain large-area longitudinal segments. Thus the tendency for the grains to slide apart along their boundary is greatly reduced. This unique morphology is achieved through special doping and mechanical processing procedures which were developed phenomenologically, beginning with Coolidge's processing experiments (Coolidge 1913, 1965) and followed by the doping process developed by Pacz. The work of Pacz (1922) led to the term "218" which has since been applied to tungsten wire having this unique grain morphology. The important microstructural feature for high-temperature mechanical strength is the linear array of fine bubbles of potassium less than about 80 nm in diameter parallel to the wire axis. The strings of bubbles are responsible for the formation and stabilization of the large-area interlocking grain boundaries (see Fig. 2), which are the basis for the high-temperature mechanical strength. The potassium bubbles are the highest-temperature barriers to grain boundary migration

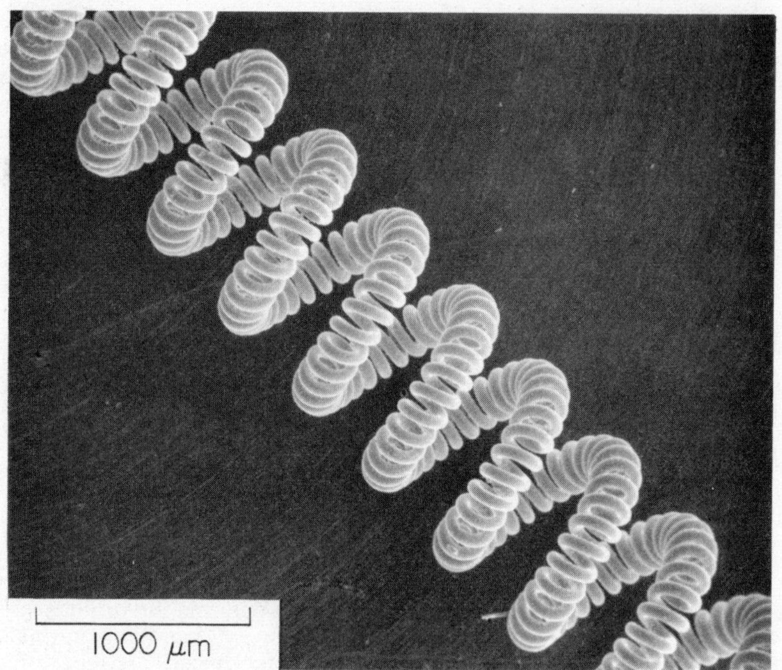

Figure 1
Scanning electron micrograph of a coiled-coil filament

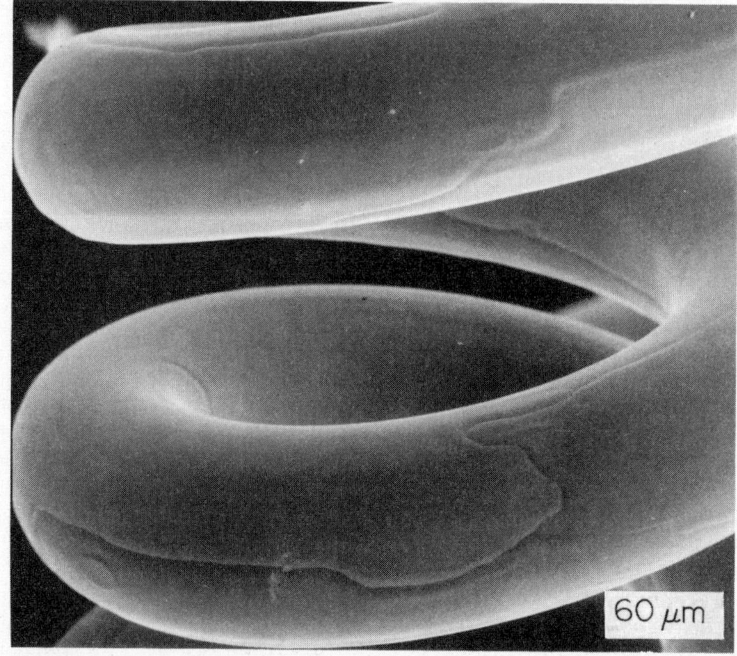

Figure 2
Scanning electron micrograph of longitudinal interlocking grain boundaries in a filament burned for 5 h at 2900 K

known. Without the bubbles, the wire would develop a bamboo-like grain morphology with all of the grain boundaries normal to the wire axis; failure by sliding apart along these transverse grain boundaries would ensue within minutes of light-up of the filament.

4. *Ease of Fabrication*

Filaments are made into intricate shapes to pack a long length of wire (up to 1 m) into a small space which increases the efficacy of the lamp and enables focusing of the light. Most lamp filaments are coiled into tight helical shapes, and often the primary coil is coiled into a double or even a triple helix. An example of a coiled-coil filament is shown in Fig. 1. Annual worldwide production of over 10^{10} filament coils requires a wire that is ductile at room temperature since up to 30% strain must be accommodated in the outer layers of the wire during coiling. Although polycrystalline tungsten is normally brittle at room temperature, Coolidge's development of a process to make wire that is ductile at room temperature has permitted mass production of tungsten filaments (Coolidge 1913, 1965).

5. *Filament Processing*

The processes used to make tungsten lamp wire have two main goals:

(a) to obtain high-temperature strength by control of the grain morphology, and

(b) to provide room temperature ductility for easy fabrication.

A combination of the Coolidge process and the doping process developed by Pacz (1922) provides for both goals and is described here in some detail. The process is all the more remarkable in that it was developed at a time when there were no instruments with sufficient resolution to detect the fine potassium bubbles responsible for control of grain morphology. Also, dislocation mechanisms in plastic deformation were unknown, so the ductilizing process could not have been understood in modern terms. The understanding of these microstructural phenomena has only developed since about the 1960s and the role of the potassium bubbles as high-temperature barriers is now known.

Tungsten filaments consist of essentially pure tungsten with a potassium concentration in the range 50–100 wt ppm (178–355 at. ppm). While this appears to be a small quantity, it is actually a huge supersaturation since potassium is insoluble in tungsten. Other dopant or impurity levels are also in the ppm range as discussed by Pugh (1973).

The processing of lamp filament wire and filaments takes place in several stages.

(a) WO_3 is extracted from the ore and dissolved in acid.

(b) It is purified through ammonium paratungstate salt precipitation.

(c) "Pure tungsten blue oxide" is produced (Shujie and Ho-yi 1987).

(d) The oxide powder is doped with potassium–aluminum silicate compound.

(e) It is reduced in hydrogen to metallic tungsten powder.

(f) Ingots are mechanically pressed.

(g) The ingots are sintered at approximately 3200 K. A large fraction of the dopants escape as gas but a sufficient concentration of potassium remains trapped in the ingot to form strings of bubbles in the filament wire (Das and Radcliffe 1968, Moon and Koo 1971, Snow 1972).

(h) They are mechanically processed by rolling, swaging and wire drawing in 30–60 reduction steps depending on the final wire diameter. The temperatures of deformation range from as high as 2200 K for ingot rolling to about 800 K in the final stages of wire drawing. A delicately balanced deformation and temperature processing schedule is a critical element of filament processing, as it ensures the following properties.

    (i) A deformation microstructure consisting of fibers or subgrains of the order of 0.1–0.3 μm in width (Fig. 3). The fibers are aligned with their ⟨011⟩ crystallographic axes parallel to the wire axis as a typical "wire texture." The subgrains contain a sufficient but not excessive density of mobile dislocations that make the wire ductile at room temperature.

    (ii) An effective dispersion of the dopant as fine ribbons which break up upon annealing into long strings of very fine bubbles (Fig. 3).

(i) The tungsten wire is coiled on mandrels (usually molybdenum wire) at or near room temperature.

(j) Some of the cold work is recovered by heating the coiled filament on the mandrel at a temperature of 1300–1700 K to prevent springback of the filament as the mandrel is etched away.

(k) The finished filaments are clamped between lead wires and mounted in glass or quartz envelopes which are then evacuated, filled with inert gas and sealed.

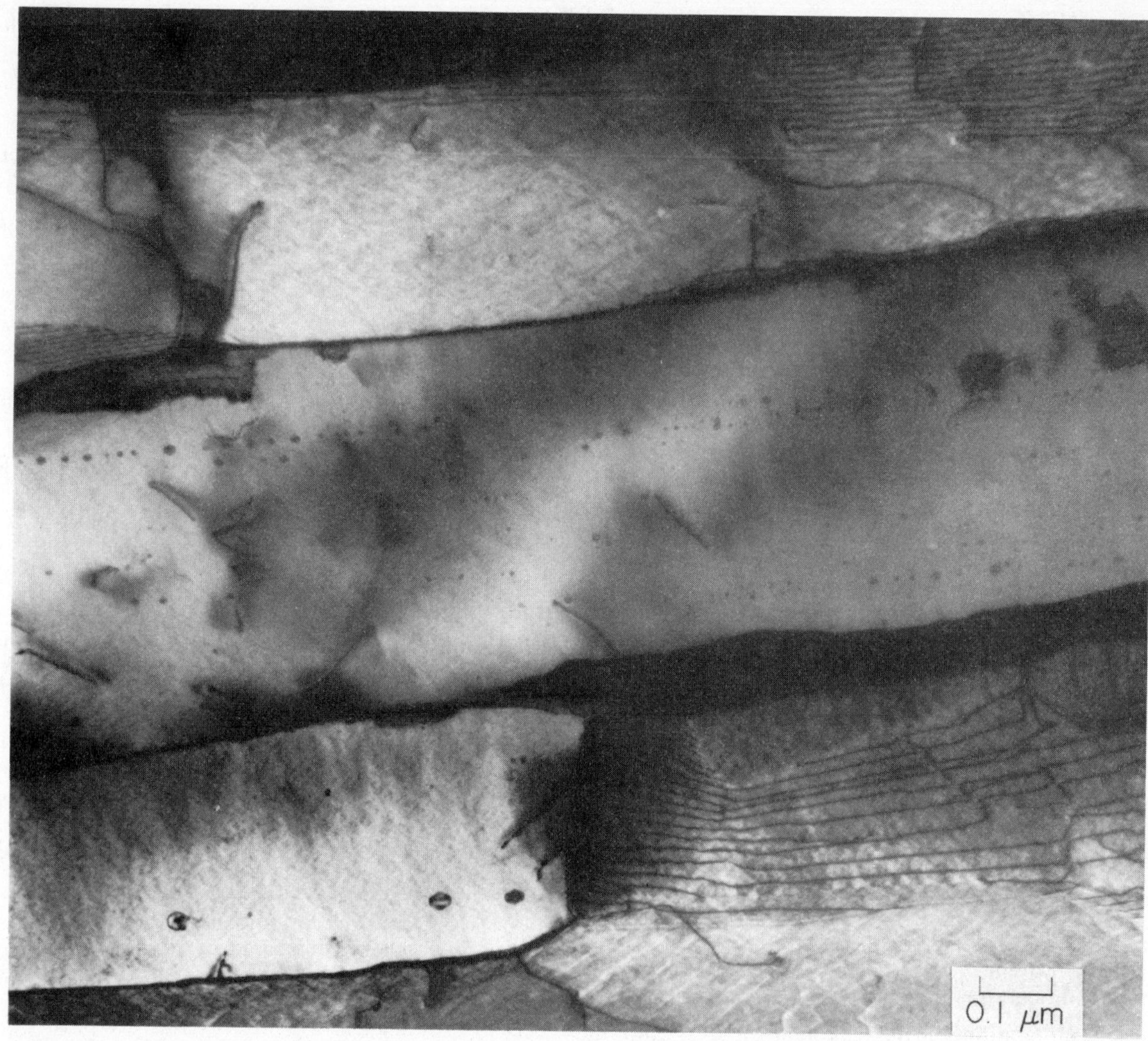

Figure 3
Transmission electron micrograph of 0.006 cm diameter doped wire heated for 1 min at 1890 K to reveal strings of potassium bubbles: the wire has not recrystallized

(l) The filaments may then be recrystallized in the lamp. The heating rate to the recrystallization temperature and the time at temperature are controlled to provide the desired grain morphology in the filament (Walter 1986).

Tungsten lamp filament wire is unique in two respects. First, in the wire there is the low-melting-point element, potassium, dispersed as strings of very fine bubbles which impede the lateral motion of grain boundaries during recrystallization to provide interlocked grains with large longitudinal grain-boundary area. Second, the brittle polycrystalline tungsten is made to be ductile by thermomechanical processing to wire.

The reason potassium provides stable barriers to grain boundaries is that the potassium atom is very large compared with the tungsten atom ($4.55\times10^{-5}\,m^3\,mol^{-1}$ for potassium compared with $9.5\times10^{-6}\,m^3\,mol^{-1}$ for tungsten) and is, therefore, insoluble in tungsten. Once incorporated as bubbles in the tungsten, potassium does not diffuse and the bubbles resist coarsening. Thus, the bubbles remain fine and retain their ability to interfere with the lateral migration of the grain boundaries at very high temperatures such as the filament operating temperatures which are 75–90% of the melting temperature of tungsten. Most of the aluminum and silicon, introduced during doping, is evaporated from the ingot during sintering; the remainder, in solution in the tungsten, has no significant role in filament performance.

The tungsten lamp wire achieves its sag resistance as a result of the interlocking large-area grain boundaries. This morphology is stabilized at high temperatures by strings of bubbles parallel to the wire axis which interfere with the migration of longitudinal grain boundaries while allowing grain growth to occur with little restraint parallel to the bubble strings. Thus, grain boundary sliding, a common failure mechanism of filaments with the "bamboo structure," is alleviated. The interaction between the bubbles and the grain boundaries is known as "Zener drag" (Zener 1949). Pinning of grain boundaries by potassium bubbles in tungsten wire has been discussed in a number of papers (Das and Radcliffe 1968, Pugh 1973). The magnitude of the drag force per unit area exerted by a string of bubbles on the boundary is proportional to the volume fraction of bubbles and to the grain-boundary energy and is inversely proportional to the bubble radius; the smaller the bubbles and the greater their number, the greater the restraint to grain-boundary motion.

On the basis of typical dopant concentrations and recrystallized grain sizes in tungsten lamp wire, Warlimont et al. (1975) calculated that the bubble diameter should be less than 60 nm to efficiently restrain grain-boundary migration. In commercial filament wire this is indeed the case (Briant and Walter 1988). The details of the doping procedure and the thermomechanical processing are important factors in determining the distribution of the potassium bubbles in the wire and the ease of processing of the wire. The finer and more numerous are the pockets of dopant in the tungsten powder particle, the closer together are the strings of bubbles and the more numerous are the bubbles resulting in more effective control of the grain morphology and the high-temperature properties of the wire (Walter and Koch 1990). The inhibition of grain-boundary migration by bubbles has been recognized to be of general validity in a variety of materials and for a variety of bubble-forming elements (Middleton et al. 1949, Ells 1963, Bhatia and Cahn 1978, Ullmaier 1983). Although bubbles form barriers to dislocation motion (Scattergood and Bacon 1982), as well as to grain-boundary migration, the use of bubbles to control microstructure and properties has been limited in commercial application to tungsten and molybdenum wire in which the fine bubble distributions are obtained by large-strain deformation (Stephens 1983).

As mentioned previously, the second unique aspect of tungsten lamp wire is its ductility at room temperature. Normally, polycrystalline tungsten is brittle at temperatures below 500–820 K depending on the impurity content. Single crystals are ductile at and below room temperature (Smithells 1952, Stephens 1983); the grain boundaries suffer brittle fracture. However, after drawing, the wire can be ductile at temperatures below room temperature. There are three reasons for the ductility of the drawn wire.

First, transverse grain boundaries which would experience tensile stresses are essentially eliminated. It is possible to split the wire along the subgrain or fiber boundaries. Under such loading conditions, the ductility is nil. Second, the subgrains are only 0.1–0.3 μm in diameter which has a limiting effect on the length of dislocation pile-ups. Thus, there is a limit to the stress concentration that could lead to fracture along the subgrain boundaries. Third, a suitable density of mobile dislocations is contained in the subgrains; the difficulty of nucleation and multiplication of dislocations is removed, and the wire will deform uniformly during plastic straining.

The dopant bubbles aid in preventing dislocations from annealing out, and raise the recrystallization temperature to about 2300 K (Fig. 4) but the wire is embrittled by annealing at temperatures above 1900 K (Fig. 5). Undoped wire, on the other hand, experiences complete dislocation recovery at 1200 K and recrystallization at about 1400 K and is, therefore, embrittled more readily than is the doped wire (Schultz 1964, Welsch et al. 1980). Although the arguments are qualitative in nature, a substantial body of literature exists that supports these interpretations.

In summary, the high-temperature sag or creep resistance of tungsten lamp filaments is determined

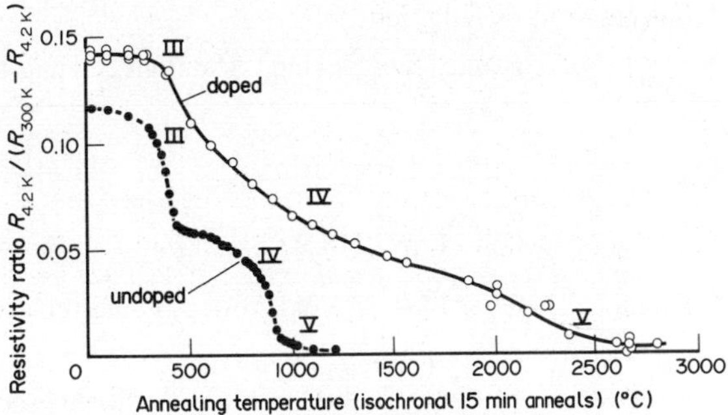

Figure 4
Plot of recovery of resistivity ratio showing the annealing stages in doped tungsten wire (Welsch et al. 1980) and undoped tungsten wire (Schultz 1964): Stage III, point defect recovery; stage IV, dislocation recovery and migration of uninhibited subgrain or grain boundaries; stage V, grain growth to complete recrystallization

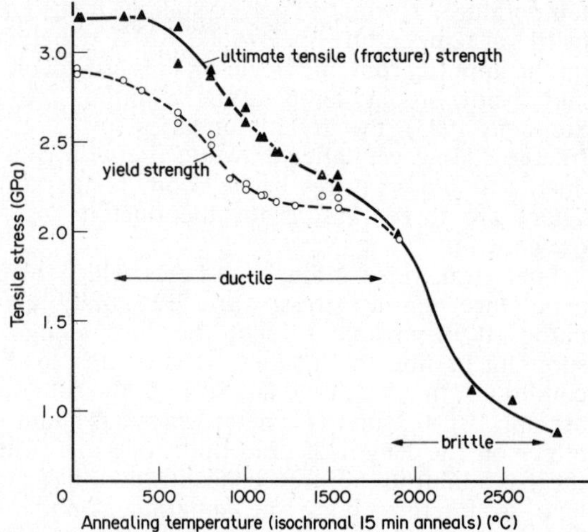

Figure 5
Room temperature tensile strength of as-drawn doped tungsten wire; change of strength and ductility with annealing (Welsch et al. 1980)

by the potassium dopant concentration and its effective distribution as fine bubbles in long strings parallel to the wire axis. The details of the powder doping process and of the thermomechanical processing control the final distribution. The ductility of the wire is the result of the fine subgrain or fiber structure of the drawn wire and of the mobile dislocations in the subgrains which are related to the details of wire drawing and intermediate anneals. Beginning with doped tungsten powder there are typically 50–100 thermal-mechanical processing steps involved in the production of lamp filaments. The parameters in these processing steps are finely tuned by experience and quality control methods to enable a gradual progression in the evolution of the desired microstructure in the filament.

See also: Incandescent Lamp Materials; Tungsten Production

Bibliography

Bhatia M L, Cahn R W 1978 Recrystallization of porous copper. *Proc. R. Soc. London, Ser. A* 362: 341–60

Briant C L, Walter J L 1988 Void growth in tungsten wire. *Acta Metall.* 36: 2503–14

Coolidge W D 1913 US Patent 1,082,933

Coolidge W D 1965 The development of ductile tungsten. In: Smith C S (ed.) 1965 *The Sorby Centennial Symposium on the History of Metallurgy*. Gordon and Breach, New York, pp 443–9

Das G, Radcliffe S V 1968 Internal void formation in thin foils of annealed tungsten wire. *Trans. Met. Soc. AIME* 242: 2191–8

Ells C E 1963 The effect of helium on recrystallization and grain boundary movement in aluminum. *Acta Metall.* 11: 87–96

Langmuir I 1913 The vapor pressure of metallic tungsten (in German). *Phys. Z.* 14: 1273–80

Middleton A B, Pfeil L A, Rhodes E C 1949 Pure platinum, of high recrystallization temperature, produced by powder metallurgy. *J. Inst. Met.* 75: 595–608

Moon D M, Koo R C 1971 Mechanisms and kinetics of bubble formation in doped tungsten. *Met. Trans.* 2: 2115–22

Pacz A 1922 US Patent 1,410,499

Pink E, Bartha L 1989 *Metallurgy of Doped/Non-sag Tungsten*. Elsevier, New York

Pugh J W 1973 On the short time creep rupture properties of lamp wire. *Met. Trans.* 4: 533–8

Scattergood R O, Bacon D J 1982 The strengthening effect of voids. *Acta Metall.* 30: 1665–77

Schultz H 1964 The recovery of electrical resistivity of cold-worked tungsten (in German). *Acta Metall.* 12: 649–64

Shujie L, Ho-Yi L 1987 The control of phases of tungsten blue oxide and their effect on the particle size of tungsten powder. *Int. J. Refract. Hard Met.* 6: 35–9

Smithells C J 1952 *Tungsten*. Chapman and Hall, London, pp. 183–5

Snow D B 1972 Dopant observation in thin foils of annealed tungsten wire. *Met. Trans.* 3: 2553–4

Stephens J R 1983 NASA Tech. Report D1581, Washington, DC

Ullmaier H 1983 Helium in metals. *Radiat. Eff.* 78: 1–10

Walter J L 1986 Failure of tungsten ribbon filaments by formation and growth of hot spots. *J. Appl. Phys.* 60: 3343–55

Walter J L, Koch E F 1990 The relationship of microstructure to mechanical properties of aluminium/silicon/potassium doped tungsten wire. *J. Mater. Sci.* 25 (in press)

Warlimont H, Necker G, Schultz H 1975 On the recrystallization of doped tungsten wire. *Z. Metallkd.* 66: 279–86

Weast R (ed.) 1970 *Handbook of Chemistry and Physics*, 51st edn. CRC Press, Boca Raton, FL. p. D-145

Welsch G, Young B J, Hehemann R F 1980 Recovery and recrystallization of doped tungsten. In: Haasen P, Gerold V, Kostorz G (eds.) 1980 *Strength of Metals and Alloys*. Pergamon, Oxford, pp 1693–8

Zener C, cited by Smith C S 1949 Grains, phases and interfaces: An interpretation of microstructure. *Trans. AIME* 175: 15–51

G. Welsch
[Case Western Reserve University, Cleveland, Ohio, USA]

J. L. Walter
[General Electric Corporate Research and Development, Schenectady, New York, USA]

Injection Molding of Polymers: Computer Simulation

The design and commissioning of molds for polymers has traditionally been a trial and error procedure. Considerable time is required to test and, if needed, modify the mold before production may begin. A desire to minimize this time and reduce retooling costs has led to the development of computer software for simulating the injection molding process. Software is commercially available for analyzing both the mold filling and the cooling phases of the injection molding process. Both types of analysis generally use a three-dimensional representation of the mold cavity. In the case of cooling analysis, a representation of the cooling line locations is also used. These representations or models are typically generated with computer-aided design (CAD) packages. Simulation of the process allows mold and product designers to determine optimal gate positions, feed systems and processing conditions by indicating problem areas so that corrective action may be taken. In addition, the use of analysis software can minimize material costs and cycle times while improving the quality of molded components.

1. *Simulation of Mold Filling*

Analysis of mold filling is the natural starting point for analysis of the injection molding process. For modelling purposes it is convenient to consider two stages: the filling phase, during which time a mold is volumetrically filled, and the packing phase, in which volumetric contraction, arising from cooling of the melt, is compensated for by the injection of more material. With an appropriate representation of the cavity geometry and of the rheological and thermal properties of the material, flow analysis can provide insight into the many factors that determine the quality of injection molded parts.

1.1 Governing Equations

Flow analysis technology is based on the solution of the governing equations of fluid dynamics (Bird et al. 1960):

$$\frac{\partial \rho}{\partial t} = -\nabla \cdot (\rho \boldsymbol{v}) \tag{1}$$

$$\frac{\partial (\rho \boldsymbol{v})}{\partial t} = -[\nabla \cdot \rho \boldsymbol{v}\boldsymbol{v}] - [\nabla \cdot \Pi] + \rho \boldsymbol{g} \tag{2}$$

$$\frac{\partial \rho \hat{U}}{\partial t} = -(\nabla \cdot \rho \hat{U} \boldsymbol{v}) - (\nabla \cdot \boldsymbol{q}) - (\Pi : \nabla \boldsymbol{v}) \tag{3}$$

where ρ is the fluid density, $\boldsymbol{v}$ is the velocity vector of a point in the fluid, Π is the stress tensor, $\boldsymbol{g}$ is the acceleration due to gravity and $\hat{U}$ is the internal energy per unit mass. Equations (1–3) are known as the equations of continuity, motion and energy, respectively. They are valid for any fluid. A constitutive equation is required to relate the stress tensor Π to the rate of deformation tensor ∇v through the fluid viscosity. Commonly a generalized Newtonian fluid constitutive equation is employed in which viscosity varies with shear rate or shear stress. Typical empirical equations (Bird et al. 1977) include the power law model, the Carreau model, the Cross model and polynomial representations. With the exception of the power law model, each of these is capable of representing the transition to Newtonian flow at low shear rates, as observed in polymer melts.

Additional terms in the constitutive equation can account for the temperature dependence of viscosity. Several functional forms are used to represent the dependence, including exponential and reciprocal functions of temperature, and relations of the Williams–Landel–Ferry (WLF) type. Another level of complexity in material property representation arises if the pressure dependence of viscosity is considered. Viscosity is most commonly modelled as an exponential function of pressure. In addition, further data about density, heat capacity and thermal diffusivity properties are also required for a complete solution of the governing equations.

Assumptions of creeping flow, no wall slip, incompressibility, zero gravitational body forces and independence of thermal diffusivity from temperature and from cooling rate are widely used to simplify solution of the equations. In addition, the filling phase is often assumed to be adequately represented by steady-state flow at any instant in time whereas transience, which appears through the inclusion of the time derivative in the equation of motion, is essential to the solution in the packing phase.

It is important to note that any flow simulation is only as good as the underlying assumptions and the material data used. Within the guidelines set by simplification of the governing fluid dynamic equations, accurate predictions can only be made with high-quality material data.

1.2 Description of Cavity Geometry

Any flow analysis requires a description of cavity geometry as this comprises one aspect of the resistance to flow of the molten polymer during injection molding. In essence, a mold cavity is modelled by a set of connected points, lines and surfaces. The degree of complexity of the cavity model is generally directly related to the degree of complexity of the equations. Simple one-dimensional surfaces provide a 'matchstick' description of the cavity and are only capable of representing simple flows such as rectangular duct flow without edge effects, and flow through circular cross-sectional tubes and annuli.

Two-dimensional surfaces give a 'house of cards' or shell description of a mold cavity and are capable of representing two-dimensional planar flows. This is adequate for the majority of molds, particularly when combined with one-dimensional surfaces where necessary. Three-dimensional cavity descriptions enable analysis of fully three-dimensional flows but modelling and analysis are far more time consuming. Consequently, three-dimensional analyses are rarely used in practical injection molding simulations.

1.3 Solution of Flow Equations

There are primarily two numerical techniques employed to solve the set of governing flow equations; the finite-difference method and the finite-element method. The finite-difference method reduces a set of differential equations to algebraic equations by substituting for the partial derivatives with approximating finite differences. In the finite-element method (Rao 1982) the solution domain is approximated by a set of piecewise polynomial approximations. In each case, increasing the order of the approximation increases the accuracy of the solution, but the solution cost becomes greater. Ultimately, there is always a trade-off between solution accuracy and computation expense.

The cavity geometry, which defines the solution domain, is mapped into a set of intersecting grid points or connected polyhedral elements, for the finite-difference or finite-element methods, respectively. Both sets are generally referred to as meshes. Solution results in a value of the field variables, pressure and temperature; for example, at grid points, element intersections (nodes) or element centroids.

Hybrid finite-element–finite-difference methods are also employed where the cavity is modelled with two-dimensional plates or shells. A finite-element mesh is used as the basis of the calculation of pressure and temperature distribution throughout the mold. Within each finite element a finite-difference mesh is used to account for the variation of velocity, shear rate, viscosity and temperature through the cavity thickness.

Finally, some packages use so-called "fast" algorithms which avoid explicit determination of the variation of parameters through the cavity thickness. These proprietary algorithms effectively increase element viscosity as a function of mold geometry and thermal and pressure history. The net effect is a rapid determination of the cavity fill pressure and of the average temperature distribution.

1.4 Analysis Results and Display

Analysis output is similar between different packages and consists primarily of pressure, temperature, fill time, velocity, shear rate, shear stress and volumetric shrinkage. However, most of these parameters change continuously during both filling and packing phases. It is prohibitively expensive to provide the user with data for each of the major variables at every instant of time during injection, so most commercial packages record results considered to be of particular interest or provide for data storage at various user-defined stages of the analysis. Output is viewed graphically with accompanying postprocessing software or general purpose postprocessors that interface to the analysis output.

One of the most useful outputs is the distribution of fill times at each point in the cavity as such a plot shows how the mold fills. It shows which areas fill first, where opposing melt fronts meet to form weld lines, and leading and lagging sections of the flow front which could lead to air entrapment or weld line formation where it was not anticipated. Consideration of the areas that fill first allows users to avoid overpacking, a common cause of differential shrinkage and warpage. Positioning of weld lines may be of aesthetic or structural importance.

Contour plots of pressure distribution at various stages during filling provide graphic evidence of high shear stress areas, flow reversals, areas of overpack and, of course, estimates of fill pressure. When the pressure distribution is integrated over the projected area of the cavity, estimates of clamp force are obtained. Both pressure and clamp force estimates aid in the selection of an injection molding machine of adequate capacity. Coupled with skillful interpretation of the filling pattern, the molder can use the output to optimize gate positioning so as to avoid overpacking and reduce shrinkage and warpage.

Excessive heat generation, which can lead to material degradation, or premature freezing, which can lead to shrinkage and warpage, can be avoided by examining the distribution of mean temperatures during filling and packing, and altering injection conditions accordingly. In addition, plots of flow-front temperature at each point in the cavity give an indication of the quality of weld lines because the hotter the flow fronts at the point of formation of the welds, the more effective the diffusion of material across the interface becomes.

Minimizing shear stress levels during injection is critical to acrylonitrile–butadiene–styrene (ABS) moldings that are to be electroplated. The molder can experiment with molding conditions to ensure maximum stress levels are not exceeded.

Volumetric shrinkage estimates are typical packing phase analysis outputs. These have been related variously to shrinkage and distortion of the entire molding. The formation of unsightly sink marks can be observed through packing analysis and can, therefore, be avoided or minimized.

The above examples are just a few of the outputs available from current commercial packages. It is apparent that the advantage of using flow analysis in injection molding is its ability to model multiple injections, varying cavity geometry, molding conditions and material type as appropriate, without the need to build costly molds and run time-consuming mold trials. The expense of component design and manufacture, and of tool making, is reduced because prototyping can be waived. Material savings are possible because runner positioning, diameters and lengths can be optimized for pressure drop and shear heat input. Similarly, local cavity thicknesses can be easily modified to provide optimum filling performance and structural strength without overdesign of the product. The ease with which a user can say "what if?" and then answer the question is certain to spread the use of flow analysis technology through the plastics injection molding industry.

2. *Cooling Analysis*

The cooling phase represents more than two-thirds of the time required to produce an injection molded part. Improperly designed cooling systems can lead to part distortion due to differential cooling and to high levels of residual stress in the molding. Residual stresses may be so high as to cause part cracking and premature failure of the component. The designer of a mold cooling system has two main aims: minimization of the cooling time and uniform cooling.

2.1 Heat Transfer in Moldings

Heat transfer in an injection molding is illustrated in Fig. 1 where Q_p is the heat flow into the mold from the incoming melt, Q_s is the heat flow from the surface of the mold due to convection, Q_r is the radiative heat flow from the mold, Q_m is the heat flow due to conduction into the injection molding machine and Q_c is the heat flow into the coolant due to convection.

Conduction of heat into the molding machine is minimized in practice by insulation and may be ignored. Heat loss due to radiation and convection from the outer surface of the mold is significant only if the mold surface temperature is above 80 °C. The major factors are therefore conduction of heat through the plastic occupying the cavity and into the mold, and heat flow from the mold into the cooling channels.

2.2 Governing Equations

Heat from the plastic in the cavity is transferred to the mold body by conduction. It is reasonable to assume conduction takes place only in the thickness direction of the plastic and so may be represented by the one-dimensional heat-conduction equation:

$$\frac{\partial T}{\partial t} = \kappa \frac{\partial^2 T}{\partial z^2} \tag{4}$$

where $T = T(z,t)$ is the temperature at a point in the cavity, κ is the thermal diffusivity of the material and z is the thickness direction of the cavity.

Heat conduction through the mold body is described by the three-dimensional heat equation:

$$\frac{\partial T}{\partial t} = \kappa \nabla^2 T \tag{5}$$

where ∇^2 is the Laplace operator and $T = T(x, y, z, t)$.

Equations (4, 5) must be solved subject to suitable boundary conditions. These consist of the prescribed initial temperature field within the plastic or mold and the rate of heat flow across the cavity–mold interface or external surfaces of the mold. The boundary conditions may incorporate the effect of radiation or convection from the mold surface. In particular, the rate of heat flow from the mold into the cooling line is given by

$$q_c = h_c A_c (T_{cs} - \bar{T}_c) \tag{6}$$

where h_c is the heat-transfer coefficient of the cooling line, A_c is the area of the cooling line, T_{cs} is the temperature of the cooling line surface and $\bar{T}_c$ is the bulk temperature of the coolant. The value of h_c, and so the ability of the cooling line to extract heat,

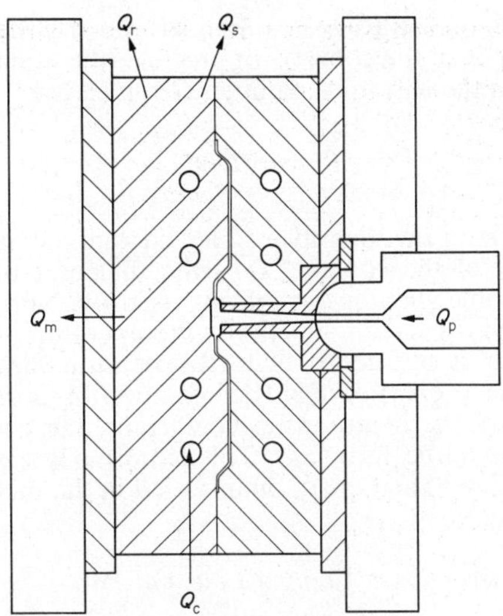

Figure 1
Heat transfer in an injection molding

depends on the Reynolds number and the thermal properties of the coolant. In practice, the dominant factor is the Reynolds number, Re given by

$$Re = \rho VD/\mu \tag{7}$$

where ρ is the density of the coolant, V is the velocity of the coolant, D is the diameter of the cooling line and μ is the coolant viscosity. Note that although ρ and μ are functions of temperature, if the flow rate is sufficiently high, temperature variation of these properties may be ignored. In order to maximize the heat-transfer coefficient, the coolant flow should be turbulent. This is normally easy to achieve with the type of pumps used in the molding industry.

2.3 Description of Cooling Line and Part Geometry

Considerable modelling time is saved if the same model for the cavity can be used for both flow and cooling analyses. Cooling lines must also be described for cooling analysis and are represented as a network of one-dimensional surfaces each of which is defined by two points. Provision to handle special configurations such as bubblers and baffles, and lines of different cross-sectional shape is often available in analysis software. The cooling line sections are meshed to form a series of one-dimensional elements. Some analysis software may also require that the outer surface of the mold be described. This will always be necessary for an analysis that takes radiative and convective losses into account.

2.4 Numerical Solution

It is important to note that the solution to the heat-transfer problem is transient in nature. As the plastic occupying the cavity cools the rate of heat transfer into the mold will vary. Despite this fact, several commercial analysis packages use a steady-state analysis. Such analyses are based on the assumption that the heat into the mold may be averaged over the molding cycle.

In order to solve Eqn (4), it is necessary to specify the initial temperature at the wall of the cavity and the initial temperature of the plastic. These may be given by the user or derived from the results of a flow analysis. Equation (4) may then be solved using finite-difference techniques.

Analysis software invariably includes cooling network calculations. The user specifies an inlet flow rate and temperature. Flow rate and temperature in each branch of the cooling system are then calculated using mass and energy conservation principles.

A very detailed solution of Eqn (5) may be obtained with a full finite-element representation of the part and the mold (Karjalainen 1987). Special elements can be used to take into account the thermal resistances between different parts of the mold; molds produced from materials with different conductivities can be handled easily. The major disadvantage is the need to mesh the interior of the mold to produce solid elements. In addition, the computer time needed can be prohibitive.

Boundary integral techniques (Brebbia et al. 1984) are also employed to solve Eqn. (5). One advantage of this approach is that no internal meshing of the mold is required. The process does, however, require the exterior mold surfaces to be modelled and meshed. One disadvantage is that it is difficult to account for molds constructed with materials of differing conductivity.

Another method involves reducing the problem to a one-dimensional approximation. A three-dimensional finite-element representation of the part is used. The reduced problem consists of a series of one-dimensional analyses where the heat from a plastic element is transferred to a single cooling line element. The distance between the elements is varied according to the heat load. Success or failure of this scheme depends on the quality of the assumptions underlying the approximation. It is necessary to consider the effect of surrounding plastic elements on the effective heat transfer. This manifests itself in the calculations as a change in the distance between the plastic and cooling line elements. For instance, in areas such as corners where the heat load is high the "effective" distance is increased. If the geometry of the part is carefully considered this scheme can rapidly produce very good results. Another advantage is that the calculated one-dimensional distances and effective conductivities can be used by flow analysis software thereby allowing temperature calculations to be based on the effect of the cooling system.

Solution by any of the above methods is iterative in the sense that after the rate of heat transfer into the cooling lines has been initially estimated, temperatures at the wall of the cavity and cooling lines are calculated, the heat transfer into the cooling lines is then calculated and the process repeats. Final convergence to the correct solution may be determined by an energy balance of heat into and out of the mold.

2.5 Analysis Results and Display

Results from cooling analysis software are displayed in a similar way to those from flow analysis with contours of the calculated variable displayed on the model. The mold temperature at ejection is an important result from cooling analysis and most graphical displays permit the user to display temperatures on either side of the plastic elements. With these results, the uniformity of cooling can be assessed. A particularly useful feature is the ability to plot the difference in temperature across the part. Areas of high temperature differential may then be seen. Minimizing temperature differences across the part will reduce warpage. Often the results from network calculations can be displayed graphically. Typical outputs include flow rate in each branch of the cooling system, the wall temperature of the cooling line and the bulk temperature of the coolant.

3. *Future Developments*

Flow analysis technology, although now widely accepted by the injection molding industry and shown to be reasonably successful for a large majority of moldings, is far from being a completed subject. There are a number of areas, particularly in relation to material behavior, where current technology is clearly inadequate.

During the filling phase, for example, material compressibility and density variation with temperature are largely ignored, although it is known that these properties can account for observed differences of 30% between molding machine pumping rate and effective volumetric flow rates within the cavity. Phenomena which reflect the viscoelastic nature of polymer melts, such as extra pressure drops in rapidly converging sections, melt fracture and distortion, orientation development and stress relaxation, are generally not considered by current commercial programs. Polymer melts also exhibit considerable tensile viscous forces, which can dominate the flow field during rapidly converging or diverging flows. This property is particularly important for accurate analysis of fiber-filled resin molding, since it plays a significant part in fiber orientation and subsequent strength and shrinkage. Adequate commercial consideration of the above phenomena depends on developments in theoretical modelling of polymer melts, material testing techniques, numerical methods and computer hardware.

Warpage of injection molded parts remains an area of concern in industry and software for its prediction is currently under development. The main causes of warpage are high temperature differentials at ejection and different amounts of shrinkage throughout the molding. Cooling analysis software can minimize high temperature differential problems but shrinkage is a more complex phenomenon (Isayev 1987). It is generally agreed that shrinkage is affected by conditions in both the filling–packing and cooling phases. While existing flow and cooling analysis software can predict the various factors that contribute to shrinkage, the interplay between them is not simple and cannot in general be interpreted by the user. Moreover, there exist no analytical models for shrinkage that may be applied to a wide range of materials.

Current research is centered on correlating values of the variables affecting shrinkage from flow and cooling analyses with measured shrinkages of test samples molded under a range of processing conditions. The results of the correlation are used to calculate material data that relate actual shrinkage to the conditions to which the material is subjected during processing. The problem is further complicated by the fact that material does not always shrink isotropically. In general, high levels of molecular orientation are induced in molded parts and the magnitude of the resulting shrinkage depends on the direction of material orientation. Consequently, it is necessary to calculate shrinkage parallel and perpendicular to the material orientation direction for each plastic element. The effects of temperature differentials are incorporated by calculating the resulting bending moment for each element. In order to predict the warped shape of the part the elemental shrinkages and bending moments are input to a finite-element stress analysis program. This calculates the deflection of each node. Results are graphically displayed and show the deflected shape of the component and the magnitudes of the deflection.

See also: Injection Molding of Thermoplastics: Mold Filling; Polymer Processing: Rheological Properties and Fluid Mechanics; Reaction Injection Molding of Polymers (Suppl. 1)

Bibliography

Bernhardt E C (ed.) 1983 *Computer Aided Engineering for Injection Molding*. Hanser, New York

Bird R B, Armstrong R C, Hassager O 1977 *Dynamics of Polymeric Liquids*, Fluid Mechanics, Vol. 1. Wiley, New York, pp. 442–51

Bird R B, Stewart W E, Lightfoot E N 1960 *Transport Phenomena*. Wiley, New York

Brebbia C A, Telles J C F, Wrobel L C 1984 *Boundary Element Techniques*. Springer, New York

Hill J M, Dewynne J N 1987 *Heat Conduction*. Blackwell Scientific, Melbourne

Isayev A I 1987 Orientation, residual stresses and volumetric effects in injection molding. In: Isayev A I (ed.) 1987 *Injection and Compression Molding Fundamentals*. Marcel Dekker, New York, pp. 227–328

Karjalainen J A 1987 Computer simulation of injection mould cooling. *Acta Univ. Ouluensis Ser. C* 43

Manzione L T (ed.) 1987 *Applications of Computer Aided Engineering in Injection Molding*. Hanser, New York

Rao S S 1982 *The Finite Element Method in Engineering*. Pergamon, Oxford

C. Friedl and P. Kennedy
[Moldflow Pty, Victoria, Australia]

Insulation Raw Materials

Insulation materials are used to control or slow heat flow. Heat flow can be reduced over a wide range of temperatures from cryogenics at 0 K to well above

Table 1
Examples of insulation raw materials, processing and products

Raw materials	Processing	Products
I Minor processing diatomite, perlite, vermiculite	mining, milling, drying, sizing, expansion or sintering	loose-fill powders, granules
II Chemical or thermal processing a. bauxite	mining, milling, solubilization, precipitation, calcining	powders, bricks, cements, refractories
b. silica, limestone, soda ash, feldspar	mining, milling, drying, blending, melting, fiberization	glass wools, rock wools, refractory wools microspheres
c. silica sand	mining, milling, drying, reaction with NaOH or Cl purification, reaction or precipitation	precipitated silicas, silica gels, aerogels, fumed silicas
d. colemanite, ulexite-probertite, kernite brines	mining, milling, washing, flotation, calcining, evaporation	borates, borax
e. limestone, silica, alumina, gypsum, iron oxide, Wollastonite	mining, milling, drying, blending, calcining, hydrothermal processing	calcium silicates, cements, bricks, blocks, pipe covering

2000 K. Insulations are made in a broad variety of types, from organics to minerals. Some minerals are used essentially as mined, others are significantly modified or composited to give improved qualities. Minerals used as insulation raw materials can be categorized by the degree of processing they require in order to be converted into useful insulation products.

The first group of minerals are used with only minor processing which could include mining, drying, size gradation, some beneficiation, and either expansion or higher-temperature sintering. Examples are perlite, vermiculite and diatomite which are all used as loose-fill insulations, cavity fill and block fill, and they can also be consolidated into structural boards, bricks, or other shapes by use of binders such as cement, sodium silicate, clays, and so forth.

The second group of minerals requires more complex chemical and/or thermal processing before they are converted into a form that can be used as insulations. These insulation raw materials are mined and frequently receive drying, size gradation and beneficiation but they then receive chemical or thermal processing. Table 1 gives, in a broad review, examples of these raw minerals, of the processing they receive and of the types of insulation usually produced. In Table 1 and throughout this article, those minerals that undergo minor processing are referred to as group I minerals and those that are chemically or thermally processed as group II minerals.

1. Sources

An abbreviated summary of the principal states in the USA and the world countries that have operating deposits of the raw materials used in insulation products is given in Table 2. The sources are listed in

Table 2
Main sources of insulation raw materials

Raw material	Main sources
Diatomite	California, Nevada, Washington, Oregon, Arizona, France, Denmark, USSR, Iceland, Spain, Mexico
Perlite	New Mexico, Arizona, California, Idaho, Colorado, Nevada, Hungary, Greece, USSR, Italy, Japan, the People's Republic of China, Bulgaria, Turkey, Mexico, Australia
Vermiculite	Montana, South Carolina, Virginia, South Africa, Japan, Brazil
Bauxite	Arkansas, Alaska, Georgia, Australia, Guinea, Jamaica, Surinam, USSR, Brazil, Ghana, India, Guyana
Silica	Illinois, Missouri, Arkansas, Oklahoma, Tennessee, Georgia, Pennsylvania, New Jersey, West Virginia, Texas, Ontario, Quebec
Limestone	Pennsylvania, Ohio, Indiana, Michigan, Texas, New York, USSR, Japan, FRG, Brazil, Mexico
Soda ash	Wyoming, California, USSR, the People's Republic of China, Bulgaria, France, UK, FRG, Japan, GDR, Romania, Poland, India
Feldspar	Italy, North Carolina, Connecticut, Georgia, California, Oklahoma, South Dakota, FRG, USSR, France, Brazil, Mexico, Finland, Norway, Sweden
Borates	California, Turkey, USSR, Argentina
Gypsum	Texas, Oklahoma, Michigan, Iowa, California, Nevada, Canada, Japan, Spain, France, Iran, USSR, the People's Republic of China, UK, Mexico
Wollastonite	New York, California, the People's Republic of China, Finland, India

an approximate order of decreasing annual production. These minerals are widely available and, therefore, only the larger producing states and countries are listed.

2. Mineralogy

As a wide variety of mineral raw materials is used in producing insulation products, the mineralogy of these materials will not be discussed here. The reader is referred to the appropriate articles on specific materials.

3. Physical Properties

Important physical properties of insulating products are summarized in Table 3. Insulating materials must endure a wide range of operating conditions. Since no one material is likely to have all the correct properties, including installation cost, the selection of a suitable insulation takes into account such properties as the melting point or maximum operating temperature, thermal conductivity at the operating condition and the necessary strength properties for the particular use. In addition, the insulation often must be chemically compatible with its environment, which may require that it have a selected chemical composition. For example, contact with stainless steel or electrical equipment often requires low sodium and low chloride content to avoid corrosion.

4. Mining and Production Techniques

The generalized production techniques for each group of minerals are given in Table 1.

The minerals in group I, the minor-processing category, are commonly open-pit mined using ripping, bulldozing and/or drilling and blasting, if necessary. The mined materials are dried, milled and classified with some beneficiation to remove tramp contaminants. Perlite and vermiculite can be "expanded" to a significantly lower density by rapid heating that converts combined water to steam and "pops" these minerals like popcorn. Diatomites are usually sintered, often with a sodium flux to improve color. The sintering process bonds smaller particles together to improve handling, increase density and remove combined water.

The minerals in group II, the chemically or thermally processed minerals, are also commonly open-pit mined with drilling and blasting. The mined materials are dried, milled and beneficiated prior to chemical and thermal processing. Since this processing varies with the subgroups under group II in Table 1, they will be discussed individually.

(*a*) *Group IIa*. Bauxite is solubilized in caustic, precipitated as aluminum hydroxide, then calcined to alumina (aluminum oxide). Alumina is used with binders, clays, bauxitic clays, etc., to make pressed, extruded, cast materials, and the like, that are recalcined to give various refractory insulations. Alumina is also used with hydraulic cements and other materials to make refractory cements to bond bricks, blocks, and so on. Aluminum metal, reduced from aluminum oxide, is used as thin foils for reflective vapor barrier insulations, often in combination with other insulations. In multiple layers and with the interlayer space evacuated, aluminum foils become effective insulations without other interlayered materials.

(*b*) *Group IIb*. Silica, soda ash, limestone, feldspar, borax and a variety of other glass-forming minerals are combined and melted into various melting point, solubility and viscosity glasses that are fiberized by drawing, flame attenuation and/or rotary spinning methods. Used as bulk fibers or formed into boards or blankets with various binders, such as phenolic resins, these fibers become effective insulating materials. Slags from other thermal-processing or various natural minerals can be fiberized, as are the synthetic glasses, to produce rock wool or mineral wool. Refractory fibers are made from even higher melting point combinations of alumina, silica and selected metal oxides.

Glass-forming minerals can also be formed into hollow bubbles or microspheres and used as insulating ingredients in composite materials or as loose-fill insulations.

(*c*) *Group IIc*. Silica gels, aerogels, precipitated silicas and fumed silicas are special categories of processed minerals that are used in specialized insulations. These are generally formed by converting silica sand to sodium silicate or to silica tetrachloride by high-temperature reaction with sodium hydroxide or with chlorine, respectively. Sodium silicate is neutralized with acid to precipitate silica which is recovered and dried to give silica gels and precipitated silicas, or dried by removal of liquid above the triple point to give aerogels. Silica tetrachloride is purified by distillation and then reacted with, or burned in, an oxygen–hydrogen flame to produce fumed silica. These synthetic silicas are used in composites as thickening agents, fillers, pozzolans, etc.; they contribute insulation characteristics owing to their low bulk density.

(*d*) *Group IId*. Boron chemicals are produced from colemanites or ulexite–probertite by mining, grinding, washing and calcining. Flotation processing is also used. Lake brines are evaporated to collect boric acids and borax. Boron minerals are used in glass-fiber compositions to produce insulations. Borates are used as fireproofing agents in organic insulations like cellulose.

Table 3
Physical property ranges for typical insulation product types

Property	Diatomite, calcined	Perlite, expanded	Vermiculite, expanded	Bauxite: Refractory brick	Bauxite: Castable cements	Bauxite: Aluminum foil	Silica: Glass fibers	Silica: Refractory fibers	Silica: Rock wool	Silica: Silicas	Limestone calcium silicate
Density (kg m^3)	128–320	32–160	48–280	560–1040	850–1250	12	46–176	64–385	16–320	10–160	192–736
Maximum operating temperature (°C)	800–900	800–900	700–1100	1000–1760	1150–1420	500	450–650	870–1540	540–980	1000	650–730
Thermal conductivity (10^{-2} W m^{-1} °C^{-1})											
−100 °C	1.1–1.7	1.1–2.3				0.0057[a]			1.7	2.3[b]	0.72[b]
0 °C		4–6	6.6				3.2	3.5	3.7–4.0	2.0	
150 °C	6.9					4.3	5.0		6.8	2.3–8.7	5.2–12.0
200 °C	7.2–10.0	7.9	10.0	13–56	21–36	6.6	6.6–7.9	7.2	7.6		5.6–13.0
400 °C	7.5–12.0	10.0	15		24–40		10	10–12	10		7.6–13.0
550 °C	8.1–13.0	11–16	15–19	16–59	26–42			13–16	13		8.9–13.0
750 °C	8.9				42–46			18–27			
1000 °C								26–39			
1100 °C				22–62				32–49			
Composition—ignited											
SiO_2	80–90	72–74	34–40	30–62	35–44		50–60	20–70	30–65	98–100	50–60
$Na_2O + K_2O$	0.8–2.0	7–10	2–8	0.1–1.0			5–20		0–20		−2
Al_2O_3	1–6	12–13	12–14	31–78	34–51	(188)	5–15	30–60	0–40		1–5
B_2O_3							5–20				
CaO	0.5–3.0	0.4–0.9	1–2	0–12	4.7–17		10–20		0–40		20–40
Cr_2O_3								0–5	0–5		
Fe_2O_3	1–3	0.7–2.0	6–13	−1–2	0.5–7.0			0–1	0–15		−2
MgO	0.5–1.0		18–24				2–6		0–40		−2
MnO								0–1	0–40		
TiO_2								0–1.5	0–10		
ZrO_2								0–20	0–10		

a Foils with separators b Vacuum

(*e*) *Group IIe*. Limestone is converted to lime by calcination. Cement is produced by calcination of mixtures of limestone, silica, alumina, gypsum, iron oxide, etc. Lime or cement may be combined with silica, expanded perlite, vermiculite, Wollastonite and other minerals and fibers, and made into various insulating materials by either hydrothermal reaction of calcium hydroxide with silica or by hydration of cement with silica ingredients. Calcium silicates are used as pipe and block insulations. Cement or concrete bricks, blocks and coverings are used in less-demanding insulation areas or as structural components. Insulation efficiency can be increased by incorporation of air bubbles, making lightweight concrete.

5. Safety

Essentially all of the raw materials for insulations have some hazards in their use, even if it is only as nuisance dusts that have low levels of hazards but require dust control, ventilation and the use of dust masks. The same would apply to the insulation products made from these materials, particularly when handled in ways to produce dust such as sawing or sanding. Insulation raw materials that contain crystalline silica above 1% need to be treated with proper care.

Fibrous insulation materials are being evaluated for specific safety hazards. Generally, safety relates to the ability of the human body to solubilize fibers. Durable fibers are more likely to cause health problems. It should be noted that asbestos has not been included in this discussion of insulation raw materials. The reason for this is that although it has been used in many forms and combinations because of its excellent insulation characteristics, the overwhelming health hazard has significantly reduced asbestos usage in the USA and is decreasing in use on a worldwide basis (see *Asbestos: Alternatives*, Suppl. 1).

High-temperature insulation uses cause particular safety problems with some materials. Long-term use at elevated temperatures can result in conversion of less-hazardous minerals to more-hazardous materials, such as cristobalite and tridymite.

6. Industrial Uses

As mentioned earlier, insulations are used to retard the flow of heat over a wide range of temperatures, from cryogenics and the very low temperatures of space to the very high temperatures of furnaces, plasmas, and so forth. Insulations are selected to minimize heat transfer at a suitable cost while giving the necessary strength and system compatibility.

Some insulations are used under high vacuums, others are used at normal pressures. Insulation characteristics involve the various methods of heat transfer conduction, convection and radiation. Gas conduction and gas convection are influenced by the concentration of gas molecules and are therefore decreased by reductions in pressure (such as in a vacuum). Solid conduction is influenced by the nature of the particle shape, size, porosity, and so on. Radiation is influenced by opacity, emissivity, adsorption and scattering. All mechanisms are usually involved and the optimum insulation must minimize all of them, particularly the mechanisms that contribute the largest percentages of the total heat transfer.

Bibliography

Dillon J B 1978 *Thermal Insulation, Recent Developments*. Noyes, Park Ridge, NJ

Kirk R E, Othmer D F 1978–1980 *Encyclopedia of Chemical Technology*, 3rd edn. Wiley, New York

Kujawa R J 1983 Insulating materials—Thermal and sound. In: Lefond S J (ed.) 1983 *Industrial Minerals and Rocks*, 5th edn. American Institute of Mining, Metallurgical and Petroleum Engineers, New York

Malloy J F 1969 *Thermal Insulation*. Van Nostrand Reinhold, New York

US Bureau of Mines 1984 Metals and minerals. *US Bureau of Mines Minerals Yearbook*. US Bureau of Mines, Washington, DC

G. Coombs
[Johns–Manville Sales Corporation, Denver, Colorado, USA]

Ion Implantation: III–V Compounds

Ion implantation involves the acceleration to a predetermined energy of a beam of ionized atoms or molecules, and the subsequent stopping of these entities in a solid target. The acceleration energies are typically between 10 keV and 400 keV, although specialized machines operate up to several MeV. The penetration depth of the ions in the target depends on their energy and mass, and on the atomic density of the target. The implanted ions have a concentration depth profile generally described by a Gaussian distribution with an average projected range R_p and a standard deviation ΔR_p. As the ions traverse the target they create displacement damage to the crystalline lattice and this requires subsequent annealing at elevated temperatures to restore the initial condition of the lattice.

Ion implantation of III–V compound semiconductors such as GaAs and InP has received renewed interest recently as an attractive method for the formation of small dimension active, contact or semi-insulating regions in device structures and circuit applications. Essentially there is no diffusion technology of *n*-type dopants in these materials, thus leading, together with the well-known advantages of implantation as a method to control the amount and location of dopants precisely, to its widespread use. In the III–V semiconductors the subsequent damage removal and dopant activation steps are somewhat more complicated than in elemental semiconductors such as silicon. Implantation in III–V semiconductors may also be used to destroy doping, in contrast to the more usual aim of creating it in selective regions.

1. Ion Stopping in III–V Materials

There are two dominant energy loss mechanisms for implanted ions in semiconductors (Ryssel and Ruge 1986):

(a) nuclear stopping, in which a part of the kinetic energy of the incoming ion is transferred to nuclei in the target via elastic collisions; and

(b) electronic stopping, in which the incoming ion undergoes inelastic collisions with bound electrons in the target material, causing excitation or ionization of the parent atoms.

The contribution from nuclear energy stopping tends to be small at the very highest implant energies because fast ions are moving past target nuclei too quickly to efficiently transfer energy to them. At intermediate energies the nuclear energy loss component increases, but falls again at the lowest energies where electron screening effects lower the effective atomic number of the target nuclei. For the most common implanted donor ion in GaAs, namely silicon, the nuclear energy loss falls from about 480 keV μm^{-1} at 20 keV initial ion energy to about 305 keV μm^{-1} at 200 keV ion energy. Conversely, over this energy range the electronic energy loss rises from about 193 keV μm^{-1} at 20 keV to 500 keV μm^{-1} at 200 keV. The total energy loss per micrometer is, therefore, roughly constant at about 700–800 keV. The projected range of silicon in GaAs is approximately 900 Å per 100 keV of initial silicon ion energy. For comparison, the range of an H^+ ion in GaAs is ~1 μm per 100 keV and that of an O^+ ion is ~1500 Å per 100 keV incident ion energy.

The ion profile $N(X)$ in GaAs or InP for an idealized stopping process is related to the average or projected range R_p, standard deviation or straggle ΔR_p and implant dose Φ (in ions cm^{-2}) by

$$N(X)=\frac{\Phi}{(2\pi)^{1/2}\,\Delta R_p}\exp\left[-\frac{(X-R_p)^2}{2\Delta R_p^2}\right]$$

and the maximum ion concentration occurs at R_p and is given by

$$N_{\max}=\frac{\Phi}{(2\pi)^{1/2}\,\Delta R_p}\simeq\frac{0.4\Phi}{\Delta R_p}$$

The ion concentration falls to ~60% of its peak value at $R_p+\Delta R_p$. It is worth noting that each implanted ion comes to rest in ~10^{-13} s (its range divided by its velocity) with the thermal spike created by ionization and excitation along its track decaying away by ~10^{-12} s after entry of the ion (Seidel 1983). Light ions such as silicon (the most common donor) and beryllium (the most common acceptor) tend to leave tracks characterized by relatively small amounts of damage. They slow down initially mainly by electronic stopping, with little displacement damage until eventually nuclear stopping becomes dominant at the end of their range. Therefore, there is little damage to the crystal except near the end of the ion range. Heavy ions like tellurium (a donor) or cadmium (an acceptor) may create damage clusters along their paths, displacing target atoms from the surface inward. If damaged areas begin to overlap with increasing ion dose, an amorphous layer can result (i.e., essentially each nucleus has been displaced from its lattice position) and no long-range order remains. The radiation-damage profile is not coincident with the ion profile and is calculated from the nuclear energy loss. The peak of the damage profile generally occurs at ~$0.75R_p$. Typically, the displacement energy of a lattice atom is ~20 eV. The number of displaced atoms for a 10 keV silicon ion in GaAs is ~400 and for a 200 keV silicon ion is ~3750. At doses near 10^{14} ions cm^{-2} the displacement disorder is approximately equal to the GaAs atom density, leading to amorphization. The critical dose for amorphization depends very strongly on the temperature of the sample during implantation. At elevated temperatures (>150 °C for InP and 180 °C for GaAs) dynamic annealing of the damage occurs, and an amorphous layer may never form.

The damage created by implantation reduces the carrier mobility in III–V materials, and creates deep level centers which trap free carriers. Therefore, the material after implantation but before annealing tends to exhibit high resistivity. This is the basis for the damage-induced isolation schemes in wide use in GaAs (Eisen 1980). Since ions require a certain energy for the production of radiation damage, the maximum of the damage distribution is always closer to the surface than that of the ion profile.

In general, to minimize both axial and planar channelling during implantation of compound semiconductor wafers, they are oriented with an appropriate azimuthal or twist direction (usually defined as the angle between the wafer flat and the direction of beam tilt) in addition to being tilted ~7° with respect to the beam direction. For large wafers the actual entry angle of the beam varies slightly as the beam is scanned across the wafer, and this can lead to spatial differences in the average range of the ions over the wafer area. The only completely effective method to eliminate channelling is to preamorphize the material, preferably by use of a lattice constituent. This is a very successful procedure for implantation in silicon, but is not appropriate for III–V materials because the regrowth of amorphous layers in all of them is singularly poor.

2. *Comparison of Implantation in III–V Compounds and Silicon*

There are a number of differences between implantation of III–V materials, and of silicon. These include:

(a) the need to avoid amorphization of III–V compounds during the implant;

(b) the creation of regions with deviations from stoichiometry resulting from the different masses of gallium and arsenide (or indium and phosphorus, etc.);

(c) the need to prevent dissociation of the III–V material during annealing;

(d) the requirement that implanted ions should occupy only one lattice site after annealing—group IV dopants, for example, display amphoteric behavior; and

(e) the rapid concentration-dependent diffusivity of *p*-type species implanted in III–V compounds.

In silicon, recoil of displaced atoms from their lattice positions is not of concern because they are all indistinguishable from each other. In III–V compounds, however, the lattice elements are distinguishable and, because they recoil unequally due to their different masses, local perturbations in stoichiometry are created. The effect is most obvious for heavy ions implanted into compound semiconductors where the lattice elements have significantly different masses (e.g., InP). Obviously the lighter element will recoil further, leading to an excess of the heavier element near the surface (shallower than R_p) and an excess of the lighter element at greater depths (between R_p and $R_p + \Delta R_p$).

Unequal recoil is related to the need not to amorphize III–V compounds during implantation. Repair of the lattice during subsequent annealing requires that the displaced atoms diffuse back to the appropriate sites, and if this reordering cannot be accomplished because the diffusion lengths are not long enough to reach the correct lattice positions, as is the case in III–V compounds, then the remnant disorder will have a significant impact on the electrical quality of the implanted layer. The regrowth of amorphous III–V compounds is a rather unsuccessful process, with the displaced lattice elements unable to diffuse quickly enough to the right positions to keep up with the growth front, leading to highly-twinned material with significant stacking-fault densities, and eventually to a complete stop of the regrowth if the initial amorphous layer was very thick ($\geqslant$2000 Å). The implant activation in regrown III–V compounds is significantly worse because of remnant disorder than if amorphization was avoided, and it is common during the implantation of heavy ions to hold the substrate at an elevated temperature (~150–300 °C). This prevents the formation of an amorphous layer because of the enhanced mobility of point defects which are able to diffuse and recombine or annihilate each other (Williams 1982). This prevention of damage accumulation is known as dynamic annealing. These results are in stark contrast to silicon in which amorphous layers regrow by a solid-phase epitaxial process around 550 °C. The crystalline quality of this regrown material and the implant activation are generally lower than if an amorphous layer was not formed.

3. *Surface Degradation*

After implantation, the implanted ions are in random positions in the sample and there is considerable lattice damage created by the stopping process. Post-implant annealing is therefore required for two reasons:

(a) to repair the disorder in the crystal, and

(b) to activate the implanted ions by causing their short-range diffusion to a lattice position.

In III–V semiconductors the annealing temperature required to move the implanted ions onto lattice sites where they are electrically active always exceeds the temperature at which the surface of the material is degraded by loss of the group V element. To preserve the integrity of the surface the annealing environment must be such that this loss is suppressed, either by providing an overpressure of the group V element, or by encapsulating the surface with a dielectric such as SiO_2 or Si_3N_4. The overpressure can basically be supplied in two ways: either through the use of a gas ambient such as AsH_3, or by placing the wafer face-to-face with another wafer of the same type. The latter technique, of course, is not ideal because any movement of the wafers relative to each other leads to scratches on the surfaces and regions which are not protected during the anneal. In principle, furnace annealing of GaAs (or InP) in an AsH_3 (or PH_3) ambient is an ideal solution, but in practice the large thermal mass of the furnace means that short anneals are not feasible and thus diffusion of the implanted dopants can occur. These large furnaces also constitute a safety hazard. Dielectric encapsulation of the surface is also far from ideal because of the different thermal expansion coefficients inducing considerable near-surface strain in the III–V compound. This can lead to a significant enhancement of the diffusivity of some implanted dopants. SiO_2 allows preferential gallium outdiffusion from the surface of GaAs, whereas Si_3N_4 is often subject to cracking and peeling during the anneal. Two promising encapsulants for GaAs are reactively sputtered AlN and spin-on phosphosilicate glass (PSG), both of which have similar expansion coefficients to GaAs. The latter is also promising for InP because adjusting the phosphorus content in the PSG enables closer matching of the thermal expansion coefficients. It is certainly true, however, that after almost two decades of study there is still no reproducible and reliable encapsulation method for annealing III–V materials and, combined with the need to restrict diffusion of some implanted dopants during annealing, this has led to considerable interest in rapid-heating methods.

4. *Rapid Thermal Annealing*

Very fast heating by lasers has proved inappropriate for III–V compounds for a number of reasons including vaporization of the surface of the wafers, the inability to anneal complete wafers and poor activation of low-dose implants. These disadvantages, and the problems with furnace annealing, have stimulated the examination of annealing methods with time scales intermediate between these two techniques (i.e., in the time regime 1–100 s). The activation of implanted dopants takes place on this time scale, and any extra time spent by the sample at high temperature can only degrade the implanted profile by diffusion. In general, the optimum annealing conditions in rapid thermal annealing (RTA) consist of a higher-temperature shorter-duration cycle than for furnace annealing. The standard surface protection method during RTA is the proximity technique, although vacuum-compatible systems capable of providing an AsH_3 or As–H_2 ambient are becoming available.

Rapid annealing gives electrical properties in the implanted layer at least as good as those achievable by furnace annealing, and has the advantage of restricting the redistribution of the normally fast-diffusing acceptor species. There remain some problems; for example, the very rapid heating and cooling rates ($>100\,°C\,s^{-1}$) can induce crystallographic slip around the edge of wafers. These are caused by the generation of thermal stresses by radiative losses at these edges, and may be largely eliminated by the use of annular guard rings into which the wafer to be annealed is placed, combined with tailoring of the temperature ramp-down cycle (Pearton et al. 1987).

Another important application of transient annealing is for the high-temperature processing of multilayer structures. For example, the extremely high ($>10^5\,cm^2\,V^{-1}\,s^{-1}$) 77 K electron mobilities in selectively doped heterostructures are drastically degraded by normal implant-activation steps in conventional furnaces. Since diffusion of dopants in these structures, even of the order of 20–100 Å, can reduce the electron mobilities, it is essential to use rapid annealing to restrict this redistribution.

5. *Activation of Dopants; Residual Defects*

The donor species in III–V materials are silicon, sulfur, selenium, tellurium, germanium and tin, while the acceptors are beryllium, magnesium, cadmium and zinc. The group IV elements silicon, tin and germanium are amphoteric, but in general display *n*-type doping. The most common implants use silicon for *n*-type doping and beryllium for *p*-type doping, because of their low masses. They therefore create less damage for room temperature implantation, and thus less annealing is required. The other dopants display better activation if they are implanted at elevated temperatures, at least in GaAs and InP. In the former it is difficult to achieve *n*-type doping levels above $\sim 4\times 10^{18}\,cm^{-3}$, while there is little diffusion of the donors (sulfur is the sole exception). By contrast, all of the acceptor species show high levels of activation, and *p*-type doping near $10^{20}\,cm^{-3}$ is possible by implantation and annealing. There is, however, marked redistribution of the acceptors during furnace annealing and loss of the dopant to the surface if the wafer is uncapped during the anneal, even for RTA. The case of dopants in InP is somewhat different. High electron concentrations are readily achievable ($>10^{19}\,cm^{-3}$) but hole concentrations above $2\times 10^{18}\,cm^{-3}$ are difficult to obtain. The annealing temperatures for optimum activation are lower than in GaAs, and typically ~750 °C is used. Once again, marked redistribution of acceptor implants occurs during furnace annealing.

Relatively little attention has been paid to the relationship between the solubility of implanted dopants in III–V compounds, and their associated electrical activity. In silicon, for example, there is a one-to-one correspondence between the occupation of a substitutional lattice position, and electrical activity. However, this is not necessarily the case in III–V compounds because of the presence of native defects which can compensate or trap charge carriers. In fact, it is clear in GaAs that substitutionality of an implanted dopant is a necessary but not sufficient condition for electrical activity.

The various stages of damage removal and dopant activation in implanted GaAs are summarized in Fig. 1. The implant damage can consist of either amorphous layers or extended crystalline defects (dislocation loops and stacking faults) depending on

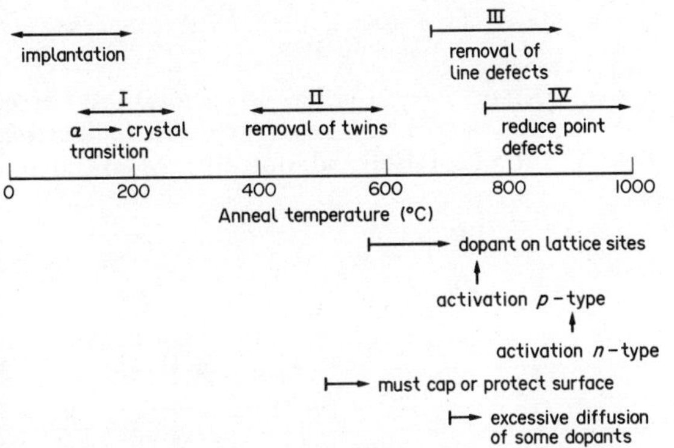

Figure 1
Summary of damage removal and dopant-activation steps in ion-implanted GaAs (after Williams 1982)

the implantation conditions. Amorphous layers recrystallize epitaxially during annealing at 150–200 °C (stage I), but the recrystallized layer is invariably highly defective, consisting of twins, stacking faults and other defects (Sadana 1985). These defects anneal out to leave only a high density of dislocation loops in the range 400–500 °C (stage II). These loops grow and annihilate above about 700 °C (stage III) and the remaining point defect clusters begin to anneal out above 750 °C (stage IV).

6. *Implant-Induced Isolation*

High resistance regions in doped III–V materials can be produced by radiation damage resulting from the implantation of ions such as H^+, B^+ or O^+. The compensation results from the trapping of free carriers by deep-level centers which are not thermally ionized at room temperature. Gallium-based materials such as GaAs, GaP and AlGaAs can be made semi-insulating ($\geqslant 10^7\ \Omega$ cm) by ion bombardment because the damage-related levels are in the middle of the bandgap. In InP and other indium-based compounds, implant bombardment is not as effective in creating high-resistivity regions. The defects created in InP tend to pin the Fermi level in the upper half of the bandgap, and therefore the resistivity of *n*-type material can only be increased to the 10^3–10^4 cm range. It is possible to make *p*-InP high resistivity, but the implant dose is critical. Above the optimum dose range (which is a function of the doping density of the material) the bombarded region becomes slightly *n*-type because of the damage-induced defects, and the resistivity is in the range 10^3–$10^4\ \Omega$ cm.

The choice of bombardment species depends on the thickness of the layer to be made resistive, and in general heavier ions are observed to have higher carrier removal rates and higher thermal stability of the compensation effect than lighter ions. A 1 MeV B^+ ion removes about 200 carriers in *n*-type GaAs, whereas a 200 keV ion creates enough damage to trap or compensate 30–100 electrons. Therefore, when implanting dopant ions, even if all the ions become electrically active, one residual defect per implanted ion is sufficient to compensate all the doping effect. Because of hopping conduction effects, the as-implanted value of resistivity in ion-bombarded material is usually not the maximum achievable, and a subsequent anneal can increase the resistivity by more than an order of magnitude.

7. *Miscellaneous*

A novel property of multilayer structures such as superlattices and heterostructures is that localized introduction of impurities by implantation or diffusion causes intermixing of the structures at considerably lower temperatures and shorter times than those required for thermal disordering. For example, in the case of a multiperiod GaAs–AlAs structure, this impurity-enhanced compositional disordering leads to a single layer of bulk AlGaAs alloy whose composition depends on the initial compositions and thicknesses of the unperturbed layers, and on the experimental conditions producing the intermixing. Selective implantation can therefore give rise to areas of different bandgaps monolithically integrated on the same substrate. This scheme has already been used to fabricate a variety of buried heterostructure and stripe geometry lasers. For implantation, the enhanced disordering disappears once the initial defects have been annealed, leaving intermixed material of good optical quality.

There are other areas such as the use of MeV ion energies to achieve thick or buried doped layers, and the application of highly focused (beam diameter ~1 μm) ion beams to produce maskless, selective area implantation, that are also gaining a lot of attention. The fundamental limitations of ion implantation, especially lateral straggle under masks, are not yet a factor in compound semiconductor devices because of their relatively primitive stage of development compared with silicon technology.

Bibliography

Eisen F H 1980 Ion implantation in III–V compounds. *Radiat. Eff.* 47: 99–114

Pearton S J, Poate J M, Sette F, Gibson J M, Jacobson D C, Williams J S 1987 Ion implantation in GaAs. *Nucl. Instrum. Methods Phys. Res. B* 19: 369–80

Ryssel H, Ruge I 1986 *Ion Implantation*. Wiley, New York, Chap. 1

Sadana D K 1985 Mechanisms of amorphization and recrystallization in ion-implanted III–V compound semiconductors. *Nucl. Instrum. Methods Phys. Res. B* 7: 375–86

Seidel T E 1983 Ion implantation. In: Sze S M (ed.) 1983 *VLSI Technology*. McGraw-Hill, New York, Chap. 6

Williams J S 1982 Compound semiconductors. In: Poate J M, Mayer J M (eds.) 1982 *Laser Annealing of Semiconductors*. Academic Press, New York, pp. 383–413

S. J. Pearton
[AT&T Bell Laboratories, Murray Hill, New Jersey, USA]

J

Junction Transient Spectroscopy

The necessity of monitoring electrically active defects in crystalline semiconductor materials has led to the development of junction transient measurement techniques. The most important of the various methods is deep-level transient spectroscopy (DLTS) which has proven to be a powerful research tool for characterizing the electrical properties of point defects in both elemental and compound semiconductors. This sensitive spectroscopy measures deep-level defect concentrations as low as $10^9\,cm^{-3}$ in the junction region of a semiconductor device. Since the processed device geometry can often be used as a sample configuration, DLTS can be a nondestructive measurement. In addition to the concentration of deep-level defects in the depletion region, DLTS can routinely determine the spatial profile of traps, the energy level for carrier emission from the defect and the majority carrier capture cross section. The spectroscopic nature of junction transient measurements has been exploited in experiments that have led to the discovery of recombination-enhanced defect diffusion, the first observation of metastable defect reactions, and the correlation of energy levels with impurities and radiation-produced point defects.

1. Junction Transient Measurements

Junction transient spectroscopies monitor the electrical effects of defects within the depletion region of a semiconductor junction. A lattice defect (such as a vacancy, an interstitial or a dislocation), an impurity (either substitutional or interstitial) or a complex of these simple defects in a crystal can introduce electronic energy states or levels into the forbidden gap of a semiconductor. Intentionally processed dopants such as arsenic or boron produce shallow states in the energy gap, approximately 0.03–0.06 eV from their related band edges. The deep states, which are measured by these spectroscopies, are those whose energy levels fall deeper in the band gap than the dopant levels. Even small concentrations of deep levels can control the carrier lifetime in semiconductors.

The application of a reverse bias to a semiconductor junction structure creates a depletion region free of mobile carriers. The junction capacitance is analogous to that of a parallel-plate capacitor with the plate separation equivalent to the width of the depleted region. The presence of occupied deep levels in the material changes this junction capacitance. All transient spectroscopies employ some method to perturb the carrier occupation of defect states and then analyze the transient signals which are a result of the return of the junction to steady state. Specifically DLTS superimposes a forward voltage pulse on a reverse-biased diode. This collapses the depletion region, filling the traps with carriers. When the pulse is removed the carriers are thermally excited from the trap to the related band and are swept out of the depletion region by the applied potential. The resulting junction capacitance exhibits a transient response with a characteristic time constant inversely proportional to the carrier emission rate of each defect level. A spectrum is generated by repeatedly pulsing the sample and correlating the resulting transients with selected instrument time constants, thus producing a unique peak for each defect as the sample temperature is ramped from carrier freeze-out (~50 K) to room temperature. Lock-in amplifiers, double box-car integrators and exponential correlators are typically employed to process analog transient signals. A block diagram for a system that employs a lock-in amplifier to monitor capacitance transients is shown in Fig. 1. Alternatively, the transient can be digitized and analyzed with software to produce a spectrum. There are numerous variations of junction spectroscopies. Carrier occupation can also be perturbed with sub-bandgap light or electron beams, and often junction current or charge transients are analyzed. Spectra can be generated as a function of incident light or instrument rate window. In admittance spectroscopy, the bias varies sinusoidally and a phase-sensitive detector is used.

Representative DLTS spectra taken on high-energy-proton-irradiated phosphorus-doped silicon are shown in Fig. 2. Each peak is identified by a thermal activation energy with E indicating electron emission from the defect to the conduction band and H indicating hole emission to the valence band. The peak height is a direct measure of the defect concentration. The instrument time constant used to produce the spectra was 1.8 ms.

2. DLTS Theory

An understanding of carrier emission and capture at a defect in a semiconductor junction depletion region is necessary to analyze the output of junction spectroscopies.

Thermal carrier emission from a deep level follows Boltzmann statistics. The emission rate e_n (s^{-1})

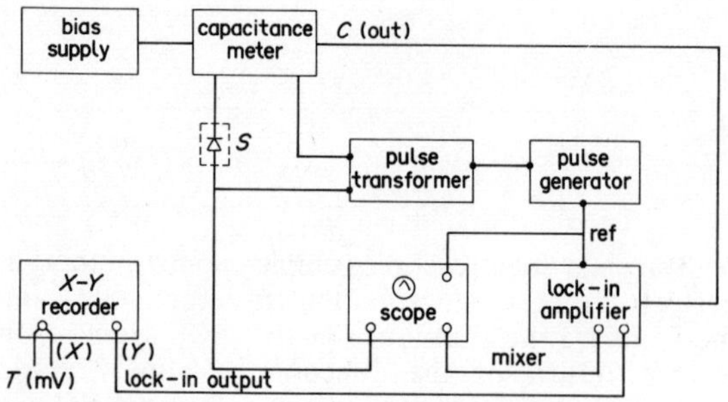

Figure 1
Block diagram of a DLTS system employing a capacitance meter and lock-in amplifier

of an electron from a defect of energy E_T (eV) is given by

$$e_n \simeq \frac{\sigma_n \langle v_n \rangle N_c}{g} \exp[-(E_c - E_T)/kT] \tag{1}$$

where σ_n is the defect state capture cross section for electrons (cm^2), $\langle v_n \rangle$ is the average thermal velocity for electrons (cm s^{-1}), E_c is the conduction band energy (eV), k is Boltzmann's constant (8.61 $\times 10^{-5}$eV K^{-1}), T is the absolute temperature (K), g is the degeneracy of the level and N_c is the effective density of states (cm^{-3}) at the conduction band edge. A similar expression can be written for hole emission from a defect to the valence band.

Experimentally, the activation energy for carrier emission from a defect to the nearest band edge E_T is obtained from the slope of an Arrhenius plot of ln $T^2 e_n^{-1}$ vs 1000/T using Eqn. (1). The emission rate e_n is the inverse of the time constant of the measured transient and must be determined independently for each experimental DLTS system. The value of T is the temperature of the spectral peak maximum.

The second process important in analyzing junction transient measurements is the capture of carriers at a defect. Defect capture of majority carriers c(s^{-1}) follows the relationship

$$c = \sigma \langle v \rangle n \tag{2}$$

where n is the majority carrier concentration (cm^{-3}), $\langle v \rangle$ is the thermal velocity of the carriers and σ is the majority carrier capture cross section (cm^2). The defect capture rate c is determined experimentally and the calculated defect capture cross section is used with the defect activation energy E_T to identify the spectral feature.

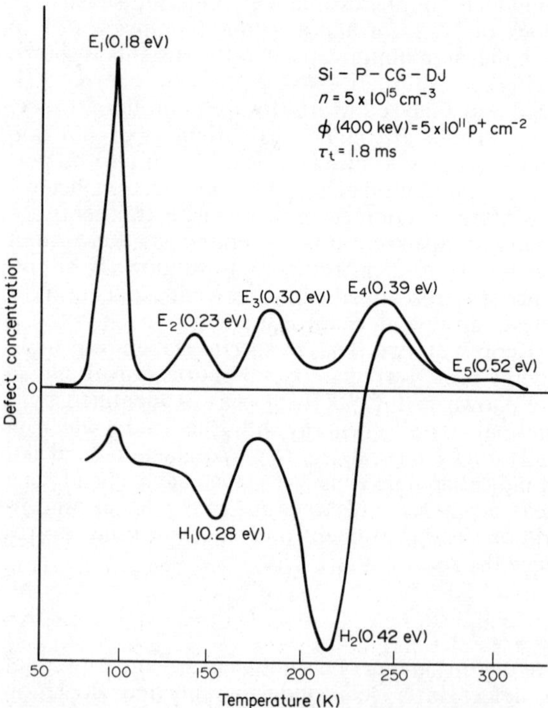

Figure 2
DLTS spectra of radiation-induced defects in phosphorus-doped silicon

3. Semiconductor Materials Characterization

Transient spectroscopy requires the filling and emptying of traps within a junction depletion region. Therefore, devices such as Schottky barriers, *p–n* junctions, metal–oxide semiconductor structures, field-effect transistors, light-emitting diodes and semiconductor lasers are proper test structures. Schottky barriers and *p–n* junctions are easily fabricated in a laboratory environment and are used to characterize bulk materials.

The spectroscopic detection limit is approximately 10^{-5} times the bulk free carrier concentration. Therefore, these methods are least effective when applied to heavily doped semiconductors.

DLTS does not provide structural or chemical information about a defect. However, extensive materials studies on defects in silicon, in conjunction with chemically specific measurements such as photoluminescence, infrared spectroscopy and electron paramagnetic resonance have resulted in the identification of over 200 DLTS peaks. A representative library is given in Table 1. The defect identity (i means interstitial, s indicates substitutional and V refers to vacancy), the temperature of the spectral peak maximum (at an instrument time constant of 1.8 ms) T(K), the thermal activation energy for carrier emission to the band edge E_T (eV) and the majority carrier capture cross section σ (cm^2) are listed. Comparisons of unknown spectra with this table do not automatically guarantee firm defect

Table 1
DLTS library of defects in silicon

Defect	T (K)	E_T (eV)	σ (10^{-18} cm^2)	Defect	T (K)	T_T (eV)	σ (10^{-18} cm^2)
Ag	184	H(0.36)	400	V	76	E(0.09)	
Ag	286	E(0.51)	100	V	79	H(0.14)	30
Au	288	E(0.53)	100	V–V	131	H(0.21)	200
Au	173	H(0.35)	1 000	V–V	241	E(0.40)	2 000
Au–Fe	170	E(0.35)	6 000	V–V	142	E(0.23)	400
Cr_i–B_s	123	H(0.28)	>5 000	V–O_i	98	E(0.17)	10 000
Cr_i	108	E(0.22)	2 000	P–V	215	E(0.44)	4 000
Cu_i–B_s	112	H(0.22)	>60 000	As–V	235	E(0.47)	
Cu	243	H(0.41)	80 000	Sb–V	224	E(0.44)	40
Fe_i	267	H(0.45)	>4 000	Sn–V	192	H(0.32)	9
Fe_i–B_s	59	H(0.10)	10 000	Sn–V	69	H(0.07)	20
Fe_i–Al_s		MH(0.13)		Al–V	282	H(0.52)	>100
Fe_i–Al_s		MH(0.20)		Al_i	203	H(0.25)	7
Fe_i–Ga_s		MH(0.14)		Al_i–Al_s		H(0.23)	
Fe_i–Ga_s		MH(0.23)		B_i	87	E(0.13)	>100
Fe_i–In_s		MH(0.15)		B_i	223	E(0.45)	700
Fe_i–In_s		MH(0.27)		B–V	190	H(0.32)	
Mn_i–B_s	251	H(0.55)		B_i–B_s		H(0.30)	800 000
Mn_i	207	H(0.25)		B_i–C_s		H(0.29)	
Mn_i	68	E(0.11)		B_i–O_i		E(0.26)	
Mn_i	216	E(0.42)		C_i	64	E(0.12)	
Mo	191	H(0.28)	1 000	C_i	165	H(0.27)	70
Ni_i–B_s	88	H(0.14)	500	C_i–O_i	206	H(0.36)	70
Ni	257	E(0.43)	100	C_i–C_s		ME(0.17)	40
Pd_s		E(0.22)	5 000	C_s–Si–C_s		ME(0.10)	
Pd–V		E(0.18)	5 000	C_i–C_s		MH(0.09)	
Pt_s	174	H(0.32)	800	C_i–C_s		MH(0.05)	
Disloc	225	E(0.38)	4 000	P_s–C_i		ME(0.30)	10
Disloc	206	H(0.35)	50	P_s–C_i		ME(0.21)	
Disloc	288	E(0.63)	>1 000	P_s–C_i		ME(0.23)	
O_D		E(0.07)		P_s–C_i		ME(0.29)	
O_D	58	E(0.15)					

identifications. Knowledge of the sample material and consideration of its processing history is crucial. Often, additional experiments are required for unambiguous defect identification.

4. DLTS Applications

DLTS research has made an extraordinary contribution to the current understanding of defect reactions in semiconductors. The phenomenon of recombination-enhanced defect reactions was first discovered in GaAs and analyzed by junction spectroscopy. The localized energy available at a defect as the result of the recombination of electrons and holes can cause enhanced defect diffusion and has proved to be the reason for enhanced degradation of injection mode semiconductor devices such as GaAs lasers.

Charge state controlled defect metastability was first investigated by DLTS. The M-center in InP was the first reported metastable defect and it exhibited different structural configurations depending on the charge state of the defect. Spectroscopic investigations of silicon, GaAs and AlGaAs have uncovered additional examples of defect metastability, and these studies have led to an understanding of configurational transformations in terms of bonding changes at the defect site.

The introduction of atomic hydrogen was initially shown by DLTS to passivate the electrical activity point defects in silicon resulting from laser annealing. Subsequent transient spectroscopy has revealed hydrogen passivation of deep and shallow levels in GaAs, GaP, GaAlAs and silicon.

Recently, engineers have begun to apply junction transient techniques to the examination of trace contamination and process-induced defects in semiconductor device manufacture. DLTS has been employed to investigate contamination introduced by processing furnaces and epitaxial reactors. DLTS feasibility studies have evaluated new processing techniques such as low-temperature, high-pressure oxidation and rapid thermal annealing.

Ion-implantation-induced traps have also been characterized by DLTS.

DLTS is also used to analyze the effect of deep levels on device performance. DLTS research has correlated the concentration of gold levels as well as vacancy–oxygen and divacancy defects with the reduction of majority-carrier lifetime in silicon devices. The presence of defects related to silver and nickel was shown to cause excessive leakage current in silicon planar diodes. Devices which exhibited changes in resistivity and minority-carrier lifetime during processing were investigated by junction transient methods revealing interstitial iron as the cause.

DLTS has matured as a semiconductor characterization technique and commercially manufactured spectrometers are now available. It appears, therefore, that DLTS will soon be utilized regularly in semiconductor device manufacturing facilities as a process monitor. Research is increasing the number of chemically identified defect levels, which adds to the potential effectiveness of DLTS in this area.

Bibliography

Benton J L, Kimerling L C 1982 Capacitance transient spectroscopy of trace contamination in silicon. *J. Electrochem. Soc.* 129(9): 2098–102

Johnson N M 1986 Deep level transient spectroscopy: Defect characterization in semiconductor devices. *Proc. Symp. Materials Research Society*, Vol. 69, pp. 75–94

Lang D V 1974 Deep level transient spectroscopy: A new method to characterize traps in semiconductors. *J. Appl. Phys.* 45: 3023–32

Miller G L, Lang D V, Kimerling L C 1977 Capacitance transient spectroscopy. *Annu. Rev. Mater. Sci.* 7: 377–448

J. L. Benton
[AT&T Bell Laboratories, Murray Hill, New Jersey, USA]

L

Laminated Veneer Lumber

Wood is a biological material with the capability to optimize itself for survival in the form of a tapered cylindrical stem with branches. Consequently, defects (e.g., knots, slope of grain) occur naturally when logs are sawn into lumber. Clear, defect-free wood is composed, by nature, of well-oriented cells with a wall structure of helically wound cellulose microfibrils (see *Wood Ultrastructure*). It has, therefore, the highest strength-to-weight ratio in tension among common structural materials. In order to utilize this excellent property of wood at a higher yield and in a greater range of dimensions, the elimination or dispersion of defects by glue jointing or glue lamination is necessary (see *Glued Joints in Wood*; *Glued Laminated Timber*).

Laminated veneer lumber (LVL), also known as parallel-laminated veneer (PLV), is one of the most suitable products for this purpose and can be processed with higher yield, and less time and labor, than glued laminated timber or plywood. LVL produced with a continuous press has been approved as an engineered material with reliable strength and stiffness.

1. *History of Development*

Development of LVL started with the production of high-strength wood aircraft members in the 1940s. Veneers were sometimes impregnated with phenolic resin to meet other requirements such as stability, hardness and screw-holding strength. In the following decades, the machinability and uniformity of mechanical properties of LVL have been appreciated by the furniture industry and used in the production of curved furniture parts.

As the supply of high-quality sawlogs diminished, the importance of LVL increased in many countries due to its higher yield potential, the introduction of automatic production methods and its adaptability to engineering end-use design. Since the beginning of the 1970s, LVL products have been used as structural members (e.g., tension chords of trusses and outer laminations of glued laminated beams) in place of lumber components because of their reliability in strength.

2. *Processing*

LVL is processed in a manner similar to plywood (see *Plywood*, Suppl. 2) but contains only parallel laminations. Rotary-cut veneers are mostly used for LVL laminations. The yield of rotary veneer has much influence on the yield of LVL. Increases in veneer yield have become necessary as the gluability and diameter of available logs diminishes. Newly developed lathes with peripheral driving devices make it possible to peel veneer down to 50 mm core diameter. It has been reported that LVL yield was at least 47% more than sawn lumber yield when green veneer yield was more than 65% of bolt volume. These percentages will increase with the use of new types of lathes.

Veneer is sometimes treated with chemicals such as preservatives and fire retardants before drying. Mixing these chemicals with the adhesive is also effective when thin veneers are used.

Normally, veneer is dried to around 5% moisture content in either conventional hot-air circulating or jet dryers. It is advantageous for LVL production that veneer can be dried with less energy and in less time than an equal volume of laminations for glued laminated timber. Press dryers are also useful in this process and it was found that the drying time required is proportional to the 1.4th power of the veneer thickness. Thinner veneers, therefore, can be dried more economically than thicker veneers. However, production of LVL with thinner veneers requires more adhesive. The balance of these factors as well as the improved properties of LVL achieved with thinner veneers should be considered in the choice of veneer thickness.

Phenolic resin or other adhesives of the same quality are used in producing LVL for structural purposes, while urea resin adhesive is generally used for nonstructural purposes.

Butt joints of veneer ends are significant defects in LVL. The joints, therefore, must be staggered and distributed as evenly as possible. The acceptability of butt joints in LVL depends on the number of laminations. Scarf joints of veneer ends are also employed in some cases. Though LVL with scarf-jointed veneers shows better strength and appearance, the process is not as simple as with butt joints. To obtain high tensile strength with a simple process, crushed lap joints (in which lap joints are staggered and high processing pressure is applied to crush overlapping veneer ends) have been used on commercial LVL products (see *Glued Joints in Wood*).

Since LVL production started as an extension or variation of the plywood process, most LVLs are hot-pressed in a multiplaten press. In this case, LVL contains no veneer end-joints but it is limited in

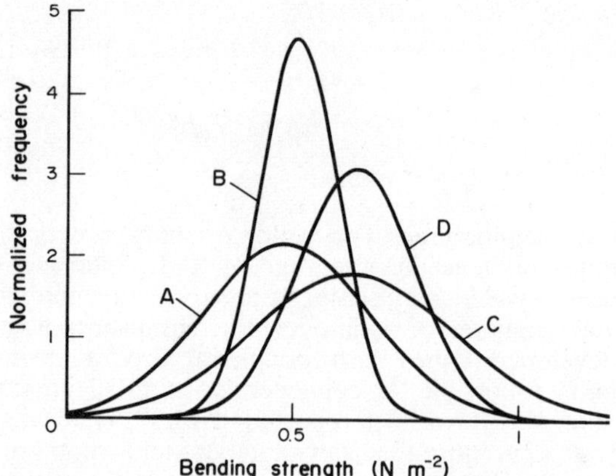

Figure 1
Comparison of bending strength between solid lumber and LVL processed from the same logs. A: solid sawn, 2 in × 4 in Siberian larch; B: 8-ply LVL, same log as A; C: solid sawn, 2 in × 4 in, southern pine; D: 6-ply LVL, same log as C

length to less than 2.5 m. These short LVL members are then scarf- or finger-jointed into longer pieces.

Modern plants for the production of structural LVL tend to employ a continuous system of veneer assembly and hot pressing in which joints are well distributed. This makes the LVL process more labor saving than either the plywood or glued laminated timber processes. A caterpillar press with electrical heating units is employed in the production of LVL (Micro-Lam) flanges with high tensile strength for commercial structural trusses. Roller-belt-type continuous presses are rather common, one of which is employed in high-efficiency production of LVL from thick veneers (Press-Lam) by utilizing residual heat of veneer drying. Other types of continuous presses employ heating devices with radiofrequency heating units or hot platens behind the belt. With these continuous presses, high yield and high efficiency production of endless LVL is feasible.

3. Mechanical Properties

Figure 1 shows how variations of strength are reduced in LVL compared to solid lumber from the same logs. The reliability in strength increases with increases in the number of laminations. The improvement in strength is higher when the quality of logs is lower. The same is true with stiffness of LVL and solid lumber.

The allowable design stress f_a can be estimated by the following approximate form:

$$f_a = (f_m - n\sigma)/2.1$$

where f_m is the average strength, σ is the standard deviation, n is a coefficient depending on the shape of the strength distribution, and the denominator 2.1 includes the effect of long-term loading, a safety factor, etc. In general, the 5% exclusion limit f_5 (the stress value below which probability of failure occurrence is 5%) is taken to be $(f_m - n\sigma)$. If the distribution is Gaussian, n is 1.645. A number of experiments with LVL showed that n should be taken as 2.0, considering safety factors in use (Koch 1973, Bodig et al. 1980).

As can be seen in Fig. 1, the estimated allowable stress of LVL will be far greater than that of solid lumber. LVL commercially manufactured from 2.5–3.0 mm thick Douglas-fir veneers (Micro-Lam) with 13 or more plies and crushed lap joints has been officially approved as a material having an allowable stress of 1509 N cm^{-2} in tension and 1921 N cm^{-2} in bending.

Butt joints are serious defects in LVL. However, their effect decreases with increases in the number of plies to just a few percent in nine-ply products and to being negligibly small for 12 or more plies. As derived from fracture mechanics theory, the tensile strength of LVL is inversely proportional to the square root of the lamination thickness. A lateral distance of 16 times the lamination thickness between adjacent butt joints prevents stress interaction (Laufenberg 1983).

The tensile strength of LVL produced by end-jointing shorter pieces is, in general, more than 90% of the strength of unjointed LVL for scarf joints with a 1:9 slope, more than 70% for finger joints of 27.5 mm length, and more than 60% for mini-finger joints of 10 mm length.

Low strength perpendicular to the long axis (i.e., cleavage and shear in the plane parallel to the fibers) is a drawback of LVL. This relates to the lathe checks of veneer created during peeling. However, the average strength increases and the variation decreases with increasing number of plies, and thus the allowable stress perpendicular to grain can be estimated to be the same value as for a solid lumber.

4. Grading and Quality Control

One of the most promising characteristics of LVL is the capability to ensure its structural performance by grading veneer according to quality and selecting the optimum combination or placement of the veneers. Visual grading of veneers which has been widely used in the plywood industry is not totally satisfactory for screening veneer used in stress-rated LVL. Mechanical methods of testing modulus of elasticity of veneer in tension are used in research, but commercial application is not feasible.

The stress-wave timing (SWT) technique is one of the most promising nondestructive methods of estimating properties of each veneer piece before it is laminated into LVL in the industrial process (Jung 1982). An ultrasonic timer with 50 kHz piezoelectric transducers, for instance, is used to determine the time required for a stress wave to travel the length of each veneer. The dynamic modulus of elasticity is computed as the product of gross mass density of veneer and square of velocity. By this method, the modulus of elasticity of LVL can be predicted with reasonable accuracy in bending applications from the measured stiffness of the veneers composing the member. However, the correlation between stiffness and ultimate strength of LVL is not always good. A grading system for commercial products employs the SWT method to categorize veneers into four stiffness classes.

5. *Applications*

The furniture and fittings industries prefer LVL because of its ease of processing, uniform properties, good adaptability to curved parts with small radius, stability and better edge appearance. Arch door frames, staircases, chairs, beds, cabinets, counter tables, wardrobes, etc., are popular applications of this material.

Uniform mechanical properties of LVL, and especially its high allowable stress in tension compared to its weight, make this material useful as the flanges of I-beams combined with plywood webs or steel-pipe lattice webs. Such I-beams are widely used as joists in house construction. Open-web trusses combining LVL-steel pipe are used as roof trusses and second-floor joists in large buildings such as factories and warehouses. These open-web trusses are also utilized for arch roofs of very large structures such as a 120 m-span football stadium and exhibition halls. The fields of application are expanding to stringers and decks of highway bridges, stringers of trailer cars, crossties, scaffold planks and to other engineered members in wood construction (Kunesh 1978).

Bibliography

Bodig J, Jayne B A 1982 *Mechanics of Wood and Wood Composites*. Van Nostrand Reinhold, New York

Jung J 1982 Properties of parallel-laminated veneer from stress-wave-tested veneers. *For. Prod. J.* 32(7): 30–5

Koch P 1973 Structural lumber laminated from 1/4-inch rotary-peeled southern pine veneer. *For. Prod. J.* 23(7): 17–25

Koch P 1985 *Utilization of Hardwoods Growing on Southern Pine Sites*, Vol. 3: Agriculture Handbook No. 605. US Government Printing Office, Washington, DC

Kunesh R H 1978 Micro-Lam: Structural laminated veneer lumber. *For. Prod. J.* 28(7): 41–4

Laufenberg T L 1983 Parallel-laminated veneer: Processing and performance research review. *For. Prod. J.* 33(9): 21–8

H. Sasaki

[Kyoto University, Kyoto, Japan]

Lignin-Based Polymers

Lignin is nature's second most abundant polymer, behind cellulose, and its annual biosynthesis rate has been estimated to be approximately 2×10^{10} t. It is an amorphous polyphenolic component common to vascular plants for which it serves as a reinforcing agent. Its function as a natural adhesive for cellulosic fibers has attracted the attention of the polymer industry, especially of manufacturers of adhesives for the wood composites market.

This article reviews sources and actual and potential uses of lignin in water-soluble and structural polymer systems.

1. *Lignin Sources for Polymers*

Scientific and technological interest in lignin has focused primarily on the role lignin plays in relation to commercial pulpmaking and papermaking processes. Isolated lignin is best known as the cell wall component that is removed by dissolution from wood during pulping (Sarkanen and Ludwig 1971, Glasser and Kelley 1987). The interest in lignin as a chemical raw material concerns those lignin fractions that can be isolated from 'spent pulping liquors', that is, from solutions in the pulping process that contain dissolved lignin. There are, in principle, two types of commercial pulping processes, and a third is currently undergoing pilot plant testing (see *Paper and Paperboard: An Overview*).

1.1 Lignin Sulfonates

Lignin from the traditional (acid) sulfite pulping process is generated as a chemically modified, sulfonated lignin derivative, which is water-soluble by virtue of having sulfonate equivalent weights in the region of 400–600 (or sulfur contents of 6–8%) (Glennie 1971, Rydholm 1965). Solubility in water is achieved by chemical modification with a hydrophilic functional group that is about ten times as polar and hydrophilic as a hydroxyl group. Chemical (hydrolytic) degradation plays an apparently less dramatic role in lignin dissolution, since the molecular weights of lignin sulfonates remain extraordinarily high. The lesser contribution of molecular fractionation to solubility in water as compared to chemical modification is also reflected in low phenolic hydroxyl contents (i.e., little hydrolytic depolymerization). Lignin sulfonates behave as inorganic, ionic polymers that are virtually insoluble in organic solvents. Certain lignin sulfonate fractions are, however, soluble in mixtures of polar organic solvents with water (e.g., water-saturated butanol). No glass transition T_g or melting T_m temperatures have been reported for lignin sulfonates, and thermal decomposition appears to occur before softening. Lignin sulfonates are generated in mixture with

Table 1
Characteristics of lignins from different sources

	Lignin sulfonates	Kraft lignin	Novel bioconversion lignin
Purity (%)	55–90	95–>99	90–>95
Sulfur (%)	5–9	2–9	0
Methoxy (%)	9–11	12–14	18–22
Phenolic OH (%)	2–4	4–6	4–6
$\bar{M} \times 10^3$ (amu)	>10	2–4	0.8–1.2
$\bar{M}_w/M_n$	~10	~10	~3
$\bar{T}_g$ (°C)		155–180	90–140
Solvent solubility	water	aq. acetone, dioxane, DMF, DMSO	aq. ethanol, acetone, chloroform, dioxane, DMF, DMSO

water-soluble carbohydrate degradation products, from which they cannot easily be separated. Industrially, fermentation of reducing C-6 (hexose) sugars and yeasting of C-5 (pentose) sugars are the most widely used 'purification' methods (McGovern and Casey 1980, Forss 1968). Less-common treatments involve ultrafiltration, reverse osmosis, electrodialysis and ion exchange separations (Adhihetty 1983, Woerner and McCarthy 1986, Dubey et al. 1965).

Lignin sulfonates can be upgraded by ion exchange, where sodium and ammonium forms are more desirable than calcium and magnesium. Quality specifications for lignin sulfonates concern ionic form, purity (as determined by methoxy and reducing sugar content), molecular weight characteristics and degree of sulfonation.

1.2 Kraft Lignin

The more modern kraft pulping process dissolves lignin from wood with aqueous mixtures of NaOH and Na_2S (Bryce 1980). This treatment involves both derivatization and molecular degradation (depolymerization), where the latter predominates. This typically results in lower molecular weights, narrower molecular weight distributions, and lignins with higher phenolic hydroxy content. Chemical depolymerization involves principally alkyl aryl ether hydrolysis. Uptake of reduced sulfur generates thiol (mercaptan) functionality, and this is the reason why kraft lignin is also sometimes called 'thio lignin'. Sulfur content reaches 2–3% (Marton 1971). Isolation of kraft lignin from spent kraft pulp liquor involves acidification with carbon dioxide and/or sulfuric acid followed by filtration of the acid-insoluble lignin. This treatment generates a 99.5% pure, completely organic, phenolic polymer with number average molecular weight of ~2500 amu; with a glass transition temperature of ~160 °C; with an ash content of <0.5%; and with limited solubility in organic solvents (aqueous acetone, pyridine, DMF, etc.). Quality specifications for kraft lignin concern sulfur, phenolic hydroxy and methoxy contents, molecular weight, and purity.

1.3 Novel Bioconversion Lignins

The separation of polysaccharides and lignin from woody plant tissues can also be achieved by treatments with such organic solvents as aqueous methanol, ethanol, butanol or phenol (Glasser and Kelley 1987, Lora and Aziz 1985), and with high-pressure steam followed by extraction with alkali or solvent (Chum et al. 1984, Wallis and Wearne 1985). Both process options are sometimes pursued in conjunction with acidic or alkaline solvent mixtures. Lignins derived from these processes are termed 'organosolv' and 'steam explosion' lignins. Both options generate lignin free of sulfur-containing functionality; with molecular weights of around 800–1000 amu; and with vastly superior solubility in organic solvents as compared to conventional pulp lignin. These, mostly hardwood-derived, bioconversion lignins have high phenolic hydroxy functionality, often high purity, and they are said to be highly reactive. The latter must probably be attributed to their greater solubility in organic solvents, their lower overall molecular weights, and their lower glass transition temperatures as compared to those of other, pulping-derived lignins. These lignins may in the future become available as uncontaminated, organic polymers (or oligomers) with good solubility and thermoplasticity. Quality specifications involve methoxy content, phenolic hydroxy content, molecular weight, molecular weight distribution, glass transition temperature and solubility in organic solvents.

Table 1 summarizes chemical and molecular characteristics of lignins from different sources.

2. Water-Soluble Polymers

Lignin sulfonates have achieved prominence among industrial surfactants, and they are the highest volume class of cationic surfactant known to industry. In the late 1970s, sales amounted to approximately 5×10^5 t per year (Chum et al. 1985). While lignin sulfonates from sulfite pulping make up the vast majority of water-soluble lignin derivatives, other types are in use and have been explored as well. Among them are sulfomethylated kraft lignins (de Groote et al. 1987). These have the advantage of better controlled surface active properties by virtue of a lower degree of sulfonation, and greater purity owing to the removal of contaminants during the isolation of parent kraft lignin. There is an approximately four-to-one price advantage in favor of the sulfite-pulping-derived raw material. Other water-soluble lignin-based polymers include those with greater concentrations of organic ionic, especially carboxylic, functionalities as a consequence of carboxylation or oxidation (Glasser and Kelley 1987, Lin 1983). The introduction of carboxyl groups has been based on reactions with dicarboxylic anhydrides such as maleic and succinic anhydride, and with chloroacetic acid. Oxidation reactions have been based on the reaction of lignin in aqueous alkali with molecular oxygen under pressure, or with hydrogen peroxide or ozone (Lin 1983). Undesirable secondary reactions often result in repolymerization of lignin leading to higher molecular weights.

A novel route to water-soluble lignin derivatives is based on graft copolymerization of lignin sulfonates with 2-propenamide using anhydrous $CaCl_2$ or cerium (Ce^{4+}) ions for initiation (Meister and Patil 1985). This reaction is illustrated in Fig. 1. Other initiation systems have been examined as well.

3. Structural Polymers

For lignin to contribute to the performance of a structural material, lignin must be soluble in a common solvent with other material components, or in the melt of a thermoplastic polymer, or it must be present as finely dispersed particles in a continuous rubber-like matrix. This solubility, miscibility or dispersibility is often the limiting factor in lignin's performance in structural materials. Three general types of structural polymers are distinguished, and these are thermoplastic, thermosetting and filled systems.

3.1 Thermoplastic Polymer Systems (Polyblends)

Thermoplastic polymer systems involve mixtures of linear polymers which undergo melt flow at elevated temperatures, with lignin or a lignin derivative in a uniform, continuous phase (Walsh et al. 1985). Upon cooling, component demixing usually occurs, and multiphase materials, polymer blends or polyblends, are formed in which the two components are separated to a greater or lesser extent. Polyblends can be produced by solution casting, in which both polymer components are dissolved in a common solvent and the solvent is removed from the mixture by (slow) evaporation; or by injection molding of a homogeneous melt mixture. The two polymer components contribute synergistically to each others' properties if some type of polymer–polymer interaction is achieved resulting in 'compatibility'. Compatibility and miscibility are governed by chemical and molecular factors, such as solubility parameter and molecular weight (Walsh et al. 1985). Thus, both chemical modification and fractionation according to molecular weight may be employed to improve superior blend behavior.

Polyblends of lignin and lignin derivatives have been studied in conjunction with polyethylene, with ethylene-vinyl acetate copolymer, with poly(methyl methacrylate), with poly(vinyl alcohol), with hydroxypropyl cellulose, and with several other thermoplastic cellulose derivatives (Ciemniecki and Glasser 1988, Rials and Glasser 1988). Results, in general, have revealed that lignin may contribute stiffness to linear, thermoplastic polymers; and that it may serve as either plasticizer or antiplasticizer, depending on degree of interaction, molecular weight parameters, T_g of the individual components, etc. In general, the degree of interaction was found to be related to the presence of polar functionality in the thermoplastic cocomponent. Thus, whereas polyethylene resulted in complete phase separation with no sign of

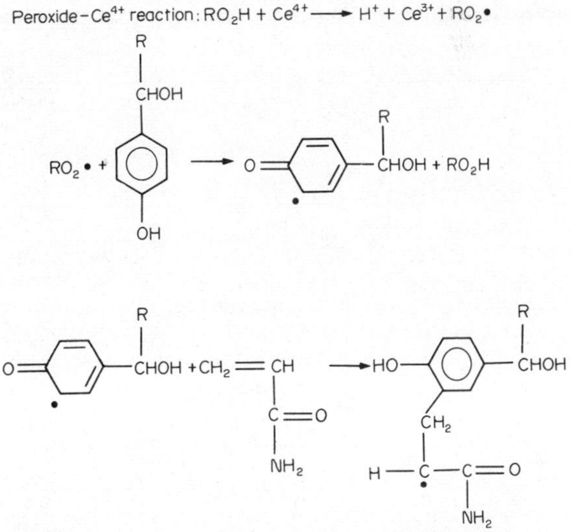

Peroxide–Ce^{4+} reaction: $RO_2H + Ce^{4+} \longrightarrow H^+ + Ce^{3+} + RO_2\bullet$

Figure 1
Schematic graft copolymerization of lignin with 2-propenamide (after Sarkanen and Ludwig 1971. © Wiley, New York. Reproduced with permission)

polymer–polymer interaction and with poor physical properties, vinyl acetate groups, carbonyl groups and alcohol groups all produced improved interaction (Glasser et al. 1988). Cellulose derivatives displayed a propensity for forming organized morphological features resembling liquid crystal mesophases (Rials and Glasser 1988).

3.2 Thermosetting Systems

The molecular weight and functionality features of lignin enhance its potential as a backbone component of network polymers. Among options for cross-linking lignins are phenol-formaldehyde, urethanes, epoxides and acrylates (Glasser 1988). Although not of industrial significance at present, all systems have been described in the recent literature.

The polyphenolic nature of lignin lends itself to association with phenol formaldehyde resins (Nimz 1983). Isolated lignin can be added to commercial phenol formaldehyde resin systems, either as powder, as alkaline solution, or as metholylated or phenolated lignin derivative. All options have been examined, and the contribution by lignin was found to depend on the method of formulation. Phenol replacement levels of up to 60% have been reported (Muller et al. 1984), but most common substitution levels which produce acceptable adhesive performance range from 25% to 30%. Phenol formaldehyde type resins with higher substitution levels require water soluble lignin derivatives, mostly lignin sulfonates, which are cross-linkable with either hydrogen peroxide or sulfuric acid (Nimz 1983). These resins cannot be used interchangeably with

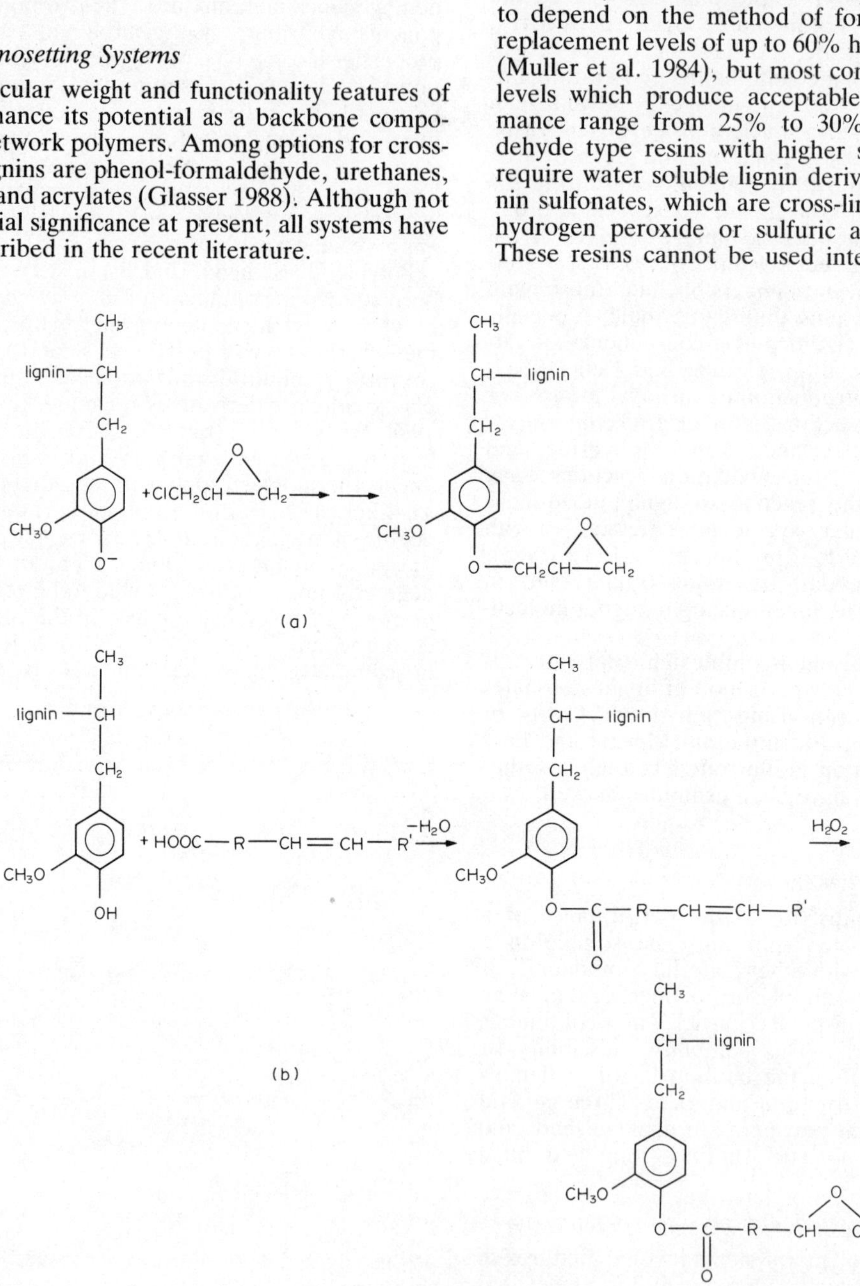

Figure 2
Lignin epoxidation: (a) with epichlorohydrin; (b) by peroxide reaction of an unsaturated ester group (after Glasser and Kelley 1987. © Wiley, New York. Reproduced with permission)

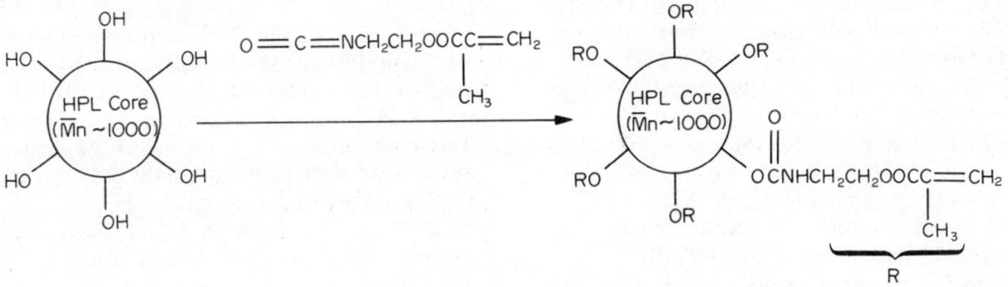

Figure 3
Schematic lignin acrylation reaction

commercial phenol formaldehyde resin formulations.

Lignin-containing polyurethane products have been described on the basis of lignin fractions soluble in low molecular weight, aliphatic glycols which are then jointly cross-linked with diisocyanates (Yoshida et al. 1987); and based on chemically modified lignin derivatives soluble in cross-linking agent or in a common solvent with the cross-linker (Saraf and Glasser 1984). In wood composites, emulsion systems involving lignin derivatives and polymeric diisocyanates or diamines have also been examined (Newman and Glasser 1985). For many applications, miscibility of lignin or lignin derivative with the cross-linking agent during gelation has been identified as the critical parameter (Glasser 1988).

Efforts to convert lignin into epoxy resin systems have concentrated on kraft lignin, phenolated kraft lignin and nonphenolic hydroxy alkyl lignin derivatives by reaction with epichlorohydrin (Tai et al. 1967); and kraft lignin containing unsaturated functionality by reaction with hydrogen peroxide (Holsopple et al. 1981). This is illustrated schematically in Fig. 2. In all cases polymeric, multifunctional epoxy derivatives are formed which can be cross-linked with conventional diamine or anhydride cross-linkers. Solubility and miscibility with a cross-linking agent become again critical performance parameters.

Acrylated lignin derivatives have been prepared from kraft lignin by reaction with acrylic acid chloride and anhydride (Naveau 1975); and by reaction of solid and liquid hydroxy alkyl lignin derivatives with isocyanate-functional methyl methacrylate derivative (Glasser et al. 1988) (Fig. 3). Lignin derivatives with acrylate functionality can be copolymerized with conventional vinyl monomers following normal initiation, such as with peroxides (Glasser et al. 1988). Reactivity ratios have been determined for certain cases, and azeotropic compositions have been established. Results suggest a preference for the formation of alternating copolymers in which copolymerization is favored over homopolymerization of any component.

Isolated lignin usually is a stiff (high modulus of elasticity), glassy polymer which contributes rigidity to polymer systems. Stiff materials typically have low impact resistance, and this has successfully been addressed in segmented materials by the use of toughness-building components (Saraf et al. 1985a, b). This approach, which has been adopted for several thermosetting polymer systems involving lignin, was made possible by the commercial availability of hydroxide, amine, vinyl terminated polyether and liquid rubber segments. Phase separation was found to be prevalent in most applications, and the desired effect of rubber toughening was easily achieved.

3.3 *Filled Systems*

Kraft lignin has been found to display surface active properties *vis-a-vis* vulcanized rubber that permit the production of particulate-filled reinforced styrene–butadiene copolymer competitive with carbon-filled tire products (Sirianni and Puddington 1972, Dimitri 1976). Careful control over isolation, post-treatment and coprecipitation methods was found to be required for lignin to achieve the desired dispersibility and reinforcing characteristic.

See also: Cellulose: Chemistry and Technology; Cellulose: Nature and Applications

Bibliography

Adhihetty T L D 1983 Utilization of spent sulfite liquor. *Cell. Chem. Technol.* 17: 395–9

Bryce J R G 1980 Alkaline pulping. In: Casey J P (ed.) 1980 *Pulp and Paper-Chemistry and Chemical Technology*, 3rd edn. Wiley, New York, pp. 377–492

Chum H L, Parker S K, Feinberg D A, Wright J D, Rice P A, Sinclair S A, Glasser W G 1985 *The Economic Contribution of Lignins to Ethanol Production from Biomass*, SERI/TR-231-2488. Solar Energy Research Institute, Golden, CO

Chum H L, Ratcliff M, Schroeder H A, Sopher D W 1984 Electrochemistry of biomass-derived materials. I. Characterization, fractionation, and reductive electrolysis of ethanol-extracted explosively-depressurized aspen lignin. *Wood Chem. Technol.* 4(4): 505–32

Ciemniecki S L, Glasser W G 1988 Multiphase materials with lignin. I. Blends of hydroxypropyl lignin with poly (methyl methacrylate). *Polymer* 29: 1021–9 de Groote R A M C, Neumann M G, Lechat J R, Curvelo A A S, Alaburda J 1987 The sulfomethylation of lignin. *Tappi J.* 70(3): 139–40

Dimitri M S 1976 Lignin reinforced polymers. US Patent 3,991,022

Dubey G A, McElhinney T R, Wiley A J 1965 Electrodialysis—Unit operation for recovery of values from spent sulfite lignor. *Tappi* 48(2): 95–8

Fross K 1967 Spent sulfite liquor—An industrial raw material. *Ind. Chem. Belge* 32 (Spec. No.): 405–10

Glasser W G 1988 Crosslinking options for lignins. In: Hemingway R W, Conner A H (eds.) 1988 *Adhesives from Renewable Resources*, ACS Symposium Series No. 385. American Chemical Society, Washington, DC

Glasser W G, Kelley S S 1987 Lignin. In: *Encyclopedia of Polymer Science and Engineering*, Vol. 8, 2nd edn. Wiley, New York, pp. 795–852

Glasser W G, Knudsen J S, Chang C-S, 1988a Multiphase materials with lignin. 3. Polyblends with ethylene-vinyl acetate copolymers. *J. Wood Chem. Tech.* 8(2), 221–34

Glasser W G, Nieh W, Kelley S S, de Oliveira W 1988b Method of producing prepolymers from hydroxyalkyl lignin derivatives. US Patent Appl. 183,213

Glennie D W 1971 Reactions in sulfite pulping. In: Sarkanen K V, Ludwig C H (eds.) 1971 *Lignins: Occurrence, Formation, Structure and Reactions*. Wiley, New York, pp. 695–768

Holsopple D B, Kurple W W, Kurple W M, Kurple K R 1981 Epoxide-lignin resins. US Patent 4,265,809

Lin S Y 1983 Lignin utilization: Potential and challenge. *Progr. Biomass Convers.* 4: 31–78

Lora J H, Aziz S 1985 Organosolv pulping—A versatile approach to wood refining. *Tappi J.* 68(8): 94–7

McGovern J N 1980 Silvichemicals. In: Casey J P (ed.) 1980 *Pulp and Paper—Chemistry and Chemical Technology*, 3rd edn. Wiley–Interscience, New York, pp. 492–503

Marton J 1971 Reactions in alkaline pulping. In: Sarkanen K V, Ludwig C H (eds.) 1971 *Lignins: Occurrence, Formation, Structure and Reactions*. Wiley, New York, pp. 639–94

Meister J J, Patil D R 1985 Solvent effects and initiation mechanisms for graft-polymerization on pine lignin. *Macromolecules* 18: 1559–64

Muller P C, Kelley S S, Glasser W G 1984 Engineering plastics from lignin. 9. Phenolic resin characterization and performance. *J. Adhes.* 17(3): 185–206

Naveau H D 1975 Methacrylic derivatives of lignin. *Cell. Chem. Technol.* 9: 71–7

Newman W H, Glasser W G 1985 Engineering plastics from lignin. 12. Synthesis and performance of lignin adhesives with isocyanates and melamine. *Holzforschung* 39(6): 345–53

Nimz H H 1983 Lignin-based wood adhesives. In: Pizzi A (ed.) 1983 *Wood Adhesives—Chemistry and Applications*. Dekker, New York, pp. 248–88

Rials T G, Glasser W G 1988 Multiphase materials with lignin. 4. Blends of hydroxypropyl cellulose with lignin. *J. Appl. Polym. Sci.* 37(8): 2399–415

Rydholm S A 1965 *Pulping Processes*. Wiley, New York

Saraf W P, Glasser W G 1984 Engineering plastics from lignin. 3. Structure property relationships in solution cast polyurethane films. *J. Appl. Polym. Sci.* 29(5): 1831–41

Saraf W P, Glasser W G, Wilkes G L 1985 Engineering plastics from lignin. 7. Structure property relationships of poly(butadiene glycol)-containing polyurethane networks. *J. Appl. Polym. Sci.* 30: 3809–23

Saraf W P, Glasser W G, Wilkes G L, McGrath J E 1985 Engineering plastics from lignin. 6. Structure property relationships of peg-containing polyurethane networks. *J. Appl. Polym. Sci.* 30: 2207–24

Sarkanen K V, Ludwig C H (eds.) 1971 *Lignins. Occurrence, Formation, Structure and Reactions*. Wiley New York

Sirianni A F, Barker C M, Barker G R, Puddington I E 1972 Lignin reinforcement of rubber. *Rubber World* 166(1): 40–1, 44–5

Tai S, Nagata M, Nakano J, Migita M, 1967 Lignin. 51. Utilization of lignin. 4. Epoxidation of thiolignin. *Mokuzai Gakkaishi* 13(3): 102–7

Wallis A F A, Wearne R H 1985 Fractionation of the polymeric components of hardwoods by autohydrolysis–explosion–extraction. *Appita* 38(6): 432–7

Walsh D J, Higgins J S, Maconnachie A (eds.) 1985 *Polymer Blends and Mixtures*. Nijhoff, The Hague

Whittaker R H, Likens G E 1975 The biosphere and man. In: Lieth H, Whittaker R H (eds.) 1975 *Primary Productivity of the Biosphere*, Ecological Studies 14. Springer, Berlin, pp. 305–28

Woerner D L, McCarthy J L 1986 The effect of manipulatable variables on fractionation by ultrafiltration. *AIChE Symp. Ser.* 82: 77–86

Yoshida H, Morck R, Kringstad K P, Hatakeyama H 1987 Kraft lignin in polyurethanes. 1. Mechanical properties of polyurethanes from a kraft lignin polyether triol polymeric MDI system. *J. Appl. Polym. Sci.* 34: 1187–98

W. G. Glasser
[Virginia Polytechnic Institute and State University, Blacksburg, Virginia, USA]

Lime

Lime is a manufactured chemical which is a calcined or burned form of limestone or dolomite. The primary lime products include quicklime, pebble lime, hard-burn dolomite or, when water is added, calcium hydroxide or slaked lime. The rather loose use of the term "lime" had led to much confusion and misunderstanding. The term is frequently, albeit erroneously, used to denote almost any kind of calcareous material or finely ground form of limestone or dolomite, as well as the true products of calcination.

Lime is usually made from high-calcium or high-magnesium limestone, generally having a minimum of 90% combined carbonate content. Normally, high-calcium limes contain less than 5% MgO. When the lime is produced from a high-magnesium limestone, the product is referred to as dolomitic lime.

1. Calcination

Calcination refers to a reaction wherein limestone or dolomite is heated to less than its melting point with a resultant change in chemistry and a weight loss. The basic chemical reaction is as follows:

$$CaCO_3 \text{ (limestone)} + \text{heat } (1000\text{–}1300\,°C) \rightleftharpoons CaO \text{ (quicklime)} + CO_2 \uparrow$$

or

$$CaCO_3 . MgCO_3 \text{ (limestone)} + \text{heat } (900\text{–}1200\,°C) \rightleftharpoons CaO . MgO \text{ (quicklime)} + 2CO_2 \uparrow$$

Calcination is strongly time-variant with the different limestones. In a very broad sense this relates to the fact that the calcination reaction starts on the exterior surfaces of the limestone and then proceeds toward the center. As the calcination reaction takes place the CO_2 released at the reaction interface must make its way through the lime to the exterior surface; therefore, because calcination is limited by gas diffusion to the surface of the partially calcined limestone, natural impurities in the stone, differences in crystallinity, grain-boundary chemistry, density variations and imperfections in the atomic lattice all play a significant role in the calcination rate. The suitability of a given limestone as a source material for lime production can be determined only after completion of adequate burn tests, designed to evaluate the various limiting parameters. When a coal-fired kiln system is considered, the entire process of calcination is made even more complex.

In the calcination reaction CO_2 is released. The result is a 44% weight loss during the complete calcination of a high-calcium limestone or a 48% weight loss for a highly dolomitic limestone. The trade term for this weight loss is "loss on ignition" (LOI).

2. Lime Production

The production (calcination) of lime requires the use of kilns which will facilitate the heating of the raw limestone such that the calcination reaction takes place. Although lime calcination is chemically a relatively simple operation, many kiln systems are used. These include vertical, rotary, rotary with preheaters, multiple-shaft regenerative and a variety of other types of kilns.

Selection of a lime kiln depends upon several factors, the most important ones being burning characteristics of the stone, fuel consumption and capital equipment cost. Other factors include customer specifications, the type of fuel to be utilized, environmental constraints, and so forth. As a result of severe energy-cost fluctuations since the early 1970s, there have been major changes in kiln selection. In the USA and Canada the emphasis has been on rotary preheater kilns. Concurrently, developments in calcining technology were generating interest in multiple-shaft regenerative kilns.

2.1 Vertical Kilns

Vertical kilns traditionally have burned only larger stone (75–300 mm) with a size ratio of approximately 1:2. However, the new designs can handle stone sizes as small as 20 mm and a widening size ratio to as much as 1:4. Fuel consumption has been a major advantage of vertical or shaft kilns, normally requiring less than 5.2 GJ t^{-1} of lime produced. Power requirements for these kilns will vary with stone size but at nominal capacity will be in the range of 54–90 MJ t^{-1}. A disadvantage of traditional vertical kilns in the current energy environment is their fuel requirement of oil, natural gas or coke; however, several recent developments in burner technology have come to the point where pulverized coal can now be utilized in the modern shaft kiln.

Vertical kilns may be of stone-masonry, reinforced-concrete or boiler-plate construction. The most widely used kiln has a refractory-lined steel shell and is usually circular in cross section. These kilns may be 2.7–7.3 m in diameter and 15–4.8 m high. Capacities vary from as low as 9.5 t d^{-1} to an excess of 700 t d^{-1}, with the larger tonnages being restricted to the newer generation of kilns.

2.2 Multiple-Shaft Kilns

Developed to burn stone of small diameter, this vertical kiln utilizes the parallel-flow calcining principle in double and triple shaft units. The shafts are interconnected in the burning zone and while one shaft is being fired the other is preheated. Fuel and combustion air are supplied to the burning shaft from above, ignite at the upper end of the burning zone, and calcine the lime in uniflow. The exhaust gases then pass into the second shaft, preheating the stone in counterflow. After 10–15 min, the shaft firing is reversed. Cooling air is blown into both shafts simultaneously. The regenerative system has experienced fuel consumption of 3.3–3.9 GJ t^{-1} of lime produced. Kilns vary in capacity from 90–550 t d^{-1} and while nominally gas or oil fired, current technology has now allowed the use of pulverized coal. These kilns have found acceptance in Europe and the USA and are now being installed in Japan.

2.3 Annular-Shaft Kilns

The burning process in this kiln is based upon counterflow for preheating, and calcining and uniflow for residual calcination. Partitions provided by an inner cylinder and by staggered bridges in the burning zone permit an even distribution of heat and a uniform flow of material down the kiln. Stone as small as 25×75 mm can be calcined to produce soft-burned lime using either oil or gas firing. Capacities vary from 90–270 t d^{-1} with a fuel consumption of 6.2 GJ t^{-1}. This type of kiln is not currently popular anywhere in the world.

2.4 Rotary Kilns

Unlike vertical kilns which operate fully charged, the rotary kiln has about 90% of its volume filled with flame and hot gases. As the kiln slowly rotates, new surfaces of stone are exposed to the hot gases but there is little passage of gases through the solids. Hence, radiation interchanged between gas, solids and refractory walls plays an important part in the overall heat transfer. Because the area of solids exposed is relatively small, a rotary kiln is less efficient than a shaft kiln. However, the rotary has the advantage of burning stone from as small as 6.5 mm to as large as 57 mm. Generally the size ratio is 1:3 in order to minimize segregation.

Rotary kilns vary greatly in size ranging from 2×25 m to 5.5×190 m with capacities of 45–945 t d^{-1}. While rotary kilns can burn a wide range of fuels, the major drawback of these kilns has been their fuel requirement of 18.6 GJ t^{-1}.

2.5 Rotary Preheater Kilns

Of the several advances which have been made in improving the heat efficiency of rotary kilns, probably the most pronounced has been the trend away from longer kilns to medium-length kilns with external preheaters. The preheater is a refractory-lined chamber located below the raw-feed bin and ahead of the kiln. Exhaust gases from the kiln are drawn counter-current to the stone preheating it to 650–900 °C. Retention time within the preheater is 1–1.5 h depending upon the system design, burning characteristics of the stone, and so on. During preheating, approximately 30% calcination is achieved before the stone is discharged to the kiln. Following preheating, the partially calcined stone is fed to the kiln at a predetermined rate by means of hydraulically actuated plungers. This type of preheater has been adapted to the new large-volume kilns and typically may be 5.2 m in diameter by 62.5 m long and have a capacity of 900–1000 t d^{-1}. This and other types of preheater kilns, including the travelling-grate variety, plus the use of heat exchangers and coolers which recover waste heat, have improved fuel efficiency to 6.9–8.5 GJ per tonne of lime.

2.6 Calcimatic Kiln

The rotary hearth calciner, the calcimatic kiln, consists of a preheater, circular hearth and cooler, all refractory lined. Like the preheater, this kiln burns small stone which is typically sized in a 1:3 ratio. The stone is carried on the hearth in a thin layer and one revolution of the hearth constitutes the calcining cycle. Numerous burners located inside and outside the hearth are used for firing, utilizing gas, fuel oil and, most recently, pulverized coal. Fuel requirements approximate 7.7 GJ t^{-1} with a wide range of burns possible because of the ease of operator control and calcining. One of the primary advantages of this type of kiln is the fact that attrition loss is negligible. This calcining system can utilize very soft limestones which would otherwise be subject to high mechanical breakage. However, their capacity is generally restricted to only 90–300 t d^{-1}.

2.7 Flash Calciners

The flash calciner is one of the most recent developments in calcination and in reality is not a kiln. This system is designed to calcine the undersized material or fines remaining after the primary kiln feed has been sized. The system includes a preheater, flash calciner and cooler. Preheating is achieved in a series of cyclones in which the heat of rising hot gases is absorbed in the counterflowing material which is in the top and heated to 980 °C. As the preheated feed drops into the furnace calciner, a forced air vortex flow pattern pulls the particles into the center of the furnace. Within the furnace the temperature is 1200 °C. The retention time is only a few seconds. The system is designed in two modes depending upon the size of feed to be used. Where stone fines are to be calcined, 60–80% of the calcination occurs in the furnace. The final calcination takes place in a fluid bed which is located between the furnace and the cooler. In those systems where −28 mesh material is fed to the calciner, no fluid-bed reactor is necessary. This system can utilize gas, oil or coal and will normally require 6.2–8.5 GJ t^{-1} of product.

3. Lime Hydration

Hydrated lime is obtained by adding water to quicklime to produce a dry, fine powder. The affinity of quicklime for moisture is then satisfied although it still retains a strong affinity for CO_2.

Hydration is the combining of calcium oxide with water in a reversible reaction to form calcium hydroxide:

$$CaO + H_2O \rightarrow Ca(OH)_2$$

As a standard condition, the reaction is exothermic for the formation of hydrate with the accepted value for the heat of hydration being 15.456 kcal g-mol^{-1} of CaO. This is a three-step reaction with the first being the disassociation of CaO into a Ca^{2+} ion and an O^{2-} ion:

$$CaO \rightarrow Ca^{2+} + O^{2-}$$

Table 1
Industrial usage of lime in the USA in 1986 as a percentage of total lime resources

Industrial use	Percentage
Steel	34.4
Water treatment and purification	10.3
Soil stabilization (includes all construction uses)	10.1
Flue-gas desulfurization	8.4
Chemical manufacturing	8.2
Pulp and paper production	7.2
Sewage treatment	6.9
Sugar refining	2.9
Refractories (dolomitic lime)	2.1
Metallurgical processing (other than steel)	2.0
Other	7.5

Source: US Bureau of Mines

Next the water ionizes into two hydroxyl ions, owing to the strong attraction that an oxygen ion exerts on a water molecule:

$$H_2O + O^{2-} \rightarrow 2(OH)^-$$

Finally the calcium and hydroxyl ions combine to form calcium hydroxide:

$$Ca^{2-} + 2(OH)^- \rightleftharpoons Ca(OH_2) + \text{heat}$$

As noted, this is a reversible reaction and hydrated lime can therefore, when heated, revert to the original oxide form.

Hydrated lime is available in two principal forms: standard, where 85% passes a No. 200 sieve and superfine or spray hydrate where 98% or more passes a No. 325 sieve. These in turn are manufactured in two types, normal (N) and special (S), each including high-calcium or dolomitic hydrate.

Type S limes are highly hydrated, generally by processes involving pressure hydration. They contain less than 8% unhydrated oxides and develop high plasticity and high water retention. Since World War Two, the demand for type S lime has grown rapidly and currently it is the chief lime for structural uses. Some lime manufacturers also add entraining catalysts to the lime to increase plasticity and durability.

4. Lime Markets

Lime has possibly the greatest number of diverse uses of any chemical or mineral commodity. Over 90% is utilized by the chemical and metallurgical process industries where it may be used as a flux, acid neutralizer, causticizing agent, flocculent, hydrolyzer, bonding agent, absorbent and/or raw material. An allied product, dead-burned dolomite, is used as a refractory material. Table 1 shows the approximate percentage uses of lime resources by industry in the USA.

Of these various markets, steel is the largest consumer industry. In this application, lime acts as a scavenger in purifying steel by fluxing out, into the molten slag, acid oxide impurities such as silica, alumina, phosphorus and sulfur.

Most steel companies use pebble or pelletized lime as flux in the basic oxygen furnace (BOF). Recent European and Japanese technology has modified the basic oxygen process. This modified process, known as O-BOP, utilizes pulverized lime.

The treatment of municipal water and industrial process water with lime is also an important market. Lime is required to remove the temporary bicarbonate hardness from the water. The high pH of 11.5 induced by the lime is of value as the secondary sterilization agent to chlorine because retention for 3–10 h at this pH will kill over 99% of the bacteria and most viruses. By introducing CO_2 into the lime-treated water, the pH is lowered to acceptable levels and most of the lime in solution is then precipitated as a carbonate sludge.

Lime is an important chemical in a number of environmental applications. The newest of these is as a primary reagent material in flue-gas desulfurization (FGD). Initially it was used in wet lime scrubbing, but more recently, attention is focusing on the so-called dry scrubber or spray dryer. FGD markets may grow significantly in the 1990s in the USA if the US Congress enacts more stringent air pollution control standards. At the other end of the environmental control spectrum, lime is utilized in acid-water treatment, sewage treatment and toxic-chemical cleanup. Overall, environmental applications may represent the largest potential growth markets for lime.

Lime is also used in a number of chemical manufacturing processes. These include the production of caustic soda, calcium carbide, bleaches and various inorganic chemicals. In addition, lime is required in the production of several important organic chemicals. Chief among these are ethylene and propylene glycols, and calcium-based organic salts such as calcium stearate, acetate, lactate and lignosulfonate.

Another primary market for lime has been in the manufacture of pulp and paper where it is universally used to causticize waste sodium carbonate. In this process it is used to regenerate sodium hydroxide; in spite of the fact that the pulp mills recover 90% to 96% of the lime employed by recalcining the dewatered calcium carbonate sludge precipitate, a substantial quantity of make-up lime continues to be used. Secondary uses of lime by the paper industry include making calcium hypochlorite bleach and the clarification of process water and color removal of waste water effluents.

The construction industry is also a major user of lime. While finding a variety of applications, the primary consumption is the area of soil stabilization. This is particularly important in areas where the subgrade soils have very high clay contents. The lime reacts with the silica and alumina in the clay soils to form complex cementing compounds which bind the soil into hard stable masses that are not sensitive to water saturation. The plasticity of the soil, as well as shrink and swell, are markedly reduced. The amount of lime used generally ranges between 3% and 5% of the dry weight of the soil. Over 80% of the lime used in this application is hydrate which is disked into the clay soils.

5. *Specifications and Analysis*

Owing to the number of limes available, their many uses and their wide variation in physical and chemical characteristics, there is no overall lime specification. Numerous specifications and buyer requirements for lime exist, not only for most of the major uses but even among individual companies or plants.

In the USA, the National Lime Association (NLA), with its member lime companies and a majority of the lime consumers, subscribe to the American Society for Testing and Materials (ASTM) specifications on lime which are promulgated by Committee C-7 on Lime. These specifications cover many uses in testing procedures, including chemical analysis, sampling, inspection, packing and marking of quicklime and lime products, and physical testing. The details of these specifications are found in ASTM Standard Section 4, volume 04.03.

Another important lime specification in the USA is the American Water Works Association (AWWA) Standard for Quick Lime and Hydrated Lime (B202-77).

See also: Calcium Aluminate Cements; Construction Materials: Crushed Stone; Construction Materials: Dimension Stone; Construction Materials: Fill and Soil; Construction Materials: Granules; Construction Materials: Industrial Minerals; Portland Cement Raw Materials (Suppl. 2); Portland Cements, Blended Cements and Mortars

Bibliography

American Society for Testing and Materials 1986 ASTM Standard, Sect. 4, Vol. 04.03. ASTM, Philadelphia, PA

American Water Works Association 1977 *Standard for Quick Lime and Hydrated Lime*, Standard B202-77. AWWA, Denver, CO

Boynton R S 1980 *Chemistry and Technology of Lime and Limestone*. Wiley, New York

Dixon T 1982 North American lime. *Ind. Miner. (London)* 177: 51–63

Freas R C, Thompson J L 1983 Lime. In: Lefond S J (ed.) 1983 *Industrial Minerals and Rocks,* 5th edn. American Institute of Mining, Metallurgical and Petroleum Engineers, New York, pp. 809–31

Gutshick K A 1987 *Lime for Industrial Uses*, STP 931. American Society for Testing and Materials, Philadelphia, PA

Mullins R C, Hatfield J D 1969 *Effects of Calcination Conditions on Properties of Lime*, STP 462. American Society for Testing and Materials, Philadelphia, PA

National Lime Association 1965 *Chemical Treatment of Sewage and Industrial Wastes*, Bulletin 215. NLA, Washington, DC

National Lime Association 1966 *Specifications for Lime and its Uses in Building*, Bulletin 322. NLA, Washington, DC

National Lime Association 1972 *Lime Stabilization Construction Manual*, Bulletin 326, 5th edn. NLA, Washington, DC

National Lime Association 1976 *Acid Neutralization with Lime*, Bulletin 216. NLA, Washington, DC

National Lime Association 1976 *Lime Handling Application and Storage in Treatment Processes*, Bulletin 213. NLA, Washington, DC

National Lime Association 1976 *Water Supply and Treatment*, Bulletin 211, 11th edn. NLA, Washington, DC

National Lime Association 1980 *Lime in Municipal Sludge Processing*, Bulletin 217. NLA, Washington, DC

National Lime Association 1981 *Chemical Lime Facts*, Bulletin 214, 4th edn. NLA, Washington, DC

Pressler J W 1986 Lime. *Minerals Yearbook 1986*, Vol. 1. US Bureau of Mines, Washington, DC

R. C. Freas
[Franklin Limestone Company, Nashville, Tennessee, USA]

M

Materials Selection in Conceptual Design: Materials Selection Charts

It is estimated that there are more than 50 000 materials available to the engineer. It is convenient to catalog these in the six broad classes shown in Fig. 1—metals, polymers, elastomers, ceramics, glasses and composites. Within a class, there is some commonality in properties, processing and pattern of usage. For example, ceramics have high moduli, polymers have low; metals can be shaped by casting and forging, composites require lay-up or special molding techniques. However, this compartmentalization has its dangers: it can lead to specialization (the metallurgist who knows nothing of polymers) and to conservative thinking ("we use steel because we have always used steel").

The range of materials available to the engineer is larger, and is growing faster, than ever before. New materials create opportunities for innovation; that is, for new products and for the evolution of existing products to give greater performance at lower cost. Markets are captured by the innovative use of new materials and lost by the failure to perceive the opportunities they present. The problem, however, is in making a sensible choice of one material from the enormous catalog available. A rational procedure for materials selection is needed.

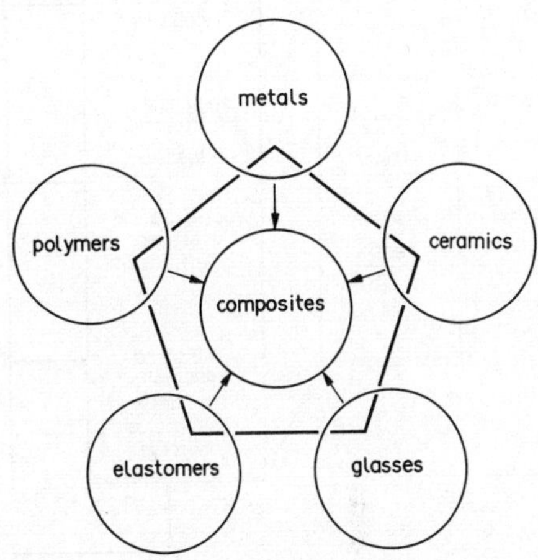

Figure 1
The menu of engineering materials. Each class has properties which occupy a particular part (or 'field') of each of the materials selection charts

1. Materials Data in the Design Process

The stages of the design process are shown, in a simplified form, in the central column of Fig. 2 (Pahl and Beitz 1984, French 1985). Initially, a market need is identified. A concept for a product which meets that need is devised. If approximate calculations (left-hand columns) show that, in principle, the concept will work, the design proceeds to the embodiment stage: a more detailed analysis, leading to a set of working drawings giving the size and layout of each component of the product, and estimates of its performance and cost. If the outcome is considered viable, the designer proceeds to the detailed design stage: full analysis (using computer methods if necessary) of critical components, preparation of detailed production drawings, specification of tolerance, precision, joining methods and finishing, and so on.

Materials selection (Dieter 1983, Crane and Charles 1984) enters at every stage of the design process, but the nature of the data for material properties needed at each stage differs greatly in its level of precision and breadth (right-hand columns). At the conceptual design stage, the designer requires approximate data for the widest possible range of materials. All options are open; for example, a polymer may be the best choice for one concept, a metal for another, even though the function is the same. This sort of data is found in low-precision tables such as those of the Fulmer Materials Optimiser (1974) and the Materials Selector (*Materials Engineering* 1988), or in materials selection charts (Ashby 1989) as will be discussed in the following sections of this article. The low level of precision is not a problem; it is perfectly adequate for the task. The problem is access; that is, how can the data be presented to give the designer the greatest freedom in considering alternatives? Materials selection charts are useful here.

Embodiment design needs data at the second level of precision and detail. The more detailed calculations involved in deciding on the scale and layout of the design require the use of more detailed compilations; that is, multivolume handbooks, such as The ASM Metals Handbook (1973), The Metals Reference Book (Smithells 1984), The Handbook of Plastics and Elastomers (Harper 1975), The

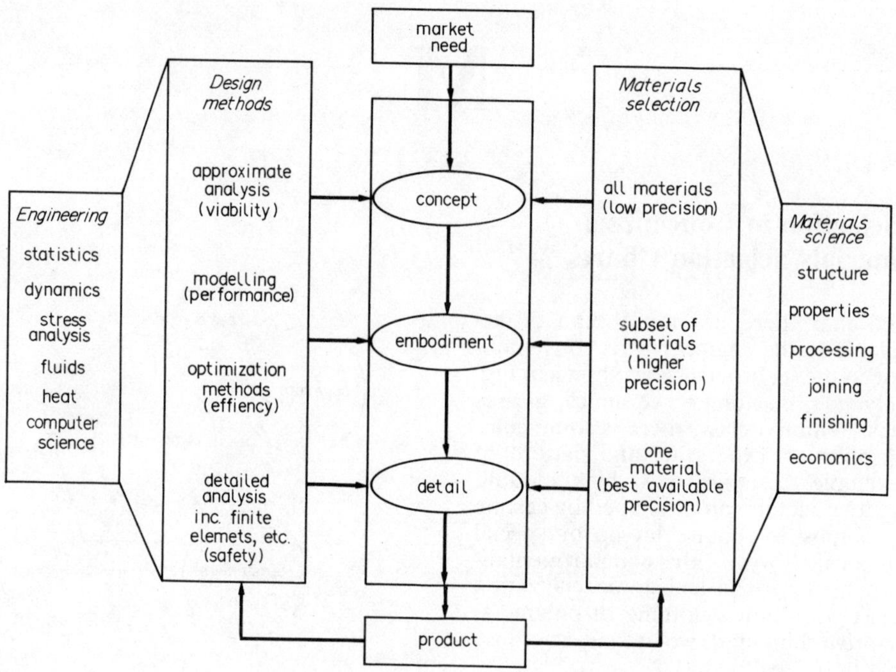

Figure 2
The design process, much simplified (central column), showing how engineering and materials science interface with each stage

Handbook of Properties of Technical and Engineering Ceramics (Morrell 1985, 1987), or computer databases. These list, plot and compare properties of a single class of materials, and allow choice at a level of detail not possible from the broader compilations that include all materials.

The final stage of detailed design requires a still higher level of precision and detail. This is best found in the data sheets issued by the material producers themselves. A given material (polyethylene, for example) can have a range of properties that derive from differences in the manufacturing processes of different producers. At the detailed design stage, a supplier should be identified and details of the properties of the product should be acquired for use in the design calculation. However, even this is not always good enough. If the component is a critical one (meaning that its failure could be disastrous) then it may be prudent to conduct in-house tests, measuring the critical property on a sample of the batch of material that will be used to make the product. The remainder of this article concerns the first level of data—the broad, low-precision compilation—and ways of presenting it that simplify the task of selection.

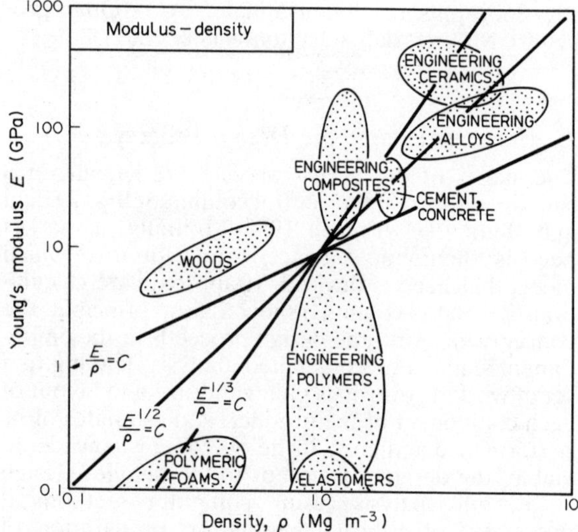

Figure 3
Schematic of a materials selection chart—Young's modulus E is plotted vs the density ρ, on logarithmic scales

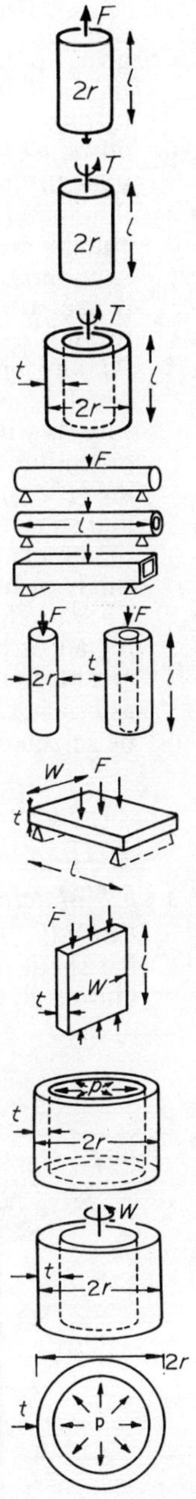

Mode of loading	Minimum weight for given: Stiffness	Ductile strength	Brittle strength
tie F, l specified r free	$\frac{E}{\rho}$	$\frac{\sigma_y}{\rho}$	$\frac{K_{IC}}{\rho}$
torsion bar T, l specified r free	$\frac{G^{1/2}}{\rho}$	$\frac{\sigma_y^{2/3}}{\rho}$	$\frac{K_{IC}^{2/3}}{\rho}$
torsion tube T, l, r specified t free	$\frac{G^{1/2}}{\rho}$	$\frac{\sigma_y^{2/3}}{\rho}$	$\frac{K_{IC}^{2/3}}{\rho}$
bending of rods and tubes F, l specified r or t free	$\frac{E^{1/2}}{\rho}$	$\frac{\sigma_y^{2/3}}{\rho}$	$\frac{K_{IC}^{2/3}}{\rho}$
buckling of slender column or tube F, l specified r or t free	$\frac{E^{1/2}}{\rho}$		
bending of plate F, l, w specified t free	$\frac{E^{1/3}}{\rho}$	$\frac{\sigma_y^{1/2}}{\rho}$	$\frac{K_{IC}^{1/2}}{\rho}$
buckling of plate F, l, w specified t free	$\frac{E^{1/3}}{\rho}$		
cylinder with internal pressure p, r specified t free	$\frac{E}{\rho}$	$\frac{\sigma_y}{\rho}$	$\frac{K_{IC}}{\rho}$
rotating cylinder w, r specified t free	$\frac{E}{\rho}$	$\frac{\sigma_y}{\rho}$	$\frac{K_{IC}}{\rho}$
sphere with internal pressure p, r specified t free	$\frac{E}{(1-\nu)\rho}$	$\frac{\sigma_y}{\rho}$	$\frac{K_{IC}}{\rho}$

Figure 4
The property combinations which determine performance in minimum weight design. For minimum cost design, ρ is replaced by ρC_R where C_R is the relative cost of the material

2. Materials Selection

As has been mentioned, it is important to start the design process with the full menu of materials in mind; failure to do so may mean a missed opportunity. The immensely wide choice is narrowed, first, by primary constraints dictated by the design and then by seeking the subset of materials which maximize the performance of the components. One way of doing this quickly and effectively is with materials selection charts.

The idea behind the charts is illustrated in Fig. 3. One material property (the Young's modulus E in this case) is plotted against another (the density ρ) on logarithmic scales. It is found that data for a given class of materials cluster together; they can be enclosed in a single "balloon." The balloon is constructed to enclose all members of the class—even those not explicitly listed on the chart. The result displays, in an accessible way, data for E and ρ for all materials.

Primary constraints in materials selection are imposed by characteristics of the design of a component which are nonnegotiable: the temperature and environment to which it is exposed, its weight, its cost and so on. If these are specified, all but a subset of materials that satisfy these constraints can be eliminated. A primary constraint corresponds to a horizontal or vertical line on the diagram: all materials to one side of the line can be rejected.

Further narrowing of the choice of materials available is achieved by seeking the combination of properties which maximize the performance of the component. For most common load-bearing components performance is limited, not by a single property, but by a combination of properties. The lightest tie rod which will carry a given axial load is that with the greatest value of σ_y/ρ (where σ_y is the yield strength). The lightest column which will support a given compressive load without buckling is that with the greatest value of $E^{1/2}/\rho$. The best material for a spring, regardless of its shape or the way it is loaded, is that with the greatest value of σ_y^2/E. Ceramics with the best thermal shock resistance are those with the largest value of $\sigma_f/E\alpha$ (where σ_f is the fracture stress and α is the thermal coefficient of expansion). There are numerous such combinations, depending on the application. Those for some simple loading geometries are listed in Fig. 4; there are many more.

The charts can be used to select materials which maximize any one of these combinations (figures of merit). Figure 3 shows E plotted against ρ on log scales. The condition

$$E/\rho = C$$

or, taking logs,

$$\log E = \log \rho + \log C$$

is a set of straight parallel lines of gradient 1, one line for each value of the constant C. The condition

$$E^{1/2}/\rho = C$$

gives another set, this time with gradient 2, and

$$E^{1/3}/\rho = C$$

gives another set, with gradient 3. It is now easy to read off the materials which are optimal for each loading geometry—assuming, of course, that nothing else (e.g., corrosion resistance) matters. If a straight line is laid parallel to the $E^{1/2}/\rho = C$ line, all the materials which lie on the line will perform equally well as a light column loaded in compression; those above the line are better, those below are worse. If the straight line is translated towards the top left corner of the diagram, the choice narrows. At any given position of the line, two materials which lie on that line are equally good, and only the subsets which remain above it are better. The same procedure, applied to the tie (E/ρ) or plate in bending ($E^{1/3}/\rho$), leads to different equivalences and optimal subsets of materials.

In mechanical design there are 12 properties that, singly or in combination, usually limit performance. They are listed in Table 1; charts exist (Ashby 1989) for all of them, combined in the ways which occur most frequently. Four of the charts will now be used as brief examples of the way they allow materials to be selected for particular applications.

3. Examples of Chart Usage

3.1 Materials for Table Legs

Consider a simple lightweight table consisting of a flat sheet of toughened glass supported on slender, unbraced, cylindrical legs (see Fig. 5). The legs must

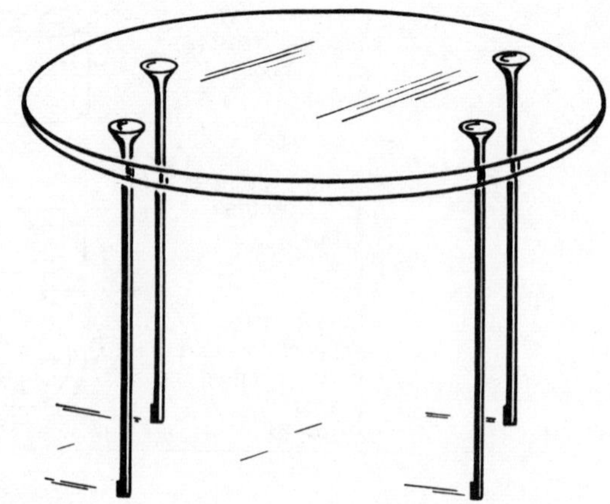

Figure 5
Lightweight table with slender, cylindrical legs

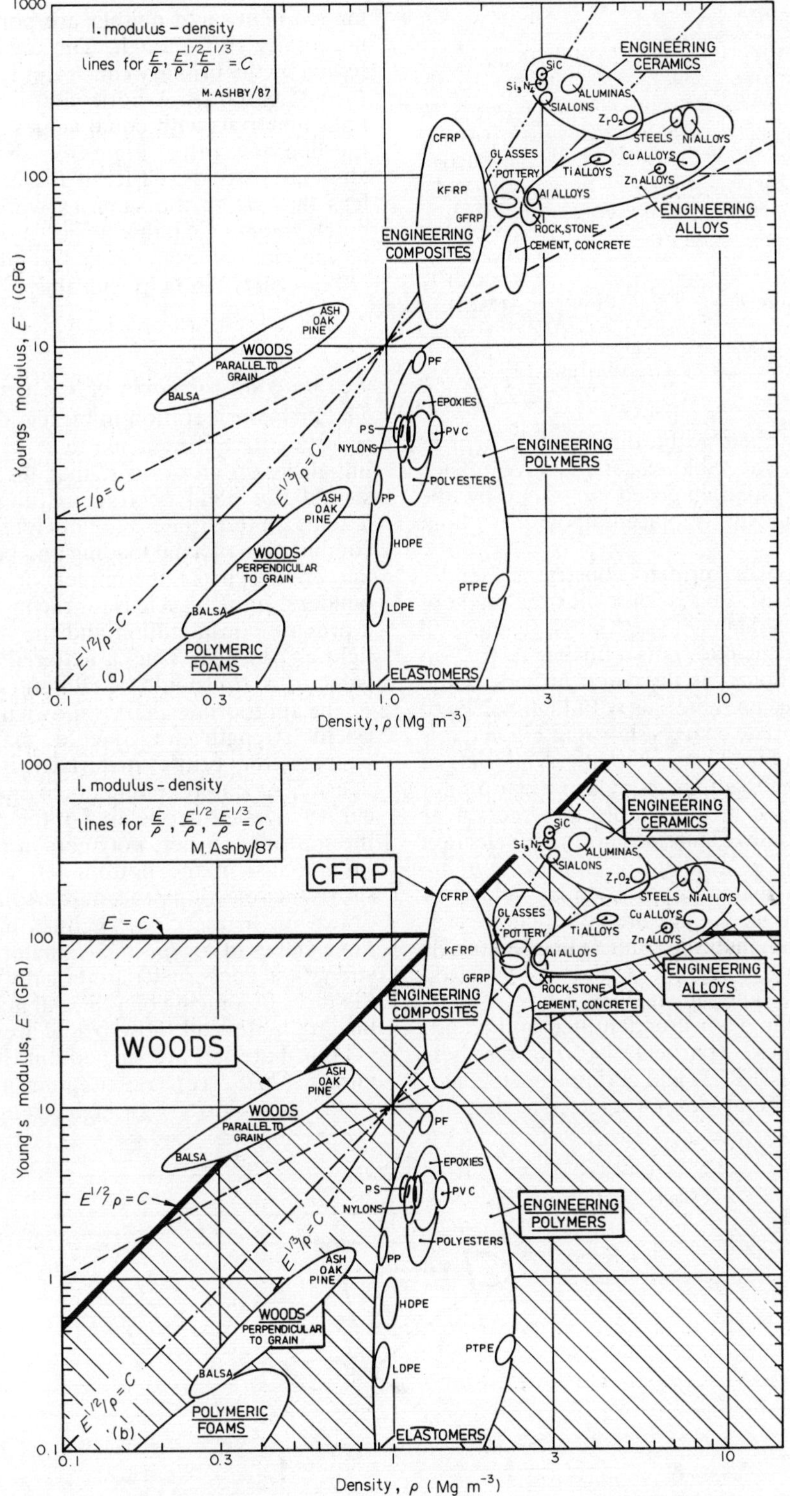

Figure 6
Materials selection chart: (a) Young's modulus vs density and (b) showing the selection of materials for light, slender columns

Table 1
Basic subset of material properties

Relative cost, C_R	(—)
Density, ρ	(Mg m^{-3})
Young's modulus, E	(GPa)
Strength, σ_y	(MPa)
Fracture toughness, K_{IC}	(MPa m$^{-1/2}$)
Damping coefficient, tan (δ)	(—)
Thermal conductivity, λ	(W mk^{-1})
Thermal diffusivity, a	(m^2 S^{-1})
Thermal expansion, α	(K^{-1})
Strength at temperature T, σ_T	(MPa)
Wear rate, W/A	(—)
Corrosion resistance	(—)

be solid (to allow them to be thin) and as light as possible (to make the table easier to move). They must support the load imposed on them by the tabletop and whatever is placed upon it without buckling.

Slenderness imposes a primary constraint. Slender columns must be stiff; that is, they must be made of a material with a high E. Lightness, while still supporting the design load, puts a further restriction on the material choice: as discussed in Sect. 2 the choice should focus on materials with high values of $E^{1/2}/\rho$. The appropriate chart is shown in Fig. 6a; it is the chart of which Fig. 3 is a schematic. Materials of a given class cluster together: metals in the top right corner, composites near the middle, polymers near the bottom, and so on. Figure 6b shows the selection procedure. A line of gradient $\frac{1}{2}$ is drawn on the diagram; it links materials with equal values of $E^{1/2}/\rho$. Materials above the line are better choices for this application than materials on the line; materials below the line are worse. The line is displaced upwards until a reasonably small selection of materials is left above it. They are identified on Fig. 6b: woods, composites (particularly carbon-fiber-reinforced plastic (CFRP)) and certain special engineering ceramics. Metals are not selected as they are far too heavy and neither are polymers as they are not nearly stiff enough. The choice is further narrowed by the primary constraint for slenderness that E must be large. A horizontal line on the diagram links materials with equal values of E; those above the line are stiffer. Figure 6b shows that this now eliminates woods. CFRP is the best choice: it gives legs that weigh the same as wooden ones but are much thinner. At this stage, other aspects of the design must be examined: strength, cost and so on. Other charts can help with this.

3.2 Materials for Forks of a Racing Bicycle

The first consideration in bicycle design (see Fig. 7) is strength. Stiffness matters, of course, but the initial design criterion is that the frame and forks should not yield or fracture in normal use. The loading on the frame is not obvious: in practice it is a combination of axial loading and bending. The loading on the forks is simpler: it is predominantly bending. If the bicycle is for racing then the weight is a primary consideration and the forks should be as light as possible. Thus a material with the greatest value of $\sigma_y^{2/3}/\rho$ should be chosen (see Fig. 4).

The appropriate chart is shown in Fig. 8a: strength (yield strength for ductile materials, crushing strength for brittle materials) is plotted against density. As before, members of one class of material cluster together in one area of the chart: metals near the top right corner, polymers in the middle, structural foams in the bottom left corner. Figure 8b shows the selection procedure. A line of gradient $\frac{2}{3}$ is drawn on to the chart; it links materials with the same values of $\sigma_y^{2/3}/\rho$, that is, materials that (as far as strength is concerned) are equally good for making the forks of a racing bicycle. All materials above the line are better; all those below are worse.

Four materials are singled out: high-strength aluminum (7075, T6) and titanium alloys are equally good; Reynolds 531 (a high-strength steel popular

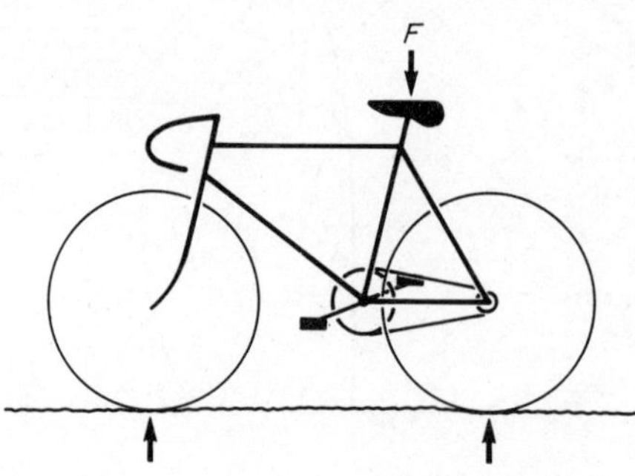

Figure 7
Bicycle with forks loaded in bending

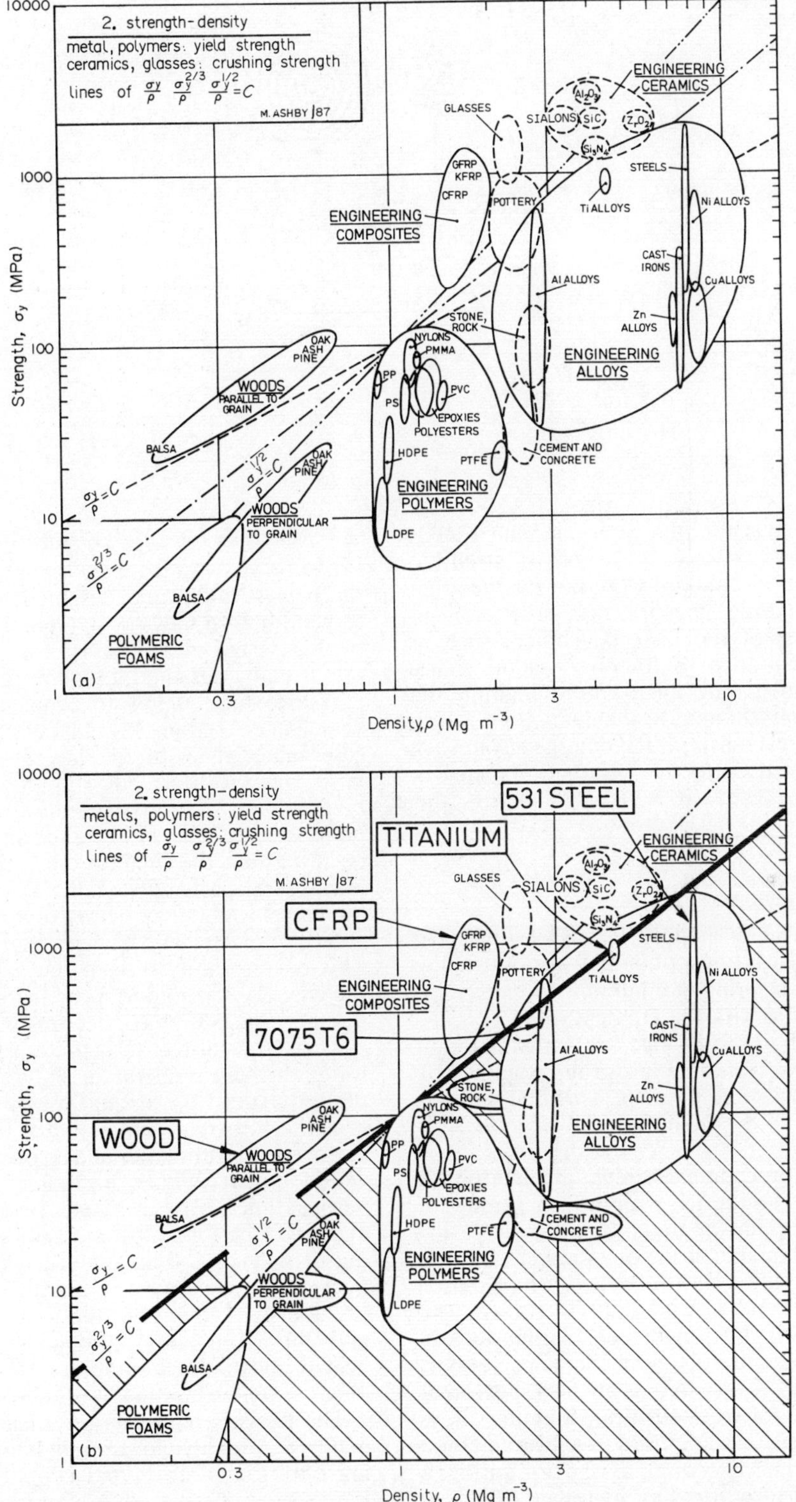

Figure 8
Materials selection chart: (a) strength against density and (b) showing the selection of materials for bicycle forks

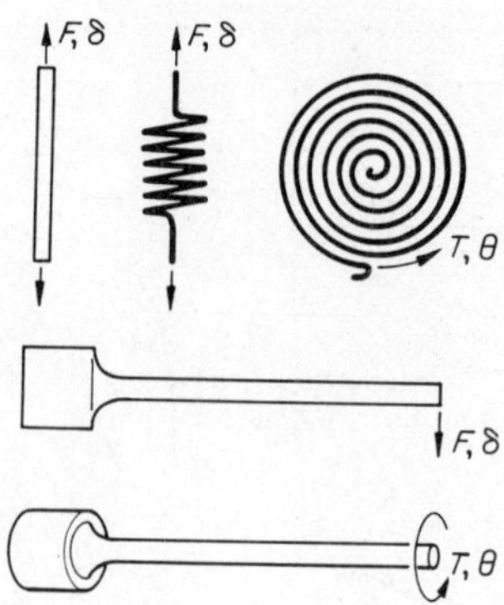

Figure 9
Springs in loading

for bicycle frames) is not quite as good; and CFRP is definitely the best choice in terms of strength. (Wood would be feasible if it were not for the great bulk required.) At this stage it is necessary to examine other aspects of the material choice, such as stiffness and resistance to fracture (again, charts exist which can help with this), and to examine the cost of fabrication (though, to the committed racing cyclist, cost is irrelevant). CFRP emerges from such an analysis as an attractive, though expensive, choice; and, of course, it is used in exactly this application.

3.3 Materials for Springs

Springs come in many shapes (see Fig. 9) and have many purposes. Regardless of their shape or use, the best material for a spring of minimum volume (e.g., for a watch) is that with the greatest value of σ_y^2/E. This result will be used in Fig. 10a with E plotted against σ_y. As always, materials group together by class, though with some overlap. This diagram has many uses; one is the identification of good materials for springs.

The selection procedure is shown in Fig. 10b. A line of slope $\frac{1}{2}$ links materials with the same values of σ_y^2/E. As the line is moved to the right (to increasing values of σ_y^2/E) a smaller selection of materials is left available. The result is shown as the unshaded area in the figure with candidate materials identified. The best choices are a high-strength steel (spring steel, in fact) and, at the other end of the line, rubber. However, certain other materials are also suggested: CFRP (now used for truck springs), titanium alloys (good but expensive), glass (used in galvanometers) and nylon (children's toys often have nylon springs). Note how the procedure has identified the best candidates from almost every class: metals, glasses, polymers and composites.

3.4 Materials for Safe Pressure Vessels

Pressure vessels, from the simplest aerosol can to the biggest boiler, are designed, for safety reasons, to yield before they break (see Fig. 11). The details of this design method vary. Small pressure vessels are usually designed to allow general yield at a pressure still too low to propagate any crack the vessel may contain ("yield before break"); in this case materials with the largest possible value of κ_{IC}/σ_y are the best choice as they will tolerate the biggest flaw. With large pressure vessels this may not be possible; instead, safe design is achieved by ensuring that the smallest crack that will propagate unstably has a length greater than the thickness of the vessel wall ("leak before break") and so the best choice of material is one with a large value of κ_{IC}^2/σ_y. This covers safety though the actual pressure that the vessel can hold is proportional to σ_y, so the designer seeks to maximize this too.

These selection criteria are most easily applied by using the chart shown in Fig. 12a which has κ_{IC} plotted against σ_y. Strong, tough materials lie towards the top right corner; hard, brittle materials in the bottom right corner and so on. The three criteria would appear as lines of gradient 1 and $\frac{1}{2}$, and as a vertical line. Taking "yield before break" as an example (see Fig. 12b), a diagonal line corresponding to $\kappa_{IC}/\sigma_y = C$ links materials with equal performance; those above the line are better. This line excludes everything but the toughest steels, copper and aluminum alloys, though some polymers are nearly above the line (pressurized lemonade and beer containers are made of these polymers).

The pressure constraint vertical line excludes all materials with a yield strength below 100 MPa, leaving only tough steels and copper alloys available. Large pressure vessels are always made of steel. Models (e.g., a model steam traction engine) are made from copper; it is favored, in such small-scale

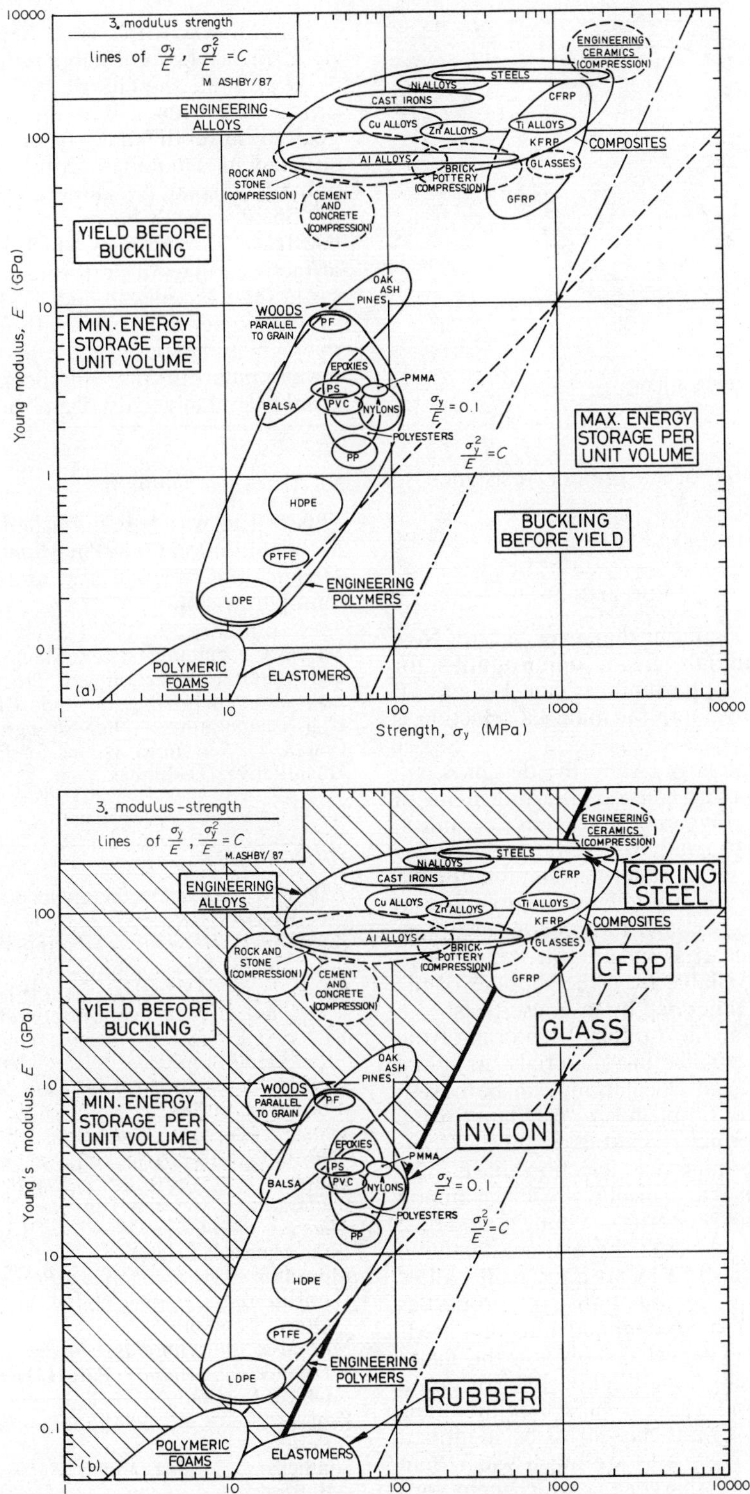

Figure 10
Materials selection chart: (a) Young's modulus against strength and (b) showing the selection of materials for springs

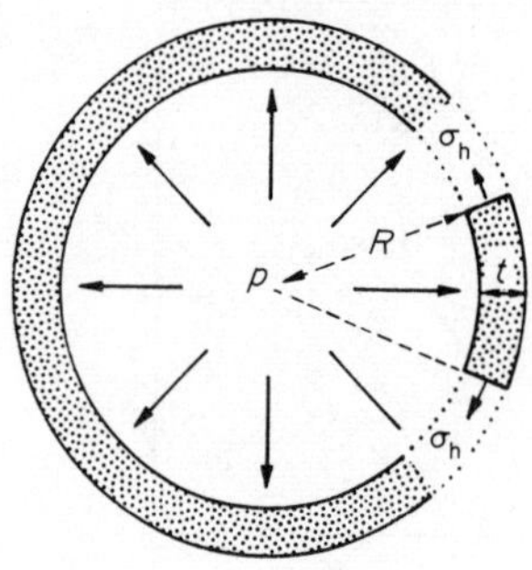

Figure 11
Pressure vessel containing a flaw

applications, because of its greater resistance to corrosion.

4. Conclusions

Materials are evolving faster than ever before. New and improved materials create opportunities for innovation. These opportunities can be missed unless a rational procedure for material selection is followed.

At the conceptual stage, while the design is still fluid, the designer must consider the full menu of materials: metals, polymers, elastomers, ceramics, glasses and numerous composites. Material data for a single class or subset of materials (suitable for the embodiment stage) are available in handbooks and computerized databases, and the precise, complete data for a single material (needed at the detailed design stage) are available from the supplier of the material or can be generated by in-house tests.

The first step is the most difficult: choosing from the vast range of engineering materials an initial subset on which design calculations can be based. One approach to this problem is to use data for the mechanical and thermal properties of all materials presented as a set of materials selection charts. The axes are chosen to display the common performance-limiting properties: Young's modulus, strength, toughness, density, thermal conductivity, wear rate and so on. The logarithmic scales allow performance-limiting combinations of properties (e.g., $E^{1/2}/\rho$ or σ_y^2/E) to be examined and compared.

The examples given in Sect. 3 show how the charts give a broad overview of material performance in a given application and allow a subset of materials (often drawn from several classes) to be identified quickly and easily. Their uses are much wider than those shown here and charts exist which help with problems of dynamics, heat transfer, thermal stress, wear and cost. They also help in finding a niche for new materials: plotted on to the charts, the applications in which the new material offers superior performance becomes apparent.

At present, the charts are hand drawn as in this article. However, it is an attractive (and attainable) goal to store the data from which they are constructed in a database coupled to an appropriate graphics display to allow charts with any combination of axes to be presented, and to be able to construct on them lines which isolate materials with attractive values of performance-limiting properties (as in the examples in Sect. 3) leading to a printout of candidate materials with their properties. A microcomputer-based system of this sort is under development in the Engineering Department of Cambridge University, Cambridge, UK.

5. Acknowledgement

This article was first published by the Institute of Metals, London (1989) in *Materials and Engineering Design: The Next Decade*, and is reprinted here by kind permission.

See also: Computer-Aided Materials Selection; Design and Materials Selection and Processing; High-Pressure Gas Containers: Safety and Efficiency (Suppl. 2); High-Temperature, High-Strength Solids: Selection (Suppl. 1); Selection Systems for Engineering Materials: Quantitative Techniques

Bibliography

Ashby M F 1989 On the engineering properties of materials. *Acta Metall.* 37: 1273–94

ASM Metals Handbook, 8th edn. 1973 American Society for Metals, Metals Park, OH

Crane F A A, Charles J A 1984 *Selection and Use of Engineering Materials*. Butterworth, London

Dieter G E 1983 *Engineering Design, A Materials and Processing Approach*. McGraw-Hill, London

French M J 1985 *Conceptual Design for Engineers*. Springer, Berlin

Fulmer Materials Optimiser 1974 Fulmer Research Institute, Stoke Poges, UK

Harper C A (ed.) 1975 *Handbook of Plastics and Elastomers*. McGraw-Hill, New York

Materials Engineering 1988 special annual issue containing the Materials Selector

Morrell R 1985 *Handbook of Properties of Technical and Engineering Ceramics*, Part I. Her Majesty's Stationery Office, London

Morrell R 1987 *Handbook of Properties of Technical and Engineering Ceramics*, Part II. Her Majesty's Stationery Office, London

Pahl G, Beitz W 1984 *Engineering Design*. Springer, Berlin

Smithells C J 1984 *Metals Reference Book*, 6th edn. Butterworth, London

M. F. Ashby
[University of Cambridge, Cambridge, UK]

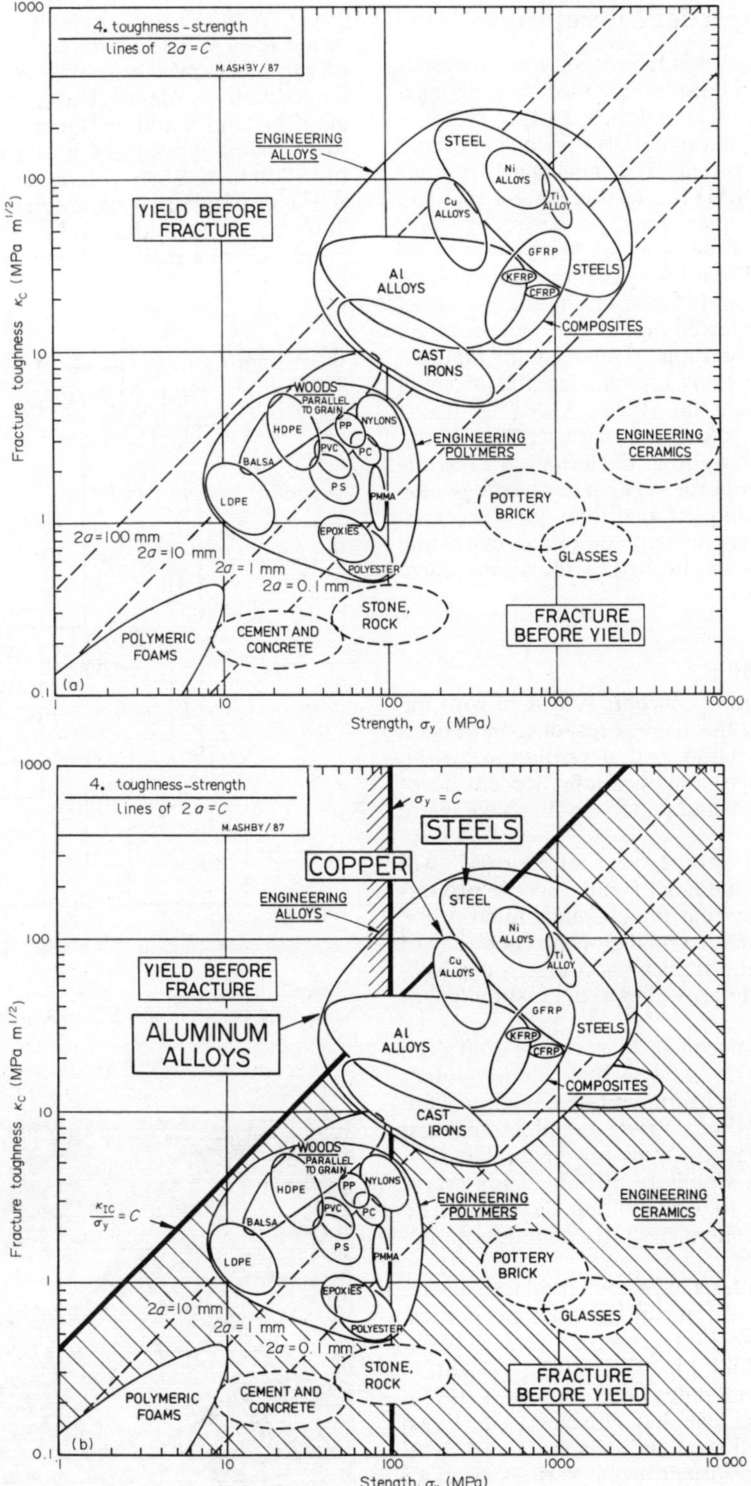

Figure 12
Materials selection chart: (a) fracture toughness against strength and (b) showing the selection of materials for a "yield before break" criterion

Mechanical Properties Microprobe

The relationships between the response of materials to imposed forces and displacements define the mechanical properties of that material. Tests to quantify these properties so that materials may be compared are numerous and varied. The mechanical properties microprobe (MPM) is a testing system that can measure some of the mechanical properties of a very small volume of material. The key to a system being designated as an MPM is the spatial resolution of the test. To qualify as an MPM the minimum volume of material that can be tested should be no greater than 1 μm in all three dimensions. Thus systems that test large areas of thin films or long lengths of small diameter filaments are not MPMs.

In terms of the volume of material tested, the MPM lies between two other devices used to determine mechanical properties. The microhardness tester is used to test volumes larger than those accessed by the MPM. Smaller contacts, sometimes consisting of only a few atoms, are probed by the atomic force microscope (AFM).

1. Physical Capabilities

An MPM is an exciting concept. A system with the ability to sample the mechanical response of a specimen with submicrometer spatial resolution has an extremely wide range of applications. Recent developments in hardware and understanding have made this goal a reality.

The mechanical properties microprobe will advance our understanding of macroscopic properties in the same way that the chemical microprobe has improved our understanding of the chemistry of the materials and the transmission electron microscope (TEM) has improved our understanding of structures.

Only one type of mechanical test can be used as an MPM. Gilman (1973) showed that a microindentation test, if properly instrumented and controlled, can perform as an MPM. During an indentation test both elastic and plastic strains are generated. For many materials the strains from both fields are of sufficient magnitude to be measured. The test can be scaled down so that submicrometer volumes of material are sampled.

Microindentation tests have been used to measure an extremely wide variety of material properties. Yield strength (Tabor 1951), creep resistance (Westbrook 1957, Chu and Li 1977), stress relaxation (Chu and Li 1980), modulus (Bulychev et al. 1975), fracture toughness (Lawn and Wilshaw 1975, Lawn and Evans 1977) and even fatigue tests (Li and Chu 1979) have been performed using various types of indentation tests. In addition, because the material being tested is close to a surface, environmental effects can be measured (Westwood and Macmillan 1973). With the proper combination of microindentation tests a nearly complete constitutive equation of the mechanical response of the sample can be mapped out. Strength (Pethica et al. 1983, Oliver et al. 1984) and modulus (Doerner and Nix 1986) are two properties that are now routinely determined from submicrometer volumes.

The experimental parameters associated with an indentation test that relate to the mechanical properties just mentioned include the force on the

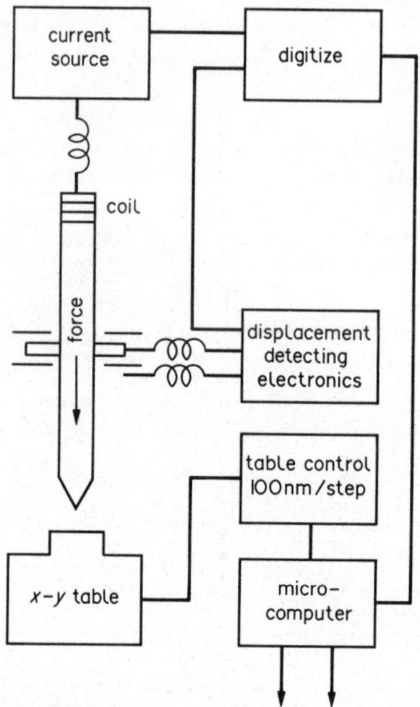

Figure 1
Schematic diagram of the Nanoindenter

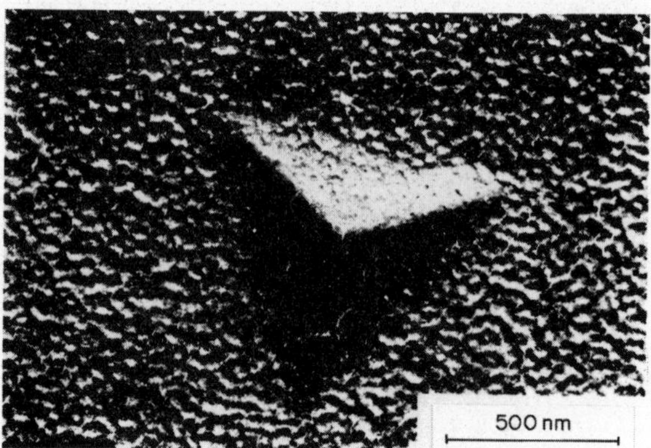

Figure 2
TEM image of a replica of a 200 nm deep indent in nickel

indenter, the geometry of the indenter, the temperature (and other important environmental parameters), the mechanical properties of the indenter and the time dependence of all the parameters. In addition, for special applications, other parameters must be added to this list (e.g., the stiffness or electrical properties of the contact).

One of the more difficult parameters to measure is the geometry of the indent. The contact area between the sample and indenter under load is the most important geometric aspect of an indentation test. The final area of a standard microhardness test is measured optically after the indenter is removed, and the assumption is made that the area does not change on unloading. Although imaging indents does give a direct measure of the most important aspect of the geometry, it becomes more difficult as the size of the indent is reduced. Submicrometer-sized indents can only be imaged using electron microscopy. The techniques used are time consuming and only yield the final size of the indent.

One geometric characteristic of the indent that is more easily measured and can be measured continuously during the entire indentation process is the displacement of the indenter after contact. This measurement provides several other distinct advantages over direct area measurement. These include the ability to sample both elastic and plastic strains, the ability to control and monitor stress and strain rates, and finally the elimination of the need for complicated, time-consuming imaging techniques. The displacement can be measured with sufficient resolution to characterize extremely small indents; however, models of the indentation process must be employed to allow the contact area to be calculated. These models have been developed and used successfully (Bulychev et al. 1975, Oliver et al. 1984).

2. *Mechanics*

A schematic of an MPM is shown in Fig. 1. The force is applied to the indenter electromagnetically.

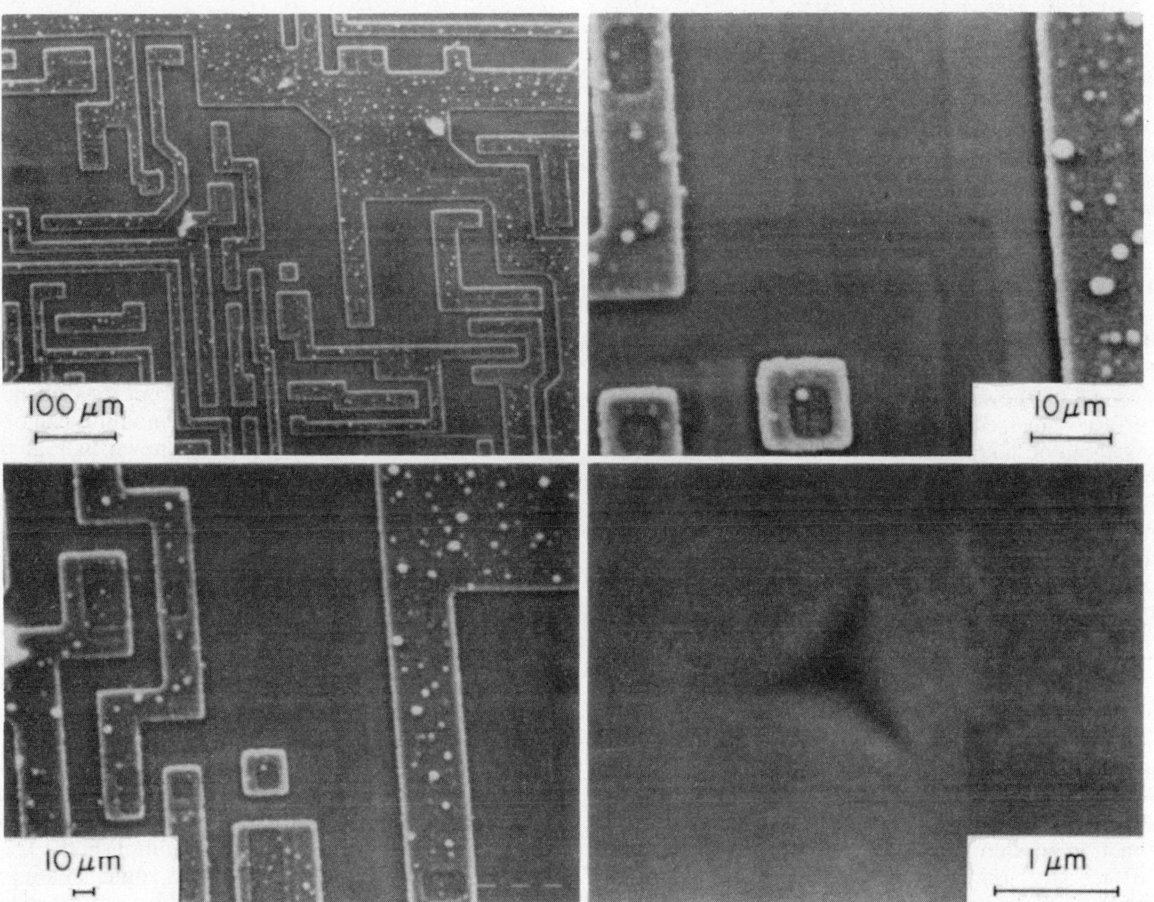

Figure 3
Two indents in a microcircuit demonstrating the fine scale of the experiment

The displacement-sensing system consists of a capacitive displacement gauge. In this case, the movable portion of the gauge is fixed to the shaft connecting the force-application system to the indenter. The whole assembly is supported by leaf springs (not shown). The sample is moved by a motorized *x–y* table. All aspects of the system are controlled by a microcomputer. The entire process is automated so that a set of indentation experiments is performed from initiation to completion without operator supervision.

The resolutions of the various components of the system depicted are carefully matched to perform as an MPM. The force application of this particular system has two ranges, 0–20 mN and 0–120 mN, with resolutions of 300 nN and 1.5 μN, respectively. The displacement sensing system has a resolution of 0.2 nm. The *x–y* table has a precision of 0.1 μm. Figures 2 and 3 show the scale of microindentations made with such a system. The image in Fig. 2 is 200 nm deep in nickel. This indent is a factor of ten larger than the smallest indents from which reasonably precise data can be obtained. The scale of the experiment relative to the scale of a microcircuit is shown in Fig. 3. The indents are clearly visible at the highest magnification.

Although a variety of indentation tests are useful, there are some experimental similarities common to all tests performed with such a system. First, the position of each indent on the surface of the specimen must be established. A combination of an optical microscope and the high accuracy *x–y* table allows this to be done with an accuracy of ±250 nm. The next step is to approach and contact the sample surface. This must be done with some care so that the kinetic energy of the indentation column does not create a significant indent when its motion is stopped by the surface. Once the indenter is on the surface, any sequence of loading and unloading steps can be taken during which the load, displacement, time and any other variables of interest are recorded. Finally, the indenter is unloaded and lifted from the surface. The *x–y* table is then moved and the process is repeated until the desired number of indentations are performed. Figure 4 shows a typical set of data from a relatively simple indentation test.

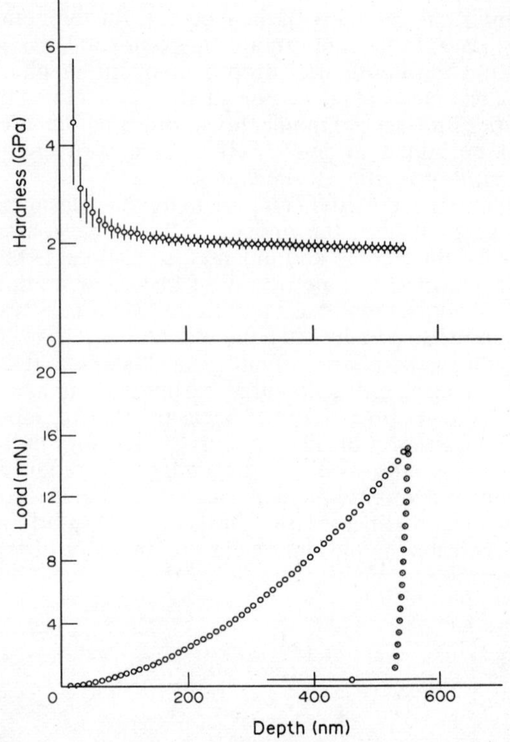

Figure 4
Typical load vs displacement and hardness vs displacement curves for electropolished nickel

3. Applications

Some examples of how the MPM has been applied to material research problems are shown in Figs. 5–11. Figure 5 shows the change in hardness as a function of depth associated with an amorphous surface layer on sapphire caused by ion bombardment. Surface treatments such as ion bombardment are being considered for high-wear applications. Fracture and wear rates associated with pin-on-disc wear tests are reduced by the presence of these layers (McHargue 1987). These treatments extend in depth only a few 100 nanometers; hence, it is very difficult to sample the mechanical properties of the treated layer. Both the hardness and modulus can be determined in such thin layers as shown in Fig. 6 (Oliver et al. 1987a). The decreased modulus shown in Fig. 6 and the decreased hardness shown in Fig. 5 help to explain the wear properties of this material. These data show how microindentation tests can be used to characterize thin films.

Another area where the MPM will have an impact is rapidly solidified materials (see *Crystalline Alloys, Rapidly Quenched*, Suppl. 1; *Rapid Quenching from the Melt: Formation of Metastable Crystalline Phases*, Suppl. 1). The product of rapid-solidification processing (i.e., powder, ribbon or splats) has at least one small dimension. Normally, to test the properties of these materials they must first be consolidated into large enough products to allow macroscopic tests to be accomplished. The consolidation step often involves some heating; hence, the as-solidified microstructure can be disturbed. Figure 7 shows a series of indentations made in the edge of a ribbon of rapidly solidified Al–Li–Be

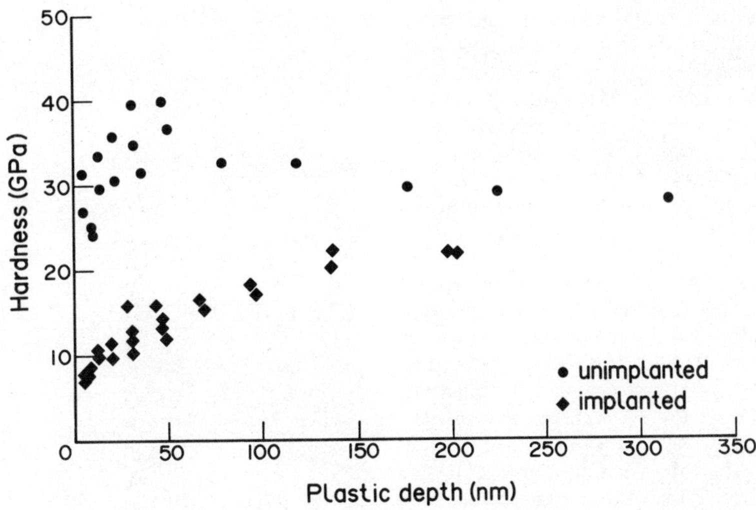

Figure 5
The change in hardness of a sapphire surface associated with an amorphous surface layer caused by ion implantation

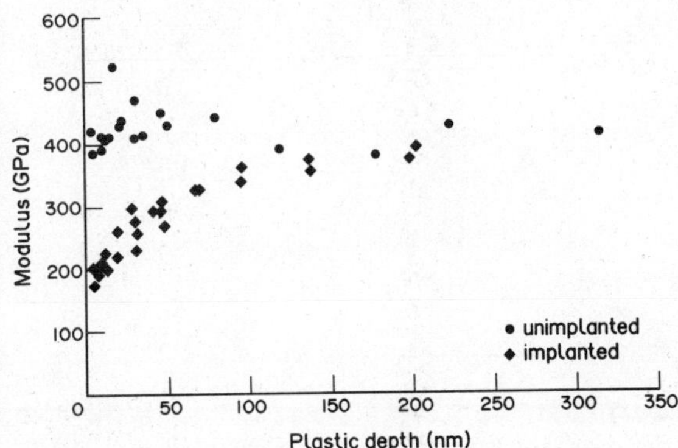

Figure 6
The change in modulus of the same ion-implanted sapphire surface shown in Fig. 5

alloy (Oliver et al. 1987b). The ribbon was approximately 50 μm thick. A study of the properties of this alloy after various annealing treatments indicates a 30% drop in strength after a 1 h anneal at 200 °C. The ability to characterize the mechanical properties of this alloy in the as-solidified state has generated further interest in processing techniques and alloy development.

In the area of welding, the MPM is being used to address the problem of weld embrittlement in duplex stainless steels. Figure 8 (David et al. 1987) shows some relatively large indents made in a specimen from a 308 austenitic stainless steel weld. The microstructure consists of an austenite matrix with ferrite islands a few micrometers in dimension. By studying the changes in the properties of both the

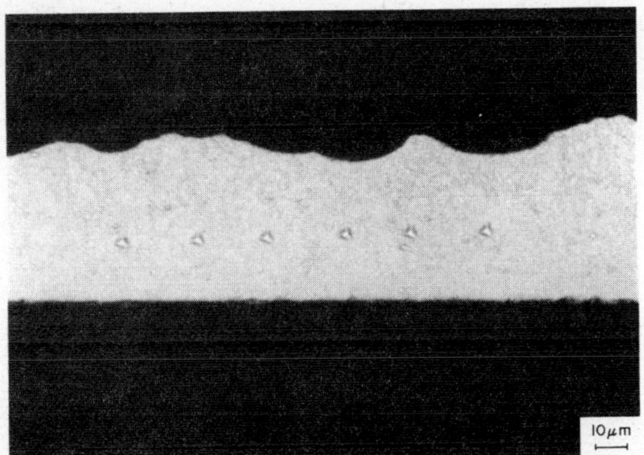

Figure 7
Indents in the edge of a rapidly solidified Al–Li–Be alloy ribbon

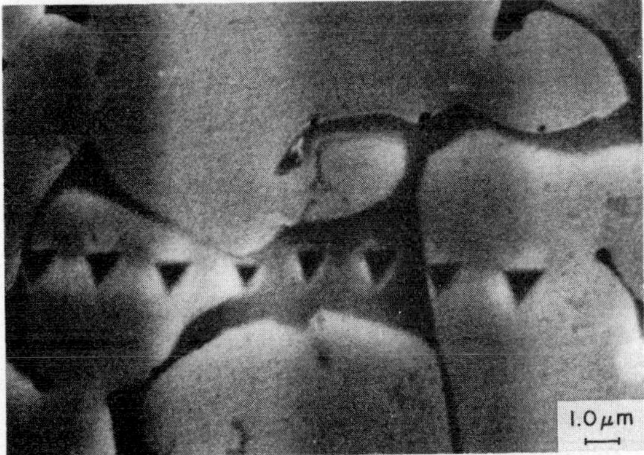

Figure 8
SEM image of indents traversing a ferrite island in a weld in a duplex stainless steel

austenite and ferrite as a function of aging treatment, the causes of the embrittling processes will be better understood.

Another example of how the MPM is applied to materials development problems involves radiation-damaged materials. Heavy ion bombardment has long been used to simulate damage structures associated with neutron irradiation. Such treatments only affect material within a few micrometers of the surface and have been extremely useful for structural studies. However, for mechanical-properties studies, macroscopic neutron-damaged specimens are typically required and are extremely expensive. Figures 9–11 (Zinkle and Oliver 1986) show how the MPM can be used to characterize the less expensive ion bombarded material. Figure 9 shows the result of indenting the damaged surface directly and comparing the resulting load-displacement curve with that same data obtained from an undamaged section of the specimen. Clearly, the effects of the damage are evident as a hardness increase near the surface. Curves of the type shown in Fig. 9 are very sensitive to surface properties; however, their interpretation is challenging owing to the uncertainty as to which region of material below the indenter is contributing most to the hardness at any given depth of penetration. Figures 10 and 11 are the result of using cross-sectional techniques to sample the surface layer of interest. The sample is prepared by plating the surface subsequent to the irradiation and then cross sectioning the resulting sandwich. A series of microindentation tests that are spatially progressed across the original interface are used to sample the material of interest in cross section. Figure 10 shows such a series, and Fig. 11 shows the resulting data.

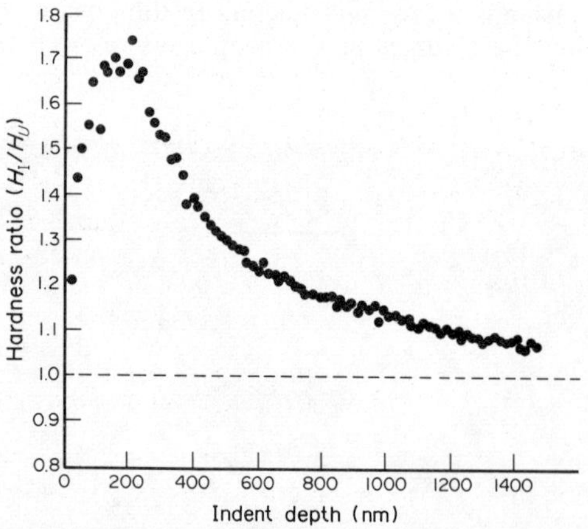

Figure 9
Hardness changes in copper due to radiation damage associated with heavy ion bombardment sampled by direct indentation of the treated surface
(4 MeV Fe^{++} + 50 at.ppm, peak damage = 15 dpa, T_{irr} = 220 °C)

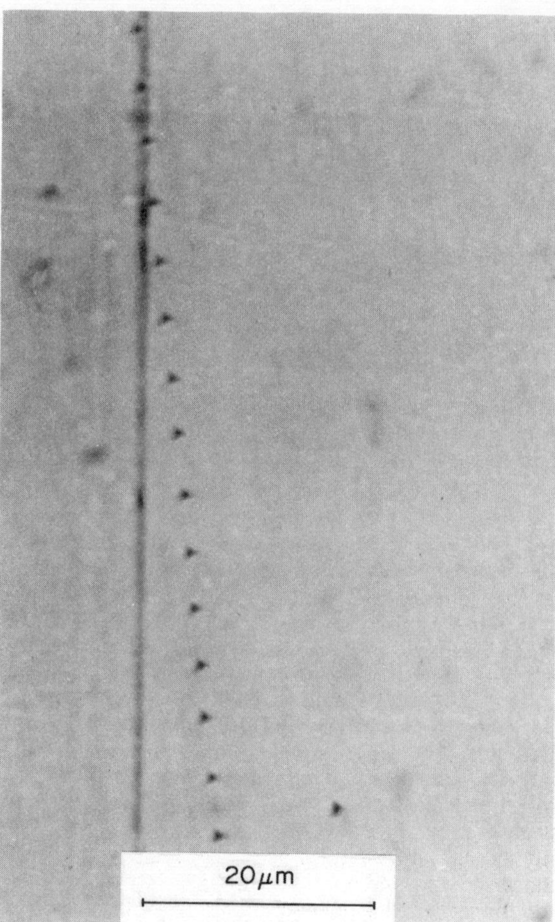

Figure 10
Series of indents approaching the original surface of a heavy ion damaged specimen that has been cross sectioned

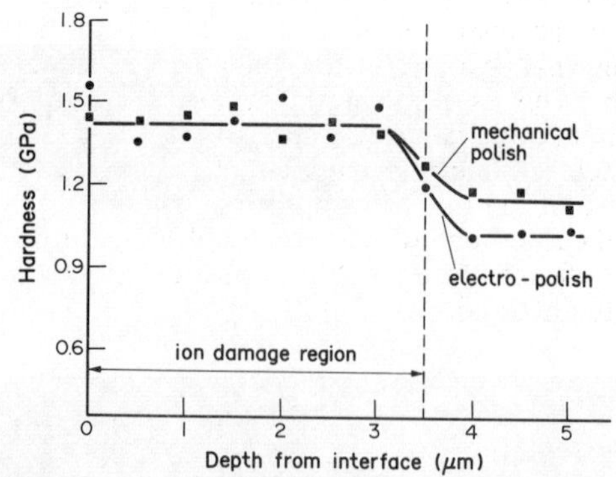

Figure 11
Results of the series of indents shown in Fig. 10: cross-section hardness of copper following 14 MeV copper-ion irradiation (indent depth = 150 nm, peak damage = 40 dpa, T_{irr} = 100 °C)

The experiment resulting in Fig. 9 is much simpler to perform; however, the results of the cross-sectional technique are more easily interpreted. Models and numerical calculations to relate the two are being pursued. If such samples and tests can be used as an initial sorting tool for characterizing resistance to radiation damage, the cost of such research could be reduced by at least an order of magnitude.

4. *Descriptive Models*

Applications like the ones just discussed provide the driving force for the second area of research related to MPMs. This is the development of models to allow researchers to deconvolute the complicated geometry of indentation tests and extract useful mechanical properties.

One example of such an effort is demonstrated by Figs. 5 and 6 which have already been discussed. The models necessary to retrieve modulus and strength as a function of depth from load-displacement data require careful consideration. The data required are obtained from an indentation test consisting of a series of loading and unloading sequences going to progressively higher maximum loads. The displacement measurements made during loading sequences, in which the size of the indent is increased plastically, contain information about both the elastic and plastic strain fields. The unloading data represents the response of the elastic field only. Using information from both sequences of data the plastic and elastic components can be separated; hence, the hardness and modulus can be calculated for each point at which an unloading sequence begins (Doerner and Nix 1986).

One major effort to improve the physical capabilities of the system is to add temperature to the list of variables that can be controlled. The MPMs now available are limited to ambient temperatures because the displacement measured is sensitive to the thermal expansion of the load application system. This is particularly important when displacements are being measured to a fraction of a nanometer. Clearly the interest in varying the temperature of the indentation test is high. The technical difficulties involved in accomplishing this goal are significant if the resolution of the experiment is to be maintained.

5. *The Future*

Interest in the MPM is high. The duration of that interest will depend on the success such systems have in contributing to the solution of important technological questions. The examples outlined here show the diversity of industries that can make use of the information the MPM generates. If a fraction of such cases result in real progress, the future of MPMs is assured. Such systems will become another important tool available to all material scientists.

Acknowledgments

The author would like to acknowledge the sponsoring of this work by the Division of Materials Sciences, US Department of Energy, under contract DE-AC05-840R21400 with Martin Marietta Energy Systems, Inc.

See also: Hardness Characterization

Bibliography

Bulychev S I, Alekhin V P, Shorshorov M K H, Ternovskii A P, Shnyrev G D 1975 Determining Young's modulus from the indentor penetration diagram. *Zavod. Lab.* 41(9): 1137–9

Chu S N G, Li J C M 1977 Impression creep; A new creep test. *J. Mater. Sci.* 12: 2200–8

Chu S N G, Li J C M 1980 Localized stress relaxation by impression testing. *Mater. Sci. Eng.* 45: 167–71

David S A, Vitek J M, Keiser J R, Oliver W C 1987 Nanoindentation microhardness study of low-temperature ferrite decomposition in austenitic stainless steel welds. *Weld. J.* 66(8): s235–s240

Doerner M F, Nix W D 1986 A method for interpreting the data from depth-sensing indentation instruments. *J. Mater. Res.* 1(4): 601–9

Gilman J J 1973 Hardness—A strength microprobe. In: Conrad H, Westbrook J H (eds.) 1973 *American Society for Metals*. ASM, Metals Park, OH, pp. 51–72

Lawn B R, Evans A G 1977 A model for crack initiation in elastic/plastic indentation fields. *J. Mater. Sci.* 12: 2195–9

Lawn B R, Wilshaw T R 1975 *Fracture of Brittle Solids*. Cambridge University Press, London

Li J C M, Chu S N G 1979 Impression fatigue. *Scr. Metall.* 13: 1021–6

McHargue C J 1987 The mechanical and tribiological properties of ion implanted ceramics. In: Mazzoldi P, Arnolds G W (eds.) 1987 *Ion Beam Modification of Insulators*. Elsevier, New York, pp. 223–44

Oliver W C, Hutchings R, Pethica J B 1984 Microindentation techniques in materials science. In: Blau P J, Lawn B R (eds.) 1984 American Society for Testing and Materials Special Technical Publication 889. ASTM, Philadelphia, PA, p. 303

Oliver W C, McHargue C J, Zinkle S J 1987a Thin film characterization using a mechanical properties microprobe. *Thin Solid Films* 153: 185–6

Oliver W C, Wadsworth J, Nieh T G 1987b Characterization of rapidly solidified Al–Be–Li and Al–Be ribbons. *Scr. Metall.* 21: 1429–33

Pethica J B, Hutchings R, Oliver W C 1983 Hardness measurement at penetration depths as small as 20 nm. *Phil. Mag. A* 48(A): 593–606

Tabor D 1951 *The Hardness of Metals*. Clarendon Press, Oxford
Westbrook J H 1957 Microhardness testing at high temperatures. *Proc. Am. Soc. Test. Mater.* 57: 873–95
Westwood A R C, Macmillan N H 1973 Environment-sensitive hardness of nonmetals. In: Conrad H, Westbrook J H (eds.) 1973 *American Society for Metals*. ASM, Metals Park, OH, pp. 377–417
Zinkle S J, Oliver W C 1986 Mechanical property measurements on ion-irradiated copper and Cu–Zr. *J. Nucl. Mater.* 143 (Nov): 548–52

W. C. Oliver
[Oak Ridge National Laboratory, Oak Ridge, Tennessee, USA]

Microengineering of Materials: Characterization

Targets for inertial-confinement fusion (ICF), must meet stringent requirements of dimension and composition (see *Microengineering of Materials: Laser Fusion Targets*, Suppl. 2). The methods used during and after the manufacture of these targets to determine their size, thickness, feature and defect location, and composition are described in this article.

Targets for ICF are compressed up to 1000 times by the rocket-like ablation of the outer layer of the target. The ablator layer is heated, and subsequently blown off, by a laser or a particle beam driver. The ablation layer must be of uniform thickness and density or it will burn through in the thin areas, drive the fuel asymmetrically and thus reduce the amount of fuel burned. The requirement is that fractional thickness variation be less than 0.1% (Weinstein 1982). Symmetric drive also requires homogeneous composition, because the absorption of the driver energy is highly dependent on the atomic number of the irradiated material. Defects as small as 10 nm high and 1 μm across can seed implosion instabilities that will grow and spoil the implosion symmetry. The fuel content of the target must be known, and if the fuel is either liquid or solid, the layer uniformity must be diagnosed. The techniques used to determine these and other target characteristics fall under the broad category of target characterization. Characterization methods in general belong to three categories: optical methods for transparent materials, and electron beam characterization and x-ray techniques for both opaque and transparent materials.

1. Optical Methods

1.1 Microscopes

The workhorse of the ICF target industry is the optical microscope used for assembly of ICF targets. Target builders use stereo microscopes with a working distance of at least 0.1 m to allow for free manipulation of the target without the risk of damage. A continuous zoom from approximately ×3 to ×60 allows the fabricator to keep all the aspects of the assembly in view and to focus closely on details when necessary.

Preliminary selection of target materials is generally accomplished on a binocular microscope, usually a higher quality instrument than an assembly microscope. A movable cross-hair or filar eyepiece is useful in sizing the target components. Alternatively a standard eyepiece can measure the shell diameter to about 0.5 μm. The target surface can be inspected with the same instrument by rotating it under the objective. Dark-field illumination using stain-free optics will detect surface defects and rough spots. Fine focus and a shallow depth of field allow inspection of the interior surfaces of transparent shells.

1.2 Interferometry

Interferometric techniques split light from a single source into two paths; one acts as a reference and the other interacts with the target under inspection. Upon recombination, the two beams of light no longer have the same phase because each beam has traversed a different optical path. (Optical path is defined as the product of the thickness and refractive index for each material that the light has passed through.) If the optical path difference of the two beams is within the coherence length of the source, the viewer or detector notes interference fringes that can be colored in the case of white-light interferometry or bands of high and low intensity in the case of a monochromatic source. The fringes are due to the coherent addition of the reference and target-altered wave fronts. Each fringe is a contour of points with equal phase difference between the two beams. Between successive fringes of the same color there is a wavelength of optical path difference. Electronic detection techniques can detect one-hundredth of a fringe in an interferogram. Thus the technique is extremely sensitive (5–50 nm) to optical path differences. Combining interferometry with microscopy can produce a depth resolution of about 500 nm.

(*a*) *White-light interferometry*. This has several important applications to ICF target characterization (Weinstein 1982). Its major use is in the measurement of the thickness and uniformity of transparent-target shell walls. The Jamin–Lebevedev interferometer is one of several that fit this category. Usually a microscope attachment, this interferometer displaces two images of the target and allows the operator to adjust the phase of one of the images relative to the other. The interference fringes that result when the path lengths are nearly equal are colored save for one, the black fringe of zero order. The unique black zeroth-order fringe allows absolute measurement of the optical path, which cannot be done as conveniently with monochromatic interferometry. To measure the thickness of a shell wall, the operator moves the black fringe from the background of the target image to the center of the target image using the phase control. In this way, the phase difference between rays traced through the target walls can be determined to 0.1 μm if the refractive index of the target-wall material is known. The complex general problem of the location of the fringes for light refracted at both the inner and outer surfaces of a transparent shell is reported by Stone et al. (1975).

At the same time that thickness is measured, wall-thickness uniformity can be judged by assessing the concentricity of the colored fringe system with the target surface. The eye can detect nonuniformities of about 3% by this method. A complete determination of the wall nonuniformity requires measurement in three orthogonal views. Two orthogonal views can be examined under the microscope without moving the target shell, by tilting the stage.

Bubbles or voids in the shell wall become centers of colored fringe systems and are easily detected with magnified white-light interferometry. Finally, the additional phase difference resulting from a fuel or diagnostic gas fill can be determined by comparing the optical path through the center of a filled target with the optical path through the same target before gas filling. The presence of the gas adds a term proportional to $P\,(\eta_{gas}-1)$ to the optical path difference (Powers 1982), where P is the gas pressure and η_{gas} is the refractive index of the gas. Hence, the pressure of the fill gas can be determined analytically.

(*b*) *Laser interferometry*. This can diagnose the symmetry of a liquid or frozen fuel layer inside an ICF target. In a simple arrangement, laser light illuminates the target from the back and passes through a wedge, or shearing cube (Tarvin et al. 1979), that displaces a pair of target images from one another. Light through the target in one image then interferes with light passing near the target in the other image. The reference field can be made flat to a quarter of a wave without much difficulty. The interference pattern shows circular fringes that are more-or-less concentric with the target image itself, depending on the symmetry of the fuel layer. The technique is similar to the white-light thickness measurement described above except that cryogenic targets require much longer working-distance optics. The thermal shields surrounding the cryogenic environment prevent placing instruments close to the target, making conventional double-arm interferometry difficult. Splitting the beam containing the target image with the shearing cube works well for this application.

Templates can be constructed by a computer showing the fringe system corresponding to specific fuel-layer nonuniformities (Miller and Sollid 1978). The target-chamber operator can compare the computed fringes with the physical ones to determine whether the cryogenic layer is acceptable for "shooting" or needs to be reprocessed (see *Microengineering of Materials: Laser Fusion Targets*, Suppl. 2, Sect. 1.3). For more precision, the interferogram can be digitized and analyzed by computer to extract the defect information (see Sect. 3.1 and Whitman 1982).

Holographic interferometry offers a more precise measurement of cryogenic layer uniformity (Bernat et al. 1982). With this method, a hologram of the target with the fuel in the gaseous state is made and compared with the liquid- or solid-layer target. Superposition of the real cold system with the hologram of the warmer symmetric system can be done with a double exposure or by viewing the cold system through the hologram in real time. The fringes produced are due to changes in the system between exposures. If motion can be eliminated between exposures, the fringes are due only to the change of state of the fuel. The only effect of the shell is to refract light out of the entrance pupil of the optical system; thus, a thick shell has an annulus inside the wall where there are no fringes. With thin shells, this method can show fuel-layer nonuniformities as small as 50 nm.

The outer surface of laser fusion targets has been completely characterized using sensitive interferometry and a method of manipulating the sphere through 4π rotation. Computer control of the rotation and data acquisition is essential so that the information can be gathered quickly. Two closely related instruments of this nature have been described (Cooper 1982, Monjes et al. 1982); although neither instrument made the transition to a production tool, both succeeded in their design goals of mapping a microsphere.

The 4π manipulator concept developed by Weinstein et al. (1978) places the target to be examined between two soft tips that are driven in precisely opposite directions. The tips can be coupled either by a microprocessor or mechanically by captive master spheres. Thus, when one tip moves, it rolls the spheres in contact with it. The other tip is driven by the opposite surface of the master spheres and moves in exactly the opposite direction. The target captured between the tips also rolls without translating. In this way the sphere can present 4π of surface to the interrogation beam of the interferometer. Figure 1 shows the 4π manipulator.

The interferometers for sphere surface characterization (Cooper 1982, Monjes et al. 1982) require sensitivity of better than $\lambda/50$ (~10 nm). To detect fringe shifts to this precision requires phase modulation. The phase of the reference beam of the

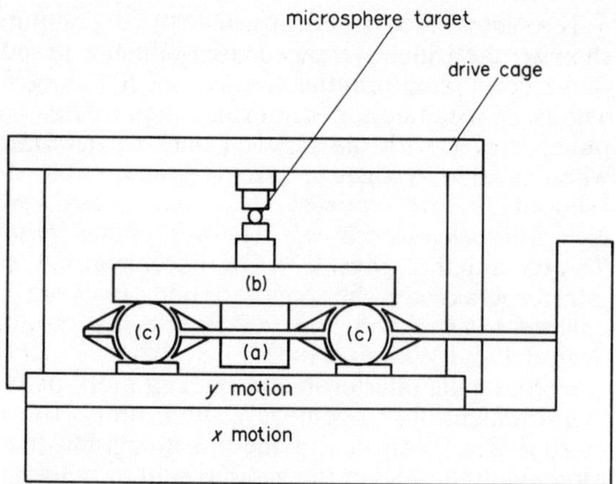

Figure 1
Schematic showing detail of 4π manipulator design: (a) magnetic loading, (b) driven plate and (c) captive master spheres

interferometer is modulated at some frequency ω_0, and the interrogating beam has some additional phase difference between the two beams. The conversion of the phase measurement to an amplitude measurement accounts for the high sensitivity. It is estimated that with a manipulator and interferometer of this type, a complete map of the surface of a 500 μm diameter sphere showing defects of 10 nm amplitude on a spatial scale of 1–10 μm takes 5 min.

2. Electron Beam Characterization of ICF Targets

Highly demanding requirements for composition, geometry and surface finish are common to the creation of targets for ICF. These requirements generally stem from the need to maintain symmetry and stability during the implosion–compression phase of each test. Targets for these experiments typically have overall dimensions less than 1 mm, intentional structures in the micrometer range and unwelcome features in the nanometer range. Such small scales required the use early on of the scanning electron microscope (SEM) and related electron beam techniques to characterize ICF targets.

2.1 Signals and Detector Types

The impact of accelerated electrons with matter results in a variety of emitted signals as enumerated in Table 1. Many SEMs have detectors capable of measuring these signals, and reaching a complete answer to an analytical question usually involves making full use of these signals and detectors. The most common use of the SEM is to produce images using the first three signals listed in Table 1. Secondary electrons are conventionally defined as those electrons emitted from the sample surface that have energies less than 50 eV. Such low energies imply that these electrons originate within a few nanometers of the surface. As a consequence, this signal has the highest spatial resolution of all the emitted signals.

Backscattered electrons are those incident electrons that undergo elastic collisions in the sample and thus retain energy near that of the original beam. The backscattered electron intensity depends on the average atomic number of the sample. This signal thus maps the composition of the surface. Resolution for elements from row four of the periodic table is approximately one-third of an atomic number.

Absorbed current consists of those electrons that are not emitted from the sample and are normally conducted to ground. The contrast in the absorbed current image is similar to the backscattered signal except inverted, bright areas in the backscattered image appear dark in absorbed current images.

The popularity of SEM images can be attributed to three basic characteristics. First, the images possess extraordinary depth of field. This is the result of the very small divergence of the electron beam in the SEM, typically less than 50 mrad. Second, high resolution results from the short wavelength of the electrons—roughly 100 000 times less than that of visible light. Third, the images have a lifelike three-dimensional quality leading to easy interpretation.

Table 1
Types of signals available in SEM and methods of detection

Emitted signal	Detector	Information conveyed
Secondary electrons	Scintillator/photomultiplier	Topography, dimensions
Backscattered electrons	Silicon diode	Topography, dimensions, atomic number
Absorbed current electrons	Current amplifier	Dimensions, atomic number
Auger electrons	Electron energy	Elemental analysis
Characteristic and bremsstrahlung x rays	Solid state detector or curved crystal spectrometer	Elemental analysis
Kossel diffraction x rays	Film	Crystallography
Cathodo-luminescence	Monochromator/photomultiplier	Chemistry, carrier lifetimes
Heat	Piezoelectric transducer	Subsurface mechanical defects

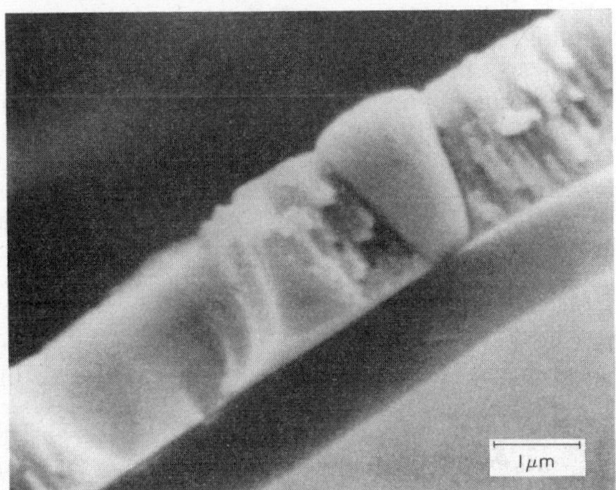

Figure 2
Magnified fracture cross section of a glass microballoon coated with gold; "cone defect" originated from contamination on glass microballoon

An application demonstrating the use of imaging is the origin of "cone-type defects". Irregular features have long been observed on the surfaces of both plastic and metallic coatings on glass substrates. The origin of these defects was expected to be contamination on the substrate that was replicated and grown in the deposited coating. The cross section of a metallic coating in Fig. 2 shows a defect of this type. Further examination showed a small surface defect on the substrate at the base of the cone defect.

2.2 Thickness Measurement

A known coating thickness is a common requirement of many multicomponent ICF targets, and several techniques are used with the SEM for determining thickness. Figure 3 shows the direct measurement of layers in the SEM image of a fractured specimen. The magnification can be determined when the specimen is mounted on a calibration standard that is then photographed under the same conditions as the specimen. This method is appropriate for films more than 100 nm thick.

Many nondestructive methods for determining thickness are based on measuring the depth distribution of x rays within the specimen (Yakowitz and Newbury 1976, Heinrich 1981, Elliott et al. 1982). These methods typically involve measuring the emitted x-ray signal from the film and dividing by the same signal measured from a bulk standard of the same composition as the film. This ratio is then converted to a mass thickness by the use of empirical or theoretical models describing the production of x rays with depth below the specimen surface. These methods cover the range from approximately 10–100 nm.

2.3 Surface Roughness Measurement

The surface roughness of targets has received great attention. A commonly stated goal for ICF targets is a surface roughness variation less than 10 nm. The current from the backscattered electron detector can be used to determine surface roughness.

The backscattered electron signal is emitted anisotropically. Thus, as the electron beam scans a specimen, the signal recorded by fixed detector will vary according to the local surface topography as well as the average atomic number of the sample. Employing two detectors, one on either side of the sample and observing the analog difference, eliminates the atomic number effect. Thus a signal proportional to the surface topography can be measured (Lebiedzick 1979). Such a system developed at Lawrence Livermore National Laboratory specifically for ICF targets showed spatial resolution of roughly 50 nm and height resolution of 10 nm.

2.4 Microradiography

The production of x rays allows specimens to be radiographed in the SEM, and the use of the electron beam to produce a point source of x rays in a thin metal foil allows magnification in the final radiograph as shown in Fig. 4. The magnification is determined by the specimen position between the x-ray source and the film. This technique has been used both to observe fine structure in coatings and to resolve small density differences in low-density materials. The ability of this technique to resolve small spatial features as well as the ease of optimizing for density resolution has resulted in the development of commercial machines designed specifically for this use, based on electron columns from SEMs (see Sect. 3.1).

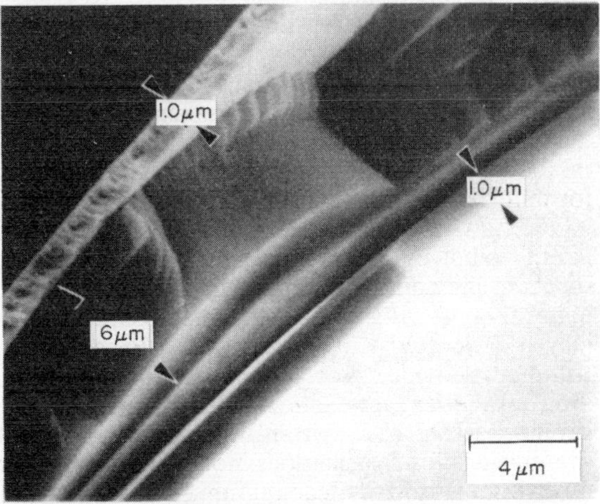

Figure 3
Measurement of coating thickness; glass microballoon coated with plastic and gold

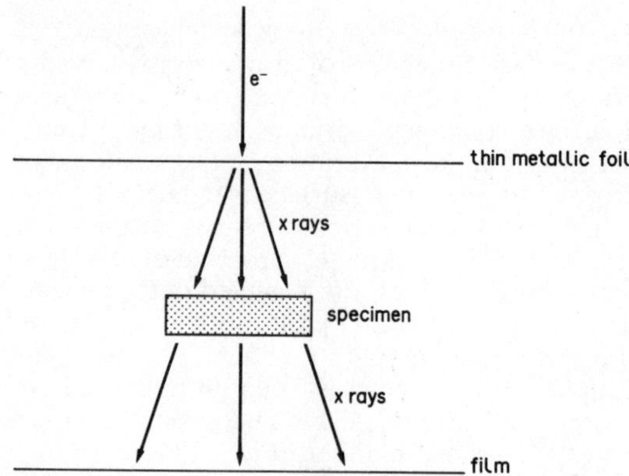

Figure 4
Schematic showing technique for producing magnified radiograph in the SEM

2.5 Quantitative Analysis

The advent of low-cost lithium-drifted silicon x-ray detectors has afforded almost all electron microscopes some capability in measuring elemental composition. The real utility of these detectors lies in converting measured x-ray intensity to elemental concentration. Three factors must be considered in this procedure (Beaman and Isasi 1972). The atomic number effect Z describes the generation of x rays in the sample by integrating the depth distribution function from the surface to the mean depth at which the average electron energy equals the excitation energy of the element being measured. The absorption factor A corrects the generated intensity for the absorption path length between the specimen and the detector. Finally, the fluorescence correction F accounts for secondary x rays generated from the continuum and characteristic x rays from other elements in the sample.

The product ZAF of these corrections gives the technique its acronym as well as a numerical value used to convert measured intensities to concentrations. Many computer codes use empirical and/or analytic models to implement this procedure. This method can provide accuracy within 1–2% relative for major constituents in a volume of about 10 μm^3.

The ZAF method makes several assumptions about sample geometry that ICF targets do not meet. In particular, the small spherical geometry of many targets can grossly affect the absorption path length of emitted x rays. Unfortunately, most ZAF routines provided by equipment manufacturers are not flexible enough to allow correction for geometry, so codes written in-house are generally preferred. This aspect does not receive sufficient attention and usually leads to unidentified errors in final results.

A variant of the ZAF technique termed $\phi(\rho z)$ has been developed more recently (Scott and Love 1983). This technique is significant for fusion targets in that it allows more accurate quantification of particles and thin films penetrated by the electron beam. The method attempts to describe mathematically the generated x-ray intensity vs mass depth from the specimen surface. The description of the x-ray depth distribution allows an alternative method to determine the absorption correction, typically the most significant of the three correction factors.

2.6 Computers

(*a*) *Data acquisition*. Operators of electron-beam instruments were quick to apply computers to the automation of their instruments. Most modern instruments use microprocessors to control instrument functions with input from front-panel operator controls. Even at modest installations there are usually additional computers controlling such functions as beam position, counting electronics, stepper motors for stage motion and spectrometer setup, and data display. Image-acquisition systems are now being added to many microscopes.

(*b*) *Data reduction*. The rapid advance in computing power now allows many tasks that were formerly run on mainframes to be performed on small dedicated systems in the electron microscope laboratory. The most obvious is the reduction of x-ray data, but other applications such as crystallography, image simulation and lens design are also used. One national bulletin board system exists solely for the dissemination of software for electron microscopy and another is planned.

(*c*) *Monte Carlo technique*. A picture of an analytical situation can be made by simulating the scattering events an electron undergoes in passing through a material. The use of random numbers to simulate individual scattering events is referred to as the Monte Carlo technique. This method has proved beneficial in the solution to several questions regarding ICF target analysis. A common application is the analysis of thin films. If the film thickness is less than the electron range in that material, quantitative analysis becomes difficult. Computer simulations of this situation permit input to ZAF routines to be corrected. Another use is the estimation of spatial resolution of x-ray analysis, which is difficult to determine experimentally. Applications involving surface effects, beam energy and imaging contrast have also been published (Heinrich 1981, Scott and Love 1983).

3. X-Ray Inspection Techniques

X-ray techniques are used on ICF targets to measure (a) coating and wall thicknesses, (b) variations in coating and wall thicknesses, (c) gas fill pressures and (d) elemental composition. These techniques are also used to measure density and thickness uniformities of bulk materials from which smaller ICF target parts are made and are equally useful for both transparent and opaque materials. Radiographic gauging, which was developed specifically for ICF target characterization, measures thicknesses and variations in coatings. Standard x-ray fluorescence techniques detect gas fill pressures and elemental composition. Standard x-ray and γ-ray gauging techniques provide maps of the density and thickness bulk samples.

An ICF target can be pictured as an onion with several layers or skins surrounding a hollow center filled with gas. The thickness of each layer is measured to 10% accuracy and the variations in thickness (deviation from the mean) of each layer to 1% accuracy. Hence, for coatings 1 μm thick, variations in thickness as small as 10 nm (100 Å) are routinely detected on an object the size of a grain of sand. With filmless radiography and microfocus x-ray sources, sensitivity will soon be 100 times better.

3.1 Characterization of Walls and Coatings Using Radiographic Gauging

The wall of the gas-containing vessel and its coatings are imaged on film by means of contact microradiography. The term "contact" here means that the targets are in contact or nearly in contact with the photographic emulsion. They are arranged in arrays as large as 150 × 150 targets on a very thin (2.5 μm thick) sheet of Mylar that lies directly on the emulsion. The term micro in microradiography indicates that the images are as small as the targets. Such sizes require that individual images in the array be enlarged by a microscope before the quality of the targets can be determined.

Figure 5 shows a diagram of a uniform spherical coating with a plot of the x-ray transmission along a

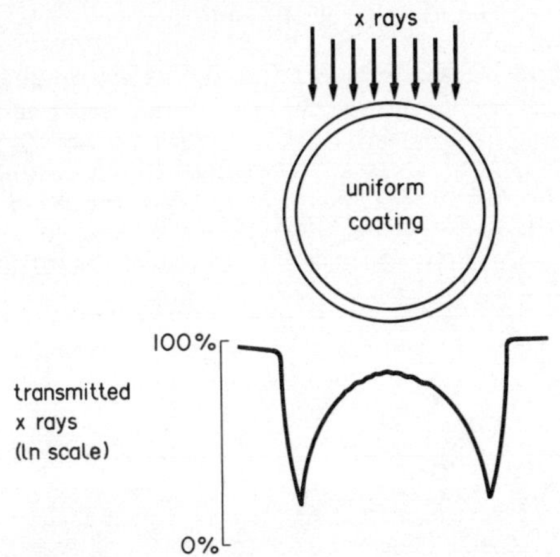

Figure 5
X-ray transmission along a diameter of a uniform spherical coating

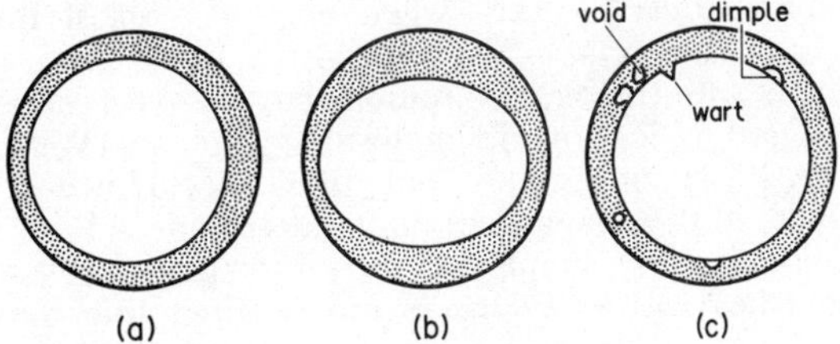

Figure 6
Three types of defects found in coatings on ICF targets: (a) concentric, (b) elliptical and (c) warts and dimples

diameter of the sphere. Parallel x rays illuminate the vessel from the top. The minimum x-ray absorption occurs at the center and increases by approximately 8000 at the edge. Hence, an exposure that is optimally exposed for the center contains much less information about the edges and vice versa. Fortunately, the exponential absorption of the x rays is moderated by the response of the photographic films so that, on an optimally exposed radiograph, the optical density is linearly proportional to the material thickness.

By optimizing the exposure for the center, Henderson et al. (1977) showed that a trace of the optical density in the central region of the image matches a theoretical model and that the match is very sensitive to coating thickness and uniformity. This method is an x-ray gauging technique that uses photographic film rather than an electronic detector. Subsequent work (Whitman and Day 1980, Whitman 1982) has shown that the symmetry of the optical density effectively discerns uniformity in the thickness of the wall or coating.

These results mean that 35% of the entire target can be characterized in a single view and that only three orthogonal views are required to completely characterize coatings. Sometimes the outermost layers of the target can interfere with the analysis of the inner ones. In such cases, as the target is built, each layer or shell is characterized before the next one is applied. Variations as small as 1% are detected in all coatings including those thinner than 1 μm. Hence, 10 nm variations are easily detected in coatings only 1 μm thick.

Two methods have evolved for computer analysis of radiographic gauging. Both methods digitize the photographic image and store it in computer memory. One method uses orthogonal diametrical scans (Singleton and Weir 1981) and the other uses a fast fourier transform of the optical-density data to extract concentric and elliptical defects (Stupin 1987a). A statistical test finds small warts and dimples. These defects are shown in Fig. 6. Concentric defects occur when the inner and outer surfaces of the shell are spherical but are not concentric.

Elliptical defects occur when one or both of the surfaces are elliptical. Concentric and elliptical defects with thickness variations larger than 1% are expected to destroy the implosion symmetry (Weinstein 1982) and so are not allowable. However, defects of this magnitude can be detected.

Wart and dimple defects include surface blemishes and voids inside the coating; thickness variations must be less than 0.1%. Current x-ray techniques do not detect defects this small, but advanced techniques are expected to do so. Radiography is more versatile than, for example, SEM characterization because it nondestructively detects warts and dimples on both surfaces of the coatings and does not require the samples to be conducting. Image analysis systems that are sensitive to thickness variations of 5–10% are used to presort images for gross concentric and elliptical defects and reduce computer processing times.

Radiographic gauging relies on a step wedge (penetrameter) to measure the absolute coating thickness by comparing the optical density in the center of the target image with that of known thicknesses of the coating material in the step wedge. The image of the penetrameter must appear on the same radiograph as the targets. In this way absolute thicknesses are determined with about 10% accuracy, although with extreme care measurements can be as accurate as 1%. However, in either case, variations as small as 1% are detected.

X-ray sources with energies between 1 kV and 50 kV are used to radiograph the thinnest freestanding shells. When possible, these sources use the characteristic K or L x-ray radiation of the anodes to produce nearly monoenergetic x-ray fluxes to improve contrast. High-energy bremsstrahlung sources are used for thicker shells. X-rays from laser-produced plasmas give extremely fine resolution using photoresists rather than film (Kim and Whitman 1987).

Advanced techniques: microfocus digital radiography and microtomography. Microfocal radiography and microtomography are advanced techniques that will improve our ability to measure wall thickness and variations in wall thickness. Microfocus x-ray sources with energies of 10–300 kV and focal spots of 10 μm detect thickness variations as small as 0.1% (Stupin 1987b, 1988, Stupin et al. 1989), and improvements in this technique are predicted to detect variations as small as 0.01%. Microtomography using synchrotron x-ray light sources, images structures only 1 μm across (Flannery et al. 1987, Kinney et al. 1988). Both methods use digital imaging; that is, images stored in computer memory, rather than photographic films.

3.2 Gas Fill Pressures

Targets are filled with gas either by diffusion or by constructing the shells in a chamber containing the fill gas. The pressure inside the shell is determined nondestructively by x-ray fluorescence. Targets filled with tritium produce bremsstrahlung radiation with an intensity proportional to the fill gas pressure and volume (Jorgensen 1982). The bremsstrahlung radiation is produced when electrons from the β decay of the tritium slow down in the gas. The electrons excite the characteristic x-ray lines of other gases in the target, and the intensities of these characteristic lines are proportional to the partial pressures of their respective gases. If tritium is absent from the fill gas, characteristic lines are proportional to the partial pressures of their respective gases. If tritium is absent from the fill gas, characteristic x-ray lines can be excited by such external sources as an x-ray generator or radioactive source. The fluorescence method is restricted to elements with atomic numbers greater than five (beryllium), and this method measures gas fill pressures with an accuracy of 10%.

See also: Microengineering of Materials: Laser Fusion Targets (Suppl. 2)

Bibliography

Beaman D R, Isasi J A 1972 *Electron Beam Microanalysis*. American Society for Testing and Materials, Philadelphia, PA

Bernat T P, Darling D H, Sanchez J J 1982 Applications of holographic interferometry to cryogenic ICF target characterization. *J. Vac. Sci. Technol.* 20: 1362–5

Clement X, Coudeville A, Eyharts P, Perrine J P, Rouillard R 1983 X-ray absorption in characterization of laser fusion targets. *J. Vac. Sci. Technol. A* 1(2): 949–51

Cooper D E 1982 A phase modulation interferometer for ICF target characterization. *J. Vac. Sci. Technol.* 20: 1075–8

Elliott N E, Anderson W E, Archuleta T A, Stupin D M 1982 Thin film thickness measurement using x-ray peak ratioing in the scanning electron microscope. *J. Vac. Sci. Technol.* 20(4): 1372–3

Feldkamp L A, Kubinski D J, Jesion G 1988 Application of high magnification to 3D x-ray computed tomography. In: Thomson D O, Chimenti D E (eds.) 1988 *Review of Progress in Quantitative Nondestructive Evaluation*, Vol. 7. Plenum, New York, pp. 381–8

Flannery B P, Deckman H W, Roberge W G, D'Amico K A 1987 Three-dimensional x-ray microtomography. *Science* 237: 1439–44

Goldstein J I, Newbury D E, Echlin P, Joy D C, Fiori G, Lipshin E 1981 *Scanning Electron Microscopy and X-Ray Microanalysis: A Text for Biologists, Materials Scientists, & Geologists*. Plenum, New York

Heinrich K F 1981 *Electron Beam X-Ray Microanalysis*. Van Nostrand Reinhold, New York

Henderson T M, Cielaszyk D E, Simms R J 1977 Microradiographic characterization of laser fusion pellets. *Rev. Sci. Instrum.* 48: 835–40

Johns W D 1967 Measurement of film thickness. In: Kaeble E F (ed.) 1967 *Handbook of X-Rays*. McGraw-Hill, New York, pp. 44-4 to 44-12

Jorgensen B S 1982 A technique for calculating DT content in glass microballoons from x ray counts. *J. Vac. Sci. Technol.* 20: 1118–19

Kim H G, Wittman M D 1987 X-ray microscopy of inertial fusion targets using a laser produced plasma as an x-ray source. *J. Vac. Sci. Technol. A* 5(4): 2781–4

Kinney J H, Johnson Q C, Saroyan R A, Nichols M C, Bonse U, Nusshardt R, Pahl R 1988 Energy-modulated x-ray microtomography. *Rev. Sci. Instrum.* 59: 196–7

Kohl J, Zentner R D, Lukens H R 1961 *Radioisotope Applications Engineering*. D. Van Nostrand, Princeton, NJ, pp. 484–5

Lambert M C 1967 Absorptiometry with polychromatic x-rays. In: Kaeble E F (ed.) 1967 *Handbook of X-Rays*. McGraw-Hill, New York, pp. 41-3 to 41-24

Lebiedzick J 1979 An automatic topographical surface reconstruction in the SEM. *Scanning* 2(4): 230–7

Miller J R, Sollid J E 1978 Interferometric measurement of cryogenic laser fusion target uniformity. *Appl. Opt.* 18: 2971–3

Monjes J A, Weinstein B W, Willenborg D L 1982 Reflection-transmission phase shift interferometer and viewing optics. *Appl. Opt.* 21: 1732–7

Powers T F 1982 Improved nonconcentricity characterization of transparent laser fusion targets by interferometry. *J. Vac. Sci. Technol.* 20: 1355–8

Scott V D, Love G 1983 *Quantitative Electron-Probe Microanalysis*. Wiley, New York

Singleton R M, Weir J T 1981 Microradiographic characterization of metal and polymer coated microspheres. *J. Vac. Sci. Technol.* 18: 1264–7

Stone R R, Gregg D W, Souers P C 1975 Nondestructive inspection of transparent microtargets for laser fusion. *J. Appl. Phys.* 17: 852–3

Stupin D M 1985 X-ray gauge measures areal density variations as small as 0.1%. *J. Vac. Sci. Technol. A* 3(3): 1266–8

Stupin D M 1987a Radiographic detection of 100 Å thickness variations in 1 μm-thick coatings on submillimeter-diameter laser fusion targets. In: Thompson D O, Chimenti D E (eds.) 1987 *Review of Progress in Quantitative Nondestructive Evaluation*, Vol. 6B. Plenum, New York, pp. 1485–94

Stupin D M 1987b Filmless radiographic detection of microscopic wires and very small areal density variations. *Mater. Eval.* 45: 1315–19

Stupin D M 1988 Near-real-time radiography detects 0.1% changes in x-ray absorption with 2 mm spatial resolution. In: Thompson D O, Chimenti D E (eds.) 1988 *Review of Progress in Quantitative Nondestructive Evaluation*, Vol. 7. Plenum, New York, pp. 1323–30

Stupin D M, Moore K R, Thomas G D, Whitman R L 1982 Automated computer analysis of x-ray radiographs greatly facilitates measurements of coating thickness variations in laser fusion targets. *J. Vac. Sci. Technol.* 20: 1071–4

Stupin D M, Stokes G H, Steven-Setchell J 1989 Near-real-time radiography detects very small wires and thickness variations. In: McGonnagle W J (ed.) 1989 *International Advances in Nondestructive Testing*. Gordon and Breach, London, pp. 323–33

Tarvin J A, Sigler R D, Busch G E 1979 Wavefront shearing interferometer for cryogenic laser-fusion targets. *Appl. Opt.* 18: 2971–3

Weinstein B W 1982 Physical measurements of inertial fusion targets. *J. Vac. Sci. Technol.* 20: 1349–54

Weinstein B W, Hendricks C D, Ward C M, Willenborg D L 1978 Simple manipulator for rotating spheres. *Rev. Sci. Instrum.* 49: 870–1

Whitman R L 1982 An efficient method to characterize laser fusion target defects more complex than concentricity. *J. Vac. Sci. Techol.* 20: 1359–61

Whitman R L, Day R H 1980 X-ray microradiographs of laser fusion targets: Improved image analysis techniques. *Appl. Opt.* 19: 1718–22

Yakowitz H, Newbury D E 1976 A simple analytical method for thin film analysis with massive pure element standards. *Scanning Electron Microsc.* 1: 150–60

L. R. Foreman, N. E. Elliott and D. M. Stupin
[Los Alamos National Laboratory, Los Alamos, New Mexico, USA]

Microengineering of Materials: Laser Fusion Targets

Inertial-confinement fusion (ICF) attempts to create an environment where a mixture of deuterium and tritium (DT) can undergo fusion on the fly. The intense laser radiation or particle flux rapidly ionizes the outer surface of the target. As the temperature rises, the plasma continues to absorb laser radiation and expand outward, and the target interior is accelerated towards zero radius by the reaction to the rocket-like ablation of the target exterior. Thus an implosive geometry is created by a driver (a laser or a particle beam) that spherically compresses the DT fuel of a fusion target core to 1000 times its liquid density for durations of the order of picoseconds. The fabrication of the targets for this work is a demanding art. The targets are less than a millimeter across, which makes them nearly too small for rotating machining methods but larger than the sizes encountered in microcircuitry. Moreover, ICF targets are more likely to have spherical or cylindrical symmetry, which complicates all the processing steps.

Targets for ICF fall into three major categories: (a) direct-drive targets, in which the illumination from a laser (or the impulse of a particle beam) implodes the fuel container by means of direct ablation of the target surface; (b) indirect-drive targets, in which a metallic shell (a hohlraum) is illuminated and the reradiation from the heated surface drives the implosion of the fuel; and (c) diagnostic targets, which we use to examine the laser–matter interaction or the extent and symmetry of the implosion. In this article it is worth remembering that currently there is insufficient power in a laboratory driver to make an ICF target ignite.

The target fabrication techniques discussed in this article apply to the manufacture of all target types.

1. Direct-Drive Targets

Direct-drive targets are the simplest to visualize. Current low-gain designs of this type consist of thick spherical shells of plastic or polymer, 300–450 μm in diameter by 30–50 μm thick, filled with 50 atm of DT gas. High-gain designs require cryogenic, that is, either liquid or solid, fuel layers inside a fuel barrier. Whichever the design, the fuel shells must meet stringent requirements on sphericity, wall uniformity, and surface finish (0.1%, 1% and <0.1 μm, respectively).

1.1 Polymer Shell Fabrication

The fuel containers are currently made of polymeric materials rather than glass because lower atomic number Z materials have better implosion properties and allow better diagnosis of the implosion. (For glass shell technology, see *Laser Fusion Targets*.) Both poly(vinyl alcohol) (PVA) and polystyrene (PS) have been used as shell materials (Campbell et al. 1986, Burnham et al. 1987). PVA is a better barrier to DT permeation (10^3 less permeable to H_2 than PS) which simplifies the target handling. PS forms shells more readily and is more radiation resistant. A combination of these two materials makes a large (400–600 μm) shell that can hold hydrogen isotopes for several days at room temperature (Burnham et al. 1987).

Polymeric shells can be made by any of four distinct techniques. Of these, those involving droplet generation and drying in 3–5 m vertical tubular furnaces give the most control over shell size and thickness. Two distinct types are currently in service. A gas-stripped droplet generator is in use at Lawrence Livermore National Laboratory (Burnham et al. 1987). In this process, a solution of PS and CH_2Cl_2 is forced through the inner capillary of a double orifice and is stripped off the tip by nitrogen gas flowing in the outer orifice. The droplets fall through a 4.5 m vertical furnace whose profile is carefully controlled at about 200 °C for the first 3.5 m and 0 °C for the last 1 m. The droplets form a skin very quickly and are then blown to size by solvent evaporating from the interior. A high yield of large thin shells instead of "raisins", or small thick shells, requires a delicate balance of furnace temperature, solvent composition (a high-volatility component such as methylene chloride is added to keep the shells from collapsing) and the flow rate of nitrogen gas up the column. Cooling the last meter of the column prevents shell collapse; it allows the shells to gain strength quickly enough to withstand the pressure drop across the shell wall created by the falling interior pressure.

In the next step of the Livermore process, the large shells are coated with PVA to provide a hydrogen barrier. PS shells made by the previous process are drawn into a glass capillary tube along with PVA solution in H_2O. The capillary is squirted into the top of the tubular drying furnace. The PVA-coated shells dry as they fall. The result is a composite shell about 500 μm in diameter and about 5–7 μm thick, 2–4 μm of which is PVA.

Analysis by interferometry of a typical 500 μm shell gives a total wall thickness (PS + PVA) of 3–8±0.3 μm and an eccentricity (the offset of the center of the bubble to the center of the exterior) of 1–2 μm. The PVA does not contribute appreciably to the eccentricity, so the composite shell has an improved wall uniformity because the overall thickness has increased.

The method used by KMSF, Inc., employs a classical Rayleigh droplet generator that uses a piezoelectric vibrator to break up a hollow stream of polymer solution into hollow droplets. This method produces a dense stream of droplets, so dense that the protoshells will catch up to one another and agglomerate as they blow larger in the furnace. The droplet stream is thinned by charging the droplets electrostatically and deflecting all but 1 in 100 or so into a catch cup. The piezoelectric-driven droplet generator makes shells that are smaller and thicker than those made by the gas-stripped generator.

A number of laboratories have tried microencapsulation as a technique for preparing polymeric shells. In the process an aqueous phase W_1 is entrained in a nonaqueous phase O, which is generally a solution of PS in dichloromethane. This mixture is then poured or dipped into a stirred, second aqueous phase W_2 in which the O/W_1 phases form spheres. The W_2 solution is heated to drive off the solvent in O, leaving a polystyrene shell containing W_1. These shells can be dried in a furnace or a vacuum chamber to create PS shells. The phase W_1 often contains PVA, which can form the barrier to hydrogen diffusion. The results of this method are mixed; some workers report difficulty in removing micrometer-sized voids in the polystyrene (Wolpert 1985, Burnham et al. 1987). The Japanese groups report that wall uniformity (5%) is suitable for compression studies and that dissolution of the PS leaves a high-quality PVA shell that is also suitable as an ICF target (Kubo and Tsubakihara 1987, Norimatsu et al. 1987).

Finally, a fourth method of producing polymer shells is spray drying, wherein dissolved polymer solution is sprayed into a drying chamber. This method is similar to a drop tower without the precise temperature control. It is not surprising then that spray drying yields shells in widely separated sizes and is unsuitable for laser-target production.

1.2 Polymer-Coating Processes

The drop towers produce shells that are generally too thin for use as fuel containers, so they are overcoated by one of several candidate polymer-coating processes. Of these, the process common to most ICF fabrication laboratories is the parylene or poly(*p*-xylene) vapor-phase-pyrolysis (VPP) process. Parylene coatings, as well as the LPP process discussed later, are conformal; that is, they replicate irregularities on the substrate rather than heal them. As the coating increases in thickness, the defects grow laterally while preserving the height of the surface irregularity (Letts et al. 1981). This process can provide thick, smooth coatings at rates of 100 Å min^{-1}, but the polymer is sticky as deposited. This makes bounce or levitated coating difficult. Jahn and Liepins at Los Alamos and Kim and workers at UR-LLE have successfully modified the VPP process with an argon plasma, which makes the coating harder and allows bouncing of the substrate shells (Jahn and Liepins 1985, Gram et al. 1986). The coater at LLE incorporates a real-time laser reflectometer to measure coating thickness.

The low-pressure plasma polymerization (LPP) process is more commonly used to augment polymer shell thickness. Known variously as glow-discharge polymerization or plasma polymerization, LPP uses a variety of precursors that are activated in a plasma discharge in the coating volume (Letts et al. 1981, Liepins et al. 1981). The plasma can be capacitively coupled or inductively coupled at either radio (MHz) or microwave (GHz) frequencies. The shell substrates can be "bounced" with a simple shaker or levitated on a molecular beam. Molecular beam levitation involves processing shells singly or in very small numbers on streams of gas drawn into the vacuum chamber. Collimated or focused hole structures channel the gas stream to support the target substrates. At the chamber pressure, the mean free path of the support gas is large compared with the target size, hence the term molecular beam levitation. This method is generally difficult to control and its low capacity is not suitable for any but very specialized coatings.

The LPP process uses the gas discharge to disassociate the feedstock vapor into radicals and ions. The fragments recombine on the substrate surface to form a brittle smooth polymer-coating. Since the substrates are bounced, the coating process is very uniform. As the shell bounces, the more heavily coated portion of the shell hangs toward the bottom, away from the flow of ions and radicals. Thus, the thinner portion of the layer receives more coating. Biasing the bouncer cup draws ions to it and their impact on the targets helps smooth the coating. Imperfections or bumps on the substrate surface grow with this conformal coating process (Letts et al. 1981). Smooth coatings require elimination of particulates on the substrate surfaces and in the feed stream. Furthermore, coater parameters must be adjusted carefully to prevent clumps of polymer from forming in the coating chamber and giving the target coating a cauliflower-like appearance.

The targets are filled with fuel (and sometimes other gases for diagnostic purposes) by permeation in much the same way as glass shells are. The shells are placed in individual "egg crates" and these are placed in an atmosphere of pressurized DT and heated. Often the final fill pressure of the shell is greater than its buckling strength. When this happens, the fill must be staged; that is, brought up to the final pressure in steps so the shells are not crushed (see *Laser Fusion Targets*). When the shells have filled, they are cooled and the fill gas is pumped away. Polymer shells do not hold gas as well as glass; they lose half their pressure in about 100 h. Thus they must be filled only a day or two before they are shot (Burnham et al. 1987).

1.3 Cryogenics

Direct-drive targets for high gain require cryogenic fuel; that is, DT that is either liquid or solid (cooled to about 20 K). These targets are more efficient because (a) they hold more fuel—liquid DT is about 1000 times denser than gas at standard temperature and pressure and solid DT is about 12% denser than the liquid; (b) the *P* work required to compress the fluid is reduced since the pressure at the target interior is less than 1 atm; and (c) the fuel is colder to begin with and thus reaches the density for ignition at a lower temperature than a gaseous target.

There are several high-gain designs for liquid targets (Sacks and Darling 1987) and solid targets (Foreman and Hoffer 1988). Liquid targets require a low-density small-pore foam sponge to hold the spherically-symmetric layer against gravity. The solid target can be made without foam by taking advantage of the heat generated by the β decay of the tritium in the fuel. A thick layer of solid DT will be warmer than a thin layer because of the heat generated by the β activity of the tritium. Thus if a layer of solid DT inside a target is nonuniform, DT will sublime from the regions where the layer is thick and condense where the layer is thin. This process proceeds to layer uniformity in about 14–300 min depending on the fuel composition and age (Hoffer and Foreman 1988).

Targets with cryogenic layers 2–3 μm thick can be processed by a fast-refreeze technique in which the cold fuel is flash-evaporated with a burst of radiation from a low-power heating laser and then allowed to condense and freeze out on the cold target walls. The method fails with thicker layers because the refreeze time lengthens and the fuel layer slumps under the effects of gravity. Uniformly thick DT liquid layers can be stabilized in targets by imposing a thermal gradient in the direction of gravity.

Cryogenic targets for laser laboratories are generally filled to about 100 atm of DT, which corresponds to about 9% of solid density. Freezing this fuel into a uniform solid layer gives a layer thickness that is about 3% of the inner target radius. Thus for a target 200 μm in diameter, the frozen fuel layer is

3 μm thick. Target designs for an ICF reactor would be about 1 mm in diameter and have about 200 μm-thick frozen layers (Sacks and Darling 1987).

Keeping a cryogenic target cold inside the target chamber is a formidable engineering task. System designs generally incorporate a retractable cold shield that contains cold helium gas and sapphire windows that allow diagnosis of the target condition, alignment of the target to chamber center and processing of the cryogenic fuel layer before shot time. The shield is snatched out of the way a few milliseconds before the laser fires to provide unrestricted access to the target. The shield collides with the shock absorbers on the target chamber wall after the shot is complete, which prevents the collision shock wave from spoiling the target alignment.

1.4 Mounting

The final step in the target fabrication process is to mount the target in the target chamber. Glass stalks are made from 2–3 mm glass tubing, 2–6 μm long, drawn to a taper with a pipette puller. The tapered portion is about 5 mm long and ends with a glass diameter of 10–20 μm. This is sufficiently thin for some applications, but most experiments require lower *Z* materials near the target. Carbon fibers 2–3 mm long and 10 μm thick are strong enough to support most targets. They are bonded with epoxy resin to the drawn glass tip, and the target itself is bonded to the carbon fiber. A hypodermic needle with a vacuum attachment mounted on a three-axis micrometer stage can position and hold the target until the bonding is complete.

The target is inserted into the center of the target chamber with some type of target fixture. Some institutions use a carousel that can accommodate a supply of targets lasting several days; others use an insertion mechanism that supplies targets individually through an air lock. The target can be aligned to the laser-focus location remotely after it is inserted into the chamber. Alternatively it can be prealigned on a locating fixture outside the chamber and then carried on the fixture to chamber center, where the fixture seats kinematically. Transfer fixtures of this type can reposition to less than 5 μm, which is about as good as can be done by remote positioning.

2. Diagnostic Targets

The bulk of target fabrication effort is directed towards making diagnostic targets that provide information about the coupling of the driver to matter and the physics of the implosion. Targets for these experiments are designed to isolate an aspect of the ICF problem. They are much more diverse than those discussed above and more challenging to build.

2.1 General Techniques: Mandrels

A central technique to all these processes involves the use of sacrificial mandrels to form target components. Made of a carefully chosen material, a mandrel is generally machined to the proper shape dimension and surface finish. It is then coated by any of the processes listed above; the innermost layer is deposited first and the other layers are added subsequently, building from the inside out.

At some suitable stage, when the layered structure has sufficient strength on its own, a suitable chemical agent or solvent dissolves the mandrel through a strategic hole or an exposed surface. The leach must be chosen carefully so it will not damage the layers already present on the prototarget. Final assembly of the target requires careful spatial positioning of all the components on a target stalk, which can be registered accurately in the target chamber.

2.2 General Techniques: Foams

Foams in fusion targets are used to simulate a fuel-filled capsule, provide hot-electron insulation or form a cushion layer (Reichelt 1985, Young 1987). ICF target designs require open-celled low-*Z* foams with densities of 10–100 kg m^{-3}; the pore size must be less than 25 μm. A number of foams can meet these specifications.

Foams made by a phase-inversion process (Young 1987, Aubert and Clough 1985) can retain some strength down to densities as low as 10 g cm^{-3}. Poly(1-4-methylpentene) (TPX) in bibenzyl is a good example of how this method works. When TPX is dissolved in bibenzyl at elevated temperature, the concentration of the TPX determines the final density of the foam in a direct way. Cooling the solution brings it first to its phase separation point, where the TPX forms a continuous phase around the separated bibenzyl. Further cooling solidifies the mixture. This is very convenient because the bibenzyl lends strength to the foam structure. At this point the material is like a crisp wax and it machines well.

The final stage in the production of a shaped foam part is the extraction of the bibenzyl. This can be done in a number of ways. Perhaps the simplest way is to dissolve the bibenzyl with methanol or isopropanol and then dry the foam. This same process can be achieved with CO_2 under pressure, which can be evaporated by releasing the pressure. Finally, the CO_2 (or other solvent) can be extracted supercritically. In this process, the temperature and pressure of the CO_2 chamber are varied so that the CO_2 liquid is taken around the critical point to the vapor phase without undergoing a phase transition and without a meniscus forming. This method is more difficult but is useful if the foam structure is very delicate and may collapse under the surface tension of the solvent. The solvent must be chosen carefully, whatever the method, because the mechanical properties

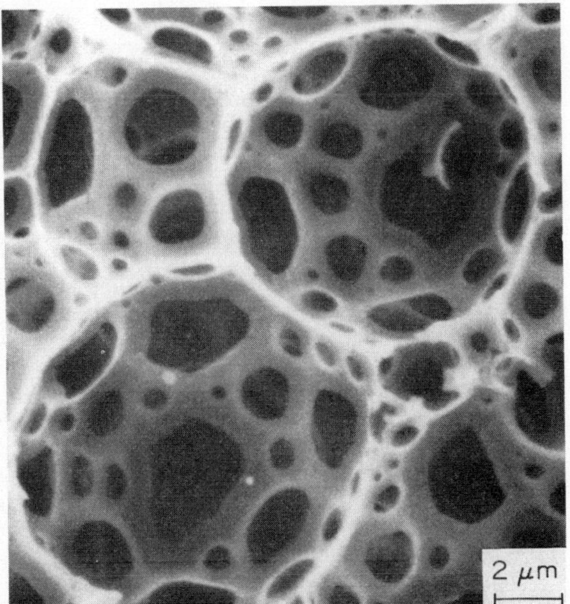

Figure 1
Scanning electron micrograph of polystyrene foam made by the emulsion process: this foam has a density of 25 kg m^{-3} and shows a spherical-void structure typical of the emulsion process

of the foam depend on residual catalysts or plasticizers left behind in the extraction process.

TPX foam with the bibenzyl extracted can be machined if the density is greater than 100 kg m^{-3}. For densities between 50 kg m^{-3} and 100 kg m^{-3}, the foam will withstand supported machining operations, such as sawing or milling, but not lathe work. Machining operations with the solid solvent present must account for the shrinkage the extraction process entails; 5% is typical, but each solvent–foam system is different.

Polystyrene foams can be made by polymerizing styrene emulsions with a water internal phase (Williams 1988). Polymerization of the styrene occurs when the emulsion is heated. The water is extracted in a vacuum oven along with other residual volatiles. The resulting foams have a distinct spherical void structure connected by smaller circular holes and can readily meet the pore size and density specifications for fusion targets. They have higher strength (at the same density) than phase-inverted TPX foam and, in the 25 g m^{-3} range, can be readily machined by conventional techniques (high-speed cutting operations are preferred). A scanning electron micrograph of this foam is shown in Fig. 1.

2.3 General Techniques: Metal Coating Processes

Metal layers are useful in diagnostic targets because they emit characteristic x rays when they ionize. With proper filtering, a metal layer can provide information about implosion timing or compression. The metal layer can be embedded in the target wall to yield heat transfer information. The two metallic coating processes in general use are described below.

(*a*) *Physical vapor deposition (PVD)*. Coating from the vapor is a process common to many industries. Typically, the material to be coated is vaporized by resistive heating or electron bombardment. The vacuum in the coating chamber is high enough to make the mean free path of the evaporating species longer than the chamber dimensions. Thus, the coating is line of sight from the coating source. Metals, salts, and even polymers can be coated by this method. The coating rate is slow, typically 0.1 μm h^{-1}. This coating method generates heat at the source, but the substrates are well removed from this heat and remain cool. Evaporative methods are well suited for polymer shells.

Sputter deposition works well for materials with high vapor pressure and substrates that can tolerate high temperatures. The coating material is biased negative, and ions formed in the chamber impact the coating cathode and eject droplets of the coating material. These droplets are hot and transfer much heat to the substrates.

Because these methods are line of sight, masks can be devised to shield areas of the substrate. Furthermore, the substrates must be rotated or bounced if an even coating is required.

(*b*) *Chemical vapor deposition (CVD)*. Coating by CVD involves either chemical reduction or thermal decomposition from a gaseous compound containing the material of interest. The material plates out on the surface of the substrate. The substrates must withstand elevated temperatures and a generally hostile environment. For example, a polymeric substrate will not survive the CVD process. More details are given in Sect. 2.4 (*b*).

2.4 Specific Examples

The targets described here illustrate how the fabrication techniques discussed above and others can be combined to produce a diagnostic target.

(*a*) *Cylinder with perturbations*. This target was designed to measure the implosion acceleration generated by laser radiation on the plastic ablator (Rhorer 1983). The aluminum layer on the inside provided x-ray emission that was recorded with a pinhole camera.

The targets were made on a copper mandrel, shown in Fig. 2, which was turned to dimension with a diamond point on an air-bearing lathe. The longitudinal surface perturbations of 25 μm wavelength and 2 μm amplitude were added with the same instrument. A version of this target with azimuthal

Figure 2
Copper mandrel for a cylindrical target: the mandrel has been machined with a diamond point to a diameter of 200 μm and a length of 750 μm. The perturbations are sinusoidal with a 25 μm wavelength and a 2 μm amplitude

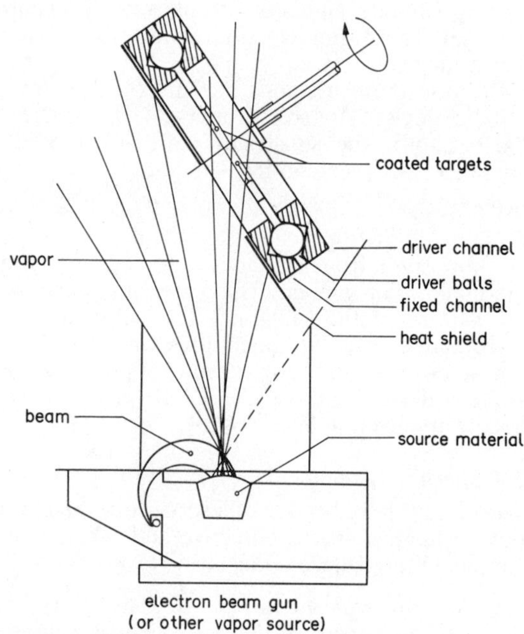

Figure 3
Dual-axis rotator for coating processes

perturbations was produced by scoring the cylindrical mandrel parallel to its axis as it was indexed in the lathe head.

Next, the machined mandrels were coated with a 2 μm-thick layer of aluminum by PVD. Even coverage was obtained by using a special rotating fixture shown in Fig. 3. Targets of arbitrary shape can be rotated by this device to receive uniform coatings. The rotator (shown in section) consists of a series of driver balls to which the targets are attached and which are themselves captured between two channels, one of which is driven by a motor external to the coating and the other of which is fixed. The balls, and hence the targets, simultaneously rotate about the axis of the channels and twist about the target axis. This two-axis motion exposes all surfaces of the target to the coating material. The exterior plastic coating was applied over the aluminum by VPP parylene in much the same way that direct-drive targets are built up except that bouncing was unnecessary. Next, the targets were faced and parted from their bases using a lensmaker's lathe. Dissolution of the mandrel left the composite cylinders 50 μm thick, 200 μm in diameter and 700 μm long. These were sparingly bonded to drawn glass capillaries using a vacuum fixture that ensured the cylinder axis would be in alignment in the pinhole camera when the target was shot.

(*b*) *Hot target*. The target shown in Fig. 4 is a 3 mm × 100 μm tungsten shell that was heated to 1800 K in the target chamber and shot at that temperature to determine the ablation products from a clean surface (Ehler and Foreman 1983). Surface contamination in the target-chamber vacuum (typically about 10^{-6} torr) occurs at the rate of about one monolayer per second. Thus, the only way to ensure that the target surface is contamination free is to heat it to at least 1800 K and boil away the low-Z materials adsorbed onto it.

The tungsten shells were made by CVD (Carroll and McCreary 1982). A schematic of the apparatus is shown in Fig. 5. Spherical molybdenum-alloy mandrels are used because their coefficient of thermal expansion closely matches that of the tungsten and because they can be acid leached without damaging the tungsten coating. The mandrels are lapped to size and surface specification in a circular track much like a thrust bearing. Several at a time are packed into a graphite reactor along with many copper beads. The reactor is heated to 450 °C and fluidized by up-flowing H_2 gas. When the bed is stabilized, WF_6 is added to the hydrogen, and the reduced tungsten coats the mandrels and beads. The rolling tumbling action in the turbulent bed produces a uniform coating; the peening of the beads creates submicrometer grain sizes and high strength. The fluidized-bed CVD process is extremely versatile. Targets requiring smooth surfaces (these did not) can be coated oversize and finish-lapped to dimension. The process works well on cylinders and other convex shapes as well as spheres. Very thin wall parts can be made (<1 μm) and these have been used in particle beam experiments.

The coated mandrels for the hot targets were drilled through along a diameter. A tungsten wire with a platinum washer on one end was threaded through the hole in the mandrel to form the mounting support and high-voltage lead. The platinum washer now at the pole of the target is melted by an Nd:YAG laser to braze the shell to the wire. Thus, the wire passes completely through the target bead as shown in Fig. 4. A close-up of the braze is shown in Fig. 6. The mandrel is removed chemically by leaching with a mixture of H_2SO_4 and HNO_3. Two vent holes are drilled into the mandrel 90° from the stalk to allow the leaching reaction to vent. The leach bath itself can be cycled between the atmospheric pressure and vacuum to allow bubbles and

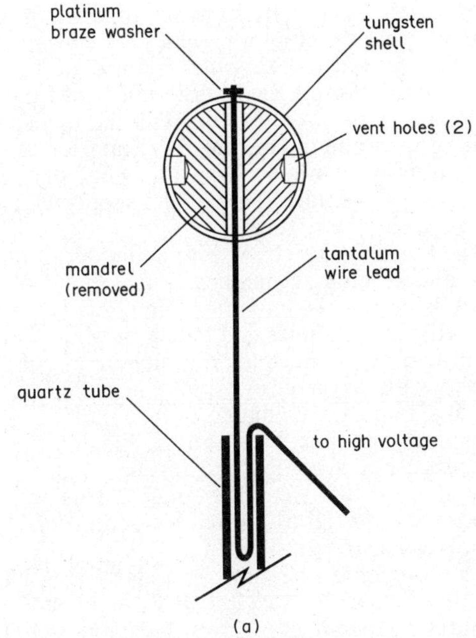

Figure 4
(a) Sectional view and (b) micrograph of hot target: the mandrel is removed by chemical dissolution through the leach holes, leaving a tungsten shell; the micrograph shows a completed 3 mm target

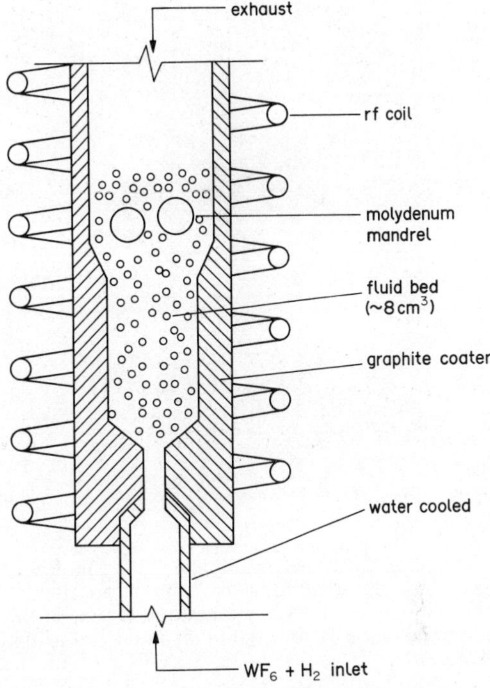

Figure 5
A fluidized-bed CVD coater: hydrogen and WF_6 react react at 450 °C in this bed to produce smooth uniform coatings

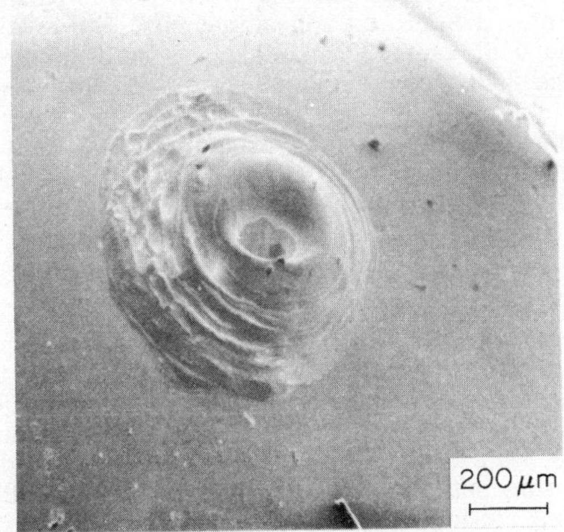

Figure 6
Platinum braze

Figure 7
Free-standing thin metallic foil target: the thin foil target, shown attached to its insertion fixture, is supported only on the top and bottom; targets of this type can be made up to 0.1 m in diameter and as thin as 0.1 μm

other reaction products to escape from the shell interior. With copper mandrels, for example, the reaction product $CuSO_4$ occupies seven times the volume of the copper metal and can burst the shell if it is not removed.

After leaching, the tungsten wire is bent double and stuffed into a quartz tube. The tubing attaches to the target insertion mechanism, which is connected to a high-voltage supply by the tail end of the wire. The target must be heated to design temperature before it is aligned to the target-chamber center. This process anneals any unresolved stresses that could cause the target to move out of alignment. The shell and wire stalk are biased 1500 V positive, and the target heats by electron impact from a thoriated tungsten wire filament oriented to avoid any incoming laser radiation or sensitive diagnostics. The quartz tube insulates the target insertion device from the target, both electrically and thermally, and provides the physical support.

(*c*) *Free-standing thin metallic foils*. Cylindrical foils have application in particle beam fusion (Duchane and Barthell 1983). Seamless aluminum or gold foils without axial supports such as the one shown in Fig. 7 can be made 0.1 m in diameter, 0.02 m long, and as thin as 0.15 μm. The metallic foils are formed by coating onto PVA film mandrels by PVD. The PVA films are formed as seamless sleeves as follows. A metal cylindrical form with an extremely smooth interior surface is dipped into a vat of PVA–water solution with some glycerin added to help film formation. As the tube is drawn out of the bath, PVA–water solution adheres to the interior in a thin film. This film is cured at 50 °C and carefully stripped from the *inside* of the form. The film is attached to two separated disks using water to make the PVA tacky. The disks are 0.020 in. (0.508 mm) undersize so they slip easily into the film sleeve. The film is washed in isopropyl alcohol to remove the glycerin plasticizer and cause it to shrink and become taut, then dried again. It stretches tightly over the cylinders and develops a wrinkle-free concave soap-bubble-like waist. The *outer* surface of the tight PVA film replicates the smooth *interior* of the form and is ready for PVD coating with the metal. After coating with aluminum, the PVA interior can then be carefully washed away, leaving the nearly pinhole-free cylindrical metal film supported at only the edges by the fixture.

The film owes its strength to a number of fabrication and coating techniques. First, the PVA substrate is extremely smooth. Unless this is true, the film will have pinholes and fall apart when the PVA is washed away. Second, the aluminum coating is pulsed with oxygen (one pulse in a 2000 Å film) which disrupts the columnar growth pattern of PVD aluminum and makes it stronger. Finally, the coating process is accomplished through a narrow slit behind which the substrate rotates. This limits the deposited material to normal incidence, which also improves its strength.

The foregoing examples were chosen to illustrate the diverse articles that the ICF target industry is required to manufacture. New techniques are being added to the usual standbys to allow more exotic designs. The following excellent reviews of the fabrication industry are provided for the interested reader: Farnum and Fries (1984), Mah et al. (1985), Reichelt (1985) and Sutton (1987).

See also: Laser Fusion Targets; Microengineering of Materials: Characterization (Suppl. 2)

Bibliography

Aubert J H, Clough R L 1985 Low-density, microcellular polystyrene foam. *Polymer* 26: 2047–54

Burnham A K, Grens J, Lilley E M 1987 Fabrication of polyvinyl alcohol coated polystyrene shells. *J. Vac. Sci. Technol.* 5: 3417–21

Campbell J H, Grens J, Poco J F, Ives B H 1986 Preparation and properties of polyvinyl alcohol microspheres, UCRL-53750. Lawrence Livermore National Laboratory, Livermore, CA

Carroll D W, McCreary W J 1982 Fabrication of thin-wall, freestanding inertial confinement fusion targets by chemical vapor deposition. *J. Vac. Sci. Technol.* 20: 1087–90

Duchane D V, Barthell B L 1983 Unbacked cylindrical metal foils of submicron thickness. *Thin Solid Films* 107: 373–8
Ehler A W, Foreman L R 1983 One-half giga electron volt tantalum ions from a CO_2 laser induced plasma. *Proc. Institute of Electrical and Electronics Engineers Conf.*, San Diego. IEEE, New York
Farnum E H, Fries R J 1984 The special materials in laser fusion targets. *Res. Mechanica* 10: 87–112
Foreman L R, Hoffer J K 1988 Solid fuel targets for the ICF reactor. *Nucl. Fusion* 28(9): 1609–12
Gram R Q, Kim H, Mason J J, Wittman W 1986 Ablation layer coating of mechanically nonsupported inertial fusion targets. *J. Vac. Sci. Technol. A* 4(3): 1145–9
Hoffer J K, Foreman L R 1988 Radioactively induced sublimation in solid tritium. *Phys. Rev. Lett.* 60: 1310–13
Jahn R K, Liepins L 1985 Di-p-xylyene polymer and method for making same. US Patent No. 4,500,562
Kubo U, Tsubakihara H 1987 Development of polyvinyl-alcohol shells overcoated with polystyrene layer for inertial confinement fusion experiments. *J. Vac. Sci. Technol. A* 5(4): 2778–80
Letts S A, Myers D W, Witt L A 1981 Ultrasmooth plasma polymerized coatings for laser fusion targets. *J. Vac. Sci. Technol.* 19: 739–42
Liepins R, Campbell M, Clements J S, Hammond J, Fries R J 1981 Plastic coating of microsphere substrates. *J. Vac. Sci. Technol.* 18: 1218–26
Mah R, Duchane D V, Young A T, Rhorer R L 1985 Target fabrication for inertial confinement fusion research. *Nucl. Inst. B* 10-1 (May): 473–7
Norimatsu T, Takagi M, Izawa Y, Nakai S, Yamanaka C 1987 Fabrication of polystyrene polyvinyl-alcohol double-layered shells by microencapsulation. *J. Vac. Sci. Technol. A* 5(4): 2785–6
Reeves G A 1980 Physical vapor deposition onto small spheres. In: *Proc. Conf. Inertial Confinement Fusion*. San Diego, CA, p. 60
Reichelt J M A 1985 Laser target fabrication activities in the United Kingdom. *J. Vac. Sci. Technol. A* 3(3): 1245–51
Rhorer R L 1983 Update on precision machining at Los Alamos. In: *Proc. SPIE* 433. San Diego, CA, 107–11
Sacks R A, Darling D H 1987 Direct drive cryogenic capsules employing DT wetted foam. *Nucl. Fusion* 27: 447–52
Sutton D W 1987 Microtarget fabrication in the United Kingdom: An overview. *J. Vac. Sci. Technol. A* 5: 2773–7
Williams J M 1988 Toroidal microstructures in water-in-oil emulsions. *Langmuir* 4(1): 44–9
Wolpert S M 1985 Production of polystyrene shells by microencapsulation. KMS Fusion annual technical report, pp. 73–7 (unpublished)
Young A T 1987 Polymer-solvent phase separation as a route to low density microcellular plastic foams. *J. Cell. Plast.* 23: 55–72
Young A T, Moreno D K, Marsters R G 1982 Preparation of multishell ICF target plastic foam cushion materials by thermally induced phase inversion processes. *J. Vac. Sci. Technol.* 20: 1094–7

L. R. Foreman
[Los Alamos National Laboratory, Los Alamos, New Mexico, USA]

Mining and Exploration: Statistical and Computer Models

Many uses of computers and statistics have developed in exploration and mining geology during the last thirty years, and particularly since the late 1970s when microcomputers have become generally available. Because words have never been as well-matched to exploration and mining geology as graphic displays of data, much of the current emphasis is on computer graphics, database development and manipulation, and man–machine interaction. Many new devices, including touch sensing, terminals and microcomputers make communication easier, quicker and more accurate than in the early days of computers.

1. Exploration

1.1 Mineral-Deposit Models

Models of mineral deposits have been employed for many years, although in the past they were given names such as hypotheses or case histories. More recently, models have been systematically developed so that they can be compared with one another and so that deposits can be assigned to them. This work has required data files stored on computers and statistical analyses implemented on computers. One collection of models contains 85 descriptive deposit models and 60 grade–tonnage models (Cox and Singer 1986). Grade–tonnage models investigate the relationship between grade and tonnage of ore; the various mathematical functions that have been proposed are statistically analyzed.

1.2 Exploration Planning

Many mining companies and some governmental agencies use computers for exploration planning. The most widely used techniques are those of operations research, including queueing, linear programming, networking and the project evaluation and research technique–critical path method (PERT–CPM) (Wignall and De Geoffroy 1987).

1.3 Resource Evaluation

Shortages of mineral commodities including petroleum have placed emphasis on resource evaluation. Because of the large amounts of data and the multivariate statistical analyses required, computers have been essential for these investigations.

Classification and evaluation of gridded areas have been of major concern in this work; later concepts have included subjective probability and probabilistic interpretations of inferences made by geologists applying scientific methodology consistently in regional evaluation (Harris 1984). Thus, the discipline imposed by the inflexibility of the computer has helped to emphasize the importance of systematic geological observation; the computer has made possible a consistent application of scientific methodology.

Resource evaluation from a governmental perspective has been extensively investigated, particularly in recent years in the USA and Canada, reflecting their large areas and well-developed science and technology (Cargill and Green 1986). A typical approach, tested in Ontario and Quebec, Canada, provided a geomathematical evaluation of copper and zinc potential. First, a database was established for geological and geophysical parameters measured in small cells. Through multivariate statistics, these measurements were compared to metal content for cells containing ore bodies and predicted metal endowment in other cells. The study demonstrated that meaningful data could be obtained from geological maps, that these data were incomplete and that effective statistical methods could be devised (Agterberg 1988).

The idea of unit regional value is that "the resources produced in some specific region may be measured in terms of the amount of resource produced and its value; if these measures are cumulated over the period of production and prorated over the area of a region, . . . they yield the unit regional weight and unit regional value of resources produced in the region" (Griffiths 1978). If these measures are related to one another and compared for relatively developed and undeveloped parts of the earth, predictions can be made about the kind and amount of mineral resources that can be developed in various regions.

The discovery-process model estimates undiscovered resources in a partially explored region based upon characteristics of the discovery process (Drew et al. 1980). Another model is a classification technique named characteristic analysis (Botbol et al. 1978). In explored areas, deposit modelling has been used successfully for resource estimation (Sinding-Larsen and Vokes 1978). All of these models have been successful for various geological situations; further work will define their generality more clearly.

1.4 Sampling Designs for Exploration

Probabilistic models for detecting ore deposits have been devised for exploration either on grids or on equally spaced traverse lines; data may be collected periodically or continuously. The designs are used most often either for drilling campaigns or for geophysical or geochemical surveys.

The models are well developed for two-dimensional situations, particularly if the ore deposits or other targets can be represented by geometrical forms, such as ellipses, susceptible to mathematical analysis. Three-dimensional models are less developed than the two-dimensional ones and apply to a narrower range of physical situations. One model explores for three-dimensional folded targets (Malmqvist and Malmqvist 1979). Another model uses the area of influence of mineralization and Bayesian statistics to select targets for univariate or multivariate data and multiple drill holes (Singer and Kouda 1988).

Statistical models are available for circles, semicircles, spheres and hemispheres. These models allow structural data associated with the host rocks and with the ore deposits themselves to be incorporated into the search models.

Dynamic program models have been devised to cope with sequential exploration, which is generally done using Bayesian statistics (Wignall and De Geoffroy 1987).

1.5 Exploration Geochemistry

Large-scale exploration geochemical surveys require the collection of thousands or tens of thousands of samples which may be analyzed for 50 or more chemical elements or other variables. Therefore, for data handling, statistical analysis and preservation of information, computers are essential.

Sampling designs and methods of statitistical analysis implemented by computers include those detailed in the previous sections. Probability graphs are widely used (Sinclair 1976) to determine *thresholds* which are cutting values separating chemical-element concentrations (or other variables) believed to be related to mineralization from those believed to be unrelated.

Besides their original variables which are generally multivariate, exploration geochemists define additional variables including ratios, sums and differences, as well as variables derived by multivariate statistical techniques including factor, cluster, regression, discriminant and cluster analysis (Howarth 1983).

Regional geochemical surveys require a systematic collection of samples, analytical work and presentation of results. Regional surveys have been completed for several small countries and extensive areas of large countries. In one survey for England and Wales, statistics were used for quality control and geological analysis; computers did the statistical analyses and processed the data for plotting (Webb et al. 1978).

1.6 Expert Systems and Artificial Intelligence

Expert systems are computer programs which store information for dealing with facts, inferences and judgements. They carry out consultations, produce maps, give advice, answer questions and explain reasoning. Thus they store expertise and judgemental knowledge for the purpose of guiding users through evaluations or diagnoses. Expert-systems software tools are becoming widely available, so a wide range of projects is now possible at reasonable cost. Expert systems have promise for storing expert knowledge, putting it into a usable form and distributing it.

In the 1960s, artificial-intelligence scientists tried without success to develop general-purpose programs. Then, in the 1970s, these scientists made a 'conceptual breakthrough' by recognizing the following principle: "To make a program intelligent, provide it with lots of high-quality, specific knowledge about some problem area" (Waterman 1986, p. 4). The resulting programs were named *expert systems*, developed through *knowledge engineering*, involving collaborative work among *knowledge engineers*, who design and build the expert systems, and *domain experts*, who have subject-matter expertise (Waterman 1986).

The first expert system in geology was PROSPECTOR; in the language of computing and information sciences it is an expert system, developed through concepts of artificial intelligence (Campbell et al. 1982).

PROSPECTOR was developed by a team of geologists and computer scientists (knowledge engineers). The geologists devised models of ore deposits (Sect. 1.1) and, with prompting from the knowledge engineers, ascribed probabilities and linkages in reasoning to them. The knowledge engineers developed inference nets to relate the models one to another and to the exploration process.

Describing PROSPECTOR, Duda et al. (1979) write that:

> The Prospector system is intended to emulate the reasoning process of an experienced exploration geologist in assessing a given prospect site or region for its likelihood of containing an ore deposit of the type represented by the model he or she designed. The performance of Prospector depends on the number of models it contains, the types of deposits modelled, and the quality and completeness of each model . . . Besides the running program, there appear to be several other benefits to this type of expert system approach. The model design process challenges the model designer to articulate, organize, and quantify his expertise. Without exception, the economic geologists who have designed Prospector models have reported that the experience aided and sharpened their own thinking on the subject matter of the model.

Expert-system applications are well suited to resource evaluation (Sect. 1.3). Many of the databases needed to make decisions are already in place, and the public, government and exploration industry need clear, rational and unbiased explanations of decisions that a good expert system gives. For resource evaluation and planning, the formulation of a rational definable mineral policy is the focus of much activity. Many methods exist for the quantitative analysis of resources. Economic modelling, environmental modelling, statistical estimation and environmental impact statements are all examples. The integration of these studies into an easily explained and understood policy is difficult and cumbersome. Expert-systems technology is clarifying the interactions of these different methods; the technology can be tested, improved and updated as problems are found or as policy changes (Koch et al. 1988).

The understanding of the object relationship, implicit in geological cross sections and maps, requires expert-system programming methods for practical implementation. Expert systems are being developed for the following applications in computer vision, remote sensing and geographic information systems: interpretation of aerial photographs; hyperspectral classification in remote-sensed imagery; image segmentation; intelligent front-ends for data integration (different sensors, ancillary data from maps and images, digitization, etc.); and analysis of data from digital terrain models (Koch et al. 1988).

The newer machines in the mineral-resource industry are controlled by electronic sensors. These measure temperature, pressure, voltage, frequency, weight, velocity and other values. Computers interpret the sensor values to regulate machines and processes. The same computers can warn operators when sensor values go beyond acceptable limits. In modern systems, numerous sensors quickly deliver large quantities of information. When things go awry, many operators have trouble interpreting the copious readings. Sensor expert advisors can be expected to give the quality of advice that an experienced operator gives on his better days. This is advantageous because advice will be available when the experienced operator is not available. In the hypothetical example of a hydrothermal production well, an expert system would be set up to continuously monitor the electronic sensors. At the first signs of trouble, the proposed sensor advisor would comment about the developments and offer advice. In contrast, experienced operators may at present get late night calls, if they can be contacted at all, when emergencies are highly developed. Although this example is hypothetical, expert sensor advisors are being used in proprietary situations today (Koch et al. 1988).

1.7 Remote Sensing

Remote sensing, in the usual sense, refers to data collection by satellite imagery or aerial photography. Because the data collected are voluminous, statistical methods are needed for sampling and analysis, and computers are needed to perform the analyses and to manage and store the data. Integration of multivariate statistical analysis, image processing and expert systems is necessary (Fabbri and Kasvand 1988). As with other methods, analysis of remote-sensing data has been done with microcomputers (Papacharalampos 1988).

1.8 Geological Mapping

In geological mapping, notes may be taken on forms suitable for keypunching for data entry into computers. Alternatively, notes can be taken directly into devices that provide machine-processible data. The disadvantage of these procedures is that they routinize geological mapping, so that data that do not fit the form may be ignored. The advantages are that data are collected systematically and language problems are minimized, as when field work is done by teams of geologists with different languages.

Once the data are in a computer they can be readily organized and statistically analyzed. By entering the data into a geographic information system (GIS) computer-generated maps can be produced to relate data collected by different geologists or on different traverses. Expert-system techniques can collate variates or rock names, and sorting and indexing become easy.

1.9 Databases

Databases for resources were first developed extensively during the 1970s. Typically, geologists and engineers wrote the specifications for the bases which were completed by computer and information scientists.

One worldwide database for economic and subeconomic ore deposits consists of more than 10 000 data bits recording grades, tonnages and geometric data (locations, strikes, dips, overburden thicknesses, etc.) (Wignall and De Geoffroy 1987). Another database for North America contains 4015 deposits; data include name, location (latitude, longitude), major and minor commodities, type of deposit, geological environment and age (Guild 1981).

The US Geological Survey maintains the computerized resources databank (CRIB), which contains some 50 000 records locating and describing mineral deposits. The US Geological Survey also has an exploration geochemical database originally compiled by the US Department of Energy for about 650 000 sites in two thirds of the USA; samples were analyzed for as many as 60 chemical elements and other variables. These are only examples of some of the many computerized databases that are now available (Peters 1987).

The current tendency is to combine databases and geographic information including topography, drainage and culture, in geographic information systems. Development of these systems has proven to be more difficult than anticipated because of statistical and computational problems.

In a somewhat different category are bibliographic databases which are required to accumulate the often scattered geological data pertinent for exploration. These are used particularly for areas in developing countries where pertinent old data, such as reports of colonial geological surveys, may be found in scientific journals of other countries. Identifying and searching these databases requires the participation of specialist librarians in cooperation with geologists who can define the parameters of the search in terms of areas and/or commodities.

Databases require combination by geographic regions, mineralization type or in other ways. One method is analysis of variance, followed by multiple comparison tests if required to establish subgroups (Wignall and De Geoffroy 1987).

2. Mining Geology

2.1 Geostatistics

Geostatistics is a specialized branch of statistics developed by Matheron (1963) and his students to analyze *regionalized variables* that are linked to one another by distance and direction (or in time). Geostatistics may be considered an extension of the weighting schemes such as inverse distance, long used in ore-reserve calculations, and of trend-surface analysis, developed in the USA and South Africa in the 1950s and 1960s (Clark 1979, David 1977, Journel and Huijbregts 1978).

2.2 Other Uses of Statistics in Ore-Reserve Estimation

While geostatistics developed so did classical and other statistics for ore-reserve estimation, as computers became increasingly available (Krige 1978). The statistical valuation of diamondiferous deposits, a difficult subject because of the extremely low grade of these deposits and their high variability, has become practicable with powerful computers to process data.

2.3 Economic Evaluation of Mineral Deposits

Current economic evaluations of mineral deposits are made with computerized models, to allow for testing various scenarios corresponding to changing conditions over the life of a mine, including the preproduction, production and shut-down stages. Among the variables considered are grade and tonnage of ore, dilution of ore by waste rock, removal of overburden, plant construction cost, cost of production (mining, milling, overheads, etc.), recovery of metals in the concentrator, preproduction development period, metal prices, revenue from sales, interest rates and income tax considerations (depreciation, depletion, etc.). By purposefully selecting variable values from statistical distributions or in other ways decision makers devise suboptimal plans (Gentry and O'Neil 1984, Peters 1987, Stermole and Stermole 1987).

3. Outlook

Today, computers and statistics are most widely used in exploration and mining geology to develop and process models that embrace the reductionist views most prevalent in twentieth-century science. While the model makers may believe that they apply "objective" rather than "subjective" methods, many fail to recognize that the methods, the commodities for which to search, the areas to be searched, the effort to be expended and other decisions are made purely or in a large part on subjective grounds. For the complexity of mineral exploration, systems models of such workers as Capra (1982) may be more appropriate.

In the future, continued rapid growth in applying statistics and computers in exploration and mining geology is anticipated. The role of Matheron's geostatistics and classical statistics will become more clearly defined. Many additional methods of data analysis, particularly those of the exploratory data-analysis school of Tukey (1977), will be employed widely and be familiar to all practitioners. Enough information will be at hand from completed exploration studies in which statistics and computers played a part to allow appraisal of the models and their refinement.

Bibliography

Agterberg F P 1988 Application of recent developments of regression analysis in regional mineral resource evaluation. In: Chung C F, Fabbri A G, Sinding-Larsen R (eds.) 1988, pp. 1–28

Botbol J M, Sinding-Larsen R, McCammon R B, Gott G B 1978 A regionalized multivariate approach to target selection in geochemical exploration. *Econ. Geol.* 73: 534–46

Campbell A N, Hollister V F, Duda R O, Hart P E 1982 Recognition of a hidden mineral deposit by an artificial intelligence program. *Science* 217: 927–9

Capra F 1982 *The Turning Point*. Bantam, New York

Cargill S M, Green S B (eds.) 1986 *Prospects for Mineral Resource Assessments on Public Lands*, US Geological Survey Circular 980. USGS, Washington, DC

Chung C F, Fabbri A G, Sinding-Larsen R (eds.) 1988 *Quantitative Analysis of Mineral and Energy Resources*. Reidel, Dordrecht

Clark I 1979 *Practical Geostatistics*. Applied Science, London

Cox D P, Singer D A 1986 *Mineral Deposit Models*. US Geological Survey Bulletin 1693. US Government Printing Office, Washington, DC

David M 1977 *Geostatistical Ore Reserve Estimation*. Elsevier, Amsterdam

Davis J C 1986 *Statistics and Data Analysis in Geology*. Wiley, New York

De Geoffroy J G, Wignall T K 1985 *Designing Optimal Strategies for Mineral Exploration*. Plenum, New York

Drew L J, Schuenemeyer J H, Root D H 1980 *Resource Appraisal and Discovery Rate Forecasting in Partially Explored Regions*. US Geological Survey Professional Paper 1138. US Government Printing Office, Washington, DC

Duda R, Gaschnig J, Hart P 1979 Model design in the PROSPECTOR consultant system for mineral exploration. In: Michie D (ed.) 1979 *Expert Systems in the Microelectronic Age*. Edinburgh University Press, Edinburgh

Fabbri A G, Kasvand T 1988 Automated integration of mineral resource data by image processing and artificial intelligence. In: Chung C F, Fabbri A G, Sinding-Larsen R (eds.) 1988, pp. 215–36

Gentry D W, O'Neil T J 1984 *Mine Investment Analysis*. American Institute of Mining, Metallurgy and Petroleum Engineers, New York

Griffiths J C 1978 Mineral resource assessment using the unit regional value concept. *J. Math. Geol.* 10: 441–72

Guild P W 1981 *Preliminary Metallogenic Map of North America*. US Geological Survey Circulars 858-A,B. USGS, Washington, DC

Harris D P 1984 *Mineral Resources Appraisal*. Clarendon, Oxford

Howarth R J (ed.) 1983 *Statistics and Data Analysis in Geochemical Prospecting*. Elsevier, Amsterdam

Journel A G, Huijbregts C 1978 *Mining Geostatistics*. Academic Press, New York

Koch G S Jr, Campbell A, Fabbri A, Lanyon W, Pereira H G, Shulman M J, Sinding-Larsen K S, Stokke P R, Toister A, Watney W L 1988 Workshop on mineral and energy resource expert system development. In: Chung C F, Fabbri A G, Sinding-Larsen R (eds.) 1988, pp. 703–13

Krige D G 1978 Lognormal-de Wijsian geostatistics for ore evaluation. *South African Institute of Mining and Metallurgy Monograph Series, Geostatistics 1*. SAIMM, Marshaltown, Transvaal

Malmqvist K, Malmqvist L 1979 A feasibility study of exploration for deep-seated sulphide ore bodies. In: O'Neil T J (ed.) 1979 *Proc. 16th Symp. Application of Computers and Operations Research in the Mineral Industry*. University of Arizona, Tucson, AZ

Matheron G 1963 Principles of geostatistics. *Econ. Geol.* 58: 1246–66

Papacharalampos D 1988 GEMS: A microcomputer-based expert system for digital image analysis. In: Chung C F, Fabbri A G, Sinding-Larsen R (eds.) 1988, pp. 529–42

Peters W C 1987 *Exploration and Mining Geology*. Wiley, New York

Ramani R V (ed.) 1986 *Proc. 19th Symp. Application of Computers and Operations in the Mineral Industry*. Society of Mining Engineers, Littleton, CO

Sinclair A J 1976 *Application of Probability Graphs in Mineral Exploration*. Special Volume No. 4. Association of Exploration Geochemists, Rexdale, Canada

Sinding-Larsen R, Vokes F M 1978 The use of deposit modelling in the assessment of potential resources as exemplified by Caledonian stratabound sulfide deposits. *J. Math. Geol.* 10: 565–80

Singer D A, Kouda R 1988 Integrating spatial and frequency information in the search for Kuroko deposits of the Hokuroku district, Japan. *Econ. Geol.* 83: 18–29

Stermole F J, Stermole J M 1987 *Economic Evaluation and Investment Decision Methods*. Investment Evaluations Corporation, Golden, CO

Tukey J W 1977 *Exploratory Data Analysis*. Addison-Wesley, Reading, MA

Waterman D A 1986 *A Guide to Expert Systems*. Addison-Wesley, Reading, MA

Webb J S, Thornton I, Thompson M, Howarth R J, Lowenstein P L 1978 *The Wolfson Geochemical Atlas of England and Wales*. Clarendon, Oxford

Wignall T K, De Geoffroy J 1987 *Statistical Models for Optimizing Mineral Exploration*. Plenum, New York

For information in addition to the bibliography, see the proceedings of the series of symposia on the application of computers and operations research in the mining industry (APCOM) that started in 1961 and have been held nearly annually since. The proceedings of the 19th symposium (Ramani 1986) lists the publishers and availability of earlier volumes in the series.

G. S. Koch Jr.
[University of Georgia, Athens, Georgia, USA]

Multifilamentary Composite Superconductor Design and Fabrication

Type II superconductors in the mixed state (i.e., the state in which most practical superconductors normally function) contain a lattice of flux lines, or fluxoids, which must be pinned on imperfections such as precipitates, dislocations, twin boundaries or grain boundaries if the material is to support a signficant resistanceless current at relatively high magnetic fields. In NbTi the flux is primarily pinned on α-titanium precipitates which are produced by repeated applications of cold work and heat treatments (Larbalestier et al. 1985). In the case of Nb_3Sn, a brittle compound which is difficult to cold work, the flux is principally pinned on grain boundaries.

Since, in practical operation, the superconductor is subjected to local fluctuations in the mechanical and electromagnetic environment, it is necessary to design practical conductors in such a way that these local perturbations do not lead to total, and perhaps catastrophic, loss of superconductivity in the device in which they are used. To help to reduce this possibility, the superconductor is surrounded by a highly thermally conductive medium such as high purity copper. In addition, in order to reduce the extent of flux motion and to ensure the proximity of the copper, it is desirable to produce the superconductor in the form of fine filaments. Therefore the structure desired is a composite one consisting of fine filaments in a copper matrix.

The design and fabrication techniques for the production of superconducting multifilamentary composites have developed along slightly different lines from those of other composites, such as those to be used in high-strength applications. The reason for this is that, in the case of the latter, at least one of the components is hard and frequently brittle. In the case of the superconductors, however, the extreme ductility and low work-hardening coefficient of niobium, niobium–titanium alloys and copper enable enormous amounts of reduction to be carried out without frequent intermediate anneals.

The same basic fabrication techniques (Hillmann 1981, Collings 1986) are used for the production of both the ductile multifilamentary niobium titanium (NbTi) and the hard intermetallic compound of niobium tin (Nb_3Sn). These two materials constitute the bulk of the commercial superconductors currently in use: NbTi for low-field and Nb_3Sn for higher field and somewhat higher temperature applications. In the case of Nb_3Sn, the two components (niobium and tin) are prevented from reacting to form the hard compound until the fabrication procedure is completed. In many cases the higher field product is fabricated into the coil required in the device before the relatively brittle Nb_3Sn is formed. These basic fabrication procedures are billet assembly, evacuation, welding, consolidation, extrusion, wire drawing, twisting, insulating and cabling. The details of the fabrication techniques employed and the design criteria used in the manufacture of multifilamentary superconductors are dependent on the device in which the conductor will be used.

1. NbTi Strands

1.1 Strands with less than 200 Filaments

These strands are made by gun drilling solid copper billets and inserting relatively large diameter NbTi rods into the holes, after carefully cleaning and etching both of the components (see Fig. 1a). The billets are evacuated, sealed by welding on a "nose" and a "tail", extruded and drawn to finished size. Unless the wire is drawn to very small dimensions, the filaments are usually of a large diameter; that is, greater than 20 μm.

Few manufacturing problems are encountered in the fabrication of this type of material when the correct production and quality control procedures are in place. Long piece lengths, which are required for magnetic resonance imaging (MRI) applications, are relatively easily achieved.

Typical applications are small and intermediate size magnets such as low-field (MRI) magnets and the ratio of copper to superconductor ranges from 1.3:1 to 7:1.

1.2 Strands with 200–2500 Filaments

When the number of strands exceeds 200, it is no longer practical to use drilled billets. One method frequently used with this number of filaments is to insert round rods into hexagonal tubes. These are stacked into 250 mm or 300 mm diameter cans and then processed in a manner similar to that illustrated in Fig. 1b. This approach is sometimes termed the "Fermi kit" approach as it was used to manufacture the strand for the Fermi tevatron dipoles and quadrupoles. Filamentary diameters down to 8 μm can be achieved with this technique in wires of diameter between 0.5 mm and 1 mm (see Fig. 2).

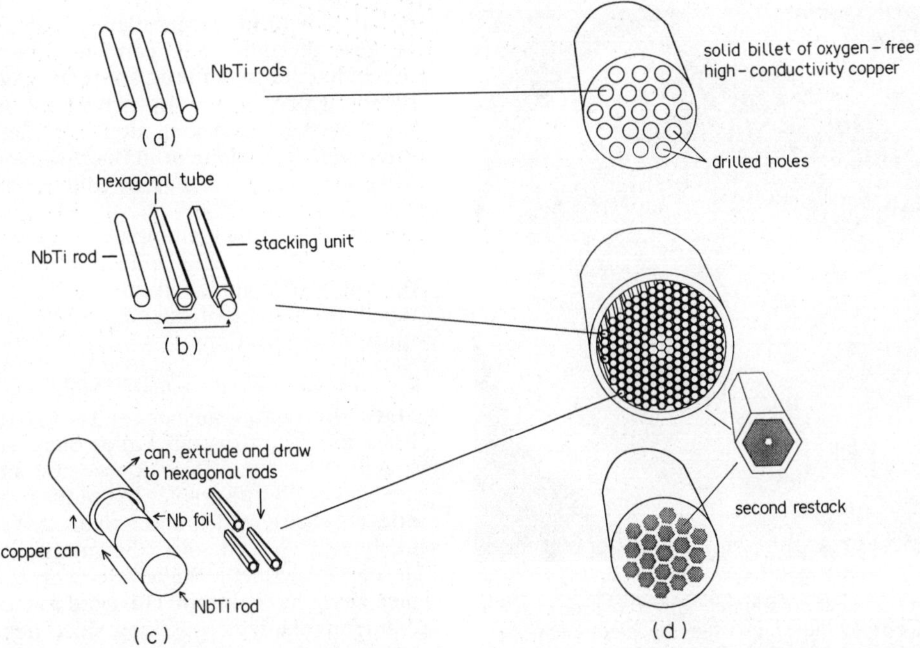

Figure 1
Fabrication methods for NbTi with the number of filaments being (a) less than 200, (b) 200–2500, (c) 2500–25 000, and (d) greater than 25 000

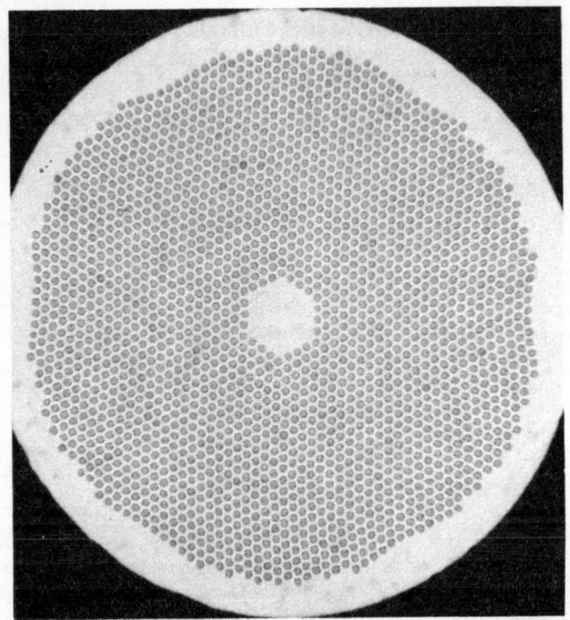

Figure 2
Cross section of a tevatron strand with approximately 2070 filaments, made by the Fermi kit technique

While the method of assembly is relatively economical, it has significant drawbacks when the capability to have smaller filament diameters combined with the highest current density (J_c) is required. The relatively thick walls of the available hexagonal tubes lead to filament 'sausaging'. The many surfaces involved at the assembly stage are difficult to clean and this may lead to piece length problems.

Conductors for a wide range of applications, including dipoles, quadrupoles and MRI magnets, have been made using this assembly technique. The values of J_c achieved can be high (~3000 A mm^{-2}), if the conductor dimensions are such that the filament diameters are greater than 15 μm. If, however, the filament diameter is below 10 μm then J_c values of 2000–2500 A mm^{-2} are more common. In the early work using this approach piece length problems were severe but these have now been reduced to tolerable levels.

1.3 Strands with 2500–25 000 Filaments

Largely in response to the increasingly demanding specifications for dipole and quadrupole magnet conductors for such proposed accelerators as the superconducting super collider (SSC) and the large hadron collider (LHC), attempts have been made to improve the values of J_c in fine filamentary NbTi wires. Considerable advances were made in the late 1980s (Gregory 1989, Kanithi et al. 1989). The best way of meeting these high J_c fine-filament

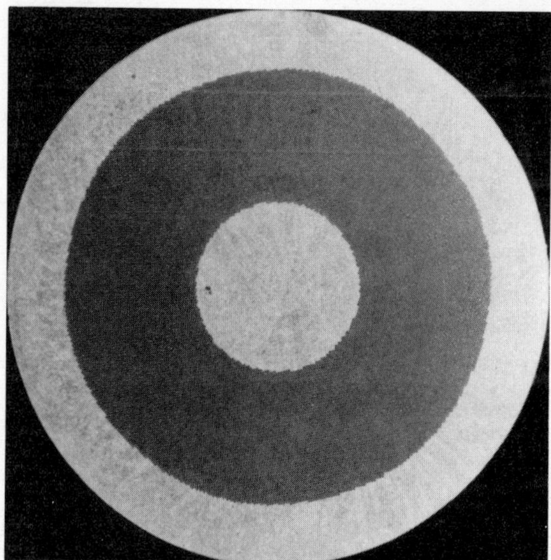

Figure 3
Cross section of an approximately 23 000 filament conductor made by the 'single stack' technique (by courtesy of Supercon Inc.)

requirements has been to first make a monofilament in which the NbTi core is surrounded by a wrap of foil of material such as niobium and then by sufficient copper (or a Cu–Mn alloy) to give the correct spacing between the filaments (see Fig. 1c). After extrusion and drawing these monofilaments are either restacked as round rods or passed through a hexagonal finishing die and then stacked. The latter approach gives the most perfect array but, in some cases, the round rods are easier and quicker to straighten and stack. The largest number of individual hexagonal monofilaments reported to have been stacked into a 300 mm diameter SSC type billet is 22 900 (Gregory et al. 1989). This makes a 0.65 mm diameter strand with 2.5 μm diameter filaments and was a prototype for an advanced outer cable conductor for the SSC. It had Cu–0.5 wt% Mn between the filaments in order to ensure that no proximity-effect coupling existed at the finished wire diameter (see Fig. 3).

This technique is now frequently used for both the inner (~7000 filaments) and the outer (~4600 filaments) 6 μm diameter filamentary material presently specified for the SSC cables. Piece length problems have been encountered during the development work and seem to be more severe as the number of filaments are increased. It now appears that, if the necessary precautions are taken, the specified piece lengths can be achieved reliably and reproducibly. While it is possible to make at least the outer conductor with 2.5 μm diameter filaments by this method, as mentioned above, the piece length and values of J_c obtained have been lower than those obtained with the larger filaments. Due to this, and because the inner SSC conductor with 2.5 μm diameter filaments requires approximately 40 000 filaments, it may be more practical and economical to use a further restacking step (see Fig. 1d) for both these very fine filament materials, as has been done for some time when making conductors for ac applications.

While the principal application for strands made by this technique is advanced accelerator magnets, the knowledge gained in this work can be applied to the development of high J_c conductors for a wide range of applications.

1.4 Strands with more than 25 000 Filaments

Since the conductors under consideration for the LHC are of an overall larger diameter than those specified for the SSC, and because small filaments are generally desirable to reduce magnetization effects, wires with greater than 25 000 strands are likely to be specified in the future, if they can be developed with the high values of J_c and the long piece lengths required. The need for submicrometer diameter filaments has long since been recognized when ac applications have been considered but the highest J_c value and long piece lengths requirements have not been stressed for these applications in the past. The approach used has been to make a subelement containing several hundred filaments by an approach such as that described in subsection 1.3, draw to a relatively large size, finish with a hexagonal die and again restack (see Fig. 1d), then extrude and draw to final size. One of the disadvantages of this approach is that the filaments on the outer layers of the second restack are not uniformly supported on all sides and can, therefore, sausage more readily than those in the centers of each group.

2. Nb_3Sn Strands

Many of the design and fabrication factors which apply to NbTi conductors must also be taken into account when Nb_3Sn strands are to be made (Suenaga and Clark 1980). There are three main commercial fabrication methods which will be discussed in this section. There is also a Dutch process based on niobium-clad $NbSn_2$ powder filaments (Elen et al. 1981) which has not yet been developed to a full commercial stage. The filament size achievable by this method appears to be significantly larger than that made by the other methods.

2.1 The Bronze Method

The process most frequently used in the past is one of "double stacking". A drilled 13.5 wt% tin bronze billet, similar to that shown in Fig. 1a, is packed with rods of pure niobium or niobium containing small percentages of titanium or tantalum. After extrusion, this is restacked into a copper can with a barrier layer of either niobium or tantalum to separate the copper stabilizer from the tin in the bronze. The principal drawback to this process is that the bronze work hardens rapidly and requires frequent annealing steps, which are expensive and tend to result in prereaction and the formation of brittle Nb_3Sn during fabrication. Although recent work using the process has overcome the technical problems, the relatively high cost of the process has led to attempts to develop alternative methods for commercial production of Nb_3Sn multifilamentary wire.

2.2 The Internal Tin Process

This process, developed by the IGC Advanced Superconductors Inc. (Schwall et al. 1983) and illustrated schematically in Fig. 4, does not require the repeated intermediate anneals. The product is free from prereaction and the method is more economical than some of the others. The process requires the production of two components: a composite tube and a stabilizer tube. The composite is made in a tubular form by a process similar to that shown in Fig. 1b and contains many filaments of niobium, or alloyed niobium, in a copper matrix. The stabilizer is made in a similar manner and contains a diffusion barrier of tantalum or niobium. After hot extrusion, the composite tube is filled with tin, or a high-tin alloy, and drawn down to a suitable size for restacking. Figure 5 shows a strand made from drawing down the restacked assembly to final size but before the reaction heat treatment used to form the Nb_3Sn has been applied.

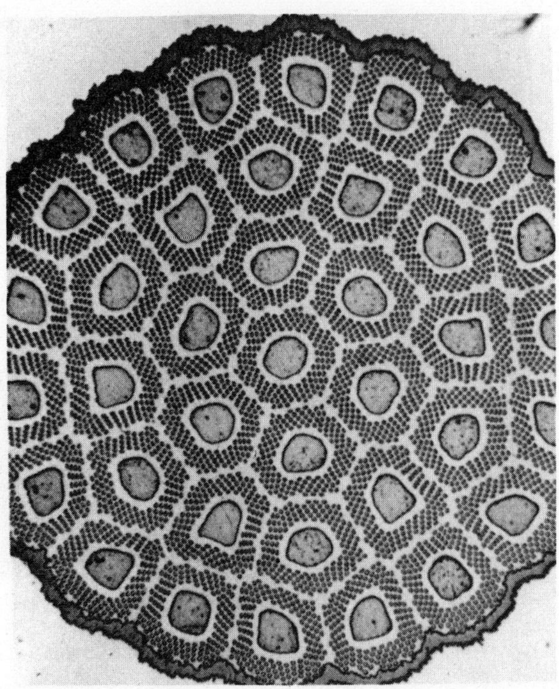

Figure 5
Cross section of internal tin Nb_3Sn conductor containing 37 subelements before reaction to form the intermetallic compound

2.3 The Modified Jelly Roll Process

The modified jelly roll (MJR) process, developed by Teledyne Wah Chang Albany (TWCA) (McDonald et al. 1983) and illustrated schematically in Fig. 6, consists of rolling expanded niobium, or niobium with small percentages of tantalum or titanium foil, interleaved with copper on a tin rod. A vanadium–niobium barrier layer separates the jelly roll configuration from the stabilizing copper. Many tin-core

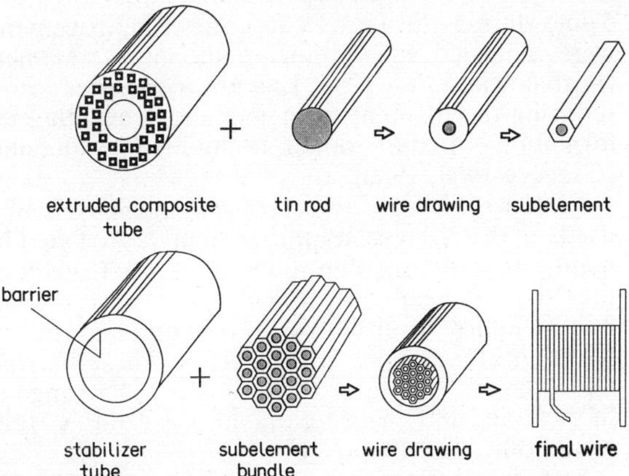

Figure 4
Fabrication steps in the internal tin process for making Nb_3Sn

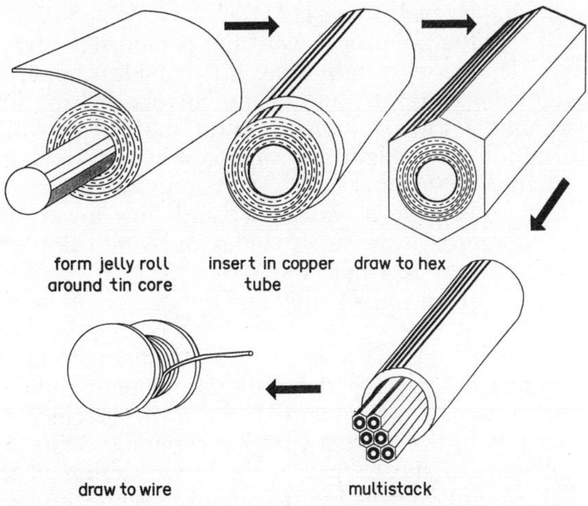

Figure 6
Fabrication steps in the MJR process for making Nb_3Sn

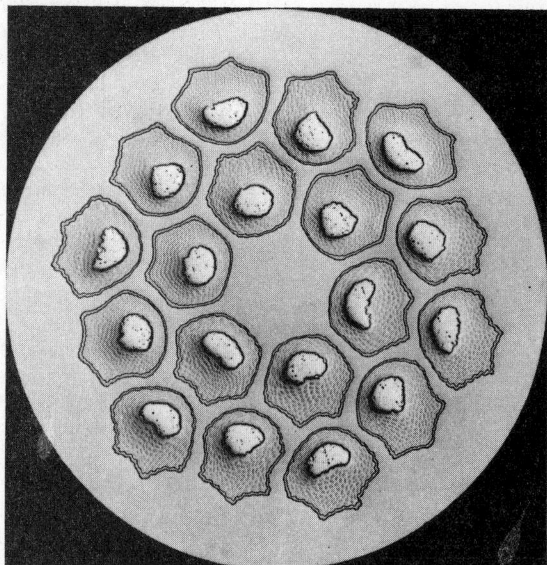

Figure 7
Cross section of MJR conductor containing 18 subelements before reaction to form the intermetallic compound (by courtesy of TWCA)

MJR units are restacked to reduce the tin diffusion distance from the core of each jelly roll and Fig. 7 shows such a multisubelement conductor.

3. Design Factors

3.1 Copper for Stability and Quench Protection

The desirability of having the filaments embedded in a material of high-conductivity copper has been alluded to in the introduction. For a more detailed explanation of the concepts of cryostability, adiabatic stability and dynamic stability see *Superconducting Magnets* (Suppl. 2).

From the point of view of the composite conductor design, it is important to provide copper in sufficient amounts, and in the correct distribution, so that a low-resistance current path is provided around any length of superconductor that has entered the normal state. The copper also conducts heat away from a "hot spot" and thus lowers the temperature of the superconductor by allowing it to return to the resistanceless state.

The type of stabilization and the degree of quench protection that the magnet designer requires will determine the ratio of copper to superconductor requested. This, together with the filament and conductor size, are frequently fixed in the specification along with the current-carrying capability. The distribution of the copper in the cross section of the wire is not specified as frequently. While uniform distribution of the copper would intuitively provide the best protection, little work has been done on the effects of copper distribution on stability.

The importance on J_c of closely spaced filaments is outlined in Sect. 3.2 and, obviously, if the total amount of copper is to remain constant that which is removed from between the filaments must be located elsewhere in the conductor cross section. Changes in this copper distribution can have profound effects on the mechanical properties. This is particularly the case in the fabrication of multifilamentary NbTi. Here the filament array becomes harder with each successive heat treatment while the copper is repeatedly annealed and remains relatively soft. These hardness differences markedly affect composite fabricability.

In the case of Nb_3Sn, the filaments are surrounded by relatively high-resistance bronze after the reaction to form the compound. It is therefore necessary to provide high-purity copper either in the center and/or on the outside of the filament array, or distributed within it. Such copper areas have to be surrounded by barrier layers to prevent their contamination by the tin in the bronze. The larger the number of discrete copper areas the more effective they will be from the point of view of providing stability but the greater the barrier volume introduced into the cross section becomes. This increase will lower the overall current-carrying capacity of the conductor.

From this section, it can be seen that the provision of copper for stability and quench protection in both NbTi and Nb_3Sn multifilamentary conductors can markedly affect mechanical behavior and thus fabricability and performance characteristics.

3.2 Improvements in Current-Carrying Capacity

The design challenge, particularly in NbTi conductors, has been to obtain the highest J_c at relatively high fields in fine filamentary material. J_c values were raised by 70% (from $1800\ \mathrm{A\,mm^{-2}}$ to $3000\ \mathrm{A\,mm^{-2}}$ at a field of 5 T) and filament sizes reduced by a factor of two or three (from 8–15 μm to 5 μm) during the late 1980s. These improvements were achieved by optimizing the heat treatment (Larbalestier et al. 1985, Li and Larbalestier 1987), reducing the filament sausaging and preventing the formation of the copper–titanium intermetallics (Gregory 1988, 1989).

NbTi with a high degree of homogeneity is now used in the designs requiring high J_c in fine filaments. In addition, thin niobium barriers surround the NbTi filaments (see Fig. 1c) to ensure that no intermetallics form between the titanium in the NbTi and the copper in the matrix. These barriers allow the manufacturer to employ a wider range of fabrication and heat-treatment procedures than would otherwise be possible.

The creation of a uniform and small spacing between the filaments, particularly in the case of filaments of small diameter, has been shown to significantly decrease sausaging (Gregory 1988). In order to achieve the small spacing the procedure of inserting round NbTi rods into copper tubes with a hexagonal exterior configuration, normally employed for the manufacture of larger filamentary material (see Fig. 1b), was abandoned in the techniques for making the finer filamentary material. A "double extrusion" (see Fig. 1c) was adopted instead. This enabled thinner copper layers to be developed on the monofilaments as well as allowing the manufacturer further cold-work and heat-treatment flexibility. The use of double stacking (as distinct from double extrusion) has been discontinued in many cases for filament numbers below 10 000, because of the desirability of maintaining a uniform filament array and a constant spacing between the individual filaments. In most billet designs, as the copper between the filaments is minimized to reduce sausaging, large areas of copper must be carefully accommodated elsewhere in the cross section, as a fixed copper-to-superconductor ratio must be maintained.

The small spacing between the filaments can lead to electrical coupling, particularly at low fields, and hence much of the advantage of subdividing the NbTi (reduced magnetization and losses under cyclic fields) is eliminated. The replacement of a small amount of low-resistivity copper between the filaments by Cu–Mn alloy has solved the coupling problem (Collings 1988, Gregory 1988).

When billets are assembled, particularly by the "single stacking" techniques, sophisticated cleaning and assembly techniques have to be employed in order to ensure that good bonding is achieved in subsequent fabrication because of the many surfaces involved. This is believed to be of particular importance in the achievement of the long piece lengths which are of great importance if economical fabrication is to be performed.

4. *Conductors for Large Devices*

Large magnets require conductors to carry currents far in excess of those that can be transported by the single strands of a small diameter wire described previously in this article and, therefore, for this type of application a multistrand cable or a composite monolith is required. The cables for the SSC dipoles are of the keystoned Rutherford-type flat cable, developed at the Rutherford laboratory in the UK. The inner and outer grades have 23 (see Fig. 8a) and 30 strands, respectively.

As has been previously mentioned for low-field MRI magnets, a single strand containing relatively few filaments (see Fig. 1a) and with a copper-to-superconductor ratio of around 7:1 is frequently used. Some of the higher field (1–2 T) MRI magnets, however, use a "wire-in-channel" type of conductor made by continuously soldering a strand that contains relatively few filaments and a low copper-to-superconductor ratio (1.35:1) into a high-purity copper channel. The overall copper-to-superconductor ratio of the resulting composite is approximately 15:1. This wire-in-channel approach (see Fig. 8b) enables a relatively high current to be obtained in a large cross-sectional conductor. In addition, long lengths of conductor can be relatively easily produced at low cost. Both these advantages are particularly valuable in MRI and many other applications.

Another technique of adding additional stabilizer to a high-current-density core is to add aluminum by extrusion cladding (Kanithi et al. 1988). Practical aluminum stabilized conductors based on continuous cladding of multifilamentary NbTi–Cu composites have been successfully developed. Many geometries can readily be fabricated to take advantage of the high-purity aluminum with a resistivity at room temperature 2000 times that at 4.2 K and these are commercially available at competitive prices. Soldering has also been used to attach an aluminum stabilizer to a normal copper-matrix superconductor. Conductors of this type are used primarily for detector magnets for accelerators.

While not many large Nb_3Sn coils have been made in the US, the notable exception is the Westinghouse LCP coil (Dresner et al. 1988). This employed one of the first cable-in-conduit conductors (CICCs), a design where the liquid helium flows

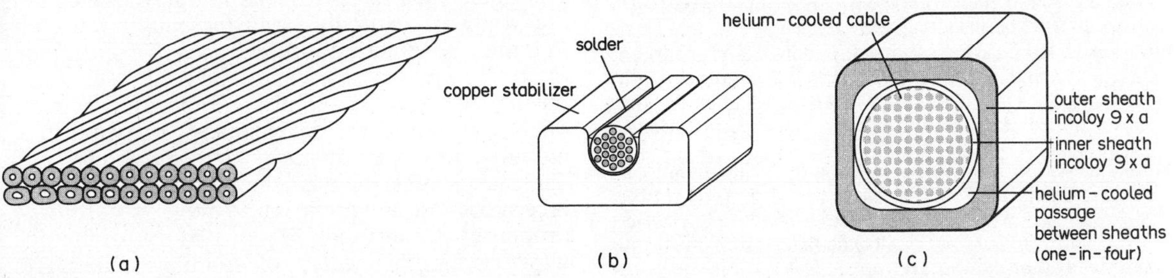

Figure 8
Methods of fabrication for large conductors for high-current applications: (a) Rutherford cable, (b) wire-in-channel conductor, and (c) CICC

through the conduit instead of the whole coil being immersed in a bath of the cryogen. A similar conductor design was used in the 12 T program at the Lawrence Livermore laboratory. Both conductors were composed of early bronze strands exhibiting low current densities. Figure 8c shows a schematic cross section of a more recent design of a CICC (Hoenig et al. 1988).

In Europe a considerable amount of work has been done, particularly in Germany, Switzerland, the Netherlands and Italy, on the design of large A-15 coils for the insert to a large test coil named "Sultan" in Switzerland and for the toroidal and poloidal coils for NET (next European torus). Many of these conductors are of the forced flow type, some with special helium conduits. In Japan the development of large-scale Nb_3Sn conductors has made considerable progress. Toroidal coils have been developed successfully for use in the test module coil in JAERI and in the demonstration poloidal coil including a Nb_3Sn pulse coil (DPC-EX), using a forced flow conductor that is currently under development.

Bibliography

Collings E W 1986 *Applied Superconductivity Applications*, Vol. 2. Plenum, New York

Collings E W 1988 Stabilizer design considerations in fine filament Cu–NbTi composites. In: Clark A F, Reed R P (eds.) 1988 *Advances in Cryogenic Engineering*, Vol. 34. Plenum, New York, pp. 867–78

Dresner L, Fehling D T, Lubell M S, Lue J W, Luton J N, McManamy T J, Shen S S, Wilson G T 1988 Stability tests of the Westinghouse coil in the international fusion superconducting magnet test facility. *IEEE Trans. Magn.* 24(2): 1779–92

Elen J D, Schinkel J W, van Wees A C A, van Beynen C A M, Hornsveld E M, Stahlie T, Veringa H J, Verkaik A 1981 Development of stabilized Nb_3Sn wire containing a reduced number of filaments. *IEEE Trans. Magn.* 17(1): 1002–5

Gregory E 1988 Recent advances in commercial multifilamentary NbTi wires in the United States. In: Reed R P, Xing Z S, Collings E W (eds.) 1988 *Cryogenic Materials '88*, Vol. 1. International Cryogenics Materials Conference Publishing, Boulder, CO, pp. 361–71

Gregory E 1989 Conventional wire and cable technology. *Proc. IEEE* 77(8): 1110–23

Gregory E, Kreilick T S, Wong J, Collings E W, Marken Jr K R, Scanlan R M, Taylor C E 1989 A conductor with uncoupled 2.5 μm diameter filaments, designed for the outer cable of SSC dipole magnets. *IEEE Trans. Magn.* 25(2): 1926–9

Hillmann H 1981 Fabrication technology of superconducting material. In: Foner S, Schwartz B B (eds.) 1981 *Superconductor Material Science, Metallurgy, Fabrications and Applications*, NATO advanced study institute series. Plenum, New York, pp. 275–388

Hoenig M O, Steeves M M, Gibson C R 1988 The selection of a 30 kA ohmic heating coil conductor. *IEEE Trans. Magn.* 24(2): 1452–4

Kanithi H C, King C G, Zeitlin B A 1989 Fine filament NbTi conductors for the SSC. *IEEE Trans. Magn.* 25(2): 1922–5

Kanithi H C, Phillips D, King C, Zeitlin B A 1988 Development and characterization of aluminum clad superconductors. *IEEE Trans. Magn.* 24(2): 1029–32

Larbalestier D C, West A W, Starch W, Warnes W, Lee P, McDonald W R, O'Larey P, Hemachalam K, Zeitlin B A, Scanlan R M, Taylor C 1985 High critical current densities in industrial-scale composites made from high homogeneity Nb 46.5 wt% Ti. *IEEE Trans. Magn.* 21(2): 269–72

Li C G, Larbalestier D C 1987 Developments of high critical current densities in niobium 46.5 wt% titanium. *Cryogenics* 27(4): 171–7

McDonald W K, Curtis C W, Scanlan R M, Larbalestier D C, Marken Jr K, Smathers D B 1983 Manufacture and evaluation of Nb_3Sn conductors fabricated by the MJR method. *IEEE Trans. Magn.* 19(3): 1124–7

Schwall R E, Ozeransky G M, Hazelton D W, Cogan S F, Rose R M 1983 Properties and performance of high current density Sn-core process MF Nb_3Sn. *IEEE Trans. Magn.* 19(3): 1135–8

Suenaga M, Clark A F 1980 *Filamentary A-15 Superconductors*. Plenum, New York

E. Gregory
[IGC Advanced Superconductors, Waterbury, Connecticut, USA]

Multilayers, Metallic

Advanced thin-film deposition techniques enable the production of multilayered materials, consisting of alternating layers of two types, in which the layer thicknesses are both very small and very precisely controlled. These materials with synthetic periodicity (in the range <1 nm to ~ 100 nm) are variously termed "multilayers," "compositionally modulated materials," or "superlattices." Multilayers offer opportunities to observe new scientific phenomena, to test existing theories and to measure materials parameters. In addition, new devices based on multilayers have considerable technological potential. The ability to design artificial multilayers may aid the study of naturally occurring multilayers, for example, long-period ordered alloys (see *Long-Period Superlattices*) and layered superconducting compounds, and also aid the study of materials unstable to the development of composition modulations (see *Spinodal Decomposition* and Tsakalakos 1984). Semiconductor superlattices have received most attention, their novel controllable properties (Chang and Giessen 1985) providing the impetus for the whole of multilayer research. Multilayers based on metals and alloys have also been widely studied; many aspects of these are treated by Barbee et al. 1988.

1. Production

Metallic multilayers are distinguished from laminated composites by their very small layer thicknesses. They are normally in thin-film form, and may consist of a few hundred layers. The first attempt to produce such a material was by DuMond and Youtz (1940) with the intention of using the multilayer as a reflector to calibrate x-ray wavelengths. Although their control of layer thickness and uniformity was not sufficiently precise to give a useful calibration, they found x-ray reflections due to the artificial layering (Sect. 3) and they noted that rapid homogenization due to interdiffusion caused the reflections to disappear. Their materials were polycrystalline copper with a layered gold content produced by evaporation. Since their work, great progress has been made in developing new deposition techniques with better control of deposition rate and of deposited structure (Barbee 1985). Metallic multilayers have been made not only by evaporation, by also by sputtering, molecular beam epitaxy (MBE), chemical (and pulsed-laser photochemical) vapor deposition and electrodeposition (see *Electrodeposited Multilayer Metallic Coatings*, Suppl. 1). The most accurate techniques can produce layers only one atom thick, and can control deposition rates to better than 0.1%. This level of precision in the dimension perpendicular to the substrate greatly exceeds the precision possible in lateral patterning of thin films by lithography. Coarser less-precisely-controlled multilayers can be produced by mechanical reduction of macroscopic laminates (Atzmon et al. 1985).

2. Classification of Structure

In this article the general term multilayer is used. The combined thickness of two successive layers is referred to as the "repeat distance." The simplest multilayers to produce are those consisting of two polycrystalline metals (which will normally have a preferred orientation with their close-packed planes parallel to the substrate). The main interest in artificial multilayers, however, concerns materials with better-defined structures. Where the layers each have the same single crystalline structure and are fully coherent (as exemplified by MBE-deposited semiconductor multilayers), the materials can be termed superlattices. (Metallic films described as single crystalline may not be truly so, but have an in-plane grain dimension much greater than the total film thickness.) Superlattices can be considered to be the result of imposing a composition modulation on a single crystalline lattice. Materials in which each layer is amorphous can also be considered to be "compositionally modulated." Multilayers with layers of two different structures are possible. When the layers are each single crystalline and display an orientation relationship with each other, the term "layered ultrathin coherent structures" (LUCS) has been used. In some cases only the crystal axes perpendicular to the interfaces are fixed; in other cases the in-plane orientation is also defined and successive layers of the same composition have identical orientation. In the latter type of multilayer, if there is good matching of atomic rows in the two structures, the interfaces can be ordered, leading to long-range coherence of the atomic planes through the thickness (Clemens and Gay 1987). If there is not good matching, the interfaces are disordered, or amorphous, and long-range coherence is lost.

3. Characterization

Complementing the production techniques are the techniques now available for the characterization of fine microstructures. Diffraction of x rays, neutrons or electrons is the most widely used technique for studying the layering. The artificial periodicity gives rise to satellite reflections around the zero-order beam. In a superlattice, or a multilayer with long-range coherence (Sect. 2), satellites also appear around higher-angle Bragg reflections. The modulation of average scattering factor contributes to all satellite intensities, and the modulation of lattice parameter to all except those around the zero-order beam (Guinier 1963). Diffraction techniques cannot provide information about local interface quality. Local structural information can come from, for example, Mössbauer spectroscopy, EXAFS (see *X-Ray Absorption Spectroscopy: EXAFS and XANES Techniques*) or transmission electron microscopy (TEM) (Stobbs 1988). (For an example of TEM showing layers edge-on, see *Electrodeposited Multilayer Metallic Coatings*, Suppl. 1.) Composition profiling of a multilayer can be achieved; for example, by Auger electron spectroscopy of the sample surface as it is ion milled, by Rutherford backscattering spectroscopy or by anodization profiling. Structural characterization during deposition can be important in controlling the process; x-ray diffraction (Spiller et al. 1980) and reflection high-energy electron diffraction (RHEED) have been used with greatest success.

4. Structural Stability

Metallic multilayers can give examples of pseudomorphism in which a nonequilibrium phase is stabilized because coherent matching to the other phase lowers the interfacial energy of the multilayer. Because of kinetic constraints, the nonequilibrium phase can persist at thicknesses greater than the critical thickness at which bulk free energies should dominate over the interfacial energies. The stabilized phase can be one expected for the material under other conditions (e.g., the body-centered-cubic (bcc) phase of zirconium stabilized in niobium–zirconium multilayers (Lowe and Geballe 1984) is a high-temperature allotrope of that element) or be quite new (e.g., a bcc phase of germanium stabilized in molybdenum–germanium multilayers (Wilson and Bienenstock 1988)). The metastable equilibrium in multilayers is complicated by the stresses in the layers, and Gibbs' phase rule and the common tangent construction do not apply (Johnson 1988). Even without nucleation barriers there may be more than one metastable state of the system.

The stability of multilayer structures is important if their properties are to be exploited. For metallic layers only a few nanometers thick, for example, lead in lead–germanium multilayers, the melting point may be measurably depressed by the interfacial energies (Willens et al. 1982). The crystallization temperatures of amorphous layers (e.g., of silicon or germanium), can be raised or lowered relative to the bulk. All of these effects may involve some diffusional mixing at the interfaces (Sevenhans et al. 1988). In addition to mixing, it is possible that the two phases in a multilayer may react. An example is the reaction of polycrystalline early and late transition-metal layers to form an amorphous alloy (see *Amorphization Reactions, Solid State*, Suppl. 1). A similar reaction may occur in transition-metal–amorphous-silicon multilayers; the rapid kinetics possible in the multilayer and the heat evolved enable the reaction to proceed explosively (Clevenger et al. 1988).

In multilayers of the compositionally modulated type, crystalline or amorphous, interdiffusion can be quantitatively analyzed (Greer and Spaepen 1985). Mostly, this is done by measurement of x-ray satellite intensities during annealing. Because of the short diffusion distances, exceptionally low diffusivities ($\geq 10^{-27}\ m^2\ s^{-1}$) can be determined (for comparison with other techniques, see *Metallic Glasses: Diffusion*, Suppl. 1), usefully permitting measurement at low temperature in metastable materials and unrelaxed structures. The technique has been applied to both metals and semiconductors in crystalline and amorphous states, in studies of homogenization and of phase separation. In the steep concentration gradients achievable, the interdiffusion coefficient $\tilde{D}$ shows a dependence on the repeat distance. Analysis of this dependence yields information on the mixing thermodynamics of the system. Strain effects on $\tilde{D}$, arising mainly from coherency, are also detectable, as are effects due to the composition dependence of $\tilde{D}$.

5. Properties and Applications

Apart from applications as samples for diffusion measurements, multilayers may be of scientific and technological interest because of special properties. The origins of these have been reviewed by Schuller (1988) and can be categorized as: thin-film effects due to limited thickness of one type of layer; interface effects from interaction with neighboring layers; coupling effects between layers of the same type across the intervening layers; and periodicity effects dependent on the overall periodicity of the multilayer. Examples of anomalous properties arising from these effects are given below. The layer thicknesses, or repeat distances, at which anomalies are found vary widely from one property to another.

5.1 X-Ray and Neutron Optics

Metallic multilayers have for some time been in use as optical elements for x rays (particularly soft x rays with wavelength $\lambda \simeq 5$ nm) and neutrons ($\lambda = 0.05$–5 nm). There is great technological potential in this area. Multilayers for x-ray reflection have been used to replace and extend the capabilities of conventional long-period crystals in spectrometers. In addition, new types of optical element with controlled reflectivity can be fabricated by extending to smaller layer thickness the principles of multilayer design which are well developed for visible light (see *Optical Thin Films: Production and Use*, Suppl. 1). These elements can be applied as mirrors (particularly for normal incidence), monochromators or polarizing mirrors. As described by Spiller et al. (1980), for x rays one layer material must have a high electron density and the other a low electron density: commonly tungsten and carbon are used, deposited in respectively microcrystalline and amorphous form. A problem for x rays (unlike visible light) is that there is severe absorption in the optically denser layers. This is alleviated by using "quasiperiodic" multilayers in which the combined two-layer thickness is kept constant, but the ratio of, for example, tungsten to carbon thickness is decreased from the bottom to the top of the multilayer. Particularly for larger wavelengths, aperiodic designs offer still better opportunities for increasing both the range of reflected wavelength and the integrated reflectivity. Advanced designs based on periodic multilayers have combined them with a deposited spacer to form a Fabry–Perot etalon (Barbee 1985) and used them etched in a grating pattern as high-efficiency dispersion elements (Barbee 1988). Linear zone plates (~1 mm thick) for focusing an x-ray point source to a line have been made by slicing multilayers (of aluminum–tantalum) perpendicular to the layers (Bionta et al. 1988).

For neutrons, periodic multilayers are used as monochromators; aperiodic multilayers of ferromagnetic and nonmagnetic materials (typically iron–germanium or iron–tungsten) are used as polarizing "supermirrors" with reflectivity over a range of angles for one spin state (Majkrzak 1989).

5.2 Magnetic Properties

Multilayers of ferromagnetic and nonmagnetic layers offer opportunities for the manipulation both of the structure and stress state of the magnetic material and for the control of interactions between the magnetic layers. They have great technological potential, particularly for the control of coercivity and anisotropy. The achievement of perpendicular anisotropy in thin films is important for high-density magnetic recording. The magnetic properties of multilayers have been reviewed by Schuller (1988). The saturation magnetization per magnetic atom in multilayers tends to decrease with decreasing thickness of the magnetic layers. The effect may be complicated by diffusional mixing at the interfaces. When this does not arise, it is consistent with the presence of "dead" (i.e., unmagnetized) layers in the magnetic material near the interfaces. Another thin-film effect is a temperature dependence of the saturation magnetization of the form expected for two-dimensional materials.

Layer thickness may also be used to control anisotropy. In, for example, palladium–cobalt multilayers, the preferred magnetization direction becomes perpendicular to the plane of the film for magnetic (cobalt) layers below a critical thickness (Garcia et al. 1985). Coupling between magnetic layers has been explored in some detail. In, for example, copper–nickel multilayers, it has been shown that the Curie temperature increases with decreasing thickness of the copper layers in the manner expected from theory. In gadolinium–yttrium multilayers with constant gadolinium thickness, the yttrium thickness can be varied to give parallel or antiparallel coupling between the layers (Majkrzak et al. 1987). In dysprosium–yttrium multilayers, there is coherent helical coupling of the magnetic layers (Hong et al. 1987). In nickel–molybdenum multilayers, the variation of the magnon (quantized spin wave) frequency with magnetic field is dependent on the thickness of the nonmagnetic molybdenum layers. The measurements are in good agreement with predictions of periodicity effects based on the development of bands for the magnons when there is coupling between the magnetic layers (Schuller 1988).

5.3 Electron Transport

The normal-state electrical resistivity of multilayers (Falco and Schuller 1985) would be expected to show anisotropy, but this has not been clearly demonstrated because of the difficulty of through-thickness measurements. Also not demonstrated, possibly because the quality of the very thin layers required has been insufficient, are anomalous transport properties arising from bandgaps due to the layering. The main effect of layering is electron scattering at the interfaces, giving a resistivity inversely proportional to the layer thickness, with an upper limit of about $0.015\ \Omega\,m$. In common with other strongly scattering materials, as the resistivity increases its temperature coefficient becomes less positive, and near the upper limit of resistivity the coefficient becomes negative. The ability to achieve a zero-valued temperature coefficient of resistivity is of significance for temperature-independent thin-film resistors. For other transport properties (e.g., thermopower, magnetoresistance and the Hall effect) only preliminary studies are available.

Superconducting interactions can be particularly well studied using artificial multilayers because the characteristic coherence length (5–100 nm) can be considerably greater than the layer thickness (Ruggiero and Beasley 1985). Multilayers can be made of either two superconducting materials, or a superconductor and a normal metal, or a superconductor and an insulator. The last category is of interest for comparison with naturally occurring layered superconductors (e.g., dichalcogenides, intercalated graphite and high transition temperature copper-oxide-based systems). In multilayers with one layer material nonsuperconducting, the critical temperature of the superconducting layers generally decreases as their thickness is reduced, owing to the proximity effect. The critical temperature also decreases as the thickness of the nonsuperconducting spacer layers is increased. The upper critical field in multilayers is anisotropic and its temperature dependence (e.g., in niobium–germanium multilayers) shows clear dimensional effects. The transition from three-dimensional to two-dimensional behavior occurs as the thickness of nonsuperconducting layers exceeds the coherence length, and this can be achieved by increasing the thickness or by lowering the temperature to decrease the coherence length. Periodicity effects are revealed in the dependence of the parallel critical field on the layer thickness, arising from the interaction between the spacing of the vortex lattice and that of the multilayer.

5.4 Mechanical Properties

One of the most remarkable properties associated with metallic multilayers is the enhanced elastic modulus for (among other modes) biaxial stretching of the thin films (Cammarata 1988). This "supermodulus effect" is found in compositionally modulated multilayers of cubic-close-packed (ccp) metals (e.g., copper–palladium) in (111) orientation. The biaxial modulus is normal, except near a particular modulation wavelength of 2–3 nm for which it shows a maximum, being up to four times greater than the modulus expected for the composite or for the homogeneous solid solution that could be formed by the elements. The effect remains controversial as the evidence is disputed by some and the origins are not agreed upon. Proposed origins are: electronic effects based on an interaction between the Fermi surface and a new Brillouin zone due to the layering; compressive and tensile strains in alternating layers due to coherency; and overall compressive strain in the multilayer due to interfacial stresses (Cammarata and Sieradski 1989). The peak in modulus has been correlated with peaks in effective interdiffusivity and in thermopower.

Epitaxial multilayers of ccp and bcc metals (e.g., nickel–molybdenum) have shown modulus dehancements, associated with expansion at the interfaces (perhaps due to disordering) perpendicular to the film plane. The effect occurs at the same repeat distance as the change from positive to negative coefficient of resistivity (Falco and Schuller 1985).

Higher yield stress (measured by indentation) is found associated with the enhanced modulus in ccp multilayers. In addition, any multilayer of materials of different stiffness will impede dislocation motion (Koehler 1970). In, for example, aluminum–copper multilayers, tensile strength varies as $d^{-1/2}$ (where d is the layer thickness) up to a limit where d (for each metal) becomes less than 70 nm; that is, less than the critical thickness for dislocation generation (Tsakalakos and Jankowski 1986). Such effects may become technologically significant: for example, 1 mm thick sheet of aluminum–transition-metal multilayer has been produced by high-rate evaporation followed by rolling. Such sheet, with aluminum-layer thickness 20–1600 nm and transition-metal thickness 0.1–21 nm, shows high strength and high temperature stability (Bickerdike et al. 1985).

5.5 Other Applications

Aluminum–transition-metal multilayers are also promising for the conducting strips on integrated circuits. These strips, commonly of deposited aluminum, suffer at high current density from electromigration damage (the growth of hillocks and voids)—a problem exacerbated by continuing miniaturization. Transition-metal layers in a multilayer constrain aluminum hillock growth and maintain electrical continuity in the presence of voids in the low-resistivity aluminum. Lifetimes are improved by a factor of up to 100, at a lower cost in increased resistivity than would be required in the alternative approach of adding solute elements (Gardner and Saraswat 1988).

With the wide possibilities for structural manipulation and novel properties offered by metallic multilayers, their further technological application can be expected. Also likely is more work on ceramic materials. Techniques for depositing metal–ceramic multilayers are being developed; these are of interest for their mechanical properties (Moustakas et al. 1988). All-ceramic multilayers have been made and hold much promise (McKee et al. 1988). For example, piezoelectric multilayers (e.g., CuCl–CuBr) are artificial ferroelectric materials because of the stresses in the layers (Wong et al. 1982).

See also: Electrodeposited Multilayer Metallic Coatings (Suppl. 1); Strained-Layer Superlattices (Suppl. 2)

Bibliography

Atzmon M, Unruh K M, Johnson W L 1985 Formation and characterization of amorphous erbium-based alloys prepared by near-isothermal rolling of elemental composites. *J. Appl. Phys.* 58: 3865–70

Barbee T W 1985 Synthesis of multilayer structures by physical vapor deposition techniques. In: Chang and Giessen 1985, pp. 313–37

Barbee T W 1988 Combined microstructure x-ray optics; Multilayer diffraction gratings. In: Barbee et al. 1988, pp. 307–14

Barbee T W, Spaepen F, Greer A L (eds.) 1988 *Multilayers: Synthesis, Properties and Non-Electronic Applications*. Materials Research Society, Pittsburgh, PA

Bickerdike R L, Clark D, Easterbrook J N, Hughes G, Mair W N, Partridge P G, Ranson H C 1985 Microstructures and tensile properties of vapour deposited aluminium alloys. Part 1: Layered microstructures. *Int. J. Rapid Solidification* 1: 305–25

Bionta R M, Jankowski A F, Makowiecki D M 1988 Fabrication and evaluation of transmissive multilayer optics for 8 keV x-rays. In: Barbee et al. 1988, pp. 257–63

Cammarata R C 1988 Elastic properties of artificially layered thin films. In: Barbee et al. 1988, pp. 315–25

Cammarata R C, Sieradzki K 1989 Effects of surface stress on the elastic moduli of thin films and superlattices. *Phys. Rev. Lett.* 62: 2005–8

Chang L L, Giessen B C (eds.) 1985 *Synthetic Modulated Structures*. Academic Press, Orlando, FL

Clemens B M, Gay J G 1987 Effect of layer thickness fluctuations on superlattice diffraction. *Phys. Rev. B* 35: 9337–40

Clevenger L A, Thompson C V, Cammarata R C, Tu K N 1988 The effect of layer thickness on the reaction kinetics of nickel/silicon multilayer films. In: Barbee et al. 1988, pp. 191–6

DuMond J, Youtz J P 1940 An x-ray method of determining rates of diffusion in the solid state. *J. Appl. Phys.* 11: 357–65

Falco C M, Schuller I K 1985 Electronic and magnetic properties of metallic superlattices. In: Chang and Giessen 1985, pp. 339–64

Garcia P F, Meinhaldt A D, Suna A 1985 Perpendicular magnetic anisotropy in Pd/Co thin film layered structures. *Appl. Phys. Lett.* 47: 178–80

Gardner D S, Saraswat K 1988 Multilayered interconnections for VLSI. In: Barbee et al. 1988, pp. 343–54

Greer A L, Spaepen F 1985 Diffusion. In: Chang and Giessen 1985, pp. 419–86

Guinier A 1963 *X-Ray Diffraction*. Freeman, San Francisco, CA

Hong M, Fleming R M, Kwo J, Schneemeyer L F, Waszczak J V, Mannaerts J P, Majkrzak C F, Gibbs D, Bohr J 1987 Synthetic magnetic rare-earth Dy–Y superlattices. *J. Appl. Phys.* 61: 4052–4

Johnson W C 1988 On the existence of multiple equilibrium states in strained-layer superlattices. In: Barbee et al. 1988, pp. 61–6

Koehler J S 1970 Attempt to design a strong solid. *Phys. Rev. B* 2: 547–51

Lowe W P, Geballe T H 1984 NbZr multilayers. I. Structure and superconductivity. *Phys. Rev. B* 29: 4961–8

McKee R A, List F A, Walker F J 1988 MBE growth of compositionally modulated ceramics. In: Barbee et al. 1988, pp. 35–40

Majkrzak C F 1989 Polarized neutron scattering methods and studies involving artificial superlattices. *Physica B* 156 (Jan): 619–26

Majkrzak C F, Cable J W, Kwo J, Hong M, McWhan D B, Yafet Y, Waszczak J V, Grimm H, Vettier C 1987 Polarized neutron diffraction studies of Gd–Y synthetic superlattices. *J. Appl. Phys.* 61: 4055–7

Moustakas T D, Koo J Y, Ozekcin A 1988 Growth and structure of tungsten carbide–transition metal superlattices. In: Barbee et al. 1988, pp. 41–6

Ruggiero S T, Beasley M R 1985 Synthetically layered superconductors. In: Chang and Giessen 1985, pp. 365–417

Schuller I K 1988 Magnetic superlattices. In: Barbee et al. 1988, pp. 335–41

Sevenhans W, Vanderstraeten H, Locquet J P, Bruyseraede Y, Homma H, Schuller I K 1988 Crystallization and melting in multilayered structures. In: Barbee et al. 1988, pp. 217–22

Spiller E, Segmüller A, Haelbich R-P 1980 The fabrication of multilayer x-ray mirrors. *Ann. N. Y. Acad. Sci.* 342: 188–98

Stobbs W M 1988 Techniques for characterising artificial layer structures using transmission electron microscopy. In: Barbee et al. 1988, pp. 121–31

Tsakalakos T (ed.) 1984 *Modulated Structure Materials*. Nijhoff, Dordrecht

Tsakalakos T, Jankowski A F 1986 Mechanical properties of composition-modulated metallic foils. *Ann. Rev. Mater. Sci.* 16: 293–313

Willens R H, Kornblit A, Testardi L R, Nakahara S 1982 Melting of Pb as it approaches a two-dimensional solid. *Phys. Rev. B* 25: 290–6

Wilson L, Bienenstock A 1988 Atomic arrangements in short period Mo–Ge multilayers determined by x-ray anomalous scattering and EXAFS. In: Barbee et al. 1988, pp. 69–78

Wong H K, Wong G K, Ketterson J B 1982 Ferroelectricity and coherent phonon generation in piezoelectric composition-modulated structures. *J. Appl. Phys.* 53: 6834–8

A. L. Greer
[University of Cambridge, Cambridge, UK]

N

Natural-Fiber-Based Composites

Natural fibers such as jute, sisal, sunhemp, banana and coir are grown in many parts of the world. Some of them have aspect ratios (ratio of length to diameter) greater than 1000 and can be easily woven. These fibers are extensively used for cordage and twine, sacks, fishnets, matting and rope, and as filling for mattresses and cushions (rubberized coir is an example). Recent reports indicate that plant-based natural fibers may be used as reinforcements in polymer composites, replacing to some extent more expensive and nonrenewable synthetic fibers such as glass.

1. Structure and Properties of Natural Fibers

Plant-based natural fibers are lignocellulosic, consisting of cellulose microfibrils in an amorphous matrix of lignin and hemicellulose. They consist of several hollow fibrils which run all along their length. Each fibril exhibits a complex layered structure, with a thin primary wall encircling a thicker secondary layer, and is similar to the structure of a single wood-pulp fiber (see *Wood Ultrastructure*).

The secondary layer is made up of three distinct layers, the middle one being by far the thickest and the most important in determining mechanical properties. In this layer parallel celullose microfibrils are wound helically around the fibrils; the angle between the fiber axis and the microfibrils is termed the microfibril angle. Natural fibers are themselves cellulose-fiber-reinforced materials in which the microfibril angle and cellulose content determine the mechanical behavior of the fiber. Stress–strain curves of these fibers have an initial portion which is approximately linear, followed by a survilinear portion to the point of maximum stress. Fibers with high microfibril angle such as coir (~45°) generally exhibit high elongation to break (~35%), whereas fibers such as sunhemp, with high cellulose content (~80%) and low microfibril angle (~10°), possess high tensile strength (~400 MPa) and low elongation (~1%).

2. Interfacial Bonding with Polymers

The overall performance of any fiber-reinforced polymer composite depends to a large extent on the fiber–matrix interface, which in turn is governed by the surface topography of the fiber and by the chemical compatibility of fiber surface and resin matrix. Natural-fiber surfaces are fairly irregular, which should in principle enhance the fiber–matrix interfacial bond. However, in many cases this advantage is likely to be offset by chemical incompatibility between the fiber surface and the resin. Some natural fibers have an outer waxy layer of fatty acids and their condensation products; coir fiber, for example, has a 3–5 μm thick waxy layer. Fatty acids, being long-chain aliphatic compounds, are not compatible with common resins such as polyester. It is thus difficult to incorporate coir fibers in polyester without introducing significant amounts of porosity in the composite. This problem may be solved by removing the outer waxy layer of the coir fiber by a simple alkali pretreatment similar to the mercerization of cotton (Prasad et al. 1982). Such treatment is not found to be a prerequisite for polyester composites reinforced with sunhemp or jute fibers.

3. Behavior of Composites under Impact and Tension

The principle advantage of natural-fiber-reinforced polymer composites stems from their ability to absorb tremendous amounts of energy during impact fracture. When a fiber-reinforced composite fractures, three types of surface are created: (a) fiber cross-sectional surfaces, (b) matrix surfaces, and (c) surfaces betwen matrix and fiber. The toughness contribution of each of these "new" surfaces will depend on the work done to create a unit area of the surface. Assuming that the fibers break randomly, some energy is absorbed while pulling the broken fibers out of the matrix after fracture. In addition, another energy-absorbing mechanism has been proposed to account for the redistribution of strain energy from fiber to matrix after a fiber breaks. The toughness of a composite is the summation of the contributions of these three different sources.

In polyester-matrix composites reinforced with unidirectionally aligned subhemp or jute fibers, the work of fracture shows a linear increase with fiber volume fraction V_f (Sanadi et al. 1986a, Roe and Ansell 1985). The work of fracture (Izod) for a 0.24 V_f sunhemp–polyester composite was reported to be 21 kJ m^{-2}; this is 15 times higher than the work of fracture for polyester resin alone. Although the high toughness of subhemp–polyester composites can be due to the fiber pullout work and the work done in creating new surfaces at the fiber–matrix interface (Sanadi et al. 1986a), one has also to consider the complex fracture mode of natural fibers as compared to the planar failure obtained in polymers reinforced with glass and carbon fibers. Since natural fibers are themselves cellulose fibril-reinforced composite materials, their fracture modes include uncoiling of fibrils, fibril pullout, plastic deformation of fibrils, fibril splitting and diversion of the crack at the fibril–fibril interface. These fracture mechanisms, which are not seen in glass-fiber-reinforced plastics (GFRP), contribute to the high toughness of natural-fiber-reinforced composites (Sanadi et al. 1986b).

Natural fibers possessing a range of tensile properties (strength, modulus and elongation) are available. High-modulus E and high-strength σ fibers such as jute, sunhemp and sisal can therefore act as reinforcements in polyester (sunhemp: $E \sim 40$ GPa, $\sigma \sim 400$ MPa; polyester: $E = 3.5$ GPa, $\sigma = 48$ MPa). For polyester reinforced with jute or subhemp fiber, both modulus and tensile strength were reported to increase linearly with V_f according to the rule of mixtures.

4. Problems and Prospects

Plant-based materials have always been important in the plastics industry. The early phenolics were modified with wood flour to lower their cost and improve processibility, and plywood is a composite of thin layers of wood joined by adhesive (see *Plywood*, Suppl. 2). However, there is considerable reluctance to use natural-fiber-reinforced composites for structural applications because it is very difficult to obtain reproducible mechanical properties with natural fibers. The tensile strengths of fiber samples obtained from different sources can sometimes vary by a factor of three.

Natural fibers are extracted from the plant by a process known as retting (see *Materials of Biological Origin: An Overview*). As this process is not well standardized, fibers with varying degrees of flaws are generally produced. Mechanical properties are also dependent on age and moisture content. Unless proper standards are developed and adopted in the extraction of natural fibers from plants, their potential in designing tough composites will be limited.

Natural-fiber-reinforced polymers absorb more moisture than do glass-fiber-reinforced plastics. This could be a serious drawback if these composites were to be used in outdoor applications where they are frequently exposed to rain. This difficulty may be minimized if the outer surface of the composite is protected by applying a gel coat or by hybridizing the outer layers with glass fibers. The natural fibers will, however, remain exposed to the moisture that diffuses through polyester resin.

On technoeconomic grounds, considering particularly specific stiffness per unit cost, composites like sunhemp–polyester (Sanadi et al. 1985) and jute–polyester (Roe and Ansell 1985) are far superior to GFRP. In applications where stiffness is important but strength is not a priority, such as a suitcase or a tabletop, the advantages of natural-fiber-reinforced polymers can be overwhelming.

See also: Natural Composite Materials; Natural Fibers in the World Economy

Bibliography

Kelly A, Macmillan N H 1986 *Strong Solids*, 3rd edn. Clarendon, Oxford

McLaughlin E C, Tait R A 1980 Failure mechanism of plant fibers. *J. Mater. Sci.* 15: 89–95

Prasad S V, Pavitran C Rohatgi P K 1982 Alkali treatment of coir fibers for coir-polyester composites. *J. Mater. Sci.* 18: 1443–54

Roe P J, Ansell M P 1985 Jute-reinforced polyester composites. *J. Mater. Sci.* 20: 4015–20

Sanadi A R, Prasad S V, Rohatgi P K 1985 Natural fibers and agro-wastes as fillers and reinforcements in polymer composites. *J. Sci. Ind. Res. (India)* 44: 437–42

Sanadi A R, Prasad S V, Rohatgi P K 1986a Sunhemp fibre reinforced polyester, Part I: Analysis of tensile and impact results. *J. Mater. Sci.* 21: 4299–304

Sanadi A R, Prasad S V, Rohatgi P K 1986b SEM observations on the origins of toughness of natural fiber-polymer composites. *J. Mater. Sci. Lett.* 5: 562–4

S. V. Prasad
[WRDC/MLBT, Dayton, Ohio, USA]

Nitrogen in Steels

Nitrogen occurs in all steels and although its solubility is generally low, it can exert large effects. Some of these effects are detrimental, often being associated with embrittlement. However, nitrogen also has beneficial effects which have led to the development of steels containing enhanced nitrogen. In many cases the beneficial effects of nitrogen are due to interactions with alloying elements that precipitate alloy nitrides or form clusters between nitrogen and the alloying elements. These interactions influence many phenomena which will be reviewed with regard to the appropriate steels in this article.

In order to understand the effects of nitrogen the much greater solubility of nitrogen compared with carbon must be appreciated. Also, many alloy nitrides have much lower solubilities than iron nitrides in both ferrite and austenite and their solubility is considerably less in ferrite than in austenite. In addition, alloy nitrides are more stable and less soluble than the corresponding carbides so that they grow more slowly and form finer particles. These differences have a fundamental influence on the effects of nitrogen on the properties of alloy steels.

1. Carbon–Manganese Steels

Carbon–manganese steels used in the as-rolled or normalized condition comprise ferrite–pearlite structures. In the formable steels that, cold rolled

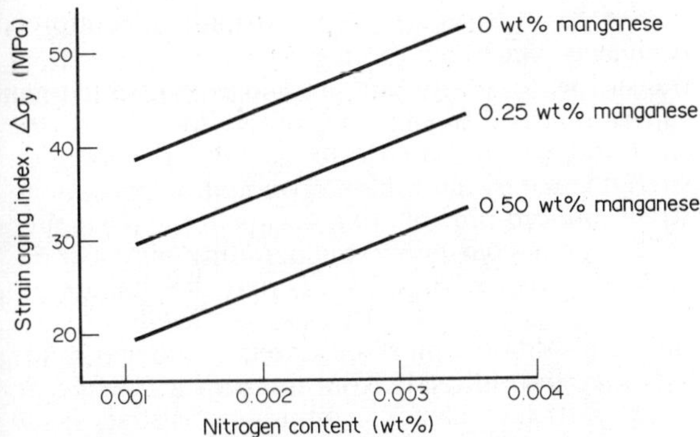

Figure 1
Effect of nitrogen and manganese content on strain aging in formable mild steels (after Duckworth and Baird)

and annealed, contain spheroidized carbides nitrogen has frequently been seen as undesirable because it introduces discontinuous yielding and strain aging, which give stretcher strains during forming. Nitrogen is predominantly responsible for strain aging and increased amounts of nitrogen accelerate strain aging and increase its intensity (see Fig. 1).

Manganese decreases strain aging by interacting with nitrogen atoms to form manganese–nitrogen clusters and also lowers nitrogen solubility. Other elements are such strong nitride formers that they form nitrides such as AlN and TiN that considerably minimize strain aging.

Aluminum-treated steels, which are stabilized against strain aging, show the advantage of nitrogen in that the precipitation of AlN results in improved deep drawability. This is achieved by preventing AlN precipitation during hot rolling and coiling, and by the use of heating rates during annealing that allow AlN to precipitate on the recovered substructure prior to recrystallization. The beneficial cube-on-corner annealing texture is thereby intensified, which increases the value of the plastic strain ratio $\bar{r}$.

In the structural carbon–manganese steels, nitrogen is used in conjunction with aluminum to precipitate AlN which refines the austenite grain size and increases the grain-coarsening temperature. This gives a much refined ferrite grain size, which results in both improved strength and improved toughness. The role of the AlN is to act as grain boundary pinning particles. For any given nitrogen content there is a maximum grain-coarsening temperature at an aluminum content for which the aluminum–nitrogen ratio equals the stoichiometric ratio for AlN. This is due to the increasing particle sizes of the undissolved AlN with increasing hyperstoichiometry and because at the stoichiometric ratio there is the largest volume fraction of fine AlN particles due to the temperature dependence of AlN solubility being a maximum. A similar effect has been observed for vanadium–nitrogen grain-refined steels. Since AlN is not readily strain induced to precipitate in austenite during hot rolling it cannot be used to retard recrystallization for low-temperature controlled rolling processes. However, an early method for recrystallization controlled rolling used increases in both aluminum and nitrogen contents in the stoichiometric ratio to preserve a fine initial austenite grain size and a refined recrystallized austenite grain size during hot rolling.

In other carbon–manganese steels used for boiler plate steels, nitrogen in solution in ferrite increases the proof stress at temperatures of 100–300 °C and this is accentuated by increasing the level of manganese and other elements with an affinity for nitrogen. The effect is removed by aluminum which precipitates the nitrogen as AlN. The phenomenon is due to dynamic strain aging from interactions between the manganese and the nitrogen. These strengthening effects are also manifested during creep testing and to obtain the required creep strength in boiler plate steels the nitrogen should be available in solution in the ferrite. Aluminum "killing" is therefore detrimental because it precipitates nitrogen as AlN, and it has been shown that the creep rate decreases with increasing nitrogen dissolved in the ferrite but not with increasing AlN. Hence, boiler plate steels are not aluminum treated and even in silicon-killed steels tempering can precipitate silicon nitride which decreases the creep strength.

2. *Microalloyed High-Strength Low-Alloy Steels*

The major property requirements for high-strength low-alloy (HSLA) steels are high yield strength, low impact transition temperature and good weldability. The high yield strength and low impact transition temperature are achieved through ferrite grain refinement, and nitrogen influences this. However, nitrogen dissolved in the ferrite increases the yield stress by 40 MPa and also increases the impact transition temperature by 70–100 °C for 0.01 wt% nitrogen content. The strengthening is small, as the solid solubility of nitrogen in ferrite is limited, but the embrittlement can be marked. Thus nitrogen dissolved in ferrite must be minimized; this is done in the microalloyed steels by additions of vanadium, niobium or titanium which form stable nitrides. These nitrides grain refine the austenite and thereby the ferrite, retard recrystallization of the austenite by strain-induced precipitation which allows the implementation of controlled rolling processes, and precipitation strengthen the ferrite under appropriate processing conditions. The detrimental effects of precipitation strengthening on toughness are less than those of nitrogen in solution.

Much work has been done on the effect of nitrogen in HSLA steels, and only a few salient points can be summarized here. In general, modern steels contain 0.008–0.010 wt% nitrogen, an enhanced level usually being 0.015–0.020 wt%. The microalloy nitrides increase in stability and decrease in solubility in the order VN, NbN, TiN. The lower the solubility, the greater is the resistance to particle coarsening. The vanadium carbides–nitrides V(CN) can be dissolved and precipitated in temperature ranges commonly used for thermomechanical and general mill processing. Consequently, enhanced nitrogen levels have usually been added to steels containing 0.10–0.20 wt% vanadium, although recently small titanium additions have been used to provide extra grain boundary pinning by very fine, stable, TiN particles, while allowing excess nitrogen to aid precipitation strengthening by V(CN). With increasing nitrogen the austenite grain size decreases and the grain-coarsening temperature increases due to the greater volume fraction of VN. This effect is not only important in normalized steels, but also in thermomechanically processed steels where a fine reheated grain size is advantageous. Being the more stable phase, VN precipitates in preference to VC particularly at higher temperatures, but in the enhanced nitrogen steels there is still sufficient nitrogen for V(CN) to cause precipitation strengthening during and after the transformation of the austenite. A feature of microalloying is for strain-induced precipitation in the austenite during rolling to inhibit recrystallization and so allow unrecrystallized austenite to be present, which gives a finer ultimate ferrite grain size. The effect can be obtained with vanadium steels, but the temperature for most rapid VN precipitation is 900 °C and, while effective controlled rolling can be carried out in this regime with vanadium–nitrogen steels, it is more usual to employ combinations of niobium and vanadium, the niobium facilitating the controlled rolling and the vanadium giving precipitation strengthening.

Controlled rolling in the unrecrystallized regime has limitations due to mill design and because strain-induced precipitation detracts from precipitation strengthening. Consequently, a concept has evolved which enables very fine recrystallized austenite to be produced by hot rolling at much higher finishing temperatures. This process is called recrystallization controlled rolling and can be used with rolling mills not designed for or suited to the more conventional low-temperature controlled rolling. In this process, a small titanium addition of about 0.01–0.02 wt% is added, which forms fine TiN particles. These TiN particles restrict austenite grain growth at the reheating temperature for rolling and maintain a fine reheated austenite grain size, which is continuously recrystallized during hot rolling to even finer austenite grains. The TiN particles prevent these very fine recrystallized austenite grains from growing. The rolling finishing temperature is above the recrystallization stop temperature so that a fully recrystallized austenite of very fine grain size is produced which on cooling to room temperature transforms to a very fine grained ferrite which has superior strength and toughness. Moreover, the steel still has the ability to be fully precipitation strengthened by interphase precipitation because no strain-induced precipitation occurs in the austenite.

The grain-coarsening temperature may be decreased by up to 300 °C by TiN particles that are too large, as can occur in large ingots, and too high a titanium addition must be avoided otherwise the TiN may form directly from the melt as ineffective large particles. Hence, continuous casting is an advantage. An enhanced nitrogen content, making the steel hypostoichiometric with respect to titanium, can be advantageous as the dependence of the TiN solvus temperature on nitrogen is then low and the decrease in titanium dissolved in the austenite minimizes TiN particle growth. Nitrogen not combined as TiN is available to form VN or V(CN) during transformation, which enhances precipitation strengthening.

The refined ferrite grain size so necessary for HSLA steels is obtained by maximizing the ratio of the austenite to ferrite grain size, which increases with increasing effective austenite grain boundary area S_v. Increasing nitrogen in vanadium steels, by decreasing the austenite grain size in normalized steels or by increasing S_v in controlled rolled steels, markedly refines the ferrite grain size. Increasing the cooling rate also refines the ferrite grain size by depressing the transformation temperature.

Precipitation strengthening results from interphase precipitation during the austenite–ferrite transformation and from precipitation in the ferrite after transformation is complete. Since nitrogen is not completely precipitated in the austenite, some is available for precipitation strengthening and the yield stress increases with increasing nitrogen content (see Fig. 2). The precipitates cause increases of 5–10 MPa in yield strength per 0.001 wt% increase of nitrogen. Controlled rolling (see *High-Strength Low-Alloy Steels; Hot Rolling*) gives less precipitation strengthening than conventional hot rolling due to the strain-induced precipitation of VN in the austenite and in the normalized condition the strengthening increases up to the stoichiometric vanadium–nitrogen ratio. These effects result from increased nitrogen contents giving finer and greater amounts of V(CN) interphase precipitation, together with a smaller intersheet spacing and a suppression of the less strengthening fibrous interphase morphology. Moreover, nitrogen extends the temperature range over which interphase precipitation occurs because it depresses the Widmanstätten ferrite temperature range and the B_s temperature (i.e. the highest temperature at which bainite forms from austenite). Hence nitrogen makes precipitation strengthening less sensitive to

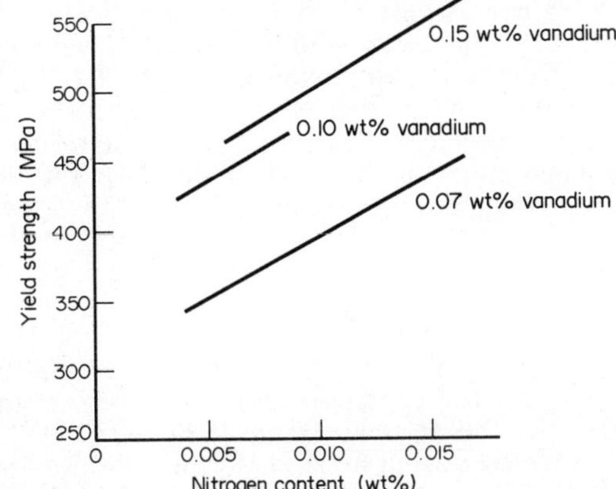

Figure 2
Effect of nitrogen content on the increase in yield stress in hot rolled vanadium steels (after Brownrigg and Ivancic)

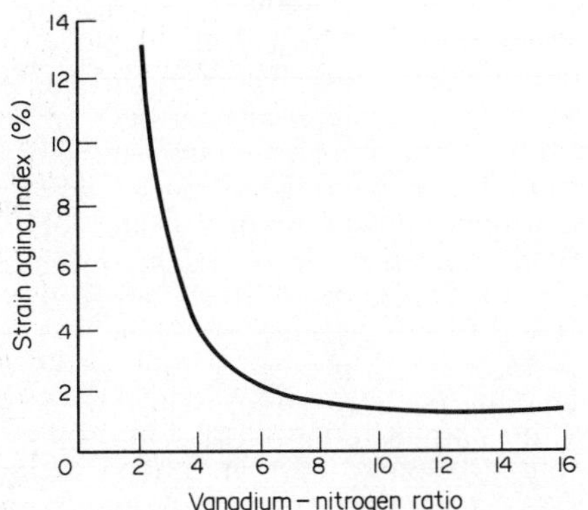

Figure 3
Effect of vanadium–nitrogen ratio on strain aging at 100 °C for 30 min (after Glover, Oldland and Voight)

cooling conditions during processing and the maximum strengthening seems to occur at approximately 600 °C. Whereas VC precipitates on ferrite grain boundaries and causes embrittlement, VN precipitates within the grains and gives less embrittlement, though any embrittlement will be offset by the finer grain size produced by the higher nitrogen content.

Increasing cooling rate increases precipitation strengthening by lowering the transformation temperature at which interphase precipitation occurs and this effect is apparent at all levels of nitrogen content. It has also been shown that the precipitation of VN so decreases the nitrogen dissolved in the ferrite, when the vanadium–nitrogen ratio exceeds stoichiometry, that the steel shows little or no strain aging (see Fig. 3). This is beneficial for reinforcing bars for earthquake-resistant structures. There has been controversy regarding the effect of nitrogen on the weldability of HSLA steels containing vanadium. It has been shown however that, in order to weld such steels satisfactorily, it is necessary to ensure that the austenite grain size in the heat-affected zone remains fine as this produces fine polygonal ferrite. It must also be ensured that much of the nitrogen is precipitated as this is less detrimental to toughness than it would be if in solid solution. Another effective method for obtaining tough welds is to use a small titanium addition which also grain refines the heat-affected zone and decreases dissolved nitrogen.

3. *Bar and Forging Steels*

For bars, hot working is usually carried out at temperatures in the recrystallization regime and microalloying is used to improve properties by precipitation strengthening rather than by grain refinement. A typical product is a concrete reinforcing bar. Microalloying has enabled the carbon content to be lowered with improved ductility, toughness and weldability, and a preferred microalloying combination is vanadium with enhanced nitrogen. Niobium steels have also been used though not with enhanced nitrogen because of the very low solubility of NbN.

The same principle of adding vanadium–nitrogen has been used with medium carbon (0.40–0.50 wt%) forging steels in order to produce equivalent properties to quenched and tempered steels, especially by using controlled cooling after hot forging. This reduces processing costs and the steels usually have improved machinability. They comprise ferrite–pearlite structures and at these higher carbon contents niobium cannot be used owing to the very low solubility of NbC. The vanadium–titanium addition refines the ferrite grains and the pearlite colony size, which is necessary for toughness, and precipitation strengthens by addition of V(CN). The precipitate is a nitrogen-rich V(CN). The tensile strength of these steels is directly proportional to the (V+5N) content (see Fig. 4) which clearly shows the effect of nitrogen content. These principles, and the use of vanadium–nitrogen additions, have also been used to produce improved rail steels.

4. *High-Carbon Wire Rod Steels*

The expensive patenting of wire rod has largely been eliminated by the use of controlled cooling by Stelmor (rod rolling followed by cooling of the coils by an air blast) or hot-water quench processes, which lower the transformation temperature and produce finer pearlite. Additional strengthening can be achieved by the precipitation of nitrogen-rich V(CN) particles in the pearlitic ferrite, using up to 0.2 wt% vanadium content with or without enhanced nitrogen content. The strengthening is maintained during wire drawing, so smaller drawing reductions or a lower carbon content can be used for the same strength of wire, both of which give better ductility. Various mechanisms of precipitation of V(CN) in the pearlitic ferrite have been proposed including interphase precipitation, precipitation on dislocations and on the pearlitic carbide lamellae, and clusters containing vanadium, chromium and possibly carbon and nitrogen. It has been concluded that interphase V(CN) precipitation in pearlitic ferrite only occurs at transformation temperatures above 600–625 °C. Work on the precipitation strengthening in vanadium-containing eutectoid steels indicates a limiting strengthening effect, giving a maximum hardness of approximately 400 HV (where HV is the Vickers hardness number) at about 0.15 wt% vanadium content for a normal nitrogen content of 0.010 wt%. For a transformation temperature of 600 °C, where maximum strengthening occurs, precipitation only occurs behind the transformation front and the nuclei may well be vanadium–carbon–nitrogen clusters. The use of enhanced nitrogen causes this limiting strength to occur at higher vanadium contents and strength increments. The consequently enhanced nitrogen contents of 0.015 wt% are useful and the increase in yield strength is of the order of 170 MPa per 0.1 wt% increase of vanadium. The enhanced nitrogen seems to be essential if less than the optimum content of 0.15 wt% vanadium is used and more rapid cooling is required to achieve the optimum transformation temperature.

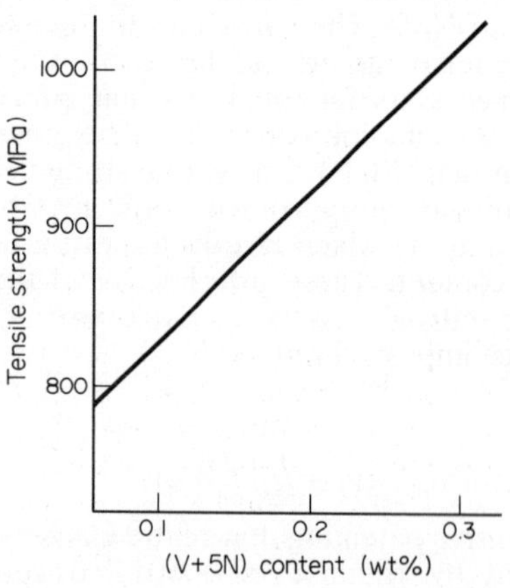

Figure 4
Effect of vanadium and nitrogen content on tensile strength of microalloyed forging steel containing 80% pearlite (After Lagneborg, Sandberg and Roberts)

The beneficial effects of the vanadium–nitrogen addition to eutectoid wire rod steels have been amply confirmed. Moreover, owing to the rapid precipitation of the V(CN) there is virtually no nitrogen dissolved in the pearlitic ferrite, so the steel is nonaging; this minimizes variability of properties due to heating during wire drawing.

5. *Quenched and Tempered Engineering Steels*

In quenched and tempered engineering steels the most important properties are the hardness of the martensite, the M_s temperature (i.e., the temperature at which martensite starts to form from austenite), the hardenability, the tempering resistance and any embrittlement effects introduced during tempering. The effect of nitrogen, especially the interaction with the alloying elements, is important.

Nitrogen has virtually identical effects to carbon on the lattice constants of both austenite and martensite. For a constant nitrogen content, the hardness of martensite is slightly less than that for the same carbon content. Nitrogen also depresses the M_s temperature slightly less than does carbon, so that for a given interstitial content nitrogen martensite will tend to suffer more autotempering and therefore be softer than carbon martensite. The effects of normal nitrogen contents on the hardness of martensite in steels are overshadowed by the carbon, but in 0.12 wt% carbon content steels an increase of 0.015 wt% of nitrogen increases the martensitic hardness by 22 HV. More significantly, however, the same amount of nitrogen causes a much larger increase in martensite hardness in the same steels containing 0.15 wt% and 0.35 wt% vanadium, namely 43 HV and 57 HV, respectively. This suggests some association between vanadium and nitrogen atoms, possibly as clusters.

Nitrogen in solution increases the hardenability of the steel as shown by continuous cooling and isothermal transformation diagrams. Nitrogen also retards the pro-eutectoid ferrite reaction and depresses the Widmanstätten ferrite and the B_s temperature. The effects have not been quantified because of the small nitrogen concentration and because nitrogen interacts markedly with other alloying elements. This latter effect has been shown by work on vanadium steels in which it has been postulated that clusters of vanadium atoms with carbon and nitrogen can segregate to the austenite grain boundaries at low austenitizing temperatures and give pronounced increases in hardenability. This is even more pronounced when the austenite grain boundaries are immobilized by nitride particles as shown by the benefit conferred by titanium–vanadium–nitrogen additions. When grain boundary pinning effects are lost, the boundaries pull free from the segregation and hardenability then decreases. This is exacerbated by thermal dispersion of the segregates at high austenitizing temperatures. It can be seen, therefore, that nitrogen plays an important, though indirect, role in hardenability.

There is little effect by nitrogen on low-temperature-tempering resistance. It might, however, be expected that nitrogen by dissolving in the precipitating V(CN) would increase secondary hardening in vanadium steels. However, no effects have been observed because of the very small increase in volume fraction of V(CN) brought about by the nitrogen. There are indications that vanadium–carbon–nitrogen clusters in the austenite can be inherited by the martensite and thus play a part in secondary harding. When the nitrogen content is increased from 0.010 wt% to 0.018 wt% in nickel–chromium steels containing vanadium, the hardness is increased by 80–100 HV on heavy tempering; this is due to the refinement of the recrystallized ferrite grains brought about by the undissolved VN particles and by VN precipitated during tempering.

Embrittlement effects during tempering are always more pronounced with increasing prior austenite grain size, and thus increasing nitrogen in the presence of vanadium, niobium or titanium is likely to decrease the impact transition temperature due to the grain refining action of the nitrides. This has been shown clearly by recent work on 3.5 wt%Ni–Cr–Mo–V steels. Increasing the nitrogen content from 0.01 wt% to 0.02 wt% decreased the impact transition temperature by up to 40 °C at the same strength. Of the usual embrittlement effects that occur during tempering, little is known of the effect of nitrogen on reversible tempering embrittlement. Nitrogen should be beneficial in reducing embrittlement due to ferrite grain growth at the highest tempering temperatures and this has been found for steels containing additions of vanadium, titanium and niobium. Alternatively, tempered martensite embrittlement has been studied and shows an effect due to vanadium–nitrogen segregation to austenite grain boundaries at low austenitizing temperatures when intergranular fracture and low toughness is observed. Maximum embrittlement is observed after austenitizing at 1025 °C, but the embrittlement decreases at higher austenitizing temperatures owing to thermal dispersion of the grain boundary segregation. Down-quenching experiments have shown the effect to be reversible. It seems that the major effect of nitrogen in quenched and tempered engineering steels resides in its interaction with strong nitride-forming alloying elements such as vanadium, titanium and niobium.

Some nitriding steels contain strong nitride-forming alloying elements such as chromium, molybdenum, vanadium and aluminum. During nitriding, alloy nitrides are precipitated and cause marked precipitation hardening, which is more pronounced with the stronger nitride-forming elements.

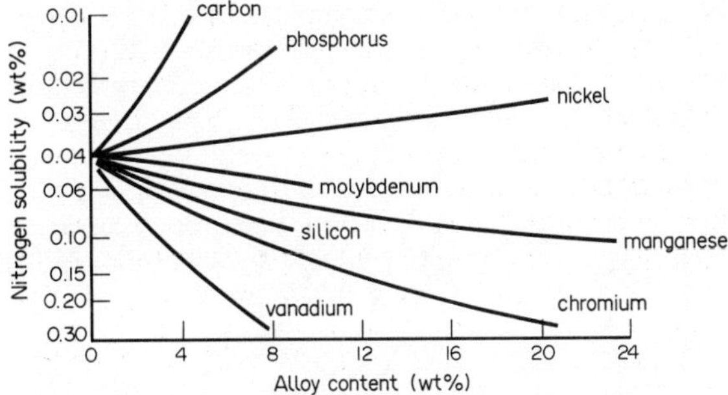

Figure 5
Effect of alloying elements on solubility of nitrogen in iron at 1600 °C (after Hilty and Kaveney)

There is also evidence that substitutional–interstitial atom clusters or zones can occur. These clusters or zones play a part in the hardening response. The effectiveness of multiple alloying may well be associated with the formation of mixed alloy–nitrogen clusters.

6. High-Chromium Transformable Steels

The 12 wt% chromium transformable steels often contain 0.10–0.15 wt% carbon. With higher chromium contents of 14–17 wt%, alloyed with nickel or manganese and other elements, the steels are used as precipitation hardening variants. The steels tend to form δ ferrite which must be minimized for maximum strength by control of the composition, the result often being predicted by use of nickel and chromium equivalents and Schaeffler-type diagrams. Nitrogen is an austenite former and in 12 wt% chromium steels has a nickel equivalent of 25 compared with 30 for carbon. The normal nitrogen content is approximately 0.02–0.04 wt% because the chromium increases nitrogen solubility (see Fig. 5).

To obtain maximum strength, it is necessary for the austenite to transform fully to martensite prior to tempering. The M_s temperature is therefore important, and nitrogen depresses the M_s by 450 °C per 1 wt% increase of nitrogen in 12 wt% chromium steels. Nitrogen also depresses the A_1 temperature (i.e., the eutectoid temperature in steels) by about 10 °C for each 0.01 wt% increase in nitrogen content and limits the maximum permissible tempering temperature. However, use of increased nitrogen contents of 0.04–0.06 wt% together with 5–7 wt% nickel have been suggested for a new class of creep-resisting rotor steel which is tempered within the critical range to produce a fine dispersion of highly stable retained austenite particles.

Nitrogen increases the tempering resistance of 12 wt% chromium steels (see Fig. 6) by changing the Cr_7C_3 into the markedly secondary hardening M_2X,

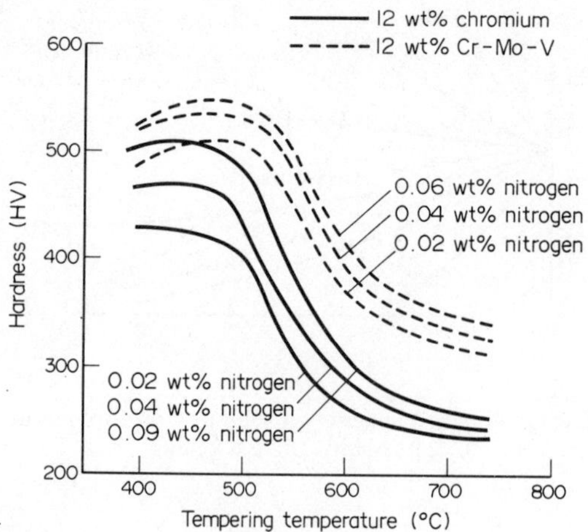

Figure 6
Effect of nitogen content on tempering characteristics (1 h) of 12 wt% chromium and 12 wt% Cr–Mo–V steels (after Irvine and Pickering)

based on $(FeCr)_2(CN)$. The beneficial effect of nitrogen is retained at all tempering temperatures and the strength increases linearly with nitrogen content after tempering at 650 °C for 1 h (see Fig. 7). The strengthening produced by nitrogen tends to lower the toughness but this can be overcome by an addition of about 0.3 wt% niobium which causes grain refinement and renders M_2X more stable, thereby improving tempering resistance (i.e., increased toughness with less loss of hardness). This has been taken advantage of in the development of better creep-resisting steels. The improved creep strength is clearly related to the nitrogen. These developments have been applied to both 10–12 wt% Cr–Mo–V and to 9 wt% Cr–Mo steels in which strength is also achieved by nitride precipitation.

7. *Ferritic Stainless Steels*

The interstitial elements carbon and nitrogen increase the strength but decrease the ductility and especially the toughness of ferritic stainless steels. However, in the presence of titanium or niobium, nitrogen can be beneficial as it forms Ti(CN) and Nb(CN) particles which inhibit grain growth. It has been suggested that introducing nitrogen from the shielding gas during welding can refine the weld metal grain size with benefit to toughness, especially in titanium steels. In general, however, deliberate additions of nitrogen to these steels are not made.

8. *Austenitic Stainless Steels*

Nitrogen has long been used as a deliberate alloying addition to austenitic stainless steels. Due to its strong austenite-forming capability, nitrogen can be used to supplement less effective austenite stabilizing elements such as maganese and it is used to control the constitution of some of the currently popular duplex stainless steels.

Nitrogen is arguably the most effective solid-solution hardening element and to obtain high nitrogen content the solubility has to be increased by manganese additions, which together with the chromium increase nitrogen solubility. The effect of dissolved nitrogen is more pronounced on the tensile

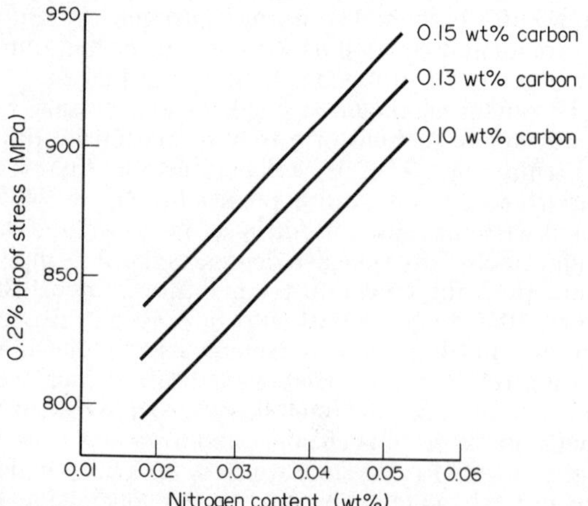

Figure 7
Effect of nitrogen content on strength of 12 wt% Cr–Mo–V steels tempered for 1 h at 650 °C (after Irvine and Pickering)

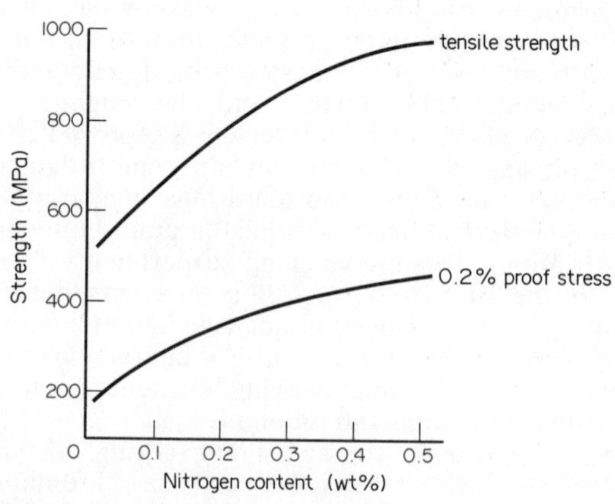

Figure 8
Effect of nitrogen content on tensile properties of austenitic stainless steels (after Pickering, Brandis and Hermann)

strength than on the proof stress (see Fig. 8) because nitrogen decreases the stacking fault energy and increases the work-hardening rate. With a high work-hardening rate, the uniform strain prior to plastic instability and the total ductility at fracture both increase. This probably explains why, despite the strengthening produced by nitrogen, the tensile ductility is also increased. There seems to be a marked interaction between nitrogen and grain size, the effect of grain size on the strength increasing as the nitrogen increases. The high work-hardening rate produced by nitrogen results in good stretch formability and the AISI 200 steels have better stretch formability than the AISI 300 steels. This is despite the fact that nitrogen depresses both the M_s and M_d (i.e., the martensite start temperature under deformation conditions) temperatures. Since nitrogen suppresses the beneficial strain-induced martensite formation, it must be concluded that the major effect controlling the increased stetch formability of nitrogen-rich austenitic steels is their high work-hardening rate.

In terms of hot workability, increased nitrogen content is generally beneficial as it eliminates δ ferrite. However, some unusual effects occur at high nitrogen contents, where there is no δ ferrite: the hot ductility decreases, possibly because of nitrogen or chromium–nitrogen clusters retarding recovery and recrystallization by segregation to grain boundaries. However, a little δ ferrite can be beneficial, as it dissolves the impurities which lead to hot intergranular cracking. Since the solidification structure, which determines the constitution with respect to δ ferrite, is controlled by nitrogen and other elements, nitrogen will indirectly affect the hot workability. Also, nitrogen with niobium causes marked strengthening during warm working, because of strain-induced nitride precipitation which can lead to loss of hot ductility.

Many nitrides precipitate in austenitic steels, the predominant one being Cr_2N. In titanium and niobium steels, TiN and NbN are formed and equations for their solubility products are available. An interesting feature in niobium steels is that the introduction of nitrogen into NbC increases its solubility, although NbN itself is much less soluble than NbC. Nitrogen also moves the stoichiometry from M_4X_3 towards MX, indicating that nitrogen occupies interstitial vacancies in the carbide. The precipitation of $M_{23}C_6$, the common carbide, is retarded by nitrogen (see Fig. 9) possibly because Cr_2N precipitation withdraws chromium from solution. It is unlikely that nitrogen enters the $M_{23}C_6$ lattice but there is clear evidence that nitrogen dissolves in the M_6C carbide.

The precipitation of $M_{23}C_6$ and Cr_2N both cause age hardening; these compounds precipitate in the matrix on dislocations and also occur as grain-boundary cellular precipitates. The higher the nitrogen content, the lower the aging temperature at which cellular precipitation occurs, this having a deleterious effect on ductility.

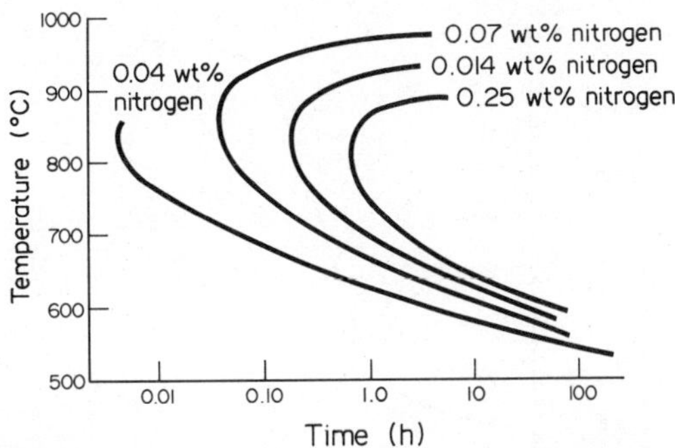

Figure 9
Effect of nitrogen content on kinetics of $M_{23}C_6$ precipitation (after Thier, Baumel and Schmidtmann)

The creep resistance of austenitic stainless steels increases with increasing nitrogen content due to enhanced precipitation. In the presence of niobium, the creep strength is also improved by nitrogen due to greater amounts of Nb(CN) precipitation, but at the expense of creep ductility. An interesting potential development employs the stability of TiN to dispersion strengthen a 25 wt%Ni–20 wt%Cr–Ti alloy. The alloy, in thin sections, is nitrided to form a fine dispersion of TiN which greatly improves the creep resistance, but diffusion considerations limit the section thickness.

Mention should be made of the effect of nitrogen on weldability because, in order to prevent hot weld metal cracking, a small amount of δ ferrite should be present. Thus care is required when welding high-nitrogen austenitic steels with filler materials of controlled ferrite content, because nitrogen pickup by the weld metal from the base plate can eliminate ferrite and allow interdendritic cracking in the weld metal.

In terms of corrosion resistance, nitrogen is important and beneficial because it promotes passivity, which is enhanced by the presence of molybdenum. Nitrogen, by extending the passive potential range, would be expected to improve general corrosion resistance but this has not always been observed because, even in the absence of sensitization, nitrogen can segregate to grain boundaries, thereby causing intergranular corrosion so that the weight loss apparently increases. Thus, nitrogen can in some environments lower general corrosion resistance. Since nitrogen retards $M_{23}C_6$ precipitation it reduces the detrimental effect of carbon on intergranular corrosion. In addition, Cr_2N precipitation does not apparently lead to such marked intergranular corrosion as does $M_{23}C_6$ precipitation. There is evidence, however, that impurities, including nitrogen, can segregate to the grain boundaries thereby promoting

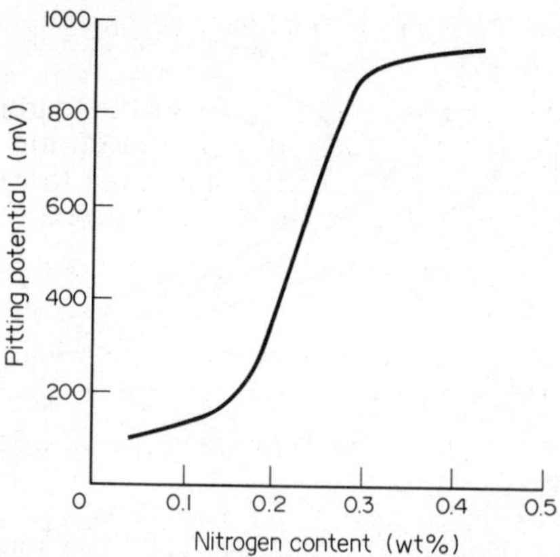

Figure 10
Effect of nitrogen on pitting potential for plain specimens of low-carbon 22 wt% Cr–20 wt% Ni–2.8 wt% Mo austenitic stainless steel (after Trueman)

intergranular corrosion even in the absence of $M_{23}C_6$.

By extending the passive range, nitrogen increases the pitting potential towards the transpassive range (see Fig. 10). Thus nitrogen improves pitting resistance, as does molybdenum, and a parameter (wt%Cr + 3.3 wt%Mo + 16 wt%N) can be used as an empirical indicator for pitting resistance. The effect of nitrogen on stress corrosion cracking in austenitic stainless steels is less clear. At the normal nitrogen levels of 0.03–0.04 wt% it seems to be detrimental in that decreasing the nitrogen seems to increase the fracture time. However, a decrease in the nitrogen content is also accompanied by a decrease in other impurities, most of which are detrimental to stress corrosion. Nevertheless, the higher stacking-fault energy brought about by the lower nitrogen content might be expected to be beneficial. A low stacking-fault energy might be the cause of poor stress-corrosion resistance at 0.15–0.20 wt% nitrogen content and still higher levels of nitrogen have little effect. High levels of nitrogen, by promoting passivity, may offset any detrimental effect of a decreased stacking-fault energy and nitrogen seems to affect the initiation rather than the propagation stage of stress corrosion cracking.

9. Conclusion

It is a clear conclusion that nitrogen is and must always be considered to be an essential alloying element. In large measures, its benefits lie in its interaction with other alloying elements present in steels and it is by understanding such interactions that the useful effects of nitrogen can be optimized and its less beneficial effects obviated. By capitalizing on its undoubted beneficial effects considerable improvements can be obtained in terms of steel properties, and this by the use of what is probably the least expensive of all alloying elements.

Bibliography

Baird J D, Jamieson A 1963 Relation between the structure and mehanical properties of metals. *Proc. National Physical Laboratory Conf.* Her Majesty's Stationery Office, London, pp. 361–9

Blickwede D J 1968 Micrometallurgy by the million. *Trans. Am. Soc. Met.* 61: 653–79

DeArdo A J, Ratz G A, Wray P J (eds.) 1982 *Thermomechanical Processing of Microalloyed Austenite*. Metallurgical Society of AIME, Warrendale, PA

Duckworth W E, Baird J D 1969 Mild steels. *J. Iron Steel Inst.* 207: 861–71

Gray J M, Ko T, Hang Shouhua Z, Baorong Wu, Xishan Xie (eds.) 1987 *HSLA Steels—Metallurgy and Applications*. American Society for Metals, Metals Park, OH

High Nitrogen Steels 1989. Institute of Metals, London

Irvine K J, Pickering F B 1965 *Metallurgical Developments in High Alloy Steels*, Special publication No. 86. Iron and Steel Institute, London, pp. 34–48

Jack D H, Jack K H 1973 Carbides and nitrides in steels. *Mater. Sci. Eng.* 11: 1–27

Korchynsky M (ed.) 1984 *HSLA Steels—Technology and Applications*. American Society for Metals, Metals Park, OH

Microalloying '75 1975. Union Carbide Corporation, New York

Peckner D, Bernstein I M (eds.) 1977 *Handbook of Stainless Steel*, McGraw-Hill, New York

Pickering F B 1978 *Physical Metallurgy and the Design of Steels*. Applied Science, London

Reed R P 1989 Nitrogen in austenitic stainless steels. *J. Min. Met. Mat. Soc.* 41(3): 16–21

Stainless Steel '77 1977. Climax Molybdenum Corporation, Ann Arbor, MI

Stainless Steels '84 1985. Institute of Metals, London

Stainless Steels '87 1988. Institute of Metals, London

F. B. Pickering
[Sheffield City Polytechnic, Sheffield, UK]

Nuclear Fuels: Burnable Poisons

A nuclear reactor system (Weinberg and Wigner 1958) normally consists of an arrangement of lattice cells containing neutron-moderating material and fuel incorporating fissile material. There are, in addition, coolant and structural materials, within which a self-sustaining chain reaction for the production, destruction and leakage of neutrons is maintained. In this condition, the system is said to be "critical." The operation of such a system at power results in the depletion, or burnup, of the fissile content of the fuel, with the resultant production of neutron-absorbing fission products and actinides such as ^{239}Pu.

The operation of such a nuclear reactor system requires that neutron-absorbing materials are used for reactor power control and shutdown. For example, many types of reactor are shut down for scheduled refuelling after one or two years of operation; the economics of operation are improved as the design time between refuelling is increased. During this operating period, the reactor core reactivity (i.e., the ability of the reactor to sustain a nuclear chain reaction by generating neutrons) decreases. This is due to the depletion of the fissile isotopes of the fuel (normally ^{235}U or ^{239}Pu) during operation, and to the buildup of fission products. Thus at the beginning of an operating cycle, the fuel must be enriched to a level that will still allow the reactor to operate (remain critical) at full power at the end of the operating cycle. This means that the reactor would be over reactive, and hence unsafe, at the beginning of the cycle if no methods were available to control the system safely.

Control is achieved by a combination of several techniques involving the insertion of neutron-absorbing materials:

(a) as movable absorbers (called control rods),

(b) as a gradually reducing level of dissolved boron in the moderator of water systems such as light water reactors (LWRs), and

(c) as fixed absorbers which themselves deplete significantly throughout the fuel cycle (called burnable poisons).

1. *Burnable Poisons*

The move to higher fuel irradiation lifetimes in pressurized water reactor (PWR) systems leads to the need for a large amount of reactivity control initially because of the increased requirement of fuel enrichment. This could be achieved in the conventional way by increasing the dissolved boron loading. It is found, however, that this leads to a positive moderator temperature coefficient; that is, an increase in reactivity with increasing moderator temperature—an unacceptable feature for system safety. The solution usually adopted is to introduce burnable poisons into the reactor.

Burnable poisons are fitted within the fuel assemblies and are designed so that at the end of the fuel cycle essentially all the absorbing material has been "burnt" away by neutron absorption processes ("burning" is the normal metaphor for depletion). Burnable poisons also have the added property that they give a degree of control over the power distributions within the reactor; for example, they have been used in advanced gas-cooled reactors (AGRs) to reduce the axial variation of power between adjacent fuel clusters.

A burnable poison must have certain characteristics. It must:

(a) have an absorption cross section sufficiently large that the material burns out during its lifetime in the reactor at a rate equal to or faster than that of the fissile material burnout;

(b) avoid a positive reactor moderator temperature coefficient;

(c) result in an absorption product that has a small absorption cross section; and

(d) be compatible with other materials over the fuel cycle.

The amount and form of burnable poison required is determined by the area between the unpoisoned reactivity vs the irradiation curve, and the required curve which is normally a constant value. This essentially defines the number of absorbing nuclei needed to absorb those neutrons in excess of those required to sustain the chain reaction (see, for example, Fig. 3).

If the cross section is very large and the absorber is contained in a concentrated form (e.g., as a single pin) then only the outermost regions of the pin will be effective, since all neutrons entering the pin surface will have been absorbed before penetrating the inner regions. This is called the self-shielding effect. However, as the poison surface is removed by neutron absorption during burnup, the effective surface of the pin moves inwards. Thus the size and shape of the burnable poison can influence its effectiveness as an absorber.

Alternatively, if the absorber is distributed uniformly throughout the fuel, or the cross section is not very large (so that self-shielding is not an important feature), then the absorber depletion rate will be almost entirely dependent on the cross section and quantity of material present, irrespective of its distribution.

2. *Burnable Poison Materials*

Among materials that have been proposed or used as burnable poisons are boron, erbium and gadolinium. Figure 1 shows the cross-section variation with neutron energy, demonstrating the common feature of very high thermal cross sections of the materials, particularly for gadolinium.

Kapil et al. (1988) describe several forms of burnable poisons that contain boron, including a discrete absorber using borosilicate glass (used extensively in PWRs) and an integral fuel absorber using fuel pellets coated with zirconium diboride. In the latter, the boride is in the form of a thin layer, avoiding self-shielding calculational problems.

Advantages of boron are that its use ensures:

(a) ease of prediction,

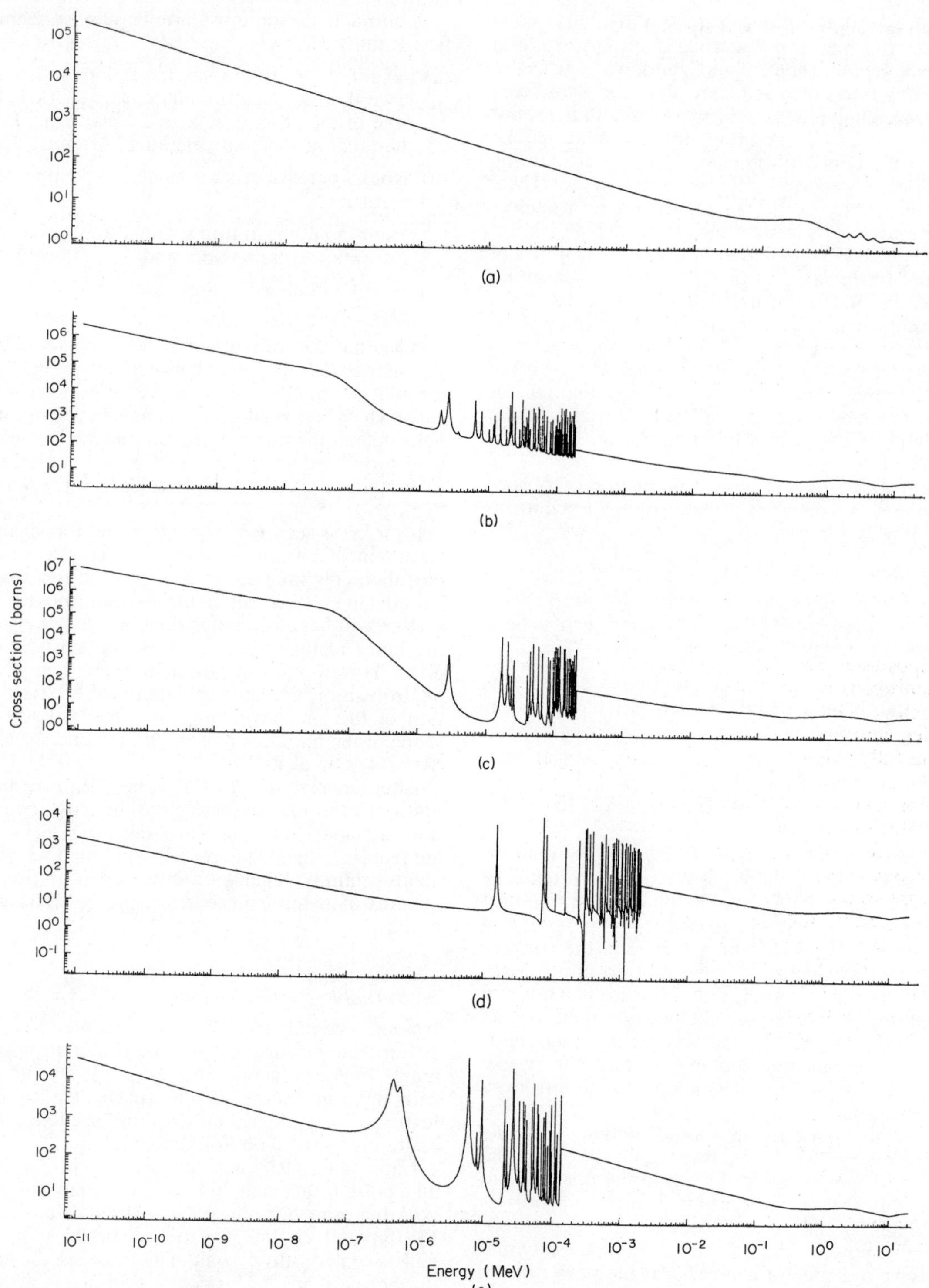

Figure 1
Total neutron cross sections for: (a) ^{10}B (note that natural boron contains ~20% ^{10}B), (b) ^{155}Gd, (c) ^{157}Gd, (d) ^{166}Er and (e) ^{167}Er (data for graphs processed from Rowlands (1985))

(b) no impact on the local enrichment,

(c) a negligible residual absorption cross section, and

(d) compatibility with all PWR fuel operations.

Boron has a cross section that varies smoothly with energy, with a peak value (for thermal neutrons) for natural boron of 700 barns. This means that for the type of pin size used in PWR fuel, self-shielding effects are unimportant. This implies that the absorption rate of any such absorber will be roughly proportional to the quanitity of the absorber that remains. The result is that it is not possible to match the reactivity effects of the change in composition of the fuel over the whole of the fuel cycle and hence achieve a flat reactivity depletion curve.

Erbium (as erbia) has been evaluated as a possible burnable poison (Beaudreau et al. 1988). The advantages claimed are that it can be mixed homogeneously with the fuel and has a cross-section variation such that it makes the moderator temperature coefficient more negative. Self-shielding effects could be important.

Gadolinium (Bailey and Crowther 1985) is particularly well suited as a burnable poison because its burnout can be designed to match approximately the change in reactivity due to the alteration in fuel composition. The core reactivity thus remains reasonably constant over the fuel cycle. Gadolinium has therefore been used extensively in the design of fuel cycles (for this reason its properties are discussed more fully below).

For gadolinium, the isotopes ^{155}Gd and ^{157}Gd both exhibit a very large thermal cross section (see Fig. 1) and this means that a gadolinium-bearing absorber can be designed to be effectively black to thermal neutrons (i.e., neutrons at $\sim$ 0.025 eV). It is this feature which allows the designer to adjust the rate of burnout by varying the geometry of the absorber and hence the self-shielding effect, as well as its composition. The complication of using gadolinium absorbers, however, lies in the need for complex computational methods to deal with the spatial depletion of the gadolinia pin.

3. *Physical Characteristics of Gadolinium Pins*

The successful introduction of any material into a reactor depends not only on its nuclear characteristics but also on its relations with other materials within the reactor. Gadolinium as gadolinia is favored for several reasons (Haas and Motte 1984, Bailey and Crowther 1985, Bairiot 1985).

(a) Mixed with UO_2 it can be incorporated into the most appropriate locations within the fuel assemblies as a pin-type feature to control both reactivity and power distributions.

(b) It can be designed to have the desired depletion rate by a judicious choice of concentration and geometry to match that of the fuel.

(c) On present evidence, gadolinium depletion does not appear to generate gaseous or volatile products that might limit the burnup accumulation in the gadolinia fuel rod.

(d) Gadolinia fuel can be routinely reprocessed.

(e) The in-reactor performance of gadolinium-loaded fuel is similar to that of UO_2 fuel.

Little (published) data are available, however, that deal with the use of gadolinium-loaded fuel, particularly with respect to its thermal and mechanical properties (Thornton et al. 1982). Problems that have been found are outlined below.

(a) It has thermal conductivity, lower than UO_2, that decreases with increasing gadolinia content. This results in the need to restrict the enrichment in order to keep the fuel within safe temperature operating limits.

(b) It has slightly lower melting point than UO_2.

(c) It has slightly higher thermal expansion coefficient than UO_2.

(d) The pellets tend to densify at a slower rate (i.e., do not sinter as easily as standard UO_2 fuel).

(e) A complex heterogeneous structure of $(U, Gd)O_2 + Gd_2O_3 + UO_2$ may exist, which can increase in $(U, Gd)O_2$ with time. This will cause the liberation of oxygen and could lead to slight oxidation of the clad inner surface.

(f) There are insufficient data to draw firm conclusions regarding fission gas release at the large concentrations required for very high burnup fuels.

Typically, the fuel pin enrichment is chosen as natural uranium compared to the standard fuel pin of some 3% ^{235}U.

The scarcity and variability of published data for gadolinia-loaded fuel is generally recognized and has in the past prompted several international experimental programs; for example, the GAIN (international experimental program on gadolinia fuel properties in PWRs) and GAP (international experimental program on gadolinia fuel evolution in PWRs) programs, both conducted by Belgonuclaire (BN), Belgium, and Centre D'Etude de L'Energie Nucléaire/Studiecentrum voor Kernenergie (CEN/SCK), Belgium. The objectives of such programs are discussed by Blanpain et al. (1987). There have also been a series of experiments in France described by Martin-Deidier et al. (1984).

4. Reactor-Physics Aspects of Gadolinium-Bearing Fuel Pins

Typically, gadolinium is dispersed within the fuel assembly as an array of gadolinia-loaded pins, with up to 10% concentration of gadolinia mixed homogeneously with UO_2. In this form the pins introduce a number of possible reactor-physics calculational problems (Jonsson 1986).

(a) The absorber-pin self-shielding is important, leading to the need to model the spatial and spectral changes within the pin accurately.

(b) The effective cross section is sufficiently large so as to cause spectrum and reactivity effects in the neighboring pins.

(c) Because of the rapid burnout of gadolinium within the pin, it is necessary to introduce small irradiation time steps of the order of 50 MW d t^{-1} into the burnup calculational route. This is to be compared with steps of 2000 MW d t^{-1} for standard fuel burnup calculations.

Gadolinium isotopes ^{155}Gd and ^{157}Gd dominate the cross section of gadolinium, with peak values of the order of 10^5 barns at thermal (0.025 eV) energies. The use of gadolinium pins therefore means that at energies below about 0.2 eV the absorber is effectively black. This gives rise to steep flux gradients into the absorber and modifies the neutron spectrum into the neighboring fuel pins. It is therefore necessary for the core analyst to use a spectrum geometry model that is as realistic as possible with current technology. This must involve the use of neutron transport theory with all the pins represented rather than the more simple diffusion theory with region smeared data. The model must therefore represent the absorber, fuel and cladding and any vacancies within the assembly designed for the purpose of holding control rods. A discussion of transport and diffusion methods of calculation for reactor lattices is given by Weinberg and Wigner (1958).

It is common practice to liken the depletion of a burnable absorber to the removal of layers from an onion; that is, the absorber is imagined as depleting from the outside inwards. This would be true if the absorber were black at all energies and there would be no need to calculate the distribution of gadolinium depletion within the absorber. In practice, while there is a strong surface effect, there is also a volume effect from absorptions occurring at epithermal neutron energies such that depletion of the inner regions continues apace with the surface depletion. Any method of calculation must therefore be able to deal with the radial variation of depletion. Of course, from a reactivity aspect it is not necessary to have a precise calculation of distribution—provided the right number of neutrons are absorbed the neutron balance within the lattice will be satisfied irrespective of position. Furthermore, if the absorption rate is incorrect at one point in time in the fuel cycle, it will tend to be compensated for by an opposite effect at a later stage so that the mean reactivity over the cycle will be approximately correct.

Among reactor-physics codes having special provision for self-shielding effects in burnable poisons are the United Kingdom Atomic Energy Authority (UKAEA) code LWRWIMS (Halsall 1982) and the Swedish code CASMO (Edenius and Ahlin 1987).

Theoretical estimates of the depletion of a burnable-poison pin for a PWR lattice have been made using LWRWIMS. The calculation considered a gadolinia-loaded pin of natural uranium containing 8 wt% of Gd_2O_3 surrounded by an array of fuel pins containing UO_2 at 3% ^{235}U enrichment.

Because of the high cross section of gadolinia and hence the rapid burnout in the outer regions, it has been found necessary to use a method of solution which completes a burnup calculation every 50 MW d t^{-1} and a transport flux solution within the poison pin with a large number of radial meshes (e.g., of the order of 50) to achieve an accurate assessment of the flux shape. With a standard non-poisoned pin, values of 2000 MW d t^{-1} and 1 mesh per pin would typically be used. This means that the calculation takes considerably more computer effort when the presence of the poison is taken into account.

Since the poison remains within the assembly for all time and because its absorptive properties vary throughout its lifetime, its effect on the spectrum of the surrounding fuel will be significant. It is therefore important to calculate the depletion of the burnable poison in a representation of the actual lattice configuration, using transport theory where possible and an adequate energy representation of the nuclear cross sections. Thus, in the example given here, a core section representing a 5×5 array of fuel pins was simulated, with the poison pin in the center.

Figure 2 demonstrates the way in which depletion of the burnable poison occurs over the range 0–14 GW d t^{-1}. The concentrations relative to the initial concentration for isotopes ^{155}Gd and ^{157}Gd are plotted to show the decrease with burnup as a function of radius across the poison pellet. It is seen that the concentrations decrease with burnup throughout the whole pellet but decrease most rapidly at the outer radii of the pellet. This demonstrates the burnout of the surface of the burnable poison. By 14 GW d t^{-1} the absorption in the pin has effectively disappeared and at this stage the pin is left with the remaining gadolinium isotopes and irradiated UO_2.

Figure 3 shows the variation of the array *k*-infinity value (a measure of the lattice reactivity) compared with that of a 5×5 array of standard UO_2 pins. The

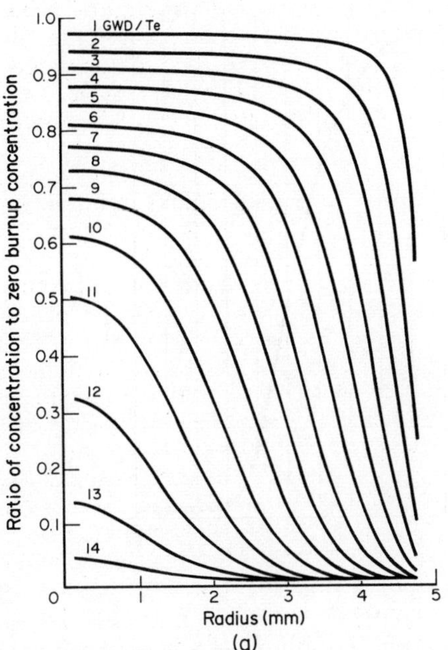

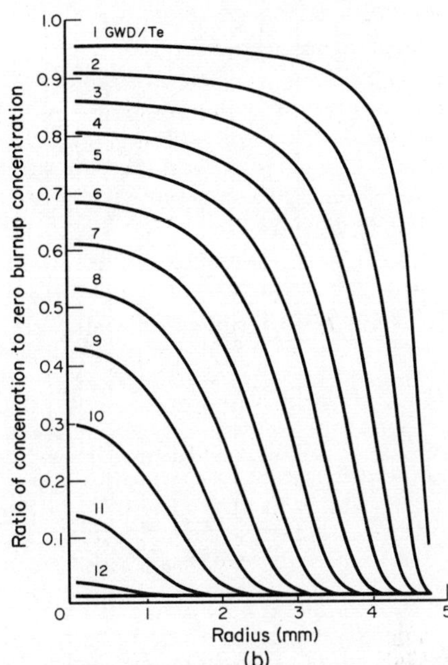

Figure 2
LWRWIMS study of burnable poison: (a) ^{155}Gd variation with radius and burnup; and (b) ^{157}Gd variation with radius and burnup

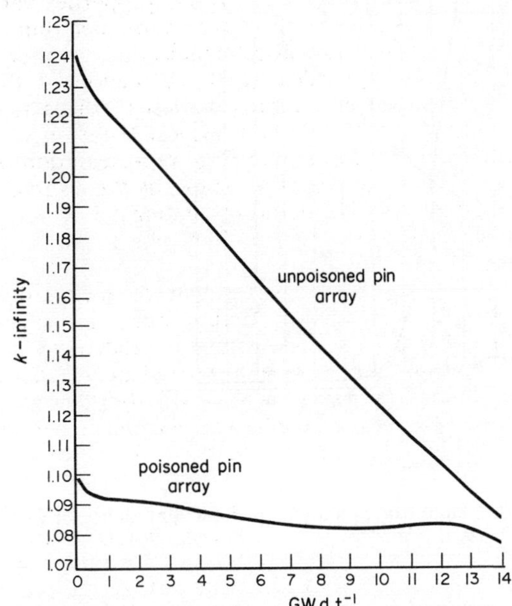

Figure 3
LWRWIMS study of burnable poison: *k*-infinity vs irradiation for unpoisoned and gadolinium poisoned fuel arrays

choice of size and gadolinium concentration for the burnable poison within the array has achieved an essentially flat curve over the range 0–12 GW d t^{-1} burnup. Clearly, in order to achieve this effect over the whole core, each assembly would need to hold burnable poison pins on a 1 in 25 basis.

Note that the two curves do not coincide at 14 GW d t^{-1}: partly because the initial UO_2 enrichment in the gadolinium-doped pin was lower than in the undoped pins, partly because of the presence of residual gadolinium isotopes in the doped fuel, and partly because of a different rate of buildup of plutonium with irradiation in the gadolinium-loaded array owing to spectrum effects by the poison pin.

5. *Experimental Measurements of Burnout of Gadolinium Isotopes*

Measurements of the depletion of isotopes ^{155}Gd and ^{157}Gd are reported by Zwicky et al. (1988) in which the measurement of local isotopic compositions were made using the secondary ion mass spectrometry (SIMS) technique. The method is nondestructive, faster and provides greater lateral resolution than other techniques; however, it is necessary to use the average-isotopic-abundance values evaluated by mass spectrometry to normalize the SIMS data. Samples irradiated under the auspices of the GAP program have been analyzed using the SIMS

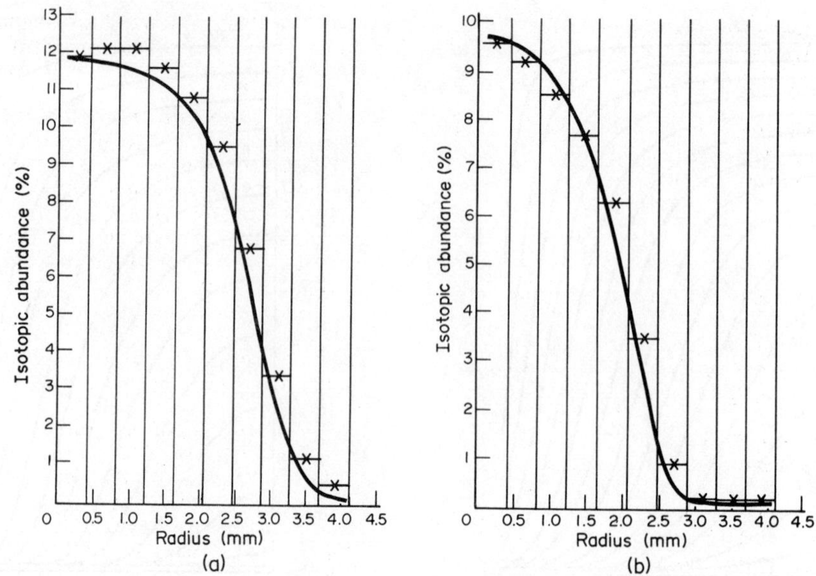

Figure 4
LWRWIMS analysis of gap experiments: comparison of measured (✳) and calculated (—) isotopic abundance of (a) ^{155}Gd with radius, and (b) ^{157}Gd with radius (local pin burnup: 3.03 GW d t^{-1})

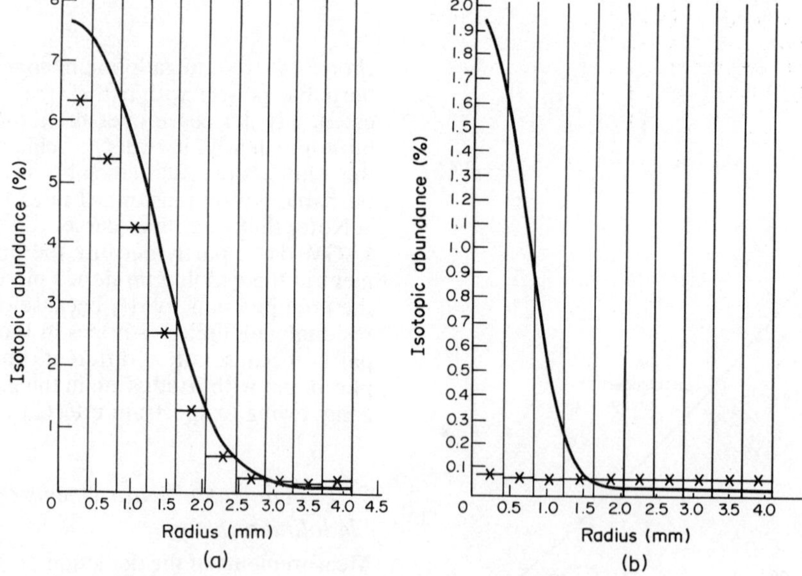

Figure 5
LWRWIMS analysis of gap experiments: comparison of measured (✳) and calculated (—) isotopic abundance of (a) ^{155}Gd with radius, and (b) ^{157}Gd with radius (local pin burnup: 5.78 GW d t^{-1})

methods and compared with theoretical estimates produced using LWRWIMS. Figure 4 shows comparisons taken from the paper by Zwicky et al. (1988) of the variation of isotopic abundance with radius at a pin burnup of 3.03 GW d t^{-1}. The results in general show reasonable agreement for ^{155}Gd and for ^{157}Gd. At higher burnup, however (corresponding to much lower concentrations of residual isotope ^{157}Gd) the agreement is poor; for example, see Fig. 5. This is probably due to deficiencies in the

LWRWIMS nuclear cross-section data for ^{157}Gd or possibly to measurement of background errors in the SIMS technique at high burnup.

See also: Control Materials for Nuclear Reactors; Fission Reactor Fuels and Materials; Oxide Nuclear Fuels; Radiation Effects on Nuclear Fuels

Bibliography

Bailey W E, Crowther R L 1985 Gadolinia performance in BWR's. *Trans. Am. Nucl. Soc.* 50 (Nov): 552–3

Bairiot H 1985 Gadolinia fuel behaviour to high burn-up. *Seminar on Performance of Fuel and Cladding Material Under Reactor Operating Conditions.* KTG, Karlsruhe, FRG

Beaudreau J J, Jonnson A, Shapiro N L 1988 Evaluation of erbia as a burnable absorber in a 24-month, low-leakage C–E core. *Trans. Am. Nucl. Soc.* 57 (Nov)

Blanpain P, Haas D, Motte F 1987 High rated and high burn-up gadolinia fuel irradiated in the BR3 17×17 PWR. *Int. Symp. Improvements in Water Reactor Fuel Technology and Utilization.* International Atomic Energy Agency, Vienna pp. 305–17

Edenius M, Ahlin A 1987 CASMO-3: New feature, benchmarking and advanced applications. *Int. Topical Meeting on Advances in Reactor Physics, Mathematics and Computation.* Paris

Haas D, Motte F 1984 Experimental programmes on gadolinia fuel utilisation in LWR's in Belgium. *International Working Group on Water Reactor Fuel Performance and Technology*. International Atomic Energy Authority, Vienna, pp. 171–81

Halsall M J 1982 LWRWIMS. A computer code for light water reactor calculations. Report AEEW-R 1498, United Kingdom Atomic Energy Authority, Winfrith, UK

Jonsson A 1986 *CRC Handbook of Nuclear Reactor Calculations*, Vol. 2. CRC Press, Boca Paton, FL

Kapil S K, Seker J R, Keller H W 1988 Achievements in Westinghouse discrete and integral fuel burnable poison usage. *Trans. Am. Nucl. Soc.* 57 (Nov)

Martin-Deider L, Santarmarina A, Doutriaua D, Roshid M, Zero S 1984 Camelon—A benchmark experiment for absorbers and burnable poisons in PWR assemblies. *Trans. Am. Nucl. Soc.* 46: 755–6

Rowlands J L, Tubbs N 1985 The joint evaluated file (JEF): A new nuclear data library for reactor calculations. In: Young P G, Brown R E, Auchampaugh G E, Lisowski P W, Stewart L (eds.) 1985 *Nuclear Data for Basic and Applied Science,* Vol. 2. Gordon and Breach, New York, pp. 1493–1509

Thornton T A, Ewanick J A, Lagedrost J F, Eldridge E A 1982 Thermal conductivity of sintered urania and gadolinia. *Trans. Am. Nucl. Soc.* 43: 348

Weinberg A M, Wigner E P 1958 *The Physical Theory of Neutron Chain Reactors.* University of Chicago Press, Chicago

Zwicky H U, Aerne E T, Bart G, Thomi H A, Sherwin J 1988 Determination of radial gadolinium burn-up in LWR fuel by secondary ion mass spectrometry (SIMS). *International Working Group on Water Reactor Fuel Performance and Technology.* International Atomic Energy Agency. Vienna, pp. 32–6

I. H. Gibson and J. Sherwin
[Winfrith Technology Centre, Winfrith, UK]

O

Oxidation of Molten Metals, Directed

The oxidation of metals, in this context used in its broadest chemical sense (i.e., including nitridation and carburization) has been considered only occasionally as a method for making ceramic bodies. Oxidation, and especially the reaction of metals and of nonoxide ceramics with oxygen, has generally been considered an undesirable feature that needs to be prevented. Only in the case of some reaction-formed ceramics, such as silicon nitride formed by the nondirected nitridation of silicon powder compacts (see *Reaction Sintering of Ceramics; Silicon Nitride*), has controlled oxidation been successfully utilized.

The oxidation of molten metals has been widely studied, though not for the sake of producing ceramics. In recent developments, however, this type of reaction has been applied in a controlled manner to the manufacture of ceramic–metal composites either as self-standing articles (Newkirk et al. 1986) or as the matrices of composites containing other reinforcing or filler materials (Newkirk et al. 1987). This novel process is the development of the Lanxide Corporation, Delaware, USA and has been tradenamed the DIMOX process. The method involves the reaction of a bulk molten metal with a gaseous oxidant to produce a solid ceramic body via a directed growth process. The growth mechanisms are not fully understood and the discussion of mechanisms in this article is therefore somewhat speculative. However, sufficient experimental data are available to render the process viable for the economic fabrication of a number of ceramic-based materials, especially the matrices for complicated fiber and particulate composites with tailored properties.

Most processing techniques for ceramic matrix composites make use of powder metallurgical technology; that is, they comprise the mixing of the reinforcement components with matrix powder, and subsequent forming and sintering, which always involve dimensional changes and their associated drawbacks. Ceramic composite techniques involving little or no shringkage, such as chemical vapor infiltration and slurry or multiple sol–gel infiltration of porous preforms, also have drawbacks. These drawbacks include high costs, high residual porosity, thermal and mechanical damage of the reinforcing components, and extremely long processing times. The directed metal oxidation process represents an alternative ceramic composite technology that overcomes many of these limitations.

The process has been applied to a number of ceramic–metal matrix systems. Of these, the

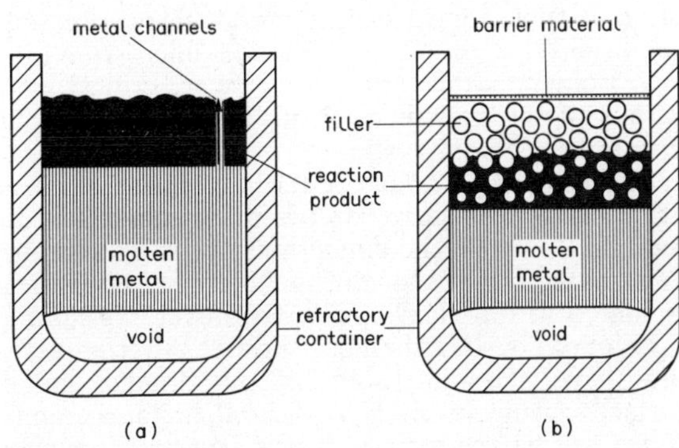

Figure 1
The directed metal oxidation process: growth into (a) free space, and (b) a filler material. The void indicates that, in this small-scale experiment, capillary forces draw the liquid metal to the growth front against the influence of gravity

Al_2O_3–Al and AlN–Al systems have been extensively investigated and are approaching commercialization as a result of successful laboratory tests and prototype component evaluations.

1. Processing

The directed metal oxidation process is based on the reaction of a molten metal with an oxidizing gas; for example, the reaction of an aluminum alloy with air to form α-Al_2O_3. In a certain critical temperature range above the melting temperature of the metal, the reaction product grows outward from the original metal pool surface either into free space (see Fig. 1a) or into a filler (see Fig. 1b). The reaction is sustained by the wicking of liquid metal through tortuous microscopic channels in the reaction product (see Fig. 2) to reach the product–gas interface. The growth proceeds until the metal supply is consumed or until the reaction front meets a barrier material which suppresses any further reaction.

The reaction product that, in the case where it "infiltrates" a porous filler, represents the matrix material is itself a ceramic–metal composite where both phases are three-dimensionally interconnected. The volume fraction of the metal may range from 0.05 to 0.3 depending on the processing conditions. Hence the mechanical properties tend to be dominated by the ceramic phase. The filler may be any material that is compatible with the reactant atmosphere and the molten metal. The filler may consist of particulates or fibers with a wide variety of possible sizes and shapes in various arrangements and volume concentrations. Some of the essential advantages of the process become apparent in the fabrication of complicated composite components; no shrinkage is involved and low temperatures are sufficient for matrix growth. The composite systems listed in Table 1 have been reported. The processing conditions for proper growth and microstructural development in the various systems are critical, and details of the growth mechanism may differ from system to system.

Forming and shaping of the composite parts is achieved by utilizing one of two phenomena. In one method the shape and dimensions of the part can be determined by the shape of the filler preform which is coated with a barrier material preventing matrix overgrowth beyond the filler surface. Alternatively, the shape and dimensions of the original metal surface can be precisely replicated in the ceramic composite part, a method that is especially suitable for intricate internal surfaces not attainable by other techniques.

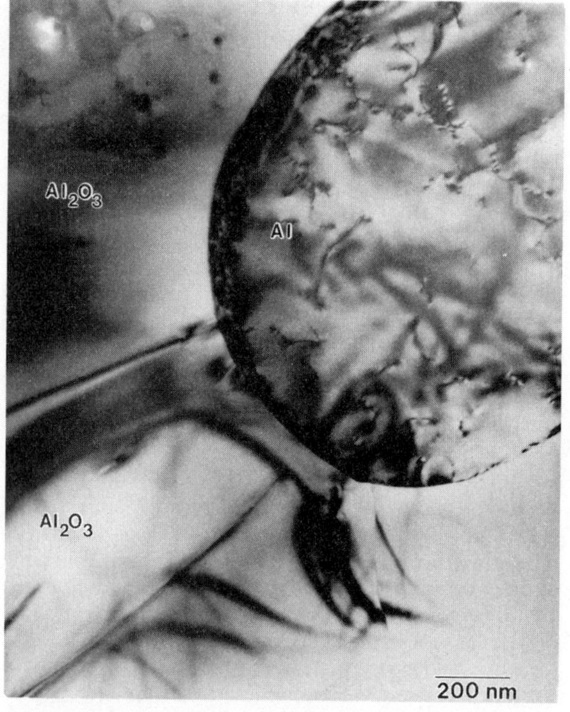

Figure 2
Bright-field transmission electron micrograph of an aluminum channel (cross section) in an Al_2O_3 matrix (© Chapman and Hall 1989: taken from Aghajanian M K et al. Properties and microstructures of Lanxide Al_2O_3–Al ceramic composite material. *J. Mater. Sci.* 24: 658–70)

Table 1
Ceramic composites made by the directed oxidation of molten metals

Matrix	Filler
Al_2O_3–Al	Al_2O_3: SiC, $BaTiO_3$
AlN–Al	AlN: Al_2O_3, B_4C, TiB_2, Si_3N_4
ZrN–Zr	ZrN: ZrB_2
TiN–Ti	TiN: TiB_2, Al_2O_3

2. Growth in the Al_2O_3–Al system

The nature of matrix growth has only been described for the Al_2O_3–Al system, although some of the features of the process may also be characteristic for other ceramic–metal systems. It is well known that oxidation of pure aluminum is usually slow due to the formation of a protective Al_2O_3 layer. In order to generate rapid and continuous growth of the oxidation product, certain dopants in the aluminum melt are necessary to initiate the growth, to prevent the formation of a protective reaction product and to improve the wettability of the growing Al_2O_3 and possibly that of the filler in respect to liquid aluminum. Alloying elements such as magnesium (1–10 wt%) have been shown to form a spinel layer (e.g., $MgAl_2O_4$) on the metal pool surface. This layer may then initiate the reaction product with its characteristic metal channel system. The same element is also active at the growth front in forming a thin (~1 μm) MgO layer on the surface of the liquid metal. Zinc as an alloying element may play a similar role to that of magnesium, suggesting that the high activity of these elements in the liquid metal may be an important factor.

The molten aluminum alloy is supplied to the surface by wicking through the microscopic channels of the Al_2O_3 product. Oxygen from the reactant atmosphere penetrates through the thin MgO layer to react with the liquid aluminum, resulting in the growth of Al_2O_3. The MgO layer appears to prevent the formation of a protective Al_2O_3 layer; it also is believed to control the growth rate which is nearly independent of the thickness of the reaction product. This constant growth rate is analogous to the constant combustion rate of candles and oil lamps (which, with appropriate level markers, were utilized in ancient times as clocks). Hence, the liquid metal (liquid wax) supply is sufficiently rapid to allow the constant growth rate at otherwise constant process parameters. Typical growth rates are in the range of 0.5–0.8 cm d^{-1} with maximum rates of 3 cm d^{-1}.

In addition, the process may require another dopant selected from the group IVa elements, such as silicon. These elements may also enhance the wettability of Al_2O_3 with respect to liquid aluminum.

Other additions to the starting aluminum alloy have been found to influence the microstructure (e.g., the fineness) of the reaction product and are useful in the tailoring of properties.

The critical temperature range in which growth takes place is determined by the threshold temperature for the formation of spinel (800–900 °C) or for the nucleation of alumina, and by a reduced reaction rate at temperatures greater than 1400 °C. A feature of the process is the reduction of less stable oxides (e.g., SiO_2) by the molten aluminum. The consequences are that the grain boundaries and filler–matrix interfaces are free of glassy phases, that SiO_2 can also be added as a dopant instead of free silicon and that the aluminum alloy phase in the channel system of the matrix may become enriched in silicon (e.g., when SiC is used as filler material, resulting in increasing silicon precipitation that may eventually reduce the reaction rate). Increasing the partial pressure of the reacting gas accelerates the growth rate moderately, the kinetics having been found to follow a $P_{O_2}^{1/4}$ rate dependence. Certain materials, such as wollastonite ($CaSiO_3$), prevent or stop the reaction. These growth-inhibiting materials are, therefore, used as shape limiters.

As indicated in Fig. 1b, the reaction product grows progressively through the filler with no displacement or disturbance of the filler configuration. Figure 3a shows a crosswoven SiC fiber composite (Nicalon) and Fig. 3b a SiC particulate composite both with an Al_2O_3–Al matrix. The filler pore spacing may be as small as to be about 1 μm; at smaller distances (e.g., when submicrometer filler particles are used) the initiation and growth of the reaction product may become difficult. The effects of size, shape, volume concentration and type of filler on growth rate, which depend on alloy content among other factors, have been little reported.

3. *Formation of Ceramic Composite Shapes*

A variety of conventional ceramic techniques (e.g., uniaxial or isostatic pressing, slip casting and injection molding) can be used to shape the filler preforms. Furthermore, all textile-type techniques developed for the manufacturing of fiber-reinforced polymers represent suitable methods for preparing fibrous filler configurations. A method for forming shapes that are used in the directed metal oxidation process is illustrated in Fig. 4, which shows the making of a hollow rotationally-symmetric component with screw threads above a spherical cavity. In this case, the filler preform can be isostatically pressed or sedimentation cast around the solid metal form. The metal–preform arrangement is then embedded in a growth barrier material (e.g., $CaSiO_3$ powder) which prevents the growth of the reaction product beyond the outer boundary of the component. Since the original shape of the metal is replicated after complete consumption of the metal (or after eliminating excess metal), the intricate internal shape of the component is determined by the parent metal shape. Other shape forming methods are related to the process shown in Fig. 1b. Thin-walled hollow parts, for example, can be made by pouring the liquid metal into the preform and, after completion of the reaction, draining the remaining liquid metal. Components with maximum dimensions of 90 cm and wall thicknesses of up to 20 cm have been fabricated. While there is, in principle, no size limit compositional changes in the alloy (e.g., increasing enrichment of precipitated silicon) may reduce the growth rate for larger thicknesses, though this will depend on the alloy-to-preform volume ratio.

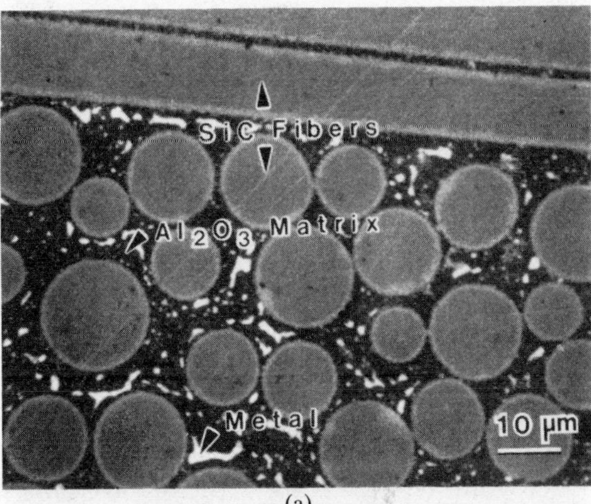

(a)

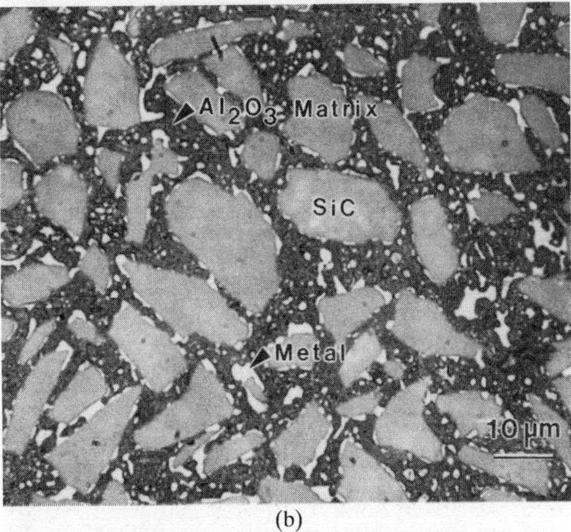

(b)

Figure 3
Al_2O_3–Al matrix growth into (a) crosswoven SiC fibers and (b) SiC particulates

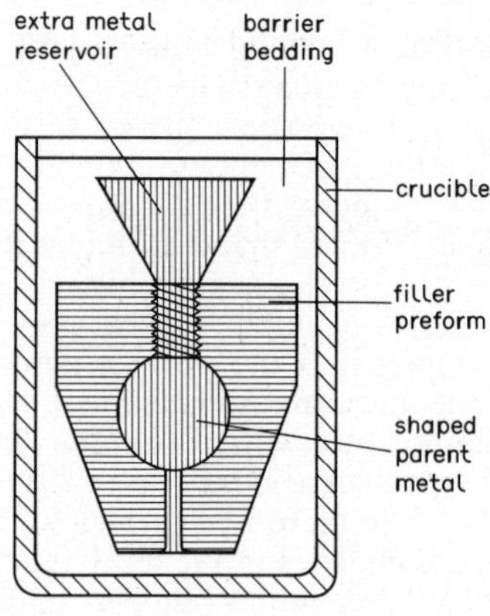

Figure 4
Shape formation by the melt oxidation process indicating the possibility of obtaining intricate internal surfaces

4. *Properties*

The properties of composites formed by the directed oxidation of molten metals depend strongly on the filler characteristics and, in some cases, on the volume fraction of the metal phase left in the channel system of the matrix. Hence, a wide property spectrum can be tailored extending, for example, from thermally conducting to insulating or from very tough to brittle. Such flexibility exists for many aspects of behavior.

Typical flexural strengths and toughnesses for the Al_2O_3–Al matrix alone in the range of 200–300 MPa and 6–11 MPa $m^{1/2}$, respectively, at densities of approximately 3500–3600 kg m^{-3}.

Composites using this matrix system uniaxially reinforced with SiC fibers have shown average strengths of 720 MPa and steady-state fracture toughnesses of 27 MPa $m^{1/2}$, ranking these materials among the toughest ceramics produced. In this case, the fracture toughness results from extensive debonding and pullout of the fibers from the grown ceramic matrix (see Fig. 5a). This fracture behavior resembles in a general way that of fiber-reinforced glass matrix composites.

Al_2O_3–Al matrix composites with either Al_2O_3 or SiC filler particles exhibit strengths of up to 550 MPa and toughnesses of up to 9 MPa $m^{1/2}$. In this case, the toughness–strength relation is controlled by the volume fraction of the metal phase which is, at least partially, determined by the growth temperature. A composite with a nominal 15 μm SiC filler, for example, exhibits strengths and toughnesses of 350 MPa and 7.8 MPa $m^{1/2}$ when grown at 900 °C, of 390 MPa and 5.4 MPa $m^{1/2}$ when grown at 1000 °C, and of 525 MPa and 4.7 MPa $m^{1/2}$ when grown at 1150 °C, respectively. Nearly 50% of the room-temperature strength can be retained at 1500 °C. Weibull moduli have been attained with values as high as 20. AlN–Al matrix strengths of 400 MPa and toughnesses of 9.5 MPa $m^{1/2}$ have been obtained. The main toughness contribution in these composites results from crack bridging by the ductile aluminum phase which necks down to sharp edges as seen in Fig. 5b. Excellent mechanical properties are also obtained with platelet fillers; for example, an AlN–Al matrix with TiB_2 platelets exhibits a strength of 512 MPa and a toughness of 10.6 MPa $m^{1/2}$.

An interesting feature of the directed metal oxidation process, which further broadens the opportunities for property tailoring, results from modifications made to the metallic constituent. For example, by adding small amounts of nickel to the starting alloy or NiO to the filler material, a nickel-aluminide-type intermetallic may replace the aluminum phase in the Al_2O_3–Al reaction product thus improving the erosion and corrosion properties. A similar modification may be obtained by annealing the composite in molten nickel (e.g., at 1525 °C for 9 h).

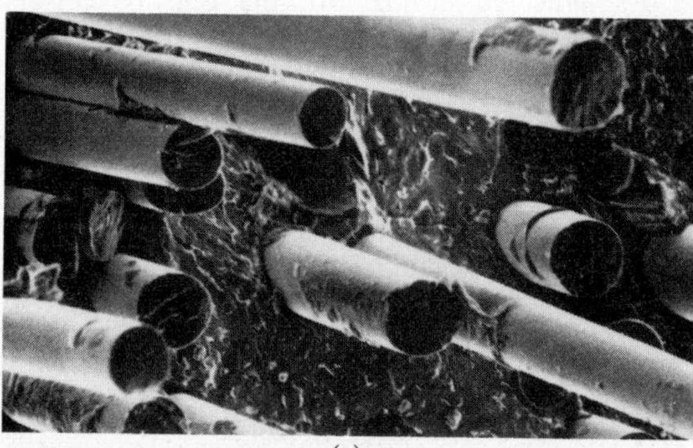

(a)

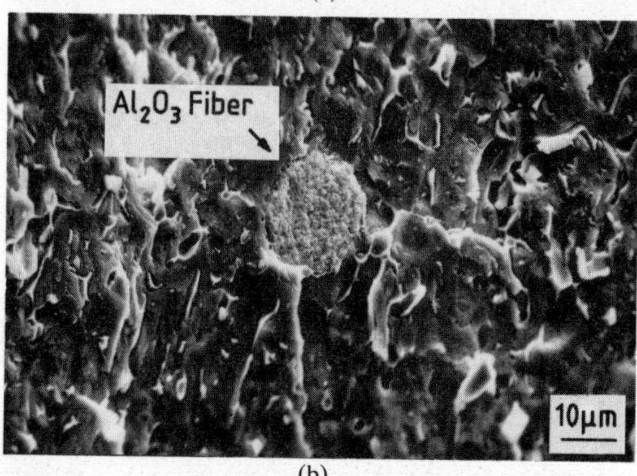

(b)

Figure 5
Fracture surfaces of (a) SiC-fiber composite and (b) Al_2O_3-fiber composite, both with an Al_2O_3–Al matrix

5. *Applications*

This new technology is being extended to a wide range of applications. These include wear parts for a variety of process industry uses; armor components; piston, gas turbine and rocket engine components (including structural, wear and thermal insulation functions); heat exchangers and heat-storage devices; and various aerospace components. Products in the areas of wear and armor are in the early commercialization stage based on successful tests and process scale up. Other product areas are in various stages of development.

Since the process involves low-cost starting materials (except when high-strength fibers are used), low processing temperatures and no densification shrinkage, it appears to be highly competitive in comparison with other conventional ceramic technologies. A special advantage may lie in the possibility of manufacturing large parts with complex shapes.

Bibliography

Newkirk M S, Lesher H D, White D R, Kennedy C R, Urquhart A W, Claar T D 1987 Preparation of Lanxide™ ceramic matrix composites: Matrix formation by the directed oxidation of molten metals. *Ceram. Eng. Sci. Proc.* 9: 879–85

Newkirk M S, Urquhart A W, Zwicker H R, Breval E 1986 Formation of Lanxide™ ceramic composite materials. *J. Mater. Res.* 1: 81–9

N. Claussen
[Technische Universität Hamburg-Harburg, Hamburg, FRG]

A. W. Urquhart
[Lanxide Corporation, Newark, Delaware, USA]

P

Permanent Magnets, Iron–Rare Earth

Permanent magnets based on the $Fe_{14}Nd_2B$ phase were introduced in 1983 and are currently the strongest magnets available. They are the latest stage of the remarkable twentieth-century progress in permanent magnets, shown in Fig. 1 in terms of maximum energy product (the most commonly used measure of magnet quality) and intrinsic coercivity (the reverse magnetic field required to reduce net magnetization to zero).

For many centuries, carbon steels were the only improvement in permanent-magnet materials over lodestones. Starting in about 1890, improved understanding of the metallurgy of alloy steels and, later, control of solid-state precipitation in iron-based alloys via heat treatment, led to substantial improvements. This path of materials development culminated in the alnicos, which retain today a large fraction of the magnet market. However, increasing availability of rare earth metals in recent decades has led to a new development path based on compounds betwen 3*d*-transition metals and rare earths. The iron–rare earth magnets represent the third "generation" of rare earth magnets, the first and second generations being based on Co_5R and

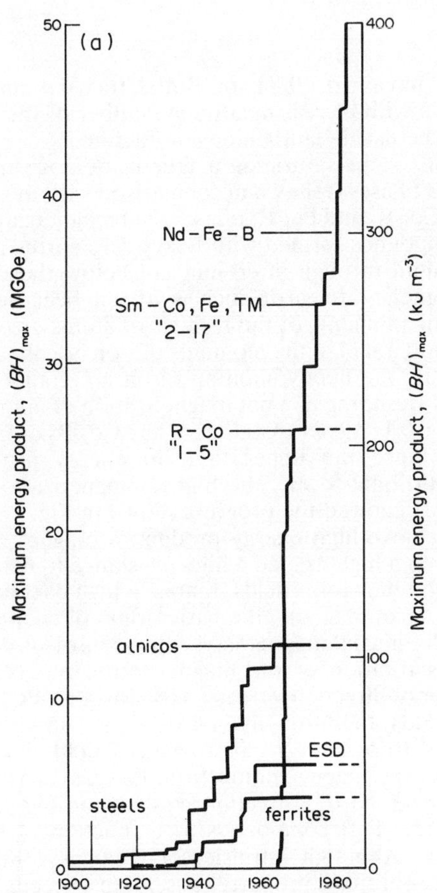

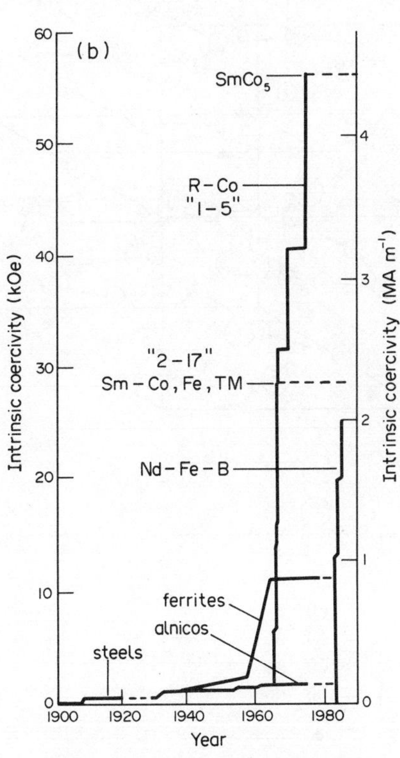

Figure 1
Chronology of magnet development since 1900: best reported laboratory values for (a) the maximum energy product and (b) intrinsic coercivity (Strnat 1988)

$(Co, Fe)_{17}R_2$ phases, respectively. As seen in Fig. 1, the Fe–Nd–B magnets have provided a fourfold increase in energy product over the alnicos—and an increase of 50 fold over carbon steels, the most popular magnet a century ago.

1. Basic Properties of $Fe_{14}R_2B$

The crystal structure of $Fe_{14}R_2B$ (Fig. 2) is complex; the tetragonal unit cell contains four formula units (68 atoms), six nonequivalent iron sites and two nonequivalent rare earth sites. There is a six-layer stacking sequence along the *c*-axis. The first and fourth layers are mirror planes containing iron, neodymium and boron atoms, and the other layers are puckered nets containing only iron atoms. Each boron atom is at the center of a trigonal prism formed by six iron atoms, three above and three below the Fe–R–B planes. (Such trigonal prisms are common to many transition metal–metalloid compounds, including FeB and Fe_3C.) For $Fe_{14}Nd_2B$, $c = 1.22$ nm and $a = 0.88$ nm. Both lattice parameters decrease with increasing atomic number of the rare earth, the usual "lanthanide contraction."

In Fig. 3, the saturation magnetization of various $Fe_{14}R_2B$ phases is shown in comparison with those of Co_5R, $Co_{17}R_2$ and $Fe_{17}R_2$ phases. Net magnetizations for compounds formed with heavy rare earths from gadolinium through ytterbium are below those for nonmagnetic rare earths such as yttrium because the magnetic moments of the rare earth atoms are coupled antiparallel to the moments of iron or cobalt. In $Fe_{14}Nd_2B$, the neodymium and iron are coupled in parallel, resulting in a net magnetization of approximately 1.6 T, higher than that of any $Co_{17}R_2$ or Co_5R phase. Since the upper limit to energy product is proportional to M_s^2, the higher magnetization of $Fe_{14}Nd_2B$ allowed the progress shown in Fig. 1.

To achieve high energy product, a magnet must have both a high M_s and a high resistance to demagnetization in reverse fields, that is, a high coercivity. In rare earth magnets, the basic origin of coercivity is a high uniaxial magnetocrystalline anisotropy that opposes rotation of the magnetization out of the low-energy direction or easy axis. In most $Fe_{14}R_2B$ compounds, the tetragonal *c*-axis is the "easy" axis. The anisotropy field H_a, the transverse field required to rotate the magnetization from the easy axis to a transverse ("hard") direction, is shown in Fig. 4 for various $Fe_{14}R_2B$ compounds as a function of temperature. Although intrinsic coercivity is a highly structure-sensitive property, it seldom exceeds 10–20% of the anisotropy field, so that anisotropy field is a good measure of the potential of a material for coercivity. For nonmagnetic rare earth ions (Y^{3+}, La^{3+}, Ce^{4+} and Lu^{3+}) and spherical ions (Gd^{3+}), the

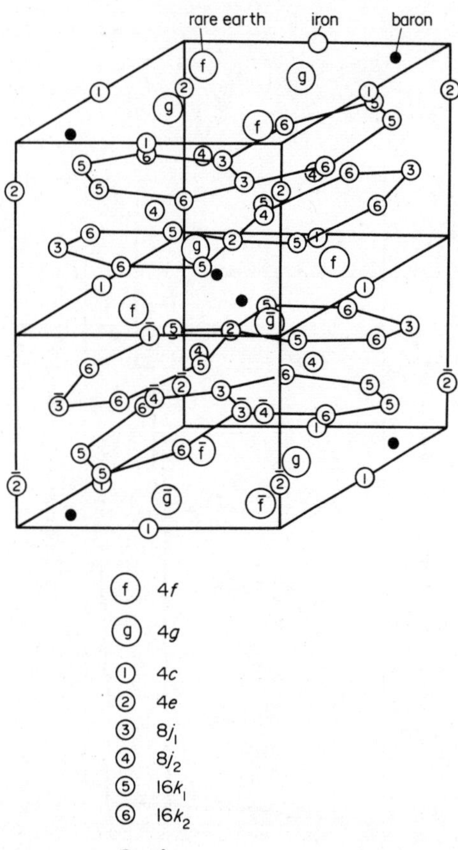

Figure 2
Schematic drawing of the crystal structure of $Fe_{14}R_2B$ (Sagawa et al. 1987)

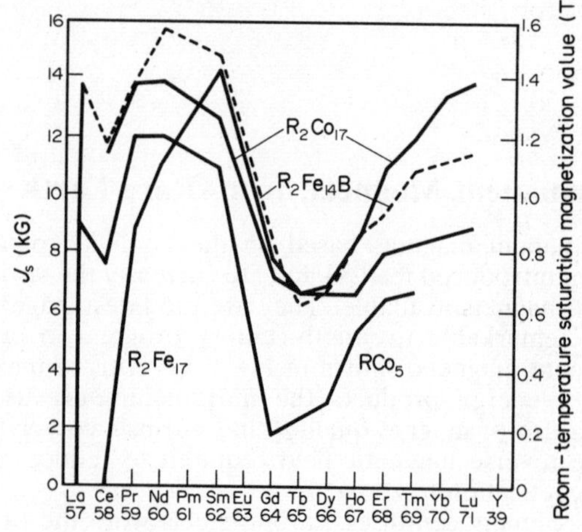

Figure 3
Room-temperature saturation magnetization values of various transition metal–rare earth (–boron) compounds (Strnat 1988)

anisotropy field of $Fe_{14}R_2B$ at room temperature is about 2 MA m^{-1}. This reflects the contribution of the iron sublattice to anisotropy. For $Fe_{14}Nd_2B$, the Nd^{3+} ions make a substantial additional contribution, yielding a total value of about 6 MA m^{-1}. Although substantially below the value of about 25 MA m^{-1} for Co_5Sm, the anisotropy field of $Fe_{14}Nd_2B$ is sufficiently high to permit the development of high coercivity. Of the other $Fe_{14}R_2B$ compounds, only $Fe_{14}Pr_2B$ has the combination of high magnetization and high anisotropy field required to make it a promising candidate for permanent magnets.

A third important parameter for permanent magnet materials is the Curie temperature T_c. As shown in Fig. 5, the $Fe_{14}R_2B$ phases have substantially higher Curie temperatures than the $Fe_{17}R_2$ phases, the only binary Fe–R phases with high magnetization. However, they remain far inferior to Co_5R and $Co_{17}R_2$. This is a major weakness that limits the temperature range of application of Fe–Nd–B magnets.

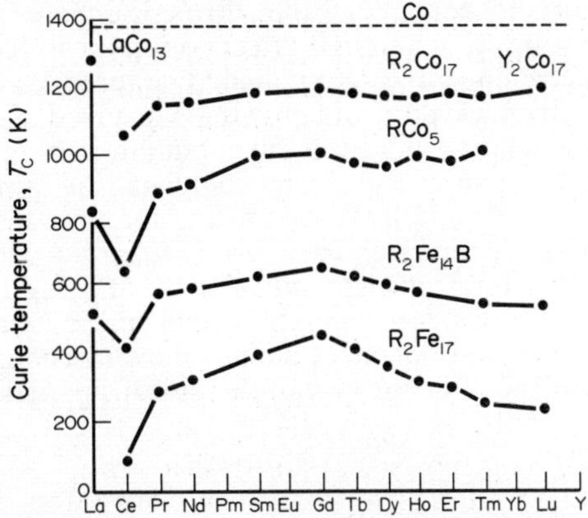

Figure 5
Curie temperatures of various transition metal–rare earth (–boron) compounds (Strnat 1988)

2. Sintered Magnets

Most Fe–Nd–B magnets are made by a process analogous to that used for Co_5R and ferrite magnets. Castings are crushed and milled to powder below 10 μm in size, sufficiently small that most particles are single crystals. Powders are then aligned in a magnetic field, pressed, sintered at about 1100 °C and rapidly cooled. A post-sintering heat treatment is usually used to enhance coercivity. To enhance densification via liquid-phase sintering, magnet compositions are more neodymium-rich and boron-rich than the stoichiometric compound, and the final microstructure consists of $Fe_{14}Nd_2B$ grains separated by a thin neodymium-rich grain-boundary phase, plus inclusions of $Fe_4Nd_{1.1}B_4$. These secondary phases are nonmagnetic at room temperature. The boron-rich phase is believed to have no beneficial effect on magnetic properties, but the neodymium-rich phase appears to enhance coercivity. On the other hand, it decreases resistance to corrosion and oxidation.

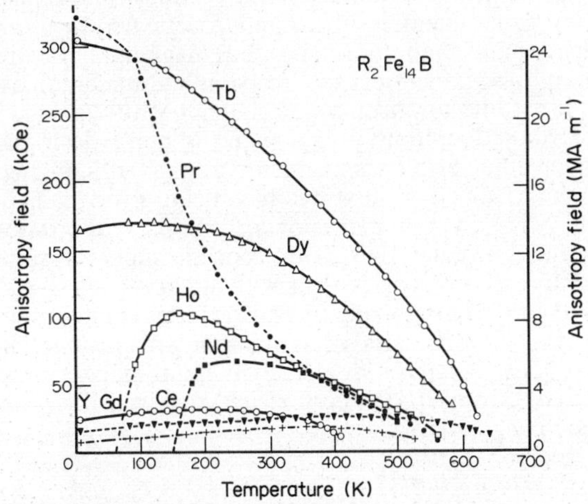

Figure 4
Anisotropy field of various $Fe_{14}R_2B$ compounds as a function of temperature (Sagawa et al. 1987)

By minimizing the presence of secondary phases, maximum energy products of up to 400 kJ m^{-3} have been reported for aligned sintered magnets, but commercial Fe–Nd–B magnets typically cover the range 200–320 kJ m^{-3}. Intrinsic coercivity approaches 1 MA m^{-1}, and remanence, the net flux density remaining after removal of the magnetizing field, is about 1.2 T.

With increasing temperature, remanence of Fe–Nd–B sintered magnets decreases by about 0.12% °C^{-1} and intrinsic coercivity by about 0.6% °C^{-1}. This decrease is much larger than for Co_5R and $(Co, Fe)_{17}R_2$ magnets, and results primarily from the lower Curie temperature. The decrease of coercivity with increasing temperature is a particular problem in many applications, since demagnetizing fields then cause irreversible losses of magnetization. Various alloying additions have been studied as a means of improving the thermal stability of Fe–Nd–B magnets. Alloying substitutions of cobalt for iron raise the Curie temperature, but also tend to lower the coercivity. As a result, cobalt additions alone do little to enhance thermal stability. Substitutions of dysprosium for neodymium raise the coercivity (Fig. 6) by increasing the anisotropy field (see Fig. 4). This improves thermal stability, but with a loss of magnetization and energy product. Numerous other alloying additions, including aluminum, gallium and magnesium, have been reported to enhance coercivity, either through enhancement of anisotropy or through effects on microstructure. Magnetization reversal in aligned sintered magnets occurs by nucleation and growth of reversed magnetic domains, with the neodymium-rich grain-boundary phase enhancing coercivity by pinning domain walls and thereby resisting the propagation of magnetic reversal from grain to grain. Subtle microstructural changes induced by alloying additions may reduce reverse-domain nucleation or enhance grain-boundary pinning, allowing coercivity to reach a higher fraction of the anisotropy field.

3. *Melt-Spun Magnets*

Commercial Fe–Nd–B magnets have also been produced by melt spinning, a process in which an alloy is rapidly solidified by directing a molten stream against a cold rotating substrate. An inert gas is used to avoid oxidation. The resulting ribbon is 15–30 μm thick, and may be either amorphous or microcrystalline depending on the substrate surface velocity, which controls the cooling rate. Optimum coercivity for Fe–Nd–B occurs at an intermediate quench rate that yields grains less than 100 nm in diameter, about 100 times finer than the grains in sintered magnets. Melt-spun ribbons are more nearly single-phase $Fe_{14}Nd_2B$ than sintered magnets, but may contain a thin neodymium-rich phase along the grain boundaries. Optimally quenched ribbon is crushed to coarse powder, blended with polymer, and formed into bonded magnets by various molding techniques. The resulting magnets have a volume fraction of $Fe_{14}Nd_2B$ of about 80%, and are isotropic, limiting the remanence to about 0.65 T and the maximum energy product to about 64 kJ m^{-3}.

Alternatively, magnets of nearly 100% $Fe_{14}Nd_2B$ can be produced by hot-pressing of melt-spun powder. Since some grain growth occurs at the pressing temperature, overquenched ribbon is used so that optimum grain size and coercivity are obtained after pressing. Such hot-pressed isotropic magnets typically have remanences approaching 0.8 T and maximum energy products of about 100 kJ m^{-3}. For both these and isotropic bonded magnets, intrinsic coercivities are about 1.3 MA m^{-1}. The relative decrease of coercivity with increasing temperature is somewhat less than for sintered magnets. Demagnetization curves for isotropic bonded and hot-pressed magnets at various temperatures are shown in Figs. 7a and 7b, respectively.

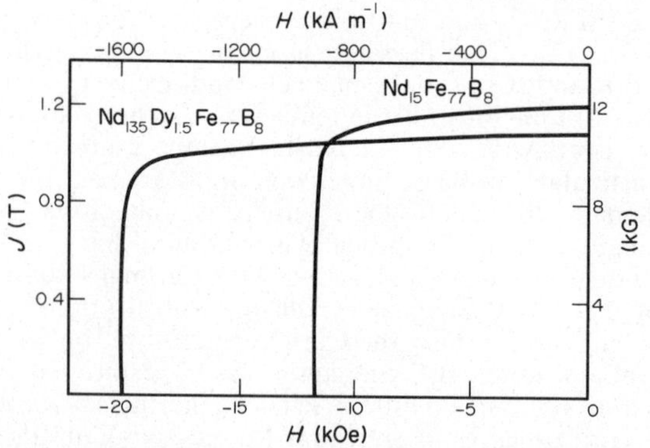

Figure 6
Magnetization vs reverse magnetic field for sintered Fe–Nd–B and Fe–Nd–Dy–B magnets, showing the enhancement in coercivity produced by dysprosium substitution (Sagawa et al. 1984)

4. *Alignment by Hot-Working*

Deformation of hot-pressed Fe–Nd–B magnets in compression at temperatures above 700 °C, called "die-upsetting," produces substantial alignment of the *c*-axis along the compression direction. After this treatment, the magnet contains platelike grains lying normal to the compression direction, typically 60 nm in thickness and 300 nm in diameter. The alignment produced by die-upsetting can increase the remanence from 0.8 T to 1.2 T (Fig. 7c). As a result, the maximum energy product is increased to over 250 kJ m^{-3}. Intrinsic coercivity, however, is decreased and becomes more temperature dependent. As with sintered magnets, alloying additions are being studied as a means of improving temperature capability.

Die-upsetting can also be used as a route to produce anisotropic bonded magnets. Die-upset magnets are reground to powder, blended with polymer, magnetically aligned and then molded into bonded magnets. As with isotropic melt-spun powder, the coercivity is related to the grain size, which is much finer than the powder size. Thus coercivity is not severely degraded by powdering or by chemical reactions with binder materials. Melt-spun magnets also contain less neodymium-rich phase than most sintered magnets, and are therefore less subject to corrosion and oxidation.

For some applications, a magnet with easy axes distributed randomly in a plane is more desirable than either a fully isotropic or a uniaxially aligned magnet. Such a texture has been produced in $Fe_{14}Nd_2B$ by hot extrusion of compacted powder through circular dies. This process imposes radial compressive forces, and, as with die-upsetting, the *c*-axes preferentially align parallel to the compression forces. Other variations on standard production methods of sintered and melt-spun magnets are described in a comprehensive review (Buschow 1988).

5. *Current Status*

Despite their high cost, rare earth magnets have found many applications where their high coercivities and energy products produce design advantages,

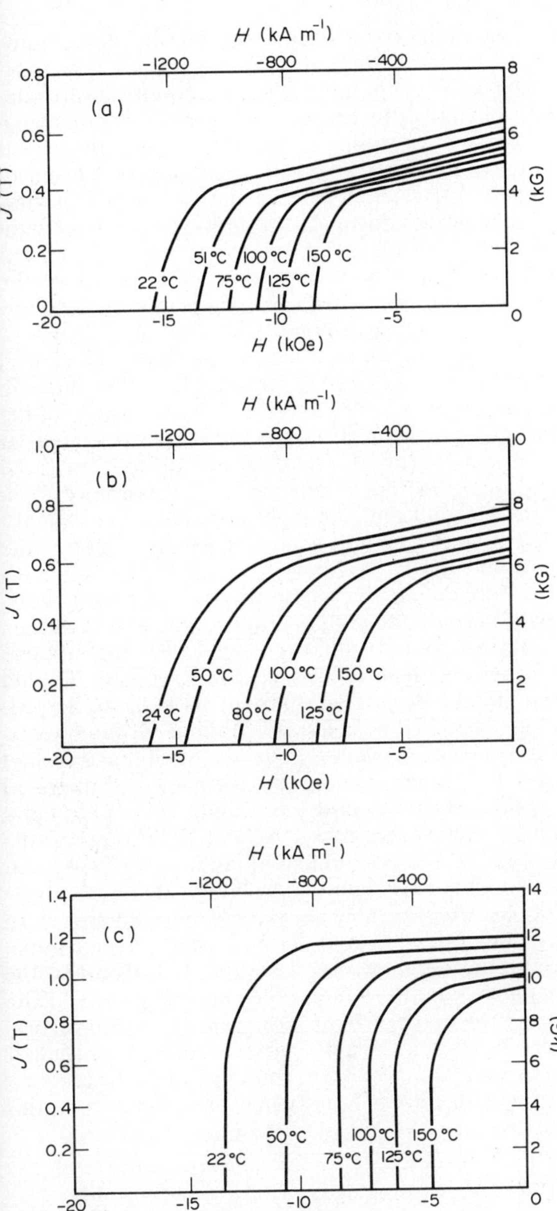

Figure 7
Magnetization vs reverse magnetic field of melt-spun Fe–Nd–B magnets at various temperatures: (a) isotropic, epoxy-resin bonded, (b) isotropic, hot-pressed, (c) anisotropic, hot-formed (Croat 1988)

including electric motors, magnetic bearings, earphones and microwave power devices. Because of higher energy products, and because iron and neodymium are more abundant than cobalt and samarium, Fe–Nd–B magnets are likely to replace Co_5Sm in applications where high thermal stability and high coercivity are not required. However, broader applicability of Fe–Nd–B will require improvements in thermal stability, coercivity and corrosion resistance. The relative importance of sintered and melt-spun magnets remains unclear, and will be determined by relative cost as well as by property differences.

Stimulated by the unexpected discovery of $Fe_{14}Nd_2B$, researchers have also begun an intensive search for other ternary and higher-order compounds with sufficiently high values of magnetization, anisotropy and Curie temperature to be useful as permanent magnets. Most attention is focused on alloy systems containing iron, cobalt and rare earths, in addition to other elements. Although higher energy product via higher magnetization is one goal of such studies, higher coercivity, better thermal stability and lower cost are equally or more important. Fe–Nd–B magnets represent the latest, (but certainly not the last) stage in the development of permanent-magnet materials.

See also: Magnetic Alloys: Cobalt–Rare Earth

Bibliography

Buschow K H J 1986 New permanent magnet materials. *Mater. Sci. Rep.* 1: 1–63

Buschow K H J 1988 Permanent magnet materials based on 3*d*-rich ternary compounds. In: Wohlfarth E P, Buschow K H J (eds.) 1988 *Ferromagnetic Materials*, Vol. 4. Elsevier, Amsterdam

Croat J J 1988 Neodymium–iron–boron permanent magnets prepared by rapid solidification. *J. Mater. Eng.* 10: 7–13

Sagawa M, Fujimura S, Yamamoto H, Matsuura Y, Hiraga K, 1984 Permanent magnet materials based on the rare earth–iron–boron tetragonal compounds. *IEEE Trans. Magnetics* 20: 1584–9

Sagawa M, Hirasawa S, Yamamoto H, Fujimura S, Matsuura Y 1987 Nd–Fe–B permanent magnetic materials. *Jpn. J. Appl. Phys.* 26: 785–800

Strnat K J 1988 Rare earth–cobalt permanent magnets. In: Wohlfarth E P, Buschow K H J (eds.) 1988 *Ferromagnetic Materials*, Vol. 4, Elsevier, Amsterdam

J. D. Livingston
[Massachusetts Institute of Technology, Cambridge, Massachusetts, USA]

Petroleum Resources: Conditions for Formation

Modern consensus on the origin of hydrocarbons states that formation takes place under exclusively organic (biogenic) conditions. Modern fauna and flora, in their nanomicroscopic to ultramicroscopic elements, reveal an actual production of hydrocarbons during the life cycle of many living taxa. Proponents of an organic origin therefore assume that hydrocarbons were present also in paleotaxa, and that these paleohydrocarbons contributed at least in part to the formation of petroleum.

Although this consensus has been strong for more than a half century, however, it has not gone unchallenged. Adherents of an inorganic (abiogenic) origin of hydrocarbons continue to present their views, thus keeping alive a theory more than a hundred years old. In the 1980s these adherents turned actively to seismology, to a carbon budget of the crust, to assumptions of massive outgassing from the earth's crust and mantle, and even to the drilling of a deep borehole in search of evidence for their theory.

1. Inorganic (Abiogenic) Origin of Hydrocarbons

Proponents of the inorganic theory of the origin of hydrocarbons accentuate the widely accepted presumption that hydrogen and carbon are present through most or all levels of the earth's crust. The presence of these same two elements in the underlying mantle also seems possible. Generation of the hydrogen may come from chemical reaction between superheated steam and mineral oxides, especially iron oxide. Catalysts to combine carbon and hydrogen deep within the crust, on the other hand, are unlikely to exist. Proponents of the inorganic theory of origin believe, however, that heat and pressure at depths many tens of kilometers below the surface of the earth may substitute for catalysts, and that the resulting hydrocarbons begin their existence at depths far below sediments, whether such sediments lie as thin columns on the craton or whether they lie as thick columns in structural foredeeps. The theory that heat and pressure can accomplish an inorganic production receives support from the presence of microscopic inclusions of hydrocarbon in diamond mined from pipes of kimberlite.

With regard to heat and pressure in the earth's crust, some new parameters which contribute to the inorganic view have appeared with the drilling of SG-3 on the Baltic Shield of the Kola Peninsula, USSR. This tremendous borehole, already more than 14 years in the drilling and by far the deepest borehole in the world, had reached a depth of some 12 000 m by 1984. Reports from the USSR mentioned gases and inflows of strongly mineralized waters at more than 11 000 m below the earth's surface. These came into the borehole despite pressures greater than 3000 bars and temperatures greater than 180 °C. Included in the gases were hydrocarbons.

Those who support the inorganic theory of origin might argue that the discovery of hydrocarbons by a single borehole in deeply buried Archean gneiss bodes well for the presence of widespread inorganic hydrocarbons in the deep crust. This borehole, however, contains 7000 m of Proterozoic sedimentary cover above the gneiss. With dolomites, siltstones and phyllites present, a generation of organic (biogenic) hydrocarbons is possible within this cover despite no more than its "cryptozoic" level of organic material. Such organic hydrocarbons may have migrated downward from the Proterozoic sedimentary beds through fractures to lodge in Archean gneiss. Such a possibility clouds the status of the "superdeep" well SG-3 as a rigid test of the inorganic (abiogenic) origin of hydrocarbons.

A deep borehole drilled near Siljan in central Sweden differs from SG-3 on the Kola Peninsula in having, as its sole objective, a definitive proof of the inorganic origin of hydrocarbons. Suspended at 6300 m, the borehole at Siljan drilled fractured granite from top to bottom, and reportedly recovered "significant" (though probably small) quantities of hydrogen and helium some 200 m above the bottom.

Opponents of inorganic theories of origin, however, note that the Siljan granite, which is more than 35 km wide, is almost ringed by fossiliferous Ordovician shales with seeps of gaseous liquid and bituminous petroleum. Opponents believe, accordingly, that if the operators continue drilling below 6300 m and discover methane, they will have proved only that organic (biogenic) methane had migrated through fractures into the granite from peripheral Ordovician source rocks. As with SG-3 on the Kola Peninsula, the borehole at Siljan in central Sweden, if deepened, may ignite considerable debate concerning its trustworthiness as a test of the origin of hydrocarbons. Insofar as the operators of Siljan claim an inorganic origin for the hydrogen and the helium in their borehole, they appear justified. The presence of these two gases through 6300 vertical meters of uninterrupted granite also justifies their assertion that these gases (but only these gases) were present in an early inorganic phase of the earth's history.

2. Organic (Biogenic) Origin of Hydrocarbons

Proponents of the organic theory of the origin of hydrocarbons agree that carbon and hydrogen were present in the primordial earth. They maintain, however, that only in an organic regime could these two elements combine to form rcognizable molecules of hydrocarbons, and eventually of petroleum. For supporting evidence they cite the exclusive association of important deposits of petroleum with sedimentary rocks. These sedimentary rocks, whether formed under marine or continental conditions, contain organic material in an original or transformed state. Up to 1988 no important deposits of petroleum had appeared in terranes of purely igneous or purely metamorphic lithofacies.

Whereas the primordial earth was cooling from a igh-temperature state, the geochemical evidence rom petroleum itself is for an origin of hydrocarbons at relatively moderate temperatures. Petroleum rotates the plane of polarized light; this ndicates that it contains molecules almost certainly f organic origin, formed at no more than moderate emperatures. Some of these molecules belong to he porphyrins, formed either from the chlorophyll f plants or the hemin of blood. Other compounds resent in petroleum also indicate derivation from iving matter; for example, cholesterol, carotene and erpenoids are derived from plants.

Modern geochemists suggest an origin of oil from rganic matter in two ways:

(a) directly from the hydrocarbons that many taxa, especially marine taxa, contain in their living cells—the contribution from this source may amount to about 10% of the total; and

(b) from decay and chemical alteration of buried organic matter—the contribution from this source, much larger than that from the first source, may amount to the remaining 90% of the total.

Evidence for the smaller contribution comes from a similarity of the hydrocarbons in petroleum to those found in plankton, kelp and in taxa even as advanced as fish. For the larger contribution, geochemists assert that burial, decay and chemical alteration produce homologues and polymers of methane. Present are all hydrocarbons that contain up to ten carbon atoms, and possibly half of all hydrocarbons that contain more than ten carbon atoms.

Before burial, photosynthesis of plants and algae converts water and carbon dioxide into oxygen and the organic saccharide glucose. Polysaccharides and a variety of complex organic compounds follow. At the end of the food chain, oxidation completes a cycle by breaking down complex organic molecules back to their original carbon dioxide and water. When burial takes place, however, the complex organic compounds escape a complete impact of the cycle, and it is these, as mentioned above, that furnish many of the hydrocarbons with ten or more carbon atoms.

In areas of incremental deposition, the depth of burial increases continually, and organic material in the buried sediment must experience continual increases of temperature and pressure. In response to these conditions the organic material must pass through a series of changes that geochemists have divided into the following three stages.

(a) *Diagenesis*. This occurs not far below the earth's surface, and with only a small increase of temperature and pressure. In this stage, reactions of organic character include decay through attack by bacteria (reactions of inorganic character also occur), the content of oxygen declines, and after a loss of methane, carbon dioxide and water the residue consists of kerogen, a solid and complex hydrocarbon.

(b) *Catagenesis*. Increasing temperature and pressure act on the residual kerogen, whose transformation during this stage, however, appears to be largely a thermal process. In addition, as in the preceding stage of diagenesis, the ratio of hydrogen to carbon declines, and theory supposes a release of oil and gas from kerogen in this stage.

(c) *Metagenesis*. Temperatures and pressures are considerably greater than in the second stage of catagenesis and are sufficient, ultimately, to expel all hydrocarbons and to convert the organic carbon to inorganic graphite.

Kerogen, described as the organic matrix of rocks, contains chemical compounds derived from carbohydrates, proteins and lipids, all three of which yield hydrocarbons under appropriate thermal conditions. Chemically, however, lipids bear the closest relations to hydrocarbons, for which reason lipids represent the most likely precursors of petroleum. In kerogen itself, slight differences in its percentages by weight of carbon, hydrogen and oxygen lead to a threefold classification.

(a) Kerogen type I—derived for the most part from algae. In this group, lipids are dominant, derivatives of oils, fats and waxes are present, there are many aliphatic chains, and there is the highest ratio of hydrogen to carbon.

(b) Kerogen type II—characterized by liptinite (exinite). Like Kerogen type I, this class contains aliphatic chains, but has derivatives of algal detritus accompanied by derivatives of zooplankton and phytoplankton, and a lesser ratio of hydrogen to carbon.

(c) Kerogen type III—this is the humic type, derived from the lignin of terrestrial woody plants, with few aliphatic chains and many aromatics, with the least ratio of hydrogen to carbon, and believed, therefore, to produce more gas than oil.

Kerogen is said to mature. Many variables, however, enter into its assumed maturation and attempts to quantify the process have taken two main forms. The first employs an index of thermal maturity (ITT) which takes into account the temperature and time of burial in each interval of temperature through each increment of 10 °C. The second approach utilizes an organic index (LOM) which is based on an assumption that maturation doubles for each increment of 10 °C. These quantifications, on the whole, correspond largely to the formation of oil and gas in catagenesis, and to the formation mainly of gas in earliest metagenesis.

3. Migration of Hydrocarbons

The rocks that serve as reservoirs for large accumulations of hydrocarbons consist for the most part of sediments deposited in or not far below the photic zone, where light, aeration and conditions of high energy would inhibit the formation of much organic material. For this reason and others, explorationists assume that oil and gas do not form in the reservoir itself, but that they migrate towards and into it from another rock, the source rock. The migration from source to reservoir can be effectively continuous, but mainly it resolves into a primary migration and a secondary migration.

3.1 Primary Migration

In the simplest form of this concept, water carries either petroleum or its precursor, in extremely low concentrations, out of the source rock as compaction drives fluid from the latter. If the source rock consists of a shale or a siltstone, its initial condition would have been that of a mud containing 70% or more of water. A constantly increasing overburden of uninterrupted sedimentation squeezes water continuously out of the compacted mud, and if organic material has survived in the original mud, some of this organic material presumably moves out together with the escaping water. Fluids migrating from the source rock may move vertically in the initial stages, but as compaction proceeds, the flat clay minerals present in the source rock assume a horizontal position, and they may impose a largely lateral motion upon migrating fluids.

The mechanism by which water carries hydrocarbons towards eventual accumulation remains largely conceptual, to a degree that the subject of primary migration has attained the status of "the last great mystery of petroleum geology." Petrophysicists believe that oil is not capable of passing through the pores of clayey mud either as small drops or as colloids. It is reasoned, consequently, that oil within the range of low molecular weights migrates in solution, and that its precursors migrate also in solution, but within the range of high molecular weights. The weak point of this model appears to be its need to convert the precursors to complete hydrocarbons by "mild cracking" after they enter the reservoir rock, presumably by "coagulation" in a changed "physical–chemical environment."

The time needed for the formation of hydrocarbons by transformation of kerogen appears to conflict, at least in part, with the model of a relatively early expulsion of hydrocarbons together with water under the impetus of simple compaction. Some petrophysicists, for this reason, have proposed that there is an increase in specific volumes and pressure within the source rock as the presumed transformation of kerogen to hydrocarbons of low molecular weight takes place. This internal pressure, developed later than the pressure of initial compaction may drive additional specific hydrocarbons out of the source rock into a primary migration. This model, while reasonable, may be insufficient to cover the whole spectrum of primary migration. A repetition or repetitions of increase in specific volumes, however, could improve considerably the expulsion of hydrocarbons and precursors from the source rock even during the comparatively late phases of its compaction. Another possibility in the late stages of compaction, especially if the pressure of the fluids in pores rises above rock pressure, is the development of microfractures. Should these microfractures connect with each other, they would offer an important pathway to primary migration.

With respect to the distance covered by hydrocarbons in primary migration, the petrophysical consensus is to the effect that the distance is short. Examples of statements concerning this factor are "only as far as needed to reach a permeable bed" or "distances covered by primary migration are commonly in the order of meters or tens of meters." The latter statement implies a strong vertical component in primary migration; that is, migration to permeable rock immediately above or below the source rock.

3.2 Secondary Migration

This phase of the movement of hydrocarbons is one that virtually by definition follows the phase of their expulsion from source rocks. The permeable beds in which secondary migration takes place are the carrier beds, precisely because they carry hydrocarbons to their ultimate accumulations. If a carrier bed remains porous without interruption in a way that does not deter the impetus of buoyancy, the length of movement of hydrocarbons through that carrier bed becomes almost limitless. Where explorationists see indications of broad and largely undeformed structures they invoke secondary migrations as long as hundred of kilometers. It should be noted that this concept of long migration is not new. Explorationists more than a half century ago referred to long "gathering grounds" as a favorable condition for prolific final entrapment of hydrocarbons.

4. Accumulation of Hydrocarbons

With the buoyancy of migrating hydrocarbons as the main motive force to propel them through carrier beds, any impediment to buoyancy terminates migration and brings about accumulation. In two dimensions, a reversal in dip of the carrier bed to form an anticline is sufficient to provide a trap in which gas can accumulate at the top and oil can accumulate below the gas. Under real subsurface conditions in three dimensions, however, the trap can accumulate and retain it hydrocarbons only if a double plunge of the structure provides complete closure. A double plunge is also necessary to retain hydrocarbons when accumulation begins with truncation or wedging at an unconformity. Faulting may replace plunge partly or wholly as an agent of closure. In trapping that is wholly stratigraphic, the agent of closure varies. Where accumulation takes place in reefs or bioherms, envelopment of these carbonate buildups by shales provides closure. Where salt penetrates carrier beds, the flanks of the salt itself convert carrier beds into reservoirs with accumulated hydrocarbons. Where a paleotopography develops on the upper surfaces of unconformities, buoyancy sends oil and gas into the high areas of this paleotopography, and accumulation occurs there under the cover of cap rock.

5. *Conclusions*

An inorganic (abiogenic) theory of the origin of hydrocarbons, now at least a century old, continues to find adherents, especially during the 1980s, despite an almost overpowering consensus for an organic origin that has gathered strength since the 1940s. Adherents to the inorganic theory maintain that hydrocarbons, especially methane, escape from the mantle and the crust in an outgassing that has continued from the time of the primordial earth to the present.

Proponents of an organic theory of the origin of hydrocarbons cite an almost exclusive association of important deposits of hydrocarbons with sedimentary rocks. They cite also the ability of oil to rotate the plane of polarized light and the presence in oil of porphyrins as indications of organic (biogenic) character. Additional evidence cited in support of this theory includes the presence of microscopic quantities of actual hydrocarbons in the cells of living taxa and the conversion of inorganic carbon dioxide and water by photosynthesis into organic glucose, the precursor of polysaccharides and other organic complex molecules.

Under incremental deposition, organic material must experience continual increases of temperature and pressure, loss of oxygen, carbon dioxide and water, and the formation of kerogen, the "organic matrix" of rocks. Kerogen itself presumably "matures," a process that theorists have quantified, and which corresponds for the most part to the release of oil and gas during the theoretical stage of "catagenesis."

The almost exclusive association of large, medium and small deposits of hydrocarbons with sedimentary rocks is certain. That such an association renders an inorganic theory of the origin of hydrocarbons untenable, is less certain. Possibly overlooked is the circumstance that for the tens of thousands of wells drilled specifically for hydrocarbons in sedimentary rocks, only one has drilled a thick body of granite from surface to total depth. The ratio of drilling in igneous rocks specifically for hydrocarbons to similar drilling in sedimentary rocks should therefore increase to even a minimum value before opponents can put forward a reasoned devaluation of the inorganic theory. An important amount of drilling in granites and gneisses, for example, might show the presence of residual methane. The polymerization of this simple aliphatic hydrocarbon to hydrocarbons of high molecular weight might seem unlikely at first glance. In theory, however, methane from the mantle and the crust could enter the sedimentary realm and become the beneficiary of pressure owing to overburden as well as the beneficiary of catalysis by clay minerals. In such a model the origin of hydrocarbons in sediments could be part inorganic and part organic.

In the organic model of recent years, details of primary migration remain purely speculative. Uncertainty exists over the relationship of the expulsion of hydrocarbons to compaction of the source rock, and over the nature of the hydrocarbons themselves at the time of expulsion. Precursors of oil and gas (protopetroleum) may be present as fluid leaves the source rock. Alternatively, oil and gas already formed may leave the source rock together with precursors. In either case, the distance to a porous carrier bed may amount to no more than a few meters. The carrier bed, by implication, is thus likely to lie immediately above the source rock. Once in the carrier bed, buoyancy becomes the motive force for the propulsion of hydrocarbons and precursors, and secondary migration has begun.

On broad epeirogenically simple forelands, secondary migration may continue along the carrier bed for distances of well over a hundred kilometers. Only when tectonic modification occurs, and when the tilt of the carrier bed reverses itself, do conditions for accumulation arise. For accumulation and consequent permanent retention of hydrocarbons, structural closure is necessary. Among such types of closure are doubly plunging anticlines, truncation or wedging, faulting, envelopment of reefs or bioherms by shale, truncation against salt and crestal paleotopography.

See also: Petroleum Resources: Oil Fields (Suppl. 2); Petroleum Resources: World Overview (Suppl. 2)

Bibliography

Anon 1987 Swedish test of Gold's gas theory inconclusive. *Pet. Rev.* October 1987: 40

Atwater G I 1980 Petroleum. In: Goetz P W (ed.) 1980 *The New Encylopaedia Brittanica, Macropaedia,* Vol. 14. Benton, Chicago, IL, pp. 164–75

Giardini A A, Melton C E 1983 A scientific explanation for the origin and location of petroleum accumulations. *J. Pet. Geol.* 6(2): 117–38

Giardini A A, Melton C E, Mitchell R S 1982 The nature of the upper 400 km of the Earth and its potential as the source for non-biogenic petroleum. *J. Pet. Geol.* 5(2): 173–90
Gold T 1979 Terrestrial sources of carbon and earth outgassing. *J. Pet. Geol.* 1(3): 3–19
Hobson G D (ed.) 1977 *Developments in Petroleum Geology—1.* Applied Science, London
Hunt J M 1979 *Petroleum Geochemistry and Geology.* Freeman, San Francisco, CA
Hunt J M 1982 Petroleum, origin of. In: Parker S P (ed.) 1982 *McGraw-Hill Encyclopedia of Science and Technology*, Vol. 10. McGraw-Hill, New York, pp. 77–8
Landes K K 1959 *Petroleum Geology.* Wiley, New York
Levorsen A I 1954 *Geology of Petroleum.* Freeman, San Francisco, CA
Selley R C 1985 *Elements of Petroleum Geology.* Freeman, New York
Shirley K 1987 Siljan project stays in cross fire. *AAPG Explorer* January 1987: 1, 12, 13
Tissot B P, Welte D H 1984 *Petroleum Formation and Occurrence.* Springer, New York
Whiteman A 1987 Gold's hypothesis. *Pet. Rev.* November 1987: 8

M. Kamen-Kaye
[Cambridge, Massachusetts, USA]

Petroleum Resources: Oil Fields

The buoyancy of gas and of oil in water led explorationists more than a century ago to a perception of the anticlinal trapping of oil and gas. In its simplest form, the anticlinal theory postulates that when a reservoir bends upwards, into roughly the shape of an inverted bowl, gas already present in pores, vugs or fractures will rise to the top, followed by oil. A century or so of exploration, not surprisingly, has brought considerable sophistication to concepts of the trapping of oil and gas. The stratigraphic parameter, in particular, has moved definitely into prominence, especially under the stimulus of seismic stratigraphy. The anticline itself, however, whether simple or whether considered together with stratigraphy, continues to strongly influence the critical decisions made in contemporary exploration.

1. Anticlinal Fields

1.1 Ghawar

The Saudi Arabian oil field of Ghawar, the largest in the world, is an anticlinal structure. Core drilling first revealed the shallow geometry of this whole anticline. Full drilling then revealed the structure at the level of the prolific "Arab D" reservoir. Consequently, the anticlinal structure of Ghawar may be relatively or virtually unfaulted, a remarkable condition in an anticline that strikes slightly east of north for at least 250 km, that covers an area of some 2300 km^2 and that originally contained a reported 70 billion barrels or more of producible reserves. Dips on the flanks of this long anticline measure as much as 12°, a phenomenon that may be due in part to compaction and draping and in part to the slow repeated rise of a crystalline Precambrian ridge.

1.2 Samotlor and Urengoy

Two fields in the USSR merit comparison with Ghawar for the structural simplicity of their anticlines and for the magnitude of their reserves. Samotlor represents oil and Urengoy represents gas. Both lie in the large west Siberian basin. At Samotlor, the structure is sufficiently periclinal to be classed as a dome, with a radius of some 25 km and with a drop from the center to the oil–water contact of only 100 m in 20–30 km (Fig. 1). The oil–water line does not cross its nearest contour around the dome, possibly because porosity and permeability are sufficient to effect a hydrodynamic equilibrium. All reservoirs at Samotlor consist of Lower Cretaceous sandstones with total initial reserves of more than 15 billion barrels of oil, and with a cumulative production of nearly 8 billion barrels by the end of 1981. The gas field of Urengoy, one of the largest in the world, occurs in an anticline with geometry notably similar to that of Ghawar. Like Ghawar, the length of Urengoy exceeds 200 km, and like Ghawar, its strike is nearly north (Fig. 2). Dips on the flanks of its anticline increase with depth, but even at a depth of 3000 m they do not exceed 300 m in 15 km. If faults are present, their throw is unlikely to exceed 50 m. Production comes mainly from Middle Cretaceous sandstones and dolomites. Initial reserves amount to an imposing 7.1×10^{12} m^3 of gas.

1.3 Painter Reservoir

In the thrust belt ("overthrust belt") of the western USA, several oil and gas fields demonstrate the ability to withstand the stress of deformations. The Painter Reservoir field, between thrust faults above and below, exhibits a tightly overturned and thrust limb, yet is one that does not affect the essentially horizontal gas–oil and oil–water contacts in the sandstone reservoir (Fig. 3). The thrust belt in which Painter Reservoir lies stretches from Canada to Mexico.

1.4 Groningen

In western Europe, the Groningen gas field, which is a large area with major reserves, belongs to the class of essentially undeformed anticlines with structural geometry moderately broken by faults. Studies of regional geology point to Groningen as a culmination on a large area that rose possibly as far back as the early Paleozoic. Groningen received gas and held it under a seal of Permian (Zechstein) evaporites. Groningen's Lower Permian reservoir, in fluviatile and eolian facies of sandstone, is broken by

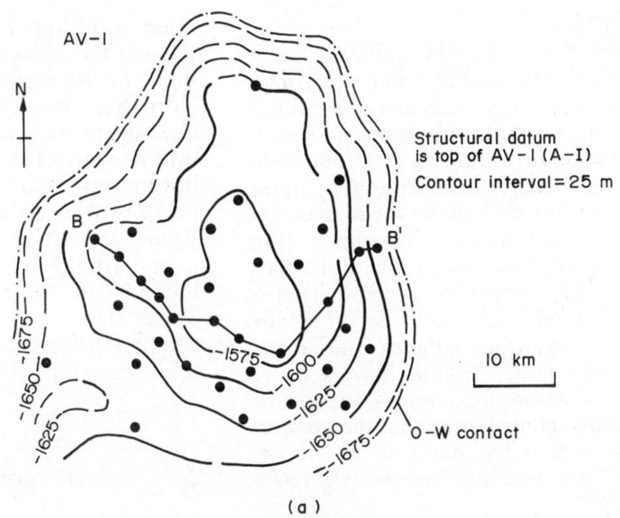

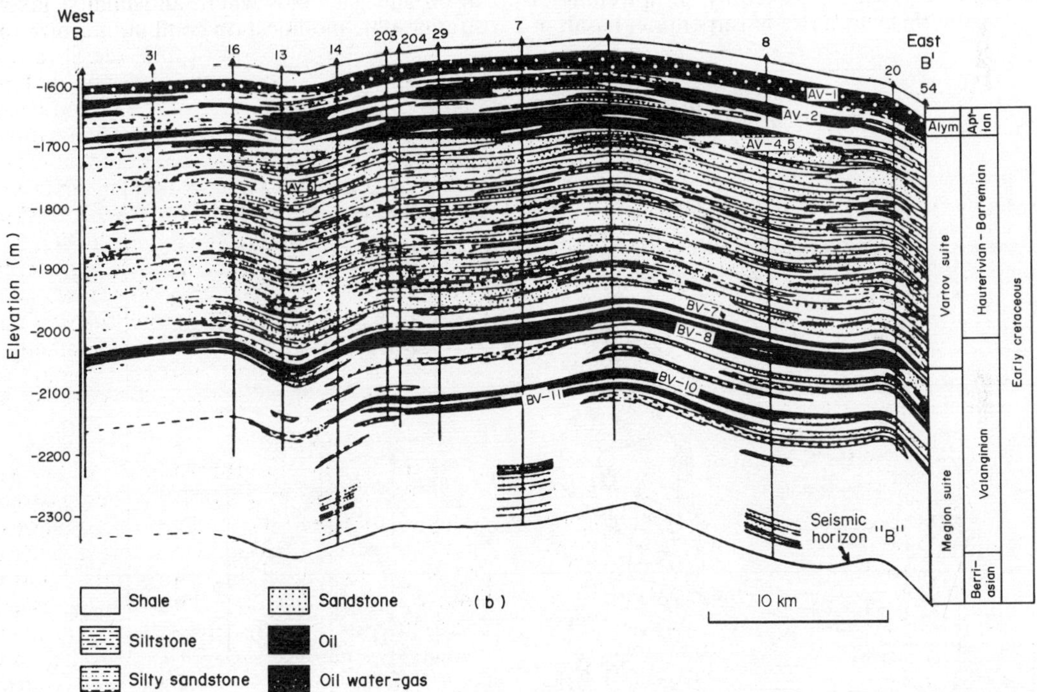

Figure 1
Samotlor oil field, west Siberian basin. Part (a) shows a contour map: contours on top produce sand; oil–water (O–W) contact is approximately 100 km vertically from center of the dome, at a distance of 20–30 km. Part (b) is a lithostratigraphic cross section of the field. (After Meyerhoff 1982. © Petroleum Exploration Society of Australia, Melbourne. Reproduced with permission)

faults that rise from underlying carboniferous strata (Fig. 4). Diagenesis of lipids and allied organic matter in carboniferous vegetal remains presumably furnished gas that rose through faults to fill the Lower Permian reservoir. Spread over an area of over 800 km^2, the initial reserves of natural gas at Groningen amount to more than 1.4×10^{12} m^3, and possibly to more than 1.7×10^{12} m^3.

1.5 Cantarell Complex

In southern Mexico, the Cantarell Complex in shelving Atlantic waters provides an example of strongly broken structure in competent beds and of abundant reserves of crude oil. Unconformities are present in the geological column (Jurassic, Cretaceous, Tertiary) but the trapping of oil is largely by updip closure against faults. Figure 5 shows the presence of both normal and reverse faults. Strike-slip faulting also may be present. Brecciation of lowest Tertiary (Paleocene) reservoirs permits the accumulation of prolific quantities of oil, but important oil is also present throughout the drilled column. Initial producible reserves of some 8 billion barrels of oil accumulated in the Cantarell Complex. These form part of the remarkable contribution that the offshore area of southern Mexico has made to the record surge in Mexico's total reserves since the late 1960s.

1.6 West Bay

Many anticlines raised or cored by an upwelling of salt, and sometimes by a rise of serpentine, basalt or shale, are important producers of oil and ga Among typical occurrences connected with salt ar those in the region of the Gulf of Mexico, in th North Sea, in the Middle East and in the USSR. A the exigencies of economics force many countries t drill deeper for oil and gas, they may find that sa has played a part in the development of structure previously regarded as normally flexed or drapec Representative of structures where the rise of salt i in a partially early and only moderately kineti phase is the West Bay field in Louisiana (Fig. 6) The acute apex of the salt breaks open an extremel complex pattern of faults that help to control loca lized reservoirs of oil and gas.

2. Anticlinal–Stratigraphic Fields

2.1 Bohai

In oil and gas fields where anticlinal structure an stratigraphic modification combine to form the tra

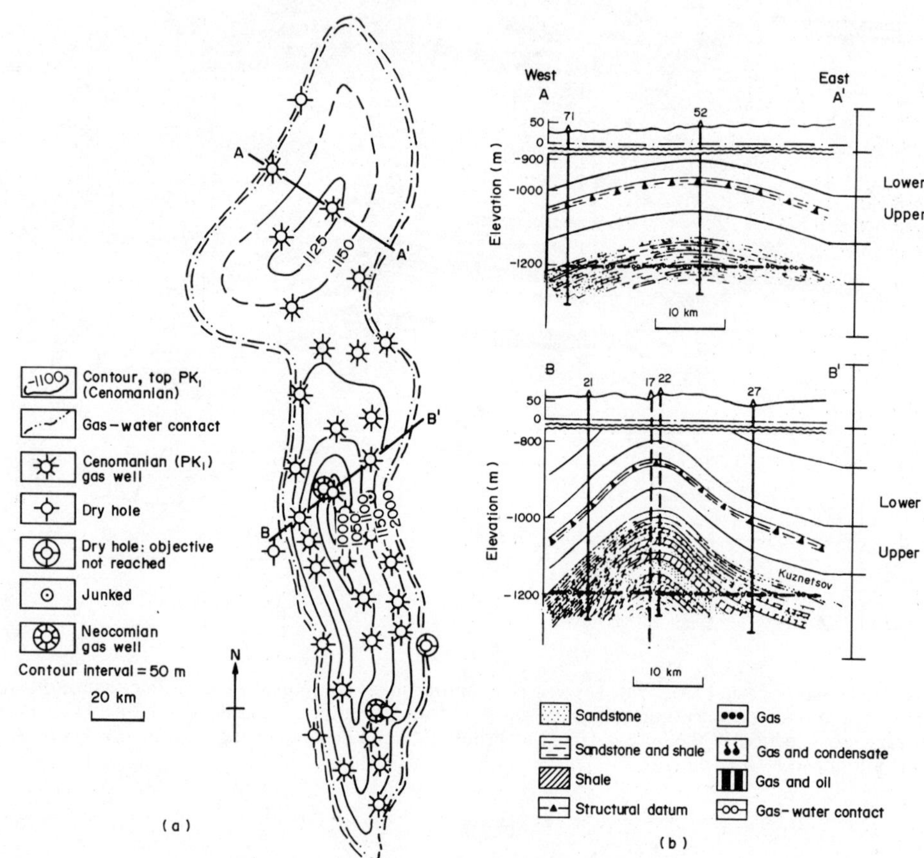

Figure 2
Urengoy gas field, west Siberian basin. Part (a) shows a contour map: contours on top produce horizon; gas–water contact is indicated. Part (b) illustrates lithostratigraphic cross sections of the field. (After Meyerhoff 1982. © Petroleum Exploration Society of Australia, Melbourne. Reproduced with permission)

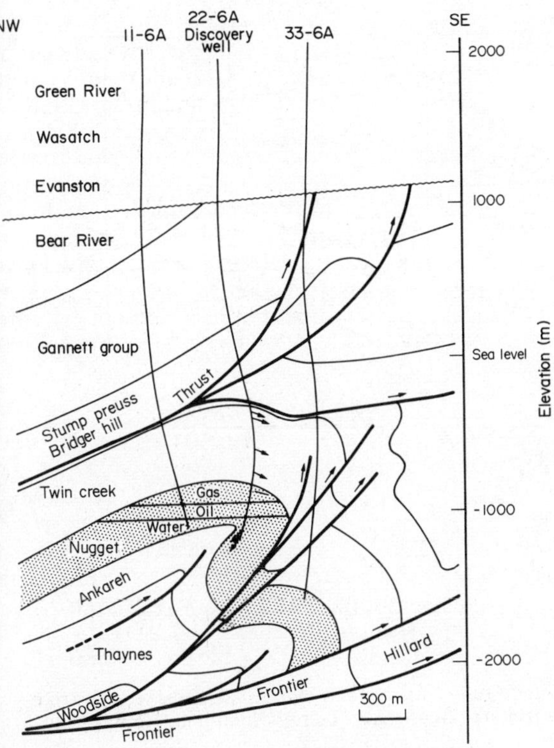

Figure 3
Painter Reservoir field, northwestern USA: horizontal gas–oil and oil–water contacts in crestal sector of overturned and thrust reservoir. (After Lamb 1980. © American Association of Petroleum Geologists, Tulsa, Oklahoma. Reproduced with permission)

for accumulation, the importance of each parameter varies from field to field. In the Bohai basin of coastal China the most prolific of its producing fields contain oil trapped in "buried hills," whose crests and/or flanks are sealed by a profound unconformity. The "central buried hill" of Fig. 7, held up by Proterozoic to Ordovician carbonates, lies under a capping of Eocene strata. This unconformity thus represents a hiatus of some 400 million years. During that long period, paleokarstic porosity was free to develop in the ancient carbonates under the unconformity, and to receive oil that drained downward and inward from Tertiary strata. As is typical in coastal China, source beds in the Tertiary column formed in a continental paleoenvironment. Those that supplied the buried hills of the Bohai basin formed mainly from Oligocene lacustrine muds. These source beds also supplied oil to penecontemporaneous Tertiary sandstone reservoirs, either above buried hills or on their flanks. Buried hills in the Bohai basin have already produced more than a billion barrels of oil.

2.2. *Prudhoe Bay*

The stratigraphic parameter has played an important part in the trapping of oil and gas at Prudhoe Bay in Alaska. The purely structural factor consists of a broad anticlinal nose with gentle plunge to the west and strong faulting to the north. This open structural condition becomes closed, however, by the eastward onlapping of Lower Cretaceous black shales across progressively eroded Jurassic, Permian, Pennsylvanian and Mississippian formations (Fig. 8).

Reservoirs do include Mississippian and Pennsylvanian limestones, but Permian and Triassic

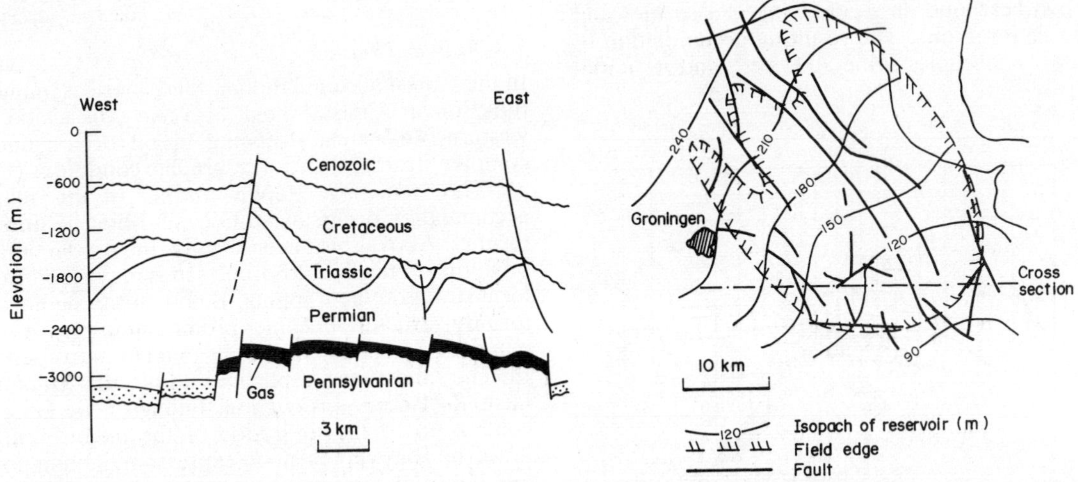

Figure 4
Groningen gas field, the Netherlands: block faulting of reservoir across broad anticline. (After Perrodon 1983. © Societé Nationale Elf–Aquitane, Pau, France. Reproduced with permission)

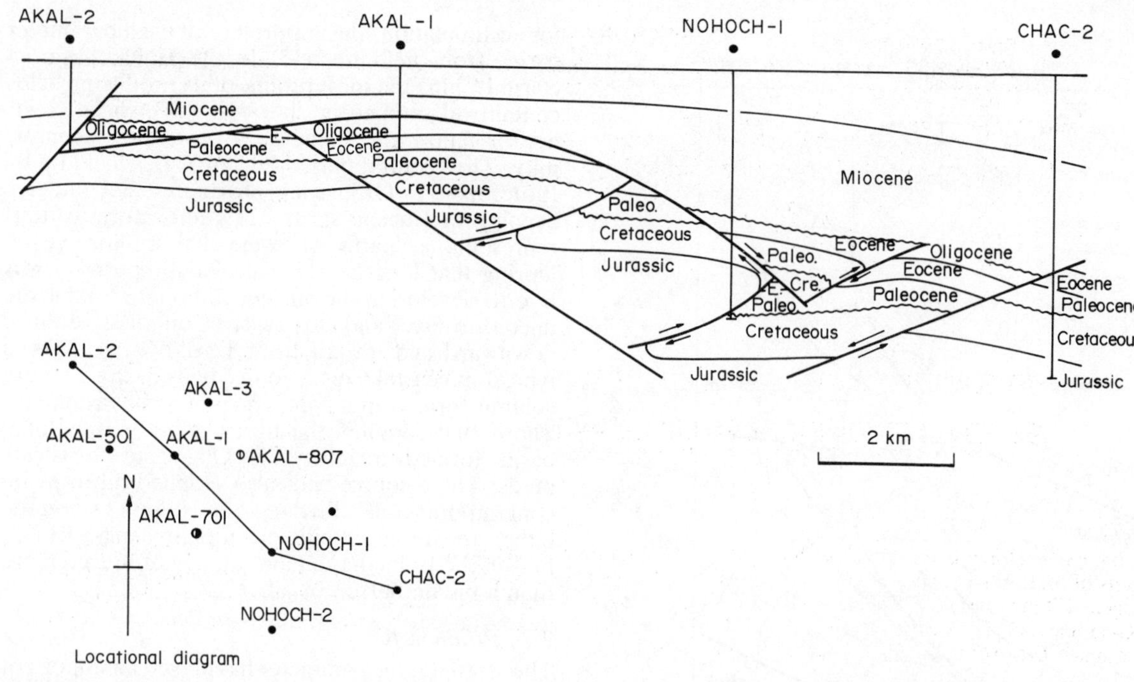

Figure 5
Cantarell Complex (marine), southern Mexico: competent carbonate reservoirs broken by normal, reverse and strike-slip faults. (After Acevedo 1980. © American Association of Petroleum Geologists, Tulsa, Oklahoma. Reproduced with permission)

sands and conglomerates in deltaic facies yield the main production at Prudhoe Bay. The Lower Cretaceous (Barremian) black shale constitutes a critical element. Not only does this formation provide stratigraphic closure for the trapping of oil and gas but also, by consensus, source material for the hydrocarbons in underlying Paleozoic and Mesozoic reservoirs. The resulting accumulation of oil and gas occupies some 2000 km^2, and its initial reserves amounted to some 10 billion barrels of oil together with probably more than 0.5×10^{12} m^3 of natural gas above its oil. Prudhoe Bay is easily the largest field in the USA and, owing to its combined reserves of oil and gas, is also the largest field in North America.

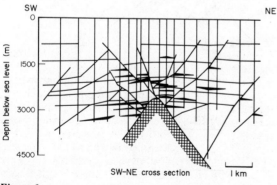

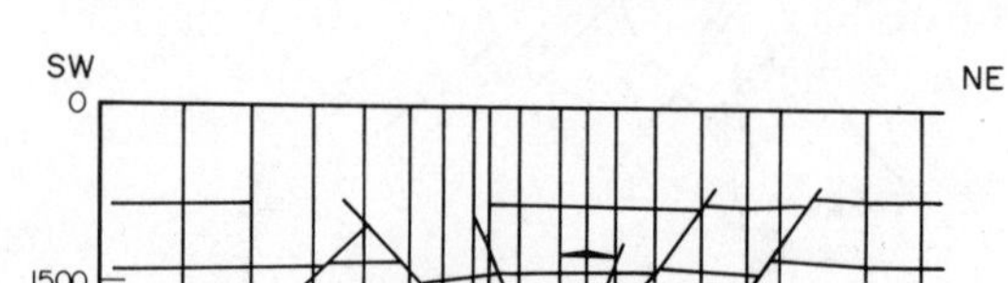

Figure 6
West Bay field, Louisiana: faults radiate from apex and flanks of diapiric salt to trap multiple reservoirs

2.3 Bolivar Coastal

In the largest accumulation in the Americas, namely the Bolivar Coastal field of Lake Maracaibo in western Venezuela, trapping of oil occurs under complex structural and stratigraphic conditions (Fig. 9). The clear stratigraphic marker of this major accumulation occurs as an Eocene–Miocene unconformity. Above this unconformity, the production of oil in the monoclinal Miocene sands may result from local stratigraphic trapping. Below this same unconformity, structure becomes prominent in the form of many high-angle normal faults that cut petroliferous Eocene sandstones. Anticlines add to the structural factor of Eocene production. Filling of the Eocene column up to the underside of the unconformity, however, may represent an expression of an important, if partial, stratigraphic control. Change in elevation of the oil–water contact from fault block to fault block may also indicate partial stratigraphic control. Production from the Eocene column is prolific, and

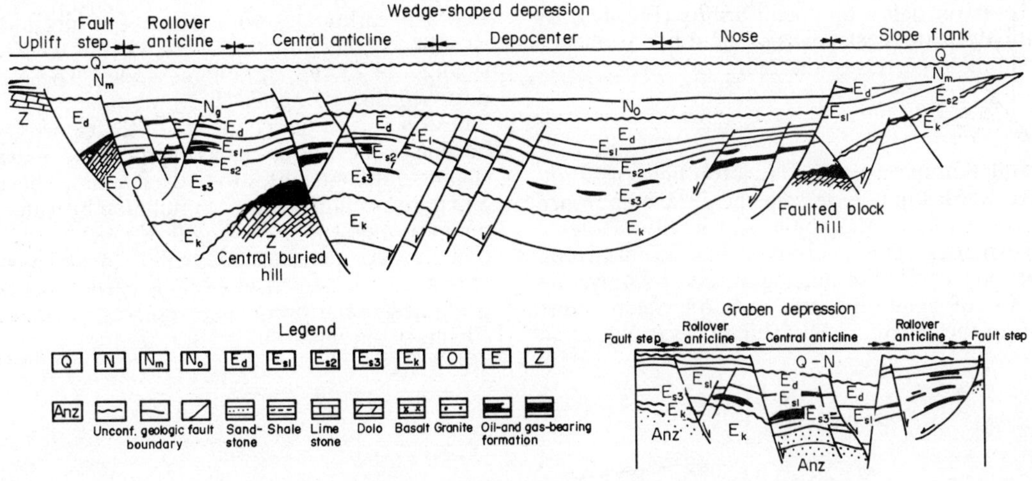

Figure 7
Bohai fields, Bohai basin, coastal China. Major accumulation in carbonates of "buried hills," and in sandstones against faults of local horsts and grabens: Q, Quaternary; N, Neogene; N_m, Neogene (Miocene); N_o, Neogene (Oligocene); E_d, Eogene (Dongying); E_{s1}, Eogene (Shahejie-1); E_{s2}, Eogene (Shahejie-2); E_{s3}, Eogene Shaheljie-3); E_k, Eogene (Kongdian); O, Oligocene; E, Eogene; Z, Lower Paleozoic; Anz, mainly Proterozoic (After Perrodon 1983. © Societé Nationale Elf–Aquitane, Pau, France. Reproduced with permission)

it forms the main contribution to an anticipated cumulative total of possibly more than 20 billion barrels of oil from the Bolivar Coastal field.

2.4 Brent

The Brent field, situated in the Viking graben of the central North Sea (UK sector), represents one of many within this graben in which a stratigraphic parameter of unconformity is present. The structural parameter of Brent and of its neighbors consists of block faulting. Figure 10 shows a paleotopography on the unconformity, and also a truncation of Jurassic and possibly latest Triassic reservoir sands against an incipient paleoscarp side. Late Jurassic and Cretaceous shales provide cover and source, both on the crest and on the flanks of the unconformity. The two main reservoirs at Brent may themselves contain partial source material for gas caps because of their vegetal material, if only in their minor partings of coal. Approximately 2 billion barrels of recoverable oil were present initially in the Brent field.

2.5 Niger Delta

In the petroliferous province of the Niger delta (west Africa) there are more than 140 oil fields. By 1986, at least two fields had produced more than 400 million barrels each; at least one field had produced more than 200 million barrels; at least seven fields had produced more than 100 million barrels each; and cumulative production from this petroliferous province as a whole had exceeded 11 billion barrels. In Fig. 11, four models illustrate the role of faulting in fields of the present Niger delta, especially the role of growth faulting. Geometry of many of the faults is listric (concave upward), and the faults themselves increase considerably in number when the crest of an underlying diapir collapses. Multiple sand reservoirs occur, many more than are shown in the models, and many are relatively thin. Minor stratigraphic trapping is probably important in a regime of Cenozoic sedimentation such as this one. Known examples include sands interpreted as simple-bar sands or point-bar sands, or sands interpreted as turbidites in the paleoprodelta. Still another type of stratigraphic trap consists of clayey paleogullies that truncate reservoirs and trap their oil.

3. Stratigraphic Fields

3.1 East Texas

The oil or gas field in which the stratigraphic parameter is total occurs only rarely. Cases often occur, on the other hand, in which the stratigraphic parameter is so important that it is justifiable to classify the field as stratigraphic. The classic example in this category is the east Texas oil field, renowned in the literature for its discovery at a random location (made by C. M. ("Dad") Joiner) in 1930. The east Texas field strikes slightly east of north for some 65 km, with an average width of some 8 km. Except for plunging local anticlines north and south, the trapping of oil has taken place by eastern truncation of the western-dipping Upper Cretaceous Woodbine

Sand reservoir below an unconformity (Fig. 12). By the late 1980s the east Texas oil field had produced some 5 billion barrels of oil.

3.2 Hugoton

Accumulation of gas in the Hugoton field of southwestern Kansas may represent the closest approach to a completely stratigraphic trap in the whole of North America. The contours in Fig. 13, drawn on top of one of the producing reservoirs, give no indication of significant structure. An "ideal" cross section of the Hugoton field (Fig. 13) indicates that Permian carbonates form the reservoirs and that closure for the trapping of gas occurs by irregular shaling-out of the carbonates updip. These carbonates consist of either drusy dolomites or oolitic limestones, and the character of their porosity may be diagenetic rather than original. The pattern of shaling-out contributes to hydrodynamic constraint, and to restriction of the accumulation by water, both undip and downdip. A length of more than 70 km, however, compensates adequately for this restraint. Hugoton, discovered in 1927 or earlier, reportedly contained initial producible reserves of some 1.5×10^{12} m^3 of gas when fully delineated.

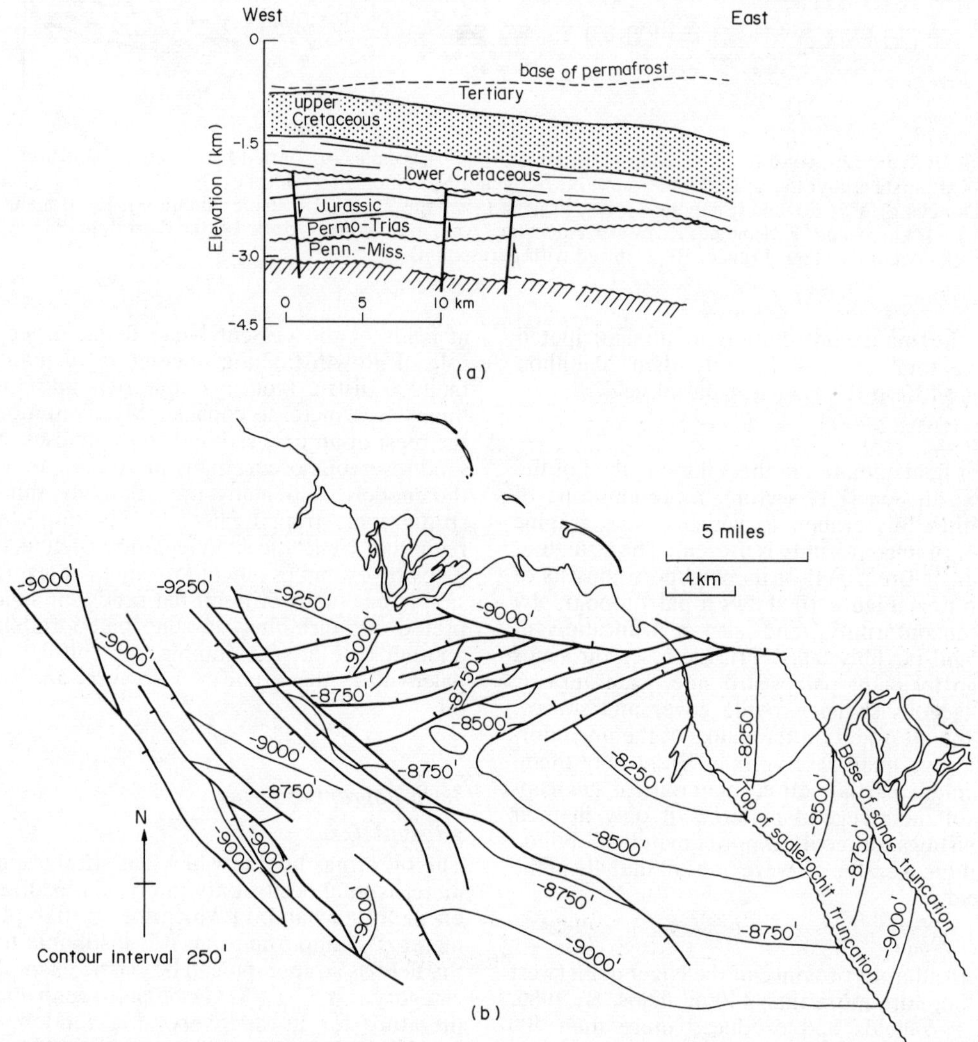

Figure 8
Prudhoe Bay field, Alaska: part (a) illustrates the eastward overlap of lower cretaceous across progressively older truncated reservoirs (after Perrodon 1983. © Societé Nationale Elf–Aquitane, Pau, France. Reproduced with permission); part (b) shows that the faulted anticline of reservoirs plunges gently westward (after Morgridge and Smith 1972. © American Association of Petroleum Geologists and Soceity of Exploration Geophysicists. Reproduced with permission)

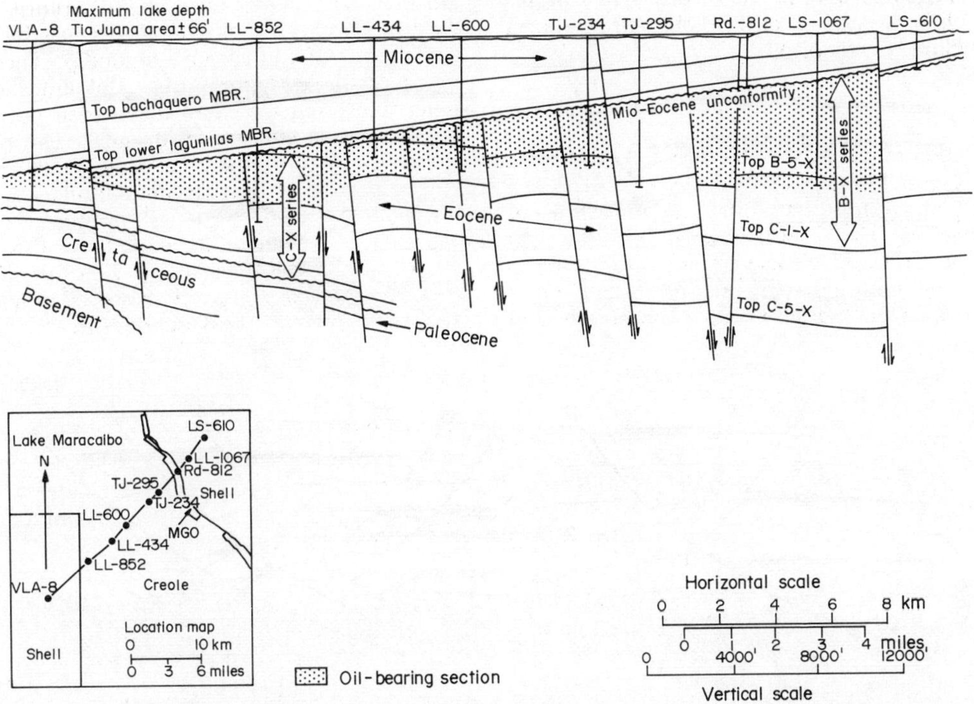

Figure 9
Bolivar Coastal field, Lake Maracaibo, western Venezuela: block faulting of main Eocene reservoirs under Miocene unconformity and Miocene overlapping reservoirs. (After Borger and Lanert 1959. © Fifth World Petroleum Congress, New York. Reproduced with permission)

4. Conclusions

Ghawar, Burgan in Kuwait, and Samotlor are spectacular examples of the trapping power of major simple anticlines. These three together have trapped nearly 150 billion barrels of oil. The anticline of Urengoy has trapped $7\times10^{12}\,m^3$ of natural gas. Hundreds of lesser structures add their production totals to this same class of simple anticlines. Among major examples of anticlines that are less simple, strongly fractured or intensely folded are the Kangan and Gach Saran anticlines in Iran, Groningen and the Cantarell complex. In all, the anticlinal concept has guided explorationists to considerable success for more than a hundred years.

The stratigraphic parameter becomes as important as the anticlinal parameter in fields such as Hassi R'Mel in Algeria and Prudhoe Bay, as well as in

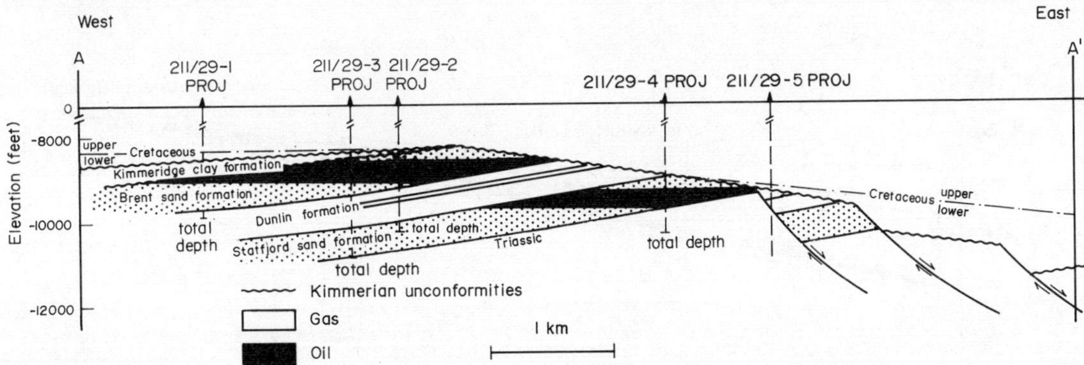

Figure 10
Brent field, North Sea. Fault block: truncation of reservoirs at underside of incipient paleoscarp on late Mesozoic unconformity. (After Bowen 1975. © Wiley, New York. Reproduced with permission)

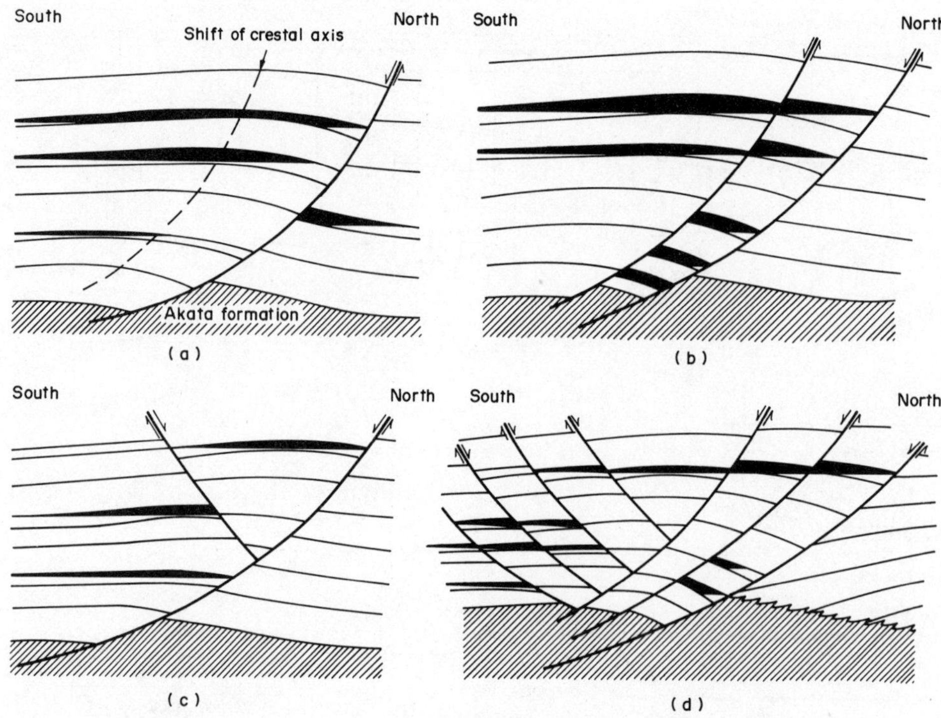

Figure 11
Niger delta fields, west Africa. Models of Cenozoic reservoirs closed by listric and antithetic faults: (a) simple rollover structure; (b) structure with multiple growth faults; (c) structure with antithetic fault; and (d) collapsed chest structure. (After Whiteman 1982. © Graham & Trotman, London. Reproduced with permission)

areas such as Bohai and the Niger delta. The more truly stratigraphic fields, as in east Texas and Hugoton, demonstrate that major accumulations can occur without the trapping power of the anticline. The anticline, either simple, relatively simple, relatively complicated or strengthened by a stratigraphic parameter, appears likely to continue strongly as a concept in the search for additional hydrocarbons. Neither land nor the well-drilled continental shelves, however, seem likely as places to repeat the huge successes of Ghawar, Burgan and Samotlor; more likely targets may consist of the continental slopes, if drilling intensifies on them in the decades to come.

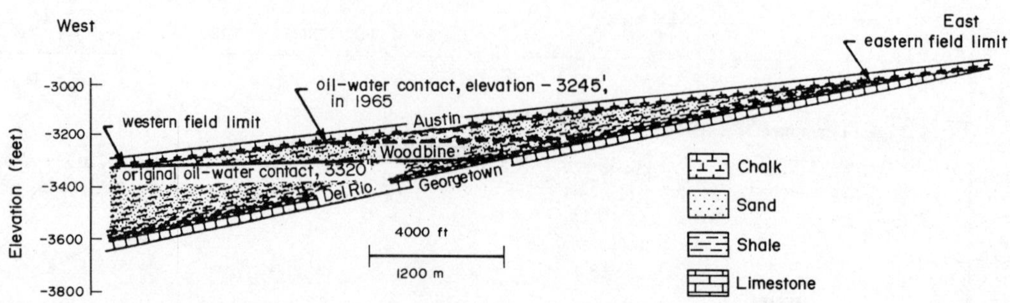

Figure 12
East Texas field: truncation of reservoir (Woodbine Sand) below onlap of Austin Chalk; western plunging anticlinal nose. (After Landes 1970. © Wiley, New York. Reproduced with permission)

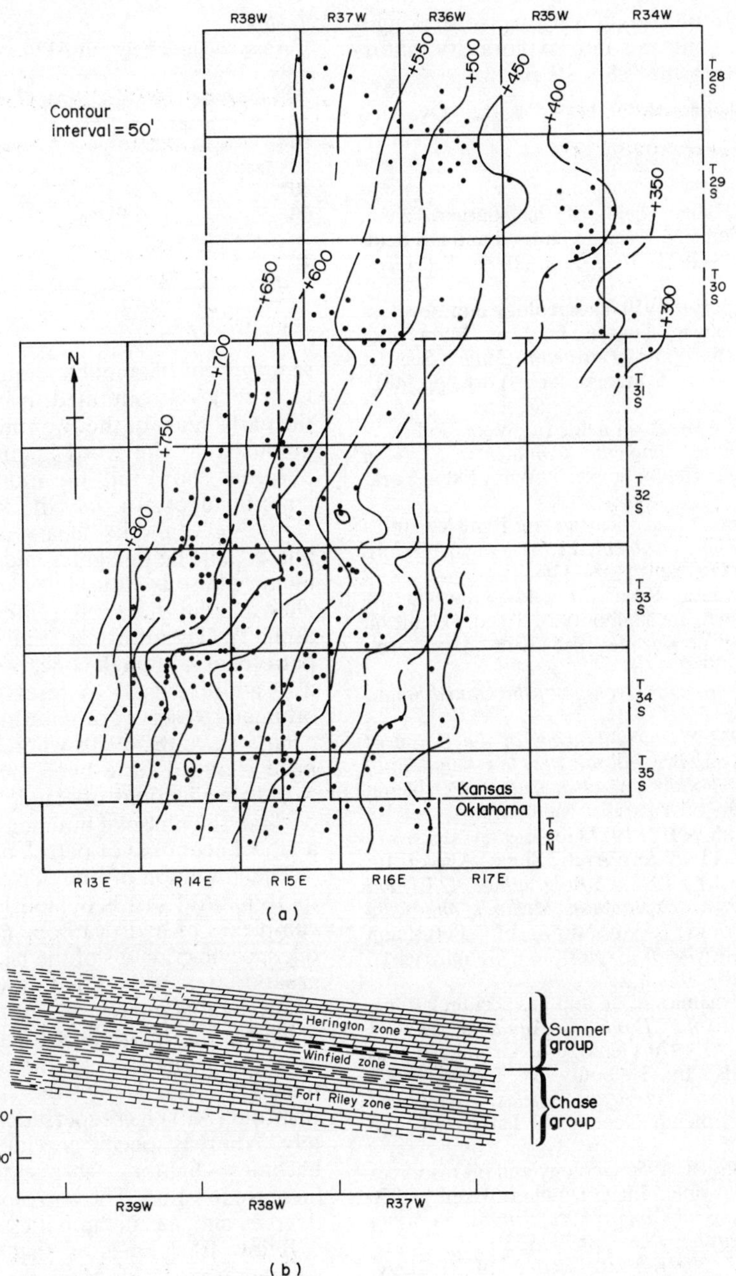

Figure 13
(a) Hugoton field, Kansas, Oklahoma: stratigraphic trap on monocline by updip shaling out of carbonates and (b) ideal cross section of Hugoton field showing clastic increase westward (After Landes 1951. © Wiley, New York. Reproduced with permission)

For the more immediate future, explorationists may increase their emphasis on the stratigraphic contribution. The resolving power of modern reflection seismography often brings into remarkably clear view the paleotopography of unconformities and the presence of traps on their undersides. For many countries, the intensive drilling of prospects with indicated stratigraphic parameters has become imperative. The results might not reverse the ongoing slide of these countries into exploratory maturity, but possibly they may slow its pace.

See also: Petroleum Resources: World Overview (Suppl. 2)

Bibliography

Acevedo J S 1980 Giant fields of the southern zone, Mexico. In: Halbouty M T (ed.) Giant oil and gas fields of the decade 1968–1978. *Mem. Am. Assoc. Pet. Geol.* 30: 339–85

Borger H D, Lenert E F 1959 The geology and development of the Bolivar Coastal field at Maracaibo, Venezuela. *Proc. 5th World Petroleum Congr.*, Sect. 1. Fifth World Petroleum Congress, New York, pp. 481–98

Bowen M J 1975 The Brent oil-field. In: Woodland A W (ed.) 1975 *Petroleum and the Continental Shelf of North-West Europe: Geology*, Vol. 1. Wiley, New York, pp. 353–61

Kuwait Oil Company Ltd 1953 Kuwait. In: Illing V C (ed.) 1953 *The Science of Petroleum*, Pt. 1, Vol. 6. Oxford University Press, Oxford, pp. 99–110

Lamb C F 1980 Painter Reservoir field—Giant in the Wyoming thrust belt. In: Halbouty M T (ed.) Giant oil and gas fields of the decade 1968–1978. *Mem. Am. Assoc. Pet. Geol.* 30: 281–8

Landes K K 1970 *Petroleum Geology of the United States*. Wiley, New York

Meyerhoff A A 1982 *Petroleum Basins of the Union of Soviet Socialist Republics and the People's Republic of China and the Politics of Petroleum*. Petroleum Exploration Society of Australia, Melbourne.

Morgridge D L, Smith W B Jr 1972 Geology and discovery of Prudhoe Bay field, eastern Arctic Slope, Alaska. In: King R E (ed.) 1972 *Stratigraphic Oil and Fields—Classification, Exploration Methods and Case Histories*. American Association of Petroleum Geologists and Society of Exploration Geophysicists, Tulsa, OK, pp. 489–501

Perrodon A 1983 Dynamics of oil and gas accumulations. *Bull. Cent. Rech. Explor. Prod. Elf–Aquitaine, Mem. 5*

Stauble A J, Milius G 1970 Geology of Groningen Gas Field, Netherlands. In: Halbouty M T (ed.) 1970 *Geology of Giant Petroleum Fields*. American Association of Petroleum Geologists, Tulsa, OK, pp. 359–69

Thralls W H, Hasson R C 1956 Geology and oil resources of eastern Saudi Arabia. In: Guzman E J (ed.) 1956 *Mexico, Int. Geological Congr. XX: Symposium Sobre Yacimientos de Petróleo y Gas*, pp. 11–32

Whiteman A 1982 *Nigeria: Its Petroleum Geology, Resources, and Potential*. Graham and Trotman, London

M. Kamen-Kaye
[Cambridge, Massachusetts, USA]

Petroleum Resources: World Overview

The two roots of the word petroleum, *petra* and *oleum*, translate as "rock oil" which was a term used in the western world before 1850 to denote natural seepages of flammable liquid hydrocarbons. The term rock oil continued to be used informally into the early years of the twentieth century, but yielded gradually to the word "petroleum." For technical purposes, however, the modern concept of petroleum includes not only the familiar liquid phase of "crude oil" (and its "heavy oil") but also two other phases, one being lighter and the other heavier. The lighter phase is natural gas and the heavier phase is almost solid bitumen (with a subphase of solid asphalt). Therefore, a study of global reserves of petroleum, although concerned with crude oil, must also address itself to reserves of natural gas and bitumen. Reserves of natural gas have risen spectacularly since the 1950s whereas reserves of bitumen have changed little over the same period; but as world production of crude oil and natural gas decline, the phase of bitumen may find more place in a global economy of petroleum.

The gradation of bitumen to crude oil, and crude oil to natural gas, is obviously a gradation of heavy complexes of hydrocarbons to light complexes. For oil, specific gravities of the heaviest phases can be as great or greater than that of water. Specific gravities of the lighter phases of crude oil can be as little as 0.74 or less. The decimal fractions of specific gravity are cumbersome to use and consequently a scale of numbers issued by the American Petroleum Institute (API) has superseded them almost completely. Whereas specific gravities decrease as crude oil becomes lighter, corresponding API degrees increase in value. The correspondence between API degrees and specific gravities are shown in Table 1.

Table 1
Correspondence between API degrees and specific gravities

API degrees	Specific gravity
60	0.74
50	0.78
40	0.83
30	0.88
20	0.93
10	1.00

Below 10 degrees on the API scale, petroleum, whether viscous or virtually solid, sinks in water. The viscous phase or the virtually solid phase of petroleum are commonly referred to as tar or bitumen.

1. World Reserves of Crude Oil

For purposes of discussing reserves, the consensus of industry and government is followed which defines "conventional crude oil" as consisting of oil lighter than 10 degrees API. Available reserves of crude oil for the world as a whole can thus be placed at approximately 500 billion (10^9) barrels, distributed as shown in Table 2.

All of the data given in Table 2 are less than those generally quoted, but in the author's opinion, many estimates of available reserves are optimistic. The predominance of the Middle East, however, is clear: the data in Table 2 show 300 billion barrels for the Middle East and 250 billion barrels for the rest of the world.

1.1 Middle East

Of the 300 billion barrels attributed to the Middle East, Saudi Arabia alone supplies an almost legendary 100 billion barrels or more. Almost as legendary is a possible remaining 50 billion barrels from a single oil field: Ghawar, the world's largest oil field, is a structure in Saudi Arabia that strikes north–south for at least 250 km, and produces oil from carbonate reservoirs at depths no greater than 2100 m. Kuwait, whose southern border is common with Saudi Arabia and which is a country covering an area that is miniscule compared to the area of Saudi Arabia, contains the world's second largest oil field. This is Burgan, a domoid anticline with an area of approximately 570 km^2 and with reserves of some 40 billion barrels. Depths to the producing reservoirs are remarkably shallow, less than 1500 m over most of this structure. These two individual oil fields in the Persian Gulf geosyncline (synclinorium) thus can supply approximately 100 billion barrels of reserves to the world's total of 500 billion barrels or more. The crude oil in such shallow horizons also gives them a tremendous economic advantage.

Elsewhere in the Middle East, Iran and Iraq each may have contained original proved reserves as great as 45 billion barrels. Since the 1930s, however, Iran has produced more than 30 billion barrels of crude oil. The corresponding figure for Iraq is more than 17 billion barrels. Iran and Iraq, for the present, draw on reserves in a relatively mature phase. The amount on which each can draw (roughly equal for both countries) could be some 25 billion barrels.

The remaining large contributor to crude-oil resources in the Middle East is Abu Dhabi, the chief sheikdom and capital of the United Arab Emirates, towards the southern end of the Persian Gulf. Abu Dhabi has produced approximately 8 billion barrels of crude oil since the mid-1960s. Some 20 billion barrels possibly remain for continued production.

Table 2
Distribution of available reserves of crude oil

Region	Barrels (× 10^9)
Middle East	300
Latin America	80
Communist bloc	50
Africa	40
North America	30
Western Europe	10
Asia-Pacific	10

1.2 Latin America

Mexico and Venezuela are the two major contributors to reserves in Latin America. Both countries are mature producers of crude oil, but the leadership in reserves has changed hands in recent years. In Venezuela, reserves of crude oil had remained in the range 15–20 billion barrels for several decades. In Mexico, the discovery of prolific reefal and reefoid carbonates in southern areas and their marine waters in the 1970s raised reserves of crude oil to an astonishing 50 billion barrels. A cumulative production of some 13 billion barrels for Mexico is attributable in large part to discoveries made before the 1970s, and therefore has made small inroads on reserves available at the end of 1987. In Venezuela, there were no extremely large additions to early discoveries. Exploitation, however, proceeded intensely, and by 1987 had resulted in a cumulative production of some 40 billion barrels, an amount exceeded by no other country in the noncommunist world except Saudi Arabia (52 billion). The net result of Venezuela's unusually large cumulative production has been to draw down available reserves of crude oil to a figure of 18 billion barrels or less. Venezuela itself, however, claims more than 30 billion.

1.3 Communist Bloc

The USSR is the only member of the communist bloc of nations whose production and reserves have attained global prominence. After nearly a century of development of crude oil and a cumulative production of some 70 billion barrels, the USSR may have reached a relatively mature stage in its total of available reserves. To take one indication of maturity, the giant Samotlor oil field was in its twenty-second year of production in 1987, and had produced some 8 billion barrels of oil from an original reserve of 17 billion barrels. Large petroliferous areas of the USSR remain to be explored, but any serious challenge to the prolific production of the Middle East may never materialize. The available reserves of the USSR may amount to some 40 billion barrels although some estimates reach 70 billion barrels.

China is the only other country in the communist bloc whose reserves merit consideration in a global perspective. Its giant oil field, Daqing, is notable not only for its production of more than 5 billion barrels of oil up to 1987 but also for its remarkable combination of continental source beds and continental reservoir beds. At least two other oil fields or oilfield areas in China have produced more than 1 billion barrels apiece. Remaining reserves, not firmly reported, may not be large. The figure of 10 billion barrels may be too high and commonly quoted figures higher than 10 billion do not appear reasonable. Large areas of northwestern China remain to be explored intensively. These northwestern areas, already productive in part, may yield sufficient oil over the next decade or two to maintain available reserves at about the present levels.

1.4 Africa

The available reserves of crude oil in Africa are low compared to the huge area of the continent. The major contributors to reserves, namely Libya, Nigeria and Algeria, have produced prolifically, but they are now in a mature phase. Reserves for Africa as a whole may have decreased to 10 billion barrels or less by the end of 1987, but most contemporary estimates show much higher figures.

1.5 North America

The two contributors to this large area, the USA and Canada, have reached a mature phase with regard to petroleum. Total discovery of crude oil in the USA amounted to some 160 billion barrels by 1987, but considerably more than a century of production has reduced available reserves to little more than 20 billion barrels. In Canada, significant production of crude oil is less than 40 years old, and available reserves amount to about 6 billion barrels.

1.6 Western Europe

Production of crude oil in western Europe remained unimportant until the end of the 1960s. Groningen in the Netherlands, a major discovery of gas with some 2×10^{12} m^3 of reserves, had indicated that large quantities of hydrocarbons existed on the southern rim of the North Sea basin. By the end of the 1960s, exploration had moved out into the North Sea itself, and by the late 1970s, initial reserves of crude oil under the waters of the North Sea had reached a total of ~10 billion barrels. Since that time, cumulative production in the UK sector of the North Sea has passed 6 billion barrels, to leave a reserve of ~5 billion barrels. Cumulative production in the Norwegian sector has amounted to 2 billion barrels, to leave a reserve of ~9 billion barrels.

1.7 Asia-Pacific

Two areas contribute to the standing of the Asia-Pacific region. The larger contribution comes from the mature petroliferous province of Indonesia–Malaysia; the smaller contribution from the more recent province of the "Bombay High" in marine waters off the western coast of India. The total amount of crude oil discovered in Indonesia–Malaysia is approximately 20 billion barrels. A cumulative production of nearly 14 billion barrels thus leaves some 6 billion barrels for available reserves. Production from the marine waters of the Bombay High, from the late 1970s to the late 1980s, amounted to about 2 billion barrels. The balance for available reserves may amount to 3 billion barrels.

Table 3
Distribution of global reserves of natural gas

Region	Reserves 10^{12} m^3
Communist bloc	42.48
Middle East	19.82
North America	7.64
Africa	4.81
Western Europe	3.96
Asia-Pacific	3.40
Latin America	3.11

2. World Reserves of Natural Gas

Since the late 1950s, discoveries of several deposits of gas, each larger than 6×10^{12} m^3, have recharged the global supply to a considerable degree. Deliverability of these gigantic accumulations may constitute a problem in some cases. In other cases, a single great field such as Urengoy has enabled the USSR to pipe natural gas as far as central and western Europe. In earlier years, discoveries like Hugoton in the USA, Groningen in the Netherlands, Hassi R'Mel in Algeria and others, had brought proved global reserves to a level of some 20×10^{12} m^3 by 1960. Discoveries made since that time may have raised that figure to as high as 85×10^{12} m^3 for the closing years of the 1980s. Data for the major geographical regions are given in Table 3.

2.1 Middle East

Deeper drilling since the 1970s has revealed gigantic accumulations of gas in carbonates of the Permian (latest Paleozoic) Khuff Formation. In Iran, the result has been the discovery of possibly 11.3×10^{12} m^3 of gas at or around Kangan near the eastern shore of the Persian Gulf. Confirmation of this discovery would mean that this one district holds some four times as much natural gas as that now available in the USA. Other reserves of natural gas from the Khuff Formation are as follows: approximately 2.8×10^{12} m^3 in Qatar; 2.8×10^{12} m^3 in Saudi Arabia; and 2.27×10^{12} m^3 in Abu Dhabi. These reserves, relatively little produced into the late 1980s, remain largely available.

2.2 Communist Bloc

Major reserves of natural gas in the communist bloc consist essentially of those within the USSR. The published figure of approximately 42.5×10^{12} m^3 may represent original reserves rather than reserves still available in the late 1980s. Even reserves which are still available, however, probably are considerably larger than those of the Middle East. At least for the near future, the USSR will continue to lead the world in reserves. One field alone has made a considerable contribution over the years. This is Urengoy, a sharply folded anticline some 200 km long, with proved ultimate reserves of some 7.1×10^{12} m^3.

To China, at least one report has assigned 0.85×10^{12} m^3 of gas, but this figure may need downward revision.

2.3 Africa

Comparatively few details are available concerning the figure of 4.8×10^{12} m^3 assigned to Africa. Data such as 10^{12} m^3 for Nigeria and 0.85×10^{12} m^3 for Libya, may include a considerable amount of gas cap; that is, gas above oil in a common reservoir. In such cases the production of oil draws down much associated gas, which is thus lost to reserves. In Algeria, the big single element of reserves is the Hassi R'Mel gas field which was mentioned in Sect. 2. Hassi R'Mel, by far the largest gas field in the whole of Africa, has been producing natural gas steadily since the 1960s. Its original reserves of some 10^{12} m^3 have probably decreased considerably. Gas fields discovered later than Hassi R'Mel may have added a figure of the order of 10^{10} m^3 to the reserves of Algeria, but a reported figure of 2.8×10^{12} m^3 for this country may be an overestimation.

2.4 North America

Hugoton, discovered in the heartland of the USA in 1918, was one of the most prominent gas fields of the world, with initial reserves of some 1.4×10^{12} m^3. There has been no successor of this stature in the USA, and reported reserves of some 5×10^{12} m^3 for the nation as a whole represent the sum of contributions from hundreds of small and medium fields. The rate of discovery of gas in the USA was of the order of 0.2×10^{12} m^3 per year in the late 1980s. Consumption during that same period, however, was rising toward 0.6×10^{12} m^3 per year. The drain on reserves is clearly a matter of concern. In Canada, where the population is relatively small, domestic consumption of natural gas is correspondingly small. Important resources, consequently are available in Canada for signfiicant delivery to the USA if the published figures of 2.6×10^{12} m^3 for "proved reserves" are correct.

2.5 Latin America

Individual fields with natural-gas reserves of the order of 10^{11} m^3 seem not to exist in Latin America. It is assumed, therefore, that fields no more than moderately large are the joint contributors to the published national "proved reserves" as follows: Argentina, 0.6×10^{12} m^3 Venezuela, 10^{12} m^3 and Mexico, 2×10^{12} m^3.

2.6 Western Europe

In the 1950s, Groningen in the Netherlands stood out as a unique phenomenon. Its initial reserves of 1.4–1.7×10^{12} m^3 of natural gas moved western Europe onto the global scene. Groningen has continued to rank high as a discovery, and has also been prominent owing to its remaining reserves, despite considerable production and distribution since the 1960s. The Netherlands is possibly capable of providing some 10^{12} m^3 of gas in the future. By the 1970s, however, Groningen was no longer alone. Both the UK on the western side of the North Sea basin and Norway on the eastern side had gained prominence in gas. Fields for the most part an order of magnitude less than Groningen in size had brought proved reserves of approximately 10^{12} m^3 to Norway and 0.6×10^{12} m^3 to the UK. Heavy domestic consumption in the UK has lowered its reserves significantly, but a recent resumption of exploration and discovery has slowed the decline. Norway, with a low domestic consumption, and with good prospects of relatively large future discoveries of natural gas, has suffered little decline of reserves.

2.7 Asia-Pacific

A steady accession of discoveries has occurred across this widespread area since the 1960s. None of these has been of phenomenal size, but many have occurred in nearshore waters as marine technology has developed. Available reports show several countries with proved reserves above 0.3×10^{12} m^3 of natural gas. Indonesia and Malaysia lead, each with approximately 0.85×10^{12} m^3 of natural gas. Pakistan, India and Australia follow, each with approximately 0.42×10^{12} m^3. The figure for Bangladesh may be around 0.3×10^{12} m^3.

3. World Reserves of Bitumen

Except for the great spread of sands in the Athabasca region of northwestern Canada, where only bitumen (tar) occurs, a distinction between bitumen and heavy oil does not appear in reports of other regions. The remarkable "Petroliferous Belt of the Orinoco" ("La Faja Petrolífera del Orinoco") in Venezuela contains many billion barrels of bitumen in the subsurface. Its potential producible reserves, however, lie much more in its heavy oils that, although viscous, might rise to the surface by the stimulation of steam at high pressure. This is the well-known *in situ* method of recovery. If *in situ* methods can succeed in the Orinoco region, a claim of one or two hundred billion barrels of reserves is within reason. Whether this method is practical may not be known for several decades. In either case, it is unlikely that bitumen in any significant amount can be moved from subsurface reservoirs to surface either in this region or elsewhere. In the USSR, available reports credit Melekess with some 100 billion barrels of bitumen, with no mention of heavy oil. Unless the bitumen of Melekess is minable at surface or below an overburden of less than 60 m its contribution to reserves is problematic.

For bitumen *sensu stricto*, only the tremendous tar sands of Athabasca in northwestern Canada make an indisputable contribution to global reserves. Two mining plants which are at work in open-pit in the Athabasca region could recover a cumulative 1.5 billion barrels of bitumen in the 1990s. The economic limit of overburden for the near future amounts to about 40 m, and within that limit it appears that a total reserve of some 35 billion barrels of bitumen exists. In the remainder of the Athabasca prospect, the volume of bitumen in place below 180 m amounts to several hundred billion barrels. Recovery of this deeper material by *in situ* methods seems remote at present.

4. Outlook for World Reserves of Petroleum

4.1 Crude Oil

Insofar as this is the age of the internal combustion engine, we must assume that crude oil will continue to play a pivotal role in the economic, social and political development of the world. With global consumption of crude oil moving toward a total of 60 million barrels a day, available reserves would decrease by more than 20 billion barrels per year. By simple arithmetic, within 25 years present reserves of crude oil would be consumed and this highlights the contrived character of a "glut of oil" in 1987.

For the western world, the economics of discovery are nothing less than hostile. Despite high technological capability, nations such as the USA, Canada, Mexico, Venezuela and the UK eventually may be forced into politically unpopular measures to sustain intensive exploration in the hope of stemming declines in their reserves. The USSR eventually may face a similarly uncomfortable necessity, possibly exacerbated if that nation's technology continues to lag.

The future of crude oil, as the general press has repeatedly stressed, lies critically with the countries of the Middle East, especially if their reserves should amount to 500 billion barrels instead of the 300 billion barrels assumed in this article. The Middle East, at either figure, is obviously less pressed than other regions, but within a few decades it will become necessary to mount a search for new reserves in this area. Whatever the ultimate result of that search might prove to be, it would play an important part in carrying the world relatively far into the twenty-first century.

4.2 Heavy Oil

This phase of petroleum, transitional between oil of natural or pumped flow and semisolid bitumen, lies in the range 10–15 API degrees. The possible importance of heavy oil in the future of petroleum stems in large measure from two main sources. First is the heavy oil of the "Petroliferous Belt of the Orinoco" in Venezuela. Claims of several hundred billion barrels of hydrocarbons in place for this belt appear reasonable. Less clear is the ratio of essentially immovable bitumen to heavy oil. Before billions of barrels of heavy oil can be forced to surface, the *in situ* method must be proved not only petrophysically but also economically feasible. The first billion barrels of heavy oil from the Orinoco region may not appear until after the first decade of the twenty-first century. In northwestern Canada, apparently large amounts of heavy oil exist and are already under limited production by moderate *in situ* methods.

4.3 Natural Gas

The shape of the future for natural gas is not unlike that for crude oil in terms of global economics, social development and politics. The unbalanced pattern of distribution follows quite closely that of crude oil. Within this pattern the only notable difference consists of an exchange of places at the head. The USSR held first place in reserves of natural gas during the late 1980s by a considerable margin. If countries in the Middle East should follow an aggressive program of deep drilling in the widespread Khuff Formation, however, the Middle East eventually could overtake the USSR. Natural-gas reserves of the order of 10^{12} m^3 in the Middle East could extend global use for a significant period.

4.4 Bitumen

As already stated, to discuss bitumen is to discuss the minable tar sands of Athabasca in northwestern Canada. In this case a sharp distinction exists within the phase of bitumen itself. The minable bitumen, with approximate reserves of 30 billion barrels, is technologically readily available for conversion to crude oil and for addition to global reserves, especially to those of North America as a whole. The several hundred billion barrels of bitumen in place below 180 m at Athabasca unfortunately are likely to resist *in situ* attempts to dislodge them even more than the recalcitrant billions of heavy oil in the Orinoco region of Venezuela. At Athabasca, in fact, the prolific volumes of bitumen below the reach of mining may prove so obdurate as never to become a relevant factor in the world's future reserves of petroleum.

5. Conclusion

Unless economic or political factors change in a way that encourages intensive and incessant programs of exploration and development, the age of petroleum in the western world could enter difficult times. The Middle East and the USSR could conceivably sustain the rest of the world and themselves on a spartan basis well into the twenty-first century. That these two regions would consent to such an arrangement, however, is highly doubtful. The viability of the western world in petroleum must rest upon its own efforts, redoubled especially in the technological sphere. The western world should be investing sums an order of magnitude greater than those of the late 1980s on research into special methods of recovery, irrespective of the fact that these methods (oil shale included) are at present largely uneconomical. As reserves decline, the uneconomical processes of today could become the mainstay of tomorrow.

See also: Petroleum Resources: Conditions for Formation (Suppl. 2); Petroleum Resources: Oil Fields (Suppl. 2)

Bibliography

Alayeto M B E, Louder L W 1974 The geology and exploration potential of the heavy oil sands of Venezuela (the Orinoco Petroleum Belt). In: Oil sands of the future. *Mem. Can. Soc. Pet. Geol.* 3: 1–18

Anon 1986 Worldwide report. *Oil Gas J.* 84(51/52): 33–73

Atwater G I 1980 Petroleum. In: Goetz P W (ed.) 1980 *The New Encyclopaedia Britannica, Micropaedia,* Vol. 14. Benton, Chicago, IL pp. 164–75

Govier G W 1974 Alberta's oil sands in the energy supply picture. In: Oil sands of the future. *Mem. Can. Soc. Pet. Geol.* 3: 35–49

Grunau H R 1984 Natural gas in major basins worldwide attributed to source rock type, thermal history and bacterial origin. *Proc. 11th World Petroleum Congr.*, Vol. 2. Wiley, Chichester, UK. pp. 229–49

Halbouty M T 1979 World ultimate reserves of crude oil. *Proc. 10th World Petroleum Congr.*, Vol. 2, *Exploration Supply and Demand*. Heyden, London, pp. 291–301

Halbouty M T 1984 Reserves of natural gas outside the communist bloc countries. *Proc. 11th World Petroleum Congr.*, Vol. 2. Wiley, Chichester, UK, pp. 281–92

Janisch A 1981 Oil sands and heavy oil: Can they ease the energy shortage? In: Meyer R F, Steele C T (eds.), Olson J C (asst. ed.) 1981 *The Future of Heavy Crude and Tar Sands*. McGraw-Hill, New York, pp. 33–41

Jardin D 1974 Cretaceous oil sands of western Canada. In: Oil sands of the future. *Mem. Can. Soc. Pet. Geol.* 3: 50–67

Masters C D, Root D H, Dietzman W D 1984 Distribution and quantitative assessment of world crude oil reserves and resources. *Proc. 11th World Petroleum Congr.*, Vol. 2. Wiley, Chichester UK, pp. 229–49

Meyer R F, Steele C T (eds.), Olson J C (asst. ed.) 1981 *The Future of Heavy Crude and Tar Sands.* McGraw-Hill, New York

Meyerhoff A A 1979 Proved and ultimate reserves of natural gas and natural gas liquids in the world. In: *Proc. 10th World Petroleum Congr.*, Vol. 2, *Exploration Supply and Demand*. Heyden, London, pp. 303–11

Meyerhoff A A 1982 *Petroleum Basins of the Union of Socialist Soviet Republics and the People's Republic of China, and the Politics of Petroleum.* Petroleum Exploration Society of Australia, Melbourne

Mossop G D, Kramers J W, Flach P D, Rotenfusser B A 1981 Geology of Alberta's oil sands and heavy oil deposits. In: Meyer R F, Steele C T (eds.), Olson J C (asst. ed.) 1981 *The Future of Heavy Crude and Tar Sands*. McGraw-Hill, New York, pp. 197–207

Root D H, Atanasi E D, Turner R M 1987 *Statistics of Petroleum Exploration in the Non-Communist World Outside the United States and Canada*, US Geological Survey Circular 981. USGS, Washington, DC

Schmerling L 1982 Petroleum. In: Parker S P (ed.) 1982 *McGraw-Hill Encyclopedia of Science and Technology*, Vol. 10. McGraw-Hill, New York, pp. 75–7

Walters E J 1974 Review of the world's major oil sand deposits. In: Oil sands of the future. *Mem. Can. Soc. Pet. Geol.* 3: 240–62

Zurrug A Y, Bois C 1984 Potential oil reserves in the Middle East and North Africa. *Proc. 11th World Petroleum Congr.*, Vol. 2. Wiley, Chichester, pp. 261–75

M. Kamen-Kaye
[Cambridge, Massachusetts, USA]

Phase Diagrams and Phase Stability: Calculations

Phase diagrams, in all their varied forms, are graphical illustrations of those mixtures of phases that are stable at a given overall composition in the presence of specific external variables, such as temperature and pressure.

The formal theory required to calculate such equilibria was established by Gibbs in the nineteenth century and involves the correspondence of such phase mixtures to the establishment of a minimum free energy condition. As long ago as 1910, work by Van Laar established that phase diagrams of markedly different topology reflect different magnitudes and signs for the interaction energies of the component species in each phase, together with the transformation energy between different geometric arrangements of the pure elements (now commonly referred to as the lattice stabilities):

$$\Delta G_\alpha(x,T)=x_i x_j \Delta G_\alpha^E+\Delta G^{\beta\to\alpha}x_j+RT(x_i\ln x_i+x_j\ln x_j) \qquad (1)$$

$$\Delta G_\beta(x,T)=x_i x_j \Delta G_\beta^E+\Delta G_i^{\alpha\to\beta}x_i+RT(x_i\ln x_i+x_j\ln x_j) \qquad (2)$$

where $\Delta G^E_{\alpha,\beta}$ are the interaction energies of components i and j in crystal structures α and β, $\Delta G_{i,j}^{\beta\to\alpha,\,\alpha\to\beta}$ are the differences in energy different allotropes (lattice stabilities) and $RT(x_i \ln x_i + x_j \ln x_j)$ is the entropy of mixing.

The temperature and composition dependence

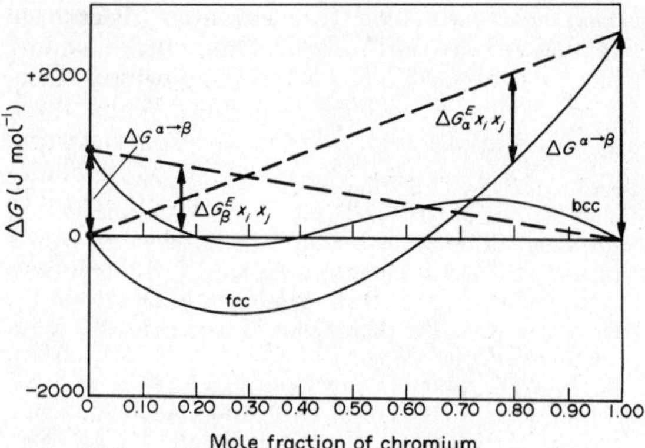

Figure 1
The effect of lattice stablilites and interaction coefficients on the free energy curves for face-centered-cubic (fcc) and body-centered-cubic (bcc) phases in the iron–chromium system at 700 K

of ΔG^E is commonly expressed as

$$\sum_{0}^{n} A_n(x_i - x_j)^n$$

The calculation of phase boundary composition is thus reduced to the mathematical problem of finding the free energy minimum for any particular combination of the variables concerned, provided the variation of free energy with temperature (and any other variables of interest) can be specified for each phase. Some of the data required to evaluate the necessary free energies (e.g., heats of formation) can be obtained directly by experiment, given a proper appreciation of the reference states. However, to perform even this initial step requires a systematic knowledge of the relevant lattice stabilities, many of which will refer to metastable phases (see Fig. 1). Since thermodynamic data are required for phases that may be experimentally inaccessible, phase diagram calculations of real systems remained largely an algebraic concept until a notable breakthrough was achieved by Kaufman and Bernstein (1970).

They realized that since the lattice stabilities of any given element have to be the same in all phase diagrams that contain that element, estimates of the missing lattice stabilities might be obtained by applying an iterative procedure to a series of interrelated phase diagrams. Once this procedure was adopted, it was found that the temperature variation of such lattice stabilities could, in many cases, be approximated by simple linear equations. These lattice stabilities in turn acted as reference states for the subsequent evaluation of missing interaction coefficients in many binary systems. With the accumulation of sufficient data, it became possible to generate empirical formulations that allowed the estimation of missing parameters, and the number of calculated diagrams increased very rapidly. However, the absence of *ab initio* calculations often raised the criticism that the revelant free energy curves were being generated by a series of arbitrary parameters. It thus became increasingly important to check all data sets for self-consistency, particularly in terms of the internal relationships that must exist between different thermodynamic properties. This led to the evolution of optimization programs, in which many separate sources of thermodynamic information (such as specific heats, melting points, heats of formation and the volume ratio of the phases) could be combined to give statistically best-fit values for all the necessary input parameters (Lukas et al. 1977). Such self-consistency also ensures that there is a reliable database that can regenerate all available thermodynamic data as well as the phase diagram (see Fig. 2). Although this article concentrates on metallic systems, analogous databases and techniques exist for salts, slags and ceramics, and advances in phase diagram calculations should be seen as having universal application.

1. Extension to Metastable Phases

The use of commercial alloys often involves heat treatments generating metastable phases that, by definition, do not appear in equilibrium phase diagrams. One of the great virtues of phase diagram calculations is the freedom to insert or omit phases, with the result that it is also possible to calculate equilibria involving metastable phases (Saunders 1989) (see Fig. 3). The accuracy with which it is

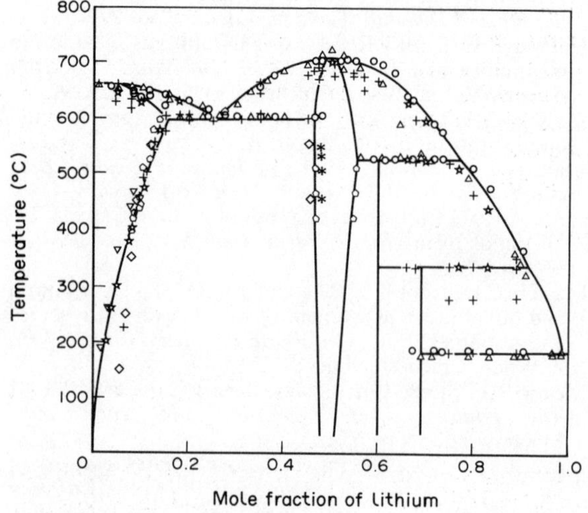

Figure 2
Calculated phase diagram of aluminum–lithium with experimental data superimposed (Saunders 1989)

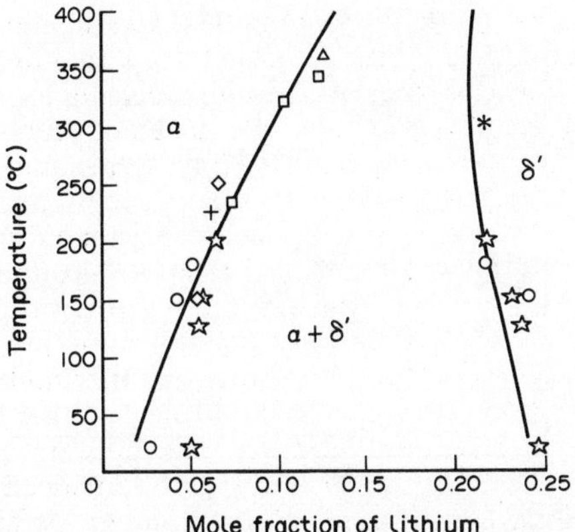

Figure 3
Comparison of experimentally determined and calculated phase boundaries between α aluminum and the metastable δ' Al_3Li phase in aluminum–lithium

possible to formulate the free energy of metastable phases may even be greater than for some of the stable phases, depending on the relative complexity of the crystal structures concerned and the experimental data available. In fact, calculations can be extended to amorphous phases and there is much more scope for realistic predictions than is generally realized (Saunders and Miodownik 1986).

2. Extension to Multicomponent Systems

Interest in calculated diagrams also increased considerably when it was found, once various binary diagrams had been calculated and had produced self-consistent results, that the same characterization parameters were capable of generating multicomponent systems by a simple expansion of the binary formalism. In the majority of cases, ternary systems can be calculated from properly characterized binary data with a minimal addition of parameters unique to the specific ternary system (see Fig. 4):

$$\Delta G_\alpha(x_i x_j x_k, T) = x_i x_j \Delta G_\alpha^{Eij} + x_j x_k \Delta G_\alpha^{Ejk} + x_k x_i \Delta G_\alpha^{Eki} + RT \sum_{x=x_i}^{x=x_k} x \ln x + x_i x_j x_k \Delta G_\alpha^{Eijk} \quad (3)$$

where ΔG^{Eijk} are ternary interaction terms which are often relatively small or zero.

This provided the major spur to the creation of an expanding database, particularly for combinations of 3*d* transition elements of industrial interest (Kaufman and Bernstein 1970, Kaufman 1972). Ternary systems, however, bring some additional choice patterns for modelling how a binary compound with a given crystal structure extends into the ternary system. One way, is to consider the corresponding stability of compounds with the same crystal structure in the other two edge-binaries. When three binary systems combine to form a ternary system, the stable compounds that occur in one of the binary systems may not have isomorphous counterparts in the other edge-binaries. It then becomes necessary to have a value for the stability of the (metastable) "counterphase" of the same crystal stucture, although this is by definition experimentally inaccessible.

This problem is in many ways analogous to that of estimating missing lattice stabilities for the elements, but as there are many more subtle variations in crystal structure for intermetallic phases, there are correspondingly fewer members of a particular crystal symmetry from which to draw guidelines. Nevertheless as databases became more extensive, empirical rules were evolved and utilized to make useful approximations, even for the introduction of parameters for ternary compounds that have no counterpart in the component binary systems.

The evolution of meaningful empirical relationships and their replacement by more fundamental expressions clearly required international collaboration. Accordingly the Calphad (calculation of phase diagrams) group was founded in 1972 to share the information necessary to the further development of this rapidly expanding technique, bringing together workers using a variety of different models and methodologies to handle an increasingly multidimensional problem. Whereas both the Calphad Group and the Calphad journal are concerned with all known methods (including first-principle calculations) relating to phase stability, the label Calphad

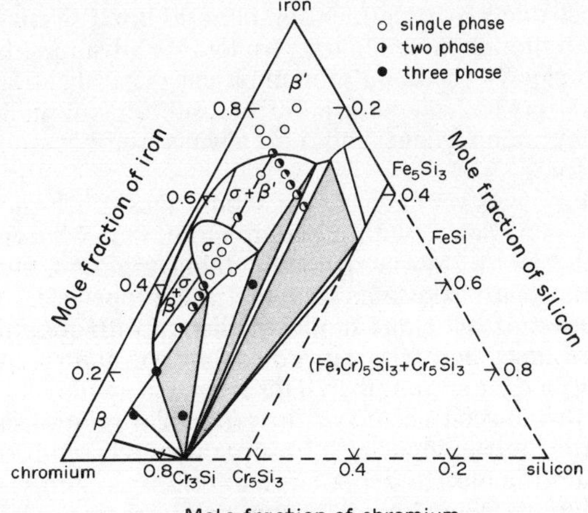

Figure 4
Comparison of calculated phase equilibria with experimental data for the chromium–iron–silicon system at 1173 K (Chart et al. 1980)

method refers to a subset of techniques that produce self-consistent results based on thermodynamic data and are capable of extension to multicomponent systems. A convenient summary of the current scope of phase diagram calculations is given by the ten year index of the Calphad journal for 1977–1986 (Bernard 1986).

3. Approaches to Modelling

Advances in modelling have taken place in three separate areas:

(a) variations in the polynomial expressions used to model known thermochemical behavior of binary systems,

(b) the extension of such expressions to cope with multicomponent systems, and

(c) the incorporation of expressions more closely related to the underlying physical reality of bonding mechanisms

The first of these is largely a matter of mathematical manipulation, and most of the systems in use are capable of mathematical transformation into each other (Tomiska 1986). However, there is obviously merit in using a system which is capable of unlimited expansion to an increasing number of components (Hillert 1986).

The increasing capability of performing multicomponent phase diagram calculations routinely, handling as many as eight components, then clearly raises a need for data relating to experimentally inaccessible phases and for reliable predictions of these missing data from relevant chemical or physical models. Chemical models incorporate well-known concepts such as chemical and magnetic ordering, nearest neighbor interactions, atomic size ratio, electronegativity, group number and bond strength. Such models have led to considerable advances, but are clearly based on semiempirical generalizations, and involve injecting a further set of parameters from independent and not always self-consistent sources.

Physical models have set themselves much more stringent goals, with a requirement that wherever possible the thermodynamic stability is consistent with elastic constants and thermal properties, the ultimate goal being a model consistent with quantum mechanics and with no adjustable constants other than the atomic number. This results in an inevitable clash between accuracy, in terms of a relationship with fundamental physical parameters, and the extension of such a formalism to ternary systems, let alone to complex multicomponent systems. It is inherently very difficult to have a simple enough overall formalism, capable of being easily differentiated to yield free energy minima in multicomponent systems that at the same time can incorporate

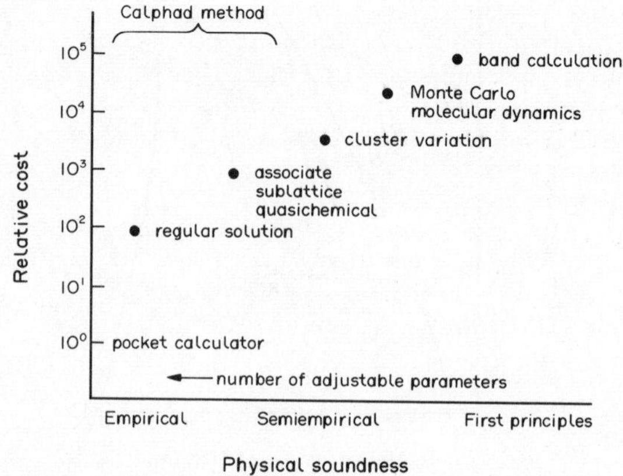

Figure 5
Overview of models and methods used to represent and calculate thermodynamic properties (Agren 1986)

quantum-mechanical calculations. This has led to the availability of a spectrum of techniques that can be selected according to user requirements, time constraints and the accuracy required in solving a particular problem (see Fig. 5).

Historically, the developments first explored in 1970–1980 were largely based on the quasichemical approach, and could be described as the addition of subroutines to established procedures.

4. The Inclusion of Chemical Ordering

One of the chief developments in the 1970–1980 period included the addition of ordering parameters to solid–solution phases, which up until that time had generally been assumed to be fully disordered. With fully disordered solutions it is easy to write an explicit expression for the entropy of mixing of any number of components, while any deviations from the fully disordered model are incorporated into the interaction terms (ΔG^E) of Eqn. (1).

By contrast, the early approaches to characterizing intermetallic phases assumed fully-ordered structures, with disorder in ternary phases being confined to specific sublattices; this device also leads to a simple explicit expression for the entropy of mixing for such phases.

Clearly a method of handling the continuous transition between ordered and disordered structures is needed. In the simplest such treatment, the Bragg–Williams approach (Inden 1974), a simple set of interactions between nearest and next nearest neighbors is assumed, and the required interaction parameters are related explicitly to the heat of solution of the disordered phase and to the ordering temperature. It is thus possible to derive these parameters from readily available experimental information in many cases, and the introduction of site occupation probabilities then allows the calculation of the free energy of a continuum of partially-ordered states. The method can be extended to ordering in ternary systems, but at present it is difficult to extend to multicomponent systems because the combinations of site occupancy become excessive. The basic equation is

$$\Delta G^{\text{ord}} = Nx_i x_j(4W_1 + 3W_2) - \frac{N}{2}(8W_1 - 6W_2)f_1^2 + 3W_2(f_2^2 + f_3^2) + \frac{NKT}{4}\sum_0^L (P_A^L \ln P_A^L + P_B^L \ln P_B^L) \quad (4)$$

where W_1 and W_2 are first and second neighbor interaction coefficients, P_A^L and P_B^L are site occupation probabilities on sublattice L, and f_1, f_2 and f_3 are functions of P_A^L.

This model also has no provision for short-range chemical order above the critical ordering temperature, so that a number of arbitrary fitting constants have to be included to match experimental information in real systems, where short-range order is generally found to be inescapably present. Consequently, a hierarchy of other techniques has been developed to include the effects of short-range order (Ackerman et al. 1989). As the name implies, the cluster variation method (CVM) considers interaction patterns in clusters of various sizes, with the tetrahedron approximation currently the most popular method but still requiring considerable computing effort. Including increasingly larger numbers of atoms ultimately requires Monte Carlo calculations (Crusius and Inden 1988), acknowledged to give the most definitive results, but requiring extensive computing time (see Fig. 5). It should be noted that, despite their mathematical rigor, all these methods require some external input in the form of interaction parameters and cannot be considered as first-principle calculations unless the relevant interaction parameters have been supplied by *ab initio* calculations.

5. *Incorporation of Magnetic Terms*

The extensive effects of magnetism were not generally appreciated until magnetic terms were quantitatively incorporated into phase diagram calculations (Miodownik 1982). For example, the magnitude of the magnetic free energy between α and γ iron is nearly ten times greater than the nonmagnetic component and of opposite sign, and so has a dominant effect on phase transformations in all ferrous alloys. Currently available methods for predicting the magnitude of possible magnetic effects require the insertion of independently available experimental or estimated inputs for the Bohr magneton number (β) and the Curie temperature (T_c), and a variety of polynomial fitting functions are used to reproduce the experimentally observed concentration dependence of these two parameters. Most of the available semiempirical formulations of magnetic free energy are based on quantum-mechanical formulae of the entropy associated with discrete magnetic spins, and all yield very similar results (Miodownik 1986). Whereas phases stabilized by chemical ordering often have very simple stoichiometries and occur in symmetrical positions in the phase diagram, the variation of β and T_c with composition can be markedly unsymmetrical, thus accounting for a number of what must otherwise be considered highly anomalous phase diagrams. In this respect, the inclusion of magnetic terms into thermodynamic calculations has much simplified the overall classification of phase diagrams from a pedagogic point of view.

By treating magnetism in terms of localized spin, it is possible to combine both chemical and magnetic ordering phenomena within the Bragg–Williams formulation (Inden 1982), although it is clear that this is an even cruder approximation than when this is used for chemical ordering alone. Increased accuracy can be produced by using mathematical fitting functions to describe the specific-heat anomalies associated with ordering effects, but such methods have the disadvantage that they are unconnected with any physical model and can therefore not be used to make predictions of magnetic effects in experimentally inaccessible metastable phases.

6. *Semiempirical Prediction of Heats of Formation*

Another area in which progress has been made in a semiempirical manner is the application of pair-potential theory to the relative stability of compounds of the same stoichiometry in different crystal structures (Machlin 1977). This has obvious applications in relation to the estimation of the energies of counterphases. The pair-potential calculations made by Machlin use the face-centered-cubic structure as a reference state and specifically exclude any *d*-band or charge-transfer contributions. The results give excellent relative stability values for different geometric variants of AB_3 compounds (and other stoichiometries), notably those that occur as mixtures between Group A and Group B elements from the periodic table. The same technique also produces a good fit with Kaufman's thermochemically-derived interaction parameters for solid solutions between transition metals (Birnie et al. 1982), but this must be considered partly fortuitous since the methodology specifically excludes strong *d*-electron bonding.

A different approach to predicting the stability of intermetallic phases was initiated by Miedema et al. (1975). This is essentially a two-parameter model based on the atomic volumes and the work function of the pure elements in the form in which they naturally exist under standard conditions of temperature and pressure. It is assumed that the macroscopic properties used to define the two ruling parameters can be applied on the atomic scale at the boundaries of modified Wigner–Seitz cells. This semiempirical method for predicting both the degree of miscibility of two elements and the stability of intermetallic compounds is surprisingly successful, but does not *a priori* take into account variations in stability due to crystal structure. (In this respect it is complementary to Machlin's model, where the emphasis is on the variation of energy with crystal structure at constant stoichiometry.)

The apparent universality of Miedema's method (de Boer et al. 1988) has caused theoretical physicists to search for the fundamental basis for its success. Miedema's model extrapolates macroscopic properties to the atomic scale without invoking any quantum-mechanical principles and relies strongly on concepts such as atomic volume and electronegativity. Using a band structure model, Pettifor (1979) has shown that there is a second-order attractive term proportional to the square of the difference in atomic energy levels, which resembles the electronegativity factor used by Miedema. However, it is not associated with the flow of charge and neither is it ionic in character. Rather it reflects the increase in band width compared to the pure constituents. A similar analysis of Miedema's second term indicates that the success of his semiempirical treatment may be due to a judicious choice of parameters that reflect certain dominant trends in the periodic table, but which are not in themselves the real causes of the observed changes. Both the methods developed by Machlin and Miedema have nevertheless proved very useful in estimating stability values for experimentally inaccessible counterphases, with the proviso that the accuracy is not intended to be better than $\pm 2.5\ \mathrm{kJ\ mol^{-1}}$.

7. *Semiempirical Methods of Predicting Entropies of Fusion*

Whereas Machlin and Miedema have concentrated exclusively on the prediction of heats of formation, the treatments of chemical and magnetic ordering have emphasized the importance of also considering the explicit calculation of entropy contributions. Although the structural disordering that occurs on melting is often used as a simple analog for the corresponding changes that occur at chemical or magnetic transition temperatures, the understanding of melting phenomena is still relatively poor, largely because it is difficult to characterize the liquid state. Although the solid-state properties of the elements are beginning to be calculated to a satisfactory level of accuracy in many cases, no calculations are available to satisfactorily predict either the melting point or the entropy of fusion of the pure elements.

This is particularly unfortunate because one of the few standard methods available for the prediction of the lattice stabilities of metastable allotropes (or counter phases) is to combine an extrapolated metastable melting point with an assumed entropy of melting for the structure in question (Miodownik 1986). Recently, the experimentally determined values of some of the entropies of fusion of the higher melting point transition metals have been revised to substantially higher values, which has had a considerable effect on the estimation of metastable lattice stabilities. Therefore it is still currently necessary to rely on semiempirical correlations to estimate the entropies of fusion of metastable phases (Saunders et al. 1988). (A hard-sphere model can be used as a basis for perturbations involving short-range orders in liquids and has achieved some success in matching experimental x-ray intensity data, but this model of the liquid state appears to be too crude as yet to produce a sufficiently accurate ΔG curve with respect to the solid phases.)

8. *First-Principle Calculations of Lattice Stabilities*

In order to calculate energy differences between different crystal structures of the same elements to the required degree of accuracy from first principles, it is necessary to make a number of simplifying assumptions. For example, the number of electrons taken into account may be restricted to the outer shell and the core electrons ignored (the "frozen core" approximation), or the number of interacting atoms may be restricted.

Another class of approximations concerns the way the quantum-mechanical interactions are handled in purely mathematical terms, as for instance with the plane-wave approximation. In the case of the transition metals, the relative importance of the *d* electrons can be used to justify the omission of certain *s–p* electron interactions. With such varied approaches, it is significant that similar results have been achieved by a number of different routes.

One of the first successful *ab initio* microscopic calculations of solid-phase transformations was produced by Yin and Cohen (1980), who achieved a resolution of 1/1000 Ry which is approximately equal to $1.4\ \mathrm{kJ\ mole^{-1}}$ for silicon. The calculations do not automatically predict the most stable structure, but they do calculate the relative energy of any particular atomic arrangement so that specification of the crystal lattice is the only external input other than the atomic number.

While Cohen and his co-workers have concentrated essentially on nontransition metals, many other groups have specialized in the transition metals, where the bonding cannot be described by interatomic pair potentials. The *d*-band contribution turns out to be by far the most dominant factor, thus accounting for the success of initial attempts to give the relative energies of face-centered-cubic, body-centered-cubic and hexagonal-close-packed structures on the basis of *d* bands alone. Many different approaches have been tried (Pettifor 1983, Skriver 1985, Watson et al. 1985) and it is very satisfying to find that they lead to very similar results. In general, state-of-the-art calculations confirm the lattice stabilities previously deduced from purely thermochemical techniques and therefore give fundamental support for the semiempirical methodology developed since the mid-1960s. However, there remain some discrepancies, whose resolution is the subject of much current activity.

9. First-Principle Calculations of Interaction Energies

Colinet et al. (1988) have recently concentrated on first-principle calculations of the effect of alloying elements that have the same crystal structure. This avoids the necessity of calculating lattice stabilities. It has become evident that, even when chosing combinations of elements such as chromium, molybdenum and tungsten which are adjacent in the periodic table, it is necessary to take into account short-range chemical order so as to obtain reasonable correlation with known thermochemical interaction energies. These workers have therefore calculated the ground-state energy of solid solutions by a combination of tight-binding and the cluster Bethe lattice method with the cluster variation method (CVM) for the evaluation of the configurational entropy. The only inputs are the bandwidth and a single site energy for each component. Figure 6 shows that the miscibility gap is correctly predicted from first principles. In the conventional thermochemical treatment this asymmetry can be easily reproduced by subregular solution models provided experimental data exists for the system, but this method clearly becomes more arbitrary when it comes to modelling metastable phases. The potential predictive capabilities of combining first-principle lattice stabilities and solid-solution interactions in the future arc sclf-evident.

The work of Mohri et al. (1988) and Sluiter et al. (1988) is representative of those groups using a combination of band calculations of atomic interactions and CVM calculations of correlation between site occupancy to obtain a free energy using no adjustable parameters. Mohri et al. used a local density functional approach to obtain interaction energies for the pure elements and for A_3B and AB stoichiometries, and then extracted the necessary pair potentials to calculate the properties of the random solution. This is an interesting variant from the conventional method of starting with the properties of a random solution and extracting the necessary parameters to allow a CVM or Monte Carlo calculation of the ordering characteristics. Since it assumes a rigid-lattice model, this method averages out the size of the constituent atoms and therefore only gives good quantitative agreement where size differences are small. Applying these principles to other binary combinations of copper, silver and gold and other more complex systems will require a refinement of this methodology before the results are of the accuracy attainable from purely thermochemical techniques, but this should not be too difficult.

However, it should be stated that because alloys of transition elements and nontransition elements have in general been treated by different methods, intermetallic compounds that involve elements from both groups (e.g., carbides and borides) have not received the theoretical attention they deserve.

10. First-Principle Calculations of Entropy

Calculations of the entropy of mixing depend solely on a knowledge of site occupancy and, therefore, no improvement can be expected from a more fundamental approach. However, a proper calculation of vibrational and electronic terms is more difficult. In principle it is certainly possible to calculate both vibrational and electronic entropy terms given the

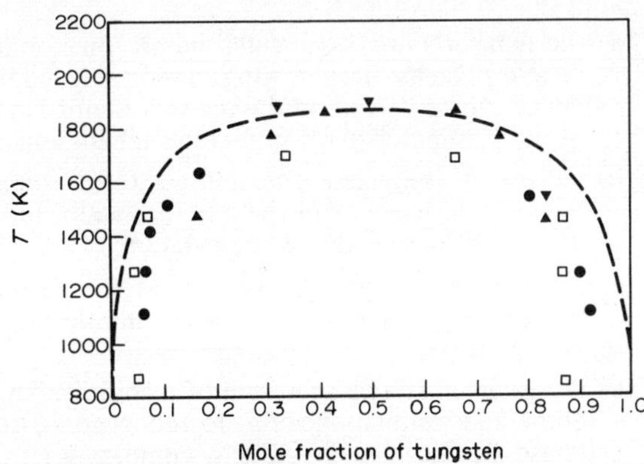

Figure 6
Predicted miscibility gap for the chromium–tungsten system vs the experimental values (Colinet et al. 1988)

band structure and the vibrational spectrum, but these will themselves be a complex function of temperature (Watson and Weinert 1984). The calculations of entropy contributions from these various sources utilize models that are not capable of being treated as independent from each other, and until a self-consistent model is evolved it will not be strictly accurate to add the various contributions. Current attempts must therefore be considered as giving only qualitative trends, but again it is encouraging that these are consistent with what has been deduced from semiempirical thermochemical routes.

11. First Principle Calculations of Magnetic Parameters

All current calculations of magnetic free energy rely on making inputs of experimentally determined values of β and T_c from the systems in question. Since such values are strongly composition dependent, the extrapolation of magnetic effects into multicomponent systems is likely to provide some exceptions to the rule that properly characterized binary systems will usually allow a good prediction of higher-order systems. It has not been possible to satisfactorily predict T_c or to include short-range magnetic order within a band-theory context, but this is under active investigation (Gyorffy et al. 1989). As with the question of structural entropy, it should eventually be possible to obtain much better input into the magnetic component of phase diagram calculations.

12. The Current Status of Phase Diagram Calculations

Summarizing these various strands of information, the most important features of modern phase diagram calculations appear to be:

(a) the convergence of results based on totally different techniques, so that increasing confidence can be placed on the physical significance of the parameters used in making calculations,

(b) the use of optimization techniques to maximize the self-consistency of available thermochemical data with phase diagram information,

(c) an increasing capability to extend the results from binary and ternary alloys to multicomponent systems,

(d) the availability of a spectrum of techniques that allow the estimation of selected features of phase equilibria with varying combinations of accuracy, time and cost, and

(e) an increasing capability to predict the free energy relationships of metastable phases, which feature strongly in many commercial alloys.

These various aspects are so interrelated that it is difficult to consider them in isolation, or place them in a particular order of importance. From a theoretical point of view, the most satisfactory aspect is the convergence of values obtained from thermochemical and *a priori* calculations. From a practical point of view, the possibility of making calculations on real multicomponent systems is a dramatic breakthrough. The capacity to handle metastable phases is an inescapable necessity in both cases. The decision as to which combination of techniques will lead to the optimum degree of accuracy and cost is a major challenge to current practitioners, both in relation to improved modelling and to the design of more flexible and user-friendly software.

Bibliography

Ackerman H, Inden G, Kikuchi R 1989 Tetrahedron approximation of CVM for BCC alloys. *Acta Metall.* 37: 1–7

Agren J 1986 Modelling of the solidification process. *Mater. Res. Soc. Bull.* September/October: 32–5

Bernard C 1986 Subject index for computer coupling of phase diagrams and thermochemistry (Vols 1 through 10 1977–86). *Calphad* 10(3/4): 255–92

Birnie D, Machlin E S, Kaufman L, Taylor K 1982 Comparison of pair potentials and thermochemical models of the heat of formation for BCC and FCC alloys. *Calphad* 6(2): 43–126

Chart T, Putland F, Dinsdale A 1980 Calculated ternary phase equilibria for the Cr–Fe–Ni–Si system. *Calphad* 4(1): 27–46

Colinet C, Bessoud A, Pasturel A 1988 Theoretical determination of thermodynamic data and phase diagrams for BCC binary transition metal alloys. *J. Phys. F* 18: 903–21

Crusius S, Inden G 1988 Order–disorder transitions in hexagonal: binary alloys: A Monte-Carlo analysis. In Komura S, Furukawa H (eds.) 1988 *Dynamics of Ordering Processes in Condensed Matter*. Plenum, New York, pp. 139–49

de Boer F R, Brown R, Mattens W C, Miedema A R, Niessen A K 1988 Cohesion in metals. In: de Boer F R, Pettifor D G (eds.) 1988 *Transition Metal Alloys*, Cohesion and Structure, Vol. 1. North-Holland, Amsterdam

Grimwall G 1986 *Thermophysical Properties of Materials*. North-Holland, Amsterdam

Gyorffy B L, Stuanton J B, Johnson D D, Pinski F J, Stocks G M 1989 Local density theory of magnetism and its interrelation with compositional order in alloys. *NATO Symp. Alloy Phase Stability*. Kluwer, Dordrecht, Netherlands, pp. 421–68

Hillert M 1986 A call for increased generality. In: Bennett L H (ed.) 1986 *Computer Modelling of Phase Diagrams*. Metallurgical Society of AIME, Warrendale, PA, pp. 1–18

Inden G 1974 Ordering and segregation reactions in BCC alloys. *Acta Metall.* 22: 945–51

Inden G 1982 Continuous transformations in phase diagrams. *Bull. Alloy Phase Diagrams* 2(4): 412–8

Kaufman L 1972 Calculation of binary phase diagrams. *National Physical Laboratory Symp. Metallurgical*

Chemistry. Her Majesty's Stationery Office, London, pp. 373–96

Kaufman L, Bernstein H 1970 *Computer Calculations of Phase Diagrams*. Academic Press, New York

Lukas H L, Henig E Th, Zimmerman B 1977 Optimisation of thermodynamic data. *Calphad* 1: 225–36

Machlin E S 1977 Modified pair-potentials and phase stability. *Calphad* 1(4): 361–77

Miedema A R, Boone R, de Boer F R 1975 On the heat of formation of solid alloys. *J. Less Common Metals* 41: 263–98

Miodownik A P 1982 The effect of magnetic transformations on phase diagrams. *Bull. Alloy Phase Diagrams* 2(4): 406–12

Miodownik A P 1986 The phase stability of the elements. In Bennett L H (ed.) 1986 *Computer Modelling of Phase Diagrams*. Metallurgical Society of AIME, Warrendale, PA, pp. 253–84

Mohri T, Terakura K, Oguchi T, Wanatabe K 1988 First principles calculation of the thermodynamic properties and phase diagrams of noble metal alloys. In: Lorimer G W (ed.) 1988 *Symp. Phase Transformations*. Institute of Metals, London, pp. 433–7

Pettifor D G 1979 Theory of the heats of formation of transition metal alloys. *Phys. Rev. Lett.* 42(13): 846–9

Pettifor D G 1983 Electron theory of metals. In: Cahn R W, Haasen P (eds.) 1983 *Physical Metallurgy*, 3rd edn. Elsevier, Amsterdam, Chap. 7

Saunders N 1989 Calculated and metastable phase equilibria in Al–Li–Zr alloys. *Z. Metallk.* 80(12): 894–903

Saunders N, Miodownik A P 1986 Thermodynamic aspects of amorphous phase transformation. *J. Mater. Res.* 1: 38–46

Saunders N, Miodownik A P, Dinsdale A T 1988 Metastable lattice stabilities for the elements. *Calphad* 12(4): 351–74

Skriver H L 1985 Crystal structure from one-electron theory. *Phys. Rev. B* 31(4): 1909–23

Sluiter M, de Fontaine D, Turchi P, Fu Zezhong F 1988 Tight-binding calculations of Ti–Rh type phase diagram. *Phys. Rev. Lett.* 60(8): 716–9

Tomiska J 1986 Interconversion of different formulations of the excess free energy. *Calphad* 10(3/4): 239–52

Watson R E, Davenport J W, Weinert M 1985 Slater orbital method for the calculation of structural energies of 5d elements. *Phys. Rev. B* 32: 4885–91

Watson R E, Weinert M 1984 Contribution to the entropy of crystals. *Phys. Rev. B* 30(4): 1641–5

Yin M T, Cohen M L 1980 Microscopic theory of the phase transformation and lattice dynamics of silicon. *Phys. Rev. Lett.* 45(12): 1004–7

A. P. Miodownik
(University of Surrey, Guildford, UK]

Phase Equilibria in Alloys: Influence of Stress

Crystalline alloys often exist in a state of stress at equilibrium. Stresses can result from many sources including external loads (such as those existing in engineering applications), internal sources such as coherency strains (that arise during the precipitation processes and which are a result of the fact that phases possess different unit cells) and epitaxial strains in thin-film systems. Just as a pressure alters phase equilibria, the crystal responds to these elastic forces by changing the volume fraction or the equilibrium composition of the phases with respect to those observed in the unstressed system, or by undergoing a transition to a new phase.

It is possible to consider at least two types of phase equilibria in crystalline solids. The first type, termed incoherent, refers to systems in which the mechanical coupling between phases occurs only through a normal force or pressure. In this case, specifying an external pressure necessarily requires an identical pressure (hydrostatic stress) in each of the phases at equilibrium. The second type, termed coherent, refers to systems in which both the displacements and tractions are continuous across phase boundaries. In a coherent system, the stresses and strains may be neither uniform nor hydrostatic at equilibrium. The characteristics of phase equilibria in incoherent systems are qualitatively similar to those observed in fluid systems, while the characteristics usually exhibited by coherent systems are qualitatively much different.

1. Stress and Strain as Thermodynamic Variables

Stress (σ_{ij}) and strain (e_{ij}), like pressure and volume in a fluid system, are conjugate thermodynamic variables that give the reversible mechanical work done by a volume element of the crystal. This work term appears in the first law of thermodynamics as

$$de = \sigma_{ij}\, d\varepsilon_{ij} \qquad (1)$$

where e is the internal energy density, usually measured with respect to the same reference state as the strains. In an alloy there are additional work terms related to the flow of heat and mass. Incorporating these work terms into Eqn. (1) and employing the second law of thermodynamics, the reversible change in the internal energy of an element of crystal is

$$de = T\, ds + \sigma_{ij}\, d\varepsilon_{ij} + \lambda_i \rho_i \qquad (2)$$

where T is the absolute temperature, s is the entropy density, ρ_i is the mass density of the ith component, and λ_i is the partial derivative of the internal energy with respect to ρ_i. For a crystalline solid, the derivatives with respect to ρ_i may not be independent owing to the existence of a lattice or may imply an implicit change in the defect concentration when the number density of a component is changed.

Equation (2) indicates how the thermodynamic state of a volume element of material depends on the various thermodynamic variables (Maxwell relations). It allows the total internal energy of the system to be written as a functional of the internal energy density. When the internal energy is minimized at constant entropy and volume, the conditions of thermodynamic equilibrium are obtained. These equilibrium conditions are in many ways similar to those of fluid systems; for example, the temperature and appropriately defined chemical potentials must be uniform at equilibrium. However, a major difference is encountered for mechanical equilibrium. Whereas the pressure must necessarily be uniform at equilibrium in a fluid or incoherent solid, the stresses in a crystal need only satisfy Cauchy's condition that the divergence of the stress tensor vanish in the absence of body forces.

2. *Qualitative Features of Coherent Equilibria*

Cauchy's condition does not require the stress state to be uniform at equilibrium. Consequently, the composition field may also be spatially nonuniform at equilibrium. This raises interesting questions as to the definition of a "phase" in stressed crystalline solids, since spatially contiguous points of the same phase in the solid occupy different points in phase space. Questions of this nature remain to be addressed.

The long-range nature of the elastic stress field also changes the qualitative nature of phase equilibria in coherent solids with respect to that of fluid or incoherent solid systems. In fluid or incoherent solid systems, the total free energy of the system is independent of the number and morphology of the phase domains and their spatial arrangement within the system when interfacial energy is neglected. One phase domain does not influence the thermodynamic state of another phase domain. This is the property that gives rise to the Gibbs phase rule and allows the equilibrium phase compositions to be determined from the common tangent construction. In coherent solids, the free energy of the system depends on the geometric arrangement of the phases as well as the morphology and number of the individual phase domains. This is because the elastic deformation depends on the morphology and spatial arrangement of the phases. The state of strain at any one point in the crystal can also be a function of the global composition field as compositional inhomogeneity can engender stress. The coupling between the stress and composition fields can then be quite complicated. Since the stress state of one phase domain influences the stress state of the other phase domains, the assumptions employed in the derivation of the Gibbs phase rule and common tangent construction are not met and these techniques may not be applicable, in general, to coherent solids.

The coupling between stress and composition, as well as the influence of the mechanical loading conditions on the nature of phase equilibria in coherent solids, can be illustrated by considering a simple two-phase binary alloy. Assume that the phases are configured as parallel plates, as depicted in Fig. 1, and are constrained to remain coherent. The isolated phases, as they would appear in their unstressed states, are shown in Fig. 1a. They possess different dimensions as the unstressed lattice parameters of the phases are different. Two different mechanical loading conditions are considered. The first consists of a set of displacement boundary conditions along the edge of the plates. The displacement boundary conditions can be imagined as having been imposed by two rigid walls, as depicted in Fig. 1b. As the displacement (strain) is fixed by the walls, the system is termed constrained. The second set of loading conditions, termed unconstrained, is one in which the system is traction free (i.e., there are no external forces (or walls) acting on the system). If the stress-free phases appear as in Fig. 1a, the α phase is always under tension and the β phase under compression when unconstrained, as depicted in Fig. 1c. For both sets of loading conditions, the elastic deformation and composition are homogeneous in each of the phases at equilibrium and the definition of a phase is unambiguous.

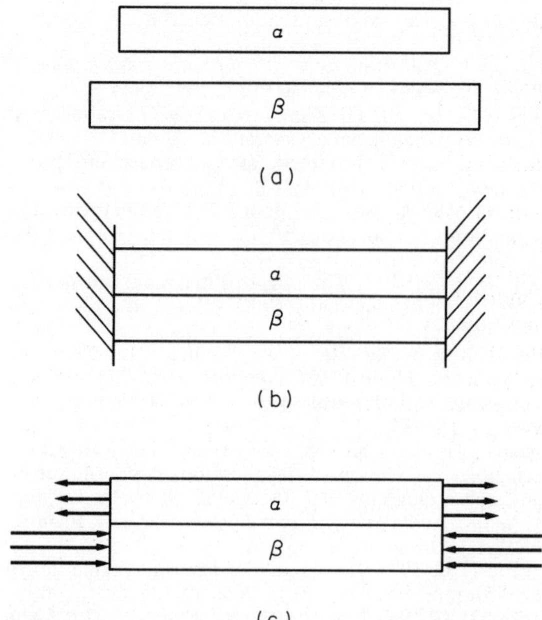

Figure 1
A two-phase binary alloy: (a) the stress-free lattice parameters of the α and β phases are different; (b) in the constrained system, the lattice parameter in the plane of the film is fixed by the imposed mechanical loading condition; (c) for the traction-free unconstrained system, the α and β phases are under tension and compression, respectively, and depend on the presence of the other phase

For the elastically constrained system, the elastic state of each phase is determined only by the displacement boundary conditions and is not influenced by the presence of the other phases. Hence, the thermodynamic state of each phase is determined once the temperature, composition of the phase, and the external displacement have been specified. As a result, the compositions of the phases (at the ends of tie lines) coincide with the phase boundaries (field lines) as they do in fluid and incoherent solid systems. The common tangent construction can be employed in determining equilibrium phase compositions as this is equivalent to satisfying the equilibrium condition for the equality of chemical (diffusion) potentials. Likewise, the Gibbs phase rule is applicable to the constrained system even though the system is coherent and nonhydrostatically stressed.

For the elastically unconstrained planar system, the elastic state of one phase depends explicitly on the presence of the other phase. For example, if the α phase is thick with respect to the β phase, the lattice parameter of the α phase within the plane will only be slightly deformed from that of its unstressed state while the lattice parameter of the β phase will be deformed so that it almost matches that of the unstressed α phase. If the phases are of approximately equal thickness, their elastic states will be determined by the relative stiffness of the phases. Thus changes in the relative thickness or volume fraction of the phases alter the elastic state and, hence, the thermodynamic state of the phases. As the thermodynamic state of one phase is influenced by the presence of the other phase, the assumptions used in deriving the Gibbs phase rule and common tangent construction are not satisfied and, therefore, are not applicable to this coherent system. In fact, it has been shown that for certain coherent systems the number of degrees of freedom of the system is independent of the number of phases present allowing, in principle, an arbitrary number of phases to coexist in equilibrium.

Changing the mechanical loading conditions alters the qualitative features of phase equilibria in coherent solids. Likewise, changing the system geometry may also change the equilibrium characteristics. In most two-phase coherent systems, the elastic state of one phase is influenced by the presence of the other, and the characteristics of phase equilibria will be different from those of fluid systems.

3. Coherent Phase Diagrams

Phase diagrams are a convenient mechanism for displaying the characteristics of phase equilibria in a limited number of coherent systems. An example is given in Fig. 2 for the InGaP pseudobinary. The system is configured as an epitaxial film on a GaP substrate. The substrate is assumed to be sufficiently thick such that the lattice parameter in the plane of the film is fixed by that of the substrate. This means that the system behaves in a manner qualitatively similar to the constrained system of Fig. 1b. The broken and solid lines of Fig. 2 delineate the limits of two-phase coexistence for an incoherent system in the absence of all stress effects and the stressed coherent system, respectively. The equilibrium composition of the phases coincides with the phase boundary. The equilibrium compositions depend strongly on the crystallographic orientation of the film because the effective elastic modulus in the plane of the film depends on its orientation.

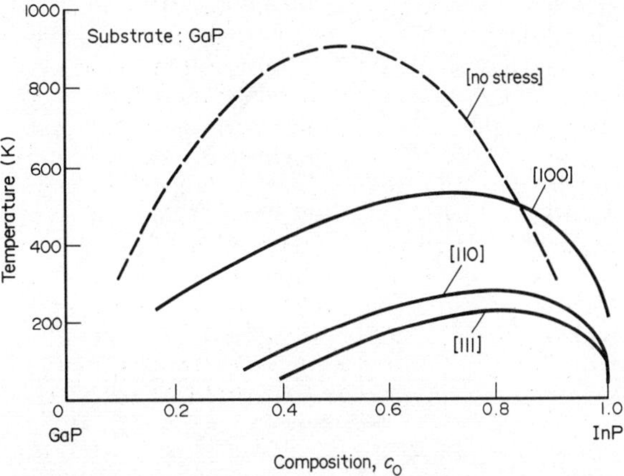

Figure 2
Calculated phase diagram for the constrained InGaP pseudobinary alloy

When the phases are configured as for the unconstrained system, phase diagrams are no longer as convenient a device for communicating information on phase equilibria. This is because a change in the volume fraction of the phases, as would accompany a change in the bulk alloy composition, results in a change in the thermodynamic state of the phase and the equilibrium phase compositions. This means that the equilibrium phase compositions are a function of the alloy composition across the two-phase field. An example of the dependence of the equilibrium phase compositions on alloy composition is illustrated in Fig. 3, for a hypothetical alloy system possessing the unconstrained planar geometry of Fig. 1. The solid line demarcates the region of two-phase coexistence for a coherent system with a consolute critical point. The broken lines indicate the equilibrium compositions of the phases for several different alloy compositions. The critical point is shifted several hundred degrees from the maximum temperature of the two-phase field. Of course, the quantitative nature of the behavior depicted in Fig. 3 depends on the thermophysical parameters of the system

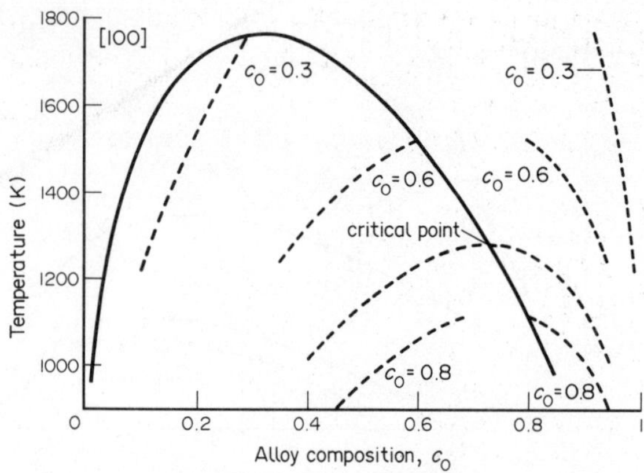

Figure 3
Calculated phase diagram for a hypothetical unconstrained system depicting the equilibrium phase compositions as a function of the alloy composition

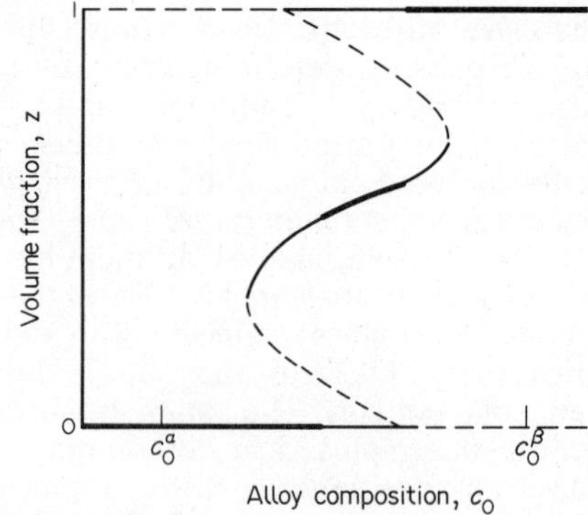

Figure 4
A schematic phase-stability diagram for an unconstrained system at constant temperature is used to depict the equilibrium volume fraction as a function of alloy composition: C_0^α and C_0^β are the phase boundaries in the absence of stress at the given temperature

employed. However, the results depicted in Fig. 3 clearly illustrate that the compositions of the phases at a particular temperature (tie lines) may be very different from the corresponding points on the phase boundaries (fields lines). Changing the system geometry from parallel plates to another geometry for which all of the equilibrium conditions can still be satisfied, to, for example, one of concentric spheres in an isotropic system, alters the equilibrium phase compositions and volume fractions because a change in the system geometry alters the stress state. The qualitative and quantitative features of a phase diagram for fully coherent phases are predicted to depend on both the mechanical loading conditions and the system geometry.

Another predicted manifestation of coherency stress on phase equilibria is the possibility that more than one stable equilbrium state may exist for a two-phase binary alloy. This means that, for a given alloy composition, temperature and pressure, more than one volume fraction and corresponding set of phase compositions gives a (relative or absolute) minimum in the free energy of the system. One way of indicating the various equilibrium states is on a phase-stability diagram for coherent systems as shown in Fig. 4. The equilibrium volume fraction z is plotted as a function of the alloy composition c_0 for an unconstrained system. The thin and thick solid lines indicate the volume fractions that give the relative and absolute minima in the system free energy, respectively, for a given alloy composition. Certain volume fractions are thermodynamically unstable corresponding to a maximum in the system free energy (dashed lines) but still satisfy all of the thermodynamic equilibrium conditions. For each equilibrium volume fraction that extremizes the system energy at a given alloy composition, there corresponds a set of equilibrium phase compositions. Owing to the potential existence of more than one stable equilibrium state, it may be possible to achieve different two-phase microstructures by changing the processing path.

Although coherency strains are well known to alter phase equilibria in solids by shifting equilibrium compositions and volume fractions and, in some cases, resulting in new phases, many of the theoretical predictions presented here have yet to be verified experimentally. Furthermore, existing theoretical analyses are applicable only to very simple systems. The possible range of the types of phase equilibria in coherent systems is not known.

Bibliography

Cahn J W 1962 Coherent fluctuations and nucleation in isotropic solids. *Acta Metall* 10: 907–13

Cahn J W, Larche F C 1984 A simple model for coherent equilibrium. *Acta Metall.* 32: 1915–24

Johnson W C 1987 On the inapplicability of Gibbs phase rule to coherent solids. *Metall. Trans.* 18A: 1903–7

Johnson W C, Chiang C S 1988 Phase equilibrium and stability of elastically stressed heteroepitaxial thin films. *J. Appl. Phys.* 64: 1155–65

Johnson W C, Voorhees PW 1987 Phase equilibrium in two-phase coherent solids. *Metall. Trans.* 18A: 1213–28

Roitburd A L 1984 Equilibrium and phase diagrams of coherent phases in solids. *Sov. Phys. Solid State* 26: 1229–33

Williams R O 1980 Long-period superlattices in the copper–gold system as two-phase mixtures. *Metall. Trans.* 11A: 247–53

W. C. Johnson
[Carnegie Mellon University, Pittsburgh, Pennsylvania, USA]

Phase Transformations at Surfaces and Interfaces

With the constant drive towards miniaturization and large-scale intergration experienced by the electronics industry, the performance of microelectronic devices depends critically on the properties of surfaces and interfaces. A modern microelectronic device is typically composed of several layers of different materials such as semiconducting silicon, noble metal gold and transition metals (e.g., tungsten, titanium and nickel) in close contact with each other. The transition region across the contact between two materials, or between two different phases of the same material, is generically referred to as an interface. The term surface is reserved for the interface between a solid and its gas phase which, for most practical purposes, can be considered as a vacuum. The interface zone can be very narrow, and essentially of atomic dimensions, or it may extend over several hundred angstroms. A common characteristic of interfaces, however, is that their physical properties change markedly across the interface zone, being generally quite different from those of the bulk materials on either side.

The importance of surfaces and interfaces in the performance of engineering materials clearly transcends the realm of microelectronic devices. For example, surfaces play a key role in friction, wear and crack formation in materials as well as in the performance of catalysts, adhesives, composites and biomaterials used for implants and prosthetics. Interfaces, however, are ubiquitous in complex engineering materials and components, since these are generally composed of a relatively fine dispersion of two or more phases. Even in single-phase materials (e.g., the ordered intermetallic compounds) interfaces between ordered domains, the so-called antiphase domain boundaries, are quite common and often play an important role in a range of mechanical properties.

Phase transformations or transitions are characterized by sudden changes in one or more long-range order parameters. For example, the order parameter in ferromagnets is the magnetization, in gas–liquid transitions is the density, and in order–disorder transitions in alloys is the difference in the probability of atomic occupancy of nonequivalent sites in the crystal. The long-range order parameter is zero in the high-symmetry phase and takes finite values in the less symmetric phase which usually occurs at low temperatures. First-order transitions are characterized by a discontinuous jump in the long-range order parameter at the transition temperature, whereas for second- or higher-order transitions the order parameter goes continuously to zero at the transition.

It is generally found that the behavior of the surface long-range order parameter may differ considerably from the behavior of the bulk long-range order parameter. In many instances it is seen that, while the bulk undergoes a first-order transition, the surface long-range order parameter goes to zero continuously in the characteristic fashion of second-order transitions. This is the phenomenon of surface-induced disorder or "wetting," whereby a layer of the disordered or high-temperature phase develops at the surface before the transition temperature is reached from below. The phenomenon of surface-induced ordering has also been reported in ferromagnetic materials. In this case the surface magnetization remains finite at temperatures above the Curie temperature of the bulk. Although it has not yet been observed experimentally, theoretical modelling has revealed that close-packed ferromagnetic surfaces may disorder via a first-order transition while the transition in the bulk is second order.

1. *Experimental Probes*

Several surface-sensitive techniques of varied degrees of sophistication are available and routinely used in the scientific and industrial laboratory. These experimental tools employ either ions, electron or photons (e.g. properly tuned x-ray radiation) to probe the structure and chemistry of surfaces. Among the techniques that use electrons as probes, scanning electron microscopy and transmission electron microscopy figure prominently since they provide a direct image of surfaces and/or interfaces at different amplifications. Low-energy electron diffraction (LEED) is extensively used to investigate the structure of surfaces and, in the case of alloys, the degree of chemical order. This technique has played an important role in establishing that most surfaces "reconstruct"; that is, the atomic positions at the outermost layer follow a pattern slightly different from the bulk. In addition, LEED has been used to investigate order–disorder transitions in alloy surfaces. A closely related experimental tool that uses spin-polarized low-energy electrons (SPLEED) has been employed to study surface ferromagnetism in several magnetic materials. Auger electron spectroscopy is used to study the chemical composition of surfaces. Using this technique, the surface composition can be determined by measuring the energy distribution of electrons that are emitted from the surface region due to the irradiation of the material with a beam of energetic electrons.

Ion beam sputtering is routinely used to remove atoms from the surface and, combined with a surface-sensitive technique such as Auger electron spectroscopy, can be used to obtain concentration profiles over relatively thick surface layers.

Low-energy ion spectroscopy (LEIS) probes the composition of the surface region over a depth of two to three layers without the need to remove material by ion sputtering. In this technique, the way in which incoming ions are scattered by surface ions of different atomic number (Rutherford scattering) provides a signature of the chemical composition of the surface and of adjacent layers. Low-energy ion spectroscopy is particularly powerful when used with LEED since both structural and compositional information are then obtained. Medium- and high-energy ion scattering are also employed to study a range of overlayers and film thicknesses. Rutherford backscattering, which analyzes the yield of protons scattered at angles close to 180°, has proved to be a powerful technique for studying the amount of structural disorder encountered near the surface of solids at high temperatures. In particular, Rutherford back-scattering has revealed that some crystalline surfaces begin to melt at temperatures lower than the melting temperature of the bulk.

Several photoelectron spectroscopic techniques are in use and they are generically based on the analysis of electrons ejected from the surface region by means of ultraviolet or soft x-ray radiation. The high-intensity tunable synchrotron source has become the instrument of choice for this type of surface study. Electronic surface states, oxidation states and the types of chemical bonding of different atomic species at surfaces can be investigated using photoelectron spectroscopy. Structural information can also be obtained by means of glancing-angle x-ray diffraction. Surface EXAFS (extended x-ray absorption fine structure) provides information on bond lengths and surface geometries through the analysis of photoemitted electrons.

2. *Surface Magnetism*

Magnetic materials are commonly used for the storage, manipulation and transfer of information. This technology is driven by the need for high-density storage and for high-speed retrieval and transfer of data. The understanding of the magnetic structure of surfaces in thin films and small particles is crucial in the development of new magnetic devices and, as such, surface magnetic transitions have been the subject of close experimental and theoretical scrutiny. In general, it has been found that surface magnetic properties can be markedly different from those in the bulk. For example, the (001) surface of the pure metal vanadium is ferromagnetic with a relatively high Curie temperature of 540 K, although vanadium itself does not display ferromagnetism in the bulk. With regard to magnetic phase transitions, the surface Curie temperature in ferromagnets may be higher than the Curie temperature of the bulk for example, in gadolinium films the Curie temperature of the (0001) close-packed surface has been found to be 22 K higher than the bulk. This phenomenon is known as surface-enhanced magnetism.

Several theoretical and experimental studies have established that the general properties, or universality class, of surface magnetic phase transitions fall in between those of two- and three-dimensional systems. In particular, the critical behavior of surfaces in thick or semi-infinite films, as characterized by the values of the critical exponents of the magnetization and magnetic susceptibility for example, is different from the critical behavior of either two- or three-dimensional systems. The nature of the surface transition as well as the value of the transition temperature relative to the bulk Curie temperature depend sensitively on the surface geometry and on the values of the surface exchange interaction. For surface exchange interactions below a critical value J_c, the surface transition is of second order, as in the bulk, and takes place at the bulk Curie temperature. Surface-enhanced magnetism is observed for values of the surface exchange interaction larger than J_c. For close-packed surface planes, such as the (111) plane in face-centered-cubic structures, the ferromagnetic transition at the surface is expected to become of first order for a finite range of surface exchange interactions. This unusual behavior has been predicted on the basis of statistical models and has not yet been reported experimentally.

3. *Wetting*

The interface between two coexisting phases is subject to profound changes as the thermodynamic parameters (e.g., temperature, pressure and composition) are varied. The phenomenon of wetting pertains to the appearance, right at the interface between two coexisting phases, of a third thermodynamically stable phase. The term wetting is borrowed from the commonly observed phenomenon of drops formation on a substrate: depending on the relative values of the gas–liquid (σ_{g-l}), liquid–substrate (σ_{l-s}) and gas-substrate (σ_{g-s}) surface tensions, the contact angle between the drop surface and the substrate may vanish, in which case the drop is said to wet the gas–substrate interface. The contact angle θ is given by

$$\cos\theta = \sigma_{g-s} - \sigma_{l-s}/\sigma_{g-l} \qquad (1)$$

Cahn (1977) pointed out that, as the bulk critical point T_c of the liquid–gas transition is approached along the coexistence line, both σ_{g-l} and the difference $\sigma_{g-s} - \sigma_{l-s}$ in Eqn. (1) will vanish. Several theoretical models indicate that the denominator in Eqn.

(1) tends to zero faster than the numerator, which implies the divergence of $\cos\theta$ at the critical point. The divergence of $\cos\theta$ is, of course, a mathematical impossibility since its absolute value must always be less than or equal to one. This simple argument led Cahn to propose the existence of a wetting transition, occurring at a temperature T_w always smaller than the bulk critical temperature T_c.

3.1 Single-Order Parameter

The so-called lattice-gas model provides a convenient and useful tool for the theoretical classification of wetting phenomena. Lattice gases are highly idealized mathematical models used to study nonsymmetry breaking first-order transitions such as the liquid–gas transition of one-component systems and the phase separation or segregation in binary alloys. These systems are characterized by the fact that they can be described by a unique long-range order parameter.

The thermodynamic behavior of the model is determined by the temperature T, the chemical potential μ (or the pressure in the case of the liquid–gas transition) and by the interaction of atoms among themselves in the bulk and with a substrate. In the present context, a substrate refers to a two-dimensional defect such as a surface, container wall or grain boundary.

The phase diagram of the lattice-gas model in temperature–chemical-potential space presents a line of first-order transitions given by $\mu = \mu_0(T)$. The line of first-order transitions ends in a critical point occurring at temperature T_c. For the case of a binary alloy, two phases of equal symmetry but different compositions coexist, at a given temperature T, when the chemical potential of the system equals $\mu_0(T)$.

In the presence of a substrate, and for sufficiently strong system–substrate interactions, the lattice-gas model predicts the existence of a sequence of first-order transitions occurring at values of chemical potentials different from the bulk coexistence chemical potential $\mu_0(T)$. This behavior is known as layering and it corresponds to the discrete segregation, plane by plane, of the minority component in the alloy as the value of $\mu_0(T)$ is approached.

For weaker values of the system–substrate interactions, the model predicts the existence of a line of second order or critical wetting. The wetting line coincides with the bulk first-order transition $\mu_0(T)$ and it extends from the wetting transition temperature T_w to the bulk critical temperature T_c. Below T_w, there is a moderate increase in segregation of the minority component to the substrate as the coexistence chemical potential $\mu_0(T)$ is approached. For temperatures higher than T_w, however, the thickness of the segregated layer increases continuously, diverging at $\mu_0(T)$.

The phenomenon of first-order wetting, whereby a macroscopic layer is formed via a first-order transition, is also possible. This phenomenon is known as prewetting. As in the case of critical wetting, the thickness of the macroscopic layer also diverges as $\mu_0(T)$ is approached.

Experiments on substrate wetting phenomena abound for one-component systems near their gas–liquid coexistence and sublimation curves. Likewise, wall and interface wetting phenomena have been studied extensively for binary-fluid mixtures using conventional contact angle measurement techniques.

The surface of a solid near its melting point also displays the characteristics of critical wetting. This phenomenon, known as surface melting, has long been suspected to occur from the observation that solids cannot be easily superheated whereas the melt can be supercooled. However, the wetting of a solid by its liquid phase has only recently been observed. These experiments were conducted for the (110) surface of lead using Rutherford backscattering spectroscopy.

Grain boundaries in polycrystalline materials are an example of solid–solid interfaces that may be wet by the liquid phase near the melting point. Several molecular dynamics, Monte Carlo and other statistical mechanics treatments of grain boundaries indicate that, for certain interatomic potentials and grain boundary orientations, strong topological disorder sets in well before reaching the melting temperature.

3.2 Multiple-Order Parameters

The phenomenon of critical wetting in systems undergoing a change of symmetry at a first-order phase transition has been reported to occur for the (001) surface of Cu_3Au crystals. Low-energy electron diffraction experiments by several researchers, using both spin polarized (SPLEED) and unpolarized (LEED) electrons, have shown conclusively that the long-range order parameter for the (001) surface of Cu_3Au vanishes continuously at or very close to the bulk transition temperature of 663 K. Furthermore, low-energy ion spectroscopy (LEIS) has been used to determine the chemical composition of the first two surface layers across the critical wetting transition. The experimental results indicate that there is considerable gold enrichment at the surface.

The theoretical modelling of surface phenomena for order–disorder transitions in alloys present several complications that are not contemplated in the simpler one-component lattice-gas model. The surface behavior depends on the lattice structure, the symmetry of the order phase, the possibility of elastic relaxation (reconstruction) of the solid surface and the interplay between at least two order parameters, namely the alloy concentration and the long-range order parameter. Accordingly, modelling of surface transition in ordering alloys is not as fully developed as that of wetting phenomena in one-component systems. Nevertheless, it is encouraging to note that the available theoretical treatments appear to be in close agreement with experiment.

4. Antiphase Domain Boundaries

The development of efficient and reliable jet engines for the aerospace industry requires materials with, among other things, high strength at elevated temperatures. Ordered alloys, such as Ni_3Al, often exhibit a substantial increase in strength with temperature and, consequently, they are attractive candidates for such applications. It has long been recognized that the interfaces between ordered domains, the so-called antiphase domain boundaries, play a significant role in the deformation behavior of ordered alloys. In these systems, antiphase domain boundaries also separate pairs of dislocations, each with a Burgers vector less than the unit translation vector of the superlattice. Consequently, the slip of one of the dislocations will disturb the symmetry of the ordered structure by the shearing of two planes, creating an antiphase domain boundary. Long-range order is restored by the second dislocation of the pair. This pair of dislocations, whose sum of Burgers vectors equals a superlattice translation vector and whose combined movement restores order, is known as a superdislocation.

The formation of films of the disordered γ phase at antiphase domain boundaries has been observed, using transmission electron microscopy, in γ' ordered intermetallic alloys based on the nickel–aluminum binary system. Critical wetting of antiphase domain boundaries is also suggested by theoretical modelling. Long-range order parameter profiles have been calculated for (111) antiphase domain boundaries near the γ–γ' order–disorder transition temperature. In these calculations, the degree of long-range order in the interface region decreases sharply relative to the bulk, indicating the onset of a disordered film at the antiphase domain boundary.

The wetting of the interface between ordered domains by the disordered phase has important scientific and technological implications. In particular, the presence of a disordered film near the antiphase domain boundary can profoundly affect the motion of superdislocations in ordered compounds and, thus, their resistance to plastic deformation or yield strength. Furthermore, as pointed out for the case of surface melting, the existence of a wetting transition in the presence of antiphase boundaries would preclude the superheating of the ordered phase beyond T_w.

See also: Surface Reactions; Surface Structure; Surfaces and Interfaces: An Overview; Surfaces and Interfaces: Composition

Bibliography

Alvarado S F, Campagna M, Fattah A, Uelhoff W 1987 Critical wetting and surface-induced continuous phase transition on $Cu_3Au(001)$. *Z. Phys. B* 66: 103–6

Cahn J W 1977 Critical point wetting. *J. Chem. Phys.* 66: 3667–72

Cahn R W, Siemers P A, Geiger J E, Hall E L 1987 The order–disorder transformation in Ni_3Al and Ni_3Al–Fe alloys–II. Phase transformations and microstructures. *Acta Metall.* 35: 2753–64

Dietrich S 1988 Wetting phenomena. In: Domb C, Lebowitz J (eds.) 1988 *Phase Transitions and Critical Phenomena*, Vol. 12. Academic Press, London

Falicov L M, Moran-Lopez J L 1986 *Magnetic Properties of Low-Dimensional Systems*. Springer, Berlin

Frenken J W M, van der Veen J F 1985 Observation of surface melting. *Phys. Rev. Lett.* 54: 134–7

Lipowsky R 1982 Critical surface phenomena at first-order bulk transitions. *Phys. Rev. Lett.* 49: 1575–8

Lipowsky R 1987 Surface critical phenomena at first-order phase transitions. *Ferroelectrics* 73: 69–81

Poate J M, Tu K N, Mayer J W 1978 *Thin Films—Interdiffusion and Reactions*. Wiley, New York

Sanchez J M, Eng S, Wu Y P, Tien J K 1987 Modeling of antiphase boundaries in $L1_2$ structures. In: Stoloff N S, Koch C C, Liu C T, Izumi O (eds.) 1987 *High Temperature Ordered Intermetallic Alloys II*, Materials Research Society Symposium Proceedings, Vol. 81. Materials Research Society, Pittsburgh, PA, pp. 57–64

Sanchez J M, Moran-Lopez J L 1987 Surface first-order phase transitions in fcc Ising ferromagnets. *Phys. Rev. Lett.* 58: 1120–2

Yamamoto M, Nehno S 1989 Phase transformations at/near the surface of ordering alloys. *Surface Sci.* 213: 502–24

J. M. Sanchez
[University of Texas at Austin,
Austin, Texas, USA]

Physical Aging of Polymers

Since the mid-1970s the usage of polymers has been increasing substantially. Even more recently, the growth rate in the class of what are often called engineering polymers (i.e., those aimed at the more high-performance applications) has been particularly significant. For such applications, the demands placed on the polymer are severe. The reason for this is that polymers suffer from a number of disadvantages in comparison with competitive materials such as metals. For example, polymers in the bulk generally have a low modulus, a limited service temperature range and exhibit time-dependent viscoelastic behavior. As a result, components made from polymers are often designed close to their limiting capabilities. Consequently, any effect that might modify the behavior of the polymer must be considered in the overall design process. One such important effect is physical aging.

The term physical aging refers in general to the phenomenon whereby the properties of materials are observed to change as a function of storage time under zero load and at constant temperature. The relevant properties may be mechanical, physical, electrical, optical or any other property which is dependent on structure. Furthermore, a dependence of properties on storage time has been observed in all classes of materials—organic and inorganic, metallic and nonmetallic. This universality of the phenomenon is remarkable, but should be interpreted with some caution; in particular, the mechanism responsible for such changes in properties cannot be the same for all materials. This article will concentrate on the physical aging of polymers as revealed through changes in their mechanical properties, with brief reference to other materials and properties.

It should be emphasized here that physical aging is distinct from chemical aging. The latter usually involves the external action of some environment (such as in oxidation, ultraviolet degradation or chemical attack) that causes irreversible changes to occur in the chemical structure of the polymer. However, physical aging is related to reversible changes that occur in the physical structure of the polymer, and which thereby have an effect on the response of the polymer to an external stimulus.

1. Characteristics

One of the earliest systematic studies of physical aging is that of Kovacs et al. (1963). Nevertheless, it is usually the extremely thorough investigation, by Struik in the 1970s, of physical aging in amorphous polymers and other materials that is considered to have led to the seminal work on the subject (Struik 1978). The important characteristic features of physical aging have been identified by Struik, and are well illustrated by reference to the small-strain creep response of an amorphous polymer, atactic polystyrene (Fig. 1).

Figure 1 shows the torsional creep compliance $J(t)$ as a function of creep time t_c, on a logarithmic scale, at the aging or test temperature of 75 °C. The several creep curves illustrated were obtained as follows. The sample was initially equilibrated above the glass transition temperature T_g of polystyrene before being cooled rapidly to the aging temperature of 75 °C, below T_g. While the sample was stored at this temperature, creep tests were performed at increasing aging times, the aging time t_a being the time

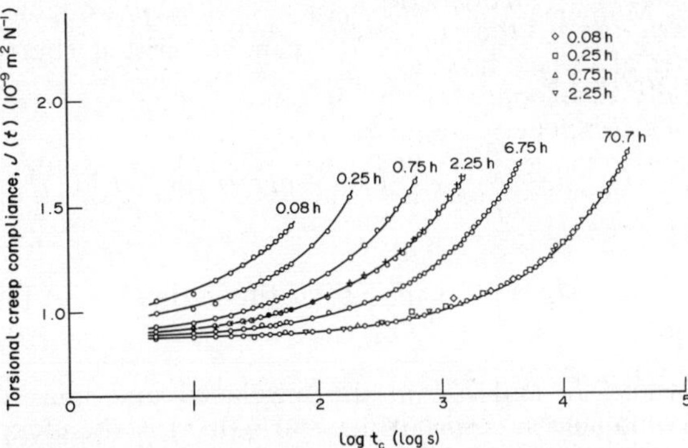

Figure 1
The torsional creep compliance $J(t)$ as a function of log(creep time) for atactic polystyrene at 75 °C (the maximum shear strain in the specimen for any creep curve was 0.025%; the data indicated by the crosses were obtained by repeating the 2.25 h aging time curve for a sample which had been aged previously for 70.7 h and then reheated above T_g; the creep curve for an aging time of 70.7 h includes data superposed by a horizontal shift from shorter aging times, as indicated in the inset)

elapsed between the first time the sample reaches the test temperature and the time of the relevant creep test. The aging time (in hours) is indicated as the parameter against each curve.

An important point to note in Fig. 1 is that the duration of the creep test is always very much less than the aging time, here by a factor of approximately five. It is an essential aspect of any study of aging that the stimulus (which here is the creep stress) must not significantly disturb the structural state of the polymer, which should be uniquely determined by the thermal history. Furthermore, there must be no significant aging taking place during the response to the stimulus, hence the restriction in Fig. 1 to limited durations of creep test. To meet these requirements, the most common experimental measurements of mechanical properties are creep at low strain and dynamic mechanical analysis.

The following characteristics of physical aging are evident from Fig. 1 and from the extensive data of Struik.

1.1 Shift of Timescale

According to Struik, the dominant effect of aging is to shift the timescale of the response of the polymer to longer times. This shift appears to occur without any change in the shape of the creep curve, so that the individual creep curves can be superposed approximately to produce a master curve. This is shown in Fig. 1 where all the creep curves have been shifted horizontally to superpose as best they can, but clearly not perfectly, onto the 70.7 h aging time curve.

Mathematically, this can be expressed as follows. The creep compliance $J(t)$ can be written (Ferry 1980):

$$J(t) = J_U + \int_{-\infty}^{\infty} [1 - \exp(-t_c/\tau)]L(\ln\tau)\, d(\ln\tau)$$
$$= J_R - \int_{-\infty}^{\infty} \exp(-t_c/\tau)L(\ln\tau)d(\ln\tau) \quad (1)$$

where J_U and J_R are the unrelaxed and relaxed compliances, respectively, and $L(\ln\tau)$ is the creep retardation time spectrum. If the whole spectrum shifts uniformly by a factor a_δ on aging, then the creep curve will simply shift along the $\log t_c$ axis by the corresponding amount $\log a_\delta$, provided that J_U and J_R remain unchanged; this last proviso is an essential requirement for horizontal shifting. The magnitude of the shift factor a_δ depends on the duration of the aging time t_a, and the relationship between them defines the double-logarithmic shift rate μ:

$$\mu = \frac{d(\log a_\delta)}{d(\log t_a)} \quad (2)$$

1.2 Reversibility

It has already been emphasized that physical and chemical aging are quite distinct: the former is reversible whereas the latter is irreversible. The reversibility of physical aging can easily be demonstrated by reference to Fig. 1. If the polymer which has previously been aged is heated above T_g, it will rapidly regain an equilibrium state. The same sample can then be cooled again to the aging temperature, whereupon the data of Fig. 1 could be reproduced as the sample ages once more. This is in fact illustrated in Fig. 1 by the creep curve for an aging time of 2.25 h, which has been repeated in this way and is indicated by the crosses. It is clear that the repeated creep curve is essentially identical to the original for the same aging time. Thus the effects of aging can be removed by reheating the polymer above the relevant value of T_g.

1.3 Temperature Range for Aging

There is a remarkable similarity between a wide range of amorphous polymers in respect of their values for the shift rate μ and of the dependence of μ on temperature. For aging temperatures in the region of T_g for each polymer, the shift rate approaches close to unity, while above T_g it falls rapidly to zero, indicative of the absence of aging in the temperature range above T_g where equilibrium is rapidly established. As the aging temperature is reduced relative to T_g, however, the shift rate decreases more slowly, approaching zero in the vicinity of the highest secondary transition (β relaxation). Therefore the important temperature range in physical aging is bounded by T_β (the temperature of the β relaxation) and T_g (the glass transition temperature).

2. Universality

The characteristic features of physical aging have been described above by reference to the creep behavior of an amorphous polymer. These same features are found, however, both in other materials and also in respect of properties other than mechanical properties. It is appropriate here to signal this apparent universality while emphasizing again that the cause of the changes in properties observed on aging is not necessarily the same for the different classes of material.

2.1 Materials

The overwhelming majority of studies of physical aging has been on amorphous polymers. Among these are included polystyrene, polyvinylacetate, polyvinylchloride, polymethylmethacrylate, polycarbonate and polysulfone, as well as a network polymer glass, epoxy resin. For semicrystalline polymers, however, since the early work of Schael (1966) attention has been concentrated principally on polypropylene. Some recent work (Struik 1987) has, however, extended the range to include low- and high-density polyethylene, polyethyleneterephthalate and polyamides, and suggests that physical aging is a much more complex phenomenon in semicrystalline polymers than it is in amorphous polymers. A major problem, in fact, with semicrystalline polymers is in separating effects due to changes in the amorphous regions, possible recrystallization and the development of internal stresses (Hutchinson and McCrum 1972).

Other amorphous materials in which aging is observed are inorganic and metallic glasses. This is entirely consistent with the many common features between all classes of amorphous materials. In fact, it was for inorganic glasses that Simon in 1930 first demonstrated the nonequilibrium nature of supercooled liquids below T_g, a concept that is fundamental to the generally accepted hypothesis for the origin of physical aging outlined below.

Interestingly, aging behavior similar to that characterized by Fig. 1 is displayed in materials for which the above concept is not immediately applicable, for example, polycrystalline metals, and in such diverse materials as sugar, cheese, bitumen and shellac (Struik 1978).

2.2 Properties

With regard to mechanical properties, the low-strain creep response such as that illustrated in Fig. 1 is particularly widely used. It is a useful approach since the creep deformation at small strains does not perturb the aging state of the material, whereas high strains can interact with the aging process to cause mechanically enhanced aging (Sternstein and Ho 1972). The use of dynamic mechanical analysis is also widespread, allowing the storage and loss moduli, G' and G'' respectively, and the loss tangent $\tan \delta$ to be determined as a function of aging time. In this case, approximate superposition of the curves of G' and G'' can be achieved by shifting along the log(frequency) axis, analogous to the shifts of creep curves along the $\log t_c$ axis. For the dependence of $\tan \delta$ on temperature, aging effects are only seen in the temperature region between T_β and T_g, with a reduction in $\tan \delta$ on aging at any temperature within this region.

The mechanical response of polymers at much larger strains is also affected by physical aging. Of particular importance here is the case of polycarbonate which, unlike amorphous polymers generally, is normally ductile below T_g. However, annealing polycarbonate in the temperature region just below its value of T_g (i.e., between approximately 90 °C and 120 °C) causes it to become embrittled, with reduction in the impact strength and breaking strain. Associated with this, and as a feature general to the aging of all polymers, the yield stress increases with aging time following a quench, implying that a greater amount of energy is necessary to initiate the molecular motions involved in the yield process.

Analogous to the changes in low-strain mechanical properties described earlier, the dielectric properties of polymers are also affected by aging. For example, the frequency at which the dielectric loss maximum occurs is observed to decrease with aging time, indicative of a shift of the dielectric relaxation spectrum to longer times, while the dielectric polarization is delayed on aging in the same way as is the mechanical creep response.

Other properties that are observed to change significantly on aging include optical properties, such as refractive index and birefringence, and properties involved in diffusion. Particularly in the latter case, the amount of gas sorbed or the rate of solvent uptake during swelling both decrease with increasing aging time. Such changes in the diffusion process have an important bearing on, for example, the dyeability of synthetic fabrics.

3. Physical Origin of Aging

The observation that physical aging occurs in many classes of material obviously has wide implications beyond the scope of this article. In confining the discussion to the aging of polymers, however, there are still some general features that require a physical interpretation. Qualitatively, and for amorphous polymers, a simple hypothesis for physical aging is generally accepted. Quantitatively, on the other hand, this simple hypothesis cannot explain some of the details of the physical aging process; nor can it easily explain the aging of semicrystalline polymers.

3.1 The Simple Hypothesis

The most commonly held view of the origin of physical aging is that the changes in mechanical and other properties that occur on aging are related to the changes in the structure or state of the polymer that occur on cooling from above to below T_g. Such structural changes are more important in amorphous rather than semicrystalline polymers since the glass transition is associated with molecular motions in the amorphous material, and this is the reason why most aging studies are carried out on amorphous polymers. The usual explanation of aging is outlined below with reference to the schematic diagram for an amorphous polymer shown in Fig. 2.

Figure 2 shows the variation in either the volume V or the enthalpy H of an amorphous polymer as a function of temperature. On cooling from equilibrium at a temperature above T_g, the polymer will, over a relatively narrow temperature interval, gradually depart from an equilibrium liquid-like state and approach a nonequilibrium glassy state. This liquid-to-glass transition is characterized by T_g, conventionally defined by the intersection of the extrapolated glassy line with that of the equilibrium

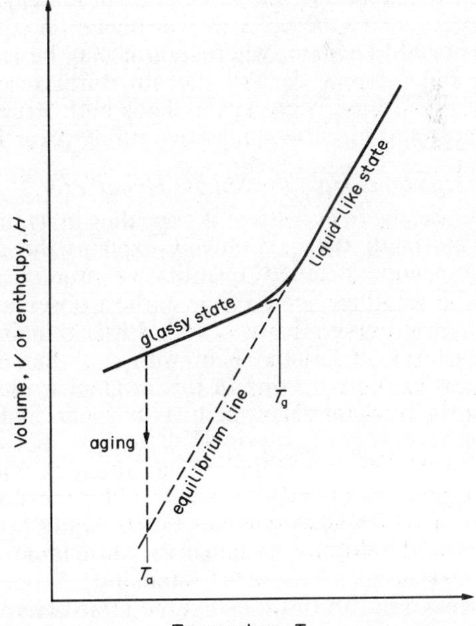

Figure 2
Schematic diagram showing the variation of volume or enthalpy on cooling an amorphous polymer from above to below the glass transition temperature

liquid, as shown in Fig. 2. However, T_g is dependent on the cooling rate, since the slower the cooling rate the more time is available for the relevant molecular motions to occur in order to accommodate to the temperature change. Hence T_g decreases with decreasing cooling rate, typically by about 2 °C per decade.

In the glassy region, below T_g, the polymer will have a nonequilibrium structure determined by the cooling rate, and which can therefore be characterized by the value of T_g pertaining to that cooling rate. This structure is often referred to as being "frozen-in." On annealing, or aging, at a temperature below T_g, however, the glass will spontaneously tend towards an equilibrium state of lower volume or enthalpy. This effect is known as structural recovery of the glass, and is widely believed to be the origin of the phenomenon of physical aging.

Certainly this hypothesis can account qualitatively for the observed phenomena. First, during structural recovery the specific volume decreases towards its equilibrium value, and hence the molecular mobility is reduced. Thus any property that depends on the molecular mobility, such as the creep compliance in Fig. 1, will be affected in such a way that the response is shifted to a longer timescale. Second, at temperatures more than 10–15 °C below T_g, the time required for structural recovery to approach equilibrium rapidly increases beyond that which is experimentally accessible. This explains why aging persists to very long times. Third, after aging, the glass may be reheated to a temperature above T_g where the polymer will regain an equilibrium liquid-like state, whereupon it may be cooled once more below T_g and the structural recovery process will then begin again. Thus both structural recovery and physical aging are entirely reversible.

3.2 Deviations from the Simple Hypothesis

While the above hypothesis is appealing in its simplicity and in that it qualitatively explains the aging phenomenon, detailed quantitative studies have found it wanting. Some of the earliest observations of physical aging (Kovacs et al. 1963) remain unsatisfactorily in disagreement with this simple hypothesis in some details. In this original work, the dynamic mechanical properties in shear and the volume recovery of polyvinylacetate were measured simultaneously. It was found that, although reasonable superposition could be achieved for the storage modulus G', the loss modulus G'' could not be well superposed to form a master curve. More important, the deviations in G'' were systematic. A possible implication of this result is that the retardation time spectrum does not remain invariant on aging, thus calling into question the usual assumption of thermorheological simplicity.

More recently, some very precise experiments (Chai and McCrum 1980, 1984, McCrum 1984) have examined the effects of aging on both an amorphous (cross-linked copolymer of acrylonitrile and butadiene) and a semicrystalline polymer (polypropylene). Systematic deviations from the predictions of the simple hypothesis have suggested two important modifications to explain why exact superposition of creep curves as a function of aging time cannot be achieved by simple horizontal shifting. The first is that the changes that take place during aging occur sequentially, the shortest retardation times in the spectrum shifting first, with the result that the distribution of retardation times changes shape. Under such circumstances, the creep curves at different aging times will not superpose exactly.

The second modification is that aging not only affects the retardation time distribution but also leads to changes in the limiting compliances J_R and J_U. In particular, following a quench from a higher temperature to the aging temperature, a nonequilibrium value of J_R–J_U is frozen in; on aging, J_R–J_U reduces towards its equilibrium value, with a consequent reduction of creep rate as a function of aging time. Even if the distribution of retardation times remained unchanged on aging, meaningful superposition of the creep curves could again not be achieved by horizontal shifting alone.

In the light of these observations, it has been suggested that it is likely that physical aging proceeds by a combination of both of the above effects: a sequential shift of the retardation time spectrum to longer times, and a diminution of the relaxation magnitude with increased elapsed time following a quench.

3.3 Semicrystalline Polymers

The simple hypothesis discussed above has been formulated with respect to an amorphous polymer, in which the process of physical aging occurs below T_g. The observation of physical aging in other materials, therefore, and in particular in semicrystalline polymers, deserves some comment. In the absence of any changes in the degree of crystallinity, the physical aging of a semicrystalline polymer below its T_g could be explained on the basis of changes occurring in the amorphous fraction of the polymer, in exactly the same way as for a fully amorphous polymer. However, it is well established that physical aging occurs, for example, in isotactic polypropylene at 40 °C, well above its value of T_g of around −15 °C.

A possible explanation for the occurrence of aging above T_g in semicrystalline polymers lies in the concept of an "extended glass transition" (Struik 1978). The fully amorphous polymer consists of molecular chains randomly arranged in a supercooled liquid-like structure with no external constraint to molecular rearrangement. However, the amorphous content of semicrystalline polymers exists in the interlamellar or interspherulitic regions, in which the molecular segments within the amorphous regions are constrained by those segments of the same molecule lying within crystalline domains. The degree of constraint will decrease from a maximum at the crystalline–amorphous interface to a minimum at the heart of the amorphous regions. The effect of such a constraint is to increase T_g, since a greater thermal energy will be required in order to allow molecular rearrangements to occur. Thus the varying degrees of constraint on molecular motion lead to an extension of the glass transition region in semicrystalline polymers such that it covers a range of temperatures from the conventional T_g of the fully amorphous material to temperatures more than 100 °C higher.

The physical aging of semicrystalline polymers is therefore clearly more complex than for amorphous polymers. Indeed, Struik (1987) has shown in a comprehensive study of a wide range of semicrystalline polymers not only that aging effects in these materials can be described using the concept of an extended glass transition but also that as many as four characteristic regions of behavior can be identified, depending on the actual aging temperature within the extended glass transition interval.

3.4 Summary

There is clear qualitative support for the simple hypothesis that physical aging, typically manifest as a change in the mechanical creep response, is related to structural recovery, measured in terms of either volume or enthalpy. However, a quantitative description of physical aging remains more elusive. The difficulty arises when experimental results are examined in detail, since it is found to be impossible to obtain exact superposition of the data. There are two possible explanations for the origins of these discrepancies: sequential relaxation and the freezing-in of nonequilibrium values for the limiting compliances.

4. Importance

The observation that physical aging occurs in both amorphous and semicrystalline polymers indicates its relevance to all thermoplastics. Since thermoplastics are most commonly formed in the melt state, by processes such as extrusion and injection molding, the importance of physical aging becomes apparent immediately in the context of the above discussion. Physical aging was described above in terms of property changes following a quench, and the melt processing of thermoplastics involves just such a quench. Thus thermoplastic components will age from the moment that they are fabricated.

Clearly this has important consequences for design, especially in terms of strength and dimensional stability. An appreciation of the changes in properties that occur on aging is essential, particularly for components that are to remain in service for a long time. Such an appreciation will be gained when the physical origin of aging is better understood, and the aging process can be quantitatively predicted.

See also: Aging of Elastomers; Degradation, Taxonomy of; Glass Transition; Polymers: An Overview of Technical Properties and Applications

Bibliography

Chai C K, McCrum N G 1980 Mechanism of physical aging in crystalline polymers. *Polymer* 21: 706–12

Chai C K, McCrum N G 1984 The freezing-in of non-equilibrium values of the limiting compliances as a mechanism of physical aging. *Polymer* 25: 291–8

Ferry J D 1980 *Viscoelastic Properties of Polymers*, 3rd edn. Wiley, New York

Hutchinson J M, McCrum N G 1972 Effect of thermal fluctuation on creep of polyethylene. *Nature* 236: 115–17

Kovacs A J, Stratton R A, Ferry J D 1963 Dynamic mechanical properties of polyvinyl acetate in shear in the glass transition temperature range. *J. Phys. Chem.* 67: 152–61

McCrum N G 1984 Sequential relaxation as the mechanism of physical aging in amorphous polymers. *Polym. Commun.* 25: 2–4

Schael G W 1966 A study of the morphology and physical properties of polypropylene films. *J. Appl. Polym. Sci.* 10: 901–15

Sternstein S S, Ho T C 1972 Biaxial stress relaxation in glassy polymers: Polymethylmethacrylate. *J. Appl. Phys.* 43: 4370–83

Struik L C E 1978 *Physical Aging in Amorphous Polymers and Other Materials*. Elsevier, Amsterdam

Struik L C E 1987 The mechanical and physical aging of semicrystalline polymers: 1 and 2. *Polymer* 28: 1521–42

J. M. Hutchinson

[Aberdeen University, Aberdeen, UK]

Plasma Spraying for Protective Coatings

Shortage of raw materials, and the need for increased lifetime and reliability are the driving forces for using surface protection on machine components. Coatings are used to produce specific surface properties that cannot be provided by the base material itself. The characteristics for choosing protective coatings have to be defined in connection with the forms of surface load, suitable deposition methods and applicable coating materials. The protective coatings are characterized by their physical properties and stability during different forms of loading.

The condition of the coating material directly prior to the coating formation classifies the deposition method: vapor (e.g., physical vapor deposition or chemical vapor deposition), solution (e.g., electroplating), solid state (e.g., sintering) or droplets

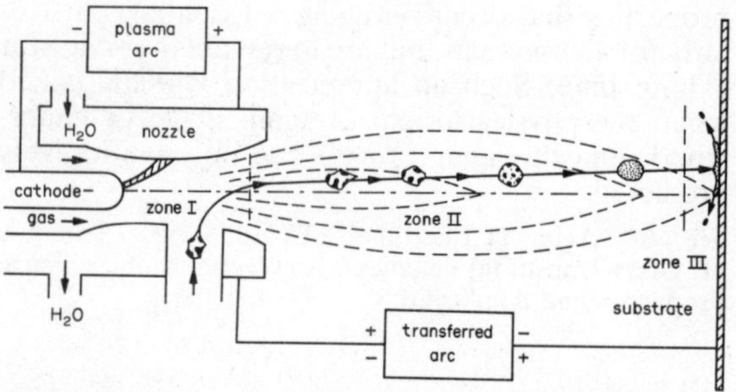

Figure 1
Schematic of plasma spraying layout with pressures of 1000 mbar for the atmosphere, 900 mbar for the inert gas and 50 mbar for the vacuum

(as in all methods of thermal spraying (GDR standard 1975)). Describing the form of the starting material and depending on the energy source that is used to produce material droplets, different methods of thermal spraying can be distinguished: wire arc spraying, powder flame spraying, flame shock spraying, laser spraying and plasma spraying. Recognition as the inventor of the thermal spray process is accorded to M U Schoop (1910), who produced the first sprayed coating, using an electric arc as an energy source for metal wire melting. Vapor deposition is preferred for the production of chemically pure thin films. Coatings with a thickness of 20 μm or more have been and remain the domain of thermal spraying due to the high deposition velocity of up to 0.2kg min^{-1} that can be attained by the liquid material droplets.

1. Technology

The high energy density within a plasma has rapidly promoted plasma spraying to a leading position among thermal spraying methods. Practically all materials avilable in powder form can be deposited as a coating. The physical conception of plasma was introduced in 1923 by Langmuir for the special gas condition in which atoms or molecules are ionized and are, together with free electrons, responsible for the electrical conductivity of the gas. This condition is sometimes referred to as the "fourth state of matter." Plasmas vary from a low density of ionization up to energy intensities of 10 W m^{-3}.

For plasma spraying, a high speed plasma torch is used to melt powder particles, as shown in Fig. 1. An electric arc is initiated by a high-frequency pulse between two water-cooled electrodes, a tungsten cathode and a nozzle-shaped anode. The tungsten cathode is thoriated to lower the electron work function. A gas is fed into the area where the arc is "burning" to stabilize and compress the arc into the anode by means of a continuous gas flow. The arc heats up, accelerates and ionizes the gas which creates a plasma flame or plasma jet. The nozzle configuration of the anode has been optimized according to the pressure relationship between the gas which is fed into the burning arc and the surroundings of the plasma flame outside the nozzle. The goal is the increase of the plasma energy density to improve the arc–gas heat transfer. The radius of the ionized part of the plasma flame is mainly a function of the arc current and should remain constant. Around the electrically conductive core of the plasma flame colder gas is flowing. Therefore, a radial drop of the electronic and thermal energy directed toward the water-cooled anode wall is ensured. Owing to the significantly lower gas temperature within the boundary layer belonging to the anode, a thermal pinch effect constricts the plasma flame. This constriction is increased additionally by a magnetic pinch, caused by a magnetic field due to conductive flux lines from the core of the plasma region to the outer area with lowered electrical conductivity.

Depending on the environmental conditions of the plasma torch, different methods of plasma spraying are distinguished: atmospheric plasma spraying (APS) or inert-gas plasma spraying (IPS) are used at normal pressure levels, and vacuum plasma spraying (VPS) is used at pressures of 20–200 mbar within a closed chamber. In the latter case, the plasma gas is additionally accelerated by its expansion into the reduced pressure environment.

Particle–plasma and particle–substrate interactions are responsible for the coating quality. Powder particles are fed into zone I of the plasma stream at various positions. Location, angle and diameter of the injection port must be varied for each individual material being sprayed. The injected particles are simultaneously accelerated and heated while passing through zone II. They travel through the extremely high temperature, enthalpy and velocity gradients of the plasma jet. Zone III denotes the part of the jet that is affected by the presence of the substrate and particle impact.

The degree of particle melting is affected by the experimental trajectories through the plasma flame. The rapidly moving particles take heat from the plasma flame by heat absorption due to ion recombination on the particle surface, supported by absorption of ultraviolet radiation. The recombinations of atoms to molecules, which also take place on the particle surface, are very effective. This is the main reason for the use of a molecular gas (N_2, H_2) as a constituent of the plasma flame, which mainly consists of argon and helium. Argon has the advantage of being easily ionized and provides a very stable arc at a low operational voltage. However, the heat content of the pure argon plasma flame is much less than can be achieved by using diatomic gas in addition. Alternatively, for a given plama energy, the temperature of the inert-gas plasma flame is much higher than that for nitrogen or hydrogen (Chang and Szekely 1982). The required flame temperature can be adjusted by the plasma arc current and by the amount of plasma gas. Using mixtures of different plasma gases, the heat content is optimized for the spray powder melting of different materials. Ion-to-molecule and atom-to-molecule recombinations, as well as ultraviolet radiation, are active at particle surfaces, just at the right place where energy for the melting is needed. Thus, it can be understood why melting temperature, heat capacity and thermal conductivity together determine the resistance to melting.

The fully molten state of the injected particles, before they impinge on the substrate surface with sufficient kinetic energy, determines the coating quality. The coating is formed by the continuous buildup of successive layers of liquid droplets flattening on impact (see Fig. 2) and this is responsible for the lamellar structure of sprayed coatings. The rapid solidification of the material after spreading determines the microstructure of the deposit. In some cases, a cooling rate of approximately 1 000 000 °C s^{-1} is sufficient to form amorphous structures.

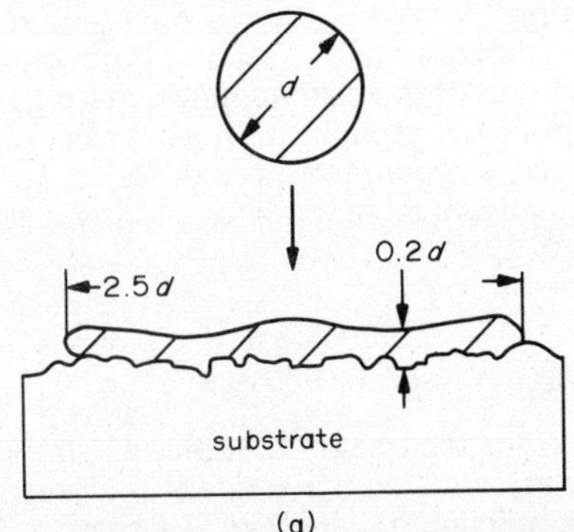

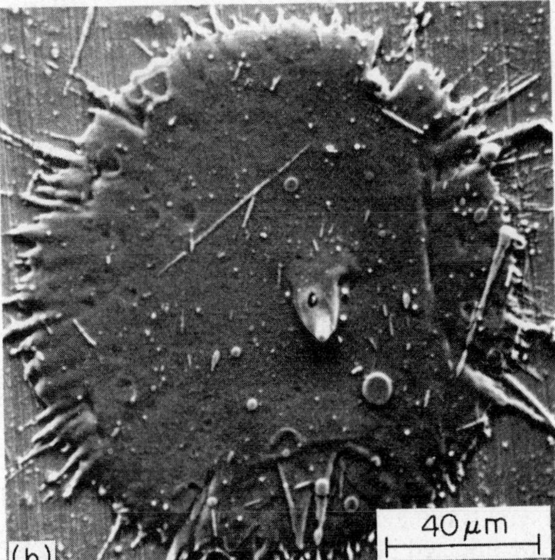

Figure 2
Liquid droplet flattening: (a) change of particle size on impact, and (b) scanning electron micrograph of a powder particle after impact

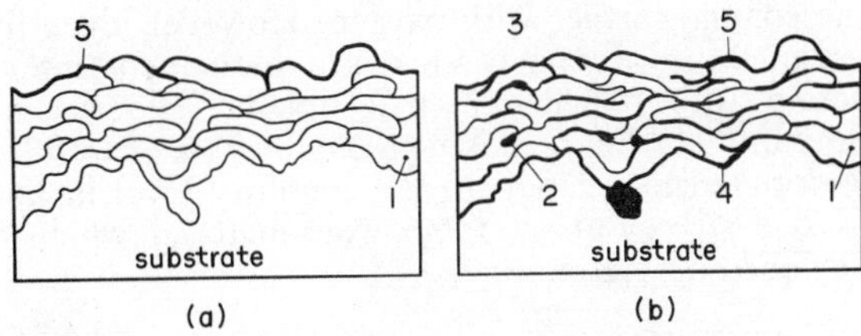

Figure 3
Differences in coating structure, produced by: (a) VPS and (b) APS, showing (1) coating lamella, (2) pore, (3) embedded oxide layer, (4) oxide layer at the substrate–coating interface, and (5) surface oxidation

2. *Process Parameters*

All of the spray parameters involved in this complex process and influencing the quality of the coating can be considered in the six interrelated groups that will be discussed in this section.

2.1 Substrate Condition

The substrate material, surface condition, roughness, cleanness and temperature prior to and during the coating do not only have an affect in the boundary layer between substrate and coating. For the VPS process, the surface can be additionally cleaned by using a transferred arc (see Fig. 3) prior to spraying (Muehlberger and Kremith 1979). This removes the oxide layer from the grit-blasted surface and results in a definite improvement in adhesive coating strength. Favorable conditions are also achieved for an interdiffusion process between the substrate and the coating material.

2.2 Flame Characteristics

The choice of plasma gases, the total gas amount, the arc and transferred arc currents, and the flame expansion determine the flame length and gas velocity and, therefore, the particle speed and the spray distance.

2.3 Plasma–Flame Interaction

Air-sprayed coatings of reactive materials always have some oxides. The surrounding air and the plasma flame strongly interact. The argon content of the flame is rapidly decreased owing to the penetration of cold air into the plasma flame. Operating the coating process in an inert gas environment limits the reaction between the molten particles and the coating surface with oxygen. However, even in vacuum the plasma torch has a permanent interaction with the flame environment. The partial pressure of the reactive gases must be carefully considered in relation to the impurity level in the used plasma gases and the powder material, which is degassed during melting.

2.4 Powder Characterization

The size of the powder particle and its surface-to-volume relationship are very important in terms of the particle–plasma interactions (see Fig. 4). Assuming that the physical properties of the spray powder are known, the particle size distribution must be matched to the demands of the spray process. The energy level required to melt the coarser particles should not be sufficient to vaporize the finer size particles prior to impact on the substrate. Such particles clearly cannot become part of the deposit. They only reduce the spray efficiency as do unmolten large particles, which have a high probability of bouncing off the substrate surface. Spray efficiency is the relationship between powder feed rate and weight gain of the substrate during spraying. Typical values are between 70% and 80%. However, the spray efficiency on small parts is dramatically reduced below this level, because of the number of spray particles that escape. Powder characterization commences with the selection of the starting material, keeps the powder manufacturing as free as possible from impurities, defines the morphology, structure and chemical composition of each particle, and ends with the selection of the required particle size distribution.

2.5 Powder Injection

The flight path of the individual particles determines the spray pattern in relation to the geometric center of the plasma flame and its geometric configuration, and the particle density distribution within the flame. They are strongly dependent on the chosen injection conditions. The center of the spray pattern can be characterized by completely molten (in some cases superheated) material. Unmolten particles are always found in the periphery. Particles that are too small may not have enough kinetic energy to enter the plasma flame. The plasma opposes particle penetration with a resistance that requires particle momentum, which is transmitted by the carrier gas. Carrier gas flow rate needs to be highest for the smallest particles. This causes high momentum transfer to bigger particles in the size distribution; these then cross through the flame and remain unmolten due to reduced dwell time. Optimum particle size distribution depends on the specific weight of the sprayed material and its resistance to melting. Both are influenced by the surface-to-volume relationship of the powder particles. Typical ranges of particle size are 5–25 μm, 10–40 μm and 20–63 μm, depending on the required properties.

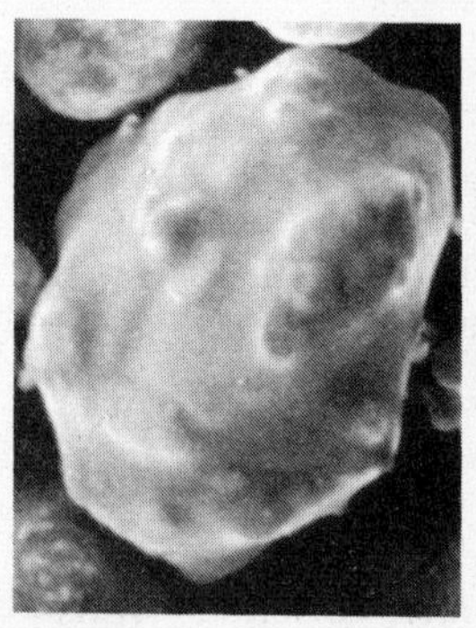

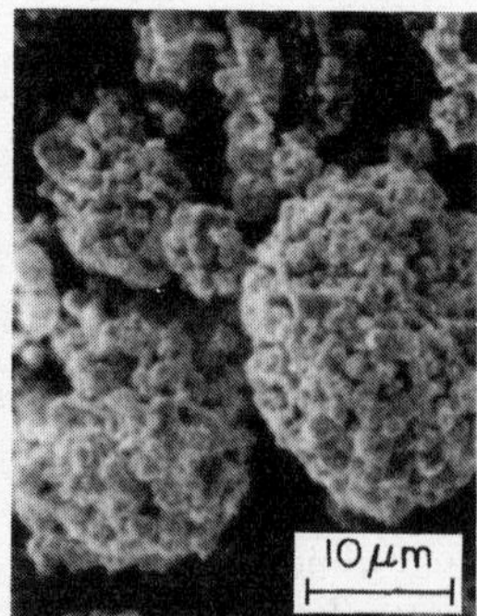

Figure 4
Scanning electron micrographs of two (WC + Co) powders with the same particle size but different surface–volume relationships

2.6 Spray Movement

The coating process is carried out using a concentrated, centered and rotation-symmetric spray stream. In order to ensure defined coating thickness distribution on complicated workpieces, robotic systems have been introduced to replace direct human involvement in the spray process (Henne and Nussbaum 1980).

3. Coating Quality

In most cases the quality of a plasma-sprayed coating is correlated with its density. Pores are not desirable and limit the lifetime of a coating. Porosity in coatings intended for corrosion protection, by permitting ingress of corrodants, cannot protect the coating–substrate interface against chemical attack. The density of plasma-sprayed coatings depends on three important factors:

(a) the momentum of the spray powder particles on impact,

(b) whether the injected powder particles are fully molten as they impinge on the substrate surface, and

(c) the amount of entrapped gas in the hollows, cavities and niches that has to be displaced or compressed during the continuous coating growth.

Under VPS conditions the partial pressure is much lower. Thus, dense coatings can be VPS sprayed (Gruner 1988) even if only a small fraction of the injected powder particles are molten as they strike the substrate surface (see Fig. 5).

Coating quality control is still carried out by using a destructive cross-sectional preparation technique despite the introduction of modern nondestructive

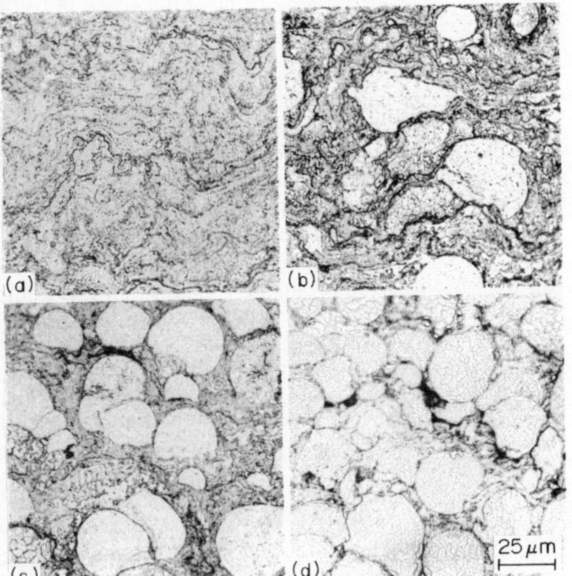

Figure 5
Density and number of unmolten particles of a VPS-sprayed cobalt–chromium alloy for the following plasma flame energies: (a) 48.6 kW, (b) 39.0 kW, (c) 28.2 kW, and (d) 19.8 kW

test methods. Metallographic evaluation requires careful preparation in order to avoid the introduction of artifacts. Many of the individual spray parameters are independent and must be optimized empirically by making their influence on particle melting visible. Diagnostics of the interactions, measurements of particle and gas velocities and temperature distribution within the plasma flame are very helpful in trying to understand this complex process. Thermal transient thermography (developed at Harwell laboratories, Harwell, UK), is one of the latest developments for nondestructive inspection of coatings for subsurface defects and other structural features (Reynolds and Wells 1984). Additional information, especially about the behavior of the individual particle on impact and its crystal structure as affected by the rapid solidification, are carried out using transmission electron microscopy (Safai and Herman 1977).

4. *Limitations*

Besides the physical limitations on the industrial use of these coatings, plasma spray technology also has certain process limitations. Only materials available in powder form and possessing a stable liquid phase can be used to produce coatings. Nevertheless, for spraying composite coatings, special structures can be produced by melting one material and building the other as solid particles into the coating.

The plasma spray process is considered to involve a line-of-sight flow of material. Therefore, only surfaces that are directly "visible" to the plasma flame can be coated. A relationship exists between the coating density and the angle of the particle impact. The more this angle diverges from the vertical, as more porosity is produced, the lower the deposition efficiency.

Plasma spray technology requires a minimum spray distance. Particles only become molten after a certain dwell time in the plasma flame. Thus, the coating of internal bores is only possible with special torches, designed to enable the inside spraying of bores down to 25 mm in diameter (Gruner and Müller 1987).

Coating and base materials can be selected and combined in a large number of ways. However, not all substrate materials mechanically resist the highly energetic plasma flame. Sufficient surface cooling during spraying is necessary and very effective if the liquefied gas cooling technique is used.

The plasma spray process itself puts no upper limit on the coating thickness; this fact is the key for the production of free-standing forms by spraying. Other technical reasons explain a thickness limitation on individual substrate materials. For example, if the temperature varies and the base material and coating exhibit different coefficients of thermal expansion this will limit coating thickness. The lowest possible coating thickness is clearly defined by the diameter of the spray powder. Using a particle size distribution of 5–20 μm, a completely closed layer is reached of approximately 10 μm thickness, assuming that numerous particles overlap after spreading on the surface.

5. *Applications*

The use of plasma spraying has resulted in new coating properties. Also, well-known substrate materials have been introduced into new areas of application by exploiting plasma-sprayed surfaces. Steady advances in processing conditions have improved the coating properties further and opened up an almost unlimited reservoir of applications for dense coatings or coatings with adjusted porosity (see Fig. 6). Tailoring the flame energy to the spray powder being used results in the ability to vary the coating structure over a wide range between very dense and extremely porous. Dense and oxide-free M–Cr–Al–Y alloy coatings (where M is cobalt, nickel or iron) (see Fig. 6a) provide the necessary resistance against hot corrosion and oxidation on superalloys. Alternatively, a thermal barrier coating, used to protect metallic parts in high-temperature applications and produced by using brittle ceramics, will only exhibit the required thermal shock resistance if the coating is produced with an adjusted internal microporosity (see Fig.

6b). The coating survives the thermal shock because the pores act as crack stoppers. For the cementless implantation of endoprostheses, very rough and porous titanium coatings (see Fig. 6c) enhance the interactions between bone and the prosthesis surface (Winkler-Gniewek et al. 1988). Large surfaces, large open cavities and pores of different sizes internected within the coating enable the bone to penetrate deeply into the coating for a rapid anchoring in order to improve the mechanical fixation.

Reactive metals can be plasma sprayed without oxidation. The VPS process provides suitable surroundings for the manufacturing of very pure coatings, whether dense or not, as the titanium coating for medical application shows. Oxide-free coatings exhibit the same ductility as do corresponding semifinished wrought products. Other physical or chemical properties can be directly correlated with the oxygen content of the coating. Reactive metals can also be APS sprayed, so long as no embedded oxide layers affect the lifetime of the coating. An example of this is the nickel–chromium–bond coating for thermal barrier applications at lower temperatures.

Figure 6
Micrographs of plasma-sprayed coatings: (a) high density, (b) adjusted microporosity, and (c) porous coating

The high energy density of the plasma flame is useful for spraying refractory metals or other materials with a high melting point. Due to their higher bond strength and improved structural stability VPS coatings are preferred in most cases. An example of a novel application is the protection of metallic surfaces against corrosion by liquid metal. Useful coatings include TiB_2, mullite and Y_2O_3, depending on the substrate material. Dense ZrO_2 coatings with very low gas penetration are a good choice for the electrolyte in solid-state fuel cells. Another use is the oxidation protection of graphite steel-melting electrodes by a dense silicon coating. Matching the spray powder to the demands of the coating process and adjusting the flame temperature and enthalpy according to the requirements of the spray materials are necessary for every new application.

Extremely interesting coatings can be fabricated by the combination of different materials that cannot be alloyed under normal conditions. Multiple injections of powders into the plasma flame or the use of blends or agglomerated materials provide composite coatings. Adjusted physical properties can be produced with preselected composition. The possible range of such metal matrix or oxide matrix composites is defined by the choice between two types of structures; that is, either both materials are molten as they impact on the substrate surface or only one material melts and acts as a binder for the other material, which is built into the coating as solid particles. For example, the distribution of alumina particles in a steel coating reaches at a level of 45 wt% Al_2O_3, the Vickers hardness of a pure alumina coating (500 HV) while retaining the coating ductility characteristic of the metal matrix (Gruner 1986). Similarly, the electrical properties can be enhanced by further increase of the alumina content. The range of 70–90 wt% of alumina within a M–Cr–Al–Y structure seems to be very important in terms of the production of thick-film resistors. Also of particular interest are metal–plastic mixtures, which open new perspectives for sliding and abradable coatings.

6. Conclusion

The use of plasma-sprayed coatings allows a component design that matches the mechanical requirements and protects the surface against environmental attacks. In-depth studies of plasma physics and coating technology ensure the reproducibility of the quality and a high efficiency in large-scale coating production using multifunctional plasma-spray systems. The protection of textile machinery parts against wear, the prevention of high-temperature corrosion and oxidation of airplane engine parts, and the improved sliding properties of piston rings are a few examples of the successful results attained with plasma-sprayed protective coatings.

See also: Metallization of Advanced Ceramics: Thermal Spraying Techniques; Metallization of Advanced Ceramics: Vacuum Deposition Techniques

Bibliography

Chang C W, Szekely J 1982 Plasma application in metals processing. *J. Metals* 2 (February): 57–64

GDR standard 1975 Thermal spraying. DIN No. 32,530

Gruner H 1986 VPS sprayed composite coatings. *Proc. 11th Int. Thermal Spraying Conf.*, Advances in Thermal Spraying. Pergamon, Toronto, pp. 73–82

Gruner H 1988 Dense oxide coatings. *Proc. Advanced Thermal Spraying Technology and Allied Coatings 88*. High Temperature Society of Japan, Osaka, pp. 265–70

Gruner H, Müller M 1987 US Patent No. 4, 661, 682

Henne R, Nussbaum H 1980 *Surfair IV*. Rencontres Techniques, Paris

Muehlberger E, Kremith R D 1979 US Patent No. 4, 328, 257

Reynolds W N, Wells G W 1984 Video-compatible thermography. *Br. J. Non-Dest. Test.* 26: 40–4

Safai S, Herman H 1977 Microstructural investigations of plasma-sprayed aluminum coatings. *Thin Solid Films* 45: 295–307

Schoop M U 1910 GDR Patent No. 233, 873

Winkler-Gniewek W, Gruner H, Stallforth H, Ungethüm M 1988 Structure and properties of VPS coatings in medical technology. *1st Plasma Technik Symp.*, Vol 3. Plasma-Technik AG, Wohlen, Switzerland

H. Gruner
[Plasma-Technik AG, Wohlen, Switzerland]

Plastics for Packaging

Approximately 35% of plastics produced in the developed countries is consumed by packaging, which is the biggest single use of these materials, amounting to US$19.9 billion in 1986. Of this quantity, 71% is accounted for by the three polyolefines: low density polyethylene (LDPE), high density polyethylene (HDPE) and polypropylene, which thus dominate. Beyond these, the next two major plastics are polystyrene which accounts for 12% and polyethylene teraphthalate (PET) at 6.4%.

The polymers used in packaging fall into two main price groups, depending on the complexity of their manufacture and the relative cost of the chemical intermediaries used. The simpler and cheaper materials are the polyolefines, followed by polyvinyl chloride (PVC) and polystyrene. Next are the nylons, PET and acrylonitrile–butadiene–styrene copolymer (ABS). The more expensive speciality polymers developed for engineering applications are not generally used for packaging. In some cases, more demanding applications can justify more expensive materials. A survey of the properties and uses of the main plastics used in packaging is given in Table 1.

The prime function of packaging is to retain all the contents of the product packed as near as possible in the state in which they leave the factory and appropriate to the expectations of the final customer. This implies protection from physical damage by external agents such as impact, corrosion, degradation by light and heat, and abuse in transit and storage; it also requires noninteraction chemically between the contents and container materials. There must also be no migration of the constituents of the packaging materials into the contents, nor absorption of the contents into the packaging material. The material used for the packaging must prevent or minimize the passage of gases such as oxygen and carbon dioxide inwards or outwards, and water vapor except in special applications.

During the processing of packaged foods and drinks, temperatures up to 125 °C for a period of up to an hour may be used to sterilize the contents. Alternatively, the pack may be subjected to deep-freeze temperatures as low as −40 °C for periods of months. An important function of packaging is to present an attractive image and to convey information about the contents. This requires the surface to be printable or capable of being labelled. It is against this background that the use of packaging should be viewed and the choice of a specific polymer and its processing selected.

Plastics are preferred to other materials by virtue of their better economics, the freedom with which they can be molded to a wide variety of shapes, and their lightness compared with metal or glass—leading to economies in transport and often savings in space. The moisture- and gas-barrier properties of plastics are not as good as those of metal and glass but are adequate for most applications, or can be enhanced by coating or by the use of special barrier polymers. If dropped or otherwise abused, plastic will perform much better than glass which also represents a safety hazard when broken. Many plastics have high impact strength and resistance to tearing or denting. Metal containers have better barrier and good impact strength but are subject to corrosion and can only be produced in a limited variety of shapes, usually cylindrical or rectangular.

Compared with paper and board containers, plastics again have the advantage of the much greater variety of shapes that can be used. Board containers are usually rectangular or cylindrical and require a lining of plastics to make them resistant to water and to render them sealable.

Table 1
Properties of plastic materials used in packaging

Material	Density kg m^{-3}	Material softening °C	Oxygen permeability[a] cm^3 mm m^{-2} d^{-1} bar^{-1} at 20 °C	Water vapor transmission g mm m^{-2} d^{-1} bar^{-1} at 20 °C	Oil/grease resistance	Solvent chemical resistance	Impact strength IZOD	Clarity	Uses in packaging
LDPE	917–932	98–115	190	0.6	good	good except nonpolar solvents	very good	fair	films, bags, coatings, moldings, bottles
LLDPE	918–960	122–124	190	0.6	good	good except nonpolar solvents	very good	fair	films, bags
HDPE	950–905	130–137	60	0.2	good	good except nonpolar solvents	very good	poor	bottles, moldings, bags, containers
Polypropylene	900–910	165–175	70	0.2	good	good except nonpolar solvents	fair–good	good	films, bags, bottles, containers, caps, tubes
PVC	1300–1580	75–105	4	1.0	good	good for nonpolar solvents	fair–good	good	films, bottles, containers
Polystyrene GP	1050	74–105	160	2	fair–good	poor	poor	good	containers, expanded foam
High impact	1030–1060	93–105	150		fair	poor	fair–very good	poor	
ABS	1020–1060	88–120	40	5	fair	poor	good–very good	poor	containers
PET	1290–1400	260–270	2.7 (amorphous) 1.4 (crystalline)	1.0	good	good	good	very good	bottles, films
Polycarbonate	1200	306–316	80	5.5	good	good	excellent	good	bottles, films
Nylon 6.6	1130–1150	255–265	2	2.5	good	good	good	fair	bottles, films
Ionomer	930–960	81–96	190	0.6	good	good	very good	good	adhesive–heatseal layer with nylon, films
			RH 0% 100%						
PVdC Extrusion Grade	1650–1720	160–172	0.06 0.06	0.04	good	fair		good	barrier layer
EVOH 30% Ethylene	1100–1200	140	0.003 0.3	1.0	very good	good		good	barrier layer
Nylon MXD6	1220	243	0.6 0.15	1.0	good	fair		good[b]	barrier layer
Acrylonitrile styrene	1130	110	0.40 0.40	2.0	good	fair		fair	bottles, films, containers

a Standard metric units; standard US units are cm^3 (10^{-3} in) (100 in^2)$^{-1}$ d^{-1} atm^{-1} at 68 °F with (US units) = 2.54 × (metric units) b Poor when crystalline

The polymers used in packaging have properties that are broadly adequate for packing most classes of product and the ease with which different shapes can be produced and linked to specific market images is a major strength compared with other materials. However, they do have a number of weaknesses. Their relatively open structure can lead to significant gas and water vapor transmission, as well as absorption of some molecular species from the contents with a resultant change in the aroma and flavor of foodstuffs and beverages. In addition, some organic solvents can be absorbed and this will lead to a reduction of strength or to swelling. Most polymers are sensitive to degradation in some degree by heat and light, particularly ultraviolet frequencies.

Plastics are inert chemically and therefore resist attack from most substances. This can make it more difficult to obtain adhesion, and special preparation of the surface of nonpolar polymers is necessary to allow the application of inks or adhesives. Such polymers are satisfactorily treated by the brief application of a gas flame or by corona electrical discharge. Thermal and light degradation have been defeated by the introduction of durable and nonmigrating thermal and ultraviolet stabilizers. Barrier deficiencies have been improved markedly, both by the development of coatings and by the addition of special barrier polymers in multilayer structures.

Nearly all the polymers used in packaging have softening and melting temperatures in the range 70–270 °C. This allows them to be processed by melting and then extruding from a screw extruder or by injection molding. In both cases, the material is cooled rapidly and at the same time shaped to its desired form, be it sheet, film, container or bottle. In some cases, extra strength is imparted to the material by stretching above the glass transition temperature in the solid state. This resultant extra degree of molecular orientation and crystallinity of the polymer also improves the gas-barrier properties.

1. Bottles

1.1 Choice of Materials

One of the most common uses of plastics in packaging is the manufacture of bottles in a wide range of sizes from 25 ml to 5 l or more. Currently some four billion plastic bottles are used annually in the UK and about 38 billion in the USA. The products packed cover an enormous range, including household cleaners, industrial chemicals, sauces, beverages, water, milk, pharmaceuticals and cosmetics. Some are dry, some water based and some contain solvents or corrosive liquids; some are toxic.

The polyolefines are the first choice for a large number of products, since for minimum cost they give bottles that are strong, flexible and, in the case of polypropylene, fairly transparent. The water vapor barrier is excellent and they are resistant to all but the most corrosive concentrated chemicals. They are however, prone to absorb essential oils from some foods, which can result in a loss of flavor and similarly a loss of aroma from cosmetics. Equally, the rather poor gas-barrier properties of the polyolefines mean that they are not used alone for oxygen-sensitive foods, nor for highly carbonated beverages where the useful storage life would be too short. For products with a short shelf life, such as milk, they are widely used.

LDPE was first used in packaging in the 1940s, mainly in the form of film. In the 1950s, bottles made from it became established for a range of chemical and household products, including the first squeeze bottles for washing-up liquids. However, the introduction of HDPE with its better rigidity led to its replacing low density material for most bottles in the 1960s because considerable weight and thus cost savings could be made. It now accounts for 55% of polymers used in bottles.

Polypropylene came into use in the 1960s and is now used for bottle manufacture on some scale, but has not replaced HDPE to any great extent because it is more difficult to process and is not any cheaper. It is mainly used where its better clarity is an advantage, as for instance in sauce bottles. Polypropylene has a somewhat higher softening temperature than the polyethylenes, which makes it better for applications where hot-fill or in-pack heat sterilization is used. It is increasingly being used for bottles to contain sterile water and other liquids for medical use.

Since these materials are so cost effective over such a wide range of packs, it has been a continual challenge to overcome some of their weaknesses and in particular their relatively poor gas barrier. This has been achieved to a considerable degree by the development of coextruded multilayer bottles incorporating an inner layer of a special polymer such as hydrolyzed ethylene vinyl acetate (EVOH) which is bonded to the polyolefine using a special adhesive polymer layer. These coextruded bottles are mainly used for sauces and ketchups which degrade or discolor when oxygen is present.

Where polyolefines do not provide adequate properties the next most commonly used material is PVC. It gives clear bottles with good gas and water vapor barrier, and better resistance to absorption of oils. PVC requires the addition of stabilizers and processing aids that must be selected for food contact applications. It is widely used for bottles to contain a range of products including bath foams, sauces, cordials and in France about half the mineral water consumed is in PVC bottles. There are now strict regulations covering the use of PVC in contact with foods.

The packaging of beers and other carbonated

beverages in plastic bottles was at a low level before the introduction of PET bottles in the 1970s. PET is a special high-molecular-weight grade of polyethylene terephthalate which is also used for the production of oriented films. It has excellent clarity and water vapor barrier and in its oriented state it has good gas barrier, about 50 times better than the polyolefines. PET bottles have a high degree of molecular orientation and thus a very high tensile modulus. This allows them to contain highly carbonated beverages within which pressures as high as 1 MPa can be experienced if they are left in direct sunlight. The loss of carbonation through the wall of the bottle is such that a useful shelf life of 26 weeks is possible with a 2 l bottle but this is halved when a 0.5 l bottle is used. This reduction is due to the higher surface-to-volume ratio in the smaller bottle.

For packaging of beers the oxygen barrier of PET bottles is not good enough so it is now normal to coat the bottles externally with a layer of polyvinylidene chloride (PVdC) polymer which is applied as an aqueous distribution and then dried in a hot-air oven.

Alternative barrier systems using an internal layer of a high-barrier polymer have recently come into use. These barrier polymers are either of the EVOH type or of special grades of aromatic nylon. They are introduced as the inner layer either by coextrusion or by coinjection with the PET.

Other polymers used to make bottles are polycarbonate and nylons which are used for their clarity and high temperature resistance. Both can be sterilized by boiling and will withstand much higher temperatures in use. They are used for such applications as baby-feeding bottles.

1.2 Production Processes

The production process used for bottle making is highly automated and nowadays is usually computer controlled. The output from the largest machines is often as high as 15 000 bottles per hour from up to 28 molds. Small units with one mold are used for low-output speciality bottle production. There are two processes: extrusion blow molding and injection blow molding, and these have been extended to include injection stretch–blow molding.

For both processes the polymer in granular form is fed to a screw extruder which has a screw with a helical flight rotating inside a heated barrel. This melts, mixes and compacts the material to the front of the screw. Coloring and processing aids can be added at the input to the extruder. Some moisture sensitive polymers such as PET and nylons must be well dried before extrusion to avoid degradation.

In extrusion blow molding, which is the process most widely used for bottle manufacture, the molten polymer is forced (usually) downwards through a tube-forming die by the pressure generated by the extruder. Two halves of a cooled bottle mold are then closed around the tube, or parison. It is severed from the emerging tube and air is introduced at pressure through either the top or bottom of the parison. After blowing and cooling, the bottle is removed from the mold and the excess material at the top and bottom is removed and recycled back into the extruder.

The injection blow process first forms the parison by injection molding using an extruder with a reciprocating screw unit to melt and inject the polymer into a multicavity mold. The parisons are then either moved on the mold core to blowing molds and blown while the material is still hot, or the parison is cooled and subsequently reheated before it is placed on a mandrel and introduced to the blowing mold. In both cases, the material is blown at a temperature above the glass transition point and usually below the melt point.

In injection stretch–blow molding, the parison is first reheated to a temperature just above the glass transition point and is then stretched longitudinally by a rod inserted through the neck while it is blown against the wall of the cooled mold. This introduces a high degree of molecular orientation into the wall of the bottle or jar. PET bottles for use with carbonated beverages are made this way and have walls with very high tensile and impact strengths. The orientation also improves the barrier of the bottle wall to gases because of the closer packing of the molecular structure. This can be further improved if the material is given increased crystallinity by heating after blowing above the glass transition temperature. However, the bottle must be restrained from shrinking by internal pressure while this is done.

1.3 Design Considerations

In all packaging applications a prime requirement in designing a pack, once the external shape has been chosen, is to develop a container that will hold, within that shape, the correct volume of contents with the appropriate fill height while minimizing the amount of plastic material used and meeting the criteria for strength, flexibility and storage. Internal pressure, whether positive or negative, which the pack may have to sustain during processing, distribution and in storage must also be considered. Significant savings are made by placing material exactly where it is required in the bottle. In the case of parison extrusion, this is done by moving the inner mandrel of the die vertically during the extrusion of each bottle length to vary the wall thickness.

Generally, a container will be rounded, avoiding flat panels, sharp angles and rectangular shapes.

2. *Pots, Tubs and Trays*

2.1 General Description

This category covers open-ended containers with a wide range of sizes in each type. The shapes are either round, rectangular or square, usually with rounded corners. Pots generally range from a capacity of 100–500 ml. Tubs range from 0.5–5 l capacity. These containers are usually closed, with either plastic lids or aluminum-foil laminates which are heat sealed to a flange. Trays are relatively shallow, rectangular in shape and are covered with a layer of film or a plastic lid.

Pots and tubs are extensively used for dairy products, soft margarine and a range of moist and dry food products. Larger tubs are used for paint and other similar home-use products as well as catering packs for food products. Trays are used for cooked and uncooked meat and fish, and a range of fresh or chilled foods including fruit and vegetables. Trays made from expanded polystyrene are used for fast foods.

2.2 Choice of Materials

The materials most widely used for the manufacture of these containers are polystyrene, polypropylene, ABS, PVC and HDPE. The polystyrene can be either "crystal" or modified with butadiene rubbers to improve impact strength. ABS and PVC have traditionally been used for greasy products because of their good resistance to stress cracking and also because of their very good thermoforming performance. HDPE is used for the manufacture of containers by injection molding. Polypropylene is increasingly replacing the other materials because of its lower cost and good physical properties.

None of these polymers has very good gas-barrier properties but the two polyolefines have very good moisture barrier. To improve oxygen barrier, a coextruded inner layer of a high-barrier polymer such as PVdC or EVOH may be incorporated. PVC combines relatively good moisture- and gas-barrier properties and is now widely used for "modified atmosphere packaging" of meats. Its use for margarine packaging is reducing as polypropylene takes over.

Improved oxygen-barrier pots and trays are increasingly being used for shelf-stable food packs. These are either filled under aseptic conditions or are sterilized by heat processing after filling and sealing. The storage life of these packs is often at least 12 months and they are used for both dairy products and ready meals.

Special grades of PET are used for making trays that are crystallized after forming to give them high temperature resistance for use with prepared meals or dishes to be cooked in ovens.

The use of polypropylene for paint containers is increasing, but only at present for water-based paints. The use of plastics for solvent-based paints is so far restricted, due to the relatively high cost of solvent-resistant polymers such as PET and the PVC copolymers specially formulated for this purpose.

2.3 Production Processes

Tubs, pots and trays are manufactured by injection molding or thermoforming. Smaller containers are mainly made by thermoforming; injection molding is used mainly for larger containers and those of higher quality. Containers made by the latter route tend to be thicker and more rigid. The process is the same as that described above for the manufacture of injection-molded parisons but the mold walls are much thinner and shorter cooling times can be used. With multicavity molds, outputs as high as 8 000 pots per hour are possible.

Thermoforming consists first of continuous extrusion of sheet through a wide-slit die onto a cooled rotating roll which is usually followed by two or three more cooling and polishing rolls. The sheet is reheated to just below or to the melt temperature and then formed into pots, tubs or trays in cooled multicavity molds using either vacuum from below or air at pressure from above to shape it into the separate cavities. It is common practice to use a plug to push the material into the cavity before the air is applied, to improve the distribution in the base and wall. The individual containers are then cut from the web in the forming position or can be conveyed forwards in the web and cut out in a separate trim press. PVC sheet is usually made by calendering between heated rolls which give a high polish to the web and produce very uniform thickness. It is otherwise handled as described above.

Foamed polystyrene sheet is made by extruding with a foaming agent that forms small bubbles in the sheet as the pressure is released after emergence from the die. The sheet is then formed in conventional thermoforming machines. It has been normal to use a chlorinated fluorocarbon as the foaming agent, but since the discovery of the damage to the ozone layer, it is now being replaced by other less harmful gases (e.g., pentane).

2.4 Design Considerations

The range of shapes that can be made by thermoforming is relatively restricted. Circular, square and rectangular containers are all made but the walls must be tapered outwards from the base for ejection from the cavity. It is normal to have a denesting feature below the flange to enable separation from a stack when the pots are placed in the filling machine.

Injection-molded pots are made in a greater variety of shapes (e.g., goblets with integral stems) but usually also require a taper for ejection from the cavity.

3. Film and Flexible Packaging

3.1 General

This class of packaging is based on polymer in the form of film, defined as material of 0.25 mm or less in thickness, either on its own or in combination with paper, aluminum film or metalized with aluminum. It can be used in preformed pouches, bags or as free film for wrapping. It can particularly be applied to other containers such as jars and tubs as a lidding material, or with trays as a covering. It is used for both consumer and industrial products. It forms 40% of the use of thermoplastic polymer in packaging in the developed countries and is the area of most rapid growth.

The term "flexible packaging" is applied to bags and pouches made of the above materials, which can be sealed to form a closed container. It accounts for more than half the use of polymer for film.

Cellophane from regenerated cellulose was the first film used for packaging and was introduced in 1929. Subsequently, PVC film appeared in the late 1930s but it was with the introduction of polyethylene film in the late 1940s that the real growth began.

3.2 Choice of Materials

Film for packaging needs to be strong, tough, puncture resistant, usually of reasonable clarity, free from creep under the tension used in wrapping and show good elastic recovery. Where foodstuffs and beverages are concerned, the consideration of oil absorption, water and gas penetrability apply as in other container forms.

The polymers used in this area are polyethylene in low density, linear low-density and high-density forms, polypropylene (oriented or unoriented), PVC, nylons, PET, ionomer and PVdC. A wide variety of combinations of these materials in film form are used to match particular requirements. LDPE, especially the linear form (LLDPE), accounts for 79% of film products. PVC film is extensively used as clingfilm.

Polyethylene film was first used on its own for the manufacture of bags and wrappings during World War II. A major advance in its use came with the introduction of coatings to improve surface properties, particularly adhesion, and thus the bonding of one film to another and to metal. This has greatly increased the range of applications by allowing combinations of film or film coating to be tailored to specific package needs. It is still the most widely used polymer in flexible packaging in both low- and high-density forms. To these has been added recently LLDPE, which has much higher impact strength and allows lower thickness and thereby strengthens its position in this field.

Polyethylene is frequently the first-choice material for film for wrapping and bags in noncritical end uses. As with bottles and jars, it provides an adequate performance for low cost for many purposes (e.g., carrier bags and retail takeaway bags). There is a wide range of form-fill bags made and used for magazine and journal transit by post. These are made on special reel-fed machines that form the bag round the item to be packed and heat seal the back and end seams before separating the unit pack from the reel.

Film is also used for holding a number of items in a collated bundle or on a pallet. The collation is wrapped round with film from one or more reels, usually in an automated process, and the ends heat sealed together. The whole is passed through a hot tunnel where the film shrinks tightly into place round the contents while the natural elasticity of the film coupled with low stress relaxation hold the pack firmly together over a prolonged period. In an alternative mode, a preformed sleeve of film is stretched and the contents inserted. When the tension on the film is relaxed, this allows them to be firmly gripped. These technologies can be usefully applied to cardboard cartons containing goods to provide a good seal, tamper-evident overwrap, to keep water off the vulnerable board containers and to provide a tough outer protection.

Polyethylene is commonly used for these shrink- or stretch-wrap applications, but the elastomeric nature can be enhanced by copolymerization of ethylene vinyl acetate (EVA). Such films will stretch over 100% and still display elastic recovery.

Reel-fed oriented polypropylene film is used in large quantities because of its clarity and strength to make bags and pouches for snack foods (e.g., potato crisps by the form-fill-seal technique described above). Oriented polypropylene of itself cannot be satisfactorily sealed, so the films are usually made by coextrusion with a lower melting copolymer with polyethylene. Heat seals can then be made without distortion of the polypropylene base film.

For many purposes, particularly when the product to be packed is food, these relatively simple films cannot provide the combination of properties required. Thus oxygen-sensitive foods require a much better barrier to the gas diffusing in from the atmosphere. Several combination systems have been developed such as polypropylene film and aluminum foil, film coated with a thin layer of PVdC or two layers of film coextruded with an inner layer of EVOH. All have their advantages and disadvantages. Thus the aluminum foil gives in principle a complete barrier but has to have an inner layer of heat sealable polymer, usually provided by an inner polyethylene layer. PVdC coating improves the barrier performance by up to 50 times that of polyolefines, but is expensive. The barrier with EVOH is excellent under dry conditions, also having the advantage over aluminum foil of allowing a transparent pack while the metal foil pack is opaque but can have a very attractive appearance. Other factors such as strength and sealability can be similarly manipulated.

The most refined pack of this type is the retort pouch which consists of an outer film of PET laminated to aluminum foil with a further inner layer of polyethylene or polypropylene laminated on the other side of the foil.

A rapidly developing use for film packaging is in "modified atmosphere packaging" for fresh foods (e.g., meats and vegetables). The packs have either to contain the produce in a specified predetermined atmosphere in the pack or to allow the entry of oxygen at a predetermined rate into a pack where its level has been reduced to a chosen degree at packing. In most cases an artificial atmosphere of oxygen, nitrogen and carbon dioxide in various proportions is used to replace the normal atmosphere at packing.

In the UK, 50% of red meat sold in supermarkets is packed in modified-atmosphere-packaging conditions using laminates of PVC and polyethylene to form the tray. The lid is of polyester, coated with PVdC and polyethylene. The polyethylene provides the seals. Combinations of nylon and ionomer or polyethylene are used for cooked meats and cheese and for cooking oil pouches in India.

3.3 Production Processes—Film Products

The majority of films used in packaging are formed by extrusion, either from a single polymer or by combination with one or more materials, using lamination, coextrusion and coating.

As with other plastic packaging products, additives are compounded with the polymer to give ultraviolet and thermal stability, antioxidant properties and, particularly with film slip additives, to allow easy separation and movement of stacked film and its products. Coloring additives are also incorporated if required.

Whether a single layer or multilayer coextruded film is being produced, the basic unit is a screw extruder, as for other products mentioned earlier. The molten polymer is forced at high pressure, typically 20–30 MPa, by the pumping action of the rotating screw into either a straight-slit flat die or a tubular ring die.

In flat-die extrusion, the molten web is first cooled by laying it on a polished cold casting roll. This is rotated at a greater speed than the extrusion speed, thus stretching and thinning the polymer sheet to a predetermined controlled thickness before winding the film up in a single layer.

In tubular-film manufacture, the molten polymer from the ring die is usually extruded vertically upwards. Air is introduced through the die into the tube to inflate the bubble of film. The tube is hauled off by driven nip rolls and then either wound up as collapsed lay-flat tube or is slit at the edges and wound as two separate rolls.

PVC can be formed into film by extrusion onto calender rolls where its thickness is reduced between the hot polished rolls. PVC film for packaging normally has to be made pliable by the addition of stabilizers and plasticizers. There is also a growing production of blown PVC for clingfilm use.

Orientation of the polymer molecules in the longitudinal and transverse directions is applied to films where increased strength and stability is required. It incidentally provides some improvement in clarity and in gas barrier. The films most often processed thus, in order of importance, are polypropylene, PET, PVC, nylon and polystyrene. Film formed in the normal way is reheated to a temperature above its glass transition and then stretched by between three and eight times in directions at right angles in the plane of the film. The stretch ratios used and their balance depend on the polymer and the properties sought from it.

4. Other Applications

There remain a number of other specific uses of plastics in packaging which can be briefly listed. These include medical and pharmaceutical packs, cosmetics and toiletries and aerosols.

Plastics have replaced glass and metal in the packaging of many medical and pharmaceutical items. Among the major uses are small bottles, jars and blister packs for ethical and proprietary pills, larger bottles and pouches for sterile fluids, intravenous feeds and plastic wrapping for disposable instruments and other one-shot items. Polyethylene, polypropylene and PVC are extensively used.

Cosmetics and toiletries are strongly influenced by the requirement to provide a strong and specific image for the product by the style, design, finish and decoration of the container. As many of these products contain oils and perfumes, the plastics used must have good resistance to these and must not absorb components of the scents. PVC, styrene-acrylonitrile and polyacetals are all used.

The main part of aerosol containers are metal or glass but the vital valve parts are plastic. The polymers used are polypropylene for the dip tube and nylon or polyacetal for the actual valve components.

Plastics are widely used for caps, closures, lidding material and overcaps. The polymers used are mainly polyethylene and polypropylene. The latter is particularly used for screw closures since these require the higher stiffness to give good thread stability.

5. Safety and Disposal

Polymers are relatively inert chemically and hence to that degree safe. However, many have added stabilizers and small quantities of low-molecular-weight oligomer which can be leached from the bulk polymer under suitable circumstances. For food contact, all additives require clearance by appropriate authorities and manufacturers that they are safe for this purpose. Oligomers and unreacted monomer are minimized by polymer manufacturing and processing.

Disposal of used plastic packaging is inherently safe and causes minimal environmental damage, whether it is by land fill or incineration. There is increasing pressure for recycling and much effort is being applied to achieving economic reuse of recovered packaging. PET bottles for example are extensively recycled in the USA.

See also: Degradation, Taxonomy of; Polymers for Bottles and Other Rigid Containers; Polymers in Flexible Packaging

Bibliography

Juran R (ed.) 1988 *Modern Plastic Encyclopedia.* McGraw-Hill, New York

Rauch R (ed.) 1987 *The Rauch Guide to the U.S. Plastics Industry. Data for 1985, 1986 and Projections to 1991.* Rauch Associates, Bridgewater, NJ

Russo J R (ed.) 1988 *Packaging Encyclopedia 1988.* Cahners, Newton, MA

J. F. E. Adams
[Metal Box, Wantage, UK]

J. McIntosh
[Maidenhead, UK]

Plywood

This article replaces the article of the same title in the Main Encyclopedia.

Plywood comprises a number of thin layers of wood called veneers which are bonded together by an adhesive. Each layer is placed so that its grain direction is at right angles to that of the adjacent layer (Fig. 1). This cross-lamination gives plywood its characteristics and makes it a versatile building material. Plywood has been used for centuries. It has been found in Egyptian tombs and was in use during the height of the Greek and Roman civilizations.

The modern plywood industry is divided into the so-called hardwood and softwood plywood industries. While the names are misnomers, they are in common use, the former referring to industry manufacturing decorative panelling and the latter referring to that which serves building construction and industrial uses. This article will be concerned primarily with the plywood manufactured for construction and for industrial uses in North America.

1. Manufacture

The fundamentals of plywood manufacture are the same for decorative and for construction grades, the primary differences being in the visual quality of the veneer faces.

1.1 Log Preparation

Logs received at the plywood plant are either stored in water or stacked in tiers in the mill yard. From there they are transported to a deck where they are debarked and cut into suitable lengths. The cut sections of logs are referred to as blocks. Debarking a log is most often accomplished by conveying it through an enclosure that contains a power-driven "ring" of scraper knives. This ring has the flexibility to adjust to varying log diameters and scrapes away the bark.

Before transforming the blocks into veneer, many manufacturers find it helpful to soften the wood fibers by steaming or soaking the blocks in hot-water vats. The block conditioning process is usually performed after debarking. The time and temperature requirements for softening the wood fibers will vary according to wood species and desired heat penetration. Block conditioning usually results in smoother and higher-quality veneer.

1.2 Conversion into Veneer

The most common method of producing veneer is by the rotary-cutting of the blocks on a lathe. Rotary

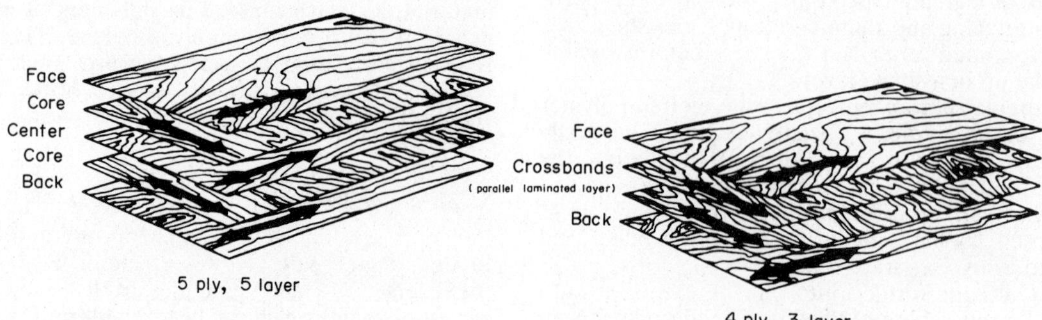

Figure 1
Ply-layer construction of plywood (courtesy of the American Plywood Association)

lathes are equipped with chucks attached to spindles and are capable of revolving the blocks against a knife which is bolted to a movable carriage. New "spindleless" lathes are being introduced which drive the block against a fixed knife with exterior power-driven rollers. Once in place, the chucks revolve the block against the knife and the process of "peeling" veneer begins. The first contact with the lathe knife produces veneer which is not full length. This is due to the fact that the blocks are never perfectly round and are usually tapered to some extent. It is therefore necessary to "round up" the block before full veneer yield is realized. Once this step is complete, the veneer comes away in a continuous sheet, in much the same manner as paper is unwound from a roll. Veneer of the highest grade and quality is usually obtained at the early cutting, knots becoming more frequent when the block diameter is reduced. A knot is a branch base embedded in the trunk of a tree (see *Wood: Macroscopic Anatomy*) and is usually considered undesirable.

All veneer has a tight and a loose side. The side which is opposite to that of the knife is the tight side because of pressure from the machinery holding it against the knife. The knife side is where the lathe checks occur and is called the loose side. The loose side is always checked to some degree and the depths of the lathe checks will depend upon the pressure exerted by the pressure bar opposite the knife (see *Machining of Wood*). Various thicknesses of veneer are peeled according to the type and grade of plywood to be produced.

Wood blocks are cut at high speed, and for this reason veneer is fed into a system of conveyors. At the opposite end of the conveyor is a machine called the clipper. Veneer coming from the lathe through the system is passed under a clipper where it is cut into prescribed widths. Defective or unusable veneer can be cut away. The clipper usually has a long knife attached to air cylinders. The knife action is a rapid up-and-down movement, cutting through the veneer quickly. Undried veneer must be cut oversized to allow for shrinkage and eventual trimming. Up to this point the wood has retained a great deal of its natural moisture and is referred to as green. Accordingly, this whole process before drying is called the green end of the mill.

1.3 Veneer Drying and Grading

The function of a veneer dryer is to reduce the moisture content of the stock to a predetermined percentage and to produce flat and pliable veneer. Most veneer dryers carry the veneer stock through the dryer by a series of rollers. The rollers operate in pairs, one above the other with the veneer between, in contact with each roll. The amount of moisture in the dryers is controlled by dampers and venting stacks. Moisture inside the dryer is essential to uniform drying, since this keeps the surface pores of the wood open. The air mixture is kept in constant circulation by powerful fans. The temperature and speed of travel through the dryer is controlled by the operator and is dependent upon the thickness of the veneer species and other important factors (see *Drying of Wood*).

Dry veneer must be graded, then stacked according to width and grade. Veneer is visually graded by individuals who have been trained to gauge the size of defects, the number of defects and the grain characteristics of various veneer pieces. Veneer grades depend upon the standard under which they are graded, but in North America follow a letter designation A, B, C and D, where A has the fewest growth characteristics and D the most.

1.4 Veneer Joining and Repair

For some grades of plywood, and especially where large panels may be necessary, pieces are cut with a specialized machine, glue is applied and the veneer strips are edge-glued together. The machine usually utilizes a series of chains which crowd the edges of the veneer tightly together, thus creating a long continuous sheet. Other methods of joining veneer may also be employed, such as splicing, stringing and stitching. Splicing is an edge-gluing process in which only two pieces of veneer are joined in a single operation. Stringing is similar to the edge-gluing process except the crowder applies string coated with a hot melt adhesive to hold the veneers temporarily together before pressing. Stitching is a variation of stringing in which industrial-type sewing machines are used to stitch rows of string through the veneer. This system has the advantage of being suitable for either green or dry stock.

The appearance of various grades of veneer may be improved by eliminating knots, pitch pockets and other defects by replacing them with sound veneer of similar color and texture. Machines can cut a patch hole in the veneer and simultaneously replace it with a sound piece of veneer. While the size and shape varies, boat-shaped and dog-bone-shaped veneer repairs are most common.

1.5 Adhesives

The most common adhesives in the plywood industry are based on phenolic resins, blood or soybeans. Other adhesives such as urea-based resorcinol, polyvinyl and melamine are used to a lesser degree for operations such as edge-gluing, panel-patching and scarfing.

Phenolic resin is synthetically produced from phenol and formaldehyde. It hardens or cures under heat and must be hot pressed. When curing, phenolic resin goes through chemical changes which make it waterproof and impervious to attack by microorganisms. Phenolic resin is used in the production of interior plywood, but is capable in certain mixtures of being subjected to permanent exterior exposure. Over 90% of all softwood plywood produced in the

USA is manufactured with this type of adhesive. Another class of phenolics is called extended resins, which have been extended with various substances to reduce the resin solids content. This procedure reduces the cost of the glueline and results in a gluebond suitable for interior or protected-use plywood.

Soybean glue is a protein type made from soybean meal. It is often blended with blood and used in both cold pressing and hot pressing. Blood glue is another protein type and made from animal blood collected at slaughter houses. The blood is spray dried and applied to the plywood and supplied in powder form. Blood and blood–soybean blends may be cold pressed or hot pressed. Protein-based glues (blood or soybean) are not waterproof and therefore are used in the production of interior-use plywood. Protein glues are not currently in common use.

The glue is applied to the veneers by a variety of techniques including roller spreaders, spraylines, curtain coaters and a more recent development, the foam glue extruder. Each of these techniques has its advantages and disadvantages, depending upon the type of manufacturing operation under consideration.

1.6 Assembly, Pressing and Panel Construction

Assembly of veneers into plywood panels takes place immediately after adhesive application. Workmanship and panel assembly must be rapid, yet careful. Speed in assembly is necessary because adhesives must be placed under pressure with the wood within certain time limits or they will dry out and become ineffective. Careful workmanship also avoids excessive gaps between veneers and veneers which lap and cause a ridge in the panel. Prior to hot pressing, many mills pre-press assembled panels. This is performed in a cold press which consists of a stationary platten and one connected to hydraulic rams. The load is held under pressure for several minutes to develop consolidation of the veneers. The purpose of pre-pressing is to allow the wet adhesive to tack the veneers together, which provides easier press loading and eliminates breaking and shifting of veneers when loaded into the hot press. The hot press comprises heated plattens with spaces between them known as openings. The number of openings is the guide to press capacity, with 20 or 30 openings being the most common, although presses with as many as 50 openings are in use. When the press is loaded, hydraulic rams push the plattens together, exerting a pressure of 1.2–1.4 MPa. The temperature of the plattens is set at a certain level, usually in the range 100–165 °C.

2. *Standards*

Construction plywood in the USA is manufactured under US Product Standard PS 1 for Construction and Industrial Plywood (US Department of Commerce 1983) or a Performance Standard such as those promulgated by the American Plywood Association. The most current edition of the Product Standard for plywood includes provisions for performance rating as well. The standard for construction and industrial plywood includes over 70 species of wood which are separated into five groups based on their mechanical properties. This grouping results in fewer grades and a simpler procedure.

Certain panel grades are usually classified as sheathing and include C-D, C-C and Structural I C-D and C-C. These panels are sold and used in the unsanded or "rough" state, and are typically used in framed construction, rough carpentry and many industrial applications. The two-letter system refers to the face and back veneer grade. A classification system provides for application as roof sheathing or subflooring. A fraction denotes the maximum allowable support spacing for rafters (in inches) and the maximum joist spacing for subfloors (in inches) under standard conditions. Thus, a 32/16 panel would be suitable over roof rafter supports at a maximum spacing of 32 in (800 mm) or as a subfloor panel over joists spaced 16 in (400 mm) on center. Common "span ratings" include 24/0, 24/16, 32/16, 40/20 and 48/24.

Many grades of panels must be sanded on the face and/or back to fulfil the requirements of their end use. Sanding most often takes place on plywood panels with A- or B-grade faces and some C "plugged" faces are sanded as well. Sanded panels are graded according to face veneer grade, such as A-B, A-C or B-C. Panels intended for concrete forming are most often sanded as well and are usually B-B panels. Panel durability for plywood is most often designated as Exterior or Exposure 1. Exterior plywood consists of a higher solids-content glueline and a minimum veneer grade of C in any ply of the panel. Such panels can be permanently exposed to exterior exposure without fear of deterioration of the glueline. Exposure 1 panels differ in the fact that D-grade veneers are allowed. Exposure 1 panels are most often made with an extended phenolic glueline and are suitable for protected exposure and short exterior exposure during construction delays.

3. *Physical Properties*

Plywood described here is assumed to be manufactured in accordance with US Product Standard PS 1, Construction and Industrial Plywood (US Department of Commerce 1983). The physical property data were collected over a period of years (O'Halloran 1975).

3.1 Effects of Moisture Content

Many of the physical properties of plywood are affected by the amount of moisture present in the wood. Wood is a hygroscopic material which nearly always contains a certain amount of water (see *Hygroscopicity and Water Sorption of Wood*). When plywood is exposed to a constant relative humidity, it will eventually reach an equilibrium moisture content (EMC). The EMC of plywood is highly dependent on relative humidity, but is essentially independent of temperature between 0 °C and 85 °C. Examples of values for plywood at 25 °C include 6% EMC at 40% relative humidity (RH), 10% EMC at 70% RH and 28% EMC at 100% RH.

Plywood exhibits greater dimensional stability than most other wood-based building products. Shrinkage of solid wood along the grain with changes in moisture content is about 2.5–5% of that across the grain (see *Shrinking and Swelling of Wood*). The tendency of individual veneers to shrink or swell crosswise is restricted by the relative longitudinal stability of the adjacent plies, aided also by the much greater stiffness of wood parallel to, as opposed to perpendicular to, the grain. The average coefficient of hygroscopic expansion or contraction in length and width for plywood panels with about the same amount of wood in parallel and perpendicular plies is about 0.002 mm mm^{-1} for each 10% change in RH. The total change from the dry state to the fiber saturation point averages about 0.2%. Thickness swelling is independent of panel size and thickness of veneers. The average coefficient of hygroscopic expansion in thickness is about 0.003 mm mm^{-1} for each 1% change in moisture content below the fiber saturation point.

The dimensional stability of panels exposed to liquid water also varies. Tests were conducted with panels exposed to wetting on one side as would be typical of a rain-delayed construction site (Bengelsdorf 1981). Such exposure to continuous wetting on one side of the panel for 14 days resulted in about 0.13% expansion across the face grain direction and 0.07% along the panel. The worst-case situation is reflected by testing from oven-dry to soaking in water under vacuum and pressure conditions. Results for a set of plywood similar to those tested on one side showed approximately 0.3% expansion across the panel face grain and 0.15% expansion along the panel. These can be considered the theoretical maximum that any panel could experience.

3.2 Thermal Properties

Heat has a number of important effects on plywood. Temperature affects the equilibrium moisture content and the rate of absorption and desorption of water. Heat below 90 °C has a limited long-term effect on the mechanical properties of wood. Very high temperatures, on the other hand, will weaken the wood.

The thermal expansion of wood is smaller than swelling due to absorption of water (see *Thermal Properties of Wood*). Because of this, thermal expansion can be neglected in cases where wood is subject to considerable swelling and shrinking. It may be of importance only in assemblies with other materials where moisture content is maintained at a relative constant level. The effect of temperature on plywood dimensions is related to the percentage of panel thickness in plies having grain perpendicular to the direction of expansion or contraction. The average coefficient of linear thermal expansion is about 6.1×10^{-6} m m^{-1} K^{-1} for a plywood panel with 60% of the plies or less running perpendicular to the direction of expansion. The coefficient of thermal expansion in panel thickness is approximately 28.8×10^{-6} m m^{-1} K^{-1}.

The thermal conductivity k of plywood is about 0.11–0.15 W m^{-1} K^{-1}, depending on species. This compares to values (in W m^{-1} K^{-1}) of 391 for copper (heat conductor), 60 for window glass and 0.04 for glass wool (heat insulator).

From an appearance and structural standpoint, unprotected plywood should not be used in temperatures exceeding 100 °C. Exposure to sustained temperatures higher than 100 °C will result in charring, weight loss and permanent strength loss.

3.3 Permeability

The permeability of plywood is different from solid wood in several ways. The veneers from which plywood is made generally contain lathe checks from the manufacturing process. These small cracks provide pathways for fluids to pass by entering through the panel edge. Typical values for untreated or uncoated plywood of 3/8 in (9.5 mm) thickness lie in the range 0.014–0.038 g m^{-2} h^{-1} mm Hg^{-1}.

Exterior-type plywood is a relatively efficient gas barrier. Gas transmission (cm^3 s^{-1} cm^{-2} cm^{-1}) for 3/8 in exterior-type plywood is as follows:

oxygen	0.000 029
carbon dioxide	0.000 026
nitrogen	0.000 021

4. Mechanical Properties

The current design document for plywood is published by the American Plywood Association (1986), and entitled *Plywood Design Specification*. The woods used to manufacture plywood under US Product Standard PS 1 are classified into five groups based on elastic modulus, and bending and other important strength properties. The 70 species are grouped according to procedures set forth in ASTM D2555 (*Establishing Clearwood Strength Values*, American Society for Testing and Materials 1987). Design stresses are presently published only for groups 1–4 since group 5 is a provisional group with little or no actual production. Currently the *Plywood Design Specification* provides for development of plywood sectional properties based on the geometry of the layup and the species, and combining those with the design stresses for the appropriate species group.

4.1 Section Properties

Plywood section properties are computed according to the concept of transformed sections to account for the difference in stiffness parallel and perpendicular to the grain of any given ply. Published data takes into account all possible manufacturing options under the appropriate standard, and consequently the resulting published value tends to reflect the minimum configuration. These "effective" section properties computed by the transformed section technique take into account the orthotropic nature of wood, the species group used in the outer and inner plies, and the manufacturing variables provided for each grade. The section properties presented are generally the minimums that can be expected.

Because of the philosophy of using minimums, section properties perpendicular to the face grain direction are usually based on a different configuration than those along the face grain direction. This compounding of minimum sections typically results in conservative designs. Information is available for optimum designs where required.

4.2 Design Stresses

Design stresses include values for each of four species groups and one of three grade stress levels. Grade stress levels are based upon the fact that bending, tension and compression design stresses depend upon the grade of the veneers. Since veneer grades A and natural C are the strongest, panels composed entirely of these grades are allowed higher design stresses than those of veneer grades B, C-plugged or D. Although grades B and C-plugged are superior in appearance to C, they rate a lower stress level because the plugs and patches which improve their appearance reduce their strength somewhat. Panel type (interior or exterior) can be important for bending, tension and compression stresses, since panel type determines the grade of the inner plies.

Stiffness and bearing strengths do not depend on either glue or veneer grade but on species group alone. Shear stresses, on the other hand, do not depend on grade, but vary with the type of glue. In addition to grade stress level, service moisture conditions are also typically presented—for dry conditions, typically moisture contents less than 16%, and for wet conditions at higher moisture contents.

Allowable stresses for plywood typically fall in the same range as for common softwood lumber, and when combined with the appropriate section property, result in an effective section capacity. Some comments on the major mechanical properties with special consideration for the nature of plywood are given below.

Bending modulus of elasticity values include an allowance for an average shear deflection of about 10%. Values for plywood bending stress assume flat panel bending as opposed to bending on edge which may be considered in a different manner. For tension or compression parallel or perpendicular to the face grain, section properties are usually adjusted so that allowable stress for the species group may be applied to the given cross-sectional area. Adjustments must be made in tension or compression when the stress is applied at an angle to the face grain.

Shear-through-the-thickness stresses are based on common structural applications such as plywood mechanically fastened to framing. Additional options include plywood panels used as the webs of I-beams. Another unique shear property is that termed rolling shear (see *Wood Strength*). Since all of the plies in plywood are at right angles to their neighbors, certain types of loads subject them to stresses which tend to make them roll, as a rolling shear stress is induced. For instance, a three-layer panel with framing glued on both faces could cause a cross-ply to roll across the lathe checks. This property must be taken into account with such applications as stressed-skin panels.

See also: Laminated Veneer Lumber (Suppl. 2)

Bibliography

American Plywood Association 1986 *Plywood Design Specification*, Form Y510. APA, Tacoma, WA

American Plywood Association 1987a *Grades and Specifications*, Form J20. APA, Tacoma, WA

American Plywood Association 1987b *303 Plywood Siding*, Form E300. APA, Tacoma, WA

American Society for Testing and Materials 1987 *Annual Book of ASTM Standards: Wood* Vol. 4(09). ASTM, Philadelphia, PA

O'Halloran M R 1975 *Plywood in Hostile Environments*, Form Z820G. APA, Tacoma, WA

Sellers T Jr 1985 *Plywood and Adhesive Technology*. Dekker, New York

US Department of Commerce 1983 *Product Standard for Construction and Industrial Plywood*, PS 1. USDC, Washington, DC (available from the American Plywood Association, Tacoma, WA)

Wood A D, Johnston W, Johnston A K, Bacon G W 1963 *Plywoods of the World: Their Development, Manufacture and Application*. Morrison and Gibb, London

M. R. O'Halloran
[American Plywood Association, Tacoma, Washington, USA]

Point Defect Equilibria in Semiconductors

The term "point defect" refers to defects of atomic dimensions in crystalline solids. Intrinsic point defects involve atoms of the host crystal only. Examples are vacancies, which are missing host atoms, and self-interstitials, consisting of squeezed-in additional host atoms. Extrinsic point defects involve atoms chemically different from the host crystal, such as unintentionally introduced impurities or atoms used for electrically doping semiconductors.

Under thermal equilibrium conditions crystals inadvertently contain a certain concentration of intrinsic point defects. This equilibrium concentration is negligibly small at room temperature, but is large enough at typical device processing temperatures to determine essential atomic materials transport phenomena. For silicon, which is the best investigated and technologically most important semiconductor, the qualitative understanding of point defects is well developed. In contrast, the quantitative knowledge of point defect parameters, required for process simulation of submicrometer devices, is far from being satisfactory. In compound semiconductors, the point defect situation is much more complex and, accordingly, the understanding of point defects and their influence on diffusion processes is in a much less advanced state.

1. Elemental Semiconductors

1.1 Thermal Equilibrium Concentrations

The thermodynamically most favorable state of a crystal under equilibrium conditions at a finite temperature is not that of an ideal crystal but one that contains a certain concentration of intrinsic point defects such as vacancies (V) or self-interstitials (I). The introduction of each point defect is locally associated with an increase in the Gibbs free energy given by

$$G_X^F = H_X^F - TS_X^F \qquad (1)$$

where X stands for V or I, T is the absolute temperature, and H_X^F and S_X^F denote the formation enthalpy and entropy of the point defect X, respectively. This increase is overcompensated by the lowering of the Gibbs free energy due to the gain in configurational entropy, leading to a thermal equilibrium concentration

$$C_X^{eq}(T) = \exp(-H_X^F/kT)\exp(S_X^F/k) \qquad (2)$$

where k is Boltzmann's constant, and $C_X^{eq}(T)$ is given in dimensionless atomic fractions. The main interest in equilibrium point defects derives from their involvement in diffusion processes (Shaw 1973, Casey and Pearson 1975, Frank et al. 1984). Establishing thermal equilibrium concentrations at a given temperature requires a sufficiently long time for the point defects to move to or from appropriate sinks and sources such as dislocations or the surfaces.

1.2 Effect of Charge States

In semiconductors, intrinsic point defects may occur in a number of different charge states, which leads to a modification of G_X^F. Provided that the concentration of intrinsic charged point defects is small enough so that they can be neglected in the overall charge balance, the equilibrium concentration of charged point defects X^r, (where r refers to the algebraic value of the charge) depends on the electron concentration n via

$$C_X^{eq}r(n)/C_X^{eq}r(n_i) = (n/n_i)^r \qquad (3)$$

For the derivation of Eqn. (3) it has been assumed that the semiconductor is nondegenerate and that $np = n_i^2$ holds, where p is the hole concentration and n_i is the intrinsic electron concentration. As expected, n-type doping enhances the concentration of negatively charged point defects and suppresses the concentration of positively charged point defects. The equilibrium concentration of neutral intrinsic point defects does not depend on the doping level. Equation (3) also holds analogously for the solubility of extrinsic charged point defects (impurities) provided that they are not determining the doping level themselves.

1.3 The Presence of Vacancies or Self-Interstitials

Whether vacancies or self-interstitials are present in higher equilibrium concentrations depends on the values of their formation enthalpies and entropies. Unfortunately, these properties are neither easy to measure nor easy to calculate (Van Vechten 1980, Lanoo and Bourgoin 1981, 1983). Quantum-mechanical calculations of these quantities (Car et al. 1985) are not accurate enough to draw any definite conclusions. For materials transport phenomena involving intrinsic point defects, it is not the concentration itself that is essential but rather its product with a mobility factor. In the case of self-diffusion, $D_I C_I^{eq}$ and $D_V C_V^{eq}$ are the relevant quantities, where D_I and D_V are the diffusivities of self-interstitials and vacancies, respectively. When considering the "dominant" point defect species under thermodynamic equilibrium conditions it must therefore be specified whether dominant refers to concentrations or to the diffusion properties of specific atoms.

Thermal equilibrium concentrations of intrinsic point defects in elemental semiconductors are much smaller than the corresponding concentrations in metals at the same temperatures normalized to the respective melting temperature. Consequently, it has not been possible to unambiguously identify whether vacancies or self-interstitials are present in higher concentrations in silicon and germanium. The relevance of vacancies or self-interstitials for self-diffusion processes can be determined via the investigation of the substitutional–interstitial diffusion of element A, predominantly dissolved substitutionally (A_s) but diffusing via a fast interstitial configuration (A_i). The interchange between A_i and A_s may be accomplished through the Frank–Turnbull mechanism (Frank and Turnbull 1956)

$$A_i + V \rightleftharpoons A_s \tag{4}$$

involving vacancies or through the kick-out mechanism (Frank et al. 1984)

$$A_i \rightleftharpoons I + A_s \tag{5}$$

involving self-interstitials. Indiffusion of such an element A (e.g., copper in germanium; gold or platinum in silicon) into dislocation-free material is dominated by the diffusion of vacancies from the surfaces or by the diffusion of self-interstitials to the surfaces. The resulting concentration profiles of A_s show drastically different shapes, which allow the determination of whether vacancies or self-interstitials are predominantly involved in self-diffusion.

1.4 Nonequilibrium Situations

Under thermal equilibrium conditions the concentrations of vacancies and self-interstitials are independent of each other. An outside perturbation of the equilibrium concentration may lead to a local equilibrium between vacancy–self-interstitial recombination and thermal generation of self-interstitial–vacancy pairs (Frenkel pairs) according to

$$V + I \leftrightharpoons 0 \tag{6}$$

where O denotes the undisturbed lattice. The actual vacancy and self-interstitial concentrations C_V and C_I are then related via

$$C_I C_V \approx C_I^{eq} C_V^{eq} \tag{7}$$

provided Eqn. (6) is the dominant annihilation and generation reaction. Equation (7) is not fulfilled for implantation or radiation-induced point defects since in this case thermal Frenkel-pair generation is no longer the dominant generation process.

1.5 Intrinsic Point Defects in Germanium

It is generally agreed upon that in germanium, vacancies are the intrinsic point defects that dominate self-diffusion and substitutional dopant diffusion under thermal equilibrium conditions (Hu 1973, Frank et al. 1984). This assumption has been checked by an investigation of the diffusion behavior of copper in germanium which diffuses via one of the substitutional–interstitial mechanisms in Eqns. (4, 5). No noticeable contribution of self-interstitials to self-diffusion has been detected (Stolwijk et al. 1985). In order to explain the diffusion of group III and group V dopants, vacancies are assumed to act as single acceptors.

1.6 Intrinsic Point Defects in Silicon

After years of controversy on the dominant intrinsic point defects under thermal equilibrium conditions, the 1968 suggestion of Seeger and Chik (1968) that both vacancies and self-interstitials are present in silicon has been verified (Frank et al. 1984, Fahey et al. 1988). The preference for vacancies or self-interstitials as "diffusion vehicles" appears to depend on the specific element. Investigation of the diffusion behavior of gold and platinum in silicon shows that both elements diffuse via the kick-out mechanism, which indicates that at least at temperatures higher than about 1000 °C self-diffusion is dominated by self-interstitials. The diffusion of the various substitutional dopants is carried partly by vacancies and partly by self-interstitials. Of the technologically important group III and group V dopants, boron and phosphorus diffusion predominantly involve self-interstitials, antimony diffusion involves mainly vacancies and arsenic diffusion involves both vacancies and self-interstitials.

The information on the relative importance of vacancies and self-interstitials was derived from experiments involving deviations from thermal equilibrium concentrations of intrinsic point defects. Surface oxidation of silicon leads to a supersaturation of self-interstitials ($C_I > C_I^{eq}$) associated with an undersaturation of vacancies ($C_V < C_V^{eq}$) according to Eqn. (7). Surface nitridation has the opposite effect. Establishing local equilibrium between vacancies and self-interstitials requires an astonishingly long time in the order of an hour at 1100 °C. Dopants preferring self-interstitials as diffusion vehicles will react with a diffusion enhancement if the self-interstitial concentration is increased, whereas dopants migrating by the vacancy mechanism will show retarded diffusion due to the decreased vacancy concentration. The diffusion of phosphorus or boron into silicon starting from a high surface concentration also leads to the generation of a self-interstitial supersaturation associated with a corresponding vacancy undersaturation. The doping dependence of group III and group V dopant diffusion may be explained in terms of neutral and negatively charged vacancies as well as in terms of neutral, and negatively and positively charged self-interstitials.

Simulation of two-dimensional diffusion processes under nonequilibrium conditions in submicrometer devices requires knowledge of the diffusivities and thermal equilibrium concentrations of intrinsic point defects (Law and Dutton 1988). Although the contribution $D_I C_I^{eq}$ of self-interstitials to self-diffusion has been measured, only estimates are available for the individual factors D_I and C_I^{eq}. The various estimates for the equilibrium concentration C_I^{eq} differ by many orders of magnitude at typical processing temperatures around 900 °C. This is also true for C_V^{eq}. The difficulty in extracting proper point defect parameters from the available experimental data on nonequilibrium point defect situations appears to be connected with the interaction of vacancies and self-interstitials via Eqn. (6).

During crystal growth the thermal equilibrium concentrations of self-interstitials and vacancies present near the melting point are lowered during cooling of the crystal by back-diffusion to the melt, by recombination and by agglomeration processes. At room temperature there usually remains a non-equilibrium concentration of intrinsic point defects and their agglomerates ("swirls") which may adversely affect further device processing of silicon wafers.

2. *Compound Semiconductors*

2.1 *Point Defect Configurations*

Compound semiconductors consisting of group III and group V elements (III–V compounds) or of group II and group VI elements (II–VI compounds) contain two sublattices each of which is formed by elements of one group. The variety of intrinsic point defects and their possible agglomerates is therefore much larger than in elemental semiconductors. Besides vacancies and self-interstitials in both sublattices, elements of one sublattice may also occupy sites on the other sublattice (anitisite defects). The classification and notation of these defects is extensively described by Kröger (1974). In addition, each of these configurations may possibly occur in various charge states. Therefore, it is hardly astonishing that the basic knowledge on equilibrium intrinsic point defects is fairly limited even in the best investigated compounds such as GaAs and InP.

2.2 *Equilibrium Concentrations and Charge Effects*

The thermal equilibrium concentrations of the basic point defects in compound semiconductors may be described by Eqn. (2), this formula also holding for elemental semiconductors. For compound semiconductors G_X^F depends on the vapor pressures of the more volatile component (Shaw 1973, Casey and Pearson 1975). An example is the case of GaAs in a pressure regime in which the vapor consists mainly of As_4 molecules. The thermal equilibrium concentrations of arsenic vacancies, gallium vacancies, arsenic self-interstitials and gallium self-interstitials depend on the As_4 vapor pressure via

$$C_{V_{As}}^{eq} \propto (1/C_{V_{Ga}}^{eq}) \propto (1/C_{I_{As}}^{eq}) \propto C_{I_{Ga}}^{eq} \propto p_{As_4}^{-1/4} \tag{8}$$

The dependence of the intrinsic point defect concentrations on the vapor pressure may be used to obtain information on the type of intrinsic point defects involved in diffusion process.

Intrinsic point defects may occur in various charge states in compound semiconductors. As long as their concentrations can be neglected in the overall charge balance, Eqn. (3) holds for the intrinsic point defects in both sublattices, as well as for the frequently occurring charged foreign interstitial atoms. In II–VI compounds the ionicity of the constituting elements is so high that intrinsic point defects generally occur in charged form only and may partly be generated to compensate for the charge of the dopants. In this case Eqn. (3) does not hold and the electron or hole concentration can no longer be treated as independent of the concentration of charged intrinsic point defects.

2.3 *The Presence of Vacancies or Self-Interstitials*

The question of which type of intrinsic point defect dominates self-diffusion and other diffusion processes has not even been answered convincingly for the best investigated III–V semiconductor compound GaAs. Although in the literature vacancies are favored (Shaw 1973, Casey and Pearson 1975) it appears likely that both vacancies and self-interstitials are involved and that their relative contributions depend on the vapor pressure of the more volatile constituent and possibly also on the doping conditions. In most III–V compounds some of the technologically most important acceptor dopants such as zinc or beryllium diffuse via a substitutional–interstitial mechanism. Including charge states of the interstitial and the substitutional forms the modified kick-out mechanism is

$$A_i^+ \rightleftharpoons A_s^- + I + 2h \tag{9}$$

where h stands for holes. An analogously extended reaction replaces the Frank–Turnbull mechanism. The distinction between the concentration profiles due to the two different substitutional–interstitial mechanisms is not as straightforward as in the case of elemental semiconductors. A detailed profile analysis as well as an analysis of the phenomenon of superlattice disordering by zinc diffusion (Laidig et al. 1981) favors the kick-out mechanism for zinc in GaAs and InP, which indicates the relevance of group III self-interstitials under *p*-doping conditions.

2.4 Nonequilibrium Effects

Nonequilibrium concentrations of intrinsic point defects in III–V compound semiconductors may be generated by the indiffusion of zinc and beryllium starting from a high surface concentration, as evidenced by the generation of diffusion-induced dislocation loops. In elemental semiconductors, equilibrium concentrations of intrinsic point defects may be established or restored either via the free surfaces or via dislocations. In compound semiconductors the path via dislocations is generally no longer considered to lead to thermodynamic equilibrium concentrations since dislocation climb involves pairs of intrinsic point defects in the two sublattices (Petroff and Kimerling 1976). Since, in addition, the equilibrium concentrations of intrinsic point defects may be much higher than in elemental semiconductors, frozen-in point defects and point defect agglomerates are more likely to cause problems in compound semiconductors. These frozen-in point defects may become mobile later on at room temperature during device operation and cause or accelerate degradation phenomena.

3. Outlook

There is a technological driving force to increase our fairly well-established qualitative knowledge of intrinsic thermal equilibrium point defects in silicon by determining point defect parameters such as thermal equilibrium concentrations and diffusivities at typical device processing temperatures. These point defect parameters will then be used in process simulation programs as well as for deciding which point defects remain in nonequilibrium after the growth of large-diameter silicon single crystals. The point defect situation in III–V compound semiconductors is much more complex and the logical research task is to unravel the qualitative picture before determining detailed point defect parameters quantitatively. Dopant-induced disordering of III–V compound superlattice structures, which appears to be a common phenomenon for most III–V compounds, measured as a function of various dopants, their concentrations and the outside vapor pressure should enable the attaining of this qualitative picture in a systematic way.

Bibliography

Car R, Kelly P J, Oshiyama A, Pantelides S 1985 Microscopic theory of impurity–defect reactions and impurity diffusion in silicon. *Phys. Rev. Lett.* 54: 360–3

Casey Jr H C, Pearson G L 1975 Diffusion in semiconductors. In: Crawford J H, Slifkin L M (eds.) 1975 *Point Defects in Solids*, Vol. 2. Plenum, New York, pp. 163–253

Fahey P M, Griffin P B, Plummer J D 1988 Point defects and dopant diffusion in silicon. *Rev. Mod. Phys.* 61(2): 289–384

Frank F C, Turnbull D 1956 Mechanism of diffusion of copper in germanium. *Phys. Rev.* 104: 617–8

Frank W, Gösele U, Mehrer H, Seeger A 1984 Diffusion in silicon and germanium. In: Murch G, Nowick A S (eds.) 1984 *Diffusion in Crystalline Solids*. Academic Press, New York, pp. 63–142

Hu S M 1973 Diffusion in silicon and germanium. In: Shaw D (ed.) 1973 *Atomic Diffusion in Semiconductors*. Plenum, New York, pp. 217–350

Kröger F A 1974 *The Chemistry of Imperfect Crystals*, Vol. 2. North-Holland, Amsterdam

Laidig W D, Holonyak Jr H, Camras M D, Hess K, Coleman J J, Dapkus P D, Bardeen J 1981 Disorder of an AlAs–GaAs superlattice by impurity diffusion. *Appl. Phys. Lett.* 38: 776–8

Lanoo M, Bourgoin J 1981 *Point Defects in Semiconductors I*. Springer, New York

Lanoo M, Bourgoin J 1983 *Point Defects in Semiconductors II*. Springer, New York

Law M E, Dutton R W 1988 Verification of analytic point defect models using SUPREM-IV. *IEEE Trans. Comput. Aided Des.* 7: 181–90

Petroff P M, Kimerling L C 1976 Dislocation climb model in compound semiconductors with zinc blende structure. *Appl. Phys. Lett.* 29: 461–3

Seeger A, Chik K P 1968 Diffusion mechanisms and point defects in silicon and germanium. *Phys. Stat. Sol.* 29: 455–542

Shaw D (ed.) 1973 *Atomic Diffusion in Semiconductors*. Plenum, New York

Stolwijk N A, Frank W, Hölzl J, Pearton S J, Haller E E 1985 The diffusion and solubility of copper in germanium. *J. Appl. Phys.* 57: 5211–19

Tan T Y, Gösele U, 1985 Point defects, diffusion processes and swirl defect formation in silicon. *Appl. Phys. A* 37: 1–17

Van Vechten J A 1980 A simple man's view of the thermochemistry of semiconductors. In: Moss T S (ed.) 1980 *Handbook on Semiconductors*, Vol. 3. North-Holland, Amsterdam, pp. 1–111

U. Gösele
[Duke University, Durham, North Carolina, USA]

Polycrystalline Silicon: Structure and Processing

Polycrystalline silicon films, usually referred to as polysilicon films, are used for a variety of functions in silicon-based integrated circuits. Lightly doped films are used for high-value load resistors, and heavily doped films are used as gate electrodes in metal-oxide-semiconductor devices, as base and emitter contacts to shallow junctions and for first-level local device interconnection. Polysilicon films are also being investigated for use as active-device

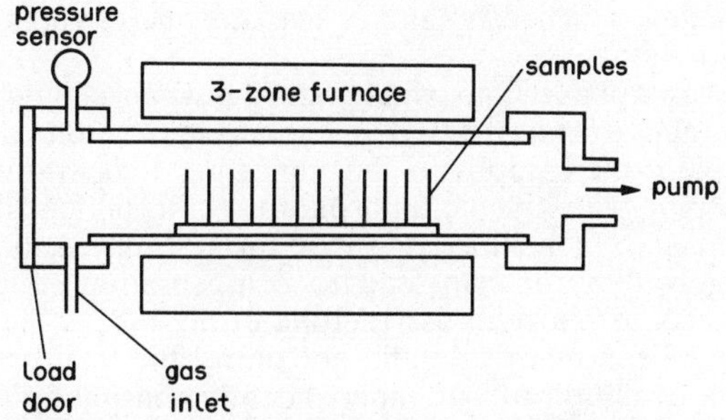

Figure 1
Schematic illustration of a hot-wall reactor for low-pressure chemical deposition (LPCVD) of polysilicon (Adams 1983)

films in thin-film transistors (especially on transparent substrates for display devices) as well as for use in solar cells and micromechanical sensors (Wong et al. 1988).

The electronic, photonic, mechanical and chemical properties of polysilicon films are strongly dependent on their composition and microstructure (e.g., grain sizes, shapes and orientations). The microstructure is affected by the choice of deposition technique and deposition conditions as well as the conditions for post-deposition processing.

1. *Chemical Vapor Deposition of Polysilicon*

Polysilicon is usually deposited by pyrolysis of silane (SiH_4) in a chemical vapor deposition (CVD) process. Silane is often also used for homoepitaxial deposition of silicon as well as deposition of silicon dioxide and silicon nitride films. Silane will decompose via the reaction

$$SiH_4 \rightarrow Si + 2H_2$$

resulting in the deposition of crystalline silicon films at rates of the order of 100 Å min^{-1} and at relatively low temperatures (600–700 °C). Low temperatures are generally desirable for applications in integrated circuits. CVD reactors can hold multiple wafers (up to about 200) which are heated either radiatively (hot-wall reactors) or via thermal conduction from a radio frequency (rf) heated graphite susceptor (cold-wall reactors). Silane is introduced, often with a carrier gas such as nitrogen or hydrogen, at either atmospheric pressure (APCVD) or at lower pressures (LPCVD). Silane is highly toxic and explosive, and must therefore be handled with great care.

A schematic diagram of a low-pressure hot-wall reactor is shown in Fig. 1. LPCVD is generally preferred because it provides economic advantages as well as improved thickness uniformity. At low pressure, convective mass transfer is less important, even when wafers are closely spaced. This allows uniform deposition on larger numbers of wafers (Rosler 1977).

The deposition rate and polysilicon microstructure are most strongly affected by the substrate temperature, the total pressure and the silane concentration. The deposition rate increases with substrate temperature, as shown in Fig. 2. Generally, temperatures between 600 °C and 700 °C are used to deposit columnar polycrystalline films (see Fig. 3a). At higher temperatures, gas phase nucleation can occur, resulting in films with poor adhesion and nonuniform thickness. At lower temperatures, fully or partially amorphous films form (Kinsbron et al. 1983). The deposition rate of polysilicon increases nonlinearly with silane concentration (see Fig. 2). The deposition rate also increases with the total pressure, which can be varied by changing the total gas flow rate or the pumping speed—the latter is preferable for reproducibility (Adams 1983). Crystalline films deposited onto amorphous SiO_2 using LPCVD at 600–650 °C have columnar grains (see Fig. 3a) with {110} fiber texture, while films deposited at 650–700 °C tend to have {100} texture. Films deposited at atmospheric pressure at 700–800 °C tend to have {111} texture (Kamins 1980).

Crystallized amorphous films generally do not have columnar grains but instead have more equiaxed grains (Fig. 3b) with sizes that can be less than or greater than the film thickness, depending on the deposition and crystallization conditions. Crystallization of amorphous silicon is often erroneously referred to as recrystallization.

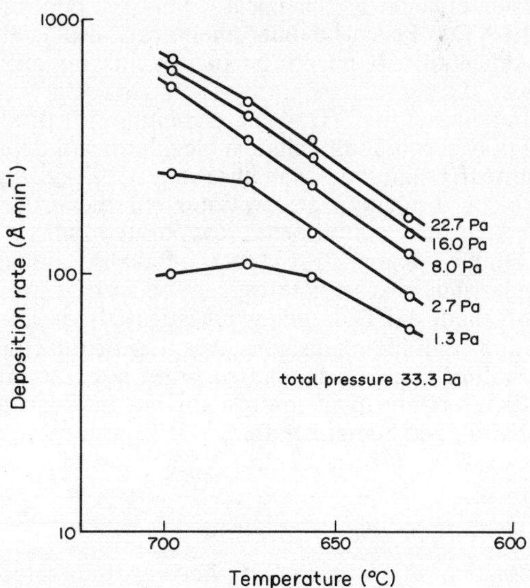

Figure 2
Arrhenius plot for silicon deposition rates for different temperatures and silane partial pressures (Adams 1983)

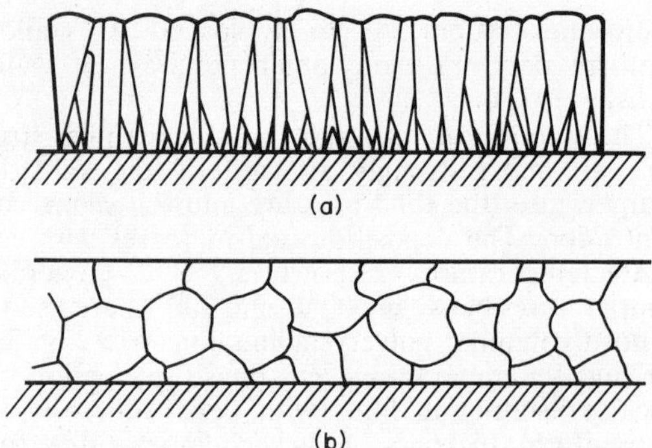

Figure 3
Schematic illustration of cross sections of films with: (a) a columnar microstructure resulting from deposition of crystalline silicon; and (b) an equiaxed microstructure resulting from crystallization of amorphous silicon

Polysilicon is often doped during deposition by adding phosphine (PH_3), arsene (AsH_3) or diborane (B_2H_6) to the silane and carrier gas. All of these gases are highly toxic. Addition of AsH_3 or PH_3 leads to decreased polysilicon growth rates while addition of B_2H_6 leads to higher growth rates (Eversteyn and Put 1973). Dopants can also affect the texture of the films and the temperature at which the amorphous-to-polycrystalline transition occurs.

The deposition rate of polysilicon can be increased and the amorphous-to-polycrystalline transition temperature can be lowered by using plasma-enhanced chemical vapor deposition (PECVD). Plasma enhancement can also lead to modification of the effects of dopants on growth rates.

In addition to CVD, other techniques for producing polysilicon films include molecular beam deposition (MBD) and bulk solidification. MBD of silicon involves deposition at low rates (of the order of $1\,\text{Å}\,s^{-1}$) of electron-beam evaporated silicon in ultrahigh vacuum ($\leqslant 10^{10}$ torr). Because impurity levels and microstructures can be very carefully controlled, MBD is being investigated for use in making thin-film transistors. Casting of bulk polycrystalline ingots can lead to grain sizes of many millimeters and is of interest for use in solar cells (Khattak and Schmid 1987).

2. *Post-Deposition Processing*

After deposition, polysilicon films are subjected to other processing steps which can lead to altered structures and properties. These include oxidation and silicide formation, as well as dopant implantation and activation. All of these processes involve heating which can, alone, lead to microstructural evolution.

Amorphous films will crystallize when heated at or above about 550 °C. The crystallization rate rapidly increases with the annealing temperature. Doping with arsenic, phosphorus or boron leads to enhanced crystallization rates. Interestingly, codoping with boron can lead to compensation of the effects of phosphorus (Lietoila et al. 1982). There have been many attempts to control the final grain size of polysilicon by independently controlling the crystal nucleation rate $\dot{N}$ and the crystal growth rate G during crystallization. If crystallization is seeded, using a single crystal of silicon, for example, spontaneous nucleation can be avoided up to grain sizes of several tens of micrometers. Without seeding, the final grain size increases with the ratio of G to $\dot{N}$. Both parameters are affected by the deposition conditions of the *amorphous* films. Amorphous-silicon films can be made by bombarding polysilicon films with silicon ions. If films are only partially amorphized in this way, polycrystalline films formed after subsequent crystallization may have grains with restricted orientations which correspond to ion channelling directions (Kung and Reif 1986).

Fully crystalline undoped films will undergo normal grain growth, at about 800 °C (Thompson 1988). Normal grain growth stops when the average in-plane grain diameter is about three times the film thickness. At higher temperatures, abnormal or secondary grain growth can lead to further increases in the grain sizes and also to the development of new restricted fiber textures. The rate of grain growth is much higher in films heavily doped with phosphorus ($\geqslant 10^{17}\,cm^{-3}$), slightly increased in films heavily doped with arsenic, and unaffected in boron-doped films. Codoping with boron can compensate the effects of phosphorus or arsenic. Oxidation, reactions with metal films to form silicides and ion bombardment can all lead to enhanced rates of grain growth. Oxygen and chlorine lead to decreased rates of grain growth.

A number of techniques involving local heating have been developed to increase the grain size of polysilicon. One such process involves scanning of resistively heated graphite strips and is known as zone-melting recrystallization (ZMR). This technique can be used to produce films with very large grains ($\geqslant 1$ mm) or, when seeded, single-crystal films. ZMR involves creation and translation of molten zones in the silicon films and might more properly be called zone melting and crystallization. Instead of using graphite strips, similar results can be obtained using lasers or electron beams.

As stated in Sect. 1, polysilicon can be doped during deposition. Alternatively, doping can be accomplished using ion implantation or by diffusion after deposition. Generally, the latter requires temperatures of the order of 1000 °C. Ion implantation can be carried out at much lower temperatures, but anneals around 1000 °C are required in order for the majority of the dopants to become electrically active. Phosphorus and arsenic tend to segregate at grain boundaries while boron does not. Dopants at grain boundaries are generally not electrically active. Boron also tends to segregate into silicon dioxide. Annealing of implanted films can therefore lead to increased carrier concentrations due to activation and, at higher temperatures, decreased carrier concentrations due to segregation.

The electronic properties of polysilicon can be improved by incorporation of hydrogen which passivates grain boundary defect states. Hydrogen can be introduced using a hydrogen plasma, by diffusion or by ion implantation (Kamins 1986). Hydrogen diffuses at appreciable rates at relatively low temperatures and will diffuse out of polysilicon at 400–500 °C.

Polysilicon can be oxidized in the same ways that single-crystal silicon is oxidized. Typically, oxidation is carried out in dry oxygen at 900–1000 °C (Adams 1983). The oxidation rate is a function of the average crystal orientation and, in heavily doped films ($\geqslant 10^{17}$ cm^{-3}), increases with phosphorus or arsenic concentration.

Polysilicon films are often reacted with metal films to form silicides for higher conductivity gates and interconnects. The desired final structure usually consists of a silicide film on top of unreacted polysilicon. The times and temperatures required for these reactions depend on the choice of metal, the desired final phase and the film thickness (Murarka 1983). During silicide formation, microstructural evolution and dopant redistribution in the polysilicon can occur.

Polysilicon can be patterned using either wet or dry etches. Etch rates can be affected by the structure and composition of the films. For example, chlorine-based dry etches react with phosphorus-doped polysilicon at significantly higher rates than the undoped silicon (Mogab and Levinstein 1980).

In summary, polysilicon films are used for a variety of functions, especially in silicon-based integrated circuits. The microstructure of polysilicon can be strongly influenced by the conditions for deposition and post-deposition processing. The structure and properties of polysilicon also depend on the concentration of dopants. The microstructure of polysilicon can be tailored for specific applications.

See also: Silicon (Semiconductor): Devices and Integrated-Circuit Processing; Silicon (Semiconductor): Preparation; Silicon (Semiconductor): Properties

Bibliography

Adams A C 1983, Dielectric and polysilicon film deposition, Chap. 3. In: Sze S M (ed.) 1983 *VLSI Technology*. McGraw-Hill, New York, pp. 93–129

Eversteyn F C, Put B H 1973 Influence of AsH_3, PH_3, and B_2H_6 on the growth rate and resistivity of polycrystalline silicon films deposited from a SiH_4–H_2 mixture. *J. Electrochem. Soc.* 120: 106–10

Kamins T I 1980 Structure and properties of LPCVD silicon films. *J. Electrochem. Soc.* 127: 686–90

Kamins T I 1986 Electrical properties of polycrystalline-silicon thin films for VLSI. In: Wittmer M, Stimmel J, Strothman M (eds.) 1986 *Materials Issues in Integrated Circuit Processing*, Materials Research Society Symposium Proceedings, Vol. 71. Materials Research Society, Pittsburgh, PA, pp. 261–72

Khattak C P, Schmid F 1987 Growth of polysilicon ingots by HEM for photovoltaic applications. In: Wang F F Y (ed.) *Materials Processing Theory and Practices*, Vol. 6. North-Holland, Amsterdam, pp. 153–84

Kinsbron E, Sternheim M, Knoell R 1983 Crystallization of amorphous-silicon films during low-pressure chemical vapor-deposition. *Appl. Phys. Lett.* 42(9): 835–7

Kung K T Y, Reif R 1986 Implant-dose dependence of grain-size and {110} texture enhancements in polycrystalline Si films by seed selection through ion channeling. *J. Appl. Phys.* 59(7): 2422–8

Lietoila A, Wakita A, Sigmon T W, Gibbons J F 1982 Epitaxial regrowth of intrinsic ^{31}P-doped and compensated (^{31}P + ^{11}B-doped) amorphous Si. *J. Appl. Phys.* 53: 4399–405

Mogab C J, Levinstein H J 1980 Anisotropic plasma etching of polysilicon. *J. Vac. Sci. Technol.* 17: 721–30

Murarka S P 1983 *Silicides for VLSI Applications*. Academic Press, Orlando

Rosler R S 1977 Low pressure CVD production processes for poly, nitride, and oxide. *Solid State Technol.* April, 63–70

Thompson C V 1988 Grain growth in polycrystalline silicon films. In: Wong C Y et al. 1988

Wong C Y, Thompson C V, Tu K-N (eds.) 1988 *Polysilicon Films and Interfaces*, Vol. 106. Materials Research Society, Pittsburgh, PA

C. V. Thompson
[Massachusetts Institute of Technology, Cambridge, Massachusetts, USA]

Polyethylene Processing to Produce Ultimate Properties

Polyethylene $(CH_2–CH_2)_n$ was discovered in 1933 and by the 1950s had become a major commodity thermoplastic. In its usual semicrystalline form, polyethylene has a room temperature Young's modulus of approximately 1 GPa and a tensile strength of 0.5 GPa, although both values depend to some extent on the molecular weight of the polymer. The long polymer chains form thin "chain-folded" platelet lamellar crystals, each of a thickness of about 10 nm. These platelets are usually organized to form centrosymmetric spherulitic crystals of the order of micrometers in diameter which give the material an overall isotropic organization, as shown in Fig. 1.

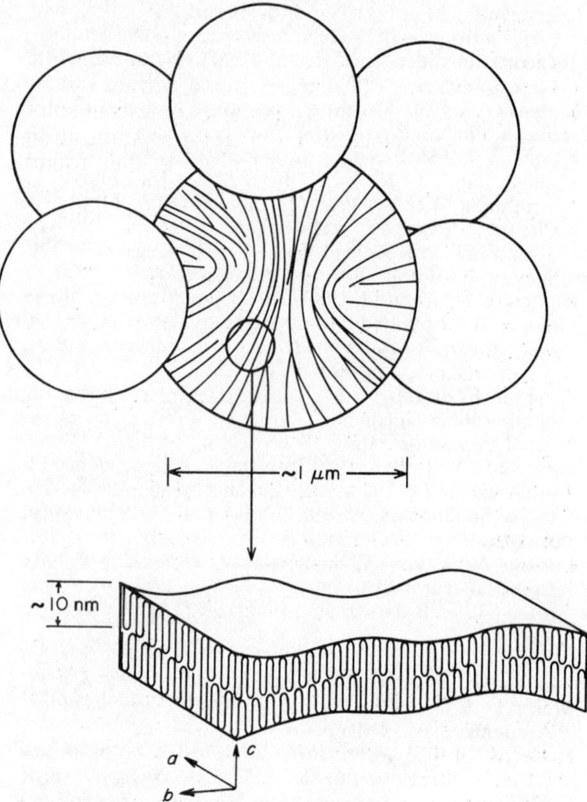

Figure 1
The structure of spherulitic melt-crystallized polyethylene

The unit cell of crystalline polyethylene was established at an early stage by Bunn (1939) and the chain configuration is shown in Fig. 2. From this diagram it can be seen that within the unit cell, the chain is in an all-trans configuration with the *c* axis aligned along the chain axis. Frank (1970) pointed out that the unit cell of crystalline polyethylene was similar to that of diamond so that if polyethylene could be persuaded to crystallize in the all-trans configuration with chains extended fully along the *c* axis, a material with significantly improved properties could be expected and in particular a room temperature Young's modulus of the order of 200–300 GPa might be anticipated. The prospect of such a high modulus from a low-cost commodity polymer generated an international effort to realize this objective and this has resulted in a number of process routes being discovered independently.

1. Solid-State Drawing

The classic route for generating anisotropy in polymers is mechanical deformation (Ward 1971). Solid-state drawing is an essential part of most synthetic fiber spinning processes and, after forming a generally isotropic fiber by melt spinning, orientation is induced by mechanically drawing the fiber at room or elevated temperature to draw ratios λ of up to ten. The draw ratio is defined as the ratio of the drawn length to the original length. By mechanically drawing, the isotropically oriented lamellar crystals of polyethylene are disrupted and aligned towards the direction of the draw axis such that the *c* crystallographic axis has a preferred orientation in the draw direction, as shown in Fig. 3.

Andrews and Ward (1970) showed that by drawing at an enhanced rate, the modulus of polyethylene could be increased from approximately 1 GPa to approximately 20 GPa. Later, Cappacio and Ward (1975) discovered that by drawing at an elevated temperature (above 80 °C) and by choosing polyethylene of a particular molecular weight distribution and crystallite morphology, it was possible to achieve even higher draw ratios of up to 20, with an associated Young's modulus of 40 GPa. It was found that there appeared to be an essentially linear relationship between the observed tensile modulus and the draw ratio, the modulus varying from 1 GPa at $\lambda = 1$ to 50 GPa at $\lambda = 25$.

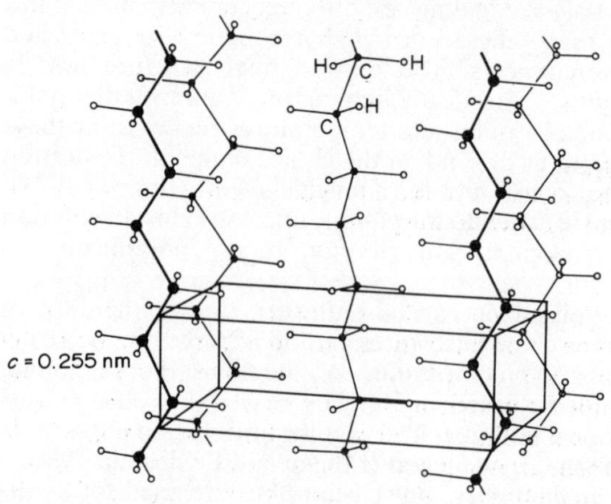

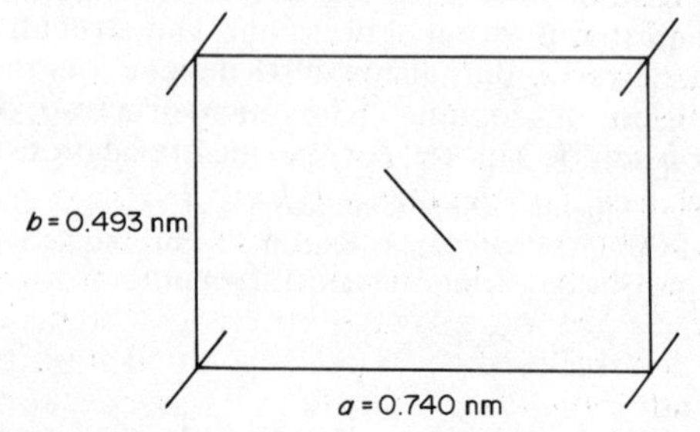

Figure 2
The unit cell of crystalline polyethylene (based on Keller 1968)

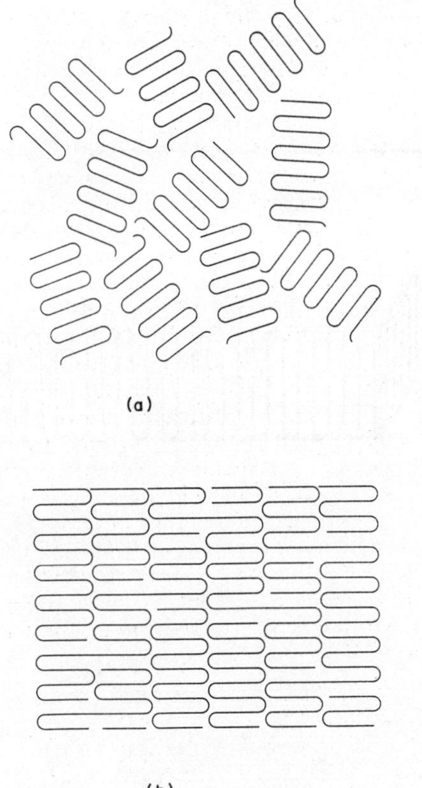

Figure 3
The alignment of crystals of polyethylene: (a) before and (b) after drawing

A variant of the solid-state drawing process is some form of ram extrusion (Ward 1985). In this case a billet of polyethylene is either pushed or pulled through a die of reduced cross section, as shown in Fig. 4. The process has many similarities to tensile drawing but in the extrusion case the dimensional changes of the material are controlled by the configuration of the die rather than by the mechanical behavior of the material.

In the case of solid-state drawing it has generally been found that drawing at high draw ratios is most readily achieved with linear polyethylenes that have an $\bar{M}_W$ below about 10^5. These are the molecular weights associated with normal melt-processable polyethylenes. The detailed mechanisms associated with the ability to generate high draw ratios in this way are not fully understood but the simplicity and elegance of the process has ensured commercial viability for this method in producing polyethylene fibers and monolithic extrusions with a tensile Young's modulus of up to about 50 GPa. This is an increase of 50 fold from the usual melt-crystallized material.

2. Solution Processing

Pioneering studies on the way in which polyethylene crystallizes from solution have been carried out by Keller (1968) for crystallization from quiescent solutions and by Pennings et al. (1970) for crystallization from mechanically stirred solutions. Keller found crystals formed thin chain-folded crystal lamella and Pennings et al. showed that stirring could induce a "shish kebab" morphology. The molecular organization associated with these two morphologies is shown in Fig. 5 and it can be seen that the core of the shish kebab polyethylene crystal has the necessary organization to have the potential for yielding a high-modulus polymer.

The first continuous length of polyethylene fiber produced by solution processing was made using a seeding technique developed by Zwijnenburg and Pennings (1975). In this process a seed crystal was placed in a flowing stream of a polyethylene solution consisting of about 0.1% w/v (weight per unit volume) of ultrahigh-molecular-weight polyethylene (UHMWPE) in an organic solvent such as xylene. At a temperature below 113 °C it was found that a continuous length of polyethylene fiber could be produced and, by using the system shown in Fig. 6, growth rates of 2 mm min^{-1} for 20 μm diameter fibers could be achieved. The level of molecular orientation in the fibers was high and tensile moduli of 22 GPa were reported.

This rather slow process was soon superseded by a more rapid and efficient surface growth route discovered by Zwijnenburg and Pennings (1976). In this method, the growing fiber was allowed to migrate towards a rotating surface and, at this point,

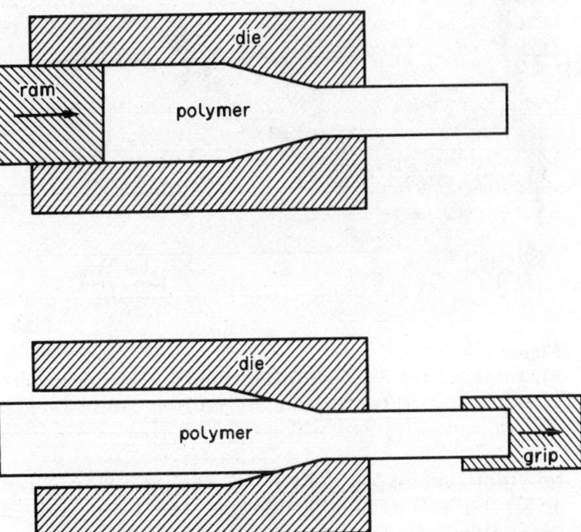

Figure 4
(a) Ram extrusion and (b) die drawing processes

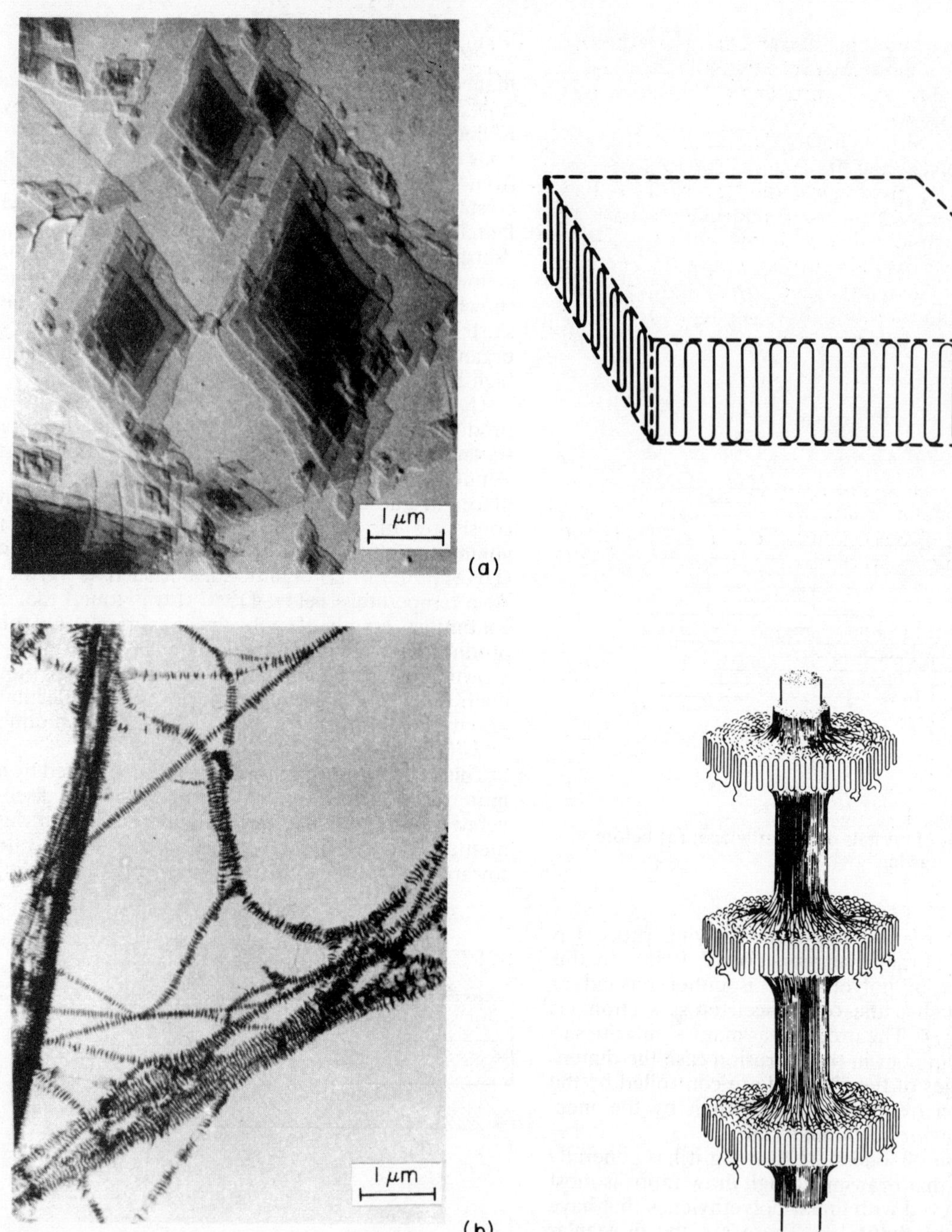

Figure 5
Micrographs and associated diagrams of polyethylene crystallized from (a) quiescent solution (chain-folded lamellar crystals), and (b) mechanically stirred solution (shish kebab crystals)

polymer chains attached to the growing fiber appeared to be caught between the fiber and an entanglement adsorption layer of polymer attached to the moving surface. The relative motion between the fiber and the surfaces causes elongation of the chains and subsequent crystallization occurs if the temperature of the solution is below 118 °C. A diagram of the system is shown in Fig. 7 and, using this type of apparatus, single fiber growth rates of up to $0.2\,\mathrm{m\,min^{-1}}$ could be achieved. Again, fiber orientation was very high and the first polyethylene with a Young's modulus exceeding 100 GPa and a tensile modulus of approximately 2 GPa was achieved in this way.

UHMWPE, with values of $\bar{M}_n$ (number average molecular weight) and $\bar{M}_w$ in excess of 10^6, is extremely difficult to process in either the melt or solid form. However, Zwijnenburg and Pennings (1975) have clearly shown that with the aid of solution processing it is possible to produce polyethylene that is both highly extended and orientated and that yields a tensile modulus that can be compared to other high-modulus fibers such as carbon fibers and Kevlar.

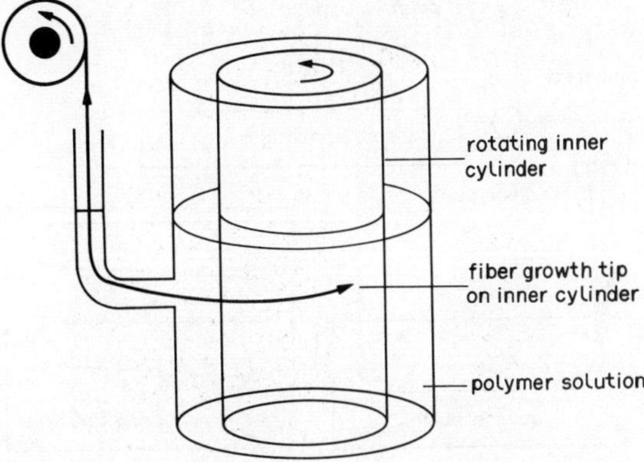

Figure 7
The "surface growth" fibrous crystal process

3. Gel Spinning

Following the success of the surface growth process, a new gel-spinning route was discovered at Dutch State Mines (DSM). Smith and Lemstra (1980) reported that they were able to gel spin UHMWPE

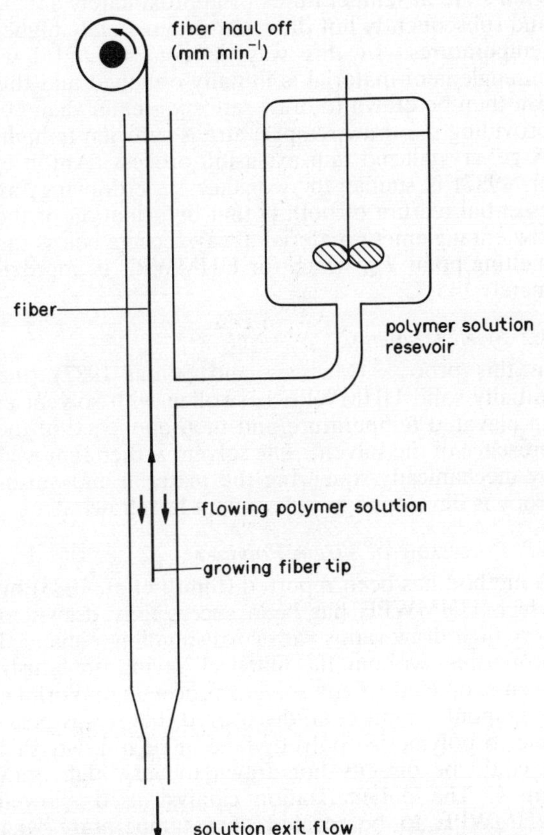

Figure 6
The "free surface" fibrous crystal growth process with 0.1–1.0% w/v UHMWPE in xylene at temperatures of 100–113 °C

fibers that could subsequently be drawn to a draw ratio λ of 31.7 which resulted in fibers with a Young's modulus of 90 GPa and a tensile strength of 3.0 GPa. Their apparatus is shown in Fig. 8 and the process route is as follows.

(a) A solution of 0.5% w/v UHMWPE in Decalin (decahydronaphthalene) is prepared under agitation at 150 °C.

(b) Gel fiber is formed by extrusion of the solution through a nozzle at 130 °C resulting in isotropically oriented fiber (~1 mm diameter).

(c) Solidification and crystallization of the fiber is achieved by quenching in a cooling bath.

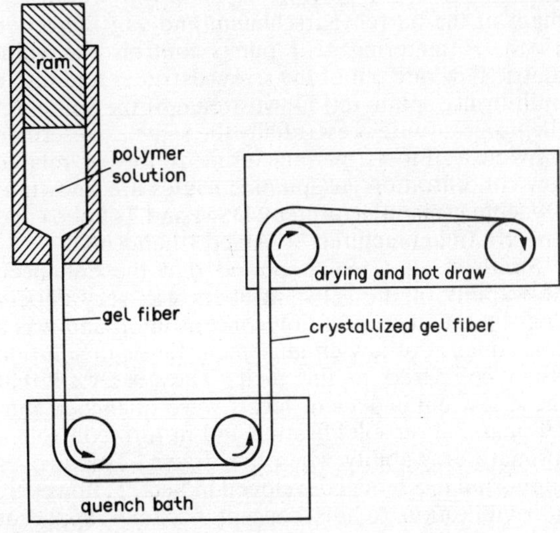

Figure 8
Batch gel-spinning apparatus

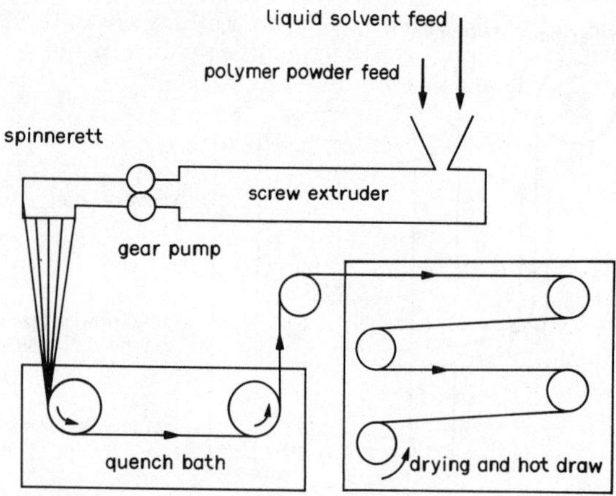

Figure 9
Continuous gel-spinning apparatus

(d) The solidified fiber is then heated to remove solvent and drawn at temperatures in excess of 90 °C to draw ratios λ of up to approximately 30.

Processing UHMWPE in this manner causes a dramatic increase in the drawability of the material. Melt-quenched UHMWPE can normally be drawn at 90 °C up to draw ratios λ of 3–4. By dissolving the polymer and subsequently crystallizing in the presence of a solvent the drawability can be increased to values of λ up to 30 with a corresponding increase in Young's modulus to 90 GPa.

Further development of this technology has enabled the polymer solution to be prepared in a continuous process as shown in Fig. 9. In this system, both solvent and polymer powder are fed into a screw extruder in the concentration range of 1–10% w/v and the polymer is dissolved within the flight of the barrel (Kirschbaum and van Dingenen 1988). A metering gear pump controls the volumetric flow rate out of the screw extruder and into a multiorifice spinneret. Downstream of the spinneret the process route is essentially the same as described previously. Fibers manufactured by this or related low-concentration gel-spinning routes are known as Dyneema (manufactured by DSM and Tohobo) and Spectra (manufactured by Allied Chemicals).

Smith et al. (1981) reported that the enhanced drawability of the gel-spun fibers was very dependent on the initial solution concentration and was a consequence of low entanglement levels in solution when compared to the melt. They asserted that these low entanglement levels were quenched into the material on solidification and in turn controlled ultimate drawability when hot drawn. This idea is supported in a model developed in Sect. 5; however, a complication to this concept becomes apparent when the third commercial polyethylene fiber known as Teknillon and manufactured by Mitsui is considered. This fiber has similar properties to Dyneema and Spectra but appears to be manufactured by the hot drawing of an isotropic fiber that has been formed by plasticizing UHMWPE with about 20% w/v of paraffin wax. The material can then be melt processed in the normal way to produce a fiber geometry and, after extraction of the paraffin, hot drawing can produce a fiber with a Young's modulus of 100 GPa.

4. Other Process Routes

There are a number of variants for producing high-modulus and high-strength polyethylene and some of these are reviewed by Barham and Keller (1985). Most, but not all, of these processes are directly related to the three routes that have been described in Sects. 1–3.

4.1 Two-Stage Superdrawing

In this method (Kanamoto et al. 1988), solution-crystallized UHMWPE is dried, coextruded in the solid-state at temperatures of approximately 110 °C and subsequently hot drawn at progressively higher temperatures. In this way, a consolidated low-entanglement material is initially obtained and this can then be drawn to draw ratios λ greater than 50, providing the draw temperature is sufficiently high. A gel-crystallized ram extrusion process (Anton et al. 1987) is similar to two-stage superdrawing; an essential feature of both is that deformation of the low-entanglement material always occurs below the melting point T_m, which for UHMWPE is approximately 145 °C.

4.2 Swell Drawing

In this process (Mackley and Solbai 1987), the initially solid UHMWPE is swollen with solvent at an elevated temperature and then quenched in the presence of the solvent. The solvent is then removed by mechanically squeezing the material and anisotropy is developed by subsequent hot drawing.

4.3 Processing of Virgin Polymer

A method has been reported (Smith et al. 1985) by which UHMWPE has been successfully drawn to very high draw ratios with corresponding enhanced properties, without the material having previously been subjected to any solvent processing. Working at Dupont, Smith et al. discovered that it was possible to polymerize polyethylene in such a way that it could be directly hot drawn to very high draw ratios. The polymerization catalyst used allowed UHMWPE to be formed at a temperature well below the melting point of the polymer and in addition it appears that a low-activity catalyst was chosen. Under these polymerization conditions the polymer was unable to become highly entangled during polymerization and, providing subsequent drawing took place below T_m, further entanglement was not possible, thereby allowing high draw ratios to be achieved in the final hot-draw process.

5. Mechanisms Associated with Ultimate Drawability

From a knowledge of the crystalline unit cell for polyethylene, modulus enhancement is clearly expected if high levels of uniaxial orientation and continuity along the chain axis are achieved. Modulus enhancement appears to be directly related to levels of chain orientation, while significant strength enhancement seems to only occur with high molecular weight variants of polyethylene.

The detailed mechanism associated with the drawing of polyethylene is complex. Drawability above 80 °C is easier for all ranges of molecular weights and this can be related to stress-induced thermal activation of the chains, either within the crystalline or the amorphous phase of the material. The drawing process appears to follow reasonably closely to that predicted by the pseudoaffine deformation assumption (see *Rubberlike Elasticity: Molecular Theory*).

The limiting drawability of polyethylene and the profound effect of solvent processing before drawing form the basis for the Smith et al. (1981) model for the deformation process. It is well known that polyethylene in the molten state exists as highly entangled chains that are themselves essentially in a random coil configuration. The molecular weight M_e of the polymer segments between these entanglement junctions can be determined from the melt plateau modulus of the polymer G'_0, given by

$$G'_0 = \frac{\rho RT}{M_e} \tag{1}$$

where ρ is the polymer density, R the universal gas constant and T the absolute temperature. If it is assumed that entanglement junctions are fixed, the distance between these junctions can be thought of in terms of the root mean square end-to-end distance l_0 of the random chain between entanglement points and this is given by $l_0 = ar^{1/2}$ where r equals the number of repeat units between entanglement junctions and a the link length of each repeat unit. r is equal to M_e/M_0 where M_0 is the equivalent molecular weight of a statistical chain segment of polyethylene and the situation is represented in Fig. 10a. Smith et al. (1981) then speculate that the entanglement level within the melt is trapped in the material during crystallization and that the subsequent drawability of the material is controlled by the drawability of this entanglement network. If the network was fully drawn, as shown in Fig. 10b, the limiting end-to-end distance between entanglement junctions would be ar yielding a maximum draw ratio of

$$\lambda_m = \frac{ar}{ar^{1/2}} = r^{1/2} = \left(\frac{M_e}{M_0}\right)^{1/2} = \left(\frac{\rho RT}{M_0 G'_0}\right)^{1/2} \tag{2}$$

For polyethylene crystallized from the melt, Eqn. (2) gives $\lambda_m \simeq 3$.

If solvent is present, the plateau modulus decreases with increasing solvent concentration and follows an equation of the form

$$G'_\phi = G'_0(1-\phi_s)^{5/3} \tag{3}$$

where ϕ_s is the volume fraction of solvent. Smith et al. use a slightly modified equation to that given in Eqn. (3); however, replacing G'_0 in Eqn. (2) by G'_ϕ from Eqn. (3) yields

$$\lambda_\phi = \left(\frac{\rho RT}{M_0 G'_0}\right)^{1/2} (1-\phi_s)^{-5/6} \tag{4}$$

where λ_ϕ is the maximum drawability after prior solvent processing.

If a λ_ϕ approximately equal to 40 is required, it can be seen from Eqn. (4) that a solvent volume fraction of 0.956 is needed, meaning that a polymer content of 4.4% volume fraction has to be used. For gel-spinning and swell-drawing behavior, Eqn. (4) holds quite well and this implies that processing with low polymer concentration appears to be essential if high drawability is to be achieved.

To some extent the virgin polymer processing route is consistent with the above concept, because here the material is polymerized well below T_m and

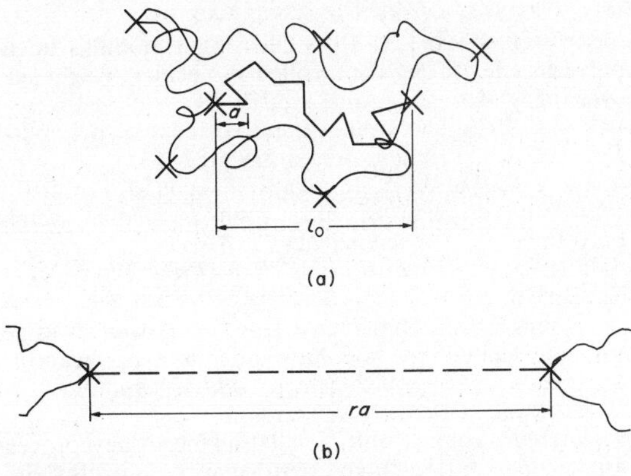

Figure 10
Molecular organization of a network: (a) undeformed, and (b) at the point of maximum deformation

in a way this deliberately attempts to produce a low entanglement state. The Mitsui process and some aspects of superdrawing, however, suggest that it is also possible to obtain high drawability using limited plasticization and/or careful control of draw temperature and of molecular weight distribution. A plausible explanation for this is that when drawing occurs close to the melting point, under certain conditions, partial disentanglement of any embedded network within the material can take place simultaneously with the development of anisotropy.

In conclusion, the new processing routes for polyethylene, and particularly UHMWPE, that have been described in this article open the way for a new and exciting chapter in the history of polyethylene. It is now possible for polyethylene to compete in the market place with other high-performance fibers such as carbon and aramid fibers, although the relatively low melting point of polyethylene at 145 °C will restrict material application areas. The potential for manufacturing high-modulus polyethylene structures other than in fiber form has been clearly demonstrated and the scope for the development of novel and more economic processing routes would still appear to be considerable.

See also: Die Extrusion of Polymers; Drawing of Polymer Fibers; High-Modulus Fibers; Polymer Fiber Formation and Processing; Texturing and Draw Texturing of Polymer Fibers

Bibliography

Andrews J A, Ward I M 1970 The cold-drawing of high density polyethylene. *J. Mater. Sci.* 5: 411–17

Anton C R, Mackley M R, Solbai S 1987 Extrusion processing of ultra high molecular weight polyethylene. *Polym. Bull.* 17: 175–9

Barham P J, Keller A 1985 Review: High strength polyethylene fibres from solution and gel spinning. *J. Mater. Sci.* 20: 2281–2302

Bunn C W 1939 The crystallographic unit cell of polyethylene. *Trans. Faraday Soc.* 35: 482–91

Cappacio G, Ward I M 1975 Ultra high modulus linear polyethylene through controlled molecular weight and drawing. *Polym. Eng. Sci.* 15: 219–24

Frank F C 1970 The strength and stiffness of polymers. *Proc. R. Soc. London, Ser. A* 319: 127–36

Kanamoto T, Tsurata A, Tanaka K, Takeda M, Porter R S 1988 Superdrawing of ultra high molecular weight polyethylene. *Macromolecules* 21: 470–7

Keller A 1968 Polymer crystals. *Rep. Prog. Phys.* 31(2): 623–705

Kirschbaum R, van Dingenen J L J 1988 Advances in gel spinning technology and Dyneema fibre applications. *Integration of Polymer Science and Technology, 3rd Rolduc Conf.* Elsevier, Amsterdam

Mackley M R, Solbai S 1987 Swell drawing: A new process for manufacturing high performance polyethylene. *Polymer* 28: 1115–20

Pennings A J, van der Mark J M A A, Kiel A M 1970 Hydrodynamically induced crystallization of polymers from solution. *Kolloid-Z. Z. Polym.* 237: 336–58

Smith P, Chanzy H D, Rotzinger B P 1985 Drawing of virgin ultrahigh molecular weight polyethylene. *Polym. Commun.* 26: 258–60

Smith P, Lemstra P J 1980 Ultra drawing of high molecular weight polyethylene cast from solution. *J. Mater. Sci.* 15: 505–14

Smith P, Lemstra P J, Booij H C 1981 Ultra drawing of high molecular weight polyethylene: The effect of concentration. *J. Polym. Sci. Polym. Phys. Ed.* 19: 877–88

Ward I M 1971 *Mechanical Properties of Polymers*. Wiley, New York

Ward I M 1985 The preparation, structure and properties of ultra high modulus flexible polymers. *Adv. Polym. Sci.* 70: 1–70

Zwijnenburg A, Pennings A J 1975 Longitudinal growth of crystals from flowing solutions II. Polyethylene crystals in Poiseuille flow. *Colloid Polym. Sci.* 253: 452–61

Zwijnenburg A, Pennings A J 1976 Longitudinal growth of crystals from flowing solutions III. Polyethylene crystals in Conette flow. *Colloid Polym. Sci.* 254: 868–81

M. R. Mackley
[University of Cambridge, Cambridge, UK]

Polymers Based on Carbon Dioxide

Chemists continually search for new ways to use abundant low-cost resources to produce commercially useful materials. Carbon dioxide copolymers, or poly(alkylene carbonates) (PACs), are a new generation of polymers currently being commercialized that represent a true breakthrough in polymer technology. In contrast to traditional thermoplastics that are petroleum-based, these new materials are derived from the copolymerization of carbon dioxide (CO_2) with one or more epoxides. In copolymers of ethylene oxide or propylene oxide, the CO_2 constitutes about half of the polymer weight. This high CO_2 content has important implications regarding economics, performance and applications.

For example, PAC polymers are potentially low cost because 50% of the polymer weight is not dependent on the price of petroleum or other high-cost petrochemicals. Additionally, in certain applications where the polymers are burned off, they form innocuous CO_2 and water (H_2O) as by-products of combustion. Similar nontoxic by-products would form if these polymers were incinerated as waste materials. The polymers burn very cleanly, leaving little residue. This makes them ideally suited for certain ceramic binder applications and for evaporative pattern casting.

Certain PAC copolymers, for example, poly(ethylene carbonate) (PEC), will biodegrade in animals, which portends possible medical uses in humans. PEC also has excellent O_2 barrier properties for food packaging. Films formed from PEC and poly(propylene carbonate) (PPC) are clear, amorphous and tough; they have good abrasion resistance and are resistant to oils, gasoline and many solvents.

These films have also been found to weather well and to resist discoloration. Polymers can be made with glass transition temperatures T_gs ranging from below 15 °C to above 100 °C, depending on the epoxide used, and can produce soft to hard polymeric materials. The materials self-adhere and adhere well to paper and wood.

Terpolymers of CO_2 made with two other epoxides can also be formed, producing another family of polymers with desirable properties not obtainable with the copolymers. In general, the PAC polymers can be custom-tailored to provide properties that are needed for a wide variety of uses. This flexibility has important commercial implications.

1. Synthesis

1.1 Copolymers

In the mid-1960s Professor Shohei Inoue and colleagues in Japan first successfully copolymerized epoxides with CO_2 using an organozinc catalyst system (Inoue et al. 1969a,b). The general equation is

$$\mathrm{RHC{-}CHR\ (epoxide,\ -O-)} + CO_2 \longrightarrow \left[\mathrm{CHR{-}CHR{-}O{-}\overset{\overset{\displaystyle O}{\|}}{C}{-}O} \right]_n \qquad (1)$$

The organozinc system is believed to involve the insertion of CO_2 into a zinc alkoxide, as shown in Eqn. (2a) and (2b) (Inoue and Yamazaki 1982):

$$\sim\mathrm{CHR{-}CHR{-}O{-}ZnX} + CO_2 \longrightarrow \sim\mathrm{CHR{-}CHR{-}O{-}\underset{\underset{\displaystyle O}{\|}}{C}{-}O{-}ZnX} \qquad (2a)$$

$$\sim\mathrm{CHR{-}CHR{-}O{-}\underset{\underset{\displaystyle O}{\|}}{C}{-}O{-}ZnX} + \mathrm{RHC{-}CHR\ (epoxide,\ -O-)} \longrightarrow \sim\mathrm{CHR{-}CHR{-}O{-}\underset{\underset{\displaystyle O}{\|}}{C}{-}O{-}CHR{-}CHR{-}O{-}ZnX} \qquad (2b)$$

The conditions for this polymerization reaction are mild, with temperatures approximately ambient and pressures of $1\text{–}50 \times 10^5$ Pa. Aliphatic polycarbonates having molecular weights in the range 30 000–200 000 are possible. In addition to ethylene and propylene oxide, other epoxides such as styrene, isobutylene, butylene and cyclohexene oxides can be reacted.

1.2 Terpolymers

Block copolymers or terpolymers, in which CO_2 is reacted with two epoxides, can also be synthesized (Santangelo et al. 1987). In this two-step reaction, an alkylene oxide is first reacted with CO_2 in an organic solvent in the presence of organometallic catalysts to form a PAC. A second alkylene oxide is then added to the initial PAC to form a block copolymer that is covalently linked, forming distinct alkylene carbonate blocks. The structural formula of the block copolymer is shown in Eqn. (3):

$$\left[\mathrm{\underset{\displaystyle CH_2}{\overset{\displaystyle R_1}{|}}{-}CH_2{-}O{-}\overset{\overset{\displaystyle O}{\|}}{C}{-}O} \right]_x {-} \left[\mathrm{\underset{\displaystyle CH}{\overset{\displaystyle R_2}{|}}{-}\underset{\displaystyle CH}{\overset{\displaystyle R_3}{|}}{-}O{-}\overset{\overset{\displaystyle O}{\|}}{C}{-}O} \right]_y \qquad (3)$$

Through block copolymerization, a range of commercially attractive molecular weights can be produced. A block copolymer of poly(cyclohexene carbonate) and PEC is but one example of the terpolymers that have been synthesized.

2. Properties

Properties of PAC polymers cover a wide range, since they are a function of the specific polymer produced. Values of T_g varying from less than 15 °C to above 100 °C are possible. Hence, polymers that range from elastomers to engineering materials can be produced. Properties of two typical potentially low-cost polymers, PEC and PPC, are shown in Table 1.

2.1 General Properties

Table 1 compares general properties of PEC and PPC. Noticeable differences are tensile strength, elongation, T_g and gas barrier properties. The low oxygen permeability of PEC, plus its good adhesion, clarity, toughness, clean by-products of combustion and potential low cost, make it ideal for food packaging applications. Also, low T_g, good oil resistance and adhesion to paper make PEC a candidate for hot-melt adhesives in the paper and board packaging industries.

2.2 Decomposition and Combustion

Most PAC copolymers decompose and burn cleanly and completely in air or nitrogen. PPC decomposes primarily into propylene carbonate, a high-boiling liquid which on further combustion decomposes into CO_2 and H_2O. Ash levels are very low with trace levels of contaminants.

When heated in air to 300 °C or nitrogen to 360 °C, PPC decomposes endothermically and completely, without tailing.

Both PEC and PPC burn with an almost invisible flame without soot or smoke. The resulting combustion products CO_2 and H_2O are environmentally attractive for many applications. Figure 1 illustrates these properties.

2.3 Chemical Resistance

Most PAC polymers are soluble in chlorinated hydrocarbons such as methylene chloride and other common solvents; for example, acetone and ethyl

Table 1
Typical properties of two PAC copolymers under development

Property	PEC	PPC	Test method[a]
Average molecular weight	50 000	50 000	GPC
Density (kg m^{-3})	1420	1320	ASTM D-792
Tensile strength (MPa at 23 °C)	4.1–4.8	48	
Elongation (%)	900–1000	9–20	ASTM D-882
Tensile modulus (MPa)	190	2070	
Melt flow index (g (10 min)$^{-1}$ at 150 °C/2, 160 g)	no flow	0.9	ASTM D-123
Hardness		79 (Shore D)	
Impact, Izod (J m^{-1})		16–32	ASTM D-256
H_2O absorption at 23 °C (%)	0.4–0.6	0.4	ASTM D-570
Refractive index	1.470	1.463	ASTM D-442
Dielectric constant (10^3 Hz)	4.32	3.00	
Loss tangent (10^3 Hz)	0.031	0.007	
Haze (%)		3.6	
Decomposition temperature (°C)	220	250	TGA
T_g(°C)	25	40	DSC
Heat of combustion (MJ kg^{-1})	9.3	10	
Permeability at 20 °C (cm^3 mm m^{-2} d^{-1} bar^{-1})[b]			
O_2[c]	0.4–0.8	3.9–9.8	
CO_2	9.1	44.1	
N_2	4.7	3.5	
H_2O[d]	1.2–5.9	1.2–7.1	

a GPC, gel permeation chromatography; TGA, thermogravimetric analysis; DSC, differential scanning calorimetry b Standard UK units; standard US units are cm^3 (10^{-3} in) (100 in^2)$^{-1}$ d^{-1} atm^{-1} at 68 °F with (US units) = 2.54 × (UK units) c For comparison, the oxygen transmission rate of poly(vinylidene chloride) with a density up to 1790 kg m^{-3} measures 0.2–0.8 cm^3 mm m^{-2} d^{-1} bar^{-1} d g mm m^{-2} d^{-1} bar^{-1}

acetate. They are also resistant to many common substances, including most alcohols, acids, water, ethylene glycol and aliphatic hydrocarbons. The solubility in common solvents permits extrusion of fibers and films and the formulation of solvent-based coatings. The resistance to many common substances increases the end-use possibilities.

2.4 Other Properties

Both PEC and PPC are amorphous, clear, easy to process and have long-term mechanical stability. They are aliphatic and free of carbon–carbon unsaturation, making them weather resistant and colorless.

These polymers adhere extremely well to paper, wood and other cellulosic materials. They can also be compounded to adhere to polyester, polyethylene, aluminum and glass. The polymers form tough printable films with good optical quality and heat-seal strength. The abrasion resistance of PPC is excellent.

PEC has been found to biologically degrade in animal studies (Kawaguchi et al. 1983). Recent preliminary testing also indicates slow biodegradability in soils for both PEC and PPC. The implications of these properties for medical applications and disposable packaging are obvious.

3. Applications

3.1 Binders for Ceramics and Powdered Metals

Because of their unique blend of properties, PAC polymers are excellent binders for ceramics and powdered metals. They decompose completely in air at temperatures at least 100 °C below that of many commercial binders, as indicated in Fig. 2. PAC binders also burn out mildly, without violent gas formation in oxidizing, reducing or inert atmospheres, thus producing less rejects due to cracking. As a result, binder burnout times can be significantly reduced. The completeness of the burnout produces strong sintered parts that are virtually contaminant free. For pure binder alone, the ash residues are typically less than 50 ppm. At the 3% binder level, residual ash in the finished part is well under 2 ppm. Ash levels of the binder alone are given in Table 2.

3.2 Evaporative Pattern (Metal) Castings

Evaporative Pattern (Metal) Casting (EPC), also called lost foam metal casting, is a technology now undergoing commercialization. The process consists of forming the finished product (e.g., automotive part) in expanded polystyrene. The polystyrene form is then embedded in sand. Molten aluminum is next poured into the polystyrene form, burning away the polymer form and leaving behind a metal casting. This process has limitations, however, since polystyrene leaves behind carbon residues and generates undesirable gases. Casting of iron parts causes additional problems, precluding its general use with iron.

Use of PAC to replace polystyrene is currently under development. PACs are advantageous because they burn cleanly and do not produce reactive carbon. This not only makes them attractive in the aluminum-casting market, but also opens up the possibilities for other casting applications, including iron and stainless steels.

3.3 Future Possibilities

The unique composition and properties of PAC polymers provide the potential for superior performance in both existing and developmental applications. They are economically and ecologically attractive, process easily and can be tailor-made for specific applications or end uses. As commercial development continues, this versatile polymer family should find applications in packaging, adhesives, castings, polyols, insulation, sealants, medical products, polymer alloys and binders.

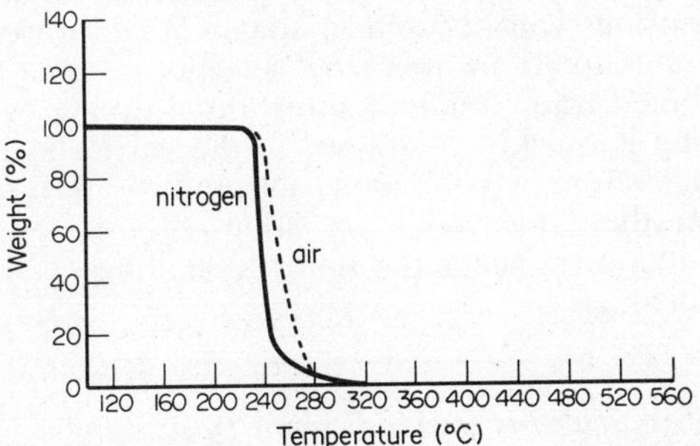

$$\left[-CH(R)-CH_2-O-C(=O)-O-\right]_n \xrightarrow{N_2 \text{ or air}} \text{(cyclic carbonate, R-substituted)}$$

major product

(a)

$$\left[-CH(R)-CH_2-O-C(=O)-O-\right]_n \xrightarrow{\text{air or } O_2} CO_2 + H_2O$$

(b)

Figure 1
PAC copolymers decompose and burn cleanly and completely—thermogravimetric analyses in both air and nitrogen show complete decomposition of PPC at low temperatures: (a) pyrolysis and (b) combustion (heating rate is 10 °C min^{-1})

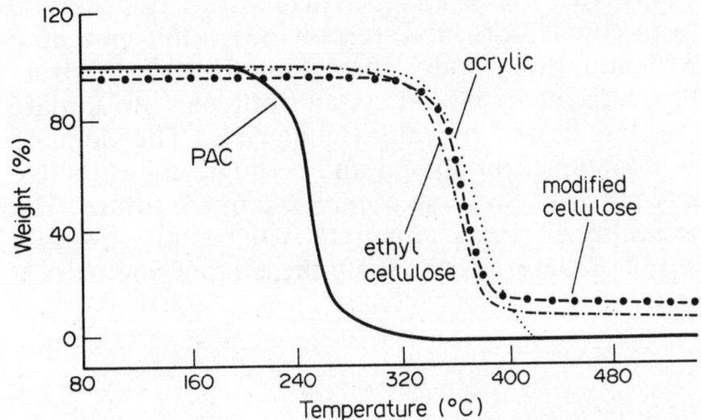

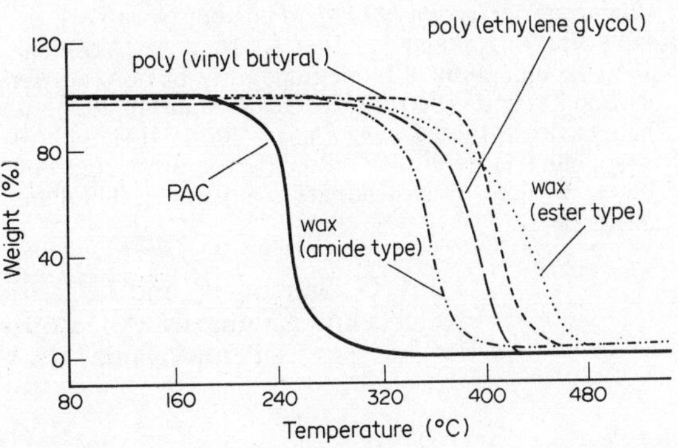

Figure 2
Thermogravimetric analysis of various binders in nitrogen illustrating the clean decomposition of a PAC binder at low temperatures

Table 2
Typical postdecomposition analyses

Element	Ash (ppm)
Sodium	<5[a]
Zinc	6.9
Iron	5.7
Lead	<5
Aluminum	<4
Calcium	2
Copper	1.8
Molybdenum	<1
Nickel	<1
Chromium	<0.5
Magnesium	<0.4
(Total)	<50

a Below limits of detection

Now that CO_2 can be reacted with epoxides to form copolymers and terpolymers, the potential exists to further modify the polymer backbone structure with other monomers and polymers to initiate new families of polymeric materials. The demand for economic, functional and ecologically attractive polymers will continue to increase in the future. The versatility of PAC polymers will permit this new family of materials to satisfy these emerging market needs.

Bibliography

Inoue S, Koinuma H, Tsuruta T 1969a *Polym. Lett.* 7: 287

Inoue S, Koinuma H, Tsuruta T 1969b *Makromol. Chem.* 130: 210

Inoue S, Yamazaki N 1982 *Organic and Bioorganic Chemistry of Carbon Dioxide*. Halsted, New York

Kawaguchi T, Nakano M, Juni K, Inoue S, Yoshida Y 1983 Examination of biodegradability of poly(ethylene carbonate) and poly(propylene carbonate) in the peritoneal cavity in rats. *Chem. Pharm. Bull.* 31(4): 1400–3

Santangelo J G et al. 1987 Process for producing novel block alkylene polycarbonate copolymers. US Patent No. 4,665,136

J. G. Santangelo and J. C. Tao
[Air Products and Chemicals, Allentown, Pennsylvania, USA]

Porcelain Enamelling Technology: Recent Advances

This article supplements the article *Porcelain Enamelling Technology* in the Main Encyclopedia.

Traditional porcelain (vitreous) enamels suffer from poor resistance to impact. This can lead to relatively high rework and scrap levels in manufacture as well as deterioration in service, particularly in domestic appliances. Enamelling has been gradually replaced by tougher paints on many products. Since the early 1960s, attempts have been made to improve the chip resistance of porcelain enamels by including metallic particles to act as crack deflectors and stoppers (US Patent No. 2,900,276; GFR Patent No. 28,29,959; UK Patent No. 1,413,713; European Patent No. 0,036,558). These enamel–metal "cermets" have been only partially successful because any increased toughness is counteracted by high levels of unpredictable contiguous porosity which leads to rough surfaces and poor corrosion performance, precluding their wide acceptance as single-coat systems. However, they have found application as catalytic self-clean oven coatings where their large surface area is advantageous.

In the early 1980s, it was recognized by TI Research that the porosity problem in aluminum-containing enamel cermets could be overcome by using low-hydroxyl frit and special processing conditions. A new genus of glass cermet, the dual phase vitreous enamels (DPVE), was developed (US Patent No. 4,555,415) that can be tailored to meet specific property requirements. These have high integrity, are tough, have improved thermal resistance and excellent adhesion. They can be applied as thin or thick single coats to a wide range of steels, irons and other materials without the necessity for stringent precleaning and pretreatment.

1. Principles of DPVE

The foaming porosity of early enamel–aluminum cermets is largely attributable to the reaction between water and aluminum which releases hydrogen during the later stages of fusing. A number of potential sources of water exist during fusing; for example, the frit itself (see *Gases in Glass*; *Water in Glass*), the fusing furnace atmosphere and the hydrated mill additions. Reduction of porosity to an acceptable level requires that the enamel frit has a low hydroxyl content, fusing is carried out in a dry atmosphere, and the slurry is prepared as a water-free, nonaqueous suspension. The third condition although desirable, is not essential and aqueous suspensions can be used in some cases. However, the mill additions must be selected so that any combined water retained after initial drying of the coating is quickly volatilized in the early stages of fusing before a continuous coverage is achieved. When these principles are adopted, low-porosity high-integrity enamel–aluminum coatings can be obtained.

2. Modifications to Traditional Processing

2.1 Preparation of Porcelain Enamel Frit

Standard enamel frits contain between 0.04 wt% and 0.1 wt% of dissolved water. The most convenient way to reduce this to the levels required for DPVE (<0.02 wt%) is to bubble a dry gas through the molten frit. Argon or dry air may be used in suitably modified melting furnaces and this adds only 1–2 h to the processing cycle. Modifications to standard frit compositions may be necessary to compensate for partial removal of the more volatile species by bubbling. Frits can be water quenched after dewatering without significant rehydration, although dry quenching is preferred.

2.2 Application of Coatings

Slurries are prepared by first ball milling the frit with mill additives (see *Porcelain Enamelling Technology*). Aluminum powder additions are made to the slurry after it has been removed from the ball mill; compositions containing up to 30 wt% of aluminum have been successfully utilized. Coatings

Table 1
Examples of DVPE slurry formulations (in parts by weight)

	Nonaqueous system	Aqueous system
Dehydrated frit	100	100
Amyl acetate	50[a]	
Cellulose nitrate	1.5[a]	
Water		50
Acid buffer		1
Xanthan gum		0.25
Sodium nitrite		0.5
Urea		1
Aluminum powder	25[b]	25[b]

a previously dissolved together b added after milling

may be prepared by either nonaqueous or aqueous routes and typical compositions of some alternatives are given in Table 1. The type and quantity of mill additions used in the aqueous route are especially critical in providing good rheological properties in the slurry without reintroducing water into the frit during fusing (standard mill additions are unsuitable). The use of aqueous slurries requires some safety precautions to be taken when aluminum is added. Due to the alkaline nature of most aqueous slurries, aluminum must never be added into the ball mill in case a violent reaction occurs. The slurry is buffered to reduce the pH to values between 7 and 8 approximately, and the aluminum blending is best done in a high-shear mixer. The shelf life of the slurries can be many months but they should be stored in vented containers. Application is possible by dipping, flow coating or spraying. Aluminum-free gloss cover coats can also be applied if required.

2.3 Fusing

When the coating is heated above its softening point (~550 °C) during fusing, diffusion of hydroxyl ions into the frit particles can occur rapidly; hence the need to ensure that mill additions to aqueous slurries allow rapid vaporization of both free and bound water. Of equal importance is the need to control the moisture content in the fusing furnace to minimize the release of hydrogen from the reaction with aluminum; the dew point should be less than about −5 °C (this is dependent on composition and fusing temperature). Standard fusing furnaces may be retrofitted with appropriate moisture control systems (UK Patent Application No. 8,622,061).

2.4 Substrates

DPVE coatings can be applied to carbon steels, gray irons, stainless steels and copper. Steels and irons that are traditionally not amenable to enamelling can be DPVE coated (e.g., hot-rolled and many Concast grades). The problem of fishscaling (see *Porcelain Enamelling Technology*) is virtually absent. This is possibly because the availability of hydrogen at the steel interface is greatly reduced due to the frit being dewatered and aluminum particles acting as sinks for hydrogen.

2.5 Process Control

Realization of the benefits of DPVE technology demands a greater degree of process control in order to maintain the water content at low levels. For example, close monitoring of the hydroxyl content of the smelted frit is required, for which customized analytical equipment has been specially developed. Although, historically, enamelling technology has tolerated wide variations in production parameters, the tighter controls demanded by DPVE technology are no greater than those employed in a wide range of metallurgical operations and are less stringent than those needed in many modern surfacing technologies.

3. Properties and Benefits of DPVE

3.1 Toughness and Adhesion

These parameters are difficult to quantify for porcelain enamels. Bending coated sheets around a mandrel gives a measure of the resistance of the enamel to cracking. DPVE coatings containing greater than 10 wt% of aluminum can be bent around a radius of approximately 10 mm before cracks become visible, whereas coatings of the same thickness containing no aluminum will crack and spall when bent around a 50 mm radius.

Adhesion can be judged from a standard enamel drop weight test. Whereas a reasonably well-bonded enamel with no aluminum additions will detach in large flakes under the area of impact partially revealing the steel, which then readily corrodes, a DPVE coating can still cover the steel when tested under identical conditions. Microscopic examination shows the heavily deformed region to be crazed on a fine scale (see Fig. 1). At a higher magnification (see Fig. 2) evidence is seen of some of the aluminum particles undergoing extensive deformation while remaining bonded to the enamel. This supports the views propounded by Krstic et al. (1981) and illustrates the role of aluminum in absorbing fracture energy. The coatings are so well bonded to the steel substrate that they can, under some conditions, delaminate in the region of the reaction zone, away from the enamel–steel interface, leaving a thin protective layer of coating. The role of aluminum in promoting very high adhesion is to accelerate the reactions involved with the adhesion-promoting oxides of nickel and cobalt. Adhesion is developed by the formation of a mechanical bond between micrometer-sized protrusions growing from the substrate, the action of the aluminum being to accelerate the decrease in FeO solubility and its reduction to produce α iron surface growths. In fact, the current model which asserts that optimum adhesion is obtained when the FeO concentration at the steel–enamel interface is maximized (see *Porcelain Enamelling Technology*) appears not to be valid (whether or not aluminum additions are made to the enamel).

This high degree of adhesion and toughness can be obtained without the need for expensive pickling and/or nickel flash treatments (although derusting and degreasing are necessary) and it is possible to carry out limited postenamelling manipulation without incurring significant property degradation (for example, bending and hole-punching).

3.2 Corrosion Resistance

For aluminum additions of 10 wt% or less, DPVE coatings have similar, excellent salt spray test resistance to standard enamels. At higher levels, white corrosion specks appear at near-surface aluminum particles but a high level of protection of the substrate is maintained. This is particularly so for damaged areas where the well-adhered, but finely crazed, coating affords a considerable degree of protection, inhibiting substrate rusting.

High-temperature oxidation protection and thermal shock resistance are key benefits of aluminum-containing enamels (Wratil 1981). The latter arises from the toughness and adhesion characteristics. Oxidation protection is possible at temperatures near the fusing temperature of the coating (~800 °C). This arises because aluminum acts as a sink for free oxygen, converting it to alumina which reacts with the glass matrix to increase its viscosity.

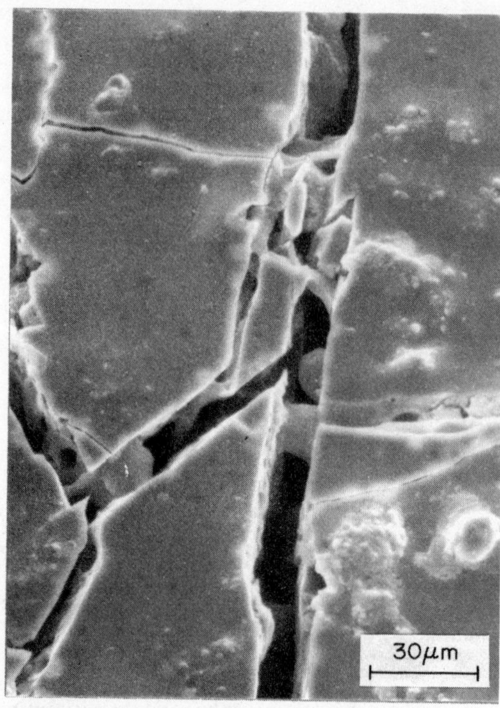

Figure 2
Enlargement of part of Fig. 1 showing an aluminum ligament bridging a crack in the enamel. Note, also, depressions on the fracture face due to aluminum particle pullout and porosity (magnification × 500)

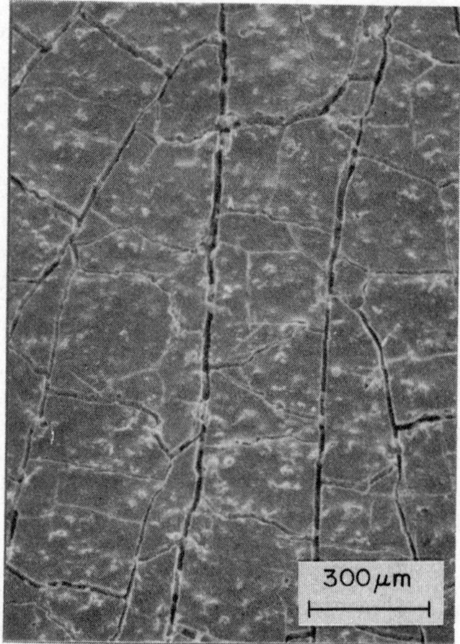

Figure 1
Scanning electron micrograph of a DPVE coating deformed in a drop weight impact test. Microcracking on a 100–300 μm scale is evident (magnification × 50)

3.3 Surface Finish and Color

Additions of aluminum reduce gloss and surface smoothness and the use of dewatered frit has eliminated the problems encountered by earlier workers of rough, blistered and highly porous finishes. Color is restricted to dark hues (e.g., browns and blacks) and, like surface finish, is dependent on the level of aluminum addition and on fusing conditions. However, a significant range of aesthetic characteristics can be tailored into the coating by changing processing conditions and coloring additives.

3.4 Cost Implications

The need to dewater and fuse under low-humidity conditions as well as the addition of aluminum powder increase enamelling costs (typically by ~30%). This is offset by, first, the fact that thinner (~75 μm) coatings can be produced compared to the conventional 100–250 μm range and, second, that much cheaper substrates can be used without the need for expensive pickling and nickel flash pretreatments. These can lead to overall cost savings of up to 20%. However, even greater savings arise from the reduced levels of scrap and rework from mechanical damage to which enamelled components are especially prone in assembly operations.

4. *Applications*

DPVE is restricted, at present, to applications in which semigloss, satin, mat or dark finishes are acceptable for single coats, or in which it is used as a ground coat. This could encompass approximately 40% of the porcelain enamelling market. DPVE is being tailored for volume production in the domestic appliance sector. Incremental applications, replacing nonenamel coatings in areas where conventional porcelain enamels have had little commercial success, are clearly possible and are being pursued. There is also scope for replacing some high-performance materials, such as stainless steels, in applications where DPVE coatings are compatible with the environment.

5. *Future Developments*

The major improvements to some of the properties of porcelain enamel, described in previous sections, have resulted from advances in technology. A detailed scientific understanding of the mechanisms controlling adhesion and toughness is needed in order that the range of applications of these enamel–metal cermets can be extended and their properties further enhanced. It would be advantageous to relax the quality control requirements by widening the tolerance bands associated with some of the processing variables; improved fundamental understanding will undoubtedly be of benefit here.

The feasibility of using DPVE as a high-adherence, tough ground coat for full gloss, traditional cover coats has been demonstrated; however, detailed quantification of the technical and cost benefits of composite coatings and the possible use of DPVE as an interlayer bonding material have yet to be carried out.

Bibliography

Krstic V V, Nicholson P S, Hoagland R G 1981 Toughening of glasses by metallic particles. *J. Am. Ceram. Soc.* 64: 499–504

Maskall K A, White D 1986 *Vitreous Enamelling, A Guide to Modern Enamelling Practice*. Pergamon, Oxford

Vargin V V 1967 *Technology of Enamels*. Maclaren, London

Wratil J 1981 High temperature enamels with outstanding qualities. *Vitr. Enamel.* 32: 42–7

Wratil J 1984 *Vitreous Enamels*. Borax Holdings, London

R. F. Price
[Cilgerran, Cardigan, UK]

N. Fletcher
[Cheswick UK, Blackpool, UK]

M. J. Stowell
[Alcan International, Banbury, UK]

Portland Cement Raw Materials

Portland cement is manufactured by firing crushed mixtures of a wide range of raw materials in kilns to form a clinker. The clinker is cooled and then finely ground with about 5 wt% gypsum ($CaSO_4 . 2H_2O$) or anhydrite ($CaSO_4$), and usually with a small amount of organic air-entraining agent to make it resistant to frost action. The most usual combination of materials fed into cement kilns is 14 wt% SiO_2, 3 wt% Al_2O_3, 3 wt% Fe_2O_3, 75 wt% $CaCO_3$, 4 wt% $MgCO_3$ and 0.6 wt% alkali (sodium and potassium oxides). Of these constituents, the first four are essential; the latter two, $MgCO_3$ and the alkalis, are tolerated as inevitable impurities.

The raw materials used are mostly natural rock but several industrial by-products may be used when available. By far the most common raw material used for the manufacture of Portland cement is limestone and most cement plants are located at their sources of limestone. The required constituents may be almost wholly within the limestone, as in the case of cement rock, or may be transported to the cement plant from some distance.

1. *Portland Cement Production*

Portland cement, because it is basic to most major construction, is produced in most countries of the world. World production for 1983 was 925 642 kt. The largest production, by the USSR, was 128 000 kt, followed by the People's Republic of China, 108 250 kt, Japan, 80 650 kt and the USA, 64 725 kt (Davis and Johnson 1983). Production of Portland cement is primarily related to the amount of heavy construction work, such as highways, dams and buildings, that is being carried out. Portland cement is produced in 117 countries and most areas underlain by limestones are geologically capable of producing the raw materials.

There are five basic types of Portland cement and all consist of similar compounds. The typical constituents from which the five primary types of Portland cement are made are summarized in Table 1, along with the critical physical characteristics of each type. The oxides of Table 1 are fired in kilns at temperatures of 1425–1650 °C, the CO_2 is driven off and the

Table 1
Typical analysis of constituents (wt%) for the five basic types of Portland cement

	Constituents (wt%)						
	SiO_2	Al_2O_3	Fe_2O_3	$CaCO_3$	$MgCO_3$	Alkalis	Critical physical characteristic
Type I	14.1	4.2	1.6	75.8	3.5	0.9	standard
Type II	14.7	3.2	2.6	74.7	4.2	0.6	moderate heat
Type III	13.7	3.6	2.0	77.3	3.0	0.4	high early strength
Type IV	16.3	3.3	3.0	72.5	4.2	0.7	low heat
Type V	12.4	1.6	1.0	76.0	3.5	0.5	sulfate resistant

oxides are combined to form a clinker which consists essentially of the compounds listed in Table 2.

Normally, Portland cements also contain 4–6 wt% gypsum or anhydrite, which is added after firing as an essential constituent to regulate the setting time of concretes and mortars. Organic air-entraining agents, to the extent of a percent or so, are also usually present in Portland cement.

A typical oxide composition of type I Portland cement with gypsum added is 63.8 wt% CaO, 20.7 wt% SiO_2, 6.2 wt% Al_2O_3, 2.5 wt% MgO, 2.4 wt% Fe_2O_3, 2.2 wt% SO_3, 1.3 wt% alkali and 1.0 wt% H_2O.

2. *Raw Materials*

The raw materials which are used in Portland cement manufacture must be inexpensive and abundant. Table 3 summarizes most materials which have been used in cement manufacture. The cement manufacturing plant is generally located at the source of $CaCO_3$, its largest constituent. Other necessary but minor constituents, such as iron oxides or calcium sulfate, may be transported to the plant from considerable distances.

Barton (1967) has given two useful summaries pertaining to raw materials used for Portland cement in the USA. These statistics are probably not very different on a worldwide scale and on this basis his tabulations, in modified form, are produced in Tables 4 and 5. Barton's statistics remain fairly accurate for the USA although certain other minor constituents such as fluorspar (CaF_2), pumicite (frothy volcanic glass), $CaCl_2$, staurolite ($HFeAl_5Si_2O_3$), fly ash (silica-rich coal ash) and diatomite (microscopic silica shells) have been used, as available and as needed. Most constituents used in Portland cement are impure, with the most common impurities also being common constituents in Portland cement. Thus most raw materials are sources of more than one required oxide.

In using materials for Portland cement manufacture, care must be taken that nonessential constituents are not excessive. Examples of the most common of such constituents are the alkalis (sodium and potassium oxides) and magnesia. The alkalis do not improve the quality of Portland cement and quantities larger than 0.6% total by weight Na_2O equivalent (percentage of $Na_2O + 0.658$ times the percentage of K_2O) tend to react with some forms of silica (opal, chalcedony, cristobalite, tridymite and finely divided quartz) to create hydrous alkali silicates which cause expansion of concrete leading to deterioration. Portland cements with higher alkali contents (2 wt% or more) are manufactured, but care must be taken to use nonreactive aggregate or to neutralize the expansive reactions in some way. Excessive magnesia in Portland cements may also create expansive reactions and therefore the amount of MgO in the final product should generally be less than about 4.0 wt%.

Any other constituent not essential to cementitious reactions is also considered to be deleterious. For example, sulfur trioxide should not be present in quantities more than 3.0–3.5 wt%. Another example is strontium (as $SrSO_4$) which may be excessive in some limestones.

2.1 *Calcium Carbonate*

The most common source of calcium carbonate is limestone, with probably over 95% of the calcium oxide sources in the world being variations of limestone or its metamorphic equivalent, marble. Rocks rich in calcium carbonate occur in nearly all countries.

Limestones result from cementation or lithification of accumulations of calcium carbonate as calcite or aragonite in the form of shells, oolites or lime

Table 2
Analysis of compounds (average wt%) in type I Portland cement clinker (after Ames and Cutcliffe 1983)

Compound	Oxide composition	wt%
Tricalcium silicate (alite)	$(CaO)_3SiO_2$	45
Dicalcium silicate (belite)	$(CaO)_2SiO_2$	27
Tricalcium aluminate	$(CaO)_3Al_2O_3$	11
Tetracalcium aluminoferrite	$(CaO)_4(Al_2O_3)(Fe_2O_3)$	8

Table 3
Sources of raw materials used in the manufacture of Portland cement clinker (after Ames and Cutcliffe 1983)

Sources	Raw materials
Calcium carbonate	limestones
	lithified limestones, chalk, marble, marl, oyster shells (reef), coquina shells, aragonite sand, slag
Silica	sand, sandstone or quartzite
	clay and claystone
	shale
	loess
	slag
	mill fines
	fly ash
Alumina	shale
	clay and mud
	loess
	slag
	fly ash
	bauxite
	alumina process waste (red mud)
	aluminum dross
	staurolite
	mill fines (granite)
	pumice or other volcanic material
Iron	iron ores
	blast furnace flue dusts
	pyrite cinder
	mill scale
	fly ash

muds with most limestones originating from varying combinations of these materials. Chalks are poorly cemented accumulations of microscopic calcium carbonate shells; marls are poorly cemented impure, often clay-rich, lime muds, usually rich in shells, and may be deposited either in fresh or marine water; and calcitic marbles are the metamorphosed and recrystallized versions of these materials.

Ideally, the easiest material to use as a raw material for Portland cement is cement rock. This is an impure limestone with a composition such that the

Table 4
Percentage of raw materials used in cement manufacture in the USA in 1965 (after Barton 1967)

Raw material	Percent
Cement rock	16.5
Limestone and shells	67.9
Marl	0.5
Sand and sandstone	1.5
Iron materials	0.6
Clay, shale and schist	9.4
Blast-furnace slag	0.8
Gypsum	2.7
Miscellaneous items	0.1

Table 5
Raw material mixes used at the 181 cement plants in the USA in 1965 (after Barton 1967)

Ingredients	Percentage of plants using mix
Cement rock or limestone only	18.8
Cement rock and limestone	7.2
Marl and limestone	0.6
Shells, marl and limestone	0.6
Limestone and/or cement rock plus clay or bauxite	29.8
Limestone or cement rock plus shale	25.4
Shells and clay	5.5
Limestone plus shale and clay	3.9
Cement rock or limestone plus slag	3.9
Limestone plus clay and slag	1.7
Limestone plus shale and slag	1.7
Shells, marl and clay	0.6
Shells, clay and marl	0.6

removal of carbon dioxide leaves an oxide composition which approximates the ideal composition of common type I or type II Portland cements without the gypsum or anhydrite. Rock of this type occurs in the Lehigh Valley of Pennsylvania where it has long been used for Portland cement manufacture. Table 6 gives three typical partial compositions of cement rock from the Lehigh Valley, along with the composition of the "average" limestone for reference.

Rock raw materials in nature are seldom compositionally homogeneous, even where cement rock is used with no additives, other than calcium sulfate and air-entraining agents. As a result, careful quarrying and blending of rock layers from within the quarry are usually needed for quality control.

Loosely consolidated materials such as oyster shells, coquina or coral-shell accumulations or oolitic aragonite (sands consisting of spherical aragonite

Table 6
Three typical partial compositions of cement rock from the Lehigh Valley, Pennsylvania, together with the composition of "average" limestone for comparison

	Composition (wt%)			
Constituent	Copley[a]	Whitehall[b]	Lehigh[c]	Average Limestone[d]
$CaCO_3$	71.8	79.3	74.6	76.03
$MgCO_3$	4.3	4.0	4.8	16.58
SiO_2	14.8	13.2	14.0	5.19
Al_2O_3	6.6	2.1	3.9	0.81
Fe_2O_3	1.6	1.1	1.3	0.54

a Copley Cement Company; Deasy et al. 1967 b Whitehall Cement Company; Deasy et al. 1967 c Lehigh Portland Cement Company; Deasy et al. 1967 d Mason 1966 grains), all occurring in shallow marine conditions, are also sources of calcium carbonate in some marine coastal areas.

Slags which result from processing iron ore to iron may be rich enough in calcium oxide to be a source of the CaO constituent. Certain slags have compositions which are relatively close to that of Portland cement and the manufacture of Portland cement from them may be a relatively easy process. A typical slag from an air-cooled blast furnace consists of 32–42 wt% SiO_2, 7–16 wt% Al_2O_3, 32–45 wt% CaO and 5–15 wt% MgO (modified from McCarl et al. 1983). A slag of this type which is low in MgO (less than 6 or 7 wt%) can be combined with high-calcium limestone or marble to approximate an ideal Portland cement composition.

2.2 *Silica*

Relatively pure silica in the form of quartz (SiO_2) occurs naturally as sand or sandstone, or their metamorphosed or recrystallized equivalent, quartzite. In addition, silica is the major constituent of clay, clay-stone, silt, shale, granites and volcanic glasses as well as loess (wind-blown dust accumulations).

In each of these latter cases alumina is the other major constituent and therefore these sedimentary materials are also sources of alumina. Somewhat arbitrarily, those sources of silica in which the silica is over four times the weight percentage of alumina will be summarized in the chemical analyses in this section. Lower-ratio materials will be summarized in Sect. 2.3. The reason for this separation is that the average Portland cement contains about four times as much silica as alumina. If the percentage of alumina in Portland cement feed is approximately correct, and silica must be raised, material in which the weight percentage of silica is over four times greater than that of alumina would have to be added so as not to raise the alumina excessively. Typical compositions of siliceous materials which may be used are given in Table 7.

Commercial waste products such as slag, mill fines and fly ash are also commonly sources of silica. Mill fines from the manufacture of granite products have the composition of granite (Table 7). According to Barton, typical fly ash consists of 44–51 wt% SiO_2, 13–26 wt% Al_2O_3, 7–15 wt% Fe_2O_3, 1.6–12 wt% CaO, 0.9–2.7 wt% MgO and 0.2–16 wt% C, and has an ignition loss of 9–17% (Clausen 1960).

Siliceous limestones, often rich in quartz sand and/or cherts, may be blended with limestones lower in silica, especially where both silica-rich and silica-poor limestone are in close proximity.

2.3 *Alumina*

Most sources of alumina are also sources of silica and several have been discussed in Sect. 2.2. Some typical compositions of alumina-rich raw materials which may be used in Portland cement manufacture are summarized in Table 8.

Table 7
Typical compositions (wt%) of common siliceous materials which may be used for Portland cement manufacture

Constituent	Granite[a]	Rhyolite[b]	Silica sand[c]	Sandstone[d]
SiO_2	70.18	72.35	99.70	78.33
TiO_2	0.39	0.25	0.02	0.25
Al_2O_3	14.47	13.98	0.08	4.77
Fe_2O_3	1.57	0.60	0.02	1.07
FeO	1.78	1.78		0.30
MnO	0.12			
MgO	0.88	0.30	0.01	1.16
CaO	1.99	1.30	0.01	5.50
Na_2O	3.48	5.04	0.01	0.45
K_2O	4.11	3.92		1.31
H_2^+	0.84	0.05	0.10	1.63
H_2^-		0.45		
P_2O_5	0.19	trace		0.08
CO_2				5.03
BaO				0.05
S				0.07
Total	100.0	99.33	99.95	100.0

a Average of 546 analyses (Rankama and Sahama 1950 p. 166) b Rhyolite obsidian, Newberry volcano, Oregon (Williams 1942) c Commercial silica sand, Silverado sandstone, Orange, California (Murphy 1975) d Average sandstone (Mason 1966)

2.4 *Iron Oxide*

Typical sources of iron oxides are the iron ores and various industrial by-products. Iron-ore minerals are limonite (approximately 85.5 wt% Fe_2O_3 and

Table 8
Average compositions (wt%) of common sources of alumina for Portland cement

Oxide	Shale[a]	Phyllite[b]	Kaolinite[c]	Bauxite[d]
SiO_2	58.38	45.20	46.77	3–16
Al_2O_3	15.47	37.02	37.79	55–70
Fe_2O_3	6.07	0.27	0.45	4–25
FeO		0.06	0.11	
TiO_2		1.26	0.02	2–35
CaO	3.12	0.52	0.13	
MgO	2.45	0.47	0.24	
Na_2O	1.31	0.36	0.05	
K_2O		0.49	1.49	
MnO			trace	
SO_3				
H_2O^-	5.02	1.55	0.61	10–20
H_2O^+		13.27		
CO_2	2.64	trace	12.18	
Total	97.71	94.69	100.21	

a Composite of 78 shales (Rankama and Sahama 1950 p. 222) b Composite of 11 phyllites (metamorphosed shales), central Norway (Rankama and Sahama 1950 p. 222) c Kaolinite, St. Austell, UK (Kerr et al. 1950) d Bauxite, range of compositions, France (Shaffer 1983)

14.5 wt% H_2O), magnetite (69.0 wt% Fe_2O_3 and 31.0 wt% FeO) and hematite (Fe_2O_3). The actual ores are highly variable in composition with varying amounts of SiO_2, Al_2O_3, CaO, MgO, P_2O_5 and MnO as the major impurities. Total impurities in iron-rich materials usable for cement manufacture are probably less than 25 wt%.

Typical industrial by-products which are sources of iron are:

(a) pyrite cinder from the manufacture of sulfuric acid from pyrite, 67.5–86.7 wt% Fe_2O_3 with the major impurities of 6.3–13 wt% SiO_2 and 2.1–2.8 wt% Al_2O_3 (Clausen, 1960);

(b) flue dust from steel manufacture, 80–90 wt% Fe_2O_3; and

(c) mill scale from the surface of iron in iron manufacturing, 93–98 wt% Fe_2O_3.

Red mud, a very fine, wet waste product resulting from the manufacture of aluminum, may also be used. It contains hydroxide of iron, along with some aluminum hydroxide and silica; the exact composition depends on the actual process employed and the nature of the raw materials.

As such small quantities of iron are usually needed, and the iron content of additives is so high, the impurities in the sources of iron are not usually a significant factor in Portland cement quality.

2.5 Calcium Sulfate

Calcium sulfate for Portland cement manufacture is obtained from gypsum and anhydrite rock. The latter is rarely pure as mined because it contains varying amounts of clay, calcite, dolomite, salt and silica. Typical gypsum or anhydrite rock as mined contains 85–98 wt% $CaSO_4.2H_2O$ or $CaSO_4$, respectively. Both are mined and available in the quantities required for Portland cement in most countries of the world. Controlled calcining of gypsum forms plaster of Paris, $CaSO_4 \cdot \frac{1}{2}H_2O$, which may then be ground for use in wallboards or plasters. Anhydrite is not as valuable as gypsum because it cannot be calcined to form plaster of Paris. Portland cement manufacturers can use somewhat less-pure calcium sulfate, and anhydrite is not a problem. Hence, they can use some less-expensive, lower grades of gypsum or anhydrites. The largest production of gypsum is from the USA, Canada, France, Iran and the USSR (Anon 1983).

3. Raw Materials for Special Portland Cements

Portland cement may be mixed with some other materials to enhance or alter the characteristics of Portland cement concretes for special purposes.

A "Portland granulated blast-furnace slag" consists of Portland cement plus 25–65 wt% very finely pulverized blast-furnace slag. Granulated blast-furnace slag results from rapid quenching of hot (1371 °C) slag in steam or water. Slag Portland cements are resistant to attack by sulfate in water and air, are not alkali reactive with silica-rich aggregates and have low permeability.

Portland–pozzolan cements contain finely divided reactive silica and are often used when Portland cements are very high in alkali and therefore tend to be reactive with silica in mineral aggregates. The 15–40 wt% of finely ground silica-rich materials may rapidly react chemically with the calcium and alkali hydroxides in Portland cement, using up the alkali and minimizing later deleterious reactions which could occur by reaction of alkalis with silica on aging. Some typical siliceous pozzolanic materials are fly ash, clay, diatomaceous earth, opaline chert and shale, tuff, and volcanic ash or pumicite.

See also: Calcium Aluminate Cements; Portland Cements, Blended Cements and Mortars

Bibliography

Ames J A, Cutcliffe W E 1983 Construction materials. Cement and cement raw materials. In: Lefond S J (ed.) 1983 *Industrial Minerals and Rocks*, 5th edn. American Institute of Mining, Metallurgical and Petroleum Engineers, New York, pp. 133–59

Anon 1983 Gypsum-optimism within a broad market. *Ind. Miner.* 193: 19–59

Barton W R 1967 Raw materials for manufacture of cement. In: Faber J H, Capp J D, Spencer J D (eds.) 1967 *Fly Ash Utilization, Proc. Edison Electric Institute–National Coal Association–Bureau of Mines Symp*. US Bureau of Mines, Washington, DC, pp. 46–51

Clausen C F 1960 Cement materials. In: Gilson J L (ed.) 1960 *Industrial Minerals and Rocks*, 3rd. edn. American Institute of Mining, Metallurgical and Petroleum Engineers, New York, pp. 203–31

Davis L L, Johnson W 1983 Cement. *Minerals Yearbook 1983*. US Bureau of Mines, Washington, DC

Deasy G F, Griess P R, Bolazik R F, Burtnett J W 1967 *The Atlas of Pennsylvania Mineral Resources*, Mineral Resources Report 50. Pennsylvania Geological Survey, Harrisburg, PA

Kerr P F, Hamilton P K, Pill R J, Wheeler G V, Lewis D R, Burkhardt W, Reno D, Taylor G L, Mielenz R C, King M E, Schieltz N C 1950 *Analytical Data on Reference Clay Materials*, Report No. 7. American Petroleum Institute, Project 49, Columbia University Preliminary

McCarl H N, Eggleston H K, Barton W R 1983 Aggregates—Slag. In: Lefond S J (ed.) 1983 *Industrial Minerals and Rocks*, 5th edn. American Institute of Mining, Metallurgical and Petroleum Engineers, New York, p. 111–31

Mason B 1966 *Principles of Geochemistry*, 3rd edn. Wiley, New York, p. 153

Murphy T D 1975 Silica and silicon. In: Lefond S J (ed.) 1975 *Industrial Minerals and Rocks*, 4th edn. American Institute of Mining, Metallurgical and Petroleum Engineers, New York, p. 153

Patterson S H, Murray H H 1983 Clays. In: Lefond S J (ed.) 1983 *Industrial Minerals and Rocks*, 5th edn. American Institute of Mining, Metallurgical and Petroleum Engineers, New York, pp. 585–651
Rankama K, Sahama T G 1950 *Geochemistry*. University of Chicago Press, Chicago, IL
Shaffer J W 1983 Bauxitic raw materials. In: Lefond S J (ed.) 1983 *Industrial Minerals and Rocks*, 5th edn. American Institution of Mining, Metallurgical and Petroleum Engineers, New York, pp. 503–27
Williams H 1942 *The Geology of Crater Lake National Park, Oregon*. Carnegie Institute, Washington, DC, No. 10, p. 149

J. R. Dunn
[Dunn Geoscience Corporation, Albany, New York, USA]

Q

Quasicrystals

The discovery of a new solid phase exhibiting five-fold rotational axes of symmetry in the reciprocal space (as observed in electron diffraction patterns) was announced by Shechtman et al. (1984). Prior to this discovery, solids were generally classified as either crystals or glasses. While crystals exhibited long-range translational periodicity, glasses had only short-range correlation; thus it was believed that long-range order was compatible only with translational periodicity of atoms and this formed the basis for conventional crystallography.

The phase discovered by Shechtman and coworkers does not fall into the category of either crystal or glass: diffraction patterns show sharp spots and their arrangement reveals icosahedral point-group symmetry (see Fig. 1), which is incompatible with translational symmetry. In addition to five-fold axes of symmetry, the patterns show three-fold and two-fold rotational axes conforming to $m\bar{3}\bar{5}$ symmetry. The arrangement of the diffraction spots in the selected area diffraction patterns which represent sections through the three-dimensional reciprocal lattice is again not periodic. They are arranged in a quasiperiodic fashion, where the distance ratios

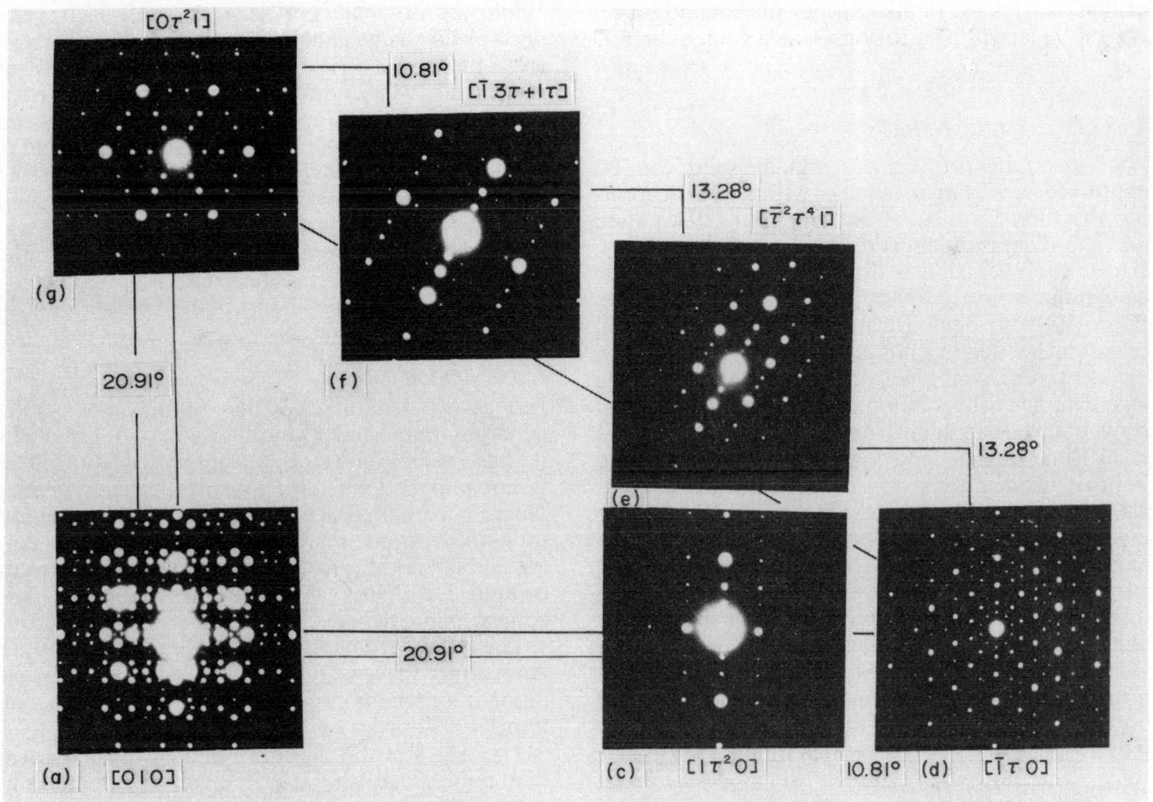

Figure 1
Complete set of electron diffraction patterns showing the icosahedral 5–3–2 symmetry of a quasicrystal; (a) [010] and (g) $[0\tau^2 1]$ show the two-fold and three-fold symmetry; (d) $[\bar{1}\tau 0]$ pattern shows a ten-fold symmetry which is the projected symmetry of five-fold axes; the angular relations correspond to those of an icosahedron

are given by τ^n, where the irrational number τ is equal to $(\sqrt{5}+1)/2$ and $n=1,2,3,\ldots$. The nonperiodic reciprocal lattice implies a nonperiodic real-space lattice; thus the advent of this phase, characterized as quasicrystals, marks a paradigmatic shift in our understanding of the solid state.

Since the original discovery, intensive research has been carried out, with over a thousand papers being published in the three years following. These investigations have tried to explore the domains of quasicrystallography and the atomic configuration of crystals. Alternative models based on icosahedral glass, large unit cells and multiple twins have also been advanced. The metallurgists have extended the systems in which quasicrystals occur and also the techniques by which they can be produced. Their morphology, growth patterns and stability have attracted attention and limited studies of properties have been undertaken.

While several significant questions still remain to be answered, it is clear that research on quasicrystals has enhanced our understanding of the interplay between order and periodicity in solids. The theories of icosahedral order running through the structure of liquids, glasses, Frank–Kasper phases and quasicrystals indeed appear to constitute a golden thread.

1. *Quasicrystallography*

Atomic arrangements in a regular solid can be expressed as a tiling problem—with atoms decorating the tiles. Grunbaum and Shepard (1986) state that the tiling problem is deceptive in its simplicity. It can easily be shown that pentagons do not tile a two-dimensional space in a periodic fashion. Penrose (1974) demonstrated that two-dimensional space can be tiled in a nonperiodic fashion by using a set of two different types of tiles. It is then possible to create five-fold rotational symmetry. It has been proved that such tilings exhibit self-similarity and local isomorphism, and Gardner (1977) made the Penrose tiles the subject of an article on mathematical recreations. Mackay (1982) studied Penrose tilings from a crystallographic point of view with the aid of a diffraction image using visible light, and coined the term "quasilattice" to describe Penrose tilings. Approaching the problem from the distinct angle of the propagation of icosahedral order from a freezing melt, Levine and Steinhardt (1984) used the term quasicrystals, the Fourier transforms of which bore startling resemblance to the electron diffraction patterns produced by Shechtman et al. (1984).

1.1 Fibonacci Tilings

To introduce the concept of quasilattices, it is instructive to begin with a one-dimensional quasilattice, owing to its simplicity. A one-dimensional quasiperiodic chain can be constructed by the use of two line segments, A and B, having different length scales (see Fig. 2). Figure 2a shows a periodic chain constructed with only one type of length scale A; the pattern repeats after every step. A quasiperiodic chain can be derived from this by a simple recursion scheme (or the replacement rule to build the next generation chain sequence), viz $A\rightarrow AB$ and $B\rightarrow A$. Figure 2b–e shows the development of the chain after every recursion step. The repeat period develops into the Fibonacci sequence of numbers $(1,2,3,5,8,\ldots)$ generated by the addition of successive numbers: the pattern remains periodic. The ratio of the number of A length scales to the number of B length scales converges to the golden mean $(\tau=(\sqrt{5}+1)/2=1.618\ldots)$ as the repeat period grows. In the limit of infinite recursions a quasiperiodic chain results with the golden mean as the ratio of the two length scales. It can be noted that the term quasiperiodic, therefore, means an ordered arrangement of objects which can be specified by a simple rule known as the recursion rule and which excludes the periodic ordered arrangement.

1.2 Penrose Tilings

While the original Penrose pattern used kites and darts as tilings, an equivalent scheme employing thin (with the angles of 36° and 144°) and fat (with angles of 72° and 108°) rhombi is convenient to use. These rhombi can be arranged to fill a two-dimensional space. The rules used to arrange these building blocks determine whether truly quasiperiodic structures are obtained. These are often called the matching rules and are illustrated in Fig. 3. The sides of the basic rhombi are arrowed by single and double arrows. The matching of the senses and the types of arrows during tiling will ensure the evolution of the truly quasiperiodic lattice.

1.3 Ammann Tilings

Just as two length scales and rhombi are used to generate one- and two-dimensional quasilattices, it has been shown by Ammann (reported in Grunbaum and Shepard (1986)) that prolate and oblate rhombohedral tiles can be used to generate three-dimensional quasilattices (Fig. 4). The matching and deflation rules for Ammann tilings are more complicated than for Penrose tilings owing to their context dependence.

The three tilings—Fibonacci, Penrose and Ammann—share features of quasiperiodicity: orientational order, minimal separation, quasiperiodic translational order and self-similarity.

The edges of the tiles are oriented in such a way that they satisfy the requirement of orientational order: all edge orientations can be related by a simple rotation operation. Minimum and maximum distances exist between any two nearest-neighbor vertices. It is possible to discern a quasiperiodic translational order by a suitable decoration of each tile by linear or planar segments in such a manner that the segments connect to form a grid whose spacings follow quasiperiodicity. Self-similarity is a property in which one can find the same environment around a lattice point, even after removing selected lattice points.

The concept of quasilattices can be further generalized to include any arbitrary rotational symmetry and quasiperiodicity. Several methods are available to construct quasilattices using arbitrary orientational order. Among these the projection from higher dimension offers powerful insights.

1.4 Projection from Higher Dimension

A periodic lattice in higher dimensions can appear to be quasiperiodic as viewed in a lower dimension. To take a specific example, a six-dimensional hypercubic lattice is strictly periodic. If the orientation of the projected space is irrational, an icosahedral quasilattice in three dimensions results: Elser (1986) has developed these ideas in an elegant fashion. Figure 5 illustrates the case of projection of a two-dimensional square lattice to yield a Fibonacci tiling in one dimension.

The projection formalism makes it clear that a single lattice parameter can be defined for the hypercubic lattice. Its projection in three dimensions will serve as a quasilattice constant. Thus it has been possible to show that all spots in the diffraction pattern of icosahedral quasicrystals in an aluminum–manganese alloy can be indexed on the basis of a quasilattice constant of 0.46 nm. It also makes it evident that six indices will be necessary to index planes and directions in six-dimensional space as well as in the projected three-dimensional space.

2. The Atomic Configuration of Quasicrystals

The question "Where are the atoms?" is central in research on quasicrystals. This remains a challenging problem, even though it has been addressed by using some of the most powerful techniques of characterization in the armory of the materials scientist. These include high-resolution electron microscopy (Hiraga et al. 1985), field-ion microscopy (Melmed and Klein 1986 (Fig. 6)), synchrotron x-ray diffraction (Bancel et al. 1985 (Fig. 7)), electron and neutron

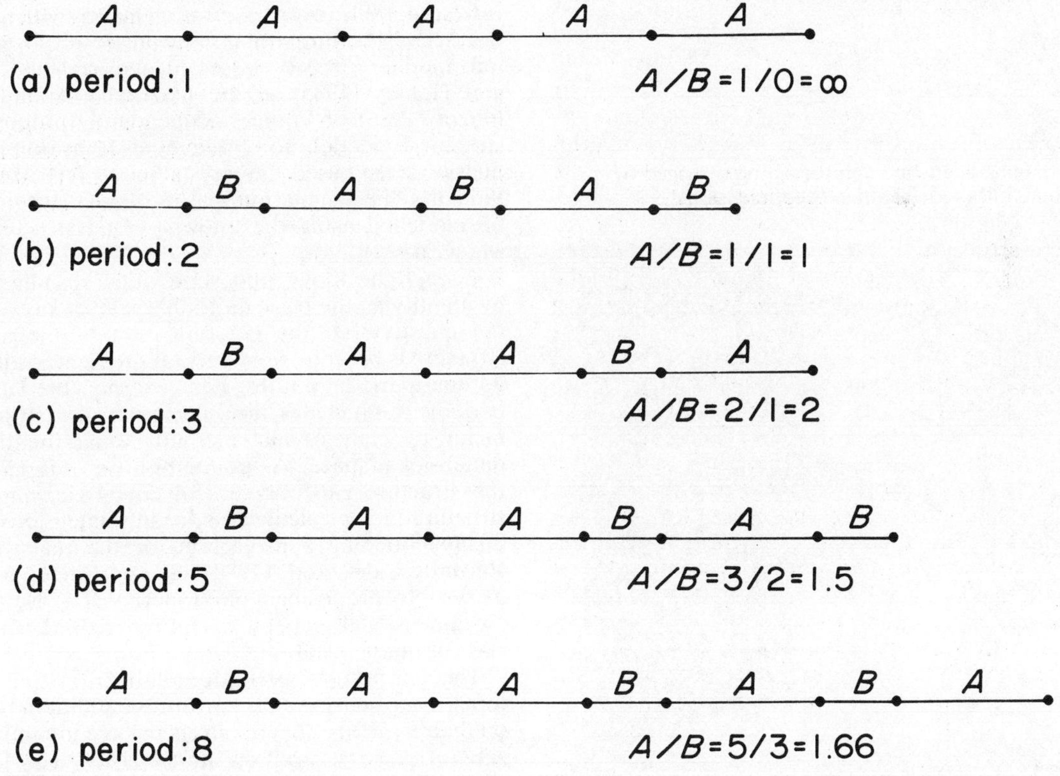

Figure 2
Diagram showing linear chains of spots separated by two distances; each successive chain can be obtained from the previous chain by a recursive substitution of $A \rightarrow AB$ and $B \rightarrow A$—with an infinite sequence this will generate a one-dimensional quasiperiodic chain

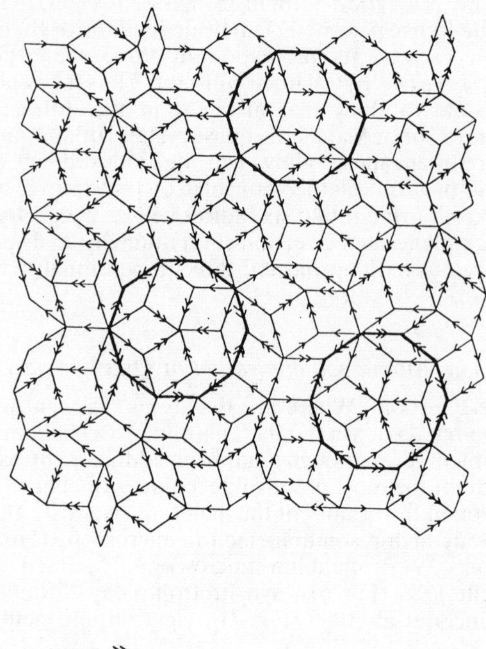

Figure 3
Two rhombi, acute and obtuse, can be arranged by matching rules to generate a two-dimensional quasiperiodic tiling

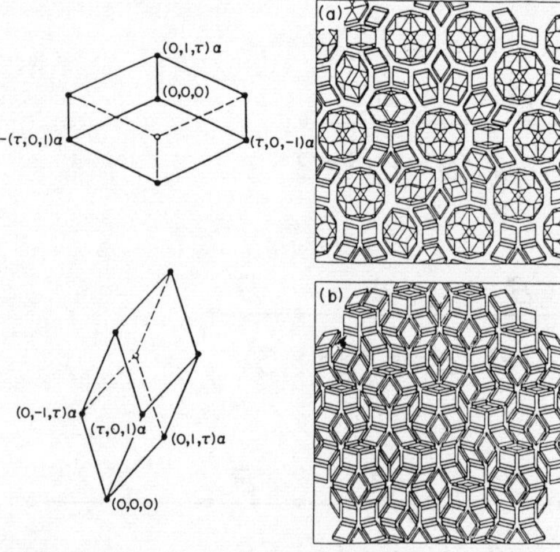

Figure 4
Perspective view of decomposition of three-dimensional quasiperiodic tiling revealing the building blocks: (a) "fat" rhombohedron and (b) "thin" rhombohedron (courtesy P. J. Steinhardt)

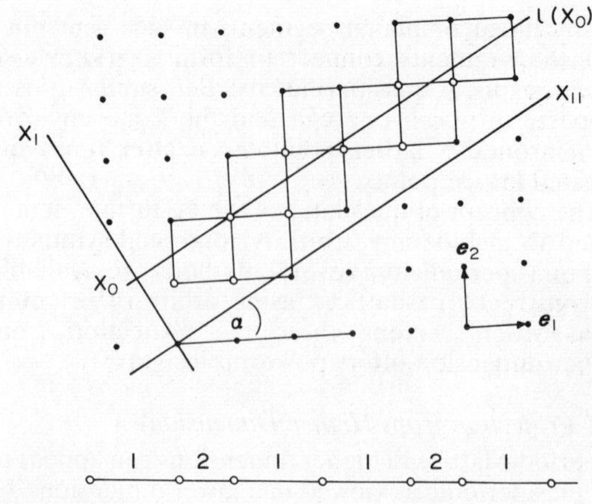

Figure 5
Generation of a quasiperiodic chain by projection from a two-dimensional periodic grid; the two parallel lines indicate the projection window; the projection angle α is irrational for a quasiperiodic chain

diffraction, extended x-ray absorption fine structure, nuclear magnetic resonance and Mössbauer techniques.

Considerable progress was achieved when two distinct crystalline structures were resolved to obtain models for two classes of quasicrystals. Elser and Henley (1985) of the USA and Audier and Guyot (1986) of France independently proposed a structural model for aluminum–transition-metal quasicrystals based on crystalline α-(Al, Mn, Si). Similarly the magnesium-based quasicrystals could be modelled using the known crystal structure of $Mg_{32}(Zn, Al)_{48}$.

Though the above models provide a starting point by identifying the basic motif that may be present in the quasicrystal, the determination of the actual structure is beset by fundamental problems owing to its quasiperiodic nature. For example, the lack of periodicity precludes the calculation of a structure factor by conventional methods using the three-dimensional space. As pointed out by Bak (1986), the structure can only be solved by carrying out structure-factor calculations in six dimensions and finally obtaining a projection for the real world. Recently Cahn et al. (1988) have performed a *tour de force* by proposing a plot in terms of a decorated six-dimensional hypercube for the α-(Al, Mn, Si) class of quasicrystal.

The fine effects associated with diffraction patterns from quasicrystals can throw additional light on their structure. These effects include breadth and shift of peaks as well as diffuse scattering. There appears to be an inherent breadth to the diffraction peaks from quasicrystals. This broadening does not scale with the reciprocal space vector and therefore the explanations applicable to crystalline materials, such as strain and grain size, are ruled out. An explanation based on phason strains has been advanced to explain these features. In addition to three elastic modes, three more modes known as phasons mode operate in quasicrystals. Physically phasons can be understood as a local rearrangement of atoms that does not alter the overall symmetry. Since the rearrangements involve a diffusional process, the phason strain will be slow to relax and can affect the diffraction patterns by broadening and shifting the diffraction spots.

In addition to the broadening discussed above, the occurrence of diffuse intensity in quasicrystalline electron diffraction patterns has been reported in aluminum-, magnesium- and titanium-base alloys. Single-crystal x-ray diffraction experiments on aluminum–copper–lithium quasicrystals also confirm this effect. The diffuse intensity develops either during slow cooling from the melt or after aging of quasicrystals. This may represent a type of ordering in quasicrystals.

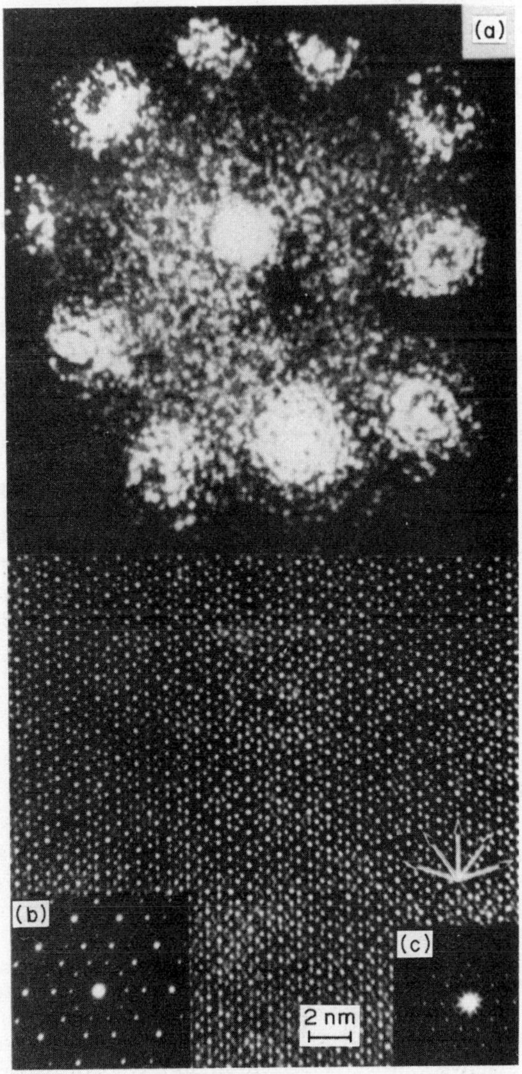

Figure 6
Icosahedral quasicrystal at atomic resolution along the five-fold axis in aluminum–manganese alloys: (a) field-ion micrograph showing the five-fold symmetry at atomic level (courtesy A. J. Melmed); (b) high-resolution transmission electron micrograph with bright dots representing atomic potentials arranged along five vectors separated by 72° (courtesy K. Hiraga); (c) high-resolution transmission electron micrograph showing five-fold symmetry at atomic level

3. Alternative Models

Although the quasicrystalline model is the most attractive model for an explanation of the intriguing electron diffraction patterns, alternative models have also been developed to explain these results. Icosahedral glass, crystals with giant unit cells decorated with icosahedral motifs, and ordered twinning are three models that merit serious scrutiny.

3.1 Icosahedral Glass

The icosahedral glass model, advanced by Shechtman and Blech consists of an assembly of icosahedra packed so as to preserve bond orientational order. Stephens and Goldman (1986) calculated the diffraction patterns from a random assembly of icosahedral units which had been grown together by sharing faces. The patterns were in excellent agreement with experimental peak positions. This model has been further developed to explain peak shifts and spot misalignments.

3.2 Giant Unit Cells

While pure metals tend to crystallize in cubic- or hexagonal-close-packed structure, alloying can lead to the formation of more complicated structures. In his classic work, Pauling (1969) emphasized the occurrence of several icosahedral motifs in metallic structures.

In $Al_{12}Mo$ alloys, the cubic unit cell has an icosahedral motif with a central molybdenum atom surrounded by 12 aluminum atoms. It is possible to add further coordination shells to develop pentagonal Frank–Kasper phases. Frank and Kasper (1958) emphasized the occurrence of CN12 icosahedral polyhedra together with CN14, CN15 and CN16 polyhedra.

Two illustrations of large unit cells will suffice. In α-(Al, Mn, Si), a central icosahedron of aluminum atoms is surrounded by a second large icosahedron, known as the Mackay icosahedron. In the case of T-$Mg_{32}(Al, Zn)_{48}$, one of the most complicated

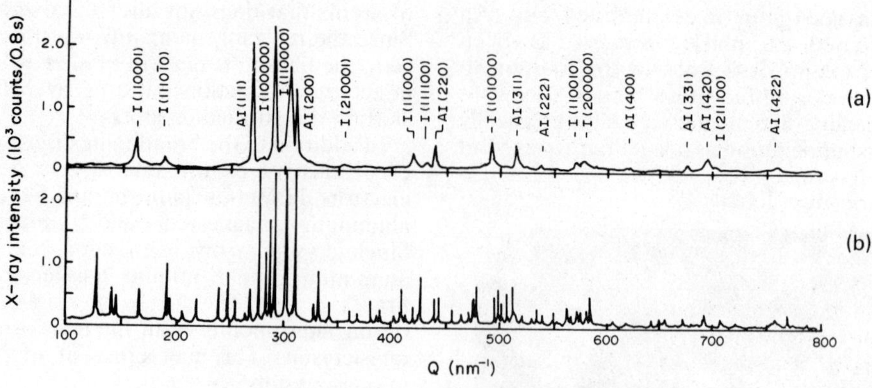

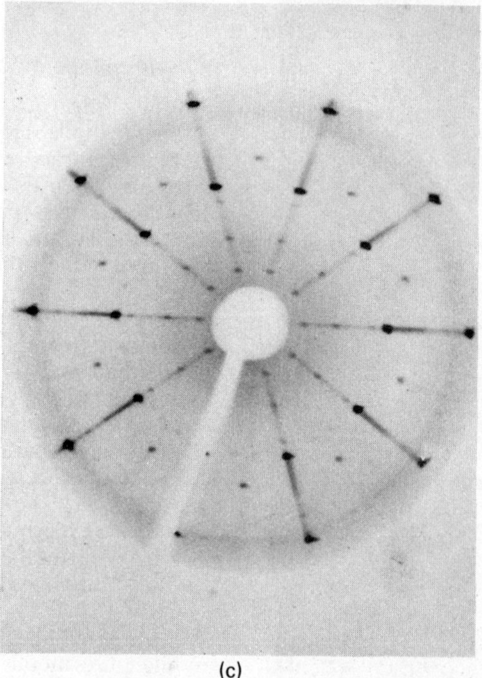

Figure 7
X-ray diffraction evidence of a quasicrystal: (a) a high-resolution powder pattern of Al–14%Mn quasicrystal using synchrotron radiation; (b) corresponding crystalline pattern (courtesy P. Heiney); (c) a precision photograph of aluminum–copper–lithium quasicrystal exhibiting the five-fold symmetry (courtesy E. S. Rajagopal)

structures in the Frank–Kasper family, the cluster has successive shells of icosahedron, pentagonal dodecahedron, rhombic triacontahedron and truncated icosahedron. The final shape of the cluster is a cubooctahedron: the cubooctahedra are packed together in a body-centered-cubic lattice such that 61% of the atoms sit at sites with icosahedral environments in this structure. The diffraction patterns from α-(Al, Mn, Si) and T-$Mg_{32}(Zn, Al)_{48}$ bear a startling similarity to those from the quasicrystals.

It is interesting to note that both the α and T phases can be developed by periodic stacking of the Ammann rhombohedra. This viewpoint was first developed by Elser and Henley (1985). It is possible to derive these structures from higher dimensional space by projection on a rational plane. By taking larger and larger unit cells, patterns can be generated that become experimentally indistinguishable from those of the corresponding quasicrystals.

3.3 Ordered Multiple Twins

Pauling (1985) proposed the existence of multiple ordered twins to explain the experimental observations relating to quasicrystals within the confines of classical crystallography. He has postulated a β–*W* structure with the motif constants of 20 interpenetrating Friauf polyhedra obtained by truncating all four corners of a tetrahedron. The crystals with this structure have 820 atoms in the unit cell; 20 tetrahedrally shaped crystals with twin relationships among them have been invoked to explain the diffraction phenomena.

This model is an extension of the giant unit cell and incorporates twinning. Multiple twins have in fact been observed in rapidly solidified foils of nickel–zirconium, aluminum–iron and aluminum–palladium alloys. However, current experimental techniques allow the establishment of twin occurrence. In the case of quasicrystals, high-resolution electron microscopy and field-ion microscopy have failed to reveal the presence of twins. Pauling's suggestion has useful applications for metastable phases produced through rapid solidification; it remains an interesting but unverified possibility as far as quasicrystals are concerned.

4. *Quasicrystal Synthesis*

4.1 Principles

Hume-Rothery evolved classical criteria based on atomic size, electronegativity and valence electrons for the formation of extended solid solutions in alloys of noble metals. Modifications of the criteria have since been applied for the occurrence of intermetallic compounds and metallic glasses. It is of interest to determine whether analogous criteria can be developed for quasicrystal-forming ability. Three approaches based on crystal chemistry, quantum structural diagrams and phase diagrams are outlined in this section.

The crystal-chemistry criterion, the most successful criterion available to date, hinges on the close structural similarity between Frank–Kasper phases with a large number of atoms in icosahedral coordination, and quasicrystals. Ramachandrarao and Sastry (1985) made use of this possibility when they outlined a basis for the synthesis of quasicrystals from $Mg_{32}(Zn, Al)_{48}$ alloy. Following a similar line of reasoning in an independent fashion, Zhang et al. (1985) succeeded in producing quasicrystals in a titanium–nickel alloy. The crystal structure of Ti_2Ni contains several atoms in icosahedral coordination and has been subsequently shown to be related to the α-(Al, Mn, Si) structure.

A second approach makes use of quantum structural diagrams to successfully isolate the crystalline compound analogy of ternary quasicrystals. Villars et al. (1986) showed that satisfactory structural classifications are obtained by the use of atomic radii, an optimized electronegativity coordinate and the average number of *s*, *p* and *d* valence electrons per atom. The quasicrystal-forming ternary alloys are referred to as PSD ternaries, based on the composition $A_{15}B_{35}C_{50}$ where A is a *p* element (such as Al, Ga, Ge or Sn), B is an *s* element (such as Li, Na or Mg) and C is a near-noble metal (such as Ni, Cu or Zn).

A third criterion stresses the occurrence of quasicrystals in systems where the phase diagram features peritectic reactions and is an echo of the criterion for glass formation in systems featuring deep eutectic reactions. As quasicrystals are metastable, they are favored to occur when undercooling bypasses the nucleation of a stable crystalline phase. This is most readily accomplished if the alloy composition is off stoichiometry of the intermetallic compound produced by peritectic reactions.

4.2 Production

While the original discovery of quasicrystals adopted the route of rapid solidification, subsequent research has shown that quasicrystals can be produced by a variety of techniques starting from liquid, glass, crystal or vapor phases. Figure 8 shows the schematic diagram of transformations among these phases.

Quasicrystals have been produced from the liquid state by rapid solidification and pressurization as well as conventional casting. Rapid-solidification techniques include melt-spinning, laser- or electron-beam processing and atomization. Rapid pressurization of an aluminum–manganese alloy by Sekhar and Rajasekharan (1986) has been shown to be a novel route for the quasicrystal synthesis. Considerable interest has been generated in the millimeter-sized particles obtained in aluminum–lithium–copper alloys by direct chill-casting from the liquid. In 1955, Hardy and Silcock reported the

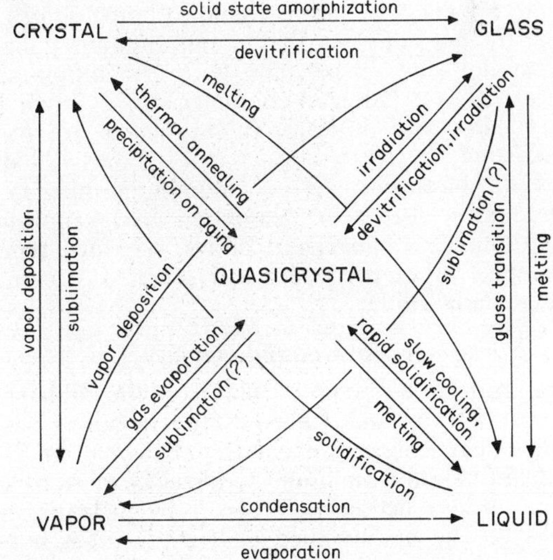

Figure 8
Schematic diagram showing phase transitions among vapor, liquid, glass, quasicrystal and crystal

Table 1
Occurrence of icosahedral quasicrystals: a selection

Base metal	Alloy system	Related crystal structure
Aluminum	Al–Cr	$Al_{45}Cr_7$
	Al–Mn	Al_4Mn, α-(Al, Mn, Si)
	Al–Mo	$Al_{12}Mo$
	Al–V	$Al_{10}V$
	Al–W	$Al_{12}W$
	Al–Ru	Al_9Ru_2
	Al–Mn–Si	α-(Al, Mn, Si)
Copper	Cu–Cd	Cu_4Cd_3
Lithium	Al–Cu–Li	Al_6CuLi_3 (T_2 phase)
	Al–Li–Cu	$Li_{32}(Al, Cu)_{48}$
	Al–Li–Zn	$Li_{32}(Al, Zn)_{48}$
Magnesium	Mg–Al–Cu	$Mg_{32}(Al, Cu)_{48}$
	Mg–Al–Zn	$Mg_{32}(Al, Zn)_{48}$
	Mg–Ga–Zn	$Mg_{32}(Ga, Zn)_{48}$
	Mg–Al–Ag	$Mg_{32}(Al, Ag)_{48}$
Titanium	Ti–Fe	Ti_2Fe
	Ti–Mn	Ti_2Mn
	Ti–Ni	Ti_2Ni
Uranium	U–Pd–Si	UPd_3

occurrence of this phase as T_2 phase and could not index its x-ray diffraction pattern. Originally it was claimed that the T_2 phase represented a stable quasicrystalline phase. Recently large quasicrystals have also been grown in gallium–magnesium–zinc and aluminum–copper–iron systems.

Annealing of glassy palladium–uranium–silicon and aluminum–copper–vanadium alloys has been found to give rise to quasicrystals. These have also been obtained as grain-boundary precipitates in heat-treated samples of aluminum–lithium–copper, aluminum–zinc–magnesium and aluminum–manganese alloys. Directed-energy processes involving ion beams and diffusion in multilayer are other solid-state routes to quasicrystal synthesis. Icosahedral phase particles of aluminum–manganese alloy have also been generated by evaporation of aluminum–manganese alloys. A few typical examples of quasicrystals and selected structures are given in Table 1.

4.3 Nucleation, Growth and Stability

The production of quasicrystals from the liquid state has attracted considerable attention as both types of phase share icosahedral arrangements of atoms. The transformation from liquid to quasicrystal occurs as a first-order reaction and proceeds by nucleation and growth. The undercooled melt experiences homogeneous nucleation of quasicrystals, often leading to ultrafine grain sizes. Occasional occurrence of heterogeneous nucleation is reported. In most cases these combinations of nucleation and growth lead to small quasicrystals. In the aluminum–lithium–copper system it has been possible to grow large quasicrystals by controlled solidification.

Quasicrystals display a rich variety in morphology (Fig. 9). Faceted shapes having icosahedral symmetry have been observed, the external morphology revealing the internal symmetry. In aluminum–manganese alloy, pentagonal dodecahedral quasicrystals are observed, while in aluminum–lithium–copper alloy a spectacular triacontahedral morphology has been documented. The dendritic growth of quasicrystals also indicates icosahedral symmetry. In magnesium-based quasicrystals, a remarkable sectoral appearance is observed. In some systems, eutectic growth of a crystal and a quasicrystal has been seen.

Transmission electron micrographs of quasicrystals very often reveal a characteristic speckle contrast. This may be evidence of local chemical inhomogeneity.

There is a substantial body of evidence to indicate that quasicrystals are metastable. They can be incorporated into metastable phase diagrams in the same manner as for metastable crystalline phases. A diagram for the aluminum–manganese system is shown in Fig. 10, with added indications of the range of formation of quasicrystals and morphological features.

Annealing of quasicrystals leads to the formation of more stable phases. Very often, orientation relationships are observed between the "parent" and the "product" phase.

5. Low-Dimensional Quasicrystals

It is possible to combine quasiperiodicity in one or two dimensions with periodicity in the remaining one or two directions. This leads to a variety of one- and two-dimensional quasicrystals.

5.1 Two-Dimensional Quasicrystals

Soon after the discovery of the icosahedral quasicrystal, the decagonal quasicrystal was independently discovered by Bendersky (1985) and Chattopadhyay et al. (1985) of the USA and India, respectively. The electron diffraction pattern shows that two-dimensional quasiperiodic layers are stacked periodically in the third dimension (Fig. 11). The morphology of the particle is rodlike and the electron micrograph shows considerable internal contrast. Convergent-beam electron diffraction has conclusively established the rotational symmetry to be tenfold. High-resolution electron microscopy indicates the presence of cylindrical domains along the decagonal axis and a possible screw axis. The point group symmetry of $10/mmm$ and the space group $P\,10_5/m$ have been proposed.

Decagonal quasicrystals form in a number of aluminum–transition-metal alloys. The binary alloys are with manganese, iron, cobalt, palladium,

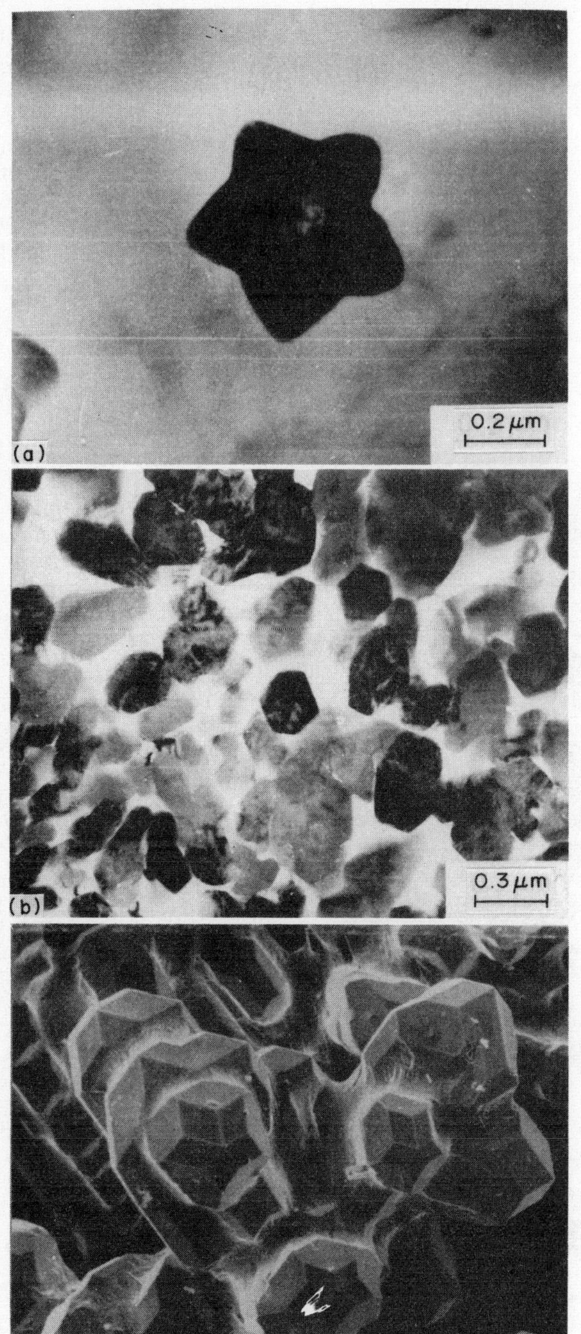

Figure 9
Series of transmission electron micrographs showing various morphologies of: (a) and (b), aluminum–manganese quasicrystals; and (c), aluminum–lithium–copper quasicrystals (courtesy N. Thangaraj, F. W. Gayle)—the aluminum–manganese quasicrystals have pentagonal dodecahedral morphology, while the aluminum–lithium–copper quasicrystals exhibit triacontahedral morphology

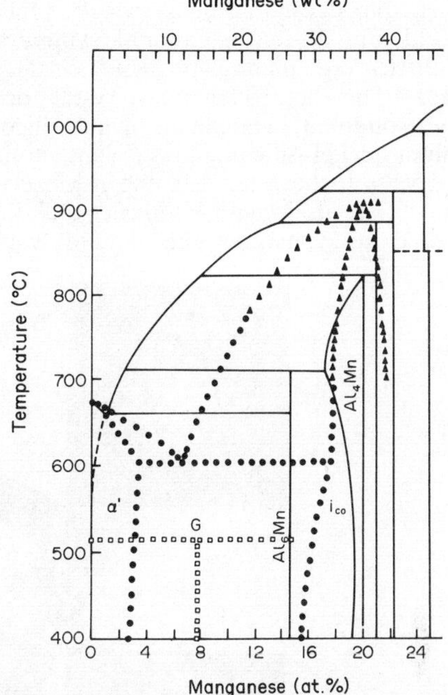

Figure 10
Metastable phase diagram of aluminum–manganese systems; the filled circles and triangles indicate quasicrystalline phase boundaries; the open squares represent a newly discovered equilibrium phase boundary (this diagram does not include the decagonal or *T* quasicrystalline phase) (courtesy Yu-Zhang)

osmium and ruthenium and the ternary alloys include aluminum–manganese–nickel, aluminum–copper–silicon and aluminum–nickel–silicon. Two distinct periodic stacking sequences have been reported to date. Aluminum–manganese, aluminum–manganese–nickel and aluminum–chromium–silicon show a six-layer repetition of 1.26 nm, while aluminum–palladium, aluminum–nickel–silicon and several other alloys show an eight-layer repetition of 1.62 nm. The aluminum–iron decagonal quasicrystal is unusual as it exhibits a different diffuse intensity distribution parallel to the decagonal axis.

In systems where both icosahedral and decagonal quasicrystals can form, the latter form at a slower cooling rate. The decagonal phase often nucleates epitaxially on the icosahedral phase with its tenfold axis parallel to one of the five-fold axes of the icosahedral phase. The close structural similarity between the two, leads to a remarkable resemblance in the electron diffraction patterns from the two phases. The electron diffraction patterns also resemble those from layered complex crystal structures such as $Al_{60}Mn_{11}Ni_4$. Considerable research effort is being directed towards discovering the rational approximants to the decagonal phase.

Apart from decagonal quasicrystals, phases with eight-fold and 12-fold rotational symmetry have been discovered mainly by Kuo et al. (1987) (Fig. 12). The octagonal quasicrystal occurs in rapidly solidified vanadium–nickel–silicon and chromium–nickel–silicon alloys (Wang et al. 1987). This appears to be related to the β-Mn crystalline structure. The dodecagonal quasicrystal has been discovered in vanadium–nickel and vanadium–nickel–silicon alloys (Chen et al. 1988) and is related to the σ phase. As mentioned in the introduction, the two length scales, a combination of which produce the icosahedral and decagonal quasiperiodic structure, are related to each other by τ, the golden mean. That they are related by τ is a consequence of the recursion rule employed to fill the space; τ is the signature for the icosahedral and decagonal symmetry. It is obvious that other length scales and

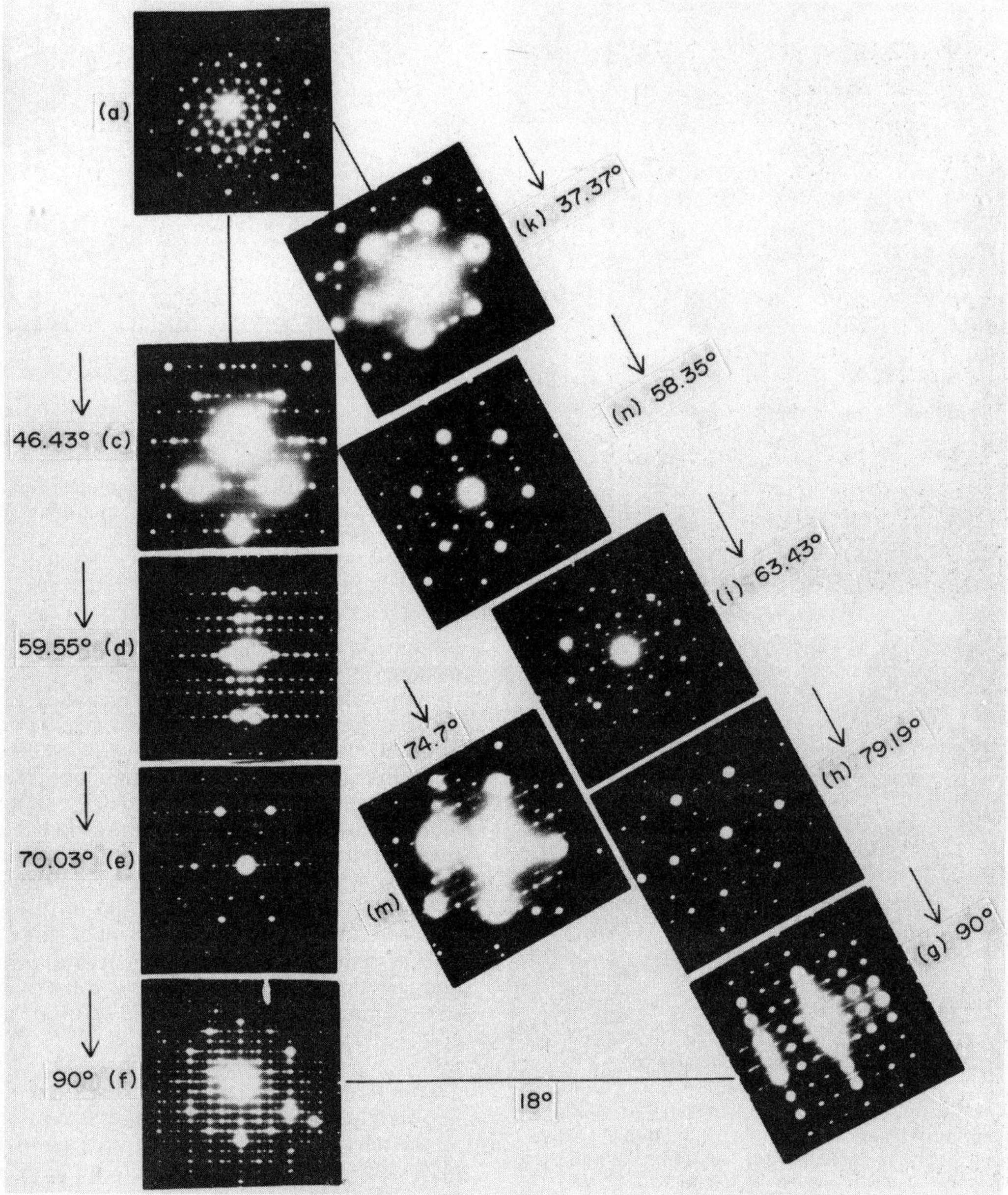

Figure 11
Complete set of diffraction patterns from a decagonal quasicrystal (note that the spots are periodic in one direction but quasiperiodic in other directions)

recursion rules can be employed to obtain the quasiperiodic arrangements, and the numbers $\sqrt{2}$ and $\sqrt{3}$ are the signature for the evolution of octagonal dodecagonal symmetry.

5.2 One-Dimensional Quasicrystals

The Fibonacci sequence can be used for the development of one-dimensional quasicrystals. The phases in rapidly solidified aluminum–copper and vapor-deposited aluminum–palladium and aluminum–zirconium alloys exhibit one-dimensional quasiperiodicity owing to the ordering of structural vacancies (Chattopadhyay et al. 1987). In addition, artificial one-dimensional quasiperiodic stacking of gallium–aluminum–arsenic layers has been accomplished by using molecular beam epitaxy. The electronic states in such one-dimensional quasicrystals are a matter of intense theoretical and practical interest.

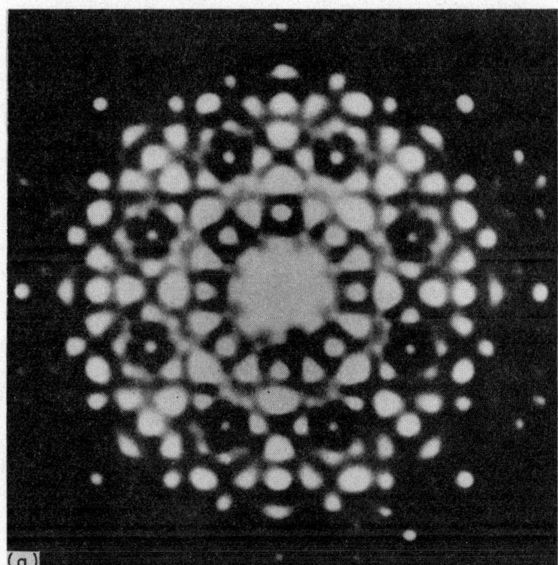

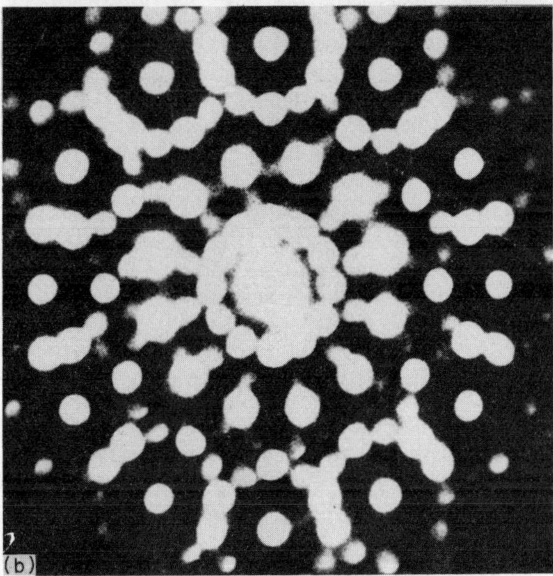

Figure 12
Electron diffraction patterns showing (a) eight-fold and (b) 12-fold rotational symmetry (courtesy K. H. Kuo)

6. Defects in Quasicrystals

Many interesting and important properties of crystals can be traced to their defects. A variety of defect configurations can be visualized in the quasicrystalline configuration, some of them unique to the quasiperiodic arrangement.

Vacancies are expected to play an important role in stabilizing the quasicrystals. Many crystalline approximants such as α-(Al, Mn, Si) and T–$Mg_{32}(Zn, Al)_{48}$ contain structural vacancies. The phason strain may also be thought of as a point defect. The decapod which occurs in Penrose tilings may act as the nucleus for the growth of a quasicrystal and can be considered as a three-dimensional aggregate of point defects.

High-resolution electron micrographs often indicate the presence of line defects. A theoretical analysis of possible dislocation-like defects has been carried out by Socolar et al. (1986). The dislocations in quasicrystals require two Burgers vectors to characterize them: the second Burgers vector describes the phason component. Since the relaxation of a phason component requires diffusional rearrangement of atoms, dislocations are generally sessile and will not act as a vehicle for plastic deformation.

Twin boundaries have been observed in quasicrystals. The addition of iron to aluminum–manganese alloy promotes twinning. Grain boundaries between quasicrystals are observed but not fully characterized; the boundary between the icosahedral and decagonal quasicrystal has been observed to be highly coherent. The interface between crystal and quasicrystal remains to be explored in detail. There is a richness of imperfections in quasicrystals and only a beginning has been made in exploring this fascinating problem.

7. Properties

Electrical, magnetic and mechanical properties of materials are very sensitive to structural changes. Since the quasicrystalline structure lacks translational periodicity, it is to be expected that this would affect these properties in a significant fashion. Studies devoted to the measurement of properties are relatively sparse compared with purely structural investigations; one reason is the nonavailability of monophase quasicrystals in several systems. These studies have tended to emphasize the similarities between quasicrystals and glasses on the one hand, and quasicrystals and Frank–Kasper phases on the other.

Bibliography

Audier M, Guyot P 1986 Al_4Mn quasicrystal: Atomic structure, diffraction data and Penrose tiling. *Phil. Mag. B* 53: L43–L51

Bak P 1986 Icosahedral crystals: Where are the atoms? *Phys. Rev. Lett.* 56: 861–4

Bancel P A, Heiny P A, Stephens P W, Goldman A J, Horn P M 1985 Structure of rapidly quenched Al–Mn. *Phys. Rev. Lett.* 54: 2422–5

Bendersky L 1985 Quasicrystal with 1-dimensional translational symmetry and a ten fold rotation axis. *Phys. Rev. Lett.* 55: 1461–3

Cahn J W, Gratias D, Mozer B 1988 A 6-D structural model for the icosahedral (Al, Si)–Mn quasicrystal. *J. Phys. (Paris)* 49: 1225–33

Cahn J W, Shechtman D, Gratias D 1986 Indexing of icosahedral quasiperiodic crystals. *J. Mater. Res.* 1: 13–26

Chattopadhyay K, Lele S, Thangaraj N, Ranganathan S 1987 Vacancy ordered phases and one-dimensional quasiperiodicity. *Acta Metall.* 35: 727–33

Chattopadhyay K, Lele S, Ranganathan S, Subbanna G N, Thangaraj N 1985 Electron microscopy of quasicrystals and related structure. *Curr. Sci.* 54: 895–903

Chen H, Li D X, Kuo K H 1988 New type of two-dimensional quasicrystals with twelve fold rotational symmetry. *Phys. Rev. Lett.* 60: 1645–8

Dubost B, Lang J M, Tanaka M, Sainfort P, Audier M 1986 Large Al–Cu–Li single quasicrystals with triacontahedral solidification morphology. *Nature (London)* 324: 48–50

Elser V 1986 The diffraction patterns of projected structure. *Acta Cryst. A* 42 (Jan): 36–43

Elser V, Henley C 1985 Crystal and quasicrystal structure in Al–Mn–Si alloys. *Phys. Rev. Lett.* 55: 2883–6

Frank F C, Kasper J S 1958 Crystal alloy structures regarded as sphere packing. *Acta Cryst.* 17: 184–90

Gardner M 1977 Extraordinary nonperiodic tiling that enriches the theory of tiles. *Sci. Am.* 236: 110–21

Gratias D, Michel L (eds.) 1986 International workshop on aperiodic crystals. *J. Phys. (Paris), Colloq.*, Suppl. 7, Vol. 47

Grunbaum B, Shepard G 1986 *Patterns and Tilings*. Freeman, New York

Hardy H K, Silcock J M 1955 The phase sections at 500 °C and 350 °C of aluminium rich Al–Cu–Li alloys. *J. Inst. Met.* 24: 423–8

Henley C L 1987 Quasicrystal order, its origin and consequences: A survey of current models. *Comments in Condensed Matter Physics* 13: 59–117

Hiraga K, Hirabayashi M, Inoe A, Masumoto T 1985 Structure of Al–Mn quasicrystals studied by high resolution electron microscopy. *J. Phys. Soc. Jpn.* 54: 4077–80

Ho T L 1986 Periodic quasicrystal. *Phys. Rev. Lett.* 56: 468–71

Jaric M V (ed.) 1988 *Introduction to Quasicrystals*. Academic Press, San Diego

Knowles K M (ed.) 1987 Quasicrystalline materials. *J. Microsc.* 146(3)

Kuo K H (ed.) 1987 Quasicrystals: Proc. Int. Workshop on Quasicrystals, Beijing, China. *Mater. Sci. Forum* 22–24

Levine D, Steinhardt P J 1984 Quasicrystals: A new class of ordered structures. *Phys. Rev. Lett.* 53: 2477–80

Mackay A C 1987 A bibliography of quasicrystals. *J. Rapid Solidification* 2: S1–S41

Mackay A L 1982 Crystallography and the Penrose pattern. *Physica A* 114: 609–13

Melmed A J, Klein R 1986 Icosahedral symmetry in a metallic phase observed by field-ion microscopy. *Phys. Rev. Lett.* 36: 1478–81

Pauling L 1969 *The Nature of the Chemical Bond*. Cornell University Press, Ithaca, NY

Pauling L 1985 Apparent icosahedral symmetry is due to directed multiple twinning of a cubic crystal. *Nature (London)* 317: 512–14

Penrose R 1974 The role of aesthetics in pure and applied mathematics research. *Inst. Math. App. Bull.* 10: 266–71

Ramachandrarao P, Sastry G V S 1985 A basis for the synthesis of quasicrystal. *Pramana* 25: L225–230

Sastry G V S, Suryanarayana C, Van Sande M, Van Tendeloo, G 1978 A new ordered phase in the Al–Pd system. *Mater. Res. Bull.* 13: 1065–70

Sekhar J A, Rajasekharan T 1986 Rapid pressurisation experiments on a liquid Al–Mn alloy. *Nature*, 320: 153–5

Shechtman D, Blech I, Gratias D, Cahn J W 1984 Metallic phase with long range orientation order and no translational symmetry. *Phys. Rev. Lett.* 53: 1951–3

Socolar J E S, Lubensky T C, Steinhardt P J 1986 Phonons, phasons, and dislocations in quasicrystals. *Phys. Rev. B* 34: 3345–60

Steinhardt P J, Ostlund S (eds.) 1987 *The Physics of Quasicrystals*. World Scientific Publishing, Singapore

Stephens P W, Goldman A I 1986 Sharp diffraction maxima from an icosahedral glass. *Phys. Rev. Lett.* 56: 1168–71

Villars P, Philips J C, Chen H S 1986 Icosahedral quasicrystals and quantum structural diagram. *Phys. Rev. Lett.* 57: 3085–8

Wang N, Chen H, Kuo K H 1987 Two dimensional quasicrystal with eight fold rotational symmetry. *Phys. Rev. Lett.* 59: 1010–13

Zhang Z, Ye H Q, Kuo K H 1985 A new icosahedral phase with *m*35 symmetry. *Phil. Mag. A* 52: L49–L52

S. Ranganathan and K. Chattopadhyay
[Indian Institute of Science, Bangalore, India]

Quenching Media: Polymeric Additives

The quenching of a heat-treatable metal is the rapid cooling of that metal from a suitably elevated temperature in order to achieve desired metallurgical properties. Traditionally, the principal quenchants were water, possibly modified by the addition of inorganic salts, and oils, either naturally occurring or refined. It is only since the early 1960s that certain water soluble organic polymers have been found to be useful in modifying the cooling characteristics of water. In order of historical use, these are polyvinyl alcohol (PVA), polyalkylene glycols (PAGs), polyvinylpyrrolidone (PVP), sodium polyacrylate and polyethyloxazoline (PEO_x). These polymer quenchants offer the heat treater a number of environmental, economic and technical advantages, the principal one being undoubtedly the elimination of the oil fire hazard. Nevertheless, insufficient initial basic understanding and experience with polymer quenchants has resulted in operational problems and cracked components; consequently, heat treaters have been reluctant to change from the tried and tested traditional quenching media. However, improved polymer products and greater understanding have led to a progressive increase in their usage since the early 1980s for the quenching of both ferrous and nonferrous materials. In 1989 polymer quenchants accounted for approximately 15% of the quenchant market, although it has been predicted that they will surpass oil in usage before the end of the twentieth century.

1. Requirements

The choice of quenching fluid and technique depends on the type of material to be quenched and the shape and thickness of the part being quenched. The main requirement of a quenching medium is that it should cool the component at the correct rate, through its section, to provide the optimum structure and properties consistent with its anticipated service performance, or to provide the correct starting conditions for subsequent treatments. The cooling rate at the center of a component is dependent on the physical properties of the material as well as the heat-extraction rate of the quenchant. Even if the surface of the component is instantaneously cooled to the temperature of the quenchant, subsurface positions cool at a slower rate. Therefore, the cooling rate at the critical position within the component is the important factor and the quenchant should be chosen to provide the correct temperature gradient to ensure that the required structure is produced at this critical depth. In some cases, it may be necessary to have a steep temperature gradient in order to obtain the required structure at the center. However, increasing the cooling rate of a quench, in addition to raising the temperature gradient, produces secondary effects that may be very important. Residual stress can be higher, and the tendency for both distortion and cracking increases. These factors must obviously be taken into account when choosing a quenchant. Ideally, a quenchant should be chosen that will produce the required structure at the slowest possible cooling rate.

Many different media have been used for quenching and some of the most popular, in order of quench severity are: brine, water, polymer quenchants, oils, molten salts and gases. Thus, it can be seen that polymer solutions are used predominantly for applications requiring rates of cooling that are between those of water and oil.

2. Advantages

Some of the advantages of polymer quenchants are as follows.

(a) The principal advantage of polymer quenchants is the elimination of the oil fire hazard.

(b) By controlling the concentration, temperature and agitation of the polymer solution, it is possible to achieve a range of cooling rates, enabling a wide variety of materials and components to be quenched.

(c) Polymer quenchants are generally supplied as concentrated solutions. Although the purchase cost of these solutions is higher than oil per unit volume, the polymer concentrates are further diluted, resulting in an overall cost saving.

(d) Heat-treatment shops are notoriously dirty places, oil spillages, sludge and scale being particular problems. Certain oil products are also known to be carcinogenic and can cause dermatitis. Polymers result in a much cleaner and safer working environment, the noxious smoke and fumes produced during oil quenching being replaced by steam.

(e) Other factors to be considered in the overall economics of use include the cost of postquench treatments, quenchant quality control, maintenance procedures, disposal, replacement of worn-out equipment, part inspection and insurance. On balance, these usually favor polymer quenchants.

Even though polymer quenchants offer all these advantages, they are by no means the answer to all the heat treater's problems. Since polymer quenchants are aqueous solutions, they generally exhibit faster cooling rates compared with oils through the critical martensite transformation temperature range. This obviously precludes the treatment of certain materials, particularly high-carbon or high-alloy steels, as well as the treatment of components with surface defects or stress raisers. Close control of quench tank conditions and more frequent monitoring is also required for the successful application of polymer products.

3. Cooling Mechanism

All polymer quenchants are believed to operate by the same fundamental mechanism. This mechanism is associated, first, with the increased viscosity of

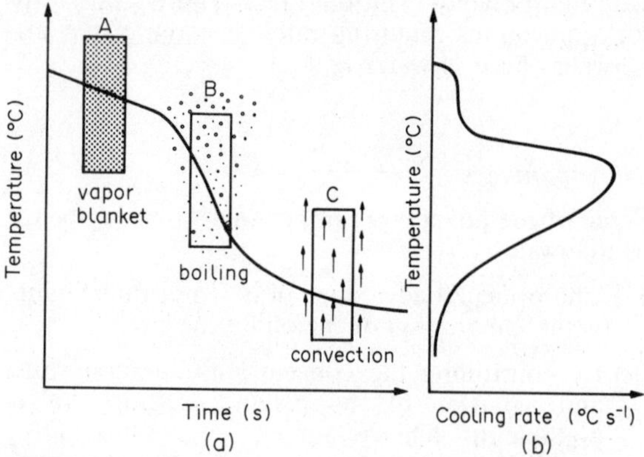

Figure 1
The three stages of cooling during quenching shown on (a) the conventional time–temperature cooling curve, and (b) the corresponding cooling-rate–temperature plot

their solutions compared with water and, second, with their ability to produce an insulating polymer-rich film around the hot metal. The most useful way of accurately describing the complex mechanism of quenching is to produce a time–temperature plot or cooling curve for the quenching liquid under controlled conditions. This forms the basis of the internationally standardized Wolfson Heat Treatment Center Engineering Group technique which uses a cylindrical test probe 12.5 mm in diameter and 60 mm long manufactured from Inconel 600, a heat-resisting nickel–chromium–iron alloy. The probe, which has a chromel–alumel thermocouple at its geometric center, is heated to 850 °C and then rapidly transferred into 2 l of quenchant. Both time–temperature and cooling-rate–temperature curves are recorded (see Fig. 1). When polymer quenchants are being evaluated, agitation of the sample is required in order to provide a more realistic test and to improve the reproducibility of the results. Quenching occurs in three distinct stages (A, B and C in Fig. 1). One or more of these stages can occur simultaneously at different positions on the hot metal surface. With polymer quenchants, the stability and hence the duration of each stage is profoundly affected by the type of polymer and its molecular weight, the concentration and temperature of the solution, and the level of agitation.

Stage A is the vapor blanket or film boiling stage, where the temperature of the metal is so high that, after the initial contact time, traditional quenchants form a thin stable film of vapor around the hot metal. This occurs when the supply of heat from the surface of the hot metal exceeds the amount required to form the maximum vapor per unit area of the metal (Leidenfrost phenomenon). Cooling is by radiation and conduction through the gaseous film and, since vapor films are poor heat conductors, the cooling rate is relatively slow during this stage. When polymers are added to water, stage A cooling is normally extended since the steam blanket is encapsulated within a polymer-rich film. The way in which this composite film forms depends on the structure of the polymer. For polymers that exhibit normal solubility in water, such as PVP and polyacrylate, localized water evaporation occurs adjacent to the hot metal surface, producing a polymer concentration gradient increasing towards the liquid–gas interface. This results in the formation of a temporary three-dimensional polymer chain entanglement network or gel. With amphipathic polymers that exhibit inverse solubility (i.e., a lower critical solution temperature), such as PAGs the polymer is actually precipitated below the boiling point of the solution. This is because the thermally labile hydrogen bonds between the water and the polymer become less numerous as the temperature increases, causing a reduction in the hydrodynamic volume of the polymer.

Stage B is the vapor transport or nucleate boiling stage. This is the fastest stage and commences when the metal has cooled to the characteristic temperature at which the vapor film is no longer stable. As the vapor blanket collapses, cool quenchant wets the metal surface and violent boiling occurs. Heat is removed from the metal very rapidly, predominantly as the latent heat of vaporization. With polymer quenchants, two processes are in competition during stage B cooling: polymer-encapsulated steam bubbles are trying to escape from the surface of the hot metal, while the surrounding polymer-rich film is attempting to restrain bubble release and thus produce a more stable film equivalent to stage A cooling. Alternate periods of nucleate boiling and stable film cooling can occur with high-molecular-weight polymers, or high-concentration solutions, where the driving force for film reformation is high. Bubble growth will tend to be limited by the inertia of the surrounding viscous polymer-rich film and the viscoelastic properties of the bulk liquid. For vaporization and bubble growth to continue, water must be diffused from the bulk solution through the polymer-rich film. The bubble growth rate will thus be reduced since mass diffusivity is of an order of magnitude smaller than thermal diffusivity. This retardation of the bubble growth rate by the viscous polymer-rich film formed at the vapor interface, combined with its insulating properties, explains the reduced stage B cooling rates experienced by polymer quenchants when compared with water.

Stage C is the liquid cooling or convection stage and starts when the surface temperature of the metal reaches the boiling point (or boiling range) of the quenching liquid. Below this point, boiling stops and slow cooling takes place by conduction and convection through the liquid. The industrial acceptance of polymer quenchants has been limited by their higher cooling rates when compared with oils at these lower temperatures. Most quench oils have an initial boiling point within the range 300–450 °C. Polymer quenchants, being aqueous solutions, have boiling points only slightly above 100 °C. This means that stage C convective cooling in polymer quenchants occurs when the surface temperature of the component has fallen much lower than that for oils. Polymer will be permanently adsorbed onto the metal surface, the amount increasing with the molecular weight of the polymer and reaching a limiting value with increased concentration.

4. *Process Variables*

4.1 *Polymer Type*

Although all polymer quenchants have some similarities, they also exhibit significant differences depending on their composition. The characteristics of polymer quenchant products of the same basic polymer type, as supplied by different manufacturers, will vary. These differences can be explained by variations in the molecular architecture of the polymer itself, and by the different levels and types of additives used. The molecular weight distribution of the polymer has a profound effect, as does its configuration and conformation in solution. Higher-molecular-weight polymers are required at lower concentrations to produce equivalent solution viscosities. Owing to the high inherent viscosity of the neat polymers, water is usually included in the commercial product to aid subsequent handling and dilution. Polymer concentrates also normally contain some form of corrosion inhibitor; defoamers, buffers and bactericides may also be included. The level of water in the as-supplied concentrates depends on the type of polymer and varies between proprietary products. Whereas the patent literature has described numerous potential polymer candidates, only five classes of product have achieved commercial prominence as quenchants in the West. These main classes will be discussed in order of historical precedence.

(*a*) *Polyvinyl alcohol* (*PVA*). Aqueous solutions of PVA were first described as quenching media in 1952 (US Patent No. 2,600,290). PVA is a solid, its solubility in water depending upon its molecular weight and the extent of residual polyvinyl acetate. For quenching applications, PVA is normally used at solution concentrations ranging from 0.05–0.30 wt%. Close control is necessary for successful quenching with this class of polymer since only slight variations in solution concentration are needed to produce significant changes in the cooling characteristics. Only a limited number of industrial PVA quench installations remain, because of major concentration control problems exacerbated by the formation of crusty cross-linked residues on components.

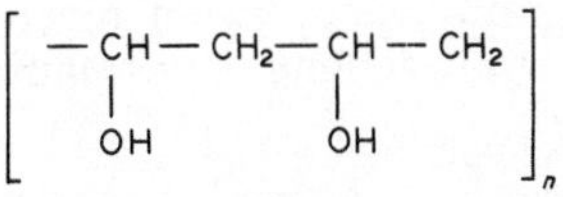

$$\left[-\underset{\displaystyle OH}{\underset{|}{CH}}-CH_2-\underset{\displaystyle OH}{\underset{|}{CH}}-CH_2 \right]_n$$

PVA

(*b*) *Polyalkylene glycols* (*PAGs*). Polyalkylene glycols or polyalkylene glycol ethers were commercially introduced as quenchants during the mid-1960s (US Patent No. 3,220,893) and hold approximately 90% of the market for polymer quenchants. These neutral or nonionic materials are formulated by the polymerization of ethylene and propylene oxides. By varying the molecular weights and the ratio of the oxides, polymers having broad applicability may be produced. PAGs exhibit inverse solubility in water; that is, they are completely soluble in water at room temperature, but insoluble at elevated temperatures. Commercial products are available with inversion points ranging from 71–88 °C. The PAGs used for quenching typically have molecular weights between 12 000 and 45 000, which is lower than for the other classes of polymer quenchant. Consequently, higher concentrations are used, resulting in less sensitivity to minor fluctuations in concentration. The inverse solubility of PAGs also enables them to be separated from certain contaminants, such as carried-over molten salts which are frequently used for heating precipitation-hardenable aluminum alloys.

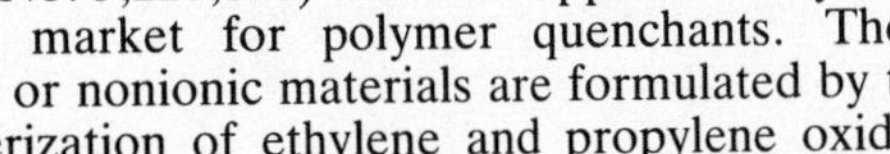

$$HO-[CH_2-CH_2-O]_n-[CH_2-\underset{\displaystyle CH_3}{\underset{|}{CH}}-O]_m-H$$

PAG

(*c*) *Polyvinylpyrrolidone* (*PVP*). This is derived from the polymerization of N-vinyl-2-pyrrolidone. This nonionic water-soluble polymer is characterized by unusual complexing and colloidal properties, and by physiological inertness. Solutions of PVP in water were first described as quenchants in 1975 (US Patent No. 3,902,929). Two molecular weights are usually employed, as denoted by the classification grades K60 and K90, derived using Fikentsher's equation. PVP products are claimed to have certain operating advantages compared with the more widely used PAGs; namely, faster stage A and stage B heat transfer, yet slower stage C convective cooling. Since PVP does not exhibit inverse solubility in water, only very small amounts of polymer film are retained on quenched parts at temperatures ranging from 30 °C to near boiling point. However, the level of drag-out is higher than would be expected from the solution viscosity, owing to chemical cross-linking adjacent to the hot metal surface such as occurs with PVA. The higher molecular weight of

PVP used for quenching, when compared with PAGs, means that it is also more susceptible to chain scission with usage. The combined influence of cross-linking and chain-scission reactions can thus result in an erratic cooling performance.

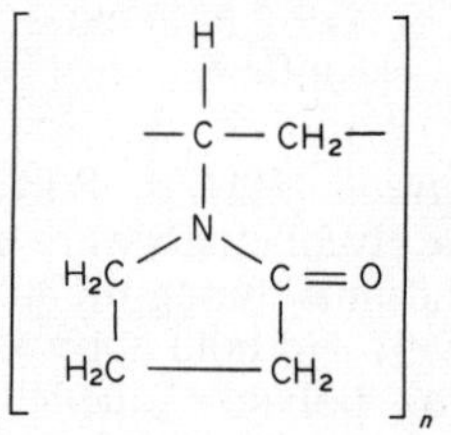

PVP

(*d*) *Sodium polyacrylate*. Polymer quenchants based on sodium polyacrylate were first patented in 1978 (US Patent No. 4,087,290). Synthesis may be via direct polymerization of sodium acrylate or by the alkaline hydrolysis of sodium polyacrylate ester. Polyacrylate products represent a class of quenchants whose structure, and hence properties, are significantly different from those of the other types. Unlike PAGs and PVP which are nonionic, sodium polyacrylate is anionic or negatively charged in solution. This type of polymer quenchant therefore acts as a polyelectrolyte. The strong polarity not only provides water solubility but also makes sodium polyacrylate a very effective aqueous thickening agent due to mutual repulsion of the polymeric ions and sodium counterions. The polyacrylate quenchants thus tend to have the most oil-like cooling characteristics for a given concentration, with extended periods of film boiling and low rates of heat transfer during the convection stage. Polyacrylate solutions are therefore used for non-martensitic quenching, patenting medium-to-high-carbon steel wire and for hardening crack-prone parts of high-hardenability steels. Applications of this kind are usually not possible with other polymer quenchants. The anionic character of sodium polyacrylate not only makes it a very effective aqueous thickening agent but also means that it is capable of being precipitated from solution by polyvalent metal cations and that their viscosity drops very rapidly with initial use with a corresponding increase in cooling rate.

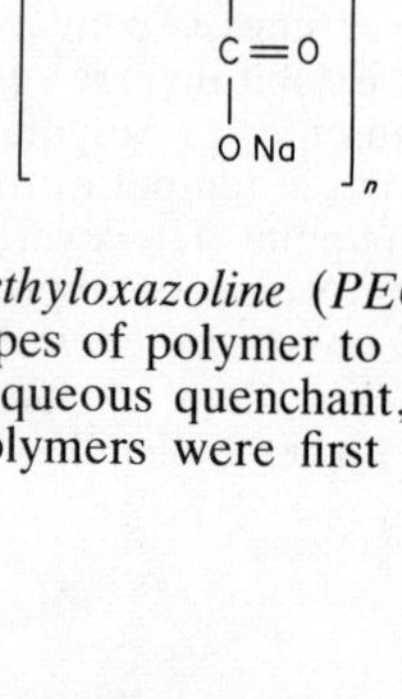

Sodium polyacrylate

(*e*) *Polyethyloxazoline* (PEO_x). One of the more recent types of polymer to be marketed as a commercial aqueous quenchant, these substituted oxazoline polymers were first patented in December 1984 (US Patent No. 4,486,246). Nonionic PEO_x solutions provide the same quenching effect as PAGs, but only one-half to one-third of the amount of polymer is required, resulting in lower bath viscosities and hence reduced drag-out. The high molecular weight (200 000–500 000) of the PEO_x solutions used for quenching means that they have relatively low cloud points (~60 °C) which may restrict their industrial usefulness. Several polymer quenchant manufacturers are therefore examining mixed polymer systems.

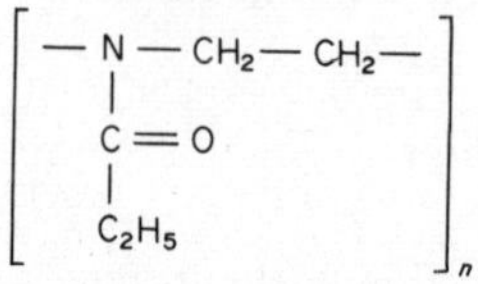

PEO_x

4.2 Polymer Solution Concentration

In general, there is a linear reduction in the rate of cooling as the polymer quenchant concentration is increased over its normal working range. Stage A is extended and the temperature at which the maximum cooling rate occurs is reduced. An increase in the concentration results in a thicker and more stable insulating polymer-rich film being formed around the hot metal. In addition, as the concentration increases, so too does the viscosity of the solutions. The more viscous the fluid surrounding the hot metal, the greater the inertia that retards the rate of bubble growth.

4.3 Polymer Solution Temperature

The maximum cooling rate for polymer quenchants is normally reduced upon raising the solution temperature. This is the opposite effect to most quench oils where raising the oil temperature between the ambient temperature and 100 °C generally reduces their viscosity sufficiently to increase the maximum rate of cooling. Raising the temperature of aqueous solutions over this range also reduces their viscosity, but this effect is outweighed by an increased contribution to the latent heat of vaporization. Raising the temperature of a polymer quenchant solution also tends to stabilize and considerably prolong stage A cooling with a corresponding reduction in both the characteristic temperature and the maximum rate of cooling.

4.4 Polymer Solution Agitation

Agitation, or externally produced fluid movement, has been established as critical for the successful use of polymer quenchants in order to prevent thermal stratification and inconsistent performance under production conditions. Forced convection produces thinner more uniform polymer-rich films around the hot metal. It also disperses the polymer-encapsulated steam bubbles and directs cooler fluid against the hot metal. The rate of cooling increases as the level of turbulence around the quenched part rises until fully turbulent conditions are established. Increased turbulence also results in enhanced mass transfer, diffusion and mixing. Therefore, insulating polymer-rich film jettisoned from the hot metal surface during nucleate boiling is more rapidly replaced. These points help to explain why agitation of polymer quenchant solutions results in faster, but more uniform, cooling.

5. *Control*

Close control of polymer quenchants is essential, since their characteristics change with usage. For example, effective concentration control is critical. Evaporation of water tends to increase the concentration of polymer quenchants, while drag-out of residual polymer on the surface of the cool metal reduces it. The normal manner of quoting polymer quenchant concentration in industry is as a volume percentage of the as-supplied concentrate. However, this figure gives no real indication of the true polymer concentration in solution. The two most popular methods of monitoring the concentration are by means of refractive index and kinematic viscosity measurements, although specific gravity measurements can also be used. Refractometer readings, although the most convenient method, are unreliable since they are influenced by contamination and take no account of polymer degradation. Kinematic viscosity measurements are more reliable but lack the superior sensitivity of the Wolfson Engineering Group test in instantly detecting subtle changes in the cooling characteristics. Contamination can also have a significant effect on the performance of polymer quenchants. Volatile contaminants generally tend to increase stage A and reduce the maximum cooling rate, while salts have the opposite effects. Salts also reduce the solubility of the polymers, especially the PAGs. During long-term use, degradation of the polymers (due to the combined influence of thermal, oxidative, mechanical and biological processes) also affects their characteristics. Aging shifts the molecular weight distribution of the polymer to lower weights. A reduction in the average molecular weight results in a corresponding reduction in the solution viscosity and an increase in the rate of cooling. Faster cooling at around 300 °C (which corresponds with the martensite start of transformation temperature for many engineering steels) increases the possibility of cracking and distortion problems.

The dilemma faced by the polymer quenchant formulator is that of trying to balance a low rate of cooling with an acceptable level of drag-out. Drag-out is associated with the viscosity of the solution: the higher the molecular weight of the polymer and hence the viscosity for a given concentration, the greater the drag-out. The charged character of sodium polyacrylate results in enhanced adsorption and higher drag-out. However, the level of the latter is reduced for each class of polymer quenchant by agitating the solution. This is particularly true for highly concentrated solutions.

See also: Aluminum Alloys: Heat Treatment; Heat-Treated Alloy Steels

Bibliography

ASM Committee on Quenching of Steel 1981 Quenching of steel. In: Masseria V, Kirkpatrick C W (eds.) 1981 "*Heat Treating*" *Metals Handbook*, 9th edn., Vol. 4. American Society for Metals, Metals Park, OH, pp. 31–69

Close D 1985 Wolfson Engineering Group quench test adopted for international standard. *Heat Treat. Met.* 12: 62

Davidson R L, Sittig M (eds.) 1968 *Water-Soluble Resins*, 2nd edn. Reinhold, New York

Foreman R W, Meszaros A 1984 Polymer quenching update. *Ind. Heat.* 51: 22–4, 29

Hilder N A 1986 Polymer quenchants—A review. *Heat Treat. Met.* 13: 15–26

Hilder N A 1987 The behaviour of polymer quenchants. *Heat. Treat. Met.* 14: 31–46

Irani R S 1989 Polymer-based quenchants—Relevance to the high pressure gas cylinder market. *Speciality Chemicals* 9: 261–6

Mueller E R 1980 A look at the polymers—Past, present and future. *Heat Treating* 12: 30–3

Segerberg S, Bodin J 1986 Polymer quenchants show benefits in metal heat treatment. *Metallurgia* 53: 425–6

N. A. Hilder
[Wolfson Heat Treatment Centre, Aston University, Birmingham, UK]

R

Rapid Thermal Processing of Semiconductors

Rapid thermal processing (RTP) or rapid thermal annealing (RTA) is a flexible furnace technology used in the pursuit of advanced microelectronic and microstructure applications. Typical heating and dwell times are several seconds for single wafers, while temperatures are in the range of approximately 400–1400 °C (the upper value is near the melting point of silicon). The limited dwell time results, in turn, in limiting the kinetics of thermal processes such as diffusion [(diffusion length)^2 ~ diffusion constant × time] and the mixing of material phases. The use of higher temperatures combined with short time cycles allows for the selective advancement of thermal processes with different activitation energies.

1. *Background*

Tungsten–halogen lamps were used in semiconductor applications as early as 1962 by Kinsel to anneal GaAs encapsulated in a quartz ampoule. This heating source was used to avoid contamination from the hot quartz walls of a standard furnace. In the mid-1970s, various workers used lasers as sources for the fast heating of semiconductors, mainly for the purpose of annealing ion-implanted crystal damage and to limit dopant diffusion.

In a broader sense, RTP covers a range of time durations from approximately 10^{-9} s to about 100 s. The shortest time regimes are implemented with pulsed lasers and electron beams. For the shortest times, a steep thermal gradient exists near the surface, with spatial and time dependencies that follow the laws of thermal diffusivity. In an intermediate time frame (10^{-5}–0.1 s) a vertical linear thermal gradient develops across the thickness of the wafer.

In both these shorter time regimes there are a number of practical limitations that preclude wide use in very-large-scale integration (VLSI) technology. There are thermal gradients which stress the silicon and introduce either residual point defects or extended dislocations. A variety of optical interference effects occur near contact-window edges, defect-free junctions are hard to obtain and annealing is limited to the area of the laser exposure. If the surface region is melted with high-energy pulses, liquid-phase epitaxial regrowth takes place and gives heavily doped shallow impurity profiles; however, in general there is poor compatibility with multiple-dielectric film and the integrated circuit device topology. Technical aspects of the shorter time RTP are addressed by Hill (1981). It is the annealing with times above about a second that result in nearly isothermal wafer heating and it is this case that holds promise for more widespread practical applications.

2. *Isothermal Annealing and Demonstrations of the RTA Advantage*

Isothermal heating of single wafers can be obtained by using arc lamps, tungsten–halogen filaments or graphite resistive elements. Although the lamp heating times can differ from 0.1–10 s, the heating rate of the wafer is mainly determined by its optical coupling (absorption and reflection), heat capacity and wafer thickness. The time to heat the wafer is typically the order of a few seconds.

$$\frac{dT}{dt} \sim (1-R) \int \frac{I_\lambda(1-\exp(a_\lambda d))}{C}\, d\lambda$$

where R is the reflectivity, C the heat capacity, a_λ the absorption coefficient, I_λ the incident energy intensity (both a_λ and I_λ are dependent on the wavelength) and d is the thickness of the wafer. The thermal conductivity/diffusivity of silicon is good, so it is difficult to obtain a very large thermal gradient *across* the thickness of a wafer, even if the wafer is heated from one side.

Near the melting point of silicon, Celler (1987) has demonstrated vertical gradients of a few degrees across the wafer for the case of silicon heated by radiation from one side.

$$\nabla T = eST^4/K$$

where e is the emissivity, $S = 5.6 \times 10^{-8}$ W m^{-2} K^{-4} is the Stefan–Boltzmann constant, K is the thermal conductivity and T is the temperature. In this case, silicon melts at the hot surface and a "mixed state" of molten "islands" is formed in a background "sea" of crystalline material. As the power is increased, the molten islands grow in size but the average temperature of the wafer system remains essentially constant as the energy goes into melting the entire surface. It is possible to hold the hotter surface at 1412 °C and the colder surface (device side) at 1405 °C for an indefinite period without any electronic feedback.

A more serious problem exists for lateral or radial uniformity. The temperature can be considerably different in the central region of the wafer than the edge, depending on radiative energy balance and the engineering of the furnace. A lateral temperature gradient can lead to stresses above the yield stress

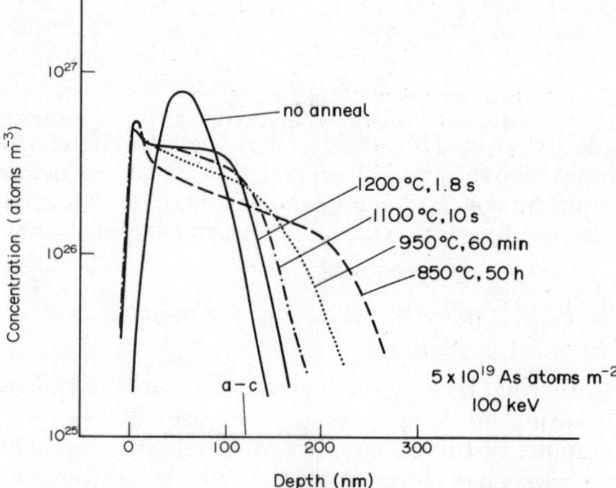

Figure 1
Concentration profiles of arsenic implanted silicon: the time was adjusted at each temperature until the residual dislocations just beyond the a–c location were removed

and result in dislocation formation or "slip." However, for the purposes of discussion here, the behavior of uniformly or isothermally heated wafers can be considered.

Whether or not temperature can be traded for time for a thermally activated process and, after some period in the kinetics, would not the same annealed physical state be obtained, is a question that might be asked. If only a single thermally activated process is being considered this can be done, but it cannot if there are two or more competing processes with different activation energies. At higher temperatures, the higher-energy process occurs relatively more quickly. For implanted dopants in silicon, with the attendant damage, the activation energy for damage removal is 5 eV and that for dopant diffusion is approximately 3.5 eV. Thus the damage can be removed at high temperatures while limiting the diffusion. This concept has been illustrated by the work of Seidel et al. (1985) for arsenic implants, and that by Sadana et al. (1983) for boron implants. The concept is not limited to implant activation applications; there are other possibilities (e.g., in metallurgical applications) where phase formation with limited diffusion is an issue.

In Fig. 1, dopant profiles are plotted for the cases where dislocations are first removed as time is increased at each temperature. For example, dislocations still exist at 8 s but not at 10 s at 1100 °C. The dislocations at issue are the localized dislocation defect arrays associated with the initial damage for arsenic ions stopping in the crystalline material just beyond the amorphous–crystalline (a–c) interface depth position. Clearly the defect-free profiles are more limited in their diffusion for the higher temperature cases. In Fig. 2, boron profiles are compared for RTA and standard-furnace anneals. The anneal conditions are adjusted so the junction depths are the same but the damage is different. The RTA case has a broadened peak, while the furnace case has a pronounced peak and residual dislocation damage.

Another demonstration case is the limited diffusion of silicon in aluminum when the contact reaction sintering between silicon and aluminum is carried out at approximately 450 °C for a few seconds (Pai et al. 1985). Also, thin thermal nitridizations are done at very high temperatures, where the thermal diffusivity of nitrogen through the thin underlying oxide and the accumulation of nitrogen at the silicon dioxide–silicon interface is limited. The activation energy for nitride formation is greater than the energy for diffusion (Chang et al. 1985).

3. *Inventory of Applications*

This section discusses a variety of applications which address issues other than the utilization of the "RTA advantage" where multiple competing thermally activated processes are at issue. Process integration issues may drive the application.

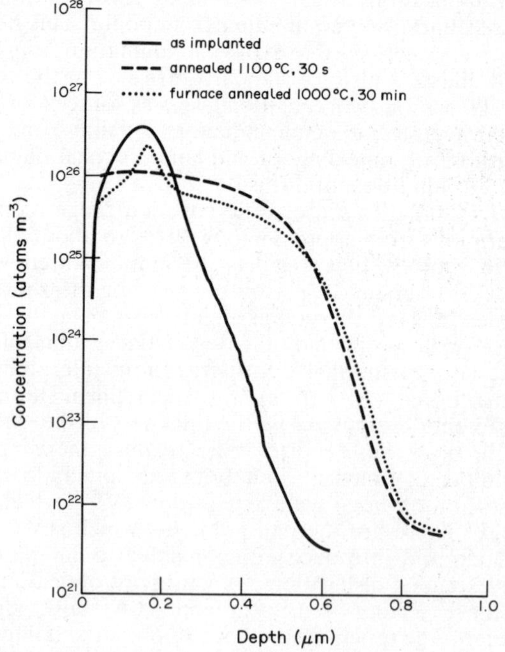

Figure 2
Concentration profiles for boron implanted into crystalline silicon: the anneals were done at different temperatures and times so that the junction depths were made the same; the standard-furnace annealed sample has a peaked impurity distribution and severe residual dislocation damage

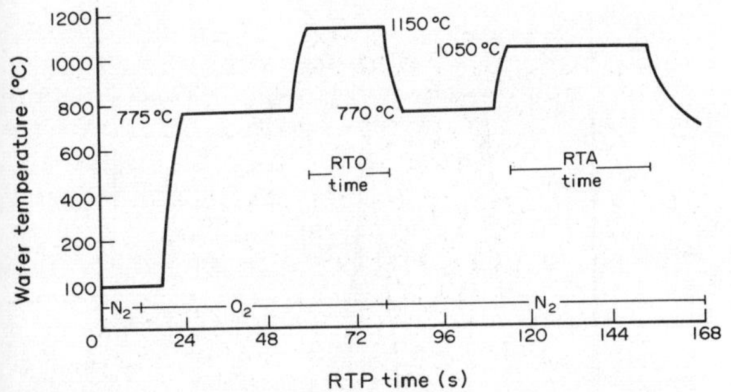

Figure 3
Wafer temperature plotted against anneal cycle time, for a typical rapid oxidation/anneal (N_2) or thermal nitridation (NH_3) sequence

3.1 Rapid Thermal Oxidation and Nitridation

It is possible to form thin gate oxides with quite high dielectric breakdown strength and excellent uniformity using the RTP approach. The oxide quality is at least as good as that obtained using conventional furnaces, the uniformity is approximately 2%, except within about 3 mm of the edge of the wafer, and most importantly the process lends itself to integration with the next serial steps.

The next step may be rapid thermal nitridation (RTN) (see Nulman 1988), as shown in Fig. 3, and/or deposition of chemical vapor deposition (CVD) polysilicon, and/or the deposition of CVD metal for silicides. The RTP concept is potentially important because of this process flexibility and the process integration. The chambers in which the processes are being carried out have high ambient integrity, and process ambients can be changed and equilibrated in times comparable with the process time.

3.2 Silicide Formation

The ability to react silicides in short times at elevated temperatures and controlled ambients, favor the use of RTA for certain silicides. Titanium is extremely reactive with oxygen, so low oxygen is required for the controlled reaction with silicon to proceed. The reaction proceeds by the diffusion of silicon into the metal. This causes the silicide to form on the sidewalls of oxide-cut topology in addition to forming TiSi in junction-contact regions. (In standard furnaces, oxygen can backstream into the wafer area.) TiSi reactions are an important example of a process which is better done using a high ambient integrity RTP/RTA approach (Praminik et al. 1985).

3.3 Shallow-Junction Science

The idea that shallow junctions can be produced by low-energy implantation into crystalline silicon is frustrated by the variability of ion-channelling effects. Implanted boron dopant ions undergo wide-angle scattering into the open axes and planar channels of the crystal target, creating deep penetrating tails on the profile of the implanted distribution. Furthermore, as the boron ions come to rest they may create damage clusters which lead to low-temperature enhanced thermal diffusion. The channelled tail of boron exhibits thermal diffusion enhanced by several orders of magnitude (Michel 1986) as shown in Fig. 4. Here the enhanced diffusion has been reduced by a high-temperature RTA step. The mechanisms for the enhanced diffusion are under research; it is possible that the boron and local damage result in an interstitial-like enhanced-diffusion process, while it is also possible that point defects from sources physically removed from the channel tail region (e.g., near the peak or maximum damaged regions) play a role. Interstitial-silicon point defects are known to replace substitutional boron dopant and enhance the boron diffusivity as long as the boron is interstitial.

The channelled tail is very sensitive to the incident angle of the ion beam with respect to the crystallographic axes of the target silicon. Thus, there are two reasons why very shallow junctions cannot be made with any great degree of control: channelling phenomena—including orientation control—and enhanced thermal diffusion. A thin (amorphous) oxide on the surface only spreads the angles of the incoming beam and does not prevent the ion-channelling phenomena once the ions have entered the crystal.

Another approach has been the use of preamorphized silicon by the preimplantation of an "inert" damaging species, such as silicon implanted into silicon (Liu and Oldham 1983). When sufficient damage occurs, the crystal disorganizes into an amorphous state and no channelling is possible for

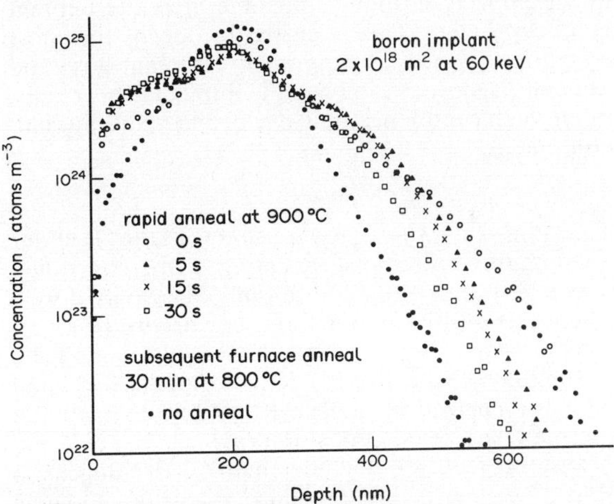

Figure 4
Boron SIMS profiles with a 900 °C RTP preanneal, before an 800 °C 30 min furnace anneal

subsequent ions implanted into this material. However, there is dislocation damage just beyond the a–c interface which should be annealed out for low-leakage junctions to be obtained.

The implanted distribution can be contained entirely within the thickness of the preamorphized layer and then controllably diffused into crystalline material. The dopant impurities may be diffused beyond the a–c disorder and this gives quality junctions with a depth of about 0.1 μm.

The implantation of dopants into filmlike layers above the crystalline silicon followed by outdiffusion into the crystalline silicon can also be considered. In this manner there may be shallow controlled junction profiles, without channelling and ion-damage effects in the crystalline silicon. The actual junction technology "process-package" is dependent on many issues. The specific application and process compatibility with the entire integrated circuit (IC) process needs to be considered. Issues such as contact resistance, carrier multiplication from high-electric-field effects, and glass flow for Na^+ passivation are issues that are just as important as the shallowness of the junction and the control of the channel lengths of metal-oxide-semiconductor (MOS) transistors.

3.4 Other Applications: Gettering and Implant Qualification

Gettering of impurities and defects away from the active regions of devices has been important for obtaining large-area wafer perfection, which allows the advance of the level of complexity in IC technology. The ability to release impurities from dislocations via the dissolution of dislocations seems to be a promising use of RTA (Sparks et al. 1986).

The use of implantation in manufacture often requires the assessment of the dose and uniformity of the implant as soon as possible after the implant has been made. This is currently one of the most common uses of RTP equipment. Implant dose and dopant diagnostics can be made immediately, resulting in high confidence in wafer product and implant equipment.

3.5 CVD with RTP

Thin films have been grown using CVD/RTP under low-pressure conditions, resulting in the controlled deposition of extremely thin films of heavily doped silicon layers by Gibbons and coworkers (Fig. 5). This work represents an important evolution in RTP technology. Basically, the temperature is switched high for deposition and switched low to stop the thermally activated deposition cycle.

System features include many advantages: a closed system with ambient integrity allowing for purging, *in situ* cleans, and fast gas switching and control, a cold wall which promotes furnace cleanliness, and independently controlled temperature—time and process gas flow—time recipes. The flexibility of the approach far exceeds that of conventional batch-loaded furnaces, where access, ambient control and speed of temperature changes are limited.

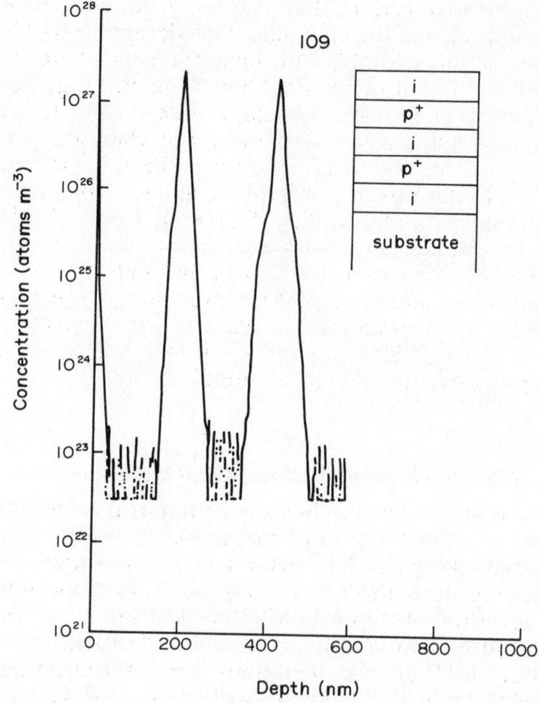

Figure 5
Concentration profiles for boron, formed using thermally switched RTP/CVD or "limited reaction processing"

The ability to perform CVD now includes the following films:

(a) silicon epitaxy, with very thin doped layers;

(b) doped and undoped SiO_2 glass;

(c) polysilicon, doped and undoped; and

(d) tungsten deposition.

Other metal silicides (titanium, cobalt and copper) can be pursued. Various III–V compounds can be deposited (Vook et al. 1987). In addition, as new chemicals are developed, thin-film new superconductors may be deposited using RTP/CVD, where it is important to limit the diffusion of impurities from the substrate.

4. Process Integration

As the capability for multiprocess deposition tools develops, processes can be anticipated that are further integrated for microelectronic manufacture.

Facility architectures that enhance yield, cycle time and throughput are emerging. One design concept is to place process tools with small footprints that are used serially in the process flow sequence together within a single clean envelope area, and connect them by automated wafer-transfer mechanisms. An example facility architecture has been proposed (Seidel 1988a, b). It is referred to as a self-contained automatic robotic facility (SCARF). Considerable improvements in cycle-time may be obtained, especially if sensor technology is applied to the control and monitoring of the processes. An example layout of a multiprocess SCARF with deposition and etching sectors is shown in Fig. 6. Various robots serve the individual modules.

5. *Equipment and Operational Issues*

The basic heating mechanism for wafers includes the overlap of wafer absorption and the incident "black body" spectrum. The wafer system, depending on the films near the surface and the doping (infrared free carrier absorption) will have a different optical response to the incoming radiation. The success of future RTP furnace operations is dependent on the ability to heat the wafers with very good lateral-temperature *uniformity*, and the ability to measure the *temperature*. Knowledge of the temperature has been a concern since the earliest days of the technology.

Thermocouples are good for experimental purposes but are destructive for IC wafers. Optical pyrometry is an excellent approach for the case of a wafer without dielectric layers, but most of the important applications use wafers with multilayered dielectric and metal structures. More fundamental methods for measuring the temperature are needed; the measurement of the lattice constant, expansion coefficient or the linewidth of temperature-sensitive optical emissions would be examples of more direct temperature metrology.

Other equipment issues are: monitoring the gas flow and gas chemistry (fluorescence); pressure, cold-wall temperature and cleanliness control; power–time controlling algorithms; and temperature and gas delivery uniformity (i.e., deposition uniformity) (Gelpey 1988). The design of the wafer-edge environment is crucial for the control of a lateral or radial temperature uniformity, although some control is also afforded by modifying the distribution of incident radiation. More energy can be delivered to the edge region of the wafer than the center. Process control using flexible recipes are available.

Finally, as the degree of process integration and complexity increases, the issue of equipment reliability and cleanliness must be addressed. The requirements for process tool *reliability* become crucial.

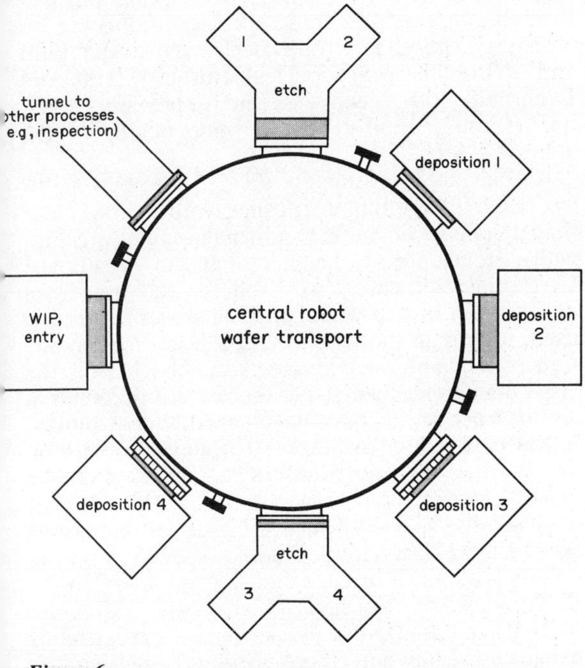

Figure 6
Schematic of an advanced integrated process system; the modules and central chamber are self-contained

Bibliography

Celler G K 1987 Thermal processing in a thermal gradient. *Solid State Technol.* 30(3): 93

Chang C C, Kamgar A, Kahng D 1985 High-temperature rapid thermal nitridation of silicon dioxide for future VLSI applications. *IEEE Electron Devices* 6(9): 476–8

Gelpey J C 1988 Recent developments in RTP equipment. *J. Electrochem. Soc.* 135(3): C125

Gibbons J F, Reynolds S, Gronet C, Vook D, King C, Opyd W, Wilson S, Nauka C, Reid G, Hull R 1987 *Mater. Res. Soc.* 92: 281

Hill C 1981 Beam processing in silicon device technology. In: Gibbons J F, Hess L D, Sigmon T W (eds.) 1981 *Laser and Electron Beam Solid Interaction and Material Processing*. North-Holland, New York p. 361

Kinsel T S, Seidel T E 1962 Heat treatment of *n*-type GaAs by radiant energy. *J. Appl. Phys.* 33: 767

Liu T M, Oldham W L 1983 Channeling effect of low-energy boron implant in (100) silicon. *IEEE Electron Devices* 4(3): 59–62

Michel A 1986 Rapid thermal processing. In: Sedgwick T O, Seidel T E, Tsaur B-Y (eds.) 1986 *Proc. Symp. Materials Research Society*, Vol. 52. MRS, Pittsburgh, PA, p. 3

Nulman J 1988 Rapid thermal processing for thin silicon dielectrics: Growth and applications. *J. Electrochem. Soc.* 135(3): C123

Pai C S, Caberos E, Lau S S, Seidel T E, Suni I 1985 Rapid thermal annealing of Al–Si contacts. *Appl. Phys. Lett.* 46(7): 652–4

Pramanik et al. 1985 Formation of TiSi by rapid thermal processing. *Semicond. Int.* May: 94

Sadana D K, Shatas S C, Gat A 1983 Heatpulse annealing of ion-implanted silicon: Structural characterization by transmission electron microscopy. *Inst. Phys. Conf. Ser.* 67: 143

Seidel T E 1988a Rapid thermal processing impact in microelectronics. *J. Electrochem. Soc.* 135(3): C124

Seidel T E 1988b Sputtering. *Semicond. Int.* May: 79

Seidel T E, Lischner D J, Pai C S, Knoll R B, Maher D M, Jacobsen D C 1985 A review of rapid thermal annealing (RTA) of B, BF_2 and As ions implanted into silicon. *Nucl. Instrum. Methods* B7(8): 251–60

Sparks D R, Chapman R G, Alvi N S 1986 Anomalous diffusion and gettering of transition-metals in silicon. *Appl. Phys. Lett.* 49(9): 525–7

Streetman B G, Dodabalapur A 1988 Rapid thermal processing of III–V materials. *J. Electrochem. Soc.* 135(3): C124

Vook D W, Reynolds S, Gibbons J F 1987 Growth of GaAs by metalorganic chemical vapor-deposition using thermally decomposed trimethylarsenic. *Appl. Phys. Lett.* 50 (19): 1386–7

T. E. Seidel
[Seidel Consultants, Cardiff, California, USA]

Rattan and Cane

Rattan and cane are terms that can be used interchangeably for the stems of climbing palms belonging to the subfamily Calamoideae of the family Palmae or Arecaceae (Uhl and Dransfield 1987). They enter world trade as raw material for furniture manufacture, and are used locally in producing countries for a wide range of purposes, most importantly for weaving basketware. In common usage there is frequently confusion between rattan or cane and bamboo (see *Bamboo*), objects referred to as "bamboo" furniture in reality being made from rattan, and some so-called rattan furniture being constructed from the leaf stalks of palms such as raphia (*Raphia* spp.) or buri (*Corypha* spp.) or from bamboo. All rattan stems are solid and enter trade either whole or variously split and cored. The center of the natural distribution of rattans lies in India, Sri Lanka, Southeast Asia and the Malay archipelago, although there are outliers in West Africa, southern China, the west Pacific and Australia. In tropical America there is a group of climbing palms belonging to the genus *Desmoncus*, analogous with the Old World rattans but belonging to a quite different group of palms (subfamily Arecoideae: tribe Cocoeae). Although these New World climbing palms may be used locally for basketware or even cheap furniture, they are not used extensively as yet and seem to have limited potential. Most rattan entering world trade is collected from wild stands from tropical rain forests. With widescale forest clearance and overexploitation, present wild stocks of rattan are being rapidly depleted and trade protectionist measures have been introduced by some rattan-producing countries to protect local supplies and to increase the value of the exported product. However, these measures have tended to increase the pressure on wild stocks elsewhere. Forest departments and other agencies within the rattan-producing countries are now investigating the possibilities of rattan cultivation, by building on the experience of villagers in Central Kalimantan, Indonesia, where two species of rattan have been successfully cultivated for over a century (Dransfield 1979).

1. Botany and Distribution

There are at least 568 different species of rattan, belonging to the following 13 genera (in decreasing order of size): *Calamus* (about 370 species; West Africa, India and China to Australia and Fiji, with the greatest number of species in Borneo), *Daemonorops* (about 115 species; South China to New Guinea), *Korthalsia* (about 26 species; Indochina to New Guinea), *Plectocomia* (about 16 species; Himalayas to Borneo and the Philippines), *Eremospatha* (about twelve species; West Africa), *Laccosperma* (about seven species; West Africa), *Ceratolobus* (six species; West Malaysia to Java), *Plectocomiopsis* (five species; Thailand to Borneo), *Oncocalamus* (five species; West Africa), *Pogonotium* (three species; West Malaysia and Borneo), *Myrialepis* (one species; Indochina to Sumatra), *Calospatha* (one species; West Malaysia) and *Retispatha* (one species; Borneo) (Uhl and Dransfield 1987). Most of the important, good-quality canes entering world trade belong to the genus *Calamus*.

Rattans are conveniently divided into two major size classes: large-diameter canes, with stems greater than 18 mm in diameter, and small-diameter canes, with stems less than 18 mm in diameter. Large-diameter canes are used for the frames of items of furniture, while small-diameter canes are used, usually in the split or cored state, for binding frameworks and weaving seats and backs. Of the large diameter canes a few species are preeminent and because of this have been grossly overexploited in the past few years. In West Malaysia, Sumatra, South Thailand and South Kalimantan, *Calamus manan* (rotan manau or manau cane) is the premier quality large species, which, because of scarcity, is now being replaced in the trade by species of poorer quality such as *C. ornatus* and *C. scipionum*. In the Philippines, the most favored large cane is *Calamus merrillii* (palasan), but stocks have been virtually exhausted. Elsewhere there are fine large-diameter canes in Sulawesi, New Guinea and Sri Lanka which may be amenable to cultivation, but their botany and in some instances even their scientific identity are not known. Amongst the small diameter canes, *Calamus caesius* (sega) (South Thailand, West Malaysia, Sumatra, Borneo and Palawan) and *C. trachycoleus* (irit) (Kalimantan) are preeminent, but many more species enter the trade.

Rattans are mostly high climbing, many reaching the forest canopy. After an initial period of establishment growth during which the full stem diameter is built up, the stem grows upwards, not increasing further in diameter with age. The plant may consist of a single unbranched stem, which when harvested is killed, or it may produce suckers at the base, in which case the clump can be harvested continually. Very rarely the stems branch aerially in the forest canopy. Some species, for example, *C. manan*, can grow to immense lengths; the longest ever recorded was about 185 m.

No reliable studies have been made on the actual or projected production of cane from wild stands of rattan, and exploitation has never been carefully controlled despite the existence of regulations. Because they are classed as minor forest products, there has been a tendency for rattans to receive rather little research attention from forest departments, yet as a forest product they have great significance to people living near forests as a supplementary source of cash, particularly in periods of agricultural scarcity.

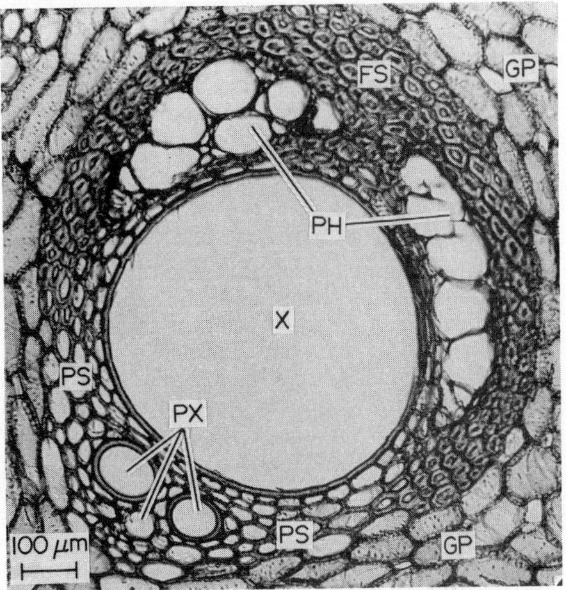

Figure 1
Vascular bundle of *Daemonorops angustifolia*, representing a common bundle type (*Calamus, Calospatha, Ceratobolus, Daemonorops, Korthalsia*). GP is the ground tissue, FS the fiber sheath, PH the phloem fields, X the metaxylem, PX the protoxylem and PS the parenchyma sheath (courtesy of Professor Walter Liese, University of Hamburg)

2. *Physical Properties*

Rattan stems usually have elongate internodes and the nodes are clearly marked. The diameter of the internodes and indeed the stem itself does not usually vary significantly along their lengths. The cross section is usually more-or-less circular; however in several species of *Calamus* (e.g., *C. scipionum*), the cross-sectional outline is circular except for a subtriangular protrusion on one side, marking the position of the vascular supply to the climbing whips or inflorescences. The position of this ridge changes from one internode to the next, corresponding to the phyllotactic spiral of the living plant. This unevenness of the bare cane has implications for utilization, such canes being less favored than those with an even cross-sectional outline. The surface of the bare cane is usually smooth and frequently lustrous, pale yellow-brown to ivory-colored. In species of *Korthalsia*, however, the cane surface is reddish-brown and rough with incompletely separated remains of the leafsheaths. Such red canes, despite being extremely durable, are not favored by the trade, although they are extensively used for varying the texture and appearance of fine local basketware. The outer layer may be two or more cells thick. The cortex of the rattan stem includes scattered vascular bundles embedded in ground tissue. The anatomy of the stem can show both generic and specific differences. The structure of a typical vascular bundle is shown in Fig. 1.

The value of rattan in the furniture industry depends on the high degree of flexibility coupled with great strength and fine appearance. Not all canes have the same properties. The finest canes have a rather uniform cross-sectional structure with an even distribution of vascular bundles and even and not excessive lignification of the ground tissue. In some species of *Plectocomia*, starch is accumulated in the central part of the stem which is also rather soft. Cane from *Plectocomia* is generally worthless, being difficult to bend without causing distortion and being particularly prone to fungal and insect attack because of the accumulations of starch. Some species of *Plectocomiopsis* and *Eremospatha* have stems which are more-or-less triangular in cross section. These are also worthless, being impossible to bend without the cane splitting. In many of the better species of *Calamus* the outer layer of the cane, or epidermis, is heavily impregnated with silica. This silica layer gives a high luster to the surface of the cane. For some purposes such as the production of fine split cane (chair cane) the silica layer is removed by a process called "runti" or "lunti" which involves twisting the harvested stems in such a way that the silica layer snaps off in thin flakes. For other purposes the silica layer is carefully preserved and protected.

Studies on the physical properties of rattans are in their infancy. Tensile strength and other mechanical properties have been investigated for some species but in some instances the results of the studies are of little value because the scientific identity of the canes was not critically established, a problem which occurs in many aspects of rattan research. Goh (1982) investigated the mechanical properties of Rotan manau (*Calamus manan*) and found that, as in timber, the strength of green material increased upon drying. Air-dry material (14.4% moisture content) had a density of 750 kg m^{-3} and compression strength parallel to grain of 30 MPa, about one-half the value one would expect of timber of the same density (see *Wood Strength*).

3. Processing

Processing of rattan is usually rather simple. Small-diameter canes newly harvested are cut into lengths of about 9 m and bent into two; they are bundled into loads sufficient for one man and carried out of the forest. If they are not to be processed immediately they may be stored under water for a short period. Initial processing of such small canes involves cleaning the cane surface of remains of the leaf sheath and the removal of dirt. This is often carried out manually by rubbing the cane surface with a handful of sand while continually washing the cane. Small canes may also be "runtied" at this point, the silica layer being removed by twisting the cane through closely placed rollers or bamboos. Canes are then sun-dried and additionally may be smoked over burning sulfur to prevent fungal attack. Fungi, particularly those responsible for "blue stain," may have a marked deleterious effect on harvested cane if it is not processed quickly. After drying and smoking, the canes are graded and packaged for shipment to a secondary processor or manufacturer. Secondary processing involves the splitting of the cane to produce skin (chair cane) and core. A wide range of machines are available for such processing. Usually they are designed in such a way that when a cane is passed into the machine over rollers, a die peels off the skin in four to eight strips while at the same time removing circular sectioned lengths of core from the remaining central portion, the number and diameter of the core strips being dependent on the die which is selected for the diameter range of the rattan itself. Strips of skin (chair cane) are extensively used for weaving the backs of both cane and wooden furniture. Core is used in usually cheaper quality woven furniture and in handicrafts.

Large-diameter cane is usually cut into lengths of about 3 m as it is harvested, the resulting "sticks" being bundled and carried out of the forest. Processing of such large-diameter rattans is not usually carried out by the collectors themselves, but by a rattan merchant. The problem of blue-stain fungus attack is a particular problem with large-diameter canes, as the time which elapses between harvesting and delivery to or collection by the rattan merchant may be prolonged. Processing of large canes usually involves treating the sticks in heated oil or kerosene. Diesel, coconut or other oil may be used on its own or in combination with kerosene. The oil is often heated in tanks made from old oil drums and the cane immersed in the heated oil for short periods. The treatment seems to act by removing excess water, by bringing natural gums in the rattan to the cane surface and by effectively killing the rattan itself together with any fungi on the cut surfaces. After treatment with oil the sticks are further sun-dried before being graded, bundled and packaged ready for shipment to the manufacturer.

4. Trading

Traditionally, rattan harvested from the forest is bought from the collectors by middlemen who sell to small rattan merchants. The rattan merchants are responsible for grading and processing and finally export. In the past rattan was for the most part exported to Singapore, Hong Kong or Cebu (Philippines), all three ports acting in an entrepreneurial role with some small-scale manufacturing, resulting in the reexport of manufactured items or of processed cane. China has been a major importer of rattan and also a major exporter of rattan handicrafts. Recently, there have been changes to this age-old pattern. A few rattan cooperatives have been organized, removing middle-men from the local trade and thus increasing the price payable to the collectors, while governments have become increasingly aware of the advantages of adding to the value of the exported cane by carrying out processing and manufacture locally. To encourage this export, duties on raw cane have been raised or the export of raw cane banned. If such bans and prohibitive taxes are effective, then the future of rattan manufacturing businesses outside the producing countries will be bleak.

5. Commercial Uses

Almost all large-diameter rattan in trade enters the furniture industry. In contrast, some small-diameter rattan is used commercially in the production of high-quality matting, mostly for Japanese markets, some for handicraft manufacture (the value of which is considerable), and some for the furniture industry. The use of small-diameter canes for reinforcing concrete has been investigated but the potential for this use seems limited, despite the apparent resilience of rattan when thus used (Baharin 1978).

6. The Future for Rattan

Rattan appears at the moment to have an uncertain future. The most serious problem facing the industry is the supply of raw material. With widescale destruction of forest and overexploitation of the canes themselves, a world shortage of rattan seems unavoidable. Techniques for cultivating small-diameter rattan are available for two species, *Calamus caesius* and *C. trachycoleus*, but very few new plantations have actually been established. Furthermore, cultivation of large-diameter canes has not progressed beyond the experimental phase. Even if plantations were established now they would be unlikely to come into bearing before ten years have elapsed, so that the shortage of rattan is likely to become more severe. In the past, manufacturing has usually taken place in importing countries rather than the producing countries themselves. Producers are now erecting trade barriers with the intention of encouraging the manufacture of rattan furniture by the producing country. However, insufficient expertise in furniture production exists in the producing countries. Considerable changes in the rattan trade worldwide can be expected as all these factors take effect. One major concern is that disruptions to the traditional trading patterns may have the effect of destroying the market entirely, before cultivation can be initiated and the long-term supply assured.

Rattan has been the subject of two recent international symposia, in Malaysia in 1984 (Wong and Manokaran 1985) and in Thailand in 1987, besides numerous local symposia. There is a rattan information center at the Forest Research Institute Malaysia, Kepong, which acts as a clearing house for the Asian region, publishing a bimonthly bulletin and a bibliography of research papers (Kong-On and Manokaran 1986). There is thus much research interest in rattan but, so far, few results of significance to the silviculture of rattans and the long-term supply.

See also: Bamboo

Bibliography

Baharin bin Puteh 1978 A study of rattan and its mechanical properties. Thesis, Mara Institute of Technology, Kuala Lumpur

Dransfield J 1979 A manual of the rattans of the Malay Peninsula. *Malay. For. Rec.* 29

Goh S C 1982 Testing of Rotan manau: Strength and machining properties. *Malay. For.* 45(2): 275–7

Kong-On H K, Manokaran N 1986 *Rattan: A Bibliography*. Rattan Information Centre, Forest Research Institute Malaysia, Kepong

Uhl N W, Dransfield J 1987 *Genera Palmarum, A Classification of Palms based on the work of H E Moore Jr.* Bailey Hortorium and International Palm Society, Kansas

Wong K M, Manokaran N 1985 *Proc. of the Rattan Seminar 2–4 October 1984, Kuala Lumpur, Malaysia*. Rattan Information Centre, Forest Research Institute Malaysia, Kepong

J. Dransfield
[Royal Botanic Gardens, Kew, Richmond, UK]

Reflection Electron Microscopy

The physical and chemical properties of surfaces are of fundamental importance in many aspects of materials science, from corrosion to the growth of novel semiconductor devices, and they have been examined by a wide variety of techniques over many years. Most of these techniques give averaged results from large areas of the sample and the effects of inhomogeneities are either ignored or, at best, implied indirectly from the data. However, it is often such defects in the form of dislocations and monatomic steps that have a crucial influence on the chemical and physical changes that occur at surfaces. In order to examine directly the role played by defects, techniques with high spatial resolution and surface sensitivity are required.

In recent years various types of microscopy have been developed that can image surfaces with the required spatial resolution and one of the most flexible of these is reflection electron microscopy (REM). Although REM was attempted soon after the invention of the electron microscope it was only after Yagi and his co-workers (Yagi 1987) succeeded in making clean flat crystalline silicon surfaces in an electron microscope that monatomic steps, dislocations and dynamic surface processes were observed. The ability to observe nucleation and growth processes and to determine the effect that monatomic steps have on these changes demonstrated to surface scientists the value of real-space imaging.

1. Imaging and Image Interpretation

For REM imaging in a conventional transmission electron microscope (see *Transmission Electron Microscopy*), the sample must be flat and crystalline. To achieve surface sensitivity the electrons should not penetrate far below the surface and this is achieved by having the incident parallel beam of electrons impinging on the surface at a glancing angle. With this geometry the electrons reflected from the surface give rise to a reflection high-energy electron diffraction (RHEED) pattern (see Fig. 1) in the back focal plane of the objective lens. In order to produce a strong signal the incident beam is tilted to

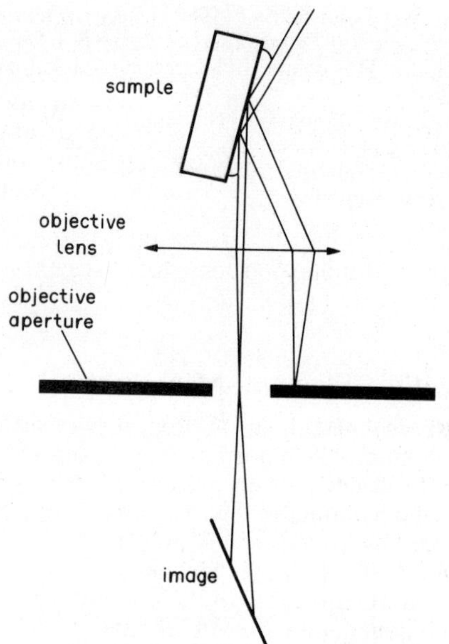

Figure 1
REM image formation

make a Bragg angle with the surface planes and the specular spot is allowed through the objective aperture and used for imaging whereas all the other diffraction spots are cut out. The intensity of the image can be further enhanced by azimuthal rotation of the specimen to a position where a surface resonance is excited (this can be considered as a diffraction condition where one beam travels parallel to, and just below, the surface). The specimen and incident beam are tilted so that the specular beam travels down the center of the imaging lens to reduce spherical aberration. It can be seen from Fig. 1 that REM is analogous to dark field imaging in transmission electron microscopy (see *Transmission Electron Microscopy*).

The geometry employed in REM has two inherent problems. First, there is considerable foreshortening of the image (by a factor of $1/\sin\theta \simeq 20$–50) reducing the resolution in the direction parallel to the beam but on good, flat surfaces this is not usually a problem since the defects of interest are often well separated. Second, only a thin strip of the image is in focus but, again, this is often not a major problem for near perfect surfaces where the RHEED pattern is composed of sharp spots. The depth of field is dependent on the angular spread of the electrons being imaged. If the diffraction spot is very sharp then large areas of the surface can be imaged. For rough or amorphous surfaces the spread of electrons is defined by the objective aperture and usually very little information can be gained about the sample.

Defects on a surface are made visible because of the presence of diffraction and phase contrast (see Fig. 2). Changes of the diffraction condition across the sample will result in changes of intensity across the image. For a specimen oriented to the Bragg condition the distortion of atomic planes around a defect will appear dark in the image. This distortion is due to the strain field around the emergence of a screw dislocation or along a step. The contrast from a screw dislocation in a sample that is not exactly at the Bragg condition can be worked out in a similar way, with one of the wings around the dislocation being closer to the Bragg condition and therefore brighter and the other being further from the Bragg condition and consequently darker. These arguments are based on the column approximation, where the intensity of any region is considered to be equal to that from a perfect surface whose orientation is the same as the region under consideration. This intensity information can be obtained from a rocking curve calculated for a perfect crystal and, in general, the full widths at half maximum of the Bragg peaks are in the range 10^{-3}–10^{-4} rad so that surface distortions as small as 10^{-4} rad can be detected. Diffraction contrast can also show domains of surface reconstruction or adsorption in cases where the different domains contribute different intensities to certain diffraction spots.

Most of an REM image is out of focus and so electrons, having different optical path lengths due to distortions around defects, can interfere constructively or destructively giving rise to dark–bright fringes around steps and screw dislocations and this phase contrast is often dominant in REM images. If two diffraction spots that are parallel to the edge of the sample are allowed through the objective aperture they can interfere to give lattice fringes (see

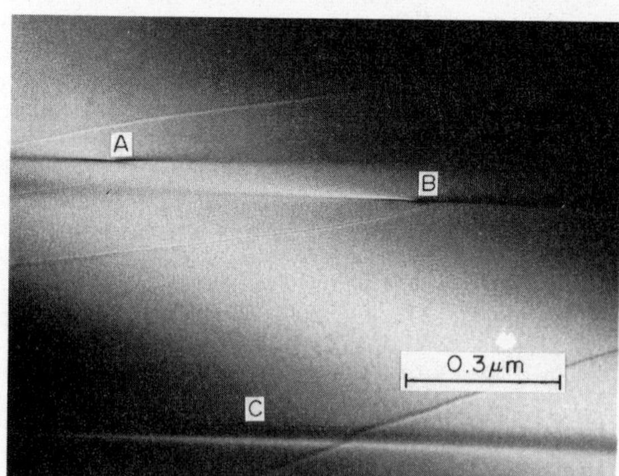

Figure 2
Monatomic steps terminate at the points where screw dislocations emerge at the surface. At A the dislocation is in focus and at the Bragg condition, at B the dislocation is slightly off the Bragg condition and at C the black–white fringe is due to phase contrast from an out-of-focus dislocation

Transmission Electron Microscopy) that are useful for detecting defects such as out-of-phase boundaries on reconstructed surfaces (see *Surface Defects*).

Although the explanations of contrast mentioned above are very useful and easy to understand they do not fully take into account the interaction of the electrons with the sample. In reality the electrons can travel considerable distances in the crystal and, for a complete description of the contrast, dynamical diffraction calculations need to be employed. The electrons can lose energy by the creation of plasmons, phonons and inner shell excitations so inelastic scattering must also be considered.

2. Observations

One of the first uses of REM was to study the phase change on a silicon (111) surface from 1×1 to 7×7 (see *Surface Defects*; *Surface Structure*). Here it was noticed that nucleation occurs preferentially on the top side of the steps. When gold is deposited on the silicon (111) 7×7 structure the nucleation of the adsorbate structure occurs on the top of the steps but when gold is deposited on a platinum (111) monolayer islands of gold nucleate at the bottom of the steps. In the latter case the growth rate depends on the width of the lower terraces indicating that the adatoms do not "fall down" the steps. Results like these obtained by REM show clearly that when the terrace width is narrower than the diffusion distance the surface dynamic processes are controlled by the steps. These facts should be borne in mind by all surface scientists since even a sample cut to within $\frac{1}{2}°$ of a low-index plane will have steps approximately every 25 nm and this is without allowing for small undulations that occur during preparation!

The controlled oxidation of surfaces is another area of research that can be examined by REM very effectively. When a silicon (111) 7×7 surface is oxidized at 920 K it is noted that the steps move back and between the steps small hollows appear. These topographic changes are caused by the formation and diffusion of vacancies. The vacancies are created by the sublimation of SiO. (Similar looking hollows have also been observed on noble metal surfaces that have been prepared by high-temperature annealing.)

The nucleation and growth of three-dimensional islands of Cu_2O on copper is also affected by the presence of steps and, conversely, the presence of these islands can affect the movement of the steps.

3. Scanning REM and Surface Analyses

REM can also be performed using a scanning transmission electron microscope (STEM) and due to the reciprocity theorem similar surface images are obtained (see *Analytical Electron Microscopy*). The disadvantage of STEM is the lower signal so that larger apertures are required. Larger apertures tend to decrease the depth of field but it is possible to compensate for this by adjusting the focus electronically during each scan. This dynamic focusing is especially useful for imaging rough or amorphous surfaces since the criterion that sharp diffraction spots are required to give a useful depth of focus is now unnecessary. The scanning form of imaging also allows the foreshortening to be reduced by adjusting the shape of the raster on the specimen. It is also possible to stop the probe on any feature of interest on the surface and use the various analytical facilities of the STEM to investigate small inhomogeneities. These facilities include microdiffraction, Auger electron spectroscopy (see *Auger Electron Spectroscopy*) and electron energy-loss spectroscopy (see *Analytical Electron Microscopy*). For example, small particles deposited on surfaces can be characterized by their crystallography, chemical composition and electronic properties.

4. Future Developments

REM has demonstrated its potential as a tool for elucidating surface structure and the mechanisms of surface reactions. Although it does not possess the spatial resolution of scanning tunnelling microscopy (see *Scanning Tunnelling Microscopy and Spectroscopy*, Suppl. 2) it is in many ways a more flexible way of observing surface changes and in several laboratories there are plans for building a tunnelling microscope inside an electron microscope so that features of interest observed by REM can be examined in more detail with the STM. Many REM studies have used samples that are easily prepared (e.g., faceted gold and platinum crystals or cleaved semiconductors) but in future samples of greater commercial and academic interest should be examined. This requires dedicated ultrahigh vacuum microscopes with better cleaning facilities that will allow the preparation of more reactive materials. The ability to perform in situ reactions will, in general, require a reaction cell to be built around the specimen position. With these developments it will be possible to pinpoint the effects that surface inhomogeneities have on many important surface reactions.

Although the main image features can now be interpreted, further work is still required to give a complete theoretical description of the more subtle image details. Due to the complex electron–specimen interaction this is not a trivial problem but good progress has been made.

Bibliography

Cowley J M 1986 Electron microscopy of surface structure. *Prog. Surf. Sci.* 21: 209–50

Dobson P J, Larsen P K (eds.) 1988 *Reflection High Energy Electron Diffraction and Reflection Electron Imaging of Surfaces*. Plenum, New York

Halliday J S 1965 Reflection electron microscopy. In: Kay D H (ed.) 1965 *Techniques for Electron Microscopy*, 2nd edn. Blackwell, Oxford

Smith D J 1986 High resolution electron microscopy in surface science. In: Vanselow R, Howe R (eds.) 1986 *Chemistry and Physics of Solid Surfaces VI*. Springer, Berlin, pp. 413–34

Yagi K 1987 Reflection electron microscopy. *J. Appl. Crystallogr*. 20: 147–60

R. H. Milne
[University of Strathclyde, Glasgow, UK]

S

Scanning Tunnelling Microscopy and Spectroscopy

Scanning tunnelling microscopy (STM) is a technique that uses tunnelling electrons to produce three-dimensional real-space images of the surfaces of materials. The technique is as simple in concept as stylus profilometry (see *Surface Roughness*), yet it provides capabilities not available with any other measurement method. In less than five years, STM developed from a difficult measurement performed in a handful of laboratories around the world into a widely accepted method of surface characterization. The excitement over this surface-analysis method, and its consequent rapid development, stem from the fact that information is obtained about atomic structure and bonding in solids at unprecedented levels of spatial resolution. Three-dimensional atomic images can be produced of semiconductors, semimetals and metals. Imaging and spectroscopic techniques have been used to investigate surface reconstructions, adsorbate reactions and charge density wave formation, making an immediate and dramatic impact in many of the traditional areas of surface science. As the technique becomes more accessible, it is being applied to issues of importance in materials science, such as diffusion, nucleation processes, fracture, fatigue, grain-boundary structure, and so on. In parallel developments, a number of techniques based on the scanning technology of STM (often referred to as near field or scanning probe microscopies) can now measure atomic forces, magnetic forces, thermal gradients, and photon emission with spatial resolution similar to STM.

1. *Fundamental Principles of Imaging*

STM is the most recent of a long list of devices based on electron tunnelling that includes: the field-ion microscope, Esaki diodes, Zener diodes, Josephson junctions and inelastic electron tunnelling spectroscopy. It was the earlier development of these tunnelling devices that provided a theoretical framework that could be almost directly applied to the STM at its invention. To summarize this theory, consider electron tunnelling in the general case.

An infinite potential energy barrier exists between two planar surfaces separated by a large distance; however, the barrier becomes finite when the two surfaces are in close proximity. When a bias is applied between two surfaces within some nanometers of each other, a statistically significant number of electrons with energies lower than that of the barrier wil negotiate the energy barrier, producing a "tunnelling current." Although classically forbidden, electron tunnelling can be explained within the wave–particle dualism concept of quantum mechanics. The equation for current resulting from electrons tunnelling between two planar electrodes in the limit of low temperature and low applied bias is:

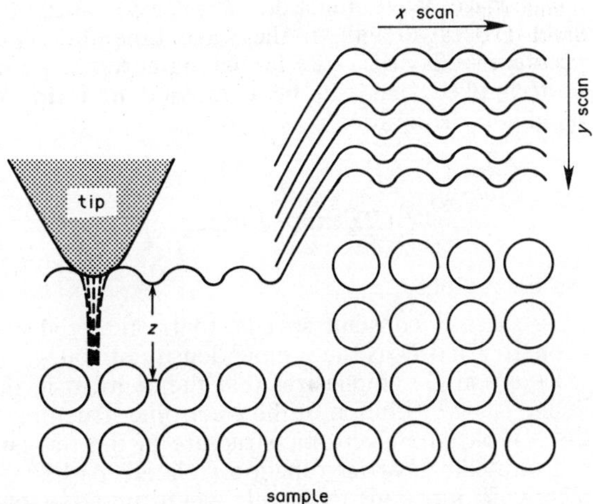

Figure 1
Schematic of tunnelling between the tip and sample in the STM configuration: in constant current imaging the sample–tip separation z is held constant as the tip is scanned over the sample surface; a series of scans produces a real space image of constant charge density

$$I=\frac{2\pi e}{\hbar}\sum f(E_s)[1-f(E_t+eV)]|\mathbf{M}_{s,t}|^2\delta(E_s-E_t) \quad (1)$$

where $\mathbf{M}_{s,t}$ is the tunnelling matrix, $f(E)$ the Fermi function, e the charge of an electron, $\hbar=h/2\pi$ where h is Planck's constant, and the subscripts s and t refer to the electrodes (later specified as sample and tip, respectively). Tunnelling, as it occurs in the STM configuration, is described in Fig. 1. One electrode is a metallic point probe (the tip), the other a planar semiconductor or metal (the sample). Tersoff and Hamann (1985) have taken the geometry of the STM and the electronic structure of the tip into consideration and evaluated the tunnelling matrix in Eqn. (1) to produce a current equation specific to the STM. Assuming a spherical wave function for the tip structure and using reasonable values for the materials parameters pertaining to the tip, they obtain for a constant applied bias:

$$I=\frac{32\pi^5 e^2}{\hbar}V\phi^2 D_t(E_f)R^2\frac{\exp(2kR)}{k^4}\int \psi^2 d(E_s-E) \quad (2)$$

in the limit of low temperature and applied bias, (V), where D_t is the density of states of the tip; ϕ, the tunnelling barrier, is on first approximation $(\phi_s+\phi_t)/2$; R is the radius of curvature of the tip; ψ the surface wave function; $k=(2m\phi)^{1/2}/h$ is the inverse decay length of the wave function. Since $\int\psi^2\propto\exp(-2kZ\phi^{1/2})$, the tunnelling current at constant applied bias can be expressed in terms of sample–tip separation z as:

$$I=CD_s\exp(-1.025z\phi^{1/2}) \tag{3}$$

where C is a constant specific to tip material and geometry and D_s is the sample density of states.

Equation (1) emphasizes that the contrast in the image is a convolution of the electronic structure of the sample, the electronic structure of the tip, and the tunnelling barrier function (related to the surface work function). It is only when the latter two contributions are constant that the variation of contrast in the image can be attributed to electronic properties of the sample. Fortunately, these conditions can be easily met on many materials; therefore, Eqn. (2), in which these assumptions are implicit, is a reasonable approximation.

To produce images, the STM can be operated in two modes. In "constant current" imaging, a feedback mechanism is enabled which maintains a constant current while a constant bias is applied between the sample and tip. As seen in Eqns. (2) and (3), these conditions require a constant sample–tip separation. As the tip is scanned over the sample, the vertical position of the tip is altered to maintain that constant separation. The signal required to alter the vertical tip position constructs the image, which represents a constant charge density contour. Altering the level of the current or the applied bias produces contours of different charge densities. An alternative imaging mode is "constant height" operation, in which constant height and constant applied bias are simultaneously maintained. The variation in current is then monitored to produce the image as the tip scans the sample surface. There are advantages inherent in both modes of operation; the former produces contrast directly related to electron charge density profiles, while the latter provides for faster scan rates not being limited by the response time of the vertical driver. Image data will be displayed differently, depending on the information required. As an example, Fig. 2 illustrates the line scan, plan view and three-dimensional rendering of an atomic resolution constant current image of the basal plane of highly oriented pyrolytic graphite. Atomic resolution images are possible only under optimized sample and tip conditions. Larger sample–tip separations and blunt tips have the

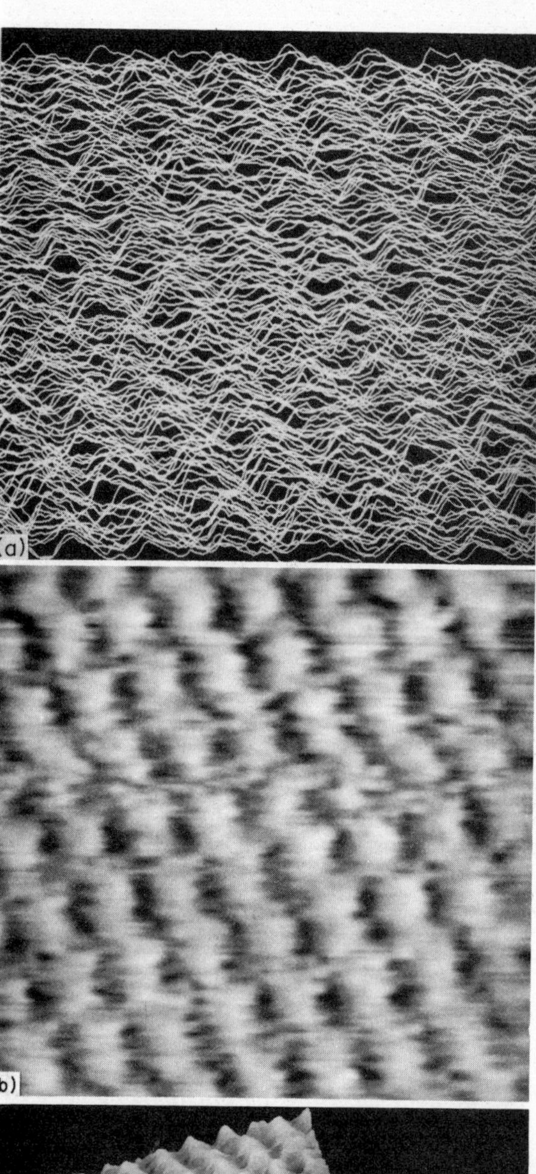

Figure 2
Comparison of image display modes showing (a) a line scan, (b) plan view and (c) a three-dimensional rendering of a constant current image of highly oriented pyrolytic graphite acquired in air

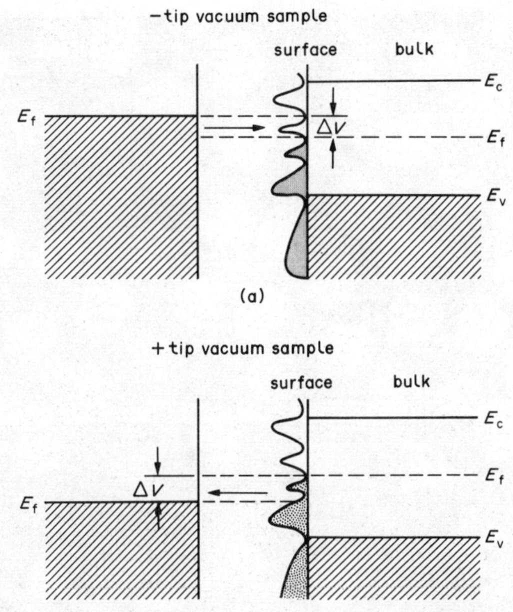

Figure 3
Idealized representation of the electronic structure of the sample and tip: E_f is the Fermi energy, E_v the valence band edge, and E_c the conduction band edge; the tip is modelled as a metal left, the sample as a semiconductor right; states on the sample surface may be due to surface reactions or extensions of midgap bulk states; a bias ΔV between the sample and tip displaces the Fermi levels with respect to each other, allowing electrons within the energy window to traverse the vacuum gap

effect of smearing the localized structure and produce topographic images with somewhat lower resolution.

2. *Tunnelling Spectroscopy*

Tunnelling spectroscopy (TS) performed in the STM provides information about the electronic structure of the sample by probing the sample density of states as a function of energy. The concept is similar to traditional sandwich junction spectroscopy (inelastic electron tunnelling spectroscopy, Hansma (1982)), the primary differences being that a vacuum gap exists between the two electrodes rather than an insulator and only elastic processes have been monitored to date. There are also two methods of performing TS. The first, referred to as point spectroscopy, involves moving the tip to a feature of interest, disengaging the feedback mechanism, modulating the tip bias and recording the resulting variation in current. The idealized band structure in Fig. 3 schematically illustrates current flow in this situation. When a bias is applied between the sample and tip, the Fermi levels of the two materials are displaced with respect to each other allowing electrons within the energy window to traverse the vacuum gap. As the bias is ramped, either positively or negatively, the current varies in response to the changing electron density in the energy window. Ramping in both directions probes both the occupied and unoccupied energy states, with the magnitude of the current at a specific voltage directly related to the density of states of the sample at that energy. Since the constant, C, in Eqn. (3) is a linear function of voltage, spectroscopy data can be related to the sample electronic structure D_s in the form of conductance:

$$I/V = C^* D_s \exp(-1.025 z\phi^{1/2}) \tag{4}$$

and voltage derivatives thereof. The second method of obtaining spectroscopic information involves simultaneously collecting images at various biases. This can be accomplished by modulating the bias at a high frequency (with respect to the time constant of the feedback controller) and recording the current at several discrete values of applied bias during a scan or by using the same sample and hold method of point spectroscopy at every image point. The simultaneous collection of multiple images is sometimes referred to as current imaging tunnelling spectroscopy, CITS. A comparison of the periodic contrast in such images with atomic charge superposition and total energy calculations identifies the spatial position of energy states (i.e., bonds) with respect to the lattice.

STM has been used extensively in the characterization of the silicon (111) surface, to the point that the (7×7) reconstruction of this surface is now universally associated with the technique. This familiar structure can be used to illustrate both point spectroscopy and CITS of semiconductors. An example of point spectroscopy is given in Fig. 4 which is taken from a study of aluminum adsorption on silicon (111) by Hamers and Demuth (1988). The electronic structure of the silicon (111) 7×7 surface determined by TS is compared with that determined by ultraviolet photoemission and inverse photoemission. Occupied states at −1.6 eV, −0.85 eV and −0.1 eV, and unoccupied states at +0.5 eV and +1.6 eV, are characteristic of the stacking fault adatom structure of the 7×7 reconstruction. These states are evident in TS and in photoemission results. The excellent agreement between the energies of the states identified in the TS spectrum and those determined from photoemission results illustrates the accuracy with which tunnelling spectra characterize surface electronic structures. The measurements average over an area larger than a unit cell and each state corresponds to the energy of a particular bond. For example, the state at −0.8 eV arises from the dangling bond, whereas the state at −1.6 eV corresponds to the silicon back bond. Naturally, the bonds are localized at different positions with respect to the atomic lattice; therefore, images of these states have different geometries but should have the same translation as the unit cell. Figure 5 is a schematic representation of current images that would be produced by CITS at +0.7 eV and −1.8 eV. Note that the features in the images appear quite different but do have the lateral translation of the unit cell. Image simulations based on theoretical calculations are necessary to interpret atomic resolution images; that is, to relate the features in the image to bond types and atom positions.

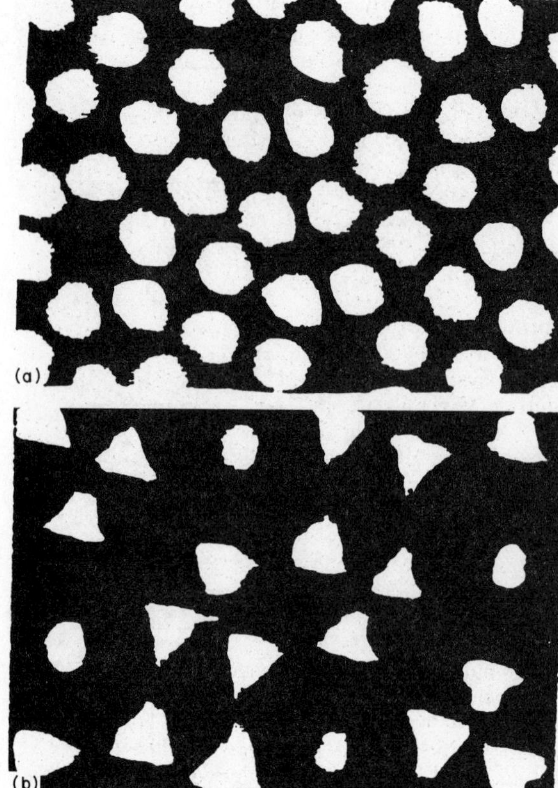

Figure 5
Schematic representation of CITS images that would be obtained at (a) +0.5 V and (b) −1.6 V applied bias on the silicon (111) 7×7 surface: the unit cell of the reconstruction is outlined in white (note that the bonding states occupied by electrons (a) and those not occupied (b) are in different positions within the unit cell)

3. Microscope Design Considerations

The heart of a scanning tunnelling microscope consists of a scanner and a sample–tip approach mechanism. Scanning tunnelling microscopes are divided into two classes based on the configuration of the mechanism that controls the tip. The scanners are made of piezoelectric elements which expand on application of an electric field, hence the tip can be scanned by ramping the voltage applied to the different piezoelectric elements. Figure 6 schematically illustrates a scanner made of three orthogonal bars and one made of a cylinder. Each type of scanner has advantages. The orthogonal arrangement is more linear by virtue of geometry and, consequently, requires no compensation to ensure a square scan area. The cylindrical scanners are more stable with respect to mechanical vibrations, as indicated by the mechanical resonance of 12–20 Hz compared with 2–7 Hz for the orthogonal scanner, but are intrinsically nonlinear and require hardware or software compensation during scanning. Scanners are controlled through simple voltage amplifiers and a standard feedback circuit by a PC size computer. The second important aspect of the microscope design is the mechanism for bringing the sample and

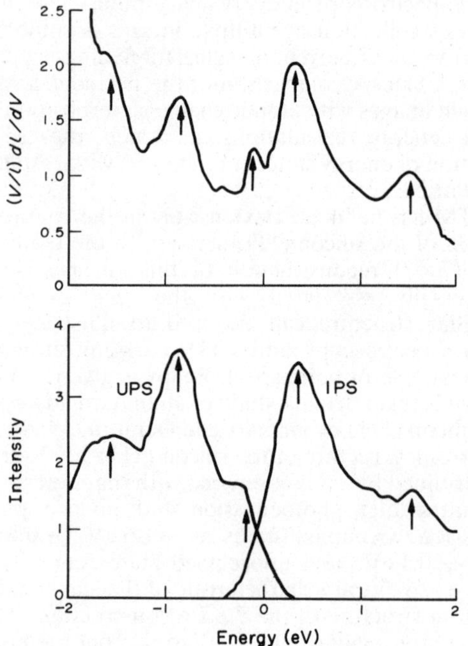

Figure 4
Comparison of density of states by TS point spectroscopy with photoemission and inverse photoemission spectra; agreement between various methods verifies the accuracy of TS data (redrawn after Tromp et al. 1985))

tip together. As tunnelling occurs only in close proximity to the surface (within 3 nm) the approach system must be capable of precise reproducible incremental steps of less than 3 nm in order to avoid tip-to-sample contact. One method of sample–tip approach is based on differential springs or levers, where the sensitivity (i.e., the smallest step) is determined by the ratio of spring constants, or lengths of levers, respectively. Electrostrictive walkers, referred to as "louses," are also used. Recently, stepper motors have been developed with enough precision to provide controlled sample–tip approach; however, these usually have the disadvantage of not being compatible with ultrahigh-vacuum (UHV) systems. A general design consideration is whether the microscope will be compatible with UHV systems. Such systems provide an environment in which to carry out controlled chemistry on atomically clean samples. Under these circumstances, images and TS can often be directly interpreted. Compatibility with UHV imposes restrictions on material choices for some of the components. Commercial microscopes are currently available for use in ambient and in UHV.

The primary obstacle to attaining atomic resolution is that imposed by electrical noise and mechanical vibration. Tunnelling currents are on the order of 0.1–10 nA and the sample–tip junction acts like an antenna to ambient electric fields; therefore, the microscope must be effectively shielded and electrically isolated from external circuits. Atomic resolution requires stability of 0.01 nm with respect to mechanical vibrations. Mechanical noise attenuation is achieved by combining a series of damping strategies. A table floating on pneumatic (air) legs or

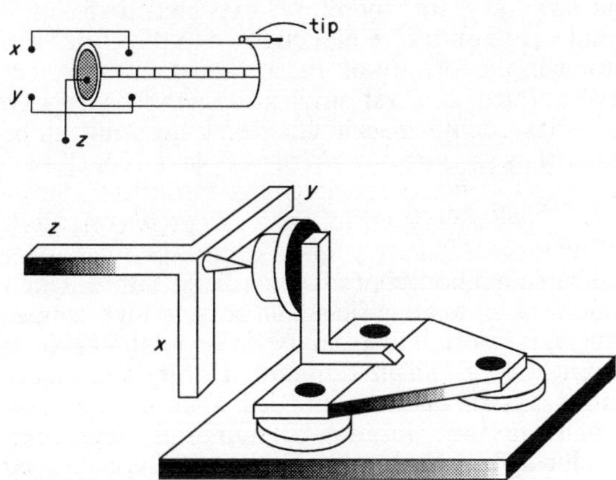

Figure 6
Schematic illustration of orthogonal and cylindrical arrangements for the piezoelectric elements which move the STM tip: *x*, *y* and *z* refer to the direction of motion when the controlling voltage is applied as shown

suspended by long springs will damp low-frequency vibrations (0.1–100 Hz). Alternating stacks of metal and rubber, or any two materials with widely differing elastic constants, damp midrange frequencies (50–300 Hz). Higher-frequency vibrations can be eliminated by electrical filtering.

4. Sample and Tip Preparation

No special preparation is required for STM analysis in air; however, under these conditions optimum resolution is usually not attained. The exception is cleaved surfaces of highly oriented pyrolitic graphite, which are often successfully imaged at atomic resolution in air, as was illustrated in Fig. 6. On materials not conducive to atomic imaging in air, imaging at levels somewhat lower than atomic resolution proves, nevertheless, to be useful for a large number of applications. Atomic resolution of semiconductors such as silicon, gallium arsenide and germanium requires that clean surfaces be produced in UHV conditions. This is achieved by *in situ* cleaving to expose the desired surface, or by a series of heating and ion-etching steps. The specific conditions for the sample depend on the structure expected on that particular surface. Similarly, metal sample surfaces can be cleaned *in situ* by heating and ion milling.

It has been both theoretically and experimentally demonstrated that the geometry and chemistry of the atoms at the end of the STM tip affect the image. A clean sharp tip is critical to attaining optimum resolution and unambiguous image interpretation. STM tips can be made of a variety of metals, the choice depending on the application. Tungsten tips are favored for imaging because of their stiffness. They are formed by mechanical sharpening with a standard grinding medium such as 600 grit SiC paper, or by electrochemically etching, using NaOH or KOH in a dc or ac electric field. In order to perform TS, further *in situ* cleaning is necessary to remove the native oxide on the tungsten, otherwise the spectra will include a contribution from the oxide–metal interface on the tip that will interfere with determination of the electronic structure of the sample. Other refractory metals are often used to avoid this problem, even though they are not as stiff as tungsten, the most common being gold, platinum and platinum alloys. These tips are formed simply by cutting with precision wire cutters. For use in UHV and to optimize resolution, atmospheric contaminates are removed from the tip by:

(a) backing the tip out of tunnelling range, applying a high voltage (20–300 V) to induce an emission current for a period of a few seconds to several minutes;

(b) *in situ* electron or ion bombardment; or

(c) *in situ* heating to relatively high temperatures.

5. Applications in Materials Science

STM has been used extensively in the study of silicon, germanium, gallium arsenide, gold, palladium and graphite (Feenstra 1988). It has been useful, not only in determining the geometric and electronic structures of various surfaces of these materials, but also in monitoring structure and bonding during adsorbate reactions. Although it is for the analysis of these surfaces that STM is widely known, it has been applied to a much wider range of materials. Five studies that illustrate the various measurements possible in STM and their relevance to materials investigations are summarized in this section. These examples emphasize several important points in the application of STM. While there is no doubt that the technique is powerful when operated at atomic resolution, valuable information is often gained at resolutions of 0.4–0.5 nm, which is orders of magnitude higher than the spatial resolution achieved by other surface characterization methods but not quite atomic STM imaging. Tunnelling spectroscopy is as powerful as STM imaging, perhaps even more so, and barrier height measurements and local capacitance measurements may have a similar impact on future investigations. The digital nature of topographic images allows complex mathematical analysis of surface structure with relative ease. As the descriptions of these studies are necessarily brief, the reader is referred to the original papers for complete discussions of experimental details.

5.1 Structure of Defects on Silicon

A large fraction of atomic level STM analysis to date has involved conventional semiconductor materials, particularly silicon. The directional bonding in these materials strongly affects the structure of surfaces. An ideally terminated surface would contain unsatisfied bonds in a system that favors tetrahedral coordination. This is such a high-energy situation that these dangling bonds interact with each other to reduce the surface energy and can, thereby, produce a surface atomic structure with a periodicity larger than that of the bulk. When the relaxation results in a periodic structure, the surfaces are said to be "reconstructed" (Zangwill 1988). Although large areas are routinely produced with homogeneously reconstructed surfaces, vacancy defects and domain boundaries are invariably observed. The use of atomic resolution TS to probe the electronic structure of these defects directly is illustrated in Fig. 7 (Hamers 1989). I–V curves acquired at different lateral positions near a vacancy type defect are shown with the constant current image of the surface (acquired at -1.5 eV). Away from the defect (Fig. 7) the *I–V* curves indicate the presence of a surface bandgap by the sharp turn-on of tunnelling current near -0.45 eV and $+0.25$ eV. Near the defect the sharp edges disappear and directly over the defect

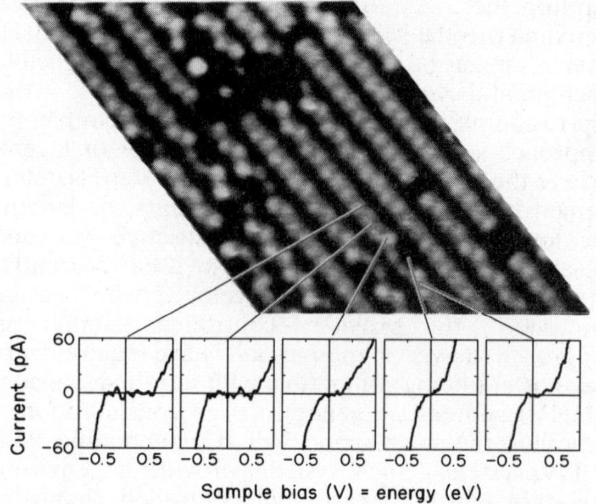

Figure 7
Constant current image of reconstructed silicon (001) surface with several defects and spatially resolved *I–V* spectra showing the variation in the surface bandgap with lateral position (courtesy of Hamers 1989)

(Fig. 7) the tunnelling curve exhibits an exponential increase in current both above and below the Fermi level. This behavior demonstrates that the defect causes a variation in the local density of states; that is, a high density of states exists at the Fermi level due to the presence of a vacancy. The spatially dependent measurements also provide a direct measure of the lateral extent of the wavefunctions associated with defects and impurities, demonstrating that the influence of the defect extends over 1.5 nm from its core. This local electronic structure is directly related to chemical reactivity. These vacancy sites are, therefore, expected to be more conducive, or less conducive, to interaction with atoms in the vicinity of the surface than the defect-free surface. Several studies along this line have investigated site-specific chemistry on semiconductor surfaces.

5.2 Growth Facets on Silicon Carbide Single Crystals

A simple method of producing silicon carbide (SiC), that involves heating silica and coke to high temperatures, is referred to as the Acheson method after its inventor (see *Silicon Carbide*). During the process single crystals nucleate on seed particles and grow, usually in the presence of numerous impurities, producing highly doped mixed-modification crystals. The most common impurities are aluminum and nitrogen which induce *n*- and *p*-type conductivity, respectively, and can be compensating when present simultaneously. These crystals provide an opportunity to study growth mechanisms. The STM image in Fig. 8 characterizes the surface of a nominally 6H modification crystal parallel to the basal plane. Growth facets are observed with the vertical dimension of one unit cell and a lateral dimension of 2–3 unit cells. TS data are also shown as they are acquired in the form of *I–V* curves, as conductance which is the first derivative, and as normalized conductance which is the logarithmic derivative of these curves (Fig. 9). Three midgap states are observed in the tunnelling normalized conductance spectrum; at +0.3 eV, −0.45 eV and −0.80 eV. Although the peak intensities vary, the energies of the peaks are reproducible. These energies correspond to one occupied and two unoccupied states, and are consistent with those expected from aluminum and nitrogen doping.

5.3 *Adsorbate Effects on the Electronic Structure of Zinc Oxide*

Zinc oxide (ZnO), the most ionic of the wurtzite semiconductors, finds major applications in catalysis and as variable resistors. It is a metal excess defect structure exhibiting *n*-type conductivity with a bandgap of 3.34 eV. The structure of internal interfaces in polycrystalline ZnO and the surface reactivity (i.e., effectiveness in catalyzing reactions) depend strongly on the atomic and electronic structure of the original surfaces. The adsorption of molecular species in the atmosphere onto ZnO surfaces is known to affect its electronic structure and thus surface reactivity (Gopel 1985). Although STM can characterize just such reactions, its application to oxides was only recently demonstrated (Bonnell and Clarke 1988). Imaging, spectroscopy and tunnel barrier height measurements were used to monitor these surface reactions at atomic levels of spatial resolution in an STM. It was found that stable imaging in air resulted only when tunnelling out of the ZnO (i.e., sample biased negatively with respect to the tip) when imaged in air. The facets on this surface, shown in Fig. 10, range from 1–7 nm. Rectification evident in the tunnelling spectrum is typical of Schottky barrier formation in *n*-type semiconductors. Increased conductance at lower energies is evidence of the valence band edge, but the conduction band edge is not observed. These results demonstrate that, when a judicious choice of imaging conditions is made, imaging of oxides even in the presence of a Schottky barrier at the sample–tip junction is possible. In contrast to the results in air, the surface is in flat band condition when examined in a vacuum of 10^{-9} torr, as indicated by the tunnelling spectrum, where both the valence and conduction band edges are detected. Under these conditions atomic resolution is possible, as shown by the image of a ZnO (10$\bar{1}$0) nonpolar surface in Fig. 10. This dramatic difference in surface electronic structure of ZnO from that which occurs in air, is a direct result of changes in surface chemistry. By measuring the change in tunnelling current that results from a variation in sample–tip separation (*I* vs *z* in Eqn. (3)), an "apparent" tunnel barrier height can be determined. This barrier height is, in principle, related to the sample surface work function. The magnitude of the change in local surface work function of ZnO determined in the STM (not shown here) is consistent with macroscopic measurements made in desorption studies and Auger analysis of the same surfaces verifies the change in surface chemistry (Fig. 11). This example illustrates the variety of measurements possible in STM which can be exploited to study local surface chemistry.

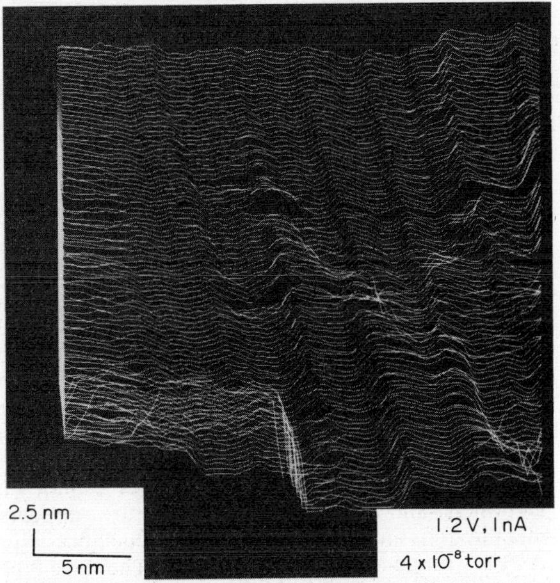

Figure 8
Constant current image of 25 nm square area of SiC observed in moderate vacuum (10^{-7} torr): facets resulting from the crystal growth mechanism are identified

5.4 *Conductive Polymers*

In the last decade, a new class of polymeric materials has been developed having electrical conductivities comparable to that of copper. In the polyaniline family of polymers, the doping mechanism is such that both the number of protons and the electrons can be varied independently by controlled chemical reaction. In the "emeraldine" oxidation state of polyaniline, the number of protons is systematically altered while the number of electrons remains constant on doping (by reaction with a protonic acid). The models invoked to describe transport properties in conductive polymers rely on the presence of localized midgap states due to variations in bond order. A necessary feature of the proposed conduction models for this material is that the electronic

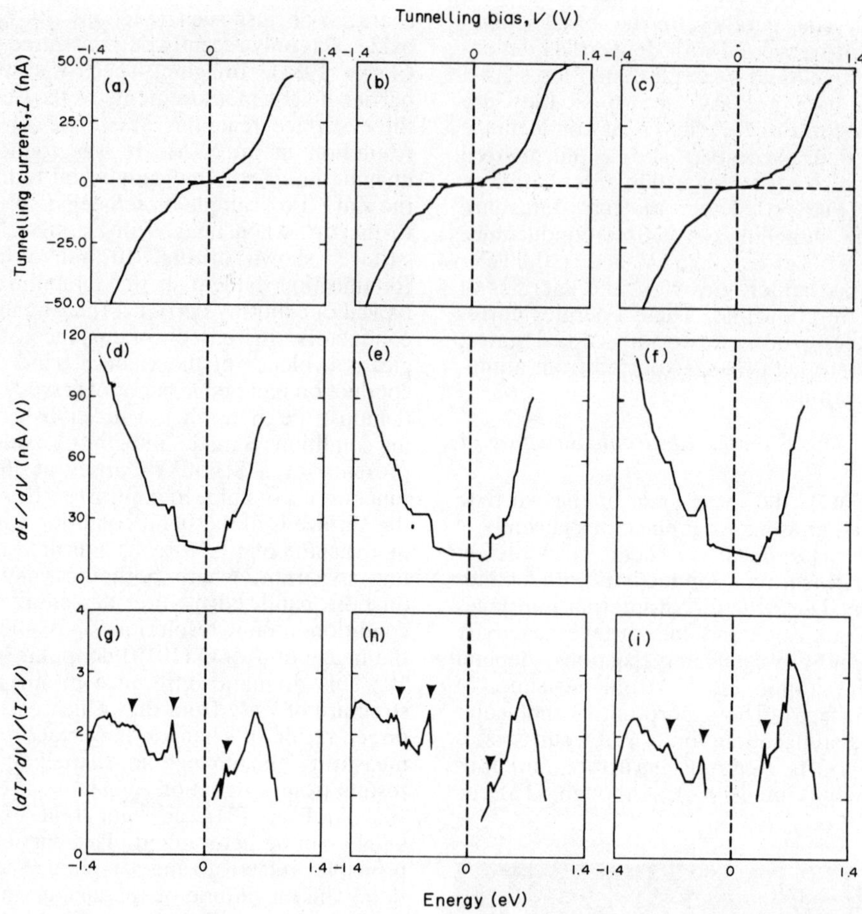

Figure 9
Tunnelling spectra from the SiC surface shown in Fig. 8: (a)–(c) show the current–voltage data as they are acquired; spectra are often displayed as conductance (d)–(f) or normalized conductance (g)–(i); spectra were acquired at three sample–tip separations (increasing from left to right) and the positions of the peaks marked by the arrows are consistent at all separations

structure of partially reacted emeraldine varies spatially, with the scale of variation estimated at tens of nanometers. While measurements of electrical behavior provide indirect evidence for this spatial variation of electronic structure, TS is the only method for direct measurement. Figure 12 illustrates the topographic structure of spin-coated emeraldine hydrochloride films (Bonnell and Angelopoulos 1989). The topographic structure is related to processing conditions and, due to its scale, could not be detected in extensive scanning electron microscopy (SEM) analysis. The material is amorphous; therefore, information about atomic bonding is not expected from the STM images. The different types of electronic structure found in fully and partially doped emeraldine hydrochloride are also compared in Fig. 13. The first type of behavior appears semiconducting; the position of the valence (v) and conduction (c) band edges is indicated by a decrease on conductance near -2.1 eV and an increase in conductance near 2.2 eV, respectively. The bandgap determined by TS (4.3 eV) agrees reasonably with that determined by optical methods (4.1 eV). In the second type of region, the band edges cannot be identified and the electronic structure appears metallic (continuously increasing conductance over ± 3 eV with no measurable bandgap). The third type of region also has a bandgap of about 4.3 eV and a significant state occupation appears to be continuous from the top of the valence band to about -1.0 eV and an additional unoccupied state is clearly resolved about 1.0 eV below the conduction band

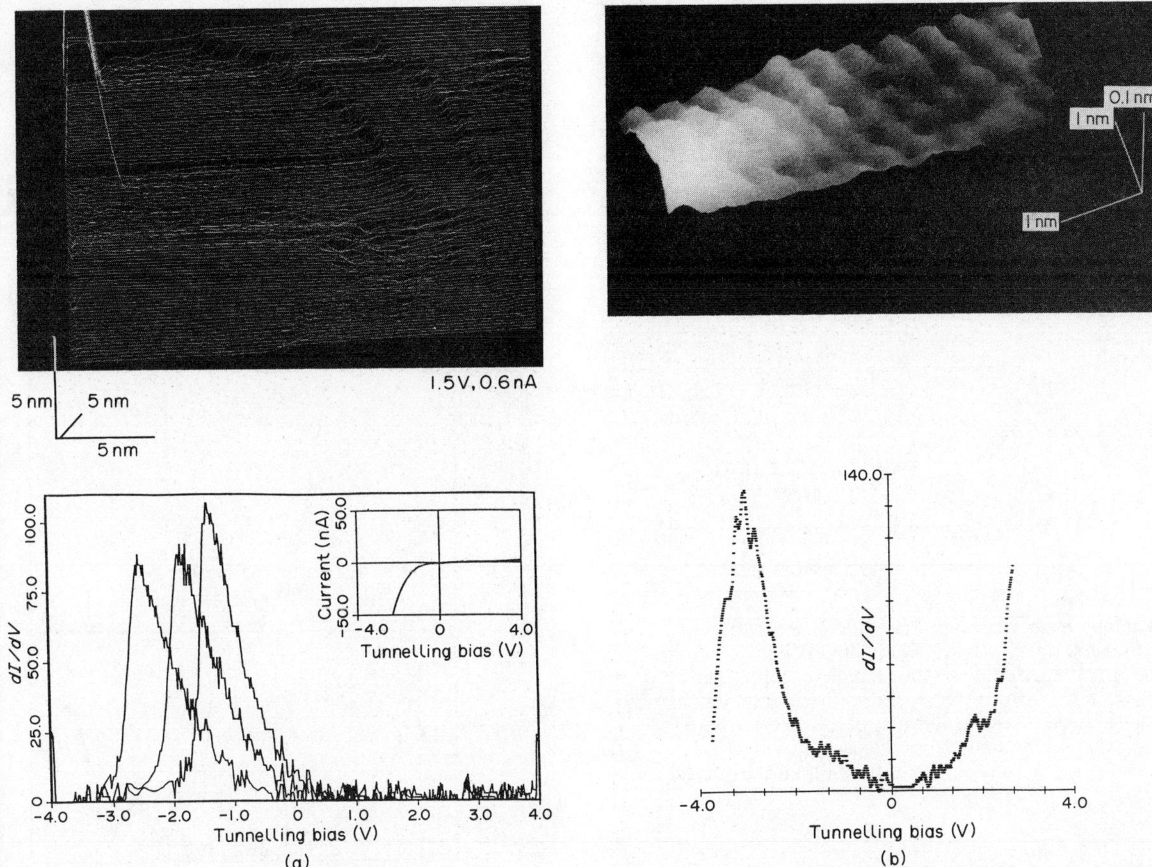

Figure 10
A comparison of the resolution of topographic imaging of ZnO in (a) air and (b) UHV with the electronic structure determined by tunnelling spectroscopy; Schottky barrier formation occurs in air as indicated by the variation in the position of the valence band edge at three different sample–tip separations and the rectified *I–V* characteristics shown in the inset; flat band conditions result in UHV, indicated by the observation of both the valence and conduction band edges

(both marked d). These results provide the first direct measurement of spatial variation of electronic structure in a conducting polymer in support of electron transport models for these systems.

5.5 Commercial Applications

In addition to making an impact as a research tool, STM is beginning to be exploited in commercial applications such as process control and calibration. Nondestructive evaluation of the smallest structures currently manufactured is not possible with other techniques. An example is illustrated in the STM images of Fig. 14 which show a diffraction grating and an integrated circuit. The diffraction grating acts as a prism to disperse light and is an internal component of infrared spectrometers. SEM lacks the vertical resolution and interferometry lacks the lateral resolution to characterize the topography of the diffraction grating; however, Fig. 14a illustrates how easily the surface structure of the top and sides of this 1 μm grating is characterized with STM. A second example of STM use in developing manufacturing processes is given in the image of an integrated circuit (Fig. 14b). STM can be used to characterize patterns produced by standard lithography processes which are used in the production of electronic devices such as computers. These materials are not necessarily conductive; consequently, deposition of a thin film of metal is necessary to provide a conductive surface for imaging. The three-dimensional information obtained in the STM analysis of integrated circuits cannot be obtained with other methods of process evaluation currently in use. Important to the potential of STM in commercial applications is that these images were acquired in air, thus avoiding the complications, and extra analysis time, associated with vacuum systems.

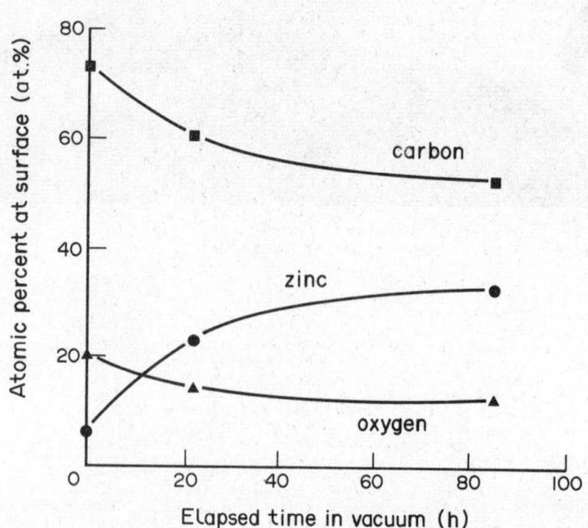

Figure 11
Change in surface chemistry due to desorption of atmospheric adsorbates when ZnO is introduced into an UHV environment: the change in "apparent" surface work function, as determined by the changing slope of the *I*/*z* measurement performed in the STM, and the Auger analysis are consistent with the desorption of oxygen and carbon dioxide from the ZnO

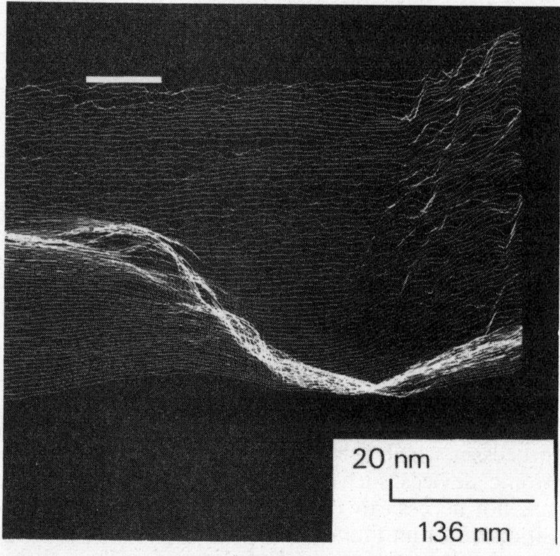

Figure 12
Topographic image of partially protonated emeraldine hydrochloride, indicating the surface structure that results from the melt spinning process of film deposition

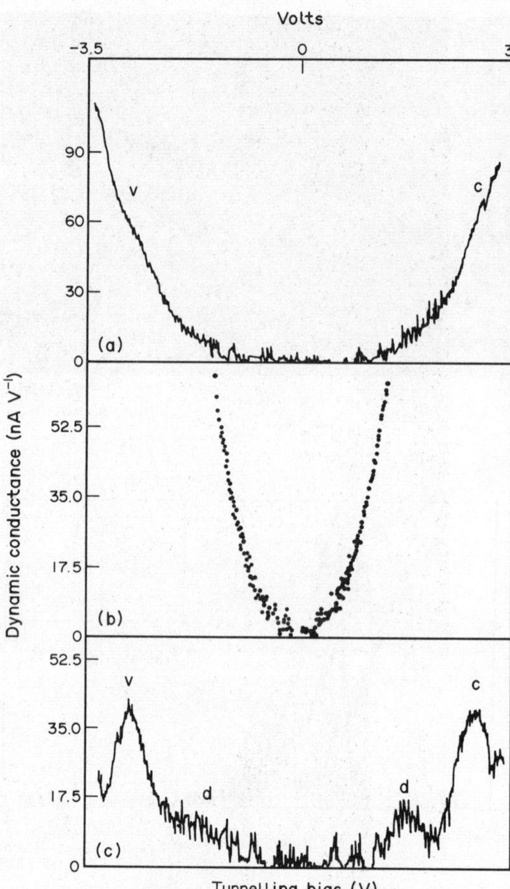

Figure 13
Three types of electronic structure detected in emeraldine hydrochloride films: (a) a semiconductor-like structure with no midgap defect states, (b) a metallic-like structure; and (c) a structure consisting of midgap states between the valence and conduction bands

6. Related Techniques

The scanning technology of STM is now being exploited to measure quantities other than tunnelling current and local densities of states. The most developed of these new "near field" or "scanning probe" techniques is atomic force microscopy (AFM), which records interatomic forces between the sample and tip. The tip is attached to a cantilever with a spring constant of about 1 N m^{-1} and resonant frequency of 1–10 kHz. As the tip is scanned over the sample, deflection of the cantilever due to the forces between atoms on the surface with those on the tip is monitored to produce a topographic image corresponding to a constant force profile. Forces detected range from 10^{-6} N to 10^{-9} N. This image is

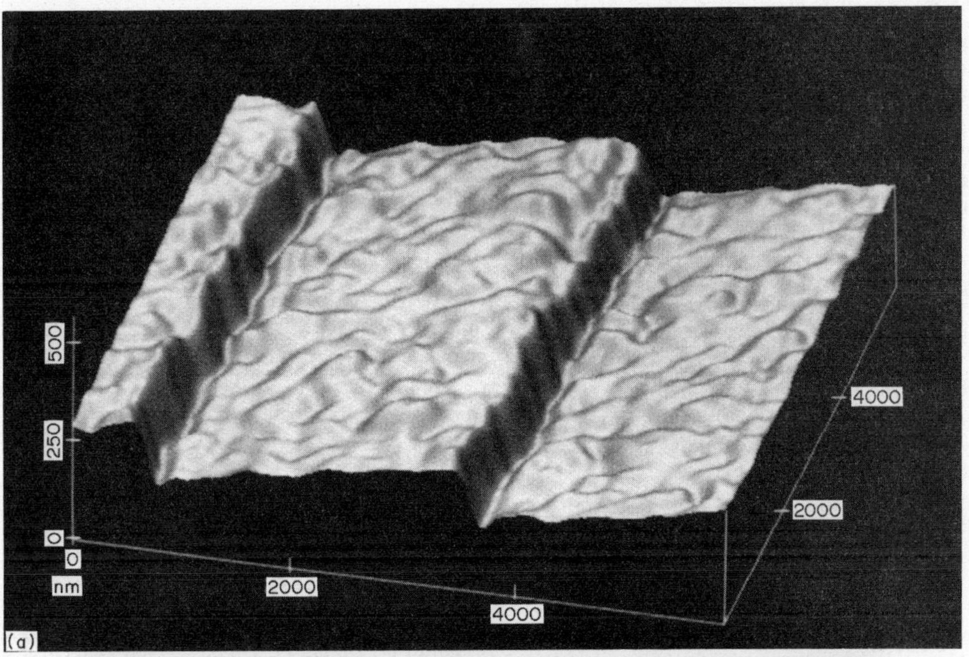

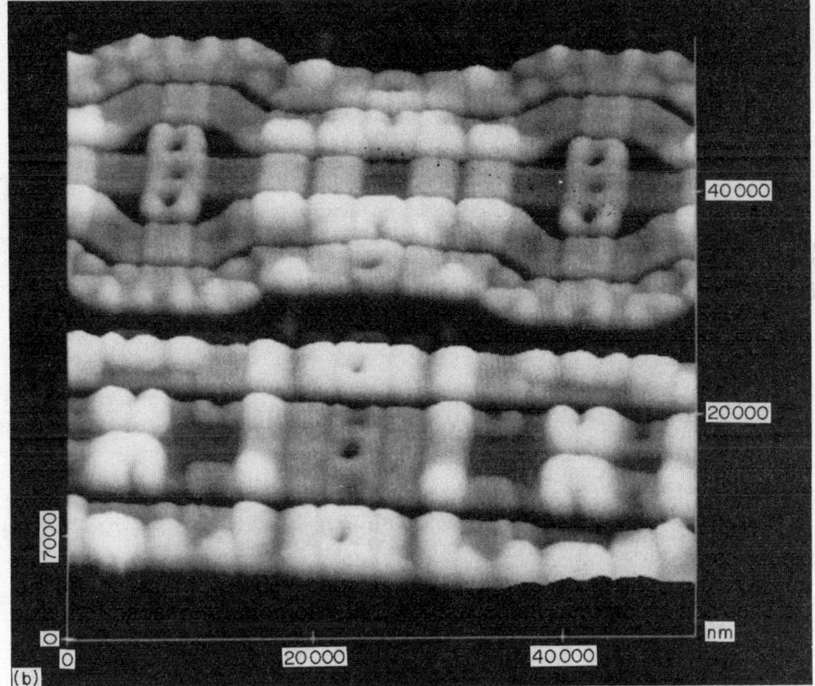

Figure 14
Constant current images: (a) a 5 μm scan of a 1 μm diffraction grating and (b) a 50 μm scan of an integrated circuit coated with a thin layer of gold—images were acquired on a Nanoscope in air (courtesy of Digital Instruments)

analogous to the constant current image in STM. Sensitive detection of the cantilever deflection is required to optimize the spatial resolution of AFM. Several methods of force detection have been employed including: electron tunnelling, capacitance, interferometry, and deflection of a laser beam reflected off the cantilever and fiber optic interferometry. The last method is preferred for its simplicity and vacuum compatibility. Atomic-resolution imaging has been demonstrated on inorganic insulators such as boron nitride and organic nonconductors such as amino acids, as well as graphite and molybdenum disilicide. One of the advantages of AFM is that it extends topographic imaging to insulators and biological materials.

Other techniques based on the new scanning probe technology of STM include magnetic force imaging, thermal gradient imaging, surface modification such as nanolithography, near field optical imaging, phonon detection, ballistic electron emission microscopy and capacitance microscopy. It remains to be seen which of these new measurements will become standard characterization techniques.

See also: Investigation and Characterization of Materials: An Overview; Mechanical Properties Microprobe (Suppl. 2)

Bibliography

Bardeen J 1961 Tunneling from a many particle point of view. *Phys. Rev. Lett.* 6: 57–9

Bell L D, Kaiser W J 1988 Observation of interface band structure by ballistic-electron-emission microscopy. *Phys. Rev. Lett.* 61: 2368–71

Binnig G, Quate C F, Gerber Ch 1986 Atomic force microscopy. *Phys. Rev. Lett.* 56: 930–3

Binnig G, Rohrer H 1982 Scanning tunnelling microscope. *Helv. Phys. Acta* 55: 726

Binnig G, Rohrer H 1987 Scanning tunnelling microscopy—From birth to adolescence. *Rev. Mod. Phys.* 59: 615–25

Binnig G, Rohrer H, Gerber Ch, Weibel E 1982 Surface studies by scanning tunneling microscopy. *Phys. Rev. Lett.* 49: 57–61

Binnig G, Rohrer H, Gerber Ch, Weibel E 1983 7×7 reconstruction on Si resolved in real space. *Phys. Rev. Lett.* 50: 120

Bonnell D A 1988 Characterization of carbides by STM. *J. Mater. Sci. Eng.* A 105/106: 55–63

Bonnell D A (ed.) 1990 *Scanning Tunneling Microscopy: Theory and Practice*. VCH Pub, New York

Bonnell D A, Angelopoulos M 1989 Spatially localized electronic structure in polyaniline by STM. IBM Research Report

Bonnell D A, Clarke D R 1988 Scanning tunneling microscopy and spectroscopy of ceramics: SiC and ZnO. *J. Am. Cer. Soc.* 71: 629–37

Feenstra R M (ed.) 1988 Proceedings of the Second International Conference on Scanning Tunneling Microscopy. *J. Vac. Sci. Technol.* A 6: 259–556

Garcia N (ed.) 1987 Proceedings of the First International Conference on Scanning Tunneling Microscopy. *Surf. Sci.* 181: 1–412

Gopel W 1985 Chemisorption and charge transfer at ionic semiconductor surfaces: implications in designing gas sensors. *Prog. Surf. Sci.* 20: 9–102

Hamers R J 1989 Atomic resolution surface spectroscopy with the scanning tunneling microscope. *Ann. Rev. Phys. Chem.* 40: 531–59

Hamers R J, Demuth J 1988 Atomic structure of Si (111) (3×3) Al. *J. Vac. Sci. Technol.* A6: 512–16

Hansma P K (ed.) 1982 *Tunneling Spectroscopy—Capabilities, Applications and New Techniques*. Plenum Press, New York

Hansma P K, Elings V B, Marti O, Bracker C E 1988 Scanning tunneling microscopy and force microscopy: Application to biology and technology. *Science* 242: 209–42

Hansma P K, Tersoff J 1987 Scanning tunneling microscopy. *J. Appl. Phys.* 61: R1–R23

Himpsel F, Fauster Th 1984 Probing valence states with photoemission and inverse photoemission. *J. Vac. Sci. Technol.* A1: 111

Lang N 1986 Theory of single atom imaging in the STM. *Phys. Rev. Lett.* 56: 1164–8

Lang N 1987 Apparent size of an atom in the scanning tunneling microscope. *Phys. Rev. Lett.* 58: 45–8

Martin Y, Williams C C, Wickramasinghe H K 1988 The techniques for microcharacterization of materials. *Scanning Microsc.* 2: 3–8

Quate C F 1986 Vacuum tunneling: A new technique for microscopy, *Phys. Today* 39(8): 26–33

Smith D, Binnig G, Quate C 1986 Detection of phonons with a STM. *Appl. Phys. Lett.* 49: 1641–3

Tersoff J, Hamann D R 1985 Theory of scanning tunneling microscopy. *Phys. Rev. B* 31: 805–13

Tromp R M, Hamers R J, Demuth J E 1985 Si (001) dimer structure observed with STM. *Phys. Rev. Lett.* 55: 1303–7

Wickramasinghe H K 1989 Scanned-probe microscopes. *Sci. Am.* 261: 74–81

Williams C C, Wickramasinghe H K 1986 Scanning thermal profiler. *Appl. Phys. Lett.* 49: 1587–9

Wolfe E L 1986 *Electron Tunnelling Spectroscopy*. Oxford University Press, Oxford

Zangwill A 1988 *Physics at Surfaces*. Cambridge University Press, Cambridge

D. A. Bonnell
[University of Pennsylvania, Philadelphia, Pennsylvania, USA]

Segmental Orientation in Elastomers

Chain segments in an undistorted isotropic amorphous elastomer are randomly oriented in all directions. Macroscopic distortion of the elastomer results in an anisotropic distribution of the chain segment orientation. A given segment of a chain orients owing to the deformation of the chain under macroscopic deformation and the distortion of the immediate environment of the segment in which it is embedded. The former is referred to as the intramolecular and the latter as the intermolecular factor contributing to segmental orientation. The extent of intramolecular and intermolecular contributions to segmental orientation depends on the type and state of the polymeric system under investigation. In a cross-linked rubber that is highly swollen with a suitable diluent, segmental orientation due to a macroscopic distortion results essentially from intramolecular contributions. In the absence of diluent, intermolecular interactions in the local environment are equally significant. In the glassy state, segmental orientation depends predominantly on the state of the local environment. For brevity, orientation under uniaxial stretching only will be considered in this article.

1. Orientation Distribution Function

Figure 1 depicts the instantaneous configuration of a chain relative to a laboratory-fixed coordinate system, $Oxyz$. $\boldsymbol{r}$ denotes the instantaneous end-to-end vector for the chain. The z axis is assumed to be the direction of macroscopic extension. $\boldsymbol{m}_i$ represents a vector rigidly embedded in the segment i of the chain. α denotes the angle between the direction of stretch and the vector $\boldsymbol{m}_i$. Instantaneously, each vector $\boldsymbol{m}_i$ of the chains of the elastomer will show different orientations relative to the z axis. The orientation distribution function $f(\alpha)$ obtained in the elastomer may be expanded (Bower 1981, Monnerie 1983) in terms of Legendre polynomials as

$$f(\alpha)=\frac{1}{2\pi}\sum_{l=0}^{\infty}\left(l+\frac{1}{2}\right)\langle P_l(\cos\alpha)\rangle P_l(\cos\alpha) \qquad (1)$$

where $P_l\,(\cos\alpha)$ is the lth Legendre polynomial and $\langle P_l\,(\cos\alpha)\rangle$ is the average over the distribution. For

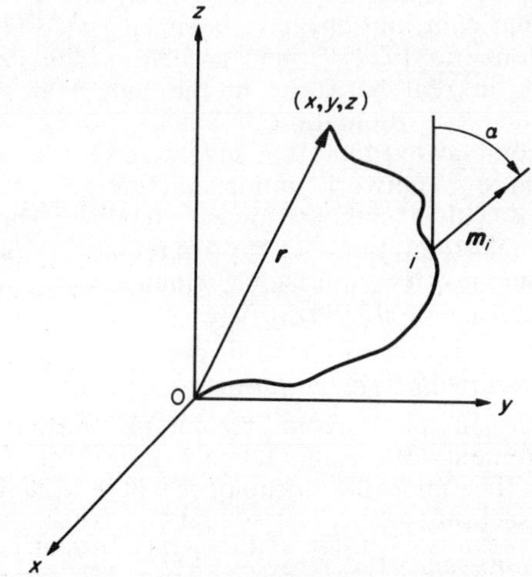

Figure 1
The instantaneous configuration of a chain

uniaxial distribution, the two averages $\langle P_2\,(\cos\alpha)\rangle$ and $\langle P_4\,(\cos\alpha)\rangle$ are commonly employed. They are given as

$$\left.\begin{aligned}\langle P_2(\cos\alpha)\rangle &= (3\langle\cos^2\alpha\rangle-1)/2\\ \langle P_4(\cos\alpha)\rangle &= (35\langle\cos^4\alpha\rangle-30\langle\cos^2\alpha\rangle+3)/8\end{aligned}\right\} \qquad (2)$$

where

$$\langle\cos^n\alpha\rangle=\int_0^{\pi} f(\alpha)\cos^n\alpha\sin\alpha\,d\alpha \qquad (3)$$

The simplest characterization of orientation rests on the determination of the orientation function, $\langle P_2(\cos\alpha)\rangle$. Various experimental techniques exist, however, for the measurement of the different averages $\langle P_n(\cos\alpha)\rangle$ (Samuels 1974, Ward 1975).

2. Segmental Orientation in the Single Chain

Characterization of segmental orientation in a deformed elastomer is made possible by first considering a network chain with fixed end-to-end vector $\boldsymbol{r}$ as shown in Fig. 1. The two ends of the chain denote cross-links which are fixed in space following the assumption of fixed $\boldsymbol{r}$. The angle α takes on different values as the chain undergoes configurational transitions, subject to the constancy of $\boldsymbol{r}$. The average $\overline{\cos^2\alpha}$ for all configurations of the chain at constant $\boldsymbol{r}$ is given (Nagai 1964, Flory 1969) as

$$\overline{\cos^2\alpha}=\frac{1}{3}\left\{1+2D_0\left[\frac{z^2}{\langle z^2\rangle_0}-\frac{1}{2}\left(\frac{x^2}{\langle x^2\rangle_0}+\frac{y^2}{\langle y^2\rangle_0}\right)\right]\right\} \qquad (4)$$

where

$$D_0=\frac{3\langle r^2\cos^2\Phi\rangle_0/\langle r^2\rangle_0-1}{10} \qquad (5)$$

Here, Φ indicates the angle between the chain vector $\boldsymbol{r}$ and $\boldsymbol{m}_i$. The averages $\langle\,\rangle_0$ shown in Eqns. (4, 5) are for a free chain. $\langle x^2\rangle_0$, $\langle y^2\rangle_0$ and $\langle z^2\rangle_0$ denote the mean square components of the free chain. Thus, Eqn. (4) gives the average value of a quantity for a chain with two ends fixed in terms of the average for a free chain. Calculating averages for a free chain is much simpler and therefore Eqn. (4) is a convenient approximation to $\cos^2 \alpha$.

3. Segmental Orientation in the Deformed Network

Equation (4) gives the value of $\cos^2 \alpha$ averaged over all configurations of a single network chain with fixed ends. Averaging Eqn. (4) over all chains of the network leads to

$$\langle \cos^2 \alpha \rangle = \frac{1}{3}\left\{1 + 2D_0\left[\frac{\langle z^2 \rangle}{\langle z^2 \rangle_0} - \frac{1}{2}\left(\frac{\langle x^2 \rangle}{\langle x^2 \rangle_0} + \frac{\langle y^2 \rangle}{\langle y^2 \rangle_0}\right)\right]\right\} \quad (6)$$

Here $\langle x^2 \rangle$, $\langle y^2 \rangle$ and $\langle z^2 \rangle$ show the mean square x, y and z components of chain vectors in the deformed state of the network. Thus, the ratios $\langle x^2 \rangle / \langle x^2 \rangle_0$, $\langle y^2 \rangle / \langle y^2 \rangle_0$ and $\langle z^2 \rangle / \langle z^2 \rangle_0$ may be regarded as averaged measures of molecular deformation along x, y and z directions. In general, we may define

$$\left.\begin{aligned} \Lambda_x^2 &= \langle x^2 \rangle / \langle x^2 \rangle_0 \\ \Lambda_y^2 &= \langle y^2 \rangle / \langle y^2 \rangle_0 \\ \Lambda_z^2 &= \langle z^2 \rangle / \langle z^2 \rangle_0 \end{aligned}\right\} \quad (7)$$

where Λ_x^2, Λ_y^2 and Λ_z^2 are respectively the x, y and z components of the molecular deformation tensor. Substituting Eqn. (7) in Eqn. (6) and rearranging leads to the orientation function $S \equiv \langle P_2(\cos \alpha) \rangle$ (Mark and Erman 1988) as

$$S \equiv \langle P_2(\cos \alpha) \rangle = D_0[\Lambda_z^2 - \tfrac{1}{2}(\Lambda_x^2 + \Lambda_y^2)] \quad (8)$$

Equation (8) relates the orientation function to the state of deformation of the molecules. If the junctions of the network are assumed to deform affinely with macroscopic strain, then

$$\Lambda_z^2 = \lambda^2, \qquad \Lambda_x^2 = \Lambda_y^2 = \lambda^{-1} \quad (9)$$

where λ is the extension ratio defined as the ratio of the final to the undeformed length along the direction of extension. A network where Eqn. (9) is obeyed is called an affine network (Mark and Erman 1988). However, only the mean positions of junctions transform affinely with macroscopic strain. The instantaneous fluctuations of junctions are independent of macroscopic strain. The relationship of microscopic to macroscopic deformation in the uniaxial extension case is obtained for the phantom network (Mark and Erman 1988) as

$$\left.\begin{aligned} \Lambda_z^2 &= (1 - 2/\phi)\lambda^2 + 2/\phi \\ \Lambda_x^2 = \Lambda_y^2 &= (1 - 2/\phi)\lambda^{-1} + 2/\phi \end{aligned}\right\} \quad (10)$$

where ϕ is the average junction functionality.

Substituting Eqns. (9, 10) into Eqn. (8) leads to the expression for the orientation function

$$\begin{aligned} S &= D_0(\lambda^2 - \lambda^{-1}) && \text{Affine network} \\ &= D_0(1 - 2/\phi)(\lambda^2 - \lambda^{-1}) && \text{Phantom network} \end{aligned} \quad (11)$$

The affine and phantom network models constitute two extreme cases that have received wide recognition (Mark and Erman 1988) in the treatment of the molecular theory of rubber elasticity. The strain dependence of S is the same for the affine and the phantom networks as observed from Eqn. (11). According to this equation, the orientation function is conveniently separated into two factors. The first factor D_0 reflects the effect of the molecular constitution of the chains on orientation. The evaluation of D_0 according to Eqn. (5) requires detailed statistical calculations (Flory 1969). In the limiting case of a very long chain, D_0 equates to $1/5N$ where N is the number of Kuhn segments in the chain. In the literature, this value of D_0 is commonly employed together with the affine network approximation:

$$S = (1/5N)(\lambda^2 - \lambda^{-1}) \quad (12)$$

It should be noted that segmental orientation in real networks deviates from the affine and phantom network predictions in two important respects. First, the expression for D_0 given by Eqn. (5) is derived on the assumption of intramolecular contributions to segmental orientation only. However, in the bulk state, intermolecular orientational contributions to segmental orientation are significant and the coefficient D_0 has to be modified. Work in this field is not conclusive. Various modifications in D_0 due to intermolecular contributions have been suggested (Jarry and Monnerie 1979, Erman and Monnerie 1985). Second, in real networks in the bulk state, the dependence of orientation on macroscopic strain is more complicated than that suggested by the affine or phantom network models. Dilution of the network with a suitable solvent, however, brings segmental orientation into close agreement with the phantom network expression (Erman and Monnerie 1985, Queslel et al. 1985).

4. Experimental Techniques

Advances in spectroscopic techniques allow direct measurement of segmental orientation in elastomers. The important techniques will be described in this section.

4.1 Fluorescence Polarization

In this technique, specific locations along the elastomer chains are labelled with fluorescent molecules.

Fluorescent molecules have the property of re-emitting, in the form of visible light, part of the energy acquired by the absorption of luminous radiation. When absorbing light of a suitable wavelength, a fluorescent molecule behaves as an electric dipole oscillator with a fixed orientation with respect to the chain. The dipole during absorption is termed an absorption transition moment $\boldsymbol{M}_0$. In the same way, fluorescent emission is represented by an emission transition moment $\boldsymbol{M}$. In Fig. 2, the directions of $\boldsymbol{M}_0$ and $\boldsymbol{M}$ are indicated relative to the direction of deformation, the z axis, by the angles α_0 and α, respectively. The angles relative to the x axis made by the projections of $\boldsymbol{M}_0$ and $\boldsymbol{M}$ to the xy plane are β_0 and β, respectively. Thus the angles α_0, α, β_0 and β represent the spherical polar angles $\Omega_0 = (\alpha_0, \beta_0)$ at time t_0 and $\Omega = (\alpha, \beta)$ at time t in the laboratory-fixed reference frame. The directions $\boldsymbol{P}$ and $\boldsymbol{A}$ in Fig. 2 indicate the directions of the polarizer and analyzer.

After illuminating the sample by a linearly polarized short pulse of light at t_0, the intensity i emitted at time $t_0 + n$ for the $\boldsymbol{P}$ and $\boldsymbol{A}$ directions of polarizer and analyzer is given by

$$i(\boldsymbol{P}, \boldsymbol{A}, t_0 + u) = K \int\int N(\Omega_0, t_0)\, P(\Omega, t_0 + u/\Omega_0, t_0) \times \cos^2(\boldsymbol{P}, \boldsymbol{M}_0) \cos^2(\boldsymbol{A}, \boldsymbol{M}) \times \exp(-u/\tau)\, d\Omega_0\, d\Omega \tag{13}$$

where K is an instrumental constant, $N(\Omega_0, t_0)$ is the orientation distribution function of $\boldsymbol{M}_0$ at time t_0, $P(\Omega, t_0 + u/\Omega_0, t_0)$ is the conditional probability density of finding at position Ω at time $(t_0 + u)$ a vector $\boldsymbol{M}$ which was at position Ω_0 at time t_0. The arguments $(\boldsymbol{P}, \boldsymbol{M}_0)$ and $(\boldsymbol{A}, \boldsymbol{M})$ denote the angles between the polarizer and $\boldsymbol{M}_0$, and analyzer and $\boldsymbol{M}$, respectively. τ is the mean lifetime of the excited state, usually called the fluorescence lifetime. The most frequent values of τ range from 1 to 100 ns. The time t_0 in Eqn. (13) corresponds to the macroscopic evolution of the sample, usually in the order of seconds in a rheological experiment. Thus the t_0 dependence of the sample may safely be ignored within the time scale ($\tau \sim 10^{-8}$ s) of fluorescence measurements.

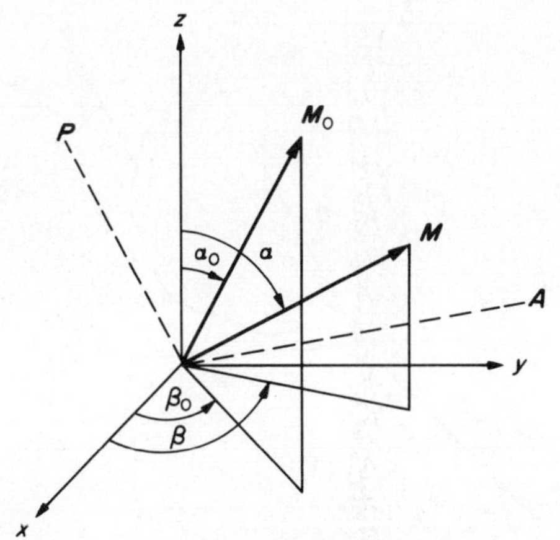

Figure 2
The direction of the dipole oscillator relative to the laboratory-fixed frame

It has been shown that the following five functions can be determined experimentally by the fluorescence polarization technique (Jarry and Monnerie 1978):

$$\left.\begin{aligned}
G_{20}^{(0)} &= \frac{1}{2}\langle 3\cos^2\alpha_0 - 1\rangle \\
G_{02}^{(0)} &= \frac{1}{2}\langle 3\cos^2\alpha - 1\rangle \\
G_{22}^{(0)} &= \frac{1}{4}\langle (3\cos^2\alpha_0 - 1)\rangle\langle 3\cos^2\alpha - 1\rangle \\
G_{22}^{(1)} &= \frac{9}{16}\langle \sin\alpha_0 \cos\alpha_0 \sin\alpha \cos\alpha \cos(\beta - \beta_0)\rangle \\
G_{22}^{(2)} &= \frac{9}{64}\langle \sin^2\alpha_0 \sin^2\alpha \cos 2(\beta - \beta_0)\rangle
\end{aligned}\right\} \tag{14}$$

If the system is frozen then $\alpha = \alpha_0$ and $\beta = \beta_0$ and the determination of $G_{20}^{(0)} (= G_{02}^{(0)})$ and $G_{22}^{(m)}$ leads to the characterization of the averages $\langle\cos^2\alpha\rangle$ and $\langle\cos^4\alpha\rangle$. In a uniaxially mobile system, the determination of $G_{20}^{(0)} (= G_{02}^{(0)})$ leads to the characterization of $\langle\cos^2\alpha\rangle$. The remaining functions $G_{22}^{(m)}$ yield information on the orientation-mobility behavior of the deformed elastomer. More detailed information about the technique of fluorescence polarization is given by Monnerie (1983).

Results of measurements of $G_{20}^{(0)}$ for a polyisoprene network at 25 °C by the fluorescence technique are shown in Fig. 3 as a function of extension ratio λ (Monnerie and Jarry 1981). The three sets of data are obtained on networks of different molecular weight M_c between cross-link.

4.2 Infrared Dichroism

The technique of infrared dichroism (Monnerie 1983) is based on the selective absorption of polarized radiation by bonds of a polymer molecule. The

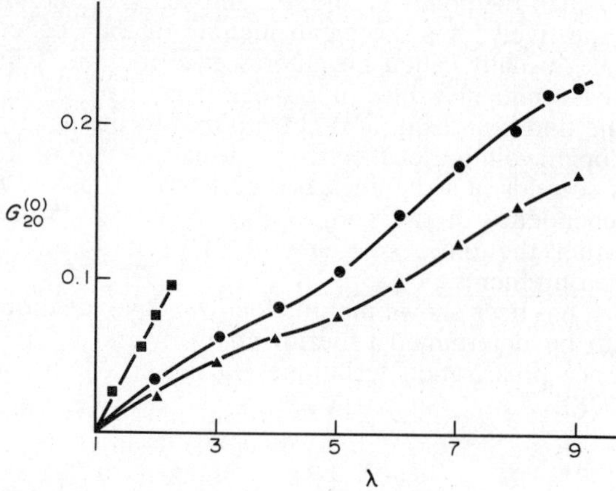

Figure 3
$G_{20}^{(0)}$ vs extension ratio λ. The three sets of data are for different cross-link densities, with $M_c = 1.1 \times 10^4$ (■), $M_c = 2.1 \times 10^4$ (●) and $M_c = 4.2 \times 10^4$ (▲)

absorption in the infrared region deals with vibrational modes of the bonds such as bond stretching and bending. Associated with each mode is an absorbing frequency and a transition dipole moment $\boldsymbol{\mu}$. The transition moment $\boldsymbol{\mu}$ has a definite orientation in the molecule. The intensity of an infrared absorption band depends upon the angle between the electric vector $\boldsymbol{E}$ of incident radiation and the transition moment $\boldsymbol{m}$. The individual absorbance a is given according to

$$a = \log_{10} I_0/I = |\boldsymbol{\mu}|^2(\boldsymbol{\mu} \, . \, \boldsymbol{E})^2 \quad (15)$$

where I_0 and I are the incident and transmitted intensities.

In Fig. 4, the transition moment $\boldsymbol{\mu}$ in a chain is shown relative to a laboratory-fixed coordinate frame. The z axis identifies the direction of stretch and $\boldsymbol{r}$ is the end-to-end vector of the chain making an angle of θ with the direction of stretch. The transition moment makes an angle of Φ, with the chain vector $\boldsymbol{r}$.

In incident radiation polarized along the z axis parallel to the direction of stretch, the absorbance $a_\parallel$ for a single $\boldsymbol{\mu}$ will be

$$a_\parallel = |\boldsymbol{\mu}|^2 \left(\cos^2\Phi \cos^2\theta + \tfrac{1}{2}\sin^2\Phi \sin^2\theta\right) \quad (16)$$

Similarly, for incident radiation perpendicular to the z axis, the absorbance $a_\perp$ will be

$$a_\perp = |\boldsymbol{\mu}|^2\left[\tfrac{1}{2}\cos^2\Phi \sin^2\theta + \tfrac{1}{4}\sin^2\Phi(1+\cos^2\theta)\right] \quad (17)$$

For a set of molecules whose end-to-end vectors are characterized by an orientation distribution function $f(\theta)$ relative to the z axis, the total absorbance from polarized radiation in parallel and perpendicular directions will be

$$\left.\begin{aligned} A_\parallel &= \int_0^\pi a\, f(\theta) \sin\theta \, d\theta \\ A_\perp &= \int_0^\pi a_\perp f(\theta) \sin\theta \, d\theta \end{aligned}\right\} \quad (18)$$

The dichroic ratio or dichroism R is defined as

$$R = A_\parallel / A_\perp \quad (19)$$

For a set of molecules perfectly aligned along the z axis, the dichroism is

$$R_0 = 2\cot^2\alpha \quad (20)$$

Substituting Eqns. (18, 20) into Eqn. (19) leads to

$$\begin{aligned} R &= \frac{\int_0^\pi [1+(R_0-1)\cos^2\theta] f(\theta)\sin\theta\, d\theta}{\int_0^\pi [1+\frac{1}{2}(R_0-1)\sin^2\theta] f(\theta)\sin\theta\, d\theta} \\ &= \frac{1+(R_0-1)\langle\cos^2\theta\rangle}{1+\frac{1}{2}(R_0-1)(1-\langle\cos^2\theta\rangle)} \end{aligned} \quad (21)$$

The orientation function $\langle P_2(\cos\theta)\rangle$ is obtained from Eqn. (21) as

$$\langle P_2(\cos\theta)\rangle = \tfrac{1}{2}\langle 3\cos^2\theta - 1\rangle = \left(\frac{R-1}{R+2}\right)\left(\frac{R_0+2}{R_0-1}\right) \quad (22)$$

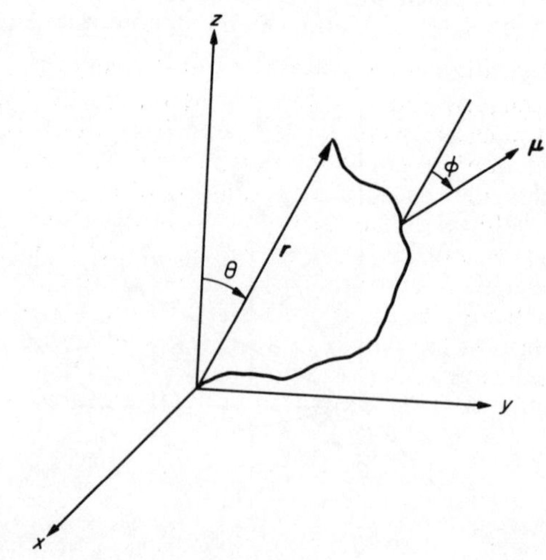

Figure 4
The transition moment $\boldsymbol{\mu}$ with the instantaneous chain end-to-end vector $\boldsymbol{r}$

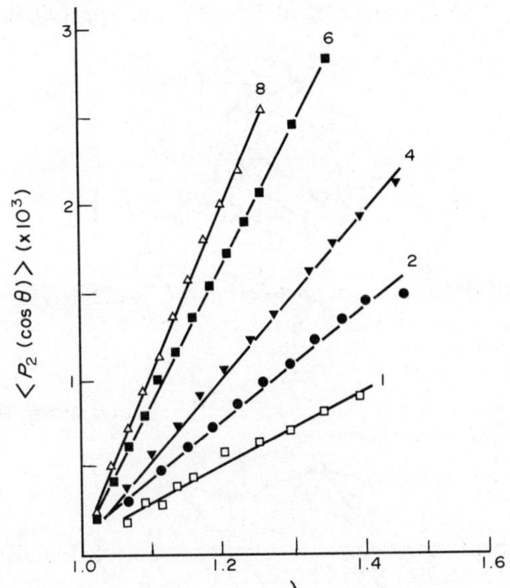

Figure 5
DNMR measurements of the orientation function $\langle P_2(\cos\theta)\rangle$ vs extension ratio λ. Samples 1 to 8 are numbered in increasing order of network cross-link density

The technique of infrared dichroism for characterizing the orientation function of deformed polymers has found wide application (Jasse and Koenig 1979). A discussion of the technique and its application to measure orientation in uniaxially stretched polyisoprene networks is given by Amram et al. (1986, 1988).

4.3 Deuterium Nuclear Magnetic Resonance

Characterization of segmental orientation by deuterium nuclear magnetic resonance (DNMR) is based on deuterium labelling of a fraction of the network chains and measuring the splitting of NMR lines in the deformed state of the network (Dubault et al. 1984, Samulski 1985). Labelling of the molecules is achieved by exchanging specific hydrogens for deuterons. In the deformed state, the anisotropic motion of the labelled molecule results in the form of splitting of the NMR spectra into a doublet whose spacing $\Delta\nu$ may be written in frequency units as

$$\Delta\nu = \tfrac{3}{2}\nu_q P_2(\cos\theta) \tag{23}$$

where ν_q is the static quadrupolar coupling constant (~200 kHz) and θ is the angle between the carbon–deuteron bond and the direction of the applied magnetic field which may be arranged to coincide with the direction of stretch.

Typical results of measurements on uniaxially deformed polybutadiene networks are shown in Fig. 5 (Dubault et al. 1984). The orientation function $\langle P_2(\cos\theta)\rangle$ is plotted as a function of extension ratio. The five sets of data refer to different degrees of cross-linking.

4.4 Other Experimental Techniques

Birefringence (Treloar 1975) and wide-angle x-ray scattering (Mitchell 1984) are two other techniques by which segmental orientation may be characterized. One shortcoming of the birefringence technique when compared with the fluorescence polarization, infrared dichroism and deuterium NMR techniques is that the polarizabilities of segments are affected by the neighboring molecules, thereby precluding any quantitative characterization of orientation.

Bibliography

Amram B, Bokobza L, Monnerie L, Queslel J P 1988 Fourier-transform infrared dichroism study of molecular orientation in high cis-1,4-polybutadiene. *Polymer* 29: 1155–60

Amram B, Bokobza L, Queslel J P, Monnerie L 1986 Fourier-transform infrared dichroism study of molecular orientation in synthetic high cis-1,4-polyisoprene and in natural rubber. *Polymer* 27: 877–82

Bower D I 1981 Orientation distribution functions for uniaxially oriented polymers. *J. Polym. Sci., Polym. Phys. Ed.* 19: 93–107

Dubault A, Deloche B, Herz J 1984 Effect of crosslinking density on the orientational order generated in strained network: A deuterium magnetic resonance study. *Polymer* 25: 1405–10

Erman B, Monnerie L 1985 Theory of segmental orientation in amorphous polymer networks. *Macromolecules* 18: 1985–91

Flory P J 1969 *Statistical Mechanics of Chain Molecules*. Wiley, New York

Jarry J P, Monnerie L 1978 Orientation and molecular dynamics in uniaxial polymers. I. Theory of fluorescence polarization. *J. Polym. Sci., Polym. Phys. Ed.* 16: 443–55

Jarry J P, Monnerie L 1979 Effects of a nematic-like interaction in rubber elasticity theory. *Macromolecules* 12: 316–20

Jasse B, Koenig J L 1979 Orientational measurements in polymers using vibrational spectroscopy. *J. Macromol. Sci., C* 17: 61–135

Mark J E, Erman B 1988 *Rubberlike Elasticity: A Molecular Primer*. Wiley, New York

Mitchell G R 1984 A wide-angle x-ray study of the development of molecular orientation in cross-linked natural rubber. *Polymer* 25: 1562–72

Monnerie L 1983 An experimental approach to the molecular viscoelasticity of bulk polymers by spectroscopic techniques: Neutron scattering, infrared dichroism and fluorescence polarization. *Faraday Symp. Chem. Soc.* 18: 57–81

Monnerie L, Jarry J P 1981 Solid state polymers as studied by fluorescence polarization: Mobility and orientation measurements. *Ann. N.Y. Acad. Sci.* 366: 328–40
Nagai K 1964 Photoelasticity of cross-linked amorphous polyethylene. *J. Chem. Phys.* 40: 2818–26
Queslel J P, Erman B, Monnerie L 1985 Experimental determination of segmental orientation in polyisoprene network by fluorescence polarization and comparison with theory. *Macromolecules* 18: 1991–4
Samuels R J 1974 Structured polymer properties. Wiley, New York
Samulski E T 1985 Investigations of polymer chains in oriented fluid phases with deuterium nuclear magnetic resonance. *Polymer* 26: 177–89
Treloar L R G 1975 *The Physics of Rubber Elasticity*, 3rd edn. Clarendon, Oxford
Ward I M 1975 *Structure and Properties of Oriented Polymers*. Applied Science, London

B. Erman
[Bogazici University, Istanbul, Turkey]

L. Monnerie
[Ecole Supérieure de Physique et de Chimie Industrielles, Paris, France]

Semiconductors: Mechanical Properties

The subject of the mechanical properties of semiconductors includes three areas of study. The first is that of elastic properties such as the elastic constants and how they depend on the chemical bond. The second is the study of plasticity and the third is the study of fracture properties. Since the bonding in semiconductors is different from that in metals, the study of the mechanical properties of semiconductors has mostly been one of scientific curiosity. However, in recent times the situation has changed; because of the immense practical importance of semiconductors for the electronic industry, their mechanical properties have significant technological implications as well. Mechanical properties have a role to play in every step of the device fabrication process. Starting from the growth of defect-free single-crystal substrates, through device processing (where the thermal stresses between dielectric films or metallic films and substrates are introduced) to device testing (where residual stresses introduced by the processing steps limit device reliability), the interrelations between stresses and strains in the material and the ambient conditions during the various device-fabrication steps have to be examined closely.

1. Bonding and Crystal Structure

Elemental semiconductors such as silicon and germanium crystallize in the diamond cubic structure.

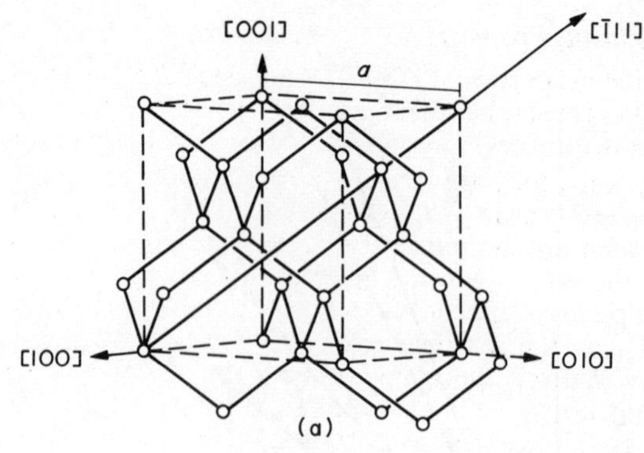

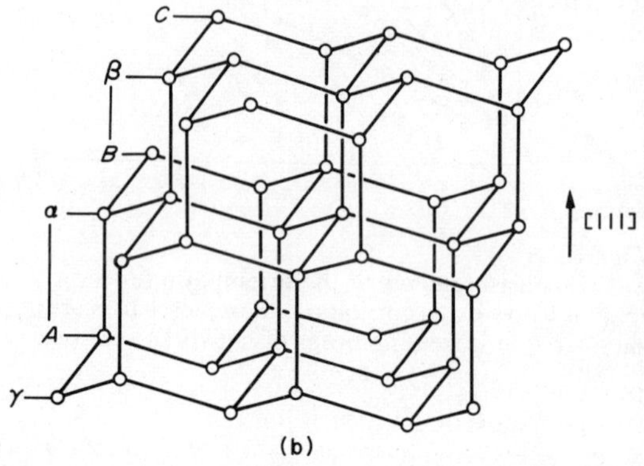

Figure 1
(a) The cubic unit cell of the diamond lattice and (b) stacking of {111} planes

The III–V compound semiconductors (e.g., GaAs, InP) crystallize in the cubic zinc blende (sphalerite) structure. Some of the II–VI compound semiconductors (e.g., CdTe, ZnSe) also crystallize in the sphalerite, while some (e.g., CdS, ZnO) crystallize in the hexagonal zinc blende (wurtzite) structure. The space lattice of the diamond cubic and sphalerite structure semiconductors is the face-centered-cubic (fcc) lattice. Since the diamond cubic structure consists of two interpenetrating fcc lattices, the close packed {111} planes of the fcc lattice become pairs of {111} planes in it as shown in Fig. 1. In the sphalerite lattice, because of the presence of the two types of atoms there is a polarity in the stacking of the {111} planes. If $ABCABC$ is the stacking order of the {111} planes in the first fcc lattice and $\alpha\beta\gamma\alpha\beta\gamma$ that in the second, then αB, βC and γA are closely spaced pairs of planes connected by three times as many bonds as there are between the three-times-wider-spaced $A\alpha$, $B\beta$, $C\gamma$ pairs of planes. In the wurtzite structure, the stacking of the tetrahedra is in the sequence $ABAB$, as in the hexagonal-close-packed lattice.

2. Elastic Properties

Elastic properties of materials are described by their elastic constants. Elastic stiffness is a measure of how the forces responsible for cohesion of solids vary when they are stretched or sheared. In general, cohesion results from electrostatic interactions between atomic particles. According to Coulomb's law the cohesive force between oppositely charged particles is proportional to e^2/r^2 where e is the charge and r is interparticle distance. Since crystals are close packed, the cohesive force acts over an area proportional to r and hence cohesive stress, or elastic stiffness, is proportional to e^2/r^4. The elastic stiffnesses C_{ijkl} are then obtained by differentiating the energy density of a crystal E with respect to elastic strain (Nye 1957):

$$C_{ijkl} = \left(\frac{\partial^2 E}{\partial \varepsilon_{ij} \partial \varepsilon_{kl}}\right)_{r=r_0} \tag{1}$$

where r_0 is the equilibrium spacing.

2.1 Elastic Constants

The usual starting point for elasticity theory is Hooke's law which states that stress $\boldsymbol{\sigma}$ is proportional to strain $\boldsymbol{\varepsilon}$ for sufficiently small strain and the constant of proportionality is the elastic constant. This linear relation between $\boldsymbol{\sigma}$ and $\boldsymbol{\varepsilon}$ is expressed in a mathematical form as

$$\sigma_{ij} = \sum_{k,l} C_{ijkl} \varepsilon_{kl} \tag{2}$$

where the stresses σ_{ij} and strains ε_{kl} are components of second-rank tensors and the elastic stiffnesses C_{ijkl} are components of a fourth-rank tensor. Equation (2) is usually expressed in its covariant form

$$\varepsilon_{ij} = \sum_{k,l} S_{ijkl} \sigma_{kl} \tag{3}$$

where $\mathbf{S}$ is the elastic compliance tensor and is the reciprocal of $\mathbf{C}$. The elastic stiffness tensor $\mathbf{C}$ has dimensions of force per unit area or energy per unit volume and $\mathbf{S}$ has reciprocal dimensions. It is customary to denote the components of these tensors in the abbreviated forms as C_{ij} and S_{ij}, respectively. The relations between the C_{ij} and the S_{ij} in a cubic crystal are

$$\left.\begin{aligned} C_{11} &= (S_{11}+S_{12})/S \\ C_{12} &= -S_{12}/S \\ C_{44} &= 1/S_{44} \end{aligned}\right\} \tag{4}$$

where $S=(S_{11}-S_{12})(S_{11}+2S_{12})$. In a hexagonal crystal the relations are

$$\left.\begin{aligned} C_{11}+C_{12} &= S_{33}/S \\ C_{11}-C_{12} &= 1/(S_{11}-S_{12}) \\ C_{13} &= -S_{13}/S \\ C_{33} &= (S_{11}+S_{12})/S \\ C_{44} &= 1/S_{44} \end{aligned}\right\} \tag{5}$$

where $S=S_{33}(S_{11}+S_{12})-2S_{13}^2$.

Because of general symmetry of matrices and that of the crystal lattice, the number of independent components of $\mathbf{C}$ and $\mathbf{S}$ is reduced from 81 to five for hexagonal crystals and to three for cubic crystals. From the point of view of practical utility, the elastic constant of interest is Young's modulus which relates the longitudinal strain produced when a crystal bar is loaded in single longitudinal tension or compression.

Another elastic property of interest is the Poisson's ratio ν which is the negative of the ratio of strains perpendicular and parallel to the stress axis. The values of E and ν can be obtained in terms of the compliance constants for an arbitrary crystallographic direction (Nye 1957). For the biaxial plane stress conditions associated with thin films, often encountered in the fabrication of semiconductor devices, longitudinal stresses and strains parallel to the film–substrate interface are related by $E/(1-\nu)$. Table 1 lists the elastic constants together with the density, lattice parameter and thermal expansion coefficient for silicon, germanium and some of the important III–V and II–VI compound semiconductors.

3. Dislocations and Plastic Deformation

Plastic deformation or mechanical transport occurs as a result of motion of line (linear) imperfections, called dislocations, within the solid. Dislocations are characterized by their Burgers vectors $\boldsymbol{b}$ which are lattice vectors. In the cubic semiconductors the common types of dislocations are the screw dislocations and 60° dislocations. In the former, $\boldsymbol{b}$ is parallel to the dislocation axis and in the latter, the angle between $\boldsymbol{b}$ and dislocation axis is 60°.

In crystals, plastic deformation, also known as "glide" or "slip," occurs easily in the close-packed planes and in the close-packed directions in the lattice. The glide plane contains the glide direction and together they form the glide system. In both the diamond and the sphalerite structures, the {111} planes and the [110] directions form the glide system. In the wurtzite structure, the {0001} basal plane and the $[11\bar{2}0]$ direction form the glide system. That these are the primary glide systems in semiconductors has been experimentally verified.

3.1 Peierls Force and Low-Temperature Brittleness

The Peierls force is a frictional force opposing the glide motion of dislocations in the crystal lattice. It is the minimum force that has to be overcome before glide occurs. In covalently bonded semiconductors, the Peierls force is high because dislocation glide involves severing of the strong directional bonding forces between the atoms. As a result, semiconductors are brittle at room temperature. They become progressively more ductile at elevated temperatures as the severing of the covalent bonds is aided appreciably by lattice vibrations and dislocation glide occurs. In general, glide is observed at temperatures at or above $0.5T_m$ where T_m is the melting point of the semiconductor in kelvin.

3.2 Critical Resolved Shear Stress

Since the volume change of the crystal in plastic deformation is not large, the stress component that causes glide is the shear stress acting on the glide plane and in the glide direction. In a compression (or tension) test, which is the standard mode of plastically deforming semiconductors, the resolved shear stress τ is given by

$$\tau = (P/A_0)\cos\phi\cos\lambda \qquad (6)$$

where P is the applied load, A_0 is the cross-sectional area of the specimen, ϕ is the angle between the normal to the glide plane and the axis of the applied stress and λ is the angle between the glide direction and the applied stress axis. τ is the critical resolved shear stress which has to be exceeded for the beginning of plastic flow in a certain glide system. The $\cos\phi\cos\lambda$ term in Eqn. (6) is called the Schmid factor. When there are several crystallographically equivalent glide systems, the one with the maximum resolved shear stress acting on it will operate first. Figure 2 shows the critical resolved shear stress (CRSS) as a function of temperature for silicon, GaAs and CdTe. As expected, with increasing temperature the Peierls barrier is easily overcome and CRSS decreases.

3.3 Glide Planes

Although the primary slip plane in the cubic semiconductors is the {111} plane, an important question arises as to whether shear occurs between αB or $B\beta$ planes (Fig. 1b). Dislocations whose axis lies in the $B\beta$ planes are referred to as the shuffle set dislocations and those whose axis lies in the αB planes are referred to as the glide set dislocations. At present it appears that the most plausible model for glide in semiconductors is that it occurs between the closely spaced αB planes (glide set) (Hirsch 1981).

Table 1
Elastic compliance, density, lattice parameter and thermal-expansion coefficient for silicon, germanium and some important III–V and II–VI compound semiconductors

Cubic semiconductors		Elastic compliance (10^{-11} m^2 N^{-1}) S_{11}	S_{12}	S_{44}	S_{13}	S_{33}	Density (kg m^{-3})	Lattice parameter (nm)	Coefficient of thermal expansion (10^{-6} K^{-1})
IV	Si	0.768	−0.214	1.256			2329.0	0.54307	2.56
	Ge	0.979	−0.267	1.497			5323.4	0.56574	5.90
III–V	GaP	0.973	−0.298	1.419			4138.0	0.54512	4.50
	GaAs	1.176	−0.365	1.684			5316.1	0.56532	6.86
	GaSb	1.582	−0.495	2.314			5613.7	0.60959	7.75
	InP	1.650	−0.594	2.170			4810.0	0.58687	4.75
	InAs	1.945	−0.685	2.525			5667.0	0.60583	4.52
	InSb	2.443	−0.863	3.311			5774.7	0.64794	5.37
II–VI	ZnS	2.030	−0.790	2.231			4083.0	0.54060	6.70
	ZnSe	2.260	−0.850	2.270			5263.0	0.56687	7.80
	ZnTe	2.40	−0.870	3.200			5636.0	0.61037	8.30
	CdTe	4.254	−1.734	5.015			6060.0	0.64810	4.80
	HgTe	4.720	−1.950	4.880			8095.0	0.64610	4.00
Hexagonal semiconductors									
II–VI	ZnO	0.786	−0.343	2.357	−0.221	0.694	5675.3	a 3.2520	∥c 2.920
								c 5.2130	∥a 4.750
	CdS	2.069	−0.999	6.649	−0.581	1.697	4820.0	4.1368	2.400
								6.7163	4.600
	CdSe	2.338	−1.122	7.595	−0.572	1.735	5810.0	4.2999	1.500
								7.0109	2.200

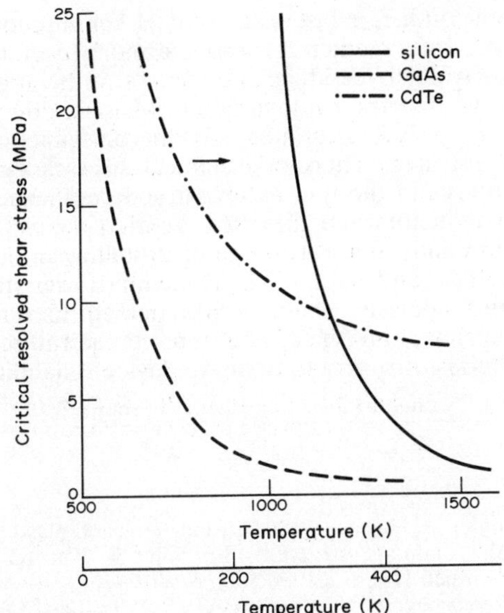

Figure 2
Critical resolved shear stress (MPa) vs temperature (K) for silicon (Schroter et al. 1983), GaAs (Guruswamy et al. 1987) and CdTe (Gutmanas et al. 1979) (note the different temperature scale for CdTe)

4. Stress–Strain Curve and Yield Point

The stress–strain curves of semiconductors exhibit a pronounced yield point which is the result of multiplication of the initial few dislocations. The values of the upper and lower yield stresses and strains depend on temperature, strain rate, machine compliance, orientation and initial dislocation density of the crystal. The temperature dependence and strain-rate dependence of the yield stresses are expressed as (Alexander and Haasen 1968)

$$\tau_y = C_y \dot{\varepsilon}^{1/n} \exp(E_y/kT) \qquad (7)$$

where $\dot{\varepsilon}$ is the shear strain rate and the constants C_y and n and the activation energy E_y are different for the upper and lower yield stresses. In certain temperature and stress ranges, n and E_y can be related to the stress dependence of dislocation velocity and the activation energy of dislocation motion, respectively. Stress–strain curves from silicon and GaAs obtained at two different temperatures are shown in Fig. 3, illustrating the upper and lower yield points and their temperature dependence (Sumino 1987a, b). Beyond the yield region, the stress–strain curves of semiconductors are very similar to that of many materials. This similarity underscores the point that the various dislocation interactions that are responsible for the different stages of deformation are more-or-less independent of structure.

4.1 Impurity Hardening

It is well documented that the dislocation velocity and consequently the macroscopic plasticity of semiconductors can be profoundly affected by the addition of donors and acceptors. In silicon and germanium, addition of donors decreases the CRSS or softens the lattice. In compounds such as GaAs, addition of donors such as silicon and tellurium at concentrations at or above $10^{15}\,mm^{-3}$ increases the CRSS or hardens the lattice. Addition of acceptors such as zinc at the $10^{15}\,mm^{-3}$ level produces a mild softening of the lattice. In the case of InP, addition of 10^{15}–$10^{16}\,mm^{-3}$ sulfur and germanium (donors) hardens the lattice. The doping effects on dislocation velocity and plasticity are mostly ascribed to electronic effects rather than to metallurgical effects (Hirsch 1981). In some special cases, metallurgical effects are supposed to be responsible for hardening. Notable examples are the hardening of silicon by the addition of oxygen and nitrogen (Sumino 1987b) and that of GaAs by the addition of indium (Guruswamy et al. 1987).

5. Applications

The mechanical properties in semiconductor-device design first become important when considering the growth of dislocation-free crystals. The primary cause for the source of dislocations in large-diameter semiconductor crystals which are grown from the melt by the Czochralski technique, is the plastic deformation caused by thermal stresses due to inhomogeneous temperature distribution in the growing crystal. A fundamental description of the thermal-stress-induced generation of dislocations in semiconductors has been obtained in terms of a quasisteady-state heat-transfer–thermal-stress model (Jordan et al.

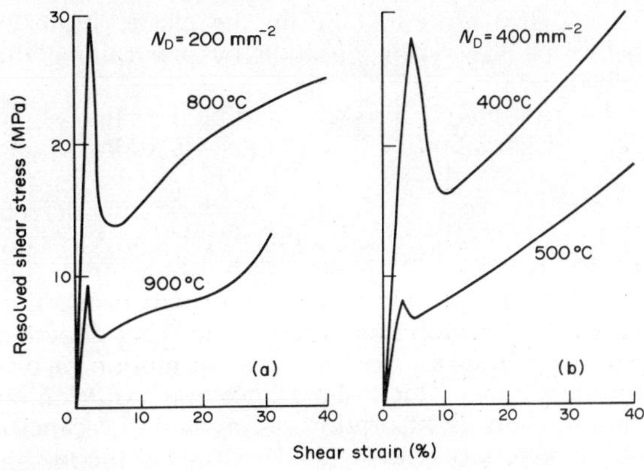

Figure 3
Stress–strain curves for (a) silicon and (b) GaAs single crystals (Sumino 1987a, b): N_D denotes the initial dislocation density in the crystals

1980). According to this model it would be possible to grow large-diameter dislocation-free silicon single crystals but not those of the compound semiconductors, owing to the higher CRSS and thermal conductivity of silicon compared with those of the compounds, consistent with the experimental evidence.

The fabrication of many electronic and photonic devices involves epitaxial growth of layers of different composition whose lattice parameters are not the same as that of the substrate. Such structures consisting of different semiconductor layers are called heterostructures. Examples of heterostructures are ternary $Al_xGa_{1-x}As$ alloy semiconductor grown on GaAs substrates, quaternary $Ga_xIn_{1-x}As_yP_{1-y}$ alloy semiconductor grown on InP substrates, Ge_xSi_{1-x} alloys on silicon and GaAs grown on silicon. The heterostructures of the III–V compounds are widely used in the fabrication of optoelectronic devices required for the optical communication systems. When such heterostructures are made, elastic strains are generated at the heteroboundaries due to the differences in the lattice parameter and thermal-expansion coefficients of the constituents. The lattice mismatch between the layers can be totally accommodated by uniform elastic strain if the layers are below certain critical values. This approach has been used to make a variety of interesting periodically layered crystalline structures known as strained-layer superlattices. These structures have opened up avenues for exploring new material systems and their physical properties, which are influenced by large built-in strains. When the thicknesses of the layers exceed the critical values, the strain is accommodated by the creation of dislocations known as "misfit" dislocations. Since misfit dislocations have deleterious effects on device quality, the growth of thick layers requires ingenious schemes of growing graded composition layers between the substrate and the mismatched layer to take up the elastic strain or achieving stress compensation by prestraining the substrate.

Dislocations are also known to be the cause of the rapid degradation of $Ga_{1-x}Al_xAs$–GaAs lasers and light-emitting diodes (LEDs). This occurs by the growth of what are known as dark line defects (DLDs) in the device area, one of the sources of the DLDs being the threading dislocations from the substrate. The DLDs, which are actually dislocation networks, develop mostly along the [100] direction and are caused by nonconservative motion of dislocations out of their slip planes (i.e., dislocation climb) involving absorption or emission of vacancies and interstitials. The detailed analysis of the microstructure of DLDs in degraded lasers and the conditions for their formation and growth, reducing dislocation density in the substrates and minimizing process-induced stresses, have led to the fabrication of semiconductor lasers for optical communication systems with predicted room-temperature lifetimes of nearly 4000 years!

Device fabrication invariably consists of depositing various dielectric and metal films on semiconductor substrates. Thermomechanical stresses due to differences in the thermal-expansion coefficients of semiconductors and dielectric or metal films often develop and can be a source of problems in device fabrication and packaging. Modelling these strains via thermoelastic theories has helped to choose appropriate film thicknesses and temperatures of film deposition, and to improve device reliability.

See also: Strained-Layer Superlattices (Suppl. 2)

Bibliography

Alexander H, Haasen P 1968 Dislocations and plastic flow in the diamond structure. In: Seitz F, Turnbull D, Ehrenreich H (eds.) 1968 *Solid State Physics: Advances in Research and Applications*, Vol. 22. Academic Press, New York, pp. 28–156

Guruswamy S, Rai R S, Faber K T, Hirth J P 1987 Deformation behavior of undoped and In-doped GaAs in the temperature range 700–1100 °C. *J. Appl. Phys.* 62: 4130–4

Gutmanas E Y, Travitzky N, Plitt U, Haasen P 1979 The mechanical behavior of CdTe. *Scr. Metall.* 13: 293–7

Hirsch P B 1981 Electronic and mechanical properties of dislocations in semiconductors. In: Narayan J, Tan T Y (eds.) 1981 *Defects in Semiconductors: Materials Research Society Symposia Proceedings*, Vol. 2. Elsevier, New York, pp. 25–71

Jordan A S, Caruso R, Von Neida A R 1980 A thermoelastic analysis of dislocation generation in pulled GaAs crystals. *Bell Syst. Tech. J.* 59: 593–637

Nye J F 1957 *Physical Properties of Crystals*. Clarendon Press, Oxford, Chap. 8

Schroter W, Brion H G, Siethoff H 1983 Yield point and dislocation mobility in Si and Ge. *J. Appl. Phys.* 54: 1816–20

Sumino K 1987a Dislocations in GaAs crystals. In: Chikawa J, Sumino K, Wada K (eds.) 1987 *Defects and Properties of Semiconductors—Defect Engineering*. KTK Scientific Publishers, Tokyo, pp. 3–24

Sumino K 1987b Interaction of dislocations with impurities in Si. In: Chikawa J, Sumino K, Wada K (eds.) 1987 *Defects and Properties of Semiconductors—Defect Engineering*. KTK Scientific Publishers, Tokyo, pp. 227–59

A. S. Jordan and V. Swaminathan
[AT&T Bell Laboratories, Murray Hill, New Jersey, USA]

Silicon Carbide Fibers

This article replaces the article of the same title in the Main Encyclopedia.

Silicon carbide fibers have a high structural stability, even at high temperatures. This is indicated by an extreme resistance to oxidation combined with good high-temperature strength, which makes them useful as fiber reinforcement in high-temperature composite materials.

Two different types of silicon carbide fibers exist: substrate-based fibers and fine ceramic fibers. Substrate-based fibers generally have a tungsten filament (SiC/W fiber) or a carbon filament (SiC/C fiber) as the substrate. The thickness of the filament lies in the range 100–150 μm. Fine ceramic fibers are based on silicon carbide, have a diameter of around 15 μm and are produced by the pyrolysis of a polycarbosilane precursor.

1. *Preparation, Microstructure and Morphology of SiC on a Substrate*

The substrate-based fiber is produced in a chemical vapor deposition process in the same type of reaction chamber as that used in boron-fiber production (see *Boron Fibers*) but with multiple injection points for the reactant gases. Various carbon-containing silanes have been used as reactants. In a typical process, with CH_3SiCl_3 as the reactant, SiC is deposited on a tungsten core.

$$CH_3SiCl_3(g) \rightarrow SiC(s) + 3HCl(g)$$

The substrate-based SiC fiber consists of a nearly unreacted core surrounded by a mantle of β-SiC microcrystallites in a preferred orientation where the (111) planes are parallel to the fiber axis. The earliest fibers made in this way had surfaces which were under slight tensile stress, making these fibers relatively sensitive to surface defects. This was overcome by depositing a thin layer of carbon onto the finished fiber, which considerably increased abrasion resistance. Prolonged use at high temperatures (above 1000 °C) produces an increasing reaction between the SiC sheath and the tungsten core, giving rise to α-W_2C and W_5Si_3, and this eventually limits the use of the fiber. The SiC/W fiber exhibits nodules which are smaller than the amorphous boron nodules in boron fibers.

Silicon carbide fibers are now produced on a carbon-filament substrate which has a diameter of 33 μm. The carbon filament is potentially cheaper than tungsten wire and it has been found that faster SiC filament production is possible using this method. The factors which prevent the use of carbon-filament substrates in boron-fiber production do not apply to SiC fiber manufacture.

The SiC/C fiber has a relatively smooth surface (compared with that of the B/W fiber). The density of a 100 μm-thick SiC/W fiber grown on a 12.5 μm-thick tungsten filament is $3.35\ g\ cm^{-3}$, while the density of a SiC/C fiber is somewhat lower ($\sim 3.2\ g\ cm^{-3}$).

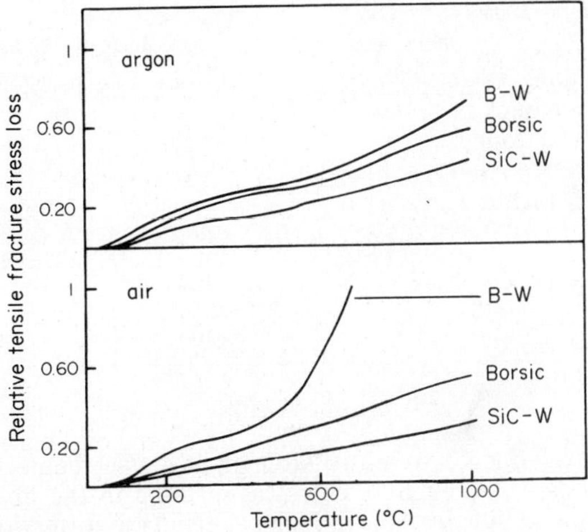

Figure 1
Relative fracture-stress loss at room temperature of different fiber types after heating in air and argon for nine minutes

As mentioned above, sensitivity to surface abrasion was overcome by depositing a layer of pyrolytic carbon on the SiC fiber surface. This has the disadvantage, however, of reducing interfacial bonding, particularly in light alloys. To overcome this difficulty SiC fibers on a carbon-fiber core are produced with a surface layer of pyrolytic carbon which itself is coated with silicon carbide. These fibers are given the designation SCS and basically there are three types: SCS-2, SCS-8 and SCS-6. The type SCS-6 has a thicker final SiC layer which makes it suitable for reinforcing titanium, and shows no degradation after five hours at 900 °C when embedded in a Ti-(6Al-4V) matrix, as demonstrated by Whatley and Wawner (1985).

2. *Properties and Chemical Compatibility*

The room-temperature axial tensile fracture-stress distribution of SiC/W fibers contains a broad maximum with a mean value in the range 3–$4\ GN\ m^{-2}$. Low fracture stresses are caused by surface flaws. At high fracture stresses the fracture is initiated in the core–mantle interface or in the core itself.

The fracture stress of SiC fibers is reduced at higher temperatures. Figure 1 and Table 1 show the reduction of the fracture stress of SiC fibers at relatively low temperatures and short-time exposure conditions and at high temperatures and long-time exposure conditions, respectively. In air, for instance, the SiC/W fiber loses only 20% of its strength after nine minutes' exposure at 700 °C, while a SiC-coated boron fiber (Borsic) loses 25% of its strength and a B/W fiber is completely degraded.

Table 1
Room-temperature fracture stresses of substrate-based SiC fibers after heating in air and argon at 1200 °C for 48 h (after Lindley and Jones 1975)

Fiber	Fiber diameter (μm)	Substrate diameter (μm)	Tensile fracture stress (GPa)		
			Initially	After exposure to air	After exposure to argon
SiC/W	100	12.5	2.99	0.66	0.73
SiC/W	150	12.5	4.00	0.92	0.74
SiC/C	100	33.0	3.99	1.48	0.81

The reduction in tensile strength at higher temperatures is caused by a chemical reaction in the fiber core–mantle interface which results in the formation of α-W_2C and W_5Si_3.

SiC fiber-reinforced composites are mostly used at temperatures above 350 °C. This means that strength and chemical compatibility with metal matrices at high temperatures are of the highest importance. At 400 °C and 800 °C the tensile fracture stress of SiC/W fibers is 90% and 75% of the room-temperature value, respectively. The axial Young's modulus of the SiC/W fibers decreases linearly with increasing temperature from 420 GPa at room temperature to 390 GPa at 600 °C. Above 600 °C the modulus decreases somewhat more rapidly.

Below 350 °C, SiC-coated boron fibers are superior to SiC fibers as reinforcement for titanium. Above 350 °C, titanium-SiC fiber composites have favorable properties in comparison with other composite materials. At high temperatures, however, an interfacial reaction between titanium and the SiC fiber occurs: Ti_3Si, Ti_5Si_3 and TiC_{1-x} are formed in the reaction zone.

3. *Preparation of Fine SiC Fibers*

The manufacture of silicon carbide fibers using a polycarbosilane precursor fiber was first described by Yajima et al. (1976). Polydimethylsilane is made by reacting sodium with dichlorodimethyl silane.

$$n\mathrm{SiCl_2(CH_3)_2} \xrightarrow{\mathrm{Na}} \mathrm{Si}\!\left(\mathrm{(CH_3)_2}\right)_n$$

This is then heated in an autoclave at a pressure of about 10 MPa, resulting in a reorganization of the polymer and the introduction of Si–C into the chain.

$$-\!\left(\mathrm{Si(CH_3)}\right)\!-_n \rightarrow \left(\begin{array}{c}\mathrm{CH_3}\\ |\\ -\mathrm{Si\text{-}CH_2}-\\ |\\ \mathrm{H}\end{array}\right)_n$$

The structure of the polycarbosilane is not linear but consists of cycles of six atoms arranged in a similar manner to the cubic structure of β-SiC.

$$\begin{array}{l}\mathrm{Si}\\ |\quad\diagdown\\ |\qquad \mathrm{C{-}Si}\\ |\qquad\quad \diagdown\\ \mathrm{C{-}Si{-}C}\end{array}$$

The molecular weight of the polymer is, however, low ($M \simeq 1500$) when compared with values of the order of hundreds of thousands for polymers drawn into textile fibers. The polymer is, therefore, more like a paste and is thus extremely difficult to draw into filaments. In addition the methyl (CH_3) groups are not included in the Si–C–Si chain so that during pyrolysis the hydrogen is driven off leaving a residue of carbon.

After synthesis and drawing, the fibers are subjected to heat treatment in air at 200 °C in order to achieve cross-linking of the structure. During this stage, some of the silicon bonds to the oxygen giving Si–O–Si although the alternative Si–O–C can also be formed. This oxidation makes the fibers infusible but has the drawback of introducing oxygen into the polymer which remains after pyrolysis. Ceramic fiber is obtained by a slow increase in temperature, in an inert atmosphere, up to 1300 °C. The fiber which is obtained by this method contains mostly SiC but also significant amounts of free carbon and excess silicon and oxygen probably combined as SiO_2. This route has been adopted by Nippon Carbon to produce a fiber called Nicalon which contains approximately 65 wt% of microcrystalline silicon carbide and which has a diameter of around 15 μm.

A similar process used by Ube Chemicals leads to an amorphous fiber, produced commercially under the name Tyranno, which contains silicon, titanium, carbon and oxygen. The Tyranno fiber is made by first producing a cross-linking organometallic polymer, polytitanocarbosilane. This is synthesized, as described by Yamamura et al. (1987), by dechlorination of dimethyldichlorosilane mixed with titanium alkoxide, heated to 340 °C in N_2 gas and polymerized. Again the molecular weight is low ($M \simeq 1500$). The general structure of the precursor fiber produced is as follows:

$$-\underset{\substack{|\\ H}}{\overset{\substack{CH_3\\ |}}{Si}}-CH_2- \qquad -\underset{\substack{|\\ CH_3}}{\overset{\substack{CH_3\\ |}}{Si}}-CH_3$$

$$-\underset{\substack{|\\ O\\ |\\ RO-Ti-OR\\ |\\ O}}{\overset{\substack{CH_3\\ |}}{Si}}-CH_2 \qquad R = C_nH_{2n+1}$$

traces of other compounds are also found in the fiber.

A ceramic fiber is obtained by heating in N_2 gas up to around 1300 °C. The Tyranno fiber has a diameter of between 8 μm and 12 μm.

Both the Nicalon and Tyranno fibers belong to the new family of fine ceramic fibers and offer the possibility of reinforcing materials for use at high temperatures. Nicalon has been available in relatively small quantities since about 1982 whereas Tyranno is still at a small pilot plant stage.

4. *Microstructure of Fine Ceramic Fibers*

Of the two fibers mentioned above, only Nicalon has been available sufficiently long enough for detailed studies of its structure to be made.

The only elements detected in Nicalon fibers are silicon, carbon and oxygen. Electron microprobe measurements of the intensity of x-ray emission characteristics of the different elements has revealed the distribution of the elements across the fiber diameter. The resolution of this technique is about 1 μm^3 and has shown a uniform distribution of silicon, carbon and oxygen across the fiber. The presence of oxygen across the diameter shows that it was introduced during the oxidation stage, although a fine layer of SiO_2 could also exist on the surface but would not be detectable by the technique employed.

X-ray diffraction studies of the Nicalon fiber reveal only one diffraction peak corresponding to microcrystalline particles of β-SiC having a size of approximately 1.7 nm. Such small crystals have been observed by Simon and Bunsell (1984a) using dark-field transmission electron microscopy. Therefore, the structure of the excess silicon, oxygen and carbon in the fiber identified by electron microprobe analysis is assumed not to be crystalline.

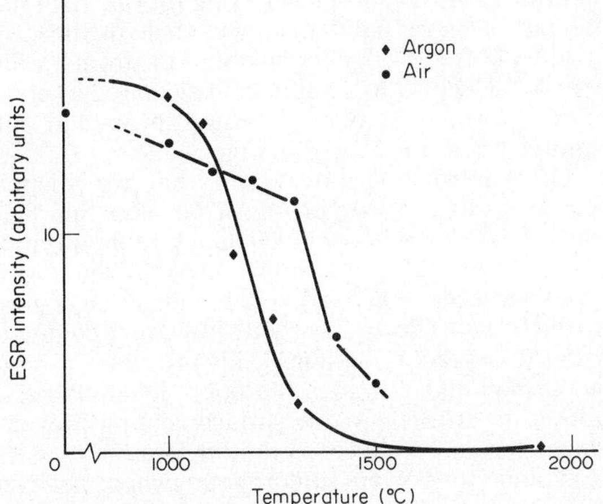

Figure 2
The reduction in the paramagnetic electron spin resonance (ESR) signal obtained from free carbon in Nicalon fibers, after heat treatment for one hour in air and in argon, due to the carbon combining with the oxygen in the structure

A study by x-ray central-beam scattering has allowed segregations of carbon to be identified (as the electron density of carbon is different from that of SiC and SiO_2) and reveals an average size of carbon segregate of about 2 nm. The only paramagnetic phase that can be present in the fiber is free carbon and the detection of a signal by electron spin resonance (ESR) is proof of its existence. The presence of free-carbon segregates in the Nicalon fibers has important consequences in controlling creep at high temperatures.

The structure of Nicalon fibers is found to evolve on heating to high temperatures. X-ray studies reveal that the diffraction peaks narrow on heating above 1200 °C, indicating an improved organization of the structure. Nicalon fibers show an increase in SiC grain size on being heated to this temperature. Grain growth is found to be slower when the fiber is under load and, when under no load, heating can lead to shrinkage. Grain growth stabilizes with a mean grain size of 3 nm. The rate of recrystallization is not affected by the environment in which the fiber is heated; however, an oxide surface layer has been seen to develop during heating in air above 1100 °C.

Under heating conditions which lead to a stabilized grain structure, the mechanical properties of Nicalon fibers are still observed to deteriorate after stabilization. This has been shown by Simon and Bunsell (1984b) to be a consequence of internal reactions between the excess carbon and oxygen in the structure.

Changes in the carbon content of Nicalon fiber were followed during heating between 1000 °C and 1500 °C, using the ESR technique. Figure 2 shows that the intensity falls after heating for one hour and the fall is more rapid in argon than in air. The quantity of free carbon in the fiber falls from a value of about 30 mol% to a minimum calculated as about 15 mol% when the fiber is heated in air and only 5 mol% when it is heated in argon.

It is clear from the above results that two independent processes are occurring in Nicalon SiC fiber when it is heated above 1000 °C. A reorganization of the structure occurs above 1100 °C, leading to shrinkage under low load and stabilizing at a grain size of 3 nm. The free carbon, however, combines with the oxygen in the fiber. This process consumes the carbon and degrades the fiber. Degradation is slowed in air as the oxide surface-coating prevents outgassing of the oxides of carbon formed.

Tyranno fibers are amorphous when received from the manufacturer; however, heating above 1000 °C results in a crystallization of the structure. The presence of titanium, up to about 5 wt%, is said to retard crystallization and reduce reactions between the fiber and an aluminum matrix.

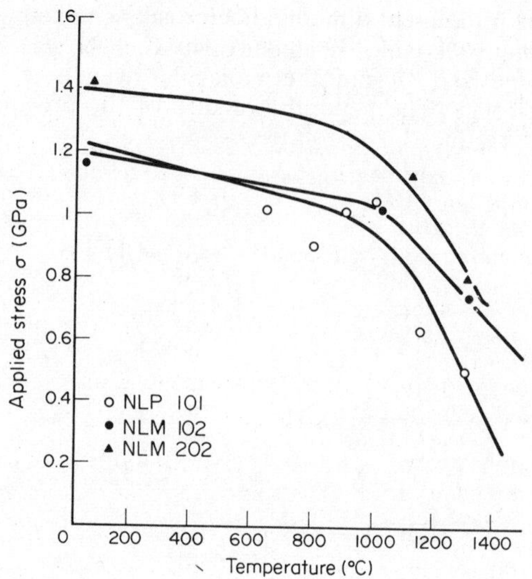

Figure 3
The strengths of three commercially available Nicalon fibers as a function of temperature

5. *Properties of Fine SiC Fibers*

The tensile strength of fibers always shows considerable scatter which can be described in statistical terms by a parameter, usually called the Weibull modulus. The higher the value of this modulus the less the scatter, so that steels can have a Weibull modulus of 60 whereas the corresponding values for ceramics are rarely above 20. Fine ceramic fibers have Weibull moduli of around four, showing that considerable scatter exists in their tensile properties. Nicalon fiber has a Young's modulus of around 190 GPa and a strength of 1.5 GPa with a density of 2.55 g cm^{-3}. Its tensile behavior is linearly elastic.

Tensile tests on Nicalon fibers in both air and argon have shown that the fibers maintain their strengths and Young's moduli up to 1000 °C. This can be seen from Fig. 3. Each point represents an average of at least 30 tests.

The fracture surface of the fibers as seen by a scanning electron microscope does not alter in appearance over the temperature range studied (up to 1400 °C) and is of a brittle nature. A smooth mirror zone of stable crack-growth is followed by the irregular fracture surface characteristic of rapid failure.

Nicalon SiC fibers have been observed to creep at temperatures above 1000 °C. At 1000 °C, creep is not observed if the applied stress is below 0.4 GPa and the time-dependent strain is less than 10^{-8} s^{-1}. Creep curves obtained are of a classical form obeying the equation

$$d\varepsilon/d\tau = A(\sigma - \sigma_0)^n \exp(-\Delta H/RT)$$

where $d\varepsilon/d\tau$ is the creep rate, σ is the applied stress, σ_0 is the creep threshold stress below which creep is not observed, ΔH is the activation energy of the mechanisms controlling creep, T is the temperature, and n and A are adjustable parameters.

The creep threshold level is found to decrease as the temperature is increased above 1000 °C so that it is practically nonexistent at 1300 °C. This threshold stress level is thought to be due to the particles of free carbon in the structure inhibiting movement. A typical activation energy at 1100–1300 °C is 490 kJ mol^{-1} although in air this value becomes rather lower.

Nicalon fibers have been reported by Favry and Bunsell (1987) to react at relatively low temperatures when in contact with molten metal so that the use of these fibers depends very much on the matrix environment in which they are embedded.

Tyranno fibers have a density of 2.4 g cm^{-3}, a room-temperature strength of 2.9 GPa and a Young's modulus of 140 GPa. The strength of Tyranno fibers falls at temperatures above 1000 °C although the presence of the titanium in its structure may retard its degradation (Okamura 1987).

See also: Boron Fibers

Bibliography

Crane R L, Krukonis V J 1975 Strength and fracture properties of silicon carbide filament. *Am. Ceram. Soc., Bull.* 54: 184–8

Favry Y, Bunsell A R 1987 Characterisation of Nicalon (SiC) reinforced aluminium wire as a function of temperature. *Compos. Sci. Technol.* 30: 85
Lindley M W, Jones B F N 1975 Thermal stability of silicon carbide fibers. *Nature (London)* 255: 474–5
Mah T I, Mendiratta M G, Katz A P, Mazdiyasni K S 1987 Recent developments in fibre-reinforced high-temperature ceramic composites. *Am. Ceram. Soc., Bull.* 66: 304
Okamura K 1987 Ceramic fibers from polymer precursors. *Composites* 18(2): 107
Simon G, Bunsell A R 1984a Mechanical and structural characterisation of the Nicalon SiC fibre. *J. Mater. Sci.* 19: 3649
Simon G, Bunsell A R 1984b Creep behaviour and structural characterisation at high temperatures of Nicalon SiC fibres. *J. Mater. Sci.* 19: 3658
Whatley W, Wawner F E 1985 Kinetics of the reaction between SiC (SCS-6) filaments and Ti-(6Al-4V) matrix. *J. Mater. Sci. Lett.* 4: 173–5
Yajima S, Hasegawa Y, Hayashi J, Limura M 1978 Synthesis of continuous silicon fibre with high tensile strength and high Young's modulus. *J. Mater. Sci.* 13: 2569–76
Yajima S, Okamura K, Hayashi J, Omori M 1976 Synthesis of continuous SiC fibers with high tensile strength. *J. Am. Ceram. Soc.* 59: 324
Yamamura T, Hurushima H, Kimoto M, Ishikawa T, Shibuya M, Iawa T 1987 Development of new continuous Si–Ti–C–O fiber with high mechanical strength and heat resistance. In: Vincenzini P (ed.) 1987 *High Tech. Ceramics*. Elsevier, Amsterdam, p. 737

A. R. Bunsell
[Ecole des Mines de Paris, Evry, France]
J.-O Carlsson
[University of Uppsala, Uppsala, Sweden]

Silicon: Preparation for Semiconductors

Silicon used for semiconductor device applications requires seven major processes to transform the ore (i.e., quartzite) to single crystal silicon. The crystals are cut into wafers from which devices are fabricated. These seven processes are:

(a) reduction of quartzite or other high-silica-content materials to metallurgical grade (MG) silicon,

(b) conversion of the MG silicon to silanes (i.e., $SiHCl_3$ or SiH_4),

(c) purification of the silanes or chlorosilanes by distillation,

(d) formation of semiconductor-grade silicon by reduction of the silanes or chlorosilanes by chemical vapor deposition (CVD) or fluidized bed processes,

(e) growth of single crystal silicon from the semiconductor-grade polysilicon,

(f) shaping and polishing in wafers, and

(g) silicon epitaxial deposition.

The MG silicon is reduced from its ore by carbon in submerged-electrode arc furnaces. The reduction process can be formulated as

$$SiO_2 + 2C \rightarrow Si + 2CO \qquad (1)$$

The MG silicon is normally 98–99% pure. This material is then purified to 99.99999% by the formation of silanes, distillation and reduction.

1. Semiconductor-Grade Polysilicon

Semiconductor (sometimes referred to as electronic-grade) silicon is the purest material ever made commercially in large quantities. Material is routinely produced with impurity concentrations of boron ≤ 0 ppba, of phosphorus ≤ 0.3 ppba and of carbon ≤ 0.5 ppma. The control of impurities is one of the most important aspects in preparing silicon for semiconductor device applications.

The demand for semiconductor silicon has increased at a 16% compound annual growth rate. Approximately 6000 t were used in 1987. Even with this very healthy growth rate the total use of semiconductor silicon is only a small fraction (~1%) of the total MG silicon produced. The major uses are as alloying additions to aluminum (60%), iron (5%), silicones (25%) and others including semiconductors (10%). From this it can readily be seen that semiconductor silicon is dependent upon other metallurgical and chemical industries.

The overall process for converting quartzite to silicon single crystals is shown in Fig. 1. The broken pieces of polycrystalline silicon (called nuggets in the

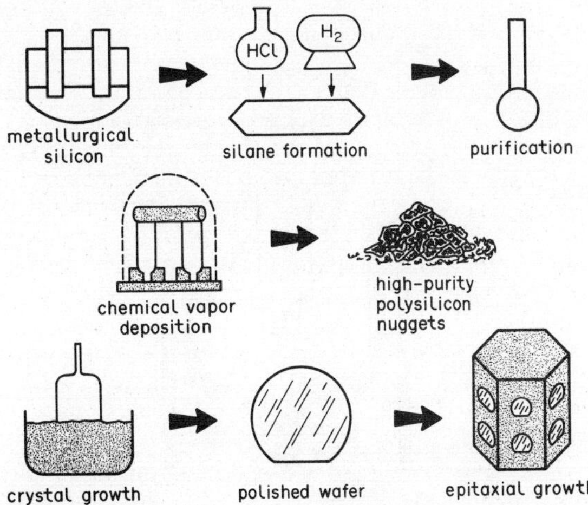

Figure 1
Conversion of quartzite to silicon single crystals

industry) are prepared by crushing the CVD deposited silicon. Fluidized bed silicon has been developed that forms beads (1 to 2 mm in diameter) of silicon that eliminate the need for the crushing process and has possibilities for use as feed stock in various proposals for continuous crystal growth.

2. Single Crystal Growth

The basic requirement of bulk semiconductor silicon for device fabrications is for a single crystal with high crystalline perfection and with the desired charge carrier concentration. Semiconductor-grade polysilicon is converted into *p*- or *n*-type single crystals through substitutional incorporation of dopants during controlled solidification from appropriately doped molten silicon. The primary melt growth techniques used for the production of cylinder-shaped ⟨111⟩ or ⟨100⟩ oriented silicon single crystals are Czochralski pulling and float zoning. Czochralski (CZ) silicon is almost exclusively used for integrated circuit fabrication (large-scale integration (LSI), very-large-scale integration (VLSI), ultralarge-scale integration (ULSI), etc.) while float zone (FZ) silicon is frequently the choice for discrete power devices.

2.1 Czochralski Growth

(*a*) *Equipment*. A modern CZ silicon grower consists of two major parts: a grower hot zone, as shown in Fig. 2, and a pull-chamber above the hot zone to provide housing for the pulled crystal. The silicon charge is contained in a fused silica crucible, supported by a rotating susceptor–pedestal assembly. The assembly is heated by a circular picket-fence-type heater surrounding the assembly. The hot zone parts are machined from high-purity, extruded or hydrostatic-pressed graphite. The entire hot zone and pull chamber are maintained in a protective argon ambient under reduced pressure (~20 torr) during crystal growth. The commercial growers can provide silicon charge sizes of 45–75 kg for growth of silicon crystals 125–200 mm in diameter.

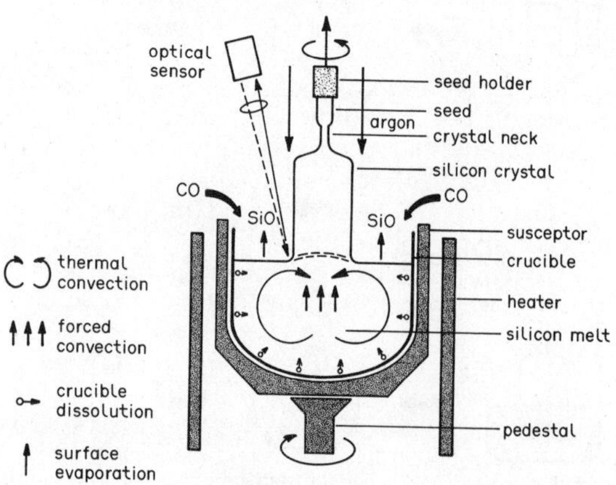

Figure 2
Grower hot zone

(*b*) *Dislocation-free growth*. CZ growth begins with melting of the initial polysilicon nuggets in a fused silica crucible. A predetermined amount of dopant is added in the initial charge or after the melting is completed. After thermal stabilization, a rotating crystal seed with desired orientation (usually ⟨100⟩ or ⟨111⟩) is lowered to make contact with the melt. The melt temperature is adjusted gradually to first cause a slight meltback of the seed and then to pull a small diameter (~3 mm) crystal until a pull speed of approximately 25 cm h^{-1} is obtained for a 3–5 cm length. This procedure establishes a dislocation-free structure for the growing seed crystal, based on a method first demonstrated by Dash. The initial growth of a small diameter crystal minimizes the radial thermal stress which is the major cause of dislocation formation. The fast pulling annihilates any existing dislocations by enhancing their movement to the periphery of the growing crystal along [110]/(111) slip systems. Once the dislocation-free feature is established, which usually displays a distinct crystallographic surface morphology, the desired crystal diameter is achieved by a combination of pull rate, temperature adjustment and crystal/crucible rotations. The constant diameter of the crystal body is maintained through an automatic diameter control (ADC) mechanism involving crystal pull rate and/or temperature change slaved to optical diameter measurements. The hot zone thermal characteristics of the grower (i.e., radial and vertical thermal gradients of the silicon melt and the crystal cooling environment) are the major factors in maintaining dislocation-free growth and cylindrical crystal shape.

(*c*) *Dopant distribution and normal freezing*. During CZ silicon growth intentionally added dopants, such as boron, arsenic, phosphorus and antimony, are subject to segregation, due to their solubility difference in solid and liquid, across the interface. The ratio of dopant concentration in the solid to that in the melt (under equilibrium conditions) is referred to as the equilibrium distribution coefficient k_0. The values of k_0 for various elements in silicon are given in Table 1. When k_0 is less than 1, as is the case for most dopants in silicon, the dopant concentration in the melt increases as growth proceeds. The axial dopant concentration of the crystal follows the normal freezing equation when complete mixing occurs:

$$C(x) = k_0 C_0 (1-x)^{k_0-1} \tag{2}$$

Table 1
Distribution coefficients k_0 for various elements in silicon

Element	k_0	Element	k_0
Aluminum	0.0020	Lithium	0.01
Arsenic	0.3	Manganese	$\sim 10^{-5}$
Gold	2.5×10^{-5}	Molybdenum	4.5×10^{-8}
Boron	0.80	Nickel	3×10^{-5}
Bismuth	7×10^{-4}	Oxygen	~ 0.3
Carbon	0.007	Phosphorus	0.35
Cobalt	8×10^{-6}	Sulfur	10^{-5}
Cesium	4×10^{-4}	Antimony	0.023
Iron	8×10^{-6}	Tin	0.016
Gallium	0.008	Tantalum	10^{-7}
Germanium	0.33	Zinc	$\sim 10^{-5}$
Indium	4×10^{-4}		

where $C(x)$ is the dopant concentration of solid formed after fraction x has solidified and C_0 is the initial dopant concentration of the melt. When growing silicon involving volatile dopants with small values of k_0, excessive segregation is reduced under reduced pressure; that is, the axial dopant distribution would deviate from Eqn. (1).

In practice, the melt is not in complete mixing conditions. The dopant concentration tends to build up near the interface, resulting in higher incorporation than that predicted by k_0. An effective distribution coefficient k is used for Eqn. (2):

$$k = k_0\{[k_0 + (1-k_0)\exp(-f\delta/D)]\}^{-1} \qquad (3)$$

where f is the growth rate, δ is the effective thickness of the boundary layer at the interface and D is the diffusivity in the liquid. Thus, the effective distribution coefficient takes into account the growing conditions related to melt convection and growth rate. A high dopant concentration at the interface can cause interface instability, caused by constitutional supercooling, resulting in high dopant nonuniformity and loss of dislocation-free structure. Constitutional supercooling usually happens before the concentration reaches the solid solubility limit.

(*d*) *Microsegregation*. The existence of a finite non-central-symmetrical thermal distribution in the melt and thermal convection driven temperature fluctuations give rise to growth rate variations. As a result, dopant concentration fluctuations occur along the growth axis, forming microscopic dopant inhomogeneities (striations). Thermal asymmetry gives rise to periodic fluctuations while those due to thermal convections are mostly random. Turbulent thermal convection and the resulting dopant microsegregation can be effectively suppressed by growing the crystal under an externally applied vertical or horizontal magnetic field.

(*e*) *Microdefects*. In the growth of dislocation-free silicon crystals, small lattice imperfections may be formed. Microdefects range from dislocation loops of a fraction of a micrometer to small chemical inhomogeneities undetectable with electron microscopy. The grown-in microdefects provide nucleation sites for oxidation-induced stacking faults (OSF) and oxide precipitates upon thermal treatment. The formation mechanism of microdefects is not clear. There is strong evidence to indicate that temperature fluctuations at the growth interface are the primary cause of these defects.

(*f*) *Oxygen, carbon and other impurities*. During CZ growth, silicon melt reacts with the silica crucible wall and slowly dissolves its oxygen. A great majority of the dissolved oxygen is evaporated through the melt surface in the form of SiO and only a small portion is incorporated into silicon during growth. The oxygen concentration in the melt is the result of a dynamic equilibrium between dissolution, surface evaporation, and the state of thermal and of forced convection. Typically, the interstitial oxygen concentration in a grown CZ crystal is of the order of $10^{18}\,cm^{-3}$. Interstitial oxygen forms donor complex SiO_x when annealed between 400 °C and 550 °C. Oxygen donors may be annihilated or reduced by a high-temperature anneal. Residual oxygen donors and residual impurities prevent the use of the CZ method for growth of high-resistivity silicon. Since the amount of oxygen incorporated significantly exceeds its equilibrium solubility at silicon processing temperatures (900–1200 °C) oxygen tends to precipitate and form oxide–dislocation complexes and/or bulk stacking faults. Crystalline defects created by oxygen precipitation are shown to be effective gettering centers for harmful metallic contaminants. To facilitate intrinsic gettering by oxide precipitates, significant effort has been devoted in CZ silicon growth to control the oxygen incorporation level, as well as its axial and radial uniformity.

Carbon in silicon originates from the graphite parts used in the grower hot zone. It is incorporated into silicon as a substitutional impurity. Carbon has been found to enhance oxygen precipitation during heat treatment. In growth employing high-purity polysilicon and reduced ambient pressure, carbon incorporation is reduced to a minimum (<0.2 ppma). Other electrically active metallic impurities can enter silicon melt during growth. Due to their small distribution coefficients, of the order of 10^{-6}, their incorporation via solidification is usually insignificant (<5 ppba). However, significant metallic contamination can enter the surface of grown CZ crystal in the grower via vapor-phase transport and solid-phase diffusion at high temperature. Sources of metals for evaporation are grower components exposed to high temperature.

(*g*) *Czochralski growth under applied magnetic field (MCZ)*. The application of a magnetic field across an electrically conductive melt effectively increases the viscosity of the melt and thus suppresses the thermal convection. Several effects on crystal properties can result. In general, the temperature fluctuation related growth rate variations are reduced, resulting in reduced impurity striations. The effects on impurity incorporation behavior vary widely depending upon the strength of the applied magnetic field and its direction with respect to the melt. The value of the MCZ for high-volume, large-capacity growth for VLSI/ULSI applications is not clear. The trend in VLSI/ULSI complementary metal-oxide semiconductor (CMOS) devices favors epitaxial over bulk structures for better reliability for the fabrication of latch-up free devices. In this view, MCZ material is expected to have less of an impact on epitaxial substrates. However, MCZ materials are shown to be beneficial to the fabrication of some devices in which high-resistivity or low-impurity striation/microdefects are needed.

(*h*) *Future trends*. Silicon grower capacity has been scaled up in proportion to the steady increase in crystal/wafer diameter as demanded by the economics of integrated circuit manufacturing. Figure 3 shows silicon crystal/wafer diameter as a function of date of introduction of the technology. A 200 mm diameter crystal growth from a 60 kg charge was not uncommon in 1988. Experimental work has shown that it is possible to grow dislocation-free, 300 mm diameter crystals with the proper thermal environment in which thermal stress of the growing crystal is minimized. For high volume production of large diameter crystals, the logical extension of the batch process is the continuous-feed growing system involving one or more containers. The production of high-purity silicon beads (1–2 mm in diameter) by the fluidized bed method has greatly aided the development of the continuous-feed process.

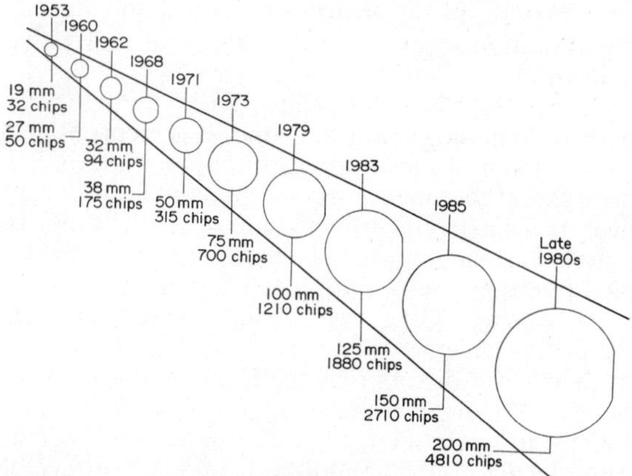

Figure 3
The increase of silicon wafer diameter and of the number of 2.5 mm × 2.5 mm chips possible on a wafer, with year

Table 2
Typical mechanical specifications for a 150 mm diameter silicon wafer

Orientation	(100)±1°
Diameter	150±0.20 mm
Thickness	675±25 μm
Fiducial mark	flats
Primary flat orientation	⟨110⟩±1°
Length	57.5±2.5 mm
Secondary flat orientation	90° cw±1°
Length	37.5±2.5 mm
Wafer identification	alpha numeric
Total thickness variation (TTV)	5 μm
Warp	⩽25 μm
Flatness—local site back reference	
Focal plane deviation	⩽0.7(±0.35) μm
Site size	20 mm × 20 mm
Percentage site in specimen	⩾90
Edge chips	0>0.25 × 0.25 mm
Edge contour	SEMI-std.
Surface scratches (number, length)	1, 1 cm
Particles	0.03 cm^{-2}>0.2 μm in size

2.2 Float Zoning

Silicon growth using the float zoning method is based on the high surface tension of silicon melt and levitation by the electromagnetic field effect of the radio frequencies in the kHz to mHz range. Since it is growth without a crucible, float zoning can produce high-resistivity silicon, in excess of 100 Ωm. Float zone technology can grow dislocation-free silicon with diameter ⩾125 mm. High resistivity, or pure, float zone silicon may be used to produce a high-resistivity phosphorus-doped crystal via neutron transmutation doping (NTD), based on the transformation of silicon isotope ^{30}Si (abundance ~3% in natural occurrence) into ^{31}P upon neutron absorption and subsequent β^- decay. The NTD process yields high dopant uniformity.

3. Wafer Process Technology

After the crystals (called ingots in the silicon industry) are grown they are converted into polished wafers using machining and chemical processes. These processes have to be controlled to meet various specifications. A typical specification for the mechanical properties of a 150 mm diameter wafer is shown in Table 2. There are also about the same number of structural, chemical and electric parameters specified on a silicon wafer used for leading edge integrated circuit technology. The mechanical dimensions are based on the needs of the various processing requirements, the capabilities of the various shaping equipment and the cost objectives. The major operations that are needed to prepare a silicon wafer suitable for integrated circuit processing or epitaxial film growth are shown in Fig. 1.

3.1 Wafer Shaping

(*a*) *Grinding*. The first operation after the ends of the crystal have been cut off is to grind the crystal to a specified diameter. Diameter control during crystal growth is not precise enough to meet the wafer specifications; therefore, crystals are grown oversize and then machined to diameter. Various grinding processes are used such as belt and wheel centerless grinding or wheel center grinding. Since silicon is a hard brittle material (1000 kg mm^{-2}), diamond abrasives are commonly used. Wheel center grinding (see Fig. 4) is a commonly used process for grinding silicon crystals to a precise diameter. The process is accomplished by passing a small quickly revolving diamond cup wheel over the crystal as it is rotated about its center. Crystals are ground slightly oversized as the final diameter is controlled by the edge contouring and etching processes.

(*b*) *Flat grinding*. Following diameter grinding, a major flat is ground on the crystal in a specified crystallographic direction. The orientation of this flat is determined by x-ray diffraction techniques. Secondary flats which are smaller than the primary flat are usually ground on the crystal to help keep various orientations and conductivity types readily identifiable. The secondary flat may be put on by surface grinding similar to that for the primary or, as is more customary, it can be put on during the edge contouring process. The primary flat is used as a mechanical locater in automated processing equipment and allows the integrated circuit to be aligned with crystallographic planes of the wafer. Although flats are the principal fiduciary marking system, some manufacturers prefer notches.

(*c*) *Slicing*. Once these operations are completed, the crystal is ready for slicing into wafers. This is an important process because it controls the bow or warp of the wafer, the initial thickness and the surface orientations (i.e., the atomic arrangement of atoms at the surface). Silicon crystals are cut into wafers with a very unique saw called an I.D. saw (see Fig. 4). The saw uses a blade with the diamonds used for cutting on the circumference of a hole in the center of the cutting blade. A typical diamond blade is fabricated from a stainless steel core (0.15–0.2 mm thick). The diamonds are attached to the stainless steel by nickel plating. The stainless steel core is tensioned in a collar similar to a drum head. This keeps the thin stainless steel blade in a flat plane and

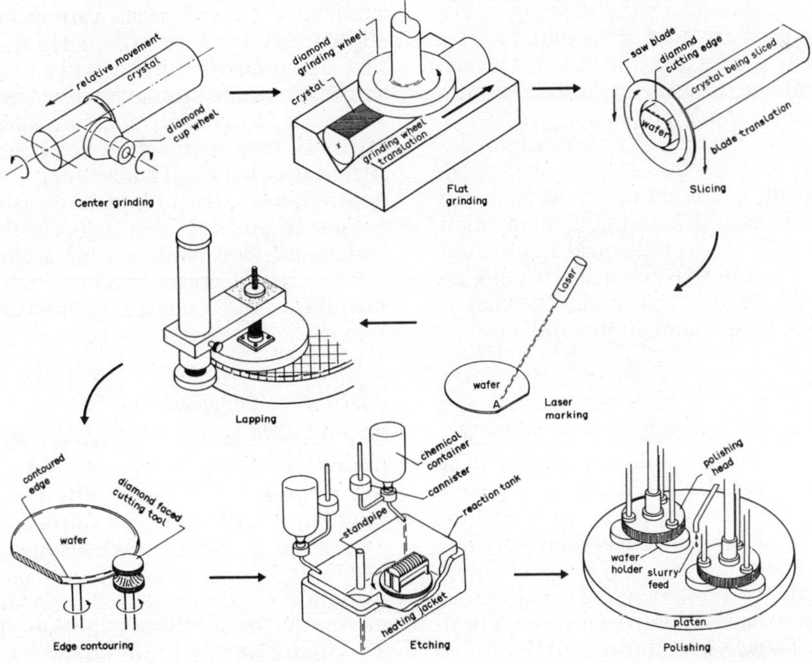

Figure 4
The wafer shaping process

also keeps the hole round. Diamond blade technology is very important in control of the tolerances of the as-cut wafer.

(*d*) *Laser marking*. After sawing, the wafers are normally inscribed with a small laser mark near the primary flat. The laser mark which is made by melting away portions of the wafer is to store information about the wafer so that it can be readily identified automatically during subsequent operations. The laser mark allows the user to be able to tell who manufactured the wafer, its conductivity type, orientation and resistivity, as well as containing a unique number.

(*e*) *Lapping*. Following this operation the wafers are normally lapped to produce the thickness uniformity and flatness requirements. This lapping process is usually accomplished with a double-sided lapping machine (see Fig. 4) using Al_2O_3 abrasives.

(*f*) *Edge contouring*. The wafer edges are contoured after lapping (see Fig. 4). This is done to prevent chipping of the wafer edges during the many (~150) process steps needed to produce an integrated circuit. Edge contouring is done with a specially shaped diamond wheel. The edge contouring machines are normally of the cam follower type which allows the edges to be contoured, the diameter to be adjusted to final dimensions and the secondary flats to be ground as the wafer rotates against the wheel.

(*g*) *Etching*. After these mechanical shaping operations, the wafers are etched to remove the damage and contamination inherent in the shaping operations. Various etchants are used in the industry. The etchants are usually acid or alkaline based. The acid etchants are usually mixtures of hydrofluoric, nitric and acetic acid. Hot alkaline etchants are usually various concentrations of sodium or potassium hydroxide. In either system the etching has to be controlled so that the planarity of the wafer established by the mechanical processing is maintained. The major differences between the acid and hydroxide etching processes are that the acid etchants are isotropic and produce exothermic reactions whereas the alkaline etchants are anisotropic and produce endothermic reactions.

3.2 Polishing

The etching process is then followed by polishing. This process is a challenge for submicrometer technology because the specular surface must be essentially defect free (i.e., no scratches, etc.) and the planarity (flatness) of the surface obtained by lapping must be maintained. Photolithographic equipment used to focus the photoengraved patterns on the wafer has a very shallow depth of focus (~1 μm). Ideally, the wafer flatness over the area to be printed should be about 25% of the focus depth. This means the area that is to be printed would have to be parallel to within ±0.25 μm to the lens focal plane forming the pattern upon the wafer. This stringent requirement is a critical issue because the metrology to measure this routinely is currently being developed. Polishing is normally accomplished by mounting the unpolished wafers into a fixture (see Fig. 4). The fixture holding the wafer is pressed against a rotating platen which is covered with a polishing pad. A colloidal silica slurry is fed continuously to the platen. Silicon wafer manufacturers use the terminology chem-mechanical polishing to describe the process, derived from what takes place during the polishing. The alkaline nature of the slurry causes a complex silica to form on the silicon wafer surface which is continuously removed by the mechanical action of the nanometer-sized particles of silica that are in the slurry and by the action of the polishing pad being pressed against the wafer. Typical operations usually use a two-step process. The first step, normally referred to as the rough or stock removal process, usually removes about 25 μm of silicon. The second step, the finishing polish, is done in a similar manner but may use other slurry concentrations, pads and pressure to produce a high specular finish with little removal of material.

3.3 Cleaning

After the wafers are polished they must be cleaned. This is an extremely important part of the process. The cleaning processes are designed to remove organic material, metallic contamination and particulates. The cleaning processes are normally done in automated systems using various aqueous mixtures of NH_4OH–H_2O_2 or H_2SO_4–H_2O_2, and HCl–H_2O_2. The ammonia or sulfuric acid solutions are used to remove organic contamination and the HCl–H_2O_2 solutions to remove metallic contaminants. These chemical treatments are followed by rinses in very-high-purity water. The cleaning processes must be controlled so that only a few particles larger than 0.2 μm in size are left on the wafer. The cleaned wafers are then packaged for use in circuit fabrication or have epitaxial deposition applied if the circuits are to be made into epitaxial layers.

4. Epitaxial Deposition

The deposition of single crystal silicon layers upon silicon substrate wafers (referred to as epitaxial deposition) has been a principal processing tool since the early days of silicon technology. This deposition process allows the device designer the flexibility of having a lightly doped region in which to fabricate the active device directly above a uniformly, or selectively, heavily doped substrate. This multilayer structure provides enhanced electrical performance superior to that of devices constructed in a uniformly doped single crystal wafer.

The epitaxial layer retains the crystal structure of the underlying substrate while allowing arbitrary vertical dopant control in the epitaxial layer as well as producing a layer with lower levels of carbon, oxygen and certain metallic impurities compared to those usually found in the substrate. Epitaxial layers are grown from thicknesses of 0.003 μm to thicknesses of greater than 100 μm with growth rates varying from 0.001 μm min^{-1} to 5 μm min^{-1} and deposition temperatures spanning the range 500–1250 °C.

Typically the silicon layer is deposited from a source gas that is either chemically reduced with H_2 or thermally decomposed at the deposition temperature. Almost all silicon chemical vapor deposition (CVD) uses one of four silicon source gases: silane (SiH_4), dichlorosilane (SiH_2Cl_2), trichlorosilane ($SiHCl_3$) or silicon tetrachloride ($SiCl_4$). These silicon source compounds are used because of their high levels of purity, low cost and ease of deposition. The electrical conductivity of the deposited layer is controlled by the introduction of an electrically active dopant provided from a gas containing either arsenic, phosphorus, or boron; typically, these gases would be arsine (AsH_3), phosphine (PH_3) or diborane (B_2H_6), respectively.

4.1 Kinetics

Epitaxial deposition can be performed over the temperature range 800–1250 °C, using the silicon source gases. This wide temperature range encompasses two separate deposition zones. In the lower temperature range, reactants reach the surface fast enough, but there is not sufficient thermal energy for complete reaction. Therefore, these deposition processes are limited by surface reaction rate and the deposition rate is exponentially dependent on the temperature. At higher temperatures there is sufficient thermal energy for complete reaction of the species and the reactions are mass transport controlled. In the mass transport region the rate limiting step is the supply of reactant species through the stagnant boundary layer just above the substrate surface or the de-adsorption of reaction products such as HCl and H_2. In the mass transport region the deposition rate increases very slowly with temperature due to the slight increase in gas phase diffusivity with temperature.

As the amount of chlorine increases in the source gas, from silane (SiH_4) to silicon tetrachloride ($SiCl_4$), the reaction goes from thermal decomposition to hydrogen reduction. As the size of the chlorine species increases more thermal energy must be supplied to the system in order to move and de-adsorb these large molecules. In practice this means that $SiCl_4$ needs a much higher deposition temperature than SiH_4 in order to achieve the same surface quality in the resultant deposited layers. Most commercial epitaxial deposition processes operate in the mass transport region to minimize thermal nonuniformity growth rate effects in the epitaxial reactor. While early epitaxial deposition processes operated at atmospheric pressure and high temperatures, more recent work has moved toward a lower deposition temperature which has been achieved by the reduction of the operating pressure from atmospheric to about 0.1 atm (76 torr). Further reductions in the pressure have resulted in SiH_4 decomposition at temperatures as low as 500 °C, while still maintaining single crystal growth without the addition of thermal energy through the use of a plasma. These extremely low temperature CVD reactions require ultralow pressures in the order of 10^{-9} torr.

Silicon epitaxial growth on a microscopic scale occurs in a lateral rather than vertical mode. Silicon-containing molecules diffuse from the gas phase to the surface of the wafer through the boundary layer and then decompose into silicon and other species. The other species will de-adsorb while the silicon atoms will migrate to two-dimensional sites where they are incorporated into the growing layer. Since the deposition process proceeds by lateral growth, high growth rates, low temperatures and high pressures result in poor quality or nonsingle crystal layers.

4.2 Dopant Incorporation

There are two sources of dopant that usually make their way into growing epitaxial layers. The first is the dopant that is intentionally added while the second source is unintentional and can come either from dopant in the substrate ("autodoping," see Fig. 5) or from the reaction chamber. The intentional dopant is produced by thermal decomposition of the hydrides mentioned previously. The dopant incorporation proceeds linearly with partial pressure at low concentrations and decreases as the concentration in the solid exceeds the intrinsic level ($\sim 2\times 10^{19}$ carriers cm^{-3}). Unintentional dopant incorporation from the substrate can occur either through solid-state diffusion or from reincorporation of dopant that is released into the gas phase and then re-adsorbed into the growing layer. While autodoping due to solid-state diffusion can be reduced by lowering the deposition temperature, the autodoping from the gas phase cannot be easily reduced. The technique used to reduce gas-phase autodoping depends on the specific dopant as well as whether the substrate is uniformly or selectively doped. For uniformly boron-doped substrates sealing the back surface of the substrate with a low diffusivity layer (e.g., SiO_2) will drastically reduce the boron autodoping. Autodoping from selectively (buried layer) arsenic-doped substrates can be reduced by lowering the deposition pressure.

The growth of sharp multilayered structures where the layer boundaries are defined by a difference in dopant level or type are difficult to obtain

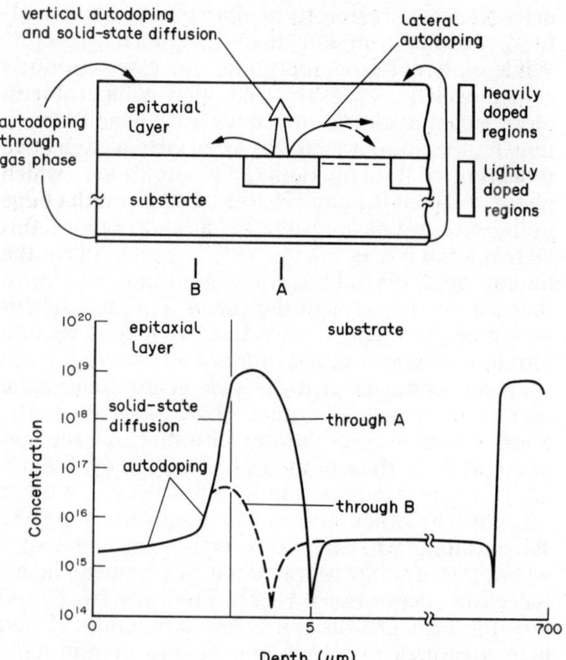

Figure 5
Variation of autodoping with dopant concentration

using epitaxial deposition. This is due to system transients in both the silicon source and the dopant species themselves. High reactant dwell time in batch reactors is common, making the elimination or the change in level of one dopant in the system a rather long process relative to a typical growth cycle which may vary from 5 min to 20 min. In many reactor designs the source and dopant gases pass over the substrates many times before they are eliminated from the system usually by dilution. Solid-state diffusion plays a significant role in dopant transport causing smearing of the interface regions at typical deposition temperatures.

4.3 Pretreatment of the Substrate

In order to produce an epitaxial layer of the highest crystal perfection with very low levels of metallic impurities anhydrous HCl has been used as a gas phase etchant in the temperature range 1100–1200 °C at etch rates in the range 0.05–1.00 μm min^{-1}. Originally HCl etching was employed to remove the damage left by the shaping processes or damage induced by the ion implantation process used to produce the selectively doped buried layer structure. Later, the HCl etch was used to reduce the level of metallic contamination found in the epitaxial layer. Often a bake in flowing H_2 is used in place of the HCl etch. This bake removes SiO_2 that is left on the surface of the water by the chemistry used during substrate cleaning or that is just the native oxide that forms on the silicon surface when exposed to air. Carbon is another contaminant usually found in deposition systems. The carbon can be eliminated by dissolution into the silicon at elevated temperatures. Some researchers have used argon sputter cleaning to remove this impurity from the substrate surface. The problem with the use of energized molecules is that they can induce structural damage into the near-surface region while removing the top layers of the silicon substrate. The use of ultraviolet radiation has also been successful in reducing the levels of carbon and other organics on the substrate surface.

Intrinsic gettering sites in the form of oxide precipitates or bulk stacking faults within the substrate (but not near the growing interface) have been found to reduce the amount of metallic contamination in the epitaxial layer. The high-energy strain fields around these defects attract the rapidly diffusing metallic species. Extrinsic gettering has also been used to reduce the contamination level in epitaxial layers. Polysilicon deposited on the rear surface of the substrate prior to deposition can act as a collection region for metallic species. Barrier layers (e.g., Si_3N_4) on the back of the substrate have also been shown to prevent transport into the substrate of certain elements coming from the graphite susceptor.

4.4 Defects and Surface Morphology

Many types of crystalline defects can be found in the epitaxial layer (see Fig. 6). Some of these defects are propagated from crystalline defects found in the substrate (e.g., edge dislocations). These defects can also be nucleated at a precipitate particle that intersects the surface prior to deposition. Another type of edge dislocation that can be generated is the "misfit" dislocation. These dislocations usually

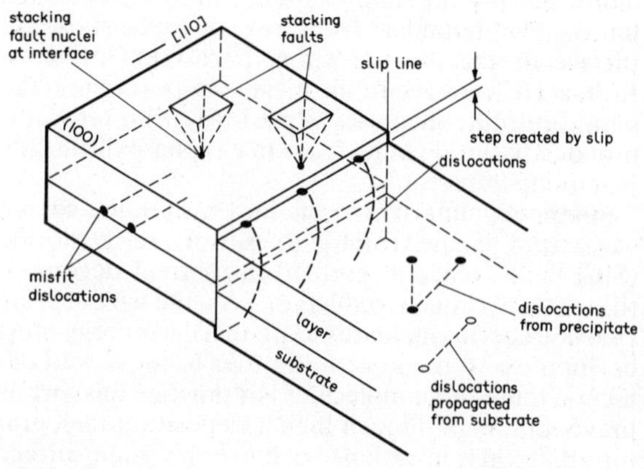

Figure 6
Epitaxial defects

occur in the plane of the interface, and are caused by a rapid change in lattice parameter; that is, a change from a heavily doped (10^{19} boron atoms cm^{-3}) substrate to a lightly doped layer. These defects depend on the layer thickness, as well as the exact doping level of the substrate. Misfit dislocations can be used as an intrinsic gettering region, and can be intentionally generated by the addition of an electrically neutral species, such as germanium, to the early stages of the growing layer.

Reducing the partial pressure of oxygen in an epitaxial reactor can be a major problem, due to the size and design of most commercial deposition systems. The same design criteria that aid in layer-thickness uniformity, can be detrimental when it comes to rapid purging of the system. Therefore, long purge or flush cycles are common in most reaction systems, leading to a total nondeposition process time of 40–70 min. The use of vacuum pumps can shorten these purge cycles, somewhat.

Stacking faults can also occur during epitaxial growth, and can be nucleated by damage to the surface of the substrate, by submicrometer particles or by residual films left on the surface. Another defect found on epitaxial layers is the tripyramid or hillock. This defect is not crystallographic in nature, but is a raised faceted area on the surface of the wafer, resulting from localized enhanced growth. These defects are usually associated with handling damage or residual films left on the surface of the substrate. Tripyramids or hillocks can also be caused by a high level of oxidant in the system. Since the higher the deposition temperature, the more rapidly the SiO_2 is reduced, these defects can be eliminated or reduced by raising the temperature. In modern robotically loaded commercial deposition equipment, the level of these defects is usually below $0.5\,cm^{-2}$.

epitaxial growth. Slip is the displacement of one part of the crystal lattice relative to another, and is accompanied by the insertion of edge dislocations along the slip line, to minimize the strain in the lattice. The density of dislocations is proportional to the misorientation between the two lattice regions. Epitaxial slip is usually caused by thermally induced strain, resulting from radial temperature nonuniformities across the wafer. In radio frequency heated reactors, where the wafer is usually at a lower temperature than the susceptor, warpage of the wafer due to radial nonuniformities can cause the wafer to bow away from the susceptor, increasing the radial temperature gradient even more. This radial situation can be rectified by putting a spherical cavity in the susceptor, which allows susceptor contact with the wafer edges but not the center, thereby preventing radial edge losses by preferentially heating the edges of the wafer. Radiantly heated reactors, where the wafer is hotter than the susceptor, also minimize the amount of slip. Heavily doped *n*-type (100) substrates are more susceptible to slip than *p*+ wafers.

The deposition of epitaxial layers over selectively doped substrates (buried layers), presents problems other than dopant control. In order to align the next mask level to the buried layer pattern, selective oxidation is used to develop a small step, which presents a problem in nonplanar growth, at the edge of the implanted region. The propagation of this step to the top of the layer is influenced by the growth rate, temperature, silicon source gas, pressure and orientation of the surface. The trace of this buried layer step can be both shifted (moved in a lateral direction) or distorted (rounded or smeared) resulting in incorrect placement of the next mask level or the inability to align the next level at all. The correct choices of deposition conditions and substrate orientation can minimize these problems.

4.5 Other Deposition Processes

In order to provide epitaxial materials that meet the challenge of submicrometer device design, novel departures from the CVD and uniform layer epitaxial growth processes have been devised, as well as innovations that result in a reduction in the total time–temperature product of the epitaxial process. Molecular beam epitaxy (MBE) involves the evaporation of silicon and the appropriate dopant(s) directly to the surface of a heated substrate in an ultrahigh-vacuum environment ($<1\times10^{-8}$ torr). MBE also requires the same type of perfect substrate surface required for CVD epitaxy and therefore a high-temperature bake (>1000 °C) or a lower temperature (~800 °C) sputter clean is required as a pretreatment. Deposition temperatures are usually in the range 400–800 °C and growth rates are low so that MBE is used for thin layers requiring very rapid changes in dopant concentration or type.

Lateral separation of adjacent regions in VLSI circuits is limited by the isolation procedures used. Selective epitaxial growth (SEG) (see Fig. 7) allows the deposition of single crystal silicon into regions defined in a SiO_2 layer. The problem with depositing silicon selectively in these oxide holes is that the silicon tends to nucleate and grow concurrently on top of the masking oxide layer as a polycrystalline deposit. It was initially noted that very low growth rates in the $SiCl_4$–H_2 system would result in selective growth in the holes but not on the oxide. Later it was shown that the SiH_4–H_2–Cl and SiH_2Cl_2–H_2–HCl systems can also produce selective growth by controlling the chlorine-to-silicon ratio. As this ratio increases the process becomes more selective while the effect of local lateral geometry increasingly affects the growth rate (higher growth rate with higher oxide-to-silicon ratio). The SEG process suffers from another defect, that of faceting. (311) facet planes become evident along the [110] directions for (100) oriented layers greater in thickness

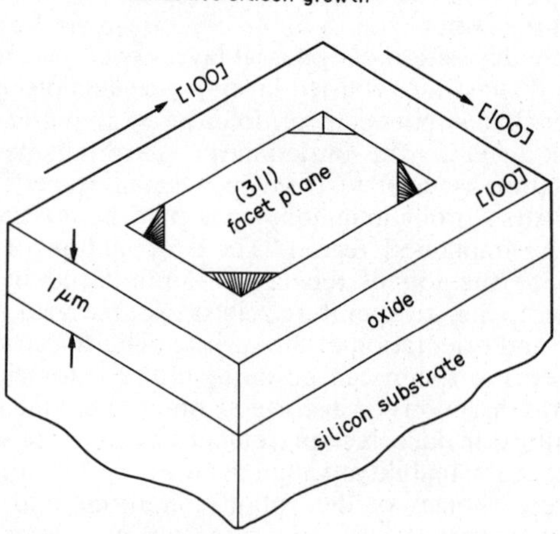

Figure 7
Selective epitaxial growth

than about 0.5 μm. These facets limit the amount of planar silicon surface area produced in the windows and also complicates coverage of these selective regions with subsequent layers. Most SEG layers are produced in the temperature range 850–950 °C with a pressure range of 10–50 torr.

An offshoot of the SEG technology involves the SEG growth as a starting point and then, when the growth has proceeded above the surface of the masking oxide, altering the conditions of the deposition process in order to promote rapid lateral growth while decreasing the vertical growth rate. This process is called epitaxial lateral overgrowth (ELO). The ELO process can provide a fully isolated structure by oxidizing away the silicon in the seed windows leaving only the single crystal silicon on top of the oxide. Problems with this process involve the achievement of rapid lateral-to-vertical growth rates and poor crystalline quality over the masking oxide.

Most epitaxial deposition processes are controlled by switching the reactant gases on and off. In the limited reaction processing (LRP) system the reactant gases are introduced into the reaction chamber at low temperatures and the wafer is very rapidly heated, with quartz–halogen lamps, into the deposition range at rates above 200 °C s^{-1}. This rapid thermal ramp bypasses the region of polycrystalline growth and results in good quality single crystal layers. The main advantage of LRP is the ability to produce rapid changes in dopant concentration making it a competitive process with MBE for thin layers that vary rapidly in dopant concentration over short distances. Problems with this technology involve slipping of the wafer edges due to radiative losses and due to uneven heating, and difficulty in controlling and measuring the wafer temperature.

4.6 Deposition Equipment

There is a large variety of epitaxial deposition equipment (see Fig. 8) available. These reactors use three heating sources: radiant heating from quartz–halogen lamps, 10 kH to 200 kH induction heating and resistance heating. Some reactors employ multiple types of heat sources. Batch system reactors have a capacity from about 8 to 50 150 mm diameter wafers run^{-1}. Process cycle times vary from about 40 min to 80 min. Two major reactors account for the bulk of the commercial systems. These are the radiantly heated barrel reactor produced by Applied Materials (Fig. 8a) and the radio frequency heated pancake reactor produced by Gemini Research (Fig. 8b). Modern reactors all use silicon-carbide-coated graphite susceptors and quartz reaction chambers. Newer reactors (Figs. 8c–e) have entered the market in order to lower the cost while increasing the quality of the epitaxial wafer. These use robotic loading and unloading to reduce particulates on the

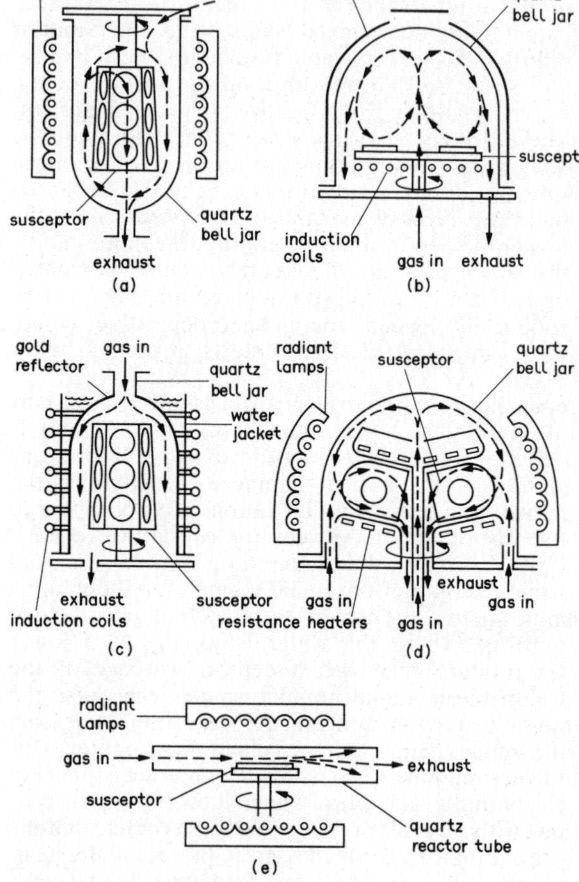

Figure 8
Epitaxial deposition equipment

wafers and minimize damage-induced defects. They also lower deposition costs by approximately a factor of two over existing equipment at the 150 mm diameter wafer level either by increasing throughput or by lowering power and gas usage. Modern reactors are large and expensive, and can exceed 5 m^2 of floor space without power supply, gas handling or pumping equipment. Prices range from US$700 000 to US$2 000 000 plus installation which may be as high as 25% of the reactor cost.

4.7 Future Directions

Silicon epitaxial deposition will continue to be a major processing technology for the production of VLSI and ULSI devices. The trend toward the increasing use of epitaxy in CMOS devices will persist while continuing to reduce the wafer cost. The lowering of the defect and contamination levels, SEG and attempts to improve dimensional control (minimization of flatness degradation) will be the major areas of study.

Bibliography

Cullen G W (ed.) 1987 *Chemical Vapor Deposition*. Electrochemical Society, Princeton, NJ

Grossman L D, Baker J A 1977 In: Huff H R, Sirtl E (eds.) 1977 *Semiconductor Silicon 1977*. Electrochemical Society, Princeton, NJ, p. 18

Jayant Baliga B (ed.) 1986 *Epitaxial Silicon Technology*. Academic Press, London

Keller W, Muhlbauer A 1981 *Floating-Zone Silicon*. Marcel Dekker,

McCormick J R 1986 In: Huff H R, Abe T (eds.) 1986 *Semiconductor Silicon 1986*. Electrochemical Society, Princeton, NJ, p. 43

Pfann W G 1965 *Zone Melting*. Wiley, New York

Sze S M (ed.) 1988 *VLSI Technology*, 2nd edn. McGraw-Hill, New York, pp. 55–140

Zulehner W, Huber D 1982 *Czochralski-Grown Silicon*, Crystals, Vol. 8. Springer, Berlin, p. 1

K. E. Benson
[AT&T Bell Laboratories, Allentown, Pennsylvania, USA]

Slag Raw Materials

The production of metal by melting requires metal ore, fluxing material (usually limestone to assist the melting process) and coke. The yield consists of molten metal and molten slag which normally floats on the metal because of its lower specific gravity. The molten slag is removed from the production area for disposal. The term slag as used in this article refers to the by-product material from the production of iron and steel.

The earliest use of slag as a construction material dates back to Roman times, but it is only since the 1920s that slag has gained wider use as a construction aggregate. Prior to about 1920 slag was a waste product that represented a waste disposal problem. Only after World War II did blast-furnace slag begin to be used increasingly as a construction aggregate. In 1985 the US construction industry used nearly 17.1 Mt of iron and steel slag with over 70% coming from the production of iron. An additional 1.8 Mt were used outside of the construction industry.

The benefits of using slag are threefold. First, the use avoids environmental concerns associated with large-scale disposal problems; second, it eliminates the costs to the metal producers that are associated with the disposal of large volumes of waste material; and third, it supplies an additional source of construction raw materials.

1. Distribution

The bulk of slag production in the USA is located around the Great Lakes in Detroit, Chicago, Gary, Cleveland and Buffalo. Other ironmaking and steelmaking centers include Pittsburgh, Philadelphia, Baltimore, Birmingham and Bethlehem. Besides these major steel production areas, there are several locations where steel is produced at "mini mills," for example, in Florida, Louisiana, New Jersey, Oklahoma and Washington State.

Imported steel has had a major impact on the production and availability of slag in the USA. In 1980 there were 56 plants producing iron in blast furnaces. By 1985 there were 38 domestic blast furnaces. During the same period, the number of open-hearth steel furnaces decreased from 11 to 5; basic-oxygen steel furnaces increased in number from 16 to 22 and the number of electric steelmaking furnaces grew from 19 to 40.

2. Production

Slag that is used by the construction industry comes mainly either from the production of iron (where it is known as blast-furnace slag) or from the production of steel (where it is referred to as steel slag). In both cases the furnace yields two products: molten metal and slag. Slag comes from the furnace at a temperature of around 1500 °C and resembles molten lava. Blast-furnace slag supplies about three times as much material as steel slag.

2.1 Blast-Furnace Slag

The blast furnace is loaded with iron ore, flux stone (limestone and/or dolomite) and coke for fuel. The flux stone serves two purposes: initially it reduces the temperature necessary to melt the ore; then, as the iron begins to melt, the slag (molten limestone) serves as a scavenger to collect (in solution) the impurities from the ore and from the coke. The slag consists essentially of silica and alumina impurities from the iron ore, and lime and magnesia from the flux stone. Since slag is lighter than iron, it floats on the molten metal. The molten slag is removed from the blast furnace and undergoes one of three cooling processes (air-cooled, granulated or expanded slag), each of which yields a different product.

(*a*) *Air-cooled blast-furnace slag*. The great majority of blast-furnace (iron) slag is dumped into slag pits and allowed to air-cool. In 1985 air-cooled slag accounted for about 90% of all blast-furnace slag in the USA. The resulting product is a vesicular, fine-grained, mainly crystalline material. After cooling, it undergoes crushing and screening similar to other crushed-stone materials that are used for construction aggregates. It is used for road-base material, coarse aggregates in asphalt and Portland cement concrete, railroad ballast and roofing aggregate.

(*b*) *Granulated blast-furnace slag*. Granulated slag is produced by quenching the molten material in water. The rapid cooling results in a material which is glassy compared with the more crystalline nature of the air-cooled variety. The granulation process yields a material that requires little or no crushing. It has cementitious properties which result in it being used as a stabilized base material beneath pavements, around bridge abutments and in building foundations. Ground granulated slag is also used as a partial replacement for cement in concrete construction. The manufacture of cement clinker from granulated slag requires less energy than the corresponding process using conventional cement-making raw materials, and the resulting cements have a better resistance to corrosion by seawater and by sulfate exposure.

(*c*) *Expanded blast-furnace slag*. Expanded slag is the material that results from treating the slag with smaller quantities of water than are required for granulation. Two common methods of expanding involve pouring the slag either into a stream of water or onto a water-cooled rotating paddle wheel. The first method yields an angular product and the latter method a more spherical product (referred to as pelletized). Both products are more vesicular than air-cooled slag because expanding techniques increase the air content in the particles. The main use of expanded slag is as a lightweight concrete aggregate. In addition to a weight saving, expanded slag also offers excellent insulation and fire-resistance properties.

2.2 Steel-Furnace Slag

Steel is produced in one of three types of furnace: open-hearth furnaces, basic-oxygen furnaces (BOF) and electric-arc furnaces. All three use scrap steel as a partial source of metal, and fluxing material. The open-hearth furnace uses limestone for flux and molten iron from the blast furnace supplements the scrap steel as the source of metal. The BOF process is much faster than the open-hearth process (45 min vs 6 h to produce 300 t of steel). Scrap steel and molten iron supply the metal, but this process differs from open-hearth melting in that oxygen is "blown" into the furnace to speed the melting process, and lime (not limestone) is usually added as a flux. Electric furnaces use scrap steel and limestone in the production of steel. Their advantages are a smaller capital investment and reduced air pollution.

Steel slag differs significantly from blast-furnace slag. It contains more iron, lime and other oxides than blast-furnace slag, as shown in Table 1. Unlike blast-furnace slag, which can undergo three cooling processes, steel slag undergoes air-cooling only, which results in a material that is darker colored and denser than blast-furnace slag mainly because it contains more iron. Steel slag does not have as many uses as blast-furnace slag because of its different weight and chemistry. Its main uses are as a highway base material, railroad ballast and some fill applications.

Table 1
Chemical composition (in %) of iron and steel slag

	Blast furnace slag	Basic-oxygen furnace slag	Electric-arc furnace slag
SiO_2	36.0	14.1	16.9
CaO	39.0	44.4	41.2
MgO	12.0	7.4	10.2
Fe_2O_3	0.5		
FeO		22.4	18.1
(Fe)		(20.7)	(13.6)
Al_2O_3	10.0	1.9	8.1
MnO	0.4	5.7	3.6
S	1.4	0.2	0.2
P_2O_5		0.8	0.6
P		0.3	0.4

3. Properties

Properties vary between the three types of blast-furnace slag and also between blast-furnace slag and steel-furnace slag. The chemical and physical properties govern, to a large extent, the end uses for which slag may be a suitable raw material. The alteration of properties may prove to be a goal for future research efforts.

3.1 Chemical Properties

The chemical characteristics of slag are controlled by the requirements of the metal production. Since the metal product is of primary importance, the technology required to produce the iron and steel dictates the composition of the slag. The use of slag as a construction material came about as a result of marketing efforts, and not due to any change in the chemical characteristics of the by-product.

Table 1 presents the average compositions for blast-furnace slag, BOF steel slag and electric-arc steel slag. Steel slag often contains free or unreacted lime. The presence of free lime, coupled with the higher iron content, creates two problems. It interferes with desired chemical reactions when used with Portland cement, and hydration of the free lime causes expansion. Therefore steel slag is not used as a concrete aggregate and the expansion problem means it is unsuitable as a structural fill material (beneath buildings).

3.2 Physical Properties

The specific gravity of blast-furnace slag ranges from 2.0 to 2.5 and depends mainly on the amount of void space (vesicules) in the particles, and the chemical composition. The unit weight of blast-furnace slag depends in part on the cooling method. Air-cooled blast-furnace slag has a loose unit weight of 1120–1360 kg m^{-3} influenced by the gradation of the particles. Granulated slag ranges from 830 kg m^{-3} to 950 kg m^{-3}. The density of expanded slag is 560–880 kg m^{-3} for coarse particle sizes and 800–1000 kg m^{-3} for fine particle sizes.

Blast-furnace slag is very durable as a construction aggregate material. It resists weathering, shows very little degradation due to freezing and thawing, and has a low coefficient of thermal expansion (about $1 \times 10^{-5}\,°C^{-1}$ from ambient temperature to melting temperature). Its resistance to polishing under traffic conditions makes it desirable as a high-friction aggregate for asphalt pavements.

The specific gravity of steel slag is about 30–50% greater than that of blast-furnace slag because of its higher iron content. The unit weight of steel slag ranges from 1660 kg m^{-3} to 2000 kg m^{-3}. Its surface texture and weight make it suitable as a road-base material and railroad ballast. Some is also used for asphalt paving as it has the properties necessary for a high-friction aggregate.

3.3 Mineralogical Properties

The mineral content of both iron and steel slags depends on the chemical composition of the slag and its cooling history. Crystallization proceeds more slowly in slags with a high silica content than in those which contain more lime and magnesia. These properties are analogous to those of acidic and basic igneous systems.

Little or no crystallization takes place in expanded or granulated blast-furnace slag because of the rapid cooling. The following minerals have been identified in air-cooled blast-furnace slags.

Akermanite $Ca_2(MgFeAl)_2O_7$
Pseudowollastonite $CaOSiO_2$
Periclase MgO
Anorthite $CaAl_2Si_2O_8$
Forsterite Mg_2SiO_4
Merwinite $Ca_3Mg(SiO_4)_2$
Gehlenite $Ca_2Al_2SiO_7$
Calcium orthosilicate Ca_2SiO_4
Olivine $(MgFe)_2SiO_4$
Monticellite $CaMgSiO_4$
Pyroxene XO (Si_2O_6)
Oldhamite CaS
Manganous sulfide MnS
Ferrous sulfide FeS

Although air-cooling is slower than water-quenching, it is still relatively fast and the resulting crystal size rarely exceeds 10 mm. The presence of glass, even in air-cooled slag, testifies that crystallization is incomplete.

Calcium orthosilicate is undesirable because its low-temperature phase is unstable. When the lime content of the slag is more than 10% greater than the silica content, calcium orthosilicate may form. Partial substitution of dolomite for limestone in the flux diminishes the likelihood that calcium orthosilicate will form in the cooled slag.

Steel slags undergo air-cooling like most blast-furnace slag. However, steel slag is more fluid, and cooling produces a higher glass content than is found in blast-furnace slags. The mineral content of steel slags has not been as widely published as that of blast-furnace slag.

4. Uses

Slag is used outside the construction market as a raw material for the manufacture of glass (amber quality) and for glass wool, soil conditioner, roofing material and sewage treatment. For a detailed description of the use of slag in the construction industry see *Slag Utilization*.

While iron and steel slags are the most commonly used, slag from the processing of other metals can be used in some instances. The production of copper, nickel, zinc and ferroalloys all yield slag. Some of this material has been used for the same range of construction materials as iron and steel slags. Production data for other slags are not reported because the volumes produced are much less than those of iron and steel slags.

In 1985 the estimated world production of blast-furnace slag was 113.4 Mt and 51.3 Mt of steel slag was produced. In several countries, for example Canada, slag is still thought of as a waste product rather than a resource. As a result, the data for international slag production and consumption are incomplete.

Uses of slag in other countries are generally not as diverse as in the USA. The manufacture of cement is apparently the major use of slag in France, the

FRG, Italy, Japan and South Africa. However, the uses within the UK parallel the US consumption patterns.

See also: Construction Materials: Crushed Stone; Construction Materials: Fill and Soil; Construction Materials: Granules; Slag Utilization

Bibliography

Eggleston H K 1968 *Slag—The All-Purpose Construction Aggregate*. National Slag Association, Washington, DC

Hogan W T 1984 *Steel in the United States: Restructuring to Compete*. Lexington Books, Lexington, MA

McCarl H N, Eggleston H K, Barton W R 1983 Construction materials, aggregates—slag. In: *Industrial Minerals and Rocks*, 5th edn. Society of Mining Engineers, Littleton, CO, pp. 111–31

Mickelson D P 1985 Slag—Iron and steel. In: *Minerals Yearbook*. US Bureau of Mines, Washington, DC

National Slag Association 1982 *Processed Blast Furnace Slag—The All-Purpose Construction Aggregate*. National Slag Association, Washington, DC

Seagall R T 1970 Slag—Michigan's All-Purpose Construction Aggregate. In: *Proc. 6th Forum on the Geology of Industrial Minerals*, Misc. Pub. #1. Michigan Geological Survey, pp. 117–26

H. L. Bourne
[Independent Consulting Geologist,
Northville, Michigan, USA]

Specialty Papers from Advanced Fibers

Twentieth-century paper machines are sophisticated and large, many producing hundreds of thousands of tonnes of paper per year. The process can handle virtually any man-made mineral or organic fiber, but its use is often ruled out because of size or volume considerations because many advanced or new fibers are only produced or needed in small quantities. Perhaps the most well-known new fiber is carbon fiber, the total world production of which could be processed on one conventional paper machine, still leaving it underutilized. However, the volume gap has been successfully bridged so that a new product form is available to engineers, designers and manufacturers working with advanced fibers.

1. Manufacture

Four routes are possible to cope with increasing production volumes.

(a) Most paper mills have laboratory or hand sheet-making facilities which can produce nonwoven sheets for initial evaluations.

(b) It is not uncommon for large paper mills or research-and-development centers of paper groups to have laboratory or pilot-plant scale machines for running trials of grades prior to running on a full-sized machine. These machines can and do run commercial orders of advanced fibers.

(c) A few companies have recognized the potential of the new fibers now available and have designed and built specialized small-capacity machines. These machines have been specifically designed to cope with the special problems of processing high-modulus fibers in short production runs (e.g., 250 m^2). As the volume requirement increases, it has proved possible to economically process a number of fibers on older paper machines whose capacity is uneconomic with conventional cellulosic grades.

1.1 Principles of Manufacture

Papermaking can be defined as the formation of a nonwoven sheet from individual fibers using water as the processing medium (see *Papermaking: Alternative Wet End Processes*). Cellulose (woodpulp) has a special property: following mechanical action in the presence of water and after subsequent drying, the fibers bind to each other on account of hydrogen bonding (see *Hydrogen Bonding in Paper: Theory*, Suppl. 1). Most man-made fibers do not have this property and therefore adhesives or binders have to be added to bind the fibers to one another to produce a nonwoven sheet. The process can now be considered in three basic steps.

First, water is used to disperse fiber bundles or tows (textile fibers) to obtain individual fibers suspended in water. A significant part of the art is in modifying the water in some way to ensure that dispersion is complete and that high-modulus fibers are either not shortened in length or, if desired, only shortened in a controllable way. Binders, most commonly thermoplastic or thermosetting polymers, can be added at this stage or after sheet forming.

Second, the suspended fibers are subsequently laid down or "formed" continuously to produce the nonwoven sheet whose characteristics should be random orientation and an even or equal distribution. The device used in this process is called a "former" and designs are normally proprietary, although the principles are widely known. At least one design has proved capable of laying down dried areal weights in a range from 10 g m^{-2} to over 2000 g m^{-2}.

Third, having formed the sheet from water it is then necessary to remove it by vacuum pressing and heat drying. Depending on the binder type chosen, drying temperatures have to be controlled in the range from 100 °C to over 250 °C.

The results of these manufacturing principles are many products, outlined in the subsequent text;

however, these should only be regarded as the forerunners of a vast range of specialty papers (nonwoven) that will be produced in the future.

2. *Papers for Thermal Insulation*

A wide range of inorganic fibrous materials are available in a suitable form for papermaking. These are most widely used for heat-insulation applications. The most well established of these products was asbestos paper; however, because of the health problems it caused, it has been almost entirely replaced by alternative materials. Ranked in order of temperature resistance, candidate materials for thermal insulation papers are as follows:

(a) glass (up to 600 °C) in various fiber forms and glass chemistries;

(b) rockfibers (up to 900 °C) usually in the form of "rockwools";

(c) aluminosilicate (ceramic) (1260 °C), the major asbestos-replacement fiber;

(d) high zirconia content, ceramic (up to 1400 °C);

(e) pure alumina (up to 1600 °C); and

(f) polycrystalline mullite (up to 1600 °C).

Ceramics with temperature resistance of up to 1400 °C are approximately twice the cost of those with temperature resistance of up to 1260 °C. Moreover, those with temperature resistances of up to 1600 °C are approximately 20 times the cost of the fiber resistant to up to 1260 °C. This limits these products to highly specialized applications where the premium price can be justified.

Papers are often the most efficient form in which these materials can be produced; typical thermal conductivity values are in the range 0.05–0.08 $W\,m^{-1}\,K^{-1}$ at ambient temperatures. These papers can be made with a variety of binder systems to give finished products that are flexible (paper) or stiff (board). Binders are usually selected for strength, compressibility and clean burning: in most cases the sheet strength is lost when the binder is destroyed, typically at temperatures of 200–300 °C. In some cases, paper can be produced with inorganic binders where sheet strength is maintained at higher temperature.

The following are some applications for thermal insulating paper.

(a) *Automotive silencers*. Papers based on rockwools, glasses and ceramic fibers are used for wrapping automotive silencers, particularly those that are wool free, where no insulating fibers are contained inside the silencers. Papers are also used to enclose catalytic convertors to give a gas-tight seal between the convertor and the outer wall of the silencer.

(b) *High temperature seals*. Papers are cut to form gaskets for use in high-temperature pipelines.

(c) *Nonferrous metals*. Papers are used for lining channels and vessels in the nonferrous-metal industries, particularly in the aluminum industry.

(d) *Glassware and ceramics*. Papers are used in the glassware and ceramic industries as a support for cooling finished pieces; papers with exceptionally smooth surfaces are used in glass-bending operations.

(e) *"White goods."* Papers are used in a variety of domestic appliances (known as white goods) such as cooker doors, electric toasters and storage heaters.

(f) *Furnace linings*. Papers are used to finish furnace linings, by forming a seal between firebricks or ceramic blankets.

3. *Papers for Battery Separators*

Materials that are used in batteries to separate the positive and negative plates are often produced by a papermaking process. The major requirements for a battery separator are:

(a) it should resist attack by the electrolyte;

(b) it should absorb and transport the electrolyte throughout the battery;

(c) it should prevent any form of contact between the electrodes;

(d) it should be electrically insulating; and

(e) it must allow ionic transport between the electrodes.

3.1 Lead–Acid Batteries

The latest generation of lead–acid batteries use a technique known as "recombination," in which the gases evolved at the electrodes recombine inside the battery. This means that the batteries are completely maintenance free. In this type of battery a separator of pure glass fiber is used. The separator is made from microfibers with diameters as low as 0.8 μm; these provide a sheet with sufficient strength without the need for any binder materials. Strength is achieved purely by physical entanglement. The small size of the fibers used also produces numerous micropores which act as capillaries to distribute the acid throughout the battery.

3.2 Alkaline–Manganese Batteries

The electrolyte in this type of cell is potassium hydroxide: because of the alkaline nature of the electrolyte, glass is not a suitable material. These separators are usually produced from polyvinyl alcohol fibers, sometimes mixed with viscose. Polyvinyl alcohol can be used as the sheet binder as well as the component fiber.

3.3 Lithium Oxyhalide Cells

In this case the electrolyte is thionyl chloride. Separators are based on glass microfiber and are similar to those used in lead–acid batteries although usually of low overall weight and thickness because of the size and shape of the batteries.

3.4 Fuel Cells

Although not strictly batteries, these operate on similar electrochemical principles but have the reacting species continuously fed to the cell. In many cases, lightweight papers of high-purity carbon fiber, bonded with high-char-yield phenolic resins, are used as a catalyst support material.

4. Filter Papers

Specialty papers are extensively used in a variety of filtration and separation products. The ability to closely control pore size (by fiber selection and/or processing variables) and the high ratios of surface area to weight that can be achieved, mean that papers are often the most efficient forms of filter media available.

4.1 Liquid Filtration

In the applications where a lower filtration efficiency is acceptable, papers based on traditional cellulosic fibers predominate. These are usually loaded with a polymeric resin to improve rigidity and pleating characteristics. A typical example is automotive oil filters. In more specialized applications, particularly where greater filtration efficiency is required, microfine glass filters can be used. This type of filter is more often used in industrial applications and in high-performance hydraulic systems.

4.2 Air Filtration

The demands for air with very high cleanliness by the electronic and nuclear industries and by medical establishments are satisfied by filters produced from papers made of very fine diameter glass fibers. These filters have efficiency ratings of 99.9–99.999% when tested with a standard 3 μm aerosol of sodium chloride solution. They are known as high-efficiency particulate air (HEPA) filters.

Lower-grade filters of microfine glass fibers can be introduced as prefilters in order to extend the life of the higher-efficiency main filters. Alternatively, they can be used in applications such as the removal of tobacco smoke from air-conditioning systems.

4.3 Other Filters

Novel fiber types offer the ability to produce new varieties of filter. One example is the activated-carbon filter—papers produced from activated-carbon fibers or impregnated with activated carbon, are used for the removal of odors, colors and bacteria. A second example is the ion exchange filter—fibers containing ion-exchange groups can be used to remove specific ions; in particular, precious metals can be recovered by this type of filter.

5. Papers for Use in Reinforced Plastics (Composites)

Because of their isotropic nature, papers are not usually suited to providing the main reinforcements for composites. However, papers produced from advanced fibers have found a variety of applications in reinforced plastics.

5.1 Glass

Glass fibers are the major reinforcement used in composites; glass-fiber tissues being the most widely used type of paper composites. Glass surfacing tissues are widely used to stabilize the resin-rich outer layer (gel coat) of reinforced plastic components. This is particularly important in chemical plants, where the stability of the gel coat and the resistance of the surfacing tissues are thought to be the major factors determining the ability to resist chemical attack. For this reason glasses with enhanced chemical resistance are used for the production of surfacing tissues.

Glass tissues are widely used as the only reinforcement in thin flexible composites such as flooring and roofing materials. Glass tissues treated with a vermiculite clay can be used to provide fire-barrier properties to composites. Untreated glass would be melted by the heat of the burning resin, but the vermiculite maintains a sheet form after the resin has burnt away and hence protects the underlying reinforcements.

5.2 Carbon Fibers

Carbon-fiber papers are mainly used to confer electrically conductive properties to composites; these electrically conductive composites are used for the following applications.

(*a*) *Static protection*. Papers produced from either 100% carbon fiber or from mixtures of carbon and nonconductive fibers can produce reinforced plastics with surface resistivities from less than 10 Ω □ to about 10^9 Ω □. These composites are used in areas where static discharge is a hazard; examples are bench tops for microelectronics production, tanks for the storage of volatile chemicals and piping and ducting for oil rigs.

(*b*) *Reflectivity*. High-conductivity papers are used to provide the electrically reflective layer in composite satellite dishes. The use of the carbon-fiber layer offers equivalent reception to metal-based dishes but avoids the problems of corrosion and distortion because of environmental effects. A related use is to give a radar "signature" to nonconductive items (e.g., inflatable rubber craft): the carbon-fiber paper reflects radar signals and hence provides radar visibility.

(*c*) *Electromagnetic interference (EMI) shielding*. In this case the conductive tissue is used to provide a barrier to the emission of undesirable EMI. Typical areas for use are in computer housing and cable wrappings. This is extended to the shielding of complete rooms or buildings (architectural shielding) in areas such as embassies or defense establishments. In all cases, the conductive layer has a dual role: in addition to preventing EMI being emitted from the contents, it also shields against malfunctions caused by EMI from external sources.

(*d*) *Resistive heating*. By passing an electric current through an electrically conductive layer, with controlled resistivity, a heating effect can be produced. This has found applications ranging from the deicing of aircraft propellors to the provision of heated floors for animal husbandry. Some other applications for carbon-fiber papers are to provide surfacing tissues with superior chemical resistance for chemical plants used in particularly harsh environments or as the final surfacing layer in reinforced-plastic x-ray tables where only carbon fiber has the necessary x-ray transparency.

5.3 Aramid Fibers

Papers produced from Nomex (R) aramid fiber are used in the manufacture of honeycomb: a lightweight structure of high strength, widely used in the noncritical parts of aircraft. These papers also have many electrical applications, their resistance to elevated temperatures and their high dielectric strengths making them particularly suited to high-voltage applications.

Papers produced from Kevlar (R) aramid fibers are used as a lightweight high-strength base for supporting syntactic foams. Aramid tissue is also used to improve the low-speed impact resistance of composites; for example, in reinforced-plastic road signs.

See also: Bonding Nonwovens; Nonwoven Fabrics; Porous Materials: Structure and Properties

Bibliography

Battista O A (ed.) 1964 *Synthetic Fibres in Papermaking*. Wiley, New York

Coppin P 1988 Nonwovens in advanced composites. *American and Canadian Chemical Societies Congr.* Toronto, June

Culpin B, Hayman J A 1986 Transport and wetting phenomena in recombination separator systems. *International Power Sources Symp.* IPSS Committee, Leatherhead, UK

Fegley N L 1985 Advanced speciality nonwovens via the wet laid process. *Technical Association of the Pulp and Paper Industry Nonwovens Symp*. TAPPI, Appleton, WI

Fry F H et al. 1985 Important factors in glass web manufacture. *Technical Association of the Pulp and Paper Industry Nonwovens Symp*. TAPPI, Appleton, WI

Hoon S R, Shelton A, Tanner B K 1985 Time-dependent resistivity in carbon fiber sheets. *J. Mater. Sci.* 20: 3311–19

Koenig A R 1984 Processed mineral fiber in mats and papers. *TAPPI J.* 67(10)

Merriman E A 1981 Kevlar (R) aramid pulp for papermaking. *Technical Association of the Pulp and Paper Industry Nonwovens Seminar*. TAPPI, Appleton, WI

Merriman E A 1984 Aramid pulp processing and properties for industrial papers. *TAPPI J.* 67(8)

Quick J R, Mate Z 1982 Conductive fiber mats as EMI shield for SMC. *Society of the Plastics Industry RP/C Institute Conf.* SPI, Washington, DC

Seager N J, Waghorne T 1988 Consideration in the design and manufacture of glassfibre separator materials for use in lithium oxyhalide cells. *33rd Int. Power Sources Symp.* IPSS Committee, Leatherhead, UK

Scheffel N B 1986 Glass fibre for paper processing and paper properties. *Paper Industry Research Association Seminar*, February 26. PIRA, Leatherhead, UK

Smith R 1984 Ceramic fibres in Nonwovens. *Technical Association of the Pulp and Paper Industry Fibres Short Course*. TAPPI, Appleton, WI

Trauve J 1981 Developments in the wet laid sector—New applications. *European Disposable and Nonwovens Association Index 81 Congr.* EDANA, Geneva

Walker N J 1985 Papers—A new dimension in carbon fibre materials. *Carbon Fibres*, Vol. 3, Plastics and Rubber Institute, London

Walker N J 1987 Veils, mats and tissues for non-structural applications. *Proc. 6th Int. Conf. Composite Materials*. Elsevier, Amsterdam

White C F, Barden M J 1979 The use of a uniflow cylinder mould machine in the manufacture of speciality paper webs. *Proc. EUCEPA 18th Int. Conf.* EUCEPA, London

White C F, Moore G K 1987 Ceramic and mineral wool fibers in wet web forming processes. *TAPPI J.* 70(12)

Williams W G 1985 Carbon Fiber for high performance applications. *Technical Association of the Pulp and Paper Industry Nonwovens Symp*. TAPPI, Appleton, WI

N. Walker and M. Allsop
[Technical Fibre Products, Kendal, UK]

Steels for Rolling Bearings

Rolling bearings comprise two major families, namely ball bearings and roller bearings, which generally consist of an inner raceway, an outer raceway and a complement of balls or rollers that are kept apart by a cage or separator. Roller bearings have very high radial load-carrying capacity

Table 1
Chemical composition of SAE 52100 bearing steel (wt%)

Carbon	0.98–1.10
Chromium	1.35–1.60
Manganese	0.25–0.45
Silicon	0.15–0.35
Sulfur	<0.025
Phosphorus	<0.025
Molybdenum	<0.10
Nickel	<0.25
Copper	<0.35

while ball bearings have lower radial carrying capacity but have the advantage of being able to support axial loads.

Rolling bearings are generally very reliable and allow very accurate rotational motion with minimum friction. They are found in all applications where rotational motion occurs, a range of standard bearings being produced for general applications such as electric motors and gearboxes, and special bearings being produced for applications such as gas turbine aeroengines and steel rolling mill support bearings.

Major features in achieving a consistent high level of performance are the bearing design, the high quality and reliability of the steel from which the bearings are made, and the high integrity of the heat treatment employed.

1. Steels for Standard Bearings

1.1 Steel Type

Most of the bearings produced throughout the world are manufactured from a 1 wt%C–1.5 wt%Cr steel whose composition is given in Table 1. It is known in different countries as, for example, SAE 52100, EN31, 535A99 and 100Cr6.

Throughout the history of the use of this steel for the manufacture of rolling bearings, the quench and temper heat treatment employed has been designed to produce a hard martensitic structure throughout the section so as to provide the maximum resistance to deformation and thus maximum load-bearing capacity, good wear resistance and high fatigue life.

1.2 Steel Making

Standard bearings are made from steel melted in electric arc furnaces and then refined to high levels of steel cleanliness in secondary steelmaking plants, which usually incorporate vacuum degassing and other process steps designed to remove nonmetallic impurities and to reduce gas contents to a minimum.

Steel casting is generally into ingot moulds followed by rolling down to billet, although continuous casting is being employed on an increasing scale.

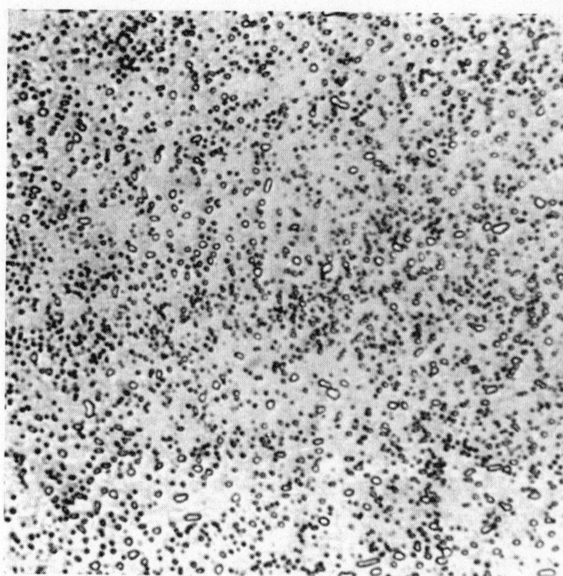

Figure 1
SAE 52100, spheroidize-annealed structure (magnification ×1000)

The product forms used in the manufacture of bearing components are tubes for machining bearing rings, wires and rods for rolling elements and forging stocks for forging bearing rings.

1.3 Annealing for Machining

Before the steel can be machined it must be spheroidize annealed, which involves heating to the austenitic state and then very slowly cooling back to the ferrite state at approximately 680 °C, followed by air cooling. Figure 1 shows a typical structure of spheroidize-annealed bearing steel, which consists of an essentially carbon-free ferrite matrix with a dispersion of spheroidal carbides embedded in it. Any residual pearlite in the structure at this stage will worsen machinability.

Some parts are cold formed, notably balls which are cold headed from parted-off slugs. Again, a spheroidize-annealed structure is required to provide maximum formability. A degree of cold working in the tube or bar is generally advantageous in improving machinability and cropping behavior.

1.4 Hardening Heat Treatment

After machining or forming to produce the bearing component, a hardening heat treatment is required to produce the desired final properties. This involves heating the components at typically 840 °C and quenching into oil. Tempering in the range 150–250 °C results in a final hardness in excess of 700 HV30 (59 Rc).

Figure 2
SAE 52100, fully hardened and tempered structure (magnification ×1000)

During the austenitizing phase at 840 °C, the carbon in the original carbides diffuses into the austenite matrix, building up the carbon content of the matrix and at the same time causing the carbide particles to shrink. After quenching, a correctly heat-treated structure consists of a martensite matrix with a dispersion of residual carbides. A small amount of retained austenite will also be present, but will be difficult to detect by metallography. Figure 2 shows a typical heat-treated structure.

There are several problems that can occur if the heat treatment is not carried out correctly.

(a) Austenitizing at too low a temperature will result in low hardness on quenching, because of too little carbon being diffused from the carbides into the matrix. The residual carbides will still be profuse.

(b) Austenitizing at too high a temperature will allow excessive carbon to diffuse into the matrix, which has the effect of lowering the martensite M_s and M_f temperatures to the point where excessive retained austenite is present on quenching. Retained austenite can produce serious service problems, notably poor dimensional stability, due to its progressive transformation to martensite with time.

(c) Quenching into a slow-quenching oil or quenching a thick section can slow down the cooling rate to the point where a mixed microstructure of martensite–bainite–pearlite can form. Lower hardness would result, giving lower load-carrying capacity, and bearing life would be reduced. To compensate for slower quenching rates in thicker sections, higher hardenability variants of the carbon–chromium steel that contain increased levels of manganese and molybdenum are used. Selection of the correct steel grade is very important and must be matched to the section size and the cooling rate of the quenchant.

(d) Correct control of furnace atmospheres is important since decarburization of the parts during austenitizing can result in cracking during quenching.

2. *Rolling Contact Fatigue*

In service, bearing life is usually limited by incorrect fitting, corrosion, ingress of dirt, inadequate lubrication and numerous other external conditions, often beyond the control of the bearing manufacturer. Most bearing companies offer advice and recommendations to avoid early failure due to such conditions, but even under well-lubricated conditions, correctly selected, installed and maintained bearings will fail eventually by some form of fatigue of the contacting surfaces.

Well-lubricated bearings can be expected to exceed what is generally referred to as the catalog life, that is, the life calculated using criteria in the standards of the American Society of Mechanical Engineers and of the International Organization for Standardization. Bearing life is expressed in terms of an L_{10} factor which is the life at which 10% of a group of ostensibly identical bearings running under the same conditions have failed; it is usually obtained by applying Weibull statistics to test data (see *Selection Systems for Engineering Materials: Quantitative Techniques*).

The stress system experienced by a bearing during running is described in Hertzian theory and in a modified theory taking account of elasto hydrodynamic lubrication (EHL). As a bearing rotates (see Fig. 3) a given point beneath the raceway surface is

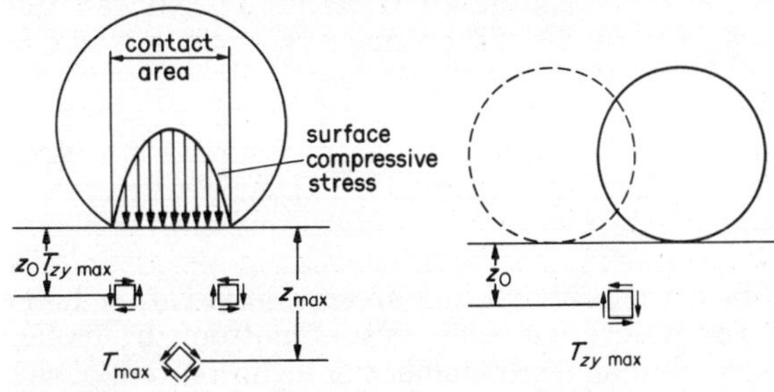

Figure 3
The stresses at a rolling contact

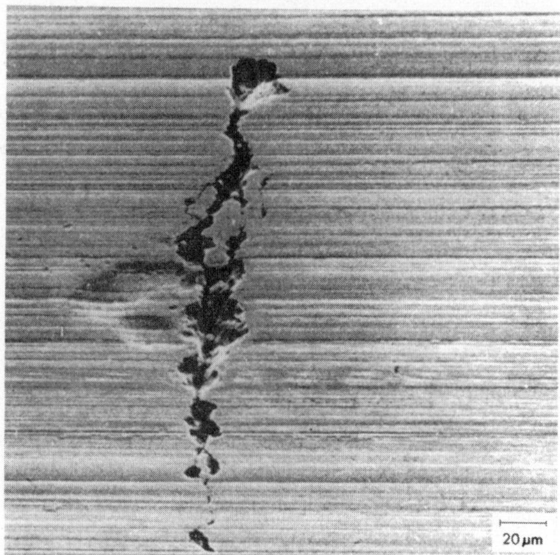

Figure 4
The emergence of a subsurface fatigue crack at the surface

subject to a buildup of stresses as the rolling element approaches that point and a falloff as it passes. Subsequent contacts repeat the process, resulting in a constant load–unload situation, the most severe feature of which is considered to be a reversing shear stress whose value is at a maximum just beneath the surface. This fluctuating stress eventually gives rise to fatigue cracking which leads to surface breakup, hence the concept of rolling contact fatigue.

The mode of fatigue failure under good lubrication conditions consists of subsurface crack nucleation in the region of peak reversing shear stress, followed by propagation of the crack to the surface (see Fig. 4). Eventually this crack propagation results in the formation of a pit at the surface (see Fig. 5). At this point the bearing becomes rough running, with a high level of vibration. Continual running after this point quickly results in a spalling of the surface around the original pit. Debris from the spalling can also cause secondary damage to other bearings or gears in the vicinity.

3. *Effect of Nonmetallic Inclusions*

3.1 *Inclusion Types*

During the steelmaking process great care is taken to remove nonmetallic inclusions from the molten steel, but a certain number of inclusions are always present after casting. Subsequent rolling elongates these inclusions into stringers. The four main types of inclusions that have been identified and designated are sulfides (type A), alumina (type B), silica (type C) and globular oxide (type D).

3.2 *Fatigue Initiation at Inclusions*

The initial nucleation of the fatigue crack is greatly influenced by the presence of nonmetallic inclusions left over from the steelmaking process. Oxides are particularly deleterious, because not only are they brittle, but they are also surrounded by tensile stresses due to differential coefficients of thermal expansion, which add to the reversing and shear stress field and help to initiate failure. Figure 6, which is a section through a typical rolling contact fatigue spall, illustrates this fact, failure in this case being initiated at an alumina inclusion trail. However, sulfide inclusions are considered much less deleterious and are even beneficial in increasing rolling contact fatigue life. One reason put forward for this is that sulfide inclusions can encapsulate oxide inclusions and reduce the tensile stresses around these inclusions. This is especially the case with steels deoxidized with calcium silicide.

Steel cleanliness is thus of paramount importance in bearing steels owing to its significant effect on rolling contact fatigue initiation. Accordingly, international bearing steel specifications require tightly controlled levels of inclusions to which steelmakers have to conform.

4. *Inclusion Content Rating*

The inclusion content in bearing steels is usually determined by a metallographic examination of samples from different positions in a cast. The inclusions present in the steels are compared against standard charts that have numerical ratings related to the

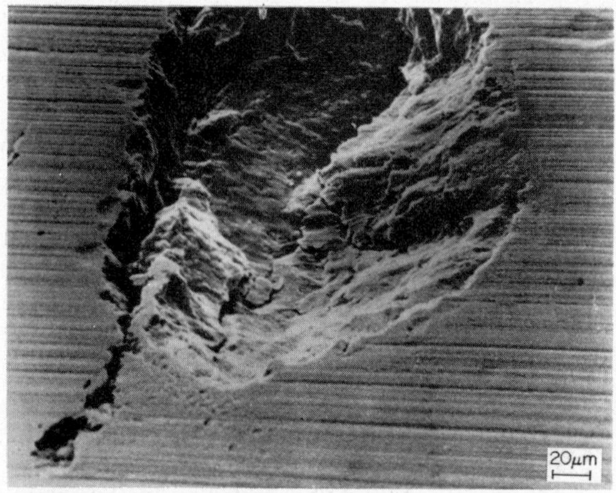

Figure 5
Typical rolling contact fatigue spall

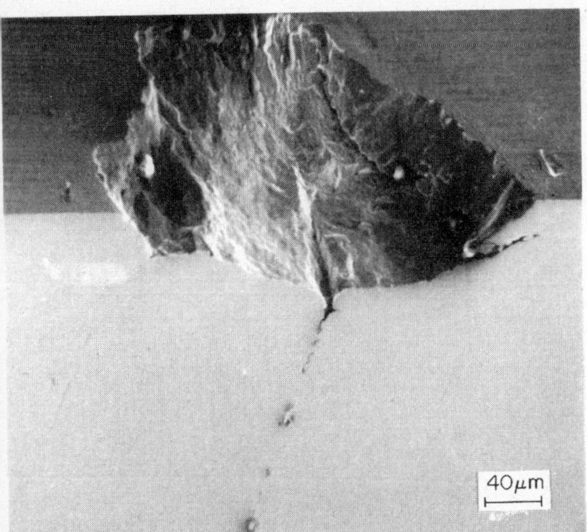

Figure 6
Section through a fatigue spall, showing association with a B-type alumina inclusion trail

severity of the inclusions. The most widespread technique is the Jernkontoret (JK) method, which is described in the E45 specification of the American Society for Testing and Materials, Philadelphia, USA. In this method, 100 fields from each sample are examined and the worst field of each type of inclusion is recorded, comparing in each case against the standard charts. The inclusions considered are sulfides (type A), alumina (type B), silicate (type C) and globular oxide (type D). The values obtained are compared against the specified maximum permitted values and a comparative rating for that cast established.

To complement the JK worst field count, a more qualitative method known as the SAM count has been developed which considers both the JK rating of inclusions and their frequency, noting all inclusions in the 100 fields examined.

From this analysis, a numerical inclusion rating for each sample is obtained and the total area of samples examined is recorded. The SAM count equates the numerical inclusion rating to an area examined of 645 mm^2. This count is generally carried out on the B-type (alumina) and D-type (globular oxide) inclusions only, since experience has shown these inclusions to have the major influence on rolling contact fatigue. A direct relationship exists between the rolling contact fatigue life and the "SAM B" and "SAM D" which allows bearing manufacturers to specify the inclusion content of their steel in terms of maximum permissible SAM counts.

The imposition of tighter material specifications and the ability of steelmakers to meet those standards has recently enabled bearing manufacturers to increase the catalog dynamic capacity by a factor of 30% for ball bearings. The designer now has the option of offering longer bearing life to the customer or alternatively of offering a smaller bearing having capacity equal to previously designed larger bearings.

5. *Higher Grade Steels*

For applications where the load and fatigue requirements are very severe, even the best of the electric arc and vacuum degassed steels may not necessarily be good enough. For these applications, steels that have been secondary melted by the electro-flux refining (EFR) or the vacuum arc refining (VAR) routes can be specified. The permissible inclusion content of these steels is very low, which produces rolling contact fatigue properties well above that of standard electric arc melted steel. However, the cost of these steels is also much higher than that of standard steels.

6. *Carburizing Steels*

Although the SAE 52100-type steel is used extensively in bearing applications worldwide, carburizing steels are also used in many applications. Often the application would involve a certain degree of shock loading which requires a higher level of toughness than quenched and tempered SAE 52100. Typical examples would be in off-highway earth moving equipment or steel rolling mill backup bearings. For these applications, lower-carbon steels in the range 0.1–0.3 wt% are used; these are subsequently carburized to produce a component with a hard surface capable of supporting contact loads and a soft tough core capable of withstanding shock without fracture. The composition of a typical carburizing bearing steel based on SAE 4720 is given in Table 2.

The steel cleanliness of these lower-carbon steels is generally not as good as the higher-carbon SAE 52100 steel due to less carbon deoxidation during the

Table 2
Chemical composition of SAE 4720 type bearing steel (wt%)

Carbon	0.16–0.21
Silicon	0.15–0.35
Manganese	0.60–0.80
Nickel	0.90–1.20
Chromium	0.35–0.55
Molybdenum	0.20–0.30
Sulfur	<0.025
Phosphorus	<0.025
Copper	<0.35

steelmaking process. However, in the finish heat-treated condition, the carburized case contains a high compressive stress which resists cracking. The net result is that the rolling contact fatigue life of commercial low-carbon and high-carbon bearings is very similar.

7. *Special Heat Treatments*

Conventional rolling contact fatigue failure occurs when operating conditions are such as to exclude failures arising from corrosion, dirt, poor lubrication and so on. However, there are many applications where these factors cannot be eliminated and special precautions must be taken to ensure satisfactory bearing life.

A typical example is bearings operating in conveyor rollers for coal mining applications, where the bearings may be subject to the ingress of coal dust after seal failure and often operate in very wet conditions, both of which markedly reduce their operating lives. A further application relates to the operation of bearings in fire-resistant fluids containing water, which are being used on a wider scale in environments where oil is seen as an unacceptable fire hazard.

The mode of failure is changed to one of abrasive wear of the balls and raceways and, on occasions, catastrophic breakup of the bearing rings, which can lead to seizure and generation of very high temperatures with the inherent risk of fire. Hydrogen embrittlement of the hard martensite microstructure is the likely cause of ring fracture.

Modifying the heat treatment of the bearing rings to produce lower bainite, which involves quenching into molten salt at 200–300 °C and holding for a while (see *Steels: Physical Metallurgy Principles*), produces a much tougher structure which is far more resistant to hydrogen embrittlement. The lower-bainite microstructure is somewhat softer than the conventional martensite structure, being typically 650–700 HV30 (56.5–59.0 Rc). Catastrophic break-up of the bearing rings in the water-containing environments mentioned is very rare and the failure mode is generally progressive wear. Tempering back of martensite to the same level of hardness does not produce the same properties. Lower-bainite heat treatment is also often used for bearings running in dirty oil conditions. In this case failure is initiated at debris indents on the bearing raceway surface and the higher toughness of the lower-bainite structure suppresses this type of failure mode.

8. *Corrosion-Resistant Bearing Steels*

For applications where dirt is not necessarily a problem but where both a highly corrosive environment exists and excessive bearing wear cannot be tolerated, then stainless steels must be used. Low-carbon austenitic steels are not suitable for bearings because their low hardness is inadequate to support contact loads; thus, the higher-carbon chromium stainless steels have to be used, in the fully martensite-hardened condition. Type A1S1 440C containing 1 wt% carbon and 18 wt% chromium is one of the most common hardenable stainless steels used in bearings of this type. The full composition of A1S1 440C is given in Table 3.

Table 3
Chemical composition of AISI 440C-type bearing steel (wt%)

Carbon	0.95–1.10
Chromium	16.0–18.0
Molybdenum	0.40–0.65
Manganese	0.50–0.80
Sulfur	<0.015
Phosphorus	<0.025
Silicon	<0.50
Nickel	<0.50
Vanadium	<0.15

In the heat treatment of 440C material, the choice of tempering temperature is very important since there is a complex interaction between hardness, dimensional stability and corrosion resistance. For operation at ambient temperatures where maximum corrosion resistance is needed, tempering in the range 125–150 °C is satisfactory. However, if operating temperatures exceed 150 °C, dimensional instability becomes a problem owing to the transformation of metastable austenite, retained in the structure during heat treatment, which can cause rings to come loose on shafts and so on. The use of higher tempering temperatures of up to 500 °C removes the retained austenite and produces good stability. However, during this process hardness drops lowering the loading capacity and, of more importance, the corrosion resistance deteriorates as precipitation of temper-induced chromium carbide removes chromium from the matrix so that it no longer aids in resisting attack. Thus a choice must be made between maximum corrosion resistance or maximum dimensional stability. Usually the low tempering temperature is adopted for optimum corrosion resistance and a subzero treatment is carried out to reduce the retained austenite content and make the bearing as stable as possible. However, maximum stability is only achieved when higher tempering temperatures are adopted.

For applications where lubrication in the usual sense is completely absent it is unwise to use the same material for both raceways and balls because this can give rise to galling problems. For example, in one application where unlubricated bearings support the fuel rod feed mechanism of a nuclear reactor in an environment comprising mainly hot water and steam, satisfactory performance has been obtained by using wear-resistant cobalt-based material for the balls in combination with stainless steel raceways.

9. *High-Temperature Bearing Steels*

The bearing material requirements of gas turbine engines have required a significant amount of steel development since the 1950s as engine temperatures, speeds and loads have increased.

Throughout the history of piston aeroengines, the SAE 52100-type steel composition had been used for all bearings. As gas turbine engines were developed in the 1950s, bearing operating temperatures increased to the extent that a steel with a higher temperature capability was required. Development in the UK and the USA followed a similar pattern although the steels that were eventually selected for aeroengine bearings in the two countries varied considerably.

9.1 UK Steel Developments

The first major engine requiring a higher-temperature bearing steel was the Rolls Royce Conway engine developed in the mid-1950s. The high-temperature steels most readily available in the UK at that time were the tungsten-containing high-speed tool steels of which T1 had a high hardness and resistance to tempering. Air-melted T1 was thus selected for the bearing rings in some of the hotter bearings in the Conway engine. Balls were still manufactured from SAE 52100 since the technology did not exist to manufacture tool steel balls.

The next major development in the UK was in the early 1960s when vacuum arc remelting (VAR) was introduced for T1 to improve cleanliness levels and to improve fatigue life. This material was used for bearings in the Rolls Royce Spey engine and for the first time balls were made in T1 steel. VAR ingot sizes were restricted to 150–200 mm in order to control segregation which is a severe problem with this steel composition.

A further development of T1 occurred in the late 1960s when the Rolls Royce RB211 was developed. The larger bearings used in this engine required a larger remelted ingot and the VAR process was replaced by the electro-flux refining (EFR) process, enabling the making of a 300 mm ingot. The final development of T1 was introduced in the early 1980s when vacuum induction melting (VIM) replaced the previous initial air-melt stage, producing VIMEFR quality. The rolling contact fatigue life of VIMEFR T1 steel is at least 25 times that of the air-melted SAE 52100 steel available in the 1950s.

9.2 US Steel Developments

The situation in the USA followed a similar pattern in that it was recognized in the mid-1950s that a high-temperature alloy was needed to replace SAE 52100 in critical high-temperature applications. Molybdenum-containing tool steels were popular steels used in the USA in the hardened state for tooling applications and a woodworking tool steel initially designated MV-1 was developed for bearing applications. This steel was later redesignated M50 and was introduced into aerobearings in 1957 in the air-melted condition. Its fatigue life was approximately twice that of air-melted SAE 52100. In 1962, vacuum arc remelted VAR M50 was introduced with a fatigue life four times that of air-melted SAE 52100, followed by VIMVAR M50 in 1970 which exhibited a fatigue life of at least eight times that of air-melted SAE 52100. In fact, operationally verified material life factors of 20–40 are now the rule rather than the exception for VIMVAR M50 and other factors, such as a requirement introduced in 1972 for conforming grain flow in M50 rings, also contribute towards improved performance.

The chemical compositions of T1 and M50 tool steels are shown in Table 4.

9.3 Heat Treatment of Bearing Tool Steels

The operating temperatures of bearings in gas turbine engines are of the order of 220 °C and during engine shutdown, soak back of heat can raise the bearing temperature significantly higher than this. To withstand these conditions, both T1 and M50 are triple tempered in the region of 550 °C, after quenching from very high carefully specified temperatures in the range 1100–1230 °C. A deep freeze operation is also carried out during the tempering operations to reduce the retained austenite content to a minimum and to generate the secondary hardening due to fine carbide precipitation. In this way, high hardness and excellent dimensional stability are produced.

The hardening cycle is usually carried out in molten salt baths, but vacuum hardening is also used, especially with M50 steel. Austenitizing times are approximately 1–2 min for T1 and 15 min for M50.

Table 4
Chemical composition of T1 and M50 bearing tool steels (wt%)

	T1	M50
Carbon	0.70–0.80	0.80–0.85
Chromium	4.00–5.00	4.00–4.25
Tungsten	17.50–19.00	
Molybdenum		4.00–4.50
Vanadium	1.00–1.50	0.90–1.10

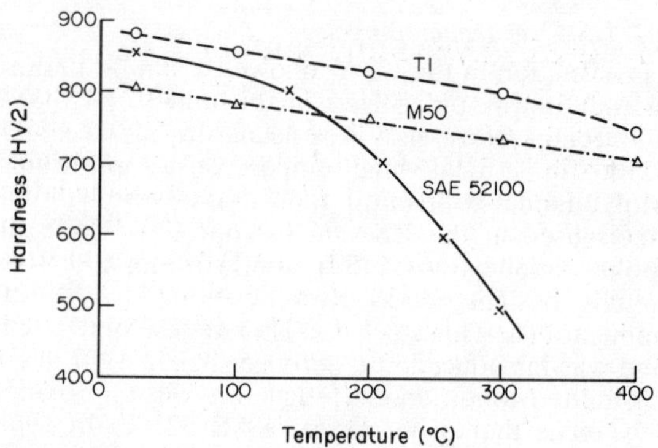

Figure 7
Hot hardness of bearing steels

Figure 7 illustrates the improvement in hot hardness of T1 and M50 over the standard SAE 52100. M50 is inherently softer than T1, but maintains its hardness in a similar manner to T1, while SAE 52100 softens considerably at temperatures over 200 °C. Holding components at these temperatures for many hundreds of hours again shows the significant advantages of the tool steel.

9.4 Current Status

The current status of the two main tool steels is that EFR T1 and VIMVAR M50 have been found to have equivalent life in rolling contact fatigue tests. Fatigue life of these steels in general is so good that conventional subsurface initiated rolling contact fatigue failures are very rare and the life-determining factors have moved from the classical fatigue failure to more insidious surface distress modes of failure. Surface distress is an umbrella term covering a variety of related failure modes. Micropitting, surface fatigue, corrosion, scoring, debris damage, contamination, plastic deformation, smearing and even process-induced damage are all inherent in, or contribute to, surface distress. With gas turbine bearings, this type of failure mode accounts for 70% of all bearing failures.

9.5 Future Developments

Currently, gas turbine bearings operate up to approximately 2.4×10^6 *DN* where *D* is the bore diameter in millimeters and *N* the revolutions per minute, but in the next generation of aeroengines operating speeds are likely to increase to the extent that values will reach 3×10^6 *DN*. Several problems start to manifest themselves under these conditions, the most important of which is the very high stress developed in rotating rings which in combination with the low fracture toughness and low stress intensity factor of the current through-hardened steels can result in catastrophic fracture and breakup of bearing rings. Fracture toughness values of bearing tool steels are approximately 16–18 MPa $m^{-1/2}$ which means that cracks of the order of a few millimeters produced around a developing rolling contact fatigue spall would exceed the critical defect size in a high hoop stress situation and could initiate rapid fracture.

Table 5
Chemical composition of RBD and M50 NiL (wt%)

	RBD	M50 NiL
Carbon	0.17–0.22	0.11–0.15
Chromium	2.75–3.25	4.00–4.25
Tungsten	9.50–10.50	
Molybdenum		4.00–4.50
Vanadium	0.35–0.50	1.20–1.40
Nickel	0.50–0.90	3.20–3.60

To cater for this requirement, high-temperature carburizing steels that have the required core toughness and the necessary surface hardness to support the contact stresses have been developed in both the UK and USA. The UK steel is known as RBD and the US steel as M50 NiL, and their chemical compositions are given in Table 5. Both these steels have been successfully used in engine mainshaft applications to overcome cracking problems. The secret behind the ability of these steels to survive under a high hoop stress situation is that when they have been correctly carburized, high residual compressive stresses are set up in the case which balance the hoop stress. Figure 8 shows a typical residual stress profile generated in M50 NiL after heat treatment prior to grinding. Typical hoop stress levels in a bearing ring at 3×10^6 *DN* would be of the order of 300 MPa, which would be balanced by the compressive residual stress in the case.

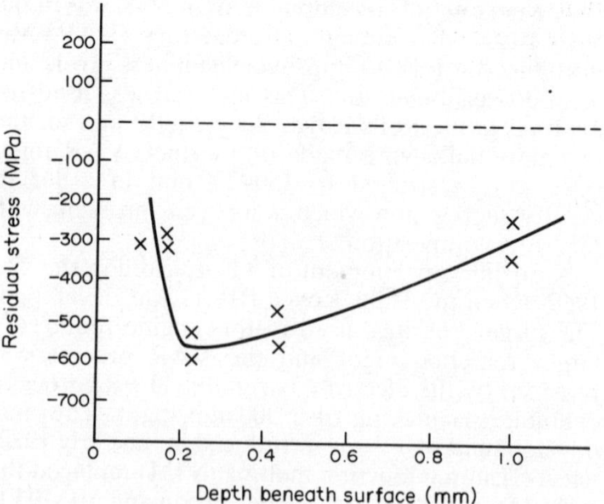

Figure 8
The residual stress profile in the case of fully heat-treated M50 NiL

The other major problem associated with high *DN* operation is the centrifugal loading effect of the ball complement on the outer raceway, especially with the high-density T1 steel. Use of a lower-density material reduces this effect dramatically. Accordingly, much work has been carried out on the performance of ceramic materials, especially silicon nitride, produced by the hot-pressing route. The rolling contact fatigue strength of this material is several times that of the best VIMVAR or VIMEFR bearing tool steels. Ceramic balls are also under consideration for use in high-speed high-precision machine tool bearings since they result in a stiffer cooler-running bearing.

The surface distress mode of failure in bearings is being addressed by the evaluation of various surface treatments, which promote a scuff-resistant hard-wearing surface. Techniques such as electroplating, plasma treatment and ion implantation are under consideration.

10. Conclusion

For standard bearings the usual choice of steel is SAE 52100, made to the most exacting standards to ensure high rolling contact fatigue performance. Enhanced fatigue performance can be gained by the use of re-refined steels. Carburizing steels are used for applications where heavy stock loading is present and corrosion-resistant stainless steels are used in severe corrosive environments. For water-containing environments where dirt is often present, the use of bainite heat-treated SAE 52100 improves lifetime and avoids catastrophic failure. Re-refined high-speed tool steels are necessary for the very demanding requirements of gas turbine aeroengines to provide the temperature and fatigue capability that is mandatory in these applications. Materials such as ceramics and special surface treatments can also play an important role in this application.

See also: Quenching Media: Polymeric Additives (Suppl. 2); Tool and Bearing Steels; Tribology

Bibliography

Hoo J J C (ed.) 1975 *Bearing Steels: The Rating of Non-Metallic Inclusions*, ASTM STP 575. American Society for Testing and Materials, Philadelphia, PA

Hoo J J C (ed.) 1982 *Rolling Contact Fatigue Testing of Bearing Steels*, ASTM STP 771. American Society for Testing and Materials, Philadelphia, PA

Hoo J J C (ed.) 1988 *The Effect of Steel Manufacturing Processes on the Quality of Bearing Steels*, ASTM STP 987. American Society for Testing and Materials, Philadelphia, PA

Landberg G, Palmgren A 1947 *Acta Polytech., Mech. Eng.* 1(3)

J. M. Hampshire
[RHP Industrial Bearings, Newark, UK]

Strained-Layer Superlattices

Strained-layer superlattices (SLS) are structures comprising alternating layers of two or more materials with different equilibrium lattice constants, grown in a structure with a common in-plane lattice constant (see Fig. 1). For sufficiently thin layers, the lattice mismatch is accommodated by uniform strain in the layers. Under these conditions, no misfit dislocations are generated at the interfaces, so high-quality crystalline materials are yielded. The alternating layers superimpose a periodic potential on the structure to yield a new electronic band structure, and the strain provides an independent variable that can be used to tune the electrical properties. These artifically structured materials are an extremely exciting recent advance, providing a new field of semiconductor research as well as permitting new solid-state electronic and optoelectronic devices.

Optical and electrical applications require high-quality dislocation-free structures. Dislocations are generated in non-lattice-matched layers when the strain exceeds a critical value. Since the strain energy is proportional to the thickness of the layer, before the advent of SLSs, the requirement that materials be lattice matched limited studies of superlattice structures and heterojunctions to AlAs and GaAs compounds, including alloys of (Al,Ga)As, where the lattice mismatch is less than 0.2%. The

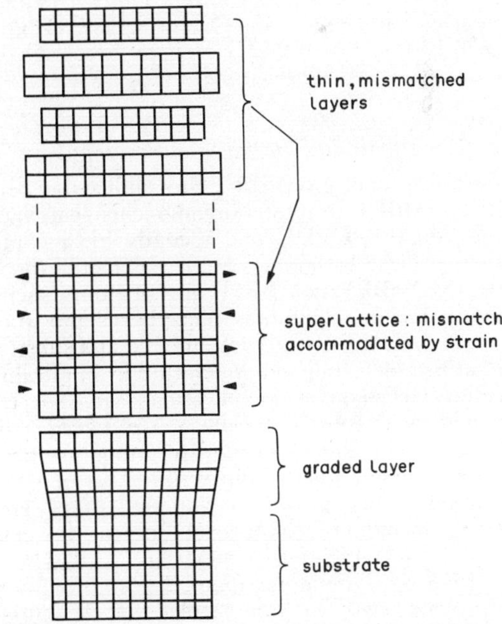

Figure 1
Schematic illustration of the elements and formation of a strained-layer superlattice

unprecedented freedom in the choice of layer materials and thicknesses available with SLSs offers a great variety of structures with a corresponding variety of materials properties. Of greatest interest for device applications are the structural, electrical and optical properties.

1. History

Superlattices consisting of adjacent layers of different materials were first proposed by Esaki and Tsu (1970) who predicted that new electronic effects would occur for superlattice layer thicknesses comparable with the electron mean free path. Frank and van der Merwe (1949) considered accommodation of lattice mismatch between crystals with different equilibrium lattice constants. They calculated that, under certain conditions, the lattice mismatch could be accommodated entirely by elastic strain without the generation of misfit dislocations.

Osbourn (1982) performed the first theoretical calculations of the electronic properties of SLSs to determine the band structure of selected systems vs lattice spacing and layer thickness. Shortly thereafter, Biefeld (1982) grew high quality SLSs in the GaAsP–GaP system, and Osbourn's theoretical predictions were confirmed by measurements of the optical properties. These original studies showed that the electronic properties of compound semiconductor SLSs are of interest and led to a large-scale effort in materials science, physics and device research in SLSs which continues to grow at an increasing rate (Peercy and Osbourn 1987, Osbourn et al. 1989).

2. SLS Growth and Stability

SLSs are usually grown by either molecular beam epitaxy (MBE) or metal-organic chemical vapor deposition (MOCVD), and recently by a combination of these techniques known as metal-organic MBE (MOMBE). In MBE, beams of atoms, such as gallium, arsenic, aluminum, phosphorus and so on, that comprise the crystal, are directed at a substrate held at a predetermined temperature in an ultrahigh vacuum (UHV) environment. Each atomic specie is supplied by evaporation from a high-purity solid, and the composition of the resulting crystal is controlled by the relative fluxes of the various types of atoms. This growth technique offers great control, including atomic-level control in many cases, and very-high-purity materials.

In MOCVD, the atoms that will form the crystal are provided by organic molecules that decompose thermally on contact with the heated substrate. MOCVD typically offers much faster crystal growth rates and higher throughput than does MBE. In MOMBE, gas sources are combined with the high vacuum of MBE to give increased growth rates compared with MBE and increased control compared with MOCVD. Dopant atoms, such as silicon, sulfur, phosphorus and boron, that are used to control the electrical properties, are introduced during growth. These growth techniques offer depth control of the order of nanometers over the composition and dopant location in the final structures.

SLSs can be grown with positive and negative strains in alternating layers, and thus no net forces, or with essentially all of the strains confined to one of the materials to yield a strain energy that increases with total SLS thickness. A typical structure designed to have no net strain consists of three principal regions: a bulk crystalline substrate, a buffer layer comprising an alloy with a graded composition (so that the lattice constant varies from that of the substrate to the average lattice constant of the SLS) and the SLS. This buffer layer provides a transition from the crystalline lattice to the SLS. The SLS consists of many alternating thin (<30 nm) single-crystal layers with equilibrium lattice constants that are alternately larger and smaller than the average constant. Each layer of the SLS is elastically strained, and the alternating strains deflect dislocations that originate in the graded layer to the edge of the wafer. The dislocations are thus confined to the first few layers of the SLS to produce dislocation-free material in the upper layers.

The amount of strain that can be accommodated in equilibrium is determined by the energy balance between misfit dislocations and strain energy. The thinner the layer, the more strain it can accommodate. Very thin layers (of the order of tens of nanometers) can accommodate misfits of several atomic percent. In this case, the lattice spacings adjust in the plane with an accompanying change in the lattice spacing perpendicular to the layer. Thus crystals with cubic equilibrium structures undergo a tetragonal distortion, which can have major implications for the electronic and optical properties.

The stability limits and structural quality of SLSs have been examined by a variety of techniques. Perhaps the most accurate measurements of layer composition and strain are obtained by high-resolution x-ray scattering and the most direct measurements of structural quality are obtained by cross section transmission electron microscopy (TEM), or, for direct-gap materials, microphotoluminescence. A TEM micrograph illustrating the uniformity and high structural quality of an SLS structure with 20 nm layer thickness is shown in Fig. 2.

Strained-layer thicknesses greater than the equilibrium thickness can be achieved under certain growth conditions. This ability to form metastable SLSs is demonstrated by MBE growth of SLSs in the silicon–germanium system (Bean et al. 1984). The observation of such metastable layers is leading to theoretical treatments of the limits of metastability and strain relaxation. The existence of such metastable structures greatly expands the parameter space available for the development of SLSs.

3. Electrical and Optical Properties

A fundamental property of semiconductors is the energy bandgap between the maximum in the valence band and the minimum in the conduction band. This bandgap determines the energy (or wavelength) at which the material absorbs or emits light; materials with a direct bandgap, in which photons directly induce transitions from the valence band to the conduction band, absorb or emit light much more efficiently than do those with indirect bandgaps, which require electrons to change the momentum by creation or annihilation of a lattice vibration (phonon) in making the transition. Direct-bandgap materials are important for optical and optoelectronic applications. Both silicon and germanium have indirect bandgaps in the near infrared spectral region, whereas compound semiconductors exist with direct bandgaps throughout the visible and infrared spectral regions.

The only binary lattice-matched III–V alloy system available with a direct bandgap is (Al,Ga)As which has a direct bandgap in the energy range 1.42–1.95 eV. Therefore, an important feature of SLSs is that they provide structures with bandgaps throughout the visible and infrared spectral regions. Furthermore, for very thin layers, the additional periodicity imposed on the structures breaks the selection rules for bulk alloys to permit direct transitions in normally indirect bandgap systems. Since the electronic energy levels can be altered by both strains and quantum confinement effects, SLSs offer unprecedented freedom in tailoring materials for electronic and optoelectronic applications.

Figure 2
Cross-section transmission electron micrograph of an $In_{0.2}Ga_{0.8}As$–GaAs SLS; the darker bands are the InGaAs regions. The layers are approximately 20 nm thick

The bandgap can be measured directly by optical absorption, or in the case of direct-gap semiconductors, by luminescence or photocurrent. Such measurements have verified theoretical calculations of the effects of strain and layer width (quantum size effects) in SLSs and demonstrated the ability to tailor the bandgap at a given composition (lattice constant). For example, the bandgap for GaAsP can be varied continuously between 1.4 eV and 2.1 eV (wavelengths 0.89–0.59 μm) at a lattice constant of 0.555 nm.

For electronic device applications, transport properties are paramount. The electronic properties are controlled locally by incorporation of dopants which provide mobile electrons to the conduction band or holes to the valence band. One of the most important properties of semiconductors for electronic device applications is high carrier mobility (the ratio of the electron or hole velocity to the imposed electric field). Mobilities are controlled by two parameters: scattering and effective mass. Mobilities are decreased by scattering from impurities and lattice vibrations or by increases in effective mass of the charge carrier. In SLSs, the mobilities are anisotropic: carriers can have low masses and high mobilities in the direction parallel to the thin SLS layers, whereas motion perpendicular to the layers is characterized by large effective masses and low mobilities.

Mobilities in the plane of the layers can be much higher than in bulk materials for two reasons. First, dopant atoms which supply the carriers can be placed in a different region of the sample than the region where the carriers move. Such "modulation doping" separates the charge carriers spatially from the charged impurities which contributed the doping. For carrier motion confined to the plane, scattering of the mobile carriers by the charged impurities will be reduced, increasing the mobility. The highest mobility reported to date using this modulation doping technique is $50\,000\ mm^2\,V^{-1}\,s^{-1}$ for the (Ga,Al)As system compared with typical mobilities of $1500\ mm^2\,V^{-1}\,s^{-1}$ for bulk GaAs.

Hole mobilities in compound semiconductor SLSs can also be increased by strain. In bulk GaAs, for example, the valence band maximum is doubly degenerate, containing both light and heavy holes.

The heavy hole bands have the largest density of carriers and dominates hole transport. With suitable strain, however, this degeneracy can be lifted in a manner that allows preferential population of bands with small in-plane effective masses which dominate the transport parallel to the layers. Measurements in the InGaAs SLS system have yielded hole masses as low as $0.11m_e$ compared with the bulk heavy hole mass of about $0.5m_e$ to yield comparable electron and hole mobilities.

The ability to tailor the electrical and optical properties in SLS systems offers new possibilities for device applications. The extremely high electron mobilities offer the potential for very-high-speed devices, and the ability to develop compound semiconductor structures in which electrons and holes have comparable mobilities offers the possibility for complementary logic analogous to the complementary metal-oxide-semiconductor (CMOS) technology that is so important in silicon technology.

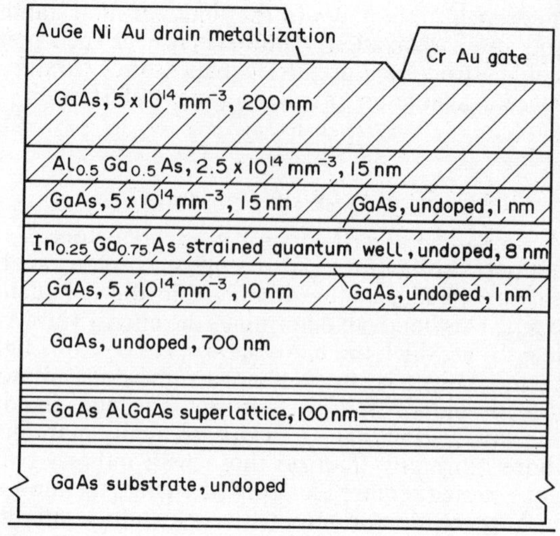

Figure 3
Schematic illustration of an *n*-channel strained quantum well transistor (SQWFET): only one-half of the device is shown; the structure is symmetric about the right side of the figure; the $In_{0.25}Ga_{0.75}As$ strained quantum well (SQW) is modulation doped from the silicon impurities on both sides of the SQW

4. SLS Devices

The electrical and optical properties of SLSs are being explored for optoelectronic and high-speed-device applications. A variety of optical detectors, light-emitting diodes and lasers have been demonstrated. The ability to tailor the bandgap over wide ranges using SLSs is being exploited to produce high-quality LEDs at various wavelengths. Such wavelength selection is also important for detectors because it permits the sensitivities optimized at a selected wavelength. SLS systems of interest for near infrared (InGaAs, SiGe), far infrared (InAsSb) and visible GaAsP, ZnSe–ZnTe, CdMnTe) wavelength ranges are currently under study.

Very recent optical devices developed in these systems included monolithic epitaxial surface emitting lasers and bistable optical switches. The same structure exhibits laser action and optical bistability. Such structures promise new freedom in chip-to-chip communications, and optical computing.

Strained-layer systems provide similar advantages for electronic devices. A particularly promising structure for field-effect transistors (FETs) is the strained quantum well (SQW) structure. A SQWFET is illustrated in Fig. 3. The thin (~8 nm) active region of InGaAs is modulation doped from donors located in the GaAs cladding layers. This device demonstrated excellent intrinsic performance even in the prototype stage, and the high-speed promise was realized by scaling this structure to submicrometer dimensions.

A related device that has the possibility for great impact is the *p*-SQWFET. This device relies on the strain-induced splitting of the valence band maxima to provide low-mass holes. Hole transport in bulk compound semiconductors is dominated by heavy holes; as a result, *p*-channel devices (e.g., from bulk GaAs) are slow. However, *p*-SQWFETs exhibit excellent performance. The high-speed capabilities of these devices offers the promise of integration with high-performance *n*-channel devices for complementary logic in compound semiconductors.

5. Summary

High-quality superlattices can be grown from semiconductor materials with lattice mismatches up to several percent. These SLSs are providing previously unexploited freedom in the choice of materials and the ability to tailor the resulting electrical and optical properties. Such artificially structured materials not only open up entirely new areas of materials science, but bandgap engineering is permitting the development of novel devices with previously unanticipated performance.

See also: Multilayers, Metallic (Suppl. 2)

Bibliography

Bean J C, Feldman L C, Fiory A T, Nakahara S, Robinson I K 1984 Ge_xSi_{1-x}/Si strained-layer superlattice grown by molecular beam epitaxy. *J. Vac. Sci. Technol. A* 2: 436

Biefield R M 1982 The preparation of device quality gallium phosphide by metal organic chemical vapor deposition. *J. Crystal Growth* 56: 382

Esaki L, Tsu R 1970 Superlattice and negative differential conductivity in semiconductors. *IBM J. Res. Dev.* 61–5
Frank F C, van der Merwe J H 1949 One-dimensional dislocations. II. Misfitting monolayers and oriented overgrowth. *Proc. R. Soc. London, Ser. A* 198: 216
Osbourn G C 1982 Strained-layer superlattices from lattice mismatched materials. *J. Appl. Phys.* 53: 1586
Osbourn G C, Gourley P L, Fritz I J, Biefield R M, Dawson L R, Zipperian T E 1989 Principles and applications of semiconductor strained-layer superlattices. In: Willardson R K, Beer A C 1989 *Semiconductors and Semimetals*, Vol. 24. Academic Press, New York, Chap. 8
Peercy P S, Osbourn G C 1987 Strained-layer superlattices. *J. Met.* 39: 14–18

P. S. Peercy
[Sandia National Laboratories, Albuquerque, New Mexico, USA]

Stringed Instruments: Wood Selection

Wood used for string instruments, such as pianos, members of the violin family and guitars, is restricted to particular species. These are different for each instrument and for each of its parts, selection being based on long-standing experience. Furthermore, only wood of the highest quality and appearance is chosen. In other words, wood for stringed instruments is selected empirically. The yield per log is presently very low; it is forecast that in the near future acquisition of suitable wood will become more difficult. Therefore, it is necessary to change the empirical selection method into a scientific one. Much research has been done to clarify scientific aspects of wood used for these instruments. In this article, basic principles and selection methods are described for wood for soundboards of pianos and sound boxes of violin-family instruments.

1. *Wood Used for Stringed Instruments*

Species of wood used for the soundboards of pianos and instruments of the violin family have been restricted to the genus *Picea*, namely European spruce (*Picea abies*), Sitka spruce (*Picea sitchensis*), and Akaezomatsu or Glehn's spruce (*Picea glehnii*). Similarly, backs of violins and related instruments are restricted to the genus *Acer*, namely sugar maple (*Acer saccharum*), Sycamore maple (*Acer pseudoplatanus*), Norway maple (*Acer platanoides*), and Itaya maple (*Acer mono*). In addition to species, wood used for instruments has also been restricted by tree age and trunk position; that is, only the lower part of trunks with a tree age of 100–300 years is acceptable. Boards are usually quarter-sawn; that is, their face is in the longitudinal/radial (LR) plane. Soundboards are graded empirically, based on the extent of various defects (sloping grain, annual ring width, indented rings, color, reaction wood, sapwood, decay, knots, resin streaks, resin pockets, checks and injuries). Boards of acceptable grade have grain which is parallel to their length, a straight, uniform annual ring width of not more than 3 mm and only slight defects of the type mentioned above. The backs have grain which is parallel to their length and usually with a flame or curl pattern.

2. *Quality Evaluation and Physical Properties*

Wood is selected on the basis of the physical properties which are important for use. For the functions of soundboards, the physical properties most desirable are generally lightness, high elasticity, speed of propagation of vibration, length of duration of sound, and bigness of sound. These properties are related to physical quantities, namely specific gravity γ, Young's modulus E, sound velocity v which is proportional to $(E\rho^{-1})^{1/2}$ (where ρ is density), internal friction Q^{-1} which is proportional to internal loss energy, and sound radiation damping ν which is proportional to $v\rho^{-1}$. Consequently, it can be summarized that wood having low γ, high $E\gamma^{-1}$ and low Q^{-1} is suitable for soundboards. Backs participate in the vibration of sound boxes of violins, and thus they must also have suitable physical properties. As flexural vibration is excited in soundboards and backs, their physical properties are important not only in the longitudinal (L) direction but also in the radial (R) direction.

Characteristically, spruce has higher $E\gamma^{-1}$, lower Q^{-1} and higher $v\rho^{-1}$ than wood of other species in the L direction. Furthermore, it has been found that wood for higher-graded instruments has higher $E\gamma^{-1}$ and lower Q^{-1}. Thus, spruce wood for soundboards has the most desirable physical properties in the L direction. As frequency increases, the proportion of deformation caused by shear as compared to bending increases, so that the value of Q^{-1} increases in the high-frequency range. In the L direction, the value of spruce wood is lower in the low-frequency range and higher in the high-frequency range when compared to hardwoods. Spruce wood shows a higher dependence of Q^{-1} on frequency because its $E_{\mathrm{L}}G_{\mathrm{LT}}^{-1}$ value (where G_{LT} is the shear modulus in the LT plane), which determines the contribution of shear relative to flexural deformation, is higher. In wood the values of $E\gamma^{-1}$ are lower and values of Q^{-1} are higher in the R than in the L direction. Nothing noteworthy can be observed in these values in the R direction for spruce wood. Maple and other hardwoods are fairly inferior to spruce with respect to physical properties in the L direction, but slightly superior in the R direction. Furthermore, maple does not have any special characteristics in its physical properties compared with other hardwoods. Consequently, maple wood has been used for violin backs because of its beautiful appearance and its characteristic figure, rather than its physical properties.

3. Physical Properties and Structure

The relationship between physical properties which are important to soundboards and wood structure has been investigated by several researchers. In conifers, tracheids make up most of the constituent cells. From this it can be shown theoretically that $E\gamma^{-1}$ of wood in the L direction is approximately proportional to the average Young's modulus of the cell wall in the L direction. Therefore, differences in $E\gamma^{-1}$ values of coniferous woods in the L direction can be attributed to differences in cell-wall structure. For conifers it has been shown experimentally that wood of species having higher $E\gamma^{-1}$ values have a smaller microfibril angle ϕ (which is a measure of the orientation of semicrystalline cellulose) and a higher degree of crystallinity α. These relationships can also be shown by numerical analysis of cell-wall models. Also, for spruce wood it has been shown experimentally that $E\gamma^{-1}$ decreases and Q^{-1} increases with increasing grain angle which corresponds to increasing values of ϕ. This explains why spruce boards having grain which is straight and parallel over their length have been used for soundboards. Reaction wood, which forms on the lower side of leaning stems in conifers, characteristically has darker color, a higher value of ϕ of the middle layer in the secondary wall, S_2, and a higher lignin content in comparison with normal wood. In the L direction severe reaction wood has lower $E\gamma^{-1}$ and higher Q^{-1} than acceptable wood. However, light reaction wood is comparable to acceptable wood in its physical properties. As annual rings become wider, that is, as the proportion of latewood becomes lower, $E\gamma^{-1}$ decreases and Q^{-1} increases in the L direction. Therefore, wood having wide annual rings is inferior. The color of wood has hardly any influence on its physical properties, and therefore colored wood traditionally was not used only because of its appearance. Sapwood, which also has not been used, has high values of both $E\gamma^{-1}$ and Q^{-1}. The Q^{-1} value of sapwood is high because of its high equilibrium moisture content. Wood containing severely indented rings and resin streaks has inferior physical properties. However, wood containing slightly indented rings, and small knots and pitch pockets, is almost comparable to acceptable wood. Therefore, these kinds of wood have been regarded as inferior not because of their physical properties but because of their appearance.

4. Nondestructive Selection Methods

4.1 Soundboards of Pianos

In order to select wood suitable for piano soundboards more easily, physical quantities which can be used as selection criteria have been investigated. The parameters v, $E\gamma^{-1}$, Q^{-1} and $v\rho^{-1}$ discussed above can also be used as criteria. It has been found that there is high correlation between $Q^{-1}(E\gamma^{-1})^{-1}$ and $E\gamma^{-1}$at any moisture content regardless of species, the three principal grain directions or grain angle. There is also a high correlation between $(QE)^{-1}$ and $E\gamma^{-1}$. If a material is subjected to sinusoidal stress with constant amplitude, the parameter $(QE)^{-1}$ is proportional to the energy per cycle dissipated as heat. Consequently, by measuring v or both E and γ, wood for piano soundboards can be selected easily. By selecting wood having higher $E\gamma^{-1}$ and low γ, wood having high $v\rho^{-1}$ (i.e., high efficiency of sound radiation) can be obtained. In addition, the parameters $v\rho^{-1}Q$, $\bar{R}_{mn}$ and EG^{-1} have been proposed as criteria. The parameter $v\rho^{-1}Q$ combines $v\rho^{-1}$ and Q^{-1}. $\bar{R}_{mn}$ is the average value of the sound pressure level in each mode produced by a board subjected to an oscillating force with constant amplitude. Therefore, $\bar{R}_{mn}$ is also a parameter related to the efficiency of sound radiation. Spruce wood has an especially high $\bar{R}_{mn}$ value. The quantity EG^{-1} is related to the slope of the envelope curve of a sound spectrum and to the magnitude of a resonance point. Spruce wood has higher values of this quantity in the L and R directions than wood of other species.

4.2 Sound Boxes of Violin Family Instruments

The relationships between the vibrational characteristics of a finished violin and those of its parts as free plates are very complex. Therefore, the selection of wood for the soundboards and backs is also complex. According to the studies done so far, in order to make violins with fine tone and playing qualities, it is most desirable (but often quite difficult) to have the free plate modes 1, 2 and 5 of a finished violin soundboard lie in a harmonic series, with mode 5 having a large amplitude and a frequency near 370 Hz, and to have the frequencies of modes 2 and 5 of the soundboard match those of the back. The shape, arching contours and thickness distributions of the soundboard and back plates are crucial in achieving these relations. The values of E, G, Q^{-1}, γ and v of the wood for soundboards and backs are very important to the sound of fine instruments. For the soundboards it is necessary to select wood having high $E\gamma^{-1}$ not only in the L but also in the R direction. The above may also be said of the other members of the violin family.

In order to determine these wood parameters instantly and automatically, measuring methods which pass the sound of wood being tapped through an FFT (fast Fourier transform) analyzer connected to a personal computer are being developed. Further improvements in selection methods of wood for instruments will require clarification of the relationship between the quality of finished stringed instruments and the physical properties of their components as free plates.

See also: Acoustic Properties of Wood

Bibliography

Holz D 1973 Untersuchungen an Resonanzholz. *Holztechnologie* 14: 195–202

Hutchins C M 1981 The acoustics of violin plates. *Sci. Am.* 245: 126–35

Kollmann F P, Côté W A 1968 *Principles of Wood Science and Technology 1*. Springer, Berlin

Meinel H 1957 Regarding the sound quality of violins and a scientific basis for violin construction. *J. Acoust. Soc. Am.* 29: 817–22

Ono T, Norimoto M 1983 Study on Young's modulus and internal friction of wood in relation to the evaluation of wood for musical instruments. *Jpn. J. Appl. Phys.* 22: 611–14

Ono T, Norimoto M 1984 On physical criteria for the selection of wood for soundboards of musical instruments. *Rheol. Acta* 23: 652–6

Ono T, Norimoto M 1985 Anisotropy of dynamic Young's modulus and internal friction in wood. *Jpn. J. Appl. Phys.* 24: 960–4

Yankovskii B A 1967 Dissimilarity of the acoustic parameters of unseasoned and aged wood. *Sov. Phys. Acoust. (Engl. Transl.)* 13: 125–7

T. Ono

[Gifu University, Gifu City, Japan]

Superconducting Device Fabrication

Although rapid progress in the development of a thin-film technology for high-critical-temperature superconducting materials is bringing superconducting devices closer, the fabrication techniques available are still primitive.

Device technology for low-critical-temperature materials has, however, progressed steadily. In the early 1980s, it became clear that the lead-alloy junctions that had been used in Josephson integrated circuits during the 1970s were not practical, because the characteristics of the lead-alloy junctions were unstable. Much work has since gone into replacing lead-alloy electrodes with refractory materials such as niobium and NbN.

In 1983, an excellent Josephson junction with niobium electrodes was made by Gurvitch et al. (1983). It had a tunnel barrier of aluminum oxide (Nb/AlO_x/Nb). Since then, niobium junctions have been applied to many logic and memory circuits (Kotani et al. 1988, Hasuo and Imamura 1989, Suzuki et al. 1989). It is now possible to fabricate a circuit with tens of thousands of junctions (Kotani et al. 1989). The most promising techniques for superconducting devices use niobium Josephson junctions.

1. Josephson Circuit Fabrication

The Josephson junction is made of a thin (a few nanometers) tunnel barrier sandwiched between two superconducting films, the base and counter electrodes. Standard Josephson integrated circuits use Nb/AlO_x/Nb Josephson junctions, molybdenum resistors, a niobium ground plane and niobium wiring usually in 9–11 layers on silicon substrates. Each layer is patterned through a photoresist mask by reactive ion etching (RIE).

A cross section of an integrated circuit with Josephson junctions is shown in Fig. 1. Josephson circuits are usually fabricated above a superconducting ground plane. Resistors that are used as loads or for damping in junction switching contact the base electrode. Via holes are made in insulators above junction counter electrodes. In some logic circuits wiring is the top layer, but usually two additional layers (insulators and control lines) are formed on the wiring.

Layer materials and thickness are listed in Table 1. Niobium, aluminum, molybdenum and SiO_2 are deposited by sputtering under the conditions in Table 2. SiO_x, used as a protective underlayer for molybdenum resistors during RIE of the niobium base electrode, is deposited by evaporating SiO in oxygen. All but the SiO_x layer are patterned by RIE. The reactive gases are listed in Table 3. Because aluminum and AlO_x are not etched by reactive CF_4 and CHF_3, the thin Al–AlO_x barrier is removed by argon sputter etching. The SiO_x layer used is patterned by liftoff.

The key process in Josephson integrated circuits is making high-quality junctions. However, several other elements are involved, including the insulation layer, the superconducting wiring, the resistors and the contacts between the two niobium electrodes.

2. Nb/AlO_x/Nb Junctions

Nb/AlO_x/Nb junctions have excellent *I–V* characteristics and are reliable. The excellent controllability and reproducibility of these junctions makes them

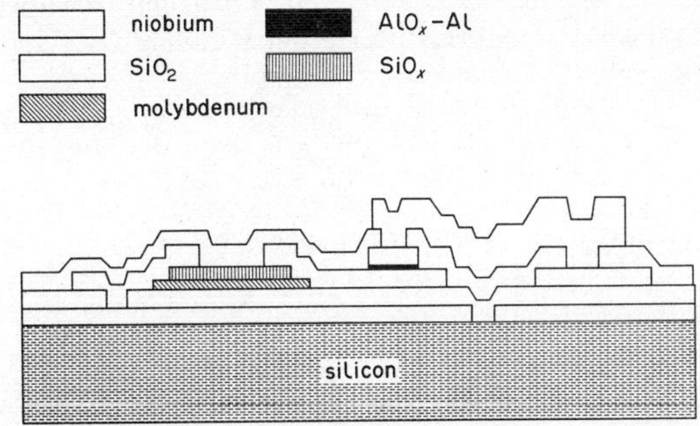

Figure 1
Cross section of integrated circuits with Nb/AlO_x/Nb junctions

Table 1
Circuit layers

Layer	Material	Thickness (nm)
Ground plane	niobium	300
Insulation	SiO_2	300
Resistor	molybdenum	100
Resistor protection	SiO or SiO_x	100
Base electrode	niobium	200
Tunnel barrier	AlO_x–Al	7
Counter electrode	niobium	100–200
Insulation	SiO_2	400
Wiring	niobium	600
Insulation	SiO_2	800
Control line	niobium	1000

Table 2
Conditions of metal and insulation layer sputtering

Material	Argon pressure (Pa)	Deposition rate (nm min^{-1})
Niobium	1.3–2.3	200
Aluminum	1.3	8
Molybdenum	0.67	130
SiO_2	1.3	8

practical. These characteristics are responsible for the stability of refractory niobium films. Josephson junction characteristics also depend on the formation of a uniform high-quality tunnel barrier on the niobium electrodes. An aluminum layer a few nanometers thick wets the niobium surface sufficiently and a uniform AlO_x layer can be formed on the aluminum layer. The discovery of the AlO_x barrier was the single most important factor in enabling the niobium junction to be put into application.

The fabrication of Josephson integrated circuits is described by Imamura et al. (1987), and Hasuo and Imamura (1989). The fabrication process is, of course, applicable to other circuits, such as superconducting quantum interference device (SQUID) chips and voltage standard chips.

Table 3
Conditions of metal and insulation layer patterning

Material	Reactive gas	Pressure (Pa)	Power density (W cm^{-2})
Niobium	CF_4(5–20% O_2)	2.7–6.7	0.10
Aluminum	argon	0.7	0.15
Molybdenum	CF_4(5% O_2)	6.7	0.10
SiO_2	CHF_3(0–15% O_2)	2.0	0.20

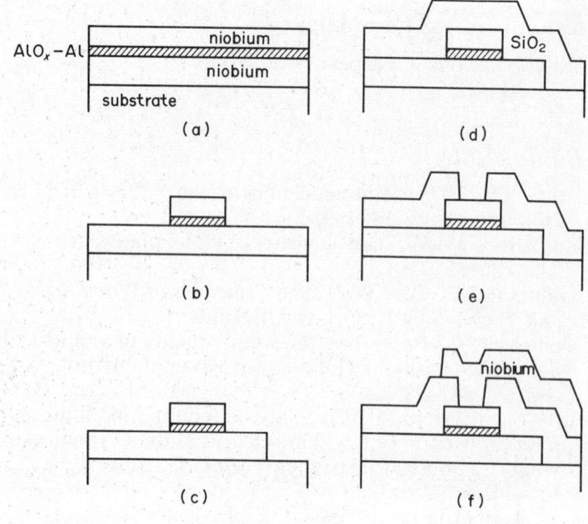

Figure 2
Nb/AlO_x/Nb junction fabrication

The Nb/AlO_x/Nb trilayer barrier is first deposited across the whole silicon substrate in the same vacuum run. The junction is defined by trimming in the next patterning process. The quality of the *I–V* characteristics depends greatly on the formation of a sharp interface between the niobium counter electrode and the AlO_x, and between the aluminum and the niobium base electrode. Optimized Nb/AlO_x/Nb trilayer deposition has been reported in detail (Morohashi and Hasuo 1987).

The Nb/AlO_x/Nb junction is usually fabricated as shown in Fig. 2. Before trilayer deposition, substrates are etched by argon sputtering to contact the niobium ground plane through contact holes. The trilayer is then deposited by sputtering on the substrate attached to a water-cooled holder. To attain good thermal contact, a copper backing plate and indium foils are used as spacers between the substrates and the holder. The base pressure is less than 10^{-5} Pa. After the base electrode and aluminum barrier are deposited, argon with 10% oxygen is introduced into the chamber to form a thin 2 nm oxide on the aluminum. The critical current density of the junction j_c depends on the pressure (usually in the range 50–130 Pa) and oxidation time (typically 30 min to 60 min). j_c ranges from 500–10 000 A cm^{-2}. For a design j_c of 2000 A cm^{-2}, junctions with j_c values in the range 1500–3000 A cm^{-2} can be reproducibly obtained. After oxidation, the niobium counter electrode is deposited on the AlO_x barrier. The trilayer is then trimmed to the device structure. The junction area is defined by patterning the niobium counter electrode. In RIE, AlO_x and aluminum act as good etch

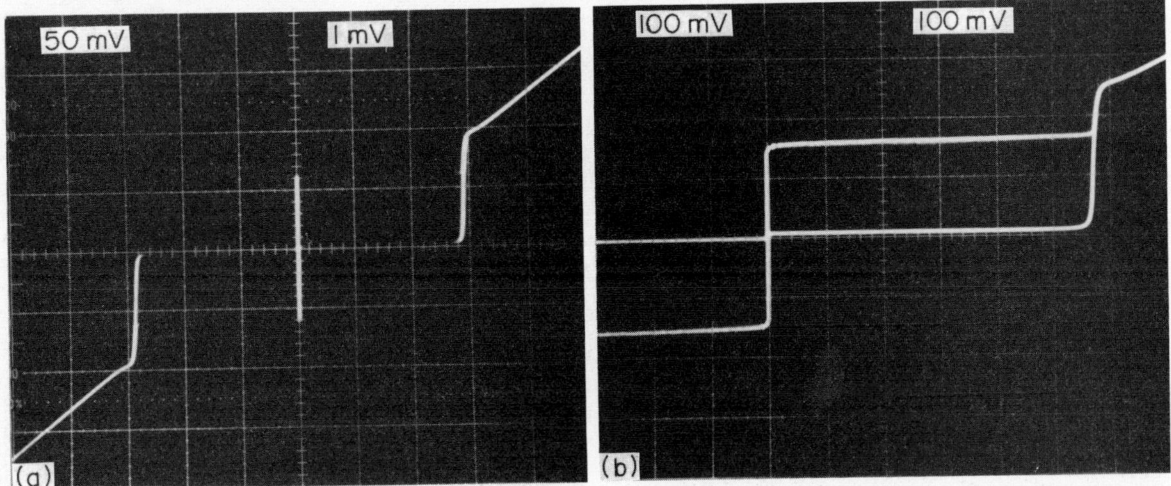

Figure 3
I–V characteristics for (a) single junction (vertical scale: 1 mA div^{-1}, horizontal scale: 1 mV div^{-1}) and (b) 200 series-connected junctions (vertical scale: 1 mA div^{-1}, horizontal scale: 100 mV div^{-1})

stops against CF_4 gas. The remaining AlO_x–Al barrier is removed by argon sputter etching. The lower niobium layer is then etched into a base electrode pattern. Junctions are covered with an SiO_2 insulation layer. Contact holes are formed in the SiO_2 by RIE, then niobium is deposited and patterned as a wiring layer.

The I–V characteristics of fabricated junctions are shown in Fig. 3. The junction size is 10 μm × 10 μm. The gap voltage is 2.9 mV. The quality parameter V_m, which is defined as the product of the critical current and the subgap resistance at 2 mV, exceeds 60 mV. Nb/AlO_x/Nb junctions are very stable in cycling between 4.2 K and 300 K, and in storage at room temperature.

3. SiO_2 Insulators

In lead-alloy Josephson circuits, SiO and SiO_x films have been used for insulation layers, but these films are not good enough for niobium circuits. The breakdown voltage between two niobium electrodes with these insulators decreases to almost zero after the counter niobium electrode is deposited. This is because of the microcracks induced by heat or stress during refractory niobium deposition. The breakdown voltage in sputtered SiO_2 exceeds 200 V even after niobium is deposited as a counter electrode. Cooling substrates during SiO_2 sputtering proved that using sputtered SiO_2 does not deteriorate the junction characteristics, which is why it is used in Josephson circuits.

4. Superconducting Niobium Wiring

As circuits become more and more densely integrated, long narrow wiring lines are needed. The superconducting critical current I_c measured for niobium wiring is shown in Fig. 4. The lines are 1.0 μm to 2.0 μm wide and 10.88 mm long. For films more than 400 nm thick, I_c depends only on the line width and not the film thickness. For films thinner than 200 nm, however, I_c changes linearly with thickness, apparently due to the supercurrent flowing in the film surface up to the London penetration depth. The current flow in Josephson logic circuits is usually

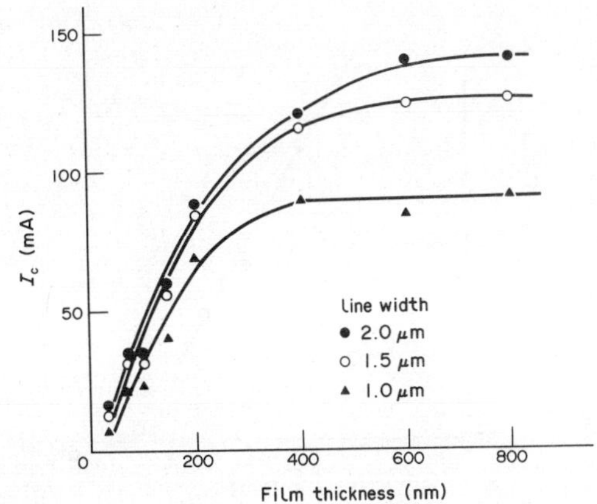

Figure 4
Superconducting critical current I_c vs film thickness niobium wiring lines

less than 1 mA. The measured I_c is more than ten times greater than the current level, so circuit malfunctions caused by normal transitions in niobium wiring cannot occur, making wiring integrity good enough for Josephson large-scale-integration circuit applications.

5. Molybdenum Resistors

Molybdenum film, deposited by sputtering, forms the resistors. Measured sheet resistance R_s is plotted as a function of thickness d in Fig. 5 (Imamura et al. 1987). R_s is proportional to $d^{-1.6}$. The spread of resistance on a wafer is only a few percent and the contact resistance between molybdenum and niobium is negligible.

6. Conclusion

Niobium Josephson junction fabrication technology has progressed remarkably in the 1980s. It has become possible to apply the technique to integrated circuits with over 10 000 junctions per chip, because of the stable, reliable, reproducible and controllable characteristics of niobium junctions.

Niobium junction fabrication will play a key role in the creation of systems with low-temperature superconducting electronics, such as multichannel SQUID systems or Josephson computers.

Bibliography

Gurvitch M, Washington M A, Huggins H A 1983 High quality refractory Josephson tunnel junctions utilizing thin aluminum layers. *Appl. Phys. Lett.* 42: 472–4

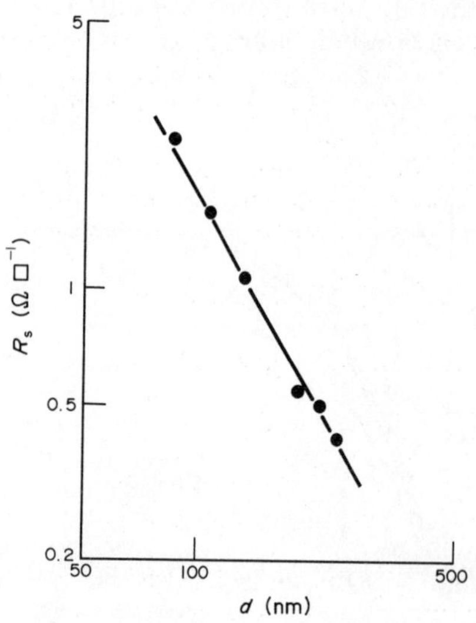

Figure 5
Sheet resistance R_s vs thickness d in molybdenum resistors

Hasuo S 1989 High-speed Josephson integrated circuit technology. *IEEE Trans. Magn.* 25: 740–9

Hasuo S, Imamura T 1989 Digital logic circuits. *IEEE Proc.* 8: 1177–93

Imamura T, Hoko H, Ohara S, Kotani S, Hasuo S 1987 Fabrication technology for Josephson integrated circuits with Nb/AlO$_x$/Nb junctions. In: Hara K (ed.) 1987 *Superconductivity Electronics*. Prentice-Hall, Englewood Cliffs, NJ, pp. 22–33

Kotani S, Fujimaki N, Imamura T, Hasuo S 1988 A Josephson 4b microprocessor. *Digest Technical Papers Int. Solid-State Circuits Conf.* IEEE, New York, pp. 150–1

Kotani S, Imamura T, Hasuo S 1989 A sub-ns clock Josephson 4b processor. *Digest Technical Paper Symp. VLSI Circuits.* IEEE, New York, pp. 23–4

Morohashi S, Hasuo S 1987 Experimental investigations and analysis for high-quality Nb/Al–AlO$_x$/Nb Josephson junctions. *J. Appl. Phys.* 61: 4835–49

Suzuki H, Fujimaki N, Tamura H, Imamura T, Hasuo S 1989 A 4K Josephson memory. *IEEE Trans. Magn.* 25: 783–8

S. Hasuo and T. Imamura
[Fujitsu Laboratories, Atsugi, Japan]

Superconducting Magnets

Superconducting magnets can be generally recognized as comprising current-carrying windings that have no electrical resistance in operation. In principle, it is possible to build a magnet from superconducting material that is a copy of a conventional winding that creates an electromagnetic field, and a wide range of magnet types should be possible. The practice is rather different, although superconducting magnets with complex shapes and weighing many tonnes have been built.

The properties of superconducting materials place constraints on the designs of magnets through the cost associated with engineering a suitable winding for the required magnetic-field profile while maintaining this in the superconducting condition. The point of considering using superconducting materials to create magnetic fields is that the field strength within any current-carrying circuit can be directly linked to the density of current in the circuit. The absence of resistance means very high current densities can be obtained, relative to conventional conductors. However, arbitrarily large current densities are not possible. The industrial exploitation of superconducting magnets has been governed by what can be reliably achieved for given capital and operating costs set against the operational circumstances of the systems using the magnetic field. Because superconductivity only occurs at relatively low temperatures, special cryogenic arrangements are necessary to cool the winding and to prevent heat generation within the windings. Many of the practical restrictions on the exploitation of superconducting magnets result from this requirement.

High-strength magnetic fields are used mainly to influence matter for experimental or practical purposes. The source of the field interaction with matter is the electric charges associated with atoms, both in their outer electron configurations and their nuclei. Physical and chemical interactions between collections of atoms making up systems, either solid, liquid or gaseous, are driven by the energy difference between the systems. The energy content of each system has a component due to the impressed magnetic field that is a characteristic of the electronic and nuclear arrangements of the system. However, the magnetic energy component will be proportional to the intensity of the impressed field although the detailed relationship can be quite complex. Therefore, many situations arise in which the higher the impressed field, the greater is the magnitude of the response of the interacting systems and, in consequence, there is a requirement for sources of high magnetic field.

1. Importance of Current Density to Field Creation

In the SI system of units, magnetic field strength $\boldsymbol{H}$ is measured as ampere-turns per meter, $\mathrm{A\,m^{-1}}$, and in the cgs system as oersteds, Oe, where $1\,\mathrm{A\,m^{-1}}$ equals $4\pi \times 10^{-3}$ Oe. The flux density $\boldsymbol{B}$ in SI units is measured in tesla, T, which are webers per square meter, $\mathrm{W\,m^{-2}}$ (the weber is a unit of flux), and in the cgs system as gauss, G, where 1 T equals 10^4 G. The flux density $\boldsymbol{B}$ is related to the impressed field $\boldsymbol{H}$ by a characteristic constant μ, termed permeability, which is unique to the medium permeated by the field.

$$\boldsymbol{B} = \mu \boldsymbol{H} \tag{1}$$

In a vacuum $\mu_0 = 4\pi \times 10^{-7}\,\mathrm{Wb\,A^{-1}\,m^{-1}}$ (or henries per meter, $\mathrm{H\,m^{-1}}$) in SI units. For other materials, tables of relative permeability μ_r are available from measurement, where

$$\mu = \mu_0 \mu_r \tag{2}$$

The relative permeability in air is almost one.

The flux density $d\boldsymbol{B}$ due to current $\boldsymbol{I}$ flowing in an element of conductor at a point P is given by

$$d\boldsymbol{B} = \mu \boldsymbol{I} \times dl \frac{\sin\theta}{4\pi r^2} \tag{3}$$

(see Fig. 1a). For any continuous path, the contribution of all the elements dl can be summed for all the points P at which it is required to know the flux

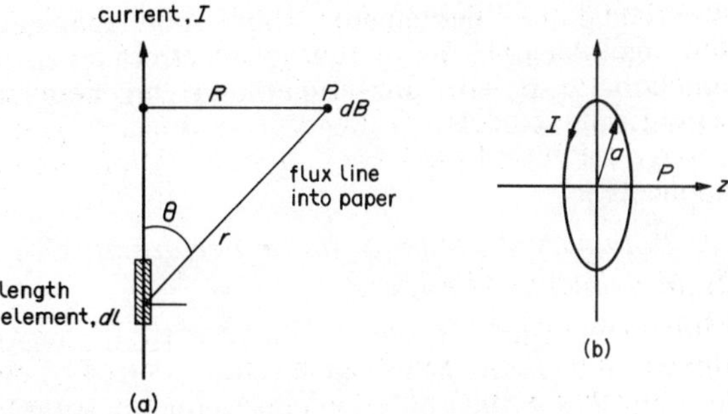

Figure 1
Relationship between magnetic field and current flow in a conductor on which a magnet design is based: (a) magnetic field at point P due to a current element $I\,dl$; and (b) dimensions of a simple loop as an elemental building unit of a solenoid

density or field strength. This may be done analytically for simple geometries, or numerically with computers for complex current paths and field zones. Thus, in principle, electromagnets can be designed to produce any required field profile. For a simple loop, the field component in the z direction on the axis at P, is given by

$$B_z = \mu I \frac{a^2}{2(a^2 + z^2)^{3/2}} \tag{4}$$

(see Fig. 1b). For a simple solenoid comprising a helical winding (like a spring) the field can be modelled by a stack of loops along the z axis; the contribution of the loops can be summed. The current density J in the conductor is the current I divided by the conductor cross-sectional area A; so J is measured in amperes per square meter. Therefore, the greater the current density in a conductor, the larger will be the flux density at the center line of the coil.

If the conductor is not superconducting, it will have a resistance to electric current and power will be dissipated. For a single loop the power W is:

$$W = 2\pi \rho a J^2 A \tag{5}$$

where a is the loop radius and ρ the resistivity in ohm meters. Using resistive conductors, as the required field is increased greater efforts to cool the conductor are needed to extract the dissipated power and eventually a practical limit is reached beyond which it will be necessary to employ a superconducting zero-resistance winding. Numbers of different ways of cooling exist and the magnet geometry plays an important part in defining "the economic" cross-over value of J favoring a superconducting magnet. For comparison of flux densities, the flux density over the earth's surface is

0.000 05 T, a permanent "horseshoe" magnet between the poles is 0.1 T, large electromagnets in industrial use 0.4 T and superconducter magnets may reach 18–20 T.

2. *Boundary Conditions on the Performance of Superconducting Magnets*

The value of field strength obtainable from a winding using superconducting materials is limited by the current density that the superconductor can support in the field of the magnet. As the field strength applied to the superconducter increases, so the maximum value of J decreases. The attainable current density in the actual winding may be less than the material is capable of under ideal conditions. This depends on the degree of success in designing the magnet structure and conductor layout to minimize the incidence of small internally generated heat pulses. Progress in the evolution of superconducting magnets can be charted by the refinement of designs to improve conductor stability in its working environment. The current density as a function of field and temperature is a fundamental characteristic of each superconducting material.

It needs to be appreciated that the specific heat of all materials tends to decrease as the temperature of the material is reduced. Liquid helium is the refrigerant used for commercial purposes, and at its boiling point (bp) (4.2 K at atmospheric pressure) specific heats are about one-thousandth of their room-temperature value. At this temperature, it requires 10^3–10^4 J m^{-3} to raise the temperature by 1 K for materials commonly used to construct superconducting magnets. Conventional superconductors only exhibit superconductivity in the temperature range 0–23 K, with each material having its own critical temperature within that range above which they will not conduct. The two most commonly used conductors are an alloy of niobium and titanium and an intermetallic compound of niobium and tin. These materials came into widespread use because reliable processes for their fabrication in composite forms were evolved; their critical temperatures are approximately 9 K and 18 K, respectively. As with increasing the applied field, increasing temperature will reduce the current density until it is zero at the critical temperature. Commercial superconducting magnets are generally operated at temperatures close to the bp of liquid helium at atmospheric pressure. This gives a working temperature range between operating conditions and the critical temperature while using a practical, if exotic, refrigeration fluid. To be economic in the use of superconductor, magnets are usually operated at current densities 25–90% of their maximum at the magnet operating temperature. This safety margin can accommodate typically between 0.1 K and 1 K temperature rise during magnet operation. Because of the small specific heat, this means that minute quantities of heat are sufficient to cause a local loss of superconductivity which may cause the whole magnet to fail.

Sources of heat that may cause the magnet winding to lose its superconductivity can be grouped into three categories: environmental, mechanical and electromagnetic. Mechanisms that give rise to heating and design solutions to minimize heat effects are reviewed next.

2.1 Environmental Sources of Heat

All superconducting magnets are housed in structures that shield the magnet from heat transferred from the environment by radiation, conduction and convection. The assembly of vessels and components is termed the cryostat and acts like the classic Dewar vessel. The size and shape of the cryostat are dictated by the dimensions, geometry and use of the magnet it contains, with various arrangements to gain access to the magnet chamber for refrigerant and to make electrical connections to the magnet. Figure 2 shows the layout of components for a medium-size magnet for magnetic resonance imaging (MRI).

A common element of cryostat design themes is to absorb most of the radiation from the room-temperature environment by boiling liquid nitrogen (bp 77 K) contained in a separate tank that is also linked to the magnet supports to thermally ground these. Reflective shields provide the magnet with its radiation environment and are either cooled by boiled-off helium gas passed through heat exchangers (negative thermal feedback) or local refrigeration engines are used to provide a low radiation temperature of 20–50 K. The whole system of vessels is evacuated and all magnet support members are made as long as possible, frequently employing low-conductivity composite materials, to minimize convection and conduction. Small magnets are nearly always immersed in a bath of helium liquid while very large magnets are sometimes built with integral cooling networks through which helium refrigerant can be pumped. Liquid helium is usually refrigerated by large liquifiers out of natural gas streams and supplied for transfer to magnets in transport dewars; typically 500–5000 l capacity. Very large magnet installations often justify their own liquifier and work with a closed cycle of helium flow. Radiation gives rise to typically 20 W m^{-2} (going from 300 K to 4.2 K) and 0.1 W m^{-2} (going from 77 K to 4.2 K). It should be noted there are many detailed design considerations placed on cryostat structures by the need to consider the working objectives of the magnet.

2.2 Mechanical Sources of Heat

When a magnet suffers a local loss of the superconducting state in the winding, the normal or "hot" zone may spread throughout the whole winding. The magnet is said to have "quenched." In practice superconducting magnets respond to the internal stress created in the winding by the electromagnetic forces with discrete movements of wire turns. The design aims to clamp turns as rigidly as possible to the position assigned to the turns during construction but, usually, very small movements are possible and the winding "settles in" to its exact operating set of positions. During settling in, bonding media between conductor turns will be distorted, possibly cracked and, because the movement of the wire turns does work, local heat pulses are generated.

Forces developed on a conductor turn per unit length are proportional to the product $\boldsymbol{H} \times \boldsymbol{I}$ and can amount to some hundreds of kilograms per meter of conductor length. The field varies in magnitude and direction within the windings, so producing a force pattern which for engineering purposes needs to be converted to a stress profile. This requires a complete knowledge of the force distribution and the effective material moduli of the components of the winding. Approximate analytical solutions can be obtained for simple geometries and finite-element methods using computers to provide solutions are employed for more complex winding shapes. In this latter approach, the solid body is divided into a series of small volumes and the stress–strain equation applied in each with boundary conditions for solution being matched at each boundary. Nevertheless, because of the small value of specific heat, it has been estimated by Wilson (1983) that the design of the winding must prevent sudden conductor movements of the order of 2 μm if the winding is to be electromagnetically efficient and reliable. The "settling in" process can give rise to the magnet quenching on a succession of runs up to field at progressively higher values of field on each run as the mechanically least-stable conductor elements "bed down." This process is referred to as "training."

2.3 Conductor Stability

If the superconductor in the magnet winding is operated at a current density less than the maximum possible in the applied field, a margin of operating

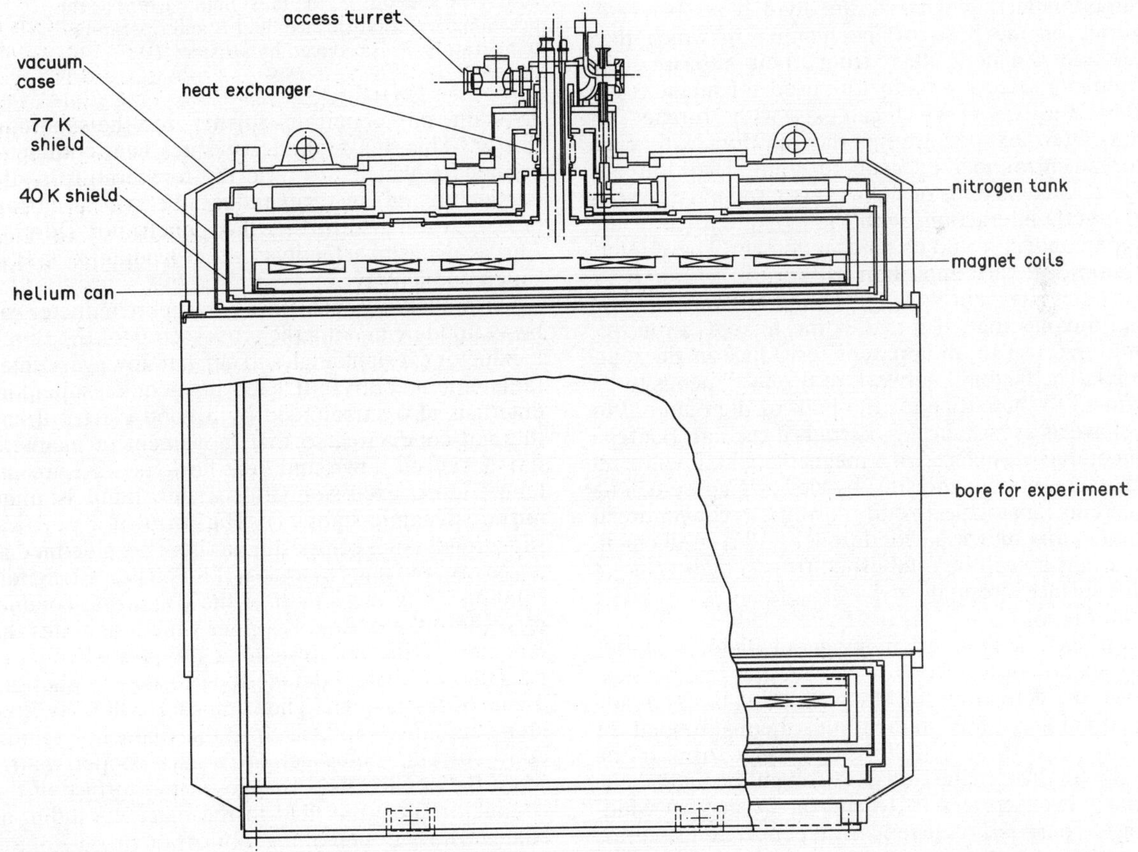

Figure 2
Principal components of a conventional superconducting MRI magnet

safety exists to accommodate disturbances. However, the resistance of superconductors in the normal state is high which results in a small normal length of conductor heating up rapidly if current is maintained in the winding. All practical superconductors are used embedded in a good conventional conductor, like copper, to provide a low-resistance current path around the normal length of superconductor. The copper also conducts heat away from the "hot zone" to the coolant and assists in lowering the temperature of the superconductor so that the winding recovers its resistanceless state after a localized discrete disturbance. Composite copper-to-superconductor volume ratios can be defined against operating current density and temperature to achieve recovery against a given size of heat pulse. An extreme is cryogenic stabilization, where the ohmic heating caused by carrying the magnet current in the copper is much less than the rate of cooling in the steady state. This latter condition requires refrigerant to be close to the composite everywhere in the winding and is only efficiently applied to large magnets.

2.4 Electromagnetic Sources of Heat

Superconducting materials are divided into two categories on the basis of the manner in which they exclude magnetic flux from their interior (the Meissner effect). Conductors used for magnets are classed as type II which can exist with "threads" of magnetic flux penetrating the interior, with each thread surrounded by a supercurrent. These threads are termed fluxoids or flux vortices and have various energetic interactions with the crystal lattice of the superconductor and its natural defects. These interactions are very important as there is a force $\boldsymbol{B}\times\boldsymbol{I}$ between a current flowing through the material and the fluxoids that, if it causes the fluxoids to move, will give rise to an apparent resistance in the material. The fluxoid "network of threads" needs to be pinned to the lattice for the bulk of the material to behave as a resistanceless carrier of current. If a type II conductor is placed in a magnetic field, it will tend to exclude magnetic flux by developing screening currents (analogous to eddy currents in conventional conductors under ac conditions) and a gradient in flux density will be established from a high value at the surface dropping towards zero at the interior (see Fig. 3).

In fact, a gradient in magnetic field is always associated with a flow of current (as can be seen from the derivations of Maxwell's equations) in any medium and a flux gradient must be maintained for transport current as well as screening current in the superconductor (the transport current being the current flowing from start to end of the magnet winding). There is no current in unpenetrated regions, however, if the current through the winding is increased, because the critical current density is

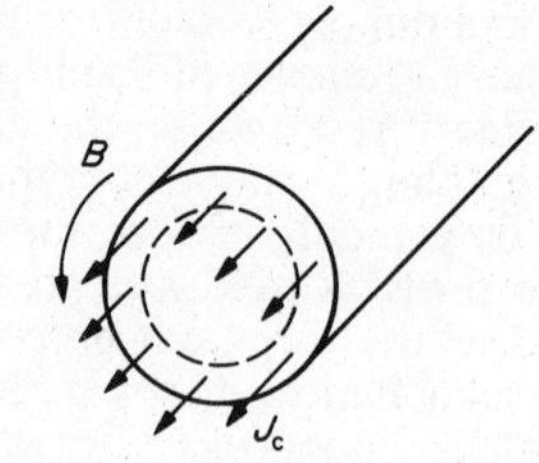

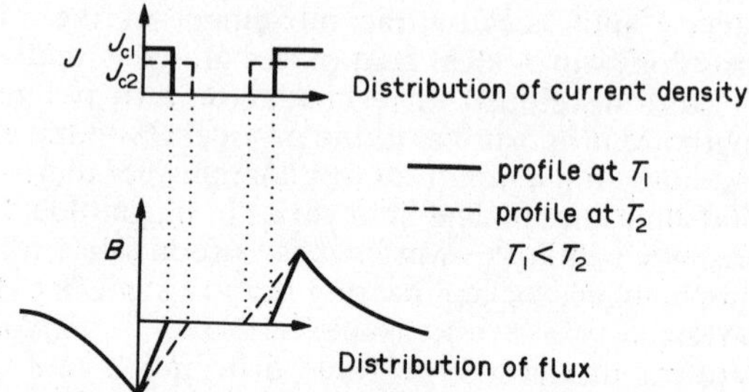

Figure 3
Superconductor carrying current and generating a "concentric self" field distribution of current and field across the section: J_c, the maximum current at the prevailing external surface field B and operating temperature T, is reduced by a rise in T

constant, flux penetrates further into the superconductor. This flux motion generates heat within the conductor, which can raise the temperature of the conductor and, in turn, reduce the critical current density leading to further flux penetration (Fig. 3). This is a positive feedback loop leading to loss of superconductivity.

Unstable flux penetration of a superconductor can be avoided by making the cross section of the superconductor (at right angles to current flow) very small (adiabatic stability). If each filament of superconductor is also surrounded by a good electrical and thermal conductor, so that movement of magnetic flux is slowed down and heat flow speeded up, the temperature excursion after a flux jump is minimized (dynamic stability). The form of a practical superconducting composite has now been defined as an array of fine filaments (10–50 μm diameter) running along the length of the composite conductor, where the matrix is copper which separates the filaments. Different designs of composite array are possible to match detailed performance requirements of the magnet. The composite will be twisted along its length to decouple the filaments against eddy currents being set up in the copper matrix when the applied field changes at the surface of the conductor (i.e., the field in the magnet winding is changed). By design of the fabrication process of the conductor, the forces pinning magnetic flux can be varied, so modifying the current field properties of a type II conductor. Optimizing conductor performance has been a major industrial activity in type II superconductor manufacture. If the field provided by the magnet is required to vary with time—an extreme case being an ac field—then losses will develop from eddy currents and flux motion hysteresis based on the mechanisms described in this section. The magnet will develop an apparent resistance across its terminals that will place an extra load on the refrigeration system and need special design responses in defining the conductor geometry. (Note that superconducting magnets are resistanceless in stable dc state but still have inductive and capacitive reactance.) Superconducting magnets can be energized, brought to a stable dc condition and a superconducting link established across their terminals. In this way, the dc field can be maintained without an external source of power and field-creating current persists in the winding to provide a very stable magnetic field. A magnet operating in this way is said to be in "persistent mode."

3. *Industrial Exploitation of Superconducting Magnets*

Initially uses of superconducting magnets were confined to laboratory physics experiments, starting in the 1950s when type II materials became available. The use of superconductivity was promoted by the availability of stable high fields. The practicality of achieving high fields (2–3 T) over several cubic meters of space, gave rise to a number of programs to provide magnets for particle physics experiments. These magnets were usually government funded and built in the laboratories where the experiments were designed. Manufacturing superconducting magnets on an industrial basis originated with the supply of instrument systems to laboratories using small magnets. A very important use of superconducting magnets has been in providing for nuclear magnetic resonance (NMR) experiments, initially for spectroscopy and, since 1980, for whole-body imaging.

Industrial superconducting magnets can cost up to US$50 per 10^{-4} T depending on size and field quality. Many of the obvious uses of superconducting magnets have not been widely accepted because of high capital costs and marginal costs associated with refrigeration. Examples of magnet applications are compact synchrotrons for x-ray lithography of chips, small cyclotrons for positron emission tomography, levitated high-speed transport, energy generation by fusion and by magnetohydrodynamics, medical imaging scanners and medical NMR diagnostic scanners. The world market for helium superconducting magnets is now some US$500 million per year, growing at a rate of 10–15% annually.

The discovery of high-temperature ceramic superconductors will be important to superconducting magnet technology in the following ways.

(a) Operation of magnets in the temperature range 50–80 K (critical temperature 100–120 K) will simplify cryostat and refrigeration systems, lowering costs but, more importantly, permitting a greater range of magnet geometries.

(b) Operating at temperatures where specific heats are nearer room-temperature values means that magnets will be more stable. It is probable that filamentary composites will not be needed and very complex stress systems will be practicable without expensive support structures. This will mean magnets can be more robust.

(c) Greater flux densities per unit cost will be possible due to the much higher field values to which ceramic conductors support current.

Bibliography

Fonar S, Schwartz B B 1974 *Superconducting Machines and Devices*. Plenum Press, New York

Lynton E A 1969 *Superconductivity*. Methuen, London

Montgomery D B 1969 *Solenoid magnet design*. Wiley, New York

Plonus M A 1978 *Applied Electromagnets*. McGraw-Hill, New York

Saint-James D, Sarma G, Thomas E J 1969 *Type II Superconductivity*. Pergamon, Oxford

Wilson M N 1983 *Superconducting Magnets*. Clarendon, Oxford

I. L. McDougall
[Oxford Instruments, Eynsham, UK]

Superconductors, Ceramic

One of the most remarkable graphs in the history of science and engineering is shown in Fig. 1, which is a plot of the critical temperature T_c of superconductors vs their year of discovery. Superconductivity was discovered in 1911 by Kamerlingh Onnes who showed that mercury became superconducting when cooled to liquid helium temperature (4 K). Between 1911 and 1986 a variety of materials were found to be superconducting (see *Superconductivity in Alloys, Compounds, Mixtures and Organic Materials*), some of which are indicated on Fig. 1, but prior to 1986 the highest known T_c was for Nb_3Ge with a T_c of 23.3 K. Between 1911 and 1986 the rate of progress in finding materials with higher values of T_c had been less than 0.3 K per year. Suddenly in 1986 the world of superconductivity changed when a new group of materials, known as high-temperature superconductors, was discovered by Bednorz and Müller, for which they were awarded the Nobel prize (the fastest Nobel prize ever to be awarded).

Bednorz and Müller (1986) originally reported work on multiphase Ba–La–Cu–O compounds with a T_c below 30 K. As work progressed in several groups, it was shown that the superconducting phase had a perovskite-like structure and that replacing barium ions with strontium pushed up the T_c to about 40 K. There was now intense excitement since the accepted BCS (Bardeen, Cooper and Schrieffer) theory of superconductivity, proposed in 1957 (see *Superconducting Materials: Theory*), predicted a maximum possible T_c of about 40 K (see *Superconductivity in Alloys, Compounds, Mixtures and Organic Materials*). In February 1987, Chu's group at Houston announced that a multiphase compound of Y–Ba–Cu–O had a T_c of 92 K and independently Zhao and his co-workers at Beijing announced a similar result. Subsequently, x ray, neutron and electron diffraction all showed that the Y–Ba–Cu–O (YBCO) phase responsible for superconductivity was perovskite-like with a composition of $Y_1Ba_2Cu_3O_{7-x}$ (known as the 1-2-3 phase). The importance of this work was twofold. First, the fact that T_c was now above the temperature of liquid nitrogen (77 K) meant that, since liquid nitrogen is much cheaper and more convenient to use than liquid helium, many more commercial applications of superconductivity might be possible. Second, the 40 K limit on T_c predicted by the BCS theory has been shattered, leading to the hope that room temperature superconductivity might be possible.

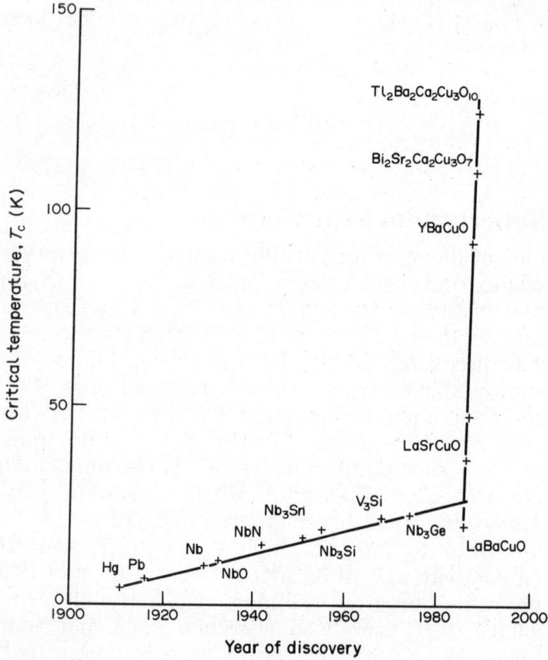

Figure 1
The critical temperature of superconductors as a function of their year of discovery

It may be worth pointing out that while Bednorz and Müller are frequently stated to have discovered ceramic superconductivity, this is not the case. For example, in 1973 superconductivity was discovered in the Li–Ti–O system ($T_c = 13.7$ K) and in 1975 in the Ba–Pb–Bi–O system ($T_c = 13$ K). The importance of the work of Bednorz and Müller was that their discovery of superconductivity in ceramics with a perovskite-like structure led directly to superconductivity above liquid nitrogen temperatures. It may also be worth pointing out that in a technology development forecast published in 1983, Japan predicted the discovery of superconductivity with T_c greater than 77 K (though they did not expect it before the twenty-first century) and in 1986, before the work of Bednorz and Müller was published, Japan predicted that such superconductors would be ceramic. Since Japanese scientists had targeted high-T_c superconductivity they were well prepared to exploit the discovery when it came and this may be the reason that the number of Japanese patents on high-T_c superconductors greatly exceeds the combined total for the rest of the world.

1. Theory of Ceramic Superconductors

There is a vast and growing literature on the theory of high-T_c superconductivity, but no clear consensus as to the mechanism. As indicated previously, standard BCS theory is probably not applicable in the high-T_c case, although some adherents of this theory believe it might be applicable with suitable modifications. However, although no detailed mechanistic theory yet exists, there is a phenomenological theory due to Ginzburg and Landau (the GL theory) which makes some extremely useful predictions for high-T_c superconductivity.

The GL theory (see *Superconducting Materials: Theory*) assumes that the superconducting electrons are paired, but does not assume a particular pairing mechanism. Experimentally it has been shown that superconducting electrons in high-T_c materials are indeed paired, as is also the case in conventional (low-T_c) superconductors, hence the GL theory should apply to all known superconductors. In conventional materials the pairing mechanism is believed to involve phonons (the BCS theory), whereas in the high-T_c materials the pairing mechanism is unknown.

If the GL theory is used in considering what happens when T_c becomes large, three important predictions can be made. First, the superconducting energy gap E_g should be large (for a definition of E_g see *Superconducting Materials: Theory*). Experimentally this has been verified: E_g for all known high-T_c superconductors is significantly larger than for conventional low-T_c superconductors. Second, the GL parameter κ should be large (see again *Superconducting Materials: Theory* for a definition of κ). Experimentally this is also found. This point is of great significance: a high value of κ implies a high value of the upper critical field H_{c2} (see *Superconducting Materials: An Overview*). In most high-T_c materials H_{c2} is extremely high (values of H_{c2} greater than 100 T have been measured before the range of existing equipment has been exceeded). This very high H_{c2} value is another reason for the potential technological importance of ceramic superconductors. Third, GL theory predicts (after some simplifications) that the coherence length ξ of the superconducting electrons is inversely proportional to T_c. Hence all high-T_c materials should have short coherence lengths for the superconducting electrons. This also has been verified experimentally; for example, for conventional superconductors ξ is approximately equal to 100 nm whereas for YBCO it is about 0.5 nm. Thus if GL theory is correct, and there is no evidence to suggest that it is not, all high-T_c superconductors (those known and those yet to be discovered) should have very short coherence lengths of typically atomic-scale dimensions. This conclusion is of fundamental importance regarding the processing of high-T_c superconductors, as will be described in Sect. 3.

2. *Superconductivity and Crystalline Defects*

Superconductors are classified as type I or type II (see *Superconducting Materials: An Overview*). Most pure elemental superconductors are type I, whereas most alloy conventional superconductors and all high-T_c ceramic superconductors are type II. The superconductors of technological interest are type II superconductors since it is these that exhibit zero resistance even in very high magnetic fields.

In type II superconductors, the transition from the superconducting state to the normal state occurs over a range of applied magnetic fields, from H_{c1} to H_{c2}. In this magnetic field range, the superconductor is in a "mixed state" (or vortex state) in which superconducting and normal regions coexist on a fine scale within the material. This occurs by having cores of normal material, called fluxoids or flux lines, passing through the superconducting material. As the magnetic field increases from H_{c1} to H_{c2} the number of fluxoids increases and their spacing decreases, hence the proportion of material in the normal state increases until at H_{c2} the material is 100% normal.

If a current flows in a type II material, this produces a Lorentz-type force on each fluxoid, causing it to move through the crystal. This movement dissipates energy, causing the crystal to heat up and if the temperature rises above T_c the superconductivity is destroyed. In a highly perfect crystal, the fluxoids move easily and the critical current density j_c is low, because the heat generated by the moving fluxoids destroys the superconductivity. For high values of j_c it is therefore important to pin the fluxoids to prevent their movement.

In conventional superconductors the coherence length ξ is approximately 100 nm, which is much greater than the widths of defects such as dislocations and grain boundaries. These defects therefore do not scatter the superconducting electron pairs and do not disrupt a flowing supercurrent. They are, however, very effective at pinning the fluxoids, thus preventing their movement through the crystal and dramatically increasing j_c. In conventional superconductors, defects such as dislocations and grain boundaries are therefore beneficial: they pin fluxoids without disrupting the supercurrent and they are often deliberately introduced (e.g., by cold working) to increase the critical current density j_c.

In high-T_c superconductors the situation is more complicated. The coherence length ξ of the electron pairs is typically 0.5 nm, which is of the same order as the width of a dislocation or a grain boundary. Such defects therefore scatter the electron pairs and reduce the critical current density. A major problem in high-T_c superconductivity is to find a crystal defect that pins the fluxoids, but that does not also appreciably disrupt the supercurrent.

3. *The Key Role of Processing in High-T_c Superconductors*

The very short coherence length of the electron pairs in all high-T_c superconductors implies that processing has a key role to play. In polycrystalline materials, grain boundaries must be sharp and contain no amorphous phases, otherwise the grain boundary width would exceed the coherence length ξ (~0.5 nm) and the electron pairs would be strongly scattered. In general, it would seem that all crystalline defects must be minimized, apart from defects (yet to be identified) that pin fluxoids without scattering the electron pairs. In addition, since the very short coherence length is comparable with the spacings of the copper–oxygen layers in YBCO, bismuth-based and thallium-based high-T_c superconductors, this implies that critical currents in a direction perpendicular to these layers (i.e., along the c axis) will be low. Single crystal studies confirm this: in YBCO and other materials high j_c is obtained in directions parallel to a–b planes but not parallel to the c axis. Hence, texturing is essential in polycrystalline materials for high critical current densities to be achieved. A further factor is strength: to obtain ceramic superconductors with sufficient strength to be wound into coils, for example, and to withstand the Lorentz force originating from a high current in a high magnetic field, the size of pores and cracks must be minimized. Various aspects of the key role of processing will now be examined in more detail. In order to be specific the Y–Ba–Cu–O system will be described, but similar considerations apply to other ceramic superconductors.

4. *The Crystal Structure of* $YBa_2Cu_3O_{7-x}$

$YBa_2Cu_3O_{7-x}$ was the first ceramic superconductor discovered with T_c greater than liquid nitrogen temperature. The highest critical current density is achieved when $x=0$, so that the composition is $YBa_2Cu_3O_7$, but the material remains superconducting for values of x down to about 0.5. X-ray and neutron diffraction identified the structure of the material to have a space group *Pmmm*, although the symmetry that is most widely observed in convergent beam electron diffraction is *Pm2m*. Initially this gave rise to considerable controversy since *Pmmm* is centrosymmetric whereas *Pm2m* does not have a center of symmetry. However, further electron diffraction work showed that the basic space group was *Pmmm*, in agreement with x-ray and neutron diffraction; the symmetry breaking observed with electrons being due to local variations in internal strain in $YBa_2Cu_3O_{7-x}$.

$YBa_2Cu_3O_7$ has an orthorhombic structure (with lattice parameters $a=0.382$ nm, $b=0.389$ nm and $c=1.168$ nm) that can be regarded as an oxygen-deficient perovskite structure with a unit cell containing three perovskite-like unit cells stacked in the c direction (see Fig. 2). The ideal tripled perovskite unit cell could in principle contain nine oxygen ions. However, oxygen vacancies occur in the yttrium plane and in the CuO_2 planes which separate the barium planes. For $YBa_2Cu_3O_{7-x}$ with x less than 0.5, if yttrium, barium and oxygen are assigned their usual valences, then simple charge considerations suggest that Cu^{2+} and Cu^{3+} ions coexist, with the Cu1 site being occupied by Cu^{3+} ions and the Cu2 site by Cu^{2+} ions. However, electron energy loss spectroscopy (EELS) and x-ray absorption spectra suggest a different interpretation; namely, that all the copper ions are essentially Cu^{2+} and that it is the oxygen ions that have the mixed valency, of O^{2-} and O^{-}.

In the production of bulk $YBa_2Cu_3O_7$, the tetragonal phase $YBa_2Cu_3O_6$ is normally formed first and annealing in oxygen then produces the tetragonal-to-orthorhombic phase transformation to $YBa_2Cu_3O_7$. $YBa_2Cu_3O_6$ is not superconducting and it is the important stoichiometry shift from O_6 to O_7 that produces the superconductivity (the onset of which starts at about $O_{6.5}$). This shift is associated with the change in the nominal oxygen valency to 1− (or alternatively with an increase in the nominal copper valency to greater than 2+) and this mixed valency appears to be associated with the superconductivity in the 1-2-3 system (and in other important systems to be discussed later). Finally, the orthorhombic (rather than tetragonal) nature of the $YBa_2Cu_3O_7$ unit cell can be seen from Fig. 2 to be due to the oxygen vacancies in the CuO plane containing the Cu1 ions. When the material transforms from the tetragonal O_6 form to the orthorhombic O_7 form, extensive twinning occurs to relieve the strain. In going from one twin to its neighbor the a and b axes of the orthorhombic unit cell are interchanged and hence the overall strain is relieved. However, this relief is only partial, as indicated by the convergent beam electron diffraction results described previously in this section and the internal strain often results in the formation of microcracks, which reduce both j_c and the strength of the material.

5. *Superconducting Thin Films*

Thin films of high-T_c superconductors are important for two reasons. First, large homogeneous bulk single crystals are extremely difficult to grow and

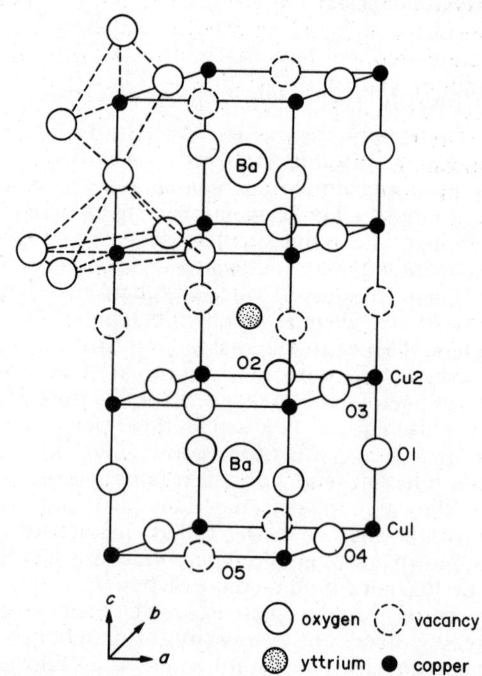

Figure 2
$YBa_2Cu_3O_7$ structure showing the three perovskite-like unit cells stacked in the c direction and the square planar and pyramidal configurations of the Cu 1 and Cu 2 sites, respectively

hence most of the important basic scientific measurements (e.g., of j_c as a function of direction) have had to be carried out on thin films. Second, there are major potential applications of superconducting thin films in electronic devices. The short coherence length in high-T_c superconductors means that grain boundaries scatter the superconducting electron pairs and act as weak links. Hence for high values of j_c it is important to minimize or eliminate grain boundaries; that is, it is important to grow single crystal thin films.

The best superconducting thin films have been grown epitaxially on substrates chosen to be closely lattice-matched to the superconductor. For example, in $Y_1Ba_2Cu_3O_{7-x}$, j_c is highest for supercurrent flow parallel to the *a–b* planes, but is low for supercurrent flow parallel to the *c* axis direction. Hence single crystal thin films of YBCO are required with the *a–b* plane parallel to the epilayer–substrate interface. Suitably closely lattice-matched substrates are (001) $SrTiO_3$ and (001) MgO, which give (001) oriented $Y_1Ba_2Cu_3O_{7-x}$ thin films as required.

There are various methods for producing superconducting YBCO thin films on a substrate. For example if $Y_1Ba_2Cu_3O_7$ is used as a target it can be sputter deposited (e.g., using argon ions) onto a substrate held at about 400 °C. The YBCO layer will be in an amorphous or polycrystalline form and will be oxygen deficient, but subsequent oxygen annealing results in epitaxial growth which nucleates at the substrate–film interface. An alternative approach is to sputter the materials directly onto the single crystal substrate held at about 650 °C. This results in a tetragonal epitaxial film of approximate composition $YBa_2Cu_3O_6$ and oxygen annealing then produces the orthorhombic superconducting $Y_1Ba_2Cu_3O_{7-x}$. Other deposition routes include laser ablation, molecular-beam epitaxy (MBE), metal organic chemical vapor deposition (MOCVD) and metal organic molecular beam epitaxy (MOMBE). Problems of thin film growth include stoichiometry control (in particular, achieving O_7 in $Y_1Ba_2Cu_3O_7$) and the maintenance of high-quality epitaxy over several square centimeters of the substrate.

As mentioned, grain boundaries act as weak links, or Josephson junctions, in high-T_c superconductors. However, to form Josephson junctions in a controlled way, as required in superconducting devices in high-T_c materials, presents a major challenge. In conventional superconductors the coherence length is about 100 nm and hence Josephson junctions need to be fabricated with a width on this scale; however, in high-T_c materials the coherence length and hence the Josephson junction width is only about 1 nm. Fabricating Josephson junctions in a controlled manner on this scale might seem an impossibility, but it has been demonstrated that an intense electron beam, focused to only 0.5 nm in diameter, locally removes the oxygen from YBCO and other oxide superconductors, thus destroying the superconductivity on a 1 nm scale and forming weak links. Since the electron beam position can be under computer control, Josephson junctions in high-T_c superconducting thin films can be made in a controlled pattern.

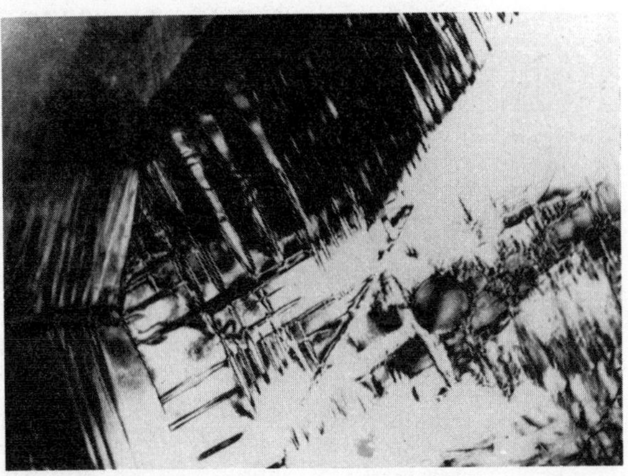

Figure 3
Microstructure of powder-processed YBaCuO material, showing the characteristic heavily twinned appearance of the orthorhombic 1-2-3 phase and a spherical particle of the 2-1-1 semiconducting phase (courtesy of D. J. Eaglesham)

6. *Bulk Superconducting* $YBa_2Cu_3O_7$

Although bulk superconducting YBaCuO can be easily made by sintering Y_2O_3, CuO and BaO, and annealing in oxygen, unless considerable care is taken the YBCO will be multiphase and have a low critical current density of about 10^6 A m^{-2}. A typical microstructure of low-quality powder-processed YBCO material is shown in Fig. 3. The 1-2-3 grains are easily identifiable by the heavy twinning. A common impurity is Y_2BaCuO_5, known as the 2-1-1 phase. This phase is green (the 1-2-3 phase being black) and often forms as spherical particles. In addition $BaCuO_5$ grains may occur. Grain boundaries in poor material frequently contain narrow strips (~1–2 nm wide) of amorphous copper-rich material plus silicates, carbonates and oxides from impurities. Since the coherence length is so short, these grain boundary phases are sufficiently wide to disrupt a flowing supercurrent.

However, by carefully controlling the annealing temperature and cooling rate, reducing the particle size and eliminating impurities it is possible to produce material that is 99% pure superconducting $Y_1Ba_2Cu_3O_{7-x}$, having 99% of grain boundaries

Figure 4
Scanning electron micrograph showing pores in typical oxygen-sintered $YBa_2Cu_3O_{7-x}$ (courtesy of J. Ringnalda)

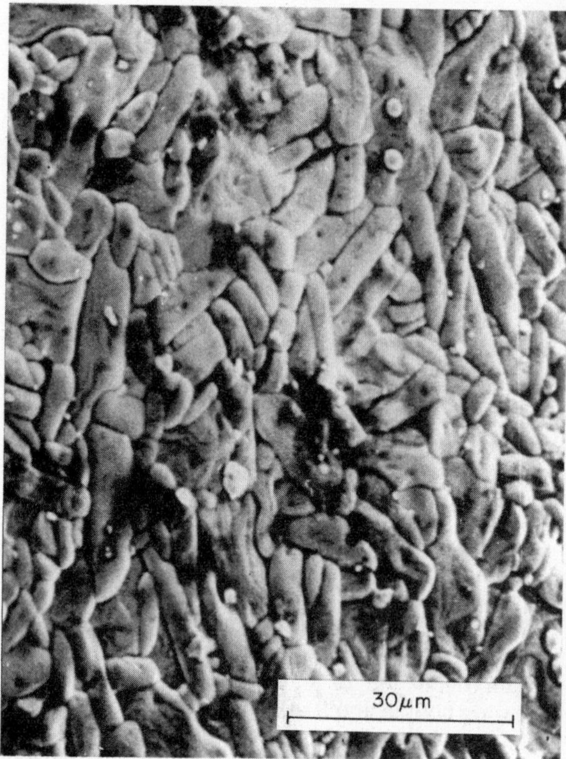

Figure 5
Scanning electron micrograph of the surface of oxygen-sintered $YBa_2Cu_3O_{7-x}$, 98% dense (courtesy of N. McN. Alford)

atomically sharp and with no traces of impurities. Typical critical currents in the pure polycrystalline materials exceed 10^7 A m^{-2}.

7. *Mechanical Properties of Ceramic Superconductors*

Again YBCO will be considered, but similar considerations apply to other ceramic materials. Poor quality sintered YBCO is only about 70% dense and contains many pores and large cracks (see Fig. 4). It is therefore very brittle (see *Advanced Ceramics: An Overview*). If the density is increased to above 85% and the crack size is reduced to below 10 μm, $YBa_2Cu_3O_{7-x}$ has sufficient strength to be fabricated into wires and coils. A good quality 1 mm diameter polycrystalline rod has a flexural strength in excess of 200 MPa and 0.2% strain to failure. As the density of the material increases, so does the critical current density j_c; for example, j_c more than doubles as the density increases from 70% to 85%. At a density of 98%, the Young's modulus is 200 GPa. Figure 5 shows a micrograph of 98% dense material.

High-density material is not only important for good mechanical properties, but is also important for increasing the lifetime of the material. If the density of YBCO is less than about 85%, the material rapidly degrades in the atmosphere and hydroxides and carbonates form because of attack by atmospheric CO_2 and H_2O. This results in j_c degrading over a period of weeks. However, if the density is approaching 100%, j_c is stable over many months. A corollary to this is that it becomes increasingly difficult to oxygenate YBCO fully to O_7 at densities above 90% because there is no easy diffusion route via pores.

Finally, it must be emphasized that many microcracks can arise from differential thermal expansion of different phases. For good mechanical properties, high-purity material is therefore essential, as is submicrometer powder starting material. Coils and shapes can then be fabricated directly using viscous processing routes.

8. *Texturing of Ceramic Superconductors*

As emphasized previously, texturing is essential to achieve high current densities in polycrystalline wires or thick films of ceramic superconductors. There are various ways of achieving texturing, such as melt texturing, rolling and solidification in high magnetic fields. There is enormous worldwide effort in texturing and the field is advancing rapidly. Rather than attempt to survey this rapidly changing field it may be useful to point out some of the difficulties involved in texturing, taking as an example melt texturing of YBCO.

Melt texturing of YBCO was the first method used to attempt texturing in any ceramic superconductor. In 1988, Jin and co-workers at AT&T Bell Laboratories reported that melt textured $YBa_2Cu_3O_{7-x}$ had a j_c of $17\times10^7\,A\,m^{-2}$ in a field of 0 T at 77 K, measured over a short length. The first problem encountered in attempting to melt texture YBCO is that it melts incongruously. Upon heating in air, stoichiometric $YBa_2Cu_3O_{7-x}$ melts at approximately 1010 °C into solid $Y_2Ba_1Cu_1O_5$ (the green semiconducting phase) and a melt of composition close to $Ba_3Cu_5O_y$. Thus we have at 1010 °C

$$2YBa_2Cu_3O_{7-x}\rightarrow \underset{\text{solid}}{Y_2BaCuO_5}+\underset{\text{liquid}}{Ba_3Cu_5O_y}$$

Upon cooling from this liquid–solid region, $YBa_2Cu_3O_{7-x}$ grains should be formed by a peritectic reaction. However, this does not happen readily and considerable temperature control is necessary. If solidification is carried out in a temperature gradient then a preferred orientation of the superconducting grains can be achieved. Slow cooling from the incongruous melt at 1010 °C usually results in large grains of $YBa_2Cu_3O_{7-x}$ containing $Y_2Ba_1Cu_1O_5$ particles. However, by carefully cooling in a temperature gradient of only $1\,K\,mm^{-1}$, growth can be nucleated at one end and large single crystals of YBCO can be grown.

9. Other Materials

The high-T_c field is continually expanding. The most significant new materials are bismuth based (Bi–Sr–Ca–Cu–O) and thallium based (Tl–Ba–Ca–Cu–O). These systems exhibit superconducting phases with a variety of values of T_c dependent upon their compositions. The highest T_c reliably reported is 125 K from a thallium-based compound $Tl_2Ba_2Ca_2Cu_3O_{10}$, but the toxicity of thallium may cause problems with widespread applications. A copper-free high-T_c superconductor Ba–K–Bi–O has been discovered showing that copper is not a necessary constituent of high-T_c ceramic superconductors. Also, whereas all the ceramic superconductors that have been mentioned in this section are essentially hole doped (i.e., they can be thought of as containing a mixed oxygen valency of O^{2-} and O^-), a high-T_c electron-doped ceramic superconductor $Nd_{2-x}Ce_xCuO_4$ has been reported. The field for discovering new high-T_c materials therefore remains wide open.

However, from an applications viewpoint the most important high-T_c materials that have been discovered are YBCO, the thallium-based compounds and, particularly, the bismuth-based compounds. All these materials are based on structures containing CuO planes and all occur in different superconducting forms in which the stacking of the CuO planes differ. It may therefore prove to be the case that most of the knowledge that has been gained from studying the YBCO system can be applied to the new families. Already small-scale commercial applications of high-T_c ceramic superconductors are being demonstrated (e.g., aerials, SQUIDS). The race to realize major practical applications has begun.

See also: Advanced Ceramics: An Overview; Ceramic Powders: Packing Characterization; Ceramics Process Engineering: An Overview; Ceramics: Relation of Microstructure to Processing History; Superconducting Materials: An Overview; Superconducting Materials: Metallurgy; Superconducting Materials: Theory; Superconductivity in Alloys, Compounds, Mixtures and Organic Materials

Bibliography

Bednorz J G, Müller K A 1986 Possible high T_c superconductivity in the Ba–La–Cu–O system. *Z. Phys. B* 64: 189–93

Clarke D R, Shaw T M, Dimos D 1989 Issues in the processing of cuprate ceramic superconductors. *J. Am. Ceram. Soc.* 77: 1103–13

Eaglesham D J, Humphreys C J, Alford N McN, Clegg W J, Harmer M A, Birchall J D 1988 High temperature superconducting ceramics. *Mater. Sci. Eng.* B1: 229–35

Evetts J E, Somekh R E 1989 The sputter deposition of superconducting ceramics. *Thin Solid Films* 174: 165–77

Friedel J 1989 The high-T_c superconductors: A conservative view. *Phys. Condens. Matter* 1: 7757–94

IBM J. Res. Dev. 33, 1989. Proceedings of the International Workshop on High Temperature Superconductivity

Physica C 153–155, 1988. Proceedings of the International Conference on High Temperature Superconductors and Materials and Mechanisms of Superconductivity

C. J. Humphreys
[University of Cambridge, Cambridge, UK]

Superconductors: Flux Pinning and the Critical Current

Large-scale applications of superconductors require both a large critical current density J_c and a high critical temperature T_c, as well as a strong upper critical field H_{c2}. Large critical currents have been achieved by the appropriate mechanical treatments during wire fabrication. These treatments yield an optimal defect morphology in the superconductor.

Defects cause local variations in the free energy of the flux-line lattice (FLL). The flux lines (or vortices) are trapped by the energy minima. Thus, defects are pinning centers which prevent the flux line motion. It is flux pinning that enables a superconductor to carry dissipation-free transport currents with densities of at least 10^2 A mm^{-2} in fields as strong as 20 T.

Although the concepts of flux pinning have been known for many years, the theoretical analysis is still a very complicated problem. Only for simple-defect systems and geometries have some successes recently been obtained. The problem consists of three issues: the elementary interactions between the FLL and various defects have to be calculated; the elastic and plastic properties of the FLL should be known; and the contributions of the individual pinning centers must be summed taking into account the distribution of pinning centers. This finally results in a prediction of the pinning force density F_p which usually is simply related to J_c by the expression

$$F_p = J_c B \qquad (1)$$

The elastic properties are treated in a separate article (see *Superconductors: Magnetic Structure*, Suppl. 2). Therefore, the discussion in this article will be concentrated on the elementary interaction and the summation problem.

1. Defect–Flux-Line Interaction

Defects alter the material properties locally, and consequently alter the superconducting parameters in their environment. These local changes couple to the periodic variations of the order parameter and the local field which are characteristic for the mixed state. A first classification of the elementary interactions discriminates between both coupling mechanisms: core interaction and magnetic interaction.

1.1 Magnetic Interaction

Examples of the magnetic interaction are the effect of surfaces parallel to the applied magnetic field (this might be the external surface as well as some large precipitate interface) and thickness variations of thin films for fields normal to the film. In the first case, the current distribution around a vortex core near the surface is forced to change in such a way that the normal component at the boundary vanishes. Theoretically, this is achieved by assuming an antivortex image at the other side of the boundary which results in an attractive surface–vortex interaction. This interaction is opposed by the repulsion due to the superconducting screening currents generated by the external field. The net effect is a potential barrier for flux entry and flux exit that is effective just above the lower critical field H_{c1}. The barrier decreases with increasing field (De Gennes 1966 pp. 76–80).

In the case of thin films with thickness variations, the vortices are trapped at the sites of smallest thickness where the line energy of the vortex is a minimum. Evidently, the typical length scale related to the magnetic interaction is the penetration depth λ. In materials with a large Ginzburg–Landau parameter κ this kind of interaction is therefore small. It generally disappears with increasing magnetic field.

1.2 Core Interaction

The coupling to the variation of the order parameter $|\Psi|^2$ (the density of Cooper pairs) is the origin of flux pinning for almost all defects; for example, dislocations, point defects, voids, grain boundaries and precipitates. Defects deviate from the surrounding material by a different density, elasticity, electron–phonon coupling or electron mean free path. The first three properties give rise to a local change in T_c, whereas the latter predominantly leads to a variation in κ. It is therefore possible to distinguish between δT_c pinning and $\delta\kappa$ pinning.

The core pin mechanism follows from the Ginzburg–Landau free energy as derived from the microscopic theory (De Gennes 1966 p. 171):

$$\mathcal{F} = \int \left(A|\Delta|^2 + \frac{B}{2}|\Delta|^4 + C|\partial\Delta|^2 + \frac{h^2}{2\mu_0} \right) d^3r \qquad (2)$$

with $A = N(0)(1-t)$, $t = T/T_c$, $B = 0.098N(0)/(kT_c)^2$, $C = 0.55\xi_0^2 N(0)\chi(\alpha)$, and $\partial = -i\nabla - (2e/\hbar)\boldsymbol{A}$. $\boldsymbol{A}$ is the vector potential related to the local field $\boldsymbol{h}$; $\Delta(\boldsymbol{r})$ is the pair potential related to the order parameter via $|\Delta|^2 = \hbar^2|\Psi|^2/(4mC)$ with m the electron mass; $\alpha = 0.882\xi_0/l_{tr}$ the impurity parameter (i.e., the ratio of the Bardeen–Cooper–Schrieffer (BCS) coherence length ξ_0 and the mean free path l_{tr}); $\chi(\alpha)$ is the Gorkov impurity function; and $N(0)$ is the density of states at the Fermi surface. The defects perturb the coefficients A, B and C, as well as the vector and pair potentials. Because $\mathcal{F}$ is supposed to be minimized with respect to Δ and $\boldsymbol{A}$, the pin energy can be in first order obtained by only taking into account the variations of the coefficients (e.g., Campbell and Evetts 1972, Thuneberg 1989).

It is clear from Eqn. (2) that the first two terms are involved in the case of T_c deviations, whereas l_{tr} fluctuations only lead to a perturbation of the third term. In the older literature, however, Eqn. (2) is usually written in reduced units which then eliminates the contribution of the third term and mixes $\delta\kappa$ effects and δT_c effects into the first and second terms. Thus, the use of reduced units incorrectly destroys the clear distinction between both pin mechanisms, which has important consequences for the temperature and field dependences of the elementary interactions (Kes 1987). These dependences play a distinguishing role in fundamental flux-pinning studies, in addition to effects related to size and concentration of pinning centers, or anisotropy and orientation effects of extended defects in monocrystalline model superconductors.

(*a*) *Electron scattering*. The theory has been worked out and reviewed by Thuneberg (1989). The concept is illustrated by deriving from Eqn. (2) the elementary pinning potential

$$\Omega_p(\boldsymbol{r}) = \tfrac{1}{2}\mu_0 H_c^2 \xi^2 g(\alpha)\xi_0 \int |\partial\psi(\boldsymbol{r}')|^2 \delta(1/l_{tr}(\boldsymbol{r}-\boldsymbol{r}'))\, d^3r' \tag{3}$$

with $g(\alpha) = 0.882\, d\ln(\chi)/d\alpha$ ($g(\alpha) = 0.85$ for $\alpha = 0$ and l_{tr}/ξ_0 for $\alpha \gg 1$), H_c the thermodynamic critical field, and $\delta(l_{tr}^{-1})$ the extra electron scattering by the defect. A reduced-order parameter has been also introduced: $|\psi|^2 = \hbar^2|\Psi|^2/(4m\xi^2\mu_0 H_c^2)$. The elementary pinning force f_p is obtained by computing the maximum value of the derivative of $\Omega_p(\boldsymbol{r})$. By the extra scattering of the electrons at the defect ξ decreases locally, so that this pin mechanism always results in an attractive interaction.

In the case of a small void (size $D < (\xi_0^{-1} + l_{tr}^{-1})^{-1}$), $\delta(l_{tr}^{-1})$ can be replaced by a δ function times the scattering cross section $\pi D^2/4$, thereby illustrating that the pinning energy is proportional to the condensation energy multiplied by the defect volume, enhanced by a factor ξ_0/D or l_{tr}/D for the "clean" or "dirty" limit, respectively. For extended defects, such as grain boundaries or large flat precipitates, parallel to the flux lines, the situation is more complicated because the integral over the entire defect should be evaluated. In addition, the scattering probability of the defect should be determined from other experiments. A special case, in which the crystal anisotropy also has to be considered, is that of the grain boundary in a bicrystal.

Regarding field and temperature dependence, two field regimes should be distinguished (see *Superconductors: Magnetic Structure*, Suppl. 2). In small fields ($B < 0.2B_{c2}$), the vortices are isolated and the order parameter changes over a distance ξ. In large fields, the vortices overlap which yields (near B_{c2})

$$|\psi|^2 \simeq (1-b)\left\{1 - \frac{1}{3}\left[\cos\left(x - \frac{y}{\sqrt{3}}\right)k_0 + \cos\left(\frac{2y}{\sqrt{3}}\right)k_0 + \cos\left(x + \frac{y}{\sqrt{3}}\right)k_0\right]\right\} \tag{4}$$

Here $k_0 = 2\pi/a_0$, $a_0 = 1.075\ (\Phi_0/B)^{1/2}$ the FLL parameter, and $b = B/B_{c2}$. In the first case, the expression for f_p contains a factor ξ^{-3} ($\propto (1-t)^{3/2}$), while in the second case a factor $(2\pi/a_0)^{-3}$ appears ($\propto B^{3/2}$). Since $a_0/\xi = 2.69b^{-1/2}$, the difference is $12.7b^{3/2}$, which is of the order of unity at $b = 0.18$. Accordingly, field and temperature dependences in both regimes are distinctly different; that is, f_p for isolated vortices should be independent of B, but it is proportional to $b^{3/2}(1-b)$ near B_{c2}. Typically, for a void ($D = 1$ nm) and an isolated vortex $f_p \simeq 0.3 \times 10^{-14}$ N, where $\mu_0 H_c = 0.1$ T and $\xi_0 = 10$ nm has been used.

(*b*) T_c *variations*. Pinning by various kinds of dislocations has been extensively studied in the past (see Campbell and Evetts 1972). The formalism used differs somewhat from the one described above. Two effects are distinguished, both related to the normal character of the vortex core: a larger density (ΔV effect) and larger elastic constants (ΔE effect). These effects, although very small, give rise to a periodic stress and elasticity field linked with the variation of $|\Psi|^2$. The pin potential arises via the coupling to the strain field around the defects and is linear or quadratic in the strain for the ΔV effect and the ΔE effect, respectively. Accordingly, these effects are referred to as first- and second-order interactions. Typical values of $\Delta V/V$ and $\Delta E/E$ are 10^{-7} and 3×10^{-5}, respectively. Some results for f_p or f_{pl} (f_p per unit length) of an isolated vortex taking again $\mu_0 H_c = 0.1$ T and $\xi_0 = 10$ nm are: edge dislocation perpendicular to a flux line $f_p \simeq 10^{-13}$ N; edge or screw dislocation parallel to a flux line $f_{pl} \simeq 5 \times 10^{-10}$ and 10^{-5} N m^{-1}, respectively; and dislocation loop ($D = 10$ nm) perpendicular to the field direction $f_p \simeq 3 \times 10^{-14}$ N. Moreover, $f_p \propto D^2$.

Precipitates can be considered as δT_c defects as well. The pin potential can be estimated to be at most $\mu_0 H_c^2 V/2$ for dielectric inclusions of volume V. For conducting precipitates, the proximity effect should be taken into account giving rise to a much lower value; no better theory exists at present. As to the field and temperature dependence of δT_c pinning, f_p now contains a factor ξ^{-1} or $(2\pi/a_0)^{-1}$ in small and large fields, respectively. This kind of mechanism may lead to attractive or repulsive interactions depending on the defect characteristics.

1.3 Concluding Remarks

From the above discussion it follows that a general formula for f_p cannot be given. For a specific material, the predominant defect structure should first be determined and then the pinning interaction estimated. Defect morphology, pinning mechanism, characteristic length scales and field orientation and regime are to be taken into account. Reasonable estimates can now be made for the most important pinning centers, although the situation for precipitates is not yet firmly settled.

In the above, only single-defect interactions have been considered. Whether this simplification is allowable depends on the concentration of pinning centers and the range of the pin interaction. The effectiveness of the pinning force will considerably decrease when the pins strongly overlap; that is, if the distance between the pins is much less than the interaction range. A good example is an amorphous superconductor. Although the defect concentration is very large, the pinning is small, since on the relevant length scale ξ the material is homogeneous. Only density or stress modulations with comparable wavelengths will be effective.

1.4 High-Temperature Superconductors

The recently discovered oxide superconductors deserve some special remarks because these materials have very short coherence lengths. This property makes them extremely sensitive to disorder, the more so since disorder tends to notably decrease T_c. If the disorder is located at grain boundaries, and if the defective layer is too thick, superconducting coherence between the grains is destroyed. Such grain boundaries act as weak links rather than as pinning centers. On the other hand, a small ξ is favorable for strong pinning as long as the defect is small enough so as to preserve the superconductivity. However, as was shown above, if the defect is too small, it will not act as a strong pin. As usual, the optimum will be $D \simeq \xi$. Moreover, it seems more effective to produce such defects in the CuO_2 planes. When the field is applied parallel to these planes, a pin mechanism may be effective related with a possible modulation of the order parameter between the layers, comparable with the situation in a multilayer. Many interesting developments are still to be expected.

2. Summing the Pinning Forces

The pinning centers in real materials generally have a random site and size distribution. In order to compute the total force on the FLL, it is usually assumed that all defects have the same size, so that only the positional distribution has to be considered. Consequently, there will be a distribution of pinning forces determined by the mutual positions of the pinning centers and the flux lines. The latter are defined by the zeros of the order parameter. In addition, it is assumed that the pins are small ($D \ll \xi$) in at least one direction perpendicular to the field.

For a perfectly uniform FLL, the effect of a random force distribution will average to zero. It can, therefore, be concluded that the deviations from perfect periodicity are essential for a net effect. The deformations of the FLL are determined by its elastic properties (see *Superconductors: Magnetic Structure*, Suppl. 2). The shear and tilt moduli c_{66} and c_{44} are most important in this respect. These moduli determine the displacements in the FLL in response to the force exerted by a specific pinning center. The effect of the other pins is to keep the FLL fixed. Such a mean-field approach is only reasonable, if the range of the FLL interaction is small compared with the distance between the pins. However, because of the electromagnetic origin, the interaction has a range λ which can be very large, especially in high-κ materials. Moreover, real pin systems are often very dense ($n_p^{1/3} \ll \lambda$, where n_p is the concentration of pinning centers) which suggests that a collective treatment of the pinning forces is more appropriate. Especially in the case of weak pins, for which the individual defect–flux-line interactions only lead to elastic deformation with strains very much less than one, the summation problem can be solved.

2.1 Concepts of Collective Pinning

The local displacements due to weak pins will be very small ($\ll a_0$), so that around each flux line a hexagonal FLL can be well defined. Over longer distances, however, the displacements may accumulate as in a random-walk process, causing a gradual break-down of the order in the FLL. This is expressed by the monotonic growth of the displacement correlation function

$$g(\boldsymbol{r}) \equiv \langle [\boldsymbol{u}(\boldsymbol{r}) - \boldsymbol{u}(0)]^2 \rangle \tag{5}$$

with distance $|\boldsymbol{r}|$. Here $\boldsymbol{u}$ is the displacement in the x–y plane (perpendicular to the field) and the origin may be located at the core of any flux line. When $g(\boldsymbol{r})$ is of order a_0^2, the positional long-range order is destroyed. However, for the summation of pinning forces, it is more relevant to know the criterion for which the pin interaction becomes unpredictable. This happens when $g(\boldsymbol{r}) = r_f^2$, where r_f is the distance in which $f(x,y)$ changes from $-f_p$ to f_p (i.e., $r_f \simeq \xi$ for $b < 0.2$ and $r_f \simeq a_0/2$ for $b > 0.2$).

The function $g(\boldsymbol{r})$ has been first derived by Larkin and Ovchinnikov for an infinitely large system in which the order breaks down in the directions both parallel and perpendicular to the field. This is called three-dimensional disorder (see the review papers by Brandt and Essmann (1987) and Kes (1987) for references). The corresponding longitudinal and perpendicular correlation lengths L_c and R_c are defined by

$$g_3(0, L_c) = g_3(R_c, 0) = r_f^2 \tag{6}$$

L_c and R_c are functions of c_{66} and c_{44} and of the pin correlation function (pin "strength") which for small point pins is given by

$$W = n_p \langle f^2 \rangle \simeq 0.5 n_p f_p^2 \quad (7)$$

The average is taken for a uniform distribution of pinning centers over a unit cell of the FLL. The relation between L_c and R_c depends in a quite natural way on the elastic moduli

$$L_c = (c_{44}/c_{66})^{1/2} R_c \quad (8)$$

It should be noted that all these quantities depend in a specific way on the temperature and field, as well as on the defect morphology.

The physical meaning of the correlation lengths is illustrated by deriving qualitative expressions for L_c and R_c. Instead of a continuous displacement field, the correlated regions with volume $V_c \simeq R_c^2 L_c$ can be considered, which are weakly coupled to each other so that they can move independently over a distance of the order of r_f. Inside V_c the FLL distortions can be ignored. The size of the correlated volume is determined by minimizing the free energy consisting of two terms: the elastic energy related to the mutual displacements of the correlated regions and the work done by the pinning centers in V_c to effect this displacement. This model readily provides the pinning force density as well, since in order to unpin the entire FLL, each correlated volume has to be unpinned independently. A statistical argument shows that the fluctuation (per unit volume) in the net effect of $n_p V_c$ pinning centers is

$$F_p = J_c B \simeq (W/V_c)^{1/2} \quad (9)$$

The situation becomes very simple for a thin-film amorphous superconductor in a perpendicular field. Since L_c is very much greater than R_c, and W is small, L_c can be easily made larger than the thickness d of the film. In this case the FLL deformations in the field direction may be ignored, so that only two-dimensional disorder in the x–y plane remains. One then has $V_c \simeq dR_{c,2D}^2$ and

$$R_{c,2D} \simeq \frac{r_f c_{66}}{(W/d)^{1/2}} \quad (10)$$

$$F_{p,2D} \simeq \frac{W}{(c_{66} r_f d)} \quad (11)$$

where d explicitly appears in the equations. The only unknown parameter is W, thereby offering a means to get information about f_p.

In the Larkin–Ovchinnikov theory, it is assumed that elastic deformations solely determine the correlation lengths. The effect of dislocations in the FLL is not considered, although their existence is known from decoration experiments. In fact, almost equivalent expressions can be derived by modelling the disorder in two dimensions by a square array of edge dislocations. This illustrates that the concept of computable correlation lengths is less settled. The statistical argument that leads to Eqn. (9) has a more general validity and can be used to obtain V_c from experiment if W is known.

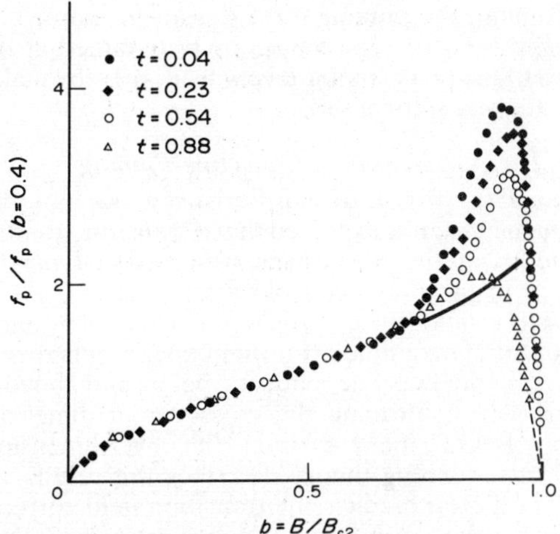

Figure 1
Typical pinning force density vs field for two-dimensional collective pinning in a thin-film amorphous superconductor: experimental data for amorphous Nb_3Ge film ($d = 1.24$ μm) at several temperatures ($T_c = 3.81$ K); the solid line represents the ideal theoretical behavior

2.2 Two-Dimensional Collective Pinning

In thin films of amorphous Nb_3Ge two-dimensional collective pinning has been observed (see Fig. 1 and Wördenweber and Kes (1989)). The drawn line represents the theory adapting W at $b = 0.4$. The field and temperature dependence of W show that the pinning is caused by quasidislocation loops. The $R_c/a_0(b)$ values computed from these data, display a dome-shaped behavior with a maximum at $b = 0.3$, typically of the order of 40, but values as large as 200 have also been determined for more homogeneous films. Both at small fields and close to H_{c2}, R_c decreases steeply, demonstrating the increase of disorder related to the softening of the FLL (i.e., the decrease of c_{66} to zero). For values of R_c/a_0 between 15 and 20 the data in Fig. 1 starts to deviate from the theoretical curve. It could be demonstrated that this peak effect is caused by the creation of edge dislocations in the FLL when, locally, the shear strength is surpassed. At the bottom of the peak effect, history effects are frequently observed. Using Eqn. (9) in the peak regime shows that at the maximum of F_p the disorder has reached saturation yielding an amorphous FLL (or vortex glass) with R_c/a_0 equal to between one and two. In the field region above the maximum, the pinning force density is simply given by $F_p \simeq W/(a_0^2 d)$. The values of W obtained at both sides of the peak compare very well, thereby making the analysis self-consistent.

2.3 Three-Dimensional Collective Pinning

From Eqn. (8), L_c/d can be calculated. For thick enough films, it is expected that a transition to three-dimensional disorder can be observed at $L_c/d = 0.5$, which in fact has been observed (Fig. 2). At the transition field. the pinning force increases by almost an order of magnitude. Further experiments reveal a different thickness dependence below and above the transition, confirming the crossover in dimensionality. Below the transition the two-dimensional collective-pinning theory describes the results very well and even predicts the transition field correctly. Above the transition, however, the three-dimensional collective-pinning expressions result in a much larger correlated volume than is determined from the data using Eqn. (9). The much too small V_c suggests that the disorder is determined by dislocations. Especially, screw dislocations created at the film surface may move freely along the flux lines, since they have no Peierls potential. This picture is justified by the observed IV curves (see Wördenweber and Kes (1989)).

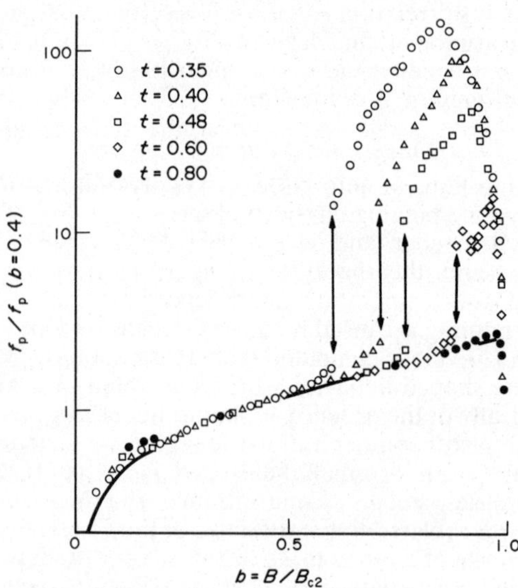

Figure 2
Typical pinning force density vs field for a thick amorphous superconducting film showing the transition from two-dimensional disorder at low fields to three-dimensional disorder at high fields: experimental data for amorphous Nb_3Ge film (17.95 μm) at several temperatures ($T_c = 4.3$ K); the transitions are indicated by the arrows; the solid line represents the two-dimensional collective-pinning theory (note the log scale for F_p)

The fact that these features are demonstrated in materials with weak pinning centers, does suggest that the "pure" three-dimensional collective-pinning situation is not stable and therefore not observable. Instead, a highly defective FLL seems to be a better picture, certainly in the case of stronger pinning centers. For the ultimate limit of disorder (i.e., an amorphous FLL) the collective-pinning theory offers an estimation for the correlation length L_c referred to as the single flux-line approximation.

$$L_c \simeq \left(\frac{B^4 a_0^6 r_f^2 (1-b)^2}{\mu_0^2 \lambda^4 W} \right)^{1/3} \tag{12}$$

This description is expected to apply close to B_{c2} where c_{66} becomes very small. In some cases, using $F_p(B) = (W/a_0^2 L_c)^{1/2}$, good agreement is found in a much larger field regime.

2.4 Real Pinning Systems

Features observed in strong pinning materials are dome-shaped $F_p(B)$ curves which, close to B_{c2}, behave either like $(1-b)$ or $(1-b)^2$. Scaling is frequently seen (i.e., $F_p(B,T) \propto f(T)\, b^p(1-b)^q$) but also "saturation" (e.g., in NbTi by Matsushita and Küpfer (1988)). (Note that these authors use a different definition of saturation.) It means that the $F_p(B)$ curves are changing in shape and increase up to a certain envelope curve when the pinning strength is enhanced. The envelope curve is dome shaped and represents the optimum pinning force achievable for the defect morphology under consideration. The nonsaturated curves typically show a much sharper peak effect closer to B_{c2}. Also, history effects are often reported in these cases.

In a pioneering paper Kramer (1973) pointed out that pin breaking gives a $(1-b)$ behavior, since it is characterized by a uniform motion of the FLL and because the elementary pinning forces decreases according to $(1-b)$ when B_{c2} is approached. On the other hand, a very inhomogeneous pinning-force distribution would lead to a shear flow of less-strong pinned regions of the FLL along strongly pinned (bundles of) flux lines. Being proportional to c_{66}, the shear flow mechanism yields a $(1-b)^2$ behavior. Although "Kramer-plots" have been very successful in describing $F_p(B)$ curves, a more profound comprehension of the mechanism for bulk materials is still needed. Recent experiments in thin films which simulated the shear flow behavior, proved that the FLL for all fields behaves like an incompressible fluid with a small, but finite shear modulus (Pruijmboom et al. 1989). It was shown that the shear modulus is considerably reduced by the effect of edge dislocations in the FLL. These results support the view that pin breaking occurs in the case of extended defects perpendicular to the flux flow direction, whereas flux-line shear would occur for smaller defects with a more inhomogeneous distribution.

In general, scaling is observed in materials that also show saturation, whereas history effects coincide with nonsaturated $F_p(B)$ curves which do not scale. In terms of the collective-pinning model it could be thought that, for intermediate pinning strengths, the FLL contains many edge and screw dislocations, dislocation loops, dislocation dipoles and other defects. Obviously, dislocations can be created or annihilated by flux-line flow giving rise to new flux-line arrangements and history effects. In the thin-film geometry with straight flux lines, how dislocation glide can make the FLL softer or more disordered is easily visualized, thereby giving it the opportunity to adapt itself more effectively to the distribution of pinning centers. For three-dimensional disorder, flux-line cutting might have to be allowed in order to model the FLL rearrangements. On the other hand, it should be realized that screw dislocations and screw-dislocation loops (causing a twist of the FLL around the field direction) can only occur by virtue of the presence of pinning centers and, in addition, that they easily glide along the flux lines. Therefore, annihilation and creation of flux-line dislocations might be more easily accomplished than dislocations in solids.

3. Pinning Models

3.1 Pin Potential

As described above, the modelling of the FLL disorder in terms of correlated volumes of greatly flexible size and shape depending on the pinning strength, can explain many experimental features. The correlated regions can be identified with the flux bundles that have to be evoked in order to describe flux-creep or flux-flow noise measurements. The flux bundles are the entities considered to be located in a pin potential which is generated by the net effect of the pinning centers within the bundles themselves. By applying an increasing driving force, which can be realized depending on the experimental geometry by either a transport current or a change of the external field, the flux bundles move up in their pin potentials, until they eventually break away, entering the flux-flow regime.

Campbell (1971) introduced this picture and showed how to derive force-displacement $F(u)$ curves from experiment. $F(u)$ curves look very similar to the stress–strain curves of solids. The slope at $u=0$ is called the Labusch parameter α. J_cB/α determines the range of the pin potential. In the case of "pure" collective pinning this ought to be r_f. However, in the presence of many flux-line dislocations this distance can be a much smaller fraction of a_0 because of the reduced flow stress in a defective lattice.

3.2 Flux Creep and Pin Potential

The picture of a flux bundle in a pin potential should be considered to represent the entire FLL. The pin potential is therefore not necessarily periodic. The question that now arises is how to describe flux creep in terms of a pin potential. A double potential well is probably most appropriate, since it contains all the ingredients: a barrier height, a hopping distance and the possibility of forward and backward hopping. Of course, this should be considered as an average potential actually representing a narrow distribution of barriers. For pure collective pinning, the hopping distance will again be equal to r_f. For a defective FLL, twice the width of the potential itself seems a better choice.

3.3 High-Temperature Superconductors

The oxide superconductors are special because of their very large anisotropy. This will make the FLL anisotropic as well. When the field is applied parallel to the crystalline *c* axis, the usual hexagonal symmetry is retained. For the field along the *a–b* plane, however, a very elongated isosceles FLL cell results, with the short basis aligned along the *c*-direction. Accordingly, the elastic moduli for this field orientation will be quite different and anisotropic. This property will make the analysis of critical currents in terms of pinning models extremely complicated. Also, in the other orientation (*H* parallel to *c*) a new phenomenon has to be taken into account; namely, it is expected from the weak coupling between the CuO_2 planes that the tilt modulus is much smaller than in conventional superconductors. This will lead to a much smaller correlation length L_c. In the most extreme case, the current vortices are considered to be entirely decoupled yielding $L_c \simeq \xi_c$. This would favor phenomena such as FLL melting and entanglement.

As is expected from the electron-scattering mechanism and has been convincingly demonstrated by decoration experiments, the twin planes in $YBa_2Cu_3O_7$ are attractive pinning centers (at least at low temperatures) for parallel flux lines. Because they have a very regular morphology, the flux lines align themselves along the twin planes as much as the average flux density allows and the flux lines in between the twin planes adapt to the other defects which will act collectively. In such a situation the summation procedure of the collective-pinning theory is not applicable, but a much simpler direct summation method should be adopted, yielding $F_p \simeq f_1/(La_0)$, in which L is the average distance between the twin planes and f_1 the elementary interaction per unit length and per flux line. This nicely exemplifies that the defect morphology predominantly determines the way in which a pinning analysis should be carried out.

Bibliography

Brandt E H, Essmann U 1987 The flux-line lattice in type-II superconductors. *Phys. Status Solidi B* 144: 13–38

Campbell A M 1971 The interaction distance between flux lines and pinning centers. *J. Phys. C.* 4: 3186–98

Campbell A M, Evetts J E 1972 Flux vortices and transport currents in type-II superconductors. *Adv. Phys.* 21: 199–428

De Gennes P G 1966 *Superconductivity of Metals and Alloys*. Benjamin, New York

Kes P H 1987 Recent developments in flux pinning. *IEEE Trans. Magn.* 23: 1160–7

Kramer E J 1973 Scaling laws for flux pinning in hard superconductors. *J. Appl. Phys.* 44: 1360–70

Matsushita T, Küpfer H 1988 Enhancement of the superconducting critical current from saturation in NbTi wire. *J. Appl. Phys.* 63: 5048–59

Pruijmboom A, Kes P H, Van Der Drift E, Radelaar S 1989 Flux-line shear studied in artificially structured superconducting double layers. *Cryogenics* 29: 232–5

Thuneberg E V 1989 Elementary pinning potentials in superconductors with anisotropic Fermi surface. *Cryogenics* 29: 236–44

Wördenweber R, Kes P H 1989 Peak effects in two- and three-dimensional collective pinning. *Cryogenics* 29: 321–7

P. H. Kes
[Leiden University, Leiden, The Netherlands]

Superconductors: Magnetic Structure

When a type II superconductor is put into a magnetic field H then at small fields ($H < H_{c1}$) the magnetic field is completely expelled from its interior (Meissner state). In large fields ($H > H_{c2}$) the superconductor becomes normal conducting and the magnetic field penetrates completely. The lower and upper critical fields of the material, H_{c1} and H_{c2}, depend on the temperature T and are approximately proportional to $1 - T^2/T_c^2$ where T_c is the superconducting transition temperature. In the Shubnikov or "mixed" state ($H_{c1} < H < H_{c2}$) magnetic flux can penetrate the superconductor in the form of flux lines or flux vortices. Flux lines are tiny current vortices that arrange to a more or less regular lattice, the flux line lattice (FLL). The existence of an FLL was predicted in 1957 by A. A. Abrikosov when he found that the phenomenological theory of superconductivity, formulated in 1950 by V. L. Ginzburg and L. D. Landau (the GL theory), had periodic solutions $\psi(x,y)$ (the complex GL or order parameter) and $B(x,y)$ (the magnetic induction) where x, y and z are Cartesian coordinates with z along the applied field H. Abrikosov interpreted this solution as an FLL, with the zeros of ψ and maxima of B defining the flux line centers. Each flux line carries one quantum of magnetic flux, $\Phi_0 = h/2e = 2.07 \times 10^{-15}$ T m^2 (h = Planck's constant, e = charge of the electron). An FLL with triangular symmetry and spacing d thus has an average induction

$$\bar{B} = \frac{2\Phi_0}{\sqrt{3}\, d^2} \tag{1}$$

For example, in niobium at $T = 4.2$ K, $\bar{B} \leqslant \mu_0 H_{c2} = 0.39$ T which yields $d \geqslant 0.08$ μm.

The existence of the FLL was confirmed first by Cribier and his co-workers in 1964 who observed Bragg reflections from the FLL in small-angle neutron scattering experiments. In 1967 Träuble and Essmann observed the FLL in an electron microscope after decorating the points where flux lines emerge from the surface of the superconductor with tiny ferromagnetic particles suspended in a helium atmosphere (see Fig. 1a). For a review of this method and of some properties of the FLL see Brandt and Essmann (1987). The FLL in the ceramic high-temperature superconductor $YBa_2Cu_3O_7$ was observed by the decoration method first by Gammel et al. (1987) (Fig. 1b).

1. Ginzburg–Landau Theory

The essential properties of the FLL may be obtained from the GL theory. Though this phenomenological theory was derived for temperatures T close to T_c, most of the GL results remain qualitatively correct at all $T < T_c$. For an exception see Sect. 4. GL introduces two characteristic lengths which both diverge as $(T_c - T)^{-1/2}$ when $T \to T_c$. The magnetic penetration depth λ is the length over which the magnetic field $\boldsymbol{B} = \nabla \boldsymbol{A}$ varies ($\boldsymbol{A}$ is the vector potential). The coherence length ξ is the length over which the complex order parameter ψ can vary. $|\psi|^2$ is proportional to the density of the superconducting electrons or Cooper pairs. $\psi(\boldsymbol{r}) = \Delta(\boldsymbol{r})/\Delta_{BCS}$ where $\Delta(\boldsymbol{r})$ is the gap function of the microscopic theory of superconductivity of Bardeen, Cooper and Schrieffer (BCS) and Δ_{BCS} is the maximum value $|\Delta|$ can attain. Thus $|\psi| \leqslant 1$.

The nearly temperature independent ratio $\lambda/\xi = \kappa$, the GL parameter, completely characterizes the type of solutions and thus the behavior of the superconductor. In particular, for $\kappa < 1/\sqrt{2}$ the superconductor is of type I and for $\kappa > 1/\sqrt{2}$ it is of type II. Lead, tin and most superconducting metals are type I superconductors. Niobium, vanadium, most alloys and the ceramic superconductors are of type II and can thus contain an FLL.

The GL equations for ψ and $\boldsymbol{A}$ and the appropriate boundary conditions of a given problem follow from a variational principle by minimizing a functional $F\{\psi, \boldsymbol{A}\}$ which is a free energy. In reduced units with length unit λ, field unit $\sqrt{2}H_c$ (H_c is the thermodynamic critical field) and energy density

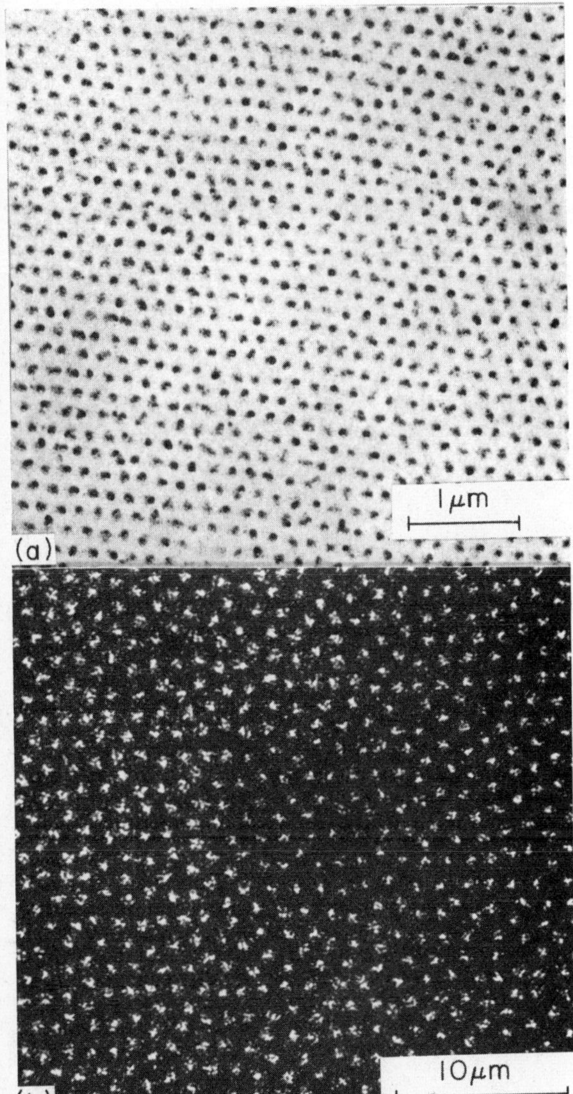

Figure 1
Flux line lattice made visible by a decoration method in the electron microscope: (a) niobium (flux lines dark), $T=1.2$ K, $\mu_0 H=98.5$ mT, $\bar{B}=82$ mT and $d=0.17$ μm (courtesy Uwe Essmann, Max-Planck Institut, Stuttgart), and (b) $YBa_2Cu_3O_7$ (flux lines light), $T=15$ K, $\mu_0 H \simeq \bar{B} \simeq 2$ mT and $d=1.2$ μm. At this low induction the FLL is strongly disordered due to pinning by material inhomogeneities and possibly by thermal fluctuations (courtesy David Bishop, AT&T Bell Laboratories)

unit H_c^2/μ_0, the free energy of a superconductor, referred to its normal conducting state at $B=0$, is

$$F=\int\left\{-|\Psi|^2+\frac{1}{2}|\psi|^4+\left|\left(\frac{\nabla}{i\kappa}-\boldsymbol{A}\right)\psi\right|^2+B^2\right\}d^3r \quad (2)$$

Note that the only material parameter in Eqn. (2) is κ. The first two terms are the superconducting condensation energy, which is a minimum when $|\psi|=1$. The third term is the kinetic energy of the supercurrent (of density $(\boldsymbol{A}-\nabla\varphi/\kappa)|\psi|^2$ where φ is the phase of ψ) plus a term $(\nabla|\psi|/\kappa)^2$ which shows that $|\psi|$ tends to be a smooth function. The last term is the energy of the magnetic field.

The applied field H, which does not appear in F, plays the role of an external pressure that forces the superconductor into some equilibrium state with average induction $\bar{B}$. The field H that establishes this state is obtained as $H=\partial F/\partial\bar{B}$. This law follows from minimization of the Gibbs free energy $G=F-\bar{B}H$. Special solutions forced by the applied field are the Meissner state ($|\psi|\equiv 1$, $B\equiv 0$, $H<H_c$ for type I and $H<H_{c1}$ for type II superconductors), the mixed or Shubnikov state with an FLL ($|\psi|^2$ and B periodic, $0<\bar{B}<\mu_0 H_{c2}$, $H_{c1}<H<H_{c2}$) and the normal conducting state ($\psi\equiv 0$, $B\equiv\mu_0 H$, $H>H_c$ for type I and $H>H_{c1}$ for type II superconductors):

$$\left.\begin{aligned} H_{c1}&=\frac{\Phi_0(\ln\kappa+\varepsilon)}{4\pi\lambda^2}\\ H_c&=\frac{\Phi_0}{2\sqrt{2}\pi\lambda\xi}\\ H_{c2}&=\frac{\Phi_0}{2\pi\xi^2}\end{aligned}\right\} \quad (3)$$

with $\varepsilon\ll 1$ for $\kappa\gg 1$ and $\varepsilon=1+\ln\sqrt{2}$ for $\kappa=1/\sqrt{2}$. For $\kappa=1/\sqrt{2}$, $H_{c1}=H_c=H_{c2}$ and, in general, $H_{c1}\simeq H_c/\sqrt{2}\kappa$, $H_{c2}=H_c\sqrt{2}\kappa$.

Type I superconductors formally exhibit $H_{c1}>H_{c2}$, thus an FLL cannot exist in them. However, in type I superconductors with a demagnetization factor $N>0$ ($N=0$ for long cylinders in parallel fields, $N=\frac{1}{3}$ for spheres and $N=1$ for thin disks in perpendicular fields) there may be an intermediate state that consists of alternating superconducting and normal conducting layers in which $B\equiv 0$ and $B\equiv\mu_0 H_c$, respectively. This state occurs in the field range $(1-N)H_c<H<H_c$. It is stable since the wall energy between these lamellae is positive; that is, the wall length tends to be as short as possible.

In contrast to type I superconductors, the wall energy in type II superconductors is negative; the walls will thus curl spontaneously until the longest possible wall (of thickness $\simeq\xi$) or finest subdivision is achieved. The result of this tendency is just the FLL: small normal vortex cores of radius $\simeq\xi$ are surrounded by a superconducting phase, each vortex carrying the smallest possible flux $\Phi=\Phi_0$. A lattice of flux lines that carry more flux, $\Phi=n\Phi_0$ with $n>1$ (n is an integer), is a solution that has higher energy than a lattice of single flux quanta. Any distorted lattice of flux lines, straight or curved, has higher energy than the perfect FLL in a homogeneous medium. In inhomogeneous media with spatially varying material parameters λ and ξ spontaneous distortion of the FLL will occur (see Sect. 4). An interesting feature of the GL equations is that, for the special case $\kappa = 1/\sqrt{2}$, any solution has the same Gibbs free energy G and is thus a possible solution for the ground state: perfect or distorted FLLs with one or more flux quanta, flux tubes (these are formally flux lines with $n \gg 1$) or lamellae, and the homogeneous states $\psi = 1$, $B = 0$ and $\psi = 0$, $B = \mu_0 H_c$.

2. Low Inductions

At low inductions $\bar{B} < 0.25\mu_0 H_{c2}$ and not too small values of $\kappa > 1.4$, the FLL may be considered as consisting of individual vortices with well-separated cores. The fields of parallel flux lines, centered at positions $\boldsymbol{r}_\nu$, then superimpose and yield

$$B(\boldsymbol{r}) \simeq \sum_\nu \frac{\Phi_0}{2\pi\lambda^2} K_0\{[(\boldsymbol{r} - \boldsymbol{r}_\nu)^2 + r_c^2]^{1/2}/\lambda\} \qquad (4)$$

where $K_0(x)$ is a modified Bessel function with limits $K_0 \simeq -\ln x$ for $x \ll 1$ and $K_0 \simeq (\pi/2x)^{1/2} \exp(-x)$ for $x \gg 1$. The isolated vortex fields in Eqn. (4) have a range λ and a core radius $r_c \simeq \xi$. In the limit $r_c \to 0$, Eqn. (4) coincides with the result of a theory established in 1935 by F. and H. London. This London theory coincides with the GL theory if one puts $|\psi| = 1$. It yields "London vortices" with zero core diameter and a logarithmic divergence of B at the vortex cores. This divergence is removed by the GL theory which provides a finite r_c. The vortex core is defined as the region where the order parameter decreases from its nearly constant value between the flux line cores:

$$|\psi(\boldsymbol{r})|^2 \simeq \prod_\nu \{1 - \exp[(\boldsymbol{r} - \boldsymbol{r}_\nu)^2/r_c^2]\} \qquad (5)$$

At low flux densities the flux lines interact with each other by a pair potential, which within the GL and London theories is

$$V(r) = \frac{\Phi_0^2}{2\pi\lambda^2\mu_0} K_0(r/\lambda) \qquad (6)$$

This is a monotonically decreasing and thus repulsive potential. The energy of the FLL is then the sum of all pair interactions, plus the self-energy J of the N vortices ($J = \Phi_0 H_{c1}$ for $B \ll \mu_0 H_{c2}$):

$$F = \tfrac{1}{2} \sum_\mu \sum_\nu V(|\boldsymbol{r}_\mu - \boldsymbol{r}_\nu|) + NJ \qquad (7)$$

Eqns. (4–7) apply to arbitrary arrangements of parallel flux lines with separations larger than $\simeq 4\xi$ and for $\kappa > 1.4$. More general approximate expressions for all separations, arbitrary κ values, and straight or curved flux lines are given by Brandt and Essmann (1987).

For a regular FLL $B(\boldsymbol{r})$ is periodic and may be written as a two-dimensional Fourier series:

$$B(\boldsymbol{r}) = \sum_K B_K \cos \boldsymbol{Kr} \qquad (8)$$

where $\boldsymbol{K} = \boldsymbol{K}_{mn}$ (m, $n = 0, \pm 1, \pm 2, \ldots$) are the reciprocal lattice vectors of the FLL. For the triangular lattice $K_{mn}^2 = (16\pi^2/3d^2)(m^2 + mn + n^2)$. At low inductions the Fourier coefficients are

$$B_K \simeq \bar{B}/(1 + K^2\lambda^2 + K^4 r_c^4) \qquad (9)$$

The term $K^4 r_c^4$ provides a cutoff at $K \simeq r_c^{-1}$ and smears the logarithmic singularity of $B(\boldsymbol{r})$ at the vortex centers over a range of radius $\simeq r_c$ (cf. Eqn. (4)).

3. High Inductions

At higher inductions the vortex cores overlap and, therefore, the vortex fields no longer superimpose linearly. In this case the shapes of $\psi(\boldsymbol{r})$ and $B(\boldsymbol{r}) - \bar{B}$ may be obtained by solving the linearized GL equations; these apply when H is close to H_{c2} since then $|\psi|^2 \ll 1$ holds. The amplitude of this solution is then obtained from the nonlinear terms. The resulting $|\psi|^2$ and B are smooth periodic functions with Fourier coefficients decreasing rapidly with increasing K (see Fig. 2):

$$|\psi(\boldsymbol{r})|^2 = \frac{1 - \bar{B}/\mu_0 H_{c2}}{1.16 - 0.16/2\kappa^2} \sum_K a_K \cos \boldsymbol{Kr} \qquad (10)$$

$$B_K = \mu_0 M a_K \quad (K \neq 0) \qquad (11)$$

where for the triangular lattice

$$a_K = (-1)^{m+mn+n} \exp[-(\pi/\sqrt{3})(m^2 + mn + n^2)] \qquad (12)$$

The magnetization $M = \bar{B}/\mu_0 - H$ is given by

$$M = -\frac{(H_{c2} - H)}{1.16(2\kappa^2 - 1)} \qquad (13)$$

Numerical computations show that the approximate analytic solutions Eqns. (10–13) apply in the large field range $0.5 < \bar{B}/\mu_0 H_{c2} < 1$.

4. Pure Superconductors at Low Temperatures

At lower temperatures ($T<T_c$) the GL results are still approximately valid for sufficiently impure superconductors; that is, for alloys with electron mean free path l smaller than the BCS coherence length ξ_0. For clean type II superconductors ($l \gg \xi_0$) at $\bar{B} \simeq \mu_0 H_{c2}$, $\psi(\mathbf{r})$ retains its GL form (Eqn. (10)) but $B(\mathbf{r})$ changes (Delrieu 1972). At $T=0.625\ T_c$ the minima and saddle points of $B(\mathbf{r})$ interchange their positions. At $T=0$, the field profile along the lines connecting neighboring flux line centers is zigzag shaped, the maximum and the three minima (per unit cell) attain conical shape, the field in the vortex centers exceeds the applied field ($B_{max}=\mu_0(H+1.8|M|)$) and the saddle points exhibit a three-fold rotational axis (see Fig. 3). This exotic behavior of the microscopic magnetic field at $T=0$ is due to the slow decrease of the Fourier coefficients $B_K \sim K^{-3}$:

$$B_K \simeq 0.363\mu_0 M(-1)^{m+mn+n}(m^2+mn+n^2)^{-3/2} \quad (K \neq 0) \quad (14)$$

The B_K for arbitrary lattice symmetry and temperatures $0 \leq T < T_c$ are given by Brandt (1988).

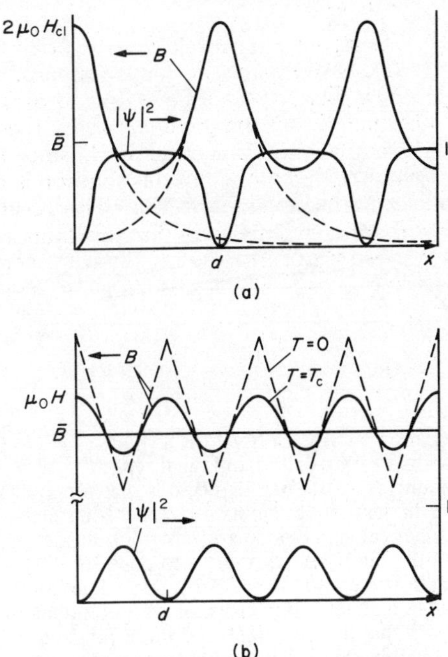

Figure 2
Profiles of magnetic field B and order parameter $|\psi|^2$ along a line connecting neighboring flux lines in pure niobium: (a) low induction $\bar{B} \simeq 0.1\mu_0 H_{c2}$ (the dashed lines give the profiles of isolated flux lines), and (b) high induction $\bar{B} \simeq 0.7 H_{c2}$ (the dashed line is the exotic field profile at zero temperature)

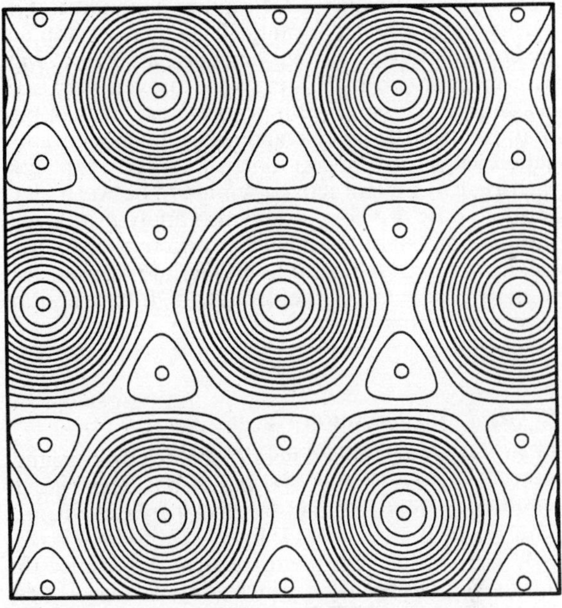

Figure 3
Lines of equal $B(x,y)$ for a type II superconductor at $\bar{B} \simeq 0.2\mu_0 H_{c2}$. The circles are centered at the flux line positions where B is at a maximum

A further modification which extends the GL picture to low temperatures is the introduction of three GL parameters $\kappa_1(T)$, $\kappa_2(T)$ and $\kappa_3(T)$ which coincide with κ at $T=T_c$ but increase slightly when T is lowered. κ_1 appears in the relation $H_{c2}=\sqrt{2}\kappa_1 H_c$, κ_2 in the slope of $M(H)$ (Eqn. (13)) and κ_3 in $H_{c1}=H_c(\ln\kappa_3+\varepsilon)/\sqrt{2}\kappa_3$. Here, again, the correction to the GL result is largest for pure superconductors. For pure niobium, which has $\kappa=0.72$, this correction has interesting consequences. At $T \simeq T_c$ its magnetization curve looks almost like that of a type I superconductor, but with decreasing T, H_{c1} decreases (in reduced units with $\sqrt{2}H_c=1$), H_{c2} increases, the slope of $-M(H)$ decreases and the area under the magnetization curve H_c^2/μ_0 stays constant. As a consequence, the magnetization curve near H_{c1} takes an S shape which, by a Maxwell construction, yields a finite jump of M for H slightly smaller than the "overheated" H_{c1}. This abrupt jump from $\bar{B}=0$ to $\bar{B}=B_0$ may be explained by an attractive interaction between the flux lines at large distances which, together with the usual repulsion at short distances d, yields a finite equilibrium distance $d_0=(1.155\Phi_0/B_0)^{1/2}$. As $T \to T_c$ the GL behavior with mere repulsive interaction (Eqn. (6)) and with $B_0=0$, results in $d_0=\infty$. Pure niobium at $T<T_c$ thus combines features of type I and type II superconductors: in the range $(1-N)H_{c1}<H<H_{c1}$ it exhibits lamellae or tubes containing an FLL with $\bar{B}=B_0$, embedded in a Meissner phase with $B=0$, or Meissner tubes embedded in an FLL.

The results of this section going beyond the GL theory were obtained from the BCS theory extended by Gor'kov to spatially varying gap parameter $\Delta(\boldsymbol{r})$. The microscopic BCS theory has proven very successful for "classical" superconductors. It is hoped that the BCS theory also applies to the new ceramic superconductors. Since it does not specify explicitly the origin of the attractive interaction between electrons, the BCS theory would also apply if it turned out that a novel mechanism, different from the classical electron–phonon interaction, causes this attraction in some of the ceramic high-temperature superconductors.

5. Ideal Magnetization Curves

In an increasing applied field H the negative magnetization $-M(H)=H-\bar{B}(H)/\mu_0$ of an ideal type II superconductor (see Fig. 4a) first increases linearly with gradient 1 since $\bar{B}=0$ for $H<H_{c1}$. At $H=H_{c1}$, $-M$ starts to decrease with vertical slope since the interaction between the penetrating vortices is very small when $d \gg \lambda$. When $-M$ has decreased to $\simeq H_{c1}/3$ it continues to decrease approximately linearly until it vanishes at $H=H_{c2}$ with slope given by Eqn. (13). This behavior applies to long cylinders in parallel fields. For different geometries the demagnetizing field of the specimens comes into play; this field is caused by the magnetized volume elements and superimposes onto the applied field H. For specimens of ellipsoidal shape with one axis parallel to H the demagnetization field is homogeneous and parallel to H and may be accounted for by the introduction of a scalar demagnetization factor N with $0 \leqslant N \leqslant 1$. The magnetization curves are then obtained by replacing in the above function $M(H)$ (which applies for $N=0$) the argument H by an effective field $H_{\text{eff}}=H-NM(H_{\text{eff}})$ and solving this implicit equation for M. This results in a shearing of the magnetization curves such that for $0<H<H'_{c1}$, where $H'_{c1}=(1-N)H_{c1}$, the slope is $1/(1-N)$ and immediately above H'_{c1} (where $-M=H_{c1}$) the slope is $-1/N$.

6. Pinning of Flux Lines

In real type II superconductors the flux lines cannot move completely freely since they are pinned by inhomogeneities of the material; that is, by crystal lattice defects, by precipitates, by a surface barrier (for flux lines parallel to the surface) or by a rough surface. Due to these inhomogeneities, the energy of a flux line depends on its position and a force is required to move a flux line away from a preferred position (a "pin"). The summation of such elementary pinning forces to an average pinning force is a complicated statistical problem into which crucially enter the elastic properties of the FLL (see Sect. 7). An ideally rigid FLL cannot be pinned by randomly positioned pins since the pinning forces average to zero when there is no correlation between the positions of the flux lines and the pins. Pinning of the FLL has important consequences on the behavior of a type II superconductor.

6.1 Current–Voltage Curves

The driving force on the FLL (i.e., the Lorentz force $\boldsymbol{j}\times\boldsymbol{B}$ exerted by a transport current or a gradient in B) must exceed a threshold value of $j_c B$ before the FLL moves. At low current densities ($j<j_c$) the FLL is pinned and the current flow is loss free. For $j>j_c$ the FLL drifts with a velocity $v \simeq (j-j_c)\bar{B}/\eta$ where $\eta \simeq \bar{B}\mu_0 H_{c2}\sigma_n$ is the viscosity of the FLL (σ_n is the

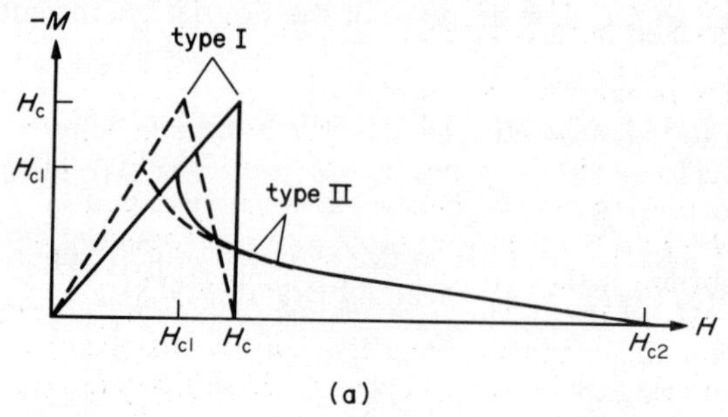

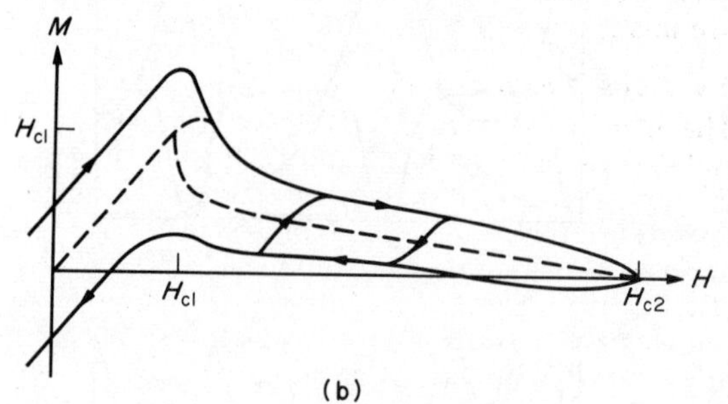

Figure 4
Magnetization curves $-M(H)$. (a) Type I superconductor ($\kappa<0.707$) and ideal type II superconductor ($\kappa>0.707$, depicted is the case $\kappa=2$) as long cylinders in parallel fields ($N=0$, solid lines) and as spherical superconductors ($N=1/3$, dashed lines); in the latter case flux starts to penetrate at $H=(2/3)H_c$ (type I) or $H=(2/3)H_{c1}$ (type II). (b) Irreversible magnetization curve of a superconducting sphere with flux line pinning (solid line), pin-free ideal reversible curve (dashed line) and virgin curve (dotted line). The depicted big hysteresis loop (symmetric about the origin) is traced when H is cycled between values $\leqslant -H_{c2}$ and values $\geqslant H_{c2}$. Smaller loops are traced when H is cycled with smaller amplitude; their initial gradient is parallel to the Meissner-state line ($-\partial M/\partial H=1$) since pinning initially keeps $\bar{B}$ constant

normal conductivity). This flux flow generates an electric field $\boldsymbol{E}=\bar{\boldsymbol{B}}\times\boldsymbol{v}$ which leads to a voltage drop along the superconductor and thus to electric losses. The current–voltage curves of a superconducting wire exhibits, therefore, a jump at $j\simeq j_c$. This jump may be rounded by an inhomogeneous j_c or by thermally activated flux creep. At sufficiently high temperatures thermally activated flux creep (TAFF) may lead to the appearance of a small voltage proportional to j at $j\ll j_c$ (Kes et al. 1989).

6.2 Flux Density Profiles

The flux density inside a real type II superconductor may be inhomogeneous and depends on its "magnetic history". $\bar{B}$ (i.e., B averaged over a few lattice cells) is higher for increasing H (lower for decreasing H) near the sample surface than near the center because a critical gradient $\partial\bar{B}/\partial x\simeq j_c\mu_0$ has to be reached before the flux lines can move.

6.3 Irreversible Magnetization

The magnetization curves of real type II superconductors exhibit hysteretic behavior (see Fig. 4b). In increasing applied field H, $\bar{B}$ is smaller, and $-M=H-\bar{B}/\mu_0>0$ is larger, than the corresponding equilibrium values. In decreasing field, $\bar{B}$ is larger and $-M$ smaller. This irreversible behavior indicates that the magnetic flux feels some resistance when entering or leaving the material. The internal field may even exceed the applied field; then M is positive (flux trapping). When H is switched off, some "remnant flux" remains in the specimen.

6.4 Stable Levitation and Suspension

The irreversibility of $M(H)$ leads to a spectacular stability of the free levitation of high-T_c superconductors above a permanent magnet or of magnets levitating above such superconductors. Free floatation is possible due to a repulsive magnetic force (Brandt 1989, Shapira et al. 1989). Depinning of flux lines leads to a frictional force which strongly damps any motion of the levitating superconductor (or of a magnet levitating above such a superconductor). A finite range of stable positions and orientations exist and, if flux pinning is sufficiently strong, even attractive forces are possible which enable free suspension of a superconductor below a magnet. In contrast, the levitation of a type I superconductor above a magnet, or vice versa, is stable only for special geometries and exhibits no damping. Levitation using only ferromagnets is always unstable.

7. Elasticity of the FLL

7.1 Uniform Deformation

When the FLL is subjected to a uniform deformation its increase in energy, the elastic energy, may be expressed in terms of three elastic moduli: for uniaxial compression (c_{11}), shear (c_{66}) and tilt (c_{44}). The modulus for isotropic compression (bulk modulus) is $c_{11}-c_{66}$. From thermodynamic considerations it is found that

$$c_{11}-c_{66}=\bar{B}^2\frac{dH}{d\bar{B}} \tag{15}$$

$$c_{44}=\bar{B}H \tag{16}$$

The shear modulus $c_{66}\ll c_{11}\simeq c_{44}\simeq\bar{B}^2/\mu_0$ cannot be obtained in this way since the magnetization curves, or $\bar{B}(H)$, are practically the same for triangular and square FLLs. At very low inductions ($\bar{B}\ll\mu_0 H_{c2}$) where the self energy J of the flux lines (defined by Eqn. (7)) does not depend on $\bar{B}$, it is possible to derive a formula that relates the shear modulus

$$c_{66}=\frac{\bar{B}}{16\Phi_0}\sum_{\nu}[R_\nu^2 V''(R_\nu)+3R_\nu V'(R_\nu)] \tag{17}$$

(R_ν are the flux line positions in an ideal triangular lattice) to $F(\bar{B})$ (Eqn. (7)) and thus to the measurable applied field $H(\bar{B})=\partial F/\partial\bar{B}$. This formula, however, is of little practical use since at $\bar{B}\ll\mu_0 H_{c2}$, or $H\simeq H_{c1}$, the measured magnetization curves are strongly influenced by pinning. From the GL equations it is found for the triangular FLL that

$$c_{66}\simeq\mu_0 H_{c2}^2\frac{b(1-b)^2}{8\kappa^2}\left(1-\frac{1}{2\kappa^2}\right)(1-0.29b) \tag{18}$$

where $b=\bar{B}/\mu_0 H_{c2}$.

For deformations varying little over an effective penetration depth $\lambda'=\lambda/(1-b)^{1/2}$ the elastic energy of the FLL is given by the "local" expression

$$\begin{aligned}U_{\text{elast}}=\frac{1}{2}\int\{&c_{11}(s_{x/x}+s_{y/y})^2\\&+c_{66}[(s_{x/y}+s_{y/x})^2+(s_{x/x}-s_{y/y})^2]\\&+c_{44}(s_{x/z}^2+s_{y/z}^2)^2\}\,d^3r\end{aligned} \tag{19}$$

where $\boldsymbol{s}(\boldsymbol{r})=(s_x;\ s_y;\ 0)$ is the flux line displacement field, and $s_{x/x}=\partial s_x/\partial x$, $s_{x/y}=\partial s_x/\partial y$ and so on are the strain components.

7.2 General Deformation

When $\boldsymbol{s}(\boldsymbol{r})$ varies considerably over λ' then the "nonlocal" theory of elasticity applies. In terms of the Fourier transformed displacement field $\tilde{\boldsymbol{s}}(\boldsymbol{k})=(\tilde{s}_x;\tilde{s}_y;\ 0)$, where $\boldsymbol{k}=(k_x;\ k_y;\ k_z)$ is the wave vector, the general expression for the elastic energy (Brandt 1986) is

$$U_{\text{elast}}=\frac{1}{2}\int_{\text{BZ}}\frac{d^3k}{8\pi^3}\sum_{\alpha,\beta}\tilde{s}_\alpha(\boldsymbol{k})\tilde{s}_\beta^*(\boldsymbol{k})\Phi_{\alpha\beta}(\boldsymbol{k}) \tag{20}$$

Here the integration is over the first Brillouin zone (BZ) of the FLL and over $-\infty<k_z<\infty$, $\Phi_{\alpha\beta}(\boldsymbol{k})$ is the elastic matrix, and α and β denote Cartesian components x and y, respectively. In general, $\tilde{s}(\boldsymbol{k})$ and $\Phi(\boldsymbol{k})$ are periodic in $\boldsymbol{k}$ space. A useful approximation to Φ, valid for small $k \ll k_B$ (πk_B^2 is the area of the BZ, $k_B^2=2b/\xi^2$) is

$$\Phi_{\alpha\beta}(\boldsymbol{k})=k_\alpha k_\beta[c_{11}(k)-c_{66}] + \delta_{\alpha\beta}[(k_x^2+k_y^2)c_{66}+k_z^2 c_{44}(k)] \quad (21)$$

For Eqn. (15)

$$c_{11}(k)-c_{66}=\mu_0 H_{c2}^2 \frac{b^2}{1+k^2/k_h^2}\frac{1-1/2\kappa^2}{1+k^2/k_\psi^2} \quad (22)$$

and for Eqn. (16)

$$c_{44}(k)=\mu_0 H_{c2}^2\left[\frac{b^2}{1+k^2/k_h^2}+\frac{b(1-b)}{2\kappa^2}\right] \quad (23)$$

where $k^2=k_x^2+k_y^2+k_z^2$, $k_h^2=(1-b)/\lambda^2$ and $k_\psi^2=2\kappa^2 k_h^2$. Eqns. (22, 23) apply if either $b\kappa^2 \gg 1$ holds or $b>0.3$; in these cases the vortex fields strongly overlap. The two terms in $c_{44}(k)$ are equal when $k^2 \simeq k_B^2$. The compression and tilt moduli are dispersive; that is, they depend on the wavelength $2\pi/k$ of the strain field. The FLL is softer with respect to short-wavelength deformations than it is with respect to uniform deformations. The ratios $c_{44}(0)/c_{44}(k) \simeq b\kappa^2/(1-b)$ and $c_{11}(0)/c_{11}(k) \simeq 2b^2\kappa^2/(1-b)^2$ are very large when $\kappa \gg 1$ ($\kappa \simeq 200$ for $YBa_2Cu_3O_7$) or when $H \simeq H_{c2}$.

7.3 Application of Nonlocal Elasticity

The nonlocality of the elastic response of the FLL has several important consequences.

(a) The flux line displacements caused by small pins become larger as compared to the result of the usual local theory in Eqn. (19). This fact makes the summation of random pinning forces more effective and yields larger critical current densities.

(b) The core structure of defects in the FLL (vacancies, interstitials, and edge and screw dislocations) is modified. This facilitates the generation and reaction of such defects, and the mutual cutting of flux lines.

(c) The thermal fluctuation of flux line positions is greatly enhanced and the "melting temperature" of the FLL is reduced.

As an example the displacement $s(0)$ of the FLL at the position of a point pinning force f and the mean square thermal displacement $\langle s^2\rangle$ of the flux line positions shall be considered. Both quantities are given by the same integral (in the continuum approximation):

$$\frac{2s(0)}{f}=\frac{\langle s^2\rangle}{k_B T} \simeq \int_{BZ}\frac{d^3k}{8\pi^3}[c_{66}(k_x^2+k_y^2)+c_{44}(k)k_z^2]^{-1} \simeq \frac{1}{\mu_0 H_{c2}^2}\frac{\kappa/\pi}{b(1-b)}\left[1+\left(\frac{b\kappa^2/2}{1-b}\right)^{1/2}\right] \quad (24)$$

The first term in the square brackets is the local result, which is obtained by replacing $c_{44}(k)$ by $c_{44}(0)$ in the integral. The second term is the nonlocal "correction" which is typically very large and even diverges at H_{c2}.

Bibliography

Brandt E H 1986 Elastic and plastic properties of the flux line lattice in type II superconductors. *Phys. Rev. B* 34: 6514–17

Brandt E H 1988 Magnetic field density of perfect and imperfect flux line lattices in type II superconductors. *J. Low Temp. Phys.* 73: 355–89

Brandt E H 1989 Levitation in physics. *Science* 243: 349–55

Brandt E H, Essmann U 1987 The flux line lattice in type II superconductors. *Phys. Status. Solidi B* 144: 13–38

Campbell A M, Evetts J E 1972 Flux vortices and transport currents in type II superconductors. *Adv. Phys.* 21: 199–428

De Gennes P G 1966 *Superconductivity of Metals and Alloys*. Benjamin, New York

Delrieu J M 1972 Singularities in the magnetic field map of type II pure superconductors. *J. Low Temp. Phys.* 6: 197–219

Dolan G J, Chandrashekhar G V, Dinger T R, Feild C, Holtzberg F 1989 Vortex structure in $YBa_2Cu_3O_7$ and evidence for intrinsic pinning. *Phys. Rev. Lett.* 62: 827–30

Essmann U, Träuble H 1971 The magnetic structure of superconductors. *Sci. Am.* 224: 75–84

Gammel P L, Bishop D J, Dolan G J, Kwo J R, Murray C A, Schneemeyer L F, Waszczak J V 1987 Observation of hexagonally correlated flux quanta in $YBa_2Cu_3O_7$. *Phys. Rev. Lett.* 60: 2592–5

Kes P H, Aarts J, van den Berg J, van der Beek C J, Mydosh J A 1989 Thermally assisted flux flow at small driving forces. *Supercond. Sci. Technol.* 1: 242–8

Shapira Y, Huong C Y, McNiff E J, Peters P N, Schwartz B B, Wu M K 1989 Magnetization and magnetic suspension of $YBa_2Cu_3O_x$–AgO ceramic superconductors. *J. Magn. & Magn. Mater.* 78: 19–30

Tinkham M 1975 *Introduction to Superconductivity*. McGraw-Hill, New York

E. H. Brandt
[Max-Planck-Institut für Metallforschung, Stuttgart, FRG]

T

Thermal Analysis: Recent Developments

This article supplements the article *Thermal Analysis* in the Main Encyclopedia.

Both the instrumentation and the application of thermal analysis are expanding rapidly. In order to update and complement the excellent earlier description of the general methods and the companion article on applications to polymers (see *Thermal Analysis of Polymers*), stress will be placed on the most important of the techniques and aspects not previously described in detail and on applications to inorganic materials—ceramics, metals and so on. In addition, the growing area of interest in the application of feedback to control the temperature program during thermoanalytical measurements is considered very briefly. It is hoped that these few examples will convince the reader of the power and versatility of thermoanalytical methods. They are conceptually very simple but extremely useful in imaginative hands.

1. Thermogravimetry and Thermomagnetometry

1.1 General and Instrumental Considerations

Thermogravimetry (TG) is essentially thermomagnetometry (TM) while the sample is subjected to a magnetic-field gradient. Consequently, the two techniques are discussed in the same section. The imposition of the field gradient allows determination of magnetic susceptibility as a function of temperature, as with a Faraday balance. More imaginatively, however, it can be used to detect the formation or disappearance of magnetic phases and the Curie temperature T_C or Néel temperature T_N of materials. Warne and Gallagher (1987) have recently reviewed TM.

In many applications of these techniques it is important to control the atmosphere over the sample during heating. Considerable efforts have also been made to operate at other than atmospheric pressure. Agarwal et al. (1987) describe the application of TG at high pressure. Far more common, however, is TG at reduced pressures. Occasionally, corrosive atmospheres are of interest and special care must be taken to preserve the integrity of the balance. This is generally accomplished by flowing an inert atmosphere through the balance chamber and subsequently blending in the corrosive component (e.g., halogen, SO_2, H_2O, etc.) prior to passing over the sample. An extreme case of balance protection and isolation is provided by the clever magnetic-coupling device described by Schubart and Knothe (1972).

1.2 Applications of TG and TM

The early applications used TG to establish the region of constant weight for the conversion of precipitates to known weighable species for gravimetric analysis. This was extended to study thermal decompositions in general, as described in the earlier articles. Related to this is the question of oxidative stability of polymers or the oxidation of alloys. It has also found use in such areas as adsorption, low-temperature nitrogen adsorption for BET surface area and porosity measurement or elevated temperature adsorption to identify various catalytically active sites on solids.

The importance of the oxygen content in high-T_C oxide superconductors has accentuated the already important area of determination of deviation from oxygen stoichiometry, particularly in electronic ceramics. For many oxides the electrical conduction, dielectric properties and magnetic behavior depend markedly on small variations in oxygen content. To follow these differences (as little as one part in 10^5) at elevated temperature, requires careful work to account for buoyancy, aerodynamic forces and the volatility of sample and container components. Figures 1 and 2 from Bracconi and Gallagher (1972) indicate the data themselves (Fig. 1) and the thermodynamic interpretation in the form of a van't Hoff Diagram (Fig. 2) for a nickel zinc ferrite. The rate of attaining constant weight is a frequent indicator of metastability. Systems that involve the precipitation or dissolution of a second phase are particularly susceptible to such effects. In Fig. 1, the change in slope of the weight-vs-temperature curves is evident at the transition between the single-phase and two-phase regions; however, the time to reach equilibrium depends markedly on the direction of crossing the phase boundary. The precipitation of the second phase requires nucleation and consequently considerably more time to reach equilibrium than the simpler process of dissolution.

The ability to superconduct at temperatures above 77 K has led to an explosion of interest in certain complex copper-based oxide systems. Most of these materials exhibit an enormous range of oxygen content while preserving the same basic structure, which adds to their scientific fascination. Figure 3 contrasts the oxygen-exchange behavior of three systems: $Ba_2YCu_3O_7$ (Gallagher et al. 1987b); $Bi_2Sr_{1.5}Ca_{1.5}Cu_2O_8$ (Grader et al. 1988); and

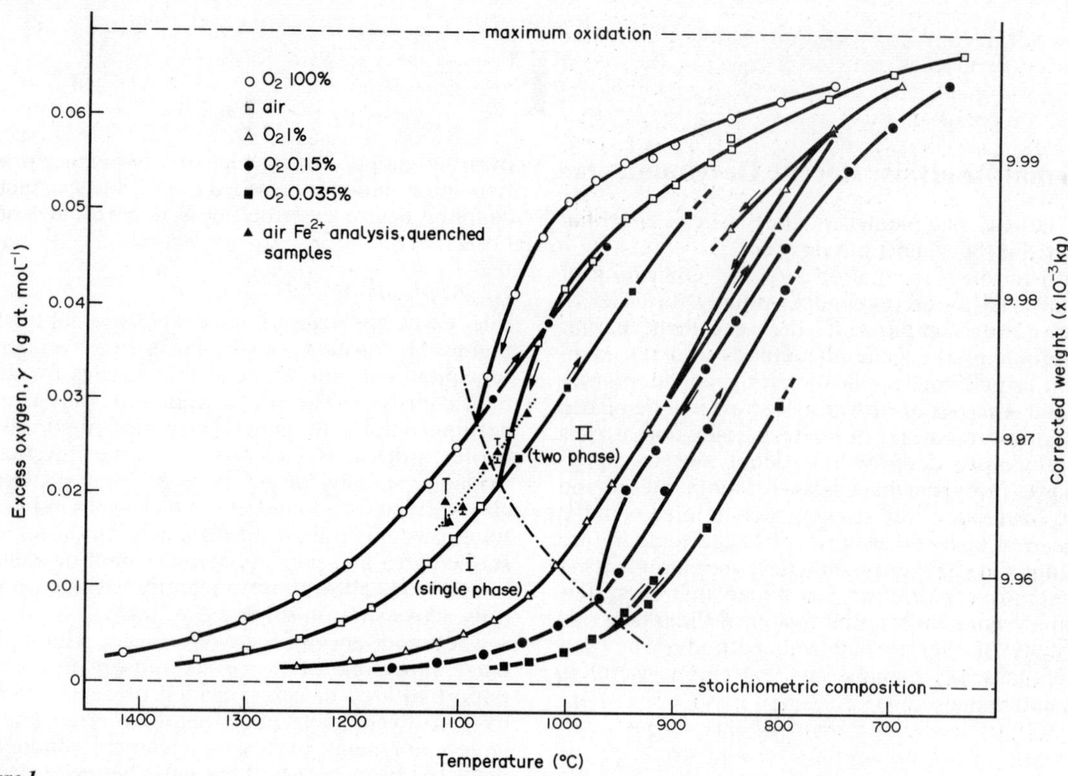

Figure 1
Excess oxygen coefficient γ and corresponding weight of $Ni_{0.685}Zn_{0.177}Fe_{2.138}O_4$ vs temperature and oxygen content of the gas phase

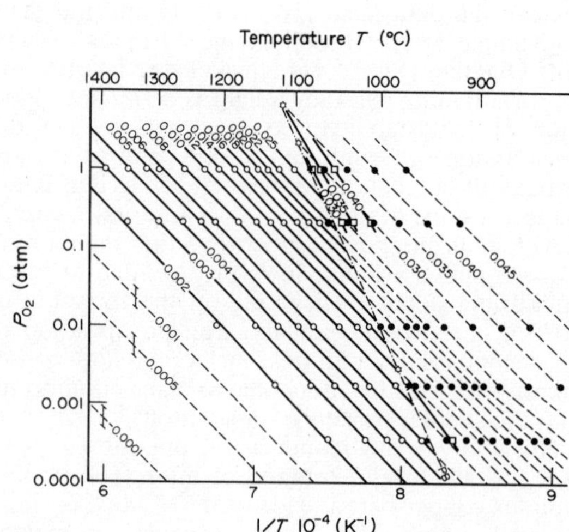

Figure 2
Isocompositional phase diagram for $Ni_{0.685}Zn_{0.1773}Fe_{2.138}O_4$ constructed with interpolated values from Fig. 1: the oxygen excess coefficient is indicated on each isocompositional line

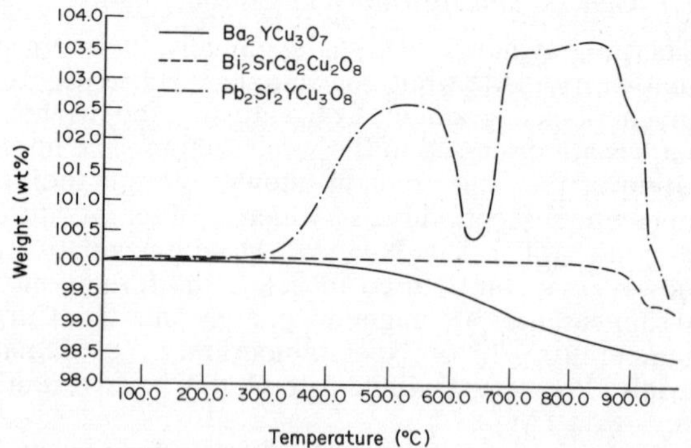

Figure 3
TG curves for several oxides heated at 1 °C min^{-1} in oxygen

$Pb_2Sr_2YCu_3O_8$ (Gallagher et al. 1989). The barium material reversibly loses O_2 above 350 °C, forming $Ba_2YCu_3O_{6.4}$ by the time 1000 °C is reached in pure oxygen. The structure transforms from orthorhombic to tetragonal symmetry during the loss of oxygen. The superconducting properties improve as the oxygen content approaches 7.00. In contrast, the lead compound reversibly gains substantial oxygen, reaching a maximum content at around 500 °C. The weight gain above 650 °C represents oxidative decomposition to new phases. Also, in contrast to the barium compound, the superconducting properties of the lead compound deteriorate with increasing oxygen content. The bismuth compound, on the other hand, is relatively stable in oxygen content and only begins to lose significant oxygen as it approaches its incongruent melting point near 900 °C.

In order to convert these weight losses to absolute values of oxygen content, it is necessary to know the precise composition at some reference point. There are a variety of analytical methods available: reduction in hydrogen on a thermobalance to yield the alkaline earth and rare earth oxides plus copper and/or lead metals is one of the most commonly used methods. It does not work for the bismuth or thallium compounds, because of the volatility of the metals.

Other ways TG is used effectively in studying these systems is to follow the decomposition of precursor compounds such as the mixed oxalates (Ozawa et al. 1988) or for assaying the initial starting compounds. The alkaline and rare earth oxides are notorious for absorbing large quantities of H_2O and CO_2 during storage and many of the soluble salts, such as nitrates and acetates, frequently used for synthesis have uncertain water contents. Degradation of the $Ba_2YCu_3O_7$ material on exposure to CO_2 and H_2O has also been investigated by TG (Gallagher et al. 1988a).

Studies involving permanent gases or materials of high volatility at relatively low temperature (e.g. H_2O) are easy to set up and control. It is more difficult to perform experiments that involve the vapor transport of ZnO, PbO, Li_2O and so on. The technique utilized by Holman (1974) is useful if suitably inert furnace tubes, sample containers and suspension materials are available. As an example, lead-deficient lead titanate ($PbTiO_3$) is suspended from a balance and surrounded by the two-phase mixture at either phase boundary. As the temperature is raised to a point for a reasonable reaction rate, the lead vapor pressure is maintained constant at the value established by the two-phase buffer material at that temperature. The change of the sample can be followed until it reaches its equilibrium value at that temperature and the partial pressure of the lead. If the composition at the phase boundary is known accurately, then the original lead deficiency can be determined. If not, then only relative values for different conditions can be established.

The use of combined TG and TM for the proximate analysis of coal is shown in Fig. 4 (Aylmer and Rowe 1982). Clearly resolved weight losses occur for moisture and volatiles in nitrogen. Switching to an oxidizing atmosphere allows combustion of the fixed carbon. After cooling to room temperature the original pyrite content is determined by a TM reduction of the iron oxide in the ash to iron metal by hydrogen. As iron metal forms, the apparent weight increases because of the strong magnetic attraction of iron. The relatively rapid and simple analysis of these five components of coal agrees very well with results obtained using the previous more cumbersome methods.

Charles (1982) provides an excellent example of the use of TM to study corrosion. A TM apparatus is shown in Fig. 5a and some results for the rate of corrosion of steel by an aqueous ammoniacal EDTA solution are presented in Fig. 5b. The oxidation of thin cobalt films or their reaction with a silicon

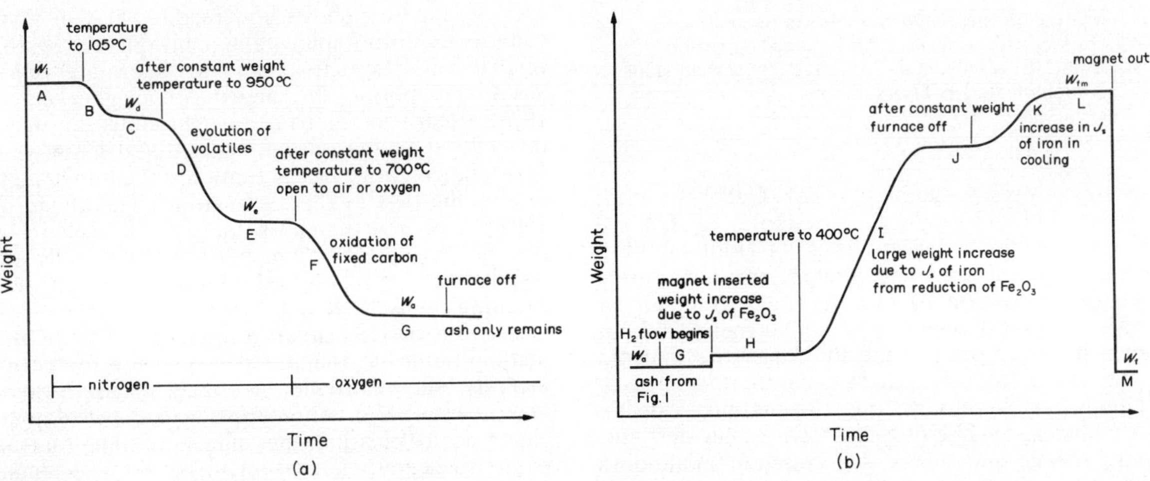

Figure 4
Proximate analysis scheme for coal analysis based on TG and TM

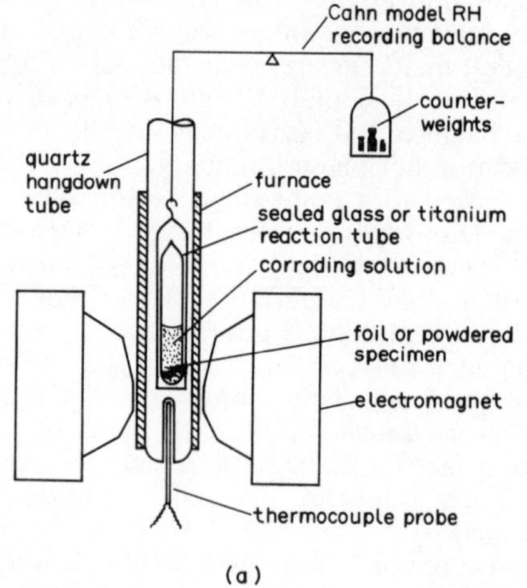

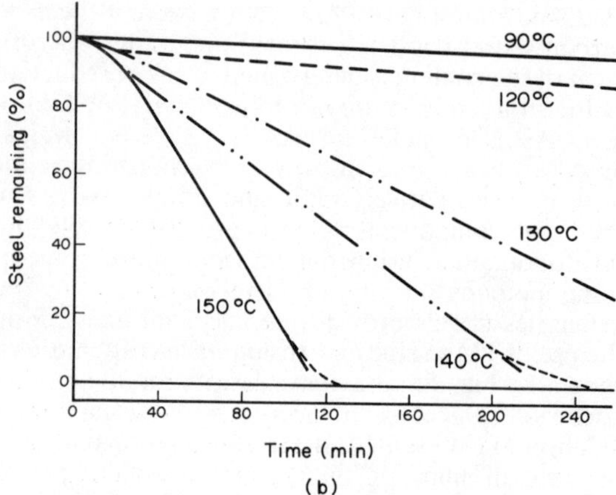

Figure 5
A TM study of the aqueous corrosion of steel: (a) TMA apparatus as used for aqueous corrosion work; and (b) corrosion of 2 mil type 285 carbon steel in 5% ammoniated EDTA solutions

substrate was also studied by TM (Gallagher et al. 1987a). Other metallurgical examples of TM are given by Haglund (1982). The use of TM to establish the presence of magnetite as an intermediate during the decomposition of siderite ($FeCO_3$) in air or oxygen has also been described. Magnetite did not form in air or oxygen and the magnetite that did form under inert conditions incorporated most of the divalent impurities (e.g., magnesium, calcium or manganese). Use of well known values of T_C for pure metals and alloys is a common method to calibrate the temperature scale during TG. Gallagher and Gyorgy (1986) described some of the factors that affect the accuracy and compared this method with the use of melting points for temperature calibration.

2. *Differential Thermal Analysis and Differential Scanning Calorimetry*

2.1 *General and Instrumental Considerations*

The scope of the terms differential thermal analysis (DTA) and differential scanning calorimetry (DSC) has been the subject of debate which is succinctly treated by Mackenzie (1980). There are two uncontroversial end members: first, DTA, in which the DT signal is measured by thermocouples inserted into the sample and reference materials or in contact with their respective containers while the sample and reference are thermally isolated; and second, DSC, in which the temperature of the sample and reference material are kept equal and the difference in the power supplied or the heat flux to each is measured. In DSC the difference in energy is directly determined. The controversial middle ground arises when a controlled heat leak or connection is made between the sample and reference materials in DTA. Boersma (1955) provided a theory that allows quantitative measurement of energy under these conditions. To the user, however, the practical consideration is whether or not the energy associated with the event of interest can or cannot be accurately determined.

2.2 *Applications of DTA and DSC*

The general tendency is to think of these techniques in terms of solid-state transitions, reactions or decompositions. However, they are equally valuable for process studies such as catalysis. Consider the oxidation of CO by oxygen, a highly exothermic reaction, releasing this heat at the surface of the catalyst. Figure 6 shows how readily DTA or DSC can be used to rapidly and conveniently screen relative catalytic activity and to determine which gases may poison this activity (Gallagher et al. 1976). Clearly in Fig. 6a copper chromite is a superior catalyst to the partially substituted lanthanum manganates, while the strontium-substituted material is the best of the latter group. The ability of 150 ppm of SO_2 to poison the catalytic activity is obvious in Fig. 6b. High pressure DSC cells are commercially available and used to evaluate hydrogenation catalysts (Kosak 1976).

The use of DSC to measure depression of the melting point has found widespread use for purity analysis, particularly for pharmaceuticals (Morros and Stewart 1976). An interesting isothermal application is to measure the induction time for the oxidative degradation of polymers and to evaluate the effectiveness of various antioxidants (Howard 1973).

A major use of DTA and DSC is in the area of phase equilibria and crystal growth. The weak second-order ferroelectric phase transition (1100–1200 °C) in $LiNbO_3$ is a strong function of the lithium deficiency in the material. Since the compound melts incongruently, it is necessary to know the composition at this melting point precisely in order to grow homogeneous crystals for use in electrooptic devices. Detecting the small change in the heat capacity by DTA as a function of composition was used to establish the congruent composition and to follow the defect (cation vacancy) equilibria in the system as it is doped with TiO_2 for device fabrication (Gallagher et al. 1988b).

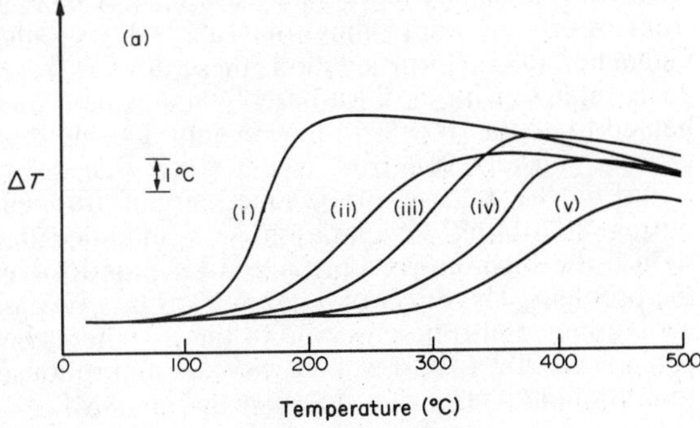

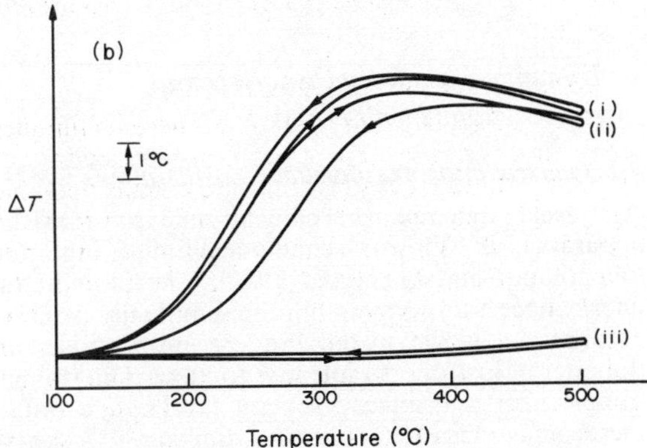

Figure 6
(a) DTA curves for oxidation of CO using catalysts with a wide variety of activities (the heating rate is 2 °C min^{-1} in a gas stream of 2% CO, 2% O_2 and 96% N_2 at 100 cm^3 min^{-1}): (i) copper chromite, 16.0 mg, 18.0 m^2 g^{-1}; (ii) $La_{0.5}Sr_{0.5}MnO_3$, 16.0 mg, 31.5 m^2 g^{-1}; (iii) $La_{0.75}K_{0.25}MnO_3$, 15.3 mg, 18.7 m^2 g^{-1}; (iv) $La_{0.7}Pb_{0.3}MnO_3$, crystals, 3000 ppm Pt, 16.4 mg, 3.1 m^2 g^{-1}; and (v) $La_{0.7}Pb_{0.3}MnO_3$, crystals, 1600 ppm Pt, 16.3 mg, 2.0 m^2 g^{-1}. (b) DTA curves for the catalytic oxidation of CO using $La_{0.5}Sr_{0.5}MnO_3$, 15.7 mg, 28.6 m^2 g^{-1} (the heating rate is 10 °C min^{-1} in a gas stream of 2% CO, 2% O_2 and 96% N_2 at 500 cm^3 min^{-1}): (i) first cycle, no SO_2; (ii) second cycle, 150 ppm SO_2 in gas stream; and (iii) third cycle, 150 ppm SO_2 in gas stream

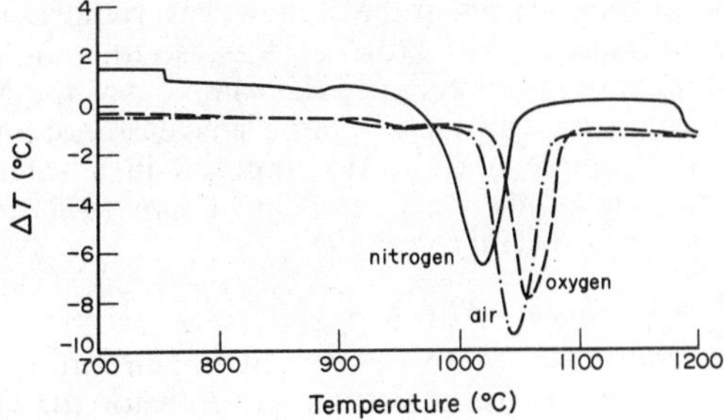

Figure 7
DTA curves of $Ba_2YCu_3O_7$ at 10 °C min^{-1} in various atmospheres

The crystal growth and sintering of the high-T_C superconducting materials is dependent on the melting temperature. In Sect. 1.2 it was shown how the oxygen content of these materials at any temperature varies with the partial pressure of oxygen. DTA has been used to show (Fig. 7) how the melting temperatures decrease with decreasing partial pressure of oxygen (Gallagher 1987).

3. *Evolved Gas Analysis*

3.1 *General Considerations and Instrumentation*

In the previously mentioned techniques, events are described in terms of their change in weight or heat content, and the chemical consequences inferred. Evolved gas detection is analogous in that the presence of volatile species is detected but not their specific nature. There are many such qualitative detectors that can be used to sense the evolution of volatile products (e.g., thermal conductivity, gas density, flame ionization, broad-band optical attenuation, etc.). Evolved gas analysis (EGA), however, utilizes techniques that are capable of specifically analyzing the evolved gas (e.g., Fourier transform infrared spectroscopy (FTIR), gas chromatography (GC), mass spectrometry (MS), etc.). Detectors sensitive to only a single species (e.g., dew-point or oxygen analyzers) are also used.

Emanation thermal analysis (ETA) is a subclass of EGA. A radioactive gas is incorporated into the sample during preparation or by subsequent implantation. The radioactivity of the gas stream passing over the sample is monitored as the temperature is raised. If the sample undergoes a phase transformation there is a sudden release of gas and a peak in the radioactivity. As the surface area and porosity of the sample changes, so does the rate of evolution. This technique is reviewed by Balek (1987).

Most often EGA is combined with TG or DTA/DSC. There is also equipment designed solely for EGA, such as pyrolysis attachments for GC or GC–MS. In the above techniques, the gas is usually at atmospheric pressure. MS, however, requires that the pressure be 10^{-4} torr or less, so that a leak system is required between the sample and the MS. Alternatively, when less volatile species are of interest the sample may be decomposed in a vacuum through heating by a laser beam (Lum 1977) or a small furnace (Gallagher 1978).

3.2 Application of EGA

The reduction of oxides by hydrogen offers an opportunity to use a specific gas detector for H_2O (Rowe et al. 1983). The small reactor package used in this research was convenient to operate between the poles of an electromagnet in order to determine the effect of an external magnetic field on the rate of reaction. The mineral components in asbestos vary considerably from one to another. A TG method had been proposed for analysis; however, EGA analysis by MS revealed the problems associated with some of the underlying assumptions (Khorami et al. 1984).

The laser heating scheme mentioned in Sect. 3.1 was used by Lum and Feinstein (1980) to study the decomposition of a Novolac epoxy used to encapsulate semiconductors. An isometric plot of the total ion current and mass ion profiles over the range of atomic masses from 12 to 122 as a function of temperature appears in Fig. 8. Some of the mass numbers due to the evolution of NO and phenol are identified, while speculation regarding other mass fragments and the mechanism of the pyrolysis was made.

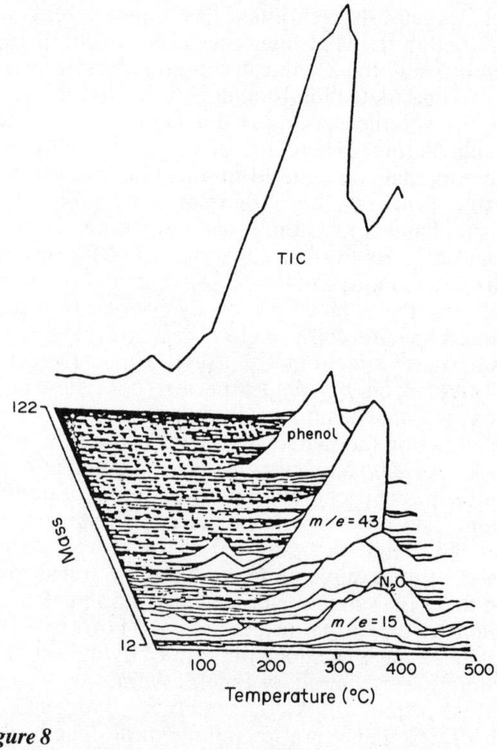

The high sensitivity of MS can be effectively used to study the very initial stages of a decomposition, such as for InP by Gallagher and Chu (1982). Similarly, the sensitivity is important for the study of thin films where large amounts of inert substrate material are involved. Kinsbron et al. (1979) showed how thin gold electrodes reacted with the underlying (GaAl)As to release arsenic while forming a gallium–gold eutectic. Duncan et al. (1988) combined EGA by MS with nuclear magnetic resonance (NMR) to characterize amorphous hydrogenated boron nitride films. Figure 9 shows MS–EGA results from sputtered thin films of $TaSi_2$ (Levy and Gallagher 1985). In production, these devices were encapsulated with a silica-based glass which was heated to about 1000 °C to flow evenly. During this process, blisters occurred which were traced by EGA to the release of occluded argon trapped during sputtering. Preparation was modified to reduce the high-temperature argon peak and solve the problem. The effect of a phase transition on gas evolution can also be observed in Fig. 9, where the peaks near 300 °C correspond with the amorphous-to-crystalline phase transition revealed by DSC.

Figure 8
Isometric plot of mass profiles of evolved gases vs temperature for Novolac epoxy

4. Dynamic Mechanical Analysis and Thermodilatometry (TD)

4.1 General and Instrumental Considerations

A typical dynamic mechanical analysis (DMA) apparatus is shown schematically in Fig. 10 (Wendlandt and Gallagher 1981). The amount of energy necessary to maintain the sample in resonant oscillation at constant amplitude is proportional to the energy lost due to internal friction. The driven sample arm is displaced several tenths of a millimeter and released to start the oscillation. A feedback mechanism from the linear transformer to the driver maintains the amplitude of oscillation. DMA is primarily used in polymer science to study relationships between structure and properties. The difference between an elastomer and a stiff plastic can be readily seen in a plot of the modulus vs temperature. The various peaks in the loss (tan δ) correlate with toughness and damping, which are of great importance in elastomers (see *Tire Adhesion: Role of Elastomer Characteristics*, Suppl. 2). Determining the frequency and temperature dependence for internal motions within polymers is valuable in determining the nature of chain segments, crystallinity, plasticizers, anisotropic properties and so on.

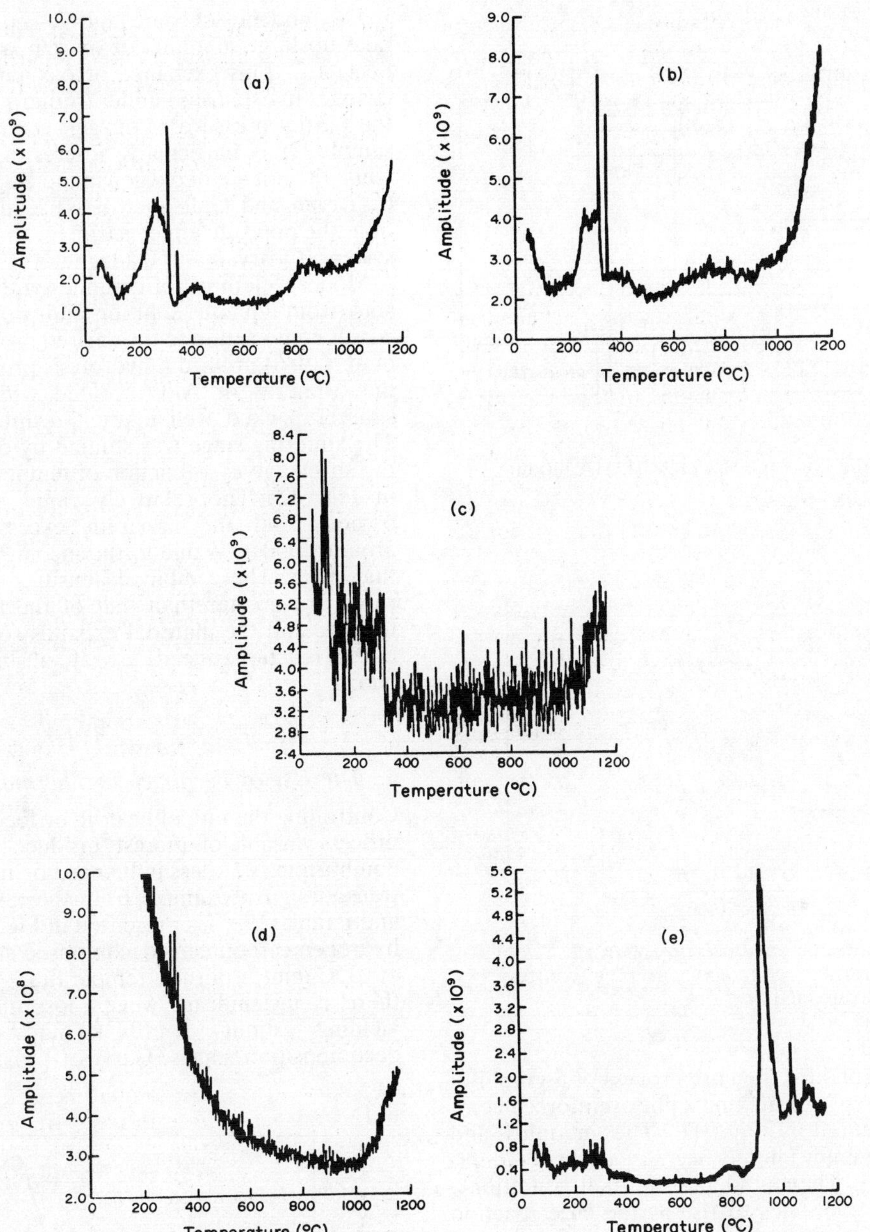

Figure 9
EGA of trapped gases in thin films of sputtered $TaSi_2$: (a) hydrogen, (b) nitrogen, (c) oxygen, (d) water vapor and (e) argon

Thermal expansion is a property of great technological importance. It determines whether parts will maintain their alignment, continue to adhere, or possibly bind as the temperature shifts. As the temperature varies, changes in length occur discontinuously at a first-order phase transition or as a change in the slope for higher-order transformations such as the glass transition. The shrinkage of a material is one of the most common methods for following the sintering process.

4.2 Applications of DMA and Thermodilatometry

Composites are currently a fertile field of study. The interrelation of time and temperature is evident in

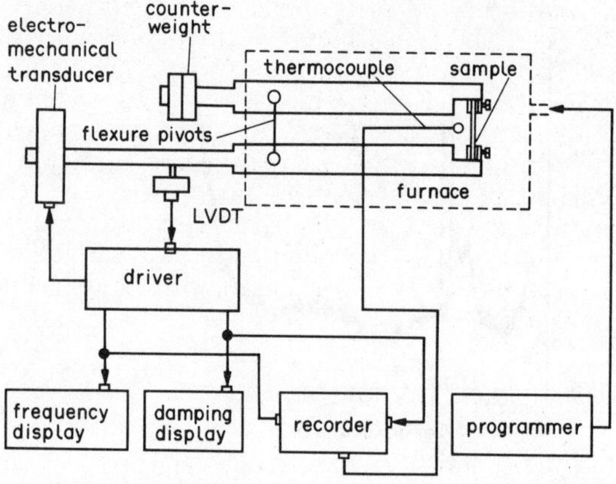

Figure 10
Schematic of the Du Pont Model 981 DMA module

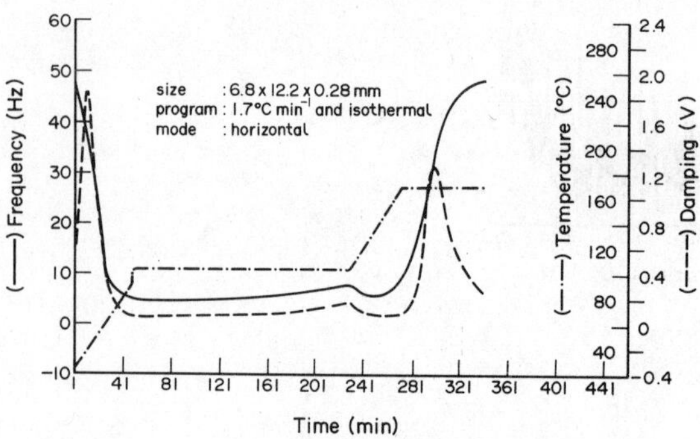

Figure 11
DMA results for a graphite-epoxy prepreg (courtesy of Dupont Instruments)

the DMA results for the cure process of a graphite-epoxy "prepreg" for making a fiber-reinforced composite, presented in Fig. 11. The modulus and resonant frequency initially decrease as the temperature is raised. Then, under isothermal conditions, the modulus continues to rise as the cure reaction proceeds.

Thermal expansion has been successful in studying several aspects of $LiNbO_3$. Since the material is used for electrooptic switching, the preservation of optical alignment within the device package is crucial. Besides providing critical expansion data for various device components, thermal expansion was used to detect subtle transitions near ambient temperature. The T_C for almost stoichiometric material; that is, only slightly lithium deficient, is where the onset of melting obscures the transition in DTA. It was, however, successfully measured by thermodilatometry (TD) (Gallagher et al. 1988b).

The tendency of some high-T_C superconductors to take up large amounts of oxygen was described in Sect. 1.2. This exchange of oxygen leads to large changes in expansion under isothermal conditions as the partial pressure of oxygen is changed over the sample. It is particularly interesting to note that, while the uptake of oxygen in the $Ba_2YCu_3O_7$ system (O'Bryan and Gallagher 1987) leads to a contraction, the opposite effect is true for the $Pb_2Sr_2YCu_3O_8$ system (O'Bryan and Gallagher 1989).

Modern methods of ceramic synthesis frequently start from aqueous solution and proceed via a precursor composition to the desired product. Okamura et al. (1986) utilized a hydrolysis process to prepare microspheres of Al_2O_3 coated with TiO_2. These spheres packed well in the presintering compact. The sintering stage was studied by dilatometry and the shrinkage as a function of temperature is shown in Fig. 12. The relatively rapid densification is obvious and the apparent excessive expansion around 1400 °C is due to the incomplete reaction at that stage. The combined densities of alumina and rutile are greater than that of the final aluminum titanate and the material expands somewhat as the rate of reaction exceeds the rate of sintering near the end.

5. *The Use of Feedback in Thermal Analysis*

Controlling the rate of heating by feedback from the process variable of interest, produce results closer to equilibrium (i.e., less influenced by mass or thermal transport). An example from the review by Paulik and Paulik (1986) is presented in Fig. 13. Potassium hydrogen carbonate is decomposed at a heating rate of 3 °C min^{-1} during temperature ranges where there is insignificant weight loss and at the rate yielding about 3×10^{-7} kg min^{-1} during the decomposition stages. Curves (i) and (ii) show the

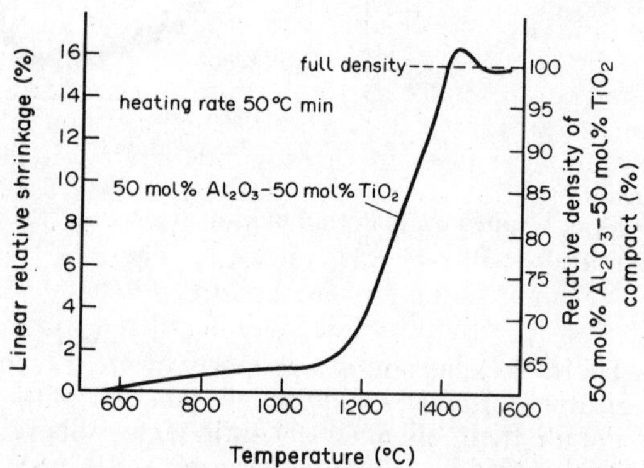

Figure 12
Sintering behavior of TiO_2-coated Al_2O_3 spheres at 50 °C min^{-1} in air

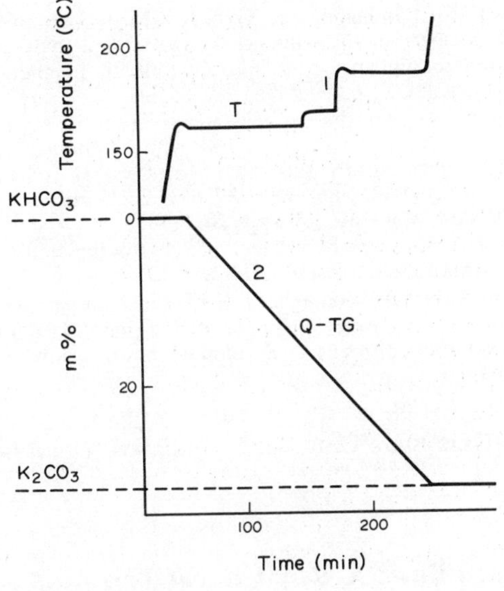

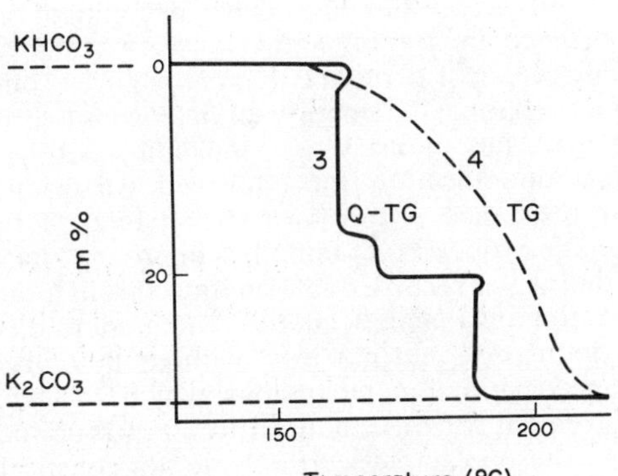

Figure 13
Dehydration of $KHCO_3$ measured by TG under quasi-isothermal (Q) conditions

temperature and weight, respectively, as a function of time, while curve (iii) shows the resulting weight vs temperature plot in comparison with the conventional TG, curve (iv). There is greatly enhanced resolution evident in the "quasiisothermal" plot. Rouquerol (1964) has demonstrated that similar benefits can be derived by using feedback from a pressure sensor to control the rate of heating. This is the analogous situation based on EGA. These methods provide a better understanding of the true processes taking place during decomposition.

This same approach can be utilized in technology to provide a better product at the expense of less energy. Huckabee and Palmour (1972) demonstrated "rate-controlled sintering" which not only allowed for a denser finer-grained sintered product but also saved energy by reducing the time–temperature needed to obtain that superior microstructure. This was achieved by predetermining the optimum rate of shrinkage that would maintain an open porosity so that a minimum of trapped pores would occur during the early stage of sintering. These trapped pores are particularly difficult to eliminate in the latter stages of sintering. The temperature vs time curve similar to curve 1 in Fig. 13) that provided this controlled shrinkage was recorded in the preliminary experiment and used for subsequent production.

Bibliography

Agarwal P K, Noh J S, Schwartz J A, Davini P 1987 Effect of surface acidity of activated carbon on hydrogen storage. *Carbon* 25: 219–26

Aylmer D, Rowe M W 1982 Thermo-magneto-gravimetric analysis of pyrite in coal and lignite. In: Miller B (ed.) 1982 *Proc. 7th Int. Conf. Thermal Imaging*, Vol. 2. Wiley, Chichester, UK, pp. 1270–5

Balek V 1987 Emanation thermal analysis: A status report. *Thermochim. Acta* 110: 222–35

Boersma S L 1955 Theory of differential analysis—Measurement and interpretation. *J. Am. Cer. Soc.* 38: 281–7

Bracconni P, Gallagher P K 1979 Phase diagram of a nickel–zinc ferrite of composition: $Ni_{0.685}Zn_{0.177}Fe_{2.138}O_{4+\gamma}$. *J. Am. Ceram. Soc.* 62: 171–6

Charles R G 1982 Thermomagnetic analysis as a means for following corrosion reactions in sealed systems. In: Miller B (ed.) 1982 *Proc. 7th Int. Conf. Thermal Imaging*, Vol. 1. Wiley, Chichester, UK, pp. 264–71

Duncan T M, Levy R A, Gallagher P K, Walsh M M 1988 Structural characterization of boron nitride films. *J. Appl. Phys.* 64: 2990–4

Gallagher P K 1978 An evolved gas analysis system. *Thermochim. Acta* 26: 175–83

Gallagher P K 1987 Characterization of $Ba_2YCu_3O_x$ as a function of oxygen partial pressure. Part I: Thermoanalytical measurements. *Adv. Ceram. Mater.* 2: 632–9

Gallagher P K, Chu S N G 1982 Kinetics of the dissociation on InP under vacuum. *J. Phys. Chem.* 86: 3246–50

Gallagher P K, Grader G S, O'Bryan H M 1988a Some effects of CO_2, CO, and H_2O upon the properties of $Ba_2YCu_3O_7$. *Mater. Res. Bull.* 23: 1491–9

Gallagher P K, Gyorgy E M 1986 Curie temperature standards for thermogravity. The effect of magnetic field strength and comparison with melting point standards using Ni and Pb. *Thermochim. Acta* 109: 19–206

Gallagher P K, Gyorgy E M, Schrey F, Hellman F 1987a Use of thermomagnetometry to follow reactions of thin films. *Thermochim. Acta* 121: 231–40

Gallagher P K, Johnson D W, Vogel E M 1976 Examples of the use of DTA for the study of catalytic activity and related phenomena. In: Rylander P N, Greenfield H (eds.) 1976 *Catalysis in Organic Synthesis*. Academic Press, New York

Gallagher P K, O'Bryan H M, Brandle C D 1988b Application of thermal analysis to study the crystal growth. *Thermochim. Acta* 133: 1–10

Gallagher P K, O'Bryan H M, Cava R J, James A C W P, Murphy D W, Rhodes W W, Krajewski J J, Peck W F,

Waszczak J V 1989 Oxidation–reduction of $Pb_2Sr_2Ln_{1-x}M_xCu_{3-y}Ag_yO_8$ *Chem. Mater.* (in press)
Gallagher P K, O'Bryan H M, Sunshine S A, Murphy D W 1987b Oxygen stoichiometry in $Ba_2YCu_3O_x$. *Mater. Res. Bull.* 22: 995–1006
Gallagher P K, Warne S S 1981 Thermomagnetometry and the thermal decomposition of siderite. *Thermochim. Acta* 43: 253–67
Grader G S, Gyorgy E M, Gallagher P K, O'Bryan H M, Johnson D W, Sunshine S A, Zahurak S M, Jin S, Sherwood R C 1988 Crystallographic, thermodynamic and transport properties of $Bi_2Sr_{3-x}Ca_xCu_2O_{8+\delta}$ superconductor. *Phys. Rev. B* 38: 757–60
Haglund B O 1982 Curie temperature of alloys, its measurement and technical importance. *J. Therm. Anal.* 25: 21–43
Holman R L 1974 Novel uses of the thermomicrobalance in the determination of nonstoichiometry in complex oxide systems. *J. Vac. Sci. Technol.* 11: 434–9
Howard J B 1973 DTA for control of stability in polyolefin wire and cable compounds. *Polym. Eng. Sci.* 13: 429–34
Huckabee M L, Palmour H 1972 Rate controlled sintering of fine grained Al_2O_3. *Am. Ceram. Soc. Bull.* 51: 574–6
Khorami J, Choquette D, Kimmerle F M, Gallagher P K 1984 Interpretation of EGA and DTG analysis of crysotile asbestos. *Thermochim. Acta* 76: 87–96
Kinsbron E, Gallagher P K, English A T 1979 Dissociation of (GaAl)As during alloying of gold contact films. *Solid-State Electron.* 22: 517–24
Kosak J 1976 Catalyst activity via thermal analysis. In: Rylander P N, Greenfield H (eds.) 1976 *Catalysis in Organic Synthesis*. Academic Press, New York, pp. 137–48
Levy R A, Gallagher P K 1985 Argon entrapment and evolution in sputtered $TaSi_2$ films. *J. Electrochem. Soc.* 132: 1986–91
Lum R M 1977 Direct analysis of polymer pyrolysis using laser microprobe techniques. *Thermochim. Acta* 18: 73–94
Lum R M, Feinstein L G 1981 Investigation of the molecular processes controlling corrosion failure mechanisms in plastic encapsulated semiconductor devices. *Microelectron. Reliab.* 21: 15–31
Mackenzie R C 1980 Differential thermal analysis and differential scanning calorimetry. Similarities and differences. *Anal. Proc. (London)* 17: 217–20
Morros S A, Stewart D 1976 Automated and computerized system for purity determination by differential scanning calorimetry. *Thermochim. Acta* 14: 13–24
O'Bryan H M, Gallagher P K 1987 Characterization of $Ba_2YCu_3O_x$ as a function of oxygen partial pressure. Part II: Dependence of the O–T transition on oxygen content. *Adv. Ceram. Mater.* 2: 642–8
Okamura H, Barringer E A, Bowen H K 1986 Preparation and sintering of monosized Al_2O_3–TiO_2 composite powder. *J. Am. Ceram. Soc.* 69: C22–C24
Ozawa T 1988 Application of thermal analysis to kinetic study of superconducting oxide formation. *Thermochim. Acta* 133: 11–16
Paulik F, Paulik J 1986 Thermoanalytical examination under quasi-isothermal–quasi-isobaric conditions. *Thermochim. Acta* 100: 23–59
Rouquerol J 1964 Thermal analysis under low pressure and constant rate of decomposition. *Bull. Soc. Chim. Fr.* 1964: 31–2
Rowe M R, Gallagher P K, Gyorgy E M 1983 Establishing the absence of any influence by an external analysis field upon the intrinsic rate of reduction of magnetite by hydrogen. *J. Chem. Phys.* 79: 3534–6
Schubart B, Knothe E 1972 A new thermogravimetric instrument for chemical analysis with weighing compartment hermetically separated. In: Gast H (ed.) 1972 *Progress in Vacuum Microbalance Techniques*, Vol. 1. Heyden, Norwich, UK, pp. 207–15
Warne S S, Gallagher P K 1987 Thermomagnetometry. *Thermochim. Acta* 110: 269–79
Wendlandt W W, Gallagher P K 1981 Instrumentation. In: Turi E A (ed.) 1981 *Thermal Characterization of Polymeric Materials*. Academic Press, New York, Chap. 1

P. K. Gallagher
[Ohio State University, Columbus, Ohio, USA]

Thin Films: X-Ray Characterization

In recent years thin films have become of major importance in several areas of science and technology. Optical component manufacturers employ surface coatings to modify optical characteristics. The materials industry uses thin film coatings and surface modification for improved corrosion and wear resistance. There is a strong interest in the magnetic properties of thin films in order to increase the density of recorded information through the use of continuous magnetic media. The modern electronics industry could not exist without its extremely advanced thin film technology which is a central requirement for the manufacture of intregrated circuits. These examples certainly do not constitute an exhaustive list of the applications of thin films in the modern world, but they do serve to illustrate the variety of areas in which they are found.

It is hardly surprising to discover that as thin films and surfaces have increased in technological importance so surface-specific analysis techniques have developed accordingly. Many of these are based on the interaction of electron or ion beams with the surface. For depth profiling, destructive methods are often the only solution. X-ray diffraction has for a long time been the definitive method for crystal structure determination. The penetration depth of x rays in solids is, however, many micrometers and conventional diffraction techniques are often of limited use in studying thinner structures and surfaces.

1. Specular Reflection of X Rays

The index of refraction n of matter at x-ray wavelengths can be expressed as

$$n = 1 - \delta - i\beta \tag{1}$$

where

$$\delta = (\lambda^2 e^2 / 2\pi mc^2) \sum_i N_i (Z_i + f_i') \tag{2}$$

and

$$\beta = (\lambda^2 e^2 / 2\pi mc^2) \sum_i N_i (f_i'') \tag{3}$$

In these equations λ is the x-ray wavelength, e and m are the charge and mass of the electron, c is the velocity of light, N_i is the number of atoms of type i in the unit cell, Z_i the corresponding atomic number, and f_i' and f_i'' are the real and imaginary resonant (or anomalous) atomic scattering factors.

For solids and liquids δ and β are of the order of 10^{-5} or less, and hence n is slightly less than unity. The index of refraction of air at x-ray wavelengths is effectively unity and, therefore, as was pointed out by Compton (1923), x rays may undergo total external reflection from smooth surfaces for incident angles ϕ below a certain critical angle ϕ_c defined by $\phi_c \simeq (2\delta)^{1/2}$. ϕ_c is typically a few tenths of a degree. Parratt (1954) developed the necessary theory to describe the reflection of x rays from a smooth homogeneous surface. The reflectivity R of the surface as a function of incident angle is given by the ratio of reflected intensity I_r to incident intensity I_0 and can be written as

$$R(\phi) = \frac{I_r}{I_0}(\phi) = \frac{(\phi - A)^2 + B^2}{(\phi + A)^2 + B^2} \tag{4}$$

where

$$2A^2 = [(\phi^2 - \phi_c^2)^2 + 4\beta^2]^{1/2} + (\phi^2 - \phi_c^2) \tag{5}$$

and

$$2B^2 = [(\phi^2 - \phi_c^2)^2 + 4\beta^2]^{1/2} + (\phi^2 - \phi_c^2) \tag{6}$$

One of the most important points to note is that the penetration of the x rays into the surface for $\phi < \phi_c$ does not drop to zero. An evanescent wave effectively probes a few nanometers into the sample surface. The penetration depth $z_{1/e}$, defined as the depth at which the electric field has dropped to 1/e of its value at the sample surface, is dependent on ϕ and for small ϕ, is given by

$$z_{1/e} \simeq \frac{\lambda}{4\pi B} \tag{7}$$

Finally, it is interesting to calculate the evanescent wave intensity at the surface. This is given by the product of the electric field E and its complex conjugate, and its dependence on ϕ is described by

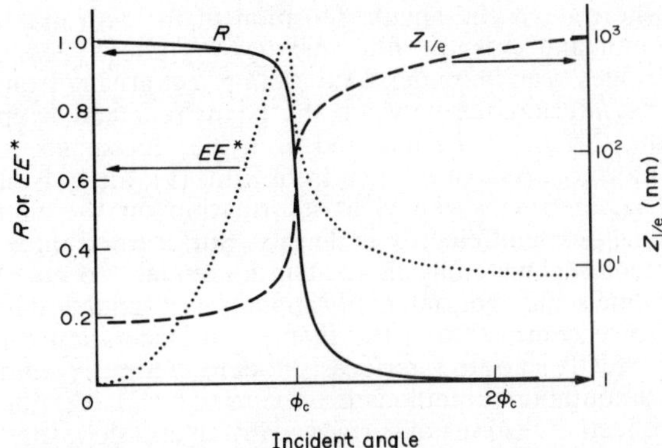

Figure 1
Typical curves of reflectivity (R), penetration depth ($z_{1/e}$) and evanescent wave intensity (EE^*) as a function of incident angle (ϕ) for x rays incident on a smooth flat surface

$$EE^*(\phi) \simeq \frac{4I\phi^2}{(\phi + A)^2 + B^2} \tag{8}$$

where I is the incident beam intensity.

Curves of $R(\phi)$, $z_{1/e}(\phi)$ and $EE^*(\phi)$ are shown in Fig. 1. It is immediately apparent that by varying the incident angle of a beam of x rays on a smooth surface the depth of penetration can be controlled in a range from a few nanometers to many micrometers. This is the basis for the various thin film analysis methods discussed in the following sections.

In theory, a perfectly smooth flat surface for which the absorption coefficient for x rays is zero will have a reflection coefficient $R(\phi)$ of unity for all values of ϕ from zero to the critical angle ϕ_c, above which $R(\phi)$ would drop sharply and tail off to zero for large ϕ. In practice this is never the case and the shape of the reflectivity curve, together with the measured value of ϕ_c, contains information on the surface density, roughness and refractive index. The effects of roughness can be taken into account in theoretical models.

For surfaces on which a thin film has been deposited an additional effect often manifests itself. Above $\phi = \phi_c$, peaks and troughs appear in the reflectivity curve. These were first reported by Kiessig (1931a,b) who derived a formula for the position of the maxima as a function of ϕ:

$$\phi_m^2 = 2\delta + \frac{(m + \frac{1}{2})\lambda^2}{4d^2} \tag{9}$$

where ϕ_m is the angular position of the mth maximum and d is the film thickness. These "Kiessig fringes" can be understood as an effect arising from the interference between the beam reflected from the surface of the film and from the film–substrate interface. As can be seen from Eqn. (9), analysis of Kiessig fringes can yield information on the film thickness and electronic density. Surface and interface roughness may also be considered in theoretical models and experimentally appears as a reduction in fringe contrast from the theoretical calculation for perfectly smooth interfaces. Modern computers and computational methods may be used to analyze the reflectivity curves of surfaces with several deposited layers. An elegant illustration of this grazing x-ray reflectometry (GXRR) technique is described by Le Boité et al. (1988) who compare it with Rutherford backscattering as an analysis technique for thin film mixing by ion beams. However, the technique is certainly not without its limitations: systems giving poor fringe contrast due to different densities between layers are unsuitable for GXRR, and the sample must be fairly large (for accurate alignment), very smooth and flat.

One further technique for thin film characterization may be mentioned at this point: total-reflection x-ray fluorescence (TXRF). This is a simple modification of the standard technique for compositional analysis, whereby the surface is illuminated with an x-ray beam and the resulting x-ray fluorescence spectrum yields information on the atomic species present. In the glancing angle geometry, a fine slit is used to collimate the beam, enabling accurate control of the incident angle. Depth profiling of composition is thus, in principle, feasible though the technique is normally only used for identifying and quantifying surface species. TXRF was originally developed for compositional analysis of trace elements deposited (e.g., from solution) onto an optically flat x-ray reflecting surface, but its application to thin film and surface analysis is obvious. Identification of interfacial atomic species is also possible.

2. Glancing Angle X-Ray Diffraction

Conventional x-ray diffractometers are often used to investigate thin film structures. The well-known Bragg–Brentano (or θ–2θ) focusing geometry suffers a major disadvantage in many instances since x-ray scattering from the substrate may overwhelm that from the thin film. This problem is partially overcome by using the Seeman–Bohlin focusing geometry, where the incident angle is kept constant and the detector is scanned around the focusing circle. Unfortunately, the minimum practical incident angle is about 5° and the diverging incident beam prevents accurate control of the penetration depth. This type of arrangement is, however, often extremely useful in thin film structure analysis and is available commercially.

In order to further reduce the angle of incidence it is necessary to abandon focusing arrangements and use a parallel beam geometry. In order to maintain angular resolution, Soller slits are employed. The resultant loss of intensity that this entails is quite serious. In addition, moving to incident angles near the critical angle for total external reflection (ϕ_c) greatly reduces the x-ray intensity on the sample surface.

Finally, below ϕ_c the depth of penetration of the incident beam may drop by several orders of magnitude, with a corresponding decrease in the intensity diffracted by the sample. It is not surprising, therefore, to find that most studies of very thin films and surfaces employ synchrotron storage rings as the x-ray source, as the intensity is high enough to study surface coatings of a fraction of a monolayer. Parallel-beam glancing angle diffractometer facilities are found at most synchrotron radiation facilities. It should be stressed, however, that conventional x-ray sources may very often be employed for effective thin film glancing angle diffraction studies using incident angles just above ϕ_c in order to enhance the surface sensitivity as much as possible while maintaining a reasonable diffracted intensity. Rotating anode sources may even be powerful enough in certain cases for very thin film and for surface studies for $\phi < \phi_c$.

Two specific techniques have been developed for glancing angle diffraction. The first, described by Lim et al. (1987), is a modification of the Seeman–Bohlin focusing geometry (see Fig. 2). Instead of a diverging incident beam, a slit has been added to allow accurate control of the angle of incidence on the sample. Diffraction from the entire illuminated section of the sample surface is collected via a Soller slit for angular resolution. In the case of synchrotron radiation facilities the incident beam will already have passed through a monochromator before the entrance slit. For conventional sources

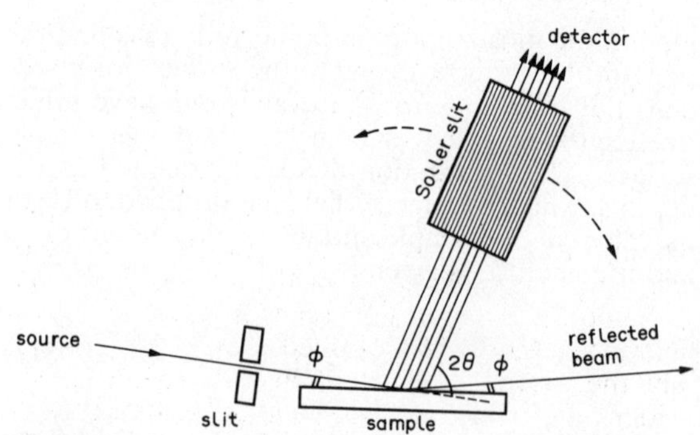

Figure 2
GIAB surface diffraction geometry

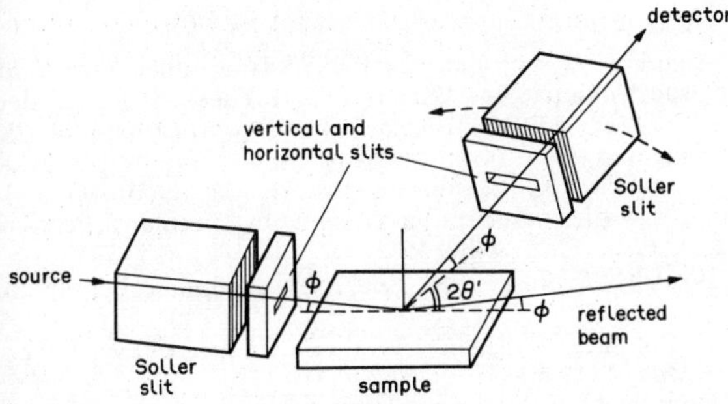

Figure 3
GIXS surface diffraction geometry

the monochromator is usually placed after the exit Soller slit. Alternatively, a solid-state semiconductor detector may be employed (these have a typical energy resolution of about 2%) to discriminate in favor of the characteristic K_α lines of the source. This arrangement is, for obvious reasons, known as the glancing incidence asymmetric Bragg (GIAB) geometry and is well suited to the study of polycrystalline or amorphous thin films since lattice information is obtained at varying angles with respect to the surface as the detector (2θ) angle changes.

The second glancing angle diffraction technique was first described by Marra et al. (1979) and has been termed glancing incidence x-ray scattering (GIXS). In this geometry both the incident beam and the detected radiation are at a glancing angle with respect to the surface (see Fig. 3). The detector is rotated round the normal to the sample surface so that only lattice planes very nearly perpendicular to the sample surface contribute to the diffracted signal. In order to maintain a reasonable angular resolution, Soller slits must be included in the incident beam as well as in the detector section. This, together with the fact that the scattered radiation suffers more absorption in the sample than in the case of GIAB, makes the technique rather unsuitable for polycrystalline or amorphous thin films. It may, however, be used as a complementary technique to GIAB in such cases as the analysis of crystallite orientation and/or strain. Where GIXS really comes into its own as a thin film analysis technique is in the study of epitaxial layers on single crystal substrates. In this case the complete surface may contribute to the diffracted intensity, thereby making the technique feasible with powerful conventional sources, even at incident angles less than ϕ_c. Sample θ detector (2θ) scans in the GIXS geometry, together with so-called "ω scans" (where only the sample is rotated) at varying angles of incidence, provide information about epitaxial relationship, crystallite size (and orientational spread) and lattice strain for epitaxial thin films.

3. *Surface X-Ray Absorption Spectroscopy*

X-ray absorption spectroscopy (XAS) is a well-known technique for structural characterization that has developed rapidly due to the increased availability of high-intensity synchrotron x-ray sources. XAS is generally divided into two regimes: x-ray absorption near edge structure (XANES) and extended x-ray absorption fine structure (EXAFS). XANES deals with the structure in the absorption coefficient close to an inner shell absorption edge of the particular atomic species under investigation in the sample, while EXAFS is concerned with the oscillations observed in the absorption coefficient up to several hundred electron volts above the absorption edge. Both furnish information about the local structure around the absorbing atomic species and are thus particularly useful in the study of amorphous solids or dilute species in host environments, where x-ray diffraction is of limited use. XANES also gives the valence state of the absorbing species.

The subject of interest is how to apply XAS to the study of thin films or surfaces. The first surface-sensitive XAS technique, known as surface EXAFS (or SEXAFS), was proposed and developed in the late 1970s. The basis of this method is that the surface photoelectron yield is closely related to the surface x-ray absorption coefficient. The small escape depth of the photoelectrons (a few nanometers) ensures surface sensitivity, so that SEXAFS is only of use in studying surfaces (e.g., absorbed atoms or monolayers) or very thin films.

By employing the glancing angle geometry described previously another range of surface and interface techniques becomes possible. Reflection EXAFS, or ReflEXAFS, was first reported by Barchewitz et al. (1978). This method (see Fig. 4) consists of reflecting an x-ray beam off the sample surface and measuring the ratio of reflected to incident intensity (I_r/I_0) as a function of the x-ray energy (E). The surface absorption spectrum may thus be found, with an effective penetration depth of a few

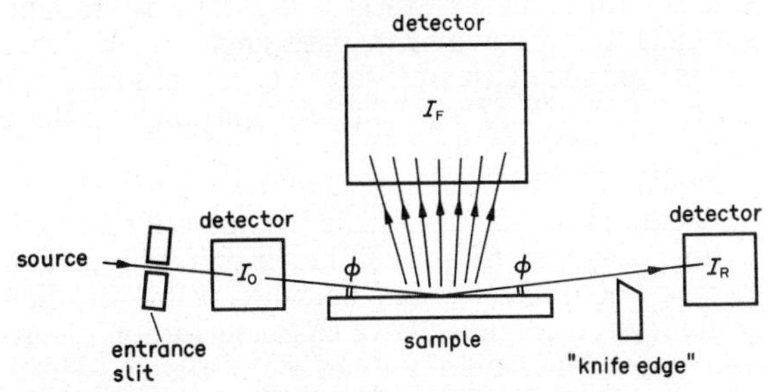

Figure 4
Experimental arrangement for ReflEXAFS and glancing angle FLEXAFS

nanometers. The possibilities of this technique do not, however, stop here. For surfaces with thin deposited films it is often possible to reflect the incident beam off the film–substrate interface. ReflEXAFS in this case will be sensitive to the complete depth of the film, plus a few nanometers of the substrate. Since XAS is "atom specific", the possibilities for studying the interfacial region are immediately apparent. The signal-to-noise ratio of such a surface-sensitive XAS technique may often be considerably enhanced by monitoring the x-ray fluorescence from the sample surface (I_F detector) instead of the reflected beam intensity (see Fig. 4). Solid-state detectors may be used here with considerable effect to further reduce unwanted background signal. Fluorescence-detected EXAFS (or FLEXAFS) at glancing angle is particularly suitable when studying the environments of atoms with low concentration and in contrast with ReflEXAFS may be employed at incident angles above ϕ_c, making it more versatile.

4. Current Situation

Characterization techniques for thin films and surfaces using x rays have developed rapidly in recent years. It should be noted, however, that these techniques are still in their infancy and are certainly not without their drawbacks. All techniques that employ the glancing angle geometry put severe requirements on the sample surface quality in terms of smoothness and flatness, and on sample size—for glancing angle work surface dimensions need to be large enough (usually several cm^2) for accurate alignment. For accurate lattice parameter measurements, GIAB diffraction data need to be corrected for refractive index effects which become important at such low incident angles. Lim et al. (1987) have discussed this and proposed an experimental procedure for determining the correction factors which are difficult to deduce theoretically for nonperfect surfaces. It was shown by Martens and Rabe (1981) that glancing angle XAS is not always as straightforward as it may appear, with angle of incidence effects producing changes both in the amplitude and phase of the EXAFS oscillations. Fortunately, these can usually be accounted for.

Much further development can certainly be expected in the various techniques described in this article, together with a continued increase in their use as more researchers become aware of their potential for nondestructive characterization of surfaces, interfaces and thin films.

See also: Ion Backscattering Analysis; X-Ray Absorption Spectroscopy: EXAFS and XANES Techniques; X-Ray Fluorescence Spectrometry; X-Ray Powder Diffraction

Bibliography

Aiginger H, Wobrauschek P 1985 Total reflectance x-ray spectrometry. In: Barrett C S, Predecki P K, Leyden D E (eds.) 1985 *Advances in X-Ray Analysis*, Vol. 28. Plenum, New York, pp. 1–10

Barchewitz R, Cremonese-Visicato M, Onori G 1978 X-ray photoabsorption of solids by specular reflection. *J. Phys. C* 11: 4439–45

Born M, Wolf E 1980 *Principles of Optics*. Pergamon, Oxford

Compton A H 1923 The total reflection of x-rays. *Philos. Mag.* 45: 1121–31

Doerner M F, Brennan S 1988 Strain distribution in thin aluminum films using x-ray depth profiling. *J. Appl. Phys.* 63(1): 126–31

Eisenberger P, Citrin P, Hewitt R, Kincaid B 1981 SEXAFS: New horizons in surface structure determination. In: Schuele D E, Hoffman R W (eds.) 1981 *CRC Critical Reviews in Solid State and Materials Science*, Vol. 10, issue 2. CRC Press, Boca Raton, FL, pp. 191–207

Fox R, Gurman S J 1980 EXAFS and surface EXAFS from measurements of x-ray reflectivity. *J. Phys. C* 13: L249–53

Heald S M, Keller E, Stern E A 1984 Fluorescence detection of surface EXAFS. *Phys. Lett. A* 103(3): 155–8

Huang T C, Toney M F, Brennan S, Rek Z 1987 Analysis of cobalt-doped iron oxide thin films by synchrotron radiation. *Thin Solid Films* 154: 439–45

Kiessig H 1931a Interference of Röntgen rays on thin layers. *Ann. Physik* 10: 769–88

Kiessig H 1931b Investigation of the total reflection of Röntgen rays. *Ann. Physik* 10: 715–68

Le Boité M G, Traverse A, Névot L, Pardo B, Corno J 1988 Grazing x-ray reflectometry and Rutherford backscattering 2. Complementary techniques for the study of thin-film mixing. *Nucl. Instrum. Methods B* 29: 653–60

Lee P A, Citrin P H, Eisenberger P, Kincaid B M 1981 Extended x-ray absorption of fine structure—Its strengths and limitations as a structural tool. *Rev. Mod. Phys.* 53: 769–806

Lim G, Parrish W, Ortiz C, Bellotto M, Hart M 1987 Grazing incidence synchrotron x-ray diffraction method for analyzing thin films. *J. Mater. Res.* 2: 471–7

Marra W C, Eisenberger P, Cho A Y 1979 X-ray total-external-reflection-Bragg diffraction: A structural study of the GaAs–Al interface. *J. Appl. Phys.* 50: 6927–33

Martens G, Rabe P 1981 The extended x-ray absorption fine structure in the reflectivity at the *K* edge of Cu. *J. Phys. C* 14: 1523–34

Parratt L G 1954 Surface studies of solids by total reflection of x-rays. *Phys. Rev.* 95: 359–69

Segmüller A 1987 Characterization of epitaxial films by grazing-incidence x-ray diffraction. *Thin Solid Films* 154: 33–42

Vineyard G H 1982 Grazing-incidence diffraction and the distorted-wave approximation for the study of surfaces. *Phys. Rev. B* 26: 4146–59

Wong J 1986 Extended x-ray absorption fine structure: A modern structural tool in materials science. *Mater. Sci. Eng.* 80: 107–28

P. N. Gibson
[Joint Research Centre—CEC, Ispra, Italy]

Timber Use: History

The tree and its wood have played a prominent role throughout history. The *International Book of Wood* points out that "man has no older or deeper debt" than that which he owes to trees and their wood (Bramwell 1976). "A culture is no better than its woods" writes W. H. Auden in *Woods*. The Bible records that Noah built the ark from gopherwood, the first ark of the covenant was made from acacia wood, and the cedars of Lebanon framed Solomon's temple. The best known of the early cultures developed along the Tigris–Euphrates, Nile, Indus and Yellow Rivers where forests were quickly depleted. Imports were soon heavily depended upon to supply people's needs for wood.

1. *Earliest Use*

Wood has been one of the most important raw materials from early Paleolithic times, both for building and for the manufacture of tools, weapons and furniture. Wood was worked early in human existence because no elaborate tools were needed. At that time, however, the quality of wood products depended more on the characteristics of the wood than on the tools available for woodworking. The availability of copper tools by about 5000 BC made possible the higher degree of craftsmanship evident in some few surviving relics of the time. This high degree of craftsmanship led to the use of wood in carefully worked form for coffers and chests for the storage of precious possessions by about 2600 BC. Closely related to this was the first coopering to make barrels of various types by about 2800 BC.

2. *The Last Millenium*

From the tenth to the eighteenth centuries in Europe, wood was the material primarily used for buildings, tools, machines, mills, carts, buckets, shoes, furniture and beer barrels, to name a few of the myriad articles of wood of the time. The first printing press was made of wood and such presses continued to be made of wood for a hundred years. Most of the essential machines and inventions to establish the machine age were developed in wood during this period. Wood played a dominant role in all industrial operations and much of the art and culture of the times. In Europe, wood use reached a high plateau around the sixteenth century. About that time, however, the availability of timber began to diminish owing to the many demands for both fuel and materials and the expansion of agriculture.

Wood use in North America reached a plateau during the middle to late nineteenth century, 150–200 years after that phase had peaked in Europe. The seemingly inexhaustible forests of colonial days were exploited, along with other natural resources, to feed a rapidly growing economy. Railroads, telegraph lines, steel mills and other industries were consuming wood at a dramatically increasing rate. During the latter half of the nineteenth century, the volume of sawnwood produced annually increased from 10 million to 85 million cubic meters, a level that has been maintained up to the present. Traditional uses of wood—for fuel, shipbuilding, construction of large buildings and bridges—were taken over by petroleum, coal, iron and steel, stone and brick. However, new uses for plywood, paper, poles, sleepers and chemicals, as well as the continuing demand for lumber, maintained a high level of timber production.

3. *Wood for Transportation*

Wood has played a key role in the transportation of people and their possessions, both as fuel and as a material, for thousands of years. Sledges made of wood were used for transport in northern Europe from 7000 BC. These were used for heavy loads such as stones and archaeologists believe that the massive stones of Stonehenge must have been moved on wood sledges placed on rollers. From this, the cart or wagon was created by putting the sledge on wheels. There are pictures of wheels that date from 3500 BC and actual wheeled vehicles found in tombs from 3000–2000 BC. In medieval times the spoked wheel became a great achievement of the joiner's art and in classical Greece both the spoked wooden wheel and the three-part solid wooden wheel were in common use. Solid wheels of planks were used on farm carts. Spoked wheels with 10–14 spokes were found in Roman forts in northern England. Rods of wood inserted into grooves and turning between the hub and the axle formed the first roller bearings.

In the nineteenth century in North America, railroads used wood for fuel, as well as for tracks, sleepers, cars, bridges, trestles, tunnel linings, sheds and stations. In the cities public transportation was mostly of wood, including horsecars, electric trolleys, cablecars, carriages and buggies. Roads made of planks laid across parallel rows of timbers embedded in the earth had come from Russia to London in the 1820s and spread to the USA during the period 1850–1857. When they were in good condition, these were the best roads in the country and more than 3000 km of such road were constructed in the mid-1800s. However, their demise soon came about as the result of excessive cost of maintenance.

Wood bridges employing advanced design concepts, laminated structural timbers and trusses were common in the nineteenth century. Covered bridges were common in North America and parts of Europe by the late nineteenth century, with the wood covering designed to protect the wood framework of the bridge itself. These were gradually replaced as the technology of wood preservation provided more economical means of wood protection and as iron, steel and concrete became the common materials for most bridge construction.

One of the first uses of wood for water transport was probably a raft or hollowed-out log. The earliest wooden ships copied the hull form of the reed boats that had been made by the Egyptians in about 4000 BC. Larger ships were built in Egypt using cedar imported from Lebanon, sometimes on a grand scale. The barge built for Queen Hatshepsut in 1500 BC to transport granite obelisks from Aswan to Thebes had a displacement of some 6800 t and required 30 oar-powered tugs to tow it. Theophrastus, a pupil of Aristotle, recorded that the shipbuilding woods in ancient Greece were silver fir, fir and cedar—silver fir for lightness, fir for decay resistance, and cedar in Syria and Phoenicia because of lack of fir. The Phoenicians dominated trade in the Mediterranean for a thousand years up to the time of Christ with their biremes, large galleys with two banks of oars. By 1000 AD, Vikings from Scandinavia were travelling mostly in wooden ships that were at least 25 m long. The wooden ship evolved somewhat in design up to the late nineteenth century. By this time, paddlewheel steamboats were cruising the major rivers of the USA and consuming large quantities of wood fuel in the process. American clipper ships were the fastest seagoing ships of the late nineteenth century, their sleek shape and large expanse of sail taking them to speeds of up to 32 km h^{-1}. However, by the 1880s iron steamships dominated the fleets of most naval powers, with wooden ships passing even in North America as the economics became unfavorable.

The discovery in 1982 of a 15 m ship sunk at some time around 1400 BC at Ulu Burun, off the coast of Turkey, with valuable cargo from around the Mediterranean, added substantially to our knowledge of the shipping, ships and wood use of the time. While most of the wood of the ship had deteriorated, enough remained of fragments, along with distribution of remains of the cargo on the ocean floor, to tell much about the size and nature of the ship. Included in the cargo were wooden tablets that served as a base for the wax used in writing and logs of wood known in Egypt as "hbny" and generally considered to be ebony. However, the latter was identified as African blackwood (*Dalbergia melanoxylon*) rather than true ebony (*Diospyros* sp.).

4. *Wood for Weapons and Tools*

Wood was also a key material for building the war devices of the ancient world. The battering ram, the scaling ladder, the tortoise and the siege tower are examples. Another example is the catapult, used to attack enemies from a safe distance. The properties of wood made it particularly suitable for such uses. High strength-to-weight ratio was a valuable and desired characteristic of wood then, just as it is now. Instruments of war such as those mentioned were essential to the expansion of Greek and Roman civilizations and of the science and technology that developed under the tutelage and guidance of the great thinkers and teachers of the time.

One of the oldest weapons, and the oldest surviving wood relic, is the pointed end of a spear made of yew from the Lower Paleolithic era found waterlogged in an English bog. Spear handles of wood were common in the Upper Paleolithic. The bow was invented during that period and there is evidence of use of the lever concept in tools and weapons as well as of hafting stones in wood to make axes and adzes. In the Mesolithic and Neolithic eras there is much evidence of new techniques for working wood, better axes and adzes, and chisels. Woodworking improved greatly with the advent of efficient carpenter tools during the Neolithic era, about 2000 BC, and still more during the Bronze age. The lathe spread from the Mediterranean to northern Europe in about 1500 BC, but was not in common use until iron cutting tools were available, by about 700 BC. The plow was in common use at this time, beginning with devices made from a single piece of wood and evolving to the use of various woods according to their characteristics.

5. *Wood in Construction*

Wood has been the most versatile and useful building material in its many forms and adaptations. Furthermore, until the relatively recent development of metallic and plastic structural materials, it was the only material from which complete structural frameworks could be fabricated readily. The type and durability of structures built at various times and places has depended on the type and quality of timber available and the conditions of use, as well as on the culture and way of life of the people concerned. Early pole structures were built in Paleolithic times from small trees growing along the rivers. The techniques used have been carried down over the ages and are very similar to those used by nomadic peoples today. However, the most significant developments in the use of wood as a building material have taken place where the culture and living conditions favored the erection of permanent structures.

In forested zones, where timber was plentiful, solid walls of tree trunks or heavy planks were built. Walls of timber houses in Neolithic Europe were frequently made of split trunks set vertically in or on the ground or a bottom sill plate, as indicated by examplesn excavated in moor settlements in

Germany. One of these also shows a beam with mortise holes. This method of wall construction continued through the Iron Age up to Norman times. A Viking fortress built in Denmark in about 1000 AD had this type of construction, as did stave churches built during that century. Later construction used similar principles, but employed vertical squared timbers or sawn planks. This "palisade" type of house construction was brought to North America by early French settlers along the Mississippi River.

Another common style of heavy timber construction, "log cabin" style with wall timbers placed horizontally rather than vertically, can be traced back nearly three thousand years. It has been most frequently used in the northern, central and mountainous areas of Europe and North America where there have been plentiful supplies of relatively large, straight trees. In this type of construction, round, squared or sawn timbers are laid horizontally and interlocked at the corners of the building. Some large farmhouses built in central Europe about 900 BC were apparently of this form. Vitruvius, a Roman architect writing at the time of Christ, referred to houses of this type built in a district of Asia Minor bordering the Black Sea. Records of the Roman period of a few centuries before and after the birth of Christ indicated considerable use of wood for roof construction, interior storage areas and lofts, and other parts of buildings.

Gothic construction in wood was introduced during the eleventh to fourteenth centuries in Europe. In this construction the roof timbers form an integral part of the frame and the principal framing members were typically large and heavy. This type of construction persisted until the seventeenth century in Europe, but variations of it were used in the USA even into the nineteenth century. Popularity of Gothic construction peaked during the last part of that century.

Wood construction has had an interesting evolution in North America because of the relatively abundant timber resource and the scattered development of much of the country. Native Indians in the forested areas of the east and northwest built homes and community houses from indigenous woods, the style depending on the culture and nature of wood available. In the east, structures were commonly made of poles covered with bark of birch or elm and palisades of vertical logs were used for protection. Along the northern Pacific coast, the Indians built houses from planks of split cedar or redwood and even had gabled roofs and decorative carvings. Even in some of the sparsely forested areas of the Great Plains and the western mountains, Indians built frames of timber and covered them with earth to make strong permanent dwellings.

Architecture of the early colonists evolved from that of their homelands, adapted to the climatic and cultural conditions of the times and materials availability. The log cabin was introduced into North America by Scandinavian immigrants in about 1638 and was adopted in the eighteenth century by Scottish–Irish immigrants. In the far northwest, explorers and settlers from Russia moved south from Alaska and built houses, forts and churches of log-cabin-type construction.

The dwellings built in New England by immigrants from England during the seventeenth and eighteenth centuries followed the pattern of the English timber frame house with wattle and daub or brick between the framing members. The classical "saltbox" type of house evolved from adaptations to that pattern to provide more protection against the severe weather—covering with oak or pine clapboards, improvement of the timber frame, and exclusion of the brick. Houses in the middle Atlantic states and the South were more likely to have exteriors of brick or stone, but elaborately decorated interior woodwork was not uncommon. The ready availability of building materials, especially wood, led to the detached or freestanding home becoming the standard dwelling in that part of the world.

Public buildings of the eighteenth century in North America followed similar trends. Churches were frequently made of wood, evolving from the simple to the ornate. Balconies were supported by wood pillars, often elaborately carved. A wooden canopy was frequently placed above the carved wood pulpit as a sounding board. Public buildings in the Georgian style frequently combined wood and brick in ways that took advantage of good features of both materials. An outstanding example is Independence Hall in Philadelphia which combined with its basic brick construction a wood superstructure for the tower clock, a wood balustrade, cedar shingles and elaborate interior woodwork.

Wood remained the principal construction material in North America well into the nineteenth century and remains so for housing today, as it does in some parts of Scandinavia and other regions where timber supplies are plentiful and the tradition of wood construction remains strong. A critical element in this continued use was the development of "balloon frame" construction in the 1830s in Chicago. This was developed to increase flexibility in building construction and to overcome the problem of lack of trained artisans needed to make the intricate connections necessary for the heavy timber and half-timber then in use. Rather than the heavy beams, posts and diagonals connected by precise and intricate joinery, the balloon frame concept employs relatively light studs held together by panels and nailed joints. Lumber and wood or other panel material can supply the necessary components inexpensively and flexibly. Following the disastrous Chicago fire of 1871, what had been done in wood was re-created in iron. Problems of fire and the need for high-rise buildings have contributed to the demise of wood-frame construction in the centers of large cities, but it is still the major form of house construction in most of the country.

6. *Wood for Furniture*

Wooden furniture appears to have been developed first in Egypt, beginning in the early Dynastic times (*c.* 3000 BC). It was rare until copper tools became available, though considerable refinement in woodworking had been made possible by the development of polished stone tools. Reeds and rushes had earlier been used for furniture and early wood furniture designs copied the light style of that material. Shortages of furniture wood in that part of the world led to early developments in economical wood use. Planks cut from the same log were laid side by side and carefully fitted together. Defects were cut out and replaced with patches or plugs. Short pieces were joined with scarf joints to make longer ones. A toilet casket of Amenhetep II (1447–1420 BC) also shows the use of precious wood as a face veneer over common wood. Mortising, tenoning and dovetailing to make strong attractive furniture joints appeared by 2600 BC. Parts of furniture were originally lashed together with linen cords or rawhide thongs. Later copper bands were used, especially on large coffins. Bronze pins or nails entered the picture in about 1440 BC.

Carvers of wood, who had evolved from carpenters or joiners, were found frequently in Europe during the middle ages. By the early thirteenth century, painting and gilding of wood had evolved into a separate craft. The joiner gradually became a complete furniture maker, while turners concentrated on turned furniture, which had been developed during Roman times and moved into northern Europe. By the sixteenth century, however, the two crafts came back together as turners worked with joiners to decorate higher-quality chairs and bed frames.

At that time furniture- and cabinet-making reached a very high level in several parts of Europe, with new designs and styles developed and executed by outstanding designers and furniture makers. As the Dark Ages passed into the Middle Ages with the Renaissance, new styles of furniture developed in Italy, England and France. Furnishings used in colonial America were at first quite simple and usually homemade. However, the rich array of woods available for furniture and other uses in the New World, together with highly skilled craftsmen, many of whom had apprenticed in Europe, soon led to high-quality manufacture in that region in styles that copied those in Europe as well as some original styles. By the latter part of the nineteenth century, the earlier small furniture shops in the USA had been largely superseded by large furniture factories that could take advantage of the new planers, mortisers, carving machines, saws, veneer lathes and veneer slicers that were becoming available.

7. *Plywood and Veneer*

The material we now know as plywood can be traced back to ancient times, even though the term "plywood" did not appear in an English language dictionary until 1927. The tomb of King Tutankhamun from 1325 BC, discovered in 1922, revealed some of the oldest examples of plywood manufacture. However, gluing of veneers to make plywood was an old art even at that time. Historians generally place the art of veneering at about 3000 BC. The purpose of decorative veneering was then, as it is largely now, to extend as far as possible the utilization of the valuable, attractive woods. Such woods had to be imported into the centers of artistic wood products manufacture in Egypt and other countries and were undoubtedly expensive, and supplies were uncertain. A tablet from Amenhotep III to King Harundaradu of Arzawa, found at Tell-el-Amarna, states "one hundred pieces of ebony I have dispatched." Ornamental woods found in Egypt in that time were limited to palms, sycamore, tamarisk and acacia. More exotic woods, such as ebony, walnut, rosewood and teak, were imported from India.

The ancient Greeks also used veneer for ornamentation, but not to the same extent as the Egyptians. The highest degree of craftsmanship and artistry was first reserved for the furnishings of temples and public buildings, but later the furnishings used by private citizens also became more luxurious.

The Romans were fond of using figured veneers on their furniture, particularly on their tables. The most expensive Roman tables were those made with the veneer of citrus trees grown in Africa. While the Romans desired comfort and utility, first consideration was given to style and quality of workmanship, with the decorative value of fine veneer being highly prized.

As compared with decorative plywood, manufactured primarily from hardwoods, softwood plywood is of relatively recent origin. It began in the early 1900s in the USA and the industry is still concentrated primarily in that country. It was developed as an alternative, and in some ways an improvement, to lumber by gluing together thin layers of knife-cut wood (cut usually on a rotary lathe) with the grain of alternate layers at right angles. Early development of the industry was in response to the need for wood panels for doors, but the industry has now expanded to produce a wide array of building panels. An exhibit at the Lewis and Clark Exposition in Portland, Oregon, in 1905, provided the interest and awareness on which was based the growth of the industry over the next decades.

The increased demands for structural wood during World War I provided the first basis for rapid growth of the softwood plywood industry. The new design and application potential of this panel product led to its being adopted for many structural applications. Exterior applications were limited, however, until the availability of synthetic resin adhesives in 1935. Those superior adhesives, coupled with improvements in manufacturing and quality control, set the stage for a further rapid expansion of the industry as a means of overcoming the declining size and quality of the timber resource. This led to further development in response to the needs of World War II and to subsequent additional expansion of the industry. Softwood plywood was made in the northwestern USA originally from Douglas fir, but in 1911 a plant was established in California to make it from ponderosa pine. Plywood made from southern pine entered the market in 1963. Several other softwoods are now also used in plywood manufacture.

Production of plywood panels has increased substantially during the past few decades in most parts of the world other than Europe. From 1964 to 1985, world production doubled from 22 to 44 million cubic meters per year. Currently, something over 40% of that is produced in the USA.

8. *Wood-Based Composites*

The development of wood-based composite materials, mostly within this century, has had a significant effect on wood use and opened new opportunities for creative products. The capability to make engineered structural products and a variety of forms and combinations with resins and other materials and the opportunity to use wood residues from other types of production provide incentives for application of this concept.

Wet-process fiberboard was developed during the latter part of the nineteenth century and commercially produced as insulating board early in the twentieth. It was based on papermaking technology. The board ranges in specific gravity from 0.16 to 0.40 and is widely used for sheathing, interior panelling, roof insulation and siding.

Wet-process hardboard was the next type of composite board to enter the picture. William H. Mason discovered the process in 1924, leading to the process and board type marketed as masonite. This is a hot-pressed board with a specific gravity of about 1.00.

Platen-pressed particleboard evolved from a series of developments in the late nineteenth and early twentieth centuries using shavings, sawdust or thin wood particles. The concept was further refined through the early part of the twentieth century, leading to commercial production after World War II. Featured in development since that time has been the concept of board made from flakes or strands of wood and suitable for structural use as well as for decorative use. This type of board, called flakeboard, waferboard or oriented strandboard has grown into competition with softwood plywood during the period since 1960. It has typically been manufactured from either softwood or medium-density hardwood, such as aspen (*Populus* sp.).

This is a field that is still evolving rapidly. Medium-density fiberboard came along in the 1960s and is used primarily as core stock in the furniture industry. Mineral-bonded building products made of excelsior (wood wool) and various mineral products, commonly cement, are suitable for a wide variety of structural products. Products molded from wood particles are relatively common in Europe and may become so in other parts of the world if the economics are favorable. World production of particleboard has increased spectacularly in recent decades, from 7.6 million cubic meters in 1964 to 43.5 million cubic meters—essentially equal to production of plywood—in 1984. Expansion of the particleboard industry has been strongest in Europe, which accounts for half of the total world production. Fiberboard production has increased relatively little over the same period, amounting to 16 million cubic meters.

9. *Wood for Paper*

Up to the middle of the nineteenth century, paper had been made primarily from rags. Their increasing cost and decreasing availability, together with a rapidly increasing demand for paper, had led to studies of other possible raw materials. The naturally fibrous structure of wood and its ready availability made wood a prime candidate and by the middle of the century three commercially feasible processes for making paper from wood had been developed. Hugh Burgess in the UK and Morris Kean in the USA developed the soda process, which could yield pulp suitable to mix with that from rags or straw in papermaking, and established a paper company in Philadelphia to use the process. Heinrich Voelter of Germany perfected the groundwood process, which became the principal method for producing woodpulp. In this process, which is still the primary method of producing pulp for newsprint, the yield is high but the strength of the paper is low. The sulfite process was perfected in Sweden and became the basis for development in the USA by Benjamin Tilghman, who had previously developed a way to grind rags mechanically as an improvement in industrial papermaking.

Paper from woodpulp greatly reduced the cost of books and papers and encouraged literacy. It also led to manufacture of paperboard, wallpaper and paper collars and bonnets. Wood is the source of pulp for most of the paper made and used in the world today. Creative new approaches to pulping and papermaking during recent years are leading to increased efficiency in use of the raw material, paper products with new use capabilities, including structural applications, and reduced environmental impacts.

See also: Timber Resources of the World

Bibliography

Bass G F 1987 Oldest known shipwreck reveals wonders of the Bronze Age. *Natl. Geogr.* 172(6): 693–733

Bramwell M (ed.) 1976 *The International Book of Wood*. Simon and Schuster, New York

Daumas M 1969–79 *A History of Technology & Invention*, Vols. 1–3. Crown, New York

Davey N 1963 *A History of Building Materials*. Phoenix House, London, pp. 32–48

Derry T K, Williams T I 1961 *A Short History of Technology*. Oxford University Press, Oxford

Hindle B (ed.) 1975 *America's Wooden Age: Aspects of Its Early Technology*. Sleepy Hollow Restorations. Tarrytown, New York

Mumford L 1963 *Technics and Civilization*. Brace and World, New York

Peters T F 1987 The rise of the skyscraper from the ashes of Chicago. *Am. Heritage Invent. Technol.* 3(2): 14–23

Singer C, Holmyard E J, Hall A R, Williams T I (eds.) 1954–59 *History of Technology*, Vols. 1–5. Clarendon Press, Oxford

Youngquist W G, Fleischer H O 1977 *Wood in American Life*: 1776–2076. Forest Products Research Society, Madison, WI

Youngs R L 1982 Every age, the age of wood. *Interdiscip. Sci. Rev.* 7(3): 211–19

R. L. Youngs
[Virginia Polytechnic Institute and State University, Blacksburg, Virginia, USA]

Tire Adhesion: Role of Elastomer Characteristics

It is taken for granted that vehicles can be driven safely under various weather conditions and on numerous types of surfaces. The reason for such confidence is due to the tires on the vehicle—the only link between the vehicle and the road surface. This results in the tire being a critical component in the design of the vehicle, having an influence on:

(a) safety,

(b) fuel consumption,

(c) vehicle handling,

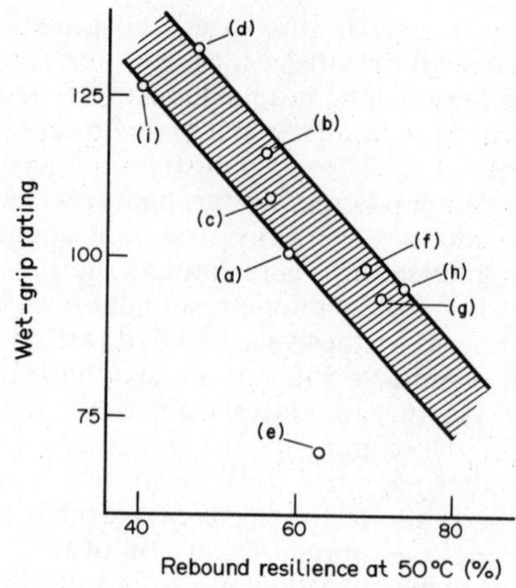

Figure 1
Wet-grip to resilience relationship for conventional polymers: (a) styrene–butadiene copolymer (23% styrene content); (b) oil-extended styrene–butadiene copolymer (23% styrene content); (c) styrene–butadiene copolymer (30% styrene content); (d) oil-extended styrene–butadiene copolymer (40% styrene content); (e) polybutadiene; (f) low *cis*-polyisoprene; (g) high *cis*-polyisoprene; (h) natural rubber; (i) butyl rubber

(d) vehicle comfort,

(e) noise generation.

Tire adhesion (or tire grip) is critical when considering safety and vehicle handling under different weather conditions. The tire tread is the only contact with the road surface and the major component of the tire tread is the elastomer used in the rubber tread compound. The tread compound also plays a significant role in the rolling resistance of the tire, which can be translated into vehicle fuel consumption. Neither tire grip (wet grip) nor tire rolling resistance (fuel consumption) should be considered in isolation; both should be taken into account in any tire development program.

1. Earlier Wet-Grip Hypotheses

There were two conventional wet-grip hypotheses generally accepted by the tire industry, one developed by Tabor (1960), the other by Bulgin et al. (1962). Figs. 1–3 (Bond, 1985) demonstrate the validity of the two hypotheses for conventional polymers.

HYPOTHESIS 1 (Tabor). *As the tread compound resilience decreases, wet grip improves.*

HYPOTHESIS 2 (Bulgin et al.). *As the glass-transition temperature of the tread polymer increases, wet grip improves.*

The wet-grip results were obtained on the variable-speed internal-drum machine situated at the University of Birmingham, UK (Lees and Williams, 1974) fitted with a Delugrip road surface (Lees et al. 1977); the resilience results were obtained on the Dunlop pendulum at 50 °C and the glass-transition temperatures were obtained using a differential scanning calorimeter.

2. Earlier Rolling Resistance Hypothesis

Collins et al. (1964) showed that the tread component to the rolling resistance of the tire is a function of the tread-compound loss modulus, E'' and of E''/E^{*2}; that is,

(a) tread bending component is proportional to E'', and

(b) general tread compression component is proportional to E''/E^{*2}.

Therefore, as E'' and E''/E^{*2} increase, the tire rolling resistance increases, where E^* is the complex modulus.

3. Relationship Between Wet Grip and Rolling Resistance

To establish how, based on conventional hypotheses, wet grip relates to rolling resistance, it was essential to determine how the tread compound resilience varies with both E'' and E''/E^{*2}. Figures 4 and 5 illustrate the relationship between resilience and E'' and E''/E^{*2}, respectively. The results show that, based on Tabor's hypothesis, as the wet grip improves, rolling resistance must increase; this was verified by the results shown in Fig. 6. Therefore, the three conclusions that can be drawn from the earlier wet-grip and rolling resistance hypotheses are:

(a) the performance of most conventional tread polymers can be explained;

(b) the performance of butyl rubber cannot be satisfactorily explained; and

(c) tread compounds and polymers developed to improve wet grip must inherently possess higher rolling-resistance characteristics and hence increase fuel consumption.

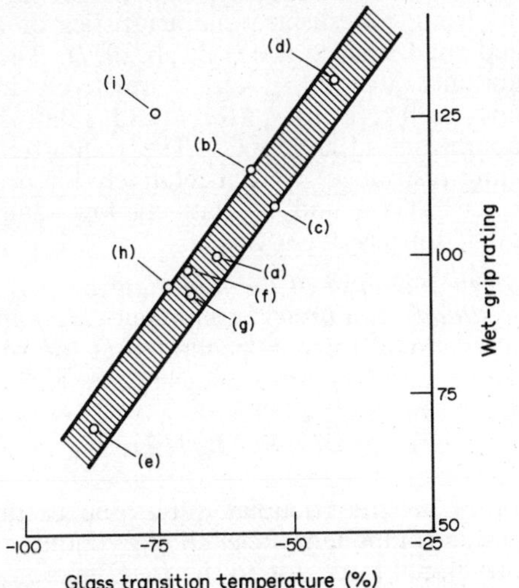

Figure 2
Wet-grip to glass-transition temperature relationship for conventional polymers (notation as Fig. 1)

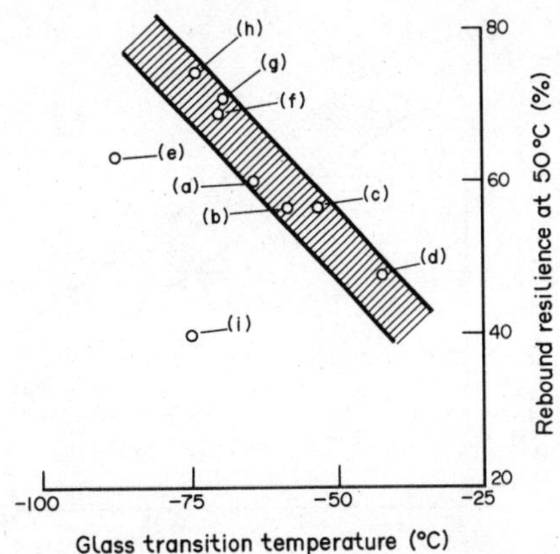

Figure 3
Resilience to glass-transition temperature relationship for conventional polymers (notation as Fig. 1)

4. New Wet-Grip/Rolling Resistance Philosophy

Today, steel-belted radial tires are the most commonly used passenger car tire. The wet-grip performance comes from the tread pattern (to

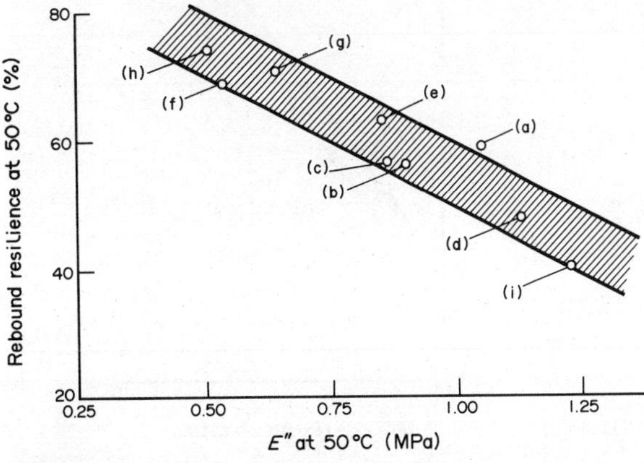

Figure 4
Resilience to E'' relationship for conventional polymers (notation as Fig. 1)

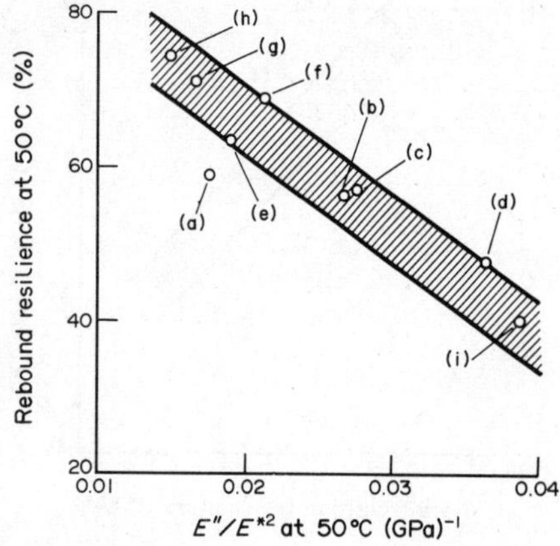

Figure 5
Resilience to E''/E^{*2} relationship for conventional polymers (notation as Fig. 1)

remove the bulk water) and the tread compound (to generate the frictional force at the tire–road-surface interface).

The rolling resistance, or temperature generation, has been shown (Williams, 1980) to be apportioned as follows: tread band, 34%; buttress region, 33%; sidewall region, 11%; and clinch region, 22%.

Figure 7 shows the various elastomeric or natural rubber compounds in a tire. The hysteresis losses can be minimized by reducing the energy absorption component of these compounds. However, it is important to recognize that any changes must not affect the other desirable properties such as comfort, structural performance, impact resistance or wear resistance. When considering the compound

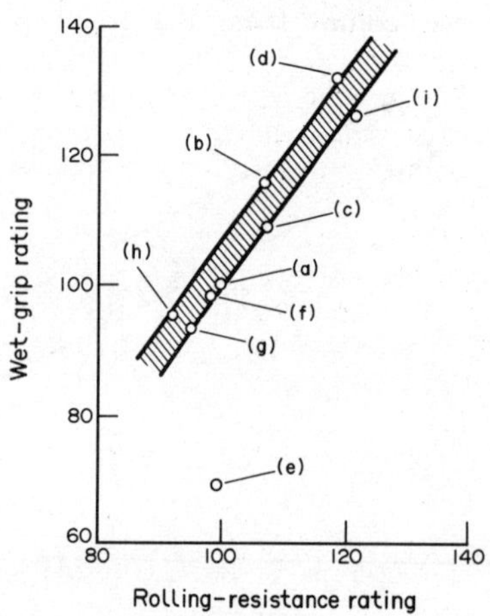

Figure 6
Wet-grip to rolling-resistance relationship for conventional polymers (notation as Fig. 1)

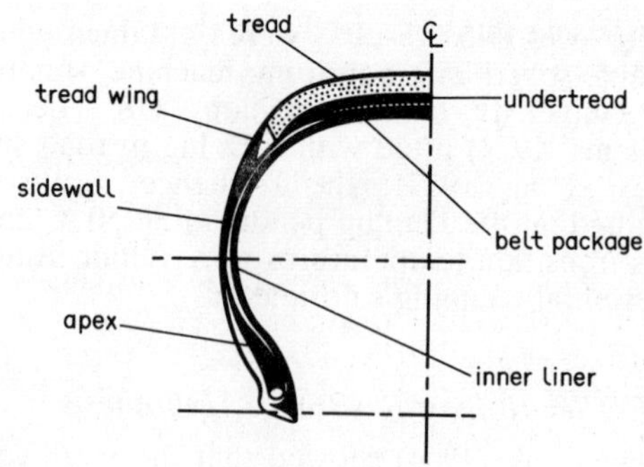

Figure 7
Tire components

aspects, the tread can be considered separately as, for one component, it makes the most significant contribution to the total rolling resistance of the tire (34%).

Therefore, a major breakthrough would be achieved if a new tread polymer, or family of polymers, could be developed which gave an improvement in wet-grip performance (safety) while reducing the rolling resistance of the tire without sacrificing the existing abrasion properties (tread life). To meet this objective, it was necessary to develop a new wet-grip/rolling-resistance philosophy.

4.1 New Philosophy

The new philosophy (Bond et al. 1984) was based on measuring the important dynamic properties of the elastomeric compound using conditions approaching those that exist at the tire–road-surface interface under wet sliding and rolling conditions. The test frequency and temperature conditions were developed using the surface characteristics of an optimized road surface (Lees et al. 1977). The sliding conditions (wet grip) were a relatively high frequency (50 kHz to 1 MHz) and relatively high temperatures (100–150 °C). The rolling conditions (rolling resistance) were a relatively low frequency (up to 120 Hz) and a relatively low temperature (≃50 °C for passenger cars).

(a) Determination of the important properties: wet-grip–sliding conditions (simplified account). The total frictional force F_t generated at the tire–road-surface interface approximates to:

$$F_t \simeq F_a + F_{mh} + F_{Bh} \qquad (1)$$

where F_a is the frictional force due to the tread adhesion component under sliding conditions, F_{mh} is the frictional force due to the tread microhysteresis component under sliding conditions and F_{Bh} is the frictional force due to the tread bulk-hysteresis component under sliding conditions 1% (Bond 1977).

Transposing the force components in Eqn. (1) into power-consumption terms, yields

$$\dot{E}_t \simeq \dot{E}_a + \dot{E}_{mh} + \dot{E}_{Bh} \quad (2)$$

where $\dot{E}$ is the rate of energy absorption and the labels correspond to those in Eqn. (1).

The bulk-hysteresis component E_{Bh} has been shown to contribute less than 1% of the total energy absorbed at the tire–road-surface interface under wet sliding conditions (Bond 1977). Therefore, E_{Bh} can be neglected, and the equation then becomes:

$$\dot{E}_t \simeq \dot{E}_a + \dot{E}_{mh} \quad (3)$$

but

$$\dot{E}_a \propto \text{loss factor (Bulgin et al. 1962)}$$
$$\dot{E}_{mh} \propto \text{loss factor (Bond 1977)}$$

Thus, the important tread-compound parameter relating wet grip is the loss factor measured at the appropriate frequencies and temperatures.

The total tread contribution to rolling resistance R_t is given by:

$$R_t = R_b + R_c + R_{tm} \quad (4)$$

where R_b is the tread-bending component to rolling resistance, R_c is the total tread-compression component to rolling resistance and R_{tm} is the tread-movement component to rolling resistance (assumed small for a well-designed tire). However,

$$R_c = R_{Gtc} + R_{Mc} + R_{mc} \quad (5)$$

where R_{Gtc} is the general tread-compression component to rolling resistance, R_{Mc} is the road-surface macrotexture tread-compression component to rolling resistance and R_{mc} is the road-surface microtexture tread-compression component to rolling resistance. Therefore,

$$R_t = R_b + R_{Gtc} + R_{Mc} + R_{mc} + R_{tm} \quad (6)$$

Now, it has been shown by Collins et al. (1964) and Bond (1977) that:

$$R_b \propto \text{loss modulus}$$

$$R_{Gtc} \propto \frac{\text{loss modulus}}{(\text{complex modulus})^2}$$

$$R_{Mc} \propto \frac{\text{loss modulus}}{(\text{complex modulus})^2}$$

$$R_{Mc} \propto \text{aggregate size}$$

$$R_{mc} \propto \frac{\text{loss modulus}}{(\text{complex modulus})^2}$$

$$R_{mc} \propto \text{microtexture level}$$

Thus, the important tread-compound parameters relating to rolling resistance are: loss modulus and loss modulus/(complex modulus)^2, measured at the appropriate frequencies and temperatures.

(*b*) *Overall tread-compound requirements.* The analysis put forward in Sect. 4.1 (*a*) has shown that for good wet grip, under sliding conditions, high values of loss factor and hence microhysteresis power consumption are required at relatively high frequencies (50 kHz to 1 MHz) and high temperatures (100–150 °C). One machine used to obtain the results was a piece of high-frequency test equipment developed specifically for the purpose (Bond 1976).

Similarly, for low rolling resistance (tread-compound component), low values of loss modulus and loss modulus/(complex modulus)^2 are required at relatively low frequencies (up to 120 Hz) and temperatures (approximately 50 °C for passenger cars). Typical machines used for obtaining these results are the rotary power loss machine (Bulgin and Hubbard 1958) and the dynamic response apparatus (Smith and Summer 1972).

4.2 New Philosophy—Advance over Earlier Hypotheses

The inclusion of the microhysteresis power-consumption term $\dot{E}_{mh}$ in the wet-grip study was the major breakthrough; this was achieved by considering the microhysteresis energy component to depend on wave propagation and not mechanical vibrations in the contact patch. The high-frequency deformation (50 kHz to 1 MHz) generated at the tire–road-surface interface due to the road-surface microtexture ($10–100 \times 10^{-6}$ m) results in both dilational and distortion waves being transmitted into the tread rubber. Under wet sliding conditions, the interface temperature ranging from 100 °C to 150 °C, it can be assumed that Poisson's ratio for the tread compound lies between 0.45 and 0.50 and hence the shear modulus G is less than 3% of the bulk modulus K. Therefore, for simplicity, it was assumed that dilational waves are transmitted through the tread perpendicular to the tire–road-surface interface, the energy transmitted by the wave propagation being absorbed by the hysteretic properties of the tread compound in the form of heat generation. Therefore, the important compound parameters involved are:

$$\text{elastic modulus} = K' + 4G'/3$$

$$\text{loss factor} = \frac{K'' + 4G''/3}{K' + 4G'/3}$$

where G' and G'' are the shear elastic and loss moduli, and K' and K'' are the bulk elastic and loss moduli.

It was considered that the waves transmitted into the tread are generated by small cylindrical pistons of cross section a_{ms} and radius r_{ms} with a displacement-amplitude equivalent to the microtexture level ξ_0, and n elements per unit area of contact. The pressure amplitude P falls off as the wave travels through the tread perpendicular to the tire–road-surface interface (i.e., in the x direction) as follows (Kinsler and Frey 1962, Bond 1976):

$$P = 2\rho_t v_t U_0 \exp(-\alpha_t x) \sin \tfrac{1}{2}k[(x^2 + r_{ms}^2)^{1/2} - x] \quad (7)$$

where ρ_t is the tread component density, v_t is the wave velocity, U_0 is the maximum velocity of displacement and r_{ms} is the radius of actual particle contact area on the microtexture scale. α_t is defined in Eqn. (11). U_0 is given by:

$$U_0 = \omega\xi_0 = 2\pi f_{ms}\,\xi_0 \quad (8)$$

where ω is the angular velocity, f_{ms} is the deformation frequency on the microtexture scale under sliding conditions and

$$k = \omega/v_t = 2\pi f_{ms}/v_t \quad (9)$$

The microhysteresis energy absorbed as a result of the microtexture deformation perpendicular to the tire–road surface at a distance x within the tread and elemental thickness δx is given by (Bond 1976):

$$E_{mh} = \frac{Ana_{ms}\alpha_t}{(\rho_t v_t)P^2\delta x} \quad (10)$$

where A is the area of contact. Also, it has been shown (Bond 1976) that

$$\alpha_t = [\pi f_{ms}(\text{loss factor})]/v_t \quad (11)$$

Therefore, from Eqns. (7) to (11) inclusive, and integrating over the full tread thickness h, the total microhysteresis energy absorbed by the tread at the tire–road-surface interface will be given by:

$$\dot{E}_{mh} = 16\pi^3 Ana_{ms}\rho_t f_{ms}^3 \xi_0^2(\text{loss factor})L \quad (12)$$

where

$$L = \int_0^h [\exp(-2\alpha_t x)] \times \sin^2\{(\pi f_{ms}/v_t)[(x^2 - r_{ms}^2)^{1/2} - x]\}dx \quad (13)$$

Figure 8 illustrates the relationship between wet-grip rating and microhysteresis power consumption obtained for the conventional polymers detailed in Fig. 1. The trend shows that as the microhysteresis power consumption increases, the wet-grip performance improves. The butyl polymer appears to be an anomaly in terms of the high microhysteresis power consumption when compared with the overall level of wet-grip performance. This anomaly was previously discussed by Bond (1977) and an explanation suggested in terms of a reduced number of adhesion bonds per unit area of contact at the tire–road-surface interface when compared with other conventional polymers.

5. *New Polymer Development*

The initial development work was carried out using blends of conventional polymers with the objective of meeting the dynamic property goals laid out in the newly developed wet-grip/rolling-resistance philosophy. Two approaches were used:

(a) blending polymers with relatively low and high glass-transition temperatures, and

(b) blending conventional tread-polymers with noncross-linking polymers, with a view to modifying either low wet-grip or high wet-grip compounds towards the desired goals.

The new philosophy could not be demonstrated using the low and high glass-transition temperature approach by blending one or two conventional tread polymers with unconventional commercially available polymers with relatively high glass-transition temperatures. The second approach led to the development of an experimental compound made up from the following blend of polymers: 40 parts polybutadiene, 30 parts natural rubber, 13 parts styrene-butadiene copolymer (23% styrene content) and 17 parts polyisobutylene (molecular weight $\bar{M}_n = 1\,200\,000$), combined with 50 parts of N326 carbon black. This compound exhibited the wet-grip

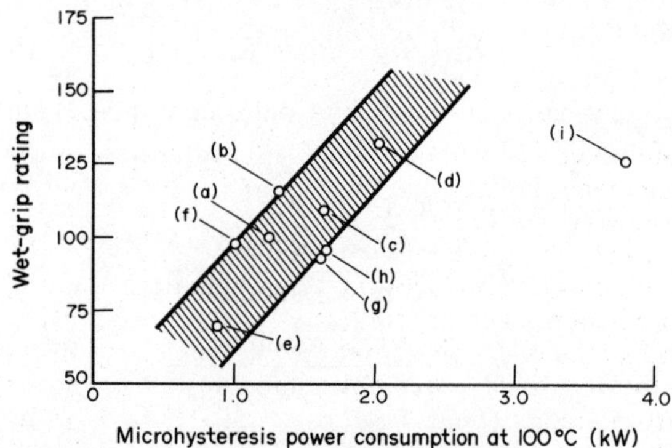

Figure 8
Wet-grip to microhysteresis power-consumption relationship for conventional polymers (notation as Fig. 1)

performance of conventional car tire tread compounds while exhibiting the rolling resistance characteristics of natural-rubber truck tread compounds, all three compounds having a similar hardness of 59 IRHD (international rubber hardness degrees). Unfortunately, the abrasion resistance of the experimental compound approached that of the natural-rubber tread, which was not acceptable for use in passenger-car tires.

Two conclusions were drawn from the work blending conventional polymers. First, tread compounds can only be developed to reduce either the rolling resistance of the tire without sacrificing wet grip or improve wet grip without increasing rolling resistance at the expense of severe reductions in abrasion resistance. Second, tread compounds cannot be developed that both reduce rolling resistance and improve wet grip while maintaining present-day levels of abrasion resistance.

Therefore, when considering the tread compound, the only successful approach to improving wet grip and reducing rolling resistance, while maintaining abrasion resistance, would be the development of a new "tailor-made" polymer which met the parameters outlined in Sect. 4.1 (*b*). The basis for the new-polymer development was the principle that the polymer was made up of two components.

(a) The element of the polymer structure that determines its low frequency or rolling-resistance characteristics.

(b) The element of the polymer structure that determines its high frequency/high temperature or wet sliding characteristics.

To ensure that any new polymer developed would be commercially feasible, two additional constraints were applied. First, the development was based on using existing monomers used by polymer manufacturers in producing polymers for tire applications. Second, any new polymer developed had to be processable on existing manufacturing equipment used throughout the tire industry.

5.1 Laboratory Studies

At an early stage, the decision was made to investigate polymers produced on a laboratory scale using butadiene and styrene monomers. The first objective was to produce a polymer that met the dynamic property requirements previously discussed. The result was a copolymer produced using the solution copolymerization technique of placing the styrene and butadiene monomers into a nonpolar solvent with diethylene glycol dimethylether, as the structure modifier, and heating to 50 °C. The polymerization reaction was then initiated by using a lithium–hydrocarbon compound. The styrene–butadiene copolymer produced had a 23% bound-styrene, 70% 1,2-butadiene content, a weight average molecular weight $\bar{M}_w$ of 450 000 and a number

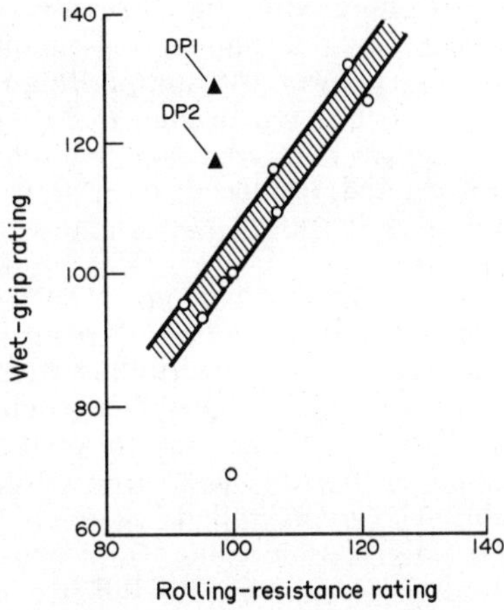

Figure 9
Wet-grip to rolling-resistance relationship obtained for the new laboratory scale polymers, DP1 and DP2, compared with the conventional polymers previously illustrated

average molecular weight $\bar{M}_n$ of 260 000. The molecular weights were obtained from gel permeation chromatography, the instrument being calibrated with polystyrene standards.

The three major characteristics of this polymer (DP1) were as follows:

(a) the fundamental characteristic of the polymer was the glass-transition temperature of −34 °C, as measured by differential scanning calorimetry;

(b) when compounded in a simple test formulation (100:50:0, polymer:N375 carbon black:free oil), the resilience of the compound, as measured on the Dunlop pendulum at 50 °C, was relatively high at 66.3%; and

(c) the microhysteresis power consumption of the simple test compound at 100 °C was relatively high at 3.0 kW.

It can be seen that the new experimental polymer (DP1) was inherently a high wet-grip polymer (relatively high glass-transition temperature) which, when used in a simple test formulation, exhibited a relatively high resilience indicating low rolling-resistance characteristics. The wet-grip and rolling-resistance results obtained in the laboratory and compared with the conventional polymers detailed in Fig. 1 can be seen in Fig. 9.

Serious difficulties were encountered processing DP1 in the laboratory, particularly in the area of compound mixing. Thus, it would not be possible to process the new polymer on existing manufacturing equipment and also difficult to manufacture it on a large scale. Therefore DP1 was not a practical commercial proposition at the time. To overcome the difficulties encountered in the laboratory, DP1 was modified to DP2 by reducing the 1,2-butadiene content to 50%. The wet-grip and rolling-resistance performance can be seen in Fig. 9. DP2 had relatively high values of glass-transition temperature (−46 °C), rebound resilience at −50 °C (66.9%) and microhysteresis power consumption at 100 °C (2.56 kW).

The new experimental polymer (DP2) was then taken a step forward to the "pilot" plant stage and tested in full-size tires. Figure 10 illustrates the full-size tire test results for the new pilot batch polymer (PB1) compared with a conventional styrene–butadiene copolymer (23% styrene content) and a conventional oil-extended styrene–butadiene copolymer (23% styrene content). Having verified the laboratory scale results on full-size tires, the pilot production level for the new polymer was increased to commercial production. The results for the commercial polymer designated SSCP901 (new polymer) are also shown in Fig. 10. It can be seen that the move from the pilot scale to the commercial scale polymer production levels resulted in a further improvement in tire performance.

The wet-grip and rolling-resistance investigation was done using 155SR13 steel-belted radial-ply tires. The wet-grip results were obtained on the internal-drum machine situated at the University of Karlsruhe in the FRG (Bond 1985) and the rolling-resistance results on the tire dynamics machine situated in the Dunlop tire research department in Birmingham, UK (Bond 1985).

To complete the laboratory analysis, Table 1 compares the new polymer characteristics with the well-

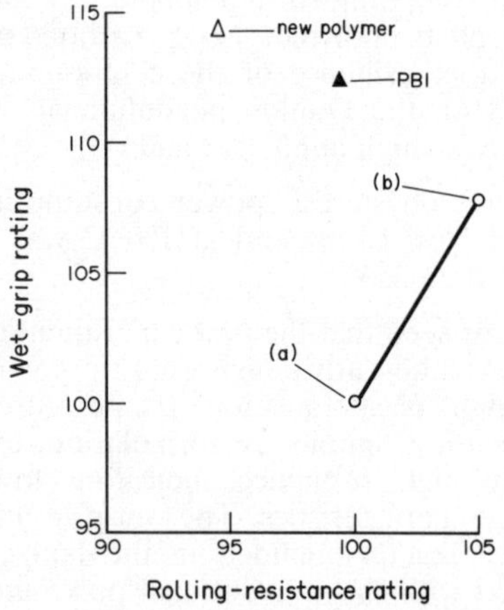

Figure 10
Wet-grip to rolling-resistance relationship obtained using 155SR13 steel-belted radial-ply tires for the pilot batch of polymer, PB1, and the new polymer compared with two conventional polymers:
(a) styrene–butadiene copolymer (23% styrene content); (b) oil-extended styrene–butadiene copolymer (23% styrene content)

established emulsion styrene–butadiene copolymer (S-1502). Table 2 compares the general compound and vulcanisate properties obtained on the laboratory formulation and Table 3, the compound, vulcanisate, wet-grip and rolling-resistance properties obtained from the simple test formulation used to evaluate the wet-grip and rolling-resistance performance of equivalent 155SR13 steel-belted radial-ply tires, the only difference being the polymer used in the tread compound.

5.2 Processability Studies

Full-scale processing trials were carried out in tire factories in the UK, FRG, USA and Japan to enable the processing characteristics of the new polymer (SSCP901) to be assessed on various types of equipment in the areas of:

(a) compound mixing (batch mixes of both the tangential and intermeshing rotor types),
(b) extrusion (hot and cold feed),
(c) tire building, and
(d) molding.

The results showed that the new polymer could be processed on existing equipment.

6. Passenger Car Tire Testing—New Polymer

Extensive test programs were carried out comparing the performance of equivalent tires with the exception of the tread compound. One series of tires used a tread compound based on the new polymer (SSCP901), and the other series of tires had used an existing practical tread compound. The closely controlled test programs included measurements of wet grip, rolling resistance, fuel consumption, temperature generation, structural performance, high-speed performance and tread wear. The results showed that tread compounds containing the new polymer, when compared with practical tread compounds in general use, produce the following improvements in tire performance:

(a) improved wet grip (+5%),
(b) reduced rolling resistance (−13%),
(c) reduced fuel consumption (−2.7%),
(d) reduced tread operating temperature,
(e) improved durability, and
(f) maintained acceptable tread life.

7. Truck Tire Testing—New Polymer

The most commonly used truck tread compounds are made from natural rubber and, on occasion, contain a proportion of synthetic polyisoprene. In

Table 1
Polymer characteristics

Characteristics	New polymer	S-1502
Stabilizer	Nonstaining hindered phenol	
Styrene content (wt%)	23.5	23.5
Cis content (%)	20	10
Trans content (%)	30	75
1,2-Butadiene content (%)	50	15
Molecular weight, $\bar{M}_n$	250 000	110 000
$\bar{M}_w$	500 000	550 000
Density (kg m^{-3})	0.93	0.93
Glass transition-temperature (DSC) (°C)	−40	−55

general, this results in a reduction of approximately 25% in wet-grip performance when compared with conventional passenger-car tread compounds. However, because of the temperatures generated in truck tires, it was essential to minimize the temperature generation characteristics of any tread compound based on the new polymer developed for use in truck tire applications. To meet the temperature generation objective, it was essential to minimize the values of E'' and E''/E^{*2} (see Sect. 4.1 (*a*)).

An experimental truck tread compound was developed based on a 70:30 new-polymer to natural-rubber blend and compared with a standard 100% natural-rubber truck tread compound. An extensive series of machine and field tests were carried out using 11.00-R-20 tires which, for vehicle tests, were fitted to selected 44 t trucks. The results demonstrated:

(a) improved wet grip (+40%),
(b) similar performance in temperature generation,
(c) similar durability performance,
(d) improved tread wear (drive axles),
(e) slight reduction in tread wear (trailer axles), and
(f) slight reduction in rib tearing.

The truck-tire evaluation indicated that, with some work to "fine tune" the experimental compound evaluated, significant advances could be made in the area of truck-tire tread-compound development.

8. *Review and Future Development*

The development of a new wet-grip/rolling-resistance philosophy enabled a new styrene–butadiene copolymer with a unique microstructure

Table 2
Laboratory, formulation, compound and vulcanisate properties

	New polymer	S-1502
Formulation (pphr)		
Polymer	100.0	100.0
N339 carbon black	45.0	45.0
Zinc oxide	5.0	5.0
Stearic acid	3.0	3.0
Santoflex 13	1.0	1.0
Santoflex 77	1.0	1.0
Santocure MOR	1.0	1.0
Sulfur	2.0	2.0
	158.0	158.0
Mooney viscosity, $M_L(1+4)$, 100 °C	80	65
Vulcanisate		
Tensile strength (MPa)	22	24
Modulus at 300% elongation (MPa)	13	14
Elongation at break (%)	425	400
Hardness (Shore A)	61	62
Tear strength, angle (kN m^{-1})	50	52
Resilience, Lu pke Pendulum at 23 °C (%)	42	52
Heat build-up (over 38 °C)	34	33

Table 3
Tread formulation, compound vulcanisate, wet grip and rolling-resistance properties

	New polymer	S-1502
Formulation (pphr)		
Polymer	100.0	100.0
Zinc oxide	2.50	2.50
Stearic acid	1.00	1.00
75% BLE	2.00	2.00
N375 carbon black	50.00	50.00
Sulfur	1.75	1.75
CBS	1.00	1.00
	158.25	158.25
Vulcanisate		
Tensile strength (MPa)	21.6	25.2
Modulus at 300% elongation (MPa)	14.3	14.2
Elongation at break (%)	390	430
Hardness (IHRD)	72.3	69.2
Resilience, Zwick at 50 °C (%)	29.9	32.8
E' at 50 °C (MPa)	7.65	7.46
E'' at 50 °C (MPa)	1.008	1.138
E''/E^{*2} at 50 °C $(MPa)^{-1}$	0.0169	0.0200
Tire Properties		
Wet-grip rating	114	100
Rolling-resistance rating	94	100

to be developed and used in tire tread compounds which, when compared with conventional passenger-car tread compounds of the day, demonstrated improved wet grip (safety), reduced rolling resistance (fuel economy) and maintained tread life. The new polymer was also evaluated in truck tread compounds with results showing that there was considerable potential for the new polymer in commercial-vehicle tread compounds.

The most important contribution from the development of the new wet-grip/rolling-resistance philosophy and the subsequent new polymer development is the fact that the new polymer should be considered as the first in a new family of polymers, tailor-made to meet defined practical applications. Using this approach, it is suggested that, in future, the tire industry should continue to make significant advances in improving tire safety while reducing rolling resistance (improving fuel economy).

See also: Abrasion of Elastomers; Automobile Tires; Fillers in Elastomers; Rubber Tires

Bibliography

Bond R 1976 The optimization of tire–road friction, Vol. 2. Ph.D. thesis, University of Birmingham, UK

Bond R 1977 Advances in the physical understanding of tire/pavement friction. Paper presented at the 111th meeting the Division of Rubber Chemistry, American Chemical Society, Washington, DC

Bond R 1985 A new tire polymer improving fuel economy and safety, Esso energy award lecture. 1984. *Proc. R. Soc. London, Ser. A* 399: 1–24

Bond R, Morton G F, Krol L H 1984 A tailor-made polymer for tire applications. *Polymer* 25: 132–40

Bulgin D, Hubbard G D 1958 Rotary power loss machine. *J. Inst. Rubber Ind.* 34(5)

Bulgin D, Hubbard G D, Walters M H 1962 Road and laboratory studies of friction of elastomers. *Proc. 4th Rubber Technology Conf.*

Collins J M, Jackson W L, Oubridge P S 1964 The relevance of elastic and loss moduli of tire components to tire energy losses. Paper presented to the Division of Rubber Chemistry, American Chemical Society, Washington, DC

Kinsler L E, Frey A R 1962 *Fundamentals of Acoustics*, 2nd edn. Wiley, New York

Lees G, Williams A R 1974 A machine for friction and wear testing of pavement surfacing materials and tyre tread compounds. *J. Inst. Rubber Ind.* 8(3): 114–20

Lees G, Williams A R, Bond R 1977 Improvements in or relating to road surfaces. UK Patent No. 1,482,920

Smith J E, Sumner E C 1972 An automatic dynamic response apparatus. *Proc. Rubber Conf.* Institution of the Rubber Industry, Brighton, UK

Tabor D 1960 Improvements in or relating to tires for road vehicles. UK Patent No. 837,849

R. Bond
[CR Industries, Elgin, Illinois, USA]

Titanium Aluminides

Titanium aluminides are a class of stoichiometric intermetallic compounds based on the elements titanium and aluminum. Three such materials exist: Ti_3Al (α_2), TiAl (γ) and $TiAl_3$ (δ). These materials are examples of ordered intermetallic compounds with a specific arrangement of the two elements in the crystal structure. This ordered arrangement of the atoms leads to some unique properties—high elastic modulus and strength, which are retained to high temperatures, and high activation energies for diffusion, which result in improved creep behavior over conventional titanium alloys. Titanium aluminides are of low density (approximately half that of nickel-based superalloys) and the aluminum-rich compositions exhibit excellent oxidation resistance. These attributes make titanium aluminides very attractive for gas turbine engines and other elevated-temperature applications.

The use of titanium aluminides has been limited by their poor ductility and fracture toughness at low temperatures. Significant progress has been made in improving these limiting properties and materials development has accelerated since the late 1970s. Commercial titanium aluminide products are available and a number of experimental components have been fabricated and tested. Alloy development efforts continue across the spectrum of titanium aluminide compositions and titanium aluminide composites reinforced by long fibers and particulates are under development.

1. Phase Equilibria

Three titanium aluminide compounds (Ti_3Al, TiAl and TiA_3) are accepted as true binary intermetallics. A fourth $TiAl_2$ has been reported, but it has been proposed that this phase is actually a ternary compound $TiAlN_2$ which forms only in nitrogen-bearing compositions. Since the solubility of nitrogen in γ and δ titanium aluminides is very low, it is likely that small amounts of nitrogen could lead to the formation of this phase. $Ti_3Al(\alpha_2)$ has a hexagonal DO_{19} crystal structure, and TiAl (γ) and $TiAl_3$ (δ) have tetragonal $L1_0$ and DO_{22} structures, respectively (see Fig. 1 and Table 1).

The titanium–aluminum phase diagram has been the subject of considerable experimental and theoretical effort. A version is shown in Fig. 2. This phase diagram is considered accurate for all but the region between the α_2 and γ phases at temperatures above the eutectoid transformation. More recent data indicate that the α phase extends to the liquidus and is a constituent of one of the two peritectic reactions that are observed within the 40–60 at.% aluminum range (see Fig. 3).

2. Ti_3Al

The development of the α_2-based titanium aluminides has progressed sufficiently that at least one alloy (Ti–24 at.%Al–11 at.% Nb) is commercially available in various product forms. Manufacturing technologies have been developed and turbine engine and other structural components have been fabricated and tested. Based on this database, α_2 titanium aluminides are likely to be incorporated into future turbine engines.

Although more oxidation resistant than conventional titanium alloys, Ti_3Al does not form a protective oxide in oxygen-containing environments and exhibits poor creep properties at temperatures higher than 700 °C. Alloying development efforts have, therefore, been directed toward increasing room temperature ductility and improving high-temperature properties.

Binary α_2 shows a brittle-to-ductile transition temperature of 700 °C (see Fig. 4). Below this temperature, α_2 fails by a cleavage fracture mechanism. Even above the transition temperature, a substantial proportion of the increase in measured elongation is a result of significant extension arising from grain-boundary cracking. The mode of fracture at high temperatures is a mixture of cleavage and intergranular separation with limited ductile tearing.

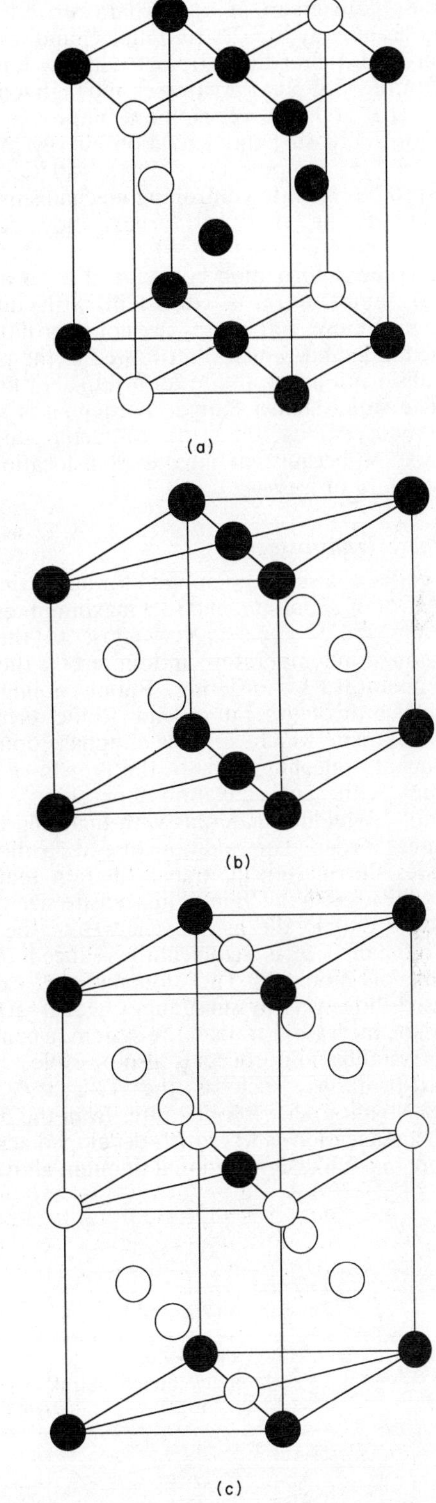

Figure 1
Crystal structures: (a) Ti_3Al, (b) TiAl, and (c) $TiAl_3$

Steady-state creep in α_2 is characterized by an activation energy of 2×10^5 J mol^{-1} and a stress exponent of 4–5 at high stresses and low temperatures, or of 2.3–2.5 at low stresses and high temperatures. The addition of niobium improves creep behavior, increasing the activation energy to 2.8×10^5 J mol^{-1} at temperatures above 650 °C. There appear to be two rate-controlling mechanisms operating in both the binary and ternary alloys depending on temperature (see Fig. 5).

The cyclic deformation behavior of α_2 as a function of temperature is consistent with ductility changes. Below 700 °C, α_2 exhibits pronounced fatigue hardening, which is attributed to the planarity of dislocation slip and to an insufficient number of active slip systems. Fatigue hardening is significantly reduced above the brittle-to-ductile transition temperature because of increased dislocation slip and recovery processes.

2.1 Phase Transformations

Ti_3Al exists as a single compound for approximately 22–39 at.% of aluminum and to a maximum temperature of 1150 °C. This is very close to the α–β transformation temperature and represents the critical temperature for ordering. Rapid cooling of β from above this temperature leads to the formation of α_2 martensite which contains antiphase domains. Subsequent annealing leads to the growth of these domains with an activation energy of 2.7×10^5 J mol^{-1}, yielding materials with high degrees of long-range order. The addition of 5 at.% niobium depresses the martensite transformation temperature and leads to a much finer martensite size. Also, it appears that in the as-quenched state the antiphase domain size is significantly reduced by the addition of niobium. The domain size can be increased, however, by subsequent heat treatment. Due to the increased strain in the system, a competitive recrystallization process is also possible.

In α_2–β alloys, such as the Ti-24 at.% Al–11 at.% Nb alloy, slow cooling rates from the β field lead to a nucleation- and growth-developed acicular structure much like conventional titanium alloys. As in the case of binary α_2, rapid cooling leads to martensitic structure of very fine antiphase domain

2.2 Deformation Characteristics

The ductility of most materials generally correlat with the number of mobile dislocations and th ease of movement at any given temperatur Examination of deformed binary α_2 indicates tha dislocation activity is mostly confined to a single sli direction, $\frac{1}{2}a[11\bar{2}0]$ and is dependent on both tem perature and strain rate, as follows:

(a) $\langle 11\bar{2}0\rangle$ $\{10\bar{1}0\}$ a dislocations on prism plane—primary system,

(b) $\langle 11\bar{2}0\rangle$ $\{0001\}$ a dislocations on basal plane—secondary system active at moderate strain rate

(c) $\langle 11\bar{2}0\rangle$ $\{2\bar{2}01\}$ a dislocations on pyramida plane—secondary system active at moderat strain rates,

(d) $\langle 11\bar{2}3\rangle$ $\{10\bar{1}0\}$ $c+a$ dislocations on prisn plane—limited to high temperature and slov strain rates only.

Some activity by c [0001] dislocations on $\{10\bar{1}0\}$ $\{11\bar{2}0\}$ and $\{3\bar{4}10\}$ has been reported at inter mediate temperatures and low strains. However this does not provide an independent slip system the strain produced by this slip system could also b achieved using suitable combinations of the primar and secondary slip systems. The complex dislocatio behavior of α_2 is in contrast to that of TiAl (γ) (Although γ is generally believed to be less ductil than α_2, it clearly shows increased numbers o mobile dislocations at similar temperatures.)

The presence of niobium in solid solution does no appear to alter the dislocation behavior relative tc the binary compound. The increase in ductility asso ciated with such additions has been credited to the reduction of slip planarity. However, the evidence is limited and there is no mechanistic explanation fo the observed improvement. The effect may be the result of microstructural refinement, acting tc shorten the dislocation pileups.

Table 1
Physical properties of titanium aluminides

	Proportion of aluminum (at.%)	Strukturbericht designation	Lattice parameter (nm) *a*	Lattice parameter (nm) *c*	Density (10^3 kg m^{-3})
Ti_3Al	22–39	DO_{19}	0.290[a]	0.464[a]	5.1
TiAl	48–70	$L1_0$	0.4008[b]	0.4070[b]	4.0
$TiAl_3$	75	DO_{22}	0.3845	0.8593	3.3

a Average at 25 at.% of aluminum b average at 46–50 at.vol of aluminum

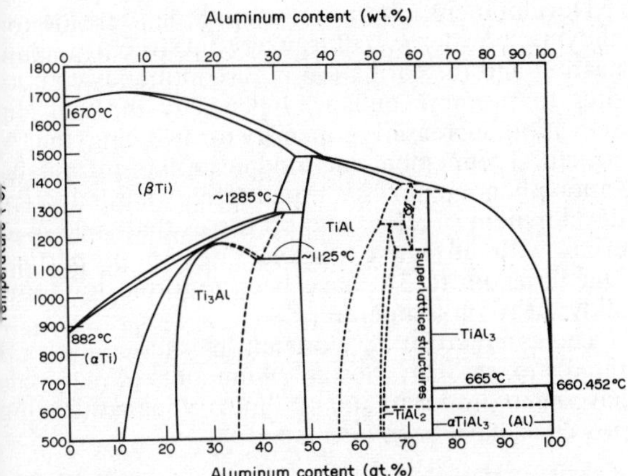

Figure 2
Phase diagram for titanium–aluminum system (Murray 1987)

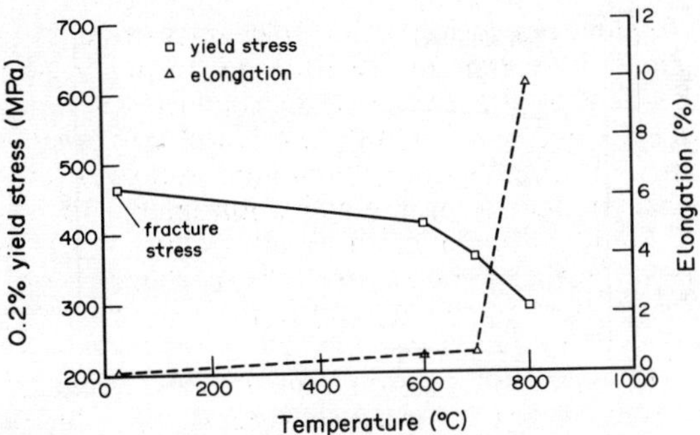

Figure 4
Tensile behavior of binary Ti_3Al as a function of temperature (Lipsitt et al. 1979)

2.3 Alloying Effects

Early development programs showed that titanium–aluminum–niobium based alloys containing near-stoichiometric amounts of aluminum and 10–15 at.% niobium showed the most promise, combining adequate room-temperature ductility and high-temperature properties. As the aluminum content of α_2 is increased above the stoichiometric ratio, toughness and ductility tend to decrease. The addition of niobium restores toughness and ductility, irrespective of aluminum content. Surprisingly, gains in ductility are realized with small additions of niobium, even at levels below those that promote retained β (disordered).

The best combination of properties has been achieved in two-phase alloys containing α_2 and β. β is retained by the addition of large amounts of niobium (~10 at.%) and appears to improve ductility by providing a ductile continuous "soft phase." The beneficial influence of the β phase can be lost if it is allowed to order to the B2 structure. Vanadium can be substituted for some of the niobium, with a concomitant decrease in density and cost. Other β stabilizers generally do not improve ductility when added as ternary alloying additions, implying that the effect of niobium extends beyond the stabilization of the β phase. Molybdenum tends to promote

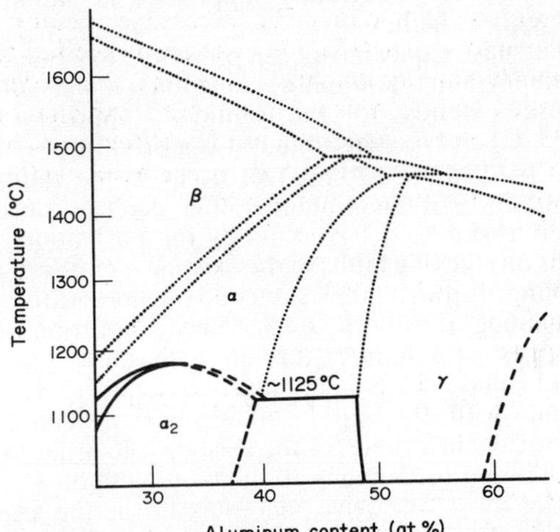

Figure 3
Phase diagram in the Ti–40 at.% Al to Ti–60 at.% Al phase field (McCullough et al. 1989)

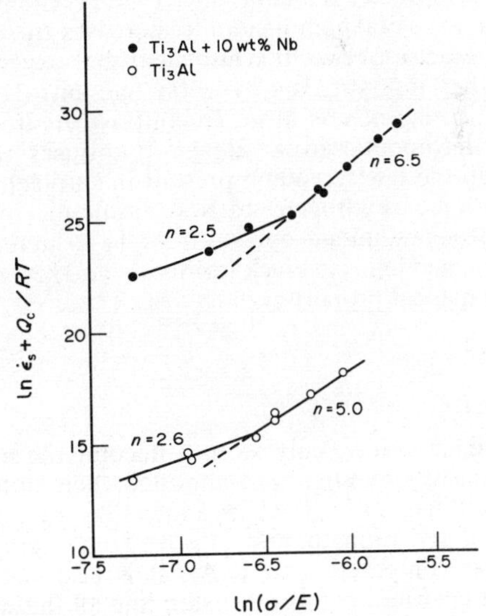

Figure 5
Temperature-compensated steady-state creep for Ti_3Al and Ti_3Al plus niobium alloys where $\dot{\varepsilon}_S$ is creep rate, σ is applied stress, E is Young's modulus, Q_c is activation energy and T is temperature (Mendiratta and Lipsitt 1980)

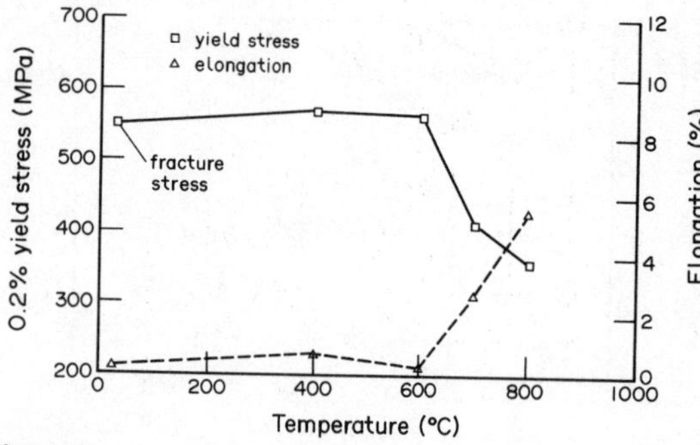

Figure 6
Tensile yield and elongation of binary TiAl (Lipsitt et al. 1975)

ordered β, particularly near the Ti_2MoAl composition. Quaternary additions of vanadium, chromium, zirconium and silver tend to induce modest increases in ductility, whereas most other transition elements have no beneficial effect.

Refinement of the microstructure generally increases ductility and fracture toughness, but is associated with a reduction in creep properties. The degree of refinement is influenced by the presence of the alloying additions, but is also a strong function of thermomechanical processing.

Ti_3Al-based alloys are quite sensitive to oxygen content. Levels above 1000 ppm lead to reduced ductility. Further reducing the oxygen content from 1000 ppm to 500 ppm generally increases the ductility by a factor of two. It is also clear that oxygen and the other interstitials of carbon and nitrogen are potent strengtheners in α_2. In contrast, hydrogen is quite detrimental to α_2 alloys. It diffuses rapidly through the lattice. When present in sufficient concentrations, hydrogen causes titanium hydride (TiH_2) to precipitate out, with an 11% increase in volume, and leads to crack formation and severe loss of mechanical properties.

3. *TiAl*

γ-based alloys have only recently become the subject of extensive investigation, although their potential was identified early. TiAl has a lower density, higher modulus and strength, and improved oxidation resistance as compared with Ti_3Al. It is also stable to much higher temperatures, extending all the way to the liquidus. The only major drawback has been its low ductility and fracture toughness at room temperature. Unlike α_2, however, γ exhibits significant plasticity at elevated temperatures. The mechanical properties of single-phase TiAl as a function of temperature are shown in Fig. 6.

Development efforts are active in the areas of alloying, powder metallurgy, crystal structure modification and the formation of discontinuous composites to promote enhanced plasticity through slip activation, increased propensity for twinning, microstructural refinement and modification of recrystallization phenomena. This broad activity has led to the development of alloys and processes that yield materials with attractive combinations of properties. Ductilities up to 3% have been reported for some alloys at room temperature.

The commercial development in γ-based alloys is rapidly expanding; substantial amounts of materials have been produced and preliminary manufacturing processes have been developed.

3.1 *Phase Transformations*

The $L1_0$ structure of γ exhibits only 2% tetragonality (i.e. $c/a = 1.02$). Consequently, efforts have been directed toward identifying alloying additions that promote a cubic structure. The assumption is that in a cubic structure an increased number of slip systems may be activated at ambient temperatures, thus improving ductility. Although alloying additions do change the lattice parameter and c/a ratio, no cubic structures have been reported for the TiAl compound to date. Efforts to correlate ductility with lattice aspect ratio have also been unsuccessful. Similarly, correlations of unit cell volume with measured ductility yield no trend.

Binary γ exists over a wide compositional range, though the exact phase boundaries are not known. The available data indicate that single-phase γ is stable at aluminum concentrations above 50 at.%. The nucleation of γ appears to be difficult and thus even relatively slow cooling rates lead to nonequilibrium amounts of α_2. During rapid solidification, γ formation is significantly suppressed at aluminum contents as high as 56 at.%.

Persuasive data have been presented to amend the titanium–aluminum phase diagram so that the α phase extends to the liquidus (approximately 1435 °C) and to suggest that it is a product of one of two peritectic reactions that occur in the range of 40–65 at.% of aluminum (see Fig. 3). The transformation of α to $\alpha+\gamma$ occurs by the nucleation of γ laths on stacking faults in the existing α grains. Upon cooling through the eutectoid temperature, the remaining α orders to α_2. The proportion of γ changes with temperature in both the $\alpha+\gamma$ and $\alpha_2+\gamma$ phase fields. This reaction sequence is consistent with the rapid solidification results: it is kinetically favorable for the existing α to order to α_2, as opposed to the nucleation and growth of γ.

The most attractive materials lie in the near-γ composition range of 45–50 at.% of aluminum. In this compositional range, the microstructure consists of a mixture of α_2 and γ, where the proportions and morphologies are a strong function of thermomechanical processing and alloy composition. In the as-cast condition, these alloys exhibit a lamellar microstructure consisting of alternating plates of α_2 and γ (see Fig. 7). These lamellae have an orientation relationship of the type $(0001)[11\bar{2}0]$ $\alpha_2//(111)[1\bar{1}0]$ γ. In this configuration there is a nearly perfect matching of the atoms across the boundary (see Fig. 8).

The lamellar structure may be transformed to a recrystallized equiaxed α_2 and γ structure by suitable thermomechanical processing (see Fig. 9). This recrystallization reaction is a discontinuous process and is stimulated by the presence of second-phase reinforcements (see Fig. 10).

3.2 Deformation Characteristics

Two types of dislocations are activated during deformation of γ at room temperature: the $a/2[110]$ dislocation and the $a/2[011]$ superdislocation, both gliding on the $\{111\}$ planes. The first dislocation is divided into two partial dislocations $a/6[211]$ and $a/6[\bar{1}12]$, separated by a stacking fault. In this case, the second partial restores the periodicity of the structure and thus movement of dislocations of this type is considered "easy." The [011] superdislocations are separated by two partials of the type $a/6[121]$ and $a/6[\bar{1}12]$, but in this case the order is not restored when the second partial passes. To restore order, a second dislocation of the correct type is required. This requirement for coupled movement makes this slip system a "hard" system.

Experimental analysis confirms the presence of the $a/2[110]$ dislocations. Consistent with a relatively high antiphase boundary energy, the separation of the partials is extremely small and often not visible. The activity of the $a/2[011]$ superdislocations is restricted, apparently by the formation of faulted dipoles which pin the partials and thereby inhibit further movement. The incidence of dipoles is reduced at temperatures above the brittle-to-ductile transition temperature of 700 °C and is eliminated at still higher temperatures, suggesting that the pinning mechanism is not operative at elevated temperatures. The reasons for this behavior are not understood but it is possible that it is caused by cross slip of the partial, or by local disorder caused by impurities or other defects.

Twinning plays an important role in the plastic deformation of TiAl. There are four types of twins available, but only two can be activated in tension. This may explain the compression–tension asymmetry observed: higher plastic strains are achievable in compression than in tension. Twinning is of the $(111)[11\bar{2}]$ type and is consistent with that found in other $L1_0$ structures. Vanadium additions increase ductility, but do not appear to change the deformation behavior of γ-based alloys. Increasing the tendency for twinning has been proposed as the reason for the increased ductility obtained with manganese additions. The suggestion that manganese stabilizes the "twin structures" may reflect a shift of the γ phase boundary to higher aluminum contents, thus expanding the $\alpha_2+\gamma$ phase field. Experimental data show that even compositions containing

Figure 7
Microstructure of a Ti–48 at.%Al alloy showing the alternating lamellae of Ti_3Al and TiAl (courtesy of R M Aikin Jr)

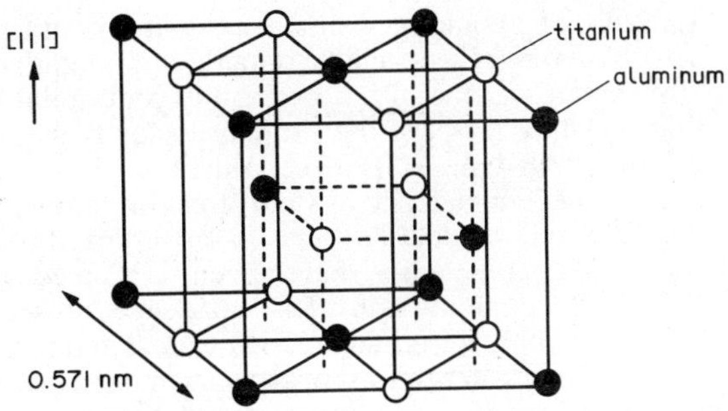

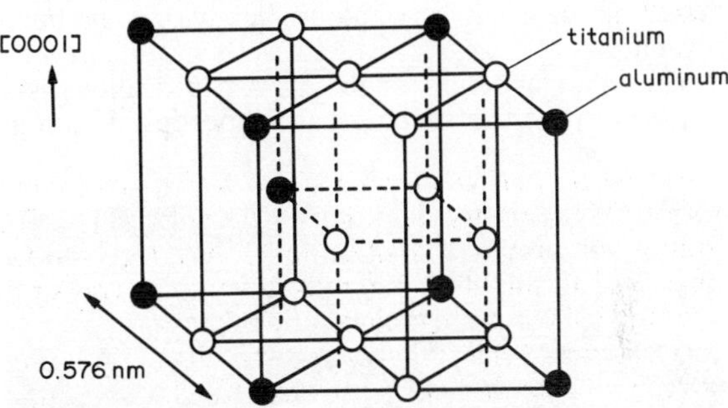

Figure 8
Orientation relationship between TiAl and TiAl lamellae (courtesy of R M Aikin Jr)

51.5 at.% of aluminum contain a small amount of Ti_3Al. Although the available data do not unequivocally support it, there is a possibility that suitable additions may increase the propensity for twinning and thus contribute to increased ductility.

A positive dependence of yield stress on temperature for single crystal TiAl has been reported and theoretical explanations have been advanced to explain the phenomenon. The incidence of such anomalous behavior remains controversial and there is convincing evidence to show that such behavior is not found in polycrystalline samples. Since γ is a product of a peritectic reaction, it is extremely difficult to grow single crystals and as yet no confirmation of such anomalous behavior has been presented.

The microstructure of the TiAl-based alloys is a key factor in their mechanical properties. Findings indicate that a significant increase in ductility can be achieved by transforming the lamellar structure to an equiaxed structure (e.g., ductility increases by 0.3–1%). This transformation is enhanced by the presence of second-phase discontinuities, such as the XD TiB_2 reinforcements (see Fig. 10). Furthermore, data for the near-γ alloys indicate that increasing the proportion of γ actually improves room temperature ductility. This is not surprising since at the aluminum compositions of interest, any α_2 will be hyperstoichiometric (~39 at.%) and thus extremely brittle. Creep behavior is also improved by this lamellar-to-equiaxed transformation in wrought products: the steady-state creep rate at 800 °C and 69 MPa is reduced from $5.5 \times 10^{-7}\,s^{-1}$ to $1.7 \times 10^{-7}\,s^{-1}$.

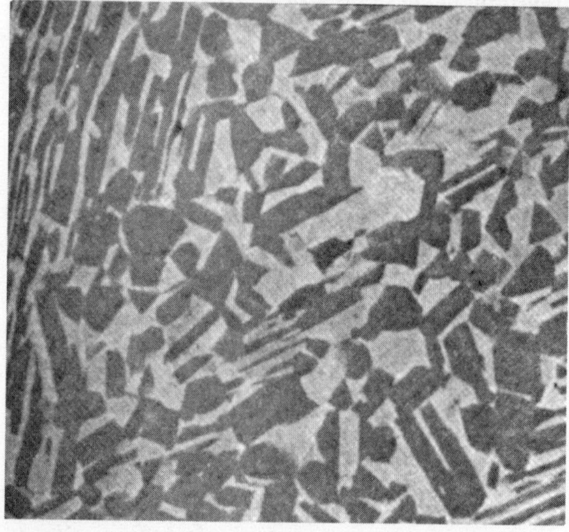

Figure 9
Transformation of the lamellar (a) to equiaxed (b) microstructure induced by suitable thermomechanical processing (courtesy of R M Aikin Jr)

3.3 Alloying Effects

The properties of γ and near-γ alloys are a strong function of the stoichiometry (see Fig. 11). This strong dependence makes it critical to accurately measure the titanium-to-aluminum ratio in any study and makes it difficult to compare materials containing different alloying additions unless the aluminum content is identical. Since the reasons for this strong dependence on stoichiometry are unknown, the analysis becomes even more difficult because the site occupation in the crystal by any alloying atom becomes an issue.

Of the alloying elements studied, the only three that appear to impart a significant increase in ductility to TiAl-based alloys are vanadium, manganese and chromium. Niobium, tungsten and tantalum have little effect, although they do improve strength, creep resistance and oxidation. (Interestingly, the three ductilizing elements all appear to significantly degrade oxidation and to some degree creep resistance.) As indicated previously in this section, one proposed mechanism for increased ductility is the propensity for twinning. However, transmission electron microscopy studies indicate that significant dislocation activity is also associated with these alloying additions. These additions may influence slip activity; for example, they may affect the mobility of the normally sessile partials.

As in the case of α_2, interstitial oxygen, nitrogen and carbon are believed to be particularly potent in strengthening (and embrittling) TiAl alloys. Boron additions appear relatively benign. In cases where boron-containing particles are present, as in composites discontinuously reinforced with titanium diboride, the transformation from the lamellar to the equiaxed morphology can be stimulated with a concomitant increase in room temperature ductility.

Nitride-reinforced alloys exhibit the best creep performance of all titanium-based alloys, whereas titanium diboride particulates only moderately improve creep performance. The mechanism of improved creep performance is unknown and may be related to high levels of interstitial nitrogen, the

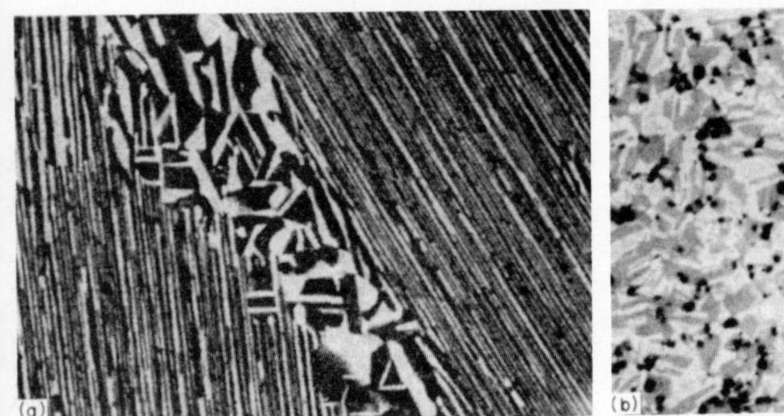

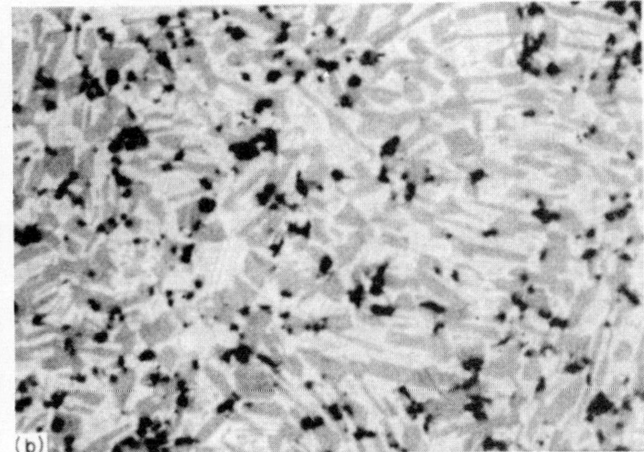

Figure 10
Second-phase TiB_2 reinforcement by the proprietary XD process induces the lamellar-to-equiaxed transformation. Following identical thermomechanical processing the discontinuously-reinforced TiAl is fully recrystallized (b) whereas the base alloy recrystallization only occurs at grain boundaries (a). (In this backscattered electron image TiB_2 appears black, TiAl grey and Ti_3Al white)

presence of the second-phase nitride particles or some other dynamic event. Early work showed that small additions of carbon also improves creep resistance. As before, this may be due to the presence of second-phase carbides or to interstitial interactive effects.

It should be noted that the amount of interstitials that can be tolerated in γ-based titanium aluminides without loss of ductility is not known; however, at least one reference indicates that reduction of oxygen from 1000 ppm to 300 ppm improves ductility by a factor of two.

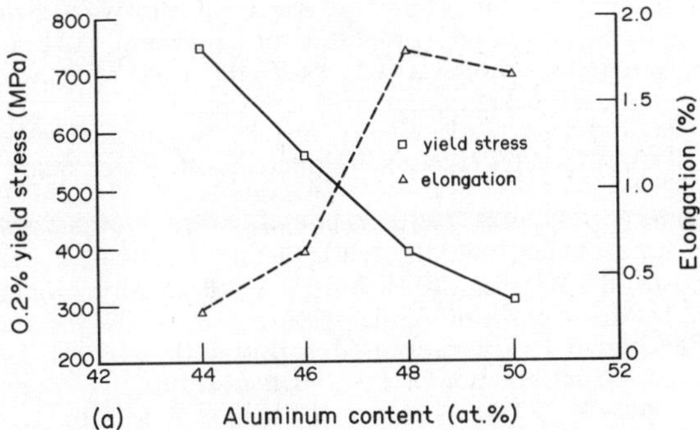

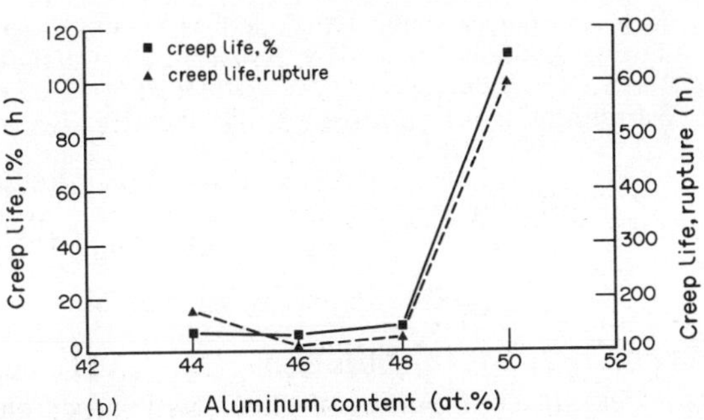

Figure 11
Tensile yield stress and elongation (a), and creep resistance (b), as a function of aluminum content in TiAl alloys (Blackburn and Smith 1979)

4. $TiAl_3$

Despite its low density of 3300 kg m^{-3} and excellent oxidation resistance, $TiAl_3$ is the least developed of the titanium aluminides. As a line compound (i.e., one with no detectable homogeneity range), it is difficult to produce and it is also extremely brittle; no tensile ductility has ever been reported. Even at moderate temperatures (330–620 °C), compressive strain is limited to less than 1%. The dominant mechanism of deformation appears to be twinning, with dislocation activity only becoming evident at the highest temperatures. Twinning is of the type (111)[11$\bar{2}$], which does not disturb the DO_{22} structure. Slip at high temperatures occurs along both the [110] and [100] directions. Of the group IVa and Va elements of the periodic table, only the addition of hafnium and zirconium to $TiAl_3$ appears to moderate the inherent brittleness of the material. Lithium induces a marginal improvement in compressive ductility; however, the mechanism responsible is unclear. The addition of boron also alleviates brittleness; this is attributed to an increased propensity for twinning.

The development of $TiAl_3$-based titanium aluminides has concentrated on expanding the compositional field and the transformation of the DO_{22} structure to the cubic $L1_2$. Ternary additions of

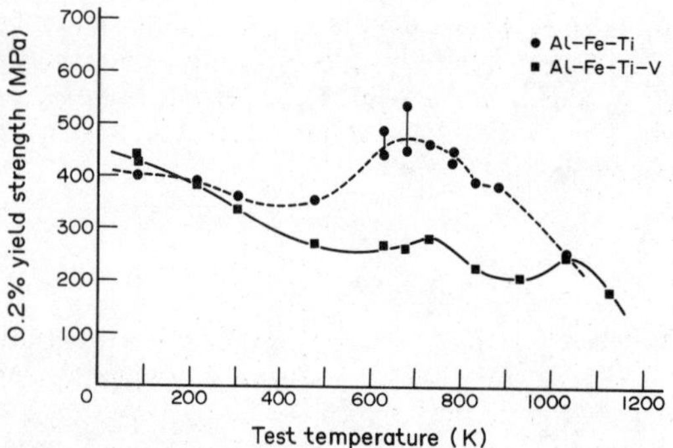

Figure 12
Compressive yield strength versus test temperature for ternary Al–Fe–Ti and quaternary Al–Fe–Ti–V $L1_2$ (τ) alloys (Kumar and Pickens 1988)

copper, iron and nickel induce this structural change. The resultant compounds (often referred to as τ aluminides) have the compositions Al_5CuTi_2, $Al_{22}Fe_3Ti_8$ and $Al_{67}Ni_8Ti_{25}$, respectively. These compounds have a relatively broad range of stability, which enables them to be produced by conventional ingot-based techniques. Noteworthy features include the positive dependence of yield stress on temperature (similar to Ni_3Al) and the presence of two peaks in the strength–temperature curve. Minor additions of vanadium reduce the strength of the τ aluminides and induce a shift in the second peak temperature by about 150 °C (see Fig. 12). Although there appears to be no measurable room temperature ductility in the $L1_2$ structures, they are significantly more resistant to crack propagation and show clear evidence of plastic deformation under compression. Further efforts in alloy development will undoubtedly endeavor to improve the ductility and the damage tolerance of this class of titanium aluminides.

See also: Aluminides for Structural Use (Suppl. 1); Titanium: Alloying; Titanium and Titanium Alloys: Selection; Titanium: Properties

Bibliography

Blackburn B J 1967 The ordering transformation in titanium aluminum alloys containing up to 25 at pt aluminum. *Trans. Metall. Soc. AIME* 239: 1200–8

Blackburn M J, Smith M P 1979 The understanding and exploitation of alloys based on the compound TiAl (gamma phase). Technical report No. AFML-TR-79-4056. Air Force Wright Aeronautical Laboratories, Dayton, OH

Court S A, Lofrander J P A, Loretto M H, Fraser H L 1989 The nature of *c*-component dislocations in samples of a polycrystalline Ti_3Al-based alloy deformed at room temperature and at 300 °C. *Philos. Mag. A* 59: 379–99

Fujiwara T, Nakamura A, Hosomi M, Nishitani S R, Shirai Y, Yamaguchi M 1990 Deformation of polysynthetically twinned crystals of TiAl with a nearly stoichiometric composition. *Philos. Mag.* 57

Greenberg B A, Gornostiner Y N 1988 Possible factors affecting the brittleness of the intermetallic compound TiAl. *Scr. Metall.* 22: 853–64

Hanamura T, Vemori R, Tanino M 1988 Mechanism of plastic deformation of Mn-added TiAl $L1_0$-type intermetallic compound. *J. Mater. Res.* 3: 656–64

Kaufman M J, Kouitzer D G, Shull R D, Fraser H L 1986 An analytical electron microscopy study of the recently reported "Ti_2Al Phase" in γ-TiAl alloys. *Scr. Metall.* 20: 103–8

Kawatata T, Tadano M, Izumi O 1989 Effect of purity and second phase on ductility of TiAl. *Scr. Metall.* 22: 1725–30

Kim Y W 1989 Intermetallic alloys based on gamma titanium aluminide. *J. Metals* 14: 24–30

Kumar K S, Pickens J R 1988 Ternary low-density cubic ‹l_2 aluminides. In: Kim Y W, Griffith W M (eds.) 1988 *Dispersion Strengthened Aluminum Alloys*. The Minerals, Metals and Materials Society, Warrendale, PA, pp. 763–86

Lipsitt H A 1985 Titanium aluminides—An overview. In: Koch C C, Liu C T, Stolloff N S (eds.) 1985 *High Temperature Ordered Intermetallic Alloys*. Materials Research Society, Boston, MA, pp. 351–64

Lipsitt H A, Shechtman D, Schafrik R E 1975 The deformation and fracture of TiAl at elevated temperatures. *Metall. Trans., A* 6: 1891–996

Lipsitt H A, Shechtman D, Schafrik R E 1979 The deformation and fracture of Ti_3Al at elevated temperatures. *Metall. Trans., A* 11: 1369–75

McCullough C et al. 1989 Phase equilibria and solidification in Ti–Al alloys. *Acta Metall.* 37: 1321–36

Martin P L et al. 1980 The effects of alloying on the microstructure and properties of Ti_3Al and TiAl. In: Kimura H, Izumi O (eds.) 1980 *Titanium '80 Science and Technology*, Vol. 2. pp. 1245–54

Mendiratta M G, Lipsitt H A 1980 Steady-state creep behavior of Ti_3Al-based intermetallics. *J. Mater. Sci.* 15: 2985–90

Murray J L 1987 *Phase Diagrams of Binary Titanium Alloys*. American Society for Metals, Metals Park, OH

Sastry S M L, Lipsitt H A 1977 Cyclic deformation of Ti_3Al. *Acta Metall.* 25: 1279–88

Shechtman D, Blackburn M J, Lipsitt H A 1974 The plastic deformation of TiAl. *Metall. Trans.* 5: 1373–81

Thomas M, Vassel A, Veyssière P 1989 '*c*' slip in Ti_3Al. *Philos. Mag. A* 59: 1013–26

Yamaguchi M, Shirai Y 1988 Deformation behavior of single and polycrystalline Al_3Ti and Al_3Ti with ternary alloying additions. *Dispersion Strengthened Aluminum Alloys*. The Minerals, Metals and Materials Society, Warrendale, PA, pp. 763–86

L. Christodoulou
[Martin Marietta Laboratories, Baltimore, Maryland, USA]

Transition Metal Silicides

Silicides are compounds of silicon with other elements. Practically all elements of the periodic table form silicides. At room temperature, silicides could be solid, liquid or gas. Similarly there are conducting, insulating or semiconducting silicides. The

Table 1
Silicides of the transition metals of group IVa, Va, VIa, VIIa and VIII of the periodic table

Ti_5Si_2	V_3Si	Cr_3Si	Mn_3Si	Fe_3Si	Co_3Si	Ni_3Si
$TiSi$	V_5Si_3	Cr_5Si_3	Mn_5Si_3	Fe_3Si_2	CO_2Si	Ni_2Si
$TiSi_2$	VSi_2	$CrSi$	$MnSi$	$FeSi$	$CoSi$	Ni_5Si_2
		$CrSi_2$	$MnSi_2$	$FeSi_2$	$CoSi_2$	Ni_3Si_2
						$NiSi$
						$NiSi_2$
Zr_4Si	Nb_4Si	Mo_3Si		Ru_2Si	Rh_2Si	Pd_3Si
Zr_2Si	Nb_5Si_3	Mo_5Si_3		$RuSi$	Rh_5Si_3	Pd_2Si
Zr_3Si_2	$NbSi_2$	$MoSi_2$		Ru_2Si_3	Rh_3Si_2	$PdSi$
Zr_6Si_3					$RhSi$	
$ZrSi$					Rh_2Si_3	
$ZrSi_2$						
Hf_2Si	$Ta_{4.5}Si$	W_5Si_3	Re_3Si	$OsSi$	Ir_3Si	Pt_3Si
Hf_5Si_2	Ta_2Si	WSi_2	Re_5Si_3	$OsSi_2$	Ir_2Si	Pt_2Si
Hf_3Si_2	Ta_5Si_3		$ReSi_2$	$OsSi_3$	Ir_3Si_2	$PtSi$
$HfSi$	$TaSi_2$				$IrSi$	
$HfSi_2$					$IrSi_2$	
					$IrSi_3$	

transition metal silicides are shown in Table 1 as compounds of groups IIIa, IVa, Va, VIa, VIIa and VIII of the periodic table. These silicides, because of their low metal-like electrical resistivity and their high temperature stability, have attracted considerable attention and have become the subject of intensive research since the beginning of the 20th century. Moissan (1904), using a newly developed electric furnace, was possibly the first to carry out a systematic preparation of various transition metal silicides.

Most of the silicide research, performed from the beginning of this century through to the 1960s, used powder metallurgical techniques to produce silicides (Wehrmann 1967). These studies primarily focused on the fundamental properties such as electrical resistivity, high-temperature stability, corrosion resistance and, most important of all, the silicide crystal chemistry and metal–silicon phase diagrams. The possibility of using silicides as conductors (as Schottky barriers and contacts and as gate and interconnection metallizations) in silicon integrated circuits (SICs) has motivated thin-film silicide research. Besides the measurement of Schottky barrier heights and resistivities, the intermetallic compound formation in the metal–silicon systems, reaction and interdiffusion kinetics, stability at the SIC processing and operating temperatures, mechanical stability, oxidation and etching characteristics, and epitaxial growth on silicon have all been investigated since the early 1960s. Most of this research has centered around the use of silicide films deposited by one of several vapor-deposition techniques or those formed by metallurgical reaction of the silicon substrate with vapor-deposited metals. The film thickness has ranged from a few tens to a few hundreds of nanometers. More recently a few silicides have been grown in crystalline form (epitaxially) on crystalline silicon substrates, opening the way to several new two- and three-dimensional electronic device processing schemes.

1. *Silicides in Integrated Circuits*

The primary thrust of very large scale integration (VLSI) has been the devices that are smaller (large packing density and hence the increased complexity on the chip), faster and operate at lower power. The interest in new metallization materials systems has been aroused by the fact that, with the scaling down of the device sizes, the conductor linewidth gets narrower and the resistance contribution to the MOSFET device delay increases. Most metals, due to undesirable resistivity, stability and/or oxidizability, do not qualify for use as metallization. Transition metal silicides became potential metals in such device applications.

Table 2 lists the electrical resistivities and other properties of the most stable silicides of the transition metals. The lower number in the resistivity column represents a high-purity bulk crystalline silicide. The higher number represents the typical value obtained for a polycrystalline 100–300 nm thick silicide film. Early applications in ICs required that the existing chip fabrication process have:

(a) high temperature stability,

(b) oxidizability, and

(c) most importantly retrofitability.

Refractory silicides, $MoSi_2$, $TaSi_2$ and WSi_2, with thin-film resistivities in the range of 0.5–1 μΩ m, thus found their applications as gate and interconnection metallization materials in 256 K to 4 M bit dynamic random access memories (DRAMs) and in newly developed microprocessors.

Refractory disilicides of molybdenum, tantalum and tungsten are most conveniently formed by a codeposition process (Murarka 1983). Silicon and the metal species are produced simultaneously in the gas phase and then condensed onto the substrate. The codeposition can be carried out using:

(a) chemical vapor deposition (CVD),

(b) plasma-assisted CVD,

(c) laser-assisted CVD,

(d) coevaporation from elemental sources, and

(e) cosputtering from elemental or silicide (alloy) targets.

Cosputtering from silicide targets has been most practical in providing silicide films of reproducible

Table 2
Properties of silicides

Silicide	Resistivity ($\mu\Omega$ m)	Melting point (°C)	Type of conduction[a]	Lattice mismatch with silicon (%)	Inert ambient stability of thin films on silicon (approximate temperature) (°C)	Weight change in air of hot pressed silicide	
						at (°C)	($kg\ m^{-2}\ h^{-1} \times 10^{-3}$)
$TiSi_2$	0.13–0.25	1500	n		950	1200	+0.75
$ZrSi_2$	0.35–0.40	1650–1700	n		950–1000	1200	+105
$HfSi_2$	0.45–0.70	~1750	n		950–1000		
VSi_2	0.50–0.55	1677	n	3.1	1000	1200	+12.5
$NbSi_2$	0.50	1930		4.2	1000	1200	−135
$TaSi_2$	0.35–0.60	2200±100	n	4.4	1000	1500	+27
$CrSi_2$	~6	1490	s,p	0		1300	+25
$MoSi_2$	0.40–1.00	2030±50	p	4.1	1000	1566	−0.031
WSi_2	0.30–0.70	2165±15	p		1000	1500	−59
$FeSi_2$	>10	1212	s,n	0.9			
$CoSi_2$	0.10–0.18	1326	p	1.2	900–950		
$NiSi_2$	~0.5	1000–1280	p	0.4	850		
PtSi	0.28–0.35	1229		9.5[b]	750		
Pd_2Si	0.30–0.35	1398		2.4	700		

a n = electron, p = hole, s = semiconducting b the effective lattice mismatch is defined as square root of the projected unit cell area

characteristics. As-deposited films are a mixture of the metal and silicon atoms and need high-temperature heat treatment to produce lower-resistivity silicides. In most applications, as-deposited films are patterned in the desired configuration using a variety of dry plasma etching processes.

The continued reduction in device dimensions, however, requires further lowering of the interconnection resistance and lower contact resistance between the metallization and the substrate. At the same time, these developments have lowered the highest temperature of processing to a range that can make the use of the cobalt and titanium disilicides feasible. They are the lowest-resistivity silicides (see Table 2). $CoSi_2$ can be easily formed by reacting a cobalt-metal film with a silicon substrate or polycrystalline silicon. At the temperature of reaction, the metal does not react with an insulating SiO_2 layer. This masking effect allows the metal film to react selectively with silicon and the unreacted metal to be selectively etched from the adjoining SiO_2 surfaces. Thus the silicide is only formed on the prepatterned gate, interconnection and contacts which are isolated by the SiO_2 layers (see Fig. 1). This selective formation of the silicide has been termed a "self-aligned process," making the development of the silicide etching process, necessary for the use of codeposited refractory silicides, unnecessary. $TiSi_2$ can also be formed in this manner. The

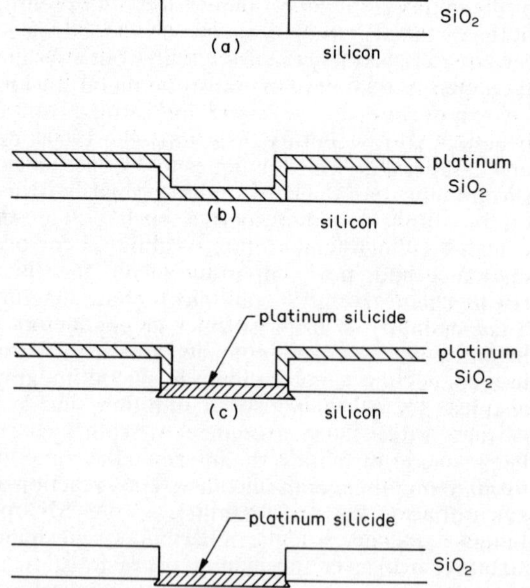

Figure 1
Simple schematic presentation of the self-aligned formation on silicon in a contact window in SiO_2 on silicon: (a) pattern oxide on silicon or polysilicon; (b) deposit platinum; (c) low-temperature sinter; and (d) etch metal

cobalt disilicide process is preferred because of simplicity and significantly less susceptibility to oxygen contamination (Murarka 1986). Both $CoSi_2$ and $TiSi_2$ can be formed simultaneously on gates and interconnections and in contact windows. These silicides can easily be used as contact metallizations, even in bipolar and high-power devices and circuits.

Besides resistivity, the Schottky barrier height (SBH) has been measured for most silicides. Group VIII metal silicides have an SBH on *n*-type silicon that is greater than half the silicon bandgap energy. Refractory silicides have an SBH on *n*-type silicon that is nearly half the bandgap energy of silicon. Information on other electrical properties of the transition metal silicides is very limited. Recently, experimental and theoretical investigations of the band structure and bonding have been reported for some silicides. These investigations find silicides to be conductors with electrons or holes as charge carriers and semiconductors (see Table 2). $CrSi_2$, $FeSi_2$, $IrSi_{1.75}$ and a few manganese silicides are reported to be semiconductors (Nicolet and Lau 1983).

2. *Epitaxial Silicides*

The possibility of making three-dimensional device structures by using epitaxial (monocrystalline) films has always intrigued scientists and engineers. Since most silicides have metal-like resistivities, the study of the epitaxial growth of the silicides on silicon and of the epitaxial growth of silicon on epitaxial silicides are aimed towards the development of such devices. Examples of such devices are the metal base transistor (Hensel et al. 1984) and the permeable base transistor (Bozler et al. 1979), first proposed in 1968 and 1979, respectively. Deposition of epitaxial metal silicide films on silicon followed by epitaxial silicon on the silicide leads to the possibility of devices with common buried metal layers. Recently, such heteroepitaxy growth has been demonstrated using $CoSi_2$ (Bean and Poate 1980, Saitoh et al. 1980, 1981). Since epitaxial disilicides offer ideal interfaces and higher temperature stability, they could possibly be used as stable contacts to shallow junctions. Table 2 also lists the calculated lattice mismatch of various silicides with silicon. Among conducting silicides, $NiSi_2$ and $CoSi_2$ offer the best possibilities for epitaxial growth. Semiconducting chromium and iron silicides also offer excellent epitaxial growth possibilities, perhaps leading to their use as heterostructure materials on silicon.

3. *Silicides as High-Temperature and Corrosion-Resistant Materials*

Table 2 lists the melting points of the most stable silicides. Refractory disilicides of group IVa, Va and VIa elements offer very high temperature stability with melting points in excess of 1450 °C. Of these, $TaSi_2$, $MoSi_2$ and WSi_2 with melting points in excess of 2000 °C have been extensively investigated for applications as protective coatings for exposure to corrosive environments at temperatures in excess of 1200 °C. In the last column of Table 2 are listed the weight changes occurring in the silicides heated to high temperatures in air (Wehrman 1967): $MoSi_2$ exhibits an excellent stability, even at 1566 °C. It is used as furnace heating element, useful to temperatures of 1600–1650 °C.

At room temperature, the refractory disilicides are resistant to salt solutions, aqueous alkalies, and single acids except hydrofluoric acid (HF). HF vigorously attacks group IVa disilicides and also dissolves group Va disilicides, although at much lower rate. $MoSi_2$ and WSi_2 are not attacked by HF unless mixed with nitric acid.

Oxidation characteristics and stability of the disilicides depend strongly on the substrate (silicon vs SiO_2) on which the silicide film is formed, and also on the silicide type. Thin-film silicides deposited on silicon or polycrystalline substrates, reveal very good stability of the disilicides during oxidation, except for group IVa disilicides. Oxidation occurs at the silicide surface leading to the formation of SiO_2 film. Silicon becomes available for oxidation, either by diffusion from the substrate through the silicide to the silicide–SiO_2 interface or by the metal ion migration inward from the silicide–SiO_2 interface releasing silicon for oxidation at this interface. Refractory silicides follow the first mechanism. Group VIII metal silicides seem to follow the second mechanism. Group IVa disilicides decompose during low-temperature oxidation leading to both metal and silicon oxidation. In the case of $TiSi_2$, at higher temperatures when silicon migration through the silicide is fast, only SiO_2 forms leaving the $TiSi_2$ on silicon intact.

Oxidation of thin-film silicides deposited on oxide, where no silicon is available for oxidation, leads to decomposition of the silicide forming SiO_2, metal oxide or even free metal depending on the metal. When the free energy of formation of the metal oxide is greater than that of SiO_2, the metal oxide is preferentially formed. Such is the case of group IVa and Va disilicides. In other cases the free energy of formation of the metal oxide is lower than that of SiO_2, and silicon oxidizes preferentially forming SiO_2 and metal-rich silicide, and eventually metal and SiO_2. On prolonged oxidation, metal also oxidizes. This behavior is exhibited by $MoSi_2$, WSi_2, PtSi and $CoSi_2$.

Addition of silicon in excess of the stoichiometric composition of the disilicide has been shown to be advantageous by providing better oxidation resistance to thin deposited films. The oxidation rate of these silicon-rich silicides on silicon or polycrystalline silicon substrates, is found to decrease with increasing silicon content (Liu et al. 1986). This process leads to a possibility of using silicon-rich refractory metal silicides for high-temperature applications.

Bibliography

Bean J C, Poate J M 1980 Silicon/metal silicide heterostructure grown by molecular beam epitaxy. *Appl. Phys. Lett.* 37: 643

Bozler C O, Alley G D, Murphy R A, Flanders R A, Lindley W T 1979 Fabrication and microwave performance of the permeable base transistors. *IEEE Int. Electron Devices* December: 384

Hensel J C, Tung R T, Poate J M, Unterwald F C 1984 Electrical transport properties of $CoSi_2$ and $NiSi_2$ thin-films. *Appl. Phys. Lett.* 44(9): 913–15

Liu R, Murarka S P, Pelleg J 1986 Effects of dopants and excess silicon on the oxidation of $TaSi_2$ polycrystalline silicon structure. *J. Appl. Phys.* 60(9): 3335–42

Moissan H 1904 *The Electric Furnace*. English translation by de Mouilpied A, Edward Arnold, London

Murarka S P 1983 *Silicides for VLSI Applications*. Academic Press, New York

Murarka S P 1986 Refractory silicides for integrated circuits. *J. Vac. Sci. Technol.* 60: 2106

Nicolet M-A, Lau S S 1983 Formation and characterization of transition-metal silicides. In: Einspruch N G, Larrabee G B (eds.) 1983 *VLSI Electronics Microstructure Science*. Academic Press, New York, p. 384

Saitoh S, Ishiwara H, Furukawa S 1980 Double heteroepitaxy in the Si(111)/$CoSi_2$/Si structure. *Appl. Phys. Lett.* 37: 203

Saitoh S, Ishiwara H, Furukawa S 1981 *Jpn. J. Appl. Phys.* Suppl. 20–1: 49

Wehrmann R 1967 Silicides. In: Campbell I E, Sherwood E M (eds.) 1967 *High-Temperature Materials and Technology*. Wiley, New York, p. 399

S. P. Murarka
[Rensselaer Polytechnic Institute, Troy, New York, USA]

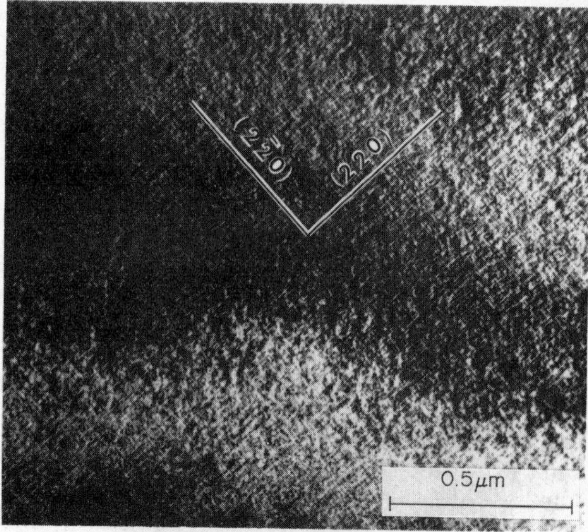

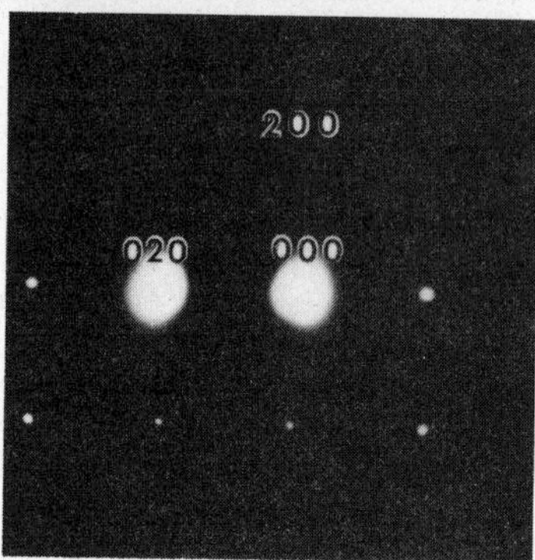

Figure 1
Typical cross-hatched appearance of tweed contrast, with two sets of striations, each parallel to traces of {110} planes in the matrix phase. This image is taken from a Cu–47.7 at.%Mn–3.5 at.% Al alloy, solution treated for 2 h at 1073 K, quenched and aged for 8 h at 673 K. The foil normal orientation is near [001], with the *g* vector 020 strongly operating

Tweed Contrast in Microstructures

The term "tweed" is a generic term for a particular sort of diffraction contrast in transmission electron microscope (TEM) images. It typically takes the form of irregular lines of contrast that lie approximately parallel to traces of {110} planes of a cubic parent phase, often forming a cross-hatched pattern. A typical example is presented in Fig. 1. The corresponding diffraction patterns show diffuse streaks along ⟨110⟩ directions.

1. Generic Character

This type of image contrast was first reported for age-hardened copper–beryllium alloys (Armitage et al. 1962). Subsequently, the tweed diffraction contrast in this alloy system was analyzed (Tanner 1966) and a very similar contrast was observed in ordered Ni_2V (Tanner 1972). The common occurrence of tweed contrast in quenched β-brass alloys was also reviewed (Delaey et al. 1972). It is evident that there are a variety of fine-scale microstructures that display virtually the same sort of diffraction contrast in TEM images.

2. Origins

There are numerous alloy systems in which tweed contrast is observed. Among these systems there are a variety of underlying causes. However, in all cases there are two basic requirements that must be satisfied in order for this sort of image contrast to be displayed. The first requirement is for an elastically anisotropic matrix phase that is "soft" with respect to certain shear distortions. (In cubic systems, these distortions are typically of the general form {110}⟨110⟩; that is, {110} planes shearing in ⟨110⟩ directions.) The second requirement is for a source of finely distributed centers of asymmetric strain.

It is with respect to this second necessity that "tweedy" systems vary. In different systems the strain centers may be Guinier–Preston zones (Armitage 1962), finely distributed precipitate particles (Delaey et al. 1972), ordered domains (Tanner 1972) or simply regions in which there is an incipient lattice transition (such as a premartensitic transition) that distorts the lattice (Schryvers et al. 1988).

With regard to the first requirement, for many cubic lattices the development of certain soft elastic constants as temperature is lowered often provides a matrix phase that is particularly susceptible to shear distortions of the type {110}⟨110⟩ (Zener 1948). In order to excite this distortion, which is based on a ⟨110⟩⟨110⟩ transverse phonon mode, the strain excitation must be asymmetric at the strain centers that are distributed in the matrix. If the strain excitation is symmetric, the strain centers are more likely to excite the longitudinal ⟨100⟩ mode, which is also typically soft. In this case, the distortions would be expected to lead to {100} contrast traces rather than the typical {110} traces of a true tweed microstructure. A well-known example of {100} cross-hatched contrast is in connection with spinodal decomposition of various cubic solid solutions.

3. Variations in Distinction

The distinction of a tweed contrast image, that is, the degree of alignment and contrast of the {110} traces, can be affected by the actual underlying microstructure and by the imaging conditions under which it is observed.

3.1 Microstructural Reasons for Contrast Variations

The distinction of a tweed contrast image is dependent on the underlying microstructure in terms of the nature of the strain centers, the magnitude of the asymmetric strain produced by each center, the distribution of the strain centers and the degree of anisotropy of the matrix phase. If any of these factors is deficient, the typical cross-hatched {110} striations will not be observed or will be weak. In cases of weak tweed contrast the image will often present a mottled appearance. Many tweedy alloys therefore show a variation in the distinction of the tweed contrast as the distribution and/or strength of the strain centers changes. This may occur, for example, as zones or precipitates develop during aging or as an alloy is cooled towards a temperature range of lattice instability, such as just above a martensite start temperature.

The distribution of the straining centers in the matrix is quite important: they must be close enough together so that the strain fields overlap and interact with one another. If the strain centers are too far apart each will exhibit its own individual pattern of local strain contrast, but thc aligned traces will not be evident.

3.2 Imaging Conditions and Contrast Variations

For a given microstructural condition, the appearance of the tweed contrast is also a strong function of the imaging conditions within the transmission electron microscope. The relevant factors with regard to this have been thoroughly reviewed and concisely presented by Robertson and Wayman (1983). For example, certain image extinctions are always observed; these are consistent with the underlying {110}⟨110⟩ shear distortions of the matrix. The basic extinction rule is that a given set of {110} striations will become invisible if, in a strong two-beam operating condition, the operating g vector for the image is perpendicular to the trace direction (within which lies the trace of the strain vector). For example, the (110) traces are visible but the $(1\bar{1}0)$ traces are invisible with $g=\bar{1}1\bar{1}$ and $g=2\bar{2}0$, whereas the opposite is true for $g=11\bar{1}$ and $g=\bar{2}\bar{2}0$. This indicates that the character of the contrast derives from strain displacements of the type {110}⟨110⟩. Also, the tweed contrast is a sensitive function of other variables in the imaging conditions such as foil orientation, degree of deviation from the Bragg condition, extinction distance and foil thickness. In general, the tweed contrast is observed to be strongest near bend contours.

4. Dynamic Image Effects: Shimmering and Flickering

In some cases, the contrast in tweedy or near-tweedy (mottled) images may appear to not only be aligned or striated, but may also appear to be shifting in space and time. There are at least two possible origins of such an effect, both having to do with excitation from the energy of the incident electron beam. In some cases, the energy of the beam may simply excite vibrations and/or local variations in orientation of the thin foil specimen (Otsuka et al. 1981). In other cases, the energy may serve to actually induce a distribution of fine-scale local shifts in crystal structure or interfaces (Perkins et al. 1988).

5. Summary

It is important to realize that the cross-hatched appearance of tweed is not an actual structure within the sample, but that it is a diffraction contrast effect in the image. However, in each case, there is an actual structure underlying the tweed contrast. This real microstructure, which can often be detected by higher-resolution techniques, provides the fine-scale distribution of strain centers which is a requisite for the tweed contrast.

Bibliography

Armitage W K, Kelly P M, Nutting J 1962 Structural changes during the aging of copper–2% beryllium alloys. *Proc. 5th Int. Cong. Electron Microscopy*. Academic Press, New York, paper K-4

Delaey L, Perkins J, Massalski T B 1972 Review: On the structure and microstructure of quenched beta-brass type alloys. *J. Mater. Sci.* 7: 1197–215

Otsuka K, Kubo H, Wayman C M 1981 Diffuse electron scattering and 'streaming' effects. *Metall. Trans. A* 12: 595–605

Perkins J, Mayes L L, Yamashita T 1988 Flickering contrast in TEM images in an aged 53Cu–45Mn–2Al alloy. *Scr. Metall.* 22: 887–92

Robertson I M, Wayman C M 1983 Tweed microstructures: I. Characterization of beta-NiAl. *Philos. Mag.* 48: 421–42

Schryvers D, Tanner L E, Van Tendeloo G 1988 Premartensitic microstructures as seen in the high resolution electron microscope: A study of a Ni–Al alloy. In: Gonis A, Stocks M (eds.) 1988 *Proc. NATO–ASI Symp. Phase Stability*. Nijhoff, Dordrecht, pp. 701–5

Tanner L E 1966 Diffraction contrast from elastic shear strains due to coherent phases. *Philos. Mag.* 14: 111–30

Tanner L E 1972 The ordering transformation in Ni_2V. *Acta Metall.* 20: 1197–227

Zener C 1948 *Elasticity and Anelasticity of Metals*. University of Chicago Press, Chicago, IL

J. Perkins
[Naval Postgraduate School, Monterey, California, USA]

Welding of Thermoplastics

The welding of metals has been performed for decades and is fairly well understood, but the welding of plastics was only undertaken in specialized applications before 1940. It is important, therefore, to understand the nature of plastics if their welding abilities are to be appreciated.

There are two types of plastics—thermosetting and thermoplastic. Thermosetting plastic is shaped by flowing under pressure, often with the application of heat, and becomes rigid at a setting temperature. Further heating will not cause softening, so it is not possible to weld thermosetting materials. Thermoplastic materials, however, need to be heated to melt or soften, and must then be cooled to be made rigid; thus they lend themselves to welding.

Thermoplastics are subdivided into amorphous and crystalline polymers. It is usual to weld like materials, but there are times when a bond to a different polymer is useful. Although it is not possible to obtain a good bond between an amorphous and a crystalline material by direct means, a hybrid method based on electromagnetic welding can achieve the same effect.

The welding of thermoplastics does have some similarities with the welding of metals, but there are also notable differences. The former are noted for their properties of thermal and electrical insulation, and relatively low melting or softening points; far less energy is therefore required to weld thermoplastics than to weld metals. Another major difference, compared with metals, is the large increase in volume of thermoplastics from ambient temperatures to their softening or melting points. With amorphous polymers this is about 2–5%, but with crystalline polymers, due to their change of state, it is in the range of 10–20%.

1. General Principles

1.1 Compatibility

An indication of which common polymers will bond to which is given in Table 1. Most of those shown in the table exist both as homopolymers and copolymers (homopolymers having all the same units in the polymer chain although different terminal units can be added to enhance thermal stability, and copolymers having chains consisting of two or more polymerizing units), while other variations are the addition of plasticizers, or other polymers (sometimes called "alloys"), to give compromises in properties. Their welding properties will be similar to those of the parent polymer. There are many

Table 1
Welding compatibility of thermoplastics

	ABS	Acetals	Acrylics	Cellulosics	Nylons (polycarbonates)	Polycarbonates	Polyethylenes	Polypropylenes	Polystyrenes	PVC rigid	PVC flexible	Thermoplastic polyesters
ABS	E[a]											
Acetals	X[b]	E										
Acrylics	G[c]	X	E									
Cellulosics	X	X	X	E								
Nylons (polyamides)	X	P[d]	X	X	E							
Polycarbonates	G	X	G	X	X	E						
Polyethylenes	X	X	X	X	X	X	E					
Polypropylenes	X	X	X	X	X	X	G	E				
Polystyrenes	E	X	G	X	X	G	X	X	E			
PVC rigid	X	X	P	X	X	P	X	X	X	E		
PVC flexible	X	X	X	X	X	X	X	X	X	P	E	
Thermoplastic polyesters	X	P	X	X	P	X	X	X	X	X	X	E

a E = excellent b G = good c P = problematic d X = not recommended

other basic polymers whose welding properties can be judged from those shown in the table.

Apart from recent developments with die-casting alloys, most metals are without fillers and are of uniform structure apart from their surfaces. Thermoplastics, however, can be very dependent on fillers and other additives for their characteristics and particularly their "engineering" properties. These additives can play a vital role, some advantageous and others not, in the welding abilities of these plastics. Glass and other fibrous fillers tend to lie with their axes parallel to the welding surface. Since there is little or no mixing between layers at the interface of the weld, the strength of the bond cannot be greater than that of the base polymer. On the other hand, the dispersion of magnetic particles is a key factor in the electromagnetic welding system (see Sect. 3.2).

Since plastics products are usually made in large numbers, any finishing operations must be kept to a minimum. This is often achieved by the design of the weld joint by providing a space into which any "spue" (excess material displaced by the welding operation) can flow without affecting the appearance of the product. Other types of welds are very dependent on the shape of the mating surfaces for their strength. Both of these conditions are made easier by the choice of the best fabrication method, and this is often injection molding because of its versatility, accuracy and consistency.

1.2 Welding Methods

The methods of melting or softening of plastic can be:

(a) direct electrical energy, such as high-frequency heating or electromagnetic heating;

(b) mechanical energy, such as spin (rotational) friction, orbital motion, linear oscillation or ultrasonic vibrations; and

(c) thermal energy, such as conducted heat, radiant heat or hot gases.

In categories (b) and (c), the prime source is almost invariably electricity.

1.3 Forms of Plastics

Plastic articles are formed by a number of methods, and these can influence the type of welding used. While most thermoplastics can be formed by most methods, certain end products that involve welding are normally made by only one or two methods (e.g., protective clothing, including some rainwear, is made from flexible PVC formed by calendering, and welded by the high-frequency method).

Since the method of manufacture of the plastic can determine the method of welding, or the fabrication method might be chosen to facilitate a welding operation, or the quality of the finished product may be adversely affected by residual fabrication strains, a brief list of the principal fabrication methods and the normal polymers used is given in Table 2. Other manufacturing methods may be widely used (e.g. wire covering) but these are not included in the list below as there is no welding interest.

1.4 Fabrication Methods

(*a*) *Injection molding*. This is the most widely used method of fabrication as the range of articles made from plastics is more varied in size and complexity than for any other material. Essentially, the material is melted or softened in a tube, from which it is forced under considerable pressure into a mold which gives the article its shape. Rapid cooling then occurs until it is rigid enough to be removed quickly from the mold. The conditions of manufacture are such that even the smallest detail, such as a dull patch on the surface of a polished mold, will be reproduced. Consistency of molding quality is now of a high order, and complicated shapes are not much higher in cost than simple ones; hence the method is particularly suitable for articles that are formed by welding. One factor that has to be borne in mind, however, is that there are residual strains in the molding, and these will be released in the area of the weld and may be a possible cause of distortion.

(*b*) *Extrusion*. This is a continuous process that lends itself to such products as pipes, films and sheet. Wire covering and profiles are produced extensively, but with little welding content.

As with injection molding, material is melted in a heated tube and forced out through a die which imparts the shape. Normally the material is cooled immediately to preserve the shape. The pressures used are much less than with injection molding, hence there are far fewer strains in the product.

(*c*) *Calendering*. In this process, hot flexible film, under little or no stress, is passed through cool rollers which can impart a pattern or a high gloss, bonding to a substrate such as paper or fabric.

(*d*) *Rotational molding*. Polymer is placed inside a hot thin-walled metal container which is rotated about all axes. As soon as a uniform section is achieved in the molten plastic, the container is cooled and the rigid shape removed. The process lends itself to short runs of large hollow articles. There is little stress in the molding.

(*e*) *Compression molding*. With stiff-flow thermoplastics, this method is used to form medium or thick sheets in a mold. Hot material is cooled at controlled rates and pressures to produce the sheet. Stresses are low and uniform, thus there is little risk of movement on heating during welding.

Table 2
Principal fabrication methods and the polymers used

Fabrication method	Polymer
Injection molding	All weldable thermoplastics
Extrusion—sections	All weldable thermoplastics
Extrusion—film	Mainly polyolefins, nylons
Calendering	Flexible PVCs
Rotational casting	Polyolefins
Compression molding	Rigid PVC and other stiff-flow materials

2. Welding Practices

2.1 Design

A number of factors have to be taken into account to obtain optimum results:

(a) application requirements (the most important factor) which include physical properties, service life and conditions (see Table 3), safety regulations (e.g., fire, toxicity, static charges), specifications, aesthetics and ergonomics;

(b) the shape of the sections;

(c) the welding method;

(d) the composition of the plastic;

(e) the strength required;

(f) the appearance of the finished article; and

(g) the possibility of postwelding operations.

Table 3
Service conditions of thermoplastic welds

Hazards	Normal conditions (short and long term)	Functional requirements
Static electrical charges	Mechanical stresses (creep, fatigue)	Length of service life
Chemicals (liquid/gases)	Temperature (maximum, average)	Dimensional tolerances required
Mechanical (vandalism)	Electrical (voltage, frequency, resistivity)	Toxicity
High thermal expansion	Chemicals (aqueous/organic)	Legal
Exposure to light	Weathering (UV degradation/change in color)	Transparency
Exposure to heat	Thermal (insulation effects)	Optical
Geographical		Appearance of surface (polished/matt, plain or decorated)
Abrasion		Noninflammability
		Bonding (to same or different materials)
		Disposal (biodegradation)

It may be necessary to change one or more of factors (b)–(d) to obtain the desired results. Here, prototyping can play an important part in evaluating designs and weld strengths before making a production commitment. It may not be necessary to make a whole assembly, instead just the component area where welding takes place, to obtain the data required. Specific designs are noted under the individual methods.

2.2 Preparation

As with metals, machining of thick sections is normal when using hot gas welding. In most other techniques, there is no necessity to undertake any preparation providing the surfaces are clean and free of lubricants (in particular, silicones). In some cases it may be advisable to scrape a mating surface to remove a small degree of oxidation which might reduce the strength of the weld.

3. Techniques

3.1 High-Frequency Heating

When an insulator is placed between electrodes it may become heated if the material has a degree of "lossiness" and the frequency is in the correct band. Most plastics are noted for their excellent electrical properties, but some will absorb high-frequency energy and become more "lossy" with an increase in temperature. This leads to poor welds, so the high-frequency type of welding is used mostly on thin flexible PVC film and foil. It has the useful property of becoming less lossy on heating, producing consistent welding. Welds can be made between layers of foil held between a plain or shaped electrode, and the table, which is earthed; or between rotating electrodes to give continuous seams. The entire apparatus must be suitably screened. The process is extensively used for the manufacture of rainwear, protective clothing, wallets, stationery folders and similar articles.

3.2 Electromagnetic Heating

When electromagnetic particles are placed in an inductive field, their temperature increases. If these particles are dispersed in a polymer, the whole of the composition becomes heated and the temperature will be conducted to the material in which it is in contact. If the temperature is high enough and the polymers compatible, then a weld will be formed on cooling.

There are a number of advantages in adding filler material to the weld:

(a) there are considerable gap-filling properties, as a result of which pressure-tight joints can be made, even when there are variations in the mating surfaces;

(b) welding takes place in all directions, with the application of light pressure;

(c) by preforming the filler to shape (e.g., by injection molding), even complicated joints can be assembled quickly and efficiently;

(d) the filler can be reheated after assembly to dissemble the joint (this can have many advantages such as replacing an expensive component if damage occurs in use—the stoppers of drums of dangerous materials can be sealed for transit and unsealed at the point of use); and

(e) suitably designed shear joints can exceed the strength of the sections being joined in fiber-filled materials where the strength is derived from the orientation of the fibers, and where other methods of welding do not maintain the fiber matrix (see Fig. 1).

There are many variables in this process and optimum results will only be achieved if consideration is given to all of them. The major factors are:

(a) most important, all aspects of the application requirements, both in manufacture and service;

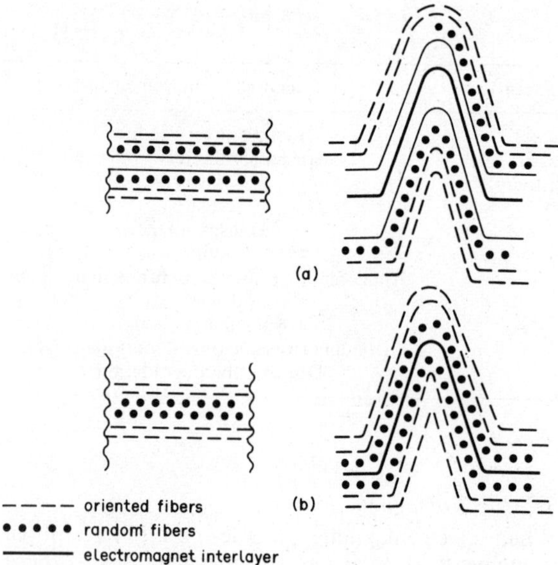

Figure 1
Schematic representation of how suitably designed shear joints can exceed the strength of the sections being joined: (a) before welding and (b) after welding

(b) component and joint design;

(c) the method of applying controlled pressure during both melting and cooling;

(d) types and grades of polymer being welded;

(e) type and grade of the matrix polymer, which need not be identical to the components being joined;

(f) thermal expansion of both the filler material and the parts of the component in the parts that melt; and the corresponding contractions on cooling;

(g) type, size and content of the electromagnetic filler, typically an iron oxide of micrometer size;

(h) design of the induction coil, which is usually water cooled to allow output powers of up to 5 kW;

(i) location of the coil relative to the work; and

(j) frequency of the generator (typically 1–10 MHz).

Care should be taken to ensure that no metals or other electromagnetic materials are in the induction fields, particularly the mechanism that applies pressure to the joint. The whole equipment must be screened to avoid radiation interference.

4. Mechanical Welding

Mechanical welding can be divided into frictional and ultrasonic vibration methods: in both cases, heat is generated at the welding interface by mechanical means.

4.1 Friction Welding

Heat is generated by mechanical energy with pressure between the mating surfaces. There are three main methods: rotational, orbital and linear.

(*a*) *Rotational*. This is very similar to the method used for metals and is the simplest of the three methods; it can be undertaken with an ordinary lathe, as was used in the initial stages of development. One component must be circular to cater for the rotary motion. One part is held stationary while the other is rotated against it under pressure to generate heat by friction. When the required melting has taken place, the stationary part is allowed to rotate with the other part, still under pressure, until sufficient rigidity has developed to release the pressure. If trimming of any spue is required to produce a tidier appearance, it can be done immediately before the work is taken from the lathe. Originally the operations were performed manually, and were dependent on the judgment of the turner as to when the second part should be allowed to rotate. With modern machines the whole operation is automatic.

Other simple machines, such as drill presses and milling machines, can be used instead of a lathe. One part is clamped to the machine table and the other is rotated on the end of a machine spindle through a clutch, which can be released instantaneously (see Fig. 2). The cleaning of any spue may not, however, be as easy as with a lathe.

One factor to bear in mind with this type of welding of solid sections, is that the periphery has a higher linear speed than the center and for equivalent pressures has greater heating. This can be minimized by doming the rod so that heat is generated on the axis before the circumference (see Fig. 3).

(*b*) *Orbital welding*. This method is used for irregular shapes that do not lend themselves to rotational

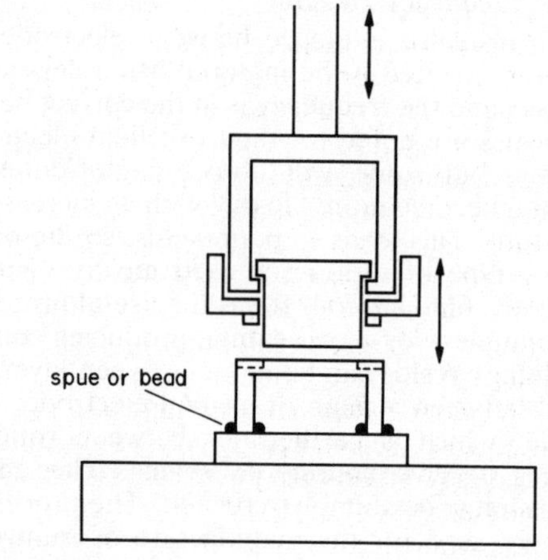

Figure 2
A drill press

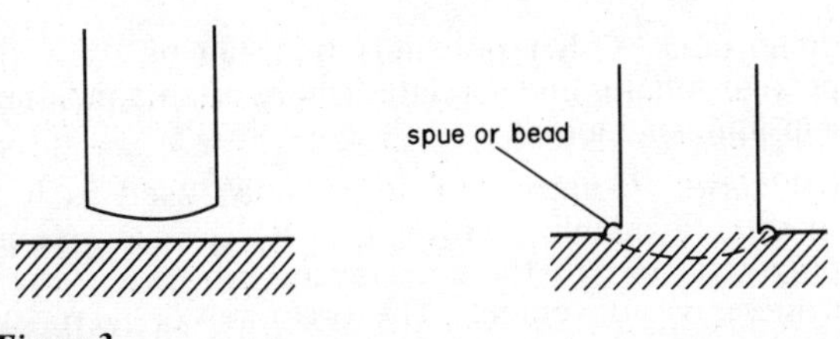

Figure 3
Doming the rod

welding. The two parts are moved relative to each other in small circles, of amplitudes typically 0.5–5 mm, again under pressure to generate the heat. The machines are designed to stop the motion so that the two parts are within register of 0.1 mm or less.

(*c*) *Linear welding*. This is similar to the two previous methods but the motion is linear. Normally the larger work piece is clamped to a table and the smaller part held by the vibrating part of the machine. Movement is achieved by electromagnets activating an element supported on leaf springs, to give motion in the plane of the weld only. When sufficient heat has been generated, the motion is stopped so that the two parts can be registered with reasonable accuracy. Pressure between the work pieces is carefully controlled throughout the whole cycle. The amplitude can be up to 5 mm and the frequency in the range of 100–240 Hz. Nearly all thermoplastics can be welded by this technique, including nylon and acetals that are not easy to bond by other means. Quite large items such as car bumpers can be handled.

4.2 Ultrasonic Welding

This technique has been established since the early 1960s and is applied mainly to rigid plastics, although small metal parts can be welded. The process is fast, clean, versatile, highly consistent, lends itself to full automation and can accept irregular joint lines. In some respects it is similar to friction welding, but with the energy generated by high-frequency (ultrasonic) compression vibrations; however, friction welds are also possible with the joint lines roughly parallel to the vibration direction. The frequencies used are in the range 20–50 kHz, with a probe in contact with the work piece to transmit the energy to the joint face. This should be as close to the horn as possible, but in some cases it is possible to transmit the energy to the joint through one work piece. The equipment consists of a high-frequency electrical generator feeding a transducer that converts the energy into mechanical movement (see Figs. 4 and 5).

Since the volume of material melted is smaller than other techniques, the cooling times are shorter. Times of welding are usually in the range 0.1–5 s

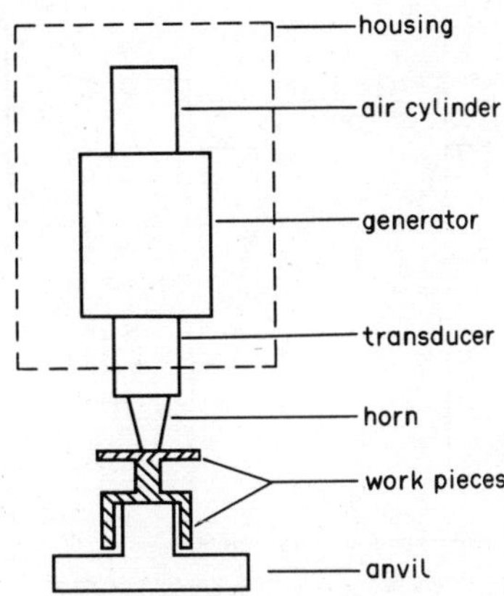

Figure 4
Ultrasonic welding machine

with the corresponding cooling times slightly longer. Light pressure is applied axially during the welding and increased during cooling. Bench machines are controlled automatically, and interlocks and guards prevent use unless two widely separated buttons are pressed simultaneously. As with other methods of welding, if appearance is important then traps are necessary to contain spue. One advantage not found in other methods is that welding virtually stops when the contact area becomes large. Thus the provision of a suitable shoulder will stop further movement, giving more consistent results. Some designs are shown in Fig. 6.

The machines are used for other techniques that join parts together but are not strictly welding, such as heat staking, threaded metal insert placing and bonding in blind holes, and the cutting of films and plastic-based fabrics.

5. *Heating by Conduction*

The oldest methods of welding plastics are those involving heating by conduction, some predating the 1940s being derived from metal-welding techniques.

5.1 Torch Welding

This method is very similar to the gas welding of metals. A V for thin and medium sheets, and a

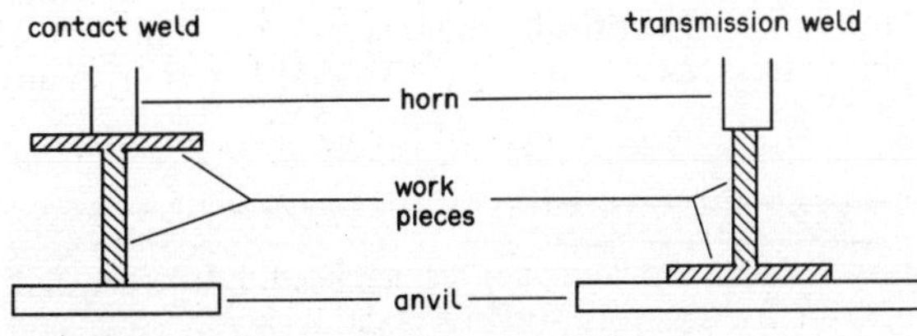

Figure 5
Ultrasonic contact and transmission welds

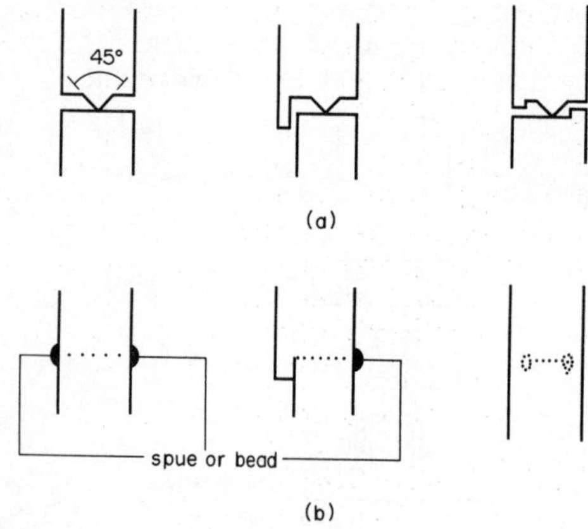

Figure 6
Designs showing traps and shoulder for an ultrasonic welding machine: (a) before welding and (b) after welding

double V for thick stock is formed on the mating surfaces of the parts to be welded, and their temperatures raised by a jet of hot gas. The tip of the welding rod is also held in the same area, so that all the surfaces soften or melt. When softening is observed, pressure is exerted on the cool part of the rod to force the softened surfaces into good contact to form a good bond. The weld is started at one end of a joint and advanced until the run is completed. The rod is given a twisting action to get as even a temperature as possible and to remove any contamination such as an oxidized skin, from both the rod and the work piece. As with metals, there is an optimum speed according to the temperature of the heating gases, the diameter of the rod, the character of the materials being joined and the skill of the operator. The method can be used for repairing articles, as well as making new products.

The similarity with metals is maintained with multiple runs, and in the case of double Vs, runs are made on alternate sides to minimize warpage (see Fig. 7).

(*a*) *Plastics*. The most widely used polymers are rigid PVC, acrylonitrile–butadiene–styrene copolymer (ABS), polyethylene and polypropylene. The end products are usually containers for water or aqueous chemicals, and are made as one-offs or in small numbers. Theoretically there is no limit to the size of the vessels, as sheets can be welded together to any dimensions. Other polymers that can be used are polycarbonates and acrylates (more often bonded by gap-filling cements).

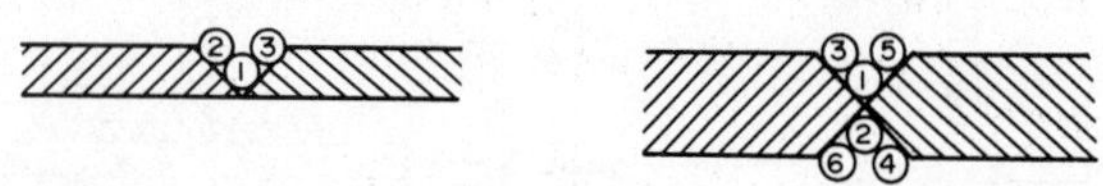

Figure 7
Single-V and double-V torch welding

(*b*) *Gases*. Compressed air is much used as it is readily available. However, it oxidizes some polymers and in these instances nitrogen from a cylinder is convenient. The pressures delivered to the torch range from 0.1 bar to 2 bar, depending on the torch aperture, the size of the work, the gas used and the polymer being welded. Heating of the gas is by a thermostatically controlled electric element in the torch.

(*c*) *Torches*. There are basically two types of torch. One is supplied with gas via a flexible tube. This allows a smaller and lighter torch with the choice of a wide range of gases; the disadvantage is that a trailing tube of some length may inhibit some sensitivity of movement. The other type is self-contained and has a built-in compressor and a thermostatically controlled heating element. This has greater portability but is bulkier and is restricted to using air as the heating gas. Each torch has an insulated handle through which the gases pass and are heated, before issuing from the nozzle. Usually the nozzle is bent through about 30–45° to direct the gas into the welding area but with a comfortable position to hold the torch. In some types of torch the filler rod is fed down an insulated guide, but this restricts the positioning of the rod for different kinds of weld.

5.2 Hot Contact Welding

There are six methods of hot contact welding: hot plate, for 1 mm and thicker sections; radiant heat; hot gas; hot knife, for film and thin stock; line sealers; and embedded metal.

(*a*) *Hot plate*. This is widely used to weld any sections that will withstand the low contact pressures necessary to counteract the shrinkage on cooling. It is applicable to most polymers, even to some dissimilar ones that have similar softening characteristics. The process can be used on moldings, extruded sections or combinations of both. The principle of the technique is to heat the mating surfaces by contact with a plate that is heated evenly and thermostatically controlled (normally by electric elements). The plate is coated with a nonstick material such as polytetrafluoroethylene (PTFE) to give a clean separation at the end of the heating period. When the heating of the plastic surfaces is sufficient, the plate is removed so that the two components can immediately be brought together to form the weld. Controlled pressure is required to form a good bond and to take up the shrinkage on cooling. There will be a bead of displaced material around the joint which, if visible, may need removal for cosmetic reasons. If it is left in place, welds can exceed the strength of the base sections owing to the greater sectional area at the weld. Contact times can range

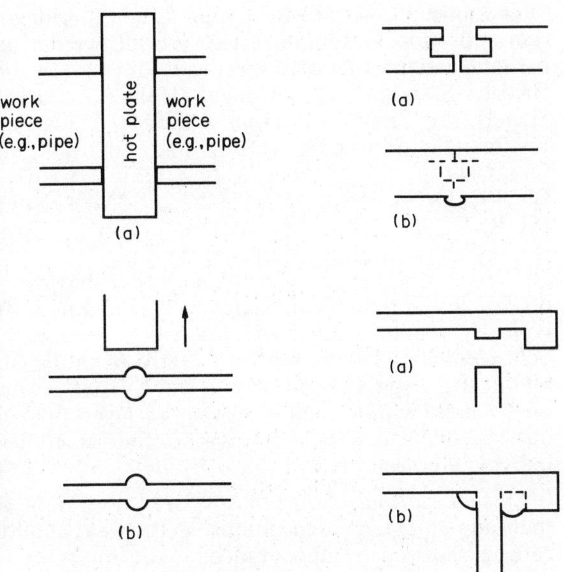

Figure 8
Variations of joint design in hot-plate welding: (a) before welding and (b) after welding

from as little as 2 s to more than 10 min. Cooling times should be somewhat longer than heating times because of the temperature gradients, and can be up to 30 min. The design of the joint depends on the compromise of appearance and strength; some variations are shown in Fig. 8.

The best reproducibility comes from fully automatic machines, but these are restricted to factory environments. However, good results can also be obtained in applications such as the laying of gas and water pipelines. These are made from extruded medium density polyethylene (MDPE), which has good rigidity as a circular section, reasonable toughness and very low permeability to gas and water, as well as good weldability.

(*b*) *Radiant heating*. Some polymers tend to stick to a hot plate in spite of nonstick coatings, and the heating has to be applied by radiation. Again plates are used, but made of suitable alloys to give constant radiation at the higher temperatures required, typically 500 °C or more. Owing to the lower rate of heat transfer, heating times are somewhat longer but cooling times will be the same as for the hot-plate method. Another disadvantage of this method is the greater risk of oxidation which gives weaker welds. Joint designs will be similar to those in the hot-plate process.

(*c*) *Hot gas*. This is similar to radiant heating, but the heat is derived from hot gases that impinge on the parts to be joined. It has the advantages of greater portability and, in some cases, greater heat control when joining surfaces of different contours (e.g., a thin-walled tube to a thick-walled pipe). The process has many of the features of torch welding.

(*d*) *Hot knife*. This is primarily a cutting operation by melting, but if two or more layers are in contact during the pass, a weld will be formed. Thus, two operations are carried out simultaneously. It is used extensively for sealing and cutting polyolefin bags made from lay-flat tubing and for small items of literature, such as mailshots and instructions with equipment, where the strength is not critical.

(*e*) *Line sealers*. This method is used also for sealing bags, but produces stronger welds and therefore is used more extensively than the hot-knife process. In many respects it is similar to hot-plate welding, but with the heat coming from the outside rather than the interface. Basically there are two straight heating elements, slightly longer than the lengths of seal required. One is above the film and the other below it, and each is covered by glass cloth which is made nonstick with PTFE or a similar material. The equipment is designed so that the two elements are adjacent to one another during welding, but can be opened to move or remove the film. An adjustable timer gives sufficient heating to melt completely through both layers of the film to make the weld, and then switches off to allow cooling and a good bond. Light pressure is applied during the heating, and slightly more during cooling. In the domestic models these pressures are applied manually, but factory models may have air or spring actuation with the whole cycle being automated.

(*f*) *Embedded metal*. The use of an electric element that can be heated and is on the face of one of the components to be welded, has been available for a number of years. It has not been widely used because the incorporation of the element adds to the expense, and possible variations in the quality of the bonds. The embedded metal may also be a disadvantage. Modern technology may be overcoming some of these objections, and there has been renewed interest.

An example is the fusing of the deck to the hull of the "Topper" sailing dinghy. In this case, the two parts were injection molded from polypropylene, and a tape comprising polypropylene fiber and heating element was placed in a groove around the periphery of the deck. The tape was spot welded at intervals to eliminate the risk of displacement during the bonding, and the hull was lowered in place for the welding operation.

The two main procedures of the embedded-metal technique depend on the method of heating the metal.

(a) The first, and possibly the oldest, has a loop of conductor (normally copper) which is heated by an inductive field (see Fig. 9). The frequency can be low, as distinct from electromagnetic welding detailed in Sect. 3.2. The metal need not be an insert in a molding but can be a separate component. The power used and the time of heating can be calculated, but a check by prototyping would be advised. Light pressure should be applied during heating to ensure a good even contact, and then increased for cooling.

(b) The second method depends on the heating of a resistive element. Originally the element was molded in with the connecting wires leading outside the molding. To make the weld, power was passed through the element for a time necessary to melt sufficient polymer on both faces. By using 12 V batteries through a timer, it was quite transportable for use on building sites for house plumbing. Extruded tubing was cut to length and connected by means of unions, tees and elbows that were injection molded with the elements molded in. Recently there has been renewed interest in this method with the extensive laying of gas and water pipelines, but the new feature is that the unions are formed by cut lengths of extruded tubing that incorporate a continuous spiral of heating element. The inside diameter of the sleeve so formed is the outside dimension of the pipe. Again, power to heat the element is easily transportable.

6. Testing

As with metals, testing can be destructive to establish the properties and the consistency of welds, or nondestructive to provide quality control. Again there are methods that parallel those used for metals, and others that are specific to plastics. The former are widely known and will be described only briefly, while those using the properties of plastics will be dealt with in more detail.

6.1 Destructive Testing

The methods used in destructive testing are the same as those for the evaluation of basic samples, such as test specimens or sections taken from moldings, extrusions, sheet or film. While absolute values are given, it is easier to quote the property as a percentage of the value of the parent material.

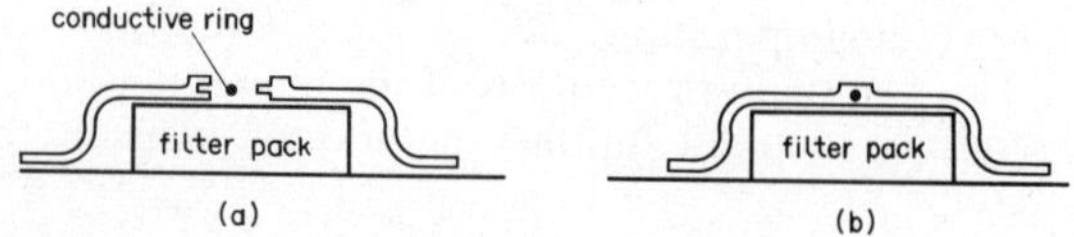

Figure 9
Encapsulation of a filter pack using inductive welding: (a) before welding and (b) after welding

The normal tests are for tensile strength, elongation and impact strength. These are all detailed in national and international specifications (e.g., by the British Standards Institution (BSI), American Society for Testing Materials (ASTM), Deutsches Institut für Normung (DIN) and International Organization for Standardization (ISO)) such as BS2782, DIN53455 and DIN53453, ASTMD638 and ISOR527. They should be used wherever possible, as they allow international comparisons.

There are tests specific to plastics that, however, do not allow numerical values to be obtained. A typical example is an examination of a section by polarized light. This shows the patterns of strains at, and in the neighborhood of, the weld. Brief details of the most widely used methods are given below, but special tests should always be considered for individual cases.

(*a*) *Visual (normal)*. Some plastics are transparent in their unpigmented compositions, so the weld quality can be assessed by appearance: if the interface is without flaws it should be of good quality. Special attention should be paid to the angles between surfaces as any notches are serious defects. Some form of impact test in comparison with a cemented joint (which should have a good fillet) will give good guidance.

(*b*) *Optical microscopy*. An examination under low powers, with and without polarized light, of sections in one or more directions and with thicknesses typically 0.05–0.5 mm, will give valuable information. With fiber-filled materials, it will show that there are seldom any fibers across the weld line, hence the strength will be only that of an unfilled composition. With polarized viewing of crystalline polymers, the size and distribution of the spherulites yields useful information.

(*c*) *Electron microscopy*. An examination, particularly of fractured surfaces, often indicates the mechanism of the failure, especially with fibrous fillers.

6.2 Nondestructive Testing

At the moment there are fewer nondestructive-testing methods applicable to plastics than there are to metals. There is the visual test in the weld area of transparent joints and the fillet area for all materials. In addition, there is the possibility of using fluorescent dye penetrants, although the solvent for the dye may attack the polymer and some plastic compositions may have some fluorescence.

(*a*) *Ultrasonics*. This technique detects lack of continuity in intended solid sections. In the past this testing has been confined to laboratories, but portable equipment is now becoming available. It is also possible (using computers) to store and correlate information gained on more than one axis and then to plot the regions and size of defects.

(*b*) *Radiography*. Low-kilovolt radiography now gives better contrast for both visual and photographic examination of welds. It is, however, a relatively expensive and a laboratory process.

(*c*) *Thermography*. Where there is a break in continuity, a drop in thermal conductivity in the area concerned can be shown. Again it is a laboratory method, and owing to the low conductivity of most compositions it is not highly sensitive.

See also: Friction Welding and Surfacing (Suppl. 2)

Bibliography

Anon 1986 Hot-plate welding of plastics. In: *Making it with Plastics*, Vol. 1(1) CMC, London, pp. 29, 31
Anon 1987a Ultrasonic welding. In: *Making it with Plastics*, Vol. 1(2) CMC, London, pp. 24, 28, 30
Anon 1987b Vibration welding. In: *Making it with Plastics*, Vol. 1(3), CMC, London, pp. 32–3, 35, 37
Anon 1988 Welding. In: *Making it with Plastics*, Vol. 2(2) CMC, London, pp. 15, 17
Baumeister M 1987 Effect of cooling on the strength of polyolefin weld seams (in Russian). *ZIS Mitt*. 29(7): 753–5
Bruehl B, Delpy U 1987 Weld line strength of glass-fiber reinforced injection moldings (in German). *Kunststoffe* 77(4): 339–42
Herrmann T 1987 Seam design in ultrasonic welding (in German). *Kunststoffe* 77(7): 673–9
Menges G, El Barbari N 1987 Material characteristics for ultrasonic welding (in German). *Kunstst.-Plast.* 34(9): 23–7
Potente H, Michel P, Ruthmann B 1987 An analysis of vibration welding (in German). *Kunststoffe* 77(7): 711–16
Watson M N Welding techniques for plastics. *Met. Mater.* 3(10): 581–5
Watson M N 1988 Ultrasonic welding of four thermoplastics. *Proc. Int. Conf. Welding and Adhesive Bonding of Plastics*. DVS, Aachen, FRG

J. Nightingale
[Cambridge Polymer Consultants, Royston, UK]

Wind Energy Systems

This article replaces the article of the same title in the Main Encyclopedia.

Humankind has been utilizing the power of the wind for many hundreds of years. There are early records and drawings of windmills dating from Persian times and throughout the Middle Ages wind was used to grind corn, to pump water and to enable worldwide travel by means of sailing ships. The gradual decline of wind energy during the industrial revolution was largely due to the availability of cheaper and more compact sources of power such as the steam engine. However, it is for exactly the same economic reasons that wind energy for power

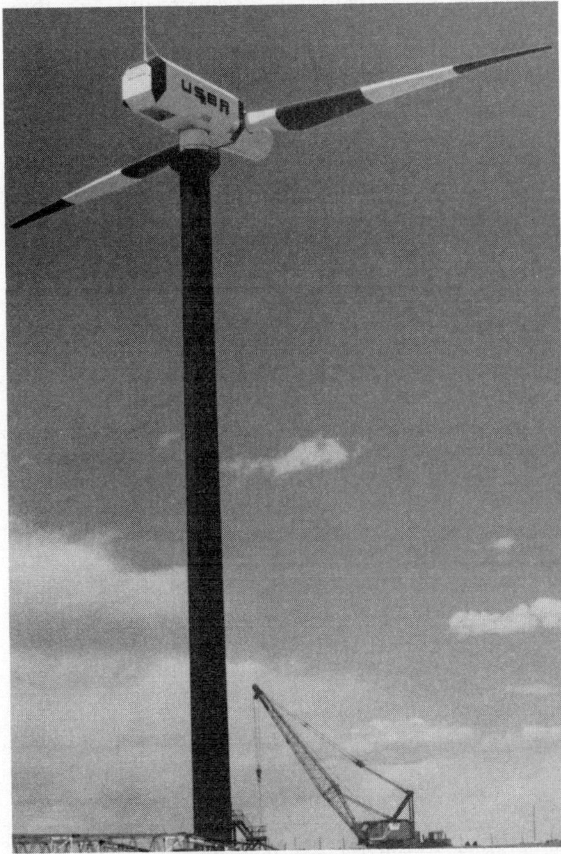

Figure 1
78 m diameter, 4 MW wind turbine built by Hamilton Standard for US Bureau of Reclamation at Medicine Bow, Wyoming, USA

generation has again become an attractive possibility, due to the way in which fossil fuel prices have gradually increased. There are already applications in which wind energy is cheaper and more convenient than any other source of power. Small wind turbines, for example, provide power for radio transmitters in remote regions, for cathodic protection on oil pipelines and for navigation buoys.

The current interest in wind energy for electricity generation can be traced back to the oil crisis of 1973–1974. The effect was to initiate government interest, principally in the USA, Sweden, the Netherlands, Denmark and the UK, in the development of machines for electricity generation. These programs generally focused on the development of large machines (50 m diameter and above, 1 MW power rating upwards). The most powerful wind turbine in the world (4 MW output, 78 m diameter), sited at Medicine Bow, Wyoming, is shown in Fig. 1. The reasoning for the development of machines of this size was that relatively few would be needed for generating large quantities of electricity and thus their environmental impact would be minimized. However, machines of this type have not generally established a good record for economy and reliability and current interest for power generation is focused on machines in the medium size range (20–35 m diameter, 200–400 kW output).

A further stimulus for the development of wind energy was provided in 1978 in the USA, principally in California, where the availability of tax credits, coupled with attractive purchase rates for electricity, led to substantial investments of private capital in "wind farms." Between 1981 and 1986 around 15 000 machines were constructed with a total capacity of around 1300 MW. Most of these machines were fairly small (50–100 kW, 15–20 m diameter) but in the late 1980s there was a trend towards larger unit sizes. The intense competition that resulted from the introduction of tax credits led to a very rapid expansion of the wind turbine industry and subsequently to a gradual improvement in engineering standards.

Although the tax credits have been withdrawn, it is still possible to take advantage of high purchase prices for electricity in California and the fall in prices for wind energy installations means that it is still an attractive proposition commercially. Wind farm construction has also taken place in Denmark and the Netherlands, as shown in Fig. 2, and there are plans for three installations in the UK.

The very early windmills were constructed primarily of wood and it is interesting to note that the technology of construction has come full circle, with many of the more recent machine blades being manufactured from wood. The material has an attractive combination of strength and lightness which is essential for harnessing a diffuse power source such as wind.

1. Principles of Operation

The basic principles of operation have remained unchanged. In essence, aerodynamic lift is used to propel a moving blade through the air and this principle is used by the vast majority of rotors having one, two, three or multiple blades, with the rotor axis horizontal or vertical. Rotors with few blades operate at tip speeds around 50–80 m s^{-1}, multiblade rotors operate at slower speeds and tend to be used for high-torque applications such as water pumping.

A few rotors make use of aerodynamic drag; these are typified by the vertical axis S-shape rotors, frequently used for advertising purposes. These require more blading and are less efficient, as well as there being relatively few applications.

The constant fluctuations in the wind mean that all rotors are subjected to loads that vary continually; hence, it is essential to make accurate assessments of the fatigue loads and to ensure that the material strength is adequate. Since the wind varies continually, so does the power and thus wind turbines are not suitable for applications where the power must be constant. However, they can be linked to battery charging circuits for remote applications, or they can be used to charge a heat store or to pump water, as in these cases the irregular nature of the supply is not crucial. It is also perfectly feasible to link wind turbines into power distribution networks, provided there are other generators available. Under these circumstances the power fluctuations become absorbed within the ever present fluctuations in a

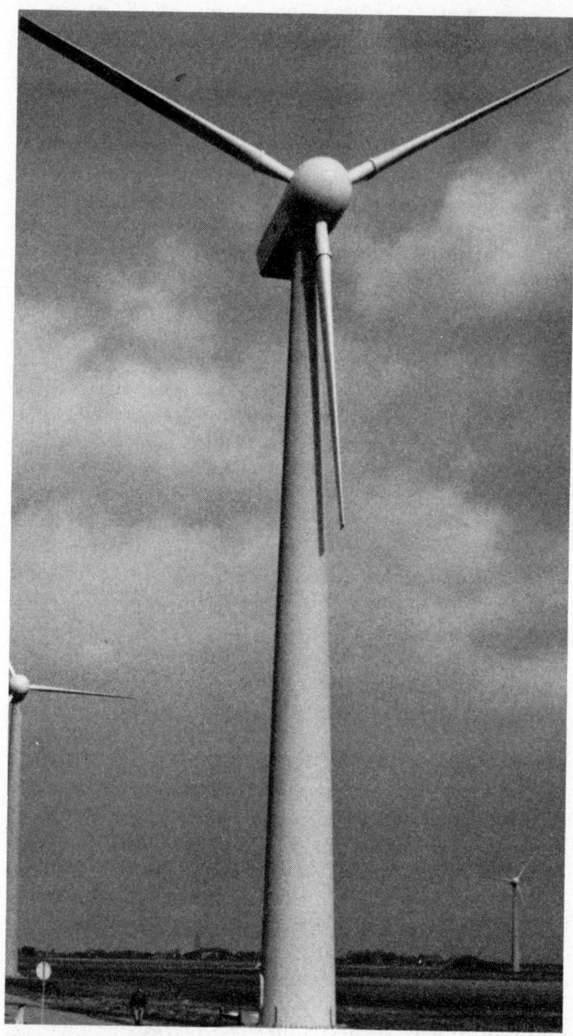

Figure 2
Wind farm comprising 18 300 kW wind turbines at Sexbierum, Netherlands

network and, again, are not of crucial importance, provided the wind turbines only supply a small proportion of the demand.

By far the majority of all electricity generating wind turbines run at constant speed and are held at this speed by the frequency of the power supply system. Synchronous generators are probably the most popular choice, but these suffer from a lack of damping which has to be supplied elsewhere in the transmission, otherwise power fluctuations and blade loads may be unacceptable. If induction generators are used damping is inherently provided, but these suffer from the disadvantage of drawing reactive power from the system and they also draw high starting currents which may cause local flicker. Suitable circuitry has to be designed to minimize these effects. Although a small amount of additional energy can be derived by operating at variable speed—the machine stays at the top of its efficiency curve—there are disadvantages in the need to resort to complex electrical or mechanical systems for coping with the variable frequency electricity. Nevertheless, there are other advantages of variable speed (or multiple-speed) operation, particularly the reduction in noise at low windspeeds. This has important implications for siting of wind turbines since care must be taken to minimize the effect on the environment.

Figure 3
28 m diameter, 300 kW wind turbine by James Howden Ltd at CEGB Wind Energy Demonstration Centre, Carmarthen Bay, UK

1.1 Power Limitation

The power in the wind varies as the cube of the wind speed. Consequently, the power available at any site varies from zero to very high levels, sometimes in a relatively short space of time. All machines therefore need to have some method of power limitation to cope with what would otherwise be excessive power generation on the relatively infrequent occasions when strong winds blow. Rating philosophy varies widely between manufacturers but most machines are designed to operate with a peak output in the range 250–500 W m^{-2} of swept area. This means, in practice, that 20 m diameter machines have ratings around 200 kW, 30 m diameter machines around 300 kW and 50 m diameter machines around 1 MW. Power limitation is accomplished in one of two ways. The most common method is to pivot all or part of the wind turbine blade about its axis so that the aerodynamic lift is reduced in high winds, thus limiting the power. This reduction of lift can be achieved either by feathering (i.e., unloading the blade by rotating it to eliminate lift) or by pushing it into stall (i.e., a situation where lift decreases sharply while drag increases). Most large machines limit power by feathering, but quite a number of small designs turn the blade into stall and there are some advantages to this technique. The second method takes advantage of the natural tendency of blades to stall as the aerofoil angle of attack increases in high winds to utilize a "fixed-pitch" rotors design. In this case there is usually some slight loss of energy, since the blade setting angles for fixed-pitch operation are governed by the need to limit power and may not result in optimum utilization of the available energy at low wind speeds. However, if fixed-pitch control is used there needs to be some auxiliary control to restrain the rotor in the event of disconnection from the power source. A rotor with "partial-span" pitch control is shown in Fig. 3.

2. Materials for Wind Turbines

Wind turbine rotor blades provide the most stimulating challenge to the designer of a wind turbine. This challenge was met in the early designs by building rotors with steel spars and glass-reinforced plastic fairings to provide the aerofoil sections. Other materials that have been investigated include aluminum, glass-reinforced plastic (to provide structural strength, not just the fairings) and, more recently, wood. Glass-reinforced plastic is by far the most popular construction material and the use of steel has declined, primarily due to the desire to reduce weight. The weight of the rotor is reflected in the weight of the transmission and the support tower and it is therefore essential to keep blade weights to the absolute minimum. This is perhaps more important than for any other engineering structure since the gross weight of a wind turbine has a strong influence on its overall cost. (Energy generation costs comprise around 60–80% capital repayment charges with the remainder being accounted for by operation and maintenance costs.) It follows that the use of lightweight, but costly, materials such as carbon-fiber-reinforced plastic has not been widespread.

A further cost factor that encourages the drive towards light blades and rotors is the tendency for blade weights to increase as the cube of the size, whereas energy yields increase as the square. This is one reason why minimum costs are found for machines around 30 m in diameter. Although there is some prospect of an upward drift in the size range of the most economic machines, this will only be achieved with materials where cost per unit weight can be reduced for very large structures. This is unlikely to be achieved soon since the very large rotor blades require specialist manufacturing techniques, workshops, transport facilities and erection procedures.

The blades of a wind turbine encounter loads from a number of sources. Roughly in order of descending severity these are (Milborrow 1982):

(a) fatigue induced by turbulent wind fluctuations in the "flapwise" direction (i.e., normal to the rotor plane),

(b) extreme gust loads when the rotor is parked,

(c) fatigue due to self weight in the "edgewise" direction (i.e., in the plane of the rotor),

(d) cyclic loads when blades operate in yaw (i.e., with the horizontal axis not parallel to the wind direction) and gyroscopic forces when the rotor head is yawed, and

(e) centrifugal tensile stresses.

There are two methods of alleviating loads and stresses in the flapwise direction, caused by the wind and its fluctuating component. The first method is coning of the rotor; i.e., the blades are permanently angled backwards downwind so that a centrifugal bending stress offsets the wind loading. The second method is "teetering" in which the blades are free to move about a central pivot point and this reduces the loads induced by wind turbulence. Teetering can naturally only be used for two-blade rotors: the teetered part of the blades move so that their axes are no longer normal to the rotor axis. Three-blade rotors are always rigidly mounted to the hub. A number of one-bladed rotors have been built, chiefly in the FRG. The advantage of these is that blade construction costs are kept to the absolute minimum, but this is offset by the need for a counter weight of some sort, together with a teetering mechanism at the rotor hub. Few power-generating machines are built with more than three blades but multibladed rotors are still built for pumping purposes where high torque is of primary importance.

The design process for wind turbine blades is somewhat complex (Madsen and Frandsen 1984), but it is possible to simplify the relevant procedures and gain an insight into the design drivers and hence the material requirements. By considering each of the blade loading mechanisms in turn, it is possible to show how material properties control rotor weights and hence assess the relative blade set weights using different materials, leading to a comparison of overall costs. A simplified analysis for horizontal axis wind turbines (Milborrow 1986) has confirmed that the fatigue properties and strength-to-weight ratios have a critical bearing on blade-set weights and enables the following conclusions to be drawn.

(a) Only large (80 m diameter and above) steel and aluminum rotors are dominated by gravitational bending stresses. In this instance, weight then increases as the fifth power of the diameter.

(b) With most other blade materials the design drivers are the flapwise bending or gust loads (in the case of wood it is the latter that tend to be the dominant factor).

(c) The use of glass-reinforced plastic or wood enables weight to be reduced by a factor of approximately two compared with steel. The use of a lighter high-strength steel structure would achieve the same object, but at considerably greater cost.

2.1 Vertical Axis Wind Turbines

The operating principles of vertical axis wind turbines demand different blade design analysis techniques. Although the cyclic loads due to reverse gravity bending are absent, there are additional cyclic aerodynamic forces due to the (roughly sinusoidal) variation in aerofoil angle of attack as the blade rotates. However, these machines do not need any mechanisms to align the rotor with the wind direction and there are no gyroscopic forces.

There are two basic types of vertical axis rotor. The Darrieus type, with curved blades (see Fig. 4), are designed so that there are no bending forces in the blades due to centrifugal forces. However, this means that it is more difficult to provide control surfaces for power limitation and for control when not connected to a load. The straight-bladed concept (see Fig. 5) overcomes the problems of control by pivoting the blades, so that they are inclined at an angle to their normal (vertical) attitude.

The need for light, strong and cheap materials is equally important when designing vertical axis rotors, since the length of blading required is greater than for horizontal axis rotors. Again, early rotors were made of steel with glass-reinforced plastic fairings though, more recently, there has been widespread use of glass-reinforced plastic for the whole blade. The drive to reduce weight has led to the use of aluminum for some rotors and there has been less interest in the use of wood, due to problems with manufacture.

Although the design process can be simplified to enable generalized conclusions about blade weights to be drawn, the actual procedures used are naturally complex and, to a certain extent, design specific. The nature of turbulent winds—which vary considerably between sites—means that the specification of rotor blades must be subject to uncertainty and it is necessary to build in some conservation to ensure adequate life. Synthesis of the turbulent fluctuations is possible, however, and it is then possible to use dynamic and aeroelastic simulations to predict rotor loads. Analytical estimates are, however, frequently cross-checked against datasets from operational machines to determine a suitable fatigue spectrum (Goodman 1985).

Since the fatigue spectrum covers a very broad level of stress ranges and mean levels, there is considerable discussion as to the most relevant method of using the data to estimate fatigue damage (Finger 1985). The traditional approach is to use Miners' rule but its validity is disputed and there are nonlinear variants with, it is claimed, better accuracy. A related question is whether the sequence of load applications influences fatigue. Answers to these questions are being sought through materials research programs both in Europe and in the USA.

Figure 4
Darrieus wind turbine

3. *Economics and Future Developments*

The cheapest machine prices in 1988 were around US\$800 kW^{-1} for wind turbines around 20–30 m diameter with output in the range 150–300 kW. Site costs and other installation charges need to be

Figure 5
25 m diameter, 130 kW vertical axis wind turbine at CEGB Wind Energy Demonstration Centre, Carmarthen Bay, UK, funded by the UK Department of Energy

added, taking installed prices to around US\$1200 kW^{-1}. Operation and maintenance costs are about 0.5–1.0 US¢ kW^{-1} and generation costs for the windiest sites lie in the range 4–7 US¢ kW^{-1}, depending on test discount rate and machine lifetime. At this level wind is economically attractive in certain locations, particularly when competing with oil-fired plants. Generation costs are higher for both smaller and larger machines but the former can supply electricity more cheaply than photovoltaics in most remote locations and are able to compete with diesel generation in many environments. In practice, wind diesel systems have been installed worldwide in large numbers to provide reliable power supplies where costs had previously been unacceptable.

Having matured rapidly during the 1980s wind energy technology is entering a period of consolidation. The market for small machines in remote applications is secure and likely to expand with further price reductions. For electricity generation in the developed nations, wind is economically attractive in the windiest regions but is not regarded as fully proven, since few machines began generating before the mid-1980s. In Europe there is the additional problem of finding sites for the large numbers of machines needed to generate significant quantities of energy. Nevertheless, the continuing pressure on conventional fuel supplies has led to the setting of targets for wind capacity in many locations and further development of windfarms is envisaged, for example, in Denmark, the Netherlands, Italy and the UK.

The development of reliable and economic machines is important but the acquisition of suitable sites for large-scale development may be a restraint, particularly in Europe. It is for this reason that the prospects for offshore wind energy have been studied in Denmark, the Netherlands, Sweden, the UK and the USA. Although windspeeds are higher the extra energy yield does not offset the increased costs. The resource is very large; for example, around the UK the energy available matches total electricity consumption. Plans for a 750 kW prototype have been announced in the UK and detailed feasibility studies have been carried out in Denmark. The prospects for offshore wind may increase if there is a continued upward drift in the size of the most economical machines which is at present around 200–300 kW. Costs of the large machines are somewhat uncertain as they have not been produced in quantity but the rapidly expanding use of wood as a blade material is tending to reduce weights and costs. The emergence of other light but strong materials at the right price would improve the prospects for both onshore and offshore wind.

Bibliography

De Gourieres D 1982 *Wind Power Plants—Theory and Design*. Pergamon, Oxford

Faddoul J R 1981 An overview of large horizontal axis wind turbine blades. *Proc. 5th Wind Energy Conf.* Solar Energy Research Institute, Golden, CO

Finger R W 1985 Methods for fatigue analysis of wind turbines. *Windpower '85 Conf.*, report No. CP-217-2902. Solar Energy Research Institute, Golden, CO

Goodman F R 1985 Fatigue-life assessment methods and application to the model WTS-4 wind turbine, report No. AP-4319. Electrical Power Research Institute, Stanford, CA

Lindley D 1988 The commercialisation of wind energy. *Euroforum New Energies Conf.* HS Stephens. Bedford, UK

Madsen P H, Frandsen S 1984 Wind—Induced failure of wind turbines. *Eng. Struct.* 6: 281–6

Milborrow D J 1982 Performance, blade loads and size limits for horizontal axis wind turbines. *Proc. 4th BWEA Conf.* British Hydromechanics Research Association, Cranfield, UK

Milborrow D J 1986 Towards lighter wind turbines. *Proc. 8th BWEA Conf.* Institution of Mechanical Engineers, London

Monroe R H 1987 Design and development of large wood/epoxy turbine blades. *Windpower '87 Conf.*, report No. CP-217-3315. Solar Energy Research Institute, Golden, CO

D. J. Milborrow and P. L. Surman
[Central Electricity Generating Board, London, UK]

Wood: Acoustic Emission and Acousto-Ultrasonic Characteristics

Acoustic emission (AE) and acousto-ultrasonics (AU) are useful techniques for determining the integrity of wood and wood-based materials. AE arises from material under stress and can be sensed with piezoelectric transducers, usually coupled to the surface of the material. Materials containing flaws or weak areas generate AE at lower stress levels than those of greater integrity. AU combines the sensitivity of AE transducers with ultrasonic transmitters to evaluate the change in energy content of a signal as it passes through the material. In general, a decrease in energy content signifies less material integrity. AE has been used with wood-based materials since the 1960s; AU, which has been in formal existence since the late 1970s, has only recently come into use. The application of AE and AU to wood-based materials has been substantially aided by developments in assessing the integrity of fiber-reinforced plastics (FRPs), which have many characteristics similar to those of wood. Like wood, FRPs require special techniques to detect internal flaws, in contrast to metals where x-ray techniques are usually sufficient.

Wood-based materials have AE characteristics similar to those of fiber-reinforced composites. Some indications of weaker (or weakened) material or structure are emissions at low stress levels, and increased numbers and/or rates of events. Another indicator of weakened material is a low Felicity ratio, which is determined from a sequence of load–unload cycles during which AE is monitored. In some materials, no AE occurs during reloading until the previous stress is reached, giving a load ratio (Felicity ratio) of unity: this is referred to as the Kaiser effect. However, many materials, such as composites, have a Felicity ratio of less than unity, not obeying the Kaiser effect.

AU differs from AE in that an active pulser is used to inject stress waves that are received by conventional AE sensors, and the resulting waveform is analyzed to determine the change in signal content from pulser to sensor. AU differs from conventional ultrasonic techniques in that more subtle flaws, such as poor-quality bonding, can be detected. In most applications, the pulsing element is also a conventional AE sensor that is energized with a narrow, high-voltage pulse, but other means of surface stimulation can be used, including mechanical impact, laser bursts and electrical discharges.

1. Acoustic Properties

Typical ultrasonic velocities for lumber and veneer are 1 km s^{-1} across the grain and 5 km s^{-1} along the grain. Reconstituted materials such as composite panels have velocities similar to solid wood across the grain. Wood-based materials are about an order of magnitude greater in attenuation than geological materials and two orders of magnitude greater than metals. Attenuation in wood is much greater across than along the grain. Because of these differences, AE sensor positioning is critical, particularly for solid wood where the grain direction is well defined.

Since attenuation increases exponentially with frequency, the usable upper frequency level for sensors on wood-based materials is about 100–200 kHz. In this frequency range, attenuation along and across the grain is about 30 dB m^{-1} and 200 dB m^{-1}, respectively. The effects of wood density and moisture content on attenuation have not been clearly determined, but appear to be of lower order than that of grain angle. Obviously, the more attenuating the material, the greater the limit on sensitivity. However, the more attenuating materials have an advantage in damping out external unwanted signals, permitting the use of higher gains to obtain the needed sensitivity. Attenuation also results from geometric spreading of the signal as it travels through the material; for planar materials, attenuation is proportional to the distance from the source, and for bulk materials, to the square of the distance. For wood-based materials, the geometric spreading may be anisotropic because of the grain direction effect.

2. Coupling of Transducers to Wood-Based Materials

Acoustic impedance (the product of velocity and density) is the determining parameter for acoustic coupling of one material to another. For wood along the grain, acoustic impedance is similar to that of metals; across the grain it is comparable to plastics and water. Coupling presents the greatest source of variability and the major impediment to on-line implementation of AE or AU in processing wood-based materials. For AE laboratory studies, coupling is generally effected using either a couplant (grease type) or by bonding the sensor to the material. Special problems in coupling to wood-based materials include porosity and roughness of the surface. Because of this, grease-type couplants are very difficult to use without them being forced into the material; this changes the pressure at the face of the sensor. Bonded coupling with contact or hot-melt adhesives can be used to overcome such problems. Bonded coupling will usually give about a 3 dB increase in efficiency over couplants because of transfer of the shear-wave component. Bonding must be used cautiously since differential expansion of the bonded interfaces can cause failure within the bond or generate stresses that might cause extraneous AE. For many AU applications, the surface must be scanned, which can be done through dry coupling using rigid or elastomeric materials, where coupling pressure must be high enough to "squeeze" out air gaps and maintain consistent local pressure on the material. For both AE and AU, most on-line uses involve materials that are moving, generally requiring a dry coupling system and adding considerable complexity in sensor–coupling design. Dry coupling introduces a loss of about 20–30 dB (at 70 kPa), which can be reduced to about 10 dB with increased pressure (300 kPa). Alternatively, but with some risk with hygroscopic materials, coupling can be achieved by using a water mist at the contact area.

3. Defect Location

Location of the origin of AE emissions can be accomplished by using multiple sensors and measuring time of arrival of the same event. With two sensors, the definition of location is limited to a plane perpendicular to the line between the sensors. The addition of more sensors permits more precise definition of the origin, assuming that the material is reasonably homogeneous and the event rate is not too high to distinguish between events. For solid wood, the relationship of velocity to grain orientation and the high, anisotropic attenuation require special precautions in sensor location and interpretation of time intervals. Some of the more nearly isotropic wood-based composites, such as certain types of fiberboard and particleboard, permit good location resolution. In contrast, plywood presents special problems from both the grain orientation of the plies and the boundaries between the plies. Oriented strandboard should behave intermediately between particleboard and plywood.

4. Applications

Although there are a limited number of examples of direct application of AE and AU in quality assessment (QA) and quality control (QC), a substantial research effort is underway that will provide the background for development of further industrial systems for nondestructive evaluation and testing. The reported research in AE and AU for wood-based materials can be placed in four major categories: fracture and/or fracture mechanics, drying (as monitored by formation of checks), decay and machining. Only the first category lacks current QC and/or QA efforts. Although about 50% of the reported research has been in fracture and/or fracture mechanics, a major area of emphasis since 1980 has been in wood drying. The most recent research has been in laboratory assessment of wood decay using both AE and AU.

4.1 Fracture and Fracture Mechanics

The first work on wood in which piezoelectric sensors were used was in the mid-1960s for general AE behavior in standard mechanical tests and crack propagation. Local failures from slow extension of intrinsic flaws produced emissions at levels as low as 5–10% of ultimate strength, with irregular "run-and-stop" flaw growth. Subsequently, it was found that longitudinal tension was characterized by early flaw growth at 5–20% of ultimate stress, with a linear increase in events to near failure. In contrast, AE from transverse tension showed a log–log response vs deformation, but for both types of tensile loading, total events to failure averaged about 300 events per cubic centimeter of specimen. Longitudinal compression produced only about 0.1% of the "AE density" (events/volume) of tensile tests. In flexure, cumulative events increased linearly with deflection to the proportional limit, continuing beyond it in a linear manner but at a much lower slope to near failure. In creep under flexural loading at 80% ultimate stress, AE accumulated stepwise from apparent redistribution of stress concentrations during irregular flaw growth. Mode I cleavage tests produced a uniform distribution of AE over the crack length, with about 8 events per millimeter and 24 events per millimeter for soaked and dry specimens, respectively.

In specimens that contain defects, the rate of emissions increases near the proportional limit. Higher emission levels of these specimens occur without corresponding changes in modulus of rupture. The critical load in fracture toughness correlates highly with cumulative AE counts. The initiation of microscopic compression lines has been postulated as the cause of early events at 40–50% ultimate load in static bending. Early AE has been used as a screening test to eliminate flawed specimens from studies.

Solid wood under tensile stress parallel to the grain produces AE in sequential slow-rate and rapid-rate periods. The slow-rate is attributed to extension of preexisting microscopic flaws in the cell walls, which has been confirmed microscopically. The cause of the rapid AE rate appears to be brittle failure of tracheids since this rate is proportional to crack velocity. AE output varies with wood species, but has not been related to gross physical or anatomical properties. Also, when the AE rate is high, the observed shearing failure, which reduces crack velocity, is low. AE has been used in static bending to distinguish the effect of compression failures in typhoon-damaged wood that cause a lower modulus of rupture (MOR) but unchanged modulus of elasticity (MOE), from the effect of knots in which both MOE and MOR are lowered. Assuming a linear relationship of total counts to the square of the load-at-failure, the slopes are quite different between clear wood and wood containing either knots or compression failures, providing a much clearer separation of the contributing source than if considering only MOR. Tension fatigue has been observed to cause an exponential increase of AE with cycles; the contribution of AE from loading and unloading, however, was not eliminated.

Clear specimens stressed in flexure generate comparably little AE until near the ultimate stress, in contrast to high activity in material containing defects. The initiation and propagation of failures can be detected by using location techniques with two or more sensors. The initiation of flaws in both laboratory and full-size specimens can be correlated with the location of natural or artificial defects.

4.2 Composites

Wood-based composite materials produce AE at substantially lower stress levels than occurs with solid wood. For example, in static bending tests, composite materials begin emissions at about 10–20% of the ultimate stress, whereas clear solid wood begins at about 40–50%. The smaller the constituent particle in the board, the higher the stress level of initial emissions. In AE monitoring during internal bond (IB) tests on composites, a clear relationship was found between one-half of the total events and ultimate stress. In particleboard with controlled resin levels, total events to failure correlated with resin level. Also, AE data correlated much better with IB values than did specimen density. Some limited data indicate a high correlation of AE with AU measured by transmission. The Felicity ratio of particleboard appears to be reasonably constant (at about 0.9) up to failure.

AE has been measured from internal microfailures during the thickness swelling of particleboard at high relative humidity. The active period of AE coincides with irreversible swelling (springback). Wood-fiber hardboard with controlled pretreatments tested in compression normal to the face produced cumulative AE that correlated well with thickness swelling from boil-swell tests, indicating that cumulative AE may be a good nondestructive predictor of dimensional stability. When the same type of materials were subjected to cyclic water-soak exposure, stress-wave factor (SWF) values decreased with increasing numbers of cycles, corresponding to increasing damage. However, undamaged boards with higher SWF had lower thickness swelling, suggesting that SWF could predict the degree of dimensional stability.

4.3 Machining

AE characteristics have been determined for cutting operations that approximate veneer peeling. The primary source of AE was found to be changes in plastic deformation in the shear zone at the tip of the cutting tool. The AE count rate, while not well correlated with cutting forces, was more sensitive to the cutting process than the rms signal. The AE output was considered potentially more important to monitor tool wear than cutting forces. This work has been extended to preliminary tests on monitoring AE from circular saw cutting, using several types of coupling attachments to the blade.

4.4 Drying and Drying Control

One of the more promising applications of AE is for monitoring and controlling the drying process of lumber (see *Drying of Wood*). AE has been used to sense the development of surface checks in both hardwoods and softwoods. The actual fracture of the surface has been directly observed simultaneously with AE emissions. By controlling the environmental conditions to prevent high rates of emissions, it is possible to dry faster with less degradation. Several investigators have demonstrated the feasibility of controlling checking during hardwood drying by using a fixed level of AE to control drying conditions. Hardwoods generally check much more readily than softwoods and consequently have AE rates about an order of magnitude greater. However, among hardwoods, there does not appear to be a good correlation between propensity to check and AE rate. AE has been shown to respond rapidly to changing surface RH conditions, independent of whether end grain or side grain is exposed. Geometry can have a large effect on AE generation, depending on the direction of moisture movement and the degree of stress development. There is some evidence that AE during drying also originates from fracture of water capillaries, analogous to AE from water stress in plants.

4.5 Biological Degradation

Several mechanical tests, including static bending and transverse compression, have shown that AE activity increases substantially in decayed wood (see *Wood: Decay During Use*). This effect also occurs within the incipient decay range, which is very difficult to detect with any other analytical technique. A substantial number of variables (including wood and fungus species) and testing configurations must be understood if the technique is to move from laboratory to field testing. Preliminary tests using AU show some promise of detection of decayed wood by nondestructive testing. Marine-borer damage has been simulated by drilled holes in wooden piling and under loading generated AE in proportion to the degree of damage. Coupling through water in which the piling was immersed was found to be more efficient that direct surface contact to the piling.

4.6 Adhesives

The characteristics of AE for adhesively bonded areas have not been well defined. There is some data that indicates lower AE levels for brittle than for flexible adhesives, even for failure at the same ultimate stress. In plywood IB testing, it was found that wood failure could be differentiated from adhesive failure by the shape of the AE vs strain curves, which were linear for adhesive and curvilinear for wood failure, respectively. Some preliminary plywood work has been done in static bending to understand the effect of voids or poorly bonded areas on AE with the objective of assessing full-size-panel integrity.

Several laboratory studies have been made on the character of AE from fingerjointed structural lumber in static bending. Although the strength of clear wood having fully cured bonds could be predicted with reasonable error at 50% and 80% of ultimate load, the presence of defects in the wood and/or fingerjoints increased the error in prediction. However, location techniques have shown that even in fingerjointed clear wood, the emissions occur predominantly from the area of the joint. In contrast, clear wood without fingerjoints has a fairly uniform distribution of emission along the length. AE has also been used in an analysis of failure modes of different furniture-joint combinations. The curing of adhesives with wood substrates has been monitored using AU. Cure time has been quantified for several types of adhesives by calculating the half-time between initial and final transmission values.

See also: Acoustic Emission; Acoustic Properties of Wood

Bibliography

Ansell M P 1982 Acoustic emission from softwoods in tension. *Wood Sci. Technol.* 16: 35–58

Beall F C 1985 Relationship of acoustic emission to internal bond strength of wood-based composite panel materials. *J. Acoust. Emission* 4(1): 19–29

Beall F C 1986 Effect of moisture conditioning on acoustic emission from particleboard. *J. Acoust. Emission* 5(2): 71–6

Beall F C 1987 Acousto-ultrasonic monitoring of glueline curing. *Wood Fiber Sci.* 19(2): 204–14

Beall F C, Wilcox W W 1987 Relationship of acoustic emission during radial compression to mass loss from decay. *For. Prod. J.* 37(4): 38–42

Beattie A G 1983 Acoustic emission, principles and instrumentation. *J. Acoust. Emission* 2(1/2): 95–128

Becker H F 1982 Acoustic emissions during wood drying. *Holz Roh- Werkst.* 40: 345–50

DeBaise G R, Porter A W, Pentoney R E 1966 Morphology and mechanics of wood fracture. *Mater. Res. Stand.* 6(10): 493–9

Dedhia D D, Wood W E 1980 Acoustic emission analysis of Douglas-fir finger joints. *Mater. Eval.* (11): 28–32

dos Reis H L M, McFarland D M 1986 On the acousto-ultrasonic characterization of wood fiber hardboard. *J. Acoust. Emission* 5(2): 67–70

Hamstad M A 1986 A review: Acoustic emission, a tool for composite-materials studies. *Exp. Mech.* 26(1): 7–13

Lemaster R L, Klamecki B E, Dornfeld D A 1982 Analysis of acoustic emission in slow speed wood cutting. *Wood Sci.* 15(2): 150–60

Niemz P, Wagner M, Theis K 1983 State and possible applications of acoustic emission analysis in wood research. *Holztechnologie* 24(2): 91–5

Noguchi M, Kitayama S, Satoyoshi K, Umetsu J 1987 Feedback control for drying Zelkova serrata using in-process acoustic emission monitoring. *For. Prod. J.* 37(1): 28–34

Sato K, Kamei N, Fushitani M, Noguchi M 1984 Discussion of tensile fracture of wood using acoustic emissions. A statistical analysis of the relationships between the characteristics of AE and fracture stress. *J. Jpn. Wood Res. Soc.* 30(8): 653–9

Vary A, Lark R F 1979 Correlation of fiber composite tensile strength with the ultrasonic stress wave factor. *J. Test. Eval.* 7(4): 185–91

F. C. Beall
[University of California, Berkeley, California, USA]

Wood: Structure, Stiffness and Strength

Wood exhibits unique structural features on virtually every dimensional scale—from molecular to tree sizes (see *Wood Constituents: Physical Nature and Structural Function*; *Wood: Chemical Composition*; *Wood: Macroscopic Anatomy*; *Wood Ultrastructure*). Glucose units are strung together to form molecular chains that are either amorphously tangled or assembled into crystalline regions; phenylpropane units are randomly linked at various locations to form an encrusting, cross-linked binding substance. Microfibrillar strands, composed of macromolecules, are helically wound in ribbon-like clusters to form layers of cell walls. The cell wall itself is composed of characteristic laminations: the primary wall (P), the secondary wall (S) and the middle lamella (M). Growth rings are distinguishable by changes in various cell geometries and the ring itself is packed with cell types arranged in unique patterns specific to the tree species. Within the cross section of the tree, zones of sapwood and heartwood can be distinguished. Reaction wood and juvenile wood zones are defined by the abnormal characteristics of the cell. Knots and the associated grain deviation patterns result from the attachment of branches to the stem of the tree. The shape of the stem itself is tapered and roughly cylindrically symmetrical. What is the reason for this compounded complexity? How do these features, cascading down through the dimensional scales, affect the mechanical properties?

The quest of the materials scientist is to relate structural features to physical and mechanical properties. Once the structure–property relations are known for man-made materials, processing can be altered to change structure, thereby obtaining the desired properties. The situation regarding structure–property relations for a biological material is different. The production process for manufacturing the material has evolved in conjunction with the evolution of the organism producing the material. The "design" of the material is intimately related to the requirements of the organism for survival. The design procedure is adaptive. Discovery of the structure–property relations for biological materials can involve finding reasons why the structure exists in relation to the function it performs. While technologists may not be able to change the essential "production process" (genetic manipulation is possible but this changes only the parameters and not the overall blueprint), they can at least utilize the material properly, exploiting its strengths and avoiding or compensating for its weaknesses.

In effect, the tree—the "manufacturer" of the wood—might be considered a factory for producing the material of which the factory is composed. The factory survives if the material performs its function under environmental duress. In addition to other functions (e.g., solar energy collection, water transport, sugar production), the tree must survive mechanical loading. It is from this perspective that scientists can understand the relationships between structure, stiffness and strength. This perspective can also carry over to the other functions (e.g., the structure of wood in relation to water transport) and a good discussion of form related to function in a tree can be found in Zimmermann and Brown (1975). A more general treatment of form versus function in biological systems is found in Thompson (1942).

1. Models

To study the relationships between structure and mechanical properties of wood, mathematical models are employed. Some are quite simple; others more complex. They all extract idealizations of the features of concern and deduce relationships that might be difficult to obtain by other means. The models are virtually indispensable tools for gaining insight into the reasons for the existence of the complex structure found in wood. On the other hand, the models are only as good as the assumptions made to employ them. Consequently, if experimental data cannot be obtained to validate model results, there is always the possibility of error.

The general purpose of most models is to establish the relationship between the properties of a particular dimensional level and the properties at another scale. In all cases, "smearing of the microstructure" occurs. In other words, for some characteristic region the material within is assumed to be a continuum. The continuum is characterized by a single parameter, or set of parameters, that remain constant within the region or vary smoothly in a specified manner. Given values for these parameters (assumed or deduced in some way) and a well defined geometry, an analysis can be performed that will yield either (a) an effective value for a large scale parameter, such as an effective Young's modulus, or (b) a distribution of some variable, a stress component for example, over some larger region. Strength can be inferred from the maximum stress and the location of failure can be identified.

In the following sections the influence of structure on stiffness and strength of wood will be discussed for three scales of dimension: the tree level (macro), the cell aggregate level (meso) and the cell wall level (micro). The distinction in scales is arbitrary but is in fact useful. For solid wood products (lumber, plywood, particleboard), the strength and stiffness of relatively large units is important. These depend on properties at the cell aggregate level, which in turn are linked to the cell wall level. Thus, through a chain of structure–property relationships the properties of common wood products can be related to the ultrastructural features of the woody substance.

2. Macroscale

Banks (1973) considered the design of the tree from an engineering point of view. Among other examples illustrating performance efficiency, the taper of the stem is shown to be compatible with the idea of economical use of material. The tree, a spruce in this case, is considered to be a cantilevered beam built into the ground. The crown is assumed to have the shape of a triangle which is responsible for transverse loading caused by wind. Wind direction, being a random variable, is considered to be the reason

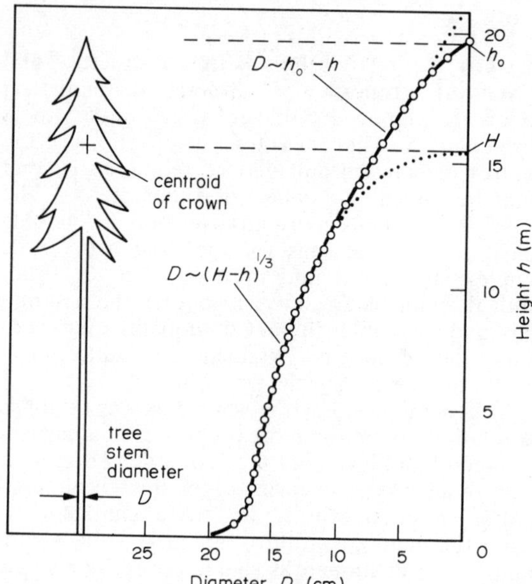

Figure 1
Taper in a 76-year-old spruce (adapted from Banks 1973)

why the tree trunk has a rotationally symmetrical moment of inertia about the axis of the stem. Assuming the tree to be composed of a homogeneous, elastic material (a continuum hypothesis) which has a unique breaking strength in bending, then the taper of the trunk (diameter versus tree height) can be deduced from another assumption: the load-bearing capacity of the stem is the same at any height level. The stem is divided into two regions: the lower and longest (ground to beginning of crown) and the upper (beginning of crown to top). The moment distribution in the lower segment is computed from a point load acting at the centroid of the crown and in the upper region it is computed from a ramp load (no load at the tip). It is found that the diameter will be proportional to the cube root of the difference between centroid height and vertical location in the lower region and will vary linearly with height in the upper zone (see Fig. 1). Some measurements of tree taper agree with this analysis.

Schniewind (1962) reviewed the older literature and found that the mechanical theory of stem form based on equal resistance to bending moments along the height dates back to 1874. A number of authors have refined the theory to take into account the nonhomogeneous nature of wood in the stem such as the change in elastic modulus with height. In fact, based on empirical evidence that wood strength in bending is exponentially related to wood density, Schniewind provides a justification for horizontal density gradients as found in many species. If a given amount of wood is added to the tree stem each year, it is more efficient from the point of view of the tree to strengthen each increment by "densifying" it, than to spread the mass out in a homogeneous manner. Moreover, the density response is apparently linked to crown development as radial density gradients clearly follow silvicultural treatments, that is, thinning, fertilizer application and tree spacing (Megraw 1985).

While it is tempting to attribute a rationale to all patterns of organization and all features found in trees, it could also be that a structure or phenomenon is not optimal but simply does not appreciably reduce the survival capability of the tree. Growth stresses may be an example. In an earlier paper Archer and Byrnes (1974) developed a model to account for the axial tension in the outer radial zones of the stem and the axial compression in the center. Extreme forms of residual stresses can cause splitting in logs when the trees are harvested. Treating the stem as a cylindrically orthotropic pole and using the "Boyd hypothesis" that each new growth increment is accompanied with a fixed growth strain state. Archer and Byrnes show that a residual stress pattern can be found which matches experimental data. The novel feature introduced in the analysis is the use of growth strains to produce the growth stresses. The growth strains are analogous to thermal or hygroelastic strains and are apparently the result of lignification of the walls in developing cells (Boyd 1973). Radial distributions of stresses, σ, in this type of material, under an axisymmetric assumption, are found to be a function of the radial coordinate r, raised to a power involving a material constant k, that is, $\sigma = Cr^{\pm(k\pm1)}$, where $k = (E_T/E_R)^{1/2}$ where E_T and E_R are the Young's moduli in the tangential and radial directions, respectively, and C is a constant. Thus, the radial position where the axial stress distribution changes from compression to tension is a function of the elastic constants. For a more detailed discussion of growth stresses see Archer (1987a).

In some species, spiral grain exists in which the wood fibers are not aligned with the axis of the stem but wind around the axis in a helical arrangement. The topic of spiral grain has been treated at length by Noskowiak (1963). Archer (1979) has modified his theory to account for this inclined principal direction of the material and has found that growth stresses are sensitive to the spiral grain angle. In addition, torsional shear stresses are also very sensitive to this parameter.

The crown of the tree is composed of branches. McMahon and Kronauer (1976) have considered the design of the branches from a mechanical perspective. They were conceived to be rectangular in cross section with width b and depth h obeying power-law tapering with branch length position s, that is, $b = b_0 s^{-\alpha}$ and $h = h_0 s^{-\beta}$. The exponents α and β are parameters whose values are predicted by two alternative design strategies. One is based on the concept of equal breaking resistance and the other on elastic similarity. In the latter case, the branch deflection under its own weight has the same relative shape for any segment. The value of β which, theoretically, produces this condition is 3/2. In contrast, the equal strength criterion yields $\beta=2$ on theoretical grounds. A plot of measured length versus diameter of oak branches asymptotically approaches an empirical relationship in which $\beta=1.50\pm0.13$. In addition, experiments involving the natural frequencies of shaking trees lead to an average value of $\beta=1.50$.

The connection between branch and trunk has not received much attention. Shigo (1985) has studied the grain configuration in the vicinity of this connection. He has found, for example, that cell orientation in the upper part of the junction is perpendicular to both stem axis and branch axis. A collar typically develops and is built up annually. The cellular arrangement appears to be related to sap and water transport as well as injury protection but must be compatible with supporting cantilevered loads acting on the branches. Lumber and veneer, whose cut surfaces sever these grain patterns, suffer significant strength losses resulting from the point-to-point change of principal material directions. Stress grades of lumber are essentially determined by the strength-reducing aspects of knots and cross grain (see *Lumber: Behavior Under Load*).

An attempt has been made by Phillips et al. (1981) to characterize the grain pattern around the knot as analogous to flow around a circular object. Using this grain pattern as a model for the material property organization of an orthotropic anisotropic elastic solid, Cramer and Goodman (1983) attempted to predict the strength of individual pieces of lumber. A finite-element approach combined with the principles of linear fracture mechanics was employed in the analysis.

3. Mesoscale

Using balsa wood to help develop a representative aggregate wood cell model, Easterling et al. (1982) have developed a model to predict the crushing strength and elastic moduli in the three principal material directions for wood. The model essentially consists of tesselated hexagonal tubes packed in stories with embedded layers of ray cells (see Fig. 2). Wood cells are represented as hexagonal tubes capped at the ends with prismatic caps. The tube walls are treated as continuous material with a density of $\rho_s=1.5\ \mathrm{g\,cm^{-3}}$. The material is also considered as a transversely orthotropic material with respect to stiffness and strength. For example, the elastic modulus in the axial direction of the tubes is

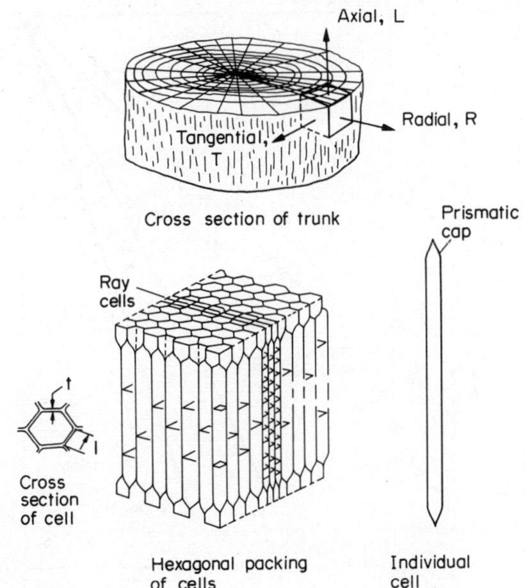

Figure 2
Cell aggregate model (adapted from Easterling et al. 1982)

considered to be 35 GPa and in the transverse direction 10 GPa.

Due to the variation of gross wood density, ρ, with cell wall dimensions (i.e., ρ/ρ_s is proportional to the ratio of cell wall thickness to hexagonal plate length, t/l) and to the stiffness of the tube structure being proportional to $(t/l)^3$, obtained from a structural analysis, the elastic modulus in the tangential direction is found to be related to the cube of gross density by $E_T=E_S\ (\rho/\rho_s)^3$.

Based on similar arrangements involving geometry, but also including ray cells, the elastic modulus in the radial direction is found to be about twice the modulus in the tangential direction. The axial stiffness of the complex was shown to be related linearly to density. A plot of normalized stiffness against normalized density reveals that the trends in the data are consistent with the theory (see Fig. 3).

Easterling et al. (1982) have also considered the crushing strength of wood using theoretical arguments based on geometry and assuming continuous cell wall material. In this case, radial and tangential crushing strengths are proportional to the square of the cell wall thickness to hexagonal wall length ratio $(t/l)^2$. The axial crushing strength is linearly related to this ratio. Again, theoretical and experimental results were consistent. Thus, structural arrangements of the units that make up wood are seen to influence the properties of the gross wood.

While the typical arrangement of tracheids in softwoods is not a perfectly hexagonal honeycomb

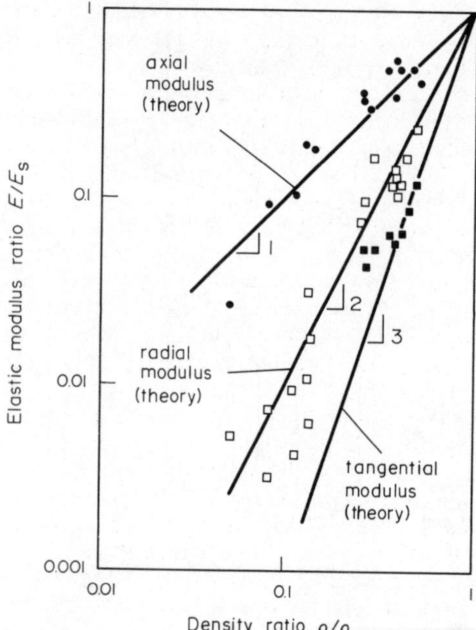

Figure 3
Normalized stiffness versus normalized density in the three orthogonal directions (adapted from Easterling et al. 1982)

structure, Gillis (1972) has noted that there are a very large number of junctures involving the intersection of three double cell walls. He was led, therefore, to develop a "triple-point element." In essence, the element is considered to be a portion of a rigid frame structure with one double wall parallel to the tangential direction and the other two inclined away from the tangential direction. The thickness of the inclined walls are the same but different from the tangentially oriented wall. All walls are considered to be composed of an isotropic material.

A strength of materials approach was taken to compute direct and bending strains resulting from forces acting through the three ends of the element. Equivalent stresses and strains were deduced from the response of the element and consequently effective stiffness moduli were determined for wood. On the basis of this analysis, Gillis was able to predict the ordering of the anisotropic elastic moduli for wood. The ordering, for the most part, was the same as that found experimentally for the few species in which all the elastic constants have been measured. Hence, the ordering of the gross anisotropic elastic moduli for wood can be deduced by using a microstructural feature which is commonly found in aggregates of cells.

Other models on the mesoscale have involved accounting for differences in earlywood and latewood by introduction of a layered system. The density of latewood may be three or more times greater than that of the earlywood in some species, which implies similar differences in mechanical properties (see *Wood Strength*). Similar layered models have also been used to model the reinforcing effect of rays in the radial direction of wood. In one case the two types of layering have been incorporated into a single model (Schniewind 1959).

4. *Microscale*

Mark (1967) has reviewed the early literature on cell wall models prior to development of his own models. Starting from the molecular level, Mark used a particular cellulose crystal structure to compute elastic modulus values for a fictitious continuum material, the framework, that deforms in a manner similar to the crystal structure. The stiffnesses of various chemical bonds (e.g., carbon–carbon, carbon–oxygen and hydrogen bonding), combined with geometrical information concerning the crystal, were used to compute equivalent elastic constants. In the direction parallel to the cellulose chains, the elastic modulus E_{FL} was found to be a little less than a measured value of 134 GPa from the literature and the experimental value was therefore adopted. Other elastic constants for the framework were calculated from the crystal structure while the constants for the lignin matrix were based on experimental values from the literature. For complete sets of elastic constants based on more recent methods of calculation see *Wood Constituents: Physical Nature and Structural Function*.

Mark then treated each cell wall layer as a filament-wound composite of framework and matrix, calculating elastic properties of each layer according to its composition. He then proceeded to construct a model for a single fiber, making use of the geometry and proportions of the helically wound cell wall layers, that is, M + P, S1, S2 and S3 (see Fig. 4). The purpose of this model was to compute internal stresses in the cell wall in order to find regions where the fiber is likely to break. His results, including a large shear stress in the S1 layer, supported experimental observations that fibers appear to rupture in the S1 layer when tested in tension.

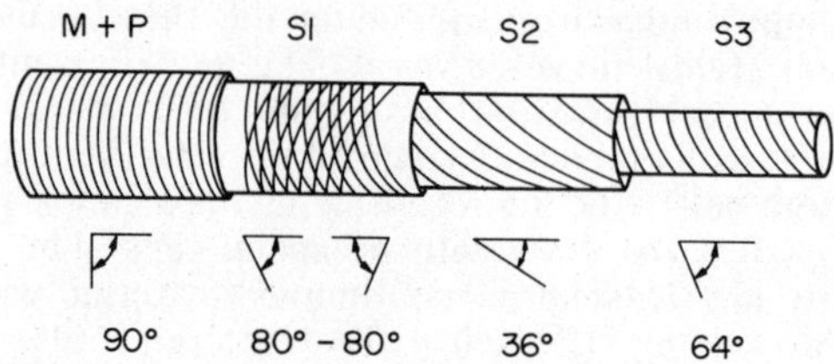

Figure 4
Microfibril angles of the layers of the radial cell wall (adapted from Mark 1967)

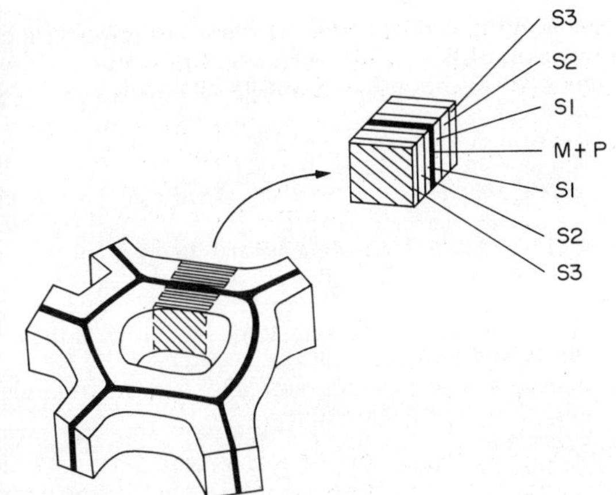

Figure 5
Rectangular flat segment of the cell wall with major layers (adapted from Schniewind and Barrett 1969)

Following Mark, there have been a series of publications dealing with the stiffness and strength of the cell wall. Cave (1968, 1969) predicted and measured the elastic modulus of the fiber in the long-axis direction as a function of microfibril angle. The microfibril wrappings were assumed to be distributed about a mean angle and the distribution function was considered to be Gaussian. In the model the concept of a balanced laminate was introduced. The adjacent walls of neighboring fibers having the same angle, but opposite inclination, prevents cell rotation when pulled axially. Only the S2 layer was considered since it is the thickest layer and tends to dominate cell behavior.

Quite independently, but similar to Cave's approach, Schniewind and Barrett (1969) developed a model for the cell wall based on the idea of a balanced laminate. The wood cell is considered to be a thin-walled tube subjected to a force along the cell axis. The wall can be broken down into many approximately flat, infinitesimal rectangular segments (see Fig. 5). Each segment is layered with orthotropic materials having appropriately inclined principal directions to represent the S1, S2, S3 and M + P layers. A shear restraint was imposed on each layer. This approach differed from Mark. The inclusion of all layers differed from Cave. The overall stiffness of the cell, when compared to Mark's work, was increased. In addition, stresses were computed in the various layers of the cell wall and it was found that shear stresses in the S1 layer were small. Compressive stresses were higher, however, and it was postulated that failure in the S1 layer might be due to elastic instability of the microfibrils.

Schniewind (1970) has followed up with more work on internal stresses in various hypothetical cell types using the shear restraint assumption and Mark and Gillis (1970) have modified Mark's earlier work to include the ideas of shear restraint as well as improvements to account for different behavior in radial and tangential wall segments.

A three-dimensional model of a wood fiber was next developed by Tang (1972). The fiber was modelled as a laminated cylinder with each layer (S1, S2, S3 and M + P) possessing cylindrical anisotropy. A stress function approach was used to calculate stresses in the layers. Rather than using shear restraint, or variations of strain restriction themes, Tang chose to match normal and shear stresses across the layer interfaces. With this boundary condition, he found large tangential stresses in the S3 layer. In more detailed analysis, Tang and Hsu (1973) have explored the influence of the constituents of the layers themselves on the internal stress distributions in the wood fiber under tension. A large number of combinations of variables related to this microstructure were explored theoretically and it was concluded that the spacings between the microfibrils were very important with respect to the elastic properties of the cell wall.

Barrett and Schniewind (1973) next used a finite-element approach to solve for stresses in the cell walls of wood. A concentric, multilayered, orthotropic cylinder model was used, with each layer of the model being represented by a group of annular elements with appropriately transformed elastic compliance tensors. A generalized plane strain condition was imposed which allowed for the overall stiffness of the fiber to be computed. The finite-element method naturally allows displacement boundary conditions, and consequently a complete shear restraint condition was imposed by setting tangential displacements at the outer surface to zero. Alternatively, an unattached, single fiber was considered by removing the constraint on outer nodal displacements. Their stress distributions differed from those of Tang (1972) substantially, a result not to be unexpected due to the differences in boundary conditions. The results, however, indicated that the two-dimensional approach utilized by the authors previously was quite adequate. The three-dimensional finite-element analysis did allow a computation of radial stresses, which can not be done with a two-dimensional analysis, but these stresses were small and considered to be inconsequential. Shear restraint, the natural condition of a fiber surrounded by its neighbors, favored the mechanical efficiency of the fiber.

Models used to explore the linkage between structure and mechanical properties of wood at any dimensional scale ultimately reflect on macroscopic mechanical behavior. Archer (1987b) has turned his attention to the cell and has used the idea of growth strain to model stress evolution in the wall of the developing cell. In this model, wood substance is laid down on the inside of a hollow cylinder, the fiber, rather than on the outside surface of a cylinder used to represent the trunk of the growing tree. The computed residual stresses based on this model indicate that the mechanical efficiency of the cell as it is developing appears to be optimized when the microfibril angle of the S2 layer is between 26° and 42°. Macroscopic properties are thus related to microstructural arrangements and a microscopic analysis reveals a rationale for macroscopic survivability.

See also: Wood: Deformation Under Load; Wood Strength

Bibliography

Archer R R 1979 On the distribution of tree growth stresses, Part III: The case of inclined grain. *Wood Sci. Technol.* 13: 67–78

Archer R R 1987a *Growth Stresses and Strains in Trees*. Springer, New York

Archer R R 1987b On the origin of growth stresses in trees, Part 1: Micromechanics of the developing cambial cell wall. *Wood Sci. Technol.* 21: 139–54

Archer R R 1987b On the origin of growth stresses in trees, Part I: Micromechanics of the developing cambial cell wall. *Wood Sci. Technol.* 21: 139–54

Banks C C 1973 The strength of trees. *J. Inst. Wood Sci.* 6(2): 44–50

Barrett J D, Schniewind A P 1973 Three-dimensional finite-element models of cylindrical wood fibers. *Wood Fiber* 5(3): 215–25

Boyd J P 1973 Tree growth stresses, Part V: Evidence of the origin in differentiation and lignification. *Wood Sci. Technol.* 7: 92–111

Cave I D 1968 The anisotropic elasticity of the plant cell wall. *Wood Sci. Technol.* 2(4): 268–78

Cave I D 1969 The longitudinal Young's modulus of Pinus radiata. *Wood Sci. Technol.* 3(1): 40–8

Cramer S M, Goodman J R 1983 Model for stress analysis and strength prediction of lumber. *Wood Fiber* 15(4): 338–49

Easterling K E, Harrysson R, Gibson L J, Ashby F R S 1982 On the mechanics of balsa and other woods. *Proc. R. Soc. Lond., Ser. A* 383: 31–41

Gillis P P 1972 Orthotropic elastic constants of wood. *Wood Sci. Technol.* 6: 138–56

McMahon T A and Kronauer R E 1976 Tree structures: Deducing the principle of mechanical design. *J. Theor. Biol.* 59: 443–66

Mark R E 1967 *Cell Wall Mechanics of Tracheids*. Yale University Press, New Haven, CT

Mark R E, Gillis P P 1970 New models of cell-wall mechanics. *Wood Fiber* 2: 79–95

Megraw R A 1985 *Wood Quality Factors in Loblolly Pine*. TAPPI Press, Atlanta, GA

Noskowiak A F 1963 Spiral grain in trees, a review. *For. Prod. J.* 13: 266–75

Phillips G E, Bodig J, Goodman J R 1981 Flow-grain analogy. *Wood Sci.* 14(2): 55–64

Schniewind A P 1959 Transverse anisotropy of wood: A function of gross anatomic structure. *For. Prod. J.* 9(10): 350–9

Schniewind A P 1962 Horizontal specific gravity variation in tree stems in relation to their support function. *For. Sci.* 8(2): 111–18

Schniewind A P 1970 Elastic behavior of the wood fiber. In: Jayne B A (ed.) 1970 *Theory and Design of Wood and Fiber Composites*. University of Washington Press, Seattle, WA

Schniewind A P, Barrett J D 1969 Cell wall model with complete shear restraint. *Wood Fiber* 1: 205–14

Shigo A L 1985 How tree branches are attached to trunks. *Can. J. Bot.* 63(8): 1391–401

Tang R C 1972 Three-dimensional analysis of elastic behavior of wood fiber. *Wood Fiber* 3(4): 210–19

Tang R C, Hsu N N 1973 Analysis of the relationship between microstructure and elastic properties of the cell wall. *Wood Fiber* 5(2): 139–51

Thompson D 1942 *On Growth and Form*. Cambridge University Press, Cambridge

Zimmermann M H, Brown C L 1975 *Trees: Structure and Function*. Springer, New York

J. A. Johnson
[University of Washington, Seattle, Washington, USA]

Wood: Surface Chemistry

Chemical interactions at the wood surface are of great importance for wood utilization, in such areas as gluing and painting. Weathering of wood is essentially a surface phenomenon. Much of the value of wood in furniture and panelling relates to the appearance of the wood surface. This article deals with the chemical nature of wood surfaces and the chemistry of changes taking place in them.

1. Nature of Wood Surfaces

The concept of a wood surface is difficult to define in a way which is both sufficiently precise and at the same time practical. Definition of the surface as a two-dimensional area is impractical since chemical surface interactions take place between three-dimensional atoms and molecules; definition of the surface as a layer of a certain specified thickness, such as length of the anhydroglucose repeat unit of cellulosic macromolecules, is not entirely satisfactory since various chemical and physical surface interactions involve surface layers of widely varying thickness. A more practical definition would be that of a surface layer of thickness involved in producing a certain specific surface effect. This definition is relative, however, since the thickness of such a layer would vary with the type of interaction in question.

The variation of the depth involved in various physical and chemical interactions represents an important consideration in studying the surface chemistry of wood as it makes it commonly impossible to arrive at the same results using efficient analytical methods.

Wood is an open-porous, composite, cellular biopolymer. Consequently, the surface of wood can be either external (i.e., artificially created and comprising the interfaces between wood and its surroundings) or internal. The internal surfaces can be either permanent and comprising the interfaces between cell walls and cell lumens, or transient, opening in response to penetration of polar liquids into cell walls and comprising the interfaces between cell-wall material and such liquids.

The area of the external wood surface is generally much smaller than the areas of the permanent and particularly of the transient internal wood surfaces. The areas of the permanent and transient internal wood surfaces have been estimated by Stamm and Millett (1941). According to their data a cube of *Pinus lambertiana* Dougl. (sugar pine) wood with an edge of 1 cm and an external surface area of 6 cm^2 (density 0.36 g cm^{-3}) will have a permanent internal surface area of about 0.11 m^2 and a transient internal surface area of about 170 m^2.

1.1 Chemical Nature of Internal Wood Surfaces

The internal wood surface is composed of the surfaces of cell lumens, including the surfaces within pit openings, and of the surfaces of transient openings in cell walls. Very little is known about the chemical makeup of the transient surfaces; most likely they are predominantly composed of amorphous carbohydrates.

We have, however, more information on the chemical composition of the tertiary cell-wall layer, or warty layer, that for all practical purposes is identical with that of the tracheid surfaces in conifers. Various scanning electron microscopy studies indicate that this layer is a combination of lignin and amorphous carbohydrates encrusted to a varying extent with extractives. The nature of the encrusting extractives is not very well known. In some cases the extractives appeared to represent sodium-hydroxide-soluble carbohydrates; in other cases they seemed to be at least partly nonpolar in nature. Most likely the nature of the extractives varies with the species.

1.2 Chemical Nature of External Wood Surfaces

Wood is composed of cellulose (linear polyanhydroglucose, 41–52% in softwoods, 37–52% in hardwoods), hemicelluloses (predominantly branched polyanhydromonosaccharides, 14–29% in softwoods, 24–39% in hardwoods) and lignin (cross-linked polymer of *p*-n-propylphenol-related units, 27–37% in softwoods, 18–32% in hardwoods), all three constituting the cellular structure of wood (see *Wood: Chemical Composition*). Wood also contains varying amounts of materials of chemically diverse nature deposited in cell lumens and cell-wall interstices (extractives, extraneous or associated materials), as well as small amounts of inorganics (generally below 1.0%). Thus the chemical composition of external wood surfaces will be influenced mainly by the factors determining the amounts, percentages and composition of the above five groups of materials in the total wood sample. These would include such factors as the taxonomic status of the wood species, conditions of tree growth, age of the tree, proportion of normal wood to reaction wood, location of the wood sample within the tree (e.g., extractive-rich heartwood vs sapwood), proportion of springwood to summerwood and density of wood.

A number of additional factors dealing specifically with the surface can substantially modify the direct relationship between the chemical composition of bulk wood and that of surface layers.

2. Conditions and Methods of Wood-Surface Formation

Cell walls of wood are constructed of cellulose microfibrils embedded in hemicelluloses and lignin. The softening points of hemicelluloses and lignin are 50–60 °C and 90–100 °C, respectively. Because of the lower mechanical strength of hemicelluloses and lignin, particularly under temperature conditions above their softening points, the formation of the surface could take place preferentially in parts of the cell wall containing more of these materials (i.e., the compound middle lamella). Thus it has been demonstrated with wood of *Picea mariana* (Mill) B.S.P. (Koran 1968) that the percentage of surface created by tangential or radial failure across cell walls decreased from 40–50% to nearly zero with a temperature increase from 0 °C to above 200 °C. Furthermore, in the case of tangential failures at above 150 °C, the fiber faces revealed mainly the primary wall structure, heavily embedded in an amorphous matrix of lignin and hemicelluloses. This suggested that under some conditions hemicelluloses and lignin might become substantially enriched at wood surfaces.

The composition of wood surfaces has been studied by electron spectroscopy for chemical analysis (ESCA) (Young et al. 1982). The results express the ratio of oxygen-to-carbon atoms at the surface. Calculated values of this ratio are 0.83 for cellulose, 0.37 for softwood lignin, and 0.10 and 0.11 for nonpolar extractives such as resin and fatty acids, respectively. In the case of pinewood an oxygen-to-carbon ratio of 0.26 was obtained, which increased to 0.42 after removal of extractives with acetone. The problem of the lower-than-expected oxygen-to-carbon ratios obtained after extraction has not as yet been satisfactorily solved. While the results could be interpreted by an enrichment of lignin at the surface, the presence of nonpolar extractives covalently bound to the surface and thus not removable with solvents represents another possibility.

3. *Redistribution of Extractives*

Redistribution of extractives during or following formation of the surface represents an important aspect of the chemistry of wood surfaces. If the surface was formed prior to the removal of water from wood (e.g., in sawmills) or if wood was wetted after drying, evaporation of the moisture at the wood surface causes the movement of water to the surface, where it evaporates leaving water-soluble extractives behind. This often results in undesirable discolorations of wood, particularly in darker woods that are rich in phenolic materials, such as *Sequoia sempervirens* (D. Don) Endl. (California redwood). The deposition of these materials can also lower the pH of the surface and interfere with gluing.

It has been well established that the surface energy and gluability of wood decrease with time of storage. This has been related to the formation of a nonpolar, lipophylic layer at the wood surface. The deposition of surprisingly small amounts of material is sufficient to substantially alter some properties of the wood surface, and even a monolayer of organic material can influence the wettability of wood (Baier et al. 1968). The studies by ESCA mentioned in Sect. 2 indicated the presence of nonpolar extractives on the surface of pinewood as the oxygen-to-carbon ratio at the surface of wood was well below the ratios expected for either cellulose or lignin and approached those of nonpolar extractives. Extraction with acetone increased this ratio which was interpreted as removal of nonpolar extractives from the wood surface. Deposition of nonpolar materials at the surface of wood can negatively influence wood bonding, particularly in the case of neutral or acidic bonding agents. Thus the Swedish regulations for the production of laminated beams of *Pinus sylvestris* L. require gluing within 24 hours after formation of the wood surfaces to be joined.

The mechanism of wood-surface inactivation is, however, controversial. Some evidence indicates that the loss of gluability is related to the migration of nonpolar extractives from the interior of the wood and their deposition on the surface, while other evidence points towards the deposition of foreign materials from the environment. Experimental results demonstrate that nonpolar, water insoluble extractives such as fatty acids, resin acids and steroids are capable of migrating and becoming deposited on the wood surface at temperatures as low as ambient and also in kiln drying. Hemingway (1969) concluded that the transport of nonpolar extractives from the inside of wood was very unlikely, except for the region closest to the surface, since the amount of the free (i.e., not glyceridically bound) saturated fatty acids such as stearic and palmitic acids was not high enough in that region to interfere with gluing by surface contamination. Hemingway favored instead air oxidation of unsaturated linoleic acid and deposition of the oxidation products on the surface.

4. *Deposition of Foreign Materials*

Foreign materials may be deposited on the surface of wood following its formation; such deposition may be purposeful or it may be incidental contamination which has been the subject of a considerable amount of work. Deposition of some materials (e.g., dust, water of condensation, rainwater, organic vapors, grime, acids and aerosols) is connected with the environment; others originate from methods of wood-surface formation; while yet others constitute intentional treatments.

4.1 *Incidental Deposition of Foreign Materials*

As mentioned, some theories of wood-surface inactivation favor the deposition of nonpolar materials from the environment over the migration of such materials from the wood interior. Thus Nguyen and Johns (1979), using contact-angle methodology, wood of *Sequoia sempervirens* containing mainly polar extractives and wood of *Pseudotsuga menziesii* containing mainly nonpolar extractives, concluded that the contribution of the polar and dispersive force components to the total surface free energy of wood is directly related to the nature of wood extractives. At the same time they concluded that inactivation of wood surfaces over time is related to environmental rather than to wood factors (i.e., not to the movement of nonpolar extractives towards the surface) since the surface free energy due to dispersive forces decreased over time to slightly less than one half, regardless of the starting proportion of the dispersive and polar force components. Still another proposed mechanistic alternative involves the sorption of atmospheric gases (Marian 1967).

Another adverse effect of surface contamination of wood is discoloration of the wood surface. Contamination of wood with compounds of iron during storage or in use (rusty nails, rusty water, dust, flying metal particles) is a common cause of surface discoloration of woods high in phenolics and similar compounds, as iron reacts with most phenolic (e.g., tannins) and tropolonic extractives under formation of dark-colored complexes. In some cases such discolorations can be eliminated by treating the discolored wood with a solution of oxalic acid in water. Other common polyvalent metal ions are generally less tinctorial, however.

Contamination of the wood surface during surface preparation is apparently less serious. While small amounts of iron from cutting surfaces can be expected in principle to become embedded in wood during sawing and related operations, such contaminations become troublesome only in special cases.

4.2 Purposeful Deposition of Foreign Materials

In order to optimize certain wood properties, wood is often treated with a variety of chemicals. These are deposited in the interior wood cavities and include wood preservatives against decay, such as:

(a) inorganic compounds of copper, zinc, arsenic and chromium, or organic compounds (e.g., creosote and pentachlorophenol);

(b) fire retardants (e.g., compounds of phosphorus and boron); and

(c) polymeric materials, which can be introduced as such or can be allowed to form *in situ* from introduced monomers.

All of the above materials change the chemistry of the wood surface according to the nature of treatment. Other treatments are connected with the deposition of such films as paints, varnishes, or adhesives on the wood surface (see *Adhesives for Wood*; *Adhesives for Wood: An Update* (Suppl. 2); *Protective Finishes and Coatings for Wood*).

5. Changes Due to Heat and Oxidation

The chemistry of wood surfaces can become modified during surface formation, wood drying, storage and use by chemical reactions triggered by heat (pyrolysis) and/or atmospheric oxygen (air oxidation). During sawing and related operations, particularly in the case of excessive saw vibrations, the temperature of the wood surface can reach levels where wood begins to decompose. Although the magnitude of the wood temperatures incurred during sawing are not well known, the temperatures of the circular saws are found to depend on the distance from the saw teeth and are generally 40–60 °C, but increasing to 100 °C and even 160 °C towards the saw teeth. The temperatures of the saw teeth are particularly high, however, reaching as high as 774 °C (Zaitsev 1968, Mote and Szymani 1977). Although the times of contact between wood and the hot metal are very short, the high temperature levels can occasionally, particularly in case of saw malfunction, result in substantial pyrolytic and oxidative changes on the surface of wood. Drying of wood particles (particleboard) or wood veneer (plywood, laminates) at elevated temperatures represents another avenue where pyrolytic and oxidative changes at the wood surface can take place.

Wood degradation at moderately elevated temperatures or shorter exposures to higher temperatures in the presence of air or oxygen includes pyrolytic and oxidative changes (i.e., transformations initiated by increased temperature and transformations due to reaction with oxygen). In these reactions the wood components (carbohydrates and lignin) change independently from each other; that is, the wood behaves like a mixture of these materials. At longer exposures to higher temperatures the combustion process sets in. This consists of autocatalytic pyrolytic decomposition coupled with the oxidation of the volatiles produced, the final products consisting of char, water and carbon dioxide.

Pyrolysis of cellulose has been the subject of intensive investigations and has been reviewed many times (Tillman 1981, Shafizadeh 1984). It begins with depolymerization by transglycosylation to yield levoglucosan and other monomeric and oligomeric sugar derivatives. Concurrently the dehydration reaction leads to the formation of the unsaturated materials. The same reactions most likely dominate the pyrolysis of hemicelluloses. Lignin pyrolysis is dominated by condensation reactions leading to the formation of ether linkages between the n-propyl sidechains and of alkyl–aryl bonds. As in the case of cellulosics, this is paralleled by dehydration reactions leading to the formation of double bonds in the sidechains (Domburg and Skripchenko 1982). Oxidation of cellulose by atmospheric oxygen apparently begins to take place at about 140 °C. It is accompanied by depolymerization and results in the formation of carbonyl and carboxyl groups, some of which decarboxylate. The process is strongly catalyzed by moisture (Tryon and Wall 1966, El-Rafie et al. 1983).

The amount of information on pyrolytic and oxidative changes on wood surfaces resulting from the history of surface preparation is meager, however. The information available is connected mainly with studies of surface inactivation in gluing processes and with dimensional stabilization of wood by exposure to moderately elevated temperatures.

Exposure of wood to temperatures moderately above 100 °C for long time periods results in loss of hygroscopicity connected with the loss of hydroxyl groups. A quantitative correlation was obtained between loss of hygroscopicity and loss of weight after heating wood samples of *Pinus taeda* L. (loblolly pine) and *Liriodendron tulipifera* L. (yellow poplar) to 200 °C for 5 min. This was explained by the formation of intramolecular epoxy groups between hydroxyls 2 and 3 of the anhydroglucose units of cellulose. Intermolecular ether linkages apparently do not form (Seborg et al. 1953, Salehuddin 1970). Heating of wood to 300 °C results in increased dimensional stabilization. This has been explained by the decomposition of hygroscopic hemicelluloses and other carbohydrates, followed by condensation and polymerization of the resulting furan-type compounds (Mitchell et al. 1953).

Changes in the chemistry of wood surfaces due to increased temperatures were studied by Chow and Mukai (1972) by exposing microsections of *Picea glauca* (Moench) Voss (white spruce) to temperatures between 100 °C and 240 °C in air and in nitrogen. Below 180 °C the changes consisted mainly of oxidation, while above 180 °C they were of mixed pyrolytic and oxidative nature. The absorption of hydroxyl in infrared spectra decreased with time at 180 °C, the color of wood darkened, and crystallinity and the degree of polymerization (DP) of cellulose decreased. Carbonyl absorption of esters and carboxyls in infrared spectra decreased first and then increased with the temperature rise. Extractives were found to catalyze the rate of the oxidation.

Changes on the wood surface due to an increase in temperature also affect the extractives. Thus phenolic extractives, particularly tannins, are likely to undergo condensation reactions and to polymerize to the water-insoluble "synthetic phlobaphenes." Volatile extractives such as monoterpenoids are likely to volatilize and resin acids are likely to isomerize by double-bond migration within their structures. Additional changes should involve unsaturated fatty acids such as linolenic acid, which is likely to transform by intramolecular double-bond migration to conjugated positions, and could ultimately oxidatively cleave into lower-molecular-weight fragments (Hemingway 1969).

6. Changes Due to Exposure to Light

Electromagnetic radiation of the visible and ultraviolet regions such as daylight and light from incandescent and fluorescent lamps changes the appearance of wood, in time, by interacting with wood constituents. Such changes are commonly noticed on wood panelling and other wooden objects by a difference in wood color between areas exposed to light and areas protected from light, such as areas covered by paintings hung on a wall. The nature of the effect (darkening or lightening, and the kind of color change) is difficult to predict, however, as it depends on the composition of the electromagnetic radiation, its intensity, temperature, moisture content of the wood, length of exposure and on the nature of the wood, particularly on the kind of extractives it contains (Kringstad 1973, Feist and Hon 1982, Hon and Chan 1982). It has been noted that exposure of lignin to light of wavelength less than 385 nm results in darkening, while exposure to light of wavelength greater than 480 nm results in lightening of the color. A similar effect has also been reported in the case of solid wood. Chemical changes involve initial formation of free radicals, and include chain scission, dehydrogenation, and dehydroxymethylation in the case of cellulose and hemicelluloses, and splitting of double bonds, formation of quinone structures, demethoxylation under formation of methanol, increased solubility (loss in Klason lignin content of wood) and polymerization in the case of lignin. The formation of free radicals most likely involves the hydroxy groups of lignin since esterification or etherification of lignin increases the light stability of wood. In the presence of oxygen and water, hydrogen peroxide and peroxy groups also form. With solid wood at 45–50 °C and 50% relative humidity, and xenon arc as a light source, a loss of lignin and hemicelluloses from the surface was noted after 75 days of exposure. Water strongly increased the rate of material loss.

Extractives are particularly prone to color changes. It has been demonstrated that the colorless flavanonols taxifolin and aromadendrin change by interaction with visible–ultraviolet light to the yellowish flavonols quercetin and kaempferol by a photooxidative reaction sequence. Concurrently a general decrease in flavonoids and an increase in vanillin-related compounds was noticed (Minemura and Umehara 1979).

7. Changes in Wood Surface Due to Weathering

Weathering of wood is a complex process, dependent on simultaneous action of several factors, such as solar radiation, moisture, temperature, air-oxygen and fungal microorganisms (mildew). Over time these agents give wood surfaces a characteristic gray color, they acquire a rough texture, checks and cracks, and become friable. The upper gray surface layer is about 125 nm thick. It is generally composed of degraded, disordered and loosely matted cellulosic fibers and contains very little if any lignin. This top layer forms mainly by direct interaction of ultraviolet radiation with wood substance, primarily with lignin under formation of free radicals. The transformations that follow occur under participation of other weathering factors and result in solubilization and disappearance of lignin. Other changes include oxidation and most likely depolymerization of cellulose; the former was demonstrated by ESCA and by infrared spectroscopy.

Below the gray layer is a brown layer (500–2500 nm) containing intermediate amounts of lignin (40–60% of normal). Since the brown layer cannot form as a result of direct interaction of wood with light, as ultraviolet light cannot penetrate that deep, it must arise as the result of either energy transfer from the surface, or as the result of migration of free radicals from the surface into the wood interior. The brown layer is underlain in turn by normal, non-weathered wood (Feist and Hon 1983).

8. Wood-Surface Modification

Certain reagents are purposely allowed to react with wood, primarily with the hydroxy groups of its constituents, in order to modify the properties of wood, particularly at its surface. Some of these reagents react with wood without the introduction of foreign molecular structures (heat, oxidation agents); in other cases foreign structures are covalently attached to the wood surface. These can include relatively small chemical units (methylation, acetylation), or larger groups including even the chains of molecular units (polymer grafting).

Treatment with physical agents includes treatment with heat, as well as with various forms of electromagnetic radiation, such as infrared, visible and ultraviolet light, and α, β and γ radiation. The respective chemistry of some of these has been discussed earlier. Such treatments, particularly in presence of air, result generally in oxidation of the surface under formation of carboxylic groups and cross-linking, an increase of the surface energy of wood, and improvements in adhesive properties.

A certain amount of information is available on the modification of wood surfaces by various types of plasma. Plasma is generally understood to be an activated gas which includes various types of reactive chemical species, such as electrons, photons, positive and negative ions, free radicals and metastables. Plasmas can be generated in a variety of ways, using any gaseous material. Reaction of plasmas with wood surfaces can radically modify surface properties. Thus oxygen plasmas, such as corona plasma, oxidize the wood surface under formation of carbonyl and carboxyl groups, and drastically increase its surface energy, while other plasmas, such as radio-frequency acetylene plasma drastically decrease the surface energy by attaching nonpolar groups to the wood surface. Other chemical groups that can be attached to wood surfaces by various plasmas include amino groups (nitrogen or ammonia plasmas), acrylic acid and styrene derivatives. The reacting wood layer is generally very thin, often making it difficult to detect the presence of these groups by spectroscopic methods.

Oxidation has been used for a long time to alter the properties of wood surfaces. The most common oxidation reagents include hydrogen peroxide and other compounds with peroxy linkages, ozone, nitric acid, nitrates, chlorates, elemental halogens, metal ions such as Fe^{3+}, metal oxides, the derived acids and their salts, such as chromates. Oxidations tend, generally, to increase the surface energy of wood. Concurrently, pH tends to change, decreasing in the cases of hydrogen peroxide, nitric acid, ozone and halogens due to the production of organic acids and occasionally also due to introduction (HNO_3) or formation (HCl, HBr) of inorganic acids. In other cases, such as oxidation by chromates and nitrates, the pH increases due to formation of salts of weaker carboxylic acids and reduction of the nitrate and chromate ions to N_2 and Cr_2O_3. As a result of oxidation, carboxylic and carbonyl groups are generally introduced into the wood. In case of oxidations by halogens or nitric acid, lignin becomes additionally nitrated or halogenated to some extent. Treatment of wood surfaces with acids, particularly if followed by heating, results in partial hydrolysis of cellulose and hemicelluloses and condensation of lignin. Under more drastic conditions the hydrolytically liberated monosaccharides, primarily the pentoses, undergo a chain of transformations leading to furfural which ultimately polymerizes to polyfurfural, a material of reduced polarity.

Treatment of wood surfaces with alkalis transforms the carboxylic groups on the wood surface into corresponding salts. This ensures a high pH of the wood surface following removal of the introduced alkali. Alkalis remove also the surface fatty acids increasing in this way the surface energy of wood.

Methylation and esterification of wood surfaces has been extensively studied in connection with attempts to dimensionally stabilize wood. The reagents included dimethylsulfate/alkali in case of methylation, and ketene, acetic anhydride, acetyl chloride and phthalic anhydride in case of esterifications. Both types of treatments generally tend to decrease polarity and surface energy of wood.

Additional surface treatments include reactions with isocyanates (methylisocyanate, phenylisocyanate, 2,4-toluene diisocyanate) forming urethane linkages, aldehydes such as formaldehyde yielding acetalic linkages and epoxides such as ethylene or propylene oxide, forming β-hydroxyether groups. The changes in the properties of the wood surface following these treatments can be easily deduced from the nature of the respective chemical transformations (see *Chemically Modified Wood*).

Still another group of surface transformations includes generally undesirable changes resulting from fire or biological decay (see *Wood: Decay During Use*).

See also: Radiation Effects on Wood (Suppl. 1); Surface Properties of Wood; Weathering of Wood; Wood: Chemical Composition

Bibliography

Baier R E, Shafrin E G, Zisman W A 1968 Adhesion: Mechanisms that assist or impede it. *Science* 162(3860): 1360–8

Chow S-Z, Mukai H N 1972 Effect of thermal degradation of cellulose on wood–polymer bonding. *Wood Sci.* 4: 202–8

Domburg G E, Skripchenko T N 1982 Process of formation of intermediate structures during thermal transformations of lignins (in Russian). *Khim. Drev.* 5: 81–8

El-Rafie M H, Khalil E M, Abdel-Hafiz S A, Hebeish A 1983 Behavior of chemically modified cottons towards thermal treatment. II: Cyanoethylated cotton. *J. Appl. Polym. Sci.* 28: 311–26

Feist W C, Hon D N-S 1983 Chemistry of weathering and protection. In: Rowell R (ed.) 1983 *The Chemistry of*

Solid Wood, ACS Advances in Chemistry Series No. 207. American Chemical Society, Washington, DC, pp. 401–51

Feist W C, Rowell R M 1982 UV degradation and accelerated weathering of chemically modified wood. In: Hon D N-S (ed.) 1982 *Graft Copolymerization of Cellulosic Fibers*, ACS Symposium Series No. 187. American Chemical Society, Washington, DC, pp. 349–70

Hemingway R W 1969 Thermal instability of fats relative to surface wettability of yellow birchwood (Betula Lutea). *Tappi* 52: 2149–55

Hon D N-S, Chan H Ch 1982 Photoinduced grafting reactions in cellulose and cellulose derivatives. In: Hon D N-S (ed.) 1982 *Graft Copolymerization of Cellulosic Fibers*, ACS Symposium Series No. 187. American Chemical Society, Washington, DC, pp. 101–18

Koran Z 1968 Electron microscopy of tangential tracheid surfaces of black spruce produced by tensile failure at various temperatures. *Svensk Papperstidning* 71: 567–76

Kringstad K P 1973 Some possible reactions in light-induced degrading of high-yield pulps rich in lignin (in German). *Das Papier* 27: 462–9

Marian J E 1967 Wood, reconstituted wood, and glued laminated structures. In: Houwink R, Salomon G (eds.) 1967 *Adhesion and Adhesives*, Vol. 2, Elsevier, Amsterdam. Chap. 14, pp. 167–280

Minemura N, Umehara K 1979 Color improvement of wood (1). Photo-induced discoloration and its control (in Japanese). *Rept. Hokkaido For. Prod. Res. Inst.* 68: 92–145

Mitchell R L, Seborg R M, Millett M A 1953 Effect of heat on the properties and chemical composition of Douglas-Fir wood and its major components. *For. Prod. Res. Soc. J.* 3(4): 38–42

Mote C D, Szymani R 1977 Principal developments in thin circular saw vibration and control research. Part 1: Vibration of circular saws. *Holz als Roh- und Werkstoff* 35: 189–96

Nguyen T, Johns W E 1979 The effects of aging and extraction on the surface free energy of Douglas fir and redwood. *Wood Sci. Technol.* 13: 29–40

Salehuddin A B M 1970 A unifying physico-chemical theory for cellulose and wood and its application in glueing. Ph.D. Thesis, North Carolina State University, pp. 1–89

Seborg R M, Tarkow H, Stamm A J 1953 Effect of heat upon the dimensional stabilization of wood. *For. Prod. Res. Soc. J.* 3(3): 59–67

Shafizadeh F 1984 The chemistry of pyrolysis and combustion. In: Rowell R (ed.) 1984 *The Chemistry of Solid Wood*, ACS Advances in Chemistry Series No. 207. American Chemical Society, Washington, DC, pp. 489–529

Stamm A J, Millett M A 1941 Internal surface of cellulosic materials. *J. Phys. Chem.* 45: 43–54

Tillman D A, Rossi A J, Kitti W D 1981 The process of wood combustion. In: Tillman D A (ed.) 1981 *Wood Combustion: Principles, Processes & Economics*. Academic Press, New York, Chap. 4, pp. 74–97

Tryon M, Wall L A 1966 Oxidation of high polymers. In: Lundberg W O (ed.) 1966 *Autoxidation and Autoxidants*, Vol. 2. Interscience, New York, pp. 963–8

Young R A, Rammon R M, Kelley S S, Gillespie G H 1982 Bond formation by wood surface reactions. Part I—Surface analysis by ESCA. *Wood Sci.* 14: 110–19

Zaitsev N A 1968 Temperature measurement on the cutting edges of the circular saws (in Russian). *Derevoobrab. Promst.* 17(4): 15

E. Zavarin
[University of California, Berkeley,
California, USA]

Z

Zirconia-Toughened Ceramics as Biomaterials

In the 1960s total hip replacement (THR) became a widely used surgical procedure due, among other reasons, to the successful use of a materials combination leading to low-friction arthroplasty: metal ball against ultrahigh molecular weight polyethylene (UHMWPE) socket. However, after a ten-year period of use, drawbacks related to UHMWPE began to appear, such as creep, wear and tissue inflammation induced by the production of polyethylene wear debris. In this respect, there was a need for the use of materials exhibiting better performances for total joint replacement. Alumina ceramic was introduced in orthopedic surgery because of both its high wear resistance and biocompatibility. Since then, high medical-grade alumina has functioned in a satisfactory manner, either in alumina–UHMWPE or in alumina–alumina combinations. However, alumina exhibits a brittle behavior, with a low fracture toughness and a low tensile strength. It is sensitive to microstructural flaws and subsequently has a low resistance to stress concentration and mechanical impact in service. This is one of the major reasons limiting the diameter of most prosthetic alumina femoral balls to 32 mm in order to avoid the occurrence of brittle fractures. In this instance, toughened ceramics including zirconia-toughened ceramics appeared particularly efficient and attractive for the highly loaded environment found in joint replacement (Pascoe et al. 1979, Christel et al. 1988).

1. Properties of Toughened Zirconia and Material Processing

Zirconium oxide ceramics have three phases: monoclinic, tetragonal and cubic. The cubic phase is stable but brittle; the tetragonal phase is tough but unstable and may transform into the monoclinic phase. Zirconia exhibits phase transformations at high temperatures. At room temperature, the stable phase of zirconium oxide has a monoclinic symmetry. During heating, it first transforms into a tetragonal phase (in the 1000–1100 °C range), then into a cubic phase (above 2000 °C). Noticeable changes in volume are associated with these transformations. During the monoclinic-to-tetragonal transformation, which occurs when zirconium oxide is heated, there is a 5% volume decrease; conversely, a 3% increase in volume is observed during the cooling process. This last retransformation into the monoclinic phase is of the same nature as the martensitic transformation occurring in steel and can be compared with it.

These phenomena are detrimental to the mechanical behavior of the zirconium oxide because the stresses induced during the phase transformations results in crack formation. This undesirable phase transformation can be inhibited by the addition of stabilizing oxides (CaO, MgO, Y_2O_3) and this process has become common practice. Accordingly, in the presence of a small amount of stabilizing additives, tetragonal particles (provided they are small enough) can be maintained in a metastable state at temperatures below the tetragonal-to-monoclinic transformation temperature. The transformation of small tetragonal grains, which should result in a volume increase, is prevented by the compressive stresses applied on these grains by their neighbors. The corresponding class of ceramics is named partially stabilized zirconia (PSZ).

Two kinds of microstructures can be generated. In the ZrO_2–MgO or ZrO_2–CaO systems, materials are sintered in the cubic state and small tetragonal precipitates are formed during cooling as a result of partial transformation of the cubic phase. In the ZrO_2–Y_2O_3 system, the extent of the stability range of the tetragonal phase, in terms of temperature and amount of yttrium oxide as shown on the phase diagram in Fig. 1, allows sintering of fully tetragonal fine-grained materials. Thus, using Y_2O_3 as a stabilizing agent, it is possible to produce a zirconium oxide ceramic made of 100% small metastable grains, leading *de facto* to a fully stabilized zirconia.

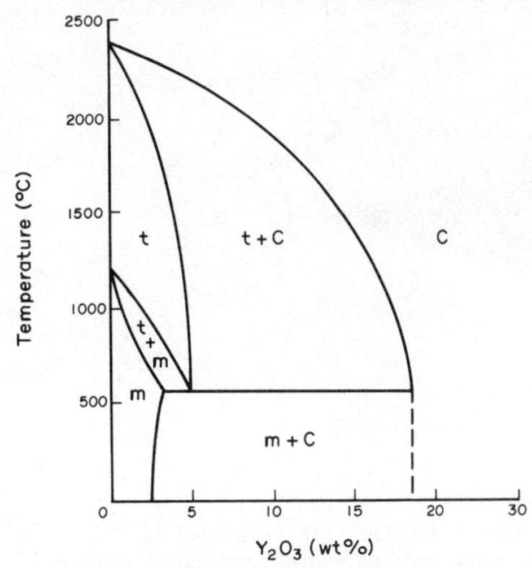

Figure 1
Y_2O_3–ZrO_2 phase diagram: the addition of less than 5% of Y_2O_3 to ZrO_2 allows the sintering of a fully tetragonal material (t = tetragonal phase; m = monoclinic phase; c = cubic phase)

The volume change related to the tetragonal-to-monoclinic phase transformation results in a prestressed material. In this respect, a propagating crack can release the stresses in the neighboring grains which then transform from the metastable state into the monoclinic phase. Because the volume of the monoclinic grains is larger than the tetragonal ones, the associated volume expansion results in compressive stresses at the edge of the crack front and extra energy is required for the crack to propagate further (Fig. 2). Thus it is believed that the main energy-absorbing mechanism is due to the martensitic-like transformation occurring at the crack tip.

Yttrium-oxide-stabilized zirconia is obtained by sintering of ultrafine zirconia powder (average particle size: 0.2 μm) with about 5 wt% Y_2O_3. This powder is highly adapted to sintering as its densification occurs in the 1400–1500 °C range, corresponding to the tetragonal phase of the ZrO_2–Y_2O_3 system. The final microstructure consists of 0.5 μm average diameter grains (Figs. 3 and 4). However, ultrafine powders are unsuitable for compacting; as a result, shaping of thick components such as hip prostheses balls can be difficult. Cold isostatic pressing is the most widely used process in the shaping of ZrO_2–Y_2O_3 ceramics. Densification can be achieved in two different manners: either pressureless sintering or sintering associated with hot isostatic pressing (HIP). This latter procedure consists of two stages: first, presintering the components without pressure up to about 95% of the theoretical density; second, removing the residual porosity in a complementary HIP process. In this procedure, complete densification occurs with only limited grain growth, resulting in improved strength.

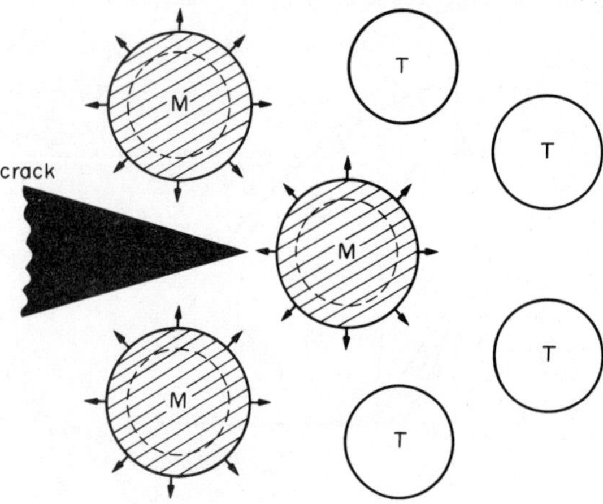

Figure 2
Diagram representing the mechanism of transformation toughening in partially-stabilized zirconia; the crack propagation induces the transformation of metastable tetragonal grain (T) into monoclinic phase (M); the monoclinic grains are larger than the tetragonal ones and stop the crack propagation

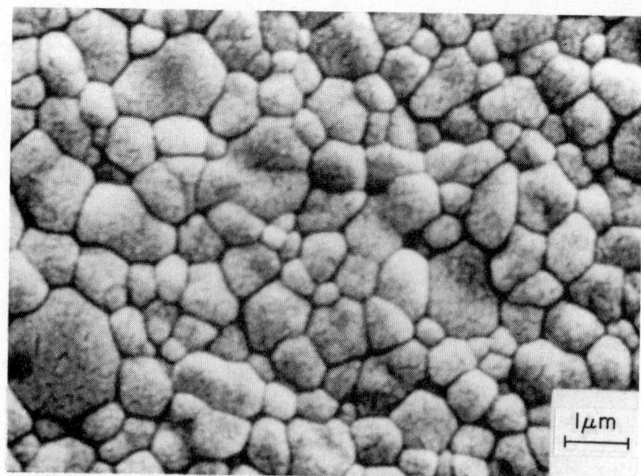

Figure 3
Microstructure of yttrium oxide–partially-stabilized zirconia as shown by scanning electron microscopy (courtesy Céramiques Techniques Desmarquest)

The materials properties of commercially available Y_2O_3–ZrO_2 ceramics, with reference to surgical-grade Al_2O_3 (International Organization for Standardization (ISO) standard ISO 130 or DIS standard DIS D13) are listed in Table 1. Accordingly, it is clear that the strength and toughness of transformation-toughened zirconia are much higher than those of alumina. In addition, the ZrO_2-based materials exhibit a lower Young's modulus, pointing to an interesting elastic deformation capability when compared with alumina (Christel et al. 1988).

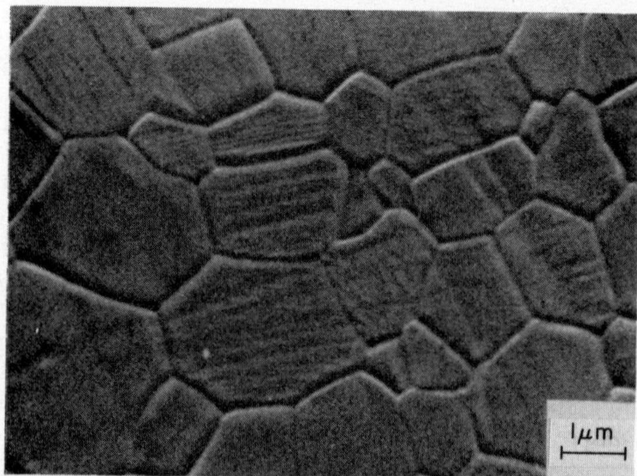

Figure 4
Microstructure of surgical-grade alumina shown for comparison with zirconia, with the same magnification as Fig. 3 (courtesy Céramiques Techniques Desmarquest)

Table 1
Comparison of several mechanical properties between surgical grade alumina (ISO requirements) and two commercially available yttrium oxide–partially-stabilized zirconia (Prozyr®, Céramiques techniques Desmarquest, France; Metoxit®, Metoxit AG, Switzerland)

Property	Alumina (ISO/DIS 130/D13 requirements)	Prozyr® YPSZ	Metoxit® YPSZ
Density (kg m^{-3})	3900	6100	*c.*6050
Average grain size (μm)	<7	<0.5	0.2
Vickers hardness (HV)	2000–3000	1000–1300	*c.*1200
Young's modulus (GPa)	380	200	150
Bending strength (MPa)	400	1200	*c.*800
Toughness, K_{Ic} (mN m$^{-3/2}$)	5–6	9–10	*c.*7

2. *Mechanical Performances of Zirconia Femoral Heads*

Tateishi and Yunoki (1987) have studied the static fracture strength of 22 mm zirconia balls fitted on 10 mm or 11 mm diameter tapers with a cross-head speed of 0.5 mm min^{-1} with reference to 28 mm alumina balls as the control. The testing was carried out at room temperature, with the taper oriented vertically. In order to avoid a direct contact between ceramic and metal cross head, polypropylene sheets were inserted at the interface.

The fracture of alumina heads occurred at loads in the range 1.5–2.5 × 10^4 N, whereas the zirconia heads broke at 4–5 × 10^4 N with the 10 mm taper and 3–4 × 10^4 N with the 11 mm taper.

Impact fracture tests were carried out by falling weights, either on the 22 mm zirconia or the 28 mm alumina heads at a constant height of 0.5 m. The alumina balls, fitted on a 13 mm, taper fractured for a 15 J impact energy; the zirconia heads, fitted with a 10 mm taper, exhibited an impact energy in the range 50–100 J. Unfortunately, the impact at fracture cannot, at present, be compared between the two ceramics because of the difference in taper size. However, it is likely that the difference will not be so great since for zirconia, increasing the taper diameter reduced the impact strength.

3. *Biodegradation of Toughened Zirconia*

The degradation of toughened zirconia can be secondary to aging, fatigue or wear occurring during its *in vivo* service.

Aging has been studied through accelerated procedures in an autoclave, combining various degrees of heat, humidity and pressure for different lengths of time (Hanninck and Garvie 1982). It has been shown that there is a strength loss of a few percent with MgO–PSZ samples boiled in saline for 1000 h. This has been related to the extension of the critical flaw size by a stress corrosion mechanism. No significant drop has been found in the fracture toughness of Y_2O_3–PSZ after γ sterilization or aging in Ringer's solution (Christel et al. 1989). *In vivo* aging of Y_2O_3–ZrO_2 samples implanted by Kumar et al. (1989) in rabbit bone or subcutaneous tissue, up to 12 months, has shown no change in bending strength of the implants. Simultaneously, a 2% transformation rate from tetragonal into monoclinic phase was measured on the same implants by x-ray diffraction analysis. It was identical for the specimens implanted in bone or subcutaneous tissue, or stored either in saline solution at room temperature or in the atmosphere. Since theoretically up to 60% of transformation occurs from tetragonal into monoclinic phase, there is no significant change in toughness of tetragonal zirconia (Sato and Shimada 1984): this 2% per year transformation rate is without consequence.

Fatigue tests have been carried out in Ringer's solution for up to 10^7 cycles on 22 mm Y_2O_3–PSZ femoral balls under an alternating load varying between 0 kN and 10 kN at a frequency of 30 Hz. The ceramic heads were placed vertically, fitted on either 10 mm or 11 mm diameter Ti_6Al_4V taper cones. Prior to the fatigue test, in order to fit the ball with the cone, the ball was preloaded by 3 kN. Under these test conditions the zirconia balls sustained 10^7 cycles without damages.

Wear tests of zirconia have been conducted using either a zirconia–UHMWPE or a zirconia–zirconia combination.

A bench test compared the deformation of UHMWPE sockets bearing with 32 mm heads made of either chromium–cobalt alloy or Y_2O_3–PSZ ceramic in saline serum at 37 °C. The heads were submitted to a bidimensional rotational movement combining a ±45° tilt of the taper cone-head assembly with a 60 revolutions per minute axial rotation under a constant load of 2 kN at a frequency of 0.5 Hz. After 10^6 cycles the UHMWPE socket deformation, bearing against a chromium–cobalt head, was 10.2×10^{-2}, whereas after 11×10^6 cycles with a Y_2O_3–PSZ ball, the plastic deformation was only 0.51×10^{-2}. Assuming that the creep of UHMWPE was identical in each test, the difference in socket deformation could be related to different wear rates in relation to the higher surface roughness of chromium–cobalt ball when compared with zirconia ceramic. In another study (Tateishi and Yunoki 1987) based on a joint simulator test to characterize the zirconia on polyethylene combination, a rotating movement at 0.125 Hz, ±10° in the frontal, transverse and sagital planes under an alternating load varying between 0 kN and 4 kN was used with pure water as a temperature-controlled lubricant at 25 °C. The socket was placed in the lower position. Measurements of frictional torques after 4×10^6 cycles showed a value of 2.6 N m for the Charnley prosthesis vs 2 N m for a 28 mm alumina ball bearing against a UHMWPE cup, and 1.5 N m for a 22 mm Y_2O_3–PSZ head and UHMWPE socket combination. However, no 28 mm Y_2O_3–PSZ head was tested.

Zirconia–zirconia tribological behavior has been compared by Sudanese et al. (1988) with alumina–alumina using a disk-on-ring test according to the ISO B474–1981 standard (angle of oscillation, ±25°; disk diameter, 30.4 mm; disk height, 4 mm; ring inside diameter, 14 mm; ring outside diameter, 20 mm; superficial pressure, 20–23 bar; frequency of oscillation, 1 Hz; and lubricant, Ringer's solution at 37 °C). The surface roughness of the Y_2O_3–PSZ disk was 0.01 μm Ra and 0.03 μm Ra for alumina. The wear rate of zirconia ($16\ mm^3\ h^{-1}$) was 5 000 times greater than alumina ($0.0033\ mm^3\ h^{-1}$). These results are difficult to extrapolate to THR because the lubrication conditions, loading and motion speed are quite different from the *in vivo* situation (Denape et al. 1984). In this test configuration, the heavy degradation of the zirconia–zirconia combination could be related to the low thermal conductivity of zirconia generating thermal shocks. Other ceramics combinations need further testing; for example, zirconia/alumina–zirconia, zirconia–alumina and zirconia–silicon nitride.

4. *Biocompatibility of Toughened Zirconia Ceramics*

The biocompatibility of pure zirconia and PSZ is not widely documented. The tissue reaction to zirconia ceramics has been studied after implantation, either in soft tissues or in bony site. The tissue reaction to plasma- or flame-sprayed ZrO_2 on a 316L stainless-steel substrate has been evaluated by Bortz and Onesto (1973) by implantation of such tubes for tracheal or vein replacement in rabbits and dogs. Neither adverse tissue response nor blood clotting have been observed. MgO–zirconia samples have been implanted by Garvie et al. (1984) in rabbits' muscles for six months, without control material. The observation of the tissue reaction, which was graded qualitatively, has led to the conclusion that the tissue response to the material was acceptable with no adverse tissue reaction. Also, no mechanical degradation of the MgO–zirconia has been observed after the six-month implantation period. Y_2O_3–ZrO_2 cylinders, differing in the methods of preparing the original powders, were implanted in the paraspinal muscles of rats for one, four and 12 weeks (Christel et al. 1989). The tissue reaction to these implants, for each implantation time, was studied by quantitative histomorphometry, in comparison with surgical-grade alumina (ISO requirements). For each cell type (macrophages, polymorphonuclear cells, giant cells, round cells or fibrocytes) the parameters of the cell distribution (amplitude, density and distances from the tissue–implant interface) and the thickness of the encapsulating membrane were computed. There was no statistical difference in any of the parameters of the distributions and membrane thickness of the cell types between Y_2O_3–ZrO_2 ceramics and alumina for each implantation time.

Implantation of Y_2O_3–zirconia cylinders in bone has been performed in rats and rabbits from two weeks up to one year, with reference to alumina. Most of the results have shown an equal amount of bone contacting the implants directly and an identical tissue reaction to both ceramics. The similarity of tissue reaction to zirconia and alumina may be explained by the fact that both materials are in the form of their most oxidated state, thus they lack the ability to leach any soluble component. In this instance, it has been shown by Soumiya (1984) when using x-ray diffraction analysis that after a four-week implantation of Y_2O_3–ZrO_2 powder and screws into the rabbit femur, no diffusion of yttrium or zirconium was detectable. However, at the time of writing there is no information on the long-term biocompatibility of bulk zirconia ceramics, nor on the tissue tolerance to submicrometer zirconia wear particles.

Bibliography

Bortz S A, Onesto E J 1973 Flame-sprayed bioceramics. *Bull. Amer. Ceram. Soc.* 52: 898

Christel P, Meunier A, Dorlot J M, Crolet J, Witvoet J, Sedel L, Boutin P 1988 Biomechanical compatibility and design of ceramic implants for orthopedic surgery. In: Ducheyne P, Lemons J (eds.) 1988 Bioceramics: Material characteristics versus in vivo behavior. *Ann. N.Y. Acad. Sci.* 523: 234–56

Christel P, Meunier A, Heller M, Torre J P, Peille C N 1989 Mechanical properties and short-term in-vivo evaluation of yttrium oxide–partially-stabilized zirconia. *J. Biomed. Mater. Res.* 23: 45–61

Denape J, Lamon J, Broussaud D 1984 Friction wear of ceramics: Theoretical and experimental studies. In: Vincezini P (ed.) 1984 *Sciences of ceramics*, Vol. 12, Elsevier, Amsterdam, pp. 529–35

Garvie R C, Urban C, Kennedy D R, McNeuer J C 1984 Biocompatibility of magnesia–partially stabilized zirconia (Mg–PSZ ceramics). *J. Mater. Sci.* 19: 3224–8

Hanninck R H J, Garvie R C 1982 Sub-eutectoid aged Mg–PSZ alloy with enhanced thermal up-shock resistance. *J. Mater. Sci.* 17: 2637–43

Kumar P, Shimizu K, Oka M, Kotoura Y, Nakayama Y, Yamamuru T, Yanagida T, Makinouchi K 1989 Biological reaction to zirconia ceramics. In: Oonishi H, Aoki H, Sawai L (eds.) 1989 *Bioceramics*. Ishiyaku Euro America Inc., Tokyo, pp. 341–6

Pascoe R T, Hughan R R, Garvie R C 1979 Strong and tough zirconia ceramics. *Sci. Sintering* 11: 185–92

Sato T, Shimada M 1984 Crystalline phase change in yttria–partially-stabilized zirconia by low temperature annealing. *J. Am. Ceram. Soc.* 67: C12–13

Soumiya M 1984 Study of zirconia ceramics. In: Oonishi H and Ooi Y (eds.) 1984 *Orthopaedic Ceramic Implants*, Vol. 4. Japanese Society of Orthopaedic Ceramic Implants, Osaka, pp. 45–9

Sudanese A, Toni A, Cattaneo G L, Ciaroni D, Greggi T, Dallari D, Galli G, Giunti A 1989 Alumina vs zirconium oxide: A comparable wear test. In: Oonishi H, Aoki H, Sawai L (eds.) 1989 *Bioceramics*. Ishiyaku Euro America Inc., Tokyo, pp. 45–9

Tateishi T, Yunoki H 1987 Research and development of advanced biocomposite materials and application to the artificial hip joint. *Bull. Mech. Eng. Lab. Jap.* 45: 1–9

P. S. Christel
[Université de Paris, Paris, France]

SYSTEMATIC OUTLINE OF THE ENCYCLOPEDIA

Cumulative Addendum for Supplementary Volumes 1 and 2

The 45 main subject areas used in the Systematic Outline of the Main Encycloedia are also used here, but they are not further subdivided since the entries in most subject areas are not sufficiently numerous to justify such subdivision. The bold number after each entry indicates the Supplementary Volume in which it appears.

Note that (R) indicates an article that replaces one with the same or a similar title appearing in the Main Encyclopedia, and (S) indicates an article that is to be read as a supplement to a similarly titled article in the Main Encyclopedia.

1. METALS PRODUCTION

2. METALS PROCESSING AND FABRICATION

3. FUNDAMENTAL PHYSICAL METALLURGY

4. APPLIED GENERAL AND NONFERROUS PHYSICAL METALLURGY

5. APPLIED FERROUS PHYSICAL METALLURGY

6. TRADITIONAL CERAMICS

7. ADVANCED CERAMICS

8. GLASSES

9. CERAMICS PROCESS ENGINEERING

10. POLYMER CHEMISTRY

11. POLYMERS: PROCESSING

12. POLYMERS: STRUCTURE, PROPERTIES AND STRUCTURE–PROPERTY RELATIONS

13. ELASTOMERS

14. POLYMERS: APPLICATIONS

15. FIBERS AND TEXTILES

16. COMPOSITE MATERIALS

17. WOOD

18. PAPER AND PAPERBOARD

19. MATERIALS OF BIOLOGICAL ORIGIN

20. INDUSTRIAL MINERALS

21. ELECTRICAL MATERIALS

22. ELECTRONIC MATERIALS

23. SUPERCONDUCTING MATERIALS

24. NUCLEAR MATERIALS

25. MATERIALS FOR OTHER ENERGY APPLICATIONS

26. MAGNETIC MATERIALS

27. OPTICAL MATERIALS

Birefringence in Crystals **1**
Float Glass Process **1**
Metal Film Deposition by Laser Pyrolysis of Gaseous Precursors **1**
Microengineering of Materials: Characterization **2**
Microengineering of Materials: Laser Fusion Targets **2**
Optical Glasses **1**
Optical Thin Films: Production and Use (S) **1**
Organically Modified Silicates—Ormosils **1**

28. BIOMEDICAL MATERIALS

Aluminum Oxide for Prosthetic Devices (S) **2**
Arteries, Synthetic **1**
Biomaterial–Blood Interactions **2**
Biomaterials: Surface Structure and Properties **2**
Heart Valve Replacement Materials **2**
Hydrogels as Biomedical Materials **1**
Medical Electrodes, Metals for **1**
Polysiloxanes as Biomedical Materials (R) **1**
Protein–Polymer Grafts: Adhesives **1**
Wound-Dressings Materials **1**
Zirconia-Toughened Ceramics as Biomaterials **2**

29. DENTAL MATERIALS

Chemical Adhesion in Dental Restoratives (R) **1**
Dental Amalgams: Further Developments (S) **2**
Dental Elastomers: An Overview **2**
Dental Gold Alloys: Age-Hardening **2**
Dental Implants (R) **1**
Denture Base Resins (R) **1**
Endodontic Materials **2**

30. BUILDING MATERIALS

Cellular Materials in Construction **2**
Ceramics: Construction Applications **2**
Electrogalvanized Steel **1**
Ice and Frozen Earth as Construction Materials **2**
Insulation Raw Materials **2**

31. MISCELLANEOUS MATERIALS

Aluminides for Structural Use **1**
Brazing Alloys by Rapid Solidification **1**
Colloidal Crystals **2**
Diamondlike and Diamond Thin Films **2**
Glassy Crystals **1**
Ice and Frozen Earth as Construction Materials **2**
Ice: Mechanical Properties **1**
Inks for Printing **1**
Inks for Writing **1**
Materials Selection in Conceptual Design: Materials Selection Charts **2**
Metal Hydrides **1**
Microengineering of Materials: Characterization **2**
Microengineering of Materials: Laser Fusion Targets **2**
Multilayers, Metallic **2**
Plastic Organic Crystals **1**
Solid-State Nuclear Track Detectors: Applications **1**
Strategic Materials: Substitution Alternatives **1**
Titanium Aluminides **2**
Water and Ice: Structure **1**
Water as a Solvent **1**
Water for Nuclear Reactors **1**
Water: Liquid Metastable States **1**
Water: Ultrapurification for Microelectronics Processing **1**

32. TECHNIQUES FOR INVESTIGATION AND CHARACTERIZATION OF MATERIALS

Art Forgeries: Scientific Detection **1**
Channelling-Enhanced Microanalysis **2**
Compton Scattering **2**
Computer Simulation of Microstructural Evolution **2**
Diffusion: Novel Measurement Methods **1**
Fractals **1**
Gamma-Ray Diffractometry **1**
Gas Sensors, Solid State **2**
High-Resolution Electron Microscopy **2**
High-Voltage Electron Microscopy **1**
Hydrogen as a Metallurgical Probe **2**
Laser Microprobe Mass Spectrometry **1**
Mechanical Properties Microprobe **2**
Microengineering of Materials: Characterization **2**
Mössbauer Spectroscopy (R) **1**
Polymers: Electron Microscopy **1**
Polymers: Light Microscopy **1**
Reflection Electron Microscopy **2**
Scanning Tunnelling Microscopy and Spectroscopy **2**
Solid-State Nuclear Track Detectors: Applications **1**
Synchrotron Radiation **1**
Thermal Analysis: Recent Developments (S) **2**
Thin Films: X-Ray Characterization **2**
Transmission Electron Microscopy: Convergent-Beam and Microdiffraction Techniques **1**
X-Ray Microanalysis, Quantitative **1**

33. NONDESTRUCTIVE EVALUATION

Art Forgeries: Scientific Detection **1**
Compton Scattering **2**
Mechanical Properties Microprobe **2**
Stereology: Image Analysis **1**
Thin Films: X-Ray Characterization **2**

34. SURFACES AND INTERFACES

35. DEGRADATION

36. SURFACE PROTECTION BY COATINGS

37. JOINING BY ADHESIVES

38. WELDING

39. SAFETY, HEALTH AND ENVIRONMENT

40. MATERIALS SCIENCE (SPECIAL TOPICS)

41. MECHANICS OF MATERIALS

42. MATERIALS ENGINEERING (SPECIAL TOPICS)

43. MINERAL RESOURCES

44. MATERIALS ECONOMICS, POLICY AND MANAGEMENT

45. HUMANISTIC ASPECTS OF MATERIALS

LIST OF CONTRIBUTORS

Contributors to *Supplementary Volume 2* are listed in alphabetical order together with their affiliations. Titles of articles that they have authored follow in alphabetical order. Where articles are coauthored, this has been indicated by an asterisk preceding the title.

Adams, J. F. E.
Metal Box Company
Research & Development
Denchworth Road
Wantage
Oxfordshire
UK
**Plastics for Packaging*

Allsop, M.
Technical Fibre Products Ltd
Burneside Mills
Kendal
Cumbria LA9 6PZ
UK
**Specialty Papers from Advanced Fibers*

Angus, J. C.
Chemical Engineering Department
Case Western Reserve University
University Circle
Cleveland, OH 44106
USA
**Diamondlike and Diamond Thin Films*

Ashby, M. F.
Department of Engineering
University of Cambridge
Trumpington Street
Cambridge CB2 1PZ
UK
Cork
Materials Selection in Conceptual Design: Materials Selection Charts

Banerjee, D.
Government of India
Ministry of Defence
Defence Metallurgical Research Laboratory
Kanchanbagh P. O.
Hyderabad 500 258
INDIA
Channelling-Enhanced Microanalysis

Beall, F. C.
University of California
Berkeley, CA 94720
USA
Wood: Acoustic Emission and Acousto-Ultrasonic Characteristics

Beardmore, P.
Engineering and Research Staff
Ford Motor Company
P O Box 2053
Dearborn, MI 48121-2053
USA
Automotive Composite Components: Fabrication

Belleville, J.
INSERM Unit 37
18 Avenue du Doyen Lepine
F-69500 Bron
FRANCE
**Biomaterial–Blood Interactions*

Benson, K. E.
AT&T Bell Laboratories
555 Union Boulevard
Allentown, PA 18103-1285
USA
Silicon: Preparation for Semiconductors

Benton, J. L.
AT&T Bell Laboratories
Room 1E-336
600 Mountain Avenue
Murray Hill, NJ 07974
USA
Junction Transient Spectroscopy

Bhadeshia, H. K. D. H.
Department of Materials Science & Metallurgy
University of Cambridge
Pembroke Street
Cambridge CB2 3QZ
UK
Bainite—The Current Situation

Bond, R.
CR Industries
900 North State Street
Elgin, IL 60123
USA
Tire Adhesion: Role of Elastomer Characteristics

Bonnell, D. A.
Department of Materials Science & Engineering
University of Pennsylvania
3231 Walnut
Philadelphia, PA 19104
USA
Scanning Tunnelling Microscopy and Spectroscopy

Bourne, H. L.
Independent Consulting Geologist
P O Box 293
Northville, MI 48167
USA
Slag Raw Materials

Bowen, D. H.
Materials Development Division
Building 47
Harwell Laboratory
United Kingdom Atomic Energy Authority
Harwell
Oxfordshire OX11 0RA
UK
Composite Materials: Applications Overview
Composite Materials: Manufacturing Overview

Brach, A. M.
Massachusetts Institute of Technology
Cambridge, MA 02139
USA
Ceramics: Construction Applications

Braden, M.
Department of Materials Science in Dentistry
The London Hospital Medical College
University of London
Turner Street
London E1 2AD
UK
Dental Elastomers: An Overview

Brandon, D. G.
Department of Materials Engineering
Technion
Israel Institute of Technology
Haifa 32000
ISRAEL
Armor, Ceramic

Brandt, E. H.
Max-Planck-Institut für Metallforschung
Heisenbergstrasse 1
D-7000 Stuttgart 80 (Busnau)
FRG
Superconductors: Magnetic Structure

Brandt, G.
AB Sandvik Coromant
S-126 80 Stockholm
SWEDEN
Ceramic Cutting Tools

Bunsell, A. R.
Centre des Materiaux
Ecole Nationale Supérieure des Mines de Paris
BP 87
91003 Evry Cedex
FRANCE
Silicon Carbide Fibers

Burggraaf, A. J.
Laboratory for Inorganic Chemistry & Materials Science
Faculty of Chemical Engineering
University of Twente
P O Box 217
7500 AE Enschede
THE NETHERLANDS
Ceramic Membranes, Porous

Carlsson, J.-O.
Institute of Chemistry
University of Uppsala
Box 532
S-751 21 Uppsala 1
Uppsala
SWEDEN
Silicon Carbide Fibers

Chattopadhyay, K.
Department of Metallurgy
Indian Institute of Science
Bangalore 560 012
INDIA
Quasicrystals

Chou, T-W.
University of Delaware
Newark, DE 19711
USA
Composite Materials: Structure–Performance Maps

Christel, P. S.
Laboratoire de Recherches Orthopédiques
Faculté de Médecine Lariboisière-Saint Louis
Université de Paris 7
10 Avenue de Verdun
F-75010 Paris
FRANCE
Zirconia-Toughened Ceramics as Biomaterials

Christodoulou, L.
Martin Marietta Laboratories
1450 South Rolling Road

Baltimore, MD 21227-0700
USA
Titanium Aluminides

Claussen, N.
Advanced Ceramics Group
Technical University Hamburg-Harburg
P O Box 90 1403
D-2100 Hamburg 90
FRG
**Oxidation of Molten Metals, Directed*

Collings, T. A.
Materials & Structures Department
X32 Building
Royal Aerospace Establishment
Farnborough
Hampshire GU14 6TD
UK
Composite Materials: Joining

Coombs, G.
Johns–Manville Sales Corporation
Research and Development Center
Ken-Caryl Ranch
Denver, CO 80217
USA
Insulation Raw Materials

Cooper, M. J.
Department of Physics
University of Warwick
Coventry CV4 7AL
UK
Compton Scattering

Damberger, H. H.
Illinois State Geological Survey
Champaign, IL 61820
USA
Coal: World Resources

Dransfield, J.
Royal Botanic Gardens
Kew
Richmond
UK
Rattan and Cane

Dunn, J. R.
Dunn Geoscience Corporation
12 Metro Park Road
Albany, NY 12205
USA
Portland Cement Raw Materials

Dyllick-Brenzinger, R.
BASF AG
D-6700 Ludwigshafen
Rhein
FRG
Heat-Sensitive Papers

Elliot, N. E.
MST Division
Mail Stop E 549
Los Alamos National Laboratory
Los Alamos, NM 87545
USA
**Microengineering of Materials: Characterization*

Eloy, R.
INSERM
Unit 37
18 Avenue du Doyen Lepine
F-69500 Bron
FRANCE
**Biomaterial–Blood Interactions*

Embury, J. D.
Materials Science & Engineering
McMaster University
Hamilton
Ontario
CANADA
Bauschinger Effect in Metals and Composites

Erman, B.
Bogazici University
Bebek
Istanbul
TURKEY
**Segmental Orientation in Elastomers*

Fair, R. B.
Microelectronics Center of North Carolina
P O Box 12889
Research Triangle Park, NC 27709
USA
Diffusion in Silicon

Fletcher, N.
Cheswick UK Ltd
Clifton Road
Blackpool

Lancashire FY4 4QF
UK
**Porcelain Enamelling Technology: Recent Advances*

Foreman, L. R.
MST Division
Mail Stop E 549
Los Alamos National Laboratory
Los Alamos, NM 87545
USA
**Microengineering of Materials: Characterization*
Microengineering of Materials: Laser Fusion Targets

Fray, D. J.
Department of Materials Science & Metallurgy
University of Cambridge
Pembroke Street
Cambridge CB2 3QZ
UK
Gas Sensors, Solid State

Freas, R. C.
Franklin Limestone Company
612 10th Avenue N
Nashville, TN 37203
USA
Lime

Friedl, C.
Leader Flow Analysis Group
Moldflow Pty. Ltd
Colchester Road
Kilsyth 3137
Victoria
AUSTRALIA
**Injection Molding of Polymers: Computer Simulation*

Gallagher, P. K.
Department of Chemistry
The Ohio State University
120 West 18th Avenue
Columbus, OH 43210-1173
USA
Thermal Analysis: Recent Developments

Gent, A. N.
University of Akron
Akron, OH 44304
USA
**Automobile Tires*

Gibson, I. H.
Atomic Energy Establishment
Winfrith Technology Centre
Dorchester
Dorset DT2 8DH
UK
**Nuclear Fuels: Burnable Poisons*

Gibson, L. J.
Massachusetts Institute of Technology
Cambridge, MA 02139
USA
Cellular Materials in Construction

Gibson, P. N.
Physics Division
Joint Research Centre—CEC
Ispra Establishment
I-21020 Ispra (Varese)
ITALY
Thin Films: X-Ray Characterization

Glasser, W. G.
Virginia Polytechnic Institute and State University
Blacksburg, VA 24061
USA
Lignin-Based Polymers

Goldsmith, D. F.
University of California
Davis, CA 95616
USA
Health Hazards in Wood Processing

Gösele, U.
Department of Mechanical Engineering & Materials Science
School of Engineering
Duke University
Durham, NC 27706
USA
Point Defect Equilibria in Semiconductors

Gray III, G. T.
MST 5
Mail Stop G730
Los Alamos National Laboratory
Los Alamos, NM 87545
USA
Deformation Twinning of Metals and Alloys

Gray, W.
Southwest Research Institute
6220 Culebra Road
P O Drawer 28510
San Antonio, TX 78284
USA
**Armor, Composite*

Greer, A. L.
Department of Materials Science & Metallurgy
University of Cambridge
Pembroke Street

Cambridge CB2 3QZ
UK
Multilayers, Metallic

Gregory, E.
Business Manager Speciality Products
Intermagnetics General Corporation
1875 Thomaston Avenue
Waterbury, CT 06704
USA
Multifilamentary Composite Superconductor Design and Fabrication

Grimes, R.
British Alcan Aluminium plc
Warwick House
Warwick Road
Solihull
West Midlands
B91 3DG
UK
Aluminum–Lithium Based Alloys

Gruner, H.
Plasma-Technik AG
Rigackerstrasse 21
CH-5610 Wohlen
SWITZERLAND
Plasma Spraying for Protective Coatings

Hampshire, J. M.
Group Quality & Research Manager
RHP Industrial Bearings
P O Box 18
Northern Road
Newark
Nottinghamshire NG24 2JF
UK
Steels for Rolling Bearings

Hancox, N. L.
Materials Development Division
Building 47
Harwell Laboratory
United Kingdom Atomic Energy Authority
Harwell
Oxfordshire OX11 0RA
UK
High-Performance Composites with Thermoplastic Matrices

Hansen, W. L.
Lawrence Berkeley Laboratory
Building 70A, Room 3353
University of California
Berkeley, CA 94720
USA
Germanium

Haresceugh, R. I.
Structural Engineering Division
British Aerospace plc
Military Aircraft Division
Warton Aerodrome
Preston PR4 1AX
UK
Composite Materials: Aerospace Applications

Hasuo, S.
Fujitsu Laboratories Ltd
10-1 Morinosato-Wakamiya
Atsugi 243-01
JAPAN
**Superconducting Device Fabrication*

Heimke, G.
Department of Bioengineering
Clemson University
301 Rhodes Engineering Research Center
Clemson, SC 29634-0905
USA
Aluminum Oxide for Prosthetic Devices

Hellawell, A.
Department of Metallurgical Engineering
Michigan Technological University
Houghton, MI 49931
USA
Aluminum–Silicon: Cast Alloys Modification

Hilder, N. A.
Esso Research Centre
Milton Hill
Abingdon
Oxfordshire OX13 6AE
UK
Quenching Media: Polymeric Additives

Humphreys, C. J.
Department of Materials Science & Metallurgy
University of Cambridge
Pembroke Street
Cambridge CB2 3QZ
UK
Superconductors, Ceramic

Hutchinson, J. M.
Department of Engineering
University of Aberdeen
Kings College
Aberdeen AB9 2UE
UK
Physical Aging of Polymers

Imamura, T.
Fujitsu Laboratories Ltd
10-1 Morinosato-Wakamiya
Atsugi 243-01
JAPAN
**Superconducting Device Fabrication*

Inoue, A. L.
Institute for Materials Research
Tohoku University
Katahira 2-1-1
Sendai 980
JAPAN
**Aluminum-Based Glassy Alloys*

Irani, R. S.
BOC Ltd
The Priestley Centre
10 Priestley Road
The Surrey Research Park
Guildford
Surrey GU2 5XY
UK
High-Pressure Gas Containers: Safety and Efficiency

Jansen, F.
Xerox Corporation
Webster Research Center
Webster, NY 14580
USA
**Diamondlike and Diamond Thin Films*

Jedynakiewicz, N. M.
The University of Liverpool School of Dentistry
P O Box 147
Pembroke Place
Liverpool L69 3BX
UK
Endodontic Materials

Johnson, J. A.
University of Washington
Seattle, WA 98105
USA
Wood: Structure, Stiffness and Strength

Johnson, W. C.
Institut für Metallforschung
Technische Universität Berlin
Hardenbergstrasse 36
D-1000 Berlin 12
FRG
Phase Equilibria in Alloys: Influence of Stress

Jordan, A. S.
AT&T Bell Laboratories
600 Mountain Avenue
Murray Hill, NJ 07974
USA
**Semiconductors: Mechanical Properties*

Jorgensen, D. B.
US Gypsum Company
101 South Wacker Drive
Chicago, IL 60606
USA
Gypsum and Anhydrite

Kamen-Kaye, M.
1 Waterhouse Street (5)
Cambridge, MA 02138
USA
Petroleum Resources: Conditions for Formation
Petroleum Resources: Oil Fields
Petroleum Resources: World Overview

Keizer, K.
Laboratory for Inorganic Chemistry & Materials Science
Faculty of Chemical Engineering
University of Twente
P O Box 217
7500 AE Enschede
THE NETHERLANDS
**Ceramic Membranes, Porous*

Kennedy, P.
Technology Transfer Manager
Moldflow Pty Ltd
Colchester Road
Kilsyth 3137
Victoria
AUSTRALIA
**Injection Molding of Polymers: Computer Simulation*

Kes, P. H.
Kamerlingh Onnes Laboratorium
Rijksuniversiteit Leiden
Nieuwsteeg 18
Postbus 9506
2300 RA Leiden
THE NETHERLANDS
Superconductors: Flux Pinning and the Critical Current

Kirchheim, R.
Max-Planck-Institut für Metallforschung
Seestrasse 92
D-7000 Stuttgart 1
FRG
Hydrogen as a Metallurgical Probe

Knowles, K. M.
Department of Materials Science and Metallurgy

University of Cambridge
Pembroke Street
Cambridge CB2 3QZ
UK
High-Resolution Electron Microscopy

Koch Jr., G. S.
University of Georgia
Athens, GA 30601
USA
Mining and Exploration: Statistical and Computer Models

Kranzmann, A.
Max-Planck-Institut für Metallforschung
Institut für Werkstoffwissenschaften
Heisenbergstrasse 5
D-700 Stuttgart 80
FRG
Aluminum Nitride

Lankford, J.
Southwest Research Institute
6220 Culebra Road
P O Drawer 28510
San Antonio, TX 78284
USA
**Armor, Composite*

Lanza, F.
Materials Division
Joint Research Centre—CEC
Ispra Establishment
I-21020 Ispra (Varese)
ITALY
High-Level Radioactive Waste Disposal: Safety

Livingston, D. J.
Livingston Associates
731 Frank Boulevard
Akron, OH 44320
USA
**Automobile Tires*

Livingston, J. D.
Massachusetts Institute of Technology
Cambridge, MA 02139
USA
Permanent Magnets, Iron–Rare Earth

Mackley, M. R.
Department of Chemical Engineering
University of Cambridge
Pembroke Street
Cambridge CB2 3RA
UK
Polyethylene Processing to Produce Ultimate Properties

Markow, M. J.
Massachusetts Institute of Technology
Cambridge, MA 02139
USA
**Ceramics: Construction Applications*

Masumoto, T.
Institute for Materials Research
Tohoku University
Katahira 2-1-1
Sendai 980
JAPAN
**Aluminum-Based Glassy Alloys*

McDougall, I. L.
Oxford Instruments Ltd
Eynsham
Oxford OX8 1TL
UK
Superconducting Magnets

McIntosh, J.
26 Harvest Hill Road
Maidenhead
Berkshire
UK
**Plastics for Packaging*

Messing, G. L.
Department of Materials Science & Engineering
Pennsylvania State University
119 Steidle Building
University Park, PA 16802
USA
Ceramic Transformations, Seeding

Milborrow, D. J.
Headquarters
National Power
Sudbury House
15 Newgate Street
London EC1A 7AU
UK
**Wind Energy Systems*

Mills, P. J.
Department of Materials Science & Engineering
University of Surrey
Guildford
Surrey GU2 5XH
UK
**Carbon-Fiber-Reinforced Plastics*

Milne, R. H.
Department of Physics and Applied Physics

University of Strathclyde
Colville Building
North Portland Street
Glasgow G1 1XN
UK
Reflection Electron Microscopy

Miodownik, A. P.
Department of Materials Science and Engineering
University of Surrey
Guildford
Surrey GU2 5XH
UK
Phase Diagrams and Phase Stability: Calculations

Monnerie, L.
Ecole Supérieure de Physique et de Chimie Industrielles de la Ville de Paris
10 rue Vauquelin
F-75231 Paris Cedex 05
FRANCE
**Segmental Orientation in Elastomers*

Morrell, J. J.
Oregon State University
Corvallis, OR 97331
USA
Fumigation of Wood

Murarka, S. P.
Department of Materials Engineering
Center for Integrated Electronics
Rensselaer Polytechnic Institute
Troy, NY 12180
USA
Transition Metal Silicides

Nicholas, E. D.
The Welding Institute
Abington Hall
Abington
Cambridge CB1 6AL
UK
Friction Welding and Surfacing

Nightingale, J.
Cambridge Polymer Consultants
Melbourne Science Park
Moat Lane
Melbourne
Royston
Hertfordshire SG8 6EJ
UK
Welding of Thermoplastics

O'Halloran, M. R.
American Plywood Association
7011 South 19th Street
Tacoma, WA 98411
USA
Plywood

Okabe, T.
Department of Dental Materials
Baylor College of Dentistry
3302 Gaston Avenue
Dallas, TX 75246
USA
Dental Amalgams: Further Developments

Oliver, W. C.
Metals & Ceramics Division
Oak Ridge National Laboratory
P O Box 2008
Oak Ridge, TN 37831
USA
Mechanical Properties Microprobe

Ono, T.
Faculty of Engineering
Gifu University
Gifu City 501-11
JAPAN
Stringed Instruments: Wood Selection

Pankove, J. I.
University of Colorado
Boulder, CO 80309-0425
USA
Hydrogenated Amorphous Silicon

Pansu, B.
Laboratoire de Physique des Solides
Bat 510
Université Paris-Sud
F-91405 Paris Cedex
FRANCE
Colloidal Crystals

Pearton, S. J.
AT&T Bell Laboratories
Murray Hill, NJ 07974
USA
Ion Implantation: III–V Compounds

Peercy, P. S.
Sandia National Laboratories
Department 1140
P O Box 5800

Albuquerque, NM 87185
USA
Strained-Layer Superlattices

Perkins, J.
Materials Science Group
Mail Code 69Ps
Naval Postgraduate School
Monterey, CA 93943-5100
USA
Tweed Contrast in Microstructures

Phillips, D. C.
Materials Development Division
Building 47
Harwell Laboratory
United Kingdom Atomic Energy Authority
Harwell
Oxfordshire OX11 0RA
UK
Fiber-Reinforced Ceramics

Pickens, J. R.
Martin Marietta Corporation
Martin Marietta Laboratories
1450 South Rolling Road
Baltimore, MD 21227
USA
Aluminum–Lithium Based Alloys: Weldability

Pickering, F. B.
Department of Engineering
Sheffield City Polytechnic
Pond Street
Sheffield S1 1WB
UK
Nitrogen in Steels

Prasad, S. V.
Council of Scientific and Industrial Research
Regional Research Laboratory
Bhopal
INDIA
Natural-Fiber-Based Composites

Price, R. F.
Cloverton House
Cilgerran
Cardigan
Dyfed SA43 3SN
UK
**Porcelain Enamelling Technology: Recent Advances*

Ranganathan, S.
Department of Metallurgy
Indian Institute of Science
Bangalore, 560 012
INDIA
**Quasicrystals*

Ratner, B. D.
NESAC/BIO
Department of Chemical Engineering
BF-10
University of Washington
Seattle, WA 98195
USA
Biomaterials: Surface Structure and Properties

Raub, Ch. J.
Forschungsinstitut für Edelmetalle und Metallchemie
Katharinenstrasse 17
D-7070 Schwäbisch Gmünd
FRG
Gold–Titanium Alloys

Reeve, K. D.
Australian Nuclear Science and Technology Organisation
Private Mail Bag 1
Menai
New South Wales 2234
AUSTRALIA
Ceramic Materials for Nuclear Waste Storage

Sanchez, J. M.
Center for Materials Science and Engineering
The University of Texas at Austin
Austin, TX 78712-1063
USA
Phase Transformations at Surfaces and Interfaces

Santangelo, J. G.
Air Products and Chemicals, Inc.
7201 Hamilton Boulevard
Allentown, PA 18195-1501
USA
**Polymers Based on Carbon Dioxide*

Sasaki, H.
Composite Wood Section
Wood Research Institute
Kyoto University
Uji
Kyoto 611
JAPAN
Laminated Veneer Lumber

Schoen, F. J.
Department of Pathology
Brigham and Women's Hospital
75 Francis Street

Boston, MA 02115
USA
Heart Valve Replacement Materials

Segal, D. L.
Materials Development Division
Building 429
Harwell Laboratory
United Kingdom Atomic Energy Authority
Harwell
Oxfordshire OX11 0RA
UK
Ceramic Powders: Chemical Preparation

Seidel, T. E.
Seidel Consultants
1165 Wales Place
Cardiff, CA 92007
USA
Rapid Thermal Processing of Semiconductors

Sherwin, J.
Atomic Energy Establishment
Winfrith Technology Centre
Dorchester
Dorset DT2 8DH
UK
**Nuclear Fuels: Burnable Poisons*

Shyam Sunder, S.
Massachusetts Institute of Technology
Cambridge, MA 02139
USA
Ice and Frozen Earth as Construction Materials

Smith, P. A.
Department of Materials Science & Engineering
University of Surrey
Guildford
Surrey GU2 5XH
UK
**Carbon-Fiber-Reinforced Plastics*

Srolovitz, D. J.
Department of Materials Science & Engineering
University of Michigan
Ann Arbor, MI 48109-2136
USA
Computer Simulation of Microstructural Evolution

Steiner, P. R.
Faculty of Forestry
University of British Columbia
2357 Main Hall
Vancouver
BC V6T 1W5
CANADA
Adhesives for Wood: An Update

Stowell, M. J.
Alcan International Ltd
Banbury Laboratory
Southam Road
Banbury
Oxfordshire OX16 7SP
UK
**Porcelain Enamelling Technology: Recent Advances*

Stupin, D. M.
MST Division
Mail Stop E 549
Los Alamos National Laboratory
Los Alamos, NM 87545
USA
**Microengineering of Materials: Characterization*

Surman, P. L.
Headquarters
National Power
Sudbury House
15 Newgate Street
London EC1A 7AU
UK
**Wind Energy Systems*

Swaminathan, V.
AT&T Bell Laboratories
600 Mountain Avenue
Murray Hill, NJ 07974
USA
**Semiconductors: Mechanical Properties*

Talreja, R.
Department of Solid Mechanics
The Technical University of Denmark
Building 404
DK-2800 Lyngby
DENMARK
Composite Materials: Fatigue

Tao, J. C.
Air Products and Chemicals, Inc.
7201 Hamilton Boulevard
Allentown, PA 18195-1501
USA
**Polymers Based on Carbon Dioxide*

Thompson, C. V.
Department of Materials Science & Engineering
Room 13-5077
Massachusetts Institute of Technology
Cambridge, MA 02139
USA
Polycrystalline Silicon: Structure and Processing

Urquhart, A. W.
Lanxide Corporation
Newark, DE 19714
USA
**Oxidation of Molten Metals, Directed*

Walker, N.
Technical Fibre Products Ltd
Burneside Mills
Kendal
Cumbria LA9 6PZ
UK
**Specialty Papers from Advanced Fibers*

Walter, J. L.
General Electric Corporate Research and Development
P O Box 8, Bldg K-1
Schenectady, NY 12301
USA
**Incandescent Lamp Filaments*

Welsch, G. E.
Case Western Reserve University
Cleveland, OH 44106
USA
**Incandescent Lamp Filaments*

Willoughby, A. F. W.
Department of Engineering Materials
Southampton University
Highfield
Southampton SO9 5NH
UK
Diffusion in Compound Semiconductors

Wohlleben, D. K.
II. Physikalisches Institut
Universität Köln
Zulpicher Strasse 77
D-5000 Köln 41
FRG
Heavy Fermion Metals: Magnetic and Superconducting Properties

Yasuda, K.
Department of Dental Materials Science
Nagasaki University School of Dentistry
Nagasaki
JAPAN
Dental Gold Alloys: Age-Hardening

Youngs, R. L.
Virginia Polytechnic Institute and State University
Blacksburg, VA 24061
USA
Timber Use: History

Zavarin, E.
University of California
Berkeley, CA 94720
USA
Wood: Surface Chemistry